FRUITS OF WARM CLIMATES

Cover images by row from top, left to right:
1. Eva Kosmas (www.evamariekosmas.com),
Roberto Verzo, David W. Eickhoff
2. Romain Guy, Eva Kosmas, Kazue Asano
3. Jeff Wright, Mitchell Hearns Bishop, Hafiz Issadeen

Cover design by Adrienne Nunez,
Echo Point Books & Media

ISBN: 978-1-62654-976-0

Published by Echo Point Books & Media
www.EchoPointBooks.com

Printed in the U.S.A.

EDITOR'S NOTE

Even in our post-exploration age, where seemingly every inch of the planet has been mapped, the tropics and subtropics continue to yield new plant life and foods. In fact, thousands of fruits exist that never make an appearance in North American or European grocery stores, leaving a large range of under or unexplored culinary opportunities. In many regions, especially rain forests, people still discover new plant life and fruits exclusive to these areas. In many cases, even residents know little or nothing about the fruits they eat (regardless of whether the fruit tastes delicious or not), and even less may be known about it scientifically. Of course this leaves—ahem—fruitful areas of discovery for many exotic and delicious foods.

In *Fruits of Warm Climates*, Morton examines the history, cultivation, uses, and nutritional value of each fruit in comprehensive detail. She further includes the varieties and relatives of each fruit to enable the reader to better understand the context of exotic fruits in relation to fruits with which they may already be familiar. With the increased attention of healthy food and globalization since original publication in 1987, many previously obscure fruits, like rambutan and mangoes, are readily found in grocery stores and on restaurant menus. However, due to her comprehensive research, Morton's guide still remains a sought after and comprehensive guide to warm climate fruits, and serves as an indispensable guide to researchers, growers, travelers and fruit enthusiasts alike.

About the Author

Julia Morton, was a self-taught and highly regarded expert on the uses of cultivated, medicinal and toxic plants. As the *New York Times* wrote of professor Morton: "To her colleagues in the field of economic botany, the study of how people use plants, Mrs. Morton was an academic's academic, a founder and developer of the Morton Collectanea, an invaluable compilation of information about plants, and the author of 10 books on the subject." She was a professor of botany at the University of Miami for four decades. She also wrote 94 scientific papers and contributed to an additional 12 books and 27 papers.

To Floridians and beyond, she was considered a "Dear Abby" for those who needed help with plants, especially harmful plants. Through letters and phone calls, lectures and even posters designed for hospital emergency rooms, Julia worked to educate the public, answering questions long after she retired.

Fruits of Warm Climates is considered Julia Morton's master work. It combines her encyclopedic knowledge of plants with her knack for reaching non-experts. For anyone interested in the topic, *Fruits of Warm Climates* is an important work.

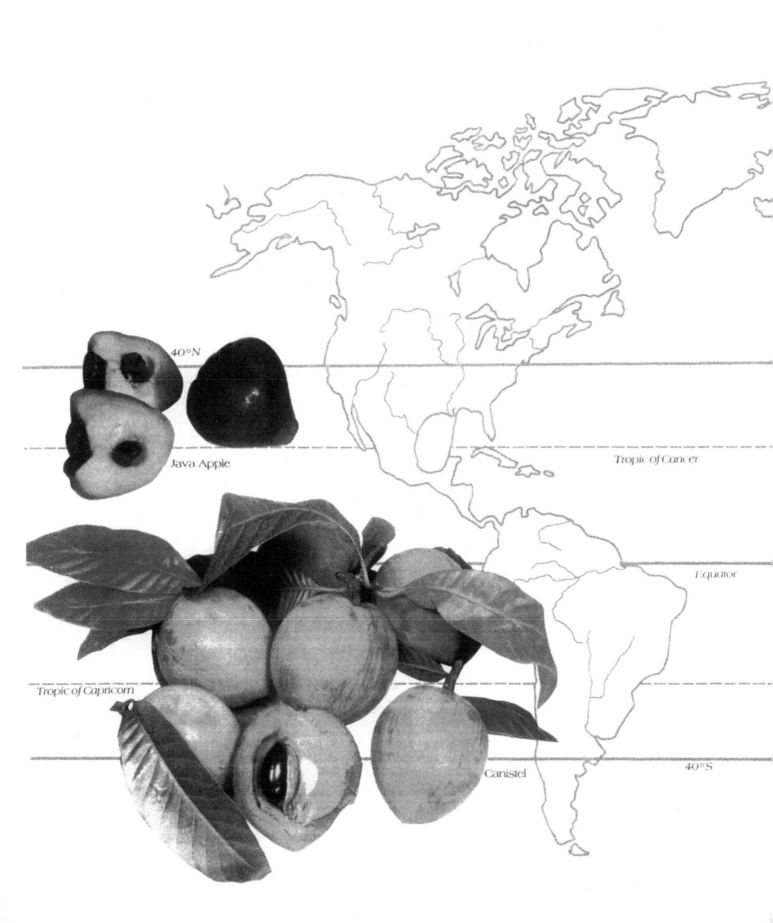

40°N

Java Apple

Tropic of Cancer

Equator

Tropic of Capricorn

Canistel

40°S

FRUITS OF WARM CLIMATES

Julia F. Morton, D. Sc., F.L.S.
Director, Morton Collectanea, University of Miami

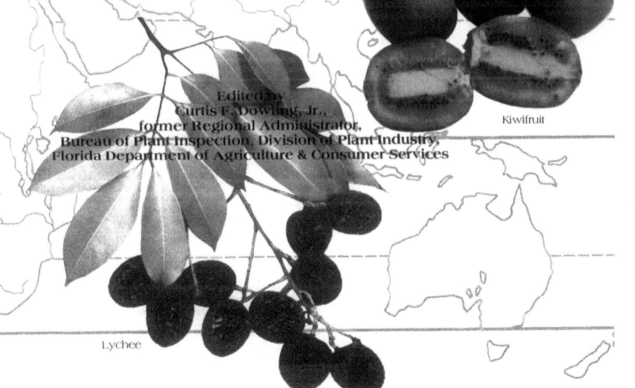

Edited by
Curtis F. Dowling, Jr.,
former Regional Administrator,
Bureau of Plant Inspection, Division of Plant Industry,
Florida Department of Agriculture & Consumer Services

Kiwifruit

Lychee

ECHO POINT BOOKS & MEDIA, LLC

Text provided and permission to reprint provided by the beneficiaries of
Julia F. Morton D. Sc., 1912-1996.

Dr. Morton was born Julia Frances McHugh on a farm in Middlebury, Vermont. Her work as a commercial artist in New York brought her in contact with Canadian botanist Kendal Morton. They married and founded the world-famed Morton Collectanea at the University of Miami. Julia Morton continued disseminating well-researched lore on economic botany for more than thirty years after the passing of her husband. Dr. Julia Morton was noted for exacting research, a generous civic spirit and a formidable personality—as exemplified by her quote:

"I don't want to suppose. I want to know."

Dedicated to

THE FLORIDA STATE HORTICULTURAL SOCIETY

Centennial
1987

ACKNOWLEDGMENTS

I owe sincere thanks to Donald Plucknett for pausing in his worldwide professional travels to give consideration to this work; to Curtis Dowling for his careful examination of the manuscript with keen concern for accuracy; to Vera Webster for her patient and talented sculpturing of words and pictures into the form of a book; to Blue Anchor, Inc., for supplying the field photographs of the kiwifruit (Figure 82); and to the United States Department of Agriculture for approval of my use of photographs represented by Figures 1, 31, 48, 52, 53, and 71.

I know readers will share my gratitude to Susan and John DuPuis for making the color reproductions possible.

FOREWORD

One of the great pleasures for visitors to the tropics or subtropics is the wide array of fruits that is available. On almost every trip, some new fruit is presented that is unknown or little known to most people, including persons knowledgeable about plants. The question that then arises — what fruit is that? — is followed by the telling of an unfamiliar local name, but little or nothing else of its origin or scientific name. So we eat the fruit, and may enjoy its taste or texture, or perhaps don't even like it very much, but still are left in the dark about the fruit and the plant on which it was grown. But it is not just the visitor who lacks knowledge about tropical fruits; very often persons residing in the tropics know little about many local fruits, and, unfortunately, have very little hope of obtaining information about them.

I believe that most persons, whether resident or visitor, would like to know more about the fruits of the tropics. They would like to know where a fruit comes from, what its other common names might be, to what it is related, how it is used, and perhaps some idea of how it is grown. Of course, some tropical or subtropical fruits are very familiar — banana, citrus, figs or dates — but what do we know of the exotic fruits of Asia, rambutan or mangosteen for example, or of the lesser-known fruits of the Amazon? And what do we know of the fruits of Africa? For most persons, the answer would be little or nothing.

This book will improve our information about tropical fruits. Dr. Julia Morton has done a great service in researching and writing *Fruits of Warm Climates*. The book contains a great deal of information on those tropical and subtropical plants, mostly trees but also some herbaceous plants, that produce fruit for mankind. In keeping with her usual attention to clarity, Dr. Morton has included much information on the history of each crop, its main cultivation requirements, food uses and nutritional value. Its coverage of most of the fruits is comprehensive and quite detailed, including very useful information on the most important varieties that are or have been used. Such information is not available for most of these crops in any other book of which I am aware.

This book is certain to be a constant companion of most persons interested in tropical horticulture. It will be an important reference for the scientist or layman who wants to learn more about the rich array of fruits that the tropics enjoys. I would also hope that it will be used widely by persons interested in economic development in tropical countries, some of which may wish to consider one or more of these fruits in their development plans. I look forward to carrying it with me on my own travels.

Donald L. Plucknett
The World Bank
Washington, D.C.

Frontispiece A.
Kendal and Julia Morton (right) with their tropical fruit exhibit at the Florida State Fair, Tampa, Florida, February 3-14, 1948. Col. William R. Grove (left), founder of Lychee Orchards, Laurel, Florida. The specimens were collected from the Univ. of Florida's Subtropical Experiment Station, Palm Lodge Tropical Grove, Kosel's Jungle Grove, Homestead; the U.S. Department of Agriculture's Plant Introduction Station, Miami, and private growers of rare fruits, including Col. Grove. Replacements were brought from Homestead midway to freshen up the display.

Frontispiece B.
Tropical fruit exhibit by Kendal and Julia Morton, Morton Collectanea University of Miami, at the Florida State Horticultural Society's Annual Meeting, November 1-3, 1948. In addition to the sources cited above, specimens were provided also from the University of Miami's Experimental Farm, at Richmond.

TABLE OF CONTENTS

(Plant families in natural order)

LIST OF COLOR PLATES

INTRODUCTION

This book is a product of the Morton Collectanea which was initiated in New York in 1933, during the Great Depression, with a view to making knowledge of foods available to all. The Collectanea was brought to the University of Miami in 1949 and has continued to grow and serve here as a reference and research center devoted to economic botany. Primary functions are the selection, acquisition, processing and maintenance of information in this broad field and its application through investigation, consultation and communication. From the inception, an area of major interest has been the fruits of warm countries, which range from those in a high state of development and of international economic importance to those which are locally utilized but have received little if any horticultural improvement.

This book is presented as a practical guide and tool for those who need to choose fruits that are of real value in the home garden or are intended to be commercial crops for domestic consumption or export. The fruits are grouped by families and the families are arranged in their natural order in the plant kingdom. Within each family, the fruits appear in their relative order of importance on a world scale. The ranking is not meant to be arbitrary but is based on weighing the evidence and recognizing that certain fruits may be locally popular because they are juicy in an arid land, or just available and abundant, or because people tend to favor what they have relished as children (the mamoncillo, for example), even though the traditional favorites often suffer by comparison with other species of more marketable nature. Then, too, a fruit may be misjudged or underestimated if one has sampled only the product of seedling trees, whereas acquaintance with superior selections may radically alter one's impression. This is particularly true of the papaya, some types being unpleasantly musky while others will appeal to any taste. Fortunately, there exist rational choices for every geographic zone and energies should be concentrated on the adaptation and protection of the few choice species that are necessary to provide year-round provender.

Following the main discussion of a particular fruit, related species of little present economic value may be appended as of casual interest, as possible rootstocks, or of potential use in breeding for pest- or disease-resistance, cold-hardiness or other desirable goals. Some may even be candidates for future cultural improvement in their own right.

Efforts have been made to ascertain the currently preferred common and botanical names and, as it is hoped that the book will be widely useful, regional names from many areas are included. If a genus is considered to be a hybrid, the generic name is preceded by "X". If a species is a hybrid, "X" appears between the genus and species (see especially the citrus fruits). In compiling the descriptions, I have tried earnestly to be accurate in providing the metric equivalents of measurements for the convenience of those accustomed to using the metric system. I realize this is risky. When writing at odd times over a period of 2½ years, it is difficult to achieve uniformity, and there may be errors in my calculations. I can only hope that flaws will be forgiven and I shall appreciate being informed of them for correction in the next edition.

Under the heading "Origin and Distribution", it has been possible in most cases to provide something of the fruit's history and development and past and current status. Statistics are given where readily available but they may vary drastically from year to year or era to era because of changes in weather, wartime demand, or other influences, and cannot be expected to be up-to-date. Some geographical regions may be referred to by names that are common in the literature but are now obsolete. I have not switched from "Ceylon" to "Sri Lanka" as it has been explained by visiting Ceylonese that the latter is only the translation of the former and not a different name. The *Flora of Ceylon* is still being published, volume by volume, with that title. To me there seems to be a parallel in the Italian application of Firenze to the city most of the world calls Florence, and the Austrian use of Wein, for Vienna.

The word "Varieties" I have employed as a general heading to embrace all kinds of variation, natural or induced by breeding, though, in a stricter sense, "variety" is a term applied only to a botanical variant, for example, the yellow passionfruit (*Passiflora edulis* var. *flavicarpa*). There are many natural variations in color, form, size, flavor, etc., that have not been given such botanical distinction. When domesticated and stabilized by vegetative reproduction and controlled cultivation, they are known as "cultivars" and usually are given names even though they may not remain entirely uniform. Individuals that display special characteristics may be selected, isolated and propagated as "strains", "lines", or "clones" and may be identified by numbers following the adopted cultivar name. Knowledge of natural varieties or cultivars may be all-important, as the right choice for a particular environment may make all the difference between success and failure. I have not hesitated to include the old with the new, as it is evident that a cultivar long abandoned in

California or Florida may be ideal for planting in Israel or Australia, and vice versa.

The work of plant-breeders never stops. There are continuous efforts to develop fruits having the ability to withstand natural enemies, adverse soil and climatic conditions, the hardships of transport and handling, and yet retain the attractiveness and quality required by the consumer. While this manuscript is being written, news of new cultivars being released for trial will come too late for inclusion. One great need, receiving insufficient attention, is for dwarf forms of large-growing fruit trees such as the avocado and mango. I am of the conviction that it is wiser to improve and redesign a species having a high ratio of acceptability than to seek untried alternatives and undue diversity and then be unprepared to confront the problems that arise when a crop enters large-scale cultivation.

The format of the text is not rigid. Where pollination is a major concern in the production of a crop (as in the date, persimmon and passionfruit), it is given a separate heading. "Climate" and "Soil" may be combined or dealt with separately depending on the amount of information available. The reader will note that the scope of the book is not limited to the tropical and subtropical zones but extends to the subtemperate to embrace the date, Japanese persimmon, hardy citrus species, the kiwifruit and others of extraordinary range, surviving and even benefitting by a degree of frost in southern Europe, North Africa, protected locations in England, central California, the Gulf States, northern Chile and Argentina and other areas far beyond the boundaries of the Tropics of Cancer and Capricorn.

Both old and modern methods of propagation and culture are covered. What is suitable in an industrialized society may not be appropriate for a developing country. Many mistakes have been made in the past trying to superimpose advanced technology over time-honored indigenous practices which may be, with slight improvements, more appropriate and quite adequate for local needs. Where a considerable body of information exists, the section on "Culture" may be subdivided into "Planting", "Fertilizing", "Irrigation", "Weed Control", and (in the case of pineapple) "Flower Induction".

Harvesting may be one of the most critical operations. It is generally held that harvesting represents the greatest single expense in crop production because of the cost and scarcity of labor in industrialized and even in semi-developed countries. This is a factor that must be given serious consideration even though all other aspects of a crop may appear favorable. Mechanized harvesting has seemed a highly desirable objective but is retarded by the cost of equipment. There are, instead, many ways in which manual labor can be assisted, and uniform ripening and easier harvesting can be achieved by the use of chemicals. In these times, when such essential crops as apples and oranges are being hedge-grown to reduce manual labor, it is only fair to advise the entrepreneur: "If you can't harvest it economically, don't grow it".

In discussions of "Pests" and "Diseases", I have endeavored to give the most commonly employed names for pests and disease organisms, though aware that they may vary in spelling and some may be synonymous.

It may seem excessive to detail food uses (even some recipes are included), but I have found that a worthy fruit (for example, the canistel) may be disdained or neglected until good methods of preparation demonstrate its true value. Even such a minor species as the Surinam cherry becomes an asset when properly treated in the kitchen.

I have emphasized food products because a perishable fruit such as the soursop that lends itself to processing has economic possibilities greater than those of its more delicious relatives (sugar apple, cherimoya, atemoya) which do not.

The data on food values are reported as they appear in the sources consulted, some of which were published many years ago. Nutritionists today may have more sophisticated ways of expressing this information. In many cases, it has seemed necessary to point out toxic or unwholesome constituents. It cannot be assumed that a fruit is completely beneficial merely because it has been consumed by certain populations over many generations. The akee, under some conditions, still causes illness and fatalities in Jamaica. The carambola may contain sufficient oxalic acid to be a renal hazard. Green bananas and green plantains, and the pomegranate and cashew apple are high in tannin, a carcinogen and antinutrient if ingested excessively.

The by-products or non-food uses of some fruit crop, or of the plant producing it, may add greatly to economic feasibility, witness the many uses of citrus residues, the importance of pineapple waste, papain from papaya, and the useful wood of the jackfruit and santol, to name a few revenue-enhancers.

Medicinal uses are set forth, not by any means in advocacy of folk-medicine, but mainly to reveal that the esteem for certain species (bael fruit, emblic, jambolan, white sapote, etc.) is based more largely on their therapeutic repute than on their food quality.

FRUITS OF WARM CLIMATES is a condensation, largely of the information in journal articles which have accumulated in the subject files of the Morton Collectanea, and to a lesser degree from literature in book form consulted primarily for historical background. Entire volumes have been written on the banana, date, fig, lychee, mango, orange and other fruit crops. I do not pretend to have digested all that has been presented in these and other books which deal in depth with chemistry, technology, pathology, entomology, handling, transport and storage and other specialized approaches to the many problems of the fruit grower, distributor and processor. These are included in the Bibliography as having contributed some details not found elsewhere and as sources of knowledge to be pursued by those active in fruit-growing or marketing.

It was impossible to key such a great number of sources to citations in the manuscript because of the continuous input of new material coming to hand over the lengthy period of writing. The Bibliography is my acknowledg-

[Left to right: Tchang-Bok Lee, John Beaman, Julia Morton, Dr. Schultes, Kendal Morton, John Freeberg and Arthur Barclay]

The research facilities of the MORTON COLLECTANEA are demonstrated to DR. RICHARD E. SCHULTES, Professor of Tropical Botany, and four graduate students of HARVARD UNIVERSITY, in the offices of the Collectanea at the University of Miami. Collectanea subject files, covering a dozen economic plants, are studied by the class in conjunction with specimens of the roots, seeds and fruits concerned, and their foliage, flowers and preserved products. Photo, June 13, 1955.

Present positions of visitors: Dr. Tchang-Bok Lee, Director, Kwanak Arboretum, College of Agriculture, Seoul National University, Suwon, Korea; Dr. John Beaman, Professor, Department of Botany and Plant Pathology, Michigan State University, East Lansing, Michigan; Dr. Richard Evans Schultes, Jeffrey Professor of Biology and Director of the Botanical Museum, Harvard University, Cambridge, Massachusetts; Dr. John Freeberg, Associate Professor of Biology, University of Massachusetts, Boston; Dr. Arthur Barclay, ret. Medicinal Plant Resources Laboratory, United States Department of Agriculture, Beltsville, Maryland.

ment of the contributions of researchers and authors whose findings have been incorporated, in abbreviated form, in this production. The alphabetical listing clearly shows who are the principal workers in this field and I trust that this will not only be a source of gratification to these leaders but also serve as a directory of advisors or consultants in the advancement of fruit culture.

The work of assembling the data in this book has been facilitated by the author's over 45 years of familiarity and experimentation with most of the fruits discussed. It was in a small market, Buchanan's, in Grand Central Station in Manhattan, that I bought my first tamarinds and sugar apples prior to World War II; in the genuine Mexican restaurant, the Xochitl, on West 46th Street, that I first ate guava paste and soursop; and in Macy's department store that I purchased a can of "Golden berries" (*Physalis peruviana*) from South Africa. Since those days, these and many others have been personally encountered and consumed, whether in the Philippines, Southeast Asia, South and Central America, southern Mexico, the Caribbean Islands, the Bahamas, or in California and Florida.

This book is dedicated, with great respect and gratitude, to the Florida State Horticultural Society, the oldest such society in this country with continuously published Proceedings. I have been a member since 1945 and have enjoyed the honor and privilege of serving the Society as President and as Chairman of the Executive Committee in the years 1979 and 1980, respectively.

Julia Frances McHugh Morton
October 17, 1986

Date

Most of the dozen or more species of the genus *Phoenix* (family Palmae) are grown as ornamental palms indoors or out. Only the common date, *P. dactylifera* L., is cultivated for its fruit. Often called the edible date, it has few alternate names except in regional dialects. To the French, it is *dattier;* in German, it is *dattel;* in Italian, *datteri,* or *dattero;* in Spanish, *datil;* and, in Dutch, *dadel.* The Portuguese word is *tâmara.*

Description

The date is an erect palm to 100 or 120 ft (30.5-36.5 m), the trunk clothed from the ground up with upward-pointing, overlapping, persistent, woody leaf bases. After the first 6 to 16 years, numerous suckers will arise around its base. The feather-like leaves, up to 20 ft (6 m) long, are composed of a spiny petiole, a stout midrib, and slender, gray-green or bluish-green pinnae 8 to 16 in (20-40 cm) long, and folded in half lengthwise. Each leaf emerges from a sheath that splits into a network of fibers remaining at the leaf base. Small fragrant flowers (the female whitish, the male waxy and cream-colored), are borne on a branched spadix divided into 25 to 150 strands 12 to 30 in (30-75 cm) long on female plants, only 6 to 9 in (15-22.5 cm) long on male plants. One large inflorescence may embrace 6,000 to 10,000 flowers. Some date palms have strands bearing both male and female flowers; others may have perfect flowers. As the fruits develop, the stalk holding the cluster may elongate 6 ft (1.8 m) while it bends over because of the weight. The fruit is oblong, 1 to 3 in (2.5-7.5 cm) long, dark-brown, reddish, or yellowish-brown when ripe with thin or thickish skin, thick, sweet flesh (astringent until fully ripe) and a single, cylindrical, slender, very hard stone grooved down one side.

Origin and Distribution

The date palm is believed to have originated in the lands around the Persian Gulf and in ancient times was especially abundant between the Nile and Euphrates rivers. Alphonse de Candolle claimed that it ranged in prehistoric times from Senegal to the basin of the Indus River in northern India, especially between latitudes 15 and 30. There is archeological evidence of cultivation in eastern Arabia in 4,000 B.C. It was much revered and regarded as a symbol of fertility, and depicted in bas relief and on coins. Literature devoted to its history and romance is voluminous. Nomads planted the date at oases in the deserts and Arabs introduced it into Spain. It has long been grown on the French Riviera, in southern Italy, Sicily and Greece, though the fruit does not reach perfection in these areas. Possibly it fares better in the Cape Verde Islands, for a program of date improvement was launched there in the late 1950's. Iraq has always led the world in date production. Presently, there are 22 million date palms in that country producing nearly 600,000 tons of dates annually. The Basra area is renowned for its cultivars of outstanding quality. The date has been traditionally a staple food in Algeria, Morocco, Tunisia, Egypt, the Sudan, Arabia and Iran. Blatter quotes the writer, Vogel, as stating: "When Abdel-Gelil besieged Suckna in 1824, he cut down no fewer than 43,000 trees, to compel the town to surrender; nevertheless, there are still at least 70,000 left."

In 1980, production in Saudi Arabia was brought to nearly a half-million tons from 11 million palms because of government subsidies, improved technology, and a royal decree that dates be included in meals in government and civic institutions and that hygienically-packed dates be regularly available in the markets. Farmers receive financial rewards for each offshoot of a high-quality date planted at a prescribed spacing. The Ministry of Agriculture has established training courses throughout the country to teach modern agricultural methods, including mechanization of all possible operations in date culture, and recognition and special roles of the many local cultivars. In West Africa, near the Sahara, only dry, sugary types can be grown.

Bonavia introduced seeds of 26 kinds of dates from the Near East into northern India and Pakistan in 1869; and, in 1909, D. Milne, the Economic Botanist for the Punjab, introduced offshoots and established the date as a cultivated crop in Pakistan. The fruits ripen well in northwestern India and at the Fruit Research Center in Saharanpur. In southern India, the climate is unfavorable for date production. A few trees around Bohol in the Philippines are said to bear an abundance of fruits of good quality. The date palm has been introduced into Australia, and into northeastern Argentina and Brazil where it may prosper in dry zones. Some dates are supplying fruits for the market on the small island of Margarita off the coast from northern Venezuela. Seed-propagated dates are found in many tropical and subtropical regions where they are valued as ornamentals but where the climate is unsuitable for fruit production.

Fig. 1: An 8-year-old 'Deglet Noor' date palm in a private garden near Indio, California. Photo'd by Avery Edwin Field, Oct. 1924. In: W.T. Swingle, *Date Growing: a new industry for Southwest States.* U.S. Dept. of Agriculture Yearbook 1926.

In November 1899, 75 plants were sent from Algiers to Jamaica. They were kept in a nursery until February 1901 and then 69 were planted at Hope Gardens. The female palms ultimately bore large bunches of fruits but they were ready to mature in October during the rainy season and, accordingly, the fruits rotted and fell. Only occasionally have date palms borne normal fruits in the Bahamas and South Florida.

Spanish explorers introduced the date into Mexico, around Sonora and Sinaloa, and Baja California. The palms were only seedlings. Still, the fruits had great appeal and were being exported from Baja California in 1837. The first date palms in California were seedlings planted by Franciscan and Jesuit missionaries in 1769. Potted offshoots from Egypt reached California in 1890 and numerous other introductions have been made into that state and into the drier parts of southern Arizona-around Tempe and Phoenix. In 1912, Paul and Wilson Popenoe purchased a total of 16,000 offshoots of selected cultivars in Algeria, eastern Arabia and Iraq and transported them to California for distribution by their father, F.O. Popenoe who was a leader in encouraging date culture in California. It became a profitable crop, especially in the Coachella Valley. There are now about a quarter of a million bearing trees in California and Arizona.

Varieties

It would be impractical to deal in depth with date cultivars here. Paul Popenoe listed 1,500 and provided descriptions of the fruit and palm, as well as the history and significance, of the most important, country by country, in 90 pages of his book, *The Date Palm*, written in 1924 but published in 1973 and readily available. In Iraq, there are presently 450 female cultivars, the most important of which are: 'Zahdi' (43% of the crop; low in price); 'Sayer' (23% of the crop and high-priced); 'Halawi' (13% of the crop and high-priced); 'Khadrawi' (6% of the crop and high-priced); also 'Khastawi', 'Brem', and 'Chipchap'. Sawaya and colleagues (1983) have reported on the sugars, tannins and vitamins in 55 major date cultivars of Saudi Arabia.

The following, with brief comments, are the dates most commonly grown:

'Barhi'—introduced into California in 1913 from Basra, Iraq; nearly cylindrical, light amber to dark-brown when ripe; soft, with thick flesh and rich flavor; of superb quality. For shipment needs refrigeration as soon as picked, then curing and special packing.

'Dayri' (the "Monastery Date")—introduced into California from convent grounds in Dayri, Iraq, in 1913; long, slender, nearly black, soft. Palm requires special care. Not grown extensively in California.

'**Deglet Noor**'—a leading date in Algeria and Tunisia; and in the latter country it is grown in inland oases and is the chief export cultivar. It was introduced into California in 1900 and now constitutes 75% of the California crop. It is semi-dry, not very sweet; keeps well; is hydrated before shipping. Much used for cooking. The palm is high-yielding but not very tolerant of rain and atmospheric humidity.

'**Halawy**' ('Halawi')—introduced into California from Iraq; soft, extremely sweet, small to medium; may shrivel during ripening unless the palm is well-watered. It is especially tolerant of humidity.

'**Hayany**' ('Hayani')—the cultivar most extensively planted in Egypt; but not exported. Introduced into California in 1901, and is sold fresh; is not easy to cure. The fruit is dark-red to nearly black; soft. The palm is one of the most cold-tolerant.

'**Khadrawy**' ('Khadrawi')—important in Iraq and Saudi Arabia, and is grown to some extent in California and Arizona. It is the cultivar most favored by Arabs but too dark in color to be popular on the American market, though it is a soft date of the highest quality. It is early-ripening; does not keep too well. This cultivar is the smallest edible date palm grown in the United States and it is fairly tolerant of rain and humidity.

'**Khastawi**' ('Khustawi'; 'Kustawy')—the leading soft date in Iraq; sirupy, small in size; prized for dessert; keeps well. The palm is large and vigorous and produces its offshoots high on the trunk in California. The fruit is resistant to humidity.

'**Maktoom**'—introduced into California from Iraq in 1902; large, red-brown; thick-skinned, soft, mealy, medium sweet; resistant to humidity.

'**Medjool**'—formerly exported from Morocco; 11 off-shoots imported into California from Bou Denib oases in French Morocco in 1927; is now marketed as a deluxe date in California; is large, soft, and luscious but ships well.

'**Saidy**' ('Saidi')—highly prized in Libya; soft, very sweet; palm is a heavy bearer; needs a very hot climate.

'**Sayer**' ('Sayir')—the most widely grown cultivar in the Old World and much exported to Europe and the Orient; dark orange-brown, of medium size, soft, sirupy, and sometimes some of the sirup is drained out and sold separately; not of high quality but the palm is one of the most tolerant of salt and other adverse factors.

'**Thoory**' ('Thuri')—popular in Algeria; does well in California. Fruit is dry; when cured is brown-red with bluish bloom with very wrinkled skin and the flesh is sometimes hard and brittle but the flavor is good, sweet and nutty. Keeps well; often carried on journeys. The palm is stout with short, stiff leaves; bears heavily, and clusters are very large; somewhat tolerant of humidity.

'**Zahdi**' ('Zahidi')—the oldest-known cultivar, consumed in great quantity in the Middle East; introduced into California about 1900. Of medium size, cylindrical, light golden-brown; semi-dry but harvested and sold in 3 stages: soft, medium-hard, and hard; very sugary; keeps well for months; much used for culinary purposes. The palm is stout, fast-growing, heavy-bearing; drought resistant; has little tolerance of high humidity.

Among the less well-known cultivars in California are:

'**Amir Hajj**'—introduced from Mandali Oasis in Iraq in 1929. The fruit is soft, with thin skin and thick flesh; of superior quality but little grown in the United States.

'**Iteema**'—offshoots from Algeria were introduced into California in 1900. The fruit is large, oblong, light amber, soft, very sweet. Much grown in Algeria but not rain resistant and little grown in California.

'Deglet Noor'

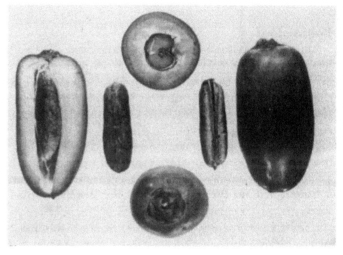

'Halawi'

'Zahdi'

Fig. 2: 'Deglet Noor', (top) a semi-soft date. 'Halawi', (center) a leading export date of Iraq. 'Zahdi', (bottom) a small date from northern Iraq. In: D.W. Albert and R.H. Hilgeman, *Date growing in Arizona*. Bull. 149, U. Arizona, Agr. Exper. Sta., Tucson, Ariz. May 1935.

'Migraf' ('Mejraf')—a very popular cultivar in Southern Yemen. Fruit is light golden-amber, large; of good quality.

In inland oases of Tunisia, in addition to the 'Deglet Noor', there is 'Ftimi' ('Alligue') which is equally subject to humidity, less productive and less disease-resistant.

'Manakhir' has a large fruit and ripens earlier but has the disadvantage that the palm produces few offshoots and its multiplication is limited.

In coastal oases, the main cultivars are 'Kenta', 'Agnioua', 'Bouhatam' and 'Lemsi' which come into season early and ripen before the fall rains. They require less heat than other cultivars. The fruits are more or less dry and the flesh firm.

In all date-growing areas, some confusion is caused if a seed from harvested fruits falls at the base of a select cultivar and the seedling springs up unnoticed among the offshoots. Such seedlings should be watched for and discarded lest they be mistakenly transplanted with the offshoots and later bear fruits of inferior quality.

Pollination

Date pollen is abundant but is not airborne very far. It has become customary to plant one male palm for every 48 or 50 females to provide pollen for artificial pollination which is an ancient practice. In Saudi Arabia and a few other areas of the Old World and in California and Arizona, the long spines on the petioles are first removed to facilitate the pollinating operation. Traditionally, a few strands of open male flowers are put upside-down in a female inflorescence while it is still upright, and a cord is bound around the latter to keep the strands in place when the cluster enlarges and bends downward. However, the pollen can be dried and will keep for 6 months at room temperature. Pollen stored for one year at 8°F (−13.33°C) has given 58% fruit set. Some has been found viable after 7 years of storage, and it is reported that pollen has been kept 14 years in Iran. There are various techniques for applying stored pollen to the female flowers. It may be dusted on by a tractor-drawn, convertible pollen/pesticide machine, or applied with a cotton pad, or sprayed on with a long-tubed applicator or other device. Lack of pollination results in small, seedless fruits. In acute shortages, pollen of another species of *Phoenix* or of some other genera may be used.

Climate

The date palm must have full sun. It cannot live in the shade. It will grow in all warm climates where the temperature rarely falls to 20°F (−6.67°C). When the palm is dormant, it can stand temperatures that low, but when in flower or fruit the mean temperature must be above 64°F (17.78°C). Commercial fruit production is possible only where there is a long, hot growing season with daily maximum temperatures of 90°F (32.22°C) and virtually no rain—less than ½ in (1.25 cm) in the ripening season. The date can tolerate long periods of drought though, for heavy bearing, it has a high water requirement. This is best supplied by periodic flooding from the rivers in North Africa and by subsurface water rather than by rain. (See remarks on irrigation under "Culture").

Soil

The date thrives in sand, sandy loam, clay and other heavy soils. It needs good drainage and aeration. It is remarkably tolerant of alkali. A moderate degree of salinity is not harmful but excessive salt will stunt growth and lower the quality of the fruit.

Propagation

Date palms grow readily from seeds if the seeds and seedlings are kept constantly wet. But seedlings are variable and take 6 to 10 years to fruit. Furthermore, 50% of the seedlings may turn out to be males. The best and common means of propagation is by transplanting the suckers, or offshoots when they are 3 to 5 years old and weigh 40 to 75 lbs (18-34 kg). They are usually separated from the parent palm as needed, but in southern Algeria suckers are often put on sale standing in tubs of water. Some offshoots are maintained in nurseries until roots are formed, though most are set directly in the field after a seasoning period of 10 to 15 days just lying on the ground, in order to lose 12 to 15% of their moisture. In parts of Egypt subject to annual flooding, very large offshoots—up to 500 lbs (226 kg) are planted to avoid water damage. In general, it is said that at least 2 offshoots can be taken from each palm annually for a period of 10 to 15 years. The potential of tissue culture for multiplication of date palms is being explored in Iraq, Saudi Arabia and in California.

Culture

In Tunisia, in former times, it was customary to plant 200 date palms per acre (500/ha). Today, optimum density is considered to be 50 per acre (120/ha) and this is about the standard in the Coachella Valley of California, but small-growing palms may be set much closer. The off-shoots, trimmed back ⅓ or ¼, leaving some of the stiff outer leaves to protect the inner ones, are usually planted 30 to 33 ft (9-10 m) apart each way. The holes should be 3 ft (0.9 m) wide and deep, prepared and enriched several months in advance, and may be encircled by a watering ditch. If the soil dries out prior to planting, the holes are filled with water at that time. In Algeria and Oman, the palms may be set much deeper in order to be closer to ground water, but this may result in drowning the palms when irrigating or they may be smothered by sandstorms.

Planting may be done at any time of year, but most often takes place in spring or fall. In Tunisia planting is done in April and May. The base is set vertically in the ground and the curving fronds will gradually assume an upright position, especially if the concave side is set to face south. Most plants will root in 2 months if the soil is kept constantly moist, while some may be delayed for a year or even several years before they show vigorous growth. Some growers expect a loss of 25% of the off-shoots. Formerly, the young plants in nursery rows were wrapped nearly to the top with old leaves, paper or burlap sacks for the first year to prevent dehydration by cold, heat or wind. But it is now held that such wrapping interferes with the proper development of the leaves.

The offshoots that survive may begin to bloom in 3 years and fruit a year later but a substantial crop is not possible before the 5th or 6th year. In 8 or 10 years, the date will attain full production and it will keep on for a century though productivity declines after 60 to 80 years and also the flowers will be too high to pollinate and the fruits too high to pick. The palm grows at the rate of 1 to 1½ ft (30-45 cm) a year and can reach 20 ft in 15 to 20 years depending on the cultivar and soil and water conditions.

In Iraq, date palms are fertilized once a year with manure at the rate of 44 lbs (20 kg) per tree. Commercial fertilizers are utilized in Saudi Arabia and the United States. Of more importance is the supply of water, a large amount being necessary and it is usually supplied by irrigation ditches. In some Old World plantations rising tides cause rivers to flood the ditches twice a day. Where this natural irrigation does not occur, the palms are watered 15 to 40 times a year. Overhead moisture (including rain) during fruit development will cause minute cracks (checking), beginning at the apex of the fruit which ultimately darkens. In California, the fruit clusters are covered with paper bags to shelter them from rain, dust, and predators.

The female inflorescences may be shortened, thinned out, or some removed entirely at pollinating time, or several weeks later when the stalk has drooped lower, in order to conserve the palm's energy for the following season. Some growers advise leaving no more than 12 bunches per palm. Many leave only 30 strands per cluster, each with about 30 fruits. Without thinning, fruits would be borne only every other year. During the pollinating operation, a grower may tie the elongating flower stalk to a palm frond to prevent breaking when later laden with fruit.

The palms are pruned twice a year, dry fronds being removed in the fall and the leaf bases may be taken off in the spring in order that their fiber may be used as a substitute for coir.

In Iraq, growth regulators have been experimentally applied to developing dates. In 'Zahdi' and 'Sayer', naphthaleneacetic acid, at 60 ppm, applied 15 to 16 weeks after pollination, improved quality and increased fruit weight by 39%. Moisture content was elevated. Ripening was delayed for 30 days or more.

In the Old World, most date plantations are intercropped with vegetables, cereals or fodder crops in the first few years and subsequently with low-growing fruit trees or grapevines. Some authorities hold that this practice distracts the grower from proper care of the dates. In mechanized plantations, intercropping is not possible inasmuch as space must be left for the mobile equipment.

Yield

Ordinarily, in palms 5 to 8 years old, the first crop will be 17.5 to 22 lbs (8-10 kg) per palm; at 13 years, 132 to 176 lbs (60-80 kg). Some improved cultivars, at high densities, have yielded over 220 lbs (100 kg) per year. 'Deglet Noor' in California may yield 4.5 to 7 tons per acre (11-17 tons/ha).

Harvesting and Ripening

Some high-quality dates are picked individually by hand, but most are harvested by cutting off the entire cluster. In North Africa, the harvesters climb the palms, use forked sticks or ropes to lower the fruit clusters, or they may pass the clusters carefully down from hand to hand. Growers in California and Saudi Arabia use various mechanized means to expedite harvesting—saddles, extension ladders, or mobile steel towers with catwalks for pickers. All fruits in a cluster and all clusters on a palm do not ripen at the same time. A number of pickings may have to be made over a period of several weeks. In the Coachella Valley, dates ripen from late September through December and there are 6 to 8 pickings per palm.

Dates go through 4 stages of development: 1) Chimri, or Kimri, stage, the first 17 weeks after pollination: green, hard, bitter, 80% moisture, 50% sugars (glucose and fructose) by dry weight; 2) Khalal stage, the next 6 weeks: become full-grown, still hard; color changes to yellow, orange or red, sugars increase, become largely sucrose; 3) Rutab stage, the next 4 weeks: half-ripe; soften, turn light-brown; some sucrose reverts to reducing sugar which gains prominence; 4) Tamar stage: ripe; the last 2 weeks; in soft dates, the sugar becomes mostly reducing sugar; semi-dry and dry dates will have nearly 50% each of sucrose and reducing sugars.

Soft dates may be picked early while they are still light-colored. Semi-dry dates may be picked as soon as they are soft and then ripened artificially at temperatures of 80° to 95°F (26.67°-35°C), depending on the cultivar. Dry dates may be left on the palm until they are fully ripe. Dry dates that have become too dehydrated and hardened on the palm are rehydrated by soaking in cold, tepid or hot water, or by exposure to steam or a humid atmosphere. Extremely dry weather will cause dates to shrivel on the palm. In the Sudan, the fruits are picked when just mature and then are ripened in jars to prevent so much loss of moisture. Rain, high humidity or cool temperatures during the maturing period may cause fruit drop or checking, splitting of the skin, darkening, black-nose, imperfect maturation, and excessive moisture content, or even rotting. Under such adverse weather conditions, as may occur in the Salt River Valley, Arizona, dates must be harvested while still immature and ripened artificially. In the Old World, there are many different methods of doing this: storing in earthen jars, placing the jars in sun hot enough to prevent spoilage, boiling the fruits in water and then sun-drying. In Australia, entire clusters are kept under cover with the cut end of the stalk in water until the fruits are fully ripe. In modern packing houses, prematurely harvested dates are ripened in controlled atmospheres, the degrees of temperature and humidity varying with the nature of the cultivar.

Where there is low atmospheric humidity outdoors and adequate sunshine, harvested dates are sun-dried whole or cut in half. For fresh shipment in California, the normally ripe, harvested fruits are carried to packing plants, weighed, inspected by agents of the United States Department of Agriculture, fumigated, cleaned, graded, packed, stored under refrigeration, and released to markets according to demand. Saudi Arabia has constructed a number of extra-modern processing plants for fumigation, washing, drying, and packing of dates prior to cold storage.

Keeping Quality

Slightly underripe 'Deglet Noor' dates will keep at 32°F (0°C) up to 10 months; fully mature, for 5 to 6 months. Freezing will extend the storage life for a much longer period. In India, sun-dried dates, buried in sand, have kept well for 1½ years and then have been devoured by worms.

Pests and Diseases

Unripe fruits are attacked by *Coccotrypes dactyliperda* which makes them fall prematurely. Ripe fruits are often infested by nitidulids—*Carpophilus hemipterus, C. multilatus (C. dimidiatus), Urophorus humeralis,* and *Heptoncus luteolus,* which cause decay. Control by insecticides is necessary to avoid serious losses. In Israel, the fruit clusters are covered with netting to protect them from such pests as *Vespa orientalis, Cadra figulilella* and *Arenipes sabella* as well as from depredations by lizards and birds.

In Pakistan, the red weevil, or Indian palm weevil, *Rhynchophorus ferrugineus,* bores into the leaf bases at the top of the trunk, causing the entire crown to wither and die. The rhinocereus beetle, or black palm beetle, *Oryctes rhinocerus,* occasionally attacks the date. Its feeding damage may provide entrance-ways for the weevil. Scale insects may infest the leaves and the trunk. They have been controlled by trimming off the heavily infested leaves, spraying the remaining ones, and treating the fire-resistant trunk with a blowtorch. Two of the most destructive scales are the Marlatt scale, *Phoenicoccus marlatti,* which attacks the thick leaf bases, and the Parlatoria scale, *Parlatoria blanchardii,* which is active in summer. The latter was the object of an eradication campaign in California and Arizona in the late 1930's. The date mite scars the fruits while they are still green.

A tineid moth and a beetle, *Lasioderma testacea,* have damaged stored dates in the Punjab. Dates held in storage are subject to invasion by the fig-moth, *Ephestia cautella,* and the Indian meal-moth, *Plodia interpunctella.*

Fusarium albedinis causes the disastrous Bayoud, or Baioudh, disease in Morocco and Algeria. It is evidenced by a progressive fading and wilting of the leaves. Over a 9-year study period of 26 resistant varieties in Morocco, Bayoud disease reduced the planting density from 364 palms per acre (900/ha) to 121 to 142 per acre (300-350/ha). It is because of this disease that 'Medjool' can no longer be grown commercially in Morocco and Algeria.

Decay of the inflorescence is caused by *Manginiella scaeltae* in humid seasons. Several brown stains will be seen on the unopened spathe and the pedicels of the opened cluster will be coated with white "down". Palm leaf pustule, small, dark-brown or black cylindrical eruptions exuding yellow spores, resulting from infestation by the fungus *Graphiola phoenicis,* is widespread but often a serious problem in Egypt. Date palm decline may be physiological or the result of a species of the fungus genus *Omphalia.* Diplodia disease is a fungus manifestation on leafstalks and offshoots and it may kill the latter if not controlled. The fungus-caused condition called "black scorch" stunts, distorts and blackens leaves and adjacent inflorescences. Other fungus diseases include pinhead spot (*Diderma effusum*), gray blight (*Pestalotia palmarum*) and spongy white rot (*Polyporus adustus*). The date, as well as its relative, *Phoenix canariensis* Hort. ex Chaub., has shown susceptibility to lethal yellowing in Florida and Texas. No commercial plantings have been affected.

Food Uses

Dry or soft dates are eaten out-of-hand, or may be seeded and stuffed, or chopped and used in a great variety of ways: on cereal, in pudding, bread, cakes, cookies, ice cream, or candy bars. The pitting may be done in factories either by crushing and sieving the fruits or, with more sophistication, by piercing the seed out, leaving the fruit whole. The calyces may be mechanically removed also. Surplus dates are made into cubes, paste, spread, powder (date sugar), jam, jelly, juice, sirup, vinegar or alcohol. Decolored and filtered date juice yields a clear invert sugar solution. Libya is the leading producer of date sirup and alcohol.

Cull fruits are dehydrated, ground and mixed with grain to form a very nutritious stockfeed. Dried dates are fed to camels, horses and dogs in the Sahara desert. In northern Nigeria, dates and peppers added to the native beer are believed to make it less intoxicating. The First International Date Conference was held in Tripoli, Libya in 1959, and led to the development of a special program under the Food and Agriculture Organization of the United Nations to promote the commercial utilization of substandard or physically defective dates.

Young leaves are cooked and eaten as a vegetable, as is the terminal bud or heart, though its removal kills the palm. In India, date seeds are roasted, ground, and used to adulterate coffee. The finely ground seeds are mixed with flour to make bread in times of scarcity.

In North Africa, Ghana and the Ivory Coast, date palms are tapped for the sweet sap which is converted into palm sugar, molasses or alcoholic beverages, but each palm should not be tapped more than 2 or 3 times. Tapping the edible date palm interferes with fruit production and it is wiser to tap *P. sylvestris,* which is not valued for its fruit, or some other of the 20 well-known palm species exploited for sugar. When the terminal bud is cut out for eating, the cavity fills with a thick, sweet fluid (called *lagbi* in India) that is drunk for refreshment but is slightly purgative. It ferments in a few hours and is highly intoxicating. Fresh spathes, by distillation, yield an aromatic fluid enjoyed by the Arabian people.

Food Value Per 100 g of Edible Portion*		
	Fresh, uncooked	*Dried*
Calories	142	274-293
Moisture	31.9-78.5 g	7.0-26.1
Protein	0.9-2.6 g	1.7-3.9 g
Fat	0.6-1.5 g	0.1-1.2 g
Carbohydrates	36.6 g	72.9-77.6 g
Fiber	2.6-4.5 g	2.0-8.5 g
Ash	0.5-2.8 g	0.5-2.7 g
Calcium	34 mg	59-103 mg
Phosphorus	350 mg	63-105 mg
Iron	6.0 mg	3.0-13.7 mg
Potassium	?	648 mg
Vitamin A (β-carotene)	110-175 mcg	15.60 mg
Thiamine	?	0.03-0.09 mg
Riboflavin	?	0.10-0.16 mg
Niacin	4.4-6.9 mg	1.4-2.2 mg
Tryptophan	?	10-17 mg
Ascorbic Acid	30 mg	0

*Based on standard analyses.

Sawaya *et al.*, in their studies of fresh dates in Saudi Arabia, reported ascorbic acid content as 1.8-14.3 mg/100 g in the Khalal stage; 1.1-6.1 in the Tamar state. They found that vitamin A ranged from 20 to 1,416 I.U. in the Khalal stage; from 0-259 I.U. in the Tamar stage. Tannin varied from 1.2 to 6.7% in the Khalal stage, 0.6 to 3.2% in the Tamar stage.

The sap contains 10% sucrose. Jaggery made from it contains 9.6% moisture, 86.1% carbohydrates, 1.5% protein, 0.3% fat, 2.6% minerals, 0.36% calcium, 0.06% phosphorus.

Other Uses

Seeds: Date seeds have been soaked in water until soft and then fed to horses, cattle, camels, sheep and goats. Dried and ground-up, they are now included in chicken-feed. They contain 7.17-9% moisture, 1.82-5.2% protein, 6.8-9.32% fat, 65.5% carbohydrates, 6.4-13.6% fiber, 0.89-1.57% ash, also sterols and estrone, and an alkali-soluble polysaccharide. The seeds contain 6 to 8% of a yellow-green, non-drying oil suitable for use in soap and cosmetic products. The fatty acids of the oil are: lauric, 8%; myristic, 4%; palmitic, 25%; stearic, 10%, oleic, 45%, linoleic, 10%; plus some caprylic and capric acid. Date seeds may also be processed chemically as a source of oxalic acid, the yield amounting to 65%. In addition, the seeds are burned to make charcoal for silver-smiths, and they are often strung in necklaces.

Leaves: In Italy, there are some groves of date palms maintained solely to supply the young leaves for religious use on Palm Sunday. In Spain, only the leaves of male palms are utilized for this purpose. In North Africa, the leaves have been commonly used for making huts. Mature leaves are made into mats, screens, baskets, crates and fans. The processed leaflets, combined with ground-up peanut shells and corn cobs, are used for making insulating board. The leaf petioles have been found to be a good source of cellulose pulp. Dried, they are used as walking sticks, brooms, fishing floats, and fuel. The midribs are made into baskets. The leaf sheaths have been prized for their scent. Fiber from the old leaf sheaths is used for various purposes including pack-saddles, rope, coarse cloth and large hats. It has been tested as material for filtering drainage pipes in Iraq, as a substitute for imported filters. Analyses of the leaves show: 0.4-0.66% nitrogen; 0.025-0.062% phosphorus; 0.33-0.66% potassium; 10-16.4% ash. There is some coumarin in the leaves and leaf sheaths.

Fruit clusters: The stripped fruit clusters are used as brooms. The fruit stalks contain 0.28-0.42% nitrogen, 0.017-0.04% phosphorus; 3.46-4.94% potassium; 7.7-9.88% ash.

Fruits: In Pakistan, a viscous, thick sirup made from the ripe fruits, is employed as a coating for leather bags and pipes to prevent leaking.

Wood: Posts and rafters for huts are fashioned of the wood from the trunk of the date palm, though this wood is lighter than that of the coconut. It is soft in the center and not very durable. That of male trees and old, unproductive females is readily available and used for aqueducts, bridges and various kinds of construction, also parts of dhows. All left-over parts of the trunk are burned for fuel.

Medicinal Uses: The fruit, because of its tannin content, is used medicinally as a detersive and astringent in intestinal troubles. In the form of an infusion, decoction, sirup or paste, is administered as a treatment for sore-throat, colds, bronchial catarrh. It is taken to relieve fever, cystisis, gonorrhea, edema, liver and abdominal troubles. And it is said to counteract alcohol intoxication.

The seed powder is an ingredient in a paste given to relieve ague.

A gum that exudes from the wounded trunk is employed in India for treating diarrhea and genito-urinary ailments. It is diuretic and demulcent. The roots are used against toothache. The pollen yields an estrogenic principle, estrone, and has a gonadotropic effect on young rats.

Pejibaye (Plates I and II)

Highly regarded today as a source of nutritious food, the pejibaye, *Bactris gasipaes* HBK. (syns. *B. speciosa* Karst.; *Guilielma gasipaes* L.H. Bailey; *G. speciosa* Mart.; *G. utilis* Orst.), family Palmae, is also called peach palm. It is known as *pejivalle* in Costa Rica; peachnut, *pewa* or *pupunha* in Trinidad; *piva* in Panama; *cachipay, chichagui, chichaguai, contaruro, chonta, chontadura, chenga, jijirre, pijiguay, pipire, pirijao, pupunha,* or *tenga* in Colombia; *bobi, cachipaes, macanilla, melocoton, pichiguao, pihiguao, pijiguao, piriguao,* or *pixabay* in Venezuela; *comer, chonta,* and *tempé* in Bolivia; *chonta dura, chonta ruru, pijuanyo, pifuayo, sara-pifuayo, pisho-guayo* in Peru; *amana,* in Surinam; *parepon* in French Guiana; *popunha* in Brazil.

Description

The palm is erect, with a single slender stem or, more often, several stems to 8 in (20 cm) thick, in a cluster; generally armed with stiff, black spines in circular rows from the base to the summit. There are occasional specimens with only a few spines. The pejibaye attains a height of 65 to 100 ft (20-30 m) and usually produces suckers freely. The leaves, with short, spiny petioles, are pinnate, about 8 to 12 ft (2.4-3.6 m) long, with many linear, pointed leaflets to 2 ft (60 cm) long and 1¼ in (3.2 cm) wide; dark-green above, pale beneath, spiny on the veins. The inflorescence, at first enclosed in a spiny spathe, is composed of slender racemes 8 to 12 in (20-30 cm) long on which the yellowish male and female flowers are mingled except for the terminal few inches where there are only male flowers.

The fruit, hanging in clusters of 50 to 100 or sometimes as many as 300, weighing 25 lbs (11 kg) or more, is yellow to orange or scarlet, yellow-and-red, or brownish at first, turning purple when fully ripe. It is ovoid, oblate, cylindrical or conical, 1 to 2 in (2.5-5 cm) long, cupped at the base by a green, leathery, 3-pointed calyx. A single stem may bear 5 or 6 clusters at a time. The skin is thin, the flesh yellow to light-orange, sweet, occasionally with a trace of bitterness, dry and mealy. Some fruits are seedless. Normally there is a single conical seed ¾ in (2 cm) long, with a hard, thin shell and a white, oily, coconut-flavored kernel. Rarely one finds 2 fused seeds.

Origin and Distribution

This useful palm is apparently indigenous to Amazonian areas of Colombia, Ecuador, Peru and Brazil, but it has been cultivated and distributed by Indians from ancient times and is so commonly naturalized as an escape that its natural boundaries are obscure. Of prehistoric introduction into Costa Rica, it is plentiful in a seemingly wild state of the Atlantic side of that country and also much cultivated. Every Indian dwelling has a patch of pejibaye palms. The palm has also been planted as partial shade for coffee. It is not as common anywhere else in Central America, though it is fairly abundant in Nicaragua, Honduras and Guatemala, and has long been grown in commercial plots in Panama to furnish fruits for local markets. In Colombia and Peru, great quantities of the fruits appear in the markets and vendors sell them along the streets. There are large stands of this palm in the Orinoco region of Venezuela and equatorial Brazil. The Indians of Colombia and Ecuador hold festivals when the pejibayes are in season, though in the latter country the fruits are valued more as feed for livestock than as food for humans.

The United States Department of Agriculture received seeds from Costa Rica in 1920 (S.P.I. #50679), but those in the first lot had lost viability. The United Fruit Company shipped whole fruits but they fermented en route and were mistakenly thrown overboard at New York, the stevedores not being aware that they were imported only for their seeds. Another shipment was made with adequate instructions and 1,000 seedlings were grown in greenhouses in Maryland and distributed. Today there are scattered specimens in southern Florida, Cuba, Puerto Rico and Trinidad. The palm was introduced into the Philippines in 1924. In the 1970's, the possibility of growing pejibayes in India was inspired by settlers of East Indian lineage in Trinidad and South America who produce and sell the fruits. In 1978, Brazilian horticulturists undertook a study to determine the feasibility of establishing pejibaye plantations in the State of Sao Paulo with a view to exploiting the fruit and the terminal bud (heart, or *palmito*). There has been much interest generated in recent years in the cultivation of the palm solely for its hearts which are of high quality. Costa Rica is a leader in this enterprise and there the hearts are being canned commercially.

Varieties

There is much variation in form, size, color and quality of the fruits. Some with longitudinal scars (*pejibaye rayado*) are considered of superior quality. These scars indicate low water content, firmness and a minimum of fiber in the flesh. In Costa Rica there are palms that bear clusters having a majority of seedless fruits. These are called *pejibaye macho* (male pejibaye) and are much prized. It has been found in surveys that only 30 to 60 palms in a seedling planting of 400 will yield highgrade fruit. As many as 100 may yield fruit of such low quality that it is not marketable for human consumption.

In recent years, germplasm collections have been initiated in Costa Rica, Panama, Colombia and Brazil, and there is a great potential for crop improvement and standardization. Spineless forms (*tapire*), especially, are being sought for breeding purposes.

Climate

The pejibaye requires a tropical climate. It is generally restricted to elevations below 6,000 ft (1,800 m). Fruiting is reduced above 5,000 ft (1,500 m). The ideal average annual temperature ranges between 64.4° and 75.2°F (18°-24°C). At low elevations with excessive rainfall, the palm cannot succeed. Optimum rainfall is 78 to 156 in (200-400 cm), rather evenly distributed the year around.

Soil

The palm does well even on poor soils but thrives best on fertile, well-drained land. In a favorable producing region of Costa Rica, the soil varies from clay loam to nearly pure clay. However, riparian, alluvial soils are deemed most desirable.

Propagation

The pejibaye is grown from seed or from suckers. Seeds can be shade-dried for a few hours, packed in moist sphagnum moss or charcoal and shipped to any part of the world. When planted, they will germinate in 3 months. Young plants must be protected from ants which will destroy the tender shoots.

Culture

The palm grows rapidly and reaches 43 ft (13 m) in 10 to 15 years. At low altitudes, seedlings begin to bear in 6 to 8 years. In cool regions, bearing may not begin until the plant is 10 to 12 years old. Productive life is said to be 50 to 75 years.

In fruit plantations, the palms are set 20 ft (6 m) apart. After a few years the suckers emerge and only 2 to 4 are allowed to remain to maturity. When they are 4 to 6 ft (1.2-1.8 m) high and about 3 in (7.5 cm) thick at the base, excess suckers are taken up, cut back severely, kept in the shade and watered until new roots are formed, and then transplanted to new locations. Weeding is done 2 or 3 times a year.

For the production of palm hearts, the spacing is closer, from 5 to 10 ft (1.5-3 m), as the terminal buds can be harvested in 2½ to 3 years. Researchers have found that an application of flurenol (10 ppm) will induce formation of lateral shoots. At 200 ppm, shoot growth is inhibited.

Season

In Colombia, the fruits of cultivated palms mature in January and February. Wild palms may bear twice a year. There are 2 crops a year in Trinidad, one without seeds, the other with seeds. In Costa Rica, the flowers appear in April, May and June in the lowlands, later in the highlands, and fruits mature from September to April.

Harvesting

Because of the spines on the stems, the fruits are knocked down with long poles or harvested with long poles equipped with cutters, unless ladders are available and the bunches can be cut intact and lowered by rope. If the bunch is dropped down, it is caught in a leaf-lined sack held by 2 men, or may land on a deep pile of banana leaves. When the palm gets too tall, the farmer usually cuts it down to obtain the fruits and the heart. If he is fortunate enough to have a number of nearly spineless palms, the spines can be trimmed off and the palm can be climbed. If all the spines are cut off the spiny trunks, the palm will die, but 5 to 8 ft (1.5-2.5 m) of trunk can be despined safely. Special gear of rope and stirrups has been devised to facilitate climbing. Then, too, if the palms have single trunks and are close enough together, the worker need climb only every other tree, using a specially-equipped pole to cut bunches from the neighboring tree. Johannessen (1966) provides details of the modes of handling the crop and the economic role of the pejibaye in the lives of Costa Rican farmers.

In the period 1948 to 1963 in Costa Rica, the harvesting cost was calculated as representing 11.4% of the total cash value of the crop. Hunter (1969) has developed data

Fig. 3: A single-stemmed pejibaye palm (*Bactris gasipaes*), photo'd by the author at Buenaventura, Colombia, in 1969.

showing that, efficiently managed, the pejibaye crop, in terms of financial return to the grower, compares favorably with maize (corn).

Yield

A palm with 4 or 5 stems may produce 150 lbs (68 kg) of fruit in a season.

Keeping Quality

Undamaged, raw fruits keep in good condition in a dry atmosphere with good air circulation for a long time, gradually dehydrating. Roughly handled and bruised fruits ferment in only 3 to 4 days. The cooked fruits, as commonly marketed, can be held for 5 or 6 days. In refrigerated storage at 35.6° to 41°F (2°-5°C), uncooked fruits can be kept for 6 weeks with a minimum of dehydration or spoilage.

Pests and Diseases

In Costa Rica, a stem borer, *Metamasius hemipterus*, sometimes penetrates the stalk of the fruit cluster, causing the fruits to rot. There have been no reports of diseases attacking the palm. Fruits injured during harvesting or transport are soon invaded by rot-inducing fungi.

Food Uses

The fruit is caustic in its natural state. It is commonly boiled; in fact, it is customary to boil the fruits for 3 hours in salted water, sometimes with fat pork added, before marketing. Boiling causes the flesh to separate easily from the seed and usually the skin as well, though in some varieties the skin adheres to the flesh even after cooking. It is only necessary to remove the skin from the cooked flesh which can then be eaten out-of-hand. The pre-boiled fruit is sometimes deep-fried or roasted and served as a snack garnished with mayonnaise or a cheese-dip. It is also mixed with cornmeal, eggs and milk and fried, and is often employed as stuffing for roasted fowl. Occasionally it is made into jam. Oven-dried fruits have been kept for 6 months and then boiled for half an hour which causes them to regain their characteristic texture and flavor. Peeled, seeded, halved fruits, canned in brine, have been exported to the United States. Dried fruits can be ground into flour for use in various dishes. A strong alcoholic drink is made by allowing the raw, sugared flesh to stand for a few days until it ferments. This is prohibited in some parts of tropical America.

Young flowers may be chopped and added to omelettes. The cooked seeds are eaten like chestnuts but are hard and considered difficult to digest.

The palm heart is excellent raw or cooked. It is served in salads or prepared with eggs and vegetables in a casserole. It is a traditional food of the Indians and its harvesting has greatly reduced the stands of wild palms.

Food Value

One average pejibaye fruit contains 1,096 calories. Analyses made in Honduras and Costa Rica show the following values for 100 g of ripe flesh and skin combined:

Food Value Per 100 g of Edible Portion	
Moisture	36.4-60.9 g
Protein	0.340-0.633 g
Fat	3.10-8.17 g
Crude Fiber	0.8-1.4 g
Ash	0.72-1.64 g
Calcium	8.9-40.4 mg
Phosphorus	33.5-55.2 mg
Iron	0.85-2.25 mg
Carotene	0.290-2.760 mg
Thiamine	0.037-0.070 mg
Riboflavin	0.099-0.154 mg
Niacin	0.667-1.945 mg
Ascorbic Acid	14.8-41.4 mg

The protein contains 7 of the 8 essential amino acids: threonine 2.5%/g/N; valine, 2.7%; methionine, 1.3%; isoleucine, 1.7%; leucine, 2.6%; phenylalanine, 1.3%; lysine, 4.6%; and 10 others. Tests for tryptophan have given negative results.

The following approximate values are shown for the seed kernels per 100 g: moisture (loss at 212°F [100°C]), 6.9%; protein (N x 6.25), 8.8%; fat, 31.3%; crude fiber, 18.2%; starch (by acid hydrolysis), 20.8%; ash, 1.9%; undetermined material, 12.1%.

Other Uses

Fruit: Excess fruits and peelings are used as feed for poultry and pigs.

Leaves: Leaflets stripped off for better visibility in harvesting are fed to livestock. The leaves have been important for thatching huts.

Sap: The trunk may be tapped for sap which is fermented into wine.

Bark: The bark is peeled off in one piece, despined and used like canvas to make a substitute for a flat spring in a crude bed or bunk.

Wood: The dark-brown wood is very hard but elastic and takes a good polish. It has been used for spears, Indian sabres, bows, arrowheads, staffs and walking-sticks. More modern uses are siding for houses, veneer and tool handles. Small pieces are fashioned into spindles and other parts used in weaving. Split trunks are used as water troughs.

Ceriman

Of the many aroids (members of the family of Araceae) that are cultivated as ornamental plants, only this one has been grown as well for its fruit. The ceriman, *Monstera deliciosa* Liebm. (syn. *Philodendron pertusum* Kunth & Bouché), is often called merely monstera and, inappropriately, false breadfruit. Because of the apertures in its leaves, some have called it Swiss-cheese plant, or hurricane plant, suggesting that the holes and slits permit the wind to pass through without damaging the foliage. Generally, in Mexico and other Latin American countries it is known as *piñanona,* or *piña anona,* but in Venezuela it is called *ojal* or *huracán;* in Colombia, *hojadillo;* in Guatemala, *harpón* or *arpón común.* In Guadeloupe it is *caroal, liane percée,* or *liane franche;* in Martinique, *siguine couleuvre;* in French Guiana, *arum du pays* or *arum troud.* In Brazil it is catalogued by a leading nursery as *ananas japonez* (Japanese pineapple).

Description

The plant is a fast-growing, stout, herbaceous vine spreading over the ground and forming extensive mats if unsupported, but climbing trees to a height of 30 ft (9 m) or more. The stems are cylindrical, heavy, 2½ to 3 in (6.25-7.5 cm) thick, rough with leaf scars, and producing numerous, long, tough aerial roots. The leathery leaves, on stiff, erect, flattened petioles to 3½ ft (105 cm) long, are oval, cordate at the base, to 3 ft (90 cm) or more in length and to 2¾ ft (82.8 cm) wide; deeply cut into 9-in (22.8 cm) strips around the margins and perforated on each side of the midrib with elliptic or oblong holes of various sizes.

Several inflorescences arise in a group from the leaf axils on tough, cylindrical stalks. The cream-colored spadix, sheltered at first by a waxy, white, calla-lily-like spathe, develops into a green compound fruit 8 to 12 in (20-30 cm) or more in length and 2 to 3½ in (5-8.75 cm) thick, suggesting an ear of corn. The thick, hard rind, made up of hexagonal plates or "scales", covers individual segments of ivory-colored, juicy, fragrant pulp much like diced pineapple. Between the segments there are thin, black particles (floral remnants). Generally there are no seeds, but sometimes, pale-green, hard seeds the size of large peas, may occur in a dozen or so of the segments.

Origin and Distribution

The ceriman is native to wet forests of southern Mexico, Guatemala and parts of Costa Rica and Panama. It was introduced into cultivation in England in 1752; reached Singapore in 1877 and India in 1878. Specimens of the fruit were exhibited by the Massachusetts Horticultural Society in 1874 and 1881. It has become familiar as an ornamental in most of the warm countries of the world and is widely used in warm and temperate regions as a potted plant indoors, —especially in conservatories and greenhouses—though it does not bloom nor fruit in confinement. In Guatemala, it is raised in pots in patios to prevent too rampant growth, as it is apt to become an aggressive nuisance.

The fruits are marketed to some extent in Queensland and, in the past, were sometimes shipped from Florida to gourmet grocers in New York and Philadelphia.

Climate

The ceriman is strictly tropical and cannot tolerate frost. It does best in semi-shade and has a high moisture requirement.

Soil

The plant grows vigorously in almost any soil, including limestone but flourishes best in well-drained, rich loam. It is not adapted to saline conditions.

Propagation

In some European nurseries, the ceriman is raised from imported seed. Rapid multiplication has been achieved through tissue culture in Denmark. Generally, propagation is by means of stem cuttings, which may be simply set in beds or pots in the ground where the vine is intended to grow. Suckers or offshoots, with or without roots, can be separated from parent plants and transplanted successfully. Mulching is desirable as well as watering until new roots have become well-established.

Culture

Suckers will fruit in 2 to 4 years; cuttings in 4 to 6 years, depending on the location, soil and attention given. Out-of-doors, the ceriman requires little care. If it is desired to expedite growth and fruiting, a complete fertilizer may be applied 3 or 4 times a year. Indoor plants need frequent repotting to accommodate the root system, and they should be set outside at least once a year in direct light.

Fig. 4: The ceriman (*Monstera deliciosa*) in flower and fruit at Palm Lodge Tropical Grove, Homestead, Fla. In: J.F. Morton, *Some Useful and Ornamental Plants of the Caribbean Gardens*, 1955.

Season

Flowering and fruiting overlap because it requires 12 to 14 months from the opening of the inflorescence to the maturity of the fruit. Therefore, there are often unopened inflorescences, immature fruits and ripening fruits together on the same plant. The current year's crop is ripening through summer and fall while the following year's crop is forming beside it.

Harvesting

The rind is always green though it assumes a lighter shade as the fruit matures. The fruit, with at least an inch (2.5 cm) of stem, should be cut from the plant when the tile-like sections of rind separate slightly at the base, making it appear somewhat bulged. At this state, the fruits have been shipped to local or distant markets. If kept at room temperature, the ceriman will ripen progressively toward the apex over a period of 5 or 6 days. The flesh should be eaten only from that portion of the fruit from which the rind segments have so loosened as to be easily flicked off. To ripen the whole fruit at one time, it should be wrapped in paper or plastic, or possibly aluminum foil, as soon as cut from the plant and kept at room temperature until the rind has loosened the entire length of the fruit. At this stage, it will be found that the flesh also falls easily away from the inedible core. Once ripened, the fruit can be kept in the refrigerator in good condition for a week or a little more. Rinsing off the floral remnants improves the appearance of the flesh, but it does cause some loss of juice.

Pests and Diseases

When grown indoors, the plants are subject to infestation by scale insects, mites and mealybugs. Outdoors, they are usually pest-free. However, in dry seasons in Florida, the lubber grasshopper (*Romalea microptera*) has rapidly consumed entire leaves, leaving only the base of the midrib and the petiole. In India, wire cages are placed around developing fruits to protect them from rats, squirrels, monkeys and other creatures.

The following diseases have been recorded in Florida: leaf spot caused by *Leptosphaeria* sp., *Macrophoma philodendri*, *Phytophthora* sp., and *Pseudomonas cichorri*; anthracnose from *Glomerella cingulata*; bacterial soft rot from infection by *Erwinia carotovora*; and root rot caused by *Pythium splendens* and *Rhizoctonia solani*.

Food Uses

Fully ripe pulp is like a blend of pineapple and banana. It may be served as dessert with a little light cream, or may be added to fruit cups, salads or ice cream. Some people cut cross-sections right through the core, creating wheel-like disks that can be held with the thumb and forefinger pinching the "hub" while the edible part is nibbled from the rim. To make a preserve, rinsed segments can be stewed for 10 minutes in a little water, a cup of sugar and a tablespoon of lime juice is then added for each 2 cups of fruit, the mixture is simmered again for 20 minutes and preserved in sterilized jars. Some cooks substitute honey for sugar.

Food Value

Philippine analyses show the following values for the edible portion: calories, 335/lb (737/kg); moisture, 77.88%; protein, 1.81%; fat, 0.2%; sugar, 16.19%; fiber 0.57%; ash, 0.85%.

Toxicity

The oxalic acid, and possibly other unidentified principles, in the unripe fruit, the floral remnants of the ripe fruit, and all parts of the plant, cause oral and skin irritation. Some sensitive individuals claim that even the ripe fruit irritates the throat. It would be well to avoid eating the ceriman in quantity until it is determined that there are no undesirable reactions. Some individuals have experienced urticaria and anaphylaxis after eating ceriman. Some children and adults have reported diarrhea and intestinal gas after consuming the flesh or products made from it.

Other Uses

The aerial roots have been used as ropes in Peru. In Mexico, they are fashioned into coarse, strong baskets.

Medicinal Uses: In Mexico, a leaf or root infusion is taken daily to relieve arthritis. A preparation of the root is employed in Martinique as a remedy for snakebite.

Fig. 5: Compound fruit of the ceriman, fully ripe, with loose segments of rind removed and flesh separated for eating. Black specks are floral remnants.

BROMELIACEAE

Pineapple

The pineapple is the leading edible member of the family Bromeliaceae which embraces about 2,000 species, mostly epiphytic and many strikingly ornamental. Now known botanically as *Ananas comosus* Merr. (syns. *A. sativus* Schult. f., *Ananassa sativa* Lindl., *Bromelia ananas* L., *B. comosa* L.), the fruit has acquired few vernacular names. It is widely called *piña* by Spanish-speaking people, *abacaxi* in the Portuguese tongue, *ananas* by the Dutch and French and the people of former French and Dutch colonies; *nanas* in southern Asia and the East Indes. In China, it is *po–lo–mah*; sometimes in Jamaica, sweet pine; in Guatemala often merely "pine".

Description

The pineapple plant is a terrestrial herb 2½ to 5 ft (.75-1.5 m) high with a spread of 3 to 4 ft (.9-1.2 m); a very short, stout stem and a rosette of waxy, straplike leaves, long-pointed, 20 to 72 in (50-180 cm) long; usually needle-tipped and generally bearing sharp, upcurved spines on the margins. The leaves may be all-green or variously striped with red, yellow or ivory down the middle or near the margins. At blooming time, the stem elongates and enlarges near the apex and puts forth a head of small purple or red flowers, each accompanied by a single red, yellowish or green bract. The stem continues to grow and acquires at its apex a compact tuft of stiff, short leaves called the "crown" or "top". Occasionally a plant may bear 2 or 3 heads, or as many as 12 fused together, instead of the normal one.

As individual fruits develop from the flowers they join together forming a cone-shaped, compound, juicy, fleshy fruit to 12 in (30 cm) or more in height, with the stem serving as the fibrous but fairly succulent core. The tough, waxy rind, made up of hexagonal units, may be dark-green, yellow, orange-yellow or reddish when the fruit is ripe. The flesh ranges from nearly white to yellow. If the flowers are pollinated, small, hard seeds may be present, but generally one finds only traces of undeveloped seeds. Since hummingbirds are the principal pollinators, these birds are prohibited in Hawaii to avoid the development of undesired seeds. Offshoots, called "slips", emerge from the stem around the base of the fruit and shoots grow in the axils of the leaves. Suckers (aerial suckers) are shoots arising from the base of the plant at ground level; those proceeding later from the stolons beneath the soil are called basal suckers or "ratoons".

Origin and Distribution

Native to southern Brazil and Paraguay (perhaps especially the Paraná-Paraguay River) area where wild relatives occur, the pineapple was apparently domesticated by the Indians and carried by them up through South and Central America to Mexico and the West Indies long before the arrival of Europeans. Christopher Columbus and his shipmates saw the pineapple for the first time on the island of Guadeloupe in 1493 and then again in Panama in 1502. Caribbean Indians placed pineapples or pineapple crowns outside the entrances to their dwellings as symbols of friendship and hospitality. Europeans adopted the motif and the fruit was represented in carvings over doorways in Spain, England, and later in New England for many years. The plant has become naturalized in Costa Rica, Guatemala, Honduras and Trinidad but the fruits of wild plants are hardly edible.

Spaniards introduced the pineapple into the Philippines and may have taken it to Hawaii and Guam early in the 16th Century. The first sizeable plantation—5 acres (2 ha)—was established in Oahu in 1885. Portuguese traders are said to have taken seeds to India from the Moluccas in 1548, and they also introduced the pineapple to the east and west coasts of Africa. The plant was growing in China in 1594 and in South Africa about 1655. It reached Europe in 1650 and fruits were being produced in Holland in 1686 but trials in England were not successful until 1712. Greenhouse culture flourished in England and France in the late 1700's. Captain Cook planted pineapples on the Society Islands, Friendly Islands and elsewhere in the South Pacific in 1777. Lutheran missionaries in Brisbane, Australia, imported plants from India in 1838. A commercial industry took form in 1924 and a modern canning plant was erected about 1946. The first plantings in Israel were made in 1938 when 200 plants were brought from South Africa. In 1939, 1350 plants were imported from the East Indies and Australia, but the climate is not a favorable one for this crop.

Over the past 100 years, the pineapple has become one of the leading commercial fruit crops of the tropics. In 1952-53, world production was close to 1,500,000 tons and reportedly nearly doubled during the next decade. Major producing areas are Hawaii, Brazil, Malaysia, Taiwan, Mexico, the Philippines, South Africa and Puerto Rico. By 1968, the total crop had risen to 3,600,000 tons, of which only 100,000 tons were shipped fresh (mainly from Mexico, Brazil and Puerto Rico) and

Fig. 6: A spiny-leaved pineapple in the Supply garden, Homestead, Fla., 1946.

925,000 tons were processed. In the period 1961-66, imports of fresh pineapples into Europe rose by 70%. Soon many new markets were opening. In 1973, the total crop was estimated at 4,000,000 tons with 2.2 million tons processed. The increased worldwide demand for canned fruit has greatly stimulated plantings in Africa and Latin America. For years, Hawaii supplied 70% of the world's canned pineapple and 85% of canned pineapple juice, but labor costs have shifted a large segment of the industry from Hawaii to the Philippines. Because production costs in Hawaii (which are 50% labor) have increased 25% or more, Dole has transferred 75% of its operation to the Philippines, where, in 1983, it employed 10,000 laborers on about 25,000, mostly rented, acres (10,117 ha).

Pineapples were first canned in Malaya by a retired sailor in 1888 and exporting from Singapore soon followed. By 1900, shipments reached a half-million cases. The industry alternately grew and declined, and then ceased entirely for 3½ years during World War II. The Malaysian Pineapple Industry Board was established in 1959. Thereafter there has been steady progress. The pineapple

was a very minor crop in Thailand until 1966 when the first large cannery was built. Others followed. Since then processing and exporting have risen rapidly. In 1977-78 many farmers switched from sugarcane to pineapple. Of the annual production of 1½ million tons, ⅛ is canned as fruit or juice.

South Africa produces 2.7 million cartons of canned pineapple yearly and exports 2.4 million. In addition, 31,000 tons of fresh pineapple are sold on the domestic market and 500,000 cartons exported yearly. As in many areas, pineapple culture existed on a small scale on the Ivory Coast until post WW II when cultural efforts were stepped up. By 1950, annual production amounted to 1800 tons. By 1972, it had risen to 200,000 tons for shipment, fresh or canned, to western Europe. Cameroun's annual production is about 6,000 tons.

In the Azores, pineapples have been grown in greenhouses for many years for export mainly to Portugal and Madeira. They are of luxury quality, carefully tended and blemish-free, graded for uniform size and well padded in each box for shipment.

As of 1971, the ten leading exporters of fresh pineapples were (in descending order): Taiwan (39,621 tons), Puerto Rico, Hawaii, Ivory Coast, Brazil, Guinea, Mexico, South Africa, Philippines and Martinique (5,000 tons). The ten leading exporters of processed pineapples were (in descending order): Hawaii, Philippines, Taiwan, South Africa, Malaysia (Singapore), Ivory Coast, Australia, Ryukyu, Mexico, Thailand (10,500,000 tons).

In Puerto Rico, the pineapple is the leading fruit crop, 95% produced, processed and marketed by the Puerto Rico Land Authority. The 1980 crop was 42,493 tons having a farm value of 6.8 million dollars.

For 250 years, pineapples have been grown in the Bahama Islands. At one time plantings on Eleuthera, Cat Island and Long Island totaled about 12,000 acres. The pineapple was a pioneer crop along the east coast of Florida and on the Keys. In 1860 fields were established on Plantation Key and Merritt's Island. And in 1876 planting material from the Keys was set out all along the central Florida east coast. Shipping to the North began in 1879. In 1910 there were 5000 to 10,000 acres stretching as far north as Ft. Pierce. There were more than a dozen families raising pineapples on Elliott's Key where an average crop was 50,000 to 75,000 dozen fruits, mostly sent by schooner to New York. When the industry was flourishing, Florida shipped to New York, Philadelphia and Baltimore one million crates of pineapples a year from the sandy ridge along the Indian River. It was believed in those days that the pineapple benefitted by closeness to salt water.

Wood-lath sheds roofed with palmetto fronds, Spanish moss or tobacco cloth were constructed to provide shade which promoted vigorous plant growth and high fruit quality. Wood-burning ovens were scattered through the sheds for frost protection in winter. Small, open boxcars operating on steam or horsepower ran on wooden rails the length of the shed to transport loads of fruit to the packing station. In open fields, plants were sheltered by palmetto fronds from mid-December to mid-March. 'Smooth Cayenne' had to be grown in sheds. It was not successful in the open. One early planter on Eden Island moved his farm to the mainland because bears ate the ripe fruits. With the coming of the railroad in 1894, pineapple growing expanded. The 1908-09 crop was 1,110,547 crates. Then Cuban competition for U.S. markets caused prices to fall and many Florida growers gave up. The ridge pineapple fields begain to fail as the humus was exhausted by cultivation. Fertilization was steadily raising the pH too high for the pineapple. World War I brought on a shortage of fertilizer, then several freezes in 1917 and 1918 devastated the industry.

In the early 1930's, the United Fruit Company supplied slips for a new field at White City but the pressure of coastal development soon reduced this to a small patch. Shortly after World War II, a plantation of 'Natal Queen' and 'Eleuthera' was established in North Miami but, after a few years, the operation was shifted inland to Sebring, in Highlands County, Central Florida, where it still produces on a small scale.

Varieties

In international trade, the numerous pineapple cultivars are grouped in four main classes: 'Smooth Cayenne', 'Red Spanish', 'Queen', and 'Abacaxi', despite much variation in the types within each class.

'Smooth Cayenne' or 'Cayenne', 'Cayena Lisa' in Spanish (often known in India, Sri Lanka, Malaysia and Thailand as 'Sarawak' or 'Kew') was selected and cultivated by Indians in Venezuela long ago and introduced from Cayenne (French Guyana) in 1820. From there it reached the Royal Botanical Gardens, Kew, England, where it was improved and distributed to Jamaica and Queensland, Australia. Because of the plant's near-freedom from spines except for the needle at the leaftip, and the size—4 to 10 lbs (1.8-4.5 kg)—cylindrical form, shallow eyes, orange rind, yellow flesh, low fiber, juiciness and rich, mildly acid flavor, it has become of greatest importance worldwide even though it is subject to disease and does not ship well. Mainly, it is prized for canning, having sufficient fiber for firm slices and cubes as well as excellent flavor.

It was the introduction of this cultivar into the Philippines from Hawaii in 1912 that upgraded the Philippine industry from the casual growing of the semi-wild type which was often seedy. There are several clones of 'Smooth Cayenne' in Hawaii which have been selected for resistance to mealybug wilt. It is the leading cultivar in Taiwan. In 1975, the Queensland Department of Primary Industries, after 20 years of breeding and testing, released a dual-purpose cultivar named the 'Queensland Cayenne'. South Africa's Pineapple Research Station, East London, after 20 years of selecting and testing of 'Smooth Cayenne' clones, has chosen 4 as superior especially for the canning industry.

'Hilo' is a variant of 'Smooth Cayenne' selected in Hawaii in 1960. The plant is more compact, the fruit is smaller, more cylindrical; produces no slips but numerous suckers. It may be the same as the 'Cayenne Lisse' strain grown in Martinique and on the Ivory Coast, the fruit of which weighs from 2 to 2¾ lbs (1-1½ kg) and has a very small crown.

'St. Michael', another strain of 'Smooth Cayenne' is the famous product of the Azores. The fruit weighs 5 to 6 lbs (2.25-2.75 kg), has a very small crown, a small core, is sweet with low acidity, and some regard it as insipid when fully ripe.

'Giant Kew', well-known in India, bears a large fruit averaging 6 lbs (2.75 kg), often up to 10 lbs (4.5 kg) and occasionally up to 22 lbs (10 kg). The core is large and its extraction results in too large a hole in canned slices.

'Charlotte Rothschild', second to 'Giant Kew' in size in India, tapers toward the crown, is orange-yellow when ripe, aromatic, very juicy. The crop comes in early. 'Baron Rothschild', a Cayenne strain grown in Guinea, has a smaller fruit 1¾ to 5 lbs (0.8-2 kg) in weight, marketed fresh.

'Perolera' (also called 'Tachirense', 'Capachera', 'Motilona', and 'Lebrija') is a 'Smooth Cayenne' type ranking second to 'Red Spanish' in importance in Venezuela. It has long been grown in Colombia. The plant is entirely smooth with no spine at the leaftip. The fruit is yellow, large—7 to 9 lbs (3-4 kg)—and cylindrical.

'Bumanguesa', of Venezuela and Colombia, is probably a mutation of 'Perolera'. The fruit is red or purple externally, cylindrical with square ends, shallow eyes, deep-yellow flesh, very slender core but has slips around the crown and too many basal slips to suit modern commercial requirements.

'Monte Lirio', of Mexico and Central America, also has smooth leaves with no terminal spine. The fruit is rounded, white-fleshed, with good aroma and flavor. Costa Rica exports fresh to Europe.

Other variants of 'Smooth Cayenne' include the 'Esmeralda' grown in Mexico and formerly in Florida for fresh, local markets; 'Typhone', of Taiwan; 'Cayenne Guadeloupe', of Guadeloupe, which is more disease-resistant than 'Smooth Cayenne'; and 'Smooth Guatemalan' and 'Palin' grown in Guatemala; also 'Piamba da Marquita' of Colombia. Some who have made efforts to classify pineapple strains have proposed grouping all smooth-leaved types under the collective name 'Maipure'. In Amazonas, Venezuela, this name is given to a large plant with smooth leaves stained with red. The fruit has 170 to 190 eyes.

Philipps Platts, a leading pineapple authority, experimented with 60 to 70 cultivars in Florida but **'Red Spanish'** proved most dependable. Despite the spininess of the plant, it still is the most popular among growers in the West Indies, Venezuela and Mexico. 'Red Spanish' constitutes 85% of all commercial planting in Puerto Rico and 75% of the production for the fresh fruit market. It is only fair for canning. The fruit is more or less round, orange-red externally, with deep eyes, and ranges from 3 to 6 lbs (1.36-2.7 kg). The flesh is pale-yellow, fibrous, with a large core, aromatic and flavorful. The fruit is hard when mature, breaks off easily and cleanly at the base in harvesting, and stands handling and transport well. It is highly resistant to fruit rot though subject to gummosis.

Two vigorous hybrids of 'Smooth Cayenne' and 'Red Spanish' were developed at the Agricultural Experiment Station of the University of Puerto Rico and released in 1970—'P.R. 1-56' and the slightly larger 'P.R. 1-67', both with good resistance to gummosis and mealybug wilt and of excellent fruit quality. 'P.R. 1-67' averages 5¾ lbs (2.5 kg), gives a high yield—32 tons per acre (79 tons/ha). The fruit is sweeter yet with more acidity than 'Red Spanish', less fibrous and good for marketing fresh and for canned juice. It was introduced into Venezuela about 1979 and is grown in the State of Lara.

'Cabezona' ('Bull Head', or 'Piña de agua') is a prominent variant (a natural tetraploid) of 'Red Spanish' long grown in Puerto Rico in the semiarid region of Lajas, to which it is well suited; also in El Salvador. The plant is large, over 3 ft (1 m) high; the leaves are gray-green. The fruit is conical but not as tall as that of 'Valera'; averages 4 to 6 lbs (1.8-2.75 kg) and may reach 18 lbs (8 kg) or more. It is orange-yellow at maturity, has few fibers and sweet-acid flesh. The stem is large and extends up into the base of the fruit and if the fruit is broken off when harvested it leaves a cavity. Consequently, it must be cut with a machete and later trimmed flush with the base in the packing house. It is marketed fresh only. It is resistant to gummosis. Platts reported that it gave a low yield and was disease-prone in Florida. There are small plantings in the States of Trujillo and Monagas, Venezuela. It has been cultivated frequently in the Philippines.

'Valera' ('Negrita', or 'Andina'), is an old cultivar originating in Puerto Rico; it is grown in the States of Lara, Merida and Trujillo in Venezuela. It is a small-to-medium plant with long, narrow, spiny, purple-green leaves. The fruit is conical-cylindrical, weighing 3½ to 5½ lbs (1.5-2.5 kg); is purple outside with white flesh.

'Valera Amarilla' is a 'Red Spanish' strain grown in the States of Lara and Trujillo in Venezuela. The fruit is broad-cylindrical and tall with a large crown; weighs 4½ to 9 lbs (2-4 kg); is yellow externally with very deep eyes, about 72 to 88 in number. The flesh is pale-yellow and very sweet in flavor.

'Valera Roja', grown in Lara, Trujillo and Merida, Venezuela, is a small-to-medium plant with cylindrical fruit 1½ to 2.2 lbs (0.6-1 kg) in weight, reddish externally, with 100 eyes. It has pale-yellow flesh.

'Castilla' is a 'Red Spanish' strain grown in Colombia and El Salvador.

'Cumanesa', supposedly a selection of 'Red Spanish', grown mainly in the State of Sucre, Venezuela, is a medium-sized plant, very spiny, producing an oblong fruit with a large crown. It is orange-yellow externally; weighs 2 to 3¾ lbs (0.9-1.70 kg), and has yellowish-white flesh.

'Morada', believed to be a variant of 'Red Spanish', is one of the less-important cultivars of Colombia and the State of Monagas, Venezuela. The plant is large, with long, narrow, purple-red leaves. The fruit is broad-cylindrical, purple-red externally, with white flesh.

'Monte Oscuro' ('Pilon'), is a large plant with broad, saw-toothed, spiny-edged leaves. The fruit is barrel-shaped, large, weighing 6.6 lbs (3 kg); has 160-180 medium-deep eyes; is yellow outside with deep-yellow, fibrous flesh. It is grown among *Mauritia* palms in the State of Monagas, Venezuela.

'Abacaxi' (also called 'White Abacaxi of Pernambuco', 'Pernambuco', 'Eleuthera', and 'English') is well known in Brazil, the Bahamas and Florida. The plant is spiny and disease-resistant. Leaves are bluish-green with red-purple tinge in the bud. The numerous suckers need thinning out. The fruit weighs 2.2 to 11 lbs (1-5 kg), is tall and straight-sided; sunburns even when erect. It is very fragrant. The flesh is white or very pale yellowish, of rich, sweet flavor, succulent and juicy with only a narrow vestige of a core. This is rated by many as the most delicious pineapple. It is too tender for commercial handling, and the yield is low. The fruit can be harvested without a knife; breaks off easily for marketing fresh.

'Sugarloaf' (also called 'Pan de Azucar') is closely related to 'Abacaxi', and much appreciated in Central and South America, Puerto Rico, Cuba and the Philippines. The leaves of the plants and crowns pull out easily and this fact gave rise to the unreliable theory that pineapple ripeness is indicated by the looseness of the leaves. The fruit is more or less conical, sometimes round; not colorful; weighs 1½ to 3 lbs (0.68-1.36 kg). Flesh is white to yellow, very sweet, juicy. This cultivar is too tender for shipping.

Among several strains of 'Sugarloaf' are 'Papelon', and 'Black Jamaica', and probably also 'Montufar' ('Sugar Slice' of Guatemala). The latter fruit is green, conical, weighs 2 to 5½ lbs (0.8-2.5 kg); has yellow, very juicy, flesh, sweet yet a little acid. This pineapple also is too tender to ship.

There are a number of tropical American cultivars not categorized as to groups, and among them are:

'Brecheche', grown to a limited extent in southern Venezuela, is a small fruit with small, spineless crown. Average weight is 1½ to 2.2 lbs (0.7-1 kg). The fruit is yellow externally. Flesh is yellow, with little fiber, small core, very fragrant, very juicy.

'Caicara', grown to a small extent in the State of Bolivar, Venezuela, is a large fruit weighing 4 to 5½ lbs (1.8-2.5 kg), with a large, spiny crown. It is cylindrical conical with deep eyes; yellow externally with white flesh, a little fiber, very juicy, with large core.

'Chocona' and 'Santa Clara' are cultivars that have been introduced into Trinidad.

'Congo Red' is a plant with bright-red, long-lasting flowers. The fruit bends over and cracks in hot, dry weather. It weighs up to 5 lbs (2.25 kg), is waxy, with yellow flesh of good flavor.

'Panare', named after the tribe of Indians that has grown it for a long time, is commercially grown to a small extent in the State of Bolivar, Venezuela. The plant is of medium size with long, spiny leaves. The fruit is bottle-shaped, small, 1 to 1½ lbs (0.45-0.70 kg), with small crown; ovate, with deep eyes; orange externally with deep-yellow flesh; slightly fragrant, with little fiber and small core.

'Santa Marta' of Colombia, is subject to cracking of the core in hot, dry weather.

In Peru, farmers still grow the old common 'Criolla' because it can be sold fresh and is not easily damaged in shipment. But modern pineapple production in that country depends on the 'Smooth Cayenne' for canning.

Minor cultivars in Colombia include: 'Amarilla de Cambao', 'Amarilla de Tocaima', 'Blanca Chocoana', 'Blanca del Atrato', 'Blanca de Valle del Cauca', 'Cimarrona', 'Española de Santander', 'Hartona', 'Jamaiqueña' and 'Manzana'.

'Cacho de Venado' is grown to a small extent in Monagas and Sucre, and 'Injerta' in Trujillo, Venezuela.

'Pearl', 'Itaparica', 'Paulista', and 'Maranhao' (or 'Amarella') are spoken of in Brazil; 'Azucaron' in El Salvador; 'Roja' in Mexico. It remains to be determined if some of these names are merely synonyms for cultivars already referred to.

'Mauritius' (also known as 'European Pine', 'Malacca Queen', 'Red Ceylon' and 'Red Malacca') is one of the 2 leading pineapple cultivars in Malaya; also important in India and Ceylon. The leaves are dark-green with broad red central stripe and red spines on the margins. The fruit is small, 3 to 5 lbs (1.36-2.25 kg), yellow externally; has a thin core and very sweet flesh. It is sold fresh and utilized for juice.

'Singapore Red' (Also called 'Red Jamaica', 'Singapore Spanish', 'Singapore Queen', 'Singapore Common') is second to 'Mauritius' in popularity. The leaves are usually all-green but sometimes have a reddish stripe near the margins; they are rarely spiny except at the tips. The fruits, cylindrical, reddish, with deep eyes, are small—3½ to 5 lbs (1.6-2.25 kg)—with slender core, fibrous, golden-yellow flesh; insipid raw but valued for canning. The plant is disease- and pest-resistant.

The related 'Green Selangor' (also called 'Selangor Green', 'Green Spanish', and 'Selassie') of Malaysia has all-green leaves prickly only at the tips. The flesh is golden-yellow, often with white dots. This cultivar is grown for canning.

'Queen' (also called 'Common Rough' in Australia) is the leading cultivar in South Africa, Queensland and the Philippines. The plant is dwarf, compact, more cold-resistant and more disease-resistant than 'Smooth Cayenne'. It matures its fruit early but suckers freely and needs thinning, and the yield is low. The fruit is conical, deep-yellow, with deep eyes; weighs 1 to 2½ lbs (0.45-1.13 kg); is less fibrous than 'Smooth Cayenne', but more fragrant; it is juicy, of fine flavor with a small, tender core. It is sold fresh and keeps well. It is only fair for canning because of its shape which makes for much waste.

'Natal Queen' of South Africa, also grown in El Salvador, produces many suckers. The fruit weighs 1½ to 2 lbs (0.75-0.9 kg).

'MacGregor', a variant of 'Natal Queen' selected in South Africa and grown also in Queensland, is a spreading, more vigorous plant with broad leaves and large suckers produced less freely. The fruit is cylindrical, medium to large, with firm flesh and flavor resembling 'Queen'.

'James Queen' (formerly 'Z') is a mutation of 'Natal Queen' that originated in South Africa. It has larger fruit with square shoulders.

'Ripley' or 'Ripley Queen', grown in Queensland, is a dwarf, compact plant with crimson tinge on leaves; takes 22 weeks from flowering to fruit maturity; is an irregular bearer. The fruit weighs 3 to 6 lbs (1.36-2.7 kg); is pale-copper externally; flesh is pale-yellow, non-fibrous, very sweet and rich. In Florida this cultivar tends to produce suckers without fruiting.

'Alexandria', a selection of 'Ripley Queen' in Queensland, is more vigorous with large suckers and fruit. The fruit is conical, tender, with 'Ripley Queen' flavor.

'Egyptian Queen' was introduced into Florida in 1870. It was popular at first, later abandoned. The fruit weighs 2 to 4 lbs (0.9-1.8 kg).

'Kallara Local' is a little-known cultivar in India. Minor strains in Thailand are 'Pattavia', 'Calcutta', 'Sri Racha', 'Intorachit' and 'Chantabun'.

In the evaluation of pineapples, the crown can be an asset or a liability. Small crowns detract from the decorative appearance of the fruit; large crowns are more attractive but hamper packing and constitute too great a proportion of inedible material from the standpoint of the purchaser.

Climate

The pineapple is a tropical or near-tropical plant limited (except in greenhouses) to low elevations between 30°N and 25°S. A temperature range of 65°-95°F (18.33-45°C) is most favorable, though the plant can tolerate cool nights for short periods. Prolonged cold retards growth, delays maturity and causes the fruit to be more acid. Altitude has an important effect on the flavor of the fruit. In Hawaii, the 'Smooth Cayenne' is cultivated from sea level up to 2,000 ft (600 m). At higher elevations the fruit is too acid. In Kenya, pineapples grown at 4500 ft (1371 m) are too sweet for canning; between 4500 and 5700 ft (1371-1738 m) the flavor is most suitable for canning; above 5700 ft (1738 m) the flavor is undesirably acid. Pineapples are grown from sea level to 7545 ft (2300 m) in Ecuador but those in the highlands are not as sweet as those of Guayaquil.

Ideally, rainfall would be about 45 in (1,143 mm), half in the spring and half in the fall; though the pineapple is drought-tolerant and will produce fruit under yearly precipitation rates ranging from 25 to 150 in (650-3,800 mm), depending on cultivar and location and degree of atmospheric humidity. The latter should range between 70 and 80 degrees.

Soil

The best soil for pineapple culture is a well-drained, sandy loam with a high content of organic matter and it should be friable for a depth of at least 2 ft (60 cm), and pH should be within a range of 4.5 to 6.5. Soils that are not sufficiently acid are treated with sulfur to achieve the desired level. If excess manganese prevents response to sulfur or iron, as in Hawaii, the plants require regular spraying with very weak sulfate of iron. The plant cannot stand waterlogging and if there is an impervious subsoil, drainage must be improved. Pure sand, red loam, clay loam and gravelly soils usually need organic enrichment. Filter presscake from sugar mills, worked into clay soils in Puerto Rico, greatly enhances plant vigor, fruit yield, number of slips and suckers.

Propagation

Crowns (or "tops"), slips (called nibs or robbers in New South Wales), suckers and ratoons have all been commonly utilized for vegetative multiplication of the pineapple. To a lesser degree, some growers have used "stumps", that is, mother plant suckers that have already fruited. Seeds are desired only in breeding programs and are usually the result of hand-pollination. The seeds are hard and slow to germinate. Treatment with sulfuric acid achieves germination in 10 days, but higher rates of germination (75-90%) and more vigorous growth of seedlings results from planting untreated seeds under intermittent mist.

The seedlings are planted when 15-18 months old and will bear fruit 16-30 months later. Vegetatively propagated plants fruit in 15-22 months.

In Queensland, tops and slips from the summer crop of 'Smooth Cayenne' are stored upside down, close together, in semi-shade, for planting in the fall. Some producers salvage the crowns from the largest grades of fruits going through the processing factory to be assured of high quality planting material.

South African experiments with 'Smooth Cayenne' have shown medium-size slips to be the best planting material. Next in order of yield were large crowns, medium-size suckers, medium-size crowns and large suckers. Medium and large suckers, however, fruited earlier. Trimming of basal leaves increased yields. Workers in Johore, Malaya, report, without specifying cultivar, that large crowns give highest yield and more slips, followed by small crowns, big slips, small slips, large and small suckers in descending order.

With the 'Red Spanish' in Puerto Rico, the utilization of large slips for planting in the first quarter of the year, medium slips during the next six months, and small slips in the final quarter, provides fruits of the maximum size over an extended period of harvest. Storage of slips until optimum planting time prevents premature bloom and diminished fruit size.

The 'Red Spanish' reaches shipping-green stage (one week before coloring begins) in Puerto Rico 150 days after natural blooming.

In South Africa the 'Queen' is grown mainly from stumps, secondly from suckers. The stumps which have fruited are detached from the mother plant as soon as possible to avoid their developing suckers of their own. In comparison with suckers, the stumps are consistently heavier in yield after the 4th crop. When suckers are used, those of medium size, approximately 18 in (45 cm) long, planted shallow and upright, yield best.

In the past, growers preferred plants that supplied abundant basal slips for planting, not recognizing the fact that such plants gave smaller fruits than those without slips or suckers. Also, breeders aim toward elimination of slips to facilitate harvesting. Because of the increased demand for planting material, a new method of mass propagation received wide attention in 1960. During the harvest, plants that have borne single-crowned, superior fruits without basal slips are selected and marked. Following harvest, these plants are cut close to the ground, the leaves are stripped off and the stems—usually 1 to 2 ft (30-60 cm) long and 3 to 4 in (7.5-10 cm) thick— are sliced lengthwise into 4 triangular strips. The strips are disinfected and placed 4 in (10 cm) apart, with exterior side upward, in beds of sterilized soil, semi-shaded and sprinkler-irrigated. Shoots emerge in 3 to 5 weeks and are large enough to transplant to the nursery in 6 to 8 weeks. 'Smooth Cayenne' yields an average of 3 shoots per slice. 'Red Spanish' and 'Natal Queen', 4 per slice.

This use of the stem is a major improvement over the former practice of allowing it to develop suckers high up after the fruit is harvested. If such suckers bear fruit *in situ* they are not strong enough to support it and collapse. They are better removed for planting, but repeated removal of suckers weakens the mother plant.

In Sri Lanka, the shortage of planting material inspired experiments at first utilizing stem cross-sections 1 in (2.5 cm) thick—15 to 24 from each stem. These sprouted in 4 weeks but plant growth was slow and fruiting was delayed for 30 months. Most of the cuttings developed a single sprout, some as many as 5, others, none at all. Accordingly, this technique was abandoned in favor of a system developed for purposes of reproducing a selected strain in Hawaii. Stems are cut into segments bearing 3 to 5 whorls of leaves. The leaves are trimmed to 4 to 5 in (10-12.5 cm) and the disinfected cuttings set upright in beds until each gives rise to one strong plantlet which is then transferred to the nursery.

The butts, or bases, of mother plants, with leaves intact, are laid end-to-end in furrows in nurseries and covered with 2 to 3 in (5-7.5 cm) of soil. Sprouting occurs in 6 to 8 weeks. The butts give an average of 6 suckers each, though some have put forth up to 25. A one-acre (0.4 ha) nursery of 25,000 butts, therefore, yields between 100,000 and 200,000 suckers.

The Pineapple Research Institute in Hawaii has also employed axillary buds at the base of crowns. Each crown segment may develop 20 plantlets. This method has been adopted in Sri Lanka for perpetuating superior strains but not for commercial cultivation because the resulting plants require 24 months or more to fruit.

In India, because of low production of slips and suckers in 'Smooth Cayenne', crown cuttings (15-16 per crown) have been adopted for propagation with 95% success, and this method is considered more economical than the utilization of butts.

Fig. 7: 'Red Spanish' (left) and 'Abacaxi' (called 'English' in the Bahamas) (right). In: K. and J. Morton, *Fifty Tropical Fruits of Nassau*, 1946.

Vegetative propagation does not assure facsimile reproduction of pineapple cultivars, as many mutations and distinct clones have occurred in spite of it.

Culture

The land should be well prepared at the outset because the pineapple is shallow-rooted and easily damaged by post-planting cultivation. Fumigation of the soil contributes to high quality and high yields.

Planting: In small plots or on very steep slopes, planting is done manually using the traditional short-handled, narrow-bladed hoe, the handle of which, 12 in (30 cm) long, is used to measure the distance between plants. Crowns are set firmly at a depth of 2 in (5 cm); slips and suckers at 3½ to 4 in (9-10 cm). Butts, after trimming and drying for several days, are laid end-to-end in furrows and covered with 4 in (10 cm) of soil.

Double-rowing has been standard practice for many years, the plantlets set 10 to 12 in (25-30 cm) apart and staggered, not opposite, in the common rows, and with 2 ft (60 cm) between the two rows. An alley 3, 5½ or 6 ft (.9, 1.6 or 1.8 m) wide is maintained between the pairs, allowing for plant populations of 17,400, 15,800 or 14,500 per acre (42,700, 37,920 or 33,800 per ha) respectively. Close spacing gives highest total crop weight—e.g., 18,000 plants/acre = 28.8 tons (43,200 plants/ha = 69.12 tons). However, various trials have shown that overcrowding has a negative effect, reducing fruit size and elongating the form undesirably, and it reduces the number of slips and suckers per plant. Density trials with 'P.R. 1-67' in Puerto Rico demonstrated that 21,360 plants per acre (51,265/ha) yielded 35.8 tons/acre (86 tons/ha) in the main crop and 18.9 tons/acre (45.43 tons/ha) in the ratoon crop, but only one slip per plant for replanting. Excessively wide spacing tends to induce multiple crowns in 'Smooth Cayenne' in Hawaii and in 'Red Spanish' in Puerto Rico.

Some plantings are mulched with bagasse. In large operations, asphalt-treated paper, or black plastic mulch is regarded as essential. It retards weeds, retains warmth in cool seasons, reduces loss of soil moisture, and can be laid by machines during the sterilization and pre-fertilization procedures. Mulch necessitates removal of basal leaves of crowns, slips and suckers and the use of a tool to punch a hole at the pre-marked planting site for the insertion of each plantlet. The mulch is usually rolled onto rounded beds 3¼ ft (1 m) wide.

Mechanical planting: Research on the potential of machines to replace the hard labor of planting pineapples was begun in Hawaii in 1945. A homemade device was first employed in Queensland in 1953. Early semi-mechanical planters were self-propelled platforms with driver and two men who made the holes in the mulch and set the plants in place. With a 2-row planter, 3 men can set 7,000 plants per hour of operation. Frequent stops are necessary to reload with planting material. With improved equipment, mechanical planting has become standard practice in large plantations everywhere. The most sophisticated machines have attachments which concurrently apply premixed fertilizer and lay a broad center strip of mulch, set the plantlets along each edge, and place a narrow strip along the outer sides. The only manual operation, apart from driving, is feeding of the plantlets to the planting unit. With this system, up to 50,000 plants have been set out per day.

Fertilization: Nitrogen is essential to the increase of fruit size and total yield. Fertilizer trials in Kenya show that a total of 420 lbs N/acre (471.7 kg/ha) in 4 equal applications during the first year is beneficial, whereas no advantage is apparent from added potassium and phosphorus. Puerto Rican studies have indicated that maximum yields are achieved by urea sprays supplying 147 lbs N/acre (151 kg/ha). In Queensland, total yield of mother plants and ratoons was increased 8% by urea spraying. Normal rate of application is 3½ gals (13.3 liters) per 1,000 plants. On acid Bayamon sandy clay in Puerto Rico, addition of magnesium to the fertilizer mix or applying it as a spray (300 lbs magnesium sulfate per acre—327 kg/ha) increased yield by 3 tons/acre (7 tons/ha). On sloping, stony clay loam high in potassium, Queensland growers obtained high yields of 'Smooth Cayenne' from side-dressings of NPK mixture 5 times a year. On poor soils, nitrogen and potassium levels of the plants may become low toward the end of the crop season. This must be anticipated early and suitable adjustments made in the application of nutrients. Potassium uptake is minimal after soil temperatures drop below 68°F (20°C). On fine sandy loam in Puerto Rico, the cultivar 'P.R. 1-67' performed best with 13-3-12 fertilizer applied at the rate of 1.5 tons/acre (3.74 tons/ha). In this experiment, 13,403 plants/acre (32,167/ha) produced 9,882 fruits/acre (23,717/ha), weighing 31.28 tons/acre (75 tons/ha). In Venezuela, 6,250 medium-size fruits per acre (15,000 fruits/ha) is considered a very good crop.

Fruit weight has been considerably increased by the addition of magnesium. In Puerto Rican trials, magnesium treatment resulted in 54% more total weight providing an average of 2.7 more tons/acre (64.8 tons/ha) than in control plots. Fruit size and total yield have been enhanced by applying chelated iron with nitrogen; also, where chlorosis is conspicuous, by accompanying nitrogen with foliar sprays of 0.10% iron and manganese.

Some growers thin out suckers and slips to promote stronger growth of those that remain.

Irrigation: Irrigation is desirable only in dry seasons and should not exceed 1 in (2.5 cm) semi-monthly.

Weed Control: Manual weeding in pineapple fields is difficult and expensive. It requires protective clothing and tends to induce soil erosion. Coir dust has been used as mulch in Sri Lanka to discourage weeds but it has a deleterious effect on the crop, delaying or preventing flowering. The use of paper or plastic mulch and timely application of approved herbicides are the best means of preventing weed competition with the pineapple crop.

Flower Induction: Pineapple flowering may be delayed or uneven, and it is highly desirable to attain uniform maturity and also to control the time of harvest in order to avoid overproduction in the peak periods. In 1874 in the Azores it was accidentally discovered that smoke would bring pineapple plants into bloom in 6 weeks. The realization that ethylene was the active ingredient in the smoke led to the development of other methods.

As far back as 1936, compressed acetylene gas, or a spray of calcium carbide solution (which generates acetylene) were employed to expedite uniform blooming. Some growers have merely deposited calcium carbide in the crown of each plant to be dissolved by rain. A more advanced method is the use of the hormone, a-naphthaleneacetic acid (ANA) or B-naphylacetic acid (BNA)—which induce formation of ethylene. In recent years, B-hydroxyethyl hydrazine (BOH) came into use. Treatment is given when the plants are 6 months old, 3 months before natural flowering time. The plants should have reached the 30-leaf stage at this age.

Spraying of a water solution of ANA on the developing fruit has increased fruit size in 'Smooth Cayenne' in Hawaii and Queensland. In West Malaysia, spraying 'Singapore Spanish' 6 weeks after flowering with Planofix, an ANA-based trade product, delayed fruit maturity, increased fruit size, weight and acidity. Similar results have been seen after hormone treatment of 'Cayenne Lisse' on the Ivory Coast.

Trials with 'Sugarloaf' in Ghana showed calcium carbide and BOH equally effective on 42- to 46-week-old plants, and Ethrel performed best on 35- to 38-week-old plants. 'Sugarloaf' seems to respond 10 days earlier than 'Red Spanish'.

Ethrel, or the more recently developed Ethephon, applied at the first sign of fruit ripening in a field will cause all the fruit to ripen simultaneously. It brings the ratoons into fruit quickly. There is a great saving in harvesting costs because it reduces the need for successive pickings.

Plants treated with naphthaleneacetic acid produce long, cylindrical, pointed fruits, maturing over an extended period of time, ripening first at the base while the apex is still unripe. Ethylene treatment results in a square-shouldered, shorter fruit maturing over a shorter period and ripening more uniformly.

In Puerto Rico, treatment in 'Cabezona' can be done to induce flowering at any time of the year.

Pests

Nematodes (*Rotylenchulus, Meloidogyne, Pratylenchus, Ditylenchus, Helicotylenchus,* and other genera) cause stunting and degeneration in pineapple plants unless soil is fumigated. In Queensland, nematicides have increased yields by 22-40%. Crop rotation has been found effective in Puerto Rico. Turning the field over to Pangola grass (*Digitaria decumbens* Stent.) or green foxtail grass (*Setaria viridis* Beauv.) for 3 years suppresses nematode populations and benefits the soil but may not be practicable unless spare land is available for pineapple culture in the interim.

Mealybugs (*Pseudococcus brevipes* and *P. neobrevipes*) attack leaf bases and cause wilt. The leaves turn orange-brown and wither due to root rot. Prevention requires spraying and dusting to control the fire ants (*Solenopsis* spp.) which carry the mealybugs from diseased to healthy plants. Control is difficult because there are many weeds and other local plants acting as mealybug hosts. Some success was achieved in Florida in combatting mealybugs with the parasitic wasp, *Hambletonia pseudococcina* Comp., though the general use of insecticides limits the activity of the wasp.

The pineapple mite, or so-called red spider (*Dolichotetranychus* (or *Stigmacus*) *floridanus* (Banks) also attacks leaf bases and is troublesome during prolonged droughts, heavily infesting the slips. The pineapple red scale (*Diaspis bromeliae*) has been a minor pest in Florida. Since 1942 this scale has spread to many pineapple districts in southeastern Queensland, with occasional serious infestations. Natural predators afford about 40% control. The palmetto beetle (*Rhynchophorus cruentatus*), which feeds on palm logs, enters the bud and lays eggs in young fruits and the fruit stalk.

The sap beetle (*Carpophilus humeralis*) is one of the main enemies of pineapple fruits in Puerto Rico, Hawaii and Malaysia and is especially attracted to fruits affected by gummosis. Populations have been diminished by sanitary procedures and growing of cultivars resistant to gummosis, and chemical control is being evaluated.

In Brazil, larvae of the large moth, *Castnia licus,* and of the butterfly, *Thecla basilides,* damage the fruit. The latter is a problem in other parts of tropical America also and in Trinidad.

Cutworms eat holes in the base of the immature fruit. Fruit fly larvae do not pupate in 'Smooth Cayenne' but new hybrids lack resistance and may require treatment.

In New South Wales, poison baits are employed to combat fruit damage by crows, rats and mice. Rats may eat the base of the stem and destroy ratoons and suckers. Rabbits in winter eat the leaves as high as they can reach.

Diseases

In Queensland, top rot and root rot are caused by the soil fungi *Phytophthora cinnamomi* and *P. nicotianae* var. *parasitica* which are most prevalent in prolonged wet weather in autumn and winter. Improved drainage helps reduce the risk and monthly spraying with fungicide gives good control. *P. cinnamomi* may also cause rot in green fruit on ratoons. These diseases are largely prevented by the use of paper or plastic mulch on raised beds.

Base rot is caused by the fungus *Ceratocystis paradoxa,* especially where drainage is poor. The imperfect form (conidial state) of this fungus, known as *Thielaviopsis paradoxa,* causes butt rot in planting material, also soft rot or breakdown of fruits during shipment and storage. If ¼-ripe 'Red Spanish' fruits are kept at temperatures between 44.6° and 46.4°F (7°-8°C) while in transit, soft rot will not develop.

Fusarium spp. in the soil are the source of wilt. Black heart is a physiological disorder not visible externally, usually occuring in winter particularly in locations where air flow is inadequate. Highest incidence in West Africa has been reported in midsummer. It begins as "endogenous brown spot" at the base of the fruitlets close to the core. Later, affected areas merge. It has been attributed to chilling or low light intensity from dense planting or cloudiness. It can be controlled by one-day heat treatment at 90° to 100°F (32°-38°C) before or after refrigerated storage. In 1974, the microorganism *Erwinia chrysanthemi* was identified in Malaya as the cause of bacterial heart rot and fruit collapse.

Yellow spot virus on leaves is transmitted by *Thrips tabaci* Lind. Black speck and water blister are mentioned among other problems of the pineapple.

A condition called Crookneck is caused by zinc deficiency. It occurs mainly in plants 12-15 months old but is also frequent in suckers. The heart leaves become curled and twisted, waxy, brittle, and light yellowish-green. Sometimes the plant bends over and grows in a nearly horizontal position. Small yellow spots appear near the edges of the leaves and eventually merge and form blisters. Later, these areas become grayish or brownish

and sunken. Treatment is usually a 1% solution of zinc sulfate. Many growers use a combined spray of 10% urea, 2% iron sulfate and 1% zinc sulfate. If burning occurs, the proportion of urea should be changed to 5%. Excessive use of urea for this or any other purpose can lead to leaf tip dieback and yellowing of older leaves due to the biuret content in urea.

Copper deficiency is evident in concave leaves with dead tips and waxiness without bloom on the underside.

Sunburn or sunscald develops when fruits fall over and expose one side to the sun, though 'Abacaxi' may sunburn even when erect. Affected fruits soon rot and become infested with pests. They must be cut as soon as noticed and safely disposed of where they will not contaminate other fruits. Dry grass, straw, excelsior or brown paper sleeves may be placed over fruits maturing in the summer to prevent sunburn.

Harvesting

It is difficult to judge when the pineapple is ready to be harvested. The grower must depend a great deal on experience. Size and color change alone are not fully reliable indicators. Conversion of starch into sugars takes place rapidly in just a few days before full maturity. In general, for the fresh fruit market, the summer crop is harvested when the eye shows a light pale-green color. At this season, sugar content and volatile flavors develop early and steadily over several weeks. The winter crop is about 30 days slower to mature, and the fruits are picked when there is a slight yellowing around the base. Even then, winter fruit tends to be more acid and have a lower sugar level than summer fruit, and the harvest period is short. Fruits for canning are allowed to attain a more advanced stage. But overripe fruits are deficient in flavor and highly perishable.

Maturity studies conducted with 'Giant Kew' in India showed that highest quality is attained when the fruit is harvested at a specific gravity of 0.98-1.02, total soluble solids of 13.8-17%, or total soluble solids/acid ratio of 20.83-27.24 with development of external yellow color. Some people judge ripeness and quality by snapping a finger against the side of the fruit. A good, ripe fruit has a dull, solid sound; immaturity and poor quality are indicated by a hollow thud.

In manual harvesting, one man cuts off or breaks off the fruits (depending on the cultivar) and tosses them to a truck or passes them to 2 other workers with baskets who convey them to boxes in which they are arranged with the stems upward for the removal of bracts and application of a 3% solution of benzoic acid on the cut stem of all fruits not intended for immediate processing. The harvested fruits must be protected from rain and dew. If moist, they must be dried before packing. All defective fruits are sorted out for use in processing.

If the work is semi-mechanized, the harvesters decrown and trim the fruits and place them on a 30-ft conveyor boom which extends across the rows and carries the fruits to a bin on a forklift which loads it onto a truck or trailer. Some conveyors take the fruits directly into the canning factory from the field. In most regions of the world, pineapples are commonly marketed with crowns intact, but there is a growing practice of removing the crowns for planting. For the fresh fruit market, a short section of stem is customarily left on to protect the base of the fruit from bruising during shipment.

Total mechanical harvesting is achieved by 2 hydraulically-operated conveyors with fingers on the top conveyor to snap off the fruit, the lower conveyor carrying it away to the decrowners. After the fruit has been conveyed away, the workers go through the field to collect the crowns (where they have been left on the tops of the plants) and place them on the conveyors for a trip to the bins which are then fork-lifted and the crowns dumped into a planting machine.

Life of plantation: In Florida, 'Abakka' fields were maintained for 2, 3, or 4 crops. Some plantings of 'Red Spanish' were prolonged for 25-26 years. In current practice, after the harvesting of the first crop, workers trim off all but 2 ratoons which will bear fruit in 15-18 months. Perhaps there may be a second or third ratoon crop. Then the field is cleared to minimize carryover of pests and diseases. The method will vary with the interest in or practicality of making use of by-products. In Malaya, fields have been cleared by cutting the plants, leaving them to dry for 12-16 weeks, then piling and burning. Spraying with kerosene or diesel fuel makes burning possible in 9 weeks. Spraying with Paraquat allows burning in 3 weeks but does not destroy the stumps which take 3-5 months to completely decay while new plants are set out between them.

Field practices will differ if pineapples are interplanted with other crops. In Malaya, pineapples have been extensively grown in young rubber plantations. In India and Sri Lanka the pineapple is often a catchcrop among coconuts. Venezuelan farmers may interplant with citrus trees or avocados.

Storage

Cold storage at a temperature of 40°F (4.44°C) and lower causes chilling injury and breakdown in pineapples. At 44.6-46.4°F (7-8°C) and above, 80-90% relative humidity and adequate air circulation, normal ripening progresses during and after storage. At best, pineapples may be stored for no more than 4-6 weeks. There is a possibility that storage life might be prolonged by dipping the fruits in a wax emulsion containing a suitable fungicide. Irradiation extends the shelf life of half-ripe pineapples by about one week.

Food Uses

In Puerto Rico and elsewhere in the Caribbean, Spaniards found the people soaking pineapple slices in salted water before eating, a practice seldom heard of today.

Field ripe fruits are best for eating fresh, and it is only necessary to remove the crown, rind, eyes and core. In Panama, very small pineapples are cut from the plant with a few inches of stem to serve as a handle, the rind is removed except at the base, and the flesh is eaten out-of-hand like corn on the cob. The flesh of larger fruits is cut up in various ways and eaten fresh, as dessert, in salads, compotes and otherwise, or cooked in pies, cakes, puddings, or as a garnish on ham, or made into sauces or preserves. Malayans utilize the pineapple in curries and various meat dishes. In the Philippines, the fermented pulp is made into a popular sweetmeat called *nata de piña*. The pineapple does not lend itself well to freezing, as it tends to develop off-flavors.

Canned pineapple is consumed throughout the world. The highest grade is the skinned, cored fruit sliced crosswise and packed in sirup. Undersize or overripe fruits are cut into "spears", chunks or cubes. Surplus pineapple juice used to be discarded after extraction of bromelain (q.v.). Today there is a growing demand for it as a beverage. Crushed pineapple, juice, nectar, concentrate, marmalade and other preserves are commercially prepared from the flesh remaining attached to the skin after the cutting and trimming of the central cylinder. All residual parts — cores, skin and fruit ends — are crushed and given a first pressing for juice to be canned as such or prepared as sirup used to fill the cans of fruit, or is utilized in confectionery and beverages, or converted into powdered pineapple extract which has various roles in the food industry. Chlorophyll from the skin and ends imparts a greenish hue that must be eliminated and the juice must be used within 20 hours as it deteriorates quickly. A second pressing yields "skin juice" which can be made into vinegar or mixed with molasses for fermentation and distillation of alcohol.

In Africa, young, tender shoots are eaten in salads. The terminal bud or "cabbage" and the inflorescences are eaten raw or cooked. Young shoots, called *"hijos de piña"* are sold on vegetable markets in Guatemala.

Food Value Per 100 g of Edible Portion*	
Moisture	81.3-91.2 g
Ether Extract	0.03-0.29 g
Crude Fiber	0.3-0.6 g
Nitrogen	0.038-0.098 g
Ash	0.21-0.49 g
Calcium	6.2-37.2 mg
Phosphorus	6.6-11.9 mg
Iron	0.27-1.05 mg
Carotene	0.003-0.055 mg
Thiamine	0.048-0.138 mg
Riboflavin	0.011-0.04 mg
Niacin	0.13-0.267 mg
Ascorbic Acid	27.0-165.2 mg
*Analyses of ripe pineapple made in Central America.	

Sugar/acid ratio and ascorbic acid content vary considerably with the cultivar. The sugar content may change from 4% to 15% during the final 2 weeks before full ripening.

Toxicity

When unripe, the pineapple is not only inedible but poisonous, irritating the throat and acting as a drastic purgative.

Excessive consumption of pineapple cores has caused the formation of fiber balls (bezoars) in the digestive tract.

Other Uses

Bromelain: The proteolytic enzyme, bromelain, or bromelin, was formerly derived from pineapple juice; now it is gained from the mature plant stems salvaged when fields are being cleared. The yield from 368 lbs (167 kg) of stem juice is 8 lbs (3.6 kg) of bromelain. The enzyme is used like papain from papaya for tenderizing meat and chill-proofing beer; is added to gelatin to increase its solubility for drinking; has been used for stabilizing latex paints and in the leather-tanning process. In modern therapy, it is employed as a digestive and for its anti-inflammatory action after surgery, and to reduce swellings in cases of physical injuries; also in the treatment of various other complaints.

Fiber: Pineapple leaves yield a strong, white, silky fiber which was extracted by Filipinos before 1591. Certain cultivars are grown especially for fiber production and their young fruits are removed to give the plant maximum vitality. The 'Perolera' is an ideal cultivar for fiber extraction because its leaves are long, wide and rigid. Chinese people in Kwantgung Province and on the island of Hainan weave the fiber into coarse textiles resembling grass cloth. It was long ago used for thread in Malacca and Borneo. In India the thread is prized by shoemakers and it was formerly used in the Celebes. In West Africa it has been used for stringing jewels and also made into capes and caps worn by tribal chiefs. The people of Guam hand-twist the fiber for making fine casting nets. They also employ the fiber for wrapping or sewing cigars. Piña cloth made on the island of Panay in the Philippines and in Taiwan is highly esteemed. In Taiwan they also make a coarse cloth for farmers' underwear.

The outer, long leaves are preferred. In the manual process, they are first decorticated by beating and rasping and stripping, and then left to ret in water to which chemicals may be added to accelerate the activity of the microorganisms which digest the unwanted tissue and separate the fibers. Retting time has been reduced from 5 days to 26 hours. The retted material is washed clean, dried in the sun and combed. In mechanical processing, the same machine can be used that extracts the fiber from sisal. Estimating 10 leaves to the lb (22 per kg), 22,000 leaves would constitute one ton and would yield 50-60 lbs (22-27 kg) of fiber.

Juice: Pineapple juice has been employed for cleaning machete and knife blades and, with sand, for scrubbing boat decks.

Animal Feed: Pineapple crowns are sometimes fed to horses if not needed for planting. Final pineapple waste from the processing factories may be dehydrated as "bran" and fed to cattle, pigs and chickens. "Bran" is also made from the stumps after bromelain extraction. Expendable plants from old fields can be processed as silage for maintaining cattle when other feed is scarce. The silage is low in protein and high in fiber and is best mixed with urea, molasses and water to improve its nutritional value.

In 1982, public concern in Hawaii was aroused by the detection of heptachlor (a carcinogen) in the milk from cows fed "green chop"—leaves from pineapple plants that had been sprayed with the chemical to control the ants that distribute mealybugs. There is supposed to be a one-year lapse to allow the heptachlor to become more dilute before sprayed plants are utilized for feed.

Folk Medicine: Pineapple juice is taken as a diuretic and to expedite labor, also as a gargle in cases of sore throat and as an antidote for seasickness. The flesh of very young (toxic) fruits is deliberately ingested to achieve abortion (a little with honey on 3 successive mornings); also to expel intestinal worms; and as a drastic treatment for venereal diseases. In Africa the dried, powdered root is a remedy for edema. The crushed rind is applied on fractures and the rind decoction with rosemary is applied on hemorrhoids. Indians in Panama use the leaf juice as a purgative, emmenagogue and vermifuge.

Ornamental Value

The pineapple fruit with crown intact is often used as a decoration and there are variegated forms of the plant universally grown for their showiness—indoors or out. Since 1963, thousands of potted, ethylene-treated pineapple plants with fruits have been shipped annually from southern Florida to northern cities as indoor ornamentals.

Banana (Plates III and IV)

The word "banana" is a general term embracing a number of species or hybrids in the genus *Musa* of the family Musaceae. Some species such as *M. Basjoo* Sieb. & Zucc. of Japan and *M. ornata* Roxb., native from Pakistan to Burma, are grown only as ornamental plants or for fiber. *M. textilis* Née of the Philippines is grown only for its fiber, prized for strong ropes and also for tissue-thin tea bags. The so-called Abyssinian banana, *Ensete ventricosum* Cheesman, formerly *E. edule* Horan, *Musa ensete* Gmel., is cultivated in Ethiopia for fiber and for the staple foods derived from the young shoot, the base of the stem, and the corm.

Most edible-fruited bananas, usually seedless, belong to the species *M. acuminata* Colla (*M. cavendishii* Lamb. ex Paxt., *M. chinensis* Sweet, *M. nana* Auth. NOT Lour., *M. zebrina* Van Houtee ex Planch.), or to the hybrid *M. X paradisiaca* L. (*M. X sapientum* L.; *M. acuminata* X *M. balbisiana* Colla).

M. balbisiana Colla of southern Asia and the East Indies, bears a seedy fruit but the plant is valued for its disease-resistance and therefore plays an important role as a "parent" in the breeding of edible bananas.

M. fehi Bertero ex Vieill. and *M. troglodytarum* L. have been applied to the group of bananas known as fehi or fe'i but taxonomists have yet to make final decisions as to the applicability of these binomials.

To the American consumer, "banana" seems a simple name for the yellow fruits so abundantly marketed for consumption raw, and "plantain" for the larger, more angular fruits intended for cooking but also edible raw when fully ripe. However, the distinction is not that clear and the terms may even be reversed. The types we call "banana" are known by similar or very different names in banana-growing areas. Spanish-speaking people say *banana china* (Paraguay), *banano enano* (Costa Rica), *cambur* or *camburi* (Colombia, Venezuela), *cachaco, colicero, cuatrofilos* (Colombia); *carapi* (Paraguay), *curro* (Panama), *guineo* (Costa Rico, Puerto Rico, El Salvador); *murrapo* (Colombia); *mampurro* (Dominican Republic); *patriota* (Panama); *plátano* (Mexico); *plátano de seda* (Peru); *plátano enano* (Cuba); *suspiro* (Dominican Republic); *zambo* (Honduras). Portuguese names in Brazil are: *banana maca, banana de Sao Tomé, banana da Prata*. In French islands or areas, the terms may be *bananier nain, bananier de Chine* (Guadeloupe), *figue, figue banane, figue naine* (Haiti). Where German is spoken, they say: *echte banane, feige*, or *feigenbaum*. In the Sudan, *baranda*.

The types Americans call "plantain", Plate IV, may be known as *banaan* (Surinam); *banano macho* (Panama); *banane* or *bananier* (Haiti, Guadeloupe, Martinique); *banane misquette* or *banane musquée*, or *pié banane* (Haiti); *bananeira de terra* (Brazil); *banano indio* (Costa Rica); *barbaro* (Mexico); *butuco* (Honduras); *parichao* (Venezuela); plantain (Guyana, Jamaica, Trinidad); *plátano* (Cuba, Puerto Rico, Dominican Republic); *plátano burro, plátano hembra* (Cuba); *plátano macho* (Cuba, Panama); *plátano de la isla* (Peru); *topocho* or *yapuru* (Venezuela); *zapolote* (Mexico). Numerous other vernacular names, according to geographical region, are provided by N.W. Simmonds in his textbook, *Bananas*.

In India, there is no distinction between bananas and plantains. All cultivars are merely rated as to whether they are best for dessert or for cooking.

Description

The banana plant, often erroneously referred to as a "tree", is a large herb, with succulent, very juicy stem (properly "pseudostem") which is a cylinder of leaf-petiole sheaths, reaching a height of 20 to 25 ft (6-7.5 m) and arising from a fleshy rhizome or corm. Suckers spring up around the main plant forming a clump or "stool", the eldest sucker replacing the main plant when it fruits and dies, and this process of succession continues indefinitely. Tender, smooth, oblong or elliptic, fleshy-stalked leaves, numbering 4 or 5 to 15, are arranged spirally. They unfurl, as the plant grows, at the rate of one per week in warm weather, and extend upward and outward, becoming as much as 9 ft (2.75 m) long and 2 ft (60 cm) wide. They may be entirely green, green with maroon splotches, or green on the upperside and red-purple beneath. The inflorescence, a transformed growing point, is a terminal spike shooting out from the heart in the tip of the stem. At first, it is a large, long-oval, tapering, purple-clad bud. As it opens, it is seen that the slim, nectar-rich, tubular, toothed, white flowers are clustered in whorled double rows along the floral stalk, each cluster covered by a thick, waxy, hoodlike bract, purple outside, deep-red within. Normally, the bract will lift from the first hand in 3 to 10 days. If the plant is weak, opening may not occur until 10 or 15 days. Female flowers occupy

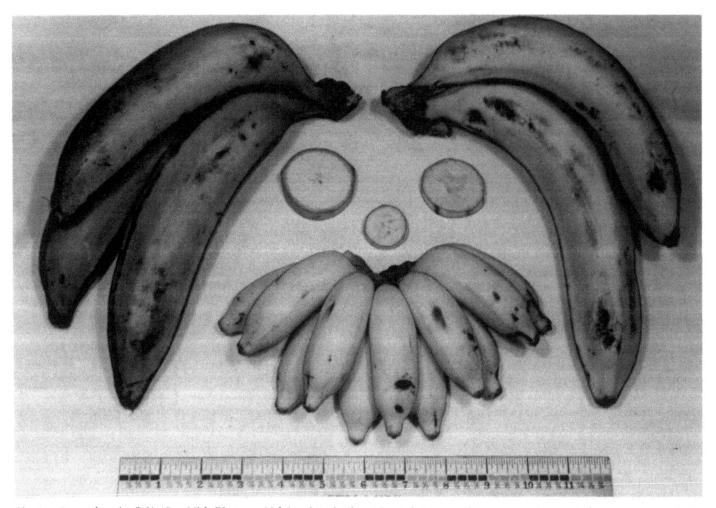

Fig. 8: Green plantains (left), 'Gros Michel' bananas (right) and 'Lady Finger' (center). In: K. and J. Morton, *Fifty Tropical Fruits of Nassau*, 1946.

the lower 5 to 15 rows; above them may be some rows of hermaphrodite or neuter flowers; male flowers are borne in the upper rows. In some types the inflorescence remains erect but generally, shortly after opening, it begins to bend downward. In about one day after the opening of the flower clusters, the male flowers and their bracts are shed, leaving most of the upper stalk naked except at the very tip where there usually remains an unopened bud containing the last-formed of the male flowers. However, there are some mutants such as 'Dwarf Cavendish' with persistent male flowers and bracts which wither and remain, filling the space between the fruits and the terminal bud.

As the young fruits develop from the female flowers, they look like slender green fingers. The bracts are soon shed and the fully grown fruits in each cluster become a "hand" of bananas, and the stalk droops with the weight until the bunch is upside-down. The number of "hands" varies with the species and variety.

The fruit (technically a "berry") turns from deep-green to yellow or red, or, in some forms, green-and white-striped, and may range from 2½ to 12 in (6.4-30 cm) in length and ¾ to 2 in (1.9-5 cm) in width, and from oblong, cylindrical and blunt to pronouncedly 3-angled, somewhat curved and hornlike. The flesh, ivory-white to yellow or salmon-yellow, may be firm, astringent, even

gummy with latex, when unripe, turning tender and slippery, or soft and mellow or rather dry and mealy or starchy when ripe. The flavor may be mild and sweet or subacid with a distinct apple tone. Wild types may be nearly filled with black, hard, rounded or angled seeds ⅛ to ⅝ in (3-16 mm) wide and have scant flesh. The common cultivated types are generally seedless with just minute vestiges of ovules visible as brown specks in the slightly hollow or faintly pithy center, especially when the fruit is overripe. Occasionally, cross-pollination by wild types will result in a number of seeds in a normally seedless variety such as 'Gros Michel', but never in the Cavendish type.

Origin and Distribution

Edible bananas originated in the Indo-Malaysian region reaching to northern Australia. They were known only by hearsay in the Mediterranean region in the 3rd Century B.C., and are believed to have been first carried to Europe in the 10th Century A.D. Early in the 16th Century, Portuguese mariners transported the plant from the West African coast to South America. The types found in cultivation in the Pacific have been traced to eastern Indonesia from where they spread to the Marquesas and by stages to Hawaii.

Bananas and plantains are today grown in every humid

tropical region and constitute the 4th largest fruit crop of the world, following the grape, citrus fruits and the apple. World production is estimated to be 28 million tons—65% from Latin America, 27% from Southeast Asia, and 7% from Africa. One-fifth of the crop is exported to Europe, Canada, the United States and Japan as fresh fruit. India is the leading banana producer in Asia. The crop from 400,000 acres (161,878 ha) is entirely for domestic consumption. Indonesia produces over 2 million tons annually, the Philippines about ½ million tons, exporting mostly to Japan. Taiwan raises over ½ million tons for export. Tropical Africa (principally the Ivory Coast and Somalia) grows nearly 9 million tons of bananas each year and exports large quantities to Europe.

Brazil is the leading banana grower in South America— about 3 million tons per year, mostly locally consumed, while Colombia and Ecuador are the leading exporters. Venezuela's crop in 1980 reached 983,000 tons. Large-scale commercial production for export to North America is concentrated in Honduras (where banana fields may cover 60 sq mi) and Panama, and, to a lesser extent, Costa Rica. In the West Indies, the Windward Islands of Martinique and Guadeloupe are the main growers and for many years have regularly exported to Europe. Green bananas are the basic food of the people of Western Samoa and large quantities are exported.

In Ghana, the plantain is a staple food but up to the late 1960's the crop was grown only in home gardens or as a shade for cacao. When the cacao trees declined, solid plantings of plantain were established in their place and in newly cleared forest land where the richness of organic matter greatly promotes growth. By 1977, Ghana was harvesting 2,204,000 tons (2,000,000 MT) annually.

The plantain is the most important starchy food of Puerto Rico and is third in monetary value among agricultural crops, being valued at $30,000,000 annually. While improved methods of culture have been adopted in recent years and production has been increased by 15% in 1980, it was still necessary to import 1,328 tons (1,207 MT) to meet local demand. Annual per capita consumption is said to be 65 lbs (29.5 kg). In the past, most of the plantains in Puerto Rico were grown on humid mountainsides. High prices have induced some farmers to develop plantations on level irrigated land formerly devoted to sugarcane.

In tropical zones of Colombia, plantains are not only an important part of the human diet but the fruits and the plants furnish indispensable feed for domestic animals as well. The total plantain area is about 1,037,820 acres (420,000 ha) with a yield of 5,500 lbs per acre (5,500) kg/ha). Mexico grows about ⅙ as much, 35% under irrigation, and the crop is valued at $1,335 US per acre ($3,300 US/ha). Venezuela has somewhat less of a crop—517,000 tons from 146,000 acres (59,000 ha) in 1980—and the Dominican Republic is fourth in order with about 114,600 acres (46,200 ha). Bananas and plantains are casually grown in some home gardens in southern Florida. There are a few small commercial plantations furnishing local markets.

Varieties

Edible bananas are classified into several main groups and subgroups. Simmonds placed first the diploid *M. acuminata* group 'Sucrier', represented in Malaya, Indonesia, the Philippines, southern India, East Africa, Burma, Thailand, the West Indies, Colombia and Brazil. The sheaths are dark-brown, the leaves yellowish and nearly free of wax. The bunches are small and the fruits small, thin-skinned and sweet. Cultivars of this group are more important in New Guinea than elsewhere.

Here belongs one of the smallest of the well-known bananas, the 'Lady Finger', also known as 'Date' or 'Fig', and, in Spanish, as 'Dedo de Dama', 'Datil', 'Niño', Bocadillo', 'Manices', 'Guineo Blanco', or 'Cambur Titiaro'. The plant reaches 25 ft (7.5 m) in height, has a slender trunk but a heavy root system that fortifies the plant against strong winds. The outer sheaths have streaks or patches of reddish-brown. The bunch consists of 10 to 14 hands each of 12 to 20 fingers. The fruit is 4 to 5 in (10-12.5 cm) long, with thin, light-yellow skin and sweet flesh. This cultivar is resistant to drought, Panama disease and the black weevil but subject to Sigatoka (leaf spot). It is common in Latin America and commercial in Queensland and New South Wales.

In second place, there is the group represented by the prominent and widely cultivated 'Gros Michel' originally from Burma, Thailand, Malaya, Indonesia and Ceylon. It was introduced into Martinique early in the 19th Century by a French naval officer and, a few years later, was taken to Jamaica; from there it was carried to Fiji, Nicaragua, Hawaii and Australia, in that sequence. It is a large, tall plant bearing long bunches of large, yellow fruits, and it was formerly the leading commercial cultivar in Central Africa, Latin America and the Caribbean, but has been phased out because of its great susceptibility to Panama disease. It has given rise to several named sports or mutants.

The Cavendish subgroup includes several important bananas:

a) The 'Dwarf Cavendish', Plate III, first known from China and widely cultivated, especially in the Canary Islands, East Africa and South Africa. The plant is from 4 to 7 ft (1.2-2.1 m) tall, with broad leaves on short petioles. It is hardy and wind-resistant. The fruit is of medium size, of good quality, but thin-skinned and must be handled and shipped with care. This cultivar is easily recognized because the male bracts and flowers are not shed.

b) The 'Giant Cavendish', also known as 'Mons Mari', 'Williams', 'Williams Hybrid', or 'Grand Naine', is of uncertain origin, closely resembles the 'Gros Michel', and has replaced the 'Dwarf' in Colombia, Australia, Martinique, in many Hawaiian plantations, and to some extent in Ecuador. It is the commercial banana of Taiwan. The plant reaches 10 to 16 ft (2.7-4.9 m). The pseudostem is splashed with dark-brown, the bunch is long and cylindrical, and the fruits are larger than those of the 'Dwarf' and not as delicate. Male bracts and flowers are shed, leaving a space between the fruits and the terminal bud.

c) 'Pisang masak hijau', or 'Bungulan', the triploid Cavendish clone of the Philippines, Indonesia and Malaya, is erroneously called 'Lacatan' in Jamaica where it replaced 'Gros

Michel' because of its immunity to Panama disease, though it is subject to Sigatoka (leaf spot). The plant is tall and slender and prone to wind injury. Its fruits ripen unevenly in winter, bruise easily and are inclined to spoil in storage. It is no longer grown commercially in Jamaica and the Windward Islands. The fruits are commonly used as cooking bananas in Jamaican households. Simmonds declares this cultivar is not the true 'Lacatan' of the Philippines. He suggested that 'Pisang masak hijau' may have been the primary source of all the members of the Cavendish group.

d) **'Robusta'**, very similar to the so-called 'Lacatan', has largely replaced that cultivar in Jamaica and the Windward Islands and the 'Gros Michel' in Central America because it is shorter, thick-stemmed, less subject to wind. It is being grown commercially also in Brazil, eastern Australia, Samoa and Fiji. It is resistant to Panama disease but prone to Sigatoka.

e) **'Valery'**, also a triploid Cavendish clone, closely resembles 'Robusta' and some believe it may be the same. However, it is being grown as a successor to 'Robusta'. It is already more widely cultivated than 'Lacatan' for export. As compared with other clones in cooking trials, it has low ratings because cooking hardens the flesh and gives it a waxy texture.

The Banana Breeding Research Scheme in Jamaica has developed a number of tetraploid banana clones with superior disease-resistance and some are equal in dessert quality to the so-called 'Lacatan' and 'Valery'.

'Bluggoe' (with many other local names) is a cooking banana especially resistant to Panama disease and Sigatoka. It bears a few distinctly separated hands of large, almost straight, starchy fruits, and is of great importance in Burma, Thailand, southern India, East Africa, the Philippines, Samoa, and Grenada.

'Ice Cream' banana of Hawaii ('Cenizo' of Central America and the West Indies; 'Krie' of the Philippines), is a relative of 'Bluggoe'. The plant grows to 10 or 15 ft (3-4.5 m), the leaf midrib is light pink, the flower stalk may be several feet long, but the bunch has only 7 to 9 hands. The fruit is 7 to 9 in (17.5-22.8 cm) long, up to 2½ in (6.25 cm) thick, 4- to 5-angled, bluish with a silvery bloom when young, pale-yellow when ripe. The flesh is white, sweetish, and is eaten raw or cooked.

'Mysore', also known as 'Fillbasket' and 'Poovan', is the most important banana type of India, constituting 70% of the total crop. It is sparingly grown in Malaya, Thailand, Ceylon and Burma. It is thought to have been introduced into Dominica in 1900 but the only place where it is of any importance in the New World is Trinidad where it is cultivated as shade for cacao. The plant is large and vigorous, immune to Panama disease and nearly so to Sigatoka; very hardy and drought-tolerant. It bears large, compact bunches of medium-sized, plump, thin-skinned, attractive, bright-yellow fruits of subacid flavor.

Other prominent commercial cultivars are **'Salembale'** and **'Rasabale'**, not suitable for canning because of starchy taste and weak flavor. **'Pachabale'** and **'Chandrabale'** are important local varieties preferred for canning. K.C. Naik described 34 cultivars as the more important among the many grown in South India.

'Silk', 'Silk Fig', or 'Apple' ('Manzana' in Spanish), is the most popular dessert banana of the tropics. It is widely distributed around the tropics and subtropics but never grown on a large scale. The plant is 10 to 12 ft (3-3.6 m) tall, only

medium in vigor, very resistant to Sigatoka but prone to Panama disease. There are only 6 to 12 hands in the bunch, each with 16 to 18 fruits. The plump bananas are 4 to 6 in (10-15 cm) long, slightly curved; astringent when unripe but pleasantly subacid when fully ripe; and apple-scented. If left on the bunch until fully developed, the thin skin splits lengthwise and breaks at the stem-end causing the fruit to fall, but it is firm and keeps well on hand in the home.

The **'Red'**, 'Red Spanish', 'Red Cuban', 'Colorado', or 'Lal Kela' banana may have originated in India, where it is frequently grown, and it has been introduced into all banana-growing regions. The plant is large, takes 18 months from planting to harvest. It is highly resistant to disease. The pseudostem, petiole, midrib and fruit peel are all purplish-red, but the latter turns to orange-yellow when the fruit is fully ripe. The bunch is compact, may contain over 100 fruits of medium size, with thick peel, and flesh of strong flavor. In the mutant called 'Green Red', the plant is variegated green-and-red, becomes 28 ft (8.5 m) tall with pseudostem to 18 in (45 cm) thick at the base. The bunch bears 4 to 7 hands, the fruits are thick, 5 to 7 in (12.5-17.5 cm) long. The purplish-red peel changes to orange-yellow and the flesh is firm, cream-colored and of good quality.

The **'Fehi'** or 'Fe'i' group, of Polynesia, is distinguished by the erect bunches and the purplish-red or reddish-yellow sap of the plants which has been used as ink and for dyeing. The plants may reach 36 ft (10.9 m) and the leaves are 20 to 30 in (50-75 cm) wide. The bunches have about 6 hands of orange or copper-colored, thick-skinned fruits which are starchy, sometimes seedy, of good flavor when boiled or roasted. These plants are often grown as ornamentals in Hawaii.

As a separate group, Simmonds places the 'I.C. 2', or **'Golden Beauty'** banana especially bred at the Imperial College of Tropical Agriculture in Trinidad in 1928 by crossing the 'Gros Michel' with a wild *Musa acuminata*. It is resistant to Panama disease and very resistant to Sigatoka. Though the bunches are small and the fruits short, they ship and ripen well and this cultivar is grown for export in Honduras and has been planted in Hawaii, Samoa and Fiji.

'Orinoco', 'Horse', 'Hog', or 'Burro', banana, a medium-tall, sturdy plant, is particularly hardy. The bunch consists of only a few hands of very thick, 3-angled fruits about 6 in (15 cm) long. The flesh has a salmon tint, is firm, edible raw when fully ripe but much better cooked—fried, baked or otherwise, as are plantains.

Trials of 5 clones of 'Giant Cavendish' and 9 other cultivars ('Robusta A', 'Robusta B', 'Cocos A', 'Cocos B', 'Golden Beauty', 'Enano Nautia', 'Enano Gigante', 'Enano' and 'Valery') were made between 1976 and 1979 at the Campo Agricola Experimental at Tecoman, Mexico. 'Enano Gigante' is the most widely grown cultivar in that region but the tests showed that 'Enano Nautia' and 'Golden Beauty' bore heavier bunches of better quality fruit, even though 'Enano Gigante' had a greater number of bunches and highest yield per ground area. 'Giant Cavendish' clones 1, 2, 3 and 4, and 'Cocos B' grew very tall, gave low yields and the fruit was of poor quality.

Among the plantains, there are many forms, some with pink, red or dark-brown leaf sheaths, some having also colored midribs or splotches on leaves or fruits. The plants are usually large, vigorous and resistant to Panama disease and Sigatoka but attacked by borers. Major subgroups are known as **'French plantain'** and **'Horn plantain'**, the former with persistent male flowers. The usually large, angled fruits are borne in few hands. All are important sources of food in southern India, East Africa, tropical America and the West Indies. The tall **'Maricongo'** and the **'Common Dwarf'** are leading commercial cultivars. A dwarf mutant is the 'Plantano enano' of Puerto Rico ('banane cochon' of Haiti). Ordinary plantains are called 'cuadrado', 'chato', and 'topocho' in Mexico. The leading commercial cultivars are 'Pelipita' and 'Saba' which are resistant to Black Sigatoka but they do not have the high culinary quality of 'Hartón', 'Dominico-Hartón', 'Currare', and 'Horn'. 'Laknau' is a fertile plantain that resembles 'Horn' but is of inferior quality. It has opened up possibilities for hybridizing and is being crossed with 'Pelipita' and 'Saba'.

Banana and plantain cultivars most often grown in Florida are the 'Dwarf Cavendish', 'Apple', and 'Orinoco' bananas and the 'Macho' plantain. The 'Red' and 'Lady Finger' bananas are very occasionally grown in sheltered locations.

There are five major collections of banana and plantain clones in the world. United Brands maintains a collection of 470 cultivars and 100 species at La Lima, Honduras.

Climate

The edible bananas are restricted to tropical or near-tropical regions, roughly the area between latitudes 30°N and 30°S. Within this band, there are varied climates with different lengths of dry season and different degrees and patterns of precipitation. A suitable banana climate is a mean temperature of 80°F (26.67°C) and mean rainfall of 4 in (10 cm) per month. There should not be more than 3 months of dry season.

Cool weather and prolonged drought retard growth. Banana plants produce only one leaf per month in winter, 4 per month in summer. If low temperatures occur just at flowering time, the bud may not be able to emerge from the stem. If fruits have already formed, maturity may be delayed several months or completely suspended. If only the leaves are destroyed, the fruits will be exposed to sunburn. Smudging, by burning dry trash covered with green clippings to create smoke, can raise the temperature 2 to 4 degrees. Flooding the field in advance of a cold snap will keep the ground warm if the chill weather is brief. In Australia, bananas are planted on sunny hillsides at elevations of 200 to 1,000 ft (60 to 300 m) to avoid the cold air that settles at lower levels. Brief frosts kill the plants to the ground but do not destroy the corm. 'Dwarf Cavendish' and the 'Red' banana are particularly sensitive to cold, whereas the dwarf cultivar 'Walha', or 'Kullen', of India is successful up to 4,000 ft (1,220 m) in the outer range of the Western Ghats. 'Vella vazhai' is ex-

Fig. 9: 'Radja' banana, introduced into Florida by Dr. J.J. Ochse about 1957.

tensively cultivated in the Lower Pulneys between 3,200 and 5,500 ft (975 and 1,616 m). A cooking banana, 'Bankel', survives winters in home gardens in northern India. In South Africa, the main banana-producing area is along the southeast coast at 3,000 ft (915 m) above sea-level with summer rainfall of 35 to 45 in (90-115 cm). The major part of the crop in East Africa is grown between 4,000 and 5,000 ft (1,220 and 1,524 m) and the total range extends from sea-level to 7,500 ft (2,286 m).

Wind is detrimental to banana plants. Light winds shred the leaves, interfering with metabolism; stronger winds may twist and distort the crown. Winds to 30 mph break the petioles; winds to 40 mph will topple a pseudo-stem that is supporting the weight of a heavy bunch unless the stem is propped, and may cause root damage in non-fruiting plants that are not blown down; winds of 60 mph or over will uproot entire plantations, especially when the soil is saturated by rain. Windbreaks are often planted around banana fields to provide some protection from cold and wind. Cyclones and hurricanes are devastating and the latter were the main reason for the shift of large-scale banana production from the West Indies to Central America, Colombia and Ecuador. Hail results from powerful convection currents in the tropics, especially in the spring, and does much damage to bananas.

Soil

The banana plant will grow and fruit under very poor conditions but will not flourish and be economically productive without deep, well-drained soil—loam, rocky sand, marl, red laterite, volcanic ash, sandy clay, even heavy clay—but not fine sand which holds water. Overhead irrigation is said to improve the tilth of heavy clay and has made possible the use of clay soils that would never have been considered for banana culture in the past. Alluvial soils of river valleys are ideal for banana-growing. Bananas prefer an acid soil but if the pH is below 5.0 lime should be applied the second year. Low pH makes bananas more susceptible to Panama disease. Where waterlogging is likely, bananas and plantains are grown on raised beds. Low, perennially wet soils require draining and dry soils require irrigation.

Propagation

Banana seeds are employed for propagation only in breeding programs. Corms are customarily used for planting and Mexican studies with 'Giant Cavendish' have shown that those over 17.5 lbs (8 kg) in weight come into bearing early and, in the first year, the bunches are longer, heavier, with more hands than those produced from smaller corms. From the second year on, the advantage disappears. Most growers prefer "bits" 2- to 4-lb (0.9-1.8 kg) sections of the corm. When corms are scarce, smaller sections—1 to 2 lbs (454-908 g) have been utilized and early fertilization applied to compensate for the smaller size. But in Queensland it is specified that "bits" of 'Dwarf Cavendish' shall not be less than 4 x 3 x 3 in (10 x 7.5 x 7.5 cm) and "bits" of 'Lady Finger' and other tall cultivars shall be not less than 5 x 5 x 3½ in (12.5 x 12.5 x 9 cm). The corm has a number of buds, or "eyes", which develop into new shoots. The two upper buds are the youngest and have a pinkish tint. These develop rapidly and become vigorous plants. To obtain the "bits", a selected, healthy banana plant, at least 7 months old but prior to fruiting, is uprooted and cut off about 4 to 5 in (10-12.5 cm) above the corm. The outer layer of leaf bases is peeled off to expose the buds, leaving just a little to protect the buds during handling and transport. The corm is split between the 2 upper buds and trimmed with square sides, removing the lower, inferior buds and any parts affected by pests or disease, usually indicated by discoloration. Then the "bits" are fumigated by immersing for 20 minutes in hot water at about 130°F (54.44°C) or in a commercial nematicide solution. Sometimes it is advisable to apply a fungicide to prevent spoilage. They should then be placed in a sanitary place (away from all diseased trash) in the shade for 48 hrs before planting.

Inasmuch as "bits" are not often available in quantity, the second choice is transplantation of suckers. These should not be too young nor too old.

The sucker first emerges as a conical shoot which opens and releases leaves that are mostly midribs with only vestiges of blade. These juvenile leaves are called "sword", "spear", or "arrow", leaves. Just before the sucker produces wide leaves resembling those of the mature plant but smaller, it has sufficient corm development to be transplanted. Sometimes suckers from old, deteriorating corms have broad leaves from the outset. These are called "water" suckers, are insubstantial, with very little vigor, and are not desirable propagating material. "Maiden" suckers that have passed the "sword"-leaved stage and have developed broad leaves must be large to be acceptably productive. In banana trials at West Bengal, India, suckers 3 to 4 months old with well-developed rhizomes proved to be the best yielders. In comparison, small, medium, or large

"sword" suckers develop thicker stems, and give much higher yields of marketable fruits per land parcel. "Bits" grow slowly at first, but in 2 years' time they catch up to plants grown from suckers or "butts" and are much more economical. "Butts" (entire corms, or rhizomes, of mature plants), called "bull-heads" in the Windward Islands, are best used to fill in vacancies in a plantation. For quick production, some farmers will use "butts" with several "sword" suckers attached. Very young suckers, called "peepers", are utilized only for establishing nurseries.

Instead of waiting for normal sucker development, multiplication has been artificially stimulated in the field by removing the soil and outer leaf sheaths covering the upper buds of the corm, packing soil around them and harvesting them when they have reached the "sword" sucker stage. A greenhouse technique involves cleaning and injuring a corm to induce callus formation from which many new plants will develop. As many as 180 plantlets have been derived from one corm in this manner.

Diseases are often spread by vegetative propagation of bananas, and this fact has stimulated efforts to create disease-free planting material on a large scale by means of tissue culture. Some commercial banana cultivars have been cultured in Hawaii. A million 'Giant Cavendish' banana plants were produced by meristem culture in Taiwan in 1983. In the field, these laboratory plantlets showed 95% survival, grew faster than suckers in the first 5 months, had bigger stems and more healthy leaves.

Rapid multiplication of 'Philippine Lacatan' and 'Grand Naine' bananas, and the Sigatoka-resistant 'Saba' and 'Pelipita' plantains by shoot-tip culture has been achieved by workers at State University of New York.

Culture

On level land where the soil is compact, deep ploughing is needed to improve aeration and water filtration, whereas on a sloping terrain minimum tillage is advised as well as contouring of rows to minimize erosion. Planting is best done at the end of the dry season and beginning of the wet season for adequate initial moisture and to avoid waterlogging of the young plants. Puerto Rico, because of its favorable climate, is able to make monthly plantings of plantains the year around in order to produce a continuous supply for processing factories. However, some consideration has been given to manipulation of planting dates to avoid a summer surplus (June-September) caused by March and May plantings and to take advantage of higher prices in winter and spring (February to April). To achieve this, it is suggested that plantings be made only in the first or second weeks of January, July, September, November and December. Generally, the banana requires 10 to 12 months from planting to harvest. Summer plantings of plantains in Puerto Rico take 14 to 16 months; winter plantings 17 to 19. In regions where there may be periods of low temperatures in winter, planting time is chosen to allow flowering and fruiting before predictable cold periods.

Spacing varies with the ultimate size of the cultivar, the

fertility of the soil, and other factors. Close planting protects plantations exposed to high winds, but results in fewer suckers, hinders disease control, and has been found to be profitable for only the first year. In subsequent years, fruits are shorter, the flesh is softer and bunches ripen prematurely. The standard practice in Puerto Rico is 500 plants of 'Maricongo' plantain per acre (1,235 plants/ha). Increasing to 800 plants/acre (1,976/ha) has increased yield by 4 tons, but elevating density to 1,300 plants/acre (3,212 plants/ha) has not shown any further increase. In Surinam, most of the plantains are grown at a density of 809 to 1,012 plants per acre (2,000-2,500/ha), but density may range from 243 to 1,780 plants per acre (600-4,400/ha).

The higher the number of plants in the field, the larger the volume of fertilizer that must be applied. The crop suffers severely from root competition, for the roots of a fully-grown banana plant may extend outward 18 ft (5.5 m). The higher the altitude, the lower the density must be because solar radiation is reduced. Too much space between plants allows excessive evaporation from uncovered soil and increases the weed problem. Growers must determine the most economical balance between sufficient light for good yields and efficient land management. Spacing distances for 'Dwarf Cavendish' range from 10 x 6 ft (3 x 1.8 m) to 15 x 12 ft (4.5 x 3.6 m). A spacing of 12 ft (3.6 m) between rows and 8 ft (2.4 m) between plants allows 450 plants per acre (1,112 plants/ha). Studies conducted with the so-called 'Lacatan' ('Pisang masak hijau') over a 3-year period in Jamaica, demonstrated the optimum density to be 680 plants per acre (2,680/ha). At closer spacings, yield increased but profits declined. Hexagonal spacing gives the maximum number of plants per area. Double- and triple-row plantings provide alleys for mechanical operations and harvesting.

Planting holes should be at least 18 in (45 cm) wide and 15 in (38 cm) deep, but may be as much as 3 ft (0.91 m) wide and 2 ft (0.6 m) deep for extra wind-resistance. They should be enriched in advance of planting. On hillsides, suckers are set with the cut surface facing downhill; the bud or "eye" of a "bit" must point uphill; so that the "follower" sucker will emerge on the uphill side where the soil is deepest. A surface cover of about 4 in (10 cm) of soil is trampled down firmly.

Weed control is essential. Geese have been installed as weeders because they do not eat the banana plants. However, they consume mostly grass and fail to eliminate certain broad-leaved weeds which still require cleaning out. Certain herbicides, including Diuron and Ametryne, have been approved for banana fields. They are applied immediately after planting but great care must be taken to minimize adverse effects on the crop. Ametryne has been shown to be relatively safe for the plants and it has a short life in the soil. The most persistent weed is *Cyperus rotundus* L. (nutgrass, yellow nutgrass, purple nutsedge, coqui or coyolillo) which decreases yields and competes with the crop for nitrogen.

In some plantations, a mulch of dry banana leaves is maintained to discourage weeds. Some growers resort to live groundcovers such as *Glycine javanica* L. (Rhodesian kudzu), *Commelina* spp., or *Zebrina pendula* Schnizl. or other creepers, but these tend to climb the banana stems and become a nuisance. Sometimes short-term crops are interplanted in young banana fields, for example, maize, eggplant, peppers, tomatoes, okra, sweetpotato, pineapple or upland rice. A space of at least 3 ft (0.91 m) must be kept clear around each banana plant. However, there are banana authorities who are opposed to interplanting.

Bananas and plantains are heavy feeders. It has been calculated that a harvest of 5 tons of fruit from an acre leaves the soil depleted by 22 lbs (10 kg) nitrogen, 4 lbs (1.8 kg) phosphorus, 55 lbs (25 kg) potash and 11 oz (312 kg) calcium. In general, it can be said that banana plants have high nitrogen and phosphorus requirements and a fertilizer formula of 8:10:8 NPK is usually suitable and normally 1 to 1½ tons/acre (1-1½ MT/ha) may be adequate. One-third of the fertilizer is worked into each planting site when most of the plants appear above ground, one-third in a circle about 1 ft (30 cm) out from each plant 2 months later, and one-third at double the distance 2 months after that. Supplementary feedings will depend on signs of deficiencies (often determined by leaf analyses) as the plantation develops. Fertilization needs vary with the soil. In Puerto Rico, most plantains are grown on humid Oxisols and Ultisols in the interior. These soils are well-drained but relatively infertile and highly acid, the pH being about 4.8. On such soils, potassium uptake may be too high and N and Mg deficiencies occur. But experts have shown that these soils respond to good fertilization practices and can be very productive. As an example, 224 lbs N per acre (224 kg/ha) applied in circular bands 1.5 ft (0.46 m) from the base of the pseudostem gives a significantly higher yield than broadcast N, and there is good response to Mg applied at time of planting and again 7 months later.

In the humid mountain regions of Puerto Rico, 250 to 325 lbs N per acre (250-325 kg/ha), 125 to 163 lbs phosphorus per acre (125-163 kg/ha), and 500 to 650 lbs potassium per acre (500-650 kg/ha) are recommended for plantains. On lowland sandy clay, phosphorus and magnesium applications appear ineffective. Applications of N at the rate of 168 to 282 lbs/acre (168-282 kg/ha) increase size and number of fruits harvested, but higher rates of N decrease yield because of the number of plants that bend over halfway or are stunted or fail to flower. Applications of 1,121 lbs N per acre (1,121 kg/ha) reduce production by 46%. Potassium at the rate of 405 to 420 lbs/acre (405-420 kg/ha) has the effect of increasing weight and number of fruits. However, there appear to be factors, possibly soil magnesium and calcium, which inhibit the uptake of potassium. One study showed that it took one year for heavy applications of K to reach down to a depth of 8 in (20 cm) where most of the roots were found in a banana plantation on clay loam. One benefit of added potassium is that it makes bananas more buoyant. In cool, dry seasons in Honduras, the fruit tissue is abnormally dense and there is a high rate of "sinkers" when hands are floated through a washing

Fig. 10: Immature banana bunch ("stem") in protective plastic cover; Hacienda Secadal, Ecuador.

As the older leaves wither and droop, they must be removed because they interfere with spraying, they shade the suckers, cause blemishes on the fruits, harbor disease, insects and other creatures, and constitute a fire hazard.

Bearing bananas require propping. This has been done with simple wooden or bamboo poles, forked poles, or two stakes fastened together to form an "X" at the top, a system much less harmful to the pseudostem. Or the plant may be tied back to pickets driven into the ground, to prevent falling with the weight of the bunch.

Various types of covering—dry banana leaves, canvas, drill cloth, sisal sacks, or burlap or so-called "Hessian" bags (made of jute), have been put over banana bunches intended for export, especially to enhance fruit development in winter and avoid blemishes. In 1955, Queensland led the trend toward adoption of tubular poly-vinyl-chloride (PVC), then the cheaper blue polyethylene covers after trials produced record bunches. At first, the transparent covering caused sunburn on the first two hands and it was found necessary to protect these with newspaper before pulling on the plastic sleeve. The use of plastic covers became standard practice not only in Australia but in Africa, India and the American tropics. In 1963, Queensland growers were turning to covers made of High Wet Strength (formaldehyde-treated) kraft paper which was already in use for garbage bags. These bags were easily stapled at the top, prevented sunburn, resisted adverse weather, and were reusable for at least another season. Some growers still prefer the burlap. It is cautioned that the cover should not be put on until the bracts have lifted from the fruits (about 21 days after "shooting") so that the young fingers will be firm enough to resist the friction of the cover.

If bunches are composed of more than 7 hands, debudding, or "de-belling"—that is, removal of the terminal male bud (which keeps on extending and growing) will result in somewhat fuller bananas, thus increasing bunch weight. The cut should be made several inches below the last hand so that the rotting tip of the severed stalk will not affect the fruits.

Harvesting

Banana bunches are harvested with a curved knife when the fruits are fully developed, that is, 75% mature, the angles are becoming less prominent and the fruits on the upper hands are changing to light-green; and the flower remnants (styles) are easily rubbed off the tips. Generally, this stage is reached 75 to 80 days after the opening of the first hand. Cutters must leave attached to the bunch about 6 to 9 in (15-18 cm) of stalk to serve as a handle for carrying. With tall cultivars, the pseudostem must be slashed partway through to cause it to bend and harvesters pull on the leaves to bring the bunch within reach. They must work in pairs to hold and remove the bunch without damaging it. In the early 1960's a "banana bender" was invented in Queensland—an 8-ft pole with a steel rod mounted at the top and shaped with a downward-pointing upper hook and an upward-pointing lower hook, the first to pull the pseudostem down after nicking

tank. Such fruits have been found deficient in potassium and increased potassium in the fertilizer has reduced the problem. Irrigation by costly overhead sprinkler systems is standard practice in large-scale banana culture in Central America. Without such equipment, irrigation basins may be necessary throughout the field and they should be able to hold at least 3 in (7.5 cm) of water. During the first 2 months, the plants should be irrigated every 7 to 10 days; older plants need irrigation only every 3 to 4 weeks in dry seasons. On heavy soils, too frequent irrigations decrease yields. For maximum root development, the water table must be between 14 and 19 in (36-48 cm) below ground level.

To preserve the original density, the plants are pruned; that is, only the most deep-seated sucker and one or more of its offshoots ("peepers") are permitted to exist beside each parent plant to serve as replacements and maintain a steady succession. All other suckers are killed to prevent competition with the pseudostem and its "followers", and a bunch of fruits will be ready for harvest every 6 to 8 months. Various methods of de-suckering have been employed: 1) wrenching by hand; 2) cutting at soil level with a banana knife; 3) cutting at soil level and filling the base with kerosene; 4) cutting at soil level and killing the underground terminal bud by thrusting in and twisting a gouging tool.

and the second to support the bent pseudostem so that the bunch can be cut at a height of about 4½ ft (1.35 m).

Formerly, entire bunches were transported to shipping points and exported with considerable loss from inevitable damage. Improved handling methods have greatly reduced bunch injuries. In modern plantations, the bunches are first rested on the padded shoulder of a harvester and then are hung on special racks or on cables operated by pulleys by means of which they can be easily conveyed to roads and by vehicle to nearby packing sheds. Where fields have been located in remote areas lacking adequate highways, transport out has been accomplished by hovercraft flying along riverbeds. In Costa Rica, when rains have prevented truck transport to railway terminals, bananas have been successfully carried in slings suspended from helicopters. Exposure to even moderate light after harvest initiates the ripening process. Therefore the fruits should be protected from light as much as possible until they reach the packing shed.

In India, studies have been made to determine the most feasible disposition of a plant from which a bunch has been harvested. It is normal for it to die and it may be left standing for 3 to 4 months to dehydrate before removal, or the top half may be removed right after harvest by means of a tool called a "mattock" (a combined axe and hoe); or the pseudostem may be cut at ground level, split open, and the tender core taken away for culinary purposes. Results indicated that the first two practices have equal effect on production, but the complete felling and removal of the pseudostem lowered the yield of the "follower" significantly. In Jamaica and elsewhere it is considered best to chop and spread as organic matter the felled pseudostem and other plant residue. This returns to the soil 404 lbs N, 101 lbs P and 1,513 lbs K from an acre of bananas (404 kg, 101 kg and 1,513 kg, respectively, from a hectare). The stump should be covered with hard-packed soil to discourage entrance of pests.

Banana plantations, if managed manually, may survive for 25 years or far longer. The commercial life of a banana "stool" is about 5 or 6 years. From the 4th year on, productivity declines and the field becomes too irregular for mechanical operations. Sanitary regulations require that the old plantings be eradicated. In the past, this has been done by digging out the plants with the mattock, or bringing in cattle to graze on them. In recent years, the old plants and the suckers that arise from the old corms are injected with herbicide until all are thoroughly killed and the field is then cleared. Where bananas or plantains are raised on cleared forest land without sophisticated maintenance practices, they become thoroughly infested with nematodes by the end of the third year and the regrowth of underbrush has begun to take over the field, so it is simply abandoned.

Fig. 11: Mature, newly harvested, banana bunches at Hacienda Secadal, Ecuador.

Yield

It is clear that many factors determine the annual yield from a banana or plantain plantation: soil and agronomic practices, the cultivar planted, spacing, the type of propagating material and the management of sucker succession. The 'Gros Michel' banana has yielded 3 to 7 tons per acre (3 to 7 MT/ha) in Central America. A 'Giant Cavendish' bunch may weigh 110 lbs (50 kg) and have a total of 363 marketable fruits. A well-filled bunch of "Dwarf Cavendish' will have no more than 150 to 200 fruits. Sword suckers of plantains have yielded 54,984 fruits per acre (135,866 fruits /ha); water suckers, 49,021 fruits per acre (121,132 fruits/ha).

With heavy fertilization, the 'Maricongo' plantain in Puerto Rico, planted at the rate of 725 per acre has produced 21,950 fruits per acre (54,238 fruits/ha); at the rate of 1,450 per acre has produced 39,080 fruits per acre (96,369 fruits/ha); in a single year.

In 1981, investigators of the earnings of plantain producers in Puerto Rico found that traditional farmers had costs of $1,568.00 per acre ($3,874.59/ha); gross income of $2,436.90 per acre ($6,021.58/ha); and net profit of $868.88 per acre ($2,146.99/ha). Those farmers who had adopted improved techniques for preparing the field, weeding and control of pests and diseases had a cost of $2,132.14 per acre ($5,268.52/ha); gross income of $4,253.26 per acre ($10,509.81/ha); and net profit of $2,121.12 per acre ($5,241.29/ha).

'Maricongo' plantains spaced at 5 x 5 ft (1.5 x 1.5 m), 1,742 plants/acre (4,303 plants/ha), have produced 33.4 tons per acre (73.5 tons/ha) over a period of 30 months.

Handling and Packing

Banana bunches were formerly padded with leaf trash which absorbed much of the sap and latex from the harvesting operation and the sites of broken-off styles, each of which can leak at least 6 drops, especially if bunches are cut early in the morning. In the 1960's, when whole bunches were being exported from the Windward Islands and Jamaica to England, they were wrapped in wadding (paperbacked layers of paper tissue) to absorb the latex, and then encased in plastic sleeves for shipment. Nowadays plastic sleeves left on the bunches help protect them during transport from the field to distant packing sheds and a cushion of banana trash on the floor and against the sides of the truck does much to reduce injury. But the plastic bags increase the problem of staining by the sap/latex which mingles with the condensation inside the bag, becomes more fluid, runs down the inside and stains the peel. When hands are cut off, additional sap/latex mixture oozes from the severed crown. Banana growers and handlers know that this substance oxidizes and makes an indelible dark-brown stain on clothing. It similarly blemishes the fruits. At packing stations, the hands are floated through water tanks to wash it off. (Sodium hydrochlorate is an effective solvent.) Some people maintain that the fruit should remain in the tank for 30 minutes until all oozing of latex ceases. At certain times of the year, up to 5% of the hands may sink to the bottom of the tank, become superficially scarred and no longer exportable. As mentioned earlier, increased potassium in fertilizer mixtures renders the bananas more buoyant and fewer hands sink. In rainy seasons, it may be necessary to apply fungicide on the cut crown surface to avoid rotting, though experiments have shown that some fungicides give an off-flavor to the fruit.

Boxing was experimented with in the late 1920's but abandoned because of various types of spoilage. Modern means of combatting the organisms that cause such problems, as well as better systems of handling and transport, quality control, and good container design, have made carton-packing not only feasible but necessary. First, the hands are graded for size and quality and then packed in layers in special ventilated cartons with plastic padding to minimize bruising.

In the past, bananas for export from Fiji to New Zealand were detached individually from the hands and packed tightly in 72-lb (33-kg) wooden boxes, with much bruising of the upper layer and of the fruits in contact with the sides. Reduction of fruit quality was found to offset the economic advantage of filling all the shipping space with fruits. Wooden boxes were abandoned and suppliers were converted to the packing of hands with cushioning material.

Controlled Ripening and Storage

At times, markets may not be able to absorb all the bananas or plantains ready for harvest. Experiments have been conducted to determine the effect of applying gibberellin, either by spraying or in the form of a lanolin paste, on the stalk just above the first hands, or by injection of a solution, powder or tablet into the stalk. In Israel, gibberellin A_4A_7, applied by any of these methods about 2 months before time of normal ripening, had the effect of delaying ripening from 10 to 19 days. If applied too early, the gibberellin treatment has no effect.

Harvested bananas allowed to ripen naturally at room temperature do not become as sweet and flavorful as those ripened artificially. Post-harvest ripening is expedited undesirably if bunches or hands are stored in unventilated polyethylene bags. As a substitute for expensive controlled-temperature storage rooms, researchers in Thailand have found that hands treated with fungicide can be stored or shipped over a period of 4 weeks in polyethylene bags if ethylene-absorbing vermiculite blocks (treated with a fresh solution of potassium permanganate) are included in the sack. The permanganate solution will be ineffective if exposed to light and oxygen. The blocks must be encased in small polyethylene bags perforated only on one side to avoid staining the fruits.

Bananas are generally ripened in storage rooms with 90 to 95% relative humidity at the outset, later reduced to 85% by ventilation: and at temperatures ranging from 58° to 75°F (14.4°-23.9°C), with 2 to 3 exposures to ethylene gas at 1:1000, or 6 hourly applications for 1 to 4 days, depending on the speed of ripening desired. The fruit must be kept cool at 56°-60°F (13.3°-15.6°C) and 80 to 85% relative humidity after removal from storage and during delivery to markets to avoid rapid spoilage. Post-ripening storage at 70°F (21°C) in air containing 10 to 100 ppm ethylene accelerates softening but the fruits will remain clear yellow and attractive with few or no superficial brown specks.

Plantains for processing in the ripe stage or marketing fresh must be stored under conditions that will provide the best quality of finished product. Puerto Rican studies have shown that uniform ripening is achieved in 4 to 5 days by storage at 56° to 72°F (13.3°-22.2°C), 95 to 100% relative humidity, and with a single exposure to ethylene gas. The initial 4% starch content is reduced to 1 to 1.74% and sugars increase by about 2%. The ripe fruit can be held another 6 days at 56°F (13.3°C) and still be acceptable for processing.

The manufacture of products from the green, still starchy, plantain is a major industry in Puerto Rico. If held at room temperature, the fruits begin to ripen 7 days after harvest and become fully ripe at the end of 2 more days. Chemically disinfected fruits stored in polyethylene bags with an ethylene absorbent (Purafil wrapped in porous paper) keep 25 days at room temperature of 85°F (29.44°C), and for 55 days under refrigeration at 55°F (12.78°C). Products of such fruits have been found to be as good as or better than those made from freshly harvested green plantains.

The potential benefits of waxing have been considered by various investigators. While it is true that waxing of pre-disinfected fruits prolongs storage life by 60% at room temperature, 78°-92°F (25.56°-33.33°C), and by 28% at 52° to 55°F (11.11°-12.78°C), there is no advantage in waxing if the fruits can be held in gas storage, a combination of waxing and gassing being no better than gassing alone. In fact, waxing may result in uneven ripening after storage.

In the mid-1960's, fumigation by ethylene dibromide (EDB) against fruit fly infestation was authorized to permit export of Hawaiian bananas to the mainland USA. The treatment accelerated ripening and it could not be applied to 'Dwarf Cavendish' without covering the bunch with opaque or semi-opaque material for at least 2 months prior to harvest. EDB is no longer approved for use on food products for marketing within the United States.

Pests

Wherever bananas and plantains are grown, nematodes are a major problem. In Queensland, bananas are attacked by various nematodes that cause rotting of the corms: spiral nematodes—*Scutellonema brachyurum*, *Helicotylenchus multicinctus* and *H. nannus*; banana root-lesion nematode, *Pratylenchus coffaea*, syn. *P. musicola;* and the burrowing nematode, *Radopholus similis* less than 1 mm long, which enters roots and corms, causing red, purple and reddish-black discoloration and providing entry for the fungus *Fusarium oxysporum*. And also prevalent is the root-knot nematode, *Meliodogyne javanica.*

Plantains in Puerto Rico are attacked by 22 species of nematodes. The most injurious is the burrowing nema-

tode and it is the cause of the common black head-toppling disease on land where plantains have been cultivated for a long time. Wherever coffee has been grown, *Pratylenchus coffaea* is the principal nematode, and where plantains have been installed on former sugar cane land, *Meliodogyne incognita* is dominant. These last two are among the three most troublesome nematodes of Surinam, the third being *Helicotylenchus* spp., especially *H. multicinctus.*

Nematicides, properly applied, will protect the crop. Otherwise, the soil must be cleared, plowed and exposed to the sun for a time before planting. Sun destroys nematodes at least in the upper several inches of earth. Some fields may be left fallow for as long as 3 years. Rotating plantains with Pangola grass (*Digitaria decumbens*) controls most of the most important species of nematodes except *Pratylenchus coffaea*. All planting material must be disinfected—corms, or parts of corms, or the bases of suckers. There are various means of accomplishing this. In Hawaii, corms are immersed in water at 122°F (50°C) for 15 minutes and soaked for 5 minutes in 1% sodium hypochlorite. In Puerto Rico, nematodes are combatted by immersing plantain corms in a solution of Nemagon for 5 minutes about 24 hours before planting and, when planting, mixing the soil in the hole with granular Dasanit (Fensulfothion) and every 6 months applying Dasanit in a ring around the pseudostem.

In Queensland, corms are immersed in hot water—131°F (55°C)—for 20 minutes or solutions of non-volatile Nemacur or Mocap. Hot water and Nemacur are equally effective but hot water has less adverse effects on plant vigor. The Australians believe that nematicidal treatment of corms must be preceded by peeling off ⅜ in (1 cm) of the outer layer (usually discolored) even though this diminishes the vigor of the planting material. However, tests with 'Maricongo' plantain corms in Puerto Rico indicate that immersing for 10 minutes in aqueous solutions of Carbofuran, Dasanit, Ethoprop, or Phenamiphos without the time-consuming and possibly detrimental peeling reduces the initial nematode populations by about 95% and all the nematicides except Carbofuran give adequate post-planting control. Carbofuran apparently does not penetrate deeply enough. The Florida spiral nematode is the most damaging nematode in Brazil and Florida, especially during hot, rainy summers. Ethoprop is the only nematicide registered for use on bananas in Florida but it is not effective against this pest. The hot water treatment must be employed.

The black weevil, *Cosmopolites sordidus*, also called banana stalk borer, banana weevil borer, or corm weevil, is the second most destructive pest of bananas and plantains. It attacks the base of the pseudostem and tunnels upward. A jelly-like sap oozes from the point of entry. It was formerly controlled by Aldrin, which is now banned. In Surinam it has been combatted by injecting pesticide into the pseudostem, or spraying the pseudostem with Monocrotophos. In Ghana, they dip planting material in a solution of Monocrotophos and apply dust of Dieldrin or Heptachlor around the base of the pseudostem. Puerto

Rican tests of several pesticides have shown that Aldicarb 10G, a nematicide-insecticide, applied at the base of plantain plants at the rate of 1 to 1½ oz (30-45 g) every 4 months, or 1 oz (30 g) every 6 months, controls both the burrowing nematode and the black weevil. Biological control of black weevil utilizing a weevil predator, *Piaesius javanus,* has not been successful.

The banana rust thrips, *Chaetanophothrips orchidii,* syn. *C. signipennis,* stains the peel, causes it to split and expose the flesh which quickly discolors. The pest is usually partially controlled by the spraying of Dieldrin around the base of the pseudostem to combat the banana weevil borer, because it pupates in the soil. Another measure has been to treat the inside of polyethylene bunch covers with insecticidal dust, especially Diazinon, before slipping them over the bunches. It is recognized that this procedure constitutes a health hazard to the workers. A great improvement is the introduction of polyethylene bags impregnated with 1% of the insecticide Dursban, eliminating the need for dusting. Bunches enclosed in these bags have been found 85% free of attack by the banana rust thrips. The bags retain their potency for at least a year in storage. Impregnated with 1 to 2% Dursban, they are equal to Diazinon in preventing banana injury by the banana fruit-scarring beetle, *Colaspis hypochlora,* also called coquito. This pest invades the bunches when the fruits are very young. It has been very troublesome in Venezuela, and at times from Guyana to Mexico. The banana scab moth, *Nacoleia octasema,* infests the inflorescence from emergence to the time half the bracts have lifted. It is a major pest in North Queensland, Malaysia and the southwest Pacific. Control may be by injection or dusting with pesticide, sometimes with lifting or removal of bracts. Corky scab of bananas in southern Queensland is caused by the banana flowers thrips, *Thrips florum,* especially in hot, dry weather. The infestation is lessened by removal of the terminal male bud which tends to harbor the pest.

Among minor enemies in Queensland is the banana spider mite, *Tetranychus lambi* which moves from beneath the leaves to the fruits in warm weather and creates dull brown specks which may become so numerous as to completely cover the peel, causing it to dehydrate and crack irregularly. The leaves of the plant will wilt. Bi-weekly sprayings of pesticide get rid of the mites.

The banana silvering thrips, *Hercinothrips bicintus,* causes silvery patches on the peel and dots them with shiny black specks of excrement. The rind-chewing caterpillar, *Barnardiella sciaphila,* usually does little damage. Two species of fruit fly—*Strumeta tryoni* and *S. musae* —occasionally attack bananas in North Queensland.

Diseases

The subject of diseases is authoritatively presented by C.W. Wardlaw in the second edition of his textbook, *Banana Diseases, including plantains and abaca,* 1972; 878 pages.

It is appropriate here only to mention the main details of those maladies which are of the greatest concern to banana and plantain growers. Sigatoka, or leaf spot, caused by the fungus *Mycosphaerella musicola* (of which the conidial stage is *Cercospora musae*) was first reported in Java in 1902, next in Fiji in 1913 where it was named after the Sigatoka Valley. It appeared in Queensland 10 years later, and in another 10 years made its appearance in the West Indies and soon spread throughout tropical America. The disease was noticed in East and West Tropical Africa in 1939 and 1940. It was discovered in Ghana in 1954 and ravaged a state farm in 1965. It is most prevalent on shallow, poorly drained soil and in areas where there is heavy dew. The first signs on the leaves are small, pale spots which enlarge to ½ in (1.25 cm), become dark purplish-black and have gray centers. When the entire plant is affected, it appears as though burned, the bunches will be of poor quality and will not mature uniformly. The fruits will be acid, the plant roots small. Control is achieved by spraying with orchard mineral oil, usually every 3 weeks, a total of 12 applications of 1½ gals per acre (14.84 liters/ha); or by systemic fungicides applied to the soil or by aerial spraying.

A much more virulent malady, Black Sigatoka, or Black Leaf Streak, caused by *Mycosphaerella fijiensis* var. *difformis,* attacked bananas in Honduras in 1969 and spread to banana plantations in Guatemala and Belize. It appeared in plantations in Honduras in 1972 where there had not been any need to spray against ordinary Sigatoka. It made headway rapidly through plantain fields in Central America to Mexico and about 10 years later was found in the Uruba region of Colombia. The disease struck Fiji in 1963 and became an epidemic. It began spreading in 1973, largely replacing ordinary Sigatoka. Surveys have revealed this previously unrecognized disease on several other South Pacific islands, in Hawaii, the Philippines, Malaysia and Taiwan. It is spread mostly by wind; kills the leaves and exposes the bunches to the sun. Cultivars which are resistant to Sigatoka have shown no resistance to Black Sigatoka. There are vigorous efforts to control the disease by fungicides or intense oil spraying. But it is not completely controlled even by spraying every 10 to 12 days—a total of 40 sprayings. The cost of control with fungicides is 3 to 4 times that of controlling ordinary Sigatoka because of the need for more frequent aerial sprayings. It is very difficult to treat properly on islands where bananas are grown mostly in scattered plantings. In Mexico where plantains are extremely important in the diet, and 65% of the production is on non-irrigated land, control efforts have elevated costs of plantain production by 145 to 168%. In the Sula Valley of Honduras, Black Sigatoka has caused annual losses of 3,000,000 boxes of bananas. The great need is for resistant cultivars of high quality.

Panama Disease or Banana Wilt, which arises from infection by the fungus, *Fusarium oxysporum* f. sp. *cubense* originates in the soil, travels to the secondary roots, enters the corm only through fresh injuries, passes into the pseudostem; then, beginning with the oldest leaves, turns them yellow first at the base, secondly along the margins, and lastly in the center. The interior leaves turn bronze and droop. The pseudostem turns brown inside. This plague has seriously affected banana production in Central America, Colombia and the Canary Islands. It started spreading in southern Taiwan in 1967 and has become the leading local banana disease. The 'Cavendish' types have been considered highly resistant but they succumb if planted on land previously occupied by 'Gros Michel'. The disease is transmitted by soil, moving agricultural vehicles or other machinery, flowing water, or by wind. It is combatted by flooding the field for 6 months. Or, if it is not too serious, by planting a cover crop. There are reportedly two races: Race #1 affects 'Gros Michel', 'Manzano', 'Sugar' and 'Lady Finger'; Race #2 attacks 'Bluggoe'. Resistant cultivars are the Jamaican 'Lacatan', 'Monte Cristo', and 'Datil' or 'Niño'. Resistant plantains are 'Maricongo', 'Enano' and 'Pelipita'.

Moko Disease, or Moko de Guineo, or Marchites bacteriana, is caused by the bacterium, *Pseudomonas solanacearum*, resulting in internal decay. It has become one of the chief diseases of banana and plantain in the western hemisphere and has seriously reduced production in the leading areas of Colombia. It attacks *Heliconia* species as well. It is transmitted by insects, machetes and other tools, plant residues, soil, and root contact with the roots of sick plants. There are said to be 4 different types transmitted by different means. Efforts at control include covering the male bud with plastic to prevent insects from visiting its mucilaginous excretion; debudding, disinfecting of cutting tools with formaldehyde in water 1:3; disinfection of planting material; disposal of infected fruits and plant parts; injection of herbicide into infected plants to hasten dehydration, and also seemingly healthy neighboring plants. If the organism is variant SFR, all adjacent plants within a radius of 16.5 ft (5 m) must be destroyed and the area not replanted for 10 to 12 months, for this variant persists in the soil that long. If it is variant B, the plants within 32.8 ft (10 m) must be injected and the area not replanted for 18 months. In either case, the soil must be kept clear of broad-leaved weeds that may serve as hosts. In Colombia, there are 12 species of weeds that serve as hosts or "carriers" but only 4 of these are themselves susceptible to the disease. Crop rotation is sometimes resorted to. The only sure defense is to plant resistant cultivars, such as the 'Pelipita' plantain.

Black-end arises from infection by the fungus *Gloeosporium musarum*, of which *Glomerella cingulata* is the perfect form. It causes anthracnose on the plant and attacks the stalk and stalk-end of the fruits forming dark, sunken lesions on the peel, soon penetrating the flesh and developing dark, watery, soft areas. In severe cases, the entire skin turns black and the flesh rots. Very young fruits shrivel and mummify. This fungus is often responsible for the rotting of bananas in storage. Immersing the green fruits in hot water, 131°F (55°C) for 2 minutes before ripening greatly reduces spoilage.

Cigar-tip rot, or Cigar-end disease, *Stachylidium (Verticillium) theobromae* begins in the flowers and extends to the tips of the fruits and turns them dark, the peel darkens, the flesh becomes fibrous. One remedy is to cut off withered flowers as soon as the fruits are formed and apply copper fungicides to the cut surfaces.

In Surinam, cucumber mosaic virus attacks plantains especially when cucumber is interplanted in the fields. Also, Chinese cabbage, Cayenne pepper and "bitter greens" (*Cestrum latifolium* Lam.) are hosts for the disease.

Cordana leaf spot (*Cordana musae*), causes oval lesions 3 in (7.5 cm) or more in length, brown with a bright-yellow border. There is progressive dying of the leaves beginning with the oldest, as in Sigatoka, with consequent undersized fruits ripening prematurely. It formerly occurred mainly in sheltered, humid regions of Queensland. Now it is seen mostly as an invader of areas affected by Sigatoka, in various geographical locations.

Bunchy top, an aphid-transmitted virus disease of banana, was unknown in Queensland until about 1913 when it was accidentally introduced in suckers brought in from abroad. In the next 10 years it spread swiftly and threatened to wipe out the banana industry. Drastic measures were taken to destroy affected plants and to protect uninvaded plantations. The disease was found in Western Samoa in 1955 and it eliminated the susceptible 'Dwarf Cavendish' from commercial plantings. A vigorous eradication and quarantine program was undertaken in 1956 and carried on to 1960. Thereafter, strict inspection and control measures continued. Other crops were provided to farmers in heavily infested areas. Leaves formed after infection are narrow, short, with upturned margins and become stiff and brittle; the leafstalks are short and unbending and remain erect, giving a "rosetted" appearance. The leaves of suckers and the 3 youngest leaves of the mother plant show yellowing and waviness of margins, and the youngest leaves will have very narrow, dark-green, usually interrupted ("dot-and-dash") lines on the underside.

Because of the seriousness of Panama disease and Bunchy Top in southern Queensland, the prospective banana planter must obtain a permit from the Queensland Department of Primary Industries. In the Southern Quarantine Area, any plant showing Bunchy Top, as well as its suckers and all plants within a 15 ft (4.6 m) radius must be killed by injecting herbicide or must be dug out completely and cut into pieces no bigger than 2 in (5 cm) wide. In restricted areas, only the immune 'Lady Finger' may be grown. In the Northern Quarantine Area, no plants may be brought in from another area and all plants within a radius of 120 ft (36.5 m) from a diseased plant must be eradicated.

Swelling and splitting of the corm and the base of the pseudostem is caused by saline irrigation water and by overfertilization during periods of drought which builds up soluble salts in the soil.

Food Uses

The ripe banana is utilized in a multitude of ways in the human diet — from simply being peeled and eaten out-of-hand to being sliced and served in fruit cups and salads, sandwiches, custards and gelatins; being mashed and incorporated into ice cream, bread, muffins, and cream pies. Ripe bananas are often sliced lengthwise, baked or broiled, and served (perhaps with a garnish of brown sugar or chopped peanuts) as an accompaniment for ham or other meats. Ripe bananas may be thinly sliced and cooked with lemon juice and sugar to make jam or sauce, stirring frequently during 20 or 30 minutes until the mixture jells. Whole, peeled bananas can be spiced by adding them to a mixture of vinegar, sugar, cloves and cinnamon which has boiled long enough to become thick, and then letting them cook for 2 minutes.

In the islands of the South Pacific, unpeeled or peeled, unripe bananas are baked whole on hot stones, or the peeled fruit may be grated or sliced, wrapped, with or without the addition of coconut cream, in banana leaves, and baked in ovens. Ripe bananas are mashed, mixed with coconut cream, scented with *Citrus* leaves, and served as a thick, fragrant beverage.

Banana puree is important as infant food and can be successfully canned by the addition of ascorbic acid to prevent discoloration. The puree is produced on a commercial scale in factories close to banana fields and packed in plastic-lined #10 cans and 55-gallon metal drums for use in baby foods, cake, pie, ice cream, cheesecake, doughnuts, milk shakes and many other products. It is also used for canning half-and-half with applesauce, and is combined with peanut butter as a spread. Banana nectar is prepared from banana puree in which a cellulose gum stabilizer is added. It is homogenized, pasteurized and canned, with or without enrichment with ascorbic acid.

Sliced ripe bananas, canned in sirup, were introduced to the food trade for commercial use in frozen tarts, pies, gelatins and other products. In 1966, the United Fruit Company built a processing plant at La Lima, Honduras, for producing canned and frozen banana puree and canned banana slices. Because of seasonal gluts and perishability and the tonnages of bananas and plantains that are not suitable for marketing or export because of overripeness or stained peel or other defects, there is tremendous interest in the development of modes of processing and preserving these fruits.

In Polynesia, there is a traditional method of preserving large quantities of bananas for years as emergency fare in case of famine. A pit is dug in the ground and lined with banana and *Heliconia* leaves. The peeled bananas are wrapped in *Heliconia* leaves, arranged in layer after layer, then banana leaves are placed on top and soil and rocks heaped over all. The pits remain unopened until the fermented food, called "masi", is needed.

In Costa Rica, ripe bananas from an entire bunch are peeled and boiled slowly for hours to make a thick sirup which is called "honey".

Green bananas, boiled in the skin, are very popular in Cuba, Puerto Rico and other Caribbean islands. In Puerto Rico, the cooked bananas are recooked briefly in a marinating sauce containing black pepper, vinegar, garlic, onions, bay leaves, olive oil and salt and left standing at room temperature for 24 hours before being eaten. Peeled, sliced green bananas are quick-frozen in Puerto Rico for later cooking. If steam-treated to facilitate peeling, the enzymes are inactivated only on the surface of the flesh and the interior, when exposed, will turn brown unless sulfited. It is more satisfactory to immerse the whole bananas in water at 200°F (93°C) for 30 minutes which wholly inactivates the enzymes. No sulfite is then needed and no browning occurs.

Much research has been conducted by food technologists at the University of Puerto Rico to determine the best procedures for canning sliced green bananas and plantains to make them readily available for cooking. Enzyme inactivation is necessary and the hot water treatment facilitates the peeling. If peeled raw, green bananas and plantains exude gummy white latex which stains materials. When canning, citric acid in a 2% brine is added, but this method of preservation has not yet met with success because of rapid detinning of the inside of the cans. The problem is not solved by using enamel-lined cans because the fruit darkens quickly after the cans are opened. Glass jars may prove to be the only suitable containers.

Through experimental work with a view to freezing peeled, blanched, sliced green bananas, it has been found that, with a pulp-to-peel ratio of less than 1:3 the fruits turn gray on exposure to air after processing and this discoloration is believed to be caused by the high iron content (4.28 p/m) of the surface layer of the flesh and its reaction to the tannin normally present in green bananas and plantains. At pulp-to-peel ratio of 1:0, the tannin level in green bananas is 241.4 mg; at 1:3, 151.0 mg, and at 1:5, 112.6 mg, per 100 g. Therefore, it is recommended that for freezing green bananas be harvested at a stage of maturity evidenced by 1:5 pulp-to-peel ratio. Such fruits have a slightly yellowish flesh, higher carotene content, and are free of off-flavors. The slices are cooked by the consumer without thawing.

Completely green plantains are 50% flesh and 50% peel. Plantains for freezing should have a pulp content of at least 60% for maximum quality in the ultimate food product, but a range of 55 to 65% is considered commercially acceptable.

Ripe plantains, held until the skin has turned mostly or wholly black, are commonly peeled, sliced diagonally and fried in olive oil, accompany the main meal daily in the majority of homes in tropical Latin America. In the Dominican Republic, a main dish is made of boiled, mashed ripe plantains mixed with beaten eggs, flour, butter, milk and cloves, and layered in a casserole with ground beef fried with Picalilli and raisins, lastly topped with grated cheese and baked until golden brown. In Guatemala, boiled plantains are usually served with honey.

Green plantains are popular sliced crosswise, fried until partially cooked, pressed into a thickness of ½ in (1.25 cm), and fried in deep fat till crisp. The product is called "tostones" and somewhat resembles French-fried potatoes. Puerto Rican "mofongo" is a ball of fried green plantain mashed with fried pork rind, seasoned with thickened stock, garlic and other condiments. It must be eaten hot before it hardens. "Mofongo" has been successfully frozen in boilable pouches. Slices of nearly ripe plantain (5% starch content) are cooked in sirup and frozen in boilable pouches. Puerto Rican plantains, shipped green to Florida, have been ripened, peeled, quartered, infused with orange juice, frozen and provided to schools for serving as luncheon dessert.

In Ghana, plantains are consumed at 5 different stages of ripeness. Fully ripe plantains are often deep fried or cooked in various dishes. A Ghanian pancake called "tatale" is made of nearly full-ripe plantains and fermented whole-meal dough of maize, seasoned with onions, ginger, pepper and salt, and fried in palm oil. "Kaklo" is the same mix but thicker and rolled into balls which are deep-fried. Because home preparation is laborious, a commercial dehydrated mix has been developed. In Ghana, green plantains are boiled and eaten in stew or mashed, together with boiled cassava, into a popular plastic product called "fufu" which is eaten with soup. Because of the great surplus of plantains in summer, technologists have developed methods for drying and storing of strips and cubes of plantain for house use in making "fufu" out of season. The cubes can also be ground into plantain flour. Use of infrared, microwave, and extrusion systems has resulted in high-quality finished products. Processing has the added advantage of keeping the peels at factories where they may be converted into useful by-products instead of their adding to the bulk of household garbage.

Banana or plantain flour, or powder, is made domestically by sun-drying slices of unripe fruits and pulverizing. Commercially, it is produced by spray-drying, or drum-drying, the mashed fruits. The flour can be mixed 50-50 with wheat flour for making cupcakes. Two popular Puerto Rican foods are "pasteles" and "alcapurias"; both are pastry stuffed with meat; the first is wrapped in plantain leaves and boiled; the latter is fried. The pastry is made of plantain flour or a mixture of plantain with cassava (*Manihot esculenta* Crantz.) or cocoyam (tanier), *Xanthosoma* spp. The plantain cultivars 'Saba', 'Tundoc' and 'Latundan' are very suitable for making flour.

Commercial production and marketing of fried green plantain and banana chips has been increasing in various parts of the world over the past 25 years and these products are commonly found in retail groceries alongside potato chips and other snack foods. 'Cariñosa' and 'Bungulan' bananas are favored for chip-making. In Puerto Rico, the plantain cultivars 'Guayamero Alto' and 'Congo Enano' are chosen for this purpose.

Dried bananas, or so-called "banana figs" are peeled firm-ripe bananas split lengthwise, sulphured, and oven-dried to a moisture content of 18 to 20%. Wrapped individually in plastic and then packed by the dozen in polyethylene bags, and encased in cartons, they can be stored for a year at room temperature — 75.2° to 86°F (24°-30°C)

and they are commonly exported. The product can be eaten as a snack or minced and used together with candied lemon peel in fruit cake and other bakery products. In India the 'Dwarf Cavendish' is preferred for drying; in the Philippines, the true 'Lacatan' or the 'Higo'.

Canadian researchers have developed a system of osmotic dehydration for sliced firm-ripe bananas and plantains, especially designed for developing countries with plentiful sugar for the solutions required.

Since the early 1960's, Brazil has produced dehydrated banana flakes for local markets and export to the USA and elsewhere in vacuum-sealed cans. The flakes are used on cereal, in baked goods, canapes, meat loaf and curries, desserts, sauces, and other products. In Israel, banana flakes have been made by steam blanching 'Dwarf Cavendish' bananas and drum-drying to 2.6% moisture. The flakes, packed in vacuum-sealed cans, keep for a year at 75.2° to 86°F (24°-30°C). At temperatures to 95°F (35°C), the flakes darken somewhat and tend to stick together. Israel has also introduced a formula for high-protein flakes made of 70% banana and 30% soybean protein and this development has been adopted in Brazil. The flakes are used by Brazilian food manufacturers in ice cream, and as fillings for cakes and other bakery products. South Africa has produced flakes of ⅔ banana and ⅓ maize meal.

In Africa, ripe bananas are made into beer and wine. The Tropical Products Institute in London has established a simple procedure for preparing an acceptable vinegar from fermented banana rejects.

The terminal male bud of the wild banana, *M. balbisiana*, is marketed in Southeast Asia. It is often boiled whole after soaking an hour in salt water, or with several changes of water to reduce astringency, and eaten as a vegetable. The male bud of cultivated bananas is considered too astringent but it is, nevertheless, sometimes similarly consumed. The flowers may be removed from the bud and prepared separately. They are used in curries in Malaya and eaten with palm oil in West Tropical Africa.

The new shoots of young plants may be cooked as greens. Banana pseudostem core constitutes about 10 to 15% of the whole and contains 1% starch, 0.68% crude fiber and 1% total ash. It is often cooked and eaten as a vegetable in India and is canned with potatoes and tomatoes in a curry sauce. Circular slices about ½ in (1.25 cm) thick are treated with citric acid and potassium metabisulphite and candied.

In India, a solution of the ash from burned leaves and pseudostems is used as salt in seasoning vegetable curries. The ash contains roughly (per 100 g): potassium, 255 mg; magnesium, 27 mg; phosphorus, 33 mg; calcium, 6.6 mg; sodium, 51 mg.

Dried green plantains, ground fine and roasted, have been used as a substitute for coffee.

Animal Feed

Reject ripe bananas, supplemented with protein, vitamins and minerals, are commonly fed to swine. Green bananas are also used for fattening hogs but, because of the dryness and astringency and bitter taste due to the tannin content, these animals do not care for them unless they are cooked, which makes the feeding costs too high for most growers. Therefore, dehydrated green banana meal has been developed and, though not equal to grain, can constitute up to 75% of the normal hog diet, 40% of the diet of gestating sows. It is not recommended for lactating sows, nor are ripe bananas, even with a 40% protein supplement.

Beef cattle are very fond of green bananas whether they are whole, chopped or sliced. Because of the fruit's deficiency in protein, urea is added at the rate of 8.8 lbs (4 kg) per ton, with a little molasses mixed in to mask the flavor. But transportation is expensive unless the cattle ranch is located near the banana fields. A minor disadvantage is that the bananas are somewhat laxative and the cattle need to be washed-down daily. With dairy cattle, it is recommended that bananas constitute no more than 20% of the feed.

In the Philippines, it has been found that meal made from dehydrated reject bananas can form 14% of total broiler rations without adverse effects. Meal made from green and ripe plantain peels has been experimentally fed to chicks in Nigeria. A flour from unpeeled plantains, developed for human consumption, was fed to chicks in a mixture of ⅔ flour and ⅓ commercial chickfeed and the birds were maintained until they reached the size of fryers. They were found thinner and lighter than those on 100% chickfeed and the gizzard lining peeled in shreds. It was assumed that these effects were the result of protein deficiency in the plantains, but they were more likely the result of the tannin content of the flour which interferes with the utilization of protein.

Leaves, pseudostems, fruit stalks and peels, after chopping, fermentation, and drying, yield a meal somewhat more nutritious than alfalfa presscake. This waste material has been considered for use as organic fertilizer in Somalia. In Malaya, pigs fed the pseudostems are less prone to liver and kidney parasites than those on other diets.

Banana peel contains beta-sitosterol, stigmasterol, campesterol, cycloeucalenol, cycloartanol, and 24-methylene cycloartanol. The major constituents are 24-methylene cycloartanol palmitate and an unidentified triterpene ketone.

Food Value Per 100 g of Edible Portion*							
Banana				Plantain			
Ripe	*Green*	*Dried*	*Flour (green)*	*Ripe*	*Ripe (cooked)*	*Green*	*Dried (green)*
Calories	65.5–111	108	298	340	110.7–156.3	77	90.5–145.9	359
Moisture	68.6–78.1 g	72.4 g	19.5–27.7 g	11.2–13.5 g	52.9–77.6 g	79.8 g	58.7–74.1 g	9.0 g
Protein	1.1–1.87 g	1.1 g	2.8–3.5 g	3.8–4.1 g	0.8–1.6 g	1.3 g	1.16–1.47 g	3.3 g
Fat	0.16–0.4 g	0.3 g	0.8–1.1 g	0.9–1.0 g	0.1–0.78 g	0.10 g	0.10–0.12 g	1.4 g
Carbohydrates	19.33–25.8 g	25.3 g	69.9 g	79.6 g	25.50–36.81 g	18.1 g	23.4–37.61 g	83.9 g
Fiber	0.33–1.07 g	1.0 g	2.1–3.0 g	3.2–4.5 g	0.30–0.42 g	0.2 g	0.40–0.48 g	1.0 g
Ash	0.60–1.48 g	0.9 g	2.1–2.8 g	3.1 g	0.63–1.40 g	0.7 g	0.63–0.83 g	2.4 g
Calcium	3.2–13.8 mg	11 mg		30–39 mg	5.0–14.2 mg		10.01–12.2 mg	50 mg
Phosphorus	16.3–50.4 mg	28 mg		93–94 mg	21.0–51.4 mg		32.5–43.2 mg	65 mg
Iron	0.4–1.50 mg	0.9 mg		2.6–2.7 mg	0.40–1.50 mg		0.56–0.87 mg	1.1 mg
B-Carotene	0.006–0.151 mg				0.11–1.32 mg		0.06–1.38 mg	45 mg
Thiamine	0.04–0.54 mg				0.04–0.11 mg		0.06–0.09 mg	0.10 mg
Riboflavin	0.05–0.067 mg				0.04–0.05 mg		0.04–0.05 mg	0.16 mg
Niacin	0.60–1.05 mg				0.48–0.70 mg		0.32–0.55 mg	1.9 mg
Ascorbic Acid	5.60–36.4 mg				18–31.2 mg		22.2–33.8 mg	1 mg
Tryptophan	17–19 mg				8–15 mg		7–10 mg	14 mg
Methionine	7–10 mg				4–8 mg		3–8 mg	
Lysine	58–76 mg				34–60 mg		37–56 mg	

*Derived from various analyses made in Cuba, Central America and Africa.

Other Uses

Banana leaves are widely used as plates and for lining cooking pits and for wrapping food for cooking or storage. A section of leaf often serves as an eye-shade. In Latin America, it is a common practice during rains to hold a banana leaf by the petiole, upside-down, over one's back as an "umbrella" or "raincoat". The leaves of the 'Fehi' banana are used for thatching, packing, and cigarette wrappers. The pseudostems have been fastened together as rafts. Split lengthwise, they serve as padding on banana-inspection turntables and as cushioning to protect the bunches ("stems") during transport in railway cars and trucks. Seat pads for benches are made of strips of dried banana pseudostems in Ecuador. In West Africa, fiber from the pseudostem is valued for fishing lines. In

Fig. 12: Banana pseudostem pad on inspection-turntable, Hacienda Secadal, Ecuador.

the Philippines, it is woven into a thin, transparent fabric called "agna" which is the principal material in some regions for women's blouses and men's shirts. It is also used for making handkerchiefs. In Ceylon, it is fashioned into soles for inexpensive shoes and used for floor coverings.

Plantain fiber is said to be superior to that from bananas. In the mid-19th Century, there was quite an active banana-fiber industry in Jamaica. Improved processes have made it possible to utilize banana fiber for many purposes such as rope, table mats and handbags. In Kerala, India, a kraft-type paper of good strength has been made from crushed, washed and dried banana pseudostems which yield 48 to 51% of unbleached pulp. A good quality paper is made by combining banana fiber with that of the betel nut husk (*Areca catechu* L.). But Australian investigators hold that the yield of banana fiber is too low for extraction to be economical. Only 1 to 4 oz (28-113 g) can be obtained from 40 to 80 lbs (18-36 kg) of green pseudostems; 132 tons of green pseudostems would yield only 1 ton of paper. Their conclusion is that the pseudostem has much greater value as organic matter chopped and left in the field.

Dried banana peel, because of its 30 to 40% tannin content, is used to blacken leather. The ash from the dried peel of bananas and plantains is rich in potash and used for making soap. That of the burned peel of unripe fruits of certain varieties is used for dyeing.

In the Philippines, the Pinatubo Negritos cut off a banana plant close to the ground, make a hollow in the top of the stump, which then fills with watery sap drunk as an emergency thirst-quencher. Central Americans obtain the sap of the red banana in the same manner and take it as an aphrodisiac.

Medicinal Uses: All parts of the banana plant have medicinal applications: the flowers in bronchitis and dysentery and on ulcers; cooked flowers are given to diabetics; the astringent plant sap in cases of hysteria, epilepsy, leprosy, fevers, hemorrhages, acute dysentery and diarrhea, and it is applied on hemorrhoids, insect and other stings and bites; young leaves are placed as poultices on burns and other skin afflictions; the astringent ashes of the unripe peel and of the leaves are taken in dysentery and diarrhea and used for treating malignant ulcers; the roots are administered in digestive disorders, dysentery and other ailments; banana seed mucilage is given in cases of catarrh and diarrhea in India.

Antifungal and antibiotic principles are found in the peel and pulp of fully ripe bananas. The antibiotic acts against *Mycobacteria*. A fungicide in the peel and pulp of green fruits is active against a fungus disease of tomato plants. Norepinephrine, dopamine, and serotonin are also present in the ripe peel and pulp. The first two elevate blood pressure; serotonin inhibits gastric secretion and stimulates the smooth muscle of the intestines.

Alleged hallucinogenic effects of the smoke of burning banana peel have been investigated scientifically and have not been confirmed.

Folklore

The banana plant because of its continuous reproduction is regarded by Hindus as a symbol of fertility and prosperity, and the leaves and fruits are deposited on doorsteps of houses where marriages are taking place. A banana plant is often installed in the corner of a rice field as a protective charm. Malay women bathe with a decoction of banana leaves for 15 days after childbirth. Early Hawaiians used a young plant as a truce flag in wars.

Fig (Plate V)

While the ancient history of the fig centers around the Mediterranean region, and it is most commonly cultivated in mild-temperate climates, it nevertheless has its place in tropical and subtropical horticulture. Botanically identified as *Ficus carica* L. (family Moraceae), it is unique in a genus embracing perhaps over 1,000 species, mostly giant "rubber trees", and mostly tropical. It is almost universally known simply as fig, common fig, or edible fig. The name is very similar in French (*figue*), German (*feige*), Italian and Portuguese (*figo*). In Spanish it is *higo* or *brevo*. Haitians give it the name, *figue France,* to distinguish it from the small, dried bananas called "figs".

Description

The fig is a tree of small dimensions, 10 to 30 ft (3-9 m) high, with numerous spreading branches and a trunk rarely more than 7 in (17.5 cm) in diameter. It contains copious milky latex. The root system is typically shallow and spreading, sometimes covering 50 ft (15 m) of ground, but in permeable soil some of the roots may descend to 20 ft (6 m). The deciduous leaves are palmate, deeply divided into 3 to 7 main lobes, these more shallowly lobed and irregularly toothed on the margins. The blade is up to 10 in (25 cm) in length and width, fairly thick, rough on the upper surface, softly hairy on the underside. What is commonly accepted as a "fruit" is technically a synconium, that is, a fleshy, hollow receptacle with a small opening at the apex partly closed by small scales. It may be obovoid, turbinate, or pear-shaped, 1 to 4 in (2.5-10 cm) long, and varies in color from yellowish-green to coppery, bronze, or dark-purple. Tiny flowers are massed on the inside wall. In the case of the common fig discussed here, the flowers are all female and need no pollination. There are 3 other types, the "Caprifig" which has male and female flowers requiring visits by a tiny wasp, *Blastophaga grossorum;* the "Smyrna" fig, needing cross-pollination by Caprifigs in order to develop normally; and the "San Pedro" fig which is intermediate, its first crop independent like the common fig, its second crop dependent on pollination. The skin of the fig is thin and tender, the fleshy wall is whitish, pale-yellow, or amber, or more or less pink, rose, red or purple; juicy and sweet when ripe, gummy with latex when unripe. Seeds may be large, medium, small or minute and range in number from 30 to 1,600 per fruit.

Origin and Distribution

The fig is believed to be indigenous to Western Asia and to have been distributed by man throughout the Mediterranean area. It has been cultivated for thousands of years, remnants of figs having been found in excavations of Neolithic sites traced to at least 5,000 B.C. As time went on, the fig-growing territory stretched from Afghanistan to southern Germany and the Canary Islands. Pliny was aware of 29 types. Figs were introduced into England some time between 1525 and 1548. It is not clear when the common fig entered China but by 1550 it was reliably reported to be in Chinese gardens. European types were taken to China, Japan, India, South Africa and Australia.

The first figs in the New World were planted in Mexico in 1560. Figs were introduced into California when the San Diego Mission was established in 1769. Later, many special varieties were received from Europe and the eastern United States where the fig reached Virginia in 1669. The Smyrna fig was brought to California in 1881-82 but it was not until 1900 that the wasp was introduced to serve as the pollinating agent and make commercial fig culture possible. From Virginia, fig culture spread to the Carolinas, Georgia, Florida, Alabama, Mississippi, Louisiana and Texas. The tree was planted in Bermuda in early times and was common around Bahamian plantations in Colonial days. It became a familiar dooryard plant in the West Indies, and at medium and low altitudes in Central America and northern South America. There are fair-sized plantations on mountainsides of Honduras and at low elevations on the Pacific side of Costa Rica. From Florida to northern South America and in India only the common fig is grown. Chile and Argentina grow the types suited to cooler zones.

In Venezuela, the fig is one of the fruits in greatest demand by fruit processors. Because of the inadequate supply, a program was launched in 1960 to encourage commercial plantings. In 1976, fresh figs were regarded as highly desirable luxuries and were selling for $6.35 to $7.25 per lb ($14-$16/kg) in Colombia. The Instituto Colombiano Agropecuario had realized some years earlier that fig growing should be encouraged and had established an experimental plantation in 1973. The results were so favorable that they circulated an advisory bulletin to farmers in 1977, including improved methods of cultivation, costs of production and potential revenue.

Varieties

There are many cultivated varieties in each class of figs. In fact, over 700 varietal names are in use but many are synonyms. Here we need only present those that are suited to warm areas and do not require pollination. Most popular among these are 'Celeste' and 'Brown Turkey', followed by 'Brunswick' and 'Marseilles', described as follows:

'Celeste'—pear-shaped, ribbed, sometimes with a short neck and slender stalk to ¾ in (2 cm) long; the eye (opening at apex) is closed; the fruit is small to medium; the skin purplish-brown or bronze tinged with purple and covered with bloom; the pulp whitish or pinkish-amber, of rich flavor and good quality; almost seedless. Main crop is heavy but of short duration. There is rarely an early, "breba", crop.

'Brown Turkey'—broad-pyriform, usually without neck; medium to large; copper-colored; pulp is whitish shading to pink or light-red; of good to very good quality; with few seeds. The tree is prolific. The main crop, beginning in mid-July, is large; the early, breba, crop is small. This cultivar is well adapted to warm climates. It is grown on all the islands of Hawaii.

'Brunswick' ('Magnolia')—leaves narrow-lobed; fruits of main crop are oblique-turbinate, mostly without neck; fruit stalk thick, often swollen; fruit of medium size; bronze or purple-brown; pulp whitish near skin, shading to pink or amber; hollow in center; of fair to good quality; nearly seedless. Ripens over a long season. Breba crop poor; large, bronze-skinned; flesh light-red; coarse.

'Marseilles' ('White Marseilles', or 'Lemon')—fruits of main crop round to oblate without neck; on slender stalks to ¼ in (6 mm) long; of medium size. Those of breba crop, turbinate with short, thick neck and short stalk; yellow-green with small green flecks; pulp white, sweet; seeds large, conspicuous. Of fair quality.

In Queensland, 'Brown Turkey', 'Adriatic', 'Genoa' and 'Purple Genoa' perform very well.

'Adriatic' ('White Adriatic', or 'Grosse Verte')—turbinate with short, thick neck and short stalk; above medium size; green to yellowish-green with red pulp; of distinctive flavor and very good quality. In early, minor, breba crop the fruits are oblique-pyriform, large, green, often tinged with purplish-red with dark-red pulp and strong flavor.

'Genoa' ('White Genoa')—pyriform or turbinate, very faintly ribbed; neck thick and short, or absent; above medium in size; skin downy, greenish-yellow; pulp greenish-white near skin, mostly amber tinged with red; hollow; of fair quality. Fruits of breba crop oblique-obovate with thick neck and short stalk; yellowish-green externally; pulp light-red; of fair to good quality.

'Purple Genoa' ('Black Genoa'; 'Black Spanish')—oblong, broad at apex, narrow at base; large; very dark-purple with thick blue bloom; pulp yellowish becoming reddish to red at the center; juicy, with sweet, rich flavor.

At Saharanpur, India, 'Brown Turkey', 'Bangalore', 'Black Ischia' and 'Lucknow' are successfully grown. Around Bombay, there is only one variety, 'Poona'.

'Black Ischia' ('Blue Ischia')—an Italian variety; main crop is elongated pear-shaped with many noticeable ribs; short neck and short to medium stalk; large, 2½ in (6.35 cm) long and 1½ in (3.8 cm) wide; dark purple-black except at the apex where it is lighter and greenish; there are many golden flecks; skin is wholly coated with thin, dark-blue bloom; eye open, with red-violet scales; pulp is violet-red, of good quality. In the breba crop, there are few ribs and mostly indistinct; the fruit is small, about 1½ in (3.8 cm) long and of the same width at the apex; the pulp is red to greenish-amber; of poor flavor. The tree is particularly ornamental and the leaves are glossy, only shallowly 3-lobed. A heavy bearer.

'Poona'-bell-shaped, of medium size, weighing about 1½ oz (42 g); thin-skinned; light-purple with red flesh, of sweet, good flavor.

We have no descriptions of 'Bangalore' and 'Lucknow'.

Climate

In southern India, 'Marseilles' flourishes on hills above 5,000 ft (1,525 m). In tropical areas generally, figs thrive between 2,600 and 5,900 ft (800-1,800 m). The tree can tolerate 10° to 20° of frost in favorable sites. It should have a dry climate with light early spring rains if it is intended for the production of fresh fruit. Rains during fruit development and ripening are detrimental to the crop, causing the fruits to split. The semi-arid tropical and subtropical regions of the world are ideal for fig-growing if means of irrigation are available. But very hot, dry spells will cause fruit-drop even if the trees are irrigated.

Soil

The fig can be grown on a wide range of soils; light sand, rich loam, heavy clay or limestone, providing there is sufficient depth and good drainage. Sandy soil that is medium-dry and contains a good deal of lime is preferred when the crop is intended for drying. Highly acid soils are unsuitable. The pH should be between 6.0 and 6.5. The tree is fairly tolerant of moderate salinity.

Propagation

Fig trees have been raised from seed, even seed extracted from commercial dried fruits. Ground- or air-layering can be done satisfactorily, and rapid mass-multiplication by tissue culture has been achieved in Greece, but the tree is commonly propagated by cuttings of mature wood 2 to 3 years of age, ½ to ¾ in (1.25-2 cm) thick and 8 to 12 in (20-30 cm) long. Planting must be done within 24 hours but, first, the upper, slanting end of the cutting should be treated with a sealant to protect it from disease, and the lower, flat, end with a root-promoting hormone. Trees of unsatisfactory varieties can be top-worked by shield- or patch-budding, or cleft- or bark-grafting.

Culture

Cuttings are raised in nursery beds and are set out in the field after 12 or 15 months. They may be spaced from 6 to 25 ft (1.8-7.5 m) apart depending on the cultivar and the fertility of the soil. A spacing of 13 x 13 ft (4x4 m) allows 260 trees/acre (625 trees/ha). In Colombia, growers are advised to set the trees at 10 x 10 ft (3x3 m) on level land, 10 x 13 ft (3x4 m) on slopes. Fruiting will commence in less than a year from planting out. Young plants will benefit from shading with palm fronds or other material until they are well established. A fertilizer formula of 10-30-10 or 10-20-20 NPK is recommended—2 oz (about 60 g) each for young plants and 1/5 lb (100 g) each for adults, plus minor elements at the rate of 1 oz (30 g) per tree every 6 months.

Fig trees are cut back severely in fall or winter, depending on whether the crop is desired the following summer or fall. Branches are often notched to induce lateral branching and increase the yield. If there are heavy rains, drainage ditches should be dug to prevent water-logging. Fig trees remain productive up to 12 or 15 years of age and thereafter the crop declines though the trees may live to a very advanced age.

Season

Fig trees usually bear 2 crops a year, the early season ("breba") fruits being inferior and frequently too acid, and only those of the second, or main, crop of actual value.

In Colombia and Venezuela, some fruits are borne throughout the year but there are 2 principal crops, one in May and June and the other in December and January.

Large-scale fig producers in California spray ethephon to speed up ripening and then wind-machines are drawn past the trees or helicopter overflights are made to hasten fruit drop, thus shortening the harvest period by as much as 10 days in order to avoid impending rain and insect attack. Proper timing of the growth regulator is crucial to fruit quality.

Harvesting and Yield

The fruits may be picked from the tree or gathered normally or by mechanical sweepers after they fall to the ground. 'Brunswick' is so tender it must be picked when slightly unripe in order to be firm enough for processing. Workers must wear gloves and protective clothing because of the latex. Harvested fruits are spread out in the shade for a day so that the latex will dry a little. Then they are transported to processing plants in wooden boxes holding 22 to 33 lbs (10-15 kg). In India, a fig tree bears 180 to 360 fruits per year. Venezuelan growers expect 132 to 176 lbs (6-8 kg) per tree.

Keeping Quality

Fresh figs are very perishable. At 40° to 43°F (4.44°-6.11°C) and 75% relative humidity, figs remain in good condition for 8 days but have a shelf-life of only 1 to 2 days when removed from storage. At 50°F (10°C) and relative humidity of 85%, figs can be kept no longer than 21 days. They remain in good condition for 30 days when stored at 32° to 35° F (0°-1.67° C). If frozen whole, they can be maintained for several months.

Pests and Diseases

Fig trees are prone to attack by nematodes (especially *Meloidogyne* spp.) and, in the tropics, have been traditionally planted close to a wall or building so that the roots can go underneath and escape damage. A heavy mulch will serve equally well. Today, control is possible with proper application of nematicides.

In India, a stem-borer, *Batocera rufomaculata*, feeds on the branches and may kill the tree. Lepidopterous pests in Venezuela include the fig borer, *Azochis gripusalis*, the larvae of which feed on the new growth, tunnel down through the trees to the roots and kill the tree. Another, called *cachudo de la higuera*, has prominently horned larvae up to 3 1/8 in (8 cm) long that can destroy a fig tree in a few days. There are also coleopterous insects of the genera *Epitrix* and *Colaspis* that perforate and severely damage the leaves and shoots. Scale insects include *Asterolecanium* sp. which attacks the bark of trees weakened by excessive humidity or prolonged drought, and the lesser enemy, *Saissetia haemispherica*.

A common and widespread problem is leaf rust caused by *Cerotelium fici*, bringing about premature leaf fall and reducing yields. It is most prevalent in rainy seasons. Leaf spot results from infection by *Cylindrocladium scoparium* or *Cercospora fici*. Fig mosaic is caused by a virus and is incurable. Affected trees must be destroyed.

The dried fruit beetle, or sour bug, *Carpophilus* spp., enters the fruit through the eye and leads to souring and smut caused by *Aspergillus niger*. This fungus may attack ripening fruits.

Food Uses

Some people peel the skin back from the stem end to expose the flesh for eating out-of-hand. The more fastidious eater holds the fruit by the stem end, cuts the fruit into quarters from the apex, spreads the sections apart and lifts the flesh from the skin with a knife blade, discarding the stem and skin. Commercially, figs are peeled by immersion for 1 minute in boiling lye water or a boiling solution of sodium bicarbonate. In warm, humid climates, figs are generally eaten fresh and raw without peeling, and they are often served with cream and sugar. Peeled or unpeeled, the fruits may be merely stewed or cooked in various ways, as in pies, puddings, cakes, bread or other bakery products, or added to ice cream mix. Homeowners preserve the whole fruits in sugar sirup or prepare them as jam, marmalade, or paste. Fig paste (with added wheat and corn flour, whey, sirup, oils and other ingredients) forms the filling for the well-known bakery product, "Fig Newton". The fruits are sometimes candied whole commercially. In Europe, western Asia, northern Africa and California, commercial canning and drying of figs are industries of great importance.

Some drying is done in Poona, India, and there is currently interest in solar-drying in Guatemala. Usually, the fruits are allowed to fully ripen and partially dehydrate on the tree, then are exposed to sulphur fumes for about a half-hour, placed out in the sun and turned daily to achieve uniform drying, and pressed flat during the 5- to 7-day process. 'Black Mission' and 'Kadota' figs are suitable for freezing whole in sirup, or sliced and layered with sugar.

Dried cull figs have been roasted and ground as a coffee substitute. In Mediterranean countries, low-grade figs are converted into alcohol. An alcoholic extract of dried figs has been used as a flavoring for liqueurs and tobacco.

Food Value Per 100 g of Edible Portion*		
	Fresh	*Dried*
Calories	80	274
Moisture	77.5-86.8 g	23.0 g
Protein	1.2-1.3 g	4.3 g
Fat	0.14-0.30 g	1.3 g
Carbohydrates	17.1-20.3 g	69.1 g
Fiber	1.2-2.2 g	5.6 g
Ash	0.48-0.85 g	2.3 g
Calcium	35-78.2 mg	126 mg
Phosphorus	22-32.9 mg	77 mg
Iron	0.6-4.09 mg	3.0 mg
Sodium	2.0 mg	34 mg
Potassium	194 mg	640 mg
Carotene	0.013-0.195 mg	—
as Vitamin A	20-270 I.U.	80 I.U.
Thiamine	0.034-0.06 mg	0.10 mg
Riboflavin	0.053-0.079 mg	0.10 mg
Niacin	0.32-0.412 mg	0.7 mg
Ascorbic Acid	12.2-17.6 mg	0 mg
Citric Acid	0.10-0.44 mg	

Note: There are small amounts of malic, boric and oxalic acids.

*According to analyses made in India, Hawaii, Central America, and by the U.S. Department of Agriculture in Washington, D.C.

Toxicity

The latex of the unripe fruits and of any part of the tree may be severely irritating to the skin if not removed promptly. It is an occupational hazard not only to fig harvesters and packers but also to workers in food industries, and to those who employ the latex to treat skin diseases.

Other Uses

Seed oil: Dried seeds contain 30% of a fixed oil containing the fatty acids: oleic, 18.99%; linoleic, 33.72%; linolenic, 32.95%; palmitic, 5.23%; stearic, 2.18%; arachidic, 1.05%. It is an edible oil and can be used as a lubricant.

Leaves: Fig leaves are used for fodder in India. They are plucked after the fruit harvest. Analyses show: moisture, 67.6%; protein, 4.3%; fat, 1.7%; crude fiber, 4.7%; ash, 5.3%; N-free extract, 16.4%; pentosans, 3.6%; carotene on a dry weight basis, 0.002%. Also present are bergaptene, stigmasterol, sitosterol, and tyrosine.

In southern France, there is some use of fig leaves as a source of perfume material called "fig-leaf absolute"—a dark-green to brownish-green, semi-solid mass or thick liquid of herbaceous-woody-mossy odor, employed in creating woodland scents.

Latex: The latex contains caoutchouc (2.4%), resin, albumin, cerin, sugar and malic acid, rennin, proteolytic enzymes, diastase, esterase, lipase, catalase, and peroxidase. It is collected at its peak of activity in early morning, dried and powdered for use in coagulating milk to make cheese and junket. From it can be isolated the protein-digesting enzyme *ficin* which is used for tenderizing meat, rendering fat, and clarifying beverages.

In tropical America, the latex is often used for washing dishes, pots and pans. It was an ingredient in some of the early commercial detergents for household use but was abandoned after many reports of irritated or inflamed hands in housewives.

Medicinal Uses: The latex is widely applied on warts, skin ulcers and sores, and taken as a purgative and vermifuge, but with considerable risk. In Latin America, figs are much employed as folk remedies. A decoction of the fruits is gargled to relieve sore throat; figs boiled in milk are repeatedly packed against swollen gums; the fruits are much used as poultices on tumors and other abnormal growths. The leaf decoction is taken as a remedy for diabetes and calcifications in the kidneys and liver. Fresh and dried figs have long been appreciated for their laxative action.

Breadfruit

One of the great food producers in its realm and widely known, at least by name, through its romanticized and dramatized history, the breadfruit, *Artocarpus altilis* Fosb. (syns. *A. communis* J.R. and G. Forst.; *A. incisus* L.f.) belongs to the mulberry family, Moraceae. The common name is almost universal, in English, or translated into Spanish as *fruta de pan* (fruit), or *arbol de pan, arbol del pan* (tree), or *pan de pobre;* into French, as *fruit à pain* (seedless), *chataignier* (with seeds), *arbre à pain* (tree); Portuguese, *fruta pão*, or *pão de massa;* Dutch, *broodvrucht* (fruit), *broodboom* (tree). In Venezuela it may be called *pan de año, pan de todo el año,* *pan de palo, pan de ñame, topán,* or *túpan;* in Guatemala and Honduras, *mazapán* (seedless), *castaña* (with seeds); in Peru, *marure;* in Yucatan, *castaño de Malabar* (with seeds); in Puerto Rico, *panapén* (seedless), *pana de pepitas* (with seeds). In Malaya and Java, it is *suku* or *sukun* (seedless); *kulur, kelur,* or *kulor* (with seeds); in Thailand, *sa-ke,* in the Philippines, *rimas* (seedless); in Hawaii, *ulu.* The type with seeds is sometimes called "breadnut", a name better limited to *Brosimum alicastrum* Swartz, an edible-seeded tree of Yucatan, Central America and nearby areas. Its Spanish name is *ramon* and the seeds, leaves and twigs are prized as stock feed.

Fig. 13: Ripe breadfruit (*Artocarpus altilis*). In: K. & J. Morton, *Fifty Tropical Fruits of Nassau*, 1946.

Description

The breadfruit tree is handsome and fast-growing, reaching 85 ft (26 m) in height, often with a clear trunk to 20 ft (6 m) becoming 2 to 6 ft (0.6-1.8 m) in width and often buttressed at the base, though some varieties may never exceed ¼ or ½ of these dimensions. There are many spreading branches, some thick with lateral foliage-bearing branchlets, others long and slender with foliage clustered only at their tips. The leaves, evergreen or deciduous depending on climatic conditions, on thick, yellow petioles to 1½ in (3.8 cm) long, are ovate, 9 to 36 in (22.8-90 cm) long, 8 to 20 in (20-50 cm) wide, entire at the base, then more or less deeply cut into 5 to 11 pointed lobes. They are bright-green and glossy on the upper surface, with conspicuous yellow veins; dull, yellowish and coated with minute, stiff hairs on the underside.

The tree bears a multitude of tiny flowers, the male densely set on a drooping, cylindrical or club-shaped spike 5 to 12 in (12.5-30 cm) long and 1 to 1½ in (2.5-3.75 cm) thick, yellowish at first and becoming brown. The female are massed in a somewhat rounded or elliptic, green, prickly head, 2½ in (6.35 cm) long and 1½ in (3.8 cm) across, which develops into the compound fruit (or syncarp), oblong, cylindrical, ovoid, rounded or pear-shaped, 3½ to 18 in (9-45 cm) in length and 2 to 12 in (5-30 cm) in diameter. The thin rind is patterned with irregular, 4- to 6-sided faces, in some "smooth" fruits level with the surface, in others conical; in some, there may rise from the center of each face a sharp, black point, or a green, pliable spine to ⅛ in (3 mm) long or longer. Some fruits may have a harsh, sandpaper-like rind. Generally the rind is green at first, turning yellowish-green, yellow or yellow-brown when ripe, though one variety is lavender.

In the green stage, the fruit is hard and the interior is white, starchy and somewhat fibrous. When fully ripe, the fruit is somewhat soft, the interior is cream-colored or yellow and pasty, also sweetly fragrant. The seeds are irregularly oval, rounded at one end, pointed at the other, about ¾ in (2 cm) long, dull-brown with darker stripes. In the center of seedless fruits there is a cylindrical or oblong core, in some types covered with hairs bearing flat, brown, abortive seeds about ⅛ in (3 mm) long. The fruit is borne singly or in clusters of 2 or 3 at the branch tips. The fruit stalk (pedicel) varies from 1 to 5 in (2.5-12.5 cm) long.

All parts of the tree, including the unripe fruit, are rich in milky, gummy latex. There are two main types: the normal, "wild" type (cultivated in some areas) with seeds and little pulp, and the "cultivated" (more widely grown) seedless type, but occasionally a few fully developed seeds are found in usually seedless cultivars. Some forms with entire leaves and with both seeds and edible pulp have been classified by Dr. F.R. Fosberg as belonging to a separate species, *A. mariannensis* Trécul. but these commonly integrate with *A. altilis* and some other botanists regard them as included in that highly variable species.

Origin and Distribution

The breadfruit is believed to be native to a vast area extending from New Guinea through the Indo-Malayan Archipelago to Western Micronesia. It is said to have been widely spread in the Pacific area by migrating Polynesians, and Hawaiians believed that it was brought from the Samoan island of Upalu to Oahu in the 12th Century A.D. It is said to have been first seen by Europeans in the Marquesas in 1595, then in Tahiti in 1606. At the beginning of the 18th Century, the early English explorers were loud in its praises, and its fame, together with several periods of famine in Jamaica between 1780 and 1786, inspired plantation owners in the British West Indies to petition King George III to import seedless breadfruit trees to provide food for their slaves.

There is good evidence that the French navigator Sonnerat in 1772 obtained the seeded breadfruit in the Philippines and brought it to the French West Indies. It seems also that some seedless and seeded breadfruit plants reached Jamaica from a French ship bound for Martinique but captured by the British in 1782. There were at least two plants of the seeded breadfruit in Jamaica in 1784 and distributions were quickly made to the other islands. There is a record of a plant having been sent from Martinique to the St. Vincent Botanical Garden before 1793. The story of Captain Bligh's first voyage to Tahiti, in 1787, and the loss of his cargo of 1,015 potted breadfruit plants on his disastrous return voyage is well known. He set out again in 1791 and delivered 5 different kinds totalling 2,126 plants to Jamaica in February 1793. On that island, the seedless breadfruit flourished and it came to be commonly planted in other islands of the West Indies, in the lowlands of Central America and northern South America. In some areas, only the seedless type is grown, in others, particularly Haiti, the seeded is more common. Jamaica is by far the leading producer of the seedless type, followed by St. Lucia. In New Guinea, only the seeded type is grown for food.

It has been suggested that the seeded breadfruit was carried by Spaniards from the Philippines to Mexico and Central America long before any reached the West Indies. On the Pacific coast of Central America, the seeded type is common and standard fare for domestic swine. On the Atlantic Coast, seedless varieties are much consumed by people of African origin. The breadfruit tree is much grown for shade in Yucatan. It is very common in the lowlands of Colombia, a popular food in the Cauca Valley, the Choco, and the San Andres Islands; mostly fed to livestock in other areas. In Guyana, in 1978, about 1,000 new breadfruit trees were being produced each year but not nearly enough to fill requests for plants. There and in Trinidad, because of many Asians in the population, both seeded and seedless breadfruits are much appreciated as a regular article of the diet; in some other areas of the Caribbean, breadfruit is regarded merely as a food for the poor for use only in emergencies. Nowadays, it is attracting the attention of gourmets and some islands are making small shipments to the United States, Canada and Europe for specialized ethnic markets. In the Palau Islands of the South Pacific, breadfruit is being outclassed by cassava and imported flour and rice. For some time breadfruit was losing ground to taro (*Colocasia esculenta* Schott.) in Hawaii, but now land for taro is limited and its culture is static.

The United States Department of Agriculture brought in breadfruit plants from the Canal Zone, Panama, in 1906 (S.P.I. #19228). For many years there have been a number of seedless breadfruit trees in Key West, Florida, and there is now at least one on Vaca Key about 50 miles to the northeast. On the mainland of Florida, the tree can be maintained outdoors for a few years with mild winters but, unless protected with plastic covering to prevent dehydration, it ultimately succumbs. A few have been kept alive in greenhouses or conservatories such as the Rare Plant House of Fairchild Tropical Garden, and the indoor garden of the Jamaica Inn on Key Biscayne.

Varieties

An unpublished report of 1921 covered 200 cultivars of breadfruit in the Marquesas. The South Pacific Commission published the results of a breadfruit survey in 1966. In it, there were described 166 named sorts from Tonga, Niue, Western and American Samoa, Papua and New Guinea, New Hebrides and Rotuma. There are 70 named varieties of seeded and seedless breadfruits in Fiji. They are locally separated into 8 classes by leaf form. The following, briefly presented, are those that are recorded as "very good". It will be noted that some varietal names are reported under more than one class.

Class I: Leaf entire, or with one or two, occasionally, three lobes.

'Koqo'—round; 4 in (10 cm) wide; seedless; does not deteriorate quickly.

'Tamaikora'—gourd-shaped (constricted around middle); to 4½ in (11.5 cm) long, 3 in (7.5 cm) wide; with many seeds. Can be eaten raw when ripe. Highly perishable. Tree to 40 or 45 ft (12-13.5 m).

Plate I
PEJIBAYE
Bactris gasipaes
(green and ripe)

Plate II
PEJIBAYE
Bactris gasipaes
(in foreground)

Plate III
DWARF CAVENDISH BANANA
Musa acuminata

Plate IV
PLANTAIN
Musa X paradisiaca

Plate V
FIG
Ficus carica

Plate VI
JACKFRUIT
Artocarpus heterophyllus

Plate VII
CHERIMOYA
Annona cherimola

Plate VIII
SUGAR APPLE
Annona squamosa

Plate IX
ATEMOYA
Annona squamosa X A. cherimola

Plate X
SOURSOP
Annona muricata

Plate XI
LOQUAT
Eriobotrya japonica

Class II: Leaf dissected at apex.

'**Temaipo**'—round; 3½ in (9 cm) long; seedless. Can be eaten raw when ripe. There is also an oblong form with many seeds.

Class III: Leaf moderately deeply dissected at apex.

'**Uto Kuro**'—round; 5 in (12.5 cm) long; does not deteriorate quickly.

Class IV: Leaf moderately deeply dissected on upper half.

'**Samoa**' ('Kasa Balavu')—round; 4 to 6 in (10-15 cm) long; seeds sparse to many.

'**Uto Yalewa**'—oblong; to 8 in (20 cm) long and 6 in (15 cm) wide; seedless.

'**Kulu Dina**'—oblong; to 16 in (40 cm) long and 13 in (33 cm) wide; seedless. Need not be peeled after cooking. Tree bears all year.

'**Sogasoga**'—oblong; to 9 in (23 cm) long and 6½ in (16.5 cm) wide; seedless.

'**Uto Dina**'—oblong; to 6 in (15 cm) long and 3 to 3½ in (7.5-9 cm) wide; seedless; need not be peeled after cooking. Tree 60 to 70 ft (18-21 m) high.

'**Buco Ni Viti**'—oblong; 11 to 14 in (28-35.5 cm) long, 6 to 7 in (15-18 cm) wide; seedless; one of the best cultivars.

'**Tamaikora**'—oblong; 7 to 9 in (18-23 cm) long, 5 to 6½ in (12.5-16.5 cm) wide; seeds sparse; pulp eaten raw when ripe. Tree to 75 or 85 ft (23-26 m) high; bears 2 crops per year.

'**Kulu Mabomabo**'—oval; 6 to 8 in (15-20 cm) long, 4 to 5½ in (10-14 cm) wide; seedless.

Class V: Leaf moderately deeply dissected; shape of leaf base variable.

'**Uto Dina**'—round; 4½ to 5 in (11.5-12.5 cm) wide; seedless. Highly recommended. Tree is 25-30 ft (7.5-9 m) tall.

'**Balekana Ni Samoa**'—round; 4 to 5 in (10-12.5 cm) long; seeds sparse. Best of all Samoan varieties. There is an oval form by the same name; seedless; deteriorates very quickly.

'**Balekana Ni Vita**'—round; 3½ to 4 in (9-10 cm) long; seedless. Does not deteriorate quickly.

'**Balekana Dina**'—oval; 6 to 8 in (15-20 cm) long, 3 to 5 in (7.5-12.5 cm) wide; seeds sparse. One of the best, especially when boiled.

'**Tabukiraro**'—round; 8 in (20 cm) long; seedless; skin sometimes eaten after cooking.

'**Sici Ni Samoa**'—oval; 5 to 6 in (12.5-15 cm) long, 3 to 3½ in (7.5-9 cm) wide; seedless. One of the highly recommended Samoan varieties.

'**Uto Me**'—oval; 5 to 6¾ in (12.5-17 cm) long, 4½ to 5 in (11.5 cm) wide; with many seeds; does not deteriorate quickly.

'**Uto Wa**'—oval; 6 to 7½ in (15-19 cm) long, 5 to 5½ in (12.5-14 cm) wide. The variety most recommended.

'**Kulu Vawiri**'—oval; 9 to 12 in (22-30 cm) long, 8 to 9 in (20-22 cm) wide; especially good when boiled.

Class VI: Leaf deeply dissected.

'**Kulu Dina**'—round; 3 to 4 in (7.5-10 cm) long; seedless. Need not be peeled after boiling. Highly recommended.

'**Balekana**'—oval; 4 in (10 cm) long, 3 in (7.5 cm) wide; of the best quality. Tree 70 to 80 ft (21-24 m) high.

'**Balekana Ni Samoa**'—round; 3 in (7.5 cm) long; seeds sparse. Best of all Samoan varieties.

'**Balekana Ni Viti**'—oblong; 5 to 6 in (12.5-15 cm) long, 3 to 4 in (7.5-10 cm) wide; seedless. The best native-type variety.

Fig. 14: Breadfruit is borne singly or in 2's or 3's at the branch tips of this handsome, large-leaved tree.

'**Uto Dina**' ('Kasa Leka')—round; 4 in (10 cm) long; seedless.

'**Uto Matala**'—round; 3 to 4 in (7.5-10 cm) long. Especially fine when boiled. Tree bears 3 times a year.

Class VII: Leaf deeply dissected; apex pointed.

'**Balekana Ni Samoa**'—round; 5 to 5½ in (12.5-14 cm) long; seeds sparse. Best of all Samoan varieties.

'**Kulu Dina**' ('Kasa Balavu')—oval; 6 to 7 in (15-18 cm) long, 4 to 5 in (10-12.5 cm) wide; seedless.

'**Uto Dina**' (Large)—oval; 8 to 9 in (20-22 cm) long, 4 to 7 in (10-18 cm) wide; seedless. Also, by the same name, a form with only moderately dissected leaves.

'**Bokasi**'—round; 4 in (10 cm) long, 3 in (7.5 cm) wide.

Class VIII: Leaf deeply dissected, wide spaces between lobes.

'**Savisavi Ni Samoa**'—oval; 4 to 5 in (10-12.5 cm) long, 3 to 3½ in (7.5-9 cm) wide. Ranks with best Samoan varieties.

'**Savisavi Ni Viti**'—oblong; 6 to 8 in (16-20 cm) long, 4 to 6 in (10-15 cm) wide; seedless; especially good when boiled.

'**Savisavi**'—round; 3 to 3½ in (7.5-9 cm) wide; especially good when boiled.

'**Balawa Ni Viti**'—oval; 6 to 7 in (15-18 cm) long, 3½ to 4 in (9-10 cm) wide; seedless.

'**Uto Kasekasei**'—round; 4 to 5 in (10-12.5 cm) long; seeds sparse.

'**Via Loa**'—oblong; 6 to 7 in (15-18 cm) long, 4 to 5 in (10-12.5 cm) wide; seedless; does not deteriorate quickly.

Koroieveibau provides a key to the 8 classes illustrated

by leaf and fruit outline sketches.

P.J. Wester, in 1928, published descriptions of 52 breadfruit cultivars of the Pacific Islands. In the book, *The Breadfruit of Tahiti*, by G.P. Wilder, there are detailed descriptions and close-up, black-and-white photographic illustrations of the foliage and fruit of 30 named varieties, and of the foliage only of one which did not have mature fruit at the time of writing. One 'Aata', an oblong fruit, is described as of poor quality and eaten by humans only when better breadfruits are scarce, but it is important as feed for pigs and horses. The tree bears heavily. Among the best are:

'Aravei'—fruit ellipsoidal; large, 8 to 12 in (10-30 cm) long, 6 to 9 in (15-22 cm) wide; rind yellowish-green with brown spots on the sunny side; rough, with sharp points which are shed on maturity. Pulp is light-yellow, dry or flaky and of delicious flavor after cooking which takes very little time. Core long, slim, with many abortive seeds.

'Havana'—fruit oval-round; the rind yellowish-green, spiny; pulp golden-yellow, moist, pasty, separates into loose flakes when cooked; very sweet with excellent flavor; core oval, large, with a row of abortive seeds. Very perishable; must be used within 2 days; cooks quickly over fire. Fruit borne in 2's and 3's. Popularly claimed to be one of the best breadfruits.

'Maohi'—fruit round; 6 in (15 cm) wide; rind bright yellow-green with patches of red-brown; rough, with spines, and often bears much exuded latex. Pulp cream-colored and smooth when cooked; of very good flavor; slow cooking, needs even heat. Core is large. Fruit is borne in 2's and 3's. Tree a heavy bearer. This is the most common breadfruit of Tahiti.

'Paea'—ellipsoidal; very large, to 11 in (28 cm) long and 9 in (22.8 cm) wide; rind yellowish-green, spiny; core oblong, thick, with a row of brown, abortive seeds; pulp bright-yellow, moist, slightly pasty, separating into flakes when cooked; agreeable but only one of its forms, 'Paea Maaroaro', is really sweet. Formerly, 'Paea' was reserved for chiefs only. Needs one hour to roast on open fire. The tree is tall, especially well-formed and elegant.

'Pei'—broad-ellipsoidal; large; rind light-green, relatively smooth; pulp light-yellow and flaky when cooked, aromatic, of sweet, delicious "fruity" flavor; cooks quickly. Ripens earlier than others. When the breadfruit crop is scant, the fruits of this cultivar are stored by burying in the ground until needed, even for a year, then taken up, wrapped in *Cordyline* leaves and boiled.

'Pucro'—fruit spherical or elongated; large; rind yellow-green with small brown spots, very rough, spiny, thin; pulp light-yellow and smooth, of excellent flavor. Cooks quickly. Highly esteemed, ranked with the very best breadfruits. There are two oblong forms, one with a large, hairy core.

'Rare'—fruit broad-ovoid; to 7 in (17.5 cm) long, rind bright-green, rough, spiny; pulp of deep-cream tone, fine-grained, smooth, flaky when cooked; of very sweet, excellent flavor. Core is small with a great many small abortive seeds. Must be cooked for about one hour. There are 3 forms that are well recognized. Fruits are borne singly on a tall, open, short-branched tree.

'Rare Aumee'—fruit round; 6½ in (16.5 cm) across; rind bright-green with red-brown splotches, fairly smooth at the base but rough at the apex; pulp deep-ivory, firm, smooth when cooked; not very sweet but of excellent flavor. Cooks quickly. Highly prized; in scarce supply because the tall, few-branched tree bears scantily.

'Rare Autia'—fruit round; 6 in (15 cm) across; rind dull-green with red-brown markings. Pulp light-yellow when cooked and separates into chunks; has excellent flavor. Core is large with small abortive seeds all around. This cultivar is so superior it was restricted to royalty and high chiefs in olden times.

'Tatara'—fruit broad-ellipsoid; very large, up to 10 lbs (4.5 kg) in weight; rind has prominent faces with long green spines; pulp light-yellow, smooth when cooked and of pleasant flavor. Core is oblong. This variety is greatly esteemed. The tree is found only in a small coastal valley where there is heavy rainfall. It is of large dimensions and high-branching and it is difficult to harvest the fruits.

'Vai Paere'—fruit is obovoid; 10 to 12 in (25-30 cm) long, 7 to 8 in (17.5-20 cm) wide; rind is yellow-green with red-brown splotches and there is a short raised point at the center of each face; pulp light-yellow, firm, smooth, a little dryish when cooked, with a slightly acid, but excellent flavor. Core is oblong, large, with a few abortive seeds attached. Fruit cooks easily. Tree is very tall, bears fruit in clusters. Grows at sea-level in fairly dry locations.

There are at least 50 cultivars on Ponape and about the same number on Truk. In Samoa, a variety known as 'Maopo', with leaves that are almost entire or sometimes very shallowly lobed, is very common and considered one of the best.

'Puou' is another choice and much-planted variety since early times. It has deeply cut leaves and nearly round fruits 6 in (15 cm) long. 'Ulu Ea', with leaves even more deeply lobed, has oblong fruits to 6⅛ in (15.5 cm) long and 5 in (12.5 cm) wide; is a longtime favorite.

In the past three decades there has been an awakening to the possibilities of increasing the food supply of tropical countries by more plantings of selected varieties of seedless breadfruit. In 1958, many appealing varieties (some early, some late in season) were collected around the South Pacific region and transferred to Western Samoa, Tahiti and Fiji for comparative trials. Two years later, plans were made to introduce Polynesian varieties into Micronesia, and propagating material of 36 Micronesian types was distributed to other areas.

Climate

The breadfruit is ultra-tropical, much tenderer than the mango tree. It has been reported that it requires a temperature range of 60° to 100°F (15.56°-37.78°C), an annual rainfall of 80 to 100 in (203-254 cm), and a relative humidity of 70 to 80%. However, in southern India, it is cultivated at sea-level and up humid slopes to an altitude of 3,500 ft (1,065 m), also in thickets in dry regions where it can be irrigated. In the "equatorial dry climate" of the Marquesas, where the breadfruit is an essential crop, there is an average rainfall of only 40 to 60 in (100-150 cm) and frequent droughts. In Central America, it is grown only below 2,000 ft (600 m).

Soil

According to many reports, the breadfruit tree must have deep, fertile, well-drained soil. But some of the best authorities on South Pacific plants point out that the seedless breadfruit does well on sandy coral soils, and seeded types grow naturally on "coraline limestone" islands in Micronesia. In New Guinea, the breadfruit tree occurs wild along waterways and on the margins of forests in the flood plain, and often in freshwater swamps. It is believed that there is great variation in the adaptability of different strains to climatic and soil conditions, and that each should be matched with its proper environment. The Tahitian 'Manitarvaka' is known to be drought-resistant. The variety 'Mai-Tarika', of the Gilbert Islands, is salt-tolerant. 'Mejwaan', a seeded variety of the Marshall Islands, is not harmed by brackish water nor salt spray and has been introduced into Western Samoa and Tahiti.

Propagation

The seeded breadfruit is always grown from seeds, which must be planted when fairly fresh as they lose viability in a few weeks. The seedless breadfruit is often propagated by transplanting suckers which spring up naturally from the roots. One can deliberately induce suckers by uncovering and injuring a root. Pruning the parent tree will increase the number of suckers, and root-pruning each sucker several times over a period of months before taking it up will contribute to its survival when transplanted. For multiplication in quantity, it is better to make root cuttings about 1 to 2½ in (2.5-6.35 cm) thick and 9 in (22 cm) long. The ends may be dipped into a solution of potassium permanganate to coagulate the latex, and the cuttings are planted close together horizontally in sand. They should be shaded and watered daily, unless it is possible to apply intermittent mist. Calluses may form in 6 weeks (though rooting time may vary from 2 to 5 months) and the cuttings are transplanted to pots, at a slant, and watered once or twice a day for several months or until the plants are 2 ft (60 cm) high. A refined method of rapid propagation uses stem cuttings taken from root shoots. In Puerto Rico, the cuttings are transplanted into plastic bags containing a mixture of soil, peat and sand, kept under mist for a week, then under 65% shade, and given liquid fertilizer and regular waterings. When the root system is well developed, they are allowed full sun until time to set out in the field.

In India, it is reported that breadfruit scions can be successfully grafted or budded onto seedlings of wild jackfruit trees.

Culture

Young breadfruit trees are planted in well-enriched holes 15 in (40 cm) deep and 3 ft (0.9 m) wide that are first prepared by burning trash in them to sterilize the soil and then insecticide is mixed with the soil to protect the roots and shoots from grubs. The trees are spaced 25 to 40 ft (7.5-12 m) apart in plantations. Usually there are about 25 trees per acre (84/ha). Those grown from root suckers will bear in 5 years and will be productive for 50 years. Some growers recommend pruning of branches that have borne fruit and would normally die back, because this practice stimulates new shoots and also tends to keep the tree from being too tall for convenient harvesting.

Standard mixtures of NPK are applied seasonally. When the trees reach bearing age, they each receive, in addition, 4.4 lbs (2 kg) superphosphate per year to increase the size and quality of the fruits.

Season

In the South Seas, the tree fruits more or less continuously, fruit in all stages of development being present on the tree the year around, but there are two or three main fruiting periods. In the Caroline Islands and the Gilbert Islands, the main ripening season is May to July or September; in the Society Islands and New Hebrides, from November to April, the secondary crop being in July and August. Breadfruits are most abundant in Hawaiian markets off and on from July to February. Flowering starts in March in northern India and fruits are ready for harvest in about 3 months. Seeded breadfruits growing in the Eastern Caroline Islands fruit only once a year but the season is 3 months long—from December to March. Seedless varieties introduced from Ponape bear 2 to 3 times a year. In the Bahamas, breadfruit is available mainly from June to November, but some fruits may mature at other times during the year.

Harvesting and Yield

Breadfruits are picked when maturity is indicated by the appearance of small drops of latex on the surface. Harvesters climb the trees and break the fruit stalk with a forked stick so that the fruit will fall. Even though this may cause some bruising or splitting, it is considered better than catching the fruits by hand because the broken pedicel leaks much latex. They are packed in cartons in which they are separated individually by dividers.

In the South Pacific, the trees yield 50 to 150 fruits per year. In southern India, normal production is 150 to 200 fruits annually. Productivity varies between wet and dry areas. In the West Indies, a conservative estimate is 25 fruits per tree. Studies in Barbados indicate a reasonable potential of 6.7 to 13.4 tons per acre (16-32 tons/ha). Much higher yields have been forecasted, but experts are skeptical and view these as unrealistic.

Keeping Quality

In Jamaica, surplus breadfruits are often kept under water until needed. Fully ripe fruits that have fallen from the tree can be wrapped in polyethylene, or put into polyethylene bags, and kept for 10 days in storage at a temperature of 53.6°F (12°C). At lower temperature, the fruit shows chilling injury. Slightly unripe fruits that have been caught by hand when knocked down can be maintained for 15 days under the same conditions. The thickness of the polyethylene is important: 38- or even 50-micrometer bags are beneficial, but not 25-micrometer.

Some Jamaican exporters partly roast the whole fruits to coagulate the latex, let them cool, and then ship them by sea to New York and Europe. Various means of preserving breadfruit for future local use are mentioned under "Food Uses", q.v.

Pests and Diseases

Soft-scales and mealybugs are found on breadfruit trees in the West Indies and ants infest branches that die back after fruiting. In southern India, the fruits on the tree are subject to soft rot. This fungus disease can be controlled by two sprays of Bordeaux mixture, one month apart. Young breadfruit trees in Trinidad have been killed by a disease caused by *Rosellinia* sp. In the Pacific Islands *Fusarium* sp. is believed to be the cause of die-back, and *Pythium* sp. is suspected in cases of root-rot. A mysterious malady, called "Pingalap disease", killed thousands of trees from 1957 to 1960 in the Gilbert and Ellice Islands, the Caroline Islands, Marshalls and Mariannas. The foliage wilts and then the branch dies back. Sometimes the whole tree is affected and killed to the roots; occasionally only half of a tree declines. The fungus, *Phytophthora palmivora*, attacks the fruit on the island of Truk. *Phomopsis*, *Dothiorella* and *Phylospora* cause stem-end rot.

Food Uses

Like the banana and plantain, the breadfruit may be eaten ripe as a fruit or underripe as a vegetable. For the latter purpose, it is picked while still starchy and is boiled or, in the traditional Pacific Island fashion, roasted in an underground oven on pre-heated rocks. Sometimes it is cored and stuffed with coconut before roasting. Malayans peel firm-ripe fruits, slice the pulp and fry it in sirup or palm sugar until it is crisp and brown. Filipinos enjoy the cooked fruit with coconut and sugar.

Fully ripe fruits, being sweeter, are baked whole with a little water in the pan. Some cooks remove the stem and core before cooking and put butter and sugar in the cavity, and serve with more of the same. Others may serve the baked fruit with butter, salt and pepper. Ripe fruits may be halved or quartered and steamed for 1 or 2 hours and seasoned in the same manner as baked fruits. The steamed fruit is sometimes sliced, rolled in flour and fried in deep fat. In Hawaii, underripe fruits are diced, boiled, and served with butter and sugar, or salt and pepper, or diced and cooked with other vegetables, bacon and milk as a chowder. In the Bahamas, breadfruit soup is made by boiling underripe chunks of breadfruit in water until the liquid begins to thicken, then adding cooked salt pork, chopped onion, white pepper and salt, stirring till thick, then adding milk and butter, straining, adding a bit of sherry and simmering until ready to serve.

The pulp scraped from soft, ripe breadfruits is combined with coconut milk (not coconut water), salt and sugar and baked to make a pudding. A more elaborate dessert is concocted of mashed ripe breadfruit, with butter, 2 beaten eggs, sugar, nutmeg, cinnamon and rosewater, a dash of sherry or brandy, blended and boiled. There are numerous other dishes peculiar to different areas. Breadfruit is also candied, or sometimes prepared as a sweet pickle.

In Micronesia, the peel is scraped off with a sharpened cowrie shell, or the fruits are peeled with a knife, cored, cut up and put into sacks or baskets, soaked in the sea for about 2 hours while being beaten or trampled; allowed to drain on shore for a few days; then packed in banana-leaf-lined boxes to ferment for a month or much longer, the leaves being changed weekly.

In Polynesia and Micronesia, a large number of fruits are baked in a native oven and left there to ferment. Over a period of a few weeks, batches are taken out as needed. In the New Hebrides, peeled breadfruits are wrapped in leaves and placed to ferment in piles of stones on open beaches where they will be flooded at high tide. In Samoa, seeded breadfruits are skinned, washed, quartered and left to ferment in a pit lined and covered with layers of banana and *Heliconia* leaves, and topped with earth and rocks. The fruits ferment for long periods, sometimes for several years, and form a pasty mass called *masi*. The seeds are squeezed out, the paste is wrapped in *Heliconia* leaves smeared with coconut cream and the product is baked for 2 hours. There is a strong, cheese-like odor, but it is much relished by the natives.

The original method of poi-making involved peeling, washing and halving the fruit, discarding the core, placing the fruits in stone pits lined with leaves of *Cordyline terminalis* Kunth, alternating the layers of fruit with old fermented poi, covering the upper layer with leaves, topping the pit with soil and rocks and leaving the contents to ferment, which acidifies and preserves the breadfruit for several years.

Modern poi is made from firm-ripe fruits, boiled whole until tender, cored, sliced, ground, pounded to a paste, kneaded with added water to thin it, strained through cloth, and eaten. If it is to be kept in the refrigerator for 2 days, only a little water is added in kneading; more is added and it is strained just before serving. Food value and digestibility are improved by mixing with poi made from taro which is rated highly as a non-allergenic food. In the Seychelles, the seedless breadfruit is cut into slices ½ in (1.25 cm) thick, dried for 4 days at 120°F (48.89°C). In some Pacific Islands, the fruits are partly roasted, then peeled, dried and formed into loaves for long-time storage. The Ceylonese dip breadfruit slices into a salt solution, then blanch them in boiling water for 5 minutes, dry them at 158°F (70°C) for 4 to 6 hours before storing. The slices will keep in good condition for 8 to 10 months. In Guam, cooked fruits may be mashed to a paste which is spread out thin, dried in the sun, and wrapped in leaves for storage. It is soaked in water to soften it for eating. This might be called "breadfruit leather". On the small Kapingamarangi Atoll in the Caroline Islands, the cooked paste is pressed into sheets 5 ft (1.5 m) long and 20 in (50 cm) wide, dried in the sun on coconut leaf mats, then rolled into cylinders, wrapped in *Pandanus* leaves and stored for at least 3 years.

The dried fruit has been made into flour and improved methods have been explored in Barbados and Brazil with a view to substituting breadfruit in part for wheat flour in breadmaking. The combination has been found more nutritious than wheat flour alone. Breadfruit flour is much richer than wheat flour in lysine and other essential amino acids. In Jamaica, the flour is boiled, sweetened, and eaten as porridge for breakfast.

Soft or overripe breadfruit is best for making chips and these are being manufactured commercially in Trinidad and Barbados. Some breadfruit is canned in Dominica and Trinidad for shipment to London and New York.

In Jamaica, Puerto Rico and the South Pacific, fallen male flower spikes are boiled, peeled and eaten as vegetables or are candied by recooking, for 2-3 hours, in sirup; then rolled in powdered sugar and sun-dried.

The seeds are boiled, steamed, roasted over a fire or in hot coals and eaten with salt. In West Africa, they are sometimes made into a purée. In Costa Rica, the cooked seeds are sold by street vendors.

Underripe fruits are cooked for feeding to pigs. Soft-ripe fruits need not be cooked and constitute a large part of the animal feed in many breadfruit-growing areas of the Old and New World. Breadfruit has been investigated as potential material for chickfeed but has been found to produce less weight gain than cassava or maize despite higher intake, and it also causes delayed maturity.

Experiments by technologists at the United States Department of Agriculture's Western Regional Research Laboratory in Berkeley, California, have demonstrated that breadfruit can be commercially dehydrated by tunnel drying or freeze-drying and the waste from these processes constitutes a highly-digestible stock feed.

Food Value Per 100 g of Edible Portion*					
	Fruit (underripe, raw)	Ripe (cooked)	Seeds (fresh)	Seeds (roasted)	Seeds (dried)
Calories	105-109				
Moisture	62.7-89.16 g	67.8 g	35.08-56.80 g	43.80 g	
Protein	1.3-2.24 g	1.34 g	5.25-13.3 g	7.72 g	13.8-19.96 g
Fat	0.1-0.86 g	0.31 g	2.59-5.59 g	3.30 g	5.1-12.79 g
Carbohydrates	21.5-29.49 g	27.82 g	30.83-44.03 g	41.61 g	15.95 g
Fiber	1.08-2.1 g	1.5 g	1.34-2.14 g	1.67 g	3.0-3.87 g
Ash	0.56-1.2 g	1.23 g	1.50-5.58 g	1.90 g	3.42-3.5 g
Calcium	0.05 mg	0.022 mg	0.11 mg	40 mg	0.12 mg
Phosphorus	0.04 mg	0.062 mg	0.35 mg	178 mg	0.37 mg
Iron	0.61-2.4 mg		3.78 mg	2.66 mg	
Carotene	0.004 mg (35-40 I.U.)				
Thiamine	0.08-0.085 mg		0.25 mg	0.32 mg	180 mcg
Riboflavin	0.033-0.07 mg		0.10 mg	0.10 mg	84 mcg
Niacin	0.506-0.92 mg		3.54 mg	2.94 mg	2.6 mg
Ascorbic Acid	15-33 mg		13.70 mg	14 mg	
Amino Acids: (According to Busson [N = 16 p. 100])					
Arginine	4.9		0.66		
Cystine	—		0.62		
Histidine	1.6		0.91		
Isoleucine	6.7		2.41		
Leucine	7.4		2.60		
Lysine	5.8				
Methionine	1.2		3.17		
Phenylalanine	8.3		1.05		
Threonine	6.8		0.78		
Tryptophan	7.0				
Valine	7.8				
Aspartic Acid	10.8				
Glutamic Acid	11.3		0.98		
Alanine	3.9		1.53		
Glycine	7.2		0.95		
Proline	6.5		0.72		
Serine	5.7		2.08		
Tyrosine			1.45		

*A composite of analyses made in Central America, Mexico, Colombia, Africa and India.
Note: There are reportedly two enzymes in the breadfruit—*papayotin* and *artocarpine*.

Negron de Bravo and colleagues in Puerto Rico show niacin content up to 8.33 mg in dried, ground seeds collected locally.
It will be seen from the above that the seedless breadfruit is low in protein, the seeds considerably higher, and therefore the seeded breadfruit is actually of more value as food.
Breadfruit flour contains 4.05% protein; 76.70% carbohydrates, and 331 calories, while cassava flour contains, 1.16% protein, 83.83% carbohydrates, and 347 calories per 100 g.

Toxicity

Most varieties of breadfruit are purgative if eaten raw. Some varieties are boiled twice and the water thrown away, to avoid unpleasant effects, while there are a few named cultivars that can be safely eaten without cooking.

The cyclopropane-containing sterol, *cycloartenol*, has been isolated from the fresh fruit. It contitutes 12% of the non-saponifiable extract.

Other Uses

Leaves: Breadfruit leaves are eagerly eaten by domestic livestock. In India, they are fed to cattle and goats; in Guam, to cattle, horses and pigs. Horses are apt to eat the bark of young trees as well, so new plantings must be protected from them.

Latex: Breadfruit latex has been used in the past as birdlime on the tips of posts to catch birds. The early Hawaiians plucked the feathers for their ceremonial cloaks, then removed the gummy substance from the birds' feet with oil from the candlenut, *Aleurites moluccana* Willd., or with sugarcane juice, and released them.

After boiling with coconut oil, the latex serves for caulking boats and, mixed with colored earth, is used as a paint for boats.

Wood: The wood is yellowish or yellow-gray with dark markings or orange speckles; light in weight; not very hard but strong, elastic and termite-resistant (except for drywood termites) and is used for construction and furniture. In Samoa, it is the standard material for house-posts and for the rounded roof-ends of native houses. The wood of the Samoan variety 'Aveloloa' which has deeply-cut leaves, is most preferred for house-building, but that of 'Puou', an ancient variety, is also utilized. In Guam and Puerto Rico the wood is used for interior partitions. Because of its lightness, the wood is in demand for surf-boards. Traditional Hawaiian drums are made from sections of breadfruit trunks 2 ft (60 cm) long and 1 ft (30 cm) in width, and these are played with the palms of the hands during Hula dances. After seasoning by burying in mud, the wood is valued for making household articles. These are rough-sanded by coral and lava, but the final smoothing is accomplished with the dried stipules of the breadfruit tree itself.

Fiber: Fiber from the bark is difficult to extract but highly durable. Malaysians fashioned it into clothing. Material for tapa cloth is obtained from the inner bark of young trees and branches. In the Philippines, it is made into harnesses for water buffalo.

Flowers: The male flower spike used to be blended with the fiber of the paper mulberry, *Broussonetia papyrifera* Vent. to make elegant loincloths. When thoroughly dry, the flower spikes also serve as tinder.

Medicinal Uses: In Trinidad and the Bahamas, a decoction of the breadfruit leaf is believed to lower blood pressure, and is also said to relieve asthma. Crushed leaves are applied on the tongue as a treatment for thrush. The leaf juice is employed as ear-drops. Ashes of burned leaves are used on skin infections. A powder of roasted leaves is employed as a remedy for enlarged spleen. The crushed fruit is poulticed on tumors to "ripen" them. Toasted flowers are rubbed on the gums around an aching tooth. The latex is used on skin diseases and is bandaged on the spine to relieve sciatica. Diluted latex is taken internally to overcome diarrhea.

Jackfruit (Plate VI)

The jackfruit, *Artocarpus heterophyllus* Lam. (syns. *A. integrifolius* Auct. NOT L. f.; *A integrifolia* L. f.; *A. integra* Merr.; *Rademachia integra* Thunb.), of the family Moraceae, is also called jak-fruit, jak, jaca, and, in Malaysia and the Philippines, *nangka;* in Thailand, *khanun;* in Cambodia, *khnor;* in Laos, *mak mi* or *may mi;* in Vietnam, *mit.* It is an excellent example of a food prized in some areas of the world and allowed to go to waste in others. O.W. Barrett wrote in 1928: "The jaks . . . are such large and interesting fruits and the trees so well-behaved that it is difficult to explain the general lack of knowledge concerning them."

Description

The tree is handsome and stately, 30 to 70 ft (9-21 m) tall, with evergreen, alternate, glossy, somewhat leathery leaves to 9 in (22.5 cm) long, oval on mature wood, sometimes oblong or deeply lobed on young shoots. All parts contain a sticky, white latex. Short, stout flowering twigs emerge from the trunk and large branches, or even from the soil-covered base of very old trees. The tree is monoecious: tiny male flowers are borne in oblong clusters 2 to 4 in (5-10 cm) in length; the female flower clusters are elliptic or rounded. Largest of all tree-borne fruits, the jackfruit may be 8 in to 3 ft (20-90 cm) long and 6 to 20 in (15-50 cm) wide, and the weight ranges from 10 to 60 or even as much as 110 lbs (4.5-20 or 50 kg). The "rind" or exterior of the compound or aggregate fruit is green or yellow when ripe and composed of numerous hard, cone-like points attached to a thick and rubbery, pale-yellow or whitish wall. The interior consists of large "bulbs" (fully developed perianths) of yellow, banana-flavored flesh, massed among narrow ribbons of thin, tough undeveloped perianths (or perigones), and a central, pithy core. Each bulb encloses a smooth, oval, light-brown "seed" (endocarp) covered by a thin white membrane (exocarp). The seed is ¾ to 1½ in (2-4 cm) long and ½ to ¾ in (1.25-2 cm) thick and is white and crisp within. There may be 100 or up to 500 seeds in a single fruit. When fully ripe, the unopened jackfruit emits a strong disagreeable odor, resembling that of decayed onions, while the pulp of the opened fruit smells of pineapple and banana.

Origin and Distribution

No one knows the jackfruit's place of origin but it is believed indigenous to the rainforests of the Western Ghats. It is cultivated at low elevations throughout India, Burma, Ceylon, southern China, Malaya, and the East Indies. It is common in the Philippines, both cultivated and naturalized. It is grown to a limited extent in Queensland and Mauritius. In Africa, it is often planted in Kenya, Uganda and former Zanzibar. Though planted in Hawaii prior to 1888, it is still rare there and in other Pacific islands, as it is in most of tropical America and the West Indies. It was introduced into northern Brazil in the mid-19th Century and is more popular there and in Surinam than elsewhere in the New World.

In 1782, plants from a captured French ship destined for Martinique were taken to Jamaica where the tree is now common, and about 100 years later, the jackfruit made its appearance in Florida, presumably imported by the Reasoner's Nursery from Ceylon. The United States Department of Agriculture's *Report on the Conditions of Tropical and Semitropical Fruits in the United States in 1887* states: "There are but few specimens in the State. Mr. Bidwell, at Orlando, has a healthy young tree, which was killed back to the ground, however, by the freeze of 1886." There are today less than a dozen bearing jackfruit trees in South Florida and these are valued mainly as curiosities. Many seeds have been planted over the years but few seedlings have survived, though the jackfruit is hardier than its close relative, the breadfruit (q.v.).

In South India, the jackfruit is a popular food ranking next to the mango and banana in total annual production. There are more than 100,000 trees in backyards and grown for shade in betelnut, coffee, pepper and cardamom plantations. The total area planted to jackfruit in all India is calculated at 14,826 acres (26,000 ha). Government horticulturists promote the planting of jackfruit trees along highways, waterways and railroads to add to the country's food supply.

There are over 11,000 acres (4,452 ha) planted to jackfruit in Ceylon, mainly for timber, with the fruit a much-appreciated by-product. The tree is commonly cultivated throughout Thailand for its fruit. Away from the Far East, the jackfruit has never gained the acceptance accorded the breadfruit (except in settlements of people of East Indian origin). This is due largely to the odor of the ripe fruit and to traditional preference for the breadfruit.

Varieties

In South India, jackfruits are classified as of two general types: 1) *Koozha chakka,* the fruits of which have small, fibrous, soft, mushy, but very sweet carpels; 2) *Koozha pazham,* more important commercially, with crisp carpels of high quality known as *Varika.* These types are apparently known in different areas by other names such as *Barka,* or *Berka* (soft, sweet and broken open with the hands), and *Kapa* or *Kapiya* (crisp and cut open with a knife). The equivalent types are known as *Kha-nun nang* (firm; best) and *Kha-nun lamoud* (soft) in Thailand; and as *Vela* (soft) and *Varaka,* or *Waraka* (firm) in

Fig. 15: A heavily fruiting jackfruit (*Artocarpus heterophyllus*) on the grounds of the old Hobson estate, Coconut Grove, Miami, Fla.

Ceylon. The *Peniwaraka,* or honey jak, has sweet pulp, and some have claimed it the best of all. The *Kuruwaraka* has small, rounded fruits. Dr. David Fairchild, writing of the honey jak in Ceylon, describes the rind as dark-green in contrast to the golden-yellow pulp when cut open for eating, but the fruits of his own tree in Coconut Grove and those of the Matheson tree which he maintained were honey jaks are definitely yellow when ripe. The *Vela* type predominates in the West Indies.

Firminger described two types: the *Khuja* (green, hard and smooth, with juicy pulp and small seeds); the *Ghila* (rough, soft, with thin pulp, not very juicy, and large seeds). Dutta says *Khujja,* or *Karcha,* has pale-brown or occasionally pale-green rind, and pulp as hard as an apple; *Ghila,* or *Ghula,* is usually light-green, occasionally brownish, and has soft pulp, sweet or acidulously sweet. He describes 8 varieties, only one with a name. This is *Hazari,* similar to *Rudrakshi,* which has a relatively smooth rind and flesh of inferior quality.

The '**Singapore**', or '**Ceylon**', jack, a remarkably early bearer producing fruit in 18 months to 2½ years from transplanting, was introduced into India from Ceylon and planted extensively in 1949. The fruit is of medium size with small, fibrous carpels which are very sweet. In addition to the summer crop (June and July), there is a second crop from October to December. In 1961, the Horticultural Research Institute at Saharanpur, India, reported the acquisition of air-layered plants of the excellent varieties, '**Safeda**', '**Khaja**', '**Bhusila**', '**Bhadaiyan**' and '**Handia**' and others. The Fruit Experimental Station at Burliar, established a collection of 54 jackfruit clones from all producing countries, and ultimately selected **'T Nagar Jack'** as the best in quality and yield. The Fruit Experimental Station at Kallar, began breeding work in 1952 with a view to developing short, compact, many-branched trees, precocious and productive, bearing large, yellow, high-quality fruits, ½ in the main season, ½ late. 'Singapore Jack' was chosen as the female parent because of its early and late crops; and, as the male parent, **'Velipala'**, a local selection from the forest having large fruits with large carpels of superior quality, and borne regularly in the main summer season. After 25 years of testing, one hybrid was rated as outstanding for precocity, fruit size, off-season as well as main-season production, and yield excelling its parents. It had not been named when reported on by Chellappan and Roche in 1982. In Assam, nurserymen have given names such as 'Mammoth', 'Everbearer', and 'Rose-scented' to preferred types.

Pollination

Horticulturists in Madras have found that hand-pollination produces fruits with more of the fully developed bulbs than does normal wind-pollination.

Climate

The jackfruit is adapted only to humid tropical and near-tropical climates. It is sensitive to frost in its early life and cannot tolerate drought. If rainfall is deficient, the tree must be irrigated. In India, it thrives in the Himalayan foothills and from sea-level to an altitude of 5,000 ft (1,500 m) in the south. It is stated that jackfruits grown above 4,000 ft (1,200 m) are of poor quality and usable only for cooking. The tree ascends to about 800 ft (244 m) in Kwangtung, China.

Fig. 16: Much white, gummy latex flows from the jackfruit stalk when the slightly underripe fruit is harvested.

Soil

The jackfruit tree flourishes in rich, deep soil of medium or open texture, sometimes on deep gravelly or laterite soil. It will grow, but more slowly and not as tall in shallow limestone. In India, they say that the tree grows tall and thin on sand, short and thick on stony land. It cannot tolerate "wet feet". If the roots touch water, the tree will not bear fruit or may die.

Propagation

Propagation is usually by seeds which can be kept no longer than a month before planting. Germination requires 3 to 8 weeks but is expedited by soaking seeds in water for 24 hours. Soaking in a 10% solution of gibberellic acid results in 100% germination. The seeds may be sown *in situ* or may be nursery-germinated and moved when no more than 4 leaves have appeared. A more advanced seedling, with its long and delicate tap root, is very difficult to transplant successfully. Budding and grafting attempts have often been unsuccessful, though Ochse considers the modified Forkert method of budding feasible. Either jackfruit or champedak (q.v.) seedlings may serve as rootstocks and the grafting may be done at any time of year. Inarching has been practiced and advocated but presents the same problem of transplanting after separation from the scion-parent. To avoid this and yet achieve consistently early bearing of fruits of known quality, air-layers produced with the aid of growth-promoting hormones are being distributed in India. In Florida cuttings of young wood have been rooted under mist. At Calcutta University, cuttings have been successfully rooted only with forced and etiolated shoots treated with indole butyric acid (preferably at 5,000 mg/1) and kept under mist. Tissue culture experiments have been conducted at the Indian Institute of Horticultural Research, Bangalore.

Culture

Soaking one-month-old seedlings in a gibberellic acid solution (25-200 ppm) enhances shoot growth. Gibberellic acid spray and paste increase root growth. In plantations, the trees are set 30 to 40 ft (9-12 m) apart. Young plantings require protection from sunscald and from grazing animals, hares, deer, etc. Seeds in the field may be eaten by rats. Firminger describes the quaint practice of raising a young seedling in a 3- to 4-ft (0.9-1.2 m) bamboo tube, then bending over and coiling the pliant stem beneath the soil, with only the tip showing. In 5 years, such a plant is said to produce large and fine fruits on the spiral underground. In Travancore, the whole fruit is buried, the many seedlings which spring up are bound together with straw and they gradually fuse into one tree which bears in 6 to 7 years. Seedlings may ordinarily take 4 to 14 years to come into bearing, though certain precocious cultivars may begin to bear in 2½ to 3½ years. The jackfruit is a fairly rapid grower, reaching 58 ft (17.5 m) in height and 28 in (70 cm) around the trunk in 20 years in Ceylon. It is said to live as long as 100 years. However, productivity declines with age. In Thailand, it is recommended that alternate rows be planted every 10 years so that 20-year-old trees may be routinely removed from the plantation and replaced by a new generation.

Little attention has been given to the tree's fertilizer requirements. Severe symptoms of manganese deficiency have been observed in India.

After harvesting, the fruiting twigs may be cut back to the trunk or branch to induce flowering the next season. In the Cachar district of Assam, production of female flowers is said to be stimulated by slashing the tree with a hatchet, the shoots emerging from the wounds; and branches are lopped every 3 to 4 years to maintain fruitfulness. On the other hand, studies at the University of Kalyani, West Bengal, showed that neither scoring nor pruning of shoots increases fruit-set and that ringing enhances fruit-set only the first year, production declining in the second year.

Season

In Asia, jackfruits ripen principally from March to June, April to September, or June to August, depending on the climatic region, with some off-season crops from September to December, or a few fruits at other times of the year. In the West Indies, I have seen many ripening in June; in Florida, the season is late summer and fall.

Harvesting

Fruits mature 3 to 8 months from flowering. In Jamaica, an "X" is sometimes cut in the apex of the fruit to speed ripening and improve flavor.

Yield

In India, a good yield is 150 large fruits per tree annually, though some trees bear as many as 250 and a fully mature tree may produce 500, these probably of medium or small size.

Storage

Jackfruits turn brown and deteriorate quickly after ripening. Cold storage trials indicate that ripe fruits can be kept for 3 to 6 weeks at 52° to 55°F (11.11°-12.78°C) and relative humidity of 85 to 95%.

Pests and Diseases

Principal insect pests in India are the shoot-borer caterpillar, *Diaphania caesalis;* mealybugs. *Nipaecoccus viridis, Pseudococcus corymbatus,* and *Ferrisia virgata,* the spittle bug, *Cosmoscarta relata,* and jack scale, *Ceroplastes rubina.* The most destructive and widespread bark borers are *Indarbela tetraonis* and *Batocera rufomaculata.* Other major pests are the stem and fruit borer, *Margaronia caecalis,* and the brown bud-weevil, *Ochyromera artocarpio.* In southern China, the larvae of the longicorn beetles, including *Apriona germarri, Pterolophia discalis, Xenolea tomenlosa asiatica,* and *Olenecamptus bilobus* seriously damage the fruit stem. The caterpillar of the leaf webbers, *Perina nuda* and *Diaphania bivitralis,* is a minor problem, as are aphids, *Greenidea artocarpi* and *Toxoptera aurantii,* and thrips, *Pseudodendrothrips dwivarna.*

Diseases of importance include pink disease, *Pellicularia (Corticium) salmonicolor,* stem rot, fruit rot and male inflorescence rot caused by *Rhizopus artocarpi,* and leafspot due to *Phomopsis artocarpina, Colletotrichum lagenarium, Septoria artocarpi,* and other fungi. Gray blight, *Pestalotia elasticola,* charcoal rot, *Ustilana zonata,* collar rot, *Rosellinia arcuata,* and rust, *Uredo artocarpi,* occur on jackfruit in some regions.

The fruits may be covered with paper sacks when very young to protect them from pests and diseases. Burkill says the bags encourage ants to swarm over the fruit and guard it from its enemies.

Fig. 17: Dried slices of peeled unripe jackfruit are commonly marketed in Southeast Asia.

Food Uses

Westerners generally will find the jackfruit most acceptable in the full-grown but unripe stage, when it has no objectionable odor and excels cooked green breadfruit and plantain. The fruit at this time is simply cut into large chunks for cooking, the only handicap being its copious gummy latex which accumulates on the knife and the hands unless they are first rubbed with salad oil. The chunks are boiled in lightly salted water until tender, when the really delicious flesh is cut from the rind and served as a vegetable, including the seeds which, if thoroughly cooked, are mealy and agreeable. The latex clinging to the pot may be removed by rubbing with oil. The flesh of the unripe fruit has been experimentally canned in brine or with curry. It may also be dried and kept in tins for a year. Cross-sections of dried, unripe jackfruit are sold in native markets in Thailand. Tender young fruits may be pickled with or without spices.

If the jackfruit is allowed to ripen, the bulbs and seeds may be extracted outdoors; or, if indoors, the odorous residue should be removed from the kitchen at once. The bulbs may then be enjoyed raw or cooked (with coconut milk or otherwise); or made into ice cream, chutney, jam, jelly, paste, "leather" or *papad,* or canned in sirup made with sugar or honey with citric acid added. The crisp types of jackfruit are preferred for canning. The canned product is more attractive than the fresh pulp and is sometimes called "vegetable meat". The ripe bulbs are mechanically pulped to make jackfruit nectar or reduced to concentrate or powder. The addition of synthetic flavoring—ethyl and *n*-butyl esters of 4-hydroxybutyric acid at 120 ppm and 100 ppm, respectively—greatly improves the flavor of the canned fruit and the nectar.

If the bulbs are boiled in milk, the latter when drained off and cooled will congeal and form a pleasant, orange-colored custard. By a method patented in India, the ripe bulbs may be dried, fried in oil and salted for eating like potato chips. Candied jackfruit pulp in boxes was being marketed in Brazil in 1917. Improved methods of preserving and candying jackfruit pulp have been devised at the Central Food Technological Research Institute, Mysore, India. Ripe bulbs, sliced and packed in sirup with added citric acid, and frozen, retain good color, flavor and texture for one year. Canned jackfruit retains quality for 63 weeks at room temperature—75° to 80°F (23.89°-26.67°C), with only 3% loss of β-carotene. When frozen, the canned pulp keeps well for 2 years.

In Malaya, where the odor of the ripe fruit is not avoided, small jackfruits are cut in half, seeded, chilled, and brought to the table filled with ice cream.

The ripe bulbs, fermented and then distilled, produce a potent liquor.

The seeds, which appeal to all tastes, may be boiled or roasted and eaten, or boiled and preserved in sirup like chestnuts. They have also been successfully canned in brine, in curry, and, like baked beans, in tomato sauce. They are often included in curried dishes. Roasted, dried seeds are ground to make a flour which is blended with wheat flour for baking.

Where large quantities of jackfruit are available, it is worthwhile to utilize the inedible portion, and the rind has been found to yield a fair jelly with citric acid. A pectin extract can be made from the peel, undeveloped perianths and core, or just from the inner rind; and this waste also yields a sirup used for tobacco curing.

Tender jackfruit leaves and young male flower clusters may be cooked and served as vegetables.

Fig. 18: Jackfruit seeds, salvaged from the ripe fruits, are sold for boiling or roasting like chestnuts.

Food Value Per 100 g of Edible Portion			
	Pulp (ripe-fresh)	*Seeds (fresh)*	*Seeds (dried)*
Calories	98		
Moisture	72.0-77.2 g	51.6-57.77 g	
Protein	1.3-1.9 g	6.6 g	
Fat	0.1-0.3 g	0.4 g	
Carbohydrates	18.9-25.4 g	38.4 g	
Fiber	1.0-1.1 g	1.5 g	
Ash	0.8-1.0 g	1.25-1.50 g	2.96%
Calcium	22 mg	0.05-0.55 mg	0.13%
Phosphorus	38 mg	0.13-0.23 mg	0.54%
Iron	0.5 mg	0.002-1.2 mg	0.005%
Sodium	2 mg		
Potassium	407 mg		
Vitamin A	540 I.U.		
Thiamine	0.03 mg		
Niacin	4 mg		
Ascorbic Acid	8-10 mg		

The pulp constitutes 25-40% of the fruit's weight.

In general, fresh seeds are considered to be high in starch, low in calcium and iron; good sources of vitamins B_1 and B_2.

Toxicity

Even in India there is some resistance to the jackfruit, attributed to the belief that overindulgence in it causes digestive ailments. Burkill declares that it is the raw, unripe fruit that is astringent and indigestible. The ripe fruit is somewhat laxative; if eaten in excess it will cause diarrhea. Raw jackfruit seeds are indigestible due to the presence of a powerful trypsin inhibitor. This element is destroyed by boiling or baking.

Other Uses

Fruit: In some areas, the jackfruit is fed to cattle. The tree is even planted in pastures so that the animals can avail themselves of the fallen fruits. Surplus jackfruit rind is considered a good stock food.

Leaves: Young leaves are readily eaten by cattle and other livestock and are said to be fattening. In India, the leaves are used as food wrappers in cooking, and they are also fastened together for use as plates.

Latex: The latex serves as birdlime, alone or mixed with *Ficus* sap and oil from *Schleichera trijuga* Willd. The heated latex is employed as a household cement for mending chinaware and earthenware, and to caulk boats and holes in buckets. The chemical constituents of the latex have been reported by Tanchico and Magpanlay. It is not a substitue for rubber but contains 82.6 to 86.4% resins which may have value in varnishes. Its bacteriolytic activity is equal to that of papaya latex.

Wood: Jackwood is an important timber in Ceylon and, to a lesser extent, in India; some is exported to Europe. It changes with age from orange or yellow to brown or dark-red; is termite-proof, fairly resistant to fungal and bacterial decay, seasons without difficulty, resembles mahogany and is superior to teak for furniture, construction, turnery, masts, oars, implements, brush backs and musical instruments. Palaces were built of jack-wood in Bali and Macassar, and the limited supply was once reserved for temples in Indochina. Its strength is 75 to 80% that of teak. Though sharp tools are needed to achieve a smooth surface, it polishes beautifully. Roots of old trees are greatly prized for carving and picture-framing. Dried branches are employed to produce fire by friction in religious ceremonies in Malabar.

From the sawdust of jackwood or chips of the heartwood, boiled with alum, there is derived a rich yellow dye commonly used for dyeing silk and the cotton robes of Buddhist priests. In Indonesia, splinters of the wood are put into the bamboo tubes collecting coconut toddy in order to impart a yellow tone to the sugar. Besides the yellow colorant, *morin,* the wood contains the colorless *cyanomaclurin* and a new yellow coloring matter, *arto-carpin,* was reported by workers in Bombay in 1955. Six

other flavonoids have been isolated at the National Chemical Laboratory, Poona.

Bark: There is only 3.3% tannin in the bark which is occasionally made into cordage or cloth.

Medicinal Uses: The Chinese consider jackfruit pulp and seeds tonic, cooling and nutritious, and to be "useful in overcoming the influence of alcohol on the system." The seed starch is given to relieve biliousness and the roasted seeds are regarded as aphrodisiac. The ash of jackfruit leaves, burned with corn and coconut shells, is used alone or mixed with coconut oil to heal ulcers. The dried latex yields artostenone, convertible to artosterone, a compound with marked androgenic action. Mixed with vinegar, the latex promotes healing of abscesses, snakebite and glandular swellings. The root is a remedy for skin diseases and asthma. An extract of the root is taken in cases of fever and diarrhea. The bark is made into poultices. Heated leaves are placed on wounds. The wood has a sedative property; its pith is said to produce abortion.

Related Species

The **Champedak,** *A. integer* Merr. (syns. *A. champeden* Spreng., *A. polyphena* Pers.), is also known as *chempedak, cempedak, sempedak, temedak* in Malaya; *cham–pa–da* in Thailand, *tjampedak* in Indonesia; *lemasa* in the Philippines. The wild form in Malaya is called *bangkong* or *baroh.* The fruit is borne by a deciduous tree, reaching about 60 ft (18 m) in cultivation, up to 100 or 150 ft (30-45.5 m) in the wild. It is easy to distinguish from the jackfruit by the long, stiff, brown hairs on young branchlets, leaves, buds and peduncles. The leaves, often 3-lobed when young, are obovate-oblong or elliptical when mature and 6 to 11 in (15-28 cm) long. The male flower spikes are only 2 in (5 cm) long and the fruit cylindrical or irregular, no more than 14 in (35.5 cm) long and 6 in (15 cm) thick, mustard-yellow to golden-brown, reticulated, warty, and highly odoriferous when ripe. In fact, it is described as having the "strongest and richest smell of any fruit in creation." The rind is thinner than that of the jackfruit and the seeds and surrounding pulp can be extracted by cutting open the base and pulling on the fruit stalk. The pulp is deep-yellow, tender, slimy, juicy and sweet. That of the wild form is thin, subacid and odorless.

The tree is native and common in the wild in Malaya up to an altitude of 4,200 ft (1,300 m) and is cultivated throughout Malaysia and by many preferred to jackfruit. It is grown from seed or budded onto self-seedlings or jackfruit or other *Artocarpus* species. Seedlings bear in 5 years. The pulp is eaten with rice and the seeds are roasted and eaten. The wood is strong and durable and yields yellow dye, and the bark is rich in tannin.

The **Lakoocha,** *A. lakoocha* Roxb., is also known as monkey jack or *lakuchi* in India; *tampang* and other similar native names in Malaya; as *lokhat* in Thailand.

The tree is 20 to 30 ft (6-9 m) tall with deciduous, large, leathery leaves, downy on the underside. Male and female flowers are borne on the same tree, the former orange-yellow, the latter reddish. The fruits are nearly round or irregular, 2 to 5 in (5 12.5 cm) wide, velvety, dull-yellow tinged with pink, with sweet-sour pulp which is occasionally eaten raw but mostly made into curries or chutney. The male flower spike, acid and astringent, is pickled.

A native of the humid sub-Himalayan region of India, up to 4,000 ft (1,200 m), also Malaya and Ceylon, it is sometimes grown for shade or for its fruit. Seedlings come into production in 5 years. A specimen was planted at the Federal Experiment Station, Mayaguez, Puerto Rico, in 1921. There was a large tree in Bermuda in 1918.

The wood, sold as *lakuch,* is heavier than that of the jackfruit, similar to teak, durable outdoors and under water, but does not polish well. It is used for piles, and in construction; for boats, furniture and cabinetwork. The bark contains 8.5% tannin and is chewed like betelnut. It yields a fiber for cordage. The wood and roots yield a dye of richer color than that obtained from the jackfruit. Both seeds and milky latex are purgative. The bark is applied on skin ailments. The fruit is believed to act as a tonic for the liver.

The **Kwai Muk,** possibly *A. lingnanensis* Merr., was introduced into Florida as *A. hypargyraea* Hance, or *A. hypargyraeus* Hance ex Benth. The tree is a slow-growing, slender, erect ornamental 20 to 50 ft (6-15 m) tall, with much milky latex and evergreen leaves 2 to 5 in (5-12.5 cm) long. Tiny male and female flowers are yellowish and borne on the same tree, the female in globular heads to 3/8 in (1 cm) long.

The fruits are more or less oblate and irregular, 1 to 2 in (2.5-5 cm) wide, with velvety, brownish, thin, tender skin and replete with latex when unripe. When ripe, the pulp is orange-red or red, soft, of agreeable subacid to acid flavor and may be seedless or contain 1 to 7 small, pale seeds. The pulp is edible raw; can be preserved in sirup or dried. Ripens from August to October in Florida.

The tree is native from Kwangtung, China, to Hong Kong, and has been introduced sparingly abroad. It was planted experimentally in Florida in 1927 and was thriving in Puerto Rico in 1929. It grows at an altitude of 500 ft (152 m) in China. Young trees are injured by brief drops in temperature to 28° to 30°F (-2.22°--1.11°C). Mature trees have endured 25° to 26°F (-3.89°--3.33°C) in Homestead, Florida; have been killed by 20°F (-6.67°C) in central Florida.

Amazon Tree-Grape

The Amazon tree-grape, *Pourouma cecropiaefolia* Mart., of the family Moraceae, is the best-known of about 50 species of *Pourouma* in Central America and tropical South America. It is known in Brazil generally as *puruma, cucura, imbauba mansa, imbauba-de-vinho, imbauba de cheiro;* in Bahia as *tararanga preta* and in Manaus as *mapati.* In Colombia it is called *puruma, caime, caimaron, caimaron silvestre, uva caimarona, camuirro, cucura, uva, sirpe, hiye* or *joyahiye.* In Peru, it is simply *uvilla.*

The tree resembles *Cecropia* spp., which are called *imbauba* in Brazil. It reaches 23 to 50 ft (7-15 m) in height. The bark is gray and marked with leaf scars. The alternate leaves, on long petioles, are nearly circular but deeply cleft into obovate-oblong-lanceolate lobes to 1 ft (30 cm) long. They are green on the upper surface, whitish or bluish-gray and velvety beneath; agreeably aromatic, like wintergreen, when crushed. The unopened inflorescence is reddish-purple, densely coated with fine white hairs. The white male and female flowers are borne on separate trees. Borne in bunches of 20 or more, the fruit is grapelike except for its wintergreen odor. It is round or round-ovate, usually 3/8 to 3/4 in (0.5-1 cm) wide, occasionally to 1½ in (4 cm). The skin is very rough to the touch, inedible but easily peeled; purple when ripe. The pulp is white, mucilaginous, juicy; of subacid, very mild flavor; and encloses 1 conical seed with fibrous, grooved coat.

The tree grows wild in the western part of Amazonas, Brazil, and adjacent areas of Ecuador and Peru. It is especially abundant in the vicinity of Iquitos. It has been cultivated since pre-Hispanic times by the Indians of southwestern Colombia and is grown by Indians and non-Indians in Brazil. Patiño says that around 1940 propagation was begun at the Estacion Agricola at Palmira, Colombia, and seeds and plants were given to the Estacion at Calima in 1945. Some trees are being grown, too, at the Estacion Agricola de Armero. There is today renewed interest in encouraging cultivation.

The tree grows on high dry land at altitudes below 1,640 ft (500 m). It may be subject to flooding every 4 or 5 years. It cannot stand prolonged drought. The seeds have short-term viability. If planted in time, they may show 86% germination. Cuttings are difficult to grow. Seedlings bear in 1 to 3 years after setting out. There may be 2 crops per year. Some trees that have been at least 3 years in the plantation have yielded 110 lbs (50 kg). The fruit is eaten raw or made into wine.

The wood is light, coarse and non-durable. It is used only for making charcoal.

Cherimoya (Plate VII)

Certainly the most esteemed of the fruits of the genus *Annona* (family Annonaceae), the cherimoya, *A. cherimola* Mill., because of its limited distribution, has acquired few colloquial names, and most are merely local variations in spelling, such as *chirimoya, cherimolia, chirimolla, cherimolier, cherimoyer.* In Venezuela, it is called *chirimorriñon;* in Brazil, *graveola, graviola,* or *grabiola;* and in Mexico, *pox or poox;* in Belize, *tukib;* in El Salvador it is sometimes known as *anona poshte;* and elsewhere merely as *anona,* or *anona blanca.* In France, it is *anone;* in Haiti, *cachiman la Chine.* Indian names in Guatemala include *pac, pap, tsummy* and *tzumux.* The name, cherimoya, is sometimes misapplied to the less-esteemed custard apple, *A. reticulata* L. In Australia it is often applied to the atemoya (a cherimoya-sugar apple hybrid).

Description

The tree is erect but low-branched and somewhat shrubby or spreading; ranging from 16 to 30 ft (5 to 9 m) in height; and its young branchlets are rusty-hairy. The leaves are briefly deciduous (just before spring flowering), alternate, 2-ranked, with minutely hairy petioles ¼ to ½ in (6 to 12.5 mm) long; ovate to elliptic or ovate-lanceolate, short blunt-pointed at the apex; slightly hairy on the upper surface, velvety on the underside; 3 to 6 in (7.5-15 cm) long, 1½ to 3½ in (3.8-8.9 cm) wide. Fragrant flowers, solitary or in groups of 2 or 3, on short, hairy stalks along the branches, have 3 outer, greenish, fleshy, oblong, downy petals to 1¼ in (3 cm) long and 3 smaller, pinkish inner petals. A compound fruit, the cherimoya is conical or somewhat heart-shaped, 4 to 8 in (10 to 20 cm) long and up to 4 in (10 cm) in width, weighing on the average 5½ to 18 oz (150-500 g) but extra large specimens may weigh 6 lbs (2.7 kg) or more. The skin, thin or thick, may be smooth with fingerprint-like markings or covered with conical or rounded protuberances. The fruit is easily broken or cut open, exposing the snow-white, juicy flesh, of pleasing aroma and delicious, subacid flavor; and containing numerous hard, brown or black, beanlike, glossy seeds, ½ to ¾ in (1.25 to 2 cm) long.

Origin and Distribution

The cherimoya is believed indigenous to the inter-andean valleys of Ecuador, Colombia and Bolivia. In Bolivia, it flourishes best around Mizque and Ayopaya, in the Department of Cochabamba, and around Luribay, Sapahaqui and Rio Abajo in the Department of La Paz. Its cultivation must have spread in ancient times to Chile and Brazil for it has become naturalized in highlands throughout these countries. Many authors include Peru as a center of origin but others assert that the fruit was unknown in Peru until after seeds were sent by P. Bernabé Cobo from Guatemala in 1629 and that thirteen years after this introduction the cherimoya was observed in cultivation and sold in the markets of Lima. The often-cited representations of the cherimoya on ancient Peruvian pottery are actually images of the soursop, *A. muricata* L. Cobo sent seeds to Mexico also in 1629. There it thrives between 4,000 and 5,000 ft (1312-1640 m) elevations.

It is commonly grown and naturalized in temperate areas of Costa Rica and other countries of Central America. In Argentina, the cherimoya is mostly grown in the Province of Tucuman. In 1757, it was carried to Spain where it remained a dooryard tree until the 1940's and 1950's when it gained importance in the Province of Granada, in the Sierra Nevada mountains, as a replacement for the many orange trees that succumbed to disease and had to be taken out. By 1953, there were 262 acres (106 ha) of cherimoyas in this region.

In 1790 the cherimoya was introduced into Hawaii by Don Francisco de Paulo Marin. It is still casually grown in the islands and naturalized in dry upland forests. In 1785, it reached Jamaica, where it is cultivated and occurs as an escape on hillsides between 3,500 and 5,000 ft (1,066-1,524 m). It found its way to Haiti sometime later. The first planting in Italy was in 1797 and it became a favored crop in the Province of Reggio Calabria. The tree has been tried several times in the Botanic Gardens, Singapore — first around 1878 — but has always failed to survive because of the tropical climate. In the Philippines, it does well in the Mountain Province at an altitude above 2,460 ft (750 m). It was introduced into India and Ceylon in 1880 and there is small-scale culture in both countries at elevations between 1,500 and 7,000 ft (457-2,134 m). The tree was planted in Madeira in 1897, then in the Canary Islands, Algiers, Egypt and, probably via Italy, in Libya, Eritrea and Somalia.

The United States Department of Agriculture imported a number of lots of cherimoya seeds from Madeira in 1907 (S.P.I. Nos. 19853, 19854, 19855, 19898, 19901, 19904, 19905).

Seeds from Mexico were planted in California in 1871. There were 9,000 trees in that state in 1936 but many of them were killed by a freeze in 1937. Several small commercial orchards were established in the 1940's. At present there may be less than 100 acres (42 ha) in the milder parts of San Diego County. Seeds, seedlings and grafted trees from California and elsewhere have been planted in Florida many times but none has done well. Any fruits produced have been of poor quality.

Varieties

In Peru, cherimoyas are classed according to degree of surface irregularity, as: **'Lisa'**, almost smooth; **'Impresa'**, with "fingerprint" depressions; **'Umbonada'**, with rounded protrusions; **'Papilonado',** or **'Tetilado',** with fleshy, nipple-like protrusions; **'Tuberculada'**, with conical protrusions having wartlike tips. At the Agricultural Experiment Station "La Molina", several named and unnamed selections collected in northern Peru are maintained and evaluated. Among the more important are: #1, **'Chavez'**, fruits up to 3.3 lbs (1½ kg); February to May; #2, **'Namas'**, fruits January to April; #3, **'Sander'**, fruits with moderate number of seeds; July and early August; #4, fruit nearly smooth, not many seeds, 1.1 to 2.2 lbs (½–1 kg), June to August; #5, nearly smooth, very sweet, 2.2 lbs (1 kg), March to June; #6, fruit with small protuberances, 1.1 to 2.2 lbs (½–1 kg), not many seeds; #7 fruit small, very sweet, many seeds, March to May; #8, fruit very sweet, 1.1 to 2.2 lbs (½–1 kg), with very few seeds, February to April.

In the Department of Antioquia, Colombia, a cultivar called **'Rio Negro'** has heart-shaped fruits weighing 1¾ to 2.2 lbs (0.8-1 kg). The cherimoyas of Mizque, Cochabamba, Bolivia, are locally famed for their size and quality. **'Concha Lisa'** and **'Bronceada'** are grown commercially in Chile. Other cultivars mentioned in Chilean literature are **'Concha Picuda'** and **'Terciopelo'**.

Dr. Ernesto Saavedra, University of Chile, after experimenting with growth regulators for 4 years, developed a super cherimoya, 4 to 6 in (10-15 cm) wide and weighing up to 4 lbs (1.8 kg); symmetrical, easy to peel and seedless, hence having 25% more flesh than an ordinary cherimoya. However, the larger fruits are subject to cracking.

The leading commercial cultivars in Spain are **'Pinchua'** (thin-skinned) and **'Basta'** (thick-skinned.)

Named cultivars in California include:

'Bays' — rounded, fingerprinted, light-green, medium to large, of excellent flavor; good bearer; early.

'Whaley' — long-conical, sometimes shouldered at the base, slightly and irregularly tuberculate, with fairly thick, downy skin. Of good flavor, but membranous sac around each seed may adhere to flesh. Bears well; grown commercially; early.

'Deliciosa' — long-conical, prominently papillate; skin thin, slightly downy; variable in flavor; only fair in quality; generally bears well but doesn't ship well; cold-resistant. Midseason.

'Booth' — short-conical, fingerprinted, medium to large; of good flavor; next to 'Deliciosa' in hardiness. Late.

'McPherson' — short-conical, fingerprinted but umbonate at the base; medium to large; of high quality; bears well. Midseason.

'Carter' — long-conical, but not shouldered; smooth or faintly fingerprinted; skin green to bronze; bears well. Late. Leaves wavy or twisted.

'Ryerson' — long-conical, smooth or fingerprinted, with thick, tough, green or yellow-green skin; of fair quality; ships well. Leaves wavy or twisted.

'White' — short-conical with rounded apex; slightly papillate to umbonate; medium to large; skin medium-thick; of good flavor; doesn't bear well near the coast.

'Chaffey' — introduced in 1940's; rounded, short, fingerprinted; of medium size; excellent quality; bears well, even without hand-pollination.

'Ott' (Patent #656) — introduced in 1940's; long-conical to heart-shaped, slightly tuberculate; of excellent flavor; ships well.

Among others that have been planted in California but considered inferior are: **'Horton', 'Golden Russet', 'Loma', 'Mira Vista', 'Sallmon'**.

Pollination

A problem with the cherimoya is inadequate natural pollination because the male and female structures of each flower do not mature simultaneously. Few insects visit the flowers. Therefore, hand-pollination is highly desirable and must be done in a 6- to 8-hour period when the stigmas are white and sticky. It has been found in Chile that in the first flowers to open the pollen grains are loaded with starch, whereas flowers that open later have more abundant pollen, no starch grains, and the pollen germinates readily. Partly-opened flowers are collected in the afternoon and kept in a paper bag overnight. The next morning the shed pollen is put, together with moist paper, in a vial and transferred by brush to the receptive stigmas. Usually only a few of the flowers on a tree are pollinated each time, the operation being repeated every 4 or 5 days in order to extend the season of ripening. The closely related *A. senegalensis* Pers., if available, is a good source of abundant pollen for pollinating the cherimoya. The pollen of the sugar apple is not satisfactory. Fruits from hand-pollinated flowers will be superior in form and size.

Climate

The cherimoya is subtropical or mild-temperate and does not succeed in the lowland tropics. It requires long days. In Colombia and Ecuador, it grows naturally at elevations between 4,600 and 6,600 ft (1,400-2,000 m) where the temperature ranges between 62.6° and 68°F (17°-20°C). In Peru, the ideal climate for the cherimoya is said to lie between 64.5° and 77°F (18°-25°C) in the summer and 64.5° and 41°F (18°-5°C) in winter. In Guatemala, naturalized trees are common between 4,000 and 8,200 ft (1,200-2,500 m) though the tree produces best between 4,000 and 5,900 ft (1,200-1,800 m) and can be grown at elevations as low as 2,950 ft (900 m). The tree cannot survive the cold in the Valle de Mexico at 7,200 ft (2,195 m). In Argentina, young trees are wrapped with dry grass or burlap during the winter. The cherimoya can tolerate light frosts. Young trees can withstand a temperature of 26°F (-3.33°C), but a few degrees lower will severely injure or kill mature trees. In February 1949, a small-scale commercial grower (B. E. Needham) in Glendora, California, reported that most of his crop was lost because of frost and snow, the cherimoya suffering more cold damage than his avocados, oranges or lemons.

The tree prefers a rather dry environment as in southern Guatemala where the rainfall is 50 in (127 cm) and there is a long dry season. It is not adaptable to northern Guatemala where the 100-inch (254-cm) rainfall is spread throughout the year.

Finally, the tree should be protected from strong winds which interfere with pollination and fruit set.

Soil

The cherimoya tree performs well on a wide range of soil types from light to heavy, but seems to do best on a medium soil of moderate fertility. In Argentina, it makes excellent growth on rockstrewn, loose, sandy loam 2 to 3 ft (0.6-0.9 m) above a gravel subsoil. The optimum pH ranges from 6.5 to 7.6. A greenhouse trial in sand has demonstrated that the first nutritional deficiency evoked in such soil is lack of calcium.

Propagation

Cherimoya seeds, if kept dry, will remain viable for several years. While the tree is traditionally grown from seed in Latin America, the tendency of seedlings to produce inferior fruits has given impetus to vegetative propagation.

Seeds for rootstocks are first soaked in water for 1 to 4 days and those that float are discarded. Then planting is done directly in the nursery row unless the soil is too cool, in which case the seeds must be placed in sand-peat seedbeds, covered with 1 in (2.5 cm) of soil and kept in a greenhouse. They will germinate in 3 to 5 weeks and when the plants are 3 to 4 in (7.5-10 cm) high, they are transplanted to pots or the nursery plot with 20 in (50 cm) between rows. When 12 to 24 months old and dormant, they are budded or grafted and then allowed to grow to 3 or 4 ft (0.9-1.2 m) high before setting out in the field. Large seedlings and old trees can be topworked by cleft-grafting. It is necessary to protect the trunk of topped trees to avoid sunburn.

The cherimoya can also be grafted onto the custard apple *(A. reticulata)*. In India this rootstock has given 90% success. Cuttings of mature wood of healthy cherimoya trees have rooted in coral sand with bottom heat in 28 days.

Fig. 19: Cherimoyas *(Annona cherimola)* from the highlands are sold at fruit stands along Venezuelan roadways.

Culture

The young trees should be spaced 25 to 30 ft (7.5-9 m) apart each way in pits 20 to 24 in (50-60 cm) wide, enriched with organic material. In Colombia, corn (maize), vegetables, ornamental foliage plants, roses or annual flowers for market are interplanted during the first few years. In Spain, the trees are originally spaced 16.5 ft (5 m) apart with the intention of later thinning them out. Thinning is not always done and around the village of Jete, where the finest cherimoyas are produced, the trees have grown so close together as to form a forest. In the early years they are interplanted with corn, beans and potatoes.

Pruning to eliminate low branches, providing a clean trunk up to 32 in (80 cm), to improve form, and open up to sunlight and pesticide control, is done preferably during dormancy. After 6 months, fertilizer (10-8-6 N,P,K) is applied at the rate of ½ lb (227 g) per tree and again 6 months later at 1 lb (454 g) per tree. In the 3rd year, the fertilizer formula is changed to 6-10-8 N,P,K and each year thereafter the amount per tree is increased by 1 lb (454 g) until the level of 5 lbs (2.27 kg) is reached. Thenceforth this amount is continued each year per tree. The fertilizer is applied in trenches 6 in (15 cm) deep and 8 in (20 cm) wide dug around each tree at a distance of 5 ft (1.5 m) from the base, at first; later, at an appropriately greater distance.

Young trees are irrigated every 15 to 20 days for the first few years except during the winter when they must be allowed to go dormant—ideally for 4 months. When the first leafbuds appear, irrigation is resumed. With bearing trees, watering is discontinued as soon as the fruits are full-grown.

In Chile, attempts to increase fruit set with chemical growth regulators have been disappointing. Spraying flowers with gibberellic acid has increased fruit set and improved form and size but induces deep cracking prior to full maturity, far beyond the normal rate of cracking in fruits from natural- or hand-pollinated flowers.

Cropping and Yield

The cherimoya begins to bear when 3½ to 5 years old and production steadily increases from the 5th to the 10th year, when there should be a yield of 25 fruits per tree—2,024 per acre (5,000 per ha). Yields of individual trees have been reported by eyewitnesses as a dozen, 85, or even 300 fruits annually. In Colombia, the average yield is 25 fruits; as many as 80 is exceptional. In Italy, trees 30 to 35 years old produce 230 to 280 fruits annually.

The fruits must be picked when full-grown but still firm and just beginning to show a slight hint of yellowish-green and perhaps a bronze cast. Bolivians judge that a fruit is at full maturity by shaking it and listening for the sound of loose seeds. Italians usually wait for the yellowish hue and the sweet aroma noticeable at a distance, picking the fruits only 24 to 28 hours prior to consumption. However, if the fruits must travel to markets in central Italy, they are harvested when the skin turns from dark-green to lighter green.

In harvesting, the fruits must be clipped from the branch so as to leave only a very short stem attached to the fruit to avoid stem-caused damage to the fruits in handling, packing and shipping.

Keeping Quality and Storage

Firm fruits should be held at a temperature of 50°F (10°C) to retard softening. When transferred to normal room temperature, they will become soft and ready to eat in 3 to 4 days. Then they can be kept chilled in the home refrigerator if not to be consumed immediately. A California grower has shipped cherimoyas ('Deliciosa' and 'Booth') packed in excelsior in 12-lb (5.5-kg) boxes to Boston and New York quite satisfactorily. And the fruit has been shipped from Madeira to London for many years.

In Bolivia, fruits for home use are wrapped in woollen cloth as soon as picked and kept at room temperature so that they can be eaten 3 days later.

Pests and Diseases

The cherimoya tree is resistant to nematodes. Very few problems have been noted in California except for infestations of mealybugs, especially at the base of the fruit, and these can be flushed off. In Colombia, on the other hand, it is said that a perfectly healthy tree is a rarity. In the Valle de Tenza, formerly an important center of production, lack of control of pests greatly reduced the plantations before 1960 when programs were launched to improve cherimoya culture here and in various other regions of the country.

Caterpillars (*Thecla* sp. and *Oiketicus kubeyi*) may defoliate the tree. A scale insect, *Conchaspis angraeci* attacks the trunk and branches. Prime enemies are reported to be fruit flies (*Anastrepha* sp.); leaf miners (*Leucoptera* sp.), particularly in the Valle de Tenza, which necessitate the collection and burning of affected leaves plus the application of systemic insecticides; and the seed borer (*Bephrata maculicollis*). The latter pest deposits eggs on the surface of the developing fruits, the larvae invade the fruit and consume the seeds, causing premature and defective ripening and rendering the fruits susceptible to fungal diseases. This pest is difficult to combat. Borers attack the tree in Argentina reducing its life span from 60 to 30 years.

The coccid, *Pseudococcus filamentosus* attacks the fruit in Hawaii, and *Aulacaspis miranda* and *Ceroputé yuccae* in Mexico. In Spain, the thin-skinned cultivar 'Pinchua' is subject to attack by the Mediterranean fruit-fly, *Ceratitis capitata*.

Stored seeds for planting are subject to attack by weevils. To avoid damping-off of young seedlings, dusting of seeds with fungicide is recommended. The tree may succumb to root-rot in clay soils or where there is too much moisture and insufficient drainage. Sooty mold may occur on leaves and fruits where ants, aphids and other insects have deposited honeydew.

Food Uses

The flesh of the ripe cherimoya is most commonly eaten out-of-hand or scooped with a spoon from the cut-

Food Value Per 100 g of Edible Portion			
Analysis of cherimoyas in Ecuador		**Colombian Analysis**	
Moisture	74.6 g	Moisture	77.1 g
Ether Extract	0.45 g	Protein	1.9 g
Crude Fiber	1.5 g	Fat	0.1 g
Nitrogen	.227 g	Carbohydrates	18.2 g
Ash	0.61 g	Fiber	2.0 g
Calcium	21.7 g	Ash	0.7 g
Phosphorus	30.2 mg	Calcium	32.0 mg
Iron	0.80 mg	Phosphorus	37.0 mg
Carotene	0.000 mg	Iron	0.5 mg
Thiamine	0.117 mg	Vitamin A (Carotene)	0.0 I.U.
Riboflavin	0.112 mg	Thiamine	0.10 mg
Niacin	1.02 mg	Riboflavin	0.14 mg
Ascorbic Acid	16.8 mg	Niacin	0.9 mg
		Ascorbic Acid	5.0 mg

open fruit. It really needs no embellishment but some people in Mexico like to add a few drops of lime juice. Occasionally it is seeded and added to fruit salads or used for making sherbet or ice cream. Colombians strain out the juice, add a slice of lemon and dilute with ice-water to make a refreshing soft drink. The fruit has been fermented to produce an alcoholic beverage.

Toxicity

The seeds, like those of other *Annona* species, are crushed and used as insecticide. Paul Allen, in his *Poisonous and Injurious Plants of Panama*, (see Bibliography), implies personal knowledge of a case of blindness resulting from "the juice of the crushed seeds coming in contact with the eyes." The seeds contain several alkaloids: caffeine, (+)-reticuline, (-)-anonaine, liriodenine, and lanuginosine.

Human ingestion of 0.15 g of the dark-yellow resin isolated from the seeds produces dilated pupils, intense photophobia, vomiting, nausea, dryness of the mouth, burning in the throat, flatulence, and other symptoms resembling the effects of atropine. A dose of 0.5 g, injected into a medium-sized dog, caused profuse vomiting.

Wilson Popenoe wrote that hogs feed on the fallen fruits in southern Ecuador where there are many cherimoya trees and few people. One wonders whether the hogs swallow the hard seeds whole and avoid injury.

The twigs possess the same alkaloids as the seeds plus michelalbine. A team of pharmacognosists in Spain and France has reported 8 alkaloids in the leaves: (+)-isoboldine, (-)-stepholidine, (+)-corytuberine, (+)-nornantenine, (+)-reticuline, (-)-anonaine, liriodenine, and lanuginosine.

Other Uses

In Jamaica, the dried flowers have been used as flavoring for snuff.

Medicinal Uses: In Mexico, rural people toast, peel and pulverize 1 or 2 seeds and take the powder with water or milk as a potent emetic and cathartic. Mixed with grease, the powder is used to kill lice and is applied on parasitic skin disorders. A decoction of the skin of the fruit is taken to relieve pneumonia.

Sugar Apple (Plate VIII)

The most widely grown of all the species of *Annona*, the sugar apple, *A. squamosa* L., has acquired various regional names: *anon* (Bolivia, Costa Rica, Cuba, Panama); *anon de azúcar, anon domestico, hanon, mocuyo* (Colombia); *anona blanca* (Honduras, Guatemala, Dominican Republic); *anona de castilla* (El Salvador); *anona de Guatemala* (Nicaragua); applebush (Grenadines); *ata, fruta do conde, fruta de condessa, frutiera de conde, pinha, araticutitaia,* or *ati* (Brazil); *ates* or *atis* (Philippines); *atte* (Gabon); *chirimoya* (Guatemala, Ecuador); *cachiman* (Argentina); *cachiman cannelle* (Haiti); *kaneelappel* (Surinam); *pomme cannelle* (Guadeloupe, French Guiana, French West Africa); *riñon* (Venezuela); *saramulla, saramuya, ahate* (Mexico); *scopappel* (Netherlands Antilles); sweetsop (Jamaica, Bahamas); *ata, luna, meba, sharifa, sarifa, sitaphal, sita pandu,* custard apple, scaly custard apple (India); *buah nona, nona, seri kaya* (Malaya) *manonah, noinah, pomme cannelle du Cap* (Thailand); *qu a na* (Vietnam); *mang cau ta* (Cambodia); *mak khbieb* (Laos); *fan–li–chi* (China).

Description

The sugar apple tree ranges from 10 to 20 ft (3-6 m) in height with open crown of irregular branches, and somewhat zigzag twigs. Deciduous leaves, alternately arranged on short, hairy petioles, are lanceolate or oblong, blunt-tipped, 2 to 6 in (5-15 cm) long and ¾ to 2 in (2-5 cm) wide; dull-green on the upperside, pale, with a bloom, below; slightly hairy when young; aromatic when crushed. Along the branch tips, opposite the leaves, the fragrant flowers are borne singly or in groups of 2 to 4. They are oblong, 1 to 1½ in (2.5-3.8 cm) long, never fully open; with 1 in (2.5 cm) long, drooping stalks, and 3 fleshy outer petals, yellow-green on the outside and pale-yellow inside with a purple or dark-red spot at the base. The 3 inner petals are merely tiny scales. The compound fruit is nearly round, ovoid, or conical; 2⅓ to 4 in (6-10 cm) long; its thick rind composed of knobby segments, pale-green, gray-green, bluish-green, or, in one form, dull, deep-pink externally (nearly always with a bloom); separating when the fruit is ripe and revealing the mass of conically segmented, creamy-white, glistening, delightfully fragrant, juicy, sweet, delicious flesh. Many of the segments enclose a single oblong-cylindric, black or dark-brown seed about ½ in (1.25 cm) long. There may be a total of 20 to 38, or perhaps more, seeds in the average fruit. Some trees, however, bear seedless fruits.

Origin and Distribution

The original home of the sugar apple is unknown. It is commonly cultivated in tropical South America, not often in Central America, very frequently in southern Mexico, the West Indies, Bahamas and Bermuda, and occasionally in southern Florida. In Jamaica, Puerto Rico, Barbados, and in dry regions of North Queensland, Australia, it has escaped from cultivation and is found wild in pastures, forests and along roadsides.

The Spaniards probably carried seeds from the New World to the Philippines and the Portuguese are assumed to have introduced the sugar apple to southern India before 1590. It was growing in Indonesia early in the 17th century and has been widely adopted in southern China, Queensland, Australia, Polynesia, Hawaii, tropical Africa, Egypt and the lowlands of Palestine. Cultivation is most extensive in India where the tree is also very common as an escape and the fruit exceedingly popular and abundant in markets. The sugar apple is one of the most important fruits in the interior of Brazil and is conspicuous in the markets of Bahia.

Cultivars

The **'Seedless Cuban'** sugar apple was introduced into Florida in 1955, has produced scant crops of slightly malformed fruits with mere vestiges of undeveloped seeds. The flavor is less appealing than that of normal fruits but it is vegetatively propagated and distributed as a novelty. Another seedless type was introduced from Brazil.

Indian horticulturists have studied the diverse wild and cultivated sugar apples of that country and recognize ten different types: **'Red'** (*A. squamosa* var. *Sangareddyii*)—red-tinted foliage and flowers, deep-pink rind, mostly non-reducing sugars, insipid, with small, blackish-pink seeds; poor quality; comes true from seed. **'Red–speckled'**—having red spots on green rind. **'Crimson'**—conspicuous red-toned foliage and flowers, deep-pink rind, pink flesh. **'Yellow'**; **'White–stemmed'**; **'Mammoth'** (*A. squamosa* var. *mammoth*)—pale-yellow petals, smooth, broad, thick, round rind segments that are light russet green; fruits lopsided, pulp soft, white, very sweet; comes true from seed. **'Balangar'**—large, with green rind having rough, warty [tuberculate], fairly thick rind segments with creamy margins; sweet; high yielding. **'Kakarlapahad'**—very high yielding. **'Washington'**—acute tuberculate rind segments, orange-yellow margins; high yielding; late in season, 20 days after others. **'Barbados'** and **'British Guiana'**—having green rind, orange-yellow margins; high-yielding; late.

Named cultivars growing at the Sabahia Experiment Station, Alexandria, Egypt, include: **'Beni Mazar'**—nearly round, large, 5¼ to 6½ oz (150-180 g); 56-60% flesh; 15-30 seeds. **'Abd El Razik'**—light-green or reddish rind; nearly round, large, maximum 8⅓ oz (236.3 g); 69.5% flesh; 14 seeds.

Climate

The sugar apple tree requires a tropical or near-tropical climate. It does not succeed in California because of the cool winters though in Israel it has survived several degrees below freezing. Generally, it does best in dry areas and it has high drought tolerance. However, in Ceylon it flourishes in the wet as well as the dry zones from sea level to 3,500 ft (1,066 m) elevation. During the blooming season, drought interferes with pollination and it is, therefore, concluded that the sugar apple should have high atmospheric humidity but no rain when flowering. In severe droughts, the tree sheds its leaves and the fruit rind hardens and will split with the advent of rain.

Soil

The sugar apple is not particular as to soil and has performed well on sand, oolitic limestone and heavy loam with good drainage. Water-logging is intolerable. The tree is shallow-rooted and doesn't need deep soil. Irrigation water containing over 300 ppm chlorine has done the tree no harm.

Propagation

Sugar apple seeds have a relatively long life, having kept well for 3 to 4 years. They germinate better a week after removal from the fruit than when perfectly fresh. Germination may take 30 days or more but can be hastened by soaking for 3 days or by scarifying. The percentage of germination is said to be better in unsoaked seeds.

While the tree is generally grown from seed, vegetative propagation is practiced where the crop is important and early fruiting is a distinct advantage.

Seedlings may be budded or grafted when one-year old. In India, selected clones grafted on *A. reticulata* seedlings have flowered within 4 months and fruited in 8 months after planting out, compared with 2 to 4 years in seedlings. The grafted trees are vigorous, the fruits less seedy and more uniform in size. *A. senegalensis* is employed as a rootstock in Egypt. *A. glabra* is suitable but less hardy. The sugar apple itself ranks next after *A. reticulata* as a rootstock. In India, budding is best done in January, March and June. Results are poor if done in July, August, November or December unless the scions are defoliated and debudded in advance and cut only after the petioles have dehisced. Side-grafting can be done only from December to May, requires much skill and the rate of success has not exceeded 58.33%. Shield-budding gives 75% success and is the only commercially feasible method.

Inarching is 100% successful. Cuttings, layers, air-layers have a low rate of success, and trees grown by these techniques have shallow root systems and cannot endure drought as well as seedlings do.

Culture

In Egypt, sugar apple trees are spaced at 10 x 10 ft (3x3 m) in order to elevate atmospheric humidity and improve pollination. Palestinian growers were spacing at 16 x 16 ft (5x5 m) but changed to 16 x 10 ft (5x3 m) as more feasible. On light soils, they apply 132 to 176 lbs (60-80 kg) manure per tree annually and they recommend the addition of nitrogen. Commercial fertilizer containing 3% N, 10% P and 10% K significantly increases flowering, fruit set and yield. Judicious pruning to improve shape and strength of tree must be done only in spring when the sap is rising, otherwise pruning may kill the tree. Irrigation during the dry season and once during ripening will increase fruit size.

Cropping and Yield

Seedlings 5 years old may yield 50 fruits per tree in late summer and fall. Older trees rarely exceed 100 fruits per tree unless hand-pollinated. With age, the fruits become smaller and it is considered best to replace the trees after 10 to 20 years. The fruits will not ripen but just turn black and dry if picked before the white, yellowish or red tint appears between the rind segments, the first signs of separation. If allowed to ripen on the tree, the fruit falls apart.

Keeping Quality

In India, mature fruits treated with 50-60 g carbide ripened in 2 days and thereafter remained in good condition only 2 days at room temperature, while those packed in straw ripened in 5-6 days and kept well for 4 days.

Storage trials in Malaya indicate that the ripening of sugar apples can be delayed by storage at temperatures between 59° and 68°F (15°-20°C) and 85-90% relative humidity, with low O_2 and C_2H_2. To speed ripening at the same temperature and relative humidity, levels of O_2 and CO_2 should be high. Storing at 39.2°F (4°C) for 5 days resulted in chilling injury.

In Egypt, of 'Beni Mazar' fruits, picked when full-grown, 115 days from set, and held at room temperature, 86% ripened in 10 days. Of 'Abd El Razik' fruits, 140 days from set, 56% were ripe in 15 days. Therefore, 'Abd El Razik' is better adapted to Upper Egypt where the climate should promote normal ripening.

Pests and Diseases

In Florida and the Caribbean, a seed borer (chalcid fly), *Bephratelloides cubensis,* infests the seeds and an associated fungus mummifies the partly grown fruits on the tree. This has discouraged many from growing the sugar apple, though in the past it was a fairly common dooryard fruit tree. Similar damage is caused by *B. maculicollis* in Colombia, Venezuela and Surinam, by *B. ruficollis* in Panama, and *B. paraguayensis* in Paraguay. The soft scale, *Philephedra* sp., attacks leaves and twigs and deposits honeydew on which sooty mold develops. Ambrosia beetles lay eggs on young stems and the larvae induce dieback during the winter.

The mealybug is the main pest in Queensland, Australia, but is easily controlled. The green tree ant is a nuisance because of the nests it makes in the tree. Bird and animal predators force Indian growers to cover the tree with netting or pick the fruits prematurely and ripen them in straw.

A serious leaf blight in India is caused by the fungus *Colletotrichum annonicola.* In 1978 a new fruit rot of sugar apple was observed in India, beginning with discoloration at one end which turns brown or black in 4 or 5 days, and 2 or 3 days later the entire fruit starts to rot. Later, the fruit is covered with gray-black mycelium and spherical bodies. The isolated fungus was identified as the *Colletotrichum* state of *Glomerella cingulata.*

Food Uses

The ripe sugar apple is usually broken open and the flesh segments enjoyed while the hard seeds are separated in the mouth and spat out. It is so luscious that it is well worth the trouble. In Malaya, the flesh is pressed through a sieve to eliminate the seeds and is then added to ice cream or blended with milk to make a cool beverage. It is never cooked.

Toxicity

The seeds are acrid and poisonous. Bark, leaves and seeds contain the alkaloid, anonaine. Six other aporphine alkaloids have been isolated from the leaves and stems: corydine, roemerine, norcorydine, norisocarydine, isocorydine and glaucine. Aporphine, norlaureline and dienone may be present also. Powdered seeds, also pounded dried fruits serve as fish poison and insecticides in India. A paste of the seed powder has been applied to the head to kill lice but must be kept away from the eyes as it is highly irritant and can cause blindness. If applied to the uterus, it induces abortion. Heat-extracted oil from the seeds has been employed against agricultural pests. Studies have shown the ether extract of the seeds to have no residual toxicity after 2 days. High concentrations are potent for 2 days and weaken steadily, all activity being lost after 8 days. In Mexico, the leaves are rubbed on floors and put in hen's nests to repel lice.

Other Uses

The **seed kernels** contain 14-49% of whitish or yellowish, non-drying oil with saponification index of 186.40. It has been proposed as a substitute for peanut oil in the manufacture of soap and can be detoxified by an alkali treatment and used for edible purposes. The **leaves** yield an excellent oil rich in terpenes and sesquiterpenes, mainly β-caryophyllene, which finds limited use in perfumes, giving a woody-spicy accent.

Fiber extracted from the bark has been employed for cordage. The **tree** serves as host for lac-excreting insects.

Medicinal Uses: In India the crushed leaves are sniffed to overcome hysteria and fainting spells; they are also applied on ulcers and wounds and a leaf decoction is taken in cases of dysentery. Throughout tropical America, a decoction of the leaves alone or with those of other plants is imbibed either as an emmenagogue, febrifuge, tonic, cold remedy, digestive, or to clarify the urine. The leaf decoction is also employed in baths to alleviate rheumatic pain. The green fruit, very astringent, is employed against diarrhea in El Salvador. In India, the crushed ripe fruit, mixed with salt, is applied on tumors. The bark and roots are both highly astringent. The bark decoction is given as a tonic and to halt diarrhea. The root, because of its strong purgative action, is administered as a drastic treatment for dysentery and other ailments.

Food Value Per 100 g of Edible Portion*	
Calories	88.9-95.7 g
Moisture	69.8-75.18 g
Fat	0.26-1.10 g
Carbohydrates**	19.16-25.19 g
Crude Fiber	1.14-2.50 g
Protein	1.53-2.38 g
Amino Acids:	
Tryptophan	9-10 mg
Methionine	7-8 mg
Lysine	54-69 mg
Minerals:	
Ash	0.55-1.34 mg
Phosphorus	23.6-55.3 mg
Calcium	19.4-44.7 mg
Iron	0.28-1.34 mg
Vitamins:	
Carotene	5-7 I.U.
Thiamine	0.100-0.13 mg
Riboflavin	0.113-0.167 mg
Niacin	0.654-0.931 mg
Ascorbic Acid	34.7-42.2 mg

*Minimum and maximum levels of constituents from analyses made in the Philippines, Central America and Cuba.

**The average sugar content is 14.58% and is about 50-50 glucose and sucrose.

Atemoya (Plate IX)

The atemoya, *Annona squamosa* X *A. cherimola*, is a hybrid of the sugar apple and cherimoya, qq.v. It was for many years mistakenly called custard apple or cherimoya in Queensland and New South Wales. The name applied in Venezuela is *chirimoriñon*.

Description

The tree closely resembles that of the cherimoya; is fast-growing; may reach 25 to 30 ft (7.5-9 m) and is short-trunked, the branches typically drooping and the lowest touching the ground. The leaves are deciduous, alternate, elliptical, leathery, less hairy than those of the cherimoya; and up to 6 in (15 cm) in length. The flowers are long-stalked, triangular, yellow, 2⅜ in (6 cm) long and 1½ to 2 in (4-5 cm) wide. The fruit is conical or heart-shaped, generally to 4 in (10 cm) long and to 3¾ in (9.5 cm) wide; some weighing as much as 5 lbs (2.25 kg); pale bluish-green or pea-green, and slightly yellowish between the areoles. The rind, ⅛ in (3 mm) thick, is composed of fused areoles more prominent and angular than those of the sugar apple, with tips that are rounded or slightly upturned; firm, pliable, and indehiscent. The fragrant flesh is snowy-white, of fine texture, almost solid, not conspicuously divided into segments, with fewer seeds than the sugar apple; sweet and subacid at the same time and resembling the cherimoya in flavor. The seeds are cylindrical, ¾ in (2 cm) long and 5/16 in (8 mm) wide; so dark a brown as to appear black; hard and smooth.

Origin and Distribution

The first cross was made by the horticulturist, P.J. Wester, at the United States Department of Agriculture's subtropical laboratory, Miami, in 1908. Seedlings were planted out in 1910. Other crosses made in 1910 fruited in 1911 and seeds were taken by Wester to the Philippines. The hybrids grew there to 7½ ft (2.3 m) high in one year, had to be moved to another location; one bloomed in 1913 and was pollinated by the custard apple, q.v. The rest of the plants fruited in 1914.

Resulting fruits were superior in quality to the sugar apple and were given the name "atemoya", a combination of "ate", an old Mexican name for sugar apple, and "moya" from cherimoya. Cuttings of 9 of the hybrids were sent by Wester to the United States Department of Agriculture in January of 1915. (S.P.I. Nos. 39808-39816), #39809 representing the hybrid tree pollinated by the custard apple. In 1917, Wester sent cuttings of #39809 under the name "cuatemoya" to the United States Department of Agriculture (S.P.I. Nos. 44671-44673). In the meantime, Edward Simmons, at the Plant Introduction Field Station, Miami, had successfully grown hybrids and they had survived an early February 1917 drop in temperature to 26.5°F (-3.10°C), showing the hardiness derived from the cherimoya. Another introduction was received from the Philippines in 1918 (S.P.I. #45571).

A few experimental growers in southern Florida maintained atemoya trees (apparently distributed by the United States Department of Agriculture) for many years while there was a general lapse of interest in this fruit. Today, there are a few small commercial plantings and the fruits are being sent to some northern fruit dealers.

In the early 1930's or 1940's, what were apparently chance hybrids between adjacent sugar apple and cherimoya orchards attracted attention in Israel and work was begun to choose and standardize the best of these for vegetative propagation.

Varieties

One of the first named selections of atemoya was the **'Page'**, so-named by Roy Page of Coral Gables who took budwood from superior atemoya trees on the property of Morrison Page in the Redlands. Perhaps the second was the **'Bradley'** which the Newcomb Nursery sold grafted onto custard apple.

An early hybrid that arose in Queensland after the introduction of cherimoya seeds from South America, was named **'Mammoth'** (or 'Pink's Prolific', or 'Pink's Mammoth') and became the basis of the commercial production of atemoyas there and on the north coast of New South Wales, though the flesh of this cultivar immediately below the rind is usually brownish and bitter. **'Island Beauty'**, a vigorous selection with excellent fruit quality was grown to a lesser extent. 'Mammoth' was introduced into Hawaii from Queensland in 1960 and grafted plants were soon being distributed by agricultural stations of the University of Hawaii in Kona and Hilo, and being sold by nurseries in Honolulu.

'African Pride' is an improved clone that originated in South Africa. It was introduced into Queensland by Langbecker Nurseries and 3,000 trees were released for commercial planting in July 1961. It was quickly adopted as a replacement for 'Mammoth' as it was free of the discoloration and bitterness next to the skin. In 1963, 6 plants of 'African Pride' were obtained from Landbecker's by private experimenters and planted at several locations in southern Florida. They began fruiting in 1965. The fruits appeared to be superior in quality to the 'Page' and 'Bradley'.

Israeli selections tried at the University of Florida's Agricultural Research and Education Center, Homestead, and the United States Department of Agriculture's Subtropical Horticulture Research Unit, Miami, are **'Geffner'**, **'Malamud'**, **'Bernitski'**, **'Kabri'** and **'Malai #1'**. Other named selections that have been grown in Florida over the years are **'Caves'**, **Chirimoriñon** A, B and C, **'Island Gem'**, **'Kaller'**, **'Lindstrom'**, **'Priestly'** and **'Stermer'**. 'Geffner' is being propagated at the AREC, Homestead; 'Priestly' by the Zill Nursery in Boynton Beach. None of the others have outstanding features; some develop hard spots in the flesh. In 'Kaller' there is frequently a black membrane around each seed-containing carpel.

'Cherimata' and **'Finny'** are Egyptian clones. 'Finny' is somewhat cylindrical, is more productive than 'Cherimata', has been grown in Egypt for many years and is considered the best for commercial production in coastal districts.

Pollination

The atemoya and other annona trees bear hermaphroditic protogynous flowers and self-pollination is rare. Atemoyas are sometimes misshapen, underdeveloped on one side, as the result of inadequate pollination. The flower, in its female stage, opens between 2 and 4 o'clock in the afternoon. Between 3 and 5 o'clock on the following afternoon, the flower converts to its male stage. In cold and humid climates it releases pollen even though it is sticky. Where the climate is hot and the humidity low at the blooming season, the carpels are short-lived and the stigmatic surface soon dries up and insects are necessary to transfer the pollen. Studies in Israel have identified the principal insect pollinators as nitidulid beetles—*Carpophilus hemipterus, C. mutilatus, Haptoncus luteolus,* and *Uroporus humeralis*. Even where these beetles are present, hand-pollination will enhance fruit-setting and this is commonly practiced in Egypt. Spraying the flowers several times with gibberellin at 1,000 ppm has increased fruit yield. The resulting fruits are seedless but smaller and less flavorful than fruits with seeds.

Climate

The atemoya is slightly hardier than the sugar apple but still is limited to tropical or near-tropical lowlands. In New South Wales, it is said to do best near the coast where rainfall and humidity are high and winters are warm. Rainy weather during the ripening season, however, may cause the fruits to split.

Soil

The tree thrives in various types of soil, from sandy loam to red basalt or heavy clay, but best growth and productivity occur in deep, rich loam of medium texture, with good organic content and a moderate amount of moisture. Good drainage is essential; waterlogging is fatal.

Propagation

Atemoyas for rootstocks are raised from seeds which germinate in about 4 weeks in seedbeds. Seedlings are transplanted to nursery rows when they are a year old and they are placed 18 in (45 cm) apart in rows 3 ft (90 cm) apart. Grafting is done in the spring, using the whip- or tongue-graft. If older trees are top-worked, it is done by cleft- or bark-grafting. Scion wood is taken from selected cultivars after the leaves have fallen. In Florida and India, the atemoya is usually grafted onto the custard apple or sugar apple. Cherimoya is used as a rootstock in Israel.

Culture

When transferred to the field at the near-dormant period, grafted plants are spaced 28 to 30 ft (8.5-9 m) apart each way and cut back to a height of 24 to 30 in (60-75 cm). Weeds are eliminated to avoid competition with the spreading, shallow root system. During the next 2 or 3 years, the trees are kept pruned to form a strong frame. Thereafter, only light pruning is done. No fertilizer is applied until after the trees are well established, since the young roots are very sensitive. A 6-10-16 formula is recommended for broadcasting over the root area, the amount gradually increased to 10 to 12 lbs (4.5-5.4 kg) annually for mature trees. Half is given in the spring a month before flowering. Irrigation during flowering and fruit-setting improves yield and fruit quality.

Season

In Florida, the atemoya ripens in the fall. In Queensland, the main blooming period is October and November and the fruits mature in April and May. If there is light fruit set in October/November, flowering may continue to February and the fruit from such late blooms may have to be picked prematurely and ripened artificially to avoid cold night temperatures, but it will not develop the highest quality.

Harvesting

The fruits must be clipped from the branch, taking care that the stalk left on the fruit does not protrude beyond the shoulders. Frequent picking is necessary to harvest the fruit at the ideal stage, that is, when creamy lines appear around the areoles showing that the spaces between them are widening. If picked too soon, the fruit will not ripen but will darken and shrivel.

Fruits colonized by mealybugs have to be cleaned by brushing or the use of compressed air before marketing. The fruits should not be wrapped because this will speed ripening, but they need to be packed in boxes with padding between layers. Because of the irregular form, the fruits must be carefully fitted together with the base of each fruit against the wall of the container and the more delicate apex inward.

Yield

The atemoya is a shy yielder, mainly for the reason mentioned under "Pollination". Trees 5 years old are expected to bear 50 fruits annually. In Queensland, commercial groves have produced 5 bushels of fruit per tree — 67 bushels per acre (165.5 bu/ha). An exceptionally large atemoya tree in Florida yielded 11 bushels of fruits in the 1972 season.

Keeping Quality

Atemoyas keep very well in cool, shady, well-ventilated storage for at least 3 weeks. The rind may darken before the interior shows any signs of spoilage. The ideal temperature for refrigerated storage is 68°F (20°C), though an acceptable temperature range is 59° to 77°F (15°-25°C). Lower temperatures cause chilling injury.

Pests and Diseases

The citrus mealybug, *Planococcus citri,* which congregates around the base of the fruit, is the most common pest, and sooty mold develops on its exudate.

In Queensland, the protective activities of the natural enemies of the mealybug are disrupted by the coastal brown ant, *Pheidole megacephala,* which carries mealybugs up the trunk and around between the fruits. Australian growers have tried sticky-banding the trunks and this has reduced the numbers of ants but not sufficiently.

The chalcid fly that lays eggs in the seeds and makes exit holes in the fruit permitting entrance of fungi, occasionally causes mummification of the atemoya. White wax, pink wax, and brown olive scales may be found on the foliage but are shed along with the leaves.

A condition called "littleleaf" is not a disease but zinc deficiency which can be corrected by foliar spraying.

Atemoyas are prone to collar rot (*Phytophthora* sp.), the first sign being an exudation of gum near the base of the trunk and on the crown roots.

Food Uses

The atemoya, preferably chilled, is one of the most delicious of fruits. It needs no seasoning. It may be simply cut in half or quartered and the flesh eaten from the

Food Value Per 100 g of Edible Portion of Ripe Fruit*	
Calories	94
Moisture	71.48-78.7 g
Protein	1.07-1.4 g
Fat	0.4-0.6 g
Carbohydrates	24 g
Fiber	0.05-2.5 g
Ash	0.4-0.75 g
Sodium	4-5 mg
Potassium	250 mg
Iron	0.3 mg
Calcium	17 mg
Magnesium	32 mg
Zinc	0.2 mg
Thiamine	0.05 mg
Riboflavin	0.07 mg
Niacin	0.8 mg
α-carotene	10 mcg
β-carotene	10 mcg
Cryptoxanthin	10 mcg
Ascorbic Acid	50 mg

*Analyses made in Florida, the Philippines and at the University of New South Wales.

"shell" with a spoon. Slices or cubes of the pulp may be added to fruit cups or salads or various dessert recipes. Some people blend the pulp with orange juice, lime juice and cream and freeze as ice cream.

Toxicity

The seeds, like those of all *Annona* species, are toxic and care should be taken to seed the pulp before it is mechanically blended.

Soursop (Plate X)

Of the 60 or more species of the genus *Annona*, family Annonaceae, the soursop, *A. muricata* L., is the most tropical, the largest-fruited, and the only one lending itself well to preserving and processing.

It is generally known in most Spanish-speaking countries as *guanabana;* in El Salvador, as *guanaba;* in Guatemala, as *huanaba;* in Mexico, often as *zapote de viejas,* or *cabeza de negro;* in Venezuela, as *catoche* or *catuche;* in Argentina, as *anona de puntitas* or *anona de broquel;* in Bolivia, *sinini;* in Brazil, *araticum do grande, graviola,* or *jaca do Pará;* in the Netherlands Antilles, *sorsaka* or *zuurzak,* the latter name also used in Surinam and Java; in French-speaking areas of the West Indies, West Africa, and Southeast Asia, especially North Vietnam, it is known as *corossol, grand corossol, corossol epineux,* or *cachiman epineux.* In Malaya it may be called *durian belanda, durian maki,* or *seri kaya belanda;* in Thailand, *thu–rian–khack.*

In 1951, Prof. Clery Salazar, who was encouraging the development of soursop products at the College of Agriculture at Mayaguez, Puerto Rico, told me that they would like to adopt an English name more appealing than the word "soursop", and not as likely as *guanabana* to be mispronounced. To date, no alternatives have been chosen.

Description

The soursop tree is low-branching and bushy but slender because of its upturned limbs, and reaches a

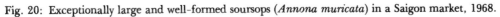

Fig. 20: Exceptionally large and well-formed soursops (*Annona muricata*) in a Saigon market, 1968.

height of 25 or 30 ft (7.5-9 m). Young branchlets are rusty-hairy. The malodorous leaves, normally evergreen, are alternate, smooth, glossy, dark-green on the upper surface, lighter beneath; oblong, elliptic or narrow-obovate, pointed at both ends, 2½ to 8 in (6.25-20 cm) long and 1 to 2½ in (2.5-6.25 cm) wide. The flowers, which are borne singly, may emerge anywhere on the trunk, branches or twigs. They are short-stalked, 1½ to 2 in (4-5 cm) long, plump, and triangular-conical, the 3 fleshy, slightly spreading, outer petals yellow-green, the 3 close-set inner petals pale-yellow.

The fruit is more or less oval or heart-shaped, sometimes irregular, lopsided or curved, due to improper carpel development or insect injury. The size ranges from 4 to 12 in (10-30 cm) long and up to 6 in (15 cm) in width, and the weight may be up to 10 or 15 lbs (4.5-6.8 kg). The fruit is compound and covered with a reticulated, leathery-appearing but tender, inedible, bitter skin from which protrude few or many stubby, or more elongated and curved, soft, pliable "spines". The tips break off easily when the fruit is fully ripe. The skin is dark-green in the immature fruit, becoming slightly yellowish-green before the mature fruit is soft to the touch. Its inner surface is cream-colored and granular and separates easily from the mass of snow-white, fibrous, juicy segments—much like flakes of raw fish—surrounding the central, soft-pithy core. In aroma, the pulp is somewhat pineapple-like, but its musky, subacid to acid flavor is unique. Most of the closely-packed segments are seedless. In each fertile segment there is a single oval, smooth, hard, black seed, ½ to ¾ in (1.25-2 cm) long; and a large fruit may contain from a few dozen to 200 or more seeds.

Origin and Distribution

Oviedo, in 1526, described the soursop as abundant in the West Indies and in northern South America. It is today found in Bermuda and the Bahamas, and both wild and cultivated, from sea-level to an altitude of 3,500 ft (1,150 m) throughout the West Indies and from southern Mexico to Peru and Argentina. It was one of the first fruit trees carried from America to the Old World Tropics where it has become widely distributed from southeastern China to Australia and the warm lowlands of eastern and western Africa. It is common in the markets of Malaya and southeast Asia. Very large, symmetrical fruits have been seen on sale in South Vietnam. It became well established at an early date in the Pacific Islands. The tree has been raised successfully but has never fruited in Israel.

In Florida, the soursop has been grown to a limited extent for possibly 110 years. Sturtevant noted that it was not included by Atwood among Florida fruits in 1867 but was listed by the American Pomological Society in 1879. A tree fruited at the home of John Fogarty of Manatee before the freeze of 1886. In the southeastern part of the state and especially on the Florida Keys, it is often planted in home gardens.

In regions where sweet fruits are preferred, as in South India and Guam, the soursop has not enjoyed great popularity. It is grown only to a limited extent in Madras. However, in the East Indies it has been acclaimed one of the best local fruits. In Honolulu, the fruit is occasionally sold but the demand exceeds the supply. The soursop is one of the most abundant fruits in the Dominican Republic and one of the most popular in Cuba, Puerto Rico, the Bahamas, Colombia and northeastern Brazil.

In 1887, Cuban soursops were selling in Key West, Florida, at 10 to 50 cents apiece. In 1920, Wilson Popenoe wrote that: "In the large cities of tropical America, there is a good demand for the fruits at all times of the year, a demand which is not adequately met at present." The island of Grenada produces particularly large and perfect soursops and regularly delivers them by boat to the market of Port-of-Spain because of the shortage in Trinidad. In Colombia, where the soursop is generally large, well-formed and of high quality, this is one of the 14 tropical fruits recommended by the Instituto Latino-americano de Mercadeo Agricola for large-scale planting and marketing. Soursops produced in small plots, none over 5 acres (2.27 ha), throughout Venezuela supply the processing plants where the frozen concentrate is packed in 6 oz (170 g) cans. In 1968, 2,266 tons (936 MT) of juice were processed in Venezuela. The strained pulp is also preserved commercially in Costa Rica. There are a few commercial soursop plantations near the south coast of Puerto Rico and several processing factories. In 1977, the Puerto Rican crop totaled 219,538 lbs (99,790 kg).

At the First International Congress of Agricultural and Food Industries of the Tropical and Subtropical Zones, held in 1964, scientists from the Research Laboratories of Nestlé Products in Vevey, Switzerland, presented an evaluation of lesser-known tropical fruits and cited the soursop, the guava and passionfruit as the 3 most promising for the European market, because of their distinctive aromatic qualities and their suitability for processing in the form of preserved pulp, nectar and jelly.

Varieties

In Puerto Rico, the wide range of forms and types of seedling soursops are roughly divided into 3 general classifications: sweet, subacid, and acid; then subdivided as round, heart-shaped, oblong or angular; and finally classed according to flesh consistency which varies from soft and juicy to firm and comparatively dry. The University of Puerto Rico's Agricultural Experiment Station at one time cataloged 14 different types of soursops in an area between Aibonito and Coamo. In El Salvador, 2 types of soursops are distinguished: *guanaba azucarón* (sweet) eaten raw and used for drinks; and *guanaba acida* (very sour), used only for drinks. In the Dominican Republic, the *guanabana dulce* (sweet soursop) is most sought after. The term "sweet" is used in a relative sense to indicate low acidity. A medium-sized, yellow-green soursop called *guanabana sin fibre* (fiberless) has been vegetatively propagated at the Agricultural Experiment Station at Santiago de las Vegas, Cuba. The foliage of this superior clone is distinctly bluish-green. In 1920, Dr. Wilson Popenoe sent to the United States Department of Agriculture, from Costa Rica, budwood of a soursop he named 'Bennett' in honor of G.S. Bennett, Agricultural Superintendent of the Costa Rican Division of the United

Fig. 21: The soursop tree may bear fruits anywhere on its trunk or branches. Multiple-stems of this tree are the result of its having been frozen to the ground more than once.

Fruit Company. He described the fruit as large and hand-some (as shown in the photograph accompanying the introduction record No. 51050) and he declared the tree to be the most productive he had seen.

Climate

The soursop is truly tropical. Young trees in exposed places in southern Florida are killed by only a few degrees of frost. The trees that survive to fruiting age on the mainland are in protected situations, close to the south side of a house and sometimes near a source of heat. Even so, there will be temporary defoliation and interruption of fruiting when the temperature drops to near freezing. In Key West, where the tropical breadfruit thrives, the soursop is perfectly at home. In Puerto Rico, the tree is said to prefer an altitude between 800 and 1,000 ft (244-300 m), with moderate humidity, plenty of sun and shelter from strong winds.

Soil

Best growth is achieved in deep, rich, well-drained, semi-dry soil, but the soursop tree can be and is commonly grown in acid and sandy soil, and in the porous, oolitic limestone of South Florida and the Bahama Islands.

Propagation

The soursop is usually grown from seeds. They should be sown in flats or containers and kept moist and shaded. Germination takes from 15 to 30 days. Selected types can be reproduced by cuttings or by shield-budding. Soursop seedlings are generally the best stock for propagation, though grafting onto custard apple (*Annona reticulata*), the mountain soursop (*A. montana*), or pond apple (*A. glabra*), is usually successful. The pond apple has a dwarfing effect. Grafts on sugar apple (*A. squamosa*) and cherimoya (*A. cherimola*) do not live for long, despite the fact that the soursop is a satisfactory rootstock for sugar apple in Ceylon and India.

Culture

In ordinary practice, seedlings, when 1 ft (30 cm) or more in height are set out in the field at the beginning of the rainy season and spaced 12 to 15 ft (3.65-4.5 m) apart, though 25 ft (7.5 m) each way has been suggested. A spacing of 20 x 25 ft (6x7.5 m) allows 87 trees per acre (215/ha). Close-spacing, 8 x 8 ft (2.4x2.4 m) is thought sufficient for small gardens in Puerto Rico. The tree grows rapidly and begins to bear in 3 to 5 years. In Queensland, well-watered trees have attained 15 to 18 ft (4.5-5.5 m) in 6 to 7 years. Mulching is recommended to avoid dehydration of the shallow, fibrous root system during dry, hot weather. If in too dry a situation, the tree will cast off all of its old leaves before new ones appear. A fertilizer mixture containing 10% phosphoric acid, 10% potash and 3% nitrogen has been advocated in Cuba and Queensland. But excellent results have been obtained in Hawaii with quarterly applications of 10-10-10 N P K — ½ lb (.225 kg) per tree the first year, 1 lb (.45 kg)/tree the 2nd year, 3 lbs (1.36 kg)/tree the 3rd year and thereafter.

Season

The soursop tends to flower and fruit more or less continuously, but in every growing area there is a principal season of ripening. In Puerto Rico, this is from March to June or September; in Queensland, it begins in April; in southern India, Mexico and Florida, it extends from June to September; in the Bahamas, it continues through October. In Hawaii, the early crop occurs from January to April; midseason crop, June to August, with peak in July; and there is a late crop in October or November.

Harvesting

The fruit is picked when full grown and still firm but slightly yellow-green. If allowed to soften on the tree, it will fall and crush. It is easily bruised and punctured and must be handled with care. Firm fruits are held a few days at room temperature. When eating ripe, they are soft enough to yield to the slight pressure of one's thumb. Having reached this stage, the fruit can be held 2 or 3 days longer in a refrigerator. The skin will blacken and become unsightly while the flesh is still unspoiled and usable. Studies of the ripening process in Hawaii have determined that the optimum stage for eating is 5 to 6 days after harvest, at the peak of ethylene production. Thereafter, the flavor is less pronounced and a faint off-odor develops. In Venezuela, the chief handicap in commercial processing is that the fruits stored on racks in a cool shed must be gone over every day to select those that are ripe and ready for juice extraction.

Yield

The soursop, unfortunately, is a shy-bearer, the usual crop being 12 to 20 or 24 fruits per tree. In Puerto Rico, production of 5,000 to 8,000 lbs per acre (roughly equal kg/ha), is considered a good yield from well-cared-for trees. A study of the first crop of 35 5-year-old trees in Hawaii showed an average of 93.6 lbs (42.5 kg) of fruits per tree. Yield was slightly lower the 2nd year. The 3rd year, the average yield was 172 lbs (78 kg) per tree. At this rate, the annual crop would be 16,000 lbs per acre (roughly equal kg/ha).

Pests & Diseases

Queensland's principal soursop pest is the mealybug which may occur in masses on the fruits. The mealybug is a common pest also in Florida, where the tree is often infested with scale insects. Sometimes it may be infected by a lace-wing bug.

The fruit is subject to attack by fruit flies — *Anastrepha suspensa*, *A. striata* and *Ceratitis capitata*. Red spiders are a problem in dry climates.

Dominguez Gil (1978 and 1983), presents an extensive list of pests of the soursop in the State of Zulia, Venezuela. The 5 most damaging are: 1) the wasp, *Bephratelloides (Bephrata) maculicollis*, the larvae of which live in the seeds and emerge from the fully-grown ripe fruit, leaving it perforated and highly perishable; 2) the moth, *Cerconota (Stenoma) anonella*, which lays its eggs in the very young fruit causing stunting and malformation; 3) *Corythucha gossipii*, which attacks the leaves; 4) *Cratosomus inaequalis*, which bores into the fruit, branches and trunk; 5) *Laspeyresia* sp., which perforates the flowers. The first 3 are among the 7 major pests of the soursop in Colombia, the other 4 being: *Toxoptera aurantii*, which affects shoots, young leaves, flowers and fruits; present but not important in Venezuela; *Aphis spiraecola*; *Empoasca* sp., attacking the leaves; and *Aconophora concolor*, damaging the flowers and fruits. Important beneficial agents preying on aphids are *Aphidius testataceipes*, *Chrysopa* sp., and *Curinus* sp. Lesser enemies of the soursop in South America include: *Talponia backeri* and *T. batesi* which damage flowers and fruits; *Horiola picta* and *H. lineolata*, feeding on flowers and young branches; *Membracis foliata*, attacking young branches, flower stalks and fruits; *Saissetia nigra*; *Escama ovalada*, on branches, flowers and fruits; *Cratosomus bombina*, a fruit borer; and *Cyclocephala signata*, affecting the flowers.

In Trinidad, the damage done to soursop flowers by *Thecla ortygnus* seriously limits the cultivation of this fruit. The sphinx caterpillar, *Cocytius antaeus antaeus* may be found feeding on soursop leaves in Puerto Rico. Bagging of soursops is necessary to protect them from *Cerconota anonella*. However, one grower in the Magdalena Valley of Colombia claims that bagged fruits are more acid than others and the flowers have to be hand-pollinated.

It has been observed in Venezuela and El Salvador that soursop trees in very humid areas often grow well but bear only a few fruits, usually of poor quality, which are apt to rot at the tip. Most of their flowers and young fruits fall because of anthracnose caused by *Collectotrichum gloeosporioides*. It has been said that soursop trees for cultivation near San Juan, Puerto Rico, should be seedlings of trees from similarly humid areas which have greater resistance to anthracnose than seedlings from dry zones. The same fungus causes damping-off of seedlings and die-back of twigs and branches. Occasionally the fungus, *Scolecotrichum* sp. ruins the leaves in Venezuela. In the East Indies, soursop trees are sometimes subject to the root-fungi, *Fomes lamaoensis* and *Diplodia* sp. and by pink disease due to *Corticum salmonicolor*.

Food Uses

Soursops of least acid flavor and least fibrous consistency are cut in sections and the flesh eaten with a spoon. The seeded pulp may be torn or cut into bits and added to fruit cups or salads, or chilled and served as dessert with sugar and a little milk or cream. For years, seeded soursop has been canned in Mexico and served in Mexican restaurants in New York and other northern cities.

Most widespread throughout the tropics is the making of refreshing soursop drinks (called *champola* in Brazil; *carato* in Puerto Rico). For this purpose, the seeded pulp may be pressed in a colander or sieve or squeezed in cheesecloth to extract the rich, creamy juice, which is then beaten with milk or water and sweetened. Or the seeded pulp may be blended with an equal amount of boiling water and then strained and sweetened. If an electric blender is to be used, one must first be careful to remove all the seeds, since they are somewhat toxic and none should be accidentally ground up in the juice.

In Puerto Rican processing factories, the hand-peeled and cored fruits are passed through a mechanical pulper having nylon brushes that press the pulp through a screen, separating it from the seeds and fiber. A soursop soft drink, containing 12 to 15% pulp, is canned in Puerto Rico and keeps well for a year or more. The juice is prepared as a carbonated bottled beverage in Guatemala, and a fermented, cider-like drink is sometimes made in the West Indies. The vacuum-concentrated juice is canned commercially in the Philippines. There soursop drinks are popular but the normal "milk" color is not. The people usually add pink or green food coloring to make the drinks more attractive. The strained pulp is said to be a delicacy mixed with wine or brandy and seasoned with nutmeg. Soursop juice, thickened with a little gelatin, makes an agreeable dessert.

In the Dominican Republic, a soursop custard is enjoyed and a confection is made by cooking soursop pulp in sugar sirup with cinnamon and lemon peel. Soursop ice cream is commonly frozen in refrigerator ice-cube trays in warm countries.

In the Bahamas, it is simply made by mashing the pulp in water, letting it stand, then straining to remove fibrous material and seeds. The liquid is then blended with sweetened condensed milk, poured into the trays and stirred several times while freezing. A richer product is made by the usual method of preparing an ice cream mix and adding strained soursop pulp just before freezing. Some Key West restaurants have always served soursop ice cream and now the influx of residents from the Caribbean and Latin American countries has created a strong demand for it. The canned pulp is imported from Central America and Puerto Rico and used in making ice cream and sherbet commercially. The pulp is used, too, for making tarts and jelly, sirup and nectar. The sirup has been bottled in Puerto Rico for local use and export. The nectar is canned in Colombia and frozen in Puerto Rico and is prepared fresh and sold in paper cartons in the Netherlands Antilles. The strained, frozen pulp is sold in plastic bags in Philippine supermarkets.

Immature soursops are cooked as vegetables or used in soup in Indonesia. They are roasted or fried in northeastern Brazil. I have boiled the half-grown fruit whole, without peeling. In an hour, the fruit is tender, its flesh

Fig. 22: Canned soursop concentrate is produced in Venezuela. On the branch at the right is a soursop flower.

off-white and mealy, with the aroma and flavor of roasted ears of green corn (maize).

Toxicity

The presence of the alkaloids anonaine and anoniine has been reported in this species. The alkaloids muricine, $C_{19}H_{21}O_4N$ (possibly des-*N*-methylisocorydine or des-*N*-methylcorydine) and muricinine, $C_{18}H_{19}O_4$ (possibly des-*N*-methylcorytuberine), are found in the bark. Muricinine is believed to be identical to reticuline. An unnamed

Food Value Per 100 g of Edible Portion*	
Calories	61.3-53.1
Moisture	82.8 g
Protein	1.00 g
Fat	0.97 g
Carbohydrates	14.63 g
Fiber	0.79 g
Ash	60 g
Calcium	10.3 mg
Phosphorus	27.7 mg
Iron	0.64 mg
Vitamin A (β-carotene)	0
Thiamine	0.11 mg
Riboflavin	0.05 mg
Niacin	1.28 mg
Ascorbic Acid	29.6 mg
Amino Acids:	
Tryptophan	11 mg
Methionine	7 mg
Lysine	60 mg

*Analyses made at the Laboratorio FIM de Nutricion, Havana, Cuba.

alkaloid occurs in the leaves and seeds. The bark is high in hydrocyanic acid. Only small amounts are found in the leaves and roots and a trace in the fruit. The seeds contain 45% of a yellow non-drying oil which is an irritant poison, causing severe eye inflammation.

Other Uses

Fruit: In the Virgin Islands, the fruit is placed as a bait in fish traps.

Seeds: When pulverized, the seeds are effective pesticides against head lice, southern army worms and pea aphids and petroleum ether and chloroform extracts are toxic to black carpet beetle larvae. The seed oil kills head lice.

Leaves: The leaf decoction is lethal to head lice and bedbugs.

Bark: The bark of the tree has been used in tanning. The bark fiber is strong but, since fruiting trees are not expendable, is resorted to only in necessity. Bark, as well as seeds and roots, has been used as fish poison.

Wood: The wood is pale, aromatic, soft, light in weight and not durable. It has been used for ox yokes because it does not cause hair loss on the neck.

In Colombia, it is deemed to be suitable for pipestems and barrelstaves. Analyses in Brazil show cellulose content of 65 to 76%, high enough to be a potential source of paper pulp.

Medicinal Uses: The juice of the ripe fruit is said to be diuretic and a remedy for haematuria and urethritis. Taken when fasting, it is believed to relieve liver ailments and leprosy. Pulverized immature fruits, which are very astringent, are decocted as a dysentery remedy. To draw out chiggers and speed healing, the flesh of an acid soursop is applied as a poultice unchanged for 3 days.

In *Materia Medica* of British Guiana, we are told to break soursop leaves in water, "squeeze a couple of limes therein, get a drunken man and rub his head well with the leaves and water and give him a little of the water to drink and he gets as sober as a judge in no time." This sobering or tranquilizing formula may not have been widely tested, but soursop leaves are regarded throughout the West Indies as having sedative or soporific properties. In the Netherlands Antilles, the leaves are put into one's pillowslip or strewn on the bed to promote a good night's sleep. An infusion of the leaves is commonly taken internally for the same purpose. It is taken as an analgesic and antispasmodic in Esmeraldas Province, Ecuador. In Africa, it is given to children with fever and they are also bathed lightly with it. A decoction of the young shoots or leaves is regarded in the West Indies as a remedy for gall bladder trouble, as well as coughs, catarrh, diarrhea, dysentery and indigestion; is said to "cool the blood," and to be able to stop vomiting and aid delivery in childbirth. The decoction is also employed in wet compresses on inflammations and swollen feet. The chewed leaves, mixed with saliva, are applied to incisions after surgery, causing proudflesh to disappear without leaving a scar. Mashed leaves are used as a poultice to alleviate eczema and other skin afflictions and rheumatism, and the sap of young leaves is put on skin eruptions.

The roots of the tree are employed as a vermifuge and the root bark as an antidote for poisoning. A tincture of the powdered seeds and bay rum is a strong emetic. Soursop flowers are believed to alleviate catarrh.

Custard Apple

Both in tree and in fruit, the custard apple, *Annona reticulata* L., is generally rated as the mediocre or "ugly duckling" species among the prominent members of this genus. Its descriptive English name has been widely misapplied to other species and to the hybrid ATEMOYA, and it is sometimes erroneously termed "sugar apple", "sweetsop" and, by Spanish-speaking people, *"anon"* or *"rinon"*; in India, *"ramphal"*; all properly applied only to *Annona squamosa*. It has, itself, acquired relatively few appropriate regional names. Most commonly employed as an alternate name in English-speaking areas is bullock's-heart or bull's-heart; in French, *coeur de boeuf;* Portuguese, *coracao de boi;* in Spanish, often merely *corazon*—all alluding to its form and external blush. The skin color is reflected in the Bolivian name, *chirimoya roia,* the Salvadoran *anona rosada,* and the Guatemalan *anona roja* or *anona colorada.* In the latter country it is also known as *anona de seso. Araticum ape* or *araticum do mato* are additional names in Brazil. Some people refer to it as Jamaica apple, or as netted custard apple, which is translated as *anona de redecilla* in Honduras and Nicaragua. *Cachiman, cachiman coeur de boeuf* and *corossol sauvage* may be heard in the French-influenced West Indies.

In the Netherlands Antilles it is *kasjoema*. This name and *boeah nona* are used in Surinam. In Cuba, it is *mamon* or *chirimoya.* Some Central Americans give it the name *anona,* or *anonillo;* Colombians, *anon pelon.* To the Carib Indians the fruit was known as *alacalyoua;* to the Aztecs, *quaultzapotl,* and to the Maya, *tsulimuy, tsulilpox, tsulipox, pox, oop,* or *op.* It is generally called in the Philippines *sarikaya;* in India *ramphal, nona* or *luvuni;* in Malaya, *nona kapri,* or *lonang;* in Thailand, *noi nong;* in Cambodia, *mo bat* or *mean bat;* in Laos, *khan tua lot;* in South Vietnam, *binh bat;* North Vietnam, *qua na.*

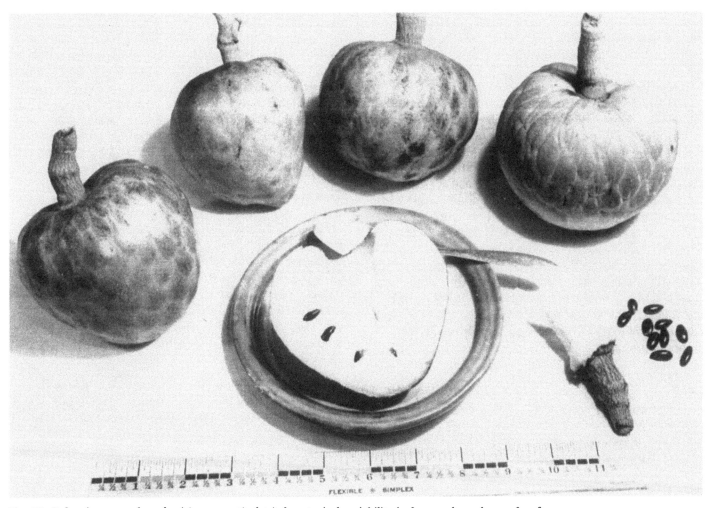

Fig. 23: Bahamian custard apples (*Annona reticulata*) show typical variability in form and roughness of surface.

Description

The custard apple tree is not especially attractive. It is erect, with a rounded or spreading crown and trunk 10 to 14 in (25-35 cm) thick. Height ranges from 15 to 35 ft (4.5-10 m). The ill-smelling leaves are deciduous, alternate, oblong or narrow-lanceolate, 4 to 8 in (10-20 cm) long, ¾ to 2 in (2-5 cm) wide, with conspicuous veins. Flowers, in drooping clusters, are fragrant, slender, with 3 outer fleshy, narrow petals ¾ to 1¼ in (2-3 cm) long; light-green externally and pale-yellow with a dark-red or purple spot on the inside at the base. The flowers never fully open.

The compound fruit, 3¼ to 6½ in (8-16 cm) in diameter, may be symmetrically heart-shaped, lopsided, or irregular; or nearly round, or oblate, with a deep or shallow depression at the base. The skin, thin but tough, may be yellow or brownish when ripe, with a pink, reddish or brownish-red blush, and faintly, moderately, or distinctly reticulated. There is a thick, cream-white layer of custardlike, somewhat granular, flesh beneath the skin surrounding the concolorous moderately juicy segments, in many of which there is a single, hard, dark-brown or black, glossy seed, oblong, smooth, less than ½ in (1.25 cm) long. Actual seed counts have been 55, 60 and 76. A pointed, fibrous, central core, attached to the thick stem, extends more than halfway through the fruit. The

flavor is sweet and agreeable though without the distinct character of the cherimoya, sugar apple, or atemoya.

Origin and Distribution

The custard apple is believed to be a native of the West Indies but it was carried in early times through Central America to southern Mexico. It has long been cultivated and naturalized as far south as Peru and Brazil. It is commonly grown in the Bahamas and occasionally in Bermuda and southern Florida.

Apparently it was introduced into tropical Africa early in the 17th century and it is grown in South Africa as a dooryard fruit tree. In India the tree is cultivated, especially around Calcutta, and runs wild in many areas. It has become fairly common on the east coast of Malaya, and more or less throughout southeast Asia and the Philippines though nowhere particularly esteemed. Eighty years ago it was reported as thoroughly naturalized in Guam. In Hawaii it is not well known.

Cultivars

No named cultivars are reported but there is considerable variation in the quality of fruit from different trees. The yellow-skinned types seem superior to the brownish, and, when well filled out, have thicker and juicier flesh.

Seeds of a purple-skinned, purple-fleshed form, from Mexico, were planted in Florida and the tree has produced fruit of unremarkable quality.

Climate

The custard apple tree needs a tropical climate but with cooler winters than those of the west coast of Malaya. It flourishes in the coastal lowlands of Ecuador; is rare above 5,000 ft (1,500 m). In Guatemala, it is nearly always found below 4,000 ft (1,220 m). In India, it does well from the plains up to an elevation of 4,000 ft (1,220 m); in Ceylon, it cannot be grown above 3,000 ft (915 m). Around Luzon in the Philippines, it is common below 2,600 ft (800 m). It is too tender for California and trees introduced into Palestine succumbed to the cold. In southern Florida the leaves are shed at the first onset of cold weather and the tree is dormant all winter. Fully grown, it has survived temperatures of 27° to 28°F (-2.78° to 2.22°C) without serious harm. This species is less drought-tolerant than the sugar apple and prefers a more humid atmosphere.

Soil

The custard apple does best in low-lying, deep, rich soil with ample moisture and good drainage. It grows to full size on oolitic limestone in southern Florida and runs wild in light sand and various other types of soil in the New and Old World tropics but is doubtless less productive in the less desirable sites.

Propagation

Seed is the usual means of propagation. Nevertheless, the tree can be multiplied by inarching, or by budding or grafting onto its own seedlings or onto soursop, sugar apple or pond apple rootstocks. Experiments in Mexico, utilizing cherimoya, ilama, soursop, custard apple, *Annona* sp. af. *lutescens* and *Rollinia jimenezii* Schlecht. as rootstocks showed best results when custard apple scions were side-grafted onto self-rootstock, soursop, or *A.* sp. af. *lutescens*. Custard apple seedlings are frequently used as rootstocks for the soursop, sugar apple and atemoya.

Culture

The tree is fast-growing and responds well to mulching, organic fertilizers and to frequent irrigation if there is dry weather during the growing period. The form of the tree may be improved by judicious pruning.

Harvesting and Yield

The custard apple has the advantage of cropping in late winter and spring when the preferred members of the genus are not in season. It is picked when it has lost all green color and ripens without splitting so that it is readily sold in local markets. If picked green, it will not color well and will be of inferior quality. The tree is naturally a fairly heavy bearer. With adequate care, a mature tree will produce 75 to 100 lbs (34–45 kg) of fruits per year. The short twigs are shed after they have borne flowers and fruits.

Pests and Diseases

The custard apple is heavily attacked by the chalcid fly. Many if not all of the fruits on a tree may be mummified before maturity. In India, the ripening fruits must be covered with bags or nets to avoid damage from fruit bats.

A dry charcoal rot was observed on the fruits in Assam in 1947. In 1957 and 1958 it made its appearance at Saharanpur. The causal fungus was identified as *Diplodia annonae*. The infection begins at the stem end of the fruit and gradually spreads until it covers the entire fruit.

Food Uses

In India, the fruit is eaten only by the lower classes, out-of-hand. In Central America, Mexico and the West Indies, the fruit is appreciated by all. When fully ripe it is soft to the touch and the stem and attached core can be easily pulled out. The flesh may be scooped from the skin and eaten as is or served with light cream and a sprinkling of sugar. Often it is pressed through a sieve and added to milk shakes, custards or ice cream. I have made a delicious sauce for cake and puddings by blending the seeded flesh with mashed banana and a little cream.

Food Value Per 100 g of Edible Portion*	
Calories	80–101
Moisture	68.3–80.1 g
Protein	1.17–2.47 g
Fat	0.5–0.6 g
Carbohydrates	20–25.2 g
Crude Fiber	0.9–6.6 g
Ash	0.5–1.11 g
Calcium	17.6–27 mg
Phosphorus	14.7–32.1 mg
Iron	0.42–1.14 mg
Carotene	0.007–0.018 mg
Thiamine	0.075–0.119 mg
Riboflavin	0.086–0.175 mg
Niacin	0.528–1.190 mg
Ascorbic Acid	15.0–44.4 mg
Nicotinic Acid	0.5 mg

*Minimum and maximum levels of constituents from analyses made in Central America, Philippines and elsewhere.

Toxicity

The seeds are so hard that they may be swallowed whole with no ill effects but the kernels are very toxic. The seeds, leaves and young fruits are insecticidal. The leaf juice kills lice. The bark contains 0.12% anonaine. Injection of an extract from the bark caused paralysis in a rear limb of an experimental toad. Sap from cut branches is acrid and irritant and can severely injure the eyes. The root bark has yielded 3 alkaloids: anonaine, liriodenine and reticuline (muricinine).

Other Uses

The leaves have been employed in tanning and they yield a blue or black dye. A fiber derived from the young twigs is superior to the bark fiber from *Annona squamosa*. Custard apple wood is yellow, rather soft, fibrous but durable, moderately close-grained, with a specific gravity of 0.650. It has been used to make yokes for oxen.

Medicinal Uses: The leaf decoction is given as a vermifuge. Crushed leaves or a paste of the flesh may be poulticed on boils, abscesses and ulcers. The unripe fruit is rich in tannin; is dried, pulverized and employed against diarrhea and dysentery. The bark is very astringent and the decoction is taken as a tonic and also as a remedy for diarrhea and dysentery. In severe cases, the leaves, bark and green fruits are all boiled together for 5 minutes in a liter of water to make an exceedingly potent decoction. Fragments of the root bark are packed around the gums to relieve toothache. The root decoction is taken as a febrifuge.

Ilama

This member of the Annonaceae was little known and the subject of much confusion until 1911, when it was investigated and fully described by W.E. Safford, of the United States Department of Agriculture's Bureau of Plant Industry, and given the botanical name of *Annona diversifolia* Safford. In Mexico, it has been called *ilama*, *izlama*, *illamatzapotl* (translated as *zapote de las viejas*, or "old woman's sapote"), *hilama*, and *papuasa*. In Guatemala, it is called *anona blanca* or *papauce;* in El Salvador, *anona blanca*.

Description

The tree may be spreading or erect, to 25 ft (7.5 m), often branching from the ground. It has aromatic, pale brownish-gray, furrowed bark and glossy, thin, elliptic to obovate or oblanceolate leaves, 2 to 6 in (5-15 cm) long. There are 1 or 2 leaflike, nearly circular, glabrous bracts, 1 to 1⅜ in (2.5-3.5 cm) long, clasping the base of the flowering branchlets. The new foliage is reddish or coppery. Solitary, long-stalked, maroon flowers, which open to the base, have small rusty-hairy sepals, narrow, blunt, minutely hairy outer petals, and stamen-like, pollen-bearing inner petals. The fruit is conical, heart-shaped, or ovoid-globose, about 6 in (15 cm) long; may weigh as much as 2 lbs (0.9 kg). Generally, the fruit is studded with more or less pronounced, triangular protuberances, though fruits on the same tree may vary from rough to fairly smooth. The rind, pale-green to deep-pink or purplish, is coated with a dense, velvety gray-white bloom. It is about ¼ in (6 mm) thick, leathery, fairly soft and granular. In green types, the flesh is white and sweet; in the pink types, it is pink-tinged near the rind and around the seeds, all-pink or even deep-rose, and tart in flavor. It is somewhat fibrous but smooth and custardy near the rind; varies from dryish to fairly juicy, and contains 25 to 80 hard, smooth, brown, cylindrical seeds, ¾ in (2 cm) long, ⅜ in (1 cm) wide, each enclosed in a close-fitting membrane easily slipped off when split.

Origin and Distribution

The ilama is native and grows wild in foothills from the southwest coast of Mexico to the Pacific coast of Guatemala and El Salvador. The earliest known record of the fruit was made by Francisco Hernandez who was sent by King Philip II of Spain in 1570 to take note of the useful products of Mexico. For many years, it was confused with either the soursop or the custard apple.

The United States Department of Agriculture introduced seeds from El Salvador in 1914 (P.I. No. 35567); from Guatemala in 1917 (P.I. No. 45548); and from Mexico in 1919, 1922 and 1923 (P.I. Nos. 46781, 55709, and 58030). One of the trees planted at the Plant Introduction Garden, Miami, Florida, bore its first fruits in 1923. Several thousand seedlings had been sent to Puerto Rico, St. Croix, various part of tropical America and Asia (including Ceylon), and the Philippines. Apparently few survived. Only in its homeland is the ilama commonly grown in dooryards, occasionally in orchards of 100 trees or more. Dr. Victor Patiño took seeds from Mexico to Colombia for planting in the Cauca Valley in 1957. In spite of early enthusiasm for this species, it is seldom mentioned in horticultural literature. In 1942, there were no more than 50 trees in southern Florida, only 3 of bearing age. In 1965, Dr. John Popenoe, Director of Fairchild Tropical Garden, brought seeds from Guatemala and raised a number of seedlings for distribution, but the tree is still quite rare in Florida. It is too tender even for southern California.

Varieties

One named cultivar, 'Imery', introduced into Florida from El Salvador and grown at the Agricultural Research and Education Center, Homestead, is large and pink-fleshed but not as flavorful as some of the white-fleshed acquisitions from Guatemala.

Climate

The ilama is strictly tropical; grows naturally not higher than 2,000 ft (610 m) in Mexico; is cultivated up to 5,000 ft (1,524 m) in El Salvador; up to 5,900 ft (1,800 m) in Guatemala. It seems to do best where there is a long dry season followed by plentiful rainfall. In areas where rainfall is scant, the tree is irrigated.

Soil

Dr. Wilson Popenoe observed that the tree was not particular as to soil but should prosper in rich, loose

Fig. 24: The ilama (*Annona diversifolia*), as grown in southern Florida, has a thick rind and dryish flesh.

loam. In Florida, it performs better on deep sand than on oolitic limestone.

Propagation

Ilama seeds, taken from ripe fruits, remain dormant for several weeks or even months and the germination rate thereafter is low. Applications of gibberellic acid at 350 ppm greatly increases germination. Higher concentrations cause malformations in the seedlings. Whip- or cleft-grafting onto custard apple (*A. reticulata*) rootstocks has been successful. Seedlings begin to bear when 3 to 5 years old.

Harvesting

The harvesting season begins in late June in Mexico and lasts only a few weeks. It extends from late July to September in Guatemala; from July to December in Florida. Traditionally, the fruits are not picked until they have begun to crack open, but they can be picked a little earlier and held up to 3 days to soften. They will not ripen if harvested too early.

Yield

The yield is typically low. In Mexico, during the normal fruiting period, some trees will have no fruits, others only

3 to 10; exceptional trees may bear as many as 85 to 100 fruits in a season.

Pests

The ilama is not as susceptible to the chalcid fly as are its more popular relatives in Florida.

Food Value Per 100 g of Edible Portion*	
Moisture	71.5 g
Protein	0.447 g
Fat	0.16 g
Fiber	1.3 g
Ash	1.37 g
Calcium	31.6 mg
Phosphorus	51.7 mg
Iron	0.70 mg
Carotene	0.011 mg
Thiamine	0.235 mg
Riboflavin	0.297 mg
Niacin	2.177 mg
Ascorbic Acid	13.6 mg

*According to analyses made in El Salvador.

Food Uses

The early plant explorers of the United States Department of Agriculture and their contacts in Mexico and Central America described the ilama as resembling the cherimoya or atemoya in flavor and expected it to be well received in this country and abroad. However, as grown in Florida, it is not as appealing as the sugar apple. There is a slightly unpleasant flavor close to the rind. The flesh is always consumed raw, either in the half-shell or, better still, shallowly scooped out, chilled, and served with a little cream and sugar to intensify the flavor, or with a dash of lime or lemon juice.

Soncoya

Among lesser species of the family Annonaceae, the soncoya, *Annona purpurea* Moc. & Sesse (syns. *A. manirote* HBK. *A. involucrata* Baill., *A. prestoli* Hemsl.) is called *cabeza de negro, cabeza de ilama, chincua, ilama,* or *ilama de Tehuantepec* in Mexico; *anona sincuya, chincuya, cabeza de muerto, sencuya, suncuyo, soncolla,* or *matacuy* in Guatemala; *guanabano torete* or *toreta* in Panama; *gallina gorda, guanabano pun,* or *matimba* in Colombia; *castiguire, manire, manirote, tiragua,* or *tucuria* in Venezuela.

Description

The tree is small to medium, to 20 or even 33 ft (6-10 m) high, with short trunk to 1½ ft (45 cm) in diameter, and spreading branches, which are rusty-woolly when young. The deciduous leaves are alternate, short-petioled, undulate, oblong-elliptic or oblong-lanceolate to oblong-obovate, 8 to 12 in (20-30 cm) long and 4 to 5½ in (10-14 cm) wide, acuminate at the apex, brown-hairy on both surfaces and with prominent veins beneath. Strong-scented flowers, which emerge with the new leaves, are solitary, fleshy, large, conical, usually enclosed at first by a pair of bracts; are held at the base by a rusty-hairy, 3-parted calyx, and have 3 very thick outer petals, brown-hairy outside, yellowish and purple-mottled within, and 3 smaller, thinner inner petals, creamy-white outside, purple inside. The fruit, thick-stalked, is ovoid or nearly round, 6 to 8 in (15-20 cm) wide, set with hard, somewhat 4-sided, conical protuberances, each tipped with a curved hook, and is coated overall with a brown felt. The pulp is agreeably aromatic, suggesting the mango; abundant, yellow or orange, soft, fibrous, of mild, agreeable flavor. Seeds are numerous, obovate, 1 to 1³⁄₁₆ in (2.5-3 cm) long, dark-brown, and each is enclosed in a thin, close-fitting membrane. The fruit carpels separate easily when ripe.

Origin and Distribution

The soncoya is native and common in coastal lowlands from southern Mexico to Panama, Colombia and Venezuela. It is grown in dooryards and the fruit is sold in local markets, though it is of mediocre quality and not popular because it is outwardly so hard. The tree was introduced into the Philippines in the early 1900's, grew well and flowered at Lamao but apparently did not set fruit for several years. It was planted at the Federal Experiment Station at Mayaguez, Puerto Rico, in 1918 and in St. Croix in 1930. Several trees have grown well and borne poorly at the Lancetilla Experimental Garden, Tela, Honduras.

Climate

The soncoya requires a hot, humid climate and it never occurs at an altitude higher than 4,000 ft (1,200 m).

Season

The fruits ripen in August in Yucatan; generally in the fall in Central America.

Food Uses

In Colombia, the pulp is eaten raw or is strained for juice, drunk as a beverage or folk remedy.

Toxicity

The seed extract destroys fleas. In Guatemala and Costa Rica, rural people believe the fruit to be unwholesome.

Medicinal Uses

In Mexico, soncoya juice is regarded as a remedy for fever and chills. Elsewhere it is given to relieve jaundice (probably because of its color). The bark decoction is effective against dysentery and a tea of the inner bark is administered in cases of edema.

Wild Custard Apple

A noteworthy, useful African member of the Annonaceae is the wild custard apple, *Annona chrysophylla* Boj. (syn. *A. senegalensis* Auct. non Pers., often cited erroneously as *A. senegalensis* Pers.). The tree is so popular in its native land that it has acquired a wealth of vernacular names: wild soursop, in several localities; *muvulu, mugosa, mbokwe, makulo, mlamote,* etc., in Kenya; *mtopetope* and *mchekwa* in Zanzibar and Pemba; *mubengeya, elipo, obwolo, ovolo,* etc., in Uganda; *aboboma, batani, bangoora, bullimbuga,* etc., in Ghana; *mposa, muroro* and *mponjela* in former Nyasaland; *dilolo, iolo,* and *malolo* in Angola; *sougni, mete, dangan, sounsoun, tangasou, dougour, ianouri, ndong, anigli* in former French West Africa.

Description

This is a sprawling shrub or a tree to 20 ft (6 m) high with smooth, silvery bark. The leaves are aromatic, deciduous, alternate, blue-green, broad-elliptic or broad-ovate, 3 to 7 in (7.5-17.5 cm) long, 1½ to 4 in (4-10 cm) wide, rounded at apex and base, blue-green above, downy, prominently veined beneath. The flowers, borne singly or in pairs in the leaf axils on stalks 1 to 1½ in (2.5-4 cm) long, are clasped by a 3-parted calyx and have 3 triangular, thick, waxy, velvety, whitish outer petals, 3 pale-yellow inner petals, and numerous stamens. Typically compound, the pineapple-scented fruit is smooth but with the carpels distinctly outlined on the surface; yellow or orange when ripe; rounded oval; 1 to 4 in (2.5-10 cm) long; fleshy; seedy.

Origin and Distribution

This species is native and common in savannas throughout tropical Africa from the Cape Verde Islands and the Nile and Upper Guinea to the Transvaal and Zululand. It is little-known outside its natural range. It was long ago introduced by the United States Department of Agriculture into Florida as a potential rootstock for related species, and into Puerto Rico in 1924 and again in 1925 and grown at the Insular Experiment Station, Rio Piedras. According to G.L. Cruz (1979), verbatim from his 1965 publication (see Bibliography), it has become well established as *araticum da areia* in Brazil, especially around Minas Gerais, Bahia and Espirito Santo, but he describes the fruit as rough-surfaced and 8 to 12 in (20-30 cm) in diameter, so he must have it confused with some other species unless there is great variation among seedlings.

Varieties

The botanical variety, *deltoides*, with elliptic to oblong-elliptic leaves, rounded to broadly deltoid at the base, is the most common form in Ghana. Eggeling mentions a variety *porpetac* in Uganda with oblong-elliptic, oval-elliptic or elliptic leaves, rounded-obtuse or broadly cuneate at the base. There is reportedly a dwarf form, the fruits of which are borne so low they touch the ground, and are of better quality than those of taller types.

Climate

The wild custard apple is limited to tropical areas up to an elevation of 5,000 ft (1,500 m) and thrives best where its roots can reach water. It remains leafless for several months in the dry season.

Food Uses

The fruit pulp is edible and said to have an apricot-like flavor. Williamson quotes an unidentified source as saying that it is one of the best of the indigenous fruits in parts of tropical Africa. It is much appreciated in the wild by shepherds. According to Irvine, the unopened flower buds are used in soup and to season native dishes; and the leaves are eaten.

Food Value

The dried leaves contain 8.2% protein.

Other Uses

Fruit: The green fruit, because of its high tannin content, is made into ink.

Leaves: Fresh leaves are employed as fodder for domestic animals. Boiled leaves serve as native perfume, and dried leaves are used as filling for mattresses.

Wood: The soft, grayish wood is fashioned into hoe handles and employed in building huts.

Ashes: The wood ashes are used in making soap and native snuff.

Bark and roots: The bark yields a yellow or brown dye. It is pounded in water and the liquid is then used as a hair dressing. A poor-quality fiber derived from the bark is made into rope for tying fences. A combination of the bark and roots serves as an insecticide, and the root has been used for homicidal purposes. The irritant, gummy sap of the bark is an adhesive for arrow poison. There are many superstitious uses of the various parts of the plant.

Medicinal Uses: Fresh fruits in quantity and dried fruits are applied on Guinea worm sores. The parched green fruits are taken to relieve diarrhea and dysentery. A tea of the young leafy twigs or of the roots is taken to alleviate pulmonary complaints. An infusion of the leaves is a popular eye lotion. Dried, powdered leaves are regarded as purgative and as a remedy for mucous diarrhea. With quantities of water, the pulverized leaves are given to horses to expel worms. Combined with the roots and bark and other materials, they are said to be effective in treating yaws in horses. The leaves also enter into a tonic for horses. Venereal diseases and intestinal disorders are treated with preparations of the roots. The bark is chewed to relieve stomachache. It is an emetic and a vermifuge and is given to overcome convulsions in children. The bark infusion, held in the mouth, relieves toothache.

In the Upper Volta, an ointment made from the bark is applied on burns. Bark and roots together will halt dysentery, expel worms, and are part of a remedy for sleeping sickness. The root infusion is employed as eye drops. Charcoal of the burned roots is applied on twitching eyelids. The root bark is considered an antidote for snakebite and is used by Nigerian medicine men as a cancer remedy. Investigations have revealed antitumor activity against sarcoma 180 ascites, and antibiotic activity. The trunk bark contains alkaloids, including 0.02% anonaine, also tannins and saponins. The leaves contain rutin, quercetin and quercetrin.

Related Species

The mountain soursop, *A. montana* Macf. (syns. *A. Marcgravii* Mart.; *A. sphaerocarpa* Splitg.; *A. Pisonis* Mart.) is also called wild soursop, *guanabana cimarrona*, *guanabana de perro*, *guanabana de loma*, *corossol zombi*, *corossolier batard*, *boszuurzak*, *araticum-ponhe* and *araticum de paca*. It grows wild from sea-level to 2,000 ft (650 m) throughout the West Indies and southward into Peru and Brazil, and is cultivated in the Philippines and rarely in Florida.

The tree somewhat resembles that of the soursop but has a more spreading crown and very glossy leaves. It is slightly hardier and bears more or less continuously. The fruit is nearly round or broad-ovoid, to 6 in (15 cm) long. Its dark-green skin is studded with numerous short, fleshy "spines". It becomes very soft and falls when ripe. The pulp is yellow, peculiarly aromatic, sour to subacid and bitter, fibrous, and contains many light-brown, plump seeds. The quality is variable but generally very poor. The fruit is generally regarded as inedible but is referred to as "edible but mediocre" in Brazil. There, the firm core attached to the base of the peduncle is pulled out and eaten as a tidbit. In southern Florida, exotic parrots eat the fruits and scatter the seeds, and a few trees are consequently occurring as escapes. The tree is of minor

Fig. 25: The scarcely-edible mountain soursop (*Annona montana*).

interest to horticulturists as an ornamental and root-stock. The wood is soft, fibrous and useful only as fuel.

The pond apple, *A. glabra* L. (syn. *A. palustris* L.), is also called alligator apple, monkey apple, custard apple, corkwood, *mamón de perro, cayur, cayuda,* and various other colloquial names. It grows wild in the Florida Everglades and in coastal swamps and marshes of the Bahamas and throughout the West Indies, in southern Mexico, Central America, and southward into Peru and Argentina; also on the coast of West Tropical Africa. It is occasionally planted in southern Florida and has been introduced into Malaya and the Philippines.

The tree may reach 45 ft (13.5 m), is rather open and spreading; may become very thick at the base; has glossy, leathery, deciduous leaves. The fruit is oval or heart-shaped, to 5 in (12.5 cm) long with thin, faintly reticulated, glossy yellow skin and salmon-colored, resinous, subacid, dryish pulp containing many light-brown, flattened-oval, longitudinally-winged seeds that float on water. When fully ripe and soft, the pulp is edible and some specimens are of fair quality and have been made into jelly or wine. The pond apple is of value as a "survival" food in extremity and of great importance as fare for wild creatures. Fishermen fashion the light, corklike wood into floats. The leaf decoction is a common, multi-purpose folk remedy in the Netherlands Antilles, Mexico and South America. Seedlings are useful as rootstock for other *Annona* species in wet soils.

Biribá

Of the approximately 65 species of the genus *Rollinia* (family Annonaceae), only a few have edible fruit and the best-known is the biribá, *R. mucosa* Baill. (syns. *R. orthopetala* A. DC.; *Annona mucosa* Jacq.; *A. sieberi* A. DC.; and possibly *R. deliciosa* Safford?). The popular Brazilian name has been widely adopted, but in that country it may also be called *biribá de Pernambuco, fruta da condessa, jaca de pobre, araticu, araticum, araticum pitaya.* In Peru, it is *anón;* in Ecuador, *chirimoya;* in Colombia, *mulato;* in Venezuela, *riñon* or *riñon de monte;* in Mexico, *anona babosa* or *zambo.* In Trinidad it is called wild sugar apple; in Guadeloupe, *cachiman morveux, cachiman cochon* or *cachiman montagne;* in Puerto Rico, *cachiman* or *anón cimarron;* in the Dominican Republic, *candongo* or *anona.*

Description

This fast-growing tree ranges from 13 to 50 ft (4–15 m) in height; has brown, hairy twigs and alternate, deciduous, oblong-elliptic or ovate-oblong leaves, pointed at the apex, rounded at the base, 4 to 10 in (10–25 cm) long, thin but somewhat leathery and hairy on the underside. The flowers, borne 1 to 3 or occasionally more together in the leaf axils, are hermaphroditic, ¾ to 1⅜ in (2–3.5 cm) wide; triangular, with 3 hairy sepals, 3 large, fleshy outer petals with upturned or horizontal wings, and 3 rudimentary inner petals. The fruit is conical to heart-shaped, or oblate; to 6 in (15 cm) in diameter; the rind yellow and composed of more or less hexagonal, conical segments, each tipped with a wart-like protrusion; nearly ⅛ in (3 mm) thick, leathery, tough and indehiscent. The pulp is white, mucilaginous, translucent, juicy, subacid to sweet. There is a slender, opaque-white core and numerous dark-brown, elliptic or obovate seeds ⅝ to ¾ in (1.6–2 cm) long.

Origin and Distribution

This species has an extensive natural range, from Peru and northern Argentina, Paraguay and Brazil and north-ward to Guyana, Venezuela, Colombia and southern Mexico; Trinidad, the Lesser Antilles including Guadeloupe, Martinique and St. Vincent; and Puerto Rico and Hispaniola. It is much cultivated around Iquitos, Peru, and Rio de Janeiro, Brazil and the fruits are marketed in abundance. It is the favorite fruit in western Amazonia.

Seeds were first introduced into the United States from Pará, Brazil, by O.W. Barrett in 1908 (S.P.I. #22512); a second time from Pará in 1910 (S.P.I. #27579) and again in 1912 (S.P.I. #27609). The United States Department of Agriculture received seeds from Rio de Janeiro in 1914 (S.P.I. #38171). P.J. Wester may have taken seeds to the Philippines where the species first fruited in 1915. Seedlings were distributed to pioneers in southern Florida but only a very few trees exist here today.

Varieties

The only named selection referred to in the literature is 'Regnard' reported by P.J. Wester in 1917 as the best variety introduced into the Philippines. A form in the western Amazon region has very pronounced points; weighs up to 8.8 lbs (4 kg).

Pollination

Brazilian scientists have found that 4 species of beetles of the family Chrysomelidae pollinate the flowers, but only 32% of the blooms set fruit. Fruiting begins 55 days after the onset of flowering.

Climate and Soil

The biribá is limited to warm lowlands, from 20° north to 30° south latitudes in tropical America. In Puerto Rico, it occurs at elevations between 500 and 2,000 ft (150–600 m). It has succumbed to temperature drops to 26.5°F (-3.10°C) in southern Florida. In Brazil, the tree grows naturally in low areas along the Amazon subject to periodic flooding and it was expected to do well in the Florida Everglades. In the Philippines it is said to flourish

Fig. 26: The biriba (*Rollinia mucosa*) is an attractive light-yellow at first.

where the rainfall is equally distributed throughout the year. Calcareous soils do not seem to be unsuitable in Florida or Puerto Rico as long as they are moist.

Season and Harvesting

In Amazonia, the tree may flower and fruit off and on during the year but the fruits are most abundant from January to June. The fruits ripen in February and March in Rio de Janeiro. In Florida, fruits have matured in November and December. In South America, the fruit is picked when still green and hard in order to transport it intact to urban markets where it gradually turns yellow and soft. When the fruit is fully ripe, handling causes the wart-like protuberances on the rind to turn brown or near-black, rendering it unattractive.

Pests and Diseases

The most important pests in Brazil are the larvae of *Cerconota anonella* (Lepidopterae) which attack fruits in the process of maturing. The borer, *Cratosomus bombina,* penetrates the bark and trunk. A stinging caterpillar, *Sabine* sp., feeds on the leaves. A white fly, *Aleurodicus cocois,* attacks foliage of young and adult plants. *Pseudococcus brevipes* and *Aspidiotus destructor* are found on the leaves and sometimes on the fruits. Black spots on the leaves are caused by the fungus *Cercospora*

Food Value Per 100 g of Edible Portion*	
Calories	80
Moisture	77.2 g
Protein	2.8 g
Lipids	0.2 g
Glycerides	19.1 g
Fiber	1.3 g
Ash	0.7 g
Calcium	24 mg
Phosphorus	26 mg
Iron	1.2 mg
Vitamin B_1	0.04 mg
Vitamin B_2	0.04 mg
Niacin	0.5 mg
Ascorbic Acid	33.0 mg

Amino Acids (mg per g of Nitrogen (N = 6.25):

Lysine	316 mg
Methionine	178 mg
Threonine	219 mg
Tryptophan	57 mg

*According to Brazilian analyses.

Fig. 27: Handling causes the conical projections on the fruit to turn black.

anonae. *Glomerella cingulata* causes dieback and fruit rot in Florida.

Food Uses

The fruit is eaten fresh and is fermented to make wine in Brazil.

Other Uses

The wood of the tree is yellow, hard, heavy, strong and is used for ribs for canoes, boat masts, boards and boxes.

Medicinal Uses: The fruit is regarded as refrigerant, analeptic and antiscorbutic. The powdered seeds are said to be a remedy for enterocolitis.

Avocado

The avocado, unflatteringly known in the past as alligator pear, midshipman's butter, vegetable butter, or sometimes as butter pear, and called by Spanish-speaking people *aguacate, cura, cupandra,* or *palta*; in Portuguese, *abacate*; in French, *avocatier*; is the only important edible fruit of the laurel family, Lauraceae. It is botanically classified in three groups: A), *Persea americana* Mill. var. *americana* (*P. gratissima* Gaertn.), West Indian Avocado; B) *P. americana* Mill. var. *drymifolia* Blake (*P. drymifolia* Schlecht. & Cham.), the Mexican Avocado; C) *P. nubigena* var. *guatemalensis* L. Wms., the Guatemalan Avocado.

Description

The avocado tree may be erect, usually to 30 ft (9 m) but sometimes to 60 ft (18 m) or more, with a trunk 12 to 24 in (30-60 cm) in diameter, (greater in very old trees) or it may be short and spreading with branches beginning close to the ground. Almost evergreen, being shed briefly in dry seasons at blooming time, the leaves are alternate, dark-green and glossy on the upper surface, whitish on the underside; variable in shape (lanceolate, elliptic, oval, ovate or obovate), 3 to 16 in (7.5-40 cm) long. Those of the Mexican race are strongly anise-scented. Small, pale-green or yellow-green flowers are borne profusely in racemes near the branch tips. They lack petals but have 2 whorls of 3 perianth lobes, more or less pubescent, and 9 stamens with 2 basal orange nectar glands. The fruit, pear-shaped, often more or less necked, oval, or nearly round, may be 3 to 13 in (7.5-33 cm) long and up to 6 in (15 cm) wide. The skin may be yellow-green, deep-green or very dark-green, reddish-purple, or so dark a purple as to appear almost black, and is sometimes speckled with tiny yellow dots, it may be smooth or pebbled, glossy or dull, thin or leathery and up to ¼ in (6 mm) thick, pliable or granular and brittle. In some fruits, immediately beneath the skin there is a thin layer of soft, bright-green flesh, but generally the flesh is entirely pale- to rich-yellow, buttery and bland or nutlike in flavor. The single seed is oblate, round, conical or ovoid, 2 to 2½ in (5-6.4 cm) long, hard and heavy, ivory in color but enclosed in two brown, thin, papery seedcoats often adhering to the flesh cavity, while the seed slips out readily. Some fruits are seedless because of lack of pollination or other factors.

Origin and Distribution

The avocado may have originated in southern Mexico but was cultivated from the Rio Grande to central Peru long before the arrival of Europeans. Thereafter, it was carried not only to the West Indies (where it was first reported in Jamaica in 1696), but to nearly all parts of the tropical and subtropical world with suitable environmental conditions. It was taken to the Philippines near the end of the 16th Century; to the Dutch East Indies by 1750 and Mauritius in 1780; was first brought to Singapore between 1830 and 1840 but has never become common in Malaya. It reached India in 1892 and is grown especially around Madras and Bangalore but has never become very popular because of the preference for sweet fruits. It was planted in Hawaii in 1825 and was common throughout the islands by 1910; it was introduced into Florida from Mexico by Dr. Henry Perrine in 1833 and into California, also from Mexico, in 1871. Vegetative propagation began in 1890 and stimulated the importation of budwood of various types, primarily to extend the season of fruiting. Some came from Hawaii in 1904 (S.P.I. Nos. 19377-19380).

Now the avocado is grown commercially not only in the United States and throughout tropical America and the larger islands of the Caribbean but in Polynesia, the Philippines, Australia, New Zealand, Madagascar, Mauritius, Madeira, the Canary Islands, Algeria, tropical Africa, South Africa, southern Spain and southern France, Sicily, Crete, Israel and Egypt.

Though the Spaniards took the avocado to Chile, probably early in the 17th Century and it was planted from the Peruvian border southward for over 1000 mi (1,600 km) actual commercial plantings were not established until California cultivars were introduced about 1930 into two areas within 100 mi (160 km) of Santiago where the industry is now centered.

The first trees were planted in Israel in 1908, but named cultivars ('Fuerte' and 'Dickinson') were not introduced until 1924. These aroused interest in the feasibility of the crop for the southern half of the coastal plain and the interior valleys, and development of the industry has steadily gone forward, except for a period in the 1960's when much planting stock was destroyed because of marketing problems. In 1979, Israel produced 33,000 tons (30,000 MT) and exported 28,600 tons (26,000 MT).

In just the last few years, New Zealand has launched a program to expand commercial production, especially in the Bay of Plenty area, with protection from wind and frost, with a view to becoming a major exporter of avocados.

California produced 265 million lbs (12,045 MT) in 1976; 486 million lbs (22,090 MT) in 1981. The Florida avocado potential is estimated at 150 million lbs (6,818 MT). Both states suffer fluctuations because of the impact of periodic freezes, droughts, high winds or other seasonal factors.

Presently, Mexico, with 150,000 acres (62,500 ha) is the leading producer—267,786 tons (243,000 MT); the Dominican Republic is second—144,362 tons (131,000 MT); U.S.A. (California and Florida combined) with 52,000 acres (21,666 ha), third—131,138 tons (119,000 MT); Brazil is fourth—128,934 tons (117,000 MT). Israel, with 16,000 acres (6,666 ha), is fifth; and South Africa sixth. Half of California's plantings are in San Diego County close to Mexico.

As an exporter, Mexico again leads, followed by California, Israel, South Africa and Florida, in that order. Nearly all of Brazil's crop is consumed domestically.

Varieties

WEST INDIAN race: Florida avocados were at first mainly of the summer-fruiting West Indian race, but these had to compete commercially with similar fruits imported from Cuba, and growers sought other cultivars maturing at a later season. This led to the development of West Indian X Guatemalan hybrids. The cessation of trade with Cuba in the early 1960's brought about a shift back to summer cultivars in new groves to fill the gap. The majority of the avocados grown in the West Indies, Bahamas and Bermuda and the tropics of the Old World are still of the West Indian race. The skin is leathery, pliable, non-granular, and the flesh low in oil. The leaves are not aromatic. The following are the most prominent of early and more recent West Indian cultivars which have played an important role in the development of the avocado industry in Florida and elsewhere. New selections appear from time to time that may have special adaptability to certain locales or conditions.

'Butler' (a USDA selection in Florida; fruited in 1909, propagated from 1914 to 1918) pear-shaped; medium-large; skin smooth; seed of medium size, tight in the cavity. Season: Aug.-Sept. No longer grown in Florida. Cultivated in Puerto Rico.

Fig. 28: West Indian avocados (*Persea americana*). The fruit cut open is a 'Hall'.

'Fuchs' ('Fuchsia') (seed of unknown origin planted in Homestead, Florida, in 1910; propagated commercially in 1926); pear-shaped to oblong, sometimes with a neck; of medium size; skin smooth; flesh pale greenish-yellow; 4 to 6% oil; seed loose. Season: early June-Aug.; a poor shipper. Tree not very productive in Florida; no longer popular in commercial groves.

'Maoz' (a seedling selected from a plot near Maoz, Israel); pear-shaped; of medium size; skin rough, leathery, violet-purple when ripe; flesh sweetish and very low in oil. Season: medium-late (Oct.). Tree is an alternate bearer but is fairly small, highly salt-tolerant; used in Israel as rootstock on either saline or calcareous soils.

'Pollock' (originated in Miami before 1896; commercially propagated in 1901); oblong to pear-shaped; very large, up to 5 lbs (2.27 kg); skin smooth; flesh green near skin, contains 3 to 5% oil; seed large, frequently loose in cavity. Season: early July to Aug. or Oct. Shy-bearing and too large but of superior quality.

'Ruehle' (a seedling of Waldin planted at the Agricultural Research and Education Center, Homestead, in 1923; first propagated in 1946); pear-shaped; of medium size, 10 to 20 oz (280-560 g); flesh low in oil (2-5%). Season begins in July in Florida; Jan. in Queensland. Heavy bearer in Florida.

'Russell' (originated in Islamorada in Florida Keys); pear-shaped at apex with long neck giving it a total length up to 13 in (32.5 cm); skin, smooth, glossy, thin, leathery; flesh of excellent quality; seed small. Season: Aug. and Sept. Tree bears well and is recommended for home gardens.

'Simmonds' (possibly from a seed of Pollock, first fruited in Miami in 1913; propagated commercially in 1921); oblong-oval to pear-shaped; large; skin smooth, light-green; flesh of good flavor, 3 to 6% oil; seed of medium size, usually tight. Season: mid-July to mid-Sept. Tree bears more regularly than Pollock but is less vigorous; sometimes sheds many of its fruits; no longer planted commercially in Florida.

'Trapp' (originated in Miami in 1894; propagated in 1901); round to pear-shaped; medium to large; skin smooth; flesh golden-yellow, green near skin, of excellent quality, 3 to 6% oil; seed large, loose in cavity. Season: medium-late (Sept. to Nov. or Dec.); a good shipper. Was prominent in Florida for 25 years despite tendency to overbloom and bear lightly some years; usually bore regularly and well.

'Waldin' (seed planted in Florida in 1909; propagated commercially in 1917); oblong to oval; medium to large; skin smooth; flesh pale to greenish-yellow, of good flavor, 5 to 10% oil; seed medium to large, tight. Season: fairly late (mid-Sept. through Oct.). Tree tends to overbear and die back; is hardy. Has been a leading commercial cultivar in central and southern Florida.

There are several Puerto Rican selections—'Alzamora', 'Avila', 'Faria', 'Garcia', 'Hernandez', 'St. Just'—and some cultivars of unknown ancestry: 'Amador', 'Galo', 'Gimenez', 'Torres', and 'Trujillo'.

GUATEMALAN race: (skin varies from thin to very thick and is granular or gritty). Among prominent early Florida and California cultivars were:

'Anaheim' (originated in California); oval to elliptical; large; skin glossy, rough, thick; flesh of fair to good flavor, up to 22% oil, but inferior to 'Fuerte', 'Nabal' and 'Benik'; is best in Mar. and Apr. in Israel, July and Aug. in Queensland. Tree slender, erect, tall, cold-sensitive; bears regularly, up to 220 lbs (100 kg) annually in Israel. Considered of poor quality and subject to disease during ripening in Queensland.

'Benik' (introduced from Guatemala to California in 1917 and from California into Israel in 1934); pear-shaped; medium to large; skin rough, purple, medium-thick; flesh of good quality, 15 to 24% oil; seed nearly round, medium. Season: Apr. to Aug. in Calif.; Jan. to Mar. in Israel; July and Aug. in Queensland. The tree begins to bear late and yields only about 116 lbs (53 kg) per year. Color is not popular on the market. Not grown in Florida.

'Dickinson' (a California selection, first propagated in 1912); oval to obovate; small to medium; skin dark-purple with large maroon dots, rough, very thick, granular, brittle; flesh of good quality; seed small to medium, tight. Season: June-Oct. in California; Feb. and Mar. in Florida; Jan. and Feb. in Puerto Rico. Tree is a moderate but regular bearer. In Israel 'Dickinson' is described as round, small to large, very thick-skinned with very large seed; of poor quality, not worth growing. It is no longer grown in Florida or California.

'Edranol' (seedling planted at Vista, California in 1927; propagated in 1932), pear-shaped; of medium size; skin olive-green, slightly rough, thin-leathery; flesh of high quality and nutty flavor, 15 to 18% oil; seed small, tight. Season: Feb. to July at Vista; Apr. to Dec. at Santa Barbara; May and June in Queensland. Disease-resistant. Rated as excellent. No longer planted in California but popular in Mexico.

'Hazzard' (seedling of 'Lyon' planted at Vista, California in 1928) pear-shaped; of medium size; skin rough, fairly thin; flesh of good quality, 15 to 34% oil; seed small. Season: Apr. to July in California, July and Aug. in Queensland where it is rated as excellent and free of external and internal diseases and discolorations in storage. The tree grows slowly, reaches only 12 to 15 ft (3.5-4.5 m), begins bearing early and is a dependable producer. Some fruits may crack if left on tree too long. More than 100 trees can be planted per acre (240 per ha).

'Itzamna' (budwood brought from Guatemala to Florida in 1916); oblong-pear-shaped; medium-large; skin rough; flesh yellow, 11% oil; seed small, tight. Season: very late (Mar. to May). May not bear well; little planted in Florida; a commercial cultivar in California and in Puerto Rico where it is a consistently heavy bearer.

'Linda' (budwood introduced into California from Guatemala in 1914; propagated in Florida in 1917); elliptical; very large; skin rough, dull-purple when ripe; flesh yellow, 10 to 14% oil; seed small, tight. Season: May to Oct. in California; late (Dec. to Feb.) in Florida. A good shipper but not popular in Florida because of size and color. Of some commercial importance in California. Tree low, spreading, vigorous and bears regularly.

'Lyon' (originated in California; propagated in 1911); broad-pear-shaped; beyond medium to large; skin somewhat rough to rough; bright-green with many small yellowish or red-brown dots; medium-thick, granular and brittle; flesh greenish near skin, of high quality; seed medium-small to medium, tight. Season: Apr. to Aug. in California. Tree comes into bearing early and bears heavily, so much so as to weaken the tree. Grown in Florida only from 1918 to 1922.

'Macarthur' (originated in 1922 at Monrovia, California); pear-shaped; large; skin thin, pliable; flesh has sweet, nutty but watery flavor, contains 13 to 16.7% oil; seed medium to large. Season: Aug. to Nov. in California; Aug. and Sept. in Queensland where it is rated as of poor quality. It is one of the 6 leading commercial cultivars in California, where it is very cold-hardy.

'Nabal' (budwood brought from Guatemala in 1917; propagated in California since 1927, in Florida from 1937; in Israel since 1934); nearly round; medium to large; skin nearly smooth, thick, granular; flesh of high quality, green near skin; 10 to 15% oil in Florida, 18 to 22% in Queensland; seed small, tight. Season: June to Sept. in California; Jan. and Feb. in Florida; Oct. and Nov. in Queensland. Tree bears well in central Florida; bears late and poorly in Israel averaging 68 lbs (31 kg) per year in alternate years. In Queensland, bears in alternate years very heavily, but is rated as of medium quality and disease-prone during prolonged ripening.

'Nimlioh' (USDA budwood brought from Guatemala in 1917; propagated commercially in 1921); elliptical; large; skin slightly rough; flesh thick; seed fairly small, tight. Season: late (Jan. and Feb.) in Florida; May to Aug. in California. Tree bears moderate crops on south coast of Puerto Rico. Abandoned in Florida in 1925 because tree found to be weak and not prolific.

'Panchoy' (a USDA introduction into Florida from Guatemala; fruited in 1919); pear-shaped to almost elliptical; medium to large; skin rough, very thick; seed of medium size, tight. Season: very late (Mar. to early Apr.) in Florida; Apr. to Aug. in California. Formerly a heavy bearer in Florida and still is on the south coast of Puerto Rico but subject to die-back. Has been commercially important in California and Hawaii.

'Pinkerton' (seedling, probably of 'Rincon', found on Pinkerton ranch in Ventura Co., California, in 1970; patented); early crop roundish; later, pear-shaped with neck; of medium size, 8 to 14 oz (227-397 g); skin medium-leathery, pliable; flesh thick, up to 10% more than in 'Hass' or 'Fuerte'; smooth-textured, of good flavor, high in oil, rated as of good quality but inferior to 'Hass' and 'Fuerte'; tends to darken in the latter part of the season; seed small, separates readily from the flesh with the coat adhering to the seed. Season: first crop, Oct. or Nov., 2nd crop, Dec. or Jan. Fruit ships well and has good shelf-life, but the neck is a disadvantage on the fresh fruit market; accordingly, the late-season fruits are sent to processing plants. The tree is of low, spreading habit; bears early and heavily; is as cold-sensitive as 'Hass'. About 1200 acres (486 ha) in California in 1984.

'Reed' (originated about 1948 on Reed property in Carlsbad, California, as a seedling, possibly of a 'Anaheim' X 'Nabal' hybrid; patented in 1960; patent now expired); round; medium to large, 8 to 18 oz (227-510 g); skin slightly rough, medium-thick, pliable; flesh cream-colored with rich, faintly nutty flavor; doesn't darken when cut; rated as excellent quality; seed small to medium, tight; coat adheres to seed. Season: July to Oct. in California; late Feb. to Apr. in New Zealand where it is one of the most promising cultivars. Tree erect, can be spaced 15 x 15 ft (4.6x4.6 m); bears early and regularly; about as cold-sensitive as 'Hass'. In 1984, about 1,000 acres (405 ha) in California.

'Schmidt' (budwood introduced into California in 1911; propagated in Florida in 1922); pear-shaped; medium to large; skin rough; flesh pale-yellow, 12 to 16% oil; seed of medium size, tight. Season: very late (Feb. and Mar.). The tree is a poor bearer and cold-sensitive and the fruit of poor keeping quality.

'Sharpless' (originated in California; propagated in 1913); slender-pear-shaped, sometimes with long neck; large to very large; skin slightly rough, greenish-purple to dark-purple with many yellowish dots, thick, granular; flesh of superior quality and flavor; seed small, tight. Very late (Oct. to Feb.) in California.

'Solano' (originated in California; propagated in 1912); obovate to oval; beyond medium to large; skin nearly smooth,

bright-green with many yellowish dots, medium-thick, granular; flesh greenish near skin, of fair quality; seed small, tight. Season: Mar. to May in California; Oct. to mid-Nov. or Dec. in Florida. A good bearer, but not grown in Florida for many years.

'Spinks' (originated in California; propagated in 1915); broad-obovate; very large; skin rough, dark-purple, thick, granular, brittle; flesh of very good quality and flavor; seed small, tight. Season: Aug. to Apr. in California. Formerly grown in central Florida.

'Taft' (originated in 1899 in California; propagated in 1912); broad-pear-shaped; medium to very large; skin faintly rough, more so at base; many yellowish dots, thick, granular but somewhat pliable; flesh of excellent quality and flavor; seed of medium size, tight. Season: May to Dec. in California; Feb. and Mar. in Florida. Poor bearer in California; fair in Florida but cold-sensitive.

'Taylor' (seed of 'Royal' planted in Florida in 1908, propagated commercially in 1914); obovate to pear-shaped, occasionally with neck; small to medium size—12 to 18 oz (340-510 g); skin rough, with many small yellow dots; fairly thin; flesh of excellent quality and flavor, 12 to 17% oil; seed of medium size, tight. Season: late (Dec. and Jan. or even to end of Mar.). The tree is cold-hardy but excessively tall and slender.

'Tonnage' (seed of 'Taylor' planted in Florida in 1916; propagated commercially in 1930); pear-shaped, medium-large; skin dark-green, rough, thick; flesh green near skin, rich in flavor, 8 to 15% oil; seed medium, fairly tight. Season: from mid-Oct. through Nov. in Florida; May to mid-Aug. in Argentina. Tree erect, fairly slender, requiring less distance between trees; is a heavy bearer. Cross-pollinated by 'Lula' and 'Collinson' in Argentina.

'Wagner' (seed of 'Royal' planted in California in 1908; propagated in Florida in 1916); rounded to obovate; small to medium; skin slightly rough; flesh light-yellow, 16 to 20% oil; seed large, tight. Season: Late (mid-Jan. to mid-Mar.). Tree lower-growing than 'Taylor', a heavy bearer, but fruit more subject to black spot than 'Taylor'. Not recommended in Florida.

'Wurtz' (originated in 1935 at Encinitas, California; cultivated in Queensland for only the past 12 or 13 years); pear-shaped, small to medium; 8 to 12 oz (226-240 g); seed large. Season: May to Sept. in Calif.; late in Queensland. Tree is small and slow-growing, bears moderately but regularly. More than 100 trees may be planted per acre (240 per ha).

GUATEMALAN X WEST INDIAN hybrids: Inasmuch as pure Guatemalan avocados proved not well adapted to Florida, Guatemalan X West Indian hybrids have come to be of utmost importance in the Florida avocado industry, representing more than half of the more than 20 major and minor commercial cultivars grown in this state today. Prominent cultivars past and present include:

'Bonita' (seed planted in Florida in 1925); obovate, slightly flattened on one side; of medium size; skin slightly rough; flesh contains 8 to 10% oil; seed of medium size. Season: late (Dec. and Jan.). Hardy in California.

'Booth 1' (seed planted in Florida in 1920); round-obovate; medium-large; skin almost smooth, medium thick, brittle; flesh pale, 8 to 12% oil; seed large and loose; Season: late (Dec. and Jan.). The tree is a heavy bearer but the fruit is of poor quality and the seed is too big.

'Booth 7' (seed planted in Florida in 1920; propagated commercially in 1935); round-obovate; of medium size; skin slightly rough, thick, brittle; flesh contains 7 to 14% oil; seed of medium size, tight. Season: late (Dec. to mid-Jan.). The fruit is commercially popular and the tree is a good bearer.

'Booth 8' (seed planted in Florida in 1920); oblong-obovate; medium-large; skin slightly rough, fairly thick, brittle; flesh contains 6 to 12% oil; seed medium-large, tight. Season: late (Nov. to mid-Dec.). Popular commercially and the tree is a heavy bearer.

'Choquette' (originated in Miami from seed planted in 1929; propagated in 1939); oval; large; skin glossy, smooth, slightly leathery; flesh of good quality, 13% oil; seed medium, tight. Season: Jan. to Mar. Tree bears heavily in alternate years.

'Collinson' (seed planted in Florida in 1915); broad-obovoid to elliptical; large; skin smooth; flesh of excellent flavor, 10 to 16% oil; seed of medium size, tight. Season: late (Nov. and Dec.). Tree doesn't produce pollen in Florida; is a heavy bearer in Puerto Rico when interplanted with other cultivars. The flesh is apt to blacken around the seed in cold storage. Cold-sensitive and unfruitful in Israel.

'Fuchs–20' (a seedling of 'Fuchs' selected in Israel); ellipsoid; medium to large; skin smooth, speckled with yellowish lenticels when ripe; flesh flavor is excellent. Season: medium-late (Oct.). Tree is vigorous but a poor bearer; seedlings vary in salt-tolerance but cuttings of resistant selections perform well in saline conditions.

'Grande' (brought to California in 1911 from Atlixco, Mexico); pear-shaped; large; skin rough, green to purplish; seed of medium size, tight. Season: late (Dec. and Jan. in Fla.; Apr. and May in Calif.). Grown in California and Puerto Rico. Tree is a heavy bearer around Mayaguez.

'Hall' (originated in Miami; of unknown parentage; fruited in 1937, propagated in 1938); pear-shaped; large; skin smooth, fairly thick; flesh deep-yellow, 12 to 16% oil; seed medium-large, tight. Season: Nov. and Dec. Heavy bearer and cold-hardy but subject to scab.

'Herman' (seed planted in Florida in 1935); obovate; skin smooth, fairly thin, flexible; flesh yellow, 10 to 14% oil; seed small. Season: fairly late (mid-Nov. to mid-Jan.). Tree a heavy bearer and hardy.

'Hickson' (seedling, fruited in Florida in 1932; propagated commercially in 1938); obovate; medium to small; skin slightly rough, thick, brittle; flesh of fair to good quality, 8 to 10% oil; seed small, tight. Season: late (Dec. and Jan.). Tree bears heavily every other year; is cold-sensitive.

'Simpson' (a sprout of 'Collinson'; fruited in Florida in 1925); obovate-elliptical; rather large; skin slightly rough and thick but not brittle; flesh pale, 10 to 14% oil; seed medium-large, tight. Season: late (mid-Nov. and December). The tree is a good bearer.

'Winslowson' (seed of 'Winslow' planted in Miami in 1911; propagated commercially in 1921); round-oblate; large; skin smooth; flesh pale, 9 to 15% oil; seed of medium size, loose. Season: late (Oct. to Dec. in Fla.; Dec. and Jan. in Puerto Rico). This hybrid is closer to the West Indian race than the Guatemalan and therefore popular in Puerto Rico. Formerly commercial in Florida but abandoned because of loose seed, overblooming, tendency to shed crop, and tree and fruit are susceptible to anthracnose.

In 1963, Puerto Rican horticulturists reported on the performance of 25 selections from 100 studied in the previous 5 years. Four of the selections preceded the establishment of the collection at the Isabela Substation of the University of Puerto Rico. One of the objectives was to identify late-maturing varieties with superior quality and yield. Of the leading 10, all are presumed to be Guatemalan X West Indian hybrids except one, 'Kanan No. 1', which is probably Guatemalan, and this and 'Melendez No. 2' are the only ones of alternate-bearing habit. 'Gripiña' Nos. 2, 5 and 12 were highly rated as, respectively, better than 'Nabal', one of the best commercial cultivars, and most attractive of all. 'Semil' Nos. 23, 31, 34, 42, 43, and 44 seemed equally desirable, with Nos. 34 and 42 noted as wind-resistant.

Puerto Rican breeders have now developed the following Guatemalan X West Indian hybrids: 'Adjuntas', 'Guatemala', 'Melendez 2', 'Gripiña 45', and 'Semil 34' and 43, as late-maturing (Nov. to Mar.), having medium oil content, rich-yellow flesh, and tight seed in order to be able to stand handling and shipment.

MEXICAN race: (skin thin and tender, clings to the flesh; flesh of high oil content, up to 30%. The foliage has a pronounced anise-like odor; the tree is more cold-resistant than those of the other races or hybrids, thriving near Puebla, Mexico, at 500 ft (1,800 m) above sea-level.

'Duke' (originated in California in 1912); elongated; rather small—5½ to 7 oz (150-200 g); flesh of good quality, 14.5% oil. Season: Sept. to Nov. in Calif.; late July or mid-Aug. to mid-Sept. in Israel. Tree is large, symmetrical and wind- and cold-resistant, and also highly resistant to root-rot, especially when grown from cuttings. It is a poor bearer in some areas of California; has borne 168 lbs (78 kg) annually from the 6th to the 15th year in Israel.

'Ganter' (originated in 1905 in California; introduced into Israel in 1943); small, about 5½ oz (150 g); of good quality, 18% oil; seed small to medium, usually loose. Season: Oct. to Dec. in Calif.; second half of Sept. in Israel. Tree is small, yields no more than 44 lbs (20 kg) per year. Poor shipper.

'Gottfried' (seed of a seedling on Key Largo planted at USDA, Miami, in 1906; distributed in 1918); pear-shaped; medium size; skin smooth, purple; flesh of excellent quality, 9 to 13% oil; seed medium. Season: Aug. to Oct. Tree prolific in California; a poor bearer in southern Florida and subject to anthracnose, but hardy and desirable for home gardens on west coast of Florida.

'Mexicola' (originated about 1910 at Pasadena, California; propagated about 1912); very small; skin black; flesh of excellent flavor; seed large. Season: Aug. to Oct. Grown only in home gardens in California. Bears early and regularly; very heat- and cold-resistant; much used as a parent in California breeding programs.

'Northrop' (seedling from C.P. Taft planted about 1900 near Tustin, California; propagated about 1911); small, 3½ to 5½ oz (100-150 g); skin nearly black; flesh of good quality, 26% oil; seed medium. Season: Oct. and Nov. in California; mid-July to mid-Sept. in Florida; mid-Sept. to mid-Oct. in Israel. Fruit does not keep well; flavor disagreeable when over-ripe. Tree bears regularly but has lower yield than 'Duke'.

'Puebla' (considered pure Mexican but some suggest may be a Mexican X Guatemalan hybrid; was found in 1911 at Atlixco near where 'Fuchs' originated). Of medium size; skin smooth, purple; flesh of good flavor; oil content nearly 20%; seed medium to large. Season: Sept. and Oct. in Florida; early to mid-winter in cool regions of California. Tree does not set fruit regularly in California or Israel and therefore is seldom planted now. Has been recommended for home gardens in Central Florida because of hardiness.

'Zutano' (hybrid, originated in 1926 at Fallbrook, California; registered in 1932); pear-shaped; medium-small; skin light-green, very thin, leathery; flesh watery, 15 to 22% oil; seed medium. Season: Dec. and Jan. in California; Apr. and May in Queensland where it is considered of poor quality, delicate to handle, and prone to disease during ripening. Tree is a good bearer. Ranks among 6 leading commercial cultivars in California, being grown where it is too cold for 'Hass'.

GUATEMALAN X MEXICAN hybrids include:

'Bacon'—Quality of flesh slightly better than 'Zutano'. Season: slightly later than 'Zutano'. Tends to be affected with end spot, an external blemish. This cultivar and 'Zutano' are the only 2 reasonably productive of 60 cultivars tried in Los Angeles and Orange Counties in California. In 1957, top-working of all the others to these 2 cold-hardy cultivars was strongly recommended. 'Bacon' is a good choice for tropical American highlands about 5,200 ft (160 m).

'Fuerte' (a natural hybrid originated at Atlixco, Mexico; introduced into California in 1911); pear-shaped; small to medium or a little larger; skin slightly rough to rough, with many small yellow dots, thin, not adherent to flesh; flesh green near skin, 12 to 17% oil; seed small, tight. Season: Jan. to Aug. in southern California; Dec. to Feb. in Israel; Apr. and May in Queensland, and New South Wales; mid-Aug. to Oct. in New Zealand. Tree is broad, very productive, but tends to bear biennially. Subject to scab and anthracnose in Florida. Formerly very popular in California (61% of all avocados shipped); now second to 'Hass' because of a trend to summer instead of winter production and marketing that began in 1972. It is the leading cultivar in Chile where it bears more dependably than in California. It is a very erratic bearer in Israel. Represents 42% of all Australian plantings. Has long been the leading avocado on the European market.

'Hass' (seed planted at La Habra Heights, Calif.; registered in 1932); pear-shaped to ovoid; of medium size; has a tendency to be undersized except in New Zealand; skin tough, leathery, dark-purple or nearly black when ripe; pebbled; fairly thin; flesh of good flavor, 18 to 22% oil, generally; up to 35% in Queensland; seed small. Season: begins in mid-Mar. in California; Nov. to Jan. in Queensland; mid-Nov. to Mar. in New Zealand; Aug. and Sept. in New South Wales. Formerly accounted for 20% of California avocados shipped; now is the leading cultivar (70% of the crop in 1984). Tree bears better than 'Nabal' in cool areas of California, but grows tall and requires topping. This is the leading cultivar in New Zealand, representing 50% of all commercial plantings; 25% in Queensland. It is second in importance to 'Fuerte' in Chile.

'Hayes' (a new hybrid in Hawaii, one parent being 'Hass'). Fruit resembles 'Hass' but is larger; skin is glossier, is pebbled, rough, thick and becomes brown-purple. Season: late (mid-Oct. to Dec. in New Zealand). Tree is erect with drooping branches and the fruit is largely sheltered by the foliage.

'Lula' (seed of 'Taft' planted in Miami in 1915); pear-shaped, sometimes with neck; medium-large; skin almost smooth; flesh pale- to greenish-yellow, 12 to 16% oil; seed large, tight. Season: medium-late (mid-Nov. and Dec.). Tree tall, bears early and heavily; cold-resistant, successful in central and southern Florida where it was formerly the leading commercial cultivar. It is the principal cultivar in Martinique for exporting to France; represents 95% of the crop.

'Rincon' (originated at Carpinteria, California); pear-shaped; small to medium; skin fairly thin, smooth, leathery; flesh buttery, contains 15 to 26.5% oil; fibers in flesh near base turn black when fruit is cut; seed of medium size. Season: Mar. and Apr. in Queensland, where it is rated as of poor quality. It is one of the 6 leading cultivars in California. Tree has a low, spreading habit.

'Ryan' (perhaps seedling of 'Amigo' found in 1927 at Whittier, California); pear-shaped; of medium size, 8 to 12 oz (226-340 g); skin medium-rough; flesh of fair quality; seed rather large. Season: May to Sept. in California; July to Oct. in Queensland. Tree large and bears regularly but not as heavily as 'Fuerte' or 'Hass' in Queensland. Important in Chile.

'Sharwil' (originated in Australia); similar to 'Fuerte' in shape but a little more oval; of medium size; skin rather rough, fairly thin; flesh rich in flavor, of high quality; 15 to 26% oil. Season: May and June in New South Wales and Queensland. Tree bears regularly but not heavily. Represents 18 to 20% of all avocados in New South Wales and Queensland. Disease-free during ripening.

'Susan' (evaluated by California Avocado Society January 2, 1975; patented but patent has now expired); pear-shaped; of medium size, averaging 8 to 10 oz (227-283 g); skin light-green, smooth, thin, peels well; flesh pale-cream-color, of bland flavor; ripens unevenly with darkening spots; has slight tendency to turn dark when cut; not attractive; of only fair quality; seed large, loose; coat adheres to seed. Season: early fall; short. Tree of medium size; grown commercially only in the San Joaquin Valley because of its cold hardiness.

Many local and introduced cultivars representing all 3 races are being grown and evaluated at the experimental station at Minas Gerais, Brazil. A large collection is also maintained in Bahia. The U.S. Department of Agriculture has an international repository of 170 clones in Miami.

In general, small to medium-sized fruits are best for commercial production and especially for metropolitan markets. Large fruits are suitable for local use especially by large families. Smooth, thin or fairly thin, pliable, green skin is preferred by the consumer. The flesh should be virtually fiberless and of agreeable flavor and, for the dieter, of low oil content. The seed must be small and tight so as not to bruise the flesh during handling and shipping. The seed coats ought to adhere to the seed and not to the cavity. The fruit should ship well and stand cold storage. The tree should be of moderate height, slender enough to permit judiciously close planting without crowding. It should bear at an early age and regularly but not so heavily as to suffer die-back, and, of course, should be disease-, insect-, and, in subtropical areas, cold-resistant. Cold-resistant cultivars stand cold-storage better than cold-sensitive cultivars.

Pollination

Many isolated avocado trees fail to fruit from lack of pollination. Commercial growers are careful to match Class A cultivars whose flowers will receive pollen in the morning with Class B cultivars that release pollen in the morning and every grower must be sure to include compatible pollinators in his grove. Bulletin 29 (1971) of the Ministry of Agriculture in Guatemala tabulates the flowering periods (varying from August to April) of 48 introduced and locally selected cultivars, and the hours of the day when each is receptive to or shedding pollen.

Climate

The West Indian race requires a tropical or near-tropical (southern Florida) climate and high atmospheric humidity especially during flowering and fruitsetting. The Guatemalan race is somewhat hardier, having arisen in subtropical highlands of tropical America, and it is successful in coastal California. The Mexican race is the hardiest and the source of most of California avocados. It is not suited to southern Florida, Puerto Rico or other areas of similar climate. Temperatures as low as 25°F (-4° C) do it little harm. In areas of strong winds, windbreaks are necessary. Wind reduces humidity, dehydrates the flowers and interferes with pollination, and also causes many fruits to fall prematurely.

Soil

The avocado tree is remarkably versatile as to soil adaptability, doing well on such diverse types as red clay, sand, volcanic loam, lateritic soils, or limestone. In Puerto Rico, it has been found healthier on nearly neutral or slightly alkaline soils than on moderately or highly acid soils. The desirable pH level is generally considered to be between 6 and 7, but, in southern Florida, avocados are grown on limestone soils ranging from 7.2 to 8.3. Mexican and Guatemalan cultivars have shown chlorosis on calcareous soils in Israel. The tree's primary requirement is good drainage. It cannot stand excessive soil moisture or even temporary water-logging. Sites with underlying hardpan must be avoided. The water table should be at least 3 ft (.9 m) below the surface. Salinity is prejudicial but certain cultivars (see 'Fuchs-20' and 'Maoz') have shown considerable salt-tolerance in Israel. Avocados grafted onto 'Fuch-20' rootstocks and irrigated with water containing 380 to 400 ppm Cl performed well in a commercial orchard. In the Rio Grande Valley of Texas, cultivars of the Mexican race must be grafted onto salt-tolerant West Indian rootstocks.

Propagation

Normally, avocado seeds lose viability within a month. 'Lula' seeds can be stored up to 5 months if placed in non-perforated polyethylene bags and kept at 40°F (4.4° C), thus indicating that it may be possible to successfully store seeds of other cultivars ripening at different seasons for later simultaneous planting. Fresh seeds germinate in 4 to 6 weeks, and many people in metropolitan areas grow avocado trees as novelty house plants by piercing the seed partway through with toothpicks on both sides to hold it on the top of a tumbler with water just covering ½ in (1.25 cm) of the base. When roots and leaves are well formed (in 2 to 6 weeks), the plant is set in potting soil. Of course, it must be given adequate light and ventilation. In nurseries, seeds that have been in contact with the soil are disinfected with hot water. Experiments with gibberellic acid and cutting of both ends of the seed with a view to achieving more uniform germination have not produced encouraging results. Seedlings will begin to bear in

4 or 5 years and the avocado tree will continue to bear for 50 years or more. Some bearing trees have been judged to be more than 100 years old.

In Australia, seeds planted in early fall germinate in 4 to 6 weeks; if planted later, they may remain dormant all winter and germinate in early spring. Seedlings should be kept in partial shade and not overwatered. While many important selections have originated from seeds, vegetative propagation is essential to early fruiting and the perpetuation of desirable cultivars. However, seedlings are grown for rootstocks.

For many years, shield-budding was commonly practiced in Florida, but this method requires considerable skill and experience and is not successful with all cultivars. Therefore, it was largely replaced by whip-, side-, or cleft-grafting, all of which make a stronger union than budding.

In the past, seedlings were grafted when 18 to 36 in (45-90 cm) high. It is now considered far better to graft when 6 to 9 in (15-23 cm) high, making the graft 1 to 3 in (2.5-7.5 cm) above ground level. West Indian rootstocks are desirable for overcoming chlorosis in avocados in Israel.

Avocado cuttings are generally difficult to root. Cuttings of West Indian cultivars will generally root only if they are taken from the tops or side shoots of young seedlings. But etiolated cuttings (new shoots) from gibberellin-treated hardwood and semi-hardwood cuttings of 'Pollock' as well as 'Lula' have been rooted with 50-60% success and, when treated with IBA, 66-83% success under mist in Trinidad. Cuttings of 'Fuchs-20' have rooted under mist with 40 to 50 or even 70% in Israel. Cuttings of 'Maoz' have rooted at the rate of 60% by a special technique developed in California. An Israeli selection, 'G.A. -13' has given 70 to 90% success in rooting cuttings under mist for the purpose of utilizing them as rootstocks in saline and high-lime situations. Air-layering is sometimes done to obtain uniform material uninfluenced by rootstock, for research on specific problems. Degree of success depends on the cultivar (those of the Mexican race rooting most quickly), and air-layering is best done in spring and early summer.

At times, mature avocado groves are top-worked to change from an unsatisfactory cultivar, or one declining in popularity, to a more profitable one, or an assortment of cultivars for different markets. In 1957, 2,700 "obsolete" avocado trees in Ventura, California, were being grafted (top-worked) to mainly 'Hass', some to 'Bacon' and 'Rincon'. This procedure may involve thousands of trees in a given region. It is done in December and January in Florida.

Inasmuch as avocado roots are sensitive to transplanting, it is now considered advisable to raise planting material in plastic bags which can be slit and set in the field without disturbing the root system.

Spacing

Spacing is determined by the habit of the cultivar and the character of the soil. In light soil, 25 x 25 ft (7.5x7.5 m) may be sufficient. In deep, rich soil, the tree makes its maximum growth and a spacing of 30 or 35 ft (9.1 or 10.7 m) may be necessary. If trees are planted so close that they will ultimately touch each other, the branches will die back. Some growers plant 10 to 15 ft (3-4.5 m) apart initially and remove every other tree at 7 to 8 years of age. If the surplus trees are not bulldozed but just cut down leaving a stump, application of herbicide may be needed to prevent regrowth. Ammonium sulfamate has been proven effective. In modernized plantings, space between rows is necessary for mechanical operations.

Holes at least 2 ft (0.6 m) deep and wide are prepared well in advance with enriched soil formed into a mound. After the young plant is put in place a mulch is beneficial, weeds should be controlled, and watering is necessary until the roots are well established. Generally small amounts of fertilizer are given every 2 months with the amount gradually increasing until fruiting begins. Bearing trees need, on the average, 3 to 4 lbs (1½-2 kg) 3 times a year, beginning when the tree is making vegetative growth. No fertilizer should be given at blooming time; one must wait until the fruits are firmly set. Nitrogen has the greatest influence on tree growth, its resistance to cold temperatures, and on fruit size and yield. Fertilizer mixes vary greatly with the type of soil. Mineral deficiencies determined by leaf analysis, are usually remedied by foliar spraying. Magnesium deficiency was formerly a serious handicap to avocado growers in Florida and Kenya. In California, zinc deficiency has been corrected by applying zinc chelates or zinc sulfate to the soil instead of spraying the foliage.

Keeping the upper soil moist has been greatly facilitated by drip irrigation, which also may carry 80% of the fertilizer requirement.

Because some cultivars tend to grow too tall for practical purposes, commercial growers cut trees back to 16 or 18 ft (4.8-5.4 m), let them grow back to 30 ft (9.1 m) and top them again. But decapitation is not a perfect remedy because the tendency of the avocado tree is to grow a new top very quickly. Recently it has been found that the growth-inhibiting chemical, TIBA (triiodobenzoic acid) slows down terminal growth and encourages lateral shoots. A system of pruning to encourage lower branching is being tried on 'Lula' in Martinique.

Avocado branches frequently need propping to avoid breaking with the weight of the developing fruits.

Some growers find it profitable to interplant bananas until the avocado trees reach bearing age.

Maturity and Harvesting

Avocados will not ripen while they are still attached to the tree, apparently because of an inhibitor in the fruit stem. Homeowners usually consider the entire crop pickable when a few mature (full grown) fruits have fallen. This is not a dependable guide because the prolonged flowering of the avocado results in fruits in varying stages of development on the tree at the same time. The largest fruits, of course, should be picked first but the problem is to determine when the largest are full grown (perfectly mature for later perfect ripening). If picked when full grown and firm, avocados will ripen in 1 to 2 weeks at

room temperature. If allowed to remain too long on the tree, the fruits may be blown down by wind and they will be bruised or broken by the fall.

Florida maturity standards for marketing have been determined by weight and time of year for each commercial cultivar so that immature fruits will not reach the market. Immature fruits do not ripen but become rubbery, shriveled and discolored. Most West Indian cultivars will ripen properly if picked when the specific gravity becomes 0.96 or lower, but 'Waldin' is fully mature when the specific gravity is still above 0.98. Guatemalan and Guatemalan X West Indian cultivars generally are harvest-mature when the specific gravity is 0.98 or lower. In California, physiological maturity of 'Bacon', 'Fuerte,' 'Hass' and 'Zutano' has been determined by measurement of length, diameter and volume, but dry weight, correlating with oil content, is considered a better maturity index. California law has, since 1925, required a minimum of 8% oil, but oil content varies greatly among cultivars and also the climatic region where the fruit is grown. Some people complain that the 8% standard is too low for some cultivars. Maximum flavor of 'Fuerte' develops when the fruit is harvested at an oil content of 16%. Therefore, a minimum dry weight standard of 21% has been recommended.

Formerly, avocados were detached by means of a forked stick and allowed to fall, but this causes much damage and loss. Nowadays harvesters usually use clippers for low-hanging fruits and for those higher up a long-handled picking pole with a sharp "V" on the metal rim to cut the stem and a strong cloth bag to catch the fruit. Gloves are worn to avoid fingernail scratches on the fruit. In California, studies have been made of the effects of hand-clipping (leaving stem on), hand-snapping (which removes the stem), tree-shaking, and limb-shaking (which removes the stem from some of the fruits). All methods are acceptable if the stem scar is waxed on stemless fruits to avoid weight loss before ripening at which time the stem detaches naturally. In Australia, some growers are using hydraulic lifts to facilitate hand-picking. A tractor fitted with a triple-decked picking platform has been adopted by some large growers in Chile. Efforts to develop dwarf avocado trees by means of sandwich interstocks from low-growing types have been going on in California since 1964.

Avocados must be handled with care and are packed and padded in single- or double-layer boxes or cartons for shipment. A special "Bruce box", holding 32 lbs (14.5 kg) is used for large fruit. The fruits may be held in position in molded trays.

Yield

It will be seen that the yield varies greatly with the cultivar, age of tree, the locale, weather and other conditions. The small tree, 'Ganter', has yielded 44 lbs (20 kg) annually; 'Nabal', 68 lbs (31 kg); 'Benik', 116 lbs (53 kg); 'Duke', 168 lbs (76 kg), and 'Anaheim', 220 lbs (100 kg). Close-planting in southern Florida provides yields averaging 11,000 lbs per acre (11,000 kg per ha) in young groves and nearly twice this amount is anticipated after the time has come to thin the planting by half.

Girdling has been tested in Florida, Australia and Israel as a means of increasing the yield of shy-bearing but popular cultivars. It must be repeated every year to be fully effective. It may decrease the yield of normally fruitful cultivars.

Marketing

Inasmuch as the avocado, outside of Latin America, has been widely regarded as a luxury fruit, large-scale marketing has been dependent on consumer education and advertising. Calavo Growers of California is an enterprising association of 2,600 avocado growers. The Mayflower Fruit Association, of which Blue Anchor is a member, packs over 60% of the avocados grown in the San Joaquin Valley. The California Avocado Commission spends millions of dollars in newspaper, magazine, television, radio and other publicity financed by grower assessments. The Florida Lime and Avocado Administrative Committees, together with the Florida Division of Marketing's Bureau of Market Expansion and Promotion, spend about ¼ million dollars annually for advertising and publicity through the Press and by means of special marketing displays and distribution of recipes. The trademarks, "Calavo" and "Flavocado" (Florida Avocado Growers Exchange), are recognized nationally and internationally.

The 8% oil standard established in California kept Florida avocados out of the California market until a court decision in 1972 outlawed the discrimination against Florida fruits which average about half the oil content of California cultivars and are advocated by growers as having better flavor and fewer calories. Calavo Growers Co-operative of California now handles 57% of the local avocado crop and 33% of the Florida crop, selling directly to the retail markets. Combined Florida and California efforts have raised the rate of regular avocado consumption in the United States from 6% in the late 1960's to over 15% today. In California, the Avocado Marketing Research Information Center was created in 1983 to gather and report information on production, foreign and domestic shipments and other activities.

Israel makes substantial investments in developing European markets for avocados and has attained the position of principal exporter to Europe. France and the United Kingdom are the chief consumers.

Storage

Ripening of avocados may be hastened by exposure to an atmosphere of at least 10 ppm ethylene 25 to 49 hours after harvest. The avocado does not respond to earlier treatment. Changes in pectinesterase activity and pectin content are being studied to measure ripening of avocados in storage. Dipping in latex has retarded decay in avocados stored at room temperature.

Avocados ship well and are sent to overseas markets under refrigeration in surface vessels. The fruits are subject to chilling injury (dark-brown or gray discoloration of the mesocarp) in refrigerated storage and degree of susceptibility varies with the cultivar and stage at harvesting and length of time in storage. Most commercial cultivars can be held safely at temperatures between 40° and 55°F (4.5°-12.8°C) for at least two weeks. The best ripening temperature after removal from storage is 60°F (15.5°C).

Removal of ethylene from controlled atmospheric storage (2% oxygen, 10% carbon dioxide) prolongs the marketable life of avocados. Reducing atmospheric pressure to subatmospheric 60 mm Hg in the refrigerated storage unit at 42.8°F (6°C) retards ripening of avocados by reducing respiration and ethylene production. Removed after 70 days, fruits have ripened normally at atmospheric pressure and 57.2°F (14°C). Experimental calcium treatments have delayed ripening and reduced internal chilling injury in storage but make the fruit externally less attractive and are, therefore, considered commercially undesirable.

'Hass' fruits dipped in fungicide 24 hours after harvest and sealed in polyethylene bags containing an ethylene absorbent (potassium permanganate on vermiculite or on aluminum silicate), have been successfully stored for 40 or 50 days at 50°F (10°C). Waxed 'Fuerte' avocados stored for 2 weeks at 41°F (5°C) and ripened at 68°F (20°C) ripened only 1 day later than non-waxed; however, waxing does reduce weight loss.

In 1965, to overcome the problem of oversupply during the harvesting season and undersupply during the off-season, California adopted liquid-nitrogen freezing of peeled or unpeeled avocado halves, which can be thawed and served as the equivalent of fresh fruits in restaurants, on airplanes and in institutions.

Pests and Diseases

Avocados have no major insect enemies in Florida but migrating cedar waxwings feed on leaves, flowers and very young fruits and the fruits are commonly attacked by squirrels, rats and mice. The avocado red mite, *Oligonychus yothersi*, is the most common predator on the leaves in some groves and not in others. Red-banded thrips, *Selenothrips rubrocinctus*, the greenhouse thrips, *Heliothrips haemorrhoidalis*, and red-spider, *Tetranychus mytilaspidis*, may feed on avocado leaves and blemish the fruits from time to time. There are several scales also which may feed on foliage, especially the Florida wax scale, *Ceroplastes floridensis*, the pyriform, or soft white, scale, *Protopulvinaria pyriformis*, Dictyospermum scale, *Chrysomphalus dictyospermi*, and the black scale, *Saissetia oleae*. Among two dozen other minor pests in Florida are the citrus mealybug, *Pseudococcus citri* and avocado mealybug, *P. nipae*. Stinkbugs may prick the fruits leaving little dents in the skin coupled with gritty areas at the same locations inside.

In California, 2 lepidopterous pests, *Amorbia cuneana* and the omnivorous looper, *Sabulodes aegrotata*, when present in large numbers, cause severe defoliation and fruit-scarring. Biological control is being achieved by release of the egg parasite, *Trichogramma platneri*, which is now commercially available to growers. Since 1949, the orange tortrix (a leaf roller), *Argyrotaenia citrana*, has been increasing as a menace to the avocado in California, the larvae feeding on twigs, terminal buds and foliage, flowers, and fruits. Since the pest requires shaded areas, it is best controlled by thinning out a close-planted grove or top-working to less susceptible cultivars.

The fruit-spotting bug, *Amblypelta nitida*, and banana spotting bug, *A. lutescens*, are important pests requiring control in Queensland. The Mediterranean fruit fly is a major hazard in Israel, but very thick-skinned fruits such as 'Anaheim' are not attacked. The Queensland fruit fly, *Dacus tryoni*, seriously damages only Mexican cultivars or Guatemalan X Mexican hybrids in Australia. In 1971, a nematode survey in Bahia, Brazil, revealed 9 genera of known or suspected parasitic nematodes associated with avocado tree decline. Israeli avocado growers are seeking and testing means of biological control of the more serious of the 3 dozen insects and mites preying on the crop in that country. In Mexico, the avocado weevil, *Heilipus lauri*, tunnels into the seeds.

The major disease of avocados in South and Central America and some islands of the West Indies, in California, Hawaii, and various other areas, is root-rot caused by the fungus, *Phytophthora cinnamomi*, which is being combatted by the use of strict sanitary procedures and resistant rootstocks, especially 'Duke'. At the University of California, Riverside, over 750 seedlings and cuttings were being tested for root-rot resistance in 1976 and 1977 and the most promising tried out for grafting compatibility with commercial cultivars. Also, soil fumigation experiments with methyl bromide and newly developed chemicals were being carried forward. The disease has been so devastating in the high rainfall areas of New South Wales and Queensland that plantings have expanded into the semi-arid Murray Valley in the hope of avoiding it. In New Zealand, it is not a problem on deep, volcanic soils, but occurs on shallow, heavier soils. It was allegedly introduced into Chile with balled trees from California and vigorous measures are being taken to control it.

Mushroom root-rot from *Clitocybe tabescens* may occasionally occur. Cercospora spot (brown spots on the leaves and fruits), caused by the fungus, *Cercospora purpurea*, may cause cracks in affected areas of the skin and thus allow entrance of the anthracnose fungus, *Colletotrichum gloeosporioides*, which invades and spoils the flesh. *Glomerella cingulata* is an important source of anthracnose in Queensland. Some cultivars are subject to scab which is readily controlled by copper sprays.

More than 30 other pathogens are variously responsible for wood rot, collar rot, dieback, leafspot, stem-end rot of fruit, branch canker, and powdery mildew. Sunblotch viroid cripples young trees and damages fruits in California and Israel. So far, it is unknown in New Zealand. Stems of young trees may be affected by sunburn, and hot, dry winds cause tipburn of leaves. The avocado tree may show copper or zinc deficiency or tipburn from an excess of mineral salts.

Food Uses

Indians in tropical America break avocados in half, add salt and eat with tortillas and a cup of coffee—as a complete meal. In North America, avocados are primarily served as salad vegetables, merely halved and garnished with seasonings, lime juice, lemon juice, vinegar, mayon-

naise or other dressings. Often the halves are stuffed with shrimp, crab or other seafood. Avocado flesh may be sliced or diced and combined with tomatoes, cucumbers or other vegetables and served as a salad. The seasoned flesh is sometimes used as a sandwich filling. Avocado, cream cheese and pineapple juice may be blended as a creamy dressing for fruit salads.

Mexican guacamole, a blend of the pureed flesh with lemon or lime juice, onion juice or powder, minced garlic, chili powder or Tabasco sauce, and salt and pepper has become a widely popular "dip" for crackers, potato chips or other snacks. The ingredients of guacamole may vary and some people add mayonnaise.

Because of its tannin content, the flesh becomes bitter if cooked. Diced avocado can be added to lemon-flavored gelatin after cooling and before it is set, and chunks of avocado may be added to hot foods such as soup, stew, chili or omelettes just before serving. In Guatemalan restaurants, a ripe avocado is placed on the table when a hot dish is served and the diner scoops out the flesh and adds it just before eating. For a "gourmet" breakfast, avocado halves are warmed in an oven at low heat, then topped with scrambled eggs and anchovies.

In Brazil, the avocado is regarded more as a true fruit than as a vegetable and is used mostly mashed in sherbet, ice cream, or milk-shakes. Avocado flesh is added to heated ice cream mixes (such as boiled custard) only after they have cooled. If mashed by hand, the fork must be a silver one to avoid discoloring the avocado. A New Zealand recipe for avocado ice cream is a blend of avocado, lemon juice, orange juice, grated orange rind, milk, cream, sugar and salt, frozen, beaten until creamy, and frozen again.

Some Oriental people in Hawaii also prefer the avocado sweetened with sugar and they combine it with fruits such as pineapple, orange, grapefruit, dates, or banana.

In Java, avocado flesh is thoroughly mixed with strong black coffee, sweetened and eaten as a dessert.

Avocado slices have been pickled and marketed in glass jars. California began marketing frozen guacamole in 1951, and a frozen avocado whip, developed at the University of Miami, was launched in 1955. To help prevent enzymatic browning of these products, it is recommended that sodium bisulfite and/or ascorbic acid be mixed in before freezing.

Avocado Oil

Oil expressed from the flesh is rich in vitamins A, B, G and E. It has a digestibility coefficient of 93.8% but has remained too costly to be utilized extensively as salad oil. The amino acid content has been reported as: palmitic, 7.0; stearic, 1.0; oleic, 79.0; linoleic, 13.0.

The oil has excellent keeping quality. Samples kept in a laboratory in Los Angeles at 40°F (4.4°C) showed only slight rancidity after 12 years. There is much interest in the oil in Italy and France. The Institut Francais de Recherches Fruitieres Outre Mer has studied the yield of oil in 25 cultivars. Joint Italian/Venezuelan studies of 5 prominent cultivars indicated that the fatty acid composition and tryglyceride structure was not influenced by

Food Value Per 100 g of Edible Portion (*Flesh*)*

Moisture	65.7–87.7 g
Ether Extract	5.13–19.80 g
Fiber	1.0–2.1 g
Nitrogen	0.130–.382 g
Ash	0.46–1.68 g
Calcium	3.6–20.4 mg
Phosphorus	20.7–64.1 mg
Iron	0.38–1.28 mg
Carotene	0.025–0.475 mg
Thiamine	0.033–0.117 mg
Riboflavin	0.065–0.176 mg
Niacin	0.999–2.220 mg
Ascorbic Acid	4.5–21.3 mg

*Analyses of West Indian, Guatemalan and Mexican avocados marketed in Central America.

Browning of the flesh of freshly cut avocado fruits is caused by polyphenol oxidase isoenzymes. Avocado halves average only 136 to 150 calories.

The avocado has a high lipid content—from 5 to 25% depending on the cultivar. Among the saturated fatty acids, myristic level may be .1%, palmitic, 7.2, 14.1 or 22.1%; stearic, 0.2, 0.6 or 1.7%. Of the unsaturated fatty acids, palmitoleic may range from 5.5 to 11.0%; oleic may be 51.9, 70.7 or 80.97%, linoleic, 9.3, 11.2 or 14.3%. Non-saponifiable represents 1.6 to 2.4%. Iodine number is 94.4. In feeding experiments which excluded animal fat, 16 patients were given $\frac{1}{2}$ to $1\frac{1}{2}$ avocados per day. Total serum cholesterol and phospholipid values in the blood began to fall in one week. Body weight did not increase. Cholesterol values did not rise and 8 patients showed decreases in total serum cholesterol and phospholipids.

Amino acids of the pulp (N = 16 p. 100) are recorded as: arginine, 3.4; cystine, 0: histidine, 1.8; isoleucine, 3.4; leucine, 5.5; lysine, 4.3; methionine, 2.1; phenylalanine, 3.5; threonine, 2.9; tryptophan, 0; tyrosine, 2.3; valine, 4.6; aspartic acid, 22.6; glutamic acid, 12.3; alanine, 6.0; glycine, 4.0; proline, 3.9; serine, 4.1.

variety. The oil is used as hair-dressing and is employed in making facial creams, hand lotions and fine soap. It is said to filter out the tanning rays of the sun, is non-allergenic and is similar to lanolin in its penetrating and skin-softening action. In Brazil, 30% of the avocado crop is processed for oil, $\frac{2}{3}$ of which is utilized in soap, $\frac{1}{3}$ in cosmetics. The pulp residue after oil extraction is usable as stockfeed.

Toxicity

Unripe avocados are said to be toxic. Two resins derived from the skin of the fruit are toxic to guinea pigs by subcutaneous and peritoneal injection. Dopamine has been found in the leaves. The leaf oil contains methyl chavicol. Not all varieties are equally toxic. Rabbits fed on leaves of 'Fuerte' and 'Nabal' died within 24 hours. Those fed on leaves of 'Mexicola' showed no adverse reactions. Ingestion of avocado leaves and/or bark has caused mastitis in cattle, horses, rabbits and goats. Large doses have been fatal to goats. Craigmill *et al.* at Davis, California, have confirmed deleterious effects on lactating goats which were allowed to graze on leaves of 'Anaheim' avocado an hour each day for 2 days. Milk was curdled and not milkable, the animals ground their teeth, necks were swollen and they coughed, but the animals would still accept the leaves on the 4th day of the experiment. By the 10th day, all but one goat were on the road to recovery. All abnormal signs had disappeared 20 days later. In another test, leaves of a Guatemalan variety were stored for 2 weeks in plastic bags and then given to 2 Nubian goats in addition to regular feed over a period of 2 days. Both suffered mastitis for 48 hours. Avocado leaves in a pool have killed the fish. Canaries have died from eating the ripe fruit.

The seeds, ground and mixed with cheese or cornmeal, have been used to poison rodents. However, tests in Hawaii did not show any ill effect on a mouse even at the rate of $\frac{1}{4}$ oz (7 g) per each 2.2 lbs (1 kg) of body weight, though the mouse refused to eat the dried, grated seed material until it was blended with cornmeal. Avocado seed extracts injected into guinea pigs have caused only a few days of hyperexcitability and anorexia. At Davis, mice given 10 to 14 g of half-and-half normal ration and either fresh or dried avocado seed died in 2 or 3 days, though one mouse given 4 times the dose of the others survived for 2 weeks.

The seed contains 13.6% tannin, 13.25% starch. Amino acids in the seed oil are reported as: capric acid, 0.6; myristic, 1.7; X, 13.5; palmitic, 23.4; X, 10.4; stearic, 8.7; oleic, 15.1; linoleic, 24.1; linolenic, 2.5%. The dried seed contains 1.33% of a yellow wax containing sterol and organic acid. The seed and the roots contain an antibiotic which prevents bacterial spoilage of food. It is the subject of two United States patents.

The bark contains 3.5% of an essential oil which has an anise odor and is made up largely of methyl chavicol with a little anethole.

Other Uses

The seed yields a milky fluid with the odor and taste of almond. Because of its tannin content, it turns red on exposure, providing an indelible red-brown or blackish ink which was used to write many documents in the days of the Spanish Conquest. These are now preserved in the archives of Popayán. The ink has also been used to mark cotton and linen textiles.

In Guatemala, the bark is boiled with dyes to set the color.

Much avocado wood is available when groves are thinned out or tall trees are topped. The sapwood is cream-colored or beige; the heartwood is pale red-brown, mottled, and dotted with small drops of gummy red sap; fine-grained; light—40 lbs per cu ft—(560-640 kg/cu m); moderately soft but brittle; not durable; susceptible to drywood termites and fungi. The wood has been utilized for construction, boards and turnery. An Australian woodworker has reported that it is suitable for carving, resembles White Beech (*Eucalyptus kirtonii*); is easy to work, and dresses and polishes beautifully. He has made it into fancy jewel boxes. It probably requires careful seasoning. A Florida experimenter made bowls of it but they cracked.

Honeybees gather a moderate amount of pollen from avocado flowers. The nectar is abundant when the weather is favorable. When unmixed by that from other sources it produces a dark, thick honey favored by those who like buckwheat honey or sugarcane sirup.

Medicinal Uses: The fruit skin is antibiotic; is employed as a vermifuge and remedy for dysentery. The leaves are chewed as a remedy for pyorrhea. Leaf poultices are applied on wounds. Heated leaves are applied on the forehead to relieve neuralgia. The leaf juice has antibiotic activity. The aqueous extract of the leaves has a prolonged hypertensive effect. The leaf decoction is taken as a remedy for diarrhea, sore throat and hemorrhage; it allegedly stimulates and regulates menstruation. It is also drunk as a stomachic. In Cuba, a decoction of the new shoots is a cough remedy. If leaves, or shoots of the purple-skinned type, are boiled, the decoction serves as an abortifacient. Sometimes a piece of the seed is boiled with the leaves to make the decoction.

The seed is cut in pieces, roasted and pulverized and given to overcome diarrhea and dysentery. The powdered seed is believed to cure dandruff. A piece of the seed, or a bit of the decoction, put into a tooth cavity may relieve toothache. An ointment made of the pulverized seed is rubbed on the face as a rubefacient—to redden the cheeks. An oil extracted from the seed has been applied on skin eruptions.

Related Species

Persea schiedeana Nees, called *coyo, coyocte, chalte, chinini, chucte, chupte, cotyo, aguacate de monte, aguacaton,* wild pear, and *yas,* grows wild in mountain forests from southern Mexico to Panama at altitudes between 4,600 and 6,200 ft (1,400-1,900 m). The tree is usually from 50 to 65 ft (15-20 m) tall, occasionally to 165 ft (50 m). Young branches are densely brown-hairy. The leaves are deciduous, obovate to oval, often cordate at the base; 5 to 12 in (12.5-30 cm) long, 2¾ to 6 in (7-15 cm) wide, white-hairy on the underside. Downy flowers, borne in densely grayish-hairy panicles, are light greenish-yellow, the perianth and stamens turning red with age. The fruit, resembling that of the avocado and equally variable, is generally pear-shaped, weighing 8 to 14 oz (227-397 g), with thick, leathery, flexible skin. Variously described as brownish-white, light-brown, pale-green, greenish-brown or dark brown, the flesh is oily with a milky juice, few to many coarse fibers, but a very appealing, avocado-coconut flavor. The seed is very large. The cotyledons, unlike those of the avocado, are pink internally.

The tree is left standing when forests are cleared and is cultivated in Veracruz and on some farms in Guatemala. The fruits from the best of the wild and cultivated trees are marketed locally. The timber is used in construction and carpentry. This species was introduced into the USA from Guatemala and Honduras in 1948 as a wilt-resistant rootstock for the avocado. It is very sensitive to frost. In 1974 it was reported to be a poor bearer in Puerto Rico.

A more distant relative is *Beilschmiedia anay* Kosterm. (*Hufelandia anay* Blake), called *anay, payta, escalalan* or *excalan,* which is native to moist, relatively low altitudes, 985 to 2,300 ft (300 to 700 m) in southern Mexico, Guatemala, Costa Rica and Colombia. Seeds were collected by Dr. Wilson Popenoe in 1917 and seedlings were set out in the Plant Introduction Garden of the U.S. Department of Agriculture, Miami.

The tree attains a height of 66 ft (20 m); the young branches are brown-hairy. Leathery leaves, broad-elliptic or broad-ovate, are 4¾ to 12 in (12-30 cm) long and 3 to 7½ in (7.5-19 cm) wide, white-hairy only on the veins. The flowers (in December and January) are fragrant, greenish, in slender panicles to 5 in (13 cm) long. The fruit is ellipsoid-pyriform, 2¾ to 6 in (7-15 cm) long, with very thin, glossy, purplish-black skin and sparse green, oily flesh similar to that of the avocado in texture and flavor. The seed is obovoid, up to 2¾ in long, with thick, purplish-yellow, red-spotted coat, and strong almond odor. In Guatemala, the fruit matures in August and September, falls while hard, and ripens in 2 or 3 days. Analyses in Guatemala show (per 100 g/flesh): moisture, 73.86 g; protein, 1.62-1.80 g; carbohydrates, 3.32-3.90 g; fat, 12.98-17.44 g; cellulose, 2.12 g; ash, 1.38 g.

Food Value Per 100 g of Edible Portion (*flesh*)*	
Moisture	76.5-77.6 g
Ether Extract	5.55-7.59 g
Fiber	1.0-1.8 g
Nitrogen	0.191-0.204 g
Ash	0.72-0.91 g
Calcium	11.4-12.5 mg
Phosphorus	35.5-36.2 mg
Iron	0.31-0.35 mg
Carotene	0.003-0.033 mg
Thiamine	0.048-0.070 mg
Riboflavin	0.067-0.089 mg
Niacin	0.598-0.718 mg
Ascorbic Acid	5.7-16.4 mg

*Analyses by Munsell *et al.*

Loquat (Plate XI)

A fruit of wide appeal, the loquat, *Eriobotrya japonica* Lindl., (syn. *Mespilus japonicus* Thunb.), of the rose family, Rosaceae, has been called Japan, or Japanese, plum and Japanese medlar. To the Italians, it is *nespola giapponese;* to French-speaking people, it is *néflier du Japon,* or *bibassier.* In the German language, it is *japanische mispel,* or *wollmispel;* in Spanish, *nispero, nispero japonés,* or *nispero del Japón;* in Portuguese, *ameixa amarella,* or *ameixa do Japao.*

Description

A tree of moderate size, the loquat may reach 20 to 30 ft (6-9 m), has a rounded crown, short trunk, and woolly new twigs. The evergreen leaves, mostly whorled at the branch tips, are elliptical-lanceolate to obovate lanceolate, 5 to 12 in (12.5-30 cm) long and 3 to 4 in (7.5-10 cm) wide; dark-green and glossy on the upper surface, whitish- or rusty-hairy beneath, thick, stiff, with conspicuous parallel, oblique veins, each usually terminating at the margin in a short, prickly point. Sweetly fragrant flowers, borne in rusty-hairy, terminal panicles of 30 to 100 blooms, are white, 5-petalled, ½ to ¾ in (1.25-2 cm) wide. The fruits, in clusters of 4 to 30, are oval, rounded or pear-shaped, 1 to 2 in (2.5-5 cm) long, with smooth or downy, yellow to orange, sometimes red-blushed, skin, and white, yellow or orange, succulent pulp, of sweet to subacid or acid flavor. There may be 1 to 10 seeds, though, ordinarily, only 3 to 5, dark-brown or light-brown, angular-ellipsoid, about ⅝ in (1.5 cm) long and 5/16 in (8 mm) thick.

Origin and Distribution

The loquat is indigenous to southeastern China and possibly southern Japan, though it may have been introduced into Japan in very early times. It is said to have been cultivated in Japan for over 1,000 years. The western world first learned of it from the botanist Kaempfer in 1690. Thunberg, who saw it in Japan in 1712, provided a more elaborate description. It was planted in the National Gardens, Paris, in 1784 and plants were taken from Canton, China, to the Royal Botanical Gardens at Kew, England, in 1787. Soon, the tree was grown on the Riviera and in Malta and French North Africa (Algeria) and the Near East and fruits were appearing on local markets. In 1818, excellent fruits were being produced in hothouses in England. The tree can be grown outdoors in the warmest locations of southern England.

Cultivation spread to India and southeast Asia, the medium altitudes of the East Indies, and Australia, New Zealand and South Africa. Chinese immigrants are assumed to have carried the loquat to Hawaii.

In the New World, it is cultivated from northern South America, Central America and Mexico to California; also, since 1867, in southern Florida and northward to the Carolinas, though it does not fruit north of Jacksonville. It was quite common as a small-fruited ornamental in California gardens in the late 1870's. The horticulturist, C.P. Taft, began seedling selection and distributed several superior types before the turn of the century, but further development was slow. Dwarfing on quince rootstocks has encouraged expansion of loquat cultivation in Israel since 1960. In the northern United States and Europe, the tree is grown in greenhouses as an ornamental, especially var. *variegata* with white and pale-green splashes on the leaves.

In India and many other areas, the tree has become naturalized, as it volunteers readily from seed. Japan is the leading producer of loquats, the annual crop amounting to 17,000 tons. Brazil has 150,000 loquat trees in the State of Sao Paulo.

Varieties

The loquat has been the subject of much horticultural improvement, increasing the size and quality of the fruit. There are said to be over 800 varieties in the Orient. T. Ikeda catalogued 46 as more or less important in Japan; over 15 have originated in Algeria through the work of L. Trabut; C.P. Taft selected and introduced at least 8 into cultivation in California; 5 or 6 have been selected in Italy; only 1 in Florida. A number of widely planted, named cultivars have been classed as either "Chinese" or "Japanese". In the Chinese group, the trees have slender leaves, the fruit is pear-shaped or nearly round with thick, orange skin and dark-orange flesh, not very juicy, subacid, but of distinct flavor. The seeds are small and numerous. The harvesting period is midseason to late and the fruits are of good keeping quality.

In the Japanese group, the tree has broad leaves, the fruit is pear-shaped or long-oval, the skin is usually pale-yellow, the flesh whitish, very juicy, acid but otherwise not very distinct in flavor. The seeds are large and there may be just a few or only one. The harvesting period is early to midseason. Keeping quality is fair to poor.

In Egypt, most loquats are of Lebanese origin. Egyptian horticulturists have selected from seedlings of 'Premier' 2 superior clones, 'Golden Ziad' and 'Maamora Golden Yellow' and have vegetatively propagated them on quince rootstocks for commercial distribution.

Some of the oldtime selections, 'Advance', 'Champagne', 'Premier', 'Success' and 'Tanaka' are no longer popular in California but are performing well in other areas. In Florida, 'Oliver' has always been the most common cultivar, though a number of others—'Advance', 'Champagne', 'Early Red', 'Pineapple', 'Premier', 'Tanaka' and 'Thales' have been more or less successful.

In the State of Sao Paulo, Brazil, 2 cultivars are raised on a commercial scale—'Precoce de Itaquera' and 'Mizuho'. In the southernmost state of the U.S.S.R., Georgia, several loquat cultivars are grown, including 'Champagne', 'Comune', 'Grossa de Sicilia', 'Premier', 'Tanaka', and 'Thales'.

The following are the cultivars most commonly described:

'Advance' (Japanese group)—A seedling selected by C.P. Taft in California in 1897. Fruit is borne in large clusters; pear-shaped to elliptic-round; of medium to large size; skin downy, yellow, thick and tough; flesh thick, cream-colored, juicy, subacid, of excellent flavor. Seeds of medium size, may be as many as 4 or 5; average is 3.20 per fruit. A late cultivar though it ripens earlier than 'Champagne' which it otherwise closely resembles. Tree is a natural dwarf, to a little over 5 ft (1.58 m); is highly resistant to pear blight. Self-infertile; a good pollinator for other cultivars. It is interplanted with 'Golden Yellow' and 'Pale Yellow' in India.

'Ahdar' (Lebanese; grown in India)—oval, of medium size; greenish-yellow with white flesh; bears moderately; late-ripening; of poor keeping quality.

'Ahmar' (Lebanese; grown in India)—pear-shaped, large, with reddish-orange skin; yellow flesh, firm, juicy; early ripening; of good keeping quality. A leading cultivar in Lebanon. Very precocious. Self-infertile.

'Akko 1' or **'Acco 1'** (of Japanese origin)—long-oval to pear-shaped, 20 to 25 g in weight; skin orange with a little russeting, thick; flesh yellow, juicy, of average flavor, and there are 3 or 4 seeds. Ripens in midseason, beginning in mid-April in Israel where it constitutes 10 to 20% of commercial plantings. Precocious and a good bearer; sets 20 to 30 fruits per cluster and requires drastic thinning, leaving about 6 fruits. Fruit is subject to sunburn. Stands harvesting and shipping well, keeps in good condition less than 2 weeks under refrigeration. This cultivar is self-fertile.

'Akko 13' or **'Acco 13'** (of Japanese origin)—pear-shaped, 20 to 25 g in weight; dark-orange, with no russeting; flesh yellow, juicy, with acid, agreeable flavor; 2 or 3 seeds. Bears from end of March through April in Israel, regularly and abundantly; constitutes 50 to 70% of commercial plantings in Israel; of good handling and keeping quality; stands transportation for 2 weeks at 32°F (0.0°C). Fruit is subject to sunburn. Needs cross-pollination.

'Asfar' (Lebanese, grown in India)—oval, smaller than 'Ahmar', with yellow skin and flesh, very juicy, of superior flavor, but very perishable.

'Blush' ('Red Blush')—Resembles 'Advance' but is very large. Was selected by C.P. Taft as being immune to blight,

but was abandoned after 'Advance' proved to be highly blight-resistant.

'Champagne' (Japanese), often misidentified as 'Early Red'. Selected and introduced into cultivation in California by C.P. Taft around 1908. Elongated pear-shaped, often oblique; small to large (depending on where it is grown); skin pale-golden to deep-yellow, thick, tough, astringent; flesh white or yellow, soft, juicy, mild and subacid to sweet; of excellent flavor. There are 3 to 5 seeds. Midseason to late. Prolific; fruits borne in large clusters. Perishable; good for preserving. Tree has long, narrow, pointed leaves; is self-infertile.

'Early Red' (Japanese); originated by Taft in 1909. Obliquely pear-shaped; medium-large; skin orange-red with white dots, thick, tough, acid; flesh orange, very juicy, sweet, of fair to excellent flavor; has 2 or 3 seeds. Earliest in season, often appearing on California markets at the end of January or in the beginning of February. Borne in compact clusters.

'Eulalia' (a seedling of 'Advance' selected by M. Payan in California in 1905)—pear-shaped to obovate-pear-shaped; skin faintly downy, orange-yellow with red blush and pale gray dots, thick, tough; flesh pinkish or orange, melting, soft, very juicy; subacid in flavor. Seeds medium in size, numerous. Early in season.

'Fire Ball' (popular in India)—ovate to ovate-elliptic; small, with yellow, thick skin; flesh white to straw-colored, thick, crisp, smooth, of mild, subacid flavor. Seeds are large; average 2.90 per fruit. Midseason. Tree is a natural dwarf to 9.5 ft (2.84 m).

'Glenorie Superb' (grown in Western Australia)—round, large, dark-orange with yellow flesh which is juicy and sweet. Somewhat late in season. Inclined to bruise during harvesting.

'Golden Red' (grown in California)—flesh pale-orange, medium-thick, smooth, melting, of subacid, agreeable taste; few seeded. Midseason.

'Golden Yellow' (grown in India)—ovate-elliptic; of medium size; skin orange-yellow; flesh pale-orange, medium-thick, soft, smooth, with subacid, mild flavor. Seeds of medium size; average 4.83 per fruit.

'Golden Ziad' (#2-6) (grown in Egypt)—dark-yellow to light-orange; up to 1½ in (3.96 cm) long; average number of seeds, 2.93-3.83 per fruit. Early. High-yielding; 50 lbs (23.5 kg) per tree.

'Herd's Mammoth' (grown in Western Australia)—long and slightly tapering at the stem end; large; yellow to orange with white to cream-colored flesh. Ripens earlier than 'Victory'. Subject to black spot; not often planted.

'Improved Golden Yellow' (grown in India)—ovate-elliptic; skin orange-yellow; flesh orange-yellow, thick, crisp, smooth, with subacid to sweet, mild flavor. Seeds large; average 3.06 per fruit. Tree to 15 ft (4.49 m). Early.

'Improved Pale Yellow' (grown in India)—flesh pale-orange or cream-colored, firm or soft, smooth, of subacid, pleasant flavor, with medium number of seeds. Midseason.

'Kusunoki' (grown in Japan)—small; early.

'Large Agra' (grown in India)—ovate-round; of medium size; skin deep-yellow; flesh yellow or pale-orange, medium thick, smooth, firm, of pleasant flavor, fairly sweet. Seeds small; average 5.10 per fruit. Midseason. Tree a medium-dwarf—to 9½ ft (2.83 m).

'Large Round' (grown in India)—ovate-round; of medium size; yellow of skin with cream-colored flesh, firm, coarse, subacid to sweet, mild. Seeds of medium size; average 4.80 per fruit. Midseason. Tree fairly tall—13 ft. (3.92 m).

'Maamora Golden Yellow' (#7-9) (grown in Egypt)—dark-yellow to light-orange; to 1½ in (3.91 cm long); seeds average

2.40 to 4.03 per fruit; late in season. High-yielding—44 lbs (20 kg) per tree.

'Mammoth' (grown in Australia; mentioned in California in 1889)—flesh orange, medium-thick, granular, coarse, of subacid, agreeable flavor. Midseason.

'Matchless' (grown in India)—pear-shaped; flesh medium-thick, pale-orange, smooth, soft, of mild, subacid flavor; medium number of seeds. Midseason.

'Mizuho' (grown in Japan)—rounded-oval; extra large (70-120 g); juicy, with agreeable, slightly acid though also sweet flavor, and with 5 or more seeds. Subject to fruit spots and sunburn.

'Mogi' (grown in Japan)—elliptical, light-yellow; small (40-50 g); Ripens in early spring. Tree is cold-sensitive. Self-fertile. Constitutes 60% of the Japanese crop of loquats.

'Obusa' (a hybrid of 'Tanaka' and 'Kusonoki', developed and grown in Japan)—deep-yellow, very large (80-100 g); of medium flavor; good keeping and shipping quality. Ripens earlier than Tanaka. Tree bears regularly and is resistant to insects and diseases, but fruit is subject to sunburn (purple stains on skin).

'Oliver' ('Olivier' X 'Tanaka'). In the past was considered the best loquat for southern Florida.

'Pale Yellow' (grown in India)—oblique-elliptic to round; light yellow, large; flesh white or cream-colored, thin, smooth, melting, of subacid to sweet flavor; seeds large; average 4.8 per fruit. Early. Tree is fairly tall—to 13 ft (4 m).

'Pineapple' (developed and introduced into cultivation in California by Taft in 1899)—round or sometimes pear-shaped; light-yellow with white flesh. Of good quality but inferior to 'Champagne'. Abandoned in California because of the weakness of the tree.

'Precoce de Itaquera' (erroneously called 'Tanaka'; grown in Brazil; believed to be a local selection of 'Mogi')—oval-pear-shaped; deep-orange; very small (25.3-29.1 g). Flesh is firm and acid-sweet. Very productive: 1,500 to 2,000 fruits per tree annually. Subject to sunburn (purple stains on skin) but less so than 'Mizuho'. Was for a long time the leading cultivar in the State of Sao Paulo but has lost ground to 'Mizuho' even though a pear-shaped fruit is preferred by consumers, because it does not keep or ship as well as the 'Mizuho', which now makes up 65% of the plantings and 'Precoce de Itaquera' 35%.

'Premier' (originated by Taft in California in 1899)—oval to oblong-pear-shaped; large; skin downy, orange-yellow to salmon-orange with large white dots; medium-thick, tough; flesh whitish, melting, juicy, subacid, of agreeable flavor; seeds average 4 or 5 per fruit. Late. Good for dooryards. Does not ship well, nor keep well.

'Safeda' (grown in India)—flesh is cream-colored, thick, smooth and melting, of subacid, excellent flavor; contains medium number of seeds. Early to midseason.

'Saint Michel' (unclassified; grown in Israel)—round but has the thin skin and white flesh of the Japanese group. Ripens late. Self-infertile.

'Swell's Enormity' (grown in Western Australia)—pear-shaped, very large; deep apricot-colored externally with flesh of the same color. Acid if harvested too early. Very late in season. Subject to sunburn in hot weather.

'Tanaka' (Chinese group; a seedling originated in Japan; young trees introduced by the United States Department of Agriculture in 1902; widely grown)—ovoid or round; large (70-80 g) in Japan; in some other areas small (30 g); skin orange or orange-yellow; flesh brownish-orange, medium thick, coarse, firm, juicy, sweet or subacid, of excellent taste. There may be 2 to 4 seeds; average 2.70 per fruit. Ripens late—

beginning the first of May, which is too late for California because of susceptibility to sunburn. The tree is of medium size—nearly 10 ft (2.98 m); precocious; bears regularly; is self-fertile to a degree. Constitutes 10% of commercial crop in Israel; 35% of the crop in Japan. Highly cold-tolerant.

'Thales', also known as 'Gold Nugget' and 'Placentia', (Chinese group; very similar to 'Tanaka' and possibly a clone. Introduced from Japan and planted at Placentia, California, between 1880 and 1900)—oblong-obovate to round, large, skin orange-yellow with numerous white dots, tough; flesh orange, thick, firm, juicy, of sweet, apricot-like flavor. There are 2 to 4 seeds. Late in season. Fruits borne only a few to a cluster; keep and ship well. Self-fertile.

'Thames Pride' (grown in India)—ovate-elliptic, of medium size or sometimes large; pale-orange or deep-yellow with cream-colored or pale-orange, juicy, coarse, somewhat granular flesh of subacid flavor; moderately seedy; average 3.20 seeds per fruit. Early in season. Tree tall, to 13½ ft (4.19 m). Bears heavily. This cultivar is grown and canned commercially.

'Tsrifin 8' (grown in Israel)—rounded pear-shaped; 25 to 30 g in weight; yellow-orange with some russeting. Of excellent quality with good acid and sugar content. Stands handling, shipping and storage well. Late—mid-April to mid-May. Precocious, bears regularly and abundantly but is subject to sunburn. Constitutes 10% of Israeli plantings.

'Victor' (originated by C.P. Taft in 1899)—oblong-pear-shaped; large; skin deep-yellow, medium-thick, tough. Flesh whitish, translucent, melting, very juicy, of sweet, mild flavor. There may be 3 to 5 seeds. Very late; too late for California. Good for canning.

'Victory' (the most popular cultivar in West Australia)—oval, large, yellow to orange, becoming amber on the sunny side. Flesh is white to cream-colored, juicy, sweet. Midseason to occasionally early.

'Wolfe', (S.E.S. #4) (a seedling of 'Advance' selected and named at the Agricultural Research and Education Center of the University of Florida in Homestead, and released in 1966)—obovoid to slightly pear-shaped; 1¾ to 2 in (4.5-5 cm) long and 1 to 1¼ in (2.5-3.2 cm) wide; yellow with fairly thick skin and pale-yellow, thick, firm, juicy flesh of excellent flavor, acid but also sweet when tree-ripe; has 1 to 5 seeds (usually 1 to 3). Tree reaches 25 ft (7.5 m) and bears well nearly every year.

Pollination

The loquat is normally pollinated by bees. Some cultivars such as 'Golden Yellow' are not self-fertile. 'Pale Yellow', 'Advance', and 'Tanaka' are partially self-fertile. In India, it has been observed that cross-pollination generally results in 10-17% increased production over self-pollination. 'Tanaka' pollinated by 'Pale Yellow' has a lower yield than when self-pollinated, indicating a degree of cross-incompatibility. Whereas, when pollinated by 'Advance', the normal yield of 'Tanaka' is nearly doubled.

When cross-pollinating for the purpose of hybridizing, only flowers of the second flush should be used, as early and late flushes have abnormal stamens, very little viable pollen, and result in poor setting and undersized fruits.

Climate

The loquat is adapted to a subtropical to mild-temperate climate. In China it grows naturally at altitudes between 3,000 and 7,000 ft (914-2,100 m). In India, it grows at all levels up to 5,000 ft (1,500 m). In Guatemala, the tree thrives and fruits well at elevations between 3,000 and 6,900 ft (900-1,200 m), but bears little or not at all at lower levels.

Well-established trees can tolerate a drop in temperature to 12°F (-11.11°C). In Japan, the killing temperature for the flower bud is 19.4°F (7°C); for the mature flower, 26.6°F (-3°C). At 25°F (-3.89°C), the seed is killed, causing the fruit to fall.

Loquats are grown on hillsides in Japan to have the benefit of good air flow. Extreme summer heat is detrimental to the crop, and dry, hot winds cause leaf scorch. Where the climate is too cool or excessively warm and moist, the tree is grown as an ornamental but will not bear fruit.

Soil

The tree grows well on a variety of soils of moderate fertility, from light sandy loam to heavy clay and even oolitic limestone, but needs good drainage.

Propagation

Generally, seeds are used for propagation only when the tree is grown for ornamental purposes or for use as rootstock. Loquat seedlings are preferred over apple, pear, quince or pyracantha rootstocks under most conditions. Quince and pyracantha may cause extreme dwarfing — to less than 8 ft (2.5 m). Quince rootstock tolerates heavier and wetter soils than loquat but is apt to put out numerous suckers. Loquat seeds remain viable for 6 months if stored in partly sealed glass jars under high humidity at room temperature, but the best temperature for storage is 40°F (5°C). They are washed and planted in flats or pots soon after removal from the fruit and the seedlings are transplanted when 6 to 7 in (15-17.5 cm) high to nursery rows. When the stem is ½ in (1.25 cm) thick at the base, the seedlings are ready to be topworked. In India, inarching is commonly practiced but budding and grafting are more popular in most other areas. Shield-budding, using 3-month-old scions, is successful. Cleft-grafting has been a common practice in Florida. Veneer-grafting in April has proved to be a superior method in Pakistan. Cuttings are not easy to root. Air-layering may be only 20% successful, though 80 to 100% of the layers root in 6 weeks if treated with 3% NAA (2-naphthoxyacetic acid).

Trees that are vegetatively propagated will begin to bear fruit in 5 years or less, as compared to 8 to 10 years in seedling trees. Old seedling trees can be converted by cutting back severely and inserting budwood of a preferred cultivar.

Culture

The rainy season is best for planting loquats. When planted on rich soil, normal size trees should be set 25 to 30 ft (7.5-9 m) apart, allowing about 83 trees per acre (200 per ha). In Brazil, a spacing of 23 x 23 ft (7x7 m) is recommended on flat land, 26 x 20 ft (8x6 m) or 26 x 16.5 ft (8x5 m) on slopes. Dwarf trees are spaced at 13 x 6.5 ft (4x2 m) in Japan and this may allow 208 per acre (500 per ha). The tree is a heavy feeder. For good fruit production the trees require ample fertilization and irrigation. In the tropics, animal manure is often used. A good formula for applications of chemical fertilizer is: 1 lb (.45 kg) 6-6-6 NPK three times a year during the period of active growth for each tree 8 to 10 ft in height. The trees should be watered at the swelling of blossoms and 2 to 3 waterings should be given during harvest-time. Thinning of flowers and young fruits in the cluster, or the clipping off of the tip of the cluster, or of entire clusters of flowers and fruits, is sometimes done to enhance fruit size. This is carefully done by hand in Japan. With the 'Tanaka' cultivar, the Japanese leave only one fruit per cluster; with the 'Mogi', two. In Taiwan, thinning is done by spraying with NAA when the flowers are fully open.

In Taiwan, because of the hazard of strong typhoons, the loquat is grown as a mini-dwarf no more than 3 ft (0.9 m) high and wide, and branch tips may be tied to the ground because branches kept at a 45° angle flower heavily. Spraying with gibberellic acid (60 ppm) at full bloom enhances fruit set and increases fruit size and weight, total reducing sugars and ascorbic acid content, reduces fruit drop, number of seeds, and acidity. Spraying the same at 300 ppm results in small, seedless fruits. There should be judicious pruning after harvest, otherwise terminal shoots become too numerous and cause a decline in vigor which may result in biennial bearing. In Brazil, the clusters are bagged to eliminate sunburn (purple staining of the skin) to which both of the leading cultivars are susceptible.

Because of the shallow root system of the loquat, great care must be taken in mechanical cultivation not to damage the roots. The growing of dwarf trees greatly reduces the labor of flower- and fruit-thinning, bagging, and, later, harvesting and pruning.

Season

Generally, the loquat tree blooms in the fall and fruits in early spring. However, in tropical climates, the tree may flower 2 or 3 times a year beginning in July and set fruit mainly from the second flowering. In Florida, ripening begins in February; in California, usually in April; in Israel, the crop ripens from March to May. In Brazil, the harvesting extends from May to October.

Harvesting

Loquats reach maturity in 90 days from full flower-opening. Determination of ripeness is not easy, but it is important because unripe fruits are excessively acid. Full development of color for each cultivar is the best guide.

The fruits are difficult to harvest because of the thick, tough stalk on each fruit which does not separate readily from the cluster, and the fruits must be picked with stalk attached to avoid tearing the skin. Clusters are cut from the branch with a sharp knife or with clippers. Whole clusters are not particularly attractive on the market, therefore the individual fruits are clipped from the cluster, the stalk is detached from each fruit and the fruits are graded for size and color to provide uniform packs. Great care is taken to avoid blemishes.

Major Japanese growers have monorail systems for conveying the picked fruits and equipment from their hillside plantations.

Yield

Dwarf loquats in Israel have produced 7 tons/ha at 3 years of age, 25 tons/ha at 7 years. Normal size trees in Brazil are expected to bear 110 lbs (50 kg) per tree, 4.17 tons per acre (10 tons/ha) when planted at a rate of 83 trees per acre (200 trees/ha). The 'Wolfe' cultivar in southern Florida has borne 100 lbs (45 kg) per tree at 5 years of age; 300 lbs (136 kg) when 15 to 20 years old.

Keeping Quality

Loquats generally will keep for 10 days at ordinary temperatures, and for 60 days in cool storage. After removal from storage, the shelf-life may be only 3 days. Treatment with the fungicide, benomyl, makes it possible to maintain loquats for one month at 60°F (15.56° C) with a minimum of decay. Other fungicides tried have proved much less effective. Cold storage of loquats in polyethylene bags alters the flavor of the fruit, promotes internal browning and the development of fungi.

Pests and Diseases

In Japan, scale insects, aphids, fruit flies and birds damage the fruits and may necessitate covering the clusters with cloth or paper bags. Laborers can attach 1,000 to 1,500 bags per day. An acre may require 62,500 bags (150,000/ha). A pole with a hook at the tip is employed to bring each branch within reach. The process is labor intensive. In Israel, wire netting is placed over trees to protect the crop from birds.

The Caribbean fruit fly *(Anastrepha suspensa)* has ruined the dooryard loquat crop for the past several years in Florida. The fruit flies, *A. striata* and *A. serpentina*, require control in Venezuela, the Mediterranean fruit fly, *Ceratitis capitata,* in Tunisia. Another fruit fly, *Dacus dorsalis,* is the major pest in India, forces the harvesting of mature fruits while they are still too hard to be penetrated, and the complete removal of all immature fruits at the same time so that they will not remain as hosts. The soil around the base of the tree must be plowed up and treated to kill the pupae. The second most important predator is the bark-eating caterpillar, *Indarbela quadrinotata.*

Minor pests include leaf-eating chafer beetles, *Adoretus duvauceli, A. lasiopygus, A. horticola* and *A. versutus;* gray weevils, *Myllocerus lactivirens* and *M. discolor* which attack the margins of the leaves. The scale insects, *Coccus viridis, Eulecanium coryli, Parlatoria oleae, P. pseudopyri, Pulvinaria psidii* and *Saissetia hemisphaerica* suck the sap from loquat leaves and branches. Carpenter bees, *Megochile anthracina,* cut holes in the leaves and take the tissue to line their mud nests. Aphids *(Aphis malvae)* suck sap from twigs and shoots and sooty mold develops on the honeydew which they excrete. Flowers are attacked by thrips *(Heliothrips* sp.). The caterpillars of the anar butterfly, *Virachola isocrates,* bore into the fruits and lay eggs on the fruits, flowers and leaves. In New Zealand, a leaf-roller caterpillar eats into the buds and flowers. In California, the main pests of loquat are the codlin moth *(Cydia pomonella)*, the green apple aphis *(Aphis pomi)* and scales.

The roots of loquat trees in India are preyed on by nematodes — *Criconemoides xenophax, Helicotylenchus* spp., *Hemicriconemoides communis, Haplolaimus* spp. and *Xiphinema insigne.*

Diseases

Pear blight *(Bacillus amylovorus)* is the major enemy of the loquat in California and has killed many trees. *Phytophthora* is responsible for crown rot and *Pseudomonas eriobotryae* causes cankers in California. Scab may occur on the bark of the trunk and larger branches. A serious disease is collar rot and root rot caused by *Diplodia natalensis. D. eriobotrya* sometimes affects the leaves. The parasitic fungus, *Monochaetia indica,* induces leaf spot in India. Leaf spot is also caused by the soil-inhabiting fungus *Schlerotium rolfsii. Spilocaeae eriobotryae* causes black spot on fruits and leaves in Italy and South Western Australia. Fleck, caused by the fungus *Fabraea maculata* is recognized by red-brown spots with whitish centers on leaves, shoots and fruit. In Florida, leaf spot may result from infection by *Pestalotia sp.* The foliage of young plants in Brazilian nurseries is damaged by the fungus *Entomosporium maculatum.* Other fungus problems of the loquat include stem-brown disease caused by *Batryosphaeria dothidee;* die-back from *Macrophoma* sp., withertip from *Collectotrichum gloeosporioides,* and twig blight and canker from *Cytospora chrysosperma.* Post-harvest fruit rot is the result of infection by *Diplodia natalensis, Pestalotia* sp. or *Aspergillus niger.*

Fig. 29: Peeled, seeded loquats *(Eriobotrya japonica)* canned in sirup in Taiwan.

Sunburn, "purple spot", is responsible for much fruit loss in hot regions with long summers. Chemical sprays have been employed to hasten fruit maturity to avoid sunburn. Various types of bags have been tried in Brazil to protect the fruit from this blemish. The best are 2- and 3-ply newspaper bags.

Food Value Per 100 g of Edible Portion*	
Calories	168
Protein	1.4 g
Fat	0.7 g
Carbohydrates	43.3 g
Calcium	70 mg
Phosphorus	126 mg
Iron	1.4 mg
Potassium	1,216 mg
Vitamin A	2,340 I.U.
Ascorbic Acid	3 mg

*Analyses reported by the Agricultural Research Service of the United States Department of Agriculture.

The fruit contains laevulose, sucrose and malic acid and lesser amounts of citric, tartaric and succinic acid. The pulp contains the carotenoids B-carotene (33%); y-carotene (6%); cryptoxanthin (22%), lutein, violaxanthin, neoxanthin (3-4% each). The peel is 5 times richer than the pulp in carotenoids which are similar to those in apricots.

Food Uses

The skin of the loquat is easily removed. Peeled and seeded fruits are eaten fresh, sometimes combined with sliced banana, orange sections and grated coconut. They are delicious simply stewed with a little sugar added. The fruits are also used in gelatin desserts or as pie-filling, or are chopped and cooked as a sauce. Loquats canned in sirup are exported from Taiwan. Some people prepare spiced loquats (with cloves, cinnamon, lemon and vinegar) in glass jars. The fruit is also made into jam and, when slightly underripe, has enough pectin to make jelly. The jelly was formerly manufactured commercially in California on a small scale.

Toxicity

A 5-year-old girl in Florida ate 4 unripe loquats, fell asleep and was difficult to awaken and seemed dazed. After about 2 hours, she was back to normal. There have been instances of poisoning in poultry from ingestion of loquat seeds. The seeds contain amygdalin (which is converted into HCN); also the lipids, sterol, β-sitosterol, triglyceride, sterolester, diglyceride and compound lipids; and fatty acids, mainly linoleic, palmitic, linolenic and oleic. There is amygdalin also in the fruit peel. The leaves possess a mixture of triterpenes, also tannin, vitamin B and ascorbic acid; in addition, there are traces of arsenic. Young leaves contain saponin. Some individuals suffer headache when too close to a loquat tree in bloom. The emanation from the flowers is sweet and penetrating.

Other Uses

Wood: The wood is pink, hard, close-grained, medium-heavy. It has been used instead of pear wood in making rulers and other drawing instruments.

Animal feed: The young branches have been lopped for fodder.

Perfume: In the 1950's, the flowers attracted the interest of the perfume industry in France and Spain and some experimental work was done in extraction of the essential oil from the flowers or leaves. The product was appealing but the yield was very small.

Medicinal Uses: The fruit is said to act as a sedative and is eaten to halt vomiting and thirst.

The flowers are regarded as having expectorant properties. An infusion of the leaves, or the dried, powdered leaves, may be taken to relieve diarrhea and depression and to counteract intoxication from consumption of alcoholic beverages. Leaf poultices are applied on swellings.

Capulin

The capulin is a true cherry and doesn't really belong with fruits of warm regions. However, it must be included here to distinguish it from the Jamaica cherry (q.v.), for the two share a number of colloquial names. *Prunus salicifolia* HBK. (syns. *P. capuli* Cav.; *P. serotina* var. *salicifolia* Koehne), of the family Rosaceae, is most often called *capulin, capuli, capoli* or *capolin,* especially in Colombia and Mexico, but in certain parts of the latter country it is known as *cerezo, detsé, detzé, taunday, jonote, puan, palman or xengua.* In Colombia it is sometimes called *cerezo criollo.* In Guatemala, it is known as *capulin, cereza, cereza común,* or wild cherry; in Bolivia, it is *capuli;* in Eucador, *capuli* or black cherry.

Description

The tree is erect, reaching 40 to 50 ft (12-15 m) in height, with a short, stout trunk to 3 ft (0.9 m) in diameter. The deciduous, alternate, aromatic leaves are lanceolate to ovate-lanceolate, 2⅜ to 7 in (6-18 cm) long, dark-green and glossy above, pale beneath; thin, finely toothed. New leaves are often rosy. Flowers, borne in slender, pendent racemes with 1 or more leaves at the base, are about ¾ in (2 cm) wide with white petals and a conspicuous tuft of yellow stamens. The aromatic fruit is round, ⅜ to ¾ in (1-2 cm) wide, with red or nearly black, rarely white or yellowish, smooth, thin, tender skin and pale-green, juicy pulp of sweet or acid, agreeable, but slightly astringent flavor. There is a single stone with a bitter kernel.

Origin and Distribution

The capulin is native and common throughout the Valley of Mexico from Sonora to Chiapas and Veracruz, and possibly also indigenous to western Guatemala. It has been cultivated since early times in these areas and other parts of Central America and in Colombia, Ecuador, Peru and Bolivia, and is extensively and abundantly naturalized. The fruit is an important food, not only of the Indians, but of all the inhabitants, and it was at times a mainstay of the invading Spaniards. Great quantities appear in the native markets, especially of El Salvador, Guatemala and Ecuador. In Guatemala, seedlings of the capulin are utilized as rootstock on which commercial cultivars of the northern cherry are grafted. The capulin is little-known in eastern South America and elsewhere in the world. It was introduced into the cool medium elevations of the Philippines in 1924.

Climate

The tree requires a subtropical to subtemperate climate. It grows naturally at elevations between 4,000 and 11,000 ft (1,200-3,400 m).

Season

In Mexico, the tree blooms from January to March and the fruits ripen in July and August. In Guatemala, flowers appear from January to May and fruits from May to September. The fruiting season in El Salvador extends from December through April.

Food Uses

The ripe fruits are eaten raw or stewed; also are preserved whole or made into jam. In Mexico they are used as filling for special tamales. With skin and seeds removed, they are mixed with milk and served with vanilla and cinnamon as dessert. Sometimes the fruits are fermented to make an alcoholic beverage.

Food Value Per 100 g of Edible Portion*	
Moisture	76.8–80.8 g
Protein	0.105–0.185 g
Fat	0.26–0.37 g
Fiber	0.1–0.7 g
Ash	0.56–0.82 g
Calcium	17.2–25.1 mg
Phosphorus	16.9–24.4 mg
Iron	0.65–0.84 mg
Carotene	0.005–0.162 mg
Thiamine	0.016–0.031 mg
Riboflavin	0.018–0.028 mg
Niacin	0.640–1.14 mg
Ascorbic Acid	22.2–32.8 mg

*According to analyses made in Guatemala and Ecuador.

Other Uses

Seeds: The seeds contain 30-38% of a yellow, semi-drying oil suitable for use in soap and paints.

Flowers: The flowers are much visited by honeybees.

Wood: The sapwood is yellow with touches of red. The heartwood is reddish-brown, fine-grained, very hard, strong, durable. It is used for furniture, interior paneling, cabinets, turnery and general carpentry. Old roots are valued for carving tobacco pipes, figurines, et cetera.

Medicinal Uses: A sirup made of the fruits is taken to alleviate respiratory troubles. The leaf decoction is given as a febrifuge and to halt diarrhea and dysentery; also applied in poultices to relieve inflammation. A leaf infusion is prescribed in Yucatan as a sedative in colic and neuralgia and as an antispasmodic. The pounded bark is employed in an eyewash.

The leaves contain essential oil, fat, resin, tannin, amygdalin, glucose, a brown pigment and mineral salts. The bark contains starch, brown pigment, amygdalin, gallic acid, fat, calcium, potassium and iron. All of these parts must be utilized cautiously because the bark, leaves or seeds in contact with water can release HCN.

Mysore Raspberry (Plate XII)

Many species of *Rubus* (family Rosaceae), especially from the warm regions of the world, have been tried in southern Florida. Only one has been truly successful here, the Mysore raspberry, *R. niveus* Thunb., (syns. *R. lasiocarpus* Hook. f. in part; *R. albescens* Roxb.; *R. mysorensis* Heyne), also called Ceylon, hill or Mahabaleshwar, raspberry in India and *pilai* in the Philippines.

Description

The plant is a large scrambling shrub growing 10 to 15 ft (3-4.5 m) high, with cylindrical, flexible stems downy when young, later purple, coated with a white bloom. It is thoroughly set with sharp, hooked thorns. The leaves, 4 to 8 in (10-20 cm) long, are composed of 5 to 9 elliptic-ovate leaflets 1 to 2½ in (2.5-6.25 cm) long, coarsely toothed, dark-green above and, on the underside, white-hairy with small, sharp spines along the rachis, petiole and midrib. Pink or red-purple, 5-petalled flowers, ½ in (1.25 cm) across, occur in lax axillary and terminal clusters. The fruit is rounded-conical, flat at the base; compound, made up of individual drupelets; red when unripe, purple-black when ripe, with a very fine bloom; ½ to ¾ in (1.25-2 cm) in diameter, juicy and of sweet, rich black-raspberry flavor. The clusters may contain as many as 2 dozen or even more. The seeds are small and not objectionable.

Origin and Distribution

The species is native to Burma and India, particularly the lower Himalayas, from Punjab to Assam, the Deccan peninsula, and the Western Ghats; and is common in the evergreen forests of Mahabaleshwar. The more hairy var. *horsfieldii* Focke extends south through Malaya to Indonesia and Bontoc and Benguet in the Philippines. From India, the Mysore raspberry was introduced into Kenya, East Africa, and has been grown in the mountains there for many years. Seeds from Kenya were obtained by F.B. Harrington of Natal, South Africa, in 1947. In 1948, he supplied seeds to the University of Florida's Agricultural Research and Education Center, Homestead. The resulting seedlings were planted out in 1949 and fruited so well the following winter that plants were distributed to many experimenters throughout south and central Florida. By 1952, many nurseries were offering the plants for sale and had difficulty filling the demand. By 1955, a major supermarket in Lake Worth was selling the fruits by the pint. In 1955, the University of Puerto Rico received planting material from Florida and established plantings in the central-western mountains of that island.

In Florida, some interest was still alive in 1965, but early enthusiasm waned as homeowners neglected their raspberry bushes, growth became too rampant, picking more and more difficult among the tangle of thorny canes, and birds competed eagerly for the crop. Many plantings were destroyed, and few remain.

Climate

This raspberry has a remarkable climatic range in Asia, from the relatively warm altitude of 1,500 ft (450 m) to the temperate environment at 10,000 ft (3,000 m). In Florida, brief drops in temperature to 35°F (1.67°C) have done the plants no harm but 20°F (-1.67°C) has killed young, tender growth, and prolonged freezing weather has killed the plants to the ground or outright.

Soil

In Florida, the plant flourishes on limestone or acid sand. In Puerto Rico it is grown on lateritic Alonso clay with a pH of 5.0. Good drainage is essential.

Propagation

The Mysore raspberry is often grown from seed but germination is slow and irregular (from 3 weeks to several months), and the seedlings are subject to damping-off. Germination can be expedited by pre-treatment with concentrated sulphuric acid. Stem cuttings root well, but the preferred method of propagation is by tip-layering. They develop plentiful roots in 3 to 4 weeks.

Culture

Florida gardeners place the plants 2½ to 4 ft (0.75-1.2 m) apart in rows 6 to 8 ft (1.8-2.4 m) apart supported by 2 or 3 strands of wire attached to end-posts. In Puerto Rico, the plants are set out in hills spaced 6 to 8 ft (1.8-2.4 m) apart each way. If taller than 18 in (45 cm), they are cut back, surrounded by 2 or 3 stakes 6 ft (1.8 m) high linked by crosswires. As the canes grow, they are loosely tied to the stakes and wires. A mulch is desirable to retain moisture and control weeds.

During the first year, in Puerto Rico, the plants are given 1 to 2 oz (28-56 g) each of ammonium sulfate quarterly. Thereafter, a 9-10-5 fertilizer formula is applied quarterly, 4 to 6 oz (113-170 g) per plant.

On Florida limestone, the recommended fertilizer consisting of 4-8-4 or 4-7-5 NPK with 3 to 4% magnesium and 30 to 40% organic nitrogen is applied every 2 to 3 weeks. And it is considered highly desirable that a mixture of zinc, copper and manganese be sprayed on the underside of the leaves 3 to 4 times per year.

Irrigation is necessary in dry seasons. Old canes should be cut to the ground at the end of the fruiting period and there should be severe pruning and thinning out in the late fall to force new growth for a winter-spring crop.

Season

The Mysore raspberry tends to bloom and fruit throughout the year but summer fruits are of poor size and quality. Therefore, the seasonal pruning has the additional purpose of preventing spring and summer flowering and allowing the first blooms to appear in December. Thus managed, the fruits are borne continuously from about February to May or June.

Harvesting

The fruits should be harvested only when they are not wet from dew or rain and when they are fully ripe and separate easily from the receptacle which remains on the plant. Gathering should be done at least 2 or 3 times a week to avoid losses by falling and spoilage. The fruits are highly perishable and should be consumed or processed as soon as possible.

Yield

In full sun, the crop is light. Where the plants receive some light shade in the afternoon, the yield is heavy. A single plant may yield 2,400 to 3,000 fruits over a 4-month period. A plot of 8 test plants in Florida produced 50 lbs (22.5 kg) in one season.

Pests and Diseases

The 2-spotted mite, *Tetranychus bimaculatus,* congregates on the underside of the leaves of shade-grown seedlings, turning them yellow. Occasionally, flower buds and fruits are attacked by the green stink bug, *Nezara viridula,* also called pumpkin or squash bug.

Anthracnose (*Elsinoe veneta*) causes spotting and scabbing of the canes toward the end of the fruiting season. Affected canes should be cut off and destroyed to prevent further infection. Damping-off of seedlings can be avoided by planting seeds in a mixture of peat moss and vermiculite, or in sphagnum moss.

Food Uses

The fruits are enjoyed fresh, alone or served with sugar and cream or ice cream. They are excellent for making pie, tarts, jam and jelly. The fresh fruits can be quick-frozen for future use.

Red Ceylon Peach (Plate XIII)

The peach, *Prunus persica* Batsch, is not ordinarily included among tropical or subtropical fruits. It is grown mainly in temperate regions of the world, including North America from the mild-temperate areas of Nova Scotia and the Ontario peninsula close to the Great Lakes, to north-central Florida, and across the Gulf States to California.

The earliest settlers in the South grew peaches, especially the highly-esteemed 'Spanish Blood'. Commercial culture began in the southern states with the introduction of other types and, by 1900, peach culture was receiving serious attention in many parts of the country. There are 5 races of the peach differing widely in their characteristics. The United States Department of Agriculture intro-

Fig. 30: The supple branches of the 'Red Ceylon Peach' (*Prunus persica*) bend to the ground when laden with fruit.

duced many cultivars of the South China race, typified by the 'Peen-to', a flat type well adapted to moderately warm climates. The peach tree has a chilling requirement of a certain number of hours at 45°F (7.22°C) from the time of leaf-fall to the emerging of new buds. This period varies with the race and cultivar from 30 to 1,000 hours. Late in the 1880's, the 'Red Ceylon', which requires no more than 50 hours of chilling, became well-established in southern Florida. In 1904, this cultivar was planted at the agricultural experiment station at Santiago de las Vegas, Cuba, and was soon being grown all around the Havana area because it was the only peach found suitable to that tropical climate and the local soils.

Description

The tree is dwarf, slender and willowy, with deciduous, alternate, slender, pointed leaves; bears pink, 5-petalled flowers on bare branches in January and February, sometimes March, and fruits heavily in April and May. The fruit is oval with a protruding knob at the apex, 2¾ in (7 cm) long and 2⅜ in (6 cm) wide; velvety, green with deep-red blush when ripe. The flesh is mainly white but a rich strawberry-red in the center; tender, juicy, and of excellent, sweet-acid flavor having a slight suggestion of bitter-almond. The stone is free, corrugated and very hard; small in proportion to the size of the fruit. Despite its unattractiveness externally and small dimensions, the 'Red Ceylon' is much-appreciated on close acquaintance. It is peeled, sliced and enjoyed fresh or stewed and can be used for various culinary purposes. The sliced fruit can be frozen in sirup and relished out-of-season as topping on cake or ice cream. In fact, one becomes so partial to this peach that the ordinary commercial peaches, though far more beautiful, seem somewhat rubbery and much less flavorful by comparison.

Other Cultivars

Two other subtropical cultivars have been successfully grown in southern Florida:

'Saharanpur'—a selection from seedlings received in 1969 from the Horticultural Research Institute, Saharanpur, India. The fruit is very similar to that of 'Red Ceylon' except that it lacks the fine red coloration in the center. The seedlings received from India were probably of the selection 'Shabati' reported by Dr. L. B. Singh as having been released in 1950 and widely distributed all over India where winter chilling requirement of 30 to 40 hours could be guaranteed.

'Okinawa'—a fruit of superior form but of inferior quality. This cultivar has been valued mainly as a rootstock because of its greater nematode-resistance.

Culture

The 'Red Ceylon' peach has been commonly propagated by seed or by grafting. The seeds may take several months to germinate unless cracked which will induce sprouting in 10 to 90 days. The tree grows rapidly and bears in 2 years from seed. It is relatively nematode-resistant and requires little care, but should receive plenty of water for good production.

Status

In the 1940's and 1950's the 'Red Ceylon' peach was being deservedly promoted as a useful fruit for home gardens. It is impractical for marketing because of the protruding tip which bruises and then spoils readily. Seedlings and grafted plants were being sold by nurseries. Unfortunately, with the advent of the Caribbean fruit fly in 1965, and its rapid spread in southern Florida, interest in peach-growing dwindled, for the peach is a major host of this pest. Marie Neal wrote that, in Hawaii, a type of peach with small fruits having whitish flesh was formerly grown from the lowlands to an altitude of 3,000 ft (900 m), but its cultivation was discouraged because of the prevalence of the Mediterranean fruit fly.

The 'Red Ceylon' and the 'Okinawa' have been used as rootstocks for peaches in central Florida, though such tender rootstocks may make the grafted tree inclined to cold-sensitivity. In 1957, a hybrid between the 'Red Ceylon' and the 'Southland' peach was developed at the University of Florida's Agricultural Experiment Station in Gainesville.

In the past 2 decades there have been continuous efforts to develop low-chilling cultivars for central Florida and also hardier types as a crop replacement for the orange in the northern part of the "Citrus Belt" where severe damage to orange trees occurred in the winter of 1962-1963 and 200,000 bearing trees were killed by freezes in December 1983 and January 1985.

Dr. Ralph Sharpe has been a leader in peach-breeding in this state for many years. Through his research and that of his colleagues, Florida now has a substantial peach industry. The low-chilling, semi-cling-stone 'Floridaprince', requiring only 150 hours below 45°F (7.22°C), was released to nurseries in late 1985.

CHRYSOBALANACEAE

Sansapote

A fruit held in rather low esteem, the sansapote, *Licania platypus* Fritsch (syn. *Moquilea platypus* Hemsl.), of the family Chrysobalanaceae, is often called *sonzapote*, *sunzapote*, *sungano*, *zapote cabelludo*, *sapote* or *sangre* in Costa Rica and El Salvador; *sonzapote* in Nicaragua; *zapote amarillo*, *zapote borracho*, *zapote cabello*, *zapote de mico*, *zapote de mono*, *mesonsapote*, *mezonzapote*, *cabeza de mico*, or *caca de niño* in Mexico; *sonza*, *sunza*, *zunza*, *chaute* or *jolobob* in Guatemala; *urraco* in Honduras; *chupa* in Colombia; monkey apple in Belize.

Description

The handsome tree is erect, stately, reaching 100 to 160 ft (30-50 m) in height; has a rounded crown of thick branches, heavily foliaged, and dark purplish or brown bark dotted with tiny white or reddish-white lenticels. It is sometimes slightly buttressed. The deciduous leaves are alternate, occasionally spiraled, elliptic- to narrow-lanceolate, pointed at both ends; 4 to 12 in (10-30 cm) long, 1¼ to 3½ in (3-9 cm) wide, with thick midrib, indented above and prominent beneath. New foliage is bronze or red-purple and very showy. The abundant, fragrant flowers, in broad terminal, branched panicles 4 to 14 in (10-35 cm) long, are small and densely hairy with recurved petals and numerous protruding stamens. Only 1 to 3 fruits develop from each panicle. The obovoid or pyriform fruit, 5 to 8 in (13-20 cm) long, 4 to 5½ in (10-14 cm) wide, has a thin, dark-brown or reddish, warty rind covered with white lenticels. The flesh, somewhat pumpkin-scented, is yellow or orange-yellow, soft, fibrous, dry or juicy and of subacid or sweet flavor. Usually there is a single rounded or ovate-oblong, flattened seed, 2⅜ to 4 in (6-10 cm) long.

Origin and Distribution

The sansapote grows wild in dense forests from southern Mexico to Panama, on both coasts, and also in northern Colombia. It is much planted as an ornamental and shade tree throughout Central America. It was introduced into the Philippines in the early 1900's and into Hawaii only about 25 years ago. In the spring of 1913, the United States Department of Agriculture received seeds from the Department of Agriculture in San José, Costa Rica (S.P.I. #34915). In November of the same year, seeds of a small-fruited type from the Pacific Coast and a large-fruited type from the Atlantic slope were received from the same source (S.P.I. #36590). Another introduction was made from Colombia in 1916 (S.P.I. #42991).

Fig. 31: The sansapote, photographed by the U.S. Department of Agriculture plant explorers, Cook, Collins and Doyle, at Nicoya, Costa Rica, in 1903. Published in Henry Pittier's *New or Noteworthy Plants from Colombia and Central America #3* (Contribution of the U.S. National Herbarium Vol. 13, Part 12), the Smithsonian Institution; 1912.

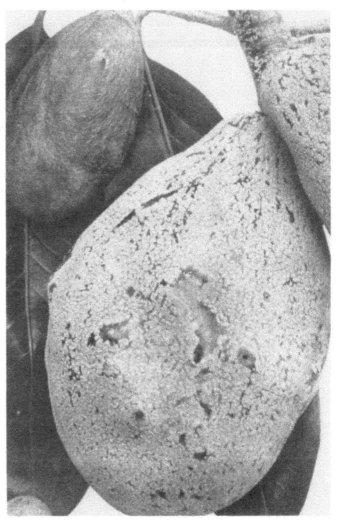

Few of the trees planted in southern Florida have survived. Several young specimens have died at the Fairchild Tropical Garden. One at the Subtropical Horticulture Research Station, Miami, has bloomed several times after rains but has not fruited. William Whitman obtained seeds from the Ministry of Agriculture, El Salvador, in 1957. One tree grew well, suffered severe hurricane damage in 1964, recovered, bloomed in late 1969 and, in the summer of 1970, produced a dozen fruits; over 100 in 1971. The fruits are not highly regarded in Central America but are sold in native markets. Tapirs and peccaries feast on those that are left on the ground.

Climate

This is a tropical species limited to low elevations — not more than 2,000 ft (600 m) above sea-level.

Season

According to Pennington, the tree blooms from July to September in Mexico and the fruits ripen from August to December. Perhaps he means of the following year. In Costa Rica and Honduras the fruit is said to take a year to develop to maturity. In Florida, one tree bloomed in November and the first fruits ripened 9 months later and the season extended from summer to fall.

Food Uses

The fruit is eaten raw when better fruits are not available. According to Standley, it has the reputation of being unwholesome, causing fever and other illnesses.

Food Value Per 100 g of Edible Portion*	
Moisture	64.6–67.4 g
Protein	0.230–0.291 g
Fat	0.26–0.49 g
Fiber	0.9–2.5 g
Ash	0.96–1.61 mg
Calcium	10.5–33.2 mg
Phosphorus	24.5–29.1 mg
Iron	0.52–1.70 mg
Carotene	0.157–0.273 mg
Thiamine	0.005–0.16 mg
Riboflavin	0.013–0.027 mg
Niacin	1.466–1.530 mg
Ascorbic Acid	11.0–35.6 mg

*According to analyses made in Costa Rica and El Salvador.

Other Uses

The sapwood is pale-yellow or light yellowish-brown; the heartwood is purplish-brown or reddish, fine-grained, very heavy and strong, suitable for fine furniture and cabinetwork, but it is not durable in contact with the ground. It is little-known inasmuch as the trees are valued and seldom felled. Related species provide timber for construction and charcoal. The seeds of *L. rigida* Benth. yield oiticica oil, much like tung oil.

Tamarind (Plate XIV)

Of all the fruit trees of the tropics, none is more widely distributed nor more appreciated as an ornamental than the tamarind, *Tamarindus indica* L. (syns. *T. occidentalis* Gaertn.; *T. officinalis* Hook.), of the family Leguminosae. Most of its colloquial names are variations on the common English term. In Spanish and Portuguese, it is *tamarindo;* in French, *tamarin, tamarinier, tamarinier des Indes,* or *tamarindier;* in Dutch and German, *tamarinde;* in Italian, *tamarandizio;* in Papiamiento of the Lesser Antilles, *tamarijn.* In the Virgin Islands, it is sometimes called *taman;* in the Philippines, *sampalok* or various other dialectal names; in Malaya, *asam jawa;* in India, it is tamarind or *ambli, imli, chinch,* etc.; in Cambodia, it is *ampil* or *khoua me;* in Laos, *mak kham;* in Thailand, *ma–kharm;* in Vietnam, *me.* The name "tamarind" with a qualifying adjective is often applied to other members of the family Leguminosae having somewhat similar foliage.

Description

The tamarind, a slow-growing, long-lived, massive tree reaches, under favorable conditions, a height of 80 or even 100 ft (24-30 m), and may attain a spread of 40 ft (12 m) and a trunk circumference of 25 ft (7.5 m). It is highly wind-resistant, with strong, supple branches, gracefully drooping at the ends, and has dark-gray, rough, fissured bark. The mass of bright-green, fine, feathery foliage is composed of pinnate leaves, 3 to 6 in (7.5-15 cm) in length, each having 10 to 20 pairs of oblong leaflets ½ to 1 in (1.25-2.5 cm) long and ⅕ to ¼ in (5-6 mm) wide, which fold at night. The leaves are normally evergreen but may be shed briefly in very dry areas during the hot season. Inconspicuous, inch-wide flowers, borne in small racemes, are 5-petalled (2 reduced to bristles), yellow with orange or red streaks. The flowerbuds are distinctly pink due to the outer color of the 4 sepals which are shed when the flower opens.

The fruits, flattish, beanlike, irregularly curved and bulged pods, are borne in great abundance along the new branches and usually vary from 2 to 7 in long and from ¾ to 1¼ in (2-3.2 cm) in diameter. Exceptionally large tamarinds have been found on individual trees. The pods may be cinnamon-brown or grayish-brown externally and, at first, are tender-skinned with green, highly acid flesh and soft, whitish, under-developed seeds. As they mature, the pods fill out somewhat and the juicy, acidulous pulp turns brown or reddish-brown. Thereafter, the skin becomes a brittle, easily-cracked shell and the pulp dehydrates naturally to a sticky paste enclosed by a few coarse strands of fiber extending lengthwise from the stalk. The 1 to 12 fully formed seeds are hard, glossy-brown, squarish in form, ⅜ to ½ in (1.1-1.25 cm) in diameter, and each is enclosed in a parchmentlike membrane.

Origin and Distribution

Native to tropical Africa, the tree grows wild throughout the Sudan and was so long ago introduced into and adopted in India that it has often been reported as indigenous there also, and it was apparently from this Asiatic country that it reached the Persians and the Arabs who called it *"tamar hindi"* (Indian date, from the date-like appearance of the dried pulp), giving rise to both its common and generic names. Unfortunately, the specific name, *"indica"*, also perpetuates the illusion of Indian origin. The fruit was well known to the ancient Egyptians and to the Greeks in the 4th Century B.C.

The tree has long been naturalized in the East Indies and the islands of the Pacific. One of the first tamarind trees in Hawaii was planted in 1797. The tamarind was certainly introduced into tropical America, Bermuda, the Bahamas, and the West Indies much earlier. In all tropical and near-tropical areas, including South Florida, it is grown as a shade and fruit tree, along roadsides and in dooryards and parks. Mexico has over 10,000 acres (4,440 ha) of tamarinds, mostly in the states of Chiapas, Colima, Guerrero, Jalisco, Oaxaca and Veracruz. In the lower Motagua Valley of Guatemala, there are so many large tamarind trees in one area that it is called "El Tamarindal". There are commercial plantings in Belize and other Central American countries and in northern Brazil. In India there are extensive tamarind orchards producing 275,500 tons (250,000 MT) annually. The pulp is marketed in northern Malaya and to some extent wherever the tree is found even if there are no plantations.

Varieties

In some regions the type with reddish flesh is distinguished from the ordinary brown-fleshed type and regarded as superior in quality. There are types of tamarinds that are sweeter than most. One in Thailand is known as 'Makham waan'. One distributed by the United States Department of Agriculture's Subtropical Horticulture Research Unit, Miami, is known as 'Manila Sweet'.

Climate

Very young trees should be protected from cold but older trees are surprisingly hardy. Wilson Popenoe wrote that a large tree was killed on the west coast of Florida (about 7.5° lat. N) by a freeze in 1884. However, no cold damage was noted in South Florida following the low temperatures of the winter of 1957-1958 which had severe effects on many mango, avocado, lychee and lime trees. Dr. Henry Nehrling reported that a tamarind tree in his garden at Gotha, Florida, though damaged by freezes, always sprouted out again from the roots. In northwestern India, the tree grows well but the fruits do not ripen. Dry weather is important during the period of fruit development. In South Malaya, where there are frequent rains at this time, the tamarind does not bear.

Soil

The tree tolerates a great diversity of soil types, from deep alluvial soil to rocky land and porous, oolitic limestone. It withstands salt spray and can be planted fairly close to the seashore.

Propagation

Tamarind seeds remain viable for months, will germinate in a week after planting. In the past, propagation has been customarily by seed sown in position, with thorny branches protecting the young seedlings. However, today, young trees are usually grown in nurseries. And there is intensified interest in vegetative propagation of selected varieties because of the commercial potential of tamarind products. The tree can be grown easily from cuttings, or by shield-budding, side-veneer grafting, or air-layering.

Culture

Nursery-grown trees are usually transplanted during the early rainy season. If kept until the second rainy season, the plants must be cut back and the taproot trimmed. Spacing may be 33 to 65 ft (10-20 m) between trees each way, depending on the fertility of the soil. With sufficient water and regular weeding, the seedlings will reach 2 ft (60 cm) the first year and 4 ft (120 cm) by the second year.

In Madagascar, seedlings have begun to bear in the 4th year; in Mexico, usually in the 5th year; but in India, there may be a delay of 10 to 14 years before fruiting. The tree bears abundantly up to an age of 50-60 years or sometimes longer, then productivity declines, though it may live another 150 years.

Season

Mexican studies reveal that the fruits begin to dehydrate 203 days after fruit-set, losing approximately ½ moisture up to the stage of full ripeness, about 245 days from fruit-set. In Florida, Central America, and the West Indies, the flowers appear in summer, the green fruits are found in December and January and ripening takes place from April through June. In Hawaii the fruits ripen in late summer and fall.

Harvesting

Tamarinds may be left on the tree for as long as 6 months after maturity so that the moisture content will be reduced to 20% or lower. Fruits for immediate processing are often harvested by pulling the pod away from the stalk which is left with the long, longitudinal fibers attached. In India, harvesters may merely shake the branches to cause mature fruits to fall and they leave the remainder to fall naturally when ripe. Pickers are not allowed to knock the fruits off with poles as this would damage developing leaves and flowers. To keep the fruit intact for marketing fresh, the stalks must be clipped from the branches so as not to damage the shell.

Yield

A mature tree may annually produce 330 to 500 lbs (150-225 kg) of fruits, of which the pulp may constitute 30 to 55%, the shells and fiber, 11 to 30%, and the seeds, 33 to 40%.

Keeping Quality

To preserve tamarinds for future use, they may be merely shelled, layered with sugar in boxes or pressed into tight balls and covered with cloth and kept in a cool, dry place. For shipment to processors, tamarinds may be shelled, layered with sugar in barrels and covered with boiling sirup. East Indians shell the fruits and sprinkle them lightly with salt as a preservative. In Java, the salted pulp is rolled into balls, steamed and sun-dried, then exposed to dew for a week before being packed in stone jars. In India, the pulp, with or without seeds and fibers may be mixed with salt (10%), pounded into blocks, wrapped in palmleaf matting, and packed in burlap sacks for marketing. To store for long periods, the blocks of pulp may be first steamed or sun-dried for several days.

Pests and Diseases

One of the major pests of the tamarind tree in India is the Oriental yellow scale, *Aonidiella orientalis.* Tamarind scale, *A. tamarindi,* and black, or olive, scale, *Saissetia oleae,* are also partial to tamarind but of less importance. Butani (1970) lists 8 other scale species that may be found on the tree, the young and adults sucking the sap of buds and flowers and accordingly reducing the crop.

The mealybug, *Planococcus lilacinus,* is a leading pest of tamarind in India, causing leaf-fall and sometimes shedding of young fruits. Another mealybug, *Nipaecoccus viridis,* is less of a menace except in South India where it is common on many fruit trees and ornamental plants. *Chionaspis acuminata-atricolor* and *Aspidiotus* spp., suck the sap of twigs and branches and the latter also feeds on young fruits. White grubs of *Holotrichia insularis* may feed on the roots of young seedlings. The nematodes, *Xiphinema citri* and *Longidorus elongatus* may affect the roots of older trees. Other predators attacking the leaves or flowers include the caterpillars, *Thosea aperiens, Thalarsodes quadraria, Stauropus alternus,* and *Laspeyresia palamedes;* the black citrus aphid, *Toxoptera aurantii,* the whitefly, *Acaudaleyrodes*

Plate XII
MYSORE RASPBERRY
Rubus niveus

Plate XIII
RED CEYLON PEACH
Prunus persica

Plate XIV
TAMARIND
Tamarindus indica

Plate XV
CAROB
Ceratonia siliqua

Plate XVI
CARAMBOLA
Averrhoa carambola

Plate XVII
BILIMBI
Averrhoa bilimbi

Plate XVIII
NAVEL GRAPEFRUIT
Citrus X paradisi

Plate XIX
TANGELO
Citrus X tangelo

Plate XX
MEYER LEMON
possibly *Citrus limon X C. reticulata*

Fig. 32: Acid-sweet pulp of the tamarind (*Tamarindus indica*) is blended with sugar as a confection, or preserved as jam or nectar. It enhances chutney and some well-known sauces.

rachispora; thrips, *Ramaswamia hiella subnudula, Scirtothrips dorsalis,* and *Haplothrips ceylonicus;* and cow bugs, *Oxyrhachis tarandus, Otinotus onerotus,* and *Laptoentrus obliquis.*

Fruit borers include larvae of the cigarette beetle, *Lasioderma serricorne,* also of *Virachola isocrates, Dichocrocis punctiferalis, Tribolium castaneum, Phycita orthoclina, Cryptophlebia (Argyroploca) illepide, Oecadarchis* sp., *Holocera pulverea, Assara albicostalis, Araecerus suturalis, Aephitobius laevigiatus,* and *Aphomia gularis.* The latter infests ripening pods on the tree and persists in the stored fruits, as do the tamarind beetle, *Pachymerus (Coryoborus) gonogra,* and tamarind seed borer, *Calandra (Sitophilus) linearis.* The rice weevil, *Sitophilus oryzae,* the rice moth, *Corcyra cepholonica,* and the fig moth, *Ephestia cautella,* infest the fruits in storage. The lesser grain borer, *Rhyzopertha dominica* bores into stored seeds.

In India, a bacterial leaf-spot may occur. Sooty mold is caused by *Meliola tamarindi.* Rots attacking the tree include saprot, *Xylaria euglossa,* brownish saprot, *Polyporus calcuttensis,* and white rot, *Trametes floccosa.* The separated pulp has good keeping quality but is subject to various molds in refrigerated storage.

Food Uses

The food uses of the tamarind are many. The tender, immature, very sour pods are cooked as seasoning with rice, fish and meats in India. The fully-grown, but still unripe fruits, called "swells" in the Bahamas, are roasted in coals until they burst and the skin is then peeled back and the sizzling pulp dipped in wood ashes and eaten. The fully ripe, fresh fruit is relished out-of-hand by children and adults, alike. The dehydrated fruits are easily recognized when picking by their comparatively light weight, hollow sound when tapped and the cracking of the shell under gentle pressure. The shell lifts readily from the pulp and the lengthwise fibers are removed by holding the stem with one hand and slipping the pulp downward with the other. The pulp is made into a variety of products. It is an important ingredient in chutneys, curries and sauces, including some brands of Worcestershire and barbecue sauce, and in a special Indian seafood pickle called "tamarind fish". Sugared tamarind pulp is often prepared as a confection. For this purpose, it is desirable to separate the pulp from the seeds without using water. If ripe, fresh, undehydrated tamarinds are available, this may be done by pressing the shelled and defibered fruits through a colander while

Fig. 33: Bahamian children hold mature but still green tamarinds in hot ashes until they sizzle, then dip the tip in the ashes and eat them. The high calcium content contributes to good teeth.

adding powdered sugar to the point where the pulp no longer sticks to the fingers. The seeded pulp is then shaped into balls and coated with powdered sugar. If the tamarinds are dehydrated, it is less laborious to layer the shelled fruits with granulated sugar in a stone crock and bake in a moderately warm oven for about 4 hours until the sugar is melted, then the mass is rubbed through a sieve, mixed with sugar to a stiff paste, and formed into patties. This sweetmeat is commonly found on the market in Jamaica, Cuba and the Dominican Republic. In Panama, the pulp may be sold in corn husks, palmleaf fiber baskets, or in plastic bags.

Tamarind ade has long been a popular drink in the Tropics and it is now bottled in carbonated form in Guatemala, Mexico, Puerto Rico and elsewhere. Formulas for the commercial production of spiced tamarind beverages have been developed by technologists in India. The simplest home method of preparing the ade is to shell the fruits, place 3 or 4 in a bottle of water, let stand for a short time, add a tablespoonful of sugar and shake vigorously. For a richer beverage, a quantity of shelled tamarinds may be covered with a hot sugar sirup and allowed to stand several days (with or without the addi-

tion of seasonings such as cloves, cinnamon, allspice, ginger, pepper or lime slices) and finally diluted as desired with ice water and strained.

In Brazil, a quantity of shelled fruits may be covered with cold water and allowed to stand 10 to 12 hours, the seeds are strained out, and a cup of sugar is added for every 2 cups of pulp; the mixture is boiled for 15 to 20 minutes and then put up in glass jars topped with paraffin. In another method, shelled tamarinds with an equal quantity of sugar may be covered with water and boiled for a few minutes until stirring shows that the pulp has loosened from the seeds, then pressed through a sieve. The strained pulp, much like apple butter in appearance, can be stored under refrigeration for use in cold drinks or as a sauce for meats and poultry, plain cakes or puddings. A foamy "tamarind shake" is made by stirring this sauce into an equal amount of dark-brown sugar and then adding a tablespoonful of the mixture to 8 ounces of a plain carbonated beverage and whipping it in an electric blender.

If twice as much water as tamarinds is used in cooking, the strained product will be a sirup rather than a sauce. Sometimes a little soda is added. Tamarind sirup is bottled for domestic use and export in Puerto Rico. In Mayaguez, street vendors sell cones of shaved ice saturated with tamarind sirup. Tamarind pulp can be made into a tart jelly, and tamarind jam is canned commercially in Costa Rica. Tamarind sherbet and ice cream are popular and refreshing. In making fruit preserves, tamarind is sometimes combined with guava, papaya or banana. Sometimes the fruit is made into wine.

Inasmuch as shelling by hand is laborious and requires 8 man-hours to produce 100 lbs (45 kg) of shelled fruits, food technologists at the University of Puerto Rico have developed a method of pulp extraction for industrial use. They found that shelling by mechanical means alone is impossible because of the high pectin and low moisture content of the pulp. Therefore, inspected and washed pods are passed through a shell-breaking grater, then fed into stainless steel tanks equipped with agitators. Water is added at the ratio of 1:1½ or 1:2 pulp/water, and the fruits are agitated for 5 to 7 minutes. The resulting mash is then passed through a screen while nylon brushes separate the shells and seeds. Next the pulp is paddled through a finer screen, pasteurized, and canned.

Young leaves and very young seedlings and flowers are cooked and eaten as greens and in curries in India. In Zimbabwe, the leaves are added to soup and the flowers are an ingredient in salads.

Tamarind seeds have been used in a limited way as emergency food. They are roasted, soaked to remove the seedcoat, then boiled or fried, or ground to a flour or starch. Roasted seeds are ground and used as a substitute for, or adulterant of, coffee. In Thailand they are sold for this purpose. In the past, the great bulk of seeds available as a by-product of processing tamarinds, has gone to waste. In 1942, two Indian scientists, T.P. Ghose and S. Krishna, announced that the decorticated kernels contained 46 to 48% of a gel-forming substance. Dr. G.R. Savur of the Pectin Manufacturing Company, Bombay, patented a process for the production of a purified product, called "jellose", "polyose", or "pectin", which has been found superior to fruit pectin in the manufacture of jellies, jams, and marmalades. It can be used in fruit preserving with or without acids and gelatinizes with sugar concentrates even in cold water or milk. It is recommended as a stabilizer in ice cream, mayonnaise and cheese and as an ingredient or agent in a number of pharmaceutical products.

Food Value Per 100 g of Edible Portion			
	Pulp (ripe)*	Leaves (young)	Flowers
Calories	115		
Moisture	28.2-52 g	70.5 g	80 g
Protein	3.10 g	5.8 g	0.45 g
Fat	0.1 g	2.1 g	1.54 g
Fiber	5.6 g	1.9 g	1.5 g
Carbohydrates	67.4 g	18.2 g	
Invert Sugars	30-41 g		
(70% glucose; 30% fructose)			
Ash	2.9 g	1.5 g	0.72 g
Calcium	35-170 mg	101 mg	35.5 mg
Magnesium	-	71 mg	
Phosphorus	54-110 mg	140 mg	45.6 mg
Iron	1.3-10.9 mg	5.2 mg	1.5 mg
Copper	-	2.09 mg	
Chlorine	-	94 mg	
Sulfur	-	63 mg	
Sodium	24 mg	-	
Potassium	375 mg	-	
Vitamin A	15 I.U.	250 mcg	0.31 mg
Thiamine	0.16 mg	0.24 mg	0.072 mg
Riboflavin	0.07 mg	0.17 mg	0.148 mg
Niacin	0.6-0.7 mg	4.1 mg	1.14 mg
Ascorbic Acid	0.7-3.0 mg	3.0 mg	13.8 mg
Oxalic Acid	-	196 mg	
Tartaric Acid	8-23.8 mg	-	
Oxalic Acid	trace only	-	

*The pulp is considered a promising source of tartaric acid, alcohol (12% yield) and pectin (2½% yield). The red pulp of some types contains the pigment, chrysanthemin.
Seeds contain approximately 63% starch, 14-18% albuminoids, and 4.5-6.5% of a semi-drying oil.

Food Value

Analyses of the pulp are many and varied. Roughly, they show the pulp to be rich in calcium, phosphorus, iron, thiamine and riboflavin and a good source of niacin. Ascorbic acid content is low except in the peel of young green fruits.

Other Uses

Fruit pulp: In West Africa, an infusion of the whole pods is added to the dye when coloring goat hides. The fruit pulp may be used as a fixative with turmeric or annatto in dyeing and has served to coagulate rubber latex. The pulp, mixed with sea water, cleans silver, copper and brass.

Leaves: The leaves are eaten by cattle and goats, and furnish fodder for silkworms—*Anaphe* sp. in India, *Hypsoides vuilletii* in West Africa. The fine silk is considered superior for embroidery.

Tamarind leaves and flowers are useful as mordants in dyeing. A yellow dye derived from the leaves colors wool red and turns indigo-dyed silk to green. Tamarind leaves in boiling water are employed to bleach the leaves of the buri palm *(Corypha elata* Roxb.) to prepare them for hat-making. The foliage is a common mulch for tobacco plantings.

Flowers: The flowers are rated as a good source of nectar for honeybees in South India. The honey is golden-yellow and slightly acid in flavor.

Seeds: The powder made from tamarind kernels has been adopted by the Indian textile industry as 300% more efficient and more economical than cornstarch for sizing and finishing cotton, jute and spun viscose, as well as having other technical advantages. It is commonly used for dressing homemade blankets. Other industrial uses include employment in color printing of textiles, paper sizing, leather treating, the manufacture of a structural plastic, a glue for wood, a stabilizer in bricks, a binder in sawdust briquettes, and a thickener in some explosives. It is exported to Japan, the United States, Canada and the United Kingdom.

Tamarind seeds yield an amber oil useful as an illuminant and as a varnish especially preferred for painting dolls and idols. The oil is said to be palatable and of culinary quality. The tannin-rich seedcoat (testa) is under investigation as having some utility as an adhesive for plywoods and in dyeing and tanning, though it is of inferior quality and gives a red hue to leather.

Wood: The sapwood of the tamarind tree is pale-yellow. The heartwood is rather small, dark purplish-brown, very hard, heavy, strong, durable and insect-resistant. It bends well and takes a good polish and, while hard to work, it is highly prized for furniture, panelling, wheels, axles, gears for mills, ploughs, planking for sides of boats, wells, mallets, knife and tool handles, rice pounders, mortars and pestles. It has at times been sold as "Madeira mahogany". Wide boards are rare, despite the trunk dimensions of old trees, since they tend to become hollow-centered. The wood is valued for fuel, especially for brick kilns, for it gives off an intense heat, and it also yields a charcoal for the manufacture of gun-powder. In Malaysia, even though the trees are seldom felled, they are frequently topped to obtain firewood. The wood ashes are employed in tanning and in de-hairing goatskins. Young stems and also slender roots of the tamarind tree are fashioned into walking-sticks.

Twigs and barks: Tamarind twigs are sometimes used as "chewsticks" and the bark of the tree as a masticatory, alone or in place of lime with betelnut. The bark contains up to 7% tannin and is often employed in tanning hides and in dyeing, and is burned to make an ink. Bark from young trees yields a low-quality fiber used for twine and string. Galls on the young branches are used in tanning.

Lac: The tamarind tree is a host for the lac insect, *Kerria lacca*, that deposits a resin on the twigs. The lac may be harvested and sold as stick-lac for the production of lacquers and varnish. If it is not seen as a useful by-product, tamarind growers trim off the resinous twigs and discard them.

Medicinal Uses: Medicinal uses of the tamarind are uncountable. The pulp has been official in the British and American and most other pharmacopoeias and some 200,000 lbs (90,000 kg) of the shelled fruits have been annually imported into the United States for the drug trade, primarily from the Lesser Antilles and Mexico. The European supply has come largely from Calcutta, Egypt and the Greater Antilles. Tamarind preparations are universally recognized as refrigerants in fevers and as laxatives and carminatives. Alone, or in combination with lime juice, honey, milk, dates, spices or camphor, the pulp is considered effective as a digestive, even for elephants, and as a remedy for biliousness and bile disorders, and as an antiscorbutic. In native practice, the pulp is applied on inflammations, is used in a gargle for sore throat and, mixed with salt, as a liniment for rheumatism. It is, further, administered to alleviate sunstroke, *Datura* poisoning, and alcoholic intoxication. In Southeast Asia, the fruit is prescribed to counteract the ill effects of overdoses of false chaulmoogra, *Hydnocarpus anthelmintica* Pierre, given in leprosy. The pulp is said to aid the restoration of sensation in cases of paralysis. In Colombia, an ointment made of tamarind pulp, butter, and other ingredients is used to rid domestic animals of vermin.

Tamarind leaves and flowers, dried or boiled, are used as poultices for swollen joints, sprains and boils. Lotions and extracts made from them are used in treating conjunctivitis, as antiseptics, as vermifuges, treatments for dysentery, jaundice, erysipelas and hemorrhoids and various other ailments. The fruit shells are burned and reduced to an alkaline ash which enters into medicinal formulas. The bark of the tree is regarded as an effective astringent, tonic and febrifuge. Fried with salt and pulverized to an ash, it is given as a remedy for indigestion and colic. A decoction is used in cases of gingivitis and asthma and eye inflammations; and lotions and poultices made from the bark are applied on open sores and caterpillar rashes. The powdered seeds are made into a paste for drawing boils and, with or without cumin seeds and palm sugar, are prescribed for chronic diarrhea and dysentery. The seedcoat, too, is astringent, and it, also, is specified for the latter disorders. An infusion of the roots is believed to have curative value in chest complaints and is an ingredient in prescriptions for leprosy.

The leaves and roots contain the glycosides: vitexin, isovitexin, orientin and isoorientin. The bark yields the alkaloid, hordenine.

Superstitions

Few plants will survive beneath a tamarind tree and there is a superstition that it is harmful to sleep or to tie a horse beneath one, probably because of the corrosive effect that fallen leaves have on fabrics in damp weather. Some African tribes venerate the tamarind tree as sacred. To certain Burmese, the tree represents the dwelling-place of the rain god and some hold the belief that the tree raises the temperature in its immediate vicinity. Hindus may marry a tamarind tree to a mango tree before eating the fruits of the latter. In Nyasaland, tamarind bark soaked with corn is given to domestic fowl in the belief that, if they stray or are stolen, it will cause them to return home. In Malaya, a little tamarind and coconut milk is placed in the mouth of an infant at birth, and the bark and fruit are given to elephants to make them wise.

Carob (Plate XV)

Non-fleshy and bean-like, the carob would not be generally regarded as a fruit, in the food-use sense, except for its sweetness. To many people it is familiar only by name as "St. John's Bread", in allusion to the "locusts" which, according to the Bible, sustained St. John the Baptist in the desert, and the "husks" which tempted the hungry Prodigal Son, though "no man gave unto him." The word "locust" was originally applied to the carob tree; later to migratory and other grasshoppers; and the name is attached to a number of other leguminous trees with pinnate leaves and oblong pods (*Gleditsia, Hymenaea, Parkia, Robinia*). The carob tree is called *carrubo* in Sicily, *carrubio* in Italy, *algarrobo* in Guatemala, *alfarrobeira* in Brazil.

Description

The tree reaches 50 to 55 ft (15-17 m) in height and at an age of 18 years may have a trunk 33 in (85 cm) in circumference. The evergreen leaves are pinnate with 6 to 10 opposite leaflets, oval, rounded at the apex, dark-green, leathery, 1 to 2½ in (2.5-6.25 cm) long. The tiny red flowers are in short, slender racemes borne in clusters along the branches—male, female or hermaphrodite on separate trees. The pod is light- to dark-brown, oblong, flattened, straight or slightly curved, with a thickened margin; 4 to 12 in (10-30 cm) long, ¾ to 1 in (1-2.5 cm) wide, glossy, tough and fibrous. It is filled with soft, semi-translucent, pale-brown pulp, scant or plentiful, and 10 to 13 flattened, very hard seeds which are loose in their cells and rattle when the pod is fully ripe and dry. The unripe pod is green, moist and very astringent; the ripe pod sweet when chewed (avoiding the seeds) but the odor of the broken pod is faintly like Limburger cheese because of its 1.3% isobutyric acid content.

Origin and Distribution

Alphonse de Candolle said that the carob "grew wild in the Levant, probably on the southern coast of Anatolia and in Syria, perhaps also in Cyrenaica. Its cultivation began within historic time. The Greeks diffused it in Greece and Italy, but it was afterwards more highly esteemed by the Arabs, who propagated it as far as Morocco and Spain. In all these countries the tree has become naturalized here and there in a less productive form . . .".

In Spain and Portugal it survives only on their Atlantic coasts. Throughout the Mediterranean region, it is grown only in the warmest areas near the coast, and the neighboring islands—Cyprus, Crete, Sicily, Sardinia and Majorca. Producers in the Bari region of Italy on the Adriatic coast have long exported the pods to Russia and central Europe. Prince Belmonte in the Province of Salerno, Italy, was a leading influence in the 19th century in the use of the carob as an ornamental and avenue tree and in the planting of thousands for reforestation of the slopes of the Appenines.

Spanish missionaries introduced the carob into Mexico and southern California. In 1856, 8,000 seedlings, from seed brought in from Spain by the United States Patent Office, were distributed in the southern states. More seeds came from Israel in 1859. Many carobs were planted in Texas, Arizona, California and a few in Florida as ornamental and street trees. Seeds privately imported from Dalmatia were planted in California in 1873.

In the Mediterranean region, peasants have virtually lived on the pods in times of famine, but the tree is valued mostly as providing great amounts of pods as feed for livestock, as it is also in the State of Campinas, Brazil. Imported pods used to be regularly sold by street vendors in the Italian section of lower New York City for chewing.

In the early 1920's, there was much promotion of carob culture in California, especially allied with the development of arid lands, and there was a flurry of activity in producing "health food" products from imported pods. Some of these products are still sold today, especially as substitutes for chocolate. Dr. J. Eliot Coit, of Vista, California, led in the study of the carob and wrote extensively on its potential improvement as a crop and its utilization.

In 1949, Dr. Walter Rittenhouse provided funds for the establishment of a 30-year test plot in northern San Diego County, where 400 local nursery seedlings and many trees grafted with Mediterranean budwood were planted and evaluated. Fruits from several thousand ornamental carob trees in California and Arizona were collected in an effort to identify superior types for human food use. Budwood of the most promising clones was supplied to horticulturists in Tunisia, Israel, Australia, South Africa, Hawaii, Mexico, Brazil and Chile.

Varieties

From more than 80 clones, 7 selections made by Coit were set out at the Citrus Research Center of the University of California for preservation. The 7 are, briefly:

'Amele'—an old commercial variety from Italy; S.P.I. #19437. Female. Pods light-brown, straight or slightly curved, 5½ to 6¼ in (14-16 cm) long, ¾ to 1 in (2-2.5 cm) wide; 53.8% sugar content under irrigation near Indio. Flavor good. Season: September at Indio; October at Vista.

'Casuda'—a very old cultivar from Spain. Female. Pod brown, mostly straight; 4¾ in (12 cm) long; ⅗ in (1.5 cm) wide; 51.7% sugar at Vista; 56.7% under irrigation at Indio. Flavor fair. Season: September at Indio; October at Vista.

'Clifford'—seedling street tree in Riverside. Hermaphrodite. Pod light-brown, slightly curved, 5⅛ in (13 cm) long, ¾ in (2 cm) wide; 52.9% sugar content. Flavor fair. Season: early October; bears regularly and heavily.

'Sfax'—from Menzel bou Zelfa, Tunisia; S.P.I. #187063. Female. Pod red-brown, straight or slightly curved; 6 in (15 cm) long, ¾ in (2 cm) wide; 56.6% sugar at Vista, 45.6% at Indio. Excellent flavor. Season: August at Indio, September at Vista. A regular, medium-heavy bearer.

'Santa Fe'—seedling from Santa Fe Springs, California. Hermaphrodite; self-fertile. Pod light-brown, slightly curved, often twisted; 7 to 7⅞ in (18-20 cm) long, ¾ in (2 cm) wide; 47.5% sugar at Vista. Excellent flavor. Season: October. Bears regular, good crops. Good for coastal foothills. Not suited to irrigated culture at Indio.

'Tantillo'—from Sicily; S.P.I. #233580. Hermaphrodite. Pod dark-brown, mostly straight; 5⅛ to 6 in (13-15 cm) long, ¾ in (2 cm) wide. Of fair flavor. Season: mid-September to mid-October. Bears heavily and regularly.

'Tylliria'—from Cyprus; their chief export variety; S.P.I. #189008. Female. Pod dark mahogany-brown, slightly curved, 6 in (15 cm) long, ¾ to 1 in (2-2.5 cm) wide; 47.4% sugar at Vista; 50.9% at Indio; 48.8% in Cyprus. Good flavor. Season: mid-August to mid-September at Indio; October at Vista. Adapted to coastal foothills. (As reported from Cyprus, seed content is 7.6 to 10.6%; pod contains 51% sugar and the seeds 49% gum).

These 7 superseded some older cultivars, including 'Bolser', 'Conejo', 'Gabriel', 'Horne', and 'Molino'; all hermaphroditic.

Other common cultivars in Cyprus are:

'Koundourka'—a tree with weeping branches; mature pods generally less than 6½ in (17 cm) long; they split readily; have 14.7% seeds with a high (58%) gum content.
'Koumbota'—a large-growing tree with "knotty" pods with low seed content. Pods contain 53% sugar; seeds, 53% gum.

Grafted types are classed as 'Imera'. The name 'Apostolika' is a general term for seedlings of fair quality. Wild types as a group are called 'Agria'.

Pollination

In a planting of female trees, one male should be included for every 25 or 30 females. In southern Europe, branches from male trees are grafted onto some of the females in an orchard instead of interplanting male trees.

Climate

The carob is slightly hardier than the sweet orange. Young trees suffer frost damage. Mature tees can endure a temperature drop to 20°F (-6.67°C). Frost during the blooming period will reduce or prevent fruit-set. The tree does best in a Mediterranean-type climate with cool, not cold, winters, mild to warm springs, and warm to hot summers with little or no rain. Temperatures in carob-growing regions of Israel may reach 104° to 122°F (40°-50°C) in summer. Ideal annual precipitation is 30 in (75 cm), but widely spaced trees will thrive with only 6 to 15 in (15-37.5 cm) without irrigation in mild climates. The pods should not be exposed to rain or heavy dew after they have turned brown and developed a high sugar content. Wet pods ferment quickly.

Soil

The tree flourishes in widely divergent soils, from rocky hillsides to deep sand or heavy loam, but must have good drainage. In Nicosia, Cyprus, a large plantation was developed by dynamiting planting holes in caprock underlaid with limestone (pH 9). The carob is not tolerant of acid or wet soils; it is extremely drought-tolerant.

Propagation

Fresh seeds germinate quickly and may be sown directly in the field. Dried, hard seeds need to be scarified or chipped and then soaked in water or dilute sulfuric or hydrochloric acid solutions until they swell. In Cyprus, seeds are planted in sand and kept wet for 6 weeks or more, periodically sifting out those that have swollen to 3 times normal size. Germination rate may be only 25%. The swollen seeds are traditionally planted in flats and when they produce the second set of leaves they are transferred to small pots. When 12 in (30 cm) tall, they are transplanted to large containers or nursery rows. A recently developed technique is to plant the seeds in 2 halves of clay drainpipes bound together or in plastic tubes packed in deep wooden boxes to accommodate the long taproot. In perhaps a year, the tubes are split and the seedlings are planted in the field in holes made with a post-hole digger. Budding is done when the stem is at least ⅜ in (1 cm) thick.

The shield-budding system is employed, or sometimes a blend of budding and grafting, in February and March in Cyprus, in April, May and June in California and Mexico. Male trees or those that bear poorly are top-worked to productive cultivars.

Culture

The carob grows slowly during the first year. Stem-elongation in young plants has been expedited by application of gibberellin (50 mg/liter monthly, or 25 mg/liter semi-monthly) for 5 months. It is necessary to cut back the taproot 6 months before transferring to the field if the plant is not grown by the tube/post-hole method. Large trees cannot be successfully transplanted.

A good spacing is 30 ft (9 m) apart each way. Most carob growers consider fertilizing unnecessary but the government of Cyprus subsidizes fertilization—so much per tree. Irrigation must be provided in very dry seasons if the tree is grown for its fruits. Budded trees begin to bear in the 6th year from planting. A carob tree may remain productive for 80 to 100 years.

Harvesting

The pods must be harvested before winter rains. They are shaken down by means of a long pole with a terminal hook to grasp the branches. Those that don't fall readily are knocked off with the pole. The pods are caught on canvas sheets laid on the ground. Then they are sun-dried for 1 or 2 days until the moisture content is reduced to 8% or below and then go through a kibbling process—crushing and grading into 4 categories: cubed, medium-kibbled, meal, and seed kernels.

Yield

At 6 years of age, a budded tree in California should yield about 5 lbs (2.25 kg). At 12 years, the crop should be 100 lbs (45 kg). Productivity increases steadily up to 25 or 30 years when the yield may average 200 lbs (90 kg). In Israel individual trees have produced 450 to 550 lbs (204-227 kg) 18 years after grafting. Some ancient trees in the Mediterranean area are reported to have borne 3,000 lbs (1,360 kg) in a season.

Pests and Diseases

In the Mediterranean area, the major pest is the carob moth, *Myelois ceratoniae*. It lays eggs on the flowers or newly-formed pods and the larvae bore into the pods and ruin them. The larvae of a midge, *Asphondylia gennadii*, cause stunting of the pods. Some of the best cultivars are resistant to these pests.

In Cyprus, the tree is subject to several scale insects: *Aspidiotus ceratoniae*, *Lecanium* sp., *Lepidosaphes* sp. and the red scale, *Aonidiella aurantii*. A beetle, *Cerambyx velutinus*, may bore holes in the trunk. Rats climb the trees, hide among the branches, gnaw the bark until the branches die. Such branches are pruned out twice a year. The only pests reported as attacking carob trees in California are scale insects, including the red scale. Ground squirrels feed on plants under 2 years of age.

Pocket gophers are very fond of carob roots, and rabbits and deer graze on the young trees.

Diseases are few. In Cyprus, deformation of young pods may be caused by the fungus *Oidium ceratoniae*. *Cercospora ceratoniae* occasionally induces leaf-spotting.

Food Uses

Apart from being chewed as a sweetmeat, carob pods are processed to a cocoa-like flour which is added to cold or heated milk for drinking. It has been combined with wheat flour in making bread or pancakes. A flour made by beating the seeded pods is high in fiber and has been utilized in breakfast foods. The finer flour is also made into confections, especially candy bars. The pods, coarsely ground and boiled in water yield a thick, honey-like sirup, or molasses.

The seeds constitute 10 to 20% of the pod. They yield a tragacanth-like gum (manogalactan), called in the trade "Tragasol", which is an important commercial stabilizer and thickener in bakery goods, ice cream, salad dressings, sauces, cheese, salami, bologna, canned meats and fish, jelly, mustard, and other food products. The seed residue after gum extraction can be made into a starch- and sugar-free flour of 60% protein content for diabetics.

In Germany, the roasted seeds have served as a substitute for coffee. In Spain, they have been mixed with coffee.

It has been demonstrated that the extracted sugars of the pod (sucrose, glucose, fructose and maltose in the ratio 5:1:1:0:7) can be utilized to produce fungal protein. Infusions of the pulp are fermented into alcoholic beverages.

Food Value Per 100 g of Carob Flour	
Calories	180
Moisture	11.2 g
Protein	4.5 g
Fat	1.4 g
Carbohydrates*	80.7 g
Fiber	7.7 g
Ash	2.2 g
Calcium	352 mg
Phosphorus	81 mg

*Sugar content may be as high as 72%.
The pods contain up to 1.5% tannins which interfere with the body's utilization of protein.

Other Uses

Pods: The pods are relished by horses, cattle, pigs, goats and rabbits. Whole pods are broken up in a hammermill in order to crush the seeds as well. Because of the tannin content, carob pods should constitute no more than 10% of total feed, otherwise they will depress growth rate. They cannot be fed to chickens. The flour is often utilized in dog biscuits. Great quantities of pods have been imported into the United States for flavoring uncured tobacco.

Seeds: The seed gum is much employed in the manufacture of cosmetics, pharmaceutical products, detergents, paint, ink, shoe polish, adhesives, sizing for textiles, photographic paper, insecticides and match heads. It is also utilized in tanning. Where rubber latex is produced, the gum is added to cause the solids to rise to the surface. It is also used for bonding paper pulp and thickening silk-screen pastes, and some derivatives are added to drilling mud. It has many other actual or potential applications. A flour made from the seeds serves as cattle feed.

Wood: The heartwood is hard and close-grained. It is prized for turnery and cabinetwork. As a fuel it burns slowly and makes excellent charcoal. It yields algarrobin, which gives textiles a light-brown hue.

Fig. 34: A rarity in southern Florida, this carob tree on the campus of the University of Miami was 15 years old when photographed in 1954. It is still bearing small fruits every year without cross-pollination.

Carambola (Plate XVI)

A curious, attractive fruit of the Oxalidaceae, the carambola, *Averrhoa carambola* L., has traveled sufficiently to have acquired a number of regional names in addition to the popular Spanish appelation which belies its Far Eastern origin. In the Orient, it is usually called *balimbing*, *belimbing*, or *belimbing manis* ("sweet belimbing"), to distinguish it from the bilimbi or *belimbing asam*, *A. bilimbi* L. In Ceylon and India, the carambola has the alternate names of *kamaranga*, *kamruk*, or other variants of the native *kamrakh*. In Vietnam, it is called *khe*, *khe ta*, or similar terms; in Kampuchea, *spu*; in Laos, *nak fuang*, or the French name, *carambolier*; in Thailand, *ma fueang*. Malayans may refer to it as *belimbing batu*, *belimbing besi*, *belimbing pessegi*, *belimbing sayur*, *belimbing saji*, *kambola*, *caramba*, or as "star fruit". Australians use the descriptive term, five corner; in Guam, it is *bilimbines*; to the Chinese, it is *yang-táo*. Early English travelers called it Chinese, or Coromandel gooseberry, or cucumber tree. In Guyana, it is five fingers; in the Dominican Republic, it is *vinagrillo*; in Haiti, *zibline*; in some of the French Antilles, *cornichon*; in El Salvador, *pepino de la India*; in Surinam, *blimbing legi* or *fransman–birambi*; Costa Rica, *tiriguro*; in Brazil, *camerunga* or *caramboleiro*, or *limas de Cayena*; in Mexico, *carambolera* or *caramboler* or *árbol de pepino*; in Trinidad, it may be called coolie tamarind. Venezuelans call it *tamarindo chino* or *tamarindo dulce*.

Description

The carambola tree is slow-growing, short-trunked with a much-branched, bushy, broad, rounded crown and reaches 20 to 30 ft (6-9 m) in height. Its deciduous leaves, spirally arranged, are alternate, imparipinnate, 6 to 10 in (15-20 cm) long, with 5 to 11 nearly opposite leaflets, ovate or ovate-oblong, 1½ to 3½ in (3.8-9 cm) long; soft, medium-green, and smooth on the upper surface, finely hairy and whitish on the underside. The leaflets are sensitive to light and more or less inclined to fold together at night or when the tree is shaken or abruptly shocked. Small clusters of red-stalked, lilac, purple-streaked, downy flowers, about ¼ in (6 mm) wide, are borne on the twigs in the axils of the leaves. The showy, oblong, longitudinally 5- to 6-angled fruits, 2½ to 6 in (6.35-15 cm) long and up to 3½ (9 cm) wide, have thin, waxy, orange-yellow skin and juicy, crisp, yellow flesh when fully ripe. Slices cut in cross-section have the form of a star. The fruit has a more or less pronounced oxalic acid odor and the flavor ranges from very sour to mildly sweetish. The so-called "sweet" types rarely contain more than 4% sugar. There may be up to 12 flat, thin, brown seeds ¼ to ½ in (6-12.5 mm) long or none at all.

Origin and Distribution

The carambola is believed to have originated in Ceylon and the Moluccas but it has been cultivated in southeast Asia and Malaysia for many centuries. It is commonly grown in the provinces of Fukien, Kuangtung and Kuangsi in southern China, in Taiwan and India. It is rather popular in the Philippines and Queensland, Australia, and moderately so in some of the South Pacific islands, particularly Tahiti, New Caledonia and Netherlands New Guinea, and in Guam and Hawaii.

There are some specimens of the tree in special collections in the Caribbean islands, Central America, tropical South America, and also in West Tropical Africa and Zanzibar. Several trees have been growing since 1935 at the Rehovoth Research Station in Israel. In many areas, it is grown more as an ornamental than for its fruits.

It was introduced into southern Florida before 1887 and was viewed mainly as a curiosity until recent years when some small groves have been established and the fruits have been used as "conversation pieces" to decorate gift shipments of citrus fruits, and also, in clear-plastic-wrapped trays, have been appearing in the produce sections of some supermarkets. One fruit-grower and shipper now has 50 acres (20 ha) planted but suggests that other prospective growers be cautious as the market may remain limited. Shipments go mainly to Vancouver, Quebec, Cleveland, and Disneyworld. Small amounts are sold locally.

Varieties

There are 2 distinct classes of carambola—the smaller, very sour type, richly flavored, with more oxalic acid; the larger, so-called "sweet" type, mild-flavored, rather bland, with less oxalic acid.

In 1935, seeds from Hawaii were planted at the University of Florida's Agricultural Research and Education Center in Homestead. A selection from the resulting seedlings was vegetatively propagated during the 1940's and 1950's and, in late 1965, was officially released under the name **'Golden Star'** and distributed to growers. The fruit is large, deeply winged, decorative, and mildly subacid to sweet. Furthermore, this cultivar shows the least minor-element deficiency in alkaline soil, and even isolated trees bear well and regularly without cross-pollination.

Several cultivars from Taiwan are being grown at the United States Department of Agriculture's Subtropical Horticulture Research Unit in Miami, including 'Mih Tao' (P.I. No. 272065) introduced in 1963, also 'Dah Pon' and 'Tean Ma' and others identified only by numbers, and 'Fwang Tung' brought from Thailand by Dr. R.J. Knight in 1973. There are certain "lines" of carambola, such as 'Newcomb', 'Thayer' and 'Arkin' being grown commercially in southern Florida. Some cultivars and seedlings bear flowers with short styles, others only flowers with long styles, a factor which affects self- and cross-pollination.

Climate

The carambola should be classed as tropical and subtropical because mature trees can tolerate freezing temperatures for short periods and sustain little damage at 27°F (-2.78°C). In Florida, the tree survives in sheltered sites as far north as St. Petersburg on the west coast and Daytona Beach on the east. It thrives up to an elevation of 4,000 ft (1,200 m) in India. In an interior valley of Israel, all trees succumbed to the prevailing hot, dry winds. The carambola needs moisture for best performance and ideally rainfall should be fairly evenly distributed all year. In Australia, it is claimed that fruit quality and flavor are best where annual rainfall is 70 in (180 cm) or somewhat more.

Soil

Not too particular as to soil, the carambola does well on sand, heavy clay or limestone, but will grow faster and bear more heavily in rich loam. It is often chlorotic on limestone. It needs good drainage; cannot stand flooding.

Propagation

The carambola is widely grown from seed though viability lasts only a few days. Only plump, fully developed seeds should be planted. In damp peat moss, they will germinate in one week in summer, require 14 to 18 days in winter. The seedlings are transplanted to containers of light sandy loam and held until time to set out. They are very tender and need good care. Seedlings are highly variable. Air-layering has been practiced and advocated. However, root formation is slow and later performance is not wholly satisfactory. Inarching is successful in India, shield-budding in the Philippines and the Forkert method in Java. Trees can be top-worked by bark-grafting, a popular technique in Java. For mass production, side-veneer grafting of mature, purplish wood, onto carambola seedlings gives best results for most workers. The rootstocks should be at least 1 year old and 3/8 to 5/7 in (1-1.5 cm) thick. One Florida farmer prefers cleft-grafting of green budwood and has 90% success. Grafted trees will fruit in 10 months from the time of planting out. Mature trees can be top-worked by bark-grafting.

Culture

The tree needs full sun. A spacing of 20 ft (6 m) has been advocated but if the trees are on good soil no less than 30 ft (9 m) should be considered. At the Research Center in Homestead, trees 8 to 10 ft (2.4-3 m) high respond well to 1 lb (0.5 kg) applications of N, P, K, Mg in the ratio of 6-6-6-3 given 3 to 4 times per year. If chlorosis occurs, it can be corrected by added iron, zinc and manganese. Some advisers recommend minor-element spraying 4 times during the year if the trees are on limestone soils. Moderate irrigation is highly desirable during dry seasons. Heavy rains during blooming season interfere with pollination and fruit production. Interplanting of different strains is usually necessary to provide cross-pollination and obtain the highest yields.

Harvesting and Yield

In India, carambolas are available in September and October and again in December and January. In Malaya, they are produced all the year. In Florida, scattered fruits are found through the year but the main crop usually matures from late summer to early winter. Some trees have fruited heavily in November and December, and again in March and April. There may even be three crops. Weather conditions account for much of the seasonal variability.

The fruits naturally fall to the ground when fully ripe. For marketing and shipping they should be hand-picked while pale-green with just a touch of yellow.

Trees that receive adequate horticultural attention have yielded 100 to 250 or even 300 lbs (45-113-136 kg) of fruit.

Keeping Quality

Carambolas have been shipped successfully without refrigeration from Florida to northern cities in avocado lugs lined and topped with excelsior. The fruits are packed solidly, stem-end down, at a 45° angle, the flanges of one fruit fitting into the "V" grooves of another. Of course, they cannot endure rough handling.

In storage trials at Winter Haven, Florida, carambolas picked when showing the first signs of yellowing kept in good condition for 4 weeks at 50°F (10°C); 3 weeks at 60°F (15.56°C); 2 weeks at 70°F (21.1°C). Waxing extends storage life and preserves the vitamin value.

Pests and Diseases

The carambola is relatively pest-free except for fruit flies. In Malaya, fruit flies (especially *Dacus dorsalis*) are so troublesome on carambolas that growers have to wrap the fruits on the tree with paper. Experimental trapping, with methyl eugenol as an attractant, has reduced fruit damage by 20%. In Florida, a small stinkbug causes superficial blemishes and a black beetle attacks overripe fruits. Reniform nematodes may cause tree decline.

Anthracnose caused by *Colletotrichum gloeosporioides* may be a problem in Florida, and leaf spot may arise from attack by *Phomopsis* sp., *Phyllosticta* sp. or *Cercospora averrhoae*. *Cercospora* leaf spot is reported also from Malaya, Ceylon, China and may occur in the Philippines as well. A substance resembling sooty mold makes many fruits unmarketable in summer.

Food Uses

Ripe carambolas are eaten out-of-hand, sliced and served in salads, or used as garnish on avocado or seafood. They are also cooked in puddings, tarts, stews and curries. In Malaya, they are often stewed with sugar and cloves, alone or combined with apples. The Chinese cook carambolas with fish. Thais boil the sliced green fruit with shrimp. Slightly underripe fruits are salted, pickled or made into jam or other preserves. In mainland China and in Taiwan, carambolas are sliced lengthwise and canned in sirup for export. In Queensland, the sweeter type is cooked green as a vegetable. Cross-sections may be covered with honey, allowed to stand overnight, and then cooked briefly and put into sterilized jars. Some cooks add raisins to give the product more character. A relish may be made of chopped unripe fruits combined with horseradish, celery, vinegar, seasonings and spices. Indian experimenters boiled horizontal slices with ¾ of their weight in sugar until very thick, with a Brix of 68°. They found that the skin became very tough, the flavor was not distinctive, and the jam was rated as only fair. Sour fruits, pricked to permit absorption of sugar and cooked in sirup, at first 33° Brix, later 72°, made an acceptable candied product though the skin was still tough. The ripe fruits are sometimes dried in Jamaica.

Carambola juice is served as a cooling beverage. In Hawaii, the juice of sour fruits is mixed with gelatin, sugar, lemon juice and boiling water to make sherbet. Filipinos often use the juice as a seasoning. The juice is bottled in India, either with added citric acid (1% by weight) and 0.05% potassium metabisulphite, or merely sterilizing the filled bottles for ½ hr in boiling water.

To make jelly, it is necessary to use unripe "sweet" types or ripe sour types and to add commercial pectin or some other fruit rich in pectin such as green papaya, together with lemon or lime juice.

The flowers are acid and are added to salads in Java; also, they are made into preserves in India. The leaves have been eaten as a substitute for sorrel.

Food Value

Ripening and storage studies were conducted at the Florida Citrus Experiment Station at Lake Alfred in 1966. They found quite a difference in the acid make-up of mature green and mature yellow carambolas. Fresh mature green fruits of 'Golden Star' were found to have a total acid content of 12.51 mg/g consisting of 5 mg oxalic, 4.37 tartaric, 1.32 citric, 1.21 malic, 0.39 a-ketoglutaric, 0.22 succinic, and a trace of fumaric. Mature yellow fruits had a total acid content of 13 mg/g, made up of 9.58 mg oxalic, 0.91 tartaric, 2.20 a-ketoglutaric, 0.31 fumaric.

In 1975, 16 carambola selections and 2 named cultivars were assayed at the United States Citrus and Subtropical Products Laboratory, Winter Haven, Florida. Preliminary taste tests ranked 'No. 17', 'No. 37', 'No. 42' and 'Tean

Food Value Per 100 g of Edible Portion*	
Calories	35.7
Moisture	89.0-91.0 g
Protein	0.38 g
Fat	0.08 g
Carbohydrates	9.38 g
Fiber	0.80-0.90 g
Ash	0.26-0.40 g
Calcium	4.4-6.0 mg
Phosphorus	15.5-21.0 mg
Iron	0.32-1.65 mg
Carotene	0.003-0.552 mg
Thiamine	0.03-0.038 mg
Riboflavin	0.019-0.03 mg
Niacin	0.294-0.38 mg
Ascorbic Acid**	26.0-53.1 mg

* According to analyses made in Cuba and Honduras.

Amino Acids: (shown in Cuban analyses)

Tryptophan	3.0 mg
Methionine	2 mg
Lysine	26 mg

Other amino acids reported by the Florida Citrus Experiment Station at Lake Alfred and expressed in micromoles per g in mature green fruits (higher) and mature yellow fruits (lower), respectively, are:

Asparagine	0.82-0.64
Threonine	0.92-0.79
Serine	3.88-2.00
Glutamic Acid	2.41-1.80
Proline	0.23-0.09
Glycine	0.20-0.10
Alanine	5.40-1.26
Valine	0.17-0.11
Isoleucine	0.03-trace
Leucine	trace
Phenylalanine	trace
Gamma Amino Bytyric Acid	0.77-0.55
Ornithine	0.11-0.13
Histidine	trace

**Analyses in India showed 10.40 mg ascorbic acid in the juice of a "sweet" variety; 15.4 mg in juice of a sour variety. Ascorbic acid content of both waxed and unwaxed fruits stored at 50°F (10°C) has been reported as 20 mg/100 ml of juice. Waxed fruits stored for 17 days at 60°F (15.56°C) had 11 mg/100 ml of juice. Unwaxed fruits had lost ascorbic acid.

Ma' as preferred. In a later test, 'Dah Pon' was ranked above 'Tean Ma'. 'No. 17' (°Brix 9.9) was described as "sweet, good and apple-like". 'No. 37' (°Brix 6.7), as "sour and sweet". 'No. 42' (°Brix 8.3), as "sour, tart and apple-like". 'Dah Pon' (°Brix 8.0), as "good and mild". 'Tean Ma' (°Brix 7.2), as "sweet, good and mild". Analyses showed that these 5 were among those with relatively high ascorbic acid content—'No. 17', 30 mg; 'Dah Pon', 30 mg; 'No. 37', 37 mg; 'No. 42', 37 mg; and 'Tean Ma',

41 mg. 'No. 40' had 43 mg and 'No. 11', 50 mg, whereas 'M-23007' had only 14 mg and 'No. 10' only 17 mg.

Oxalic acid content of the 18 selections and cultivars ranged from 0.039 mg to 0.679 mg and 4 of the preferred carambolas were in the lower range as follows: 'No. 17', 0.167; 'Dah Pon', 0.184; 'Tean Ma', 0.202; 'No. 42', 0.276 mg, but 'No. 37', with 0.461 was 3rd from the highest of all.

Puerto Rican technologists found the oxalic acid content of ripe carambolas to average 0.5 g per 100 ml of juice, the acid being mostly in the free state. They likened the juice to rhubarb juice and advised that physicians be informed of this because there are individuals who may be adversely affected by ingestion of even small amounts of oxalic acid or oxalates. Other investigators have presumed the oxalic acid in fully ripe carambolas to be precipitated as calcium oxalate or in solution as neutral salts. The health risk needs further study.

Other Uses

The acid types of carambola have been used to clean and polish metal, especially brass, as they dissolve tarnish and rust. The juice will also bleach rust stains from white cloth. Unripe fruits are used in place of a conventional mordant in dyeing.

Wood: Carambola wood is white, becoming reddish with age; close-grained, medium-hard. It has been utilized for construction and furniture.

Medicinal Uses: In India, the ripe fruit is administered to halt hemorrhages and to relieve bleeding hemorrhoids; and the dried fruit or the juice may be taken to counteract fevers. A conserve of the fruit is said to allay biliousness and diarrhea and to relieve a "hangover" from excessive indulgence in alcohol. A salve made of the fruit is employed to relieve eye afflictions. In Brazil, the carambola is recommended as a diuretic in kidney and bladder complaints, and is believed to have a beneficial effect in the treatment of eczema. In *Chinese Materia Medica* it is stated, "Its action is to quench thirst, to increase the salivary secretion, and hence to allay fever."

A decoction of combined fruit and leaves is drunk to overcome vomiting. Leaves are bound on the temples to soothe headache. Crushed leaves and shoots are poulticed on the eruptions of chicken-pox, also on ringworm.

The flowers are given as a vermifuge. In southeast Asia, the flowers are rubbed on the dermatitis caused by lacquer derived from *Rhus verniciflua* Stokes.

Burkill says that a preparation of the inner bark, with sandalwood and *Alyxia* sp., is applied on prickly heat. The roots, with sugar, are considered an antidote for poison. Hydrocyanic acid has been detected in the leaves, stems and roots.

A decoction of the crushed seeds acts as a galactagogue and emmenagogue and is mildly intoxicating. The powdered seeds serve as a sedative in cases of asthma and colic.

Bilimbi (Plate XXII)

The bilimbi, *Averrhoa bilimbi*, L., (Oxalidaceae), is closely allied to the carambola but quite different in appearance, manner of fruiting, flavor and uses. The only strictly English names are "cucumber tree" and "tree sorrel", bestowed by the British in colonial times. "Bilimbi" is the common name in India and has become widely used. In Malaya, it is called *belimbing asam, belimbing buloh, b'ling,* or *billing-billing*. In Indonesia, it is *belimbing besu, balimbing, blimbing,* or *blimbing wuluh;* in Thailand, it is *taling pling,* or *kaling pring*.

In Haiti, it is called *blimblin;* in Jamaica, *bimbling plum;* in Cuba, it is *grosella china;* in El Salvador and Nicaragua, *mimbro;* in Costa Rica, *mimbro* or *tiriguro;* in Venezuela, *vinagrillo;* in Surinam and Guyana, *birambi;* in Argentina, *pepino de Indias*. To the French it is *carambolier bilimbi,* or *cornichon des Indes.* Filipinos generally call it *kamias* but there are about a dozen other native names.

Description

The tree is attractive, long-lived, reaches 16 to 33 ft (5-10 m) in height; has a short trunk soon dividing into a number of upright branches. The leaves, very similar to those of the Otaheite gooseberry and mainly clustered at the branch tips, are alternate, imparipinnate; 12 to 24 in (30-60 cm) long, with 11 to 37 alternate or subopposite leaflets, ovate or oblong, with rounded base and pointed tip; downy; medium-green on the upper surface, pale on the underside; ¾ to 4 in (2-10 cm) long, ½ to 1⅛ in (1.2-1.25 cm) wide.

Small, fragrant, 5-petalled flowers, yellowish-green or purplish marked with dark-purple, are borne in small, hairy panicles emerging directly from the trunk and oldest, thickest branches and some twigs, as do the clusters of curious fruits. The bilimbi is ellipsoid, obovoid or nearly cylindrical, faintly 5-sided, 1½ to 4 in (4-10 cm) long; capped by a thin, star-shaped calyx at the stem-end and tipped with 5 hair-like floral remnants at the apex. The fruit is crisp when unripe, turns from bright-green to yellowish-green, ivory or nearly white when ripe and falls to the ground. The outer skin is glossy, very thin, soft and tender, and the flesh green, jelly-like, juicy and extremely acid. There may be a few (perhaps 6 or 7) flattened, disc-like seeds about ¼ in (6 mm) wide, smooth and brown.

Origin and Distribution

Perhaps a native of the Moluccas, the bilimbi is cultivated throughout Indonesia; is cultivated and semi-wild everywhere in the Philippines; is much grown in Ceylon and Burma. It is very common in Thailand, Malaya and Singapore; frequent in gardens across the plains of India, and has run wild in all the warmest areas of that country. It is much planted in Zanzibar. Introduced into Queensland about 1896, it was readily adopted and commercially distributed to growers.

In 1793, the bilimbi was carried from the island of Timor to Jamaica and, after some years, was planted in Cuba and Puerto Rico, Trinidad, the lowlands of Central America, Venezuela, Colombia, Ecuador, Surinam, Guyana and Brazil, and even in northern Argentina, and it is very popular among the Asiatic residents of those countries as it must be in Hawaii. Still it is grown only as an occasional curiosity in southern Florida.

Varieties

Bilimbis are all much the same wherever they are grown, but P.J. Wester reported that a form with sweet fruits had been discovered in the Philippines.

Climate

The bilimbi is a tropical species, more sensitive to cold than the carambola, especially when very young. In Florida, it needs protection from cold and wind. Ideally, rainfall should be rather evenly distributed throughout most of the year but there should be a 2- to 3-month dry season. The bilimbi is not found in the wettest zones of Malaya. The tree makes slow growth in shady or semi-shady situations. It should be in full sun.

Soil

While the bilimbi does best in rich, moist, but well-drained soil, it grows and fruits quite well on sand or limestone.

Propagation

Most efforts at grafting and budding have not been rewarding, though Wester had success in shield-budding, utilizing non-petioled, ripe, brown budwood cut 1½ to 2 in (3.8-5 cm) long. Air-layering has been practiced in Indonesia for many years. However, the tree is more widely grown from seed.

Bilimbi trees are vigorous and receive no special horticultural attention. It has been suggested that they would respond well to whatever cultural treatment gives good results with the carambola.

Season, Harvesting and Keeping Quality

In India as in Florida, the tree begins to flower about February and then blooms and fruits more or less continuously until December. The fruits are picked by hand, singly or in clusters. They need gentle handling because of the thin skin. They cannot be kept on hand for more than a few days.

Pests and Diseases

No pests or diseases have been reported specifically for the bilimbi.

Food Uses

The bilimbi is generally regarded as too acid for eating raw, but in Costa Rica, the green, uncooked fruits are prepared as a relish which is served with rice and beans. Sometimes it is an accompaniment for fish and meat. Ripe fruits are frequently added to curries in the Far East.

They yield 44.2% juice having a pH of 4.47, and the juice is popular for making cooling beverages on the order of lemonade.

Mainly, the bilimbi is used in place of mango to make chutney, and it is much preserved. To reduce acidity, it may be first pricked and soaked in water overnight, or soaked in salted water for a shorter time; then it is boiled with much sugar to make a jam or an acid jelly. The latter, in Malaya, is added to stewed fruits that are oversweet. Half-ripe fruits are salted, set out in the sun, and pickled in brine and can be thus kept for 3 months. A quicker pickle is made by putting the fruits and salt into boiling water. This product can be kept only 4 to 5 days.

The flowers are sometimes preserved with sugar.

Food Value Per 100 g of Edible Portion*	
Moisture	94.2-94.7 g
Protein	0.61 g
Fiber	0.6 g
Ash	0.31-0.40 g
Calcium	3.4 mg
Phosphorus	11.1 mg
Iron	1.01 mg
Carotene	0.035 mg
Thiamine	0.010 mg
Riboflavin	0.026 mg
Niacin	0.302 mg
Ascorbic Acid	15.5 mg

*According to analyses of fruits studied in Nicaragua and the Philippines.

Other Uses

Fruit: Very acid bilimbis are employed to clean the blade of a *kris* (dagger), and they serve as mordants in the preparation of an orange dye for silk fabrics. Bilimbi juice, because of its oxalic acid content, is useful for bleaching stains from the hands and rust from white cloth, and also tarnish from brass.

Wood: The wood is white, soft but tough, even-grained, and weighs 35 lbs/cu ft. It is seldom available for carpentry.

Medicinal Uses: In the Philippines, the leaves are applied as a paste or poulticed on itches, swellings of mumps and rheumatism, and on skin eruptions. Elsewhere, they are applied on bites of poisonous creatures. Malayans take the leaves fresh or fermented as a treatment for venereal disease. A leaf infusion is a remedy for coughs and is taken after childbirth as a tonic. A leaf decoction is taken to relieve rectal inflammation. A flower infusion is said to be effective against coughs and thrush.

In Java, the fruits combined with pepper are eaten to cause sweating when people are feeling "under the weather". A paste of pickled bilimbis is smeared all over the body to hasten recovery after a fever. The fruit conserve is administered as a treatment for coughs, beri-beri and biliousness. A sirup prepared from the fruit is taken as a cure for fever and inflammation and to stop rectal bleeding and alleviate internal hemorrhoids.

RUTACEAE

Sour Orange

A species of multiple uses, the sour orange (*Citrus aurantium* L.), is also known as bitter, bigarade, or Seville orange. In Spanish-speaking areas it may be called *naranja ácida, naranja agria,* or *naranja amarga.* In Arabia, it is *naranji;* in Italy, *melangolo;* in India, *khatta;* in Samoa, *moli;* in Guam, soap orange.

Description

The tree ranges in height from less than 10 ft (3 m) to 30 ft (9 m), is more erect and has a more compact crown than the sweet orange; has smooth, brown bark, green twigs, angular when young, and flexible, not very sharp, thorns from 1 in to 3⅛ in (2.5-8 cm) long. The evergreen leaves (technically single leaflets of compound leaves), are aromatic, alternate, on broad-winged petioles much longer than those of the sweet orange; usually ovate with a short point at the apex; 2½ to 5½ in (6.5-13.75 cm) long, 1½ to 4 in (3.75-10 cm) wide; minutely toothed; dark-green above, pale beneath, and dotted with tiny oil glands. The highly fragrant flowers, borne singly or in small clusters in the leaf axils, are about 1½ in (3.75 cm) wide, with 5 white, slender, straplike, recurved, widely-separated petals surrounding a tuft of up to 24 yellow stamens. From 5 to 12% of the flowers are male.

The fruit is round, oblate or oblong-oval, 2¾ to 3⅛ in (7-8 cm) wide, rough-surfaced, with a fairly thick, aromatic, bitter peel becoming bright reddish-orange on maturity and having minute, sunken oil glands. There are 10 to 12 segments with bitter walls containing strongly acid pulp and from a few to numerous seeds. The center becomes hollow when the fruit is full-grown.

Origin and Distribution

The sour orange is native to southeastern Asia. Natives of the South Sea Islands, especially Fiji, Samoa, and Guam, believe the tree to have been brought to their shores in prehistoric times. Arabs are thought to have carried it to Arabia in the 9th Century. It was reported to be growing in Sicily in 1002 A.D., and it was cultivated around Seville, Spain, at the end of the 12th Century. For 500 years, it was the only orange in Europe and it was the first orange to reach the New World. It was naturalized in Mexico by 1568 and in Brazil by 1587, and not long after it was running wild in the Cape Verde Islands, Bermuda, Jamaica, Puerto Rico and Barbados. Sir Walter Raleigh took sour orange seeds to England; they were planted in Surrey and the trees began bearing regular crops in 1595, but were killed by cold in 1739.

Spaniards introduced the sour orange into St. Augustine, Florida. It was quickly adopted by the early settlers and local Indians and, by 1763, sour oranges were being exported from St. Augustine to England. Sour orange trees can still be found in Everglades hammocks on the sites of former Indian dwellings. The first sweet orange budwood was grafted onto sour orange trees in pioneer dooryards and, from that time on, the sour orange became more widely grown as a rootstock in all citrus-producing areas of the world than for its fruit or other features. Today, the sour orange is found growing wild even in southern Georgia and from Mexico to Argentina.

It is grown in orchards or groves only in the Orient and the various other parts of the world where its special products are of commercial importance, including southern Europe and offshore islands, North Africa, the Middle East, Madras, India, West Tropical Africa, Haiti, the Dominican Republic, Brazil and Paraguay.

Varieties

There are various well-established forms of the sour orange. In the period 1818-1822, 23 varieties were described and illustrated in Europe. A prominent subspecies is the Bergamot orange, *C. aurantium* var. *bergamia* Wight & Arn., grown in the Mediterranean area since the 16th Century but commercially only in Italy. Trees grown in California and Florida under this name are actually the 'Bouquet' variety of sour orange (see below). The flowers of the Bergamot are small, sweetly fragrant; the fruits round or pear-shaped, with strongly aromatic peel and acid pulp.

The myrtle-leaved orange (*C. aurantium* var. *myrtifolia),* is a compact shrub or tree with small leaves and no thorns. It was found as a bud mutation on trunks of old sour orange trees in Florida. It is propagated and grown only on the French and Italian Riviera for its small fruits which are preserved in brine and exported for candying.

Apart from these special types, there are several groups of sour oranges, within which there are placed certain cultivars:

1) *Normal group* (large, seedy fruits)
'African', 'Brazilian', 'Rubidoux', 'Standard', 'Oklawaha' and 'Trabut'. 'Oklawaha' originated in the United States. It has large fruits rich in pectin and is prized for marmalade.

Fig. 35: The sour orange (*Citrus aurantium*) has a rough, fairly thick skin, very sour juice.

2) *Aberrant group*

'**Daidai**', or 'Taitai', popular in Japan and China. Its fruits are large with very thick peel, very acid pulp, and many seeds. The tree is somewhat dwarf and almost thornless; immune to citrus canker in the Philippines. It is prized for its flower buds which are dried and mixed with tea for their scent.

'**Goleta**' has medium-large fruits with juicy, medium-sour pulp and very few seeds. The tree is of medium size and almost thornless.

'**Bouquet**' has small, deep-orange fruits, acid, with few seeds. The tree is less than 10 ft (3 m) high and is grown as an ornamental.

3) *Bittersweet group* includes any sweet-acid forms of the sour orange introduced by Spaniards and formerly found growing in the Indian River region of Florida. These oranges are often seen in a naturalized state in the West Indies. The peel is orange-red, the pulp is darker in hue than that of the normal sour orange.

'**Paraguay**' was introduced from Paraguay in 1911. The fruit is of medium size, with sweet pulp, moderately seedy. The tree is large, thorny and hardy.

Among other forms of sour orange, there is in India a type called 'Karna', 'Khatta' or 'Id Nimbu', identified as *C. aurantium* var. *khatta* (or *C. karna* Raf.) but suspected of being a hybrid of sour orange and lemon. The fruits are typical sour oranges but the flowers are red-tinted like those of the lemon.

Two cultivars are grown as rootstocks for the sweet orange in China:

'**Vermilion Globe**' has oblate fruits containing 30 to 40 seeds. The tree has long, narrow, pointed leaves.

'**Leather-head**' has small, oblate, rough fruits with 20 seeds. The tree has elliptic, blunt leaves.

Cultivars grown especially for the production of Neroli oil in France and elsewhere, have flowers in large, more concentrated clusters than the ordinary types of sour orange. One of these, 'Riche Défouille', has unusual, wingless leaves.

Climate

The sour orange flourishes in subtropical, near-tropical climates, yet it can stand several degrees of frost for short periods. Generally it has considerable tolerance of adverse conditions. But the Bergamot orange is very sensitive to wind and extremes of drought or moisture.

Soil

Unlike its sweet relative, the sour orange does well on low, rich soils with a high water table and is adapted to a wide range of soil conditions.

Propagation

Sour orange trees volunteer readily from self-sown seeds. As generally grown for rootstock for sweet oranges, they are raised in nurseries for 1 or 2 years and then budded. Growth of the seedlings, especially in diameter, has been expedited by weekly applications of gibberellic acid to the stems, making it possible to bud them much earlier.

Culture

In the proper climatic and soil conditions, the sour orange is self-maintaining and receives only a modicum of cultural attention. It has an extraordinary ability to survive with no care at all. Some trees in Spain are said to be over 600 years old and one tree in a tub at Versailles, which, of course, must be carefully tended, was reportedly planted in the year 1421.

Pests and Diseases

The sour orange is subject to most of the pests that attack the sweet orange. In addition to its susceptibility to the disease called tristeza, the tree is liable to other viruses—crinkly leaf, gummy bark, psorosis, and xyloporosis. The Division of Plant Industry of the Florida State Department of Agriculture has recorded the following fungal problems as sometimes seen: leaf spot (*Alternaria citri, Cercospora penzigii, Mycophaerella horii, Cladosporium oxysporum,* and *Phyllosticta hesperidearum*); greasy spot (*Cercospora citri-grisea*); tar spot (*C. gigantea*); leprosis (*Cladosporium herbarum*); mushroom root rot (*Clitocybe tabescens*); anthracnose (*Colletotrichum gloeosporioides*); thread blight (*Corticium koleroga* and *C. stevensii*); gummosis and dieback (*Diaporthe citri*); foot rot and root rot (*Fusa-rium oxysporum, Macrophomia phaseolina, Phytophthora* spp.); heart rot and wood rot (*Fomes applanatus, Ganoderma sessilis, Xylaria polymorpha*), and others.

Food Uses

The normal types of sour orange are usually too sour to be enjoyed out-of-hand. In Mexico, however, sour oranges are cut in half, salted, coated with a paste of hot chili peppers, and eaten.

The greatest use of sour oranges as food is in the form of marmalade and for this purpose they have no equal. The fruits are largely exported to England and Scotland for making marmalade. Sour oranges are used primarily for marmalade in South Africa.

The juice is valued for ade and as a flavoring on fish and, in Spain, on meat during cooking. In Yucatan, it is employed like vinegar. In Egypt and elsewhere, it has been fermented to make wine.

"Bitter orange oil", expressed from the peel, is in demand for flavoring candy, ice cream, baked goods, gelatins and puddings, chewing gum, soft drinks, liqueurs and pharmaceutical products, especially if the water- or alcohol-insoluble terpenes and sesquiterpenes are removed. The oil is produced in Sicily, Spain, West Africa, the West Indies, Brazil, Mexico and Taiwan.

The essential oil derived from the dried peel of immature fruit, particularly from the selected types—'Jacmel' in Jamaica and the much more aromatic 'Curacao orange' (var. *curassaviensis*)—gives a distinctive flavor to certain liqueurs.

"Neroli oil", or "Neroli Bigarade Oil", distilled from the flowers of the sour orange, has limited use in flavoring candy, soft-drinks and liqueurs, ice cream, baked goods and chewing gum.

'Petitgrain oil', without terpenes, is used to enhance the fruit flavors (peach, apricot, gooseberry, black currant, etc.) in food products, candy, ginger ale, and various condiments.

'Orange leaf absolute' enters into soft-drinks, ice cream, baked goods and candy.

The ripe peel of the sour orange contains 2.4 to 2.8%, and the green peel up to 14%, neohesperidin dihydrochalcone which is 20 times sweeter than saccharin and 200 times sweeter than cyclamate. Potential use as a sweetener may be hampered by the limited supply of peel.

Other Uses

Soap substitute: Throughout the Pacific Island, the crushed fruit and the macerated leaves, both of which make lather in water, are used as soap for washing clothes and shampooing the hair. Safford described the common scene in Guam of women standing in a river with wooden trays on which they rub clothing with sour orange pulp, then scrub it with a corncob. He wrote: "Often the entire surface of the river where the current is sluggish is covered with decaying oranges." On the islands of Zanzibar and Pemba, the fruits are used for scouring floors and brass.

Food Value Per 100 g of Edible Portion		
	Fruit (raw)	Fruit (raw, with only superficial layer of peel removed)*
Calories	37-66	
Moisture	83-89.2 g	77.8-83.1 g
Protein	0.6-1.0 g	0.154-0.167 g
Fat	trace-0.1 g	0.05-0.07 g
Carbohydrates	9.7-15.2 g	?
Fiber	0.4 g	1.8-2.2 g
Ash	0.5 g	0.57-0.69 g
Calcium	18-50 mg	64.3-81.9 mg
Iron	0.2 mg	0.22-0.85 mg
Phosphorus	12 mg	19.6-20.4 mg
Vitamin A	290 mcg or 200 I.U.	0.055-0.07 mg
Thiamine	100 mcg	0.048-0.059 mg
Riboflavin	40 mcg	0.030-0.040 mg
Niacin	0.3 mg	0.282-0.400 mg
Ascorbic Acid	45-90 mg	55.2-103.5 mg

*Sampled in Guatemala and El Salvador.

Fig. 36: Dried peel of the locally-grown sour orange yields the essential oil that flavors "Curacao liqueur".

Perfumery: All parts of the sour orange are more aromatic than those of the sweet orange. The flowers are indispensable to the perfume industry and are famous not only for the distilled Neroli oil but also for "orange flower absolute" obtained by fat or solvent extraction. During favorable weather in southern France, 2,200 lbs (1,000 kg) of flowers will yield 36 to 53 oz (1,000-1,500 g) of oil.

Neroli oil consists of 35% terpenes (mainly dipentene, pinene and camphene), 30% 1-linalool, and 4% geraniol and nerol, 2% d-terpineol, 6% d-nerolidol, traces of decyclic aldehyde, 7% 1-linalyl acetate, 4% neryl and geranyl acetates, traces of esters of phenylacetic acid and benzoic acid, as much as 0.1% methyl anthranilate, and traces of jasmone, farnesol, and palmitic acid. Orange flower water is usually a by-product of oil production.

Petitgrain oil is distilled from the leaves, twigs and immature fruits, especially from the Bergamot orange. Both Petitgrain and the oil of the ripe peel are of great importance in formulating scents for perfumes and cosmetics. Petitgrain oil is indispensable in fancy eau-de-cologne. The seed oil is employed in soaps.

Honey: The flowers yield nectar for honeybees.

Wood: The wood is handsome, whitish to pale-yellow, very hard, fine-grained, much like boxwood. It is valued for cabinetwork and turnery. In Cuba it is fashioned into baseball bats.

Medicinal Uses: Sour orange juice is antiseptic, antibilious and hemostatic. Africans apply the cut-open orange on ulcers and yaws and areas of the body afflicted with rheumatism. In Italy, Mexico and Latin America generally, decoctions of the leaves are given for their sudorific, antispasmodic, stimulant, tonic and stomachic action. The flowers, prepared as a sirup, act as a sedative in nervous disorders and induce sleep. An infusion of the bitter bark is taken as a tonic, stimulant, febrifuge and vermifuge.

The fresh young leaves contain as much as 300 mg of ascorbic acid per 100 g. The mature leaf contains 1-stachyhydrine.

Orange

One of the most widely favored of the world's fruits, the orange, sweet orange, or round orange, was for many years known as *Citrus aurantium* var. *sinensis* L. and considered to be a form of the sour orange (q.v.). It is still not universally agreed to be a distinct species, *C. sinensis* Osbeck, but it is usually treated as though it were. One of its first recorded regional names was the Persian *narang*, from which were derived the Spanish name, *naranja*, and the Portuguese, *laranja*. In some Caribbean and Latin American areas, the fruit is called *naranja de China*, *China dulce*, or simply *China* (pronounced *cheena*).

Description

The orange tree, reaching 25 ft (7.5 m) or, with great age, up to 50 ft (15 m), has a rounded crown of slender branches. The twigs are twisted and angled when young and may bear slender, semi-flexible, bluntish spines in the leaf axils. There may be faint or conspicuous wings on the petioles of the aromatic, evergreen, alternate, elliptic to ovate, sometimes faintly toothed "leaves" — technically solitary leaflets of compound leaves. These are 2½ to 6 in (6.5-15 cm) long, 1 to 3¾ in (2.5-9.5 cm) wide. Borne singly or in clusters of 2 to 6, the sweetly fragrant white flowers, about 2 in (5 cm) wide, have a saucer-shaped, 5-pointed calyx and 5 oblong, white petals, and 20 to 25 stamens with conspicuous yellow anthers. The fruit is globose, subglobose, oblate or somewhat oval, 2½ to 3¾ in (6.5-9.5 cm) wide. Dotted with minute glands containing an essential oil, the outer rind (epicarp) is orange or yellow when ripe, the inner rind (mesocarp) is white, spongy and non-aromatic. The pulp (endocarp), yellow, orange or more or less red, consists of tightly packed membranous juice sacs enclosed in 10 to 14 wedge-shaped compartments which are readily separated as individual segments. In each segment there may be 2 to 4 irregular seeds, white externally and internally, though some types of oranges are seedless. The sweet orange differs physically from the sour orange in having a solid center.

Origin and Distribution

The orange is unknown in the wild state; is assumed to have originated in southern China, northeastern India, and perhaps southeastern Asia (formerly Indochina). It was carried to the Mediterranean area possibly by Italian traders after 1450 or by Portuguese navigators around 1500. Up to that era, citrus fruits were valued by Europeans mainly for medicinal purposes, but the orange was quickly adopted as a luscious fruit and wealthy persons grew it in private conservatories, called "orangeries". By 1646 it had been much publicized and was well known.

Spaniards undoubtedly introduced the sweet orange into South America and Mexico in the mid-1500's, and probably the French took it to Louisiana. It was from New Orleans that seeds were obtained and distributed in Florida about 1872 and many orange groves were established by grafting the sweet orange onto sour orange rootstocks. Arizona received the orange tree with the founding of missions between 1707 and 1710. The orange was brought to San Diego, California, by those who built the first mission there in 1769. An orchard was planted at the San Gabriel Mission around 1804. A commercial orchard was established in 1841 on a site that is now a part of Los Angeles. In 1781, a surgeon and naturalist on the ship, "Discovery", collected orange seeds in South Africa, grew seedlings on board and presented them to tribal chiefs in the Hawaiian Islands on arrival in 1792. In time, the orange became commonly grown throughout Hawaii, but was virtually abandoned after the advent of the Mediterranean fruit fly and the fruit is now imported from the United States mainland.

The orange has become the most commonly grown tree fruit in the world. It is an important crop in the Far East, the Union of South Africa, Australia, throughout the Mediterranean area, and subtropical areas of South America and the Caribbean. The United States leads in world production, with Florida, alone, having an annual yield of more than 200 million boxes, except when freezes occur which may reduce the crop by 20 or even 40%. California, Texas and Arizona follow in that order, with much lower production in Louisiana, Mississippi, Alabama and Georgia. Other major producers are Brazil, Spain, Japan, Mexico, Italy, India, Argentina and Egypt. In Brazil, oranges are grown everywhere in the coastal plain and in the highlands but most extensively in the States of Sao Paulo and Rio de Janeiro, where orange culture rose sharply in the years immediately following World War II and is still advancing. Mexico's citrus industry is located largely in the 4 southern states of Nuevo Leon, Tamaulipas, San Luis Potosi and Veracruz. The orange crop is over one million MT and Nuevo Leon has 20 modern packing plants, mostly with fumigation facilities. Large quantities of fresh oranges and orange juice concentrate are exported to the United States and small shipments go to East Germany, Canada and Argentina. However, overproduction has glutted domestic markets and brought down prices and returns to the farmer to such an extent that plantings have declined and growers are switching to grapefruit. Cuba's crop has become nearly ⅓ as large as that of Florida. Lesser quantities are produced in Puerto Rico, Central America (especially

Guatemala), some of the Pacific Islands, New Zealand, and West Africa, where the fruit does not acquire an appealing color but is popular for its quality and sweetness. Many named cultivars have been introduced and grown in the Philippines since 1912, but the fruit is generally of low quality because of the warm climate.

Varieties

Most of the oranges grown in California are of 2 cultivars: the 'Washington Navel' and the 'Valencia'. Florida's commercial cultivars are mainly: (early) 'Hamlin'; (mid-season) 'Pineapple'; (late) 'Valencia'.

The 'Washington Navel' (formerly known as 'Bahia') originated, perhaps as a mutant in Bahia, Brazil, before 1820. It was introduced into Florida in 1835 and several other times prior to 1870. In 1873, budded trees reached California where the fruit matures at the Christmas season. It is large but with a thick, easily removed rind; not very juicy; of excellent flavor, and seedless or nearly so. Ease of peeling and separation of segments makes this the most popular orange in the world for eating out-of-hand or in salads. Limonene content of the juice results in bitterness when pasteurized and therefore this cultivar is undesirable for processing. The tree needs a relatively cool climate and should not be grown below an elevation of 3,300 ft (1,000 m) in tropical countries. Today it is commercially grown, not only in Brazil and California, but also in Paraguay, Spain, South Africa, Australia and Japan.

'Trovita', a non-navel seedling raised in 1914-1915 at the Citrus Experiment Station in California and released in 1935, is milder in flavor and has a few seeds, but may be earlier in season, and it has been considered promising in hot, dry regions unsuitable for 'Washington Navel'. There are several other named variations such as 'Robertson Navel', 'Summer Navel', 'Texas Navel', and the externally attractive 'Thompson Navel' which was grown in California for a time but dropped because of its poor quality. Various mutants, more suitable for warmer climates, have been selected and named in Florida, including 'Dream', 'Pell', 'Summerfield', 'Surprise'—the latter being more productive than 'Washington Navel' in Florida but still not grown to any extent. 'Bahiamina' is a small version of the 'Washington Navel' developed in Brazil in the late 1940's. It follows 'Pera' and 'Natal' sweet oranges in importance in tropical Bahia.

'Valencia', or 'Valencia Late', is the most important cultivar in California, Texas and South Africa. It has been the leader in Florida until recently. In 1984, 40% of the oranges being planted in Florida were 'Valencia', 60% were 'Hamlin'. The 'Valencia' may have originated in China and it was presumably taken to Europe by Portuguese or Spanish voyagers. The well-known English nurseryman, Thomas Rivers, supplied plants from the Azores to Florida in 1870 and to California in 1876. In Florida, it was quickly appreciated and cultivated, at first labeled 'Brown' and later renamed 'Hart's Tardiff', 'Hart' and 'Hart Late' until it was recognized as identical to the 'Valencia' in California. It was not propagated for sale in California until 1916 and was slow to be adopted commercially. It is smaller than the 'Washington Navel', with a thinner, tighter rind; is far juicier and richer in flavor; nearly seedless except in Chile where the dry climate apparently allows better pollination and development of many more seeds—up to 980 in 44 lbs (20 kg). It needs a warm climate. In fact, it is the most satisfactory orange for the tropics, even though it may not develop full color in warm regions. In Colombia, the quality is good from sea-

level to 5,000 ft (1,600 m). It bears two crops a year, overlapping and giving it the great advantage of a late and long season lasting until midsummer. The fruits on the trees in spring will "regreen", lose their orange color and turn green at the stem end, but the quality is not affected. They were formerly dyed to improve market appearance but since the 1955 Food & Drug Administration ban on the synthetic dyes used on oranges, they have been colored by exposure to ethylene gas in storage. The gas removes the chlorophyll layer, revealing the orange color beneath. "Degreening" does not occur in California where 'Valencia' oranges from one growing area or another are marketed from late spring through fall.

'Lue Gim Gong' was claimed to be a hybrid of 'Valencia' and 'Mediterranean Sweet' made by a Chinese grower in 1886. 'Lue Gim Gong' was awarded the Wilder Silver Medal by the American Pomological Society in 1911 but, later on, his "hybrid" was judged to be a nucellar seedling of 'Valencia'. Propagated and distributed by Glen St. Mary Nurseries in 1912, this cultivar closely resembles 'Valencia', matures and is marketed with its parent without distinction. It is best cited as the 'Lue Gim Gong Strain' of 'Valencia'. 'Mediterranean Sweet' was introduced into Florida from Europe in 1875, was briefly popular, but is no longer grown.

Certain strains of 'Valencia' are classed as summer oranges because the fruits can be left on the trees longer without dehydrating. One is known as 'Pope', 'Pope Summer', or 'Glen Summer'. It was found in a grove of 'Pineapple' oranges near Lakeland about 1916, was propagated in 1935, and trademarked in 1938. On sour orange or sweet orange rootstocks in hammock soils, the fruit matures in April but is still in good condition on the tree in July and August.

'Rhode Red Valencia' was discovered in 1955 in a grove near Sebring, Florida, by Paul Rhode, Sr., of Winter Haven. Some budwood was put on sour orange stock which caused dwarfing and some on rough lemon which produced large, vigorous, productive trees. In 1974, 5 trees were accepted into the Citrus Budwood Registration Program but there was no budwood free of exocortis and xyloporosis viruses. The fruit equals 'Valencia' in soluble solids, excels 'Valencia' in volume of juice, is less acid, has slightly less ascorbic acid, but has a far more colorful juice due to its high content of cryptoxanthin, a precursor of vitamin A which remains nearly stable during processing.

In Cuba, 'Campbell Valencia' (a 1942 seedling similar to 'Valencia'), 'Frost Valencia' (a 1915 nucellar seedling of 'Valencia'), and 'Olinda Valencia' (a virus-free nucellar seedling of 'Valencia' discovered in California in 1939), each on 2 different rootstocks—sour orange and Cleopatra mandarin—were test-planted in 1973 and evaluated in 1982. 'Olinda Valencia' on sour orange excelled in quality and in productivity.

'Hamlin', discovered in 1879 near Glenwood, Florida, in a grove later owned by A.G. Hamlin, is small, smooth, not highly colored, seedless and juicy but the juice is pale. The fruit is of poor-to-medium quality but the tree is high-yielding and cold-tolerant. The fruit is harvested from October to December and this cultivar is now the leading early orange in Florida. On pineland and hammock soil it is budded on sour orange which gives a high solids content. On sand, it does best on rough lemon rootstock.

'Homosassa', a selected Florida seedling named in 1877, is of rich orange color, of medium size, and excellent flavor. It was formerly one of the most valued midseason oranges in Florida but it is too seedy to maintain that position. It is no longer planted except perhaps in Texas and Louisiana.

'Shamouti' ('Jaffa'; 'Khalili'; 'Khalili White')—originated as a limb sport on a 'Beledi' tree near Jaffa, Israel, in 1844; introduced into Florida about 1883; oval, medium-large; peel entirely orange when ripe; leathery, thick, easy to remove; pulp very juicy, of good quality. Constitutes 75% of the Lebanese and Israeli crops; is one of the 2 main cultivars in Syria; was formerly an important, midseason, cold-tolerant, cultivar in Florida and was grown in all other orange-growing regions of the United States. However, the tree tends to alternate-bearing, the fruit does not hold for long on the tree and is subject to the fungus, *Alternaria citri,* and it is no longer planted in this country.

'Parson Brown' was discovered in a grove owned by Parson Brown in Wester, Florida; was purchased, propagated and distributed by J.L. Carney between 1870 and 1878. It is rough-skinned, with pale juice; moderately seedy; of low-to-medium quality. It was formerly popular in Florida because of its earliness and long season (October through December), but has been largely replaced by 'Hamlin'. It is grown in Texas, Arizona and Louisiana but is not profitable in California where it matures at the same time as 'Washington Navel'. It does not develop acceptable quality in the tropics.

'Pineapple' is a seedling found in a grove near Citra, Florida. It was propagated in 1876 or 1877 under the name of 'Hickory'. It is pineapple-scented, smooth, highly colored, especially after cold spells; of rich, appealing flavor, and medium-seedy. It is the favorite midseason orange in Florida, its tendency to preharvest drop having been overcome by nutrition and spray programs. If the crop is allowed to remain too long on the tree, it may induce alternate-bearing. It is grown to some extent in Texas, rarely in California; succeeds on sour orange rootstock in low hammock land, on rough lemon in light sand. Seedless mutants of 'Pineapple' have been produced by seed irradiation. This cultivar does fairly well in tropical climates though not as well as 'Valencia'.

'Queen' is a seedling of unknown origin which was found in a grove near Bartow, Florida. Because it survived the freeze of 1894-95, it was propagated in 1900 under the name 'King' which was later changed to 'Queen'. It is much like 'Pineapple', has fewer seeds, higher soluble solids, persists on the tree better in dry spells; is high-yielding and somewhat more cold-tolerant than 'Pineapple'.

'Blood Oranges' are commonly cultivated in the Mediterranean area, especially in Italy, and also in Pakistan. They are grown very little in Florida where the red coloration rarely develops except during periods of cold weather. In California they are grown only as novelties. Among the well-known cultivars in this group are 'Egyptian', which tends to develop a small navel; 'Maltese', 'Ruby', and 'St. Michael'.

Pollination

Orange blossoms yield very little pollen and orange growers do not practice artificial pollination. However, there is evidence of self-incompatibility and need for cross-pollination in the TANGOR and TANGELO (qq.v.).

Climate

The orange is subtropical, not tropical. During the growing period, the temperature should range from 55° to 100°F (12.78°-37.78°C). In the winter dormancy, the ideal temperature range is 35° to 50°F (1.67°-10°C). Mature, dormant trees have survived 10 hours at temperatures below 25°F (-3.89°C) but fruit is damaged by freezing—30° to 26°F (-1.11°-3.33°C). Young trees may be killed outright by even brief frosts. Hardiness, however, varies with the cultivar and rootstock. Seedling orange trees of bearing age are capable of enduring more cold than budded cultivars. Prolonged cold is more injurious than short periods of freezing temperatures. In Florida, many efforts have been made to protect orange trees from winter cold, which is most damaging if preceded or accompanied by drought.

In the early days, slatted shadehouses were erected over young groves. Windbreaks have been planted on the northeast exposure. Old automobile tires have been burned in piles throughout groves. A commercially produced heater has been fueled and lit in the coldest predawn hours. Helicopters have been flown back and forth to cause movement of air, and, more recently, wind machines have been installed. Most recent, and most effective are overhead sprinklers which give maximum protection from cold damage.

Favorable annual precipitation varies from 5 to 20 in (12.5-50 cm), though oranges are frequently grown in areas receiving 40 to 60 in (100-150 cm) of rain. Benthall says that in the damp climate of Lower Bengal, the fruits lack juice and are usually very sour. California's generally dry climate contributes to more intense color in the orange peel than is seen in humid areas. Success in orange culture depends a great deal on the selection of cultivars tolerant of the weather conditions where they are to be grown.

Soil

The best soil for orange-growing in Florida is known as "Lakeland fine sand," well-drained, and often identified as high hammock or high pineland soil. There must be adequate depth for good root development. Shallow soils of high water-holding ability are avoided. In Egypt, it has been found that where the water table is too high— 30 in (78 cm) or less below the surface of the soil—root growth, vegetative vigor and fruit yield of orange trees are greatly reduced. In the alkaline soil of South Florida, neglected orange trees develop chlorosis and gradually decline. Many old groves planted in the southern part of the state to avoid cold have been totally lost. In California, the best soils for orange groves are deep loams. It is important to select the appropriate rootstock for particular soil conditions.

Propagation

While the orange will often come true from seed because of nucellar embryos, the common means of assuring the reproduction of cultivars of known quality is by budding onto appropriate rootstocks. It is believed that budding was practiced by Europeans during the 16th and 17th Centuries, but, with the realization that seedling trees were more vigorous and productive, Italian and Spanish orange growers went back to planting seeds. Fortunately, budded orange trees from Europe had been imported into Florida in 1824 and budwood from these and of others later brought in from England was utilized in topworking existing sour and sweet orange seedlings. It was soon apparent that the budded trees came into bearing earlier than seedlings, were less thorny, and matured uniformly. The sweet orange lost popularity as a rootstock because of its susceptibility to foot rot. Sour orange, resistant to foot rot, became the preferred rootstock in low hammock and flatwoods soils with high water table until the discovery of the virus disease, tristeza, in Florida orange groves in 1952. This caused many to switch from the susceptible sour orange to 'Cleopatra mandarin'. Unfortunately, trees on 'Cleopatra' stock are reduced in size, they have lower yields than those on sour orange, and acidity of the fruit is elevated.

As citrus-growing stretched southward into high pineland, rough lemon (*Citrus jambhiri*) rootstock gained favor and was found to induce more rapid and vigorous growth and earlier bearing, counterbalancing its sensitivity to cold and tendency toward foot rot. Rough lemon became the dominant rootstock in Florida until it was found to be extremely susceptible to blight and was abandoned. Sour orange has been reinstated in recent years because tristeza has been more or less dormant since the 1940's and sour orange is now the prevailing stock for 50% of the orange and grapefruit trees in the state. In second place is the 'Carrizo citrange', resistant to tristeza but subject to exocortis and also to blight though less so than rough lemon. 'Carrizo' is somewhat resistant to the burrowing nematode and gives a little higher yield than the similar rootstocks. Growers are advised to quickly replace blight-affected orange trees on rough lemon with new plants on 'Carrizo' held ready for this purpose. Because exocortis can now be detected quickly, it has become possible to utilize 'Carrizo' as a rootstock for hundreds of thousands of orange trees in Florida.

About 90% of commercial orange groves in Queensland are on rough lemon rootstock, as are 90% of the citrus trees in Jamaica. In Egypt, rough lemon rootstock has been found short-lived on heavy soils. In that country, early budding was done on citron (*Citrus medica* L.) but that stock was abandoned when sour orange was found much more desirable on the prevailing loamy-clay. Second to the sour orange rootstock is the Egyptian lime, locally considered "native" and used mainly on lighter soils.

In the tropical citrus-growing region of Bahia, Brazil, Rangpur lime (*C. X limonia* Osbeck) has been the dominant rootstock—95% in orchards and 100% in nurseries—but experiments in the past few years have shown that rough lemon and Cleopatra mandarin give better results. Also, 'Cleopatra' has good resistance to "citrus decline", whereas Rangpur is susceptible to Phytophthora root rot and exocortis.

Some oranges are budded onto the so-called "trifoliate orange" (*Poncirus trifoliata* Raf.) which tends to reduce the growth but is cold-tolerant and able to flourish on low, wet soils. It does poorly in light sand. Rootstocks capable of dwarfing orange trees may become necessary if close spacing is to be considered more advantageous. Trifoliate orange cultivar 'English Small' has successfully dwarfed 'Valencia'. 'Rusk' and 'Carrizo' ('Troyer') citranges (*P. trifoliata* X *C. sinensis*) show promise for semi-dwarfing of 'Valencia'. However, all of these are very susceptible to the exocortis virus. Alternative rootstocks include 'Swingle citrumelo' (*P. trifoliata* X *C.*

paradisi)—cold-hardy, resistant to tristeza, exocortis, xyloporosis, and the citrus nematode but not the burrowing nematode—and the 'Volkamer lemon' (*C. volkameriana*) which behaves much like rough lemon but gives very high yields of fruit of slightly better quality.

In India, the sweet lime (*C. limettioides* Tanaka) was found to be the best rootstock for their 'Mosambi' orange in wet zones with high maximum temperatures.

Cuban horticulturists are currently experimenting with various *Citrus* species as potential rootstocks to replace sour orange.

In Florida, nurseries of seedling rootstocks must be approved by the Department of Agriculture, Division of Plant Industry. The seeds must not be more than 3 to 4 weeks old unless they have been washed, dried, then mixed with sand and kept in a cool place, or put into a plastic bag and refrigerated for a few weeks at about 40°F (4.4°C). Seeds of *P. trifoliata* are planted in the fall but sour orange and 'Cleopatra mandarin' are planted in spring. Seeds are set in rows 3 to 4 ft (0.9-1.2 m) apart and will germinate in 3 weeks. When the stems reach ½ in (1.25 cm) in diameter, the seedlings are ready for budding. The budding technique most commonly used in Florida is shield-budding by the inverted "T" method, inserting the bud 2 to 3 in (5-7.5 cm) above ground level. California propagators favor the upright "T". Usually the trees are ready for transplanting after one growing season. Mature trees that have been frozen back, or that are to be converted to more suitable cultivars, may be top-worked by cleft-grafting, crown grafting, or budding of the sprouts that arise after the tree is cut off close to the ground.

It must be kept in mind that the rootstock influences not only the rate of growth, disease resistance and productivity of the cultivar but also the physical and chemical attributes of the crop. For example, 'Valencia' oranges on sour orange stock have been found to have more dry matter in the peel, pulp and juice than those on rough lemon. 'Washington Navel' oranges on rough lemon stock have had low levels of potassium in the peel, pulp and juice; and, on 'Cleopatra mandarin' stock, even lower in the pulp and juice. Trifoliate orange rootstock produces high levels of potassium throughout the fruit. In southeastern Queensland, Australia, nearly half of the oranges for processing are grown in the Near North Coast area. There, trials of 'Valencia' on rough lemon revealed that fruit quality was inferior to that in Florida; there was bitterness in the juice and only a small percentage of the fruits met the minimum standards for processing as frozen orange juice concentrate. General quality, flavor and ascorbic acid content were considerably higher on sweet orange rootstock. Trifoliate orange gave second-best results. Rootstocks affect the chemistry of the peel oil, especially the aldehyde content, and the oil content of the peel is influenced by selection of budwood. Dr. Walter T. Swingle, one of the early and renowned plant explorers of the United States Department of Agriculture, was an authority on *Citrus* and vitally interested in rootstocks. He was convinced that they were the key to the successful future of the citrus industry.

Culture

A spacing of 25 x 25 ft (7.5x7.5 m) was standard in the past. However, many orange groves today are being close-planted and "hedged" to facilitate both manual and mechanical harvesting, and between-row alleys must be wide enough to accommodate mobile machinery for fertilizing, spraying, pruning and harvesting. There are arguments against close-spacing; mainly that, as the trees grow and become more crowded, productivity declines; also that close-spacing requires expensive pruning. However, data gathered on yields of the 'Pineapple' orange on rough lemon rootstock at Lake Alfred, Florida, over an 11-year trial, showed total yields for the period as: 2,380 boxes per acre (5,880/ha) at 25 x 20 ft (7.5x6 m)—87 trees per acre (215/ha); 3,496 boxes per acre (8,639/ha) at 20 x 15 ft (6x4.5 m)—145 trees per acre (358/ha); 4,484 boxes per acre (11,079/ha) at 15 x 10 ft (4.5-3 m)—290 trees per acre (716/ha). Other examples are given under **"Yield"**.

The young trees must be carefully tended and kept weed-free for the first 2 or 3 years in the field. Citrus trees have special nutritional requirements. The soil should be tested to determine the best balance of major and minor elements to be added. In general, orange trees need to be fertilized with N P K very soon after harvesting. The balance of major nutrients has to be considered in relation to the ultimate use of the crop. For example, extra nitrogen increases the peel oil content of oranges, while extra potassium decreases it. In California, 1 lb (0.45 kg) of nitrogen per tree per year has been found sufficient to maintain high productivity. Indian scientists, after a 4-year study, concluded that sweet oranges of the best quality were produced by applications of nitrogen at the rate of 2 lbs (0.9 kg) per year for 8-year-old trees. Orange trees are watched for signs of deficiencies which may be counteracted by foliar spraying. Leaf analysis reveals what is lacking or being applied in excess.

Efforts in northern India to control spring fruit drop with growth regulators have not been successful but pre-harvest drop has been greatly reduced. Gibberellic acid at 100 to 1,000 ppm, whether applied at full bloom or small fruit stage, has significantly increased the number of 'Washington Navel' fruits harvested.

Irrigation: Irrigation of orange trees is carefully managed. Ordinarily, it is omitted in the fall in order to avoid the production of tender new growth that would be damaged in winter cold spells. It may be very desirable in the spring dry season to prevent wilting. Excessive irrigation lowers the solids content of the fruit. The deeper the soil, the better the root system and the greater the ability to withstand drought. Soils at least 4 ft (1.2 m) deep can be given 1½ in (6.25 cm) of water as needed, whereas soils only 1½ ft (45 cm) deep should receive no more than 1 in (2.5 cm) of water at a time but more frequently.

Pruning: Orange trees are self-forming and do not need to be shaped by early pruning. Removal of water sprouts from young and older trees is important. Branches that are lower than 1 ft (30 cm) from the ground should be taken off. Deadwood from any cause—adverse soil conditions, pests or diseases, nutritional deficiencies, or cold injury—should be cut out and cut surfaces over 1 in (2.5 cm) in diameter should be sealed with pruning compound. Orange trees that are close-planted and "hedged" are being mechanically pruned by special equipment. Cuban experimenters claim that this procedure is beneficial in increasing the number of new shoots and that it decreases pest and disease problems.

In Israel, the old practice of girdling has been revived. If done in winter, it will enhance the sprouting of buds in the spring. Summer girdling increases the size of the fruits.

Harvesting

In the early days of the orange industry, harvesters climbed ladders and pulled the fruits off by hand, putting them into pails or shoulder-sacks which they later emptied into 90-lb (40.8 kg) field boxes. From 1900 to 1940, they used clippers. With the erstwhile shortage and increased cost of field labor, various changes and improvements have been made in harvesting methods. Pulling is again practiced, especially with fruits destined for processing. In the United States, Federal regulations and the individual state Department of Agriculture and state Citrus Commission control the stage of maturity at which the fruits may be picked and the grading of the fruits for marketing and shipping.

In anticipation of drastic increases in the cost of conventional harvesting, various methods of wholly or partly mechanized harvesting have been explored, including limb and tree shakers and air jets. Devices developed are not being widely utilized as yet because of the investments necessary for their acquisition and the current availability of manual labor. Manual picking is less laborious now that oranges for processing can be allowed to fall on the ground instead of being placed in sacks which have to be carried down ladders. The efficiency of hand-harvesting has been enhanced also by the use of fiberglass ladders and abscission agents which make it possible to pluck the fruit with less force and consequently greater speed. Good workers who have harvested oranges at the rate of 6.5 boxes per hour are now able to pick 9.1 boxes per hour. The effectiveness of the abscission agent depends largely on the lapsed time after spray-application and the prevailing temperature and relative humidity during that period.

Yield

On the average, a 'Washington Navel' orange tree may bear approximately 100 fruits in a season. Horticulturists at the University of Puerto Rico have selected Navel orange clones and budded them onto orange seedlings for test plantings. Of 5 that were numbered 4, 5, 6, 7 and 8, numbers 5 and 7 surpassed the others in productivity, number 7 yielding 293 fruits per tree. These two clones are considered worthy of propagation and naming. It is said that very old, large orange trees in the Mediterranean area may bear 3,000 to 5,000 oranges each year.

Growers everywhere are testing high-density as a means of gaining higher yields. In Australia, 'Valencia' orange trees 6 years old, planted 1,011 to 2,023 trees per acre (2,500-5,000/ha), yielded 24 tons/acre (60 tons/ha). 'St. Ives Valencia' trees on *P. trifoliata* rootstock and inoculated in the nursery with mildly dwarfing exocortis, were planted in 1973 at densities ranging from 270 to 2,023 trees per acre (667-5,000 trees/ha). Those at 506 trees/acre (1,250/ha) yielded 55 tons/acre (135 tons/ha). Those at 1,214 to 2,023 trees/acre (3,000-5,000/ha) yielded 105 tons/acre (260 tons/ha) until after the 4th crop, when productivity began to decline.

Keeping Quality

Oranges can be stored for 3 months at 52°F (11.11°C); up to 5 months at 36° to 39°F (2.22°-3.89°C). Deterioration in market quality is primarily due to transpiration—loss of moisture in the peel and pulp. After 2 months of storage at 68°F (20°C) and relative humidity of 60 to 80%, 'Valencia' oranges have been found to have lost 9.5% of the moisture in the peel but only 2.1% of that in the pulp. The peel becomes 50% thinner, the pulp 10%. Later, the peel is very thin, dry and brittle while the pulp is still juicy. Coating the fruits with a polyethylene/wax emulsion doubles the storage life.

Pests

Oranges and other citrus fruits are commonly affected by citrus rust mites causing external blemishing and, in extreme infestations, smaller fruits, pre-mature falling and even shedding of leaves. Citrus red mites (purple mites) and Texas citrus mites, common in summer, disfigure the surface of the fruit and the foliage mainly in the winter and during droughts. Parasitic fungi (*Hirsutella thompsonii* and *Triplosporium floridana*) help to eradicate rust mites and the Texas citrus mite.

Several scale insects prey on citrus trees. The most harmful enemy is citrus snow scale infesting the woody portions of the tree. Purple scale and glover scale suck sap from the branches, twigs, leaves and fruit. Florida red scale and yellow scale induce shedding of fruit and foliage. Chaff scale may be found on the fruit, foliage and bark and produces green spots on the fruit. Cottony cushion scale often infests young trees. Maintaining populations of the Vedalia lady beetle in nurseries and groves is a fairly effective means of controlling this scale. Parasitic wasps (*Aphytis* spp.) are able to control Citrus snow scale, purple scale and Florida red scale.

California red scale (*Aonidiella aurantii*) is fairly well controlled by insect parasites in desert orchards but chemical treatment is necessary in the San Joaquin Valley when pheromone trapping of males reveals infestations. Pheromone trapping has virtually eliminated this scale in commercial groves in Arizona.

Mealybugs, prevalent in spring and early summer, form white masses underneath and between fruits in the early stages of development and may cause shedding, and their excretion of honeydew provides a base for the fungal manifestation termed sooty mold. The whitefly in its immature stage congregates on the lower side of the leaves, sucking the sap, and also excreting honeydew

leading to sooty mold. Immature whiteflies are preyed upon by the parasitic fungi, *Aschersonia* spp. and *Aegerita* sp., which are frequently mistaken for harmful pests. The citrus blackfly, *Aleurocanthus woglumi*, deposits eggs in spiral formations on the underside of the leaves. It is a serious pest in many of the citrus regions of the world. In January 1976, an inspection program was launched in Florida with the expectation that spraying could eventually be replaced with biological control utilizing the blackfly parasites, *Amitus hesperidum* and *Prospaltella opulenta*. By 1978, the parasites were credited with a 97% reduction in the blackfly population.

Aphids (plant lice) cause leaves to curl and become crinkled. The brown citrus aphid, *Toxoptera citricidus*, is the main vector of the tristeza virus. The orange dog is a large brown-and-white caterpillar, the larva of a black-and-yellow, swallowtailed butterfly. These pests damage the trees in summer and autumn.

In 1953, it was discovered that the burrowing nematode, *Radopholus similis*, was the cause of "spreading decline" in Florida and extraordinary measures costing over 21 million dollars in the next 22 years were taken to remove infested trees, treat the soil and create buffer zones to prevent spread into other groves.

Fruit flies are a constant threat to oranges and massive steps have been taken against the spread of the Mediterranean fruit fly whenever it has appeared in Florida or California. The Caribbean fruit fly is common in Florida and oranges from this state were, until 1980, fumigated with ethylene dibromide before export. When this chemical was reported to have caused cancer in experimental animals, it was banned for export or domestic use. Instead, cold treatment for 17 days at 34°F (1.1°C) has been required. Quality of 'Valencia' oranges has remained stable for only 1 week at 40°F (4.4°C) following cold treatment; has deteriorated in a further 2 weeks at 70°F (21.1°C).

Diseases

Orange and other citrus trees are subject to a great number of fungal diseases affecting the roots, the trunk and branches, the foliage and the fruits. Greasy spot, caused by *Cercospora citri-grisea*, is seen, 2 to 9 months after severe infection, as yellow-brown, blistery, oily, brown or black spots on the foliage. Severe defoliation may follow. The fungus, *Diaporthe citri*, is responsible for gummosis, melanose, dieback and stem-end rot. The fungus, *Elsinoe australis*, causes sweet orange scab which is frequently seen on oranges in South America and in Sicily and New Caledonia. *Phytophthora megasperma*, *P. palmivora* and *P. parasitica* are common causes of foot rot.

There are also viruses and viroids usually named for the syndromes they cause—crinkly leaf; gummy bark; exocortis (scaly butt) transmitted by budwood and by tools; psorosis, xyloporosis (cachexia), transmitted only by budwood. Tristeza has been a major problem in Florida in the past and still is in Brazil. Since 1953, Florida has maintained a Citrus Budwood Registration program for the production of virus-tested citrus trees. Under this program, the Etrog citron was adopted as a test plant for identifying exocortis virus in one year's time, and techniques have been developed for identifying tristeza in a few hours instead of months.

In 1984, an outbreak of citrus canker (*Xanthomonas campestris* pr. *citri* or *Phytomonas citri*) in four wholesale citrus nurseries in Florida caused widespread alarm and forced the burning of thousands of nursery plants and a search for plants that had been sold by those nurseries, in efforts to prevent the spread of this menace. The virus causes lesions on fruits, stems, and, unlike other diseases, on both sides of the leaves; induces leaf fall and premature fruit drop and, in severe cases, the death of the tree. Canker is common in various countries including India, the Philippines, the Middle East, parts of Africa and in Brazil and Argentina. The highly virulent "Oriental Strain A" was introduced into Florida in 1910 and was eradicated in Florida and the Gulf States by 1933. In anticipation of reintroduction, pathologists have gone abroad to study the disease. By January 1986, "Strain E" had been reported in 17 nurseries and over 15 million young trees had been destroyed. Eradication programs were intensified when "Oriental Strain A" reappeared on Florida's west coast in midsummer, and 5 million more trees had to be burned.

Blight, or young tree decline (YTD), is the leading cause of losses of orange trees—up to a half-million per year—in Florida, especially 'Valencia' on rough lemon, but any cultivars on any rootstocks. Sour orange rootstock seems somewhat more resistant than the others. Blight was thought to be the result of nutritional deficiencies or physiological or soil problems. But root-grafting of healthy trees onto affected trees has shown the disease to be infectious.

Experiments at Lake Alfred have shown that substantial recovery from YTD can be achieved by early treatment of an affected tree with 20 gals (76 liters) of a 1½% solution of sodium erythorbate or erythorbic acid applied to the soil, and 10 gals (38 liters) applied as a foliar spray, plus soil application of 5 to 7½ lbs (2.2–3.3 kg) of calcium chloride or calcium nitrate—about 6 ft (1.8 m) out from the base of the trunk. Foliar sprays of urea—5 lbs (2.2 kg) per 100 gals (380 liters)—with a wetter-sticker are given to encourage new growth.

Californian scientists have traced decline of the 'Navel' orange to incompatibility with trifoliate orange rootstock (especially 'Rubidoux'; rarely 'Rich 16-6'). Malformation at the union, evident in about 20 years, fully developed in 25, takes two forms—"tongue-and-groove", and "shelf-and-shoulder" distortions.

Often, abnormal aspects of leaves, occasioned by mineral deficiencies, may be mistaken for signs of disease. Exanthema is the result of copper deficiency. Mottle-leaf indicates zinc deficiency. Yellow spot signals lack of molybdenum. On the other hand, star melanose is brought about by late copper spraying. Inspection by trained entomologists and/or plant pathologists is usually necessary to determine the actual cause, or causes, of disfigurations or decline. Citrus quarantine laws are very strict with a view to preventing the introduction and spread of pests and diseases, and failure to comply with these laws can have disastrous consequences.

Food Uses

In the past, oranges were primarily eaten fresh, out-of-hand, and many are so consumed in warm climates. In Cuba, oranges are peeled by an old-fashioned apple peeler mounted on the pushcart of fruit vendors. Today, pre-peeled oranges in plastic bags are sold to motorists by Latin American street vendors in Miami. The hand-labor of peeling oranges has limited the production of sliced oranges for use by restaurants and orange-salad packers. However, a peeling machine developed by John Webb in Clearwater, Florida, is peeling 80 oranges a minute and this device, together with his successful sectioning machine, is expected to greatly expand the commercial use of fresh oranges.

In the home, oranges are commonly peeled, segmented and utilized in fruit cups, salads, gelatins and numerous other desserts, and as garnishes on cakes, meats and poultry dishes. They were also squeezed daily in the kitchen for juice but housewives are becoming less and less inclined to do this. In South America, a dozen whole, peeled oranges are boiled in 3 pints (1.41 liters) of slightly sweetened water for 20 minutes and then strained and the liquid is poured over small squares of toast and slices of lemon and served as soup.

In the past few decades, the commercial extraction of orange juice and its marketing in waxed cartons or cans has become a major industry, though now surpassed on a grand scale by the production of frozen orange concentrate to be diluted with water and served as juice. Dehydrated orange juice (orange juice powder), developed in 1963, is sold for use in food manufacturing, adding flavor, color and nutritive elements to bakery goods and many other products. Whole oranges are sliced, dried and pulverized, and the powder is added to baked goods as flavoring.

Orange slices and orange peel are candied as confections. Grated peel is much used as a flavoring and the essential oil, expressed from the outer layer of the peel, is employed commercially as a food, soft-drink and candy flavor and for other purposes. Pectin for use in fruit preserves and otherwise, is derived from the white inner layer of the peel. Finisher pulp, consisting mostly of the juice sacs after the extraction of orange juice, has become a major by-product. Dried to a moisture content of less than 10%, it has many uses as an emulsifier and binder in the food and beverage industries.

Orange wine was at one time made in Florida from fruits too affected by cold spells to be marketed. It is presently produced on a small scale in South Africa. Orange wine and brandy are made in Brazil from fruits which have been processed for peel oil and then crushed.

Food Value

The chemistry of the orange is affected by many factors. On the average, 'Valencia', 'Washington Navel', and other commercial oranges have been found to possess the values shown on the next page.

	Fruit (fresh)	Juice (fresh)*	Juice (canned, unsweetened, undiluted)	Frozen concentrate (unsweetened, undiluted)	Juice (dehydrated)	Orange Peel (raw)**
Food Value Per 100 g of Edible Portion						
Calories	47–51	40–48	223	158	380	
Moisture	86.0 g	87.2–89.6 g	42.0 g	58.2 g	1.0 g	72.5%
Protein	0.7–1.3 g	0.5–1.0 g	4.1 g	2.3 g	5.0 g	1.5 g
Fat	0.1–0.3 g	0.1–0.3 g	1.3 g	0.2 g	1.7 g	0.2 g
Carbohydrates	12.0–12.7 g	9.3–11.3 g	50.7 g	38.0 g	88.9 g	25.0 g
Fiber	0.5 g	0.1 g	0.5 g	0.2 g	0.8 g	-
Ash	0.5–0.7 g	0.4 g	1.9 g	1.3 g	3.4 g	0.8 mg
Calcium	40–43 mg	10–11 mg	51 mg	33 mg	84 mg	161 mg
Phosphorus	17–22 mg	15–19 mg	86 mg	55 mg	134 mg	21 mg
Iron	0.2–0.8 mg	0.2–0.3 mg	1.3 mg	0.4 mg	1.7 mg	0.8 mg
Sodium	1.0 mg	1.0 mg	5 mg	2 mg	8.0 mg	3.0 mg
Potassium	190–200 mg	190–208 mg	942 mg	657 mg	1,728 mg	212 mg
Vitamin A	200 I.U.	200 I.U.	960 I.U.	710 I.U.	1,680 I.U.	420 I.U.
Thiamine	0.10 mg	0.09 mg	0.39 mg	0.30 mg	0.67 mg	0.12 mg
Riboflavin	0.04 mg	0.03 mg	0.12 mg	0.05 mg	0.21 mg	0.09 mg
Niacin	0.4 mg	0.4 mg	1.7 mg	1.2 mg	2.9 mg	0.9 mg
Ascorbic Acid	45–61 mg	37–61 mg	229 mg	158 mg	359 mg	136 mg

*Volatile properties include: ethyl, *iso*amyl and phenylethyl alcohols; acetone; acetaldehyde; formic acid; esters of formic, acetic and caprylic acids; geraniol and terpineol. The juice also contains β-sitosteryl-*D*-glucoside and β-sitosterol.

**Orange Peel Oil: *d*-limonene (90%); citral; citranellal; methyl ester of anthranilic acid; decyclic aldehyde; linalool; *d-l*-terpineol; nonyl alcohol; methyl anthranilate; and traces of caprilic acid esters.

Toxicity

Persons in close proximity to orange trees in bloom may have adverse respiratory reactions. Sawdust of the wood of orange trees, formerly used for polishing jewelry, has caused asthma. Excessive contact with the volatile oils in orange peel can produce dermatitis. People who suck oranges often suffer skin irritation around the mouth. Those who peel quantities of oranges may have rash and blisters between the fingers. If they touch their faces, they are apt to have facial symptoms as well. In southern Florida, a young woman shook an orange tree in order to cause the fruit to fall. An hour later, she broke out in hives, presumably from exposure to a spray of citrus oils from the ruptured peduncles, stem-end peel, and broken leaf petioles. A similar reaction has occurred from shaking down the fruits of a lime tree in Miami. Sensitive individuals may have respiratory reactions in proximity to the volatile emanations from broken orange peel.

Other Uses

Pulp: Citrus pulp (¾ being a by-product of orange juice extraction) is highly valued as pelleted stockfeed with a protein content of 6.58 to 7.03%, and it is also being marketed as cat litter. It is a source of edible yeast, non-potable alcohol, ascorbic acid, and hesperidin.

Peel: In addition to its food uses, orange peel oil is a prized scent in perfume and soaps. Because of its 90–95% limonene content, it has a lethal effect on houseflies, fleas and fireants. Its potential as an insecticide is under investigation. It is being used in engine cleaners and in waterless hand-cleaners in heavy machinery repair shops. It is commercially produced mainly in California and Florida, followed distantly by Italy, Israel, Jamaica, South Africa, Brazil and Greece, in that order. Terpenes extracted from the outer layer of the peel are important in resins and in formulating paints for ships. Australians have reported that a shipment of platypuses sent to the United States in the 1950s was fed mass-produced worms raised on orange peel.

Seeds: Oil derived from orange and other citrus seeds is employed as a cooking oil and in soap and plastics. The high-protein seed residue is suitable for human food and an ingredient in cattlefeed, and the hulls enter into fertilizer mixtures.

Flowers and foliage: The essential oils distilled from orange flowers and foliage are important in perfume manufacturing. Some Petitgrain oil is distilled from the leaves, flowers, twigs, and small, whole, unripe fruits.

Nectar: The nectar flow is more abundant than that from any other source in the United States and is actually a nuisance to grove workers in California, more moderate in Florida. It is eagerly sought by honeybees and the delicious, light-colored honey is widely favored, though it darkens and granulates within a few months. Citrus honey constitutes 25% of all honey produced in California each year. There are efforts to time pest-control spraying to avoid adverse effects on honeybees during the period of nectar-gathering.

Wood: The wood is yellowish, close-grained and hard but prone to attack by drywood termites. It has been valued for furniture, cabinetwork, turnery and engraver's blocks. Branches are fashioned into walking-sticks. Orange wood is the source of "orange sticks" used by manicurists to push back the cuticle.

Medicinal Uses: Oranges are eaten to allay fever and catarrh. The roasted pulp is prepared as a poultice for skin diseases. The fresh peel is rubbed on acne. In the mid-1950s, the health benefits of eating peeled, whole oranges was much publicized because of its protopectin, bioflavonoids and inositol (related to vitamin B). The orange contains a significant amount of the vitamin-like glucoside, hesperidin, 75-80% of it in the albedo, rag and pulp. This principle, also rutin, and other bioflavonoids were for a while much advocated for treating capillary fragility, hemorrhages and other physiological problems, but they are no longer approved for such use in the United States.

An infusion of the immature fruit is taken to relieve stomach and intestinal complaints. The flowers are employed medicinally by the Chinese people living in Malaya. Orange flower water, made in Italy and France as a cologne, is bitter and considered antispasmodic and sedative. A decoction of the dried leaves and flowers is given in Italy as an antispasmodic, cardiac sedative, antiemetic, digestive and remedy for flatulence. The inner bark, macerated and infused in wine, is taken as a tonic and carminative. A vinous decoction of husked orange seeds is prescribed for urinary ailments in China and the juice of fresh orange leaves or a decoction of the dried leaves may be taken as a carminative or emmenagogue or applied on sores and ulcers. An orange seed extract is given as a treatment for malaria in Ecuador but it is known to cause respiratory depression and a strong contraction of the spleen.

Mandarin Orange

Mandarin is a group name for a class of oranges with thin, loose peel, which have been dubbed "kid-glove" oranges. These are treated as members of a distinct species, *Citrus reticulata* Blanco. The name "tangerine" could be applied as an alternate name to the whole group, but, in the trade, is usually confined to the types with red-orange skin. In the Philippines all mandarin oranges are called *naranjita*. Spanish-speaking people in the American tropics call them *mandarina*.

Description

The mandarin tree may be much smaller than that of the sweet orange or equal in size, depending on variety. With great age, some may reach a height of 25 ft (7.5 m) with a greater spread. The tree is usually thorny, with slender twigs, broad- or slender-lanceolate leaves having minute, rounded teeth, and narrowly-winged petioles. The flowers are borne singly or a few together in the leaf axils. The fruit is oblate, the peel bright-orange or red-orange when ripe, loose, separating easily from the segments. Seeds are small, pointed at one end, green inside.

Origin and Distribution

The mandarin orange is considered a native of southeastern Asia and the Philippines. It is most abundantly grown in Japan, southern China, India, and the East Indies, and is esteemed for home consumption in Australia. It gravitated to the western world by small steps taken by individuals interested in certain cultivars. Therefore, the history of its spread can be roughly traced in the chronology of separate introductions. Two varieties from Canton were taken to England in 1805. They were adopted into cultivation in the Mediterranean area and, by 1850, were well established in Italy. Sometime between 1840 and 1850, the 'Willow-leaf' or 'China Mandarin' was imported by the Italian Consul and planted at the Consulate in New Orleans. It was carried from there to Florida and later reached California. The 'Owari' Satsuma arrived from Japan, first in 1876 and next in 1878, and nearly a million budded trees from 1908 to 1911 for planting in the Gulf States. Six fruits of the 'King' mandarin were sent from Saigon in 1882 to a Dr. Magee at Riverside, California. The latter sent 2 seedlings to Winter Park, Florida. Seeds of the 'Oneco' mandarin were obtained from India by the nurseryman, P.W. Reasoner, in 1888. In 1892 or 1893, 2 fruits of 'Ponkan' were sent from China to J.C. Barrington of McMeskin, Florida, and seedlings from there were distributed and led to commercial propagation.

The commercial cultivation of mandarin oranges in the United States has developed mostly in Alabama, Florida and Mississippi and, to a lesser extent, in Texas, Georgia and California. Mexico has overproduced tangerines, resulting in low market value and cessation of plantings. The 1971-72 crop was 170,000 MT, of which, 8,600 MT were exported to the United States and lesser amounts to East Germany, Canada and Argentina. There is limited culture in Guatemala and some other areas of tropical America. These fruits have never been as popular in western countries as they are in the Orient, Coorg, a mountainous region of the Western Ghats, in India, is famous for its mandarin oranges. For commercial exploitation, mandarins have several disadvantages: the fruit has poor holding capacity on the tree, the peel is tender

Fig. 37: Easily-peeled Mandarin oranges (*Citrus reticulata*) are ideal for eating out-of-hand and very popular in Central America.

and therefore the fruits do not stand shipping well, and the tree has a tendency toward alternate bearing.

Climate

Mandarin oranges are much more cold-hardy than the sweet orange, and the tree is more tolerant of drought. The fruits are tender and readily damaged by cold.

Varieties

Mandarin cultivars fall into several classes:

Class I, Mandarin:

'Changsa' — brilliant orange-red; sweet, but insipid; seedy. Matures early in the fall. The tree has high cold resistance; has survived 4°F (-15.56°C) at Arlington, Texas. It is grown as an ornamental.

'Le–dar' — arose from a climbing branch discovered on an 'Ellendale Beauty' mandarin tree in Bundaberg, Queensland, Australia, about 1959. The owners, named Darrow, took bud-wood from the branch and found that it retained its climbing tendency. Commercial propagation was undertaken by Langbecker Nurseries and the name was trademarked in 1965 when over 5,000 budded trees were put on sale. The budded trees produced large fruits, of rich color and high quality, maturing a little later than the parent.

'Emperor' — believed to have originated in Australia, and a leading commercial cultivar there; oblate, large, 2½ in (6.5 cm) wide, 1¾ in (4.5 cm) high; peel pale-orange, medium thin; pulp pale-orange; 9-10 segments; seeds long, pointed, 10-16 in number. Midseason. Grown on rough lemon rootstock or, better still, on *Poncirus trifoliata*.

'Oneco' — closely related to 'Emperor'; from northwestern India; introduced into Florida by P.W. Reasoner in 1888. Oblate to faintly pear-shaped; medium to large, 2½-3½ in (6.25-9 cm) wide, 2¼-3 in (5.7-7.5 cm) high; peel orange-yellow, glossy, rough and puffy; pulp orange-yellow, of rich, sweet flavor; 5-10 seeds. Medium to late in season. Tree large and vigorous, high-yielding. Not grown commercially in the United States.

'Willow–leaf' ('China Mandarin') — oblate to rounded, of medium size, 2-2½ in (5-6.25 cm) wide, 1¾-2¼ in (4.5-5.7 cm) high; peel orange, smooth, glossy, thin; pulp orange, with 10-12 segments; very juicy, of sweet, rich flavor; 15-20 seeds. Early in season. Tree is small to medium, with very slender, willowy branches, almost thornless, and slim leaves. Reproduces true from seed. Grown mainly as an ornamental and for breeding.

Class II, Tangerine:

'Clementine' ('Algerian Tangerine') — introduced into Florida by the United States Department of Agriculture in 1909 and from Florida into California in 1914; also brought directly from the Government Experiment Station in Algeria about the same time; round to elliptical; of medium size, 2-

2⅜ in (5-6.1 cm) wide, 2-2¾ in (5-7 cm) high; peel deep orange-red, smooth, glossy, thick, loose, but scarcely puffy; pulp deep-orange with 8-12 segments; juicy, and of fine quality and flavor; 3-6 seeds of medium size, non-nucellar; season early but long, extending into the summer. Tree is of medium size, almost thornless; a shy bearer. In Spain it has been found that a single application of gibberellic acid at color-break, considerably reduces peel blemishes and permits late harvesting. 'Clementine' crossed with pollen of the 'Orlando' tangelo produced the hybrid selections, 'Robinson', 'Osceola', and 'Lee', released in 1959. The last two are no longer grown as fruit crops; only utilized in breeding programs.

'Cleopatra' ('Ponki', or 'Spice')—(now being shown as *Citrus reshni* Hort. ex Tanaka)—introduced into Florida from Jamaica before 1888; oblate, small; peel dark orange-red; pulp of good quality but seedy. Fruits too small to be of commercial value; they remain on the tree until next crop matures, adding to the attractiveness of the tree which is itself highly ornamental; much used as a rootstock in Japan and Florida.

'Dancy'—may have come from China; found in the grove of Col. G.L. Dancy at Buena Vista, Florida, and brought into cultivation in 1871 or 1872. Oblate to pear-shaped; of medium size, 2¼-3 in (5.7-7.5 cm) wide, 1½-2⅛ in (4-5.4 cm) high; peel deep orange-red to red, smooth, glossy at first but lumpy and fluted later, thin, leathery, tough; pulp dark-orange with 10-14 segments, of fine quality, richly flavored; 6-20 small seeds. In season in late fall and winter. This is the leading tangerine in the United States, mainly grown in Florida, secondarily in California, and, to a small extent, in Arizona. Tree is vigorous, cold-tolerant, bears abundantly. Alternate-bearing induced by an abnormally heavy crop, can be avoided by spraying with a chemical thinner (Ethephon) when the fruits are very young. Thinning enhances fruit size and market value. This cultivar is disease-resistant but highly susceptible to chaff scale (*Parlatoria pergandii*) which leaves green feeding marks on the fruit making it unmarketable. Control can be achieved by spring and summer or spring and fall spraying of an appropriate pesticide.

'Ponkan' ('Chinese Honey Orange')—round to oblate; large, 2¾-3³⁄₁₆ in (7-8 cm) wide; peel orange, smooth, furrowed at apex and base; medium thick; pulp salmon-orange, melting, with 9-12 segments, very juicy, aromatic, sweet, of very fine quality and with few seeds. Tree not as cold-hardy as 'Dancy', small, upright; can be maintained as a "dwarf" and in China, where the fruit is greatly prized, may be planted 900 to the acre (2,224/ha). R.C. Pitman, Jr., of Apopka, Florida, organized the Florida Ponkan Corporation in 1948, served as its President, and has continuously promoted the culture of this delicious fruit.

'Robinson'—the result of pollinating the 'Clementine' tangerine with the 'Orlando' tangelo, at the United States Department of Agriculture's Horticultural Field Station, Orlando, Florida, was introduced into cultivation in 1960. It is essentially a tangerine, has 10 to 20 seeds. Back-crossing with pollen of the 'Orlando' greatly elevates fruit-set but also results in increasing the seed count to an average of 22 per fruit. This cultivar had lost popularity with growers but the recent practice of spraying with Ethrel (a ripening agent) to speed up coloring on the tree and loosen the fruit has been such an important advance in harvesting and in reducing time in the coloring room that it has reinstated the 'Robinson' as a commerical cultivar. In 1980, the crop forecast was 1.1 million boxes, about 40% of that of 'Dancy'.

'Sunburst'—This cultivar was selected in 1967 from 15 seedlings of hybrids of 'Robinson' and 'Osceola', the latter being another 'Clementine' pollinated with 'Orlando' tangelo but still dominantly a tangerine. 'Sunburst' was propagated on several rootstocks in 1970 and released in Florida in 1979. Oblate, medium-sized, 2½-3 in (6.25-7.5 cm) wide; peel is orange to scarlet in central Florida, orange around the Indian River area; pulp in 11-15 segments with much colorful juice; seeds 10 to 20 according to degree of pollination; green inside. Matures in a favorable season: (mid-November to mid-December). Tree vigorous, thornless, early-bearing, self-infertile; needs cross-pollination for good fruit set; amenable to sour orange, rough lemon, 'Carrizo' and 'Cleopatra' rootstocks though the latter results in slightly reduced fruit size; medium cold-hardy; resistant to *Alternaria* and very tolerant of snow scale.

Class III, Satsuma (sometimes marketed as "Emerald Tangerine")

The Satsuma orange is believed to have originated in Japan about 350 years ago as a seedling of a cultivar, perhaps the variable 'Zairi'. It is highly cold-resistant; has survived 12°F (-11.11°C); is more resistant than the sweet orange to canker, gummosis, psorosis and melanose. It is budded onto *Poncirus trifoliata* in Florida, sweet orange in California. It has been found in Spain that spraying with gibberellic acid 4 to 5 weeks before commercial maturity prevents puffiness, delays ripening, and permits harvesting 2 months later than normal, but this leads to reduced yields the following year.

'Owari'—oblate to rounded or becoming pear-shaped with age; of medium size, 1½-2¾ in (4-6.1 cm) wide, 1½-2½ in (4-6.25 cm) high; peel orange, slightly rough, becoming lumpy and fluted, thin, tough; pulp orange, of rich, subacid flavor; nearly seedless, sometimes 1-4 seeds. Early but short season. Peel often remains more or less green after maturity and needs to be artificially colored in order to market before loss of flavor. Tree small, almost thornless, large-leaved, with faint or no wings on petioles; cultivated commercially in northern Florida, Alabama and other Gulf States; very little in California.

'Wase'—Discovered at several sites in Japan from before 1895; believed to be a bud sport of 'Owari'; was propagated and extensively planted in Japan before 1910; was growing in Alabama in 1917; one tree was sent to California in 1929; oblate to rounded or somewhat conical; large, 2⅓ in (5.81 cm) wide, 1¾ in (4.5 cm) high; peel orange, thin, smooth; pulp salmon-orange, melting, sweet, with 10 segments more or less. Very early in season. Tree is dwarf, slow-growing, heavy-bearing, but susceptible to pests and diseases; has been planted to a limited extent in California and southern Alabama.

'Kara' ('Owari' X 'King' tangor)—a hybrid developed at the California Citrus Experiment Station and distributed in 1935; sub-oblate or nearly round; of medium size, 2⅛-3 in (5.4-7.5 cm) wide, 2⅛-2¾ in (5.4-7 cm) high; peel deep-orange to orange-yellow, lumpy and wrinkled at apex, puffy with age, thin to medium, fairly tough; pulp deep yellow-orange, with 10-13 segments, tender, very juicy, aromatic, of rich flavor, acid until fully ripe, then sweet; usually 12-20 large seeds, at times nearly seedless. Late in season. Tree is vigorous, thornless, with large leaves, the petiole narrowly winged. Grown in coastal California.

Keeping Quality and Storage

Tangerines generally do not have good keeping quality. Commercially washed and waxed 'Dancy' tangerines show a high rate of decay if kept for 2 weeks, will totally decay if held 4 weeks, at 70°F (21°C). To prolong storage life, pads impregnated with the fungistat, diphenyl, have been placed in shipping cartons. The chemical is partly absorbed by the fruit and Federal regulations allow a residue of only 110 ppm. Storage trials have shown that washed and waxed 'Dancy' and 'Sunburst', with 2 pads per carton, absorbed more than 110 ppm in 2 weeks at 70°F (21°C). Though 'Dancy' absorbed more of the fungistat than 'Sunburst', it showed more decay. Storage of unwashed 'Dancy' fruits for 2 weeks at 39.2°F (3°C) with 1 pad per carton showed diphenyl absorption below the legal limit. Unwashed 'Sunburst' fruits with 2 pads can be stored 4 weeks without absorbing excessive diphenyl. Early-harvested tangerines are less susceptible to decay but apt to absorb an excess of diphenyl.

In the Coorg region of India, mandarins of the main crop, harvested in January/February, lose moisture and become shriveled and unmarketable in 10 days at room temperature, 69°F (20.26°C). Wax-coating extends shelf-life to 14 days. Fruits stored in perforated polyethylene bags remain marketable for 21 days at room temperature, and, whether waxed or unwaxed, held at 41°F (5°C), retain quality for 31 days.

Food Uses

Mandarin oranges of all kinds are primarily eaten out-of-hand, or the sections are utilized in fruit salads, gelatins, puddings, or on cakes. Very small types are canned in sirup.

The essential oil expressed from the peel is employed commercially in flavoring hard candy, gelatins, ice cream, chewing gum, and bakery goods. Mandarin essential oil paste is a standard flavoring for carbonated beverages. The essential oil, with terpenes and sesquiterpenes removed, is utilized in liqueurs. Petitgrain mandarin oil, distilled from the leaves, twigs and unripe fruits, has the same food applications. Tangerine oil is not suitable for flavoring purposes.

Food Value Per 100 g of Edible Portion*	
Moisture	82.6–90.2 g
Protein	0.61–0.215 g
Fat	0.05–0.32 g
Fiber	0.3–0.7 g
Ash	0.29–0.54 g
Calcium	25.0–46.8 mg
Phosphorus	11.7–23.4 mg
Iron	0.17–0.62 mg
Carotene	0.013–0.175 mg
Thiamine	0.048–0.128 mg
Riboflavin	0.014–0.041 mg
Niacin	0.199–0.38 mg
Ascorbic Acid	13.3–54.4 mg

*Analyses of tangerines made in Central America.

In 1965, the 'Dancy' tangerine was found to contain more of the decongestant synephrine than any other citrus fruit—97–152 mg/liter, plus 80 mg/100 g ascorbic acid.

Mandarin peel oil contains decylaldehyde, γ-phellandrene, ρ-cymene, linalool, terpineol, nerol, linalyl, terpenyl acetate, aldehydes, citral, citronellal, and d-limonene. Petitgrain mandarin oil contains a-pinene, dipentene, limonene, ρ-cymene, methyl anthranilate, geraniol, and methyl methylanthranilate.

Other Uses

Mandarin essential oil and Petitgrain oil and tangerine oil, and their various tinctures and essences, are valued in perfume-manufacturing, particularly in the formulation of floral compounds and colognes. They are produced mostly in Italy, Sicily and Algiers.

Tangor

Tangors are deliberate or accidental hybrids of the mandarin (*Citrus reticulata*) and the sweet orange (*C. sinensis*). The following are among the better known:

'King' ('King of Siam'); formerly identified as *Citrus nobilis* Lour.; is believed to have originated in Malaya and to have traveled from there to Japan and then to Florida; oblate to rounded; large, 2½–3¾ in (6.25–9.5 cm) wide, 2¼–3½ in (5.7–9 cm) high; peel deep orange-yellow to orange, thick, rough, lumpy; pulp dark-orange, with 10 to 12 segments, very little rag, melting, of fine quality and flavor; 5–15 or more seeds, white within. Late in season. Tree of medium size, erect, thorny to almost thornless, large-leaved, with narrowly-winged petioles; cold-resistant, very productive; may overbear and break branches. Formerly popular in Florida; of limited culti-

vation in California. No longer grown commercially in the United States. Does very well at cool elevations in Peru.

'Murcott' ('Honey Murcott'; 'Murcott Honey Orange'; 'Red'; 'Big Red'; 'Honey Bell' tangelo)—believed to have resulted from breeding work by Dr. Walter Swingle and associates at the United States Department of Agriculture nursery in the Little River district of northeast Miami. The original tree was sent to R.D. Hoyt, in Safety Harbor, about 1913 for trial. Budwood was given to his nephew, Charles Murcott Smith, who propagated several trees about 1922. This led to propagation by several nurseries beginning in 1928 under the name, 'Honey Murcott'. Large-scale production began in 1952. The fruit is oblate, of medium size, 2¾–3³⁄₁₆ in (7.0–8.0 cm) wide, 1⅘–2¹⁄₁₆ in (4.7–5.2 cm) high; peel yellow to deep-orange, glossy, smooth, faintly ribbed, thin, clings to pulp but easily removed when fresh; pulp orange, 11-12 segments, with little rag; tender, having an abundance of reddish-orange juice, with high soluble solids; flavor rich, sweet-subacid; seeds 18-24, small, white inside. Because of the thin peel, the fruit is clipped from the tree, not pulled. It stores and ships well; is in high demand as a fresh fruit, not desirable for canned juice or frozen juice concentrate because of poor processed flavor. Tree is bushy with slender branches bearing fruits near the tips where they are subject to wind and cold damage. Very productive on rough lemon rootstock. Tends to alternate bearing. In heavy-fruiting years, crop may be so heavy as to break the limbs, or the tree may collapse ('Murcott Decline'), or many branches may die back. This cultivar is subject to a virus disease known as *fovea*.

'Temple' (believed identical to the 'Magnet' of Japan)—a seedling discovered by a fruit buyer named Boyce who went to Jamaica in 1896 to purchase oranges after a severe freeze in Florida. He sent budwood to several friends in Winter Park, Florida, who later shared budwood with others. One budded tree fruiting in the grove of L.A. Hakes was brought to the attention of W.C. Temple who recommended it to H.E. Gillett, owner of Buckeye Nurseries. The latter named and propagated it and offered it for sale in 1919. It was not extensively planted until after 1940. The fruit is oblate to round, medium to large, 2⅝-3¼ in (6.6-8.25 cm) wide, 2¼-2½ in (5.7-6.25 cm) high; peel is deep-orange to red-orange, glossy, slightly rough, loose, thick, leathery; pulp orange, melting, of rich, sprightly flavor and superb quality; about 20 seeds of medium size, 25% being under-developed; green inside. Midseason. Tree not very cold-hardy, moderately thorny, bushy; most satisfactory on sour orange rootstock, and succeeds better in Florida than in California or Texas. Excessive applications of nitrogen and potassium increase acidity of the juice. For low-acid juice, low rates of nitrogen and potassium and high rates of phosphorus are necessary. Florida produced 3.3 million boxes in 1984-85 despite severe freezes.

'Umatilla' (incorrectly 'Umatilla Tangelo')—arose from pollination of the flowers of a 'Ruby' orange by 'Owari' Satsuma at Eustis, Florida, in 1911. The progeny was propagated in 1931. Much like 'King'; oblate to rounded; large, 3¼-4¾ in (8.25-12 cm) wide, 2½-2¾ in (6.25-7 cm) high; peel red-orange, smooth, glossy, medium-thick, not very loose; pulp orange, with usually 10 segments, melting, very juicy, of rich sweet-acid flavor and fine quality; 10 or more large seeds or occasionally none. Late in season; holds well on tree. Tree is slow-growing, high-yielding; leaves thick and leathery without wings. Not extensively grown but prized for gift-boxes in Florida.

'Ortanique'—believed to be a chance cross of sweet orange and tangerine; discovered in the Christiana market, Jamaica, by a Manchester man named Swaby who bought 6 fruits. Of resulting seedlings, 2 bore fruit true to type which were exhibited at an agricultural show in the early 1900's. A man named C.P. Jackson, from Mandeville, bought 2 fruits, planted 130 seeds. Some of the seedlings were very thorny. Jackson selected the least thorny, least seedy, and named the fruit—a contraction of orange, tangerine, and unique. The Citrus Growers Association took charge of the marketing for export in 1944. Fruit closely resembles 'Temple'; oblate; peel deep-orange, thin, adherent; pulp divided into 16 segments with scant rag, very juicy, of distinctive acid-sweet flavor; seedless or with few seeds; subject to bruising when freshly picked; needs special handling by harvesters and packers. Grown commercially only in Jamaica but planted to some extent on other Caribbean islands. Fresh fruits and hot-pack concentrate have been shipped to the United Kingdom and New Zealand for many years. Citrus Growers Association took charge of marketing for export in 1944. The fruit is in demand domestically and abroad and brings a premium price. The tree is budded onto pummelo rootstock; cannot tolerate excessive moisture; optimum rainfall is 55-60 in (140-150 cm) annually, half in spring, half in fall. Ideal day temperature is 70°-80° or up to 90°F (21.11°-26.67° or up to 35°C), with 55°F (12.7°C) at night. The 'Ortanique' does well in hot, dry weather on shallow bauxite soil between 2,000 and 3,000 ft (600-900 m) elevation. There is less flavor in fruits from trees grown on clay or alluvial soils or at lower elevations. On clay, the 'Ortanique' is budded on sour orange rootstock. Rough lemon rootstock produces very inferior fruit. The tree begins to bear regularly at 3 years, and a 5-year-old tree will yield 1½ to 2½ 90-lb (40.8 kg) field boxes; a 10-year-old tree, 3½ to 4½ boxes; and trees 15 to 20 years old, 4 to 5½ boxes.

The 'Ortanique' has not performed well in Florida. In South Africa, fruiting has been somewhat irregular. Horticulturists at the Citrus and Subtropical Fruit Research Institute, Nelspruit, found it to be self-incompatible. Cross-pollination with the 'Valencia' orange, 'Minneola' and 'Orlando' tangelos and 'Marsh' grapefruit greatly increases fruit-set and elevates the seed count.

Pests and Diseases

In Jamaica, the 'Ortanique' is attacked by aphids (*Aphis gossypii*), rust mite (*Phyllocoptruta oleivora*), Florida red scale (*Chrysomphalus aonidum*), purple scale (*Lepidosaphes beckii*), and occasionally the West Indian red scale (*Selanaspidus articulatus*). Frequently seen are the fruit-piercing moth (*Gonodonta* spp.) and moths of the genus *Tortrix*.

The fungus, *Sphaeropsis tumefaciens*, sometimes causes large galls or knots around new twigs. Thread blight (*Corticium stevensii*) may occur in some localities.

Pummelo

This, the largest citrus fruit, is known in the western world mainly as the principal ancestor of the grapefruit. As a luscious food, it is famous in its own right in its homeland, the Far East.

Botanically it is identified as *Citrus maxima* Merr., (*C. grandis* Osbeck; *C. decumana* L.). The common name is derived from the Dutch *pompelmoes*, which is rendered *pompelmus* or *pampelmus* in German, *pamplemousse* in French. An alternate vernacular name, shaddock, now little used, was acquired on its entry into the Western Hemisphere as related below. The current Malayan names are *limau abong, limau betawi, limau bali, limau besar, limau bol, limau jambua, Bali lemon,* and pomelo.

Description

The pummelo tree may be 16 to 50 ft (5-15 m) tall, with a somewhat crooked trunk 4 to 12 in (10-30 cm) thick, and low, irregular branches. Some forms are distinctly dwarfed. The young branchlets are angular and often densely hairy, and there are usually spines on the branchlets, old limbs and trunk. Technically compound but appearing simple, having one leaflet, the leaves are alternate, ovate, ovate-oblong, or elliptic, 2 to 8 in (5-20 cm) long, ¾ to 4¾ in (2-12 cm) wide, leathery, dull-green, glossy above, dull and minutely hairy beneath; the petiole broadly winged to occasionally nearly wingless. The flowers are fragrant, borne singly or in clusters of 2 to 10 in the leaf axils, or sometimes 10 to 15 in terminal racemes 4 to 12 in (10-30 cm) long; rachis and calyx hairy; the 4 to 5 petals, yellowish-white, ⅗ to 1⅓ in (1.5-3.5 cm) long, somewhat hairy on the outside and dotted with yellow-green glands; stamens white, prominent, in bundles of 4 to 5, anthers orange. The fruit ranges from nearly round to oblate or pear-shaped; 4 to 12 in (10-30 cm) wide; the peel, clinging or more or less easily removed, may be greenish-yellow or pale-yellow, minutely hairy, dotted with tiny green glands; ½ to ¾ in (1.25-2 cm) thick, the albedo soft, white or pink; pulp varies from greenish-yellow or pale-yellow to pink or red; is divided into 11 to 18 segments, very juicy to fairly dry; the segments are easily skinned and the sacs may adhere to each other or be loosely joined; the flavor varies from mildly sweet and bland to subacid or rather acid, sometimes with a faint touch of bitterness. Generally, there are only a few, large, yellowish-white seeds, white inside; though some fruits may be quite seedy. A pummelo cross-pollinated by another pummelo is apt to have numerous seeds; if cross-pollinated by sweet orange or mandarin orange, the progeny will not be seedy.

Origin and Distribution

The pummelo is native to southeastern Asia and all of Malaysia; grows wild on river banks in the Fiji and Friendly Islands. It may have been introduced into China around 100 B.C. It is much cultivated in southern China (Kwangtung, Kwangsi and Fukien Provinces) and especially in southern Thailand on the banks to the Tha Chine River; also in Taiwan and southernmost Japan, southern India, Malaya, Indonesia, New Guinea and Tahiti. The first seeds are believed to have been brought to the New World late in the 17th Century by a Captain Shaddock who stopped at Barbados on his way to England. By 1696, the fruit was being cultivated in Barbados and Jamaica. Dr. David Fairchild was enthusiastic about the first pummelo he tasted, aboard ship between Batavia and Singapore in 1899. In 1902, the United States Department of Agriculture obtained several plants from Thailand (S.P.I. Nos. 9017, 9018, 9019). Only one (No. 9017) survived and was planted in the agricultural greenhouse in Washington, and budwood from it was sent to Florida, California, Puerto Rico, Cuba (the Isle of Pines), and Trinidad. When the trees fruited, the flavor and general quality were inferior and aroused no enthusiasm. Other introductions were attempted in 1911 but all the plants died in transit. In 1913, a horticulturist of the Philippine Bureau of Agriculture was given the assignment of collecting the best types of pummelos in Thailand. He shipped to San Francisco one tree of a 'Bangkok' type that had been introduced into the Philippines in 1912; it was planted in the greenhouse of the Plant Introduction Garden at Chico. When it fruited several years later, the fruit was of such poor quality that it was considered useless. However, budwood was sent to Riverside and grafted onto two grapefruit trees growing on sour orange rootstock. One of the trees died but the other bore high-quality fruits which were much admired. Budwood was sent to different locations in Florida. In 1919, two trees of a superior pummelo (possibly 'Hao Phuang') from Thailand, which had been doing well in the Philippines, were shipped to the United States Quarantine Station in Bethesda, Maryland, and one of these survived. In addition, seeds from Thailand and from fruits in Chinese markets had been sent to Washington and seedlings were growing in greenhouses.

Dr. Fairchild was eager to introduce the red-fleshed type he had enjoyed in 1899. In 1926, he collected budwood at a hotel in Bandoeng and sent it, together with seeds, to the United States Department of Agriculture but they did not survive the trip. However, seeds of a cultivar in Kediri with flesh nearly as red as his ideal pummelo did reach the Citrus Quarantine Station in Bethesda, Maryland (as S.P.I. No. 67641), and the seedlings were grown there successfully.

In all the succeeding years, the pummelo has never attained significant status in this hemisphere. Generally, it is casually grown as a curiosity in private gardens in Florida and the Caribbean area, and mainly for experimental and breeding purposes at the United States Department of Agriculture's research stations in Orlando and Leesburg, Florida, and at Indio, California, and Mayaguez, Puerto Rico, and at the University of California's Citrus Experiment Station, Riverside. There are small commercial plantings in southern Mexico furnishing fruits for local markets. At least one fruit-grower in Ladylake, Florida, raises pummelos on a small commercial scale. He ships the fruits to New York's Chinatown for $3 each for Chinese New Year festivities. They must be 5 in (12.5 cm) or more in diameter.

Varieties

Professor G. Weidman Groff, in his *Culture and Varieties of Siamese Pummelos*, lists 20 named Thai cultivars, giving the date and identification number of their introduction into the United States. He describes nine. Dr. J.J. Ochse, in *Fruits and Fruitculture in the Dutch East Indies*, described 8 types commonly grown in Batavia. All have red or pink pulp; most have a more or less acid flavor, or a sweetish flavor with an astringent aftertaste. None seems to be of outstanding quality. Reuther, Webber and Batchelor, in *Citrus Industry*, Volume I, 2nd ed., describe 14 cultivars, including the best-known in Thailand, Japan, Indonesia and Tahiti and hybrids created in California. The following 22, from these and other sources, are briefly presented in alphabetical order:

'**Banpeiyu**' (believed to be the same as 'Pai Yau' of Taiwan)—originated in Malaya, introduced into Taiwan in 1920 and from there into Japan; nearly round, very large; peel pale-yellow, smooth, thick, tightly clinging; pulp pale-yellow, in 15-18 segments with thin but tough walls; firm but tender, juicy, of excellent, sweet-acid flavor; medium-late in season; keeps well for several months. Tree large, vigorous, with hairy new growth; leaves hairy beneath. Widely grown in the Orient; the leading cultivar of Japan where it attains high quality only in the warmest locations.

'**Chandler**'—a hybrid of 'Siamese Sweet' (white) and 'Siamese Pink' (acid) developed at Indio, California and released in 1961; oblate to globose; of medium size; peel smooth, at times minutely hairy, medium-thick; core small; pulp pink, fine-grained, tender, fairly juicy; segment walls thin; flavor superior to that of either parent; subacid, about 12% sugar. Seedy. Early in season; of good keeping quality.

Fig. 38: Pummelos (*Citrus maxima*) vary in form, size, color and flavor of pulp.

'Daang Ai Chaa' ('Red Bantam')—grown in Thailand; round, faintly furrowed at base and apex; peel very smooth with conspicuous oil glands; the albedo sometimes tinted with pink; pulp rich-red; the segment walls thick; pulp sacs separate easily from the walls and each other; juicy; of mild flavor, neither sweet nor acid. Tree is more or less dwarfed, with low-lying branches. Non-commercial.

'Double' (incorrectly called 'Banda Navel'; known locally as 'Lemon Banda', 'Lemon Bonting', 'Lemon pompelmoes')—grown in the Banda and Ambon Islands, the Moluccas, Batavia and Java; first reported by Rumphius in 1741; sought out and found by O.A. Reinking in 1926. He supplied budwood of various types to the Departments of Agriculture in Java, Manila, and Washington, D.C. Fruit is round, oblate or faintly pear-shaped; 6 to 8 in (15-20 cm) wide; peel smooth, up to 1 in (2.5 cm) thick; shows no evidence of deformity but, inside, there is a second, rindless fruit the size of a small orange embedded in the apex. The main fruit has 19 segments, the lesser fruit 4; pulp may be red, pink-and-white, or white; is sweet and juicy; mostly seedless, rarely with one or a few more seeds. Occasionally, under adverse conditions, there are many seeds. Fruits are borne in clusters of 5 or 6; not all on a tree will be double. Tree may be low and spreading, to 15 ft (4.5 m) or upright and 18 to 30 ft (5.5-9 m) high.

'Hirado' ('Hirado Buntan')—a chance seedling found in Nagasaki Prefecture, Japan; named and introduced into cultivation around 1910; oblate; large; peel bright-yellow, smooth, glossy, medium-thick, clings tightly; pulp pale greenish-yellow, in numerous segments with thin, tough walls; tender, medium-juicy; of good, subacid flavor, faintly bitter. Medium-early in season; of good keeping quality. Tree of fairly large size, vigorous, unusually cold-tolerant. Occupies second place as a commercial cultivar in Japan.

'Hom Bai Toey' ('Scented Toey Leaf')—grown in Thailand; nearly round, slightly depressed at apex; large, 5 1/8 in (13 cm) wide; peel yellow, smooth, nearly 5/8 in (1.5 cm) thick; pulp of peculiar aroma, white, non-juicy; of slightly bitter flavor. Non-commercial.

'Kao Lang Sat' ('White Lang Sat')—grown in Thailand; oval-pyriform without neck, faintly furrowed at both ends; 4 in (10 cm) wide; peel slightly rough, less than 3/8 in (1 cm) thick; pulp has peculiar aroma; pale pinkish, resembling that of the Langsat (q.v.); divided into 11 or 12 segments; sacs very dry and loosely packed; very sweet without a trace of acid; of inferior quality. Non-commercial.

'Kao Pan' ('Kao Panne', 'Khao Paen', 'White flat')—grown mainly in Nakhon Chaisri district, south of Bangkok, Thailand, for about 160 years; subglobose, flattened at base and apex; 4 1/2 in (11.5 cm) wide; peel light lemon-yellow, smooth, 3/8 to 3/4 in (1-2 cm) thick, tightly clinging; shrinks in storage; core is large and stringy; pulp is divided into 12-15 segments difficult to separate; walls are thick and tough, inedible; they are skinned off and the individual pulp sacs separate readily from each other and are eaten by the handful, like those of the pomegranate (q.v.). They are very juicy, of sweet, faintly acid flavor with hardly a hint of bitterness. Seeds under-developed and inconspicuous in June as grown locally; may be fully developed and numerous in November or when planted elsewhere. Considered the most delicious of Thai pummelos. Almost ever-bearing. Tree is round-topped and spreading, nearly thornless, very productive, but not vigorous and is subject to insects and diseases, especially prone to citrus canker. Non-commercial in Thailand. Air layers were sent to the United States Department of Agriculture's Date Garden, Indio, California, in 1929 and grown as 'Siamese Pink'. All produced

seedy fruits. Trees in the United States Department of Agriculture's Foundation Farm near Leesburg, Florida, bear fruits of excellent flavor.

'Kao Phuang' ('Khao Phoang'; 'White tassel')—grown in Thailand; Groff records P.J. Wester's description of a cultivar that he named 'Siam', the budwood of which was taken by H.H. Boyle from a tree in the garden of Prince Yugelar in Bangkok, and grafted onto calamondin rootstock at the Lamao experiment station, Philippines, in 1913. The trees fruited in 1916. Reinking and Groff later determined that the Prince's tree was the 'Kao Phuang'. Fruit is elongated-pear-shaped with neck; 5 in (12.5 cm) wide or more; peel greenish to yellow, smooth, glossy, 1/2 to 3/4 in (1.25-2 cm) thick, not clinging; pulp in 11-13 segments which separate readily; walls medium thick and tough, ordinarily not eaten; pulp sacs easy to separate, very juicy; flavor excellent, somewhat acid, turning nearly sweet when fully ripe, non-bitter; seeds, few; virtually none in fruits of the third season. This is the leading and perhaps the only commercial cultivar of Thailand; in great demand; considerable quantities are exported to Hong Kong, Singapore and Malaysia. Tree is of upright habit with more thorns than 'Kao Pan'; vigorous, everbearing, high-yielding. Thai growers maintain that this cultivar never attains the same quality when grown in other locations that it does in the Bang Bakok district. However, fruit produced at Indio, California, is of excellent quality.

'Kao Ruan Tia' ('White Dwarf')—grown in Thailand; bell-shaped; larger than 'Kao Phuang'; peel pale-yellow; pulp in as many as 16 segments; of excellent flavor; seeds numerous. Later in season than 'Kao Phuang'; non-commercial.

'Kao Yai' ('White Large')—native to the area east of the Chao Phraya River south of Bangkok; globose, symmetrical; very large, 5 1/2 in (14 cm) or more in diameter; peel light-yellow outside, slightly pinkish inside, exudes a little gum when cut, 1/2 to 3/4 in (1.25-2 cm) thick; pulp in 13 segments; sacs irregularly arranged, clinging tightly together; juicier and sweeter than 'Kao Phuang' but they become tough and indigestible if fruit is left too long on tree; seeds numerous and fully developed. Tree is upright, with a rounded top, large leaves, wavy-edged, with strongly winged petiole. Non-commercial.

'Khun Nok' ('Eagle'; 'Bang Khun Non'; 'Khun Hon Village')—closely allied to 'Kao Pan'; well suited to northern Thailand; fruit subglobose, much like 'Kao Pan'; 5 3/5 in (14.5 cm) wide; pulp of fine flavor and quality; seeds fully developed and numerous. Fruit stores and ships very well.

'Mato' ('Mato Butan'; 'Amoy')—Originated in China and introduced into Taiwan around 1700; obovoid to pear-shaped; peel pale-yellow; rough because of prominent oil glands, medium-thick, closely adhering to pulp; pulp white; segment walls thin, tough; sacs non-juicy, rather dry; flavor sweet. Early in season. The leading cultivar in Punan, China, and Taiwan; one of the three main cultivars in Japan.

'Nakhon' (mispelled 'Nakorn')—a seedling of 'Kao Pan' (PI 52388), introduced from Thailand in 1930 and grown by United States Department of Agriculture at Orlando, Florida, and at Foundation Farm at Leesburg; broad pear-shaped; small, 4 in (10 cm) wide; peel lemon-yellow; pulp white, of fine flavor. Midseason; remains in good condition for a long time on the tree.

'Pandan Bener'—grown in Java; oblate; peel smooth with small oil glands, thick but brittle; pulp dark-red; segment membranes thin but adherent; juice sacs solidly packed, less juicy than 'Pandan Wangi'; sweetish but somewhat astringent. Tree bears a moderate crop. Fruit is rarely attacked by the borer.

'Pandan Wangi'—grown in Java; oblate to round; peel rough because of large oil glands, fairly thick; pulp red, coarse-grained; segment walls thin, bitter, difficult to remove from the juice sacs which are fibrous, slightly juicy, but sweet. Tree vigorous, productive, pest- and disease-resistant but the fruits are heavily attacked by citrus rind borer.

'Reinking'—a selected seedling from a cross of 'Kao Phuang' and the 'Shamouti' orange made at Indio, California, but still a typical pummelo.

'Seeloompang'—grown in Java; pronouncedly oblate, flattened at both ends; peel green even when fully ripe, smooth, thin, brittle; pulp red; segment membranes non-adherent; juice sacs densely compacted, very juicy, acid-sweet and somewhat astringent. Very early in season.

'Siamese Sweet'—introduced by the United States Department of Agriculture in 1930 (CES 2240) and grown at the Citrus Research Center, Riverside, California; oblate to broad-ovoid; pulp white, with large, crisp, non-juicy sacs easily separating from each other; mild-flavored but faintly bitter. Tree is a dwarf with drooping branches and hairy new growth.

'Tahitian' ('Moanalua'; often called 'Tahitian grapefruit')—grown from seed thought to have been taken from Borneo to Tahiti; later introduced into Hawaii; a typical pummelo but with a thin peel and amber-colored, very juicy pulp. The flavor and quality are excellent and it is locally popular.

'Thong Dee' ('Khao Thongdi'; 'Golden')—grown in Thailand; oblate; large, 6 in (15 cm) wide; peel pinkish inside, 3/8 in (1 cm) thick; pulp white with light-brown streaks; pulp sacs large, separating easily from the segment walls; juicy; flavor good but inferior to that of 'Kao Pan'; seedy. Not outstanding as a shipper. Non-commercial. Tree vigorous and produces good quality fruits under unfavorable conditions. A seedling at the United States Department of Agriculture's research station in Orlando, Florida, bears fruits with pink flesh and of good quality despite having a number of seeds. In William Cooper's garden at Winter Park, a tree of this cultivar produces some seedless fruits but most are seedy. Trees at the Foundation Farm near Leesburg bear fruits of excellent flavor.

'Tresca'—a seedling from a tree in the Bahamas grew in the grove of Captain Fred Tresca in Manatee County, Florida. Discovered and propagated in 1887 by Reasoner's Nurseries, Oneco. Fruit is oblate to round, obovoid, or pear-shaped; of medium size, 4 in (10 cm) wide; peel light-yellow, smooth, thick; albedo cream-colored to white; pulp pale-orange, or pink, in 12 to 14 segments; of good flavor; very juicy; many, medium-sized seeds. Late in season. Tree of medium size; new growth hairy; very sensitive to cold. Has been grown commercially in Florida and marketed as a grapefruit. Flesh shows very little color in California.

Climate

The pummelo is tropical or near-tropical and flourishes naturally at low altitudes close to the sea. It has never performed well in New Zealand because of insufficient heat. In the prime growing region of Bang Bakok in southern Thailand, the mean temperature is 82.4°F (28°C) and mean annual rainfall is 56 in (143 cm), being heaviest from May through October and scant in January, February and March, and November and December.

Soil

It is obvious from its coastal habitat that the pummelo revels in the rich silt and sand overlying the organically enriched clay-loam of the flood plain, and that it is highly tolerant of brackish water pushed inland by high tides. On the salty mud flats, farmers dig ditches and create elevated beds of soil for planting pummelo trees. They claim that salt contributes to the flavor and juiciness of the fruits. The salt content of the water varies throughout the year but may be as high as 2.11% at times. In southern Florida and the Bahamas, the trees grow and fruit modestly on oolitic limestone. In Malaya, the tree grows well on the tailings of tin mines.

Propagation

Though the seeds of the pummelo are monoembryonic, seedlings usually differ little from their parents and therefore most pummelos in the Orient are grown from seed. The seeds can be stored for 80 days at 41°F (5°C) and 56-58% relative humidity. Only the best varieties are vegetatively propagated—traditionally by air-layering but more modernly by budding onto rootstocks of pummelo, 'King' or 'Cleopatra' mandarin, rough lemon, or Rangpur lime. In experimental work in the United States, the "T", or shield-budding, method has been found most satisfactory.

Culture

Pummelo growers in Thailand and elsewhere in southeastern Asia are primarily Chinese who dike the swampy land, dig the ditches and canals for drainage and as routes of transportation, and build the raised beds. In the 3- to 5-year period before the beds are ready for the pummelo trees, quick crops such as bananas, sugarcane and peanuts are grown on them. Water gates at intervals along the base of the dikes, allow water to flow through hollow coconut trunks and into the ditches in the dry season. Continual deepening and widening of ditches and adding of soil to the beds is necessary to counteract erosion. Coconut and betel nut palms are planted for shade for the young citrus trees but are removed at the end of 3-5 years, or sometimes not until the pummelos are 10 to 15 years old. Rice may be grown in the ditches. The pummelo trees are spaced 10 to 15 ft (3-4.5 m) apart. Some growers interplant the columnar tree, *Erythrina fusca* Lour., to shade the mature pummelos, and to help retain the soil with its extensive, fibrous root system, and enrich the soil with its falling leaves. Weeds are removed by hoeing. Night soil, of course, is the standard fertilizer in the Orient and is used on pummelos but, more commonly, paddy ash (the ash of burned rice hulls) is placed in piles under each tree to gradually seep down to the roots. The air-layered trees have a low, spreading habit and must eventually be pruned.

An analysis of production methods by farmers whose main source of income is marketing pummelos was made by agriculturists at the University of Malaya in 1974. It was concluded that labor input was excessive; fertilizer (all organic on mature trees), was under-utilized on young trees; chemicals were over-utilized on young trees and under-utilized on older trees which the farmers are inclined to neglect because most of them suffer from *Phytophthora* root rot. Pummelo trees may need nutritional sprays to correct zinc, manganese or boron deficiencies.

Harvesting and Keeping Quality

Pummelos may flower 2 to 4 times a year. In the Old World, there are usually 4 harvesting seasons. The main crop matures in November but it is said that fruits that ripen at other seasons have fewer seeds and superior quality. In Florida, the fruits ripen from November to February and there may be a small crop in the spring. In Thailand, fruits for marketing are generally picked when just beginning to turn yellow, heaped in large piles for sale. If not disposed of immediately, they are stored in dry, ventilated sheds shaded by trees. The fruits keep for long periods and ship well because of the thick peel. After 3 months, the peel will be deeply wrinkled but the pulp will be juicier and of more appealing flavor than in the fresh fruit. If stored too long, they may become bitter. Paper-wrapped fruits in ventilated crates have kept in good condition for 6 to 8 months during sea transport to Europe. According to an old Chinese Atlas, the fruits of the 'Double' pummelo, if hung in the house, will remain in good condition for a year.

Pests and Diseases

Among the leading insect pests of pummelo in the Orient are a leaf miner, *Phyllocnistis citrella;* a flea beetle which attacks the leaves; a stinging red ant (*Pheidologeton* sp.) that damages roots, twigs, leaves and trunk, sometimes girdling and killing the tree. Scale insects (*Chrysomphalus aonidum* and *C. aurantii, Coccus hesperidum, Lepidosaphes gloverii, Parlatoria brasiliensis* and *P. zizyphus, Pseudaonidia trilobitiformis,* and *Saissetia* sp.) are prevalent but are partly controlled by natural enemies — a black ant (*Dolochonderus* sp.) and a parasitic fungus, *Aschersonia aleyrodis.* The weaver ant, *Oecophylla smaragdina,* tends scale insects for their honeydew. Fruit growers in China and Southeast Asia put out chicken entrails to encourage the weaver ant to construct its long, hanging nests on citrus trees because it controls the tree borers (*Pentatomidae*) and other pests. Though beneficial, it is a nuisance at harvest time because it inflicts painful stings. The "eggs" (pupae) are commonly eaten.

In Indonesia, the fruits of one cultivar, 'Bali Merah', which has a thin rind, are so heavily attacked by the citrus rind borer and other insects that they are commonly wrapped in old banana leaves, paper or cloth when young.

Sooty mold, develops on the honeydew excreted by the scale insects. The pummelo is subject to most of the diseases that affect the orange (q.v.). But Dr. Walter Swingle, on his trip to Japan, China and the Philippines, found some varieties very resistant to canker. Most of the older trees in Malaya, as already mentioned, succumb to *Phytophthora* root rot.

Mistletoe (*Loranthus* sp.) is a great pest on pummelo trees in Asia.

Food Uses

Though there is some labor involved, it is worth the effort to peel good pummelos, skin the segments, and eat the juicy pulp. The skinned segments can be broken apart and used in salads and desserts or made into preserves. The extracted juice is an excellent beverage. The peel can be candied.

Food Value Per 100 g of Edible Portion*	
Calories	25-58
Moisture	84.82-94.1 g
Protein	0.5-0.74 g
Fat	0.2-0.56 g
Carbohydrates	6.3-12.4 g
Fiber	0.3-0.82 g
Ash	0.5-0.86 g
Calcium	21-30 mg
Phosphorus	20-27 mg
Iron	0.3-0.5 mg
Vitamin A	20 I.U.
Thiamine	0.04-0.07 mg
Riboflavin	0.02 mg
Niacin	0.3 mg
Ascorbic Acid	30-43 mg

*Analyses made in China and the United States.

Toxicity

Like that of other citrus fruits, the peel of the pummelo contains skin irritants, mainly limonene and terpene, also citral, aldehydes, geraniol, cadinene and linalool, which may cause dermatitis in individuals having excessive contact with the oil of the outer peel. Harvesters, workers in processing factories, and housewives may develop chronic conditions on the fingers and hands.

Other Uses

The flowers are highly aromatic and gathered in North Vietnam for making perfume. The wood is heavy, hard, tough, fine-grained and suitable for making tool handles.

Medicinal Uses: In the Philippines and Southeast Asia, decoctions of the leaves, flowers, and rind are given for their sedative effect in cases of epilepsy, chorea and convulsive coughing. The hot leaf decoction is applied on swellings and ulcers. The fruit juice is taken as a febrifuge. The seeds are employed against coughs, dyspepsia and lumbago. Gum that exudes from declining trees is collected and taken as a cough remedy in Brazil.

Grapefruit (Plate XVIII)

A relative newcomer to the citrus clan, the grapefruit was originally believed to be a spontaneous sport of the pummelo (q.v.). James MacFayden, in his *Flora of Jamaica*, in 1837, separated the grapefruit from the pummelo, giving it the botanical name, *Citrus paradisi* Macf. About 1948, citrus specialists began to suggest that the grapefruit was not a sport of the pummelo but an accidental hybrid between the pummelo and the orange. The botanical name has been altered to reflect this view, and it is now generally accepted as *Citrus* X *paradisi*.

When this new fruit was adopted into cultivation and the name grapefruit came into general circulation, American horticulturists viewed that title as so inappropriate that they endeavored to have it dropped in favor of "pomelo". However, it was difficult to avoid confusion with the pummelo, and the name grapefruit prevailed, and is in international use except in Spanish-speaking areas where the fruit is called *toronja*. In 1962, Florida Citrus Mutual proposed changing the name to something more appealing to consumers in order to stimulate greater sales. There were so many protests from the public against a name change that the idea was abandoned.

Description

The grapefruit tree reaches 15 to 20 ft (4.5-6 m) or even 45 ft (13.7 m) with age, has a rounded top of spreading branches; the trunk may exceed 6 in (15 cm) in diameter; that of a very old tree actually attained nearly 8 ft (2.4 m) in circumference. The twigs normally bear short, supple thorns. The evergreen leaves are ovate, 3 to 6 in (7.5-15 cm) long, and 1¾ to 3 in (4.5-7.5 cm) wide; dark-green above, lighter beneath, with minute, rounded teeth on the margins, and dotted with tiny oil glands; the petiole has broad, oblanceolate or obovate wings. The white, 4-petalled flowers, are 1¾ to 2 in (4.5-5 cm) across and borne singly or in clusters in the leaf axils. The fruit is nearly round or oblate to slightly pear-shaped, 4 to 6 in (10-15 cm) wide with smooth, finely dotted peel, up to ⅜ in (1 cm) thick, pale-lemon, sometimes blushed with pink, and aromatic outwardly; white, spongy and bitter inside. The center may be solid or semi-hollow. The pale-yellow, nearly whitish, or pink, or even deep-red pulp is in 11 to 14 segments with thin, membranous, somewhat bitter walls; very juicy, acid to sweet-acid in flavor when fully ripe. While some fruits are seedless or nearly so, there may be up to 90 white, elliptical, pointed seeds about ½ in (1.25 cm) in length. Unlike those of the pummelo, grapefruit seeds are usually polyembryonic. The number of fruits in a cluster varies greatly; a dozen is unusual but there have been as many as 20.

Origin and Distribution

The grapefruit was first described in 1750 by Griffith Hughes who called it the "forbidden fruit" of Barbados. In 1789, Patrick Browne reported it as growing in most parts of Jamaica and he referred to it as "forbidden fruit" or "smaller shaddock". In 1814, John Lunan, in *Hortus Jamaicensis*, mentions the "grapefruit" as a variety of the shaddock, but not as large; and, again, as "forbidden fruit", "a variety of the shaddock, but the fruit is much smaller, having a thin, tough, smooth, pale yellow rind". In 1824, DeTussac mentions the "forbidden fruit or smaller shaddock" of Jamaica as a variety of shaddock the size of an orange and borne in bunches. William C. Cooper, a citrus scientist (USDA, ARS, Orlando, Florida, to 1975), traveled widely observing all kinds of citrus fruits. In his book, *In Search of the Golden Apple,* he tells of the sweet orange and the grapefruit growing wild on several West Indian islands. He cites especially a fruit similar to grapefruit that is called *chadique* growing wild on the mountains of Haiti and marketed in Port-au-Prince. The leaves are like those of the grapefruit. He says that it was from the nearby Bahama Islands in 1823 that Count Odette Phillipe took grapefruit seeds to Safety Harbor near Tampa, Florida. When the seedlings fruited, their seeds were distributed around the neighborhood.

At first, the tree was grown only as a novelty in Florida and the fruit was little utilized. Even in Jamaica, the trees were often cut down. Mrs. Mary McDonald Carter of Eustis, Florida, was quoted in the *Farm and Livestock Record,* Jacksonville, in 1953, as relating that her father, John A. MacDonald, settled in Orange County in 1866. In 1870, he was attracted to a single grapefruit tree with clusters of lemon-colored fruits on the Drawdy property at Blackwater. He bought the entire crop of fruits, planted the seeds and established the first grapefruit nursery. The first grapefruit grove planted from this nursery by a man named Hill was sold in 1875 to George W. Bowen who developed it commercially. In 1881, MacDonald bought the Drawdy crop and once more raised seedlings for his nursery in Eustis. Early settlers began planting the tree and acquired a taste for the fruit. There was already a small demand in the North. New York imported 78,000 fruits from the West Indies in 1874. Florida started sending small shipments to markets in New York and Philadelphia between 1880 and 1885.

In 1898, Dr. David Fairchild was excited to learn of a grove of 2,000 grapefruit trees in the Kendall area south of Miami on the property of the Florida East Coast Railway. In 1904, he was amazed to see one tree in the dooryard of the Kennedy ranch in southern Texas where he thought the climate too cold for it. He was told that the tree had been frozen to the ground but had recovered. He predicted that a citrus industry could not be established in that region of the country. In 1928, he photographed the same tree, which had been killed back several times in the interim, but was again in fruit. By 1910, grapefruit had become an important commercial crop in the Rio Grande Valley and, to a lesser extent, in Arizona and desert valleys of California. By 1940, the United States was exporting close to 11,000,000 cases of grapefruit juice and nearly one-half million cases of canned sections. Cultivation had reached commercial proportions in Jamaica and Trinidad and spread to Brazil, South America and Israel. In 1945/46, the United States (mainly Florida) produced a record of 2,285,000 tons of grapefruit. In 1967/68, this country accounted for 70% of the world crop despite a great decline in Texas production because of severe weather. Grapefruit was moving forward by leaps and bounds. Israel, in 1967, supplied only 11% of the world crop but, by 1970, her production had increased by 300%. In 1980, Florida exported just under 10 million boxes, making grapefruit this state's most valuable export crop. Japan is the main importer and has, at times, suspended shipments to determine the safety of fungicide residues or because of discovery of larvae of the Caribbean fruit fly. Great care is taken to maintain this important trade. Other countries which had entered the grapefruit industry were Mexico, Argentina, Cyprus, Morocco and some areas of South America which raise grapefruit for local markets. In Central America, the grapefruit is not much favored because of its acidity.

In the late 1960's and early 1970's, Mexico was rapidly expanding its grapefruit plantings, especially in the states of Tamaulipas and Veracruz, to save its citrus industry in view of the decline in market value of oranges and tangerines brought on by over-production. Furthermore, there were great advantages in the lower costs of producing grapefruit without irrigation and with good biological control of pests. Now Mexico exports large quantities of grapefruit to the United States and lesser amounts to Canada and Japan. Puerto Rico formerly exported grapefruit to the United States but is no longer able to compete in the trade and has only remnants of former plantations. Cuba has planted 370,000 acres (150,000 ha) of citrus, mostly grapefruit with expectations of exporting to the Soviet Union and eastern European countries. The grapefruit is grown only in a small way in the Orient where the pummelo is cultivated. In recent years, the grapefruit has become established in India in hot regions where the sweet orange and the mandarin are prone to sunburn.

Varieties

Named varieties of grapefruit appeared in the official list of the American Pomological Society in 1897, but pioneers had selected and named favorite clones for several years before that time. The following are among the most noteworthy of old and new cultivars:

'Duncan'—the original trees were virtually identical seedlings that grew in a grove owned by a man named Snedicor

near Safety Harbor, Florida. Propagation was first undertaken by A.L. Duncan of Dunedin in 1892. The fruit is round or slightly obovate; large, 3½ to 5 in (9-12.5 cm) wide; peel is very light yellow (usually called "white"), with large oil glands, medium-thick, highly aromatic; pulp is buff, in 12-14 segments with medium-tender membranous walls, very juicy, of fine flavor; seeds medium-large, 30-50. Early to mid-season. Tree is unusually cold-hardy. This was the leading cultivar for many years in Florida and Texas and was introduced into all the grapefruit-growing areas of the world. Today, in the United States, it has largely given way to cultivars with fewer seeds, but it is being grown commercially in India. Recent seed irradiation experiments have shown that a high percentage of seedless mutants results from exposure to 20-25 krad.

'Foster' ('Foster Pink Flesh')—Originated as a branch sport of a selection called 'Walters' in the Atwood Grove near Ellenton, Florida, discovered by M.B. Foster of Manatee in 1906, and propagated for sale by the Royal Palm Nurseries. Fruit is oblate to round; medium-large, averaging 3¾ in (9.5 cm) in width; peel light-yellow blushed with pink, smooth but with large, conspicuous oil glands; albedo pink; pulp light-buff, pinkish near the center; in 13 or 14 segments with pinkish walls, tender, juicy, of good quality despite seeds, up to 50 or even more, of medium size. Medium-early in season. Not very popular; grown to a limited extent in Florida, Texas, Arizona and India. In Texas, it is more colorful, the pulp being entirely pinkish in hue.

'Marsh' ('Marsh Seedless')—one of 3 seedling trees on the property of a Mrs. Rushing near Lakeland, Florida, purchased by William Hancock in 1862. Because the fruits of this tree were seedless, C.M. Marsh took budwood from it for nursery propagation and he bought young trees previously budded by others. He sold the budded offspring and, in time, the 'Marsh' was planted more than any other cultivar. The original tree was killed by cold in the winter of 1895-96. The fruit is oblate to round, medium in size, 3½ to 4¾ in (9-12 cm) wide; peel is light-yellow, very smooth, with medium-size oil glands, mildly aromatic; pulp is buff, in 12-14 segments with tender membranes, melting, extremely juicy and rich in flavor; seeds absent or 3-8, medium-sized. Medium to late in season and holds well on the tree. Keeps well after harvest. The leading grapefruit cultivar; grown in Florida, California, Texas, Arizona, South America, Australia, South Africa, Israel and India. A local selection, presumably of a seedling 'Marsh', in Surinam is known there as 'Hooghart'. The two are almost indistinguishable.

'Oroblanco'—a triploid from a grapefruit X pummelo cross made in 1958 by geneticists R.K. Soost and J.W. Cameron of the University of California, Riverside. Patent obtained in 1981 and assigned to the University of California Board of Regents. Fruit form and size similar to 'Marsh'; peel paler and thicker; pulp paler and has larger hollow in center; sections easily skinned; tender, juicy, non-bitter; has faintly astringent after-taste before full maturity or in cooler climates; seedless. Season early: December to April at Riverside; early November through February at Landcove. Tree is vigorous, large, hardy, can tolerate temperatures down to 30°F (-1.11°C); yields medium to heavy crops and may tend to alternate bearing. Seems better adapted to California's inland citrus locations than to desert sites. Has been grown experimentally on trifoliate orange, 'Troyer' citrange, citremon 1449, Brazilian sour orange, grapefruit, sweet orange, rough lemon and 'Red' rough lemon rootstocks. The two latter have adversely affected internal quality.

Fig. 39: Grapefruit (*Citrus paradisi*): pink (left); yellow (center); and russet (right). In: K. & J. Morton, *Fifty Tropical Fruits of Nassau*, 1946.

'Paradise Navel'—a selection from the 100-year-old Nicholson citrus grove near Winter Garden, Florida; propagated and patented by W.H. Nicholson, improved and released for distribution in 1976. Fruit is oblate, smaller than a typical grapefruit. Originally very seedy, but, by budding onto various rootstocks and transferring from one rootstock to another over a period of years, there eventually emerged one tree bearing fruit without seeds. Budwood from this tree has produced uniformity of seedlessness regardless of rootstock. The fruits have been sold to local customers but no scions nor trees were sold prior to 1976.

'Redblush' (including 'Ruby', 'Ruby Red', 'Shary Red', 'Curry Red', 'Fawcett Red', 'Red Radiance', and 'Webb' ['Webb's Redblush Seedless'])—originated as sports—lower branches—growing out of 'Thompson' trees which a Texas nursery had purchased from Glen St. Mary Nursery and sold to growers in the Rio Grande Valley, and which were frozen back in 1929. All are seedless and otherwise similar to 'Thompson' but display redder color. 'Redblush' grapefruits have been extensively planted in Florida in the past few decades though the juice is not suitable for canning as it tends to turn brown with age. By 1950, 75% of Florida's grapefruit crop was of the pink or red seedless type. Under the name, 'Ruby Red', a member of this group is a standard commercial cultivar in Texas. In 1958, budwood of 'Redblush' from California was acquired by the Regional Fruit Research Station at Abohar, India, was propagated on rough lemon, and the resulting trees

performed so well and showed such disease resistance that the cultivar was recommended for growing under irrigation in the arid regions of the Punjab and Haryana, where it averages 250 fruits annually per tree. Probably includable in this group is 'Burgundy'. Its peel is not blushed but the pulp is intense red throughout the season. 'Ray Ruby' and the similar if not identical 'Henderson' are branch sports propagated in Texas and introduced into Florida in the 1970's. The peel is redder than that of 'Ruby Red' and the pulp is red though not as intense as 'Star Ruby' throughout the season. Recently, budwood of 'Ray Ruby' has become available from the Florida Department of Agriculture's Bureau of Citrus Budwood Registration in Winter Haven. 'Ray Ruby' is expected to perform better than 'Star Ruby' on standard rootstocks.

'Star Ruby'—a lower branch mutation bearing red-blushed fruits, noticed on a 'Foster' tree at San Benito, Texas, in the mid 1930's. The tree had been frozen back nearly to the bud union the previous year. Budwood from the branch was propagated by C.E. Hudson as the 'Hudson Red' but, because of its coarse texture and high number of seeds (40–60), it was not adopted commercially. Seeds were irradiated at the Texas A & I Citrus Center, Weslaco, in 1959. The seedling from one of these treated seeds was named the 'Star Ruby' and introduced into cultivation in 1971 by Richard Hensz of Texas A & I University. Several thousand trees were planted in Texas. At least 65,000 budded trees were brought into Florida in 1971 by commercial interests without proper qualifications and permits under the Division of Plant Industry. Investigation revealed a susceptibility to *Phytophthora* root rot and ringspot virus in Texas. The Florida State Agricultural Commissioner ordered the destruction of all unauthorized imported trees. About 25,000 were voluntarily destroyed by owners but the ruling was contested and the trees were placed under quarantine. Subsequently, ringspot virus was found on one of the imported trees which had already been used as a source of budwood. Infected trees from this source were found in a nursery and were destroyed together with all neighboring healthy trees. By April 1977, certified, disease-free budwood of 'Star Ruby' was made available and nearly 200,000 "budeyes" were released to growers. They were urged to make only limited plantings until more was known of this cultivar's fruiting habits. The tree tends to become more chlorotic than 'Ruby Red' when sunburned or affected by poor drainage, or high applications of herbicides and pesticides, and it is sensitive to adverse weather conditions.

'Star Ruby' has a yellow peel distinctly red-blushed and intensely red pulp and juice, 3 times more colorful than 'Ruby Red'. Though the color decreases with maturity, it is maintained throughout the season. The pulp is smooth and firmer than that of 'Ruby Red' and has a bit more sugar and acid. Furthermore, there may be no seeds or no more than nine. Some of the juice color is dissipated by heat in the pasteurization process but there is still enough for the product to be blended with white or pink grapefruit juice to provide more consumer appeal.

'Sweetie'—a grapefruit X pummelo hybrid released in 1984 by the Citrus Marketing Board in Israel, has all the features of a typical grapefruit but the flavor is sweet.

'Thompson' ('Pink Marsh')—In 1913, one branch of a 'Marsh' tree owned by W.R. Thompson, Oneco, Florida, bore pink-fleshed, seedless fruits. Propagation of budwood from the branch was undertaken by the Royal Palm Nurseries in 1924. A similar bud variation of the 'Marsh' had appeared around 1920 at Riverside, California. The fruit is oblate to round, of medium size, 2¾ to 3¾ in (7-9.5 cm) wide; peel is light-yellow, smooth, with small, inconspicuous oil glands, faintly aromatic; pulp is light- to deep-buff more or less flushed with pink, sometimes throughout, occasionally just near the center. There are 12 to 14 segments with abundant, colorless juice, and few seeds—usually 3 to 5. The color of the pulp is most intense in January and February. By late March and April it has faded to nearly amber.

'Triumph' (possibly the same as 'Royal' and 'Isle of Pines')-a seedling on the grounds of the Orange Grove Hotel in Tampa, Florida, propagated in 1884. The fruit is oblate to ellipsoid, slightly flattened at both ends; of medium size; peel light-yellow, very smooth, with oil glands of medium size; medium-thick; pulp pale, tender, juicy, only faintly bitter, the flavor having a touch of orange; the center is semi-hollow; of superior quality; 35-50 seeds. Medium-early in season, beginning in November. Grown only in dooryards in Florida, but has been widely distributed in citrus regions; does better than 'Marsh' in South Africa.

A grapefruit-like, triploid hybrid named 'Melogold' was developed by crossing a sweet pummelo with a seedy, white, tetraploid grapefruit in 1958. The fruit is larger than 'Marsh' grapefruit and its pummelo-like flavor is considered superior though it may have a trace of bitterness at the beginning and end of the season which extends from early November or December through February. 'Melogold' is grafted onto rough lemon and 'Troyer'

citrange rootstocks and is recommended for interior California, not in hot desert nor in humid coastal situations. Patent rights are held by the University of California and budwood is released only to licensed nurserymen.

Climate

The grapefruit prospers in a warm subtropical climate. Temperature differences affect the length of time from flowering to fruit maturity. At Riverside, California the period is 13 months; at warmer Brawley in the Imperial Valley of southern California, only 7 to 8 months. The fruit is lower in acidity in the Indian River region and areas of southern Florida, the lower Rio Grande Valley of Texas, and in the tropics than in cooler situations.

Humidity contributes to thinness of peel, while in arid climates the peel is thicker and rougher and, as might be expected, the juice content is lower. Low winter temperatures also result in thicker peel the following year and even affect the fruit shape.

Ideal rainfall for grapefruit is 36 to 44 in (91.4-111.7 cm) rather evenly distributed the year around.

Soil

The grapefruit is grown on a range of soil types. In the main growing area of Florida, the soil is mildly acid sand and applications of lime may be beneficial. On the east coast there are coquina shell deposits and, in the extreme southern part of the peninsula, there is little soil mixed with the prevailing oolitic limestone. Where the grapefruit is grown in California, Arizona and Texas, the soils are largely alkaline and frequent irrigation causes undesirable alkaline salts to rise to the surface. In Surinam, grapefruit is grown on clay. Successful grapefruit culture depends mainly on the choice of rootstock best adapted to each type of soil.

Salinity of the soil and in irrigation water retards water uptake by the root system and reduces yields.

Propagation

In the early years of grapefruit-growing, the customary citrus rootstocks were utilized: sour orange on heavy hammock and flatwoods soils, rough lemon on sand, though trees grafted on this stock were short-lived. In the early 1950's, sweet orange was being preferred over sour orange. In 1946, the United States Department of Agriculture, Texas A & M University, and Rio Farms, Inc., of Monte Alto, Texas, launched a cooperative program of testing grapefruit on different rootstocks. Of 13 different rootstocks utilized, 'Swingle citrumelo', 'Morton' and 'Troyer' citranges gave the best yield of large fruits. Rough lemon and 'Christian' trifoliate orange reduced acidity. 'Swingle citrumelo' was never used extensively as a rootstock until 1974 when it was released to nurserymen and growers because of its tolerance of exocortis, xyloporosis, and tristeza and resistance to foot-rot and citrus nematode, and low uptake of salts, together with its ability to support heavy crops. It is now in third place after 'Troyer' citrange and sour orange.

In the past, 'Marsh' and 'Hooghart', the commercial grapefruits of Surinam, have been grown there on sour orange rootstock, but fear of tristeza inspired a rootstock testing program. Among the stocks tried, 'King' and 'Sunki' resulted in high yield and excellent quality in contrast to rough lemon and Rangpur lime. The two latter also showed susceptibility to *Phytophthora* root rot. 'Cleopatra' lowered the yield, and trifoliate orange proved unsatisfactory in such a humid climate. In the Lower Rio Grande Valley of Texas, grapefruit trees on 'Swingle citrumelo' have grown very poorly on heavy clay as compared to those on sour orange.

Culture

In general, culture of grapefruit is similar to that of the orange, q.v., except that wider spacing is necessary.

Nutritional experiments with grapefruit have shown that excessive nitrogen results in malformed fruit, coarser texture and less juice. Lack of certain minor elements is evident in symptoms often mistaken for disease. The condition called exanthema is caused by copper deficiency; mottle leaf results from zinc deficiency.

Harvesting and Handling

In Florida, all commercial cultivars reach legal maturity in September or October if sprayed after blooming with lead arsenate to reduce acidity. Even after legal maturity the grapefruit can be "stored" on the tree for months, merely increasing in size, and extending the marketing season. The fruits can be harvested until near the end of May when they begin to fall and seeds start sprouting in the fruit. The only adverse effect of late harvesting is a corresponding reduction in the following year's crop. It has been found that spot-picking of the largest fruits partially counteracts this effect of late harvest. Fruit drop can be retarded by spraying with a combination of gibberellic acid and 2,4-D. Either of these agents or both together will reduce the germination of seeds. Germination may be inhibited for periods up to 11 weeks by cool storage at 50°F (10°C).

Grapefruits were formerly harvested by climbing the trees or using picking hooks which frequently damaged the fruit. Today, the fruits on low branches are picked by hand from the ground; higher fruits are usually harvested by workers on ladders who snap the stems or clip the fruits as required. California began utilizing a modified olive limb-shaker for harvesting grapefruit in 1972. The machines work in pairs to harvest opposite sides of each tree and the trees must be pruned to remove deadwood and to give access to 3-5 main limbs for shaking. Lower branches must be lopped off to leave a clear 2½ ft (75 cm) space for the catching frame. Mechanical harvesting causes some superficial injury. A team of 3 workers with one machine can harvest 150 to 188 field boxes—50 lbs (22.7 kg) when filled—per hour, as compared with 45 boxes per hour for 3 manual pickers. Stems are removed from the fruits before packing to avoid stem-damage.

Early in the season, when the fruits are mature but not fully colored, they are often degreened by exposure to ethylene gas. The grapefruit is remarkable for its durability, but modern practices of applying fungicide to the harvested fruit are given credit for the great reduction in marketing losses. The cull rate in New York wholesale warehouses in 1983 was found to be 1.4% (mostly fungal), as compared with 13% estimated in 1960. Retail losses in 1983 were 3.5%, and only a small proportion were the result of physical injury.

Keeping Quality

The grapefruit keeps well at 65°F (18.33°C) or higher for a week or more and for 2 or 3 weeks in the fruit/vegetable compartment of the home refrigerator. The first sign of breakdown is dehydration and collapse of the stem-end. To retard moisture loss, fruits for marketing are washed and waxed as soon as possible after harvest. When kept in prolonged storage, the grapefruit is subject to chilling injury (peel pitting) at temperatures below 50°F (10°C). The degree of injury depends on several factors: the fruits on the outside of the tree are more susceptible than the fruits that have been sheltered by foliage. The use of preharvest growth regulators tends to reduce susceptibility, as does 100% relative humidity during storage. Preconditioning at 60.8°F (16°C) for 7 days before storing at 33.8°F (1°C) prevents injury. Lowering the temperature gradually after preconditioning is also beneficial, as is sealing the fruit in polyethylene shrink-film before refrigerating.

The banning of ethylene dibromide fumigation except for export has made it necessary to resort to cold treatment as an alternative measure against fruit fly infestation for shipment to Texas, Arizona and California. The United States Department of Agriculture now requires that imported citrus fruits be kept at 32°F (0°C) for 10 days or at 36°F (2.2°C) for 16 days after the fruit has been cooled down to the specified temperature. In Israel, investigators have found that waxing with a coating containing fungicide, and holding the packed fruit for 6 days at 62.6°F (17°C) before the cold treatment, gives good protection from chilling injury and decay in storage. Cold treatment costs 5 times as much as fumigation with ethylene dibromide. Methyl bromide has been tested and proposed as an effective fumigant.

Pests and Diseases

The grapefruit is subject to most of the same pests that attack the orange, including Caribbean and Mediterranean fruit flies. In addition to the cold treatment referred to above, irradiation has been studied as a method of disinfection, but has not been authorized for citrus fruit treatment. Exposure of early-season fruit to 60 and 90 krad causes scald and rind breakdown after 28 days of storage, and mainly pitting in midseason and late fruits. Minimal injury results from exposure to 7.5, 15, and 30 krad.

The following diseases have been reported for the grapefruit tree and its fruit by the Florida Division of Plant Industry: leaf spot (*Alternaria citri, Mycosphaerella horii, Phyllosticta hesperidearum*); algal leaf spot (*Cephaleuros* virescens); greasy spot (*Cercospora citri-grisea*); tar spot (*C. gigantea*); anthracnose (*Colletotrichum gloeosporioides*); thread blight (*Corticium koleroga* and *C. stevensii*); gummosis (*Diaporthe citri*); dieback (*Diplodia natalensis*); heart rot (*Fomes applanatus, Ganoderma sessilis,* and *Xylaria polymorpha*); charcoal root rot (*Macrophomina phaseolina*); root rot (*Fusarium oxysporum*); sooty blotch (*Gloeodes pomigena*); flyspeck (*Leptothyrium pomi*); mushroom root rot (*Clitocybe tabescens*); foot rot (*Phytophthora megasperma, P. palmivora,* and *P. parasitica*); damping-off (*Rhizoctonia solani*); seedling blight (*Sclerotium rolfsii*); felt fungus, (*Septobasidium pseudopedi- cellatum*); branch knot (*Sphaeropsis tumefaciens*); leaves may be attacked by *Chaetothyricum hawaiiense,* and twigs by *Physalospora fusca.* Brown rot of fruit is caused by *Phytophthora citrophthora* and *P. terrestris;* stem-end rot, *Botryosphaeria ribis;* dry rot of fruit (*Nematospora coryli*); green mold (*Penicillium digitatum*); blue mold, (*P. italicum*); pink mold (*P. roseum*); scab (*Elsinoe fawcetti*).

The tree is highly susceptible to citrus canker and several viruses: crinkly leaf virus, psorosis, tristeza, xyloporosis, and infectious variegation. Mesophyll collapse is caused by extreme drought and dehydrating wind.

Food Uses

As a relatively new food, the grapefruit has made great advances in the past 75 years. In 1970, consumption of grapefruit was temporarily heightened by a widely promoted "grapefruit diet" plan claimed to achieve a loss of 10 lbs (4.5 kg) in 10 days and continuous gradual loss until the achievement of normal body weight. In 1983, the United States Department of Agriculture Marketing Service reported that, among fresh fruits and vegetables consumed in Metropolitan New York, grapefruit was exceeded only by potatoes, lettuce, oranges and apples.

Grapefruit is customarily a breakfast fruit, chilled, cut in half, the sections loosened from the peel and each other by a special curved knife, and the pulp spooned from the "half-shell". Some consumers sweeten it with white or brown sugar, or a bit of honey. Some add cinnamon, nutmeg or cloves. As an appetizer before dinner, grapefruit halves may be similarly sweetened, lightly broiled, and served hot, often topped with a maraschino cherry. The sections are commonly used in fruit cups or fruit salads, in gelatins or puddings and tarts. They are commercially canned in sirup. In Australia, grapefruit is commercially processed as marmalade. It may also be made into jelly.

The juice is marketed as a beverage fresh, canned, or dehydrated as powder, or concentrated and frozen. It can be made into an excellent vinegar or carefully fermented as wine.

Grapefruit peel is candied and is an important source of pectin for the preservation of other fruits. The peel oil, expressed or distilled, is commonly employed in soft-drink flavoring, after the removal of 50% of the monoterpenes. The main ingredient in the outer peel oil is nookatone. Extracted nookatone, added to grapefruit juice powder, enhances the flavor of the reconstituted juice. Naringin, extracted from the inner peel (albedo), is used as a bitter in "tonic" beverages, bitter chocolate, ice cream and ices. It is chemically converted into a sweetener about 1,500 times sweeter than sugar. After the extraction of naringin, the albedo can be reprocessed to recover pectin.

Grapefruit seed oil is dark and exceedingly bitter but, bleached and refined, it is pale-yellow, bland, much like olive oil in flavor, and can be used similarly. Because it is an unsaturated fat, its production has greatly increased since 1960.

Food Value Per 100 g of Edible Portion*			
	Pulp (raw)	Juice (raw)	Peel (candied)**
Calories	34.4-46.4	37-42	316
Moisture	87.5-91.3 g	89.2-90.4 g	17.4 g
Protein	0.5-1.0 g	0.4-0.5 g	0.4 g
Fat	0.06-0.20 g	0.1 g	0.3 g
Carbohydrates	8.07-11.5 g	8.8-10.2 g	80.6 g
Fiber	0.14-0.77 g	trace	2.3 g
Ash	0.29-0.52 g	0.2-0.3 g	1.3 g
Calcium	9.2-32.0 mg	9.0 mg	
Phosphorus	15-47.9 mg	15.0 mg	
Iron	0.24-0.70 mg	0.2 mg	
Sodium	1.0 mg	1.0 mg	
Potassium	135 mg	162 mg	
Vitamin A			
(white)	10 I.U.	10. I.U.	
(pink/red)	440 I.U.	440 I.U.	
Thiamine	0.04-0.057 mg	0.04 mg	
Riboflavin	0.01-0.02 mg	0.02 mg	
Niacin	0.157-0.29 mg	0.2 mg	
Ascorbic Acid	36-49.8 mg	36-40 mg	
Tryptophan	2 mg		
Methionine	0-1 mg		
Lysine	12-14 mg		

*According to analyses made in California, Texas, Florida, Cuba and Central America.

**Peel Oil:* 90% limonene; the volatile fraction (2-3%) consists mainly of oxygen compounds and sesquiterpenes; the waxy fraction (7-8%) consists of C_8 and C_{10} aldehydes, plus geraniol, cadinene and small amounts of citral and dimethyl arthranilate, plus acid. Also present are 9 coumarins and 0.88% 22-dihydrostigmasterol. The dried pulp and seeds contain β-sitosteryl-D-glucoside and β-sitosterol.

The glycoside 7 β-neohesperidosyl-4-(β-D-glucopyranosyl) naringenin occurs in the pulp segments. Feruloylputrescine is found in the juice and leaves. Mature grapefruit leaves contain the flavonoid, apigenin 7 β-rutinoside. Young leaves contain the 7 β-neohesperidoside and 7 β-rutinoside of naringenin.

Other Uses

Factory waste: The waste from grapefruit packing plants has long been converted into molasses for cattle.

Seed hulls: After oil extraction, the hulls can be used for soil conditioning, or, combined with the dried pulp, as cattlefeed. A detoxification process must precede the feeding of this product to pigs or poultry.

Wood: Old grapefruit trees can be salvaged for their wood. The sapwood is pale-yellow or nearly white, the heartwood yellow to brownish, hard, fine-grained, and useful for domestic purposes. Mainly, pruned branches and felled trees are cut up for firewood.

Medicinal Uses: An essence prepared from the flowers is taken to overcome insomnia, also as a stomachic, and cardiac tonic. The pulp is considered an effective aid in the treatment of urinary disorders. Leaf extractions have shown antibiotic activity.

Tangelo (Plate XIX)

Tangelos are deliberate or accidental hybrids of any mandarin orange and the grapefruit or pummelo. The first known crosses were made by Dr. Walter T. Swingle at Eustis, Florida, in 1897, and Dr. Herbert J. Webber at Riverside, California, in 1898. They are so unlike other citrus fruits that they have been set aside in a class by themselves designated *Citrus* X *tangelo* J. Ingram & H.E. Moore (*C.* X *paradisi* X *C. reticulata*).

Tangelos range from the size of a standard sweet orange to the size of a grapefruit, but are usually somewhat necked at the base. The peel is fairly loose and easily removed. The pulp is often colorful, subacid, of fine flavor and very juicy. The trees are large, more cold-tolerant than the grapefruit but not quite as hardy as the mandarin. Nucellar embryos are not uncommon in these hybrids and most of the cultivars are self-sterile, so a majority come true from seed. Tangelos are not commonly grown in California but are produced commercially and in home gardens in Florida. They are much more satisfactory on limestone in southern Florida than the sweet orange and are prized for their quality.

Among the better-known tangelo cultivars are:

'K-Early' ('Sunrise Tangelo')—a hybrid propagated by growers. It is an early-maturing cultivar of such poor quality that it gave tangelos a bad reputation. The Official Rules Affecting the Florida Citrus Industry require that it be sold only as 'K-Early Citrus Fruit'.

'Minneola'—a hybrid of 'Bowen' grapefruit and 'Dancy' tangerine; oblate, faintly necked; medium-large, 3¼ in (8.25 cm) wide, 3 in (7.5 cm) high; peel deep red-orange, thin, firm, not loose; pulp orange, with 10-12 segments, melting, sweet-acid; of fine flavor; 7-12 small seeds, green inside. Late in season. Ships well. If crop is left too long on tree, the next crop will be light. Bears better if honeybees are provided and if 'Temple' tangor is interplanted as a pollenizer, but the 'Temple' is not as cold-hardy as the 'Minneola', and the trees tend to crowd each other. The 'Minneola' needs fertile soil, irrigation and adequate nutrition. Effects to increase production of seedless fruits include spraying the blooms with gibberellic acid, or girdling during full bloom. The former reduces fruit size and the latter may induce virus outbreaks causing scaling and flaking of the bark.

'Nova'—a 'Clementine' tangerine and 'Orlando' tangelo cross made by Dr. Jack Bellows in 1942, first fruited in 1950, and released by the United States Department of Agriculture's Horticultural Field Station, Orlando, Florida, in 1964. Fruit is oblate to rounded, of medium size, 2¾ -3 in (7-7.5 cm) wide, 2½ -2¾ in (6.25-7 cm) high; peel is orange to scarlet, thin, slightly rough, leathery, easy to remove; pulp dark-orange, with about 11 segments, of good, sweet flavor; seeds numerous if cross-pollinated; polyembryonic, green inside. Early in season (mid-September to mid-December). Does very well on 'Cleopatra' rootstock. The tree resembles that of the 'Clementine' tangerine, its twigs are thornless, and it is more cold-hardy than 'Orlando'. This cultivar is self-infertile and trials have shown that 'Temple' tangor is a good pollenizer.

'Orlando' (formerly 'Lake')—result of 'Bowen' grapefruit pollinated with 'Dancy' tangerine, by Dr. Swingle in 1911. The fruit is oblate to rounded, of medium size, 3 in (7.5 cm) wide, 2¾ in (7 cm) high; peel deep-orange, slightly rough, not loose; pulp deep-orange, with 12 to 14 segments, melting, very juicy, sweet; seeds 10-12. Early in season but after 'Nova'. A good commercial fruit in Florida. Needs cross-pollination by 'Temple' tangor, or by 'Dancy' or 'Fairchild' tangerines. The presence of honeybees, even without interplanting with a pollinator tree, has greatly increased yields. 'Cleopatra' mandarin is often used as a rootstock on sandy soils, but higher yields have been obtained on sweet lime and rough lemon in Florida. In Texas, 'Orlando' is most productive on 'Swingle citrumelo', 'Morton citrange', 'Rangpur lime' and 'Cleopatra' mandarin. Fruit quality is best on 'Morton citrange', sour orange, 'Sun Cha Sha Kat', 'Keraji' and 'Kinokune' mandarins.

'Seminole'—a hybrid of 'Bowen' grapefruit and 'Dancy' tangerine; oblate, not necked; medium-large, 3¼ in (8.25 cm) wide, 2¾ in (7 cm) high; peel deep red-orange, thin, firm, almost tight but not hard to remove; pulp deep-orange with 11-13 segments, little rag, melting, of fine, subacid flavor; seeds small, 20-25, green inside. Early in season but holds well through March. Tree vigorous and high-yielding, scab-resistant; leaves with faint or no wings, tangerine-scented.

'Thornton'—a tangerine-grapefruit hybrid created by Dr. Swingle in 1899; oblate to obovate, a little rough and lumpy, puffy with age; medium-large, 3¼-3¾ in (8.25-9.5 cm) wide, 2⅞-3¼ in (7.25-8.25 cm) high; peel, light-orange, medium-thick, almost loose, easily removed; pulp pale- to deep-orange, with 10-12 segments, soft, melting, juicy, of rich subacid to sweet flavor; seeds slender, 10-25, green inside. Matures from December to March. Tree vigorous and high-yielding, large-leaved, well adapted to hot, dry regions of California. Fruit is a poor shipper.

'Ugli'—believed to be a chance hybrid between a mandarin orange and grapefruit. The discoverer, G.G.R. Sharp, owner

Fig. 40: The 'Ugli' tangelo of Jamaica is believed to be a chance hybrid between a Mandarin orange and a grapefruit.

of Trout Hall Estate, Jamaica, reported that it was found growing in a pasture around 1917. He took budwood and grafted onto sour orange, and kept on regrafting the progeny with the fewest seeds. Sharp was exporting to England and Canada in 1934 and to markets in New York City in 1942. The fruit is obovoid, compressed to nearly oblate, necked at the base, puffy; large, 4¼ to 6 in (10.8-15 cm) wide, 3¼-4½ in (8.25-11.5 cm) high; peel is light-yellow with light-green areas at apex, leathery, loose, medium-thin; albedo is thick; pulp light-orange, or apricot, divided into 12 segments with tough membranes, easily skinned; tender, melting, very juicy; of fine flavor, superior to grapefruit, only faintly bitter; seedless or with 3 or a few more medium-sized seeds, white inside. In Jamaica, matures in December and January.

In January 1942, Kendal Morton purchased fruits on the New York market, sent 2 to Dr. H. Harold Hume of the University of Florida, and 4 to Dr. H.J. Webber of the University of California, Riverside. Dr. Webber was able to examine them only at the Quarantine Station but he wrote up the description for the first edition of the book, *The Citrus Industry,* by Batchelor and Webber. He planted the seeds and reported that, of 13 seedlings, 6 had strongly mandarin-scented leaves, 3 had weak-mandarin scent, and 4 had leaf-scent reminiscent of grapefruit or sweet orange leaves. Dr. Webber passed on in 1943 before he could carry out his plans to bud 2 trees from each seedling. Dr. W.P. Betters, Associate Horticul-

turist, reported that in 1947 the 4 seedlings still in the nursery were bearing fruit, mostly in May-June; the fruits averaged 6 in (15 cm) in diameter, the peel was orange-yellow with a slight tendency to regreen in the spring, the albedo was very thick and fibrous, the flavor of the orange, juicy pulp was good but with a grapefruit tang, and there was, on the average, one seed in each segment. These trees were destroyed in 1951 because they were in the path of campus development, but budwood was taken for propagation and the new trees were beginning to bear in 1954. The 'Ugli' was considered a good fruit for home dooryards in California and was being tried as a rootstock for lemon. The 'Ugli' is little known in Florida. James McClure of Lake Placid has a few trees that bear in February. There are small groves of 'Ugli' in South Africa. In New Zealand a similar fruit has been grown since 1861 as "Poorman's orange", or "Poorman grapefruit".

'Alamoen' — a fruit rather like the 'Ugli' commonly grown from seeds in Surinam. J.B. Rorer, a Plant Pathologist in Trinidad, saw it in Surinam, considered it better-flavored than the grapefruit, and sent 3 specimens to Dr. David Fairchild in 1914. Under the introduction number 37804, seeds were planted at the United States Department of Agriculture's Garden at Chico, California. Fruits borne by the seedlings had very thick peel and very little juice. Two of the trees were sent to Dr. Fairchild and he planted them at his home, The Kampong, in Coconut Grove, Miami. They began fruiting in 1931 and the fruits were not equal in quality to those he had received from Surinam, which were much lighter in weight because of large, hollow centers. In 1944, he sent fruits to Dr. Webber who detected several points of similarity to the 'Ugli' but found the latter easier to peel and superior in quality and flavor.

Orangelo

Several sweet orange X grapefruit crosses were made by citrus breeders in California in early years and were given the name "orangelo", a combination of orange and pomelo, the original name for the grapefruit. None of the hybrids was sufficiently productive to be of horticultural value. However, the group name is the only one on record to which such hybrids can be referred. The only promising one being currently exploited is the following:

'Chironja' — This seemingly spontaneous hybrid was noticed by Carlos G. Moscoso, Fruit Specialist in Horticulture, Agricultural Extension Service, of the University of Puerto Rico, when he was interviewing citrus growers in the interior, mountainous, coffee zone of that island in November 1956. He saw a tree with large, bright-yellow fruits in contrast to the normal sweet orange and grapefruit trees grown by farmers as shade for their coffee plantations. He learned that there were several other trees of the same type on other farms in the neighborhood, some of them quite a few years old and all raised from seed and showing only slight variations in form and size, and greater variation in season of fruiting.

He described the fruit as round to pear-shaped, necked, equal to grapefruit in size; peel a brilliant yellow, slightly adherent, easy to remove; the inner peel non-bitter; pulp yellow-orange, with 9-13 segments having tender walls and much juice; the mild flavor reminiscent of both orange and grapefruit, hardly acid or bitter even when immature. The seed count ranges from 7 to 15, with an average of 11, and some fruits have as few as 2. The fruit is borne singly or in clusters. The tree, reaching 22 ft (6.7 m), has leaves that smell like and resemble those of the grapefruit except that they are usually deformed. Young shoots may have prominent thorns. Flowering and fruiting may occur throughout the year, though most trees flower mainly in late spring and early summer.

By 1969, horticulturists in Puerto Rico had evaluated 500 seedlings in a test planting and selected 12 clones, 3 being considered superior. It was observed that 7-year-old trees may produce 300 to 500 fruits over a period of one year, while a 7-year-old grapefruit tree in Puerto Rico may produce about 70.

In rootstock trials, grapefruit rootstock gave best results at the Adjuntas Agricultural Experiment Substation and sour orange at the Isabela Substation. On grapefruit rootstock, the 'Chironja' is larger than ordinary but not as sweet. A planting of seedlings was made at the Corozal Substation with simultaneous planting of grafted trees for comparison. So much variation was seen in the seedlings it was concluded that the 'Chironja' must be vegetatively propagated for uniform results. Ten clones selected from the Corozal planting were grafted onto sour orange and set out at the 3 Substations. The trees reached heavy production at 6 years of age. Yield was highest at Isabela Substation and 'Clone 2-4' had the best yield, the thinnest peel and the most seeds. 'Clone 2-3' had 11 seeds and 'Clone 3-6' had 14.

Storage tests revealed that fruit in polyethylene bags at 44.5°F (7°C) and relative humidity of 90%, maintained acceptable quality for 70 days. But fruits harvested 5 months after fruit-set and stored for periods of 30 to 55 days were of the best quality. Fruits harvested 7 months after fruit-set retained high quality for only 25 days.

The 'Chironja's' productivity makes it popular with Puerto Rican growers and it is in demand on Puerto Rican markets, mainly because it is more colorful than the grapefruit, sweeter, and easy to peel.

The fruit is cut in half and eaten with a spoon as a grapefruit is eaten, or is peeled and the sections eaten individually, or they are squeezed for juice. The sections can be canned in sirup with added citric acid to enhance the flavor. The rind can be candied successfully.

Lemon (Plate XX)

The leading acid citrus fruit, because of its very appealing color, odor and flavor, the lemon, *Citrus limon* Burm. f. (syns. *C. limonium* Risso, *C. limonia* Osbeck, *C. medica* var. *limonium* Brandis), is known in Italy as *limone;* in most Spanish-speaking areas as *limón, limón agria, limón real,* or *limón francés;* in German as *limonen;* in French as *citrónnier;* in Dutch as *citroen.* In Haiti, it is *limon France;* in Puerto Rico, *limon amarillo.* In the

Netherlands Antilles, *lamoentsji,* or *lamunchi,* are locally applied to the lime, not to the lemon as strangers suppose. The lemon is not grown there.

Several lemon-like fruits are domestically or commercially regarded as lemons wherever they are grown and, accordingly, must be discussed under this heading. These include: Rough lemon (*C. jambhiri* Lush.), Sweet lemon (*C. limetta* Risso), 'Meyer' (lemon X mandarin hybrid); 'Perrine' (lime X lemon hybrid); 'Ponderosa' (presumed lemon X citron hybrid), qq.v. under "Varieties".

Description

The true lemon tree reaches 10 to 20 ft (3-6 m) in height and usually has sharp thorns on the twigs. The alternate leaves, reddish when young, become dark-green above, light-green below; are oblong, elliptic or long-ovate, 2½ to 4½ in (6.25-11.25 cm) long, finely toothed, with slender wings on the petioles. The mildly fragrant flowers may be solitary or there may be 2 or more clustered in the leaf axils. Buds are reddish; the opened flowers have 4 or 5 petals ¾ in (2 cm) long, white on the upper surface (inside), purplish beneath (outside), and 20-40 more or less united stamens with yellow anthers. The fruit is oval with a nipple-like protuberance at the apex; 2¾ to 4¾ in (7-12 cm) long; the peel is usually light-yellow though some lemons are variegated with longitudinal stripes of green and yellow or white; it is aromatic, dotted with oil glands; ¼ to ⅜ in (6-10 mm) thick; pulp is pale-yellow, in 8 to 10 segments, juicy, acid. Some fruits are seedless, most have a few seeds, elliptic or ovate, pointed, smooth, ⅜ in (9.5 mm) long, white inside.

Origin and Distribution

The true home of the lemon is unknown, though some have linked it to northwestern India. It is supposed to have been introduced into southern Italy in 200 A.D. and to have been cultivated in Iraq and Egypt by 700 A.D. It reached Sicily before 1000 and China between 760 and 1297 A.D. Arabs distributed it widely in the Mediterranean region between 1000 and 1150 A.D. It was prized for its medicinal virtues in the palace of the Sultan of Egypt and Syria in the period 1174-1193 A.D. Christopher Columbus carried lemon seeds to Hispaniola in 1493. The Spaniards may have included lemons among the fruits they introduced to St. Augustine. They were grown in California in the years 1751-1768. Lemons were reported to be increasingly planted in northeastern Florida in 1839. Because of heavy imports from Sicily, commercial culture in Florida and California was begun soon after 1870 and grew to the point where 140,000 boxes were being shipped out of Florida alone. The small Florida industry was set back by a freeze in 1886, the susceptibility of the lemon to scab, and the unfavorable climate for curing the fruit, and also competition from California. Following the devastating freeze of 1894-95, commercial lemon culture was abandoned in Florida. Not until 1953 was interest in lemon-growing revived in Central Florida to take advantage of the demand for frozen concentrate and for natural cold-press lemon oil. At that time, Florida was importing lemons from Italy for processing. Plantings grew to 8,700 acres by 1975. Freezes caused 50% reduction by 1980. Still, in 1984, Florida exported $2 million worth of lemons.

In the meantime, Arizona had developed lemon orchards, though on a smaller scale than California. In the 1956-57 season, California produced 11 million gallons (42 million liters) of frozen lemon concentrate while Florida's output was still very small. California and Arizona became the leading sources of lemons in the western hemisphere. In recent years, California has produced nearly double the crop that can be profitably marketed fresh or processed. Foreign competition has increased and many California growers have destroyed their lemon groves or topworked the trees to oranges, but new cultural techniques making summer production possible may reverse the trend.

Guatemala has in the past 2 decades developed commercial lemon culture, primarily to produce the peel oil for its essential oil industry and secondarily for the purpose of dehydrating the fruit and preparing a powder for reconstituting into juice. Southern Mexico, too, is now a major grower of lemons, also primarily for lemon peel oil. Lemons are rarely grown for the fresh fruit market in Latin America. In South America, Argentina leads in lemon culture with Chile a distant second. Among the world's leading lemon growers and exporters are Italy, Spain, Greece, Turkey, Cyprus, Lebanon, South Africa and Australia. Lemons can be grown only at medium and high elevations in the Philippines.

Varieties

With the resumption of lemon-growing in Florida, workers at the Citrus Experiment Station, Lake Alfred, began a search for the most suitable cultivars, whether in dooryards, or in the United States Department of Agriculture planting at Orlovista, or the Lake Alfred collection. By late 1950, 200 selections had been brought together from various parts of the United States. Of these, 40 were budded onto 30-year-old grapefruit trees on rough lemon rootstock on the Minute Maid property at Avon Park. Two selections grown elsewhere were included in the studies—evaluation for thorniness, cold- and disease-susceptibility, sizes, juiciness, flavor, number of segments and seeds, yields, and quality of peel oil. The majority of the selections were judged undesirable; only a few showed promise for processing and fresh fruit marketing purposes. For processing, 'Villafranca' rated highest, followed by 'Eustis', 'Bearss', 'Perkin' and 'Avon'. Any of these, properly harvested and cured would be suitable for marketing fresh. Libby, McNeil & Libby, when planning for their lemon orchard at Babson Park, Florida, about 1948, tested varieties from all major lemon-producing areas of the world and chose 'Bearss' as rating highest in quality and quantity of juice, which was their chief concern at the time. In 1960, they added marketing of the fresh fruit and found the 'Bearss' equally desirable for this purpose.

The following are brief descriptions of most of the better known cultivars of true lemons and of lemon-like fruits that are accepted as lemons in home or commercial usage, and a few of the lesser-known.

'Armstrong' ('Armstrong Seedless')—a sport discovered in a private grove at Riverside, California, about 1909. Patented in 1936 by Armstrong Nurseries. Resembles 'Eureka' except that it usually bears seedless or near-seedless fruits. If planted among other lemon trees will occasionally have a few seeds.

'Avon'—first noticed as a budded tree in Arcadia, Florida. A budded tree propagated from the original specimen around 1934 was planted in the Alpine Grove in Avon Park; it produced heavy crops of fruits highly suitable for frozen concentrate. It, therefore, became the source of budwood for commercial propagation by Ward's Nursery beginning in 1940.

'Bearss' ('Sicily', but not the original introduction by Gen. Sanford in 1875, which has disappeared)—a seedling believed to have been planted in 1892, discovered in the Bearss grove near Lutz, Florida, about 1952. Closely resembles 'Lisbon'. It is highly susceptible to scab and greasy spot and oil spotting. The tree is vigorous and tends to produce too many water sprouts. Nevertheless, it has been propagated commercially by Libby, McNeill & Libby since 1953 because the peel is rich in oil. It constitutes 20% of Brazil's lemon/lime crop.

'Berna' ('Bernia', 'Verna', 'Vernia')—oval to broad-elliptic, with pronounced nipple, short neck; peel somewhat rough, medium-thick, becoming thinner in summer, tightly clinging. Seeds generally few or absent. Ripens mostly in winter; fruits keep well on tree until summer but become too large. Tree is vigorous, large, prolific. This is the leading cultivar of Spain and important in Algeria and Morocco. It is too much like the 'Lisbon' to be of value in California. In Florida, it has been found deficient in acid, low in juice, and too subject to scab.

'Eureka'—originated from seed taken from an Italian lemon (probably the 'Lunario') and planted in Los Angeles in 1858; selected in 1877 and budwood propagated by Thomas Garey who named it 'Garey's Eureka'. The fruit is elliptic to oblong or rarely obovate, with moderately protruding nipple at apex, a low collar at the base; peel yellow, longitudinally ridged, slightly rough because of sunken oil glands, medium-thick, tightly clinging; pulp greenish-yellow, in about 10 segments, fine-grained, tender, juicy, very acid. Fruits often borne in large terminal clusters unprotected by the foliage. Bears all year but mostly late winter, spring and early summer when the demand for lemons is high. Tree of medium size, almost thornless, early-bearing, prolific; not especially vigorous, cold-sensitive, not insect-resistant; relatively short-lived. Not suitable for Florida. Grown commercially in Israel. One of the 2 leading cultivars of California, though now being superseded by clonal selections with more vigor, e.g., 'Allen', 'Cascade', 'Cook', and 'Ross'. 'Lambert Eureka' is a chance seedling found in 1940 on the property of Horace Lambert in New South Wales. It is vigorous and productive.

'Femminello Ovale'—one of the oldest Italian varieties; short-elliptic with low, blunt nipple; slightly necked or rounded at base; of medium size; peel yellow, finely pitted, medium-smooth, medium-thick, tightly clinging; pulp in about 10 segments, tender, juicy, very acid, of excellent quality, with few, mostly undeveloped, seeds. Fruits all year but mainly in late winter and spring; ships and stores well. The tree is almost thornless, medium- to very-vigorous, but highly susceptible to *mal secco* disease. This is the leading cultivar in Italy, accounting for ¾ of the total lemon production, and ⅕ of the crop is processed as single-strength juice.

'Genoa'—introduced into California from Genoa, Italy, in 1875. Almost identical to 'Eureka'; ovoid or ovate-oblong with blunt nipple at apex; base rounded or slightly narrowed; of medium size; peel yellow, medium-thick, tightly clinging; pulp in 10-12 segments, melting, medium-juicy, with 29 to 51 seeds which are light-brown within. Tree is shrubby, nearly trunkless, spreading, very thorny, cold-hardy. Grown commercially in India, Chile and Argentina.

'Harvey'—of unknown parentage; was found by Harvey Smith on the property of George James in Clearwater, Florida. Fruit much like 'Eureka'. Tree highly cold-tolerant, compatible with several rootstocks. Commercially propagated by Glen St. Mary Nurseries Company, near Jacksonville, Florida, since 1943.

'Interdonato' ('Special')—a lemon X citron hybrid that originated on property of a Colonel Interdonato, Sicily, around 1875; oblong, cylindrical, with conical, pointed nipple at apex, short neck or collar at base; large; peel yellow, smooth, glossy, thin, tightly clinging; pulp greenish-yellow, in 8 or 9 segments, crisp, juicy, very acid, faintly bitter. Very few seeds. Earliest in season; mostly fall and early winter. Tree vigorous, usually thornless, medium-resistant to *mal secco*; of medium yield; accounts for 5% of Italy's crop.

'Lisbon' (perhaps the same as 'Portugal' in Morocco and Algeria)—originated in Portugal, possibly as a selection of 'Gallego'; reached Australia in 1824; first catalogued in Massachusetts in 1843; introduced into California about 1849 and catalogued there in 1853; introduced into California from Australia in 1874 and again in 1875. Fruit almost identical to 'Eureka'; elliptical to oblong, prominently nippled at apex, base faintly necked; peel yellow, barely rough, faintly pitted, sometimes slightly ribbed, medium-thick, tightly clinging; pulp pale greenish-yellow, in about 10 segments, fine-grained, tender, juicy, very acid, with few or no seeds. Main crop in February, second crop in May. Fruit is borne inside the canopy, sheltered from extremes of heat and cold. Tree large, vigorous, thorny, prolific, resistant to cold, heat, wind. Not well adapted to Florida. It is low-yielding and short-lived in India. Surpasses 'Eureka' in California. Has given rise to a number of clonal selections, particularly 'Frost', originated by H.B. Frost at the Citrus Research Station, Riverside, California in 1917 and released about 1950; also 'Prior Lisbon' and the more vigorous 'Monroe Lisbon'.

'Meyer'—a hybrid, possibly lemon X mandarin orange; introduced into the United States as S.P.I. #23028, by the agricultural explorer, Frank N. Meyer, who found it growing as an ornamental pot-plant near Peking, China, in 1908; obovate, elliptical or oblong, round at the base, occasionally faintly necked and furrowed or lobed; apex rounded or with short nipple; of medium size, 2¼ to 3 in (5.7-7.5 cm) wide and 2½ to 3½ in (6.25-9 cm) high; peel light-orange with numerous small oil glands, ⅛ to ¼ in (3-6 mm) thick; pulp pale orange-yellow, usually in 10 segments with tender walls, melting, juicy, moderately acid with medium lemon flavor; seeds small, 8 to 12. Tends to be everbearing but fruits mostly from December to April. Tree small, with few thorns, prolific, cold-resistant; produces few water sprouts, and is only moderately subject to greasy spot and oil spotting. It is easily and commonly grown from cuttings. Does well on sweet orange and rough lemon rootstocks; is not grafted onto sour orange because it is a carrier of a virulent strain of tristeza. Grown for home use in California; in Florida, both for home use and to some extent commercially for concentrate though the product must be enhanced by the addition of peel oil from true lemons, since that from 'Meyer' peel is deficient in flavoring properties. Has been fairly exten-

Fig. 41: Lemons: 'Ponderosa', perhaps a lemon X citron hybrid (left); 'Lisbon'-type commercial lemons (*Citrus limon*) (center); and rough lemon (*C. jambhiri*) (right).

sively planted in Texas and in Queensland, Australia, and New Zealand.

'**Monachello**' ('Moscatello') — suspected of being a lemon X citron hybrid; elliptical, with small nipple and no neck, merely tapered at apex and base; medium-small; peel yellow, smooth except for large, sunken oil glands, thin, clinging very tightly; pulp in 10 segments, tender, not very juicy, not sharply acid. Bears all year but mainly winter and spring. Tree not vigorous, slow-growing, almost thornless, with abundant, large leaves; bears medium-well, resistant to *mal secco*, and has been extensively planted in Italy in areas where the disease is common.

'**Nepali Oblong**' ('Assam', 'Pat Nebu') — originated in Assam; fruit resembles citron in some aspects; long-elliptic to oblong-obovate, with wide, short nipple; medium-large; peel greenish-yellow, smooth, glossy, medium-thick; pulp greenish-yellow in 11 segments, fine-grained, very juicy, of medium acidity, with few or no seeds. Everbearing. Tree large, vigorous, spreading, medium-thorny, prolific; foliage resembles that of the citron. Commercial in India.

'**Nepali Round**' — of Indian origin; round, without distinct nipple; juicy; seedless. Tree large, vigorous, compact, nearly thornless, medium-prolific. Successfully cultivated in South India.

'**Perrine**' — a Mexican lime X 'Genoa' lemon hybrid created by Dr. Walter Swingle and colleagues in 1909, but still a fairly typical lemon; it is lemon-shaped, with small nipple at apex,

necked at base; of medium size; peel pale lemon-yellow, smooth, slightly ridged, thin, tough; pulp pale greenish-yellow, in 10 to 12 segments having thin walls; tender, very juicy, with slightly lime-like flavor but acidity more like lemon; seeds usually 4 to 6, occasionally as many as 12, long-pointed. Everbearing. Tree cold-sensitive but less so than the lime; resistant to wither tip and scab but prone to gummosis and other bark diseases. In the early 1930's, was extensively planted in southern Florida on rough lemon rootstock, but no longer grown.

'**Ponderosa**' ('Wonder'; 'American Wonder') — a chance seedling, possibly of lemon/citron parentage, grown by George Bowman, Hagerstown, Maryland around 1886 or 1887; appeared in nursery catalogs in 1900 and 1902; obovate, lumpy and faintly ribbed, slightly necked at base; large, 3½ to 4⅛ in (9-11 cm) wide, 3½ to 4¾ in (9-12 cm) high; peel light orange-yellow, with medium-large oil glands, flush or slightly depressed; ⅜ to ½ in (1-1.25 cm) thick; pulp pale-green, in 10 to 13 segments with thick walls; juicy, acid; seeds of medium size, 30 to 40 or more, brown within. Everbearing. Tree small, moderately thorny; buds and flowers white or barely tinged with red-purple. More sensitive to cold than true lemons. Grown for home use and as a curiosity in California and Florida and in small-scale commercial plantings since 1948. Rather widely cultivated as an indoor potted plant in temperate regions.

'**Rosenberger**' — a clone found in a grove of 'Lisbon' and 'Villafranca' trees at Upland, California; was planted in the

Rosenberger orchard and gained recognition as a superior cultivar. Tree closely resembles that of 'Villafranca'. Fruit is somewhat like 'Lisbon' but is shorter and broader and less tapered at base. Tree vigorous and prolific. Became popular in California in the 1960's.

'Rough Lemon' ('Florida Rough'; 'French'; 'Mazoe'; 'Jamberi')—perhaps a lemon X citron hybrid, but has been given the botanical name of *C. jambhiri* Lush. Believed to have originated in northern India, where it grows wild; carried in 1498 or later by Portuguese explorers to southeastern Africa where it became naturalized along the Mazoe River; soon taken to Europe, and brought by Spaniards to the New World; is naturalized in the West Indies and Florida; oblate, rounded or oval, base flat to distinctly necked, apex rounded with a more or less sunken nipple; of medium size, averaging 2¾ in.(7 cm) wide, 2½ (6.25 cm) high; peel lemon-yellow to orange-yellow, rough and irregular, with large oil glands, often ribbed; ³/₁₆ to ³/₈ in (5-10 mm) thick; pulp lemon-yellow, usually in 10 segments, medium-juicy, medium-acid, with moderate lemon odor and flavor; seeds small, 10 to 15, brownish within. Reproduces true from seeds, which are 96% to 100% nucellar. Tree large, very thorny; new growth slightly tinged with red; buds and flowers with red-purple. The scant pulp and juice limit the rough lemon to home use. It is appreciated as a dooryard fruit tree in Hawaii and in other tropical and subtropical areas where better lemons are not available. The tree has been of great importance as a rootstock for the sweet orange, mandarin orange and grapefruit. It is not now used as a rootstock for lemon in Florida because of its susceptibility to "blight" (young tree decline). It is also prone to Alternaria leaf spot (*Alternaria citri*) in the nursery, to foot rot (*Phytophthora parasitica*). Incidence varies with the clone and certain clones show significant resistance. In trials at Lake Alfred, 3 atypical clones showed immunity to leaf spot, while a typical rough lemon clone, 'Nelspruit 15', from South African seed, proved highly resistant to leaf spot and also extremely cold tolerant.

'Santa Teresa'—an old tree discovered to be disease-free in a 'Femminello Ovale' orchard in Italy that had been devastated by *mal secco*. Budded trees from the original specimen were being commonly planted in the 1960's wherever the disease was prevalent in Italy.

Sweet Lemon (*C. limetta* Risso)—a general name for certain non-acid lemons or limettas, favored in the Mediterranean region. In India, they are grown in the Nilgiris, Malabar and other areas. The fruits are usually insipid, occasionally subacid or acid. The seeds are white within and the tree is large, resembling that of the orange. One cultivar, called 'Dorshapo' after the plant explorers, Dorsett, Shamel and Popenoe, who introduced it from Brazil in 1914, resembles the 'Eureka' in most respects except for the lack of acidity. Another, called 'Millsweet', apparently was introduced into California from Mexico and planted in a mission garden. It was reproduced at the old University of California Experiment Station at Pomona. Neither is of any commercial value.

'Villafranca'—believed to have originated in Sicily; introduced into Sanford, Florida, from Europe around 1875 and later into California. Closely resembles 'Eureka'; of medium size. Tree is more vigorous, larger, more densely foliaged, and more thorny than 'Eureka' but becomes thornless with age. One strain is everbearing; another fruits heavily in summer. This was the leading lemon cultivar in Florida for many years; is cultivated commercially in Israel; is low-yielding and short-lived in India. It is little grown in California but has given rise to certain selections that are of importance, particularly 'Galligan Lisbon' and 'Corona Foothill Eureka'.

Climate

Because of its more or less continuous state of growth, the lemon is more sensitive to cold than the orange and less able to recover from cold injury. The tree is defoliated at 22° to 24°F (-5.56°--4.44°C). A temperature drop to 20°F (-6.67°C) will severely damage the wood unless there has been a fortnight of near-freezing weather to slow down growth. Flowers and young fruits are killed by 29°F (-1.67°C) and nearly mature fruits are badly damaged below 28°F (-2.22°C). On the other hand, the lemon attains best quality in coastal areas with summers too cool for proper ripening of oranges and grapefruit. Therefore, the lemon has a relatively limited climatic range. In Florida, lemons are produced commercially as far north as Ft. Pierce on the East Coast and Ruskin on the West Coast. The 'Meyer' lemon, as a dooryard tree, can be grown wherever oranges thrive, even as far west as Pensacola.

The fruits are scarred and the tree readily defoliated by winds, and benefit by the protection of windbreaks.

Lemons are grown in both dry and humid atmospheres, the latter being a disadvantage mainly in the processes of curing and storing. Over a large lemon-growing region in California, annual rainfall varies from 25 to 125 cm. In long, dry periods, the lemon must be irrigated.

Soil

The lemon tree has the reputation of tolerating very infertile, very poor soil. In Florida, groves are mostly on sand. In California, excellent growth is maintained on silty clay loam of high water-holding capacity. In Guatemala, recommended soils are sand, clay and sandy-clay—deep, with high permeability and good drainage. Black soils are also suitable if not lying over calcareous subsoil. Ph should be between 5.5 and 6.5. If acidity is high, it is necessary to apply lime to achieve the optimum level.

Propagation

The rough lemon is widely grown from seed. The 'Meyer' lemon is easily reproduced by rooting large cuttings in the nursery and planting them directly in the grove. They fruit 2 to 3 years sooner than budded trees and have a long life, remaining in full production for over 30 years, perhaps much longer.

In Florida, commercial lemons have been budded onto 'rough lemon', sweet orange, and 'Cleopatra' mandarin rootstocks. More recent practices are the utilization of sour orange, Volkamer lemon (*C. volkameriana*), and alemow (*C. macrophylla* Wester, an old Philippine lemon/pummelo hybrid). The latter is employed in California on soils containing an excess of soluble salts and boron. If citranges are used as rootstocks for 'Eureka', bud union crease will kill the tree.

Culture

Lemon trees should be spaced 25 ft (7.6 m) apart each way. If crowded or "hedged", production declines. The trees must be pruned when young and kept below 10 or 12 ft (3-3.6 m) in height. They are cut back severely after 12 years or replaced. Weeds must be controlled but lemon

Fig. 42: Flowers of the lemon (*Citrus limon*) are larger and showier than those of the orange.

trees are very sensitive to herbicides.

In Florida, fertilizing may be done 3 times a year between mid-November and the end of April, at the gradually increasing rate of 4 to 10 lbs (1.8–4.5 kg) per tree up to an age of 50 years. Nitrogen and potash are given in equal amounts under normal soil conditions. A nutritional spray with copper added is applied after spring bloom. Fertilizer and irrigation programs should be varied according to the desired goal: fresh fruit marketing or processing. High nitrogen steps up yield and peel oil content but also results in more scab infection and poor curing. Potash increases acidity. Heavy irrigation increases yield and peel oil, scab infection, size of fruit and accelerates maturity.

In California, foliar spraying of urea is preferred over ground application of nitrogen which can lead to accumulations of salts and also contamination of groundwater. Leaf analyses are made to determine the nitrogen requirements of each cultivar for maximum yield. 'Eureka', in a 6-year test, showed no response to increased levels of nitrogen. In New Zealand, mature trees (15 to 20 years old) are given 25 to 30 lbs (11.3–13.6 kg) of complete mixed fertilizer annually, also heavy dressings of organic manure or mulch.

In Sicily, growers have, for over 50 years, made a practice of withholding water in summer — for 35 to 60 days — until the trees begin to wilt. Then the trees are heavily irrigated and given high nitrogen fertilizer which induces a second bloom in August or early September, producing a crop the following summer when lemons are scarce and prices are high. This system, called the "Verdelli process", was adopted on a little over 1,000 acres (405 ha) in California in 1983. Adequate bloom did not occur on sandy or shallow soils, but 80% of the plantings on gradually dehydrating, fine-textured soil bloomed well. Nearly $3 million was expected from this extra crop of summer lemons in the Central Valley and the Riverside area in 1984. New horticultural techniques are needed to overcome the handicaps of higher use of fertilizer, increased insect and fungus problems, effects of moisture stress on fruit quality, and low temperature hazard to immature fruits in winter.

In 1965, a team of California horticulturists initiated experimental trellis culture of 'Prior Lisbon' lemon on *C. macrophylla* rootstock. It was found that the labor of training, and repeated pruning either manually or by machine hedging and topping, was excessive and uneconomic.

Guatemalan and Mexican growers interplant short-term crops such as beans, cassava, yautia (*Xanthosoma*), in the rainy season, and tomatoes and peppers during the winter when the lemon trees will be irrigated and fertilized.

Harvesting and Handling

The marketability of lemons depends on the stage at which they are picked. Italian lemons for export are harvested as early as possible and are naturally "cured" in transit. In early days, California and Florida lemons were allowed to remain on the trees until they became too large. It was realized that early picking is necessary and California and Arizona adopted the practices of picking at any time after the fruits reach a 25% juice content, and using rings to gauge the commercially acceptable size, and repeated spot-picking with clippers. Mechanical picking is impossible with lemons. The fruits are highly prone to oil spotting (oleocellosis) and cannot be handled roughly nor picked wet.

Formerly, Florida lemons were picked from mid-July to October for shipping fresh, and the balance in November was harvested for processing. Lemons under 2 1/8 in (5.4 cm) are too immature to attain proper quality for marketing and fruits over 2½ in (6.25 cm) are too large. Manual spot-picking has been commonly practiced, but some producers have found it too costly, and are harvesting the entire crop at one time and grading for fresh sale or processing in the packing-house, discarding all undersized fruits. The lemons, after sorting according to color, washing and coating with a fungicide and a thin layer of wax are stored (cured) until ready for shipping.

Yield

Lemon tree yields vary considerably with the cultivar, the location and weather conditions. A yield of 3 boxes per tree is commercially satisfactory in Florida. In India, a 6-year-old tree bore 966 fruits and, at 9 years of age, had produced a total of 3,173 fruits.

Storage

Florida's climate is unfavorable for long-term curing. It has been claimed that a 10-day curing period is adequate and degreening of Florida fruit is not needed. A major producer keeps the newly harvested fruits for 48 hours at 60°F (15.56°C) and 95% humidity, then passes them through a pre-grading procedure to eliminate all that are unusable. The usable fruits are then treated with fungicide against stem-end rot and returned to the curing room. Those harvested early in the season need 3 weeks to color-up, the last may require less than a week. Finally, the fruits are washed, given a second fungicidal treatment, dried, waxed and packed.

Generally, lemons are cured at 56° to 58°F (13.33°-14.4°C) and 85-90% relative humidity. Green fruits may be held for 4 months or more, while the peel becomes yellow and thinner, the pulp juicier (6-80%) and the proportion of soluble solids higher (7-24%). Sometimes the degreening process is hastened by exposing the fruit to ethylene gas, ethephon, or silane, but this practice tends to stimulate decay, mainly through the shedding of the "button" (stem stub), the absence of which allows entry of *Diplodia natalensis*, *Phomopsis citri*, or *Alternaria* mycelium. Various auxins have been studied to determine which can be applied before storage to prevent button loss without delaying degreening. In 1982, Israeli investigators reported that decay losses from degreening procedures can be greatly reduced (from over 50% to 6.3%) by packaging the fruits in 10 micrometer-thick high-density polyethylene. This treatment makes it possible to store lemons with minimum damage for as long as 6 months.

In the past, New Zealand lemons for storage have been individually wrapped in diphenyl-treated paper after washing and dipping in a 200 ppm solution of 2,4,5-T and then waxing. The fruits were marketable after storing for 4 months at room temperature. Lemons can be kept for weeks in the home refrigerator if placed in a jar with a tight-fitting lid to prevent loss of moisture.

Lemons for export from Florida to Hawaii and Arizona must be fumigated with methyl bromide because of possible infestation by the Caribbean fruit fly. For sale within the state, other methods must be employed.

Pests and Diseases

In Southeast Asia, many species of ants attack the root system and the farmer times the opening of the water gates so as to force the ants to the surface of the beds, where he burns them with fire.

One of the 3 most serious arthropod pests of the lemon and other citrus trees in California is California red scale, *Aonidiella aurantii*. In the southern part of the state it is under biological control but it requires applications of pesticides in the San Joaquin Valley. In Florida, rust mites, purple mites and purple scale may at times be troublesome but they are all controllable with appropriate sprays.

Young lemon trees in California sometimes require protection from wild rabbits.

Diseases are the greater challenges. In Florida, the main lemon diseases are scab (*Elsinoe fawcetti*) on fruit, leaves and twigs; anthracnose of fruit (stylar-end-rot), leaves and twigs caused by both *Colletotrichum gloeosporioides* and *Glomerella cingulata;* greasy spot (*Mycosphaerella citri* or *Cercospora citri-grisea*); and gummosis (*Diaporthe citri*). The latter organism also causes melanose and dieback, and stem-end rot. Stem-end rot may also arise from attack by *Botryosphaeria ribis* and *Diplodia natalensis*.

Other lemon diseases recorded in Florida are branch knot (*Sphaeropsis tumefaciens*), damping-off (*Rhizoctonia solani*), leaf spot (*Mycosphaerella horii, Alternaria citri,* and *Catenularia* sp.; algal leaf spot or green scurf (*Cephaleuros virescens*); tar spot (*Cercospora gigantea*); felt fungus (*Septobasidium pseudopedicellatum*); charcoal root rot (*Macrophomia phaseolina*); root rot (*Fusarium oxysporum, Pythium ultimum,* and *Phytophthora parasitica;* heart rot and wood rot (*Fomes applanatus* and *Ganoderma sessilis*); crinkly leaf and exocortis viruses; and green mold (*Penicillium digitatum*); blue mold (*P. italicum*); and pink mold (*P. roseum*). In 1955, the lemon budwood certification program was begun to provide virus-free stock for growers.

Red algae infests lemon trees and causes much dieback unless controlled with copper fungicide in the summer. Zinc deficiency causes stunting of twigs, reduced flowering, premature dropping of fruit, and yellow bands along the leaf veins. Manganese deficiency is evidenced by interveinal chlorosis and subsequent necrosis, shedding of leaves, flowers and young fruit. In India, fruit cracking occurs when dry periods are followed by heavy rains. Cracking can be largely avoided by frequent light irrigation during the dry period and early picking.

Stored lemons are subject to the stem-end rots and the molds listed above. The albedo may show small dark sunken areas even though this defect is not visible externally. Cultivars differ in their ability to resist decay.

Food Uses

Slices of lemon are served as a garnish on fish or meat or with iced or hot tea, to be squeezed for the flavorful juice. In Colombia, lemon soup is made by adding slices of lemon to dry bread roll that has been sautéed in shortening until soft and then sieved. Sugar and a cup of wine are added and the mixture brought to a boil, and then served.

Lemon juice, fresh, canned, concentrated and frozen, or dehydrated and powdered, is primarily used for lemonade, in carbonated beverages, or other drinks. It is also used for making pies and tarts, as a flavoring for cakes, cookies, cake icings, puddings, sherbet, confectionery, preserves and pharmaceutical products. A few drops of lemon juice, added to cream before whipping, gives stability to the whipped cream.

Lemon peel can be candied at home and is preserved in brine and supplied to manufacturers of confectionery and baked goods. It is the source of lemon oil, pectin and citric acid. Lemon oil, often with terpenes and sesquiterpenes removed, is added to frozen or otherwise processed lemon juice to enrich the flavor. It is much employed as a flavoring for hard candies.

Food Value Per 100 g of Edible Portion*						
	Fruit (fresh, peeled)	Juice (fresh)	Juice (canned, unsweetened)	Juice (frozen, unsweetened)	Lemonade (concentrate, frozen)	Peel (raw)**
Calories	27	25	23	22	195	
Moisture	90.1 g	91.0 g	91.6 g	92.0 g	48.5 g	81.6 g
Protein	1.1 g	0.5 g	0.4 g	0.4 g	0.2 g	1.5 g
Fat	0.3 g	0.2 g	0.1 g	0.2 g	0.1 g	0.3 g
Carbohydrates	8.2 g	8.0 g	7.6 g	7.2 g	51.1 g	16.0 g
Fiber	0.4 g	trace	trace	trace	0.1 g	–
Ash	0.3 g	0.3 g	0.3 g	0.2 g	0.1 g	0.6 g
Calcium	26 mg	7 mg	7 mg	7 mg	4 mg	134 mg
Phosphorus	16 mg	10 mg	10 mg	9 mg	6 mg	12 mg
Iron	0.6 mg	0.2 mg	0.2 mg	0.3 mg	0.2 mg	0.8 mg
Sodium	2 mg	1 mg	1 mg	1 mg	0.2 mg	6 mg
Potassium	138 mg	141 mg	141 mg	141 mg	70 mg	160 mg
Vitamin A	20 I.U.	20 I.U.	20 I.U.	20 I.U.	20 I.U.	50 I.U.
Thiamine	0.04 mg	0.03 mg	0.03 mg	0.03 mg	0.02 mg	0.06 mg
Riboflavin	0.02 mg	0.01 mg	0.01 mg	0.01 mg	0.03 mg	0.08 mg
Niacin	0.1 mg	0.1 mg	0.1 mg	0.1 mg	0.3 mg	0.4 mg
Ascorbic Acid	53 mg	46 mg	42 mg	44 mg	30 mg	129 mg

*Analyses of true lemons, as marketed.

**Lemon Peel Oil* consists mainly of terpenes, particularly limonene, also gamma terpinene and beta-phellandrene. There are small amounts of sesquiterpenes and aldehydes. Among the aliphatic aldehydes are n-octyl aldehyde, n-nonyl aldehyde, and citral.

Toxicity

The thorns of the lemon tree inflict painful punctures and scratches. Lemon peel oil may cause contact dermatitis, chronic in those who handle, cut and squeeze lemons daily. Parts of the body touched by contaminated hands may show severe reactions after exposure to the sun. People that suck lemons may suffer irritation and eruptions around the mouth. The wood of lemon trees and its sawdust may induce skin reactions in sensitive woodworkers.

Other Uses

Lemon juice is valued in the home as a stain remover, and a slice of lemon dipped in salt can be used to clean copper-bottomed cooking pots. Lemon juice has been used for bleaching freckles and is incorporated into some facial cleansing creams.

Lemon peel oil is much used in furniture polishes, detergents, soaps and shampoos. It is important in perfume blending and especially in colognes.

Petitgrain oil (up to 50% citral), is distilled from the leaves, twigs and immature fruits of the lemon tree in West Africa, North Africa and Italy. With terpenes removed, it is greatly prized in colognes and floral perfumes.

Lemon peel, dehydrated, is marketed as cattlefeed.

Lemonade, when applied to potted plants, has been found to keep their flowers fresh longer than normal. But it cannot be used on chrysanthemums without turning their leaves brown.

Wood: The wood is fine-grained, compact, and easy to work. In Mexico, it is carved into chessmen, toys, small spoons, and other articles.

Medicinal Uses: Lemon juice is widely known as a diuretic, antiscorbutic, astringent, and febrifuge. In Italy, the sweetened juice is given to relieve gingivitis, stomatitis, and inflammation of the tongue. Lemon juice in hot water has been widely advocated as a daily laxative and preventive of the common cold, but daily doses have been found to erode the enamel of the teeth. Prolonged use will reduce the teeth to the level of the gums. Lemon juice and honey, or lemon juice with salt or ginger, is taken when needed as a cold remedy. It was the juice of the Mediterranean sweet lemon, not the lime, that was carried aboard British sailing ships of the 18th Century to prevent scurvy, though the sailors became known as "limeys".

Oil expressed from lemon seeds is employed medicinally. The root decoction is taken as a treatment for fever in Cuba; for gonorrhea in West Africa. An infusion of the bark or of the peel of the fruit is given to relieve colic.

Mexican Lime

Of the two acid, or sour, limes in world trade, the one longest known and most widely cultivated is the Mexican, West Indian, or Key lime, *Citrus aurantifolia* Swingle (syns. *C. acida* Roxb., *C. lima* Lunan; *C. medica* var. *ácida* Brandis; and *Limonia aurantifolia* Christm.). It is often referred to merely as "lime". In Spanish it is, *lima ácida, lima chica, lima boba, limón chiquito, limón criollo, limón sutil, limón corriente,* or *limón agria.* In French, it is *limette* or *limettier acide;* in German, *limett;* Italian, *limetta;* in Dutch, *lemmetje* or *limmetje.* In East Africa, it is *ndimu;* in the Philippines, *dalayap* or *dayap;* in Malaya, *limau asam;* in India, *nimbu, limbu, nebu, lebu* or *limun.* In Papiamento in the Netherlands Antilles it is *lamoentsji* or *lamunchi;* in Brazil, *limao galego,* or *limao miudo.* In Egypt and the Sudan it is called *limûn baladi,* or *baladi;* in Morocco, *doc.*

Description

The Mexican lime tree is exceedingly vigorous; may be shrubby or range from 6½ to 13 ft (2-4 m) high, with many slender, spreading branches, and usually has numerous, very sharp, axillary spines to ⅜ in (1 cm) long. The evergreen, alternate leaves are pleasantly aromatic, densely set; elliptic- or oblong-ovate, rounded at the base, 2 to 3 in (5-7.5 cm) long, leathery; light purplish when young, dull dark-green above, paler beneath, when mature; with minute, rounded teeth and narrowly-winged petioles. Faintly fragrant or scentless, the axillary flowers, to 2 in (5 cm) across are solitary or 2 to 7 in a raceme, and have 4 to 6 oblong, spreading petals, white but purple-tinged when fresh, and 20-25 bundled white stamens with yellow anthers. The fruit, borne singly or in 2's or 3's (or sometimes large clusters), at the twig tips, is round, obovate, or slightly elliptical, sometimes with a slight nipple at the apex; the base rounded or faintly necked; 1 to 2 in (2.5-5 cm) in diameter; peel is green and glossy when immature, pale-yellow when ripe; somewhat rough to very smooth, ¹/₁₆ to ⅛ in (1.5-3 mm) thick; the pulp is greenish-yellow in 6 to 15 segments which do not readily separate; aromatic, juicy, very acid and flavorful, with few or many small seeds, green inside.

Origin and Distribution

The Mexican lime is native to the Indo-Malayan region. It was unknown in Europe before the Crusades and it is assumed to have been carried to North Africa and the Near East by Arabs and taken by Crusaders from Palestine to Mediterranean Europe. In the mid-13th Century, it was cultivated and well-known in Italy and probably also in France. It was undoubtedly introduced into the Caribbean islands and Mexico by the Spaniards, for it was reportedly commonly grown in Haiti in 1520. It readily became naturalized in the West Indies and Mexico, There is no known record of its arrival in Florida. Dr. Henry Perrine planted limes from Yucatan on Indian Key and possibly elsewhere. In 1839, cultivation of limes in southern Florida was reported to be "increasing". The lime became a common dooryard fruit and by 1883 was being grown commercially on a small scale in Orange and Lake Counties. When pineapple culture was abandoned on the Florida Keys, because of soil depletion and the 1906 hurricane, people began planting limes as a substitute crop for the Keys and the islands off Ft. Myers on the west coast. The fruits were pickled in saltwater and shipped to Boston where they were a popular snack for school children. The little industry flourished especially between 1913 and 1923, but was demolished by the infamous hurricane of 1926. Thereafter, the lime was once again mainly a casual dooryard resource on the Keys and the southern part of the Florida mainland.

In 1953, George D. Fleming, Jr., proprietor of Key Lime Associates, at Rock Harbor, on Key Largo, was the chief producer of limes. Though he had sold several of his groves, he was developing a new one as part of a "vacation cottage colony".

Fearing that this little lime might disappear with lack of demand and the burgeoning development of the Keys, the Upper Florida Keys Chamber of Commerce launched in 1954, and again in 1959 with the help of the Upper Keys Kiwanis Club, an educational campaign to arouse interest and encourage residents to plant the lime and nurseries to propagate the tree for sale.

The Mexican lime continues to be cultivated more or less on a commercial scale in India, Egypt, Mexico, the West Indies, tropical America, and throughout the tropics of the Old World. There are 2,000,000 seedling trees near Colima, Mexico. Mexico raises this lime primarily for sale as fresh fruit but also exports juice and lime oil. New plantings are being made to elevate oil production. In 1975, Rodolfo Guillen Paiz, Chief of the Citrus and Tropical Fruit Subproject of ANACAFE in Guatemala, reported the initiation of a program to establish the Mexican lime as an all-year commercial crop for the fresh fruit market, the production of juice and lime peel oil, and, as a first step, the creation of a collection of selections as a genetic base for development of an industry, possibly in association with cattle-raising since it had been observed that cattle do little damage to the trees.

Production of Mexican limes for juice has been the major industry on the small Caribbean island of Dominica for generations. There are at least 8 factories expressing the juice which is exported largely to the United Kingdom in wooden casks after "settling" in wooden vats and clarifying. In England, it is bottled as the world-famous "Rose's Lime Juice" put out by L. Rose & Co., Ltd., or as the somewhat different product of the chief competitor, A.C. Shellingford & Co. Surplus juice, over their requirements, is sold to soft-drink manufacturers. Since 1960, Rose has produced lime juice concentrate in Dominica for export. There is also considerable export of lime oil distilled from lime juice and oil expressed from the whole fruit. Jamaica, Grenada, Trinidad and Tobago, Guyana,

and the Dominican Republic export lesser amounts of juice and oil. But the Dominican Republic has recently enlarged its plantings in order to increase its oil output. Montserrat ships only juice. Ghana is now the leading producer of lime juice and oil for L. Rose & Co., Ltd. Gambia began serious lime processing in 1967.

The Mexican lime grows wild in the warm valleys of the Himalayas and is cultivated not only in the lowlands but up to an elevation of 4,000 ft (1,200 m). It was first planted on the South Pacific island of Niue in 1930. A small commercial industry has been expanding since 1966. Some of the fruit is sold fresh but most of the crop is processed for juice and oil by the Niue Development Board Factory. These products are shipped to New Zealand, as are a good part of the peels for the manufacture of marmalade and jam. Production was crippled by a hurricane in 1979. This storm inspired a search for rootstocks that could be expected to withstand strong winds.

Varieties

There are few varieties of the Mexican lime, except for several spineless selections, inasmuch as there is no great variation in the wild or under cultivation. Some old named cultivars may not be recognized today.

'Everglade' ('Philippine Islands #2182') – a seedling of a Mexican lime pollinated by flowers of a grapefruit or pummelo, but the fruits show no grapefruit or pummelo characteristics. Introduced into Trinidad in 1922. Planted in the Citrus Experiment Station collection at Riverside, California, it showed little or no distinguishing features. It is limelike, elliptical, with fairly large nipple at apex; 1½ to 2 in (4-5 cm) wide, 1¾ to 2⅛ in (4.5-5.4 cm) high; peel light-yellow when ripe, medium-smooth, the largest oil glands slightly sunken; thin, about 1/16 in (1.5 m); pulp light-greenish, in 8 to 10 segments with tender walls; aromatic, very juicy, of excellent quality and texture; the flavor sprightly acid; seeds 2 to 10, averaging about 5. The fruits are borne in large clusters because all the flowers are perfect. Tree is highly susceptible to withertip.

'Kagzi' – the name given the Mexican lime most commonly cultivated throughout India. It is represented by numerous subtypes differing slightly in size, shape and color.

'Palmetto' – a selected seedling from a Mexican lime pollinated by the 'Sicily' lemon; first described by Dr. H.J. Webber in the United States Department Yearbook for 1905; elliptical or nearly round with small nipple at apex; small of size, 1⅜ to 1½ in (3.6-4 cm) wide; 1⅜ to 1¾ in (3.6-4.5 cm) high; peel pale-yellow when ripe, smooth, very thin, less than 1/16 in (1.5 mm); pulp light greenish-yellow, in 8 to 10 segments; tender, very juicy, of fine quality, aromatic, with sprightly acid flavor; usually 3 to 6 seeds.

'Yung' ('Spineless Mexican') – of unknown origin; was introduced into California from Mexico by George Yung around 1882.

Another spineless sport was reported in Dominica in 1892 and apparently the same was sent to the United States Department of Agriculture from Trinidad in 1910, and several thornless sports were found in lime groves near Weslaco, Texas, after a 1925 freeze. In 1967, seeds of a lime tree seen flourishing in the desert at Yuma, Arizona, were brought to southern Florida by Burt Colburn and planted. Of 50 resulting seedlings, 8 were practically thornless. Budwood from these was grafted onto rough lemon stock for distribution.

In Trinidad, hybridization was undertaken in 1925 in the hope of developing a type immune to withertip. A seedling selection from hybrids was labeled 'T-1'. The fruits were not as juicy in the green stage and a bit larger than the typical Mexican lime. Back-crossing was done to arrive at 'T-145' more closely resembling a typical Mexican lime in size.

Climate

The Mexican lime is more sensitive to cold than the lemon, and can be grown only in protected locations in California. It thrives in a warm, moist climate with annual rainfall between 80 and 150 in (203-381 mm). Nevertheless, it tolerates drought better than any other citrus fruit. When there is excessive rainfall, the tree is subject to fungus diseases.

Soil

The oolitic limestone of the Florida Keys seems perfectly acceptable to the Mexican lime. The tree grows reasonably well in a variety of other soils. In sandy locations on the Florida mainland, best growth is achieved by the periodic addition of lime to raise the pH. Otherwise there will be a lighter crop of fruits; they will be larger than normal with thicker peel and less juice. In Hawaii, this lime is cultivated in rich sandy or gravelly, well-drained soil. Porous lava soil is acceptable if there is abundant rainfall. Stiff clay soils are unsuitable. On the island of Niue, limes are grown on a thin layer of topsoil underlain with limestone. Farmers are advised to avoid breaking up the limestone too much and mixing excessive calcium with the topsoil.

Propagation

The Mexican lime is usually propagated by seed because most seeds are polyembryonic and reproduce faithfully to the parent. In some areas, root sprouts from mature trees are taken up and transplanted into groves. Sprouting may be encouraged by digging around the parent tree to sever the roots wholly or partly. Cuttings of mature wood may also serve for propagation but usually do not develop strong root systems. Selected clones have been budded onto rough lemon or sour orange. The latter is said to provide more resistance to hurricanes. Pummelo has been used in Hawaii but doesn't make a perfect union. In Indonesia, this lime has always been air-layered. In the 1940's, air-layering became popular in Florida. It was adopted in India with 100% success, using indole butyric acid to aid root development of the 'Kagzi' lime.

Culture

In pioneer days, people on the Florida Keys had unsophisticated methods of raising limes. They often sowed the seeds thickly in a pot-hole in the limestone having a bit of soil in the bottom. When the seedlings were a few inches high, they were taken up and transplanted during the rainy season into any pot-hole with enough soil to sustain them until the roots were strong enough to penetrate the porous rock. The result was irregular groves, and this practice was called "jungle" planting. Sometimes volunteer seedlings would be taken up from beneath fruiting trees and transplanted in the same manner. Later on, growers began to dynamite holes in a regular pattern in order to have uniform rows. The breaking up of the rock enhanced root development.

The trees are best set 25 ft (7.5 m) apart each way, which allows for 70 trees per acre (28/ha). Closer spacings of 15 or 20 ft (4.5-6 m) do not permit enough room for good cultural practices. For many years, the trees on the Keys were fertilized only by a mulch of cured seaweed. On the mainland, nitrogen was supplied by leguminous cover crops such as velvet bean (*Mucuna deeringiana* Merr.), beggarweed (*Desmodium canum* Sch. & Thell.), or Showy Crotalaria (*Crotalaria spectabilis* Roth.). Dade County growers came to apply commercial fertilizer, using a 2-8-10, or 2-10-10 NPK formula. Increasing potash is a means of checking growth and promoting fruiting.

Before planting, in Niue, 1 to 2 tablespoons of zinc sulphate are placed in each hole. One month later, and then every 4 months thereafter, 3½ oz (100 g) of mixed nitrogen and potassium are applied around the base. In the second year, the amount given is 18 oz (500 g) in 3 applications; in the third year, 3.3 lbs (1.5 kg); in the 4th year, 6.5 lbs (3 kg) and the 5th year and beyond, 9 lbs (4.5 kg).

Seedlings will begin to fruit in 3 to 6 years and reach full production in 8 to 10 years. The fruits ripen and fall 5 to 6 months after flowering. Trees grown from air layers or cuttings tend to fruit the first year and then cease fruiting until they have attained some growth. If the trees have been correctly pruned when young, there is no further need for pruning except to remove deadwood and watersprouts, or for the purpose of thinning the fruits to increase size.

Harvesting

On the Florida Keys, the trees produce some fruits more or less the year around, but there are two main seasons—May/June and November/December. The peak season on Niue is in April and May. The fruits may be picked while still somewhat green for home use or for the fresh fruit market, but grove workers are reluctant to pick them because of the thorniness of the tree, unless they are provided with protective gloves. If picked too soon, the peel is apt to develop a dark "rind scald". The ideal stage is when the color has changed from dark to light green, the surface is smooth and the fruit feels slightly soft to the touch. For processing, the fully ripe, yellow limes are gathered from the ground twice a week. Because of the rough ground, pioneer growers on the Keys collected the fruits with wheelbarrows pushed along boards placed over the limestone.

Storage

The Mexican lime ripens to full yellow and loses weight rapidly at normal room temperature in warm climates. In the home, the fruits can be held fresh for 2 or 3 weeks if kept in water in a closed jar. They are prone to cold injury under refrigeration at 44.6°F (7°C). A storage temperature of 48.2°F (9°C) with 85-90% relative humidity has been recommended for delaying ripening and loss of moisture. Controlled atmospheres low in oxygen and high in carbon dioxide are also effective in prolonging storage life. Experiments in the Sudan have shown that packing the fruits in polyethylene bags with an ethylene absorbent retards ripening and moisture loss and makes possible the shipping of the fruit by air freight to the United Kingdom.

In India, Mexican limes picked green were coated with wax emulsion containing the growth regulator, indole butyric acid, at 2,000 ppm and kept at room temperature of 65° to 85°F (18.33-29.44°C) and relative humidity of 60 to 90% for 17 days. On removal from storage, 75% of the fruits were marketable, while fruits left untreated and those coated with wax only were completely unmarketable.

A study in Trinidad demonstrated that Mexican limes treated with gibberellic acid, packaged in polyethylene bags to retain moisture, and stored at ambient temperature, remained in marketable condition for 65 days. Yellowing was retarded and there was no adverse effect on quality.

Pests

The Mexican lime is attacked by few pests. On the island of Niue, the most important enemy is snow scale, *Unaspis citri,* in prolonged droughts. Severe infestations cause dieback of branches; lighter attacks induce splitting of the bark which permits entry of other insects and fungi. The scale insect is transported from tree to tree by ants.

Diseases

Withertip, or lime anthracnose, (*Gleosporium limetticolum*) is a serious affliction of the Mexican lime in Florida. *Fusarium oxysporum* causes wilt of seedlings in Florida greenhouses, induces twig dieback in India, and has been identified on Mexican lime grafted onto Rangpur mandarin lime in Brazil.

When the weather is too humid, the Mexican lime is prone to attack by the fungus, *Elsinoe fawcetti,* causing scab. It is also subject to algal disease and oil spotting can be severe. In Niue, the trees are often afflicted with collar rot, caused by *Phytophthora* sp. The fungus, *Sphaeropsis tumefaciens,* causing lime knot and witches broom, has destroyed many trees in Jamaica.

In 1982, a new strain of citrus canker, *Xanthomonas campestris* pv. *citri,* was found on 20,000 trees in the state of Colima, Mexico, in a 5-sq. mile (12.8 sq. km) area. Seedlings that had been shipped from this area were destroyed and the United States Department of Agriculture culture set up the requirement that all citrus imports from Mexico would have to be accompanied by a phytosanitary certificate. Canker is a common plague of limes in India and in 1960 the Horticultural Research Institute reported that Streptomycin sulfate at 500 ppm reduced the incidence by 34%.

The fruits are attacked by decay organisms in storage, principally *Rhizopus nigricans* and *Penicillium* spp.

Food Uses

The Mexican lime, because of its special bouquet and unique flavor, is ideal for serving in half as a garnish and flavoring for fish and meats, for adding zest to cold drinks, and for making limeade. In the Bahamas, fishermen and others who spend days in their sailboats, always have with them their bottles of homemade "old sour" — lime juice and salt. Throughout Malaysia, this lime is grown mainly to flavor prepared foods and beverages. Commercially bottled lime juice is prized the world over for use in mixed alcoholic drinks. If whole limes are crushed by the screwpress process, the juice should be treated to remove some of the peel oil. It is calculated that 2,200 lbs (1 metric ton) of fruit should yield 1,058 lbs (480 kg) of juice.

Lime juice is made into sirup and sauce and pies similar to lemon pie. "Key Lime Pie" is a famous dish of the Florida Keys and southern Florida, but today is largely made from the frozen concentrate of the 'Tahiti' lime.

Mexican limes are often made into jam, jelly and marmalade. In Malaya, they are preserved in sirup. They are also pickled by first making 4 incisions in the apex, covering the fruits with salt, and later preserving them in vinegar. Before serving, the pickled fruits may be fried in coconut oil and sugar and then they are eaten as appetizers.

Pickling is done in India by quartering the fruits, layering the pieces with salt in glass or glazed clay jars, and placing in the sun for 3 to 4 days. The contents are stirred once a day. Green chili peppers, turmeric, ginger or other spices may be included at the outset. Coconut or other edible oil may be added last to enhance the keeping quality. Another method of pickling involves scraping the fruits, steeping them in lime juice, then salting and exposing to the sun.

Hard, dried limes are exported from India to Iraq for making a special beverage.

The oil derived from the Mexican lime is obtained by three different methods in the West Indies:

1) by hand-pressing in a copper bowl studded with spikes (which is called an écuelle). This method yields oil of the highest quality but it is produced in limited amounts. It is an important flavoring for hard candy.

2) by machine pressing, cold expression, of the oil from the spent half-shells after juice extraction, or simultaneously but with no contact with the juice.

3) by distillation from the oily pulp that rises to the top of tanks in which the washed, crushed fruits have been left to settle for 2 weeks to a month. This yields the highest percentage of oil. With terpenes and sesquiterpenes removed, it is extensively used in flavoring soft drinks, confectionery, ice cream, sherbet, and other food products. The settled juice is marketed for beverage manufacturing. The residue can be processed to recover citric acid.

The minced leaves are consumed in certain Javanese dishes. In the Philippines, the chopped peel is made into a sweetmeat with milk and coconut.

Food Value Per 100 g of Edible Portion*	
Moisture	88.7–93.5 g
Protein	0.070–0.112 g
Fat	0.04–0.17 g
Fiber	0.1–0.5 g
Ash	0.25–0.40 g
Calcium	4.5–33.3 mg
Phosphorus	9.3–21.0 mg
Iron	0.19–0.33 mg
Vitamin A	0.003–0.040 mg
Thiamine	0.019–0.068 mg
Riboflavin	0.011–0.023 mg
Niacin	0.14–0.25 mg
Ascorbic Acid	30.0–48.7 mg

*According to analyses made in Central America.

Other Uses

Juice: In the West Indies, the juice has been used in the process of dyeing leather. On the island of St. Johns, a cosmetic manufacturer produces a bottled Lime Moisture Lotion as a skin-conditioner.

Peel: The dehydrated peel is fed to cattle. In India, the powdered dried peel and the sludge remaining after clarifying lime juice are employed for cleaning metal.

Peel oil: The hand-pressed peel oil is mainly utilized in the perfume industry.

Twigs: In tropical Africa, lime twigs are popular chewsticks.

Medicinal Uses: Lime juice dispels the irritation and swelling of mosquito bites.

In Malaya, the juice is taken as a tonic and to relieve stomach ailments. Mixed with oil, it is given as a vermifuge. The pickled fruit, with other substances, is poulticed on the head to allay neuralgia. In India, the pickled fruit is eaten to relieve indigestion. The juice of the Mexican lime is regarded as an antiseptic, tonic, an antiscorbutic, an astringent, and as a diuretic in liver ailments, a digestive stimulant, a remedy for intestinal hemorrhage and hemorrhoids, heart palpitations, headache, convulsive cough, rheumatism, arthritis, falling hair, bad breath, and as a disinfectant for all kinds of ulcers when applied in a poultice.

The leaves are poulticed on skin diseases and on the abdomen of a new mother after childbirth. The leaves or an infusion of the crushed leaves may be applied to relieve headache. The leaf decoction is used as eye drops and to bathe a feverish patient; also as a mouth wash and gargle in cases of sore throat and thrush.

The root bark serves as a febrifuge, as does the seed kernel, ground and mixed with lime juice.

In addition, there are many purely superstitious uses of the lime in Malaya.

Tahiti Lime

This acid lime lacks the long history and wide usage that glamorize the small Mexican lime. Its identity has been in doubt and only in recent years has it been given the botanical name, *Citrus latifolia* Tan. An alternate common name is Persian lime.

Description

The Tahiti lime tree is moderately vigorous, medium to large, up to 15 or 20 ft (4.5–6 m), with nearly thornless, widespread, drooping branches. The leaves are broad-lanceolate, with winged petioles; young shoots are purplish. Flowers, borne off and on during the year but mainly in January, are slightly purple-tinged. The fruit is oval, obovate, oblong or short-elliptical, usually rounded at the base, occasionally ribbed or with a short neck; the apex is rounded with a brief nipple; 1½ to 2½ in (4–6.25 cm) wide, 2 to 3 in (5–7.5 cm) high; peel is vivid green until ripe when it becomes pale-yellow; smooth, thin, tightly clinging; pulp is light greenish-yellow when ripe, in 10 segments, tender, acid, but without the distinctive bouquet of the Mexican lime; usually seedless, rarely with one or a few seeds, especially if planted among a number of other *Citrus* species. The Tahiti lime flowers have no viable pollen.

Origin and Distribution

The origin of the Tahiti lime is unknown. It is presumed to be a hybrid of the Mexican lime and citron, or, less likely, the lemon, and it is genetically a triploid though only the normal 18 chromosomes have been reported. Dr. Groff, in a reference to *Citrus aurantifolia* in his "Culture and Varieties of Siamese Pummelos . . .", said: ". . .it is represented by a large variety known as *Manow klom* and by a small one known as *Manow yai*." One might speculate as to whether the large variety might be the female parent of the Tahiti lime. At any rate, it is believed that the Tahiti was introduced into the Mediterranean region by way of Iran (formerly called Persia). It is said that, for some centuries, a virtually identical lime called 'Sakhesli' has been cultivated on the island of Djerba off the coast of Tunisia, and that the local name means "from Sakhos", an old Arabic name for Chios, a Gre-

Fig. 43: Tahiti, or Persian lime (*Citrus latifolia*) (left); and the Mexican, or West Indian (*C. aurantifolia*) which is especially aromatic.

cian island. Portuguese traders probably carried it to Brazil, and it was apparently taken to Australia from Brazil about 1824. It reached California from Tahiti between 1850 and 1880 and had arrived in Florida by 1883. It was being grown at Lake Placid in 1897. This lime was adopted into cultivation in California but is not extensively grown there, the bulk of California's lime crop being mainly the Mexican lime. In Florida, the Tahiti quickly took the place of the more sensitive small lime and the lemon. Following World War I, the Tahiti lime became a well-established commercial crop. At first, there was market resistance, buyers viewing the Tahiti lime as a "green lemon", and, for some time, Canadians would not accept it because they were accustomed to the more flavorful Mexican lime. In the 1930's, many Florida citrus growers planted limes for extra income and, in 1949, the development of limeade concentrate provided further impetus to the Tahiti lime industry.

In 1954, Libby, McNeil & Libby topworked 100 acres (40 ha) of grapefruit trees in Florida to Tahiti lime. Production increased 60% from 1970 to 1980. In 1979, the total crop was valued at close to $9 million. Nearly 1 million bushels (250 limes per bushel) were shipped fresh and the same amount was processed. By 1980, there were approximately 8,000 acres (about 3,250 ha) of commercial groves. Five years later, Dade County shipped 110 million lbs (50 million kg) of fresh fruit worth about $14 million to the growers, from a total of 6,500 acres (2,630 ha). Florida produces 90% of the national crop, for marketing fresh and for canned lime juice, frozen lime juice, frozen lime juice concentrate, frozen limeade and powdered lime juice. The Florida Lime and Avocado Administrative Committee conducts research on production and carries on national promotional activity.

Varieties

There have been only a few named cultivars, or alleged cultivars, of the Tahiti lime:

'Bearss' ('Bearss Seedless', 'Byrum Seedless')—This was first put forward as a new variety of Tahiti lime originating in the grove of T.J. Bearss at Porterville, California, in 1895. It was described and illustrated in 1902 and cultivated and catalogued by the Fancher Creek Nursery Company in 1905. It was grown in California, Arizona and Hawaii under the name, 'Bearss', at least until the late 1940's. However, comparative studies made in California led to the decision that the 'Bearss' did not differ sufficiently from the typical Tahiti lime to be maintained as a distinct cultivar.

'Idemor' — a limb sport found around 1934 in a grove owned by G.L. Polk in Homestead, Florida, and patented in 1941 (U.S. Plant Patent #444). The fruit is smaller and more rotund than the typical Tahiti. A very similar sport has been reported from Morocco. This lime is no longer planted because of its susceptibility to virus diseases.

'Pond' — In 1914, budwood was obtained by Dr. H.J. Webber from a Tahiti lime tree in the Moanalua Gardens, in Honolulu. Budded trees bore fruits that were somewhat smaller than the typical Tahiti but otherwise much the same. The trees were somewhat lower growing. This cultivar seems to have disappeared.

'USDA 'No. 1' and 'No. 2' — selections from many seedlings grown by Dr. James Childs of the United States Department of Agriculture at the Horticultural Field Station, Orlando, Florida. They are free of exocortis and xyloporosis viruses and are available to growers through Florida's Budwood Registration Program. The fruit does not differ significantly in character from the typical Tahiti lime. The development of these virus-free clones has been a great boon to Florida's lime industry.

Climate

The Tahiti lime is hardier than the Mexican lime and better adapted to the mainland of Florida. Most of the commercial groves are in Dade County, but, with some cold protection, this lime can be grown on the east and west coasts and the central ridge as far north as Winter Haven. Even in southern Florida, drastic drops in temperature have made it necessary to protect lime groves with wind machines or overhead sprinkling.

Soil

The plantings in southern Florida are on oolitic limestone. Those further north are on deep sand. The soil must be well drained. In low land subject to standing water, lime trees are planted on elevated beds.

Propagation

The seeds of the Tahiti lime are largely monoembryonic; few seeds are available for planting; and seedlings, for the most part, are exceedingly variable. Only 10 trees of 114 seedlings grown at the Agricultural Research and Education Center of the University of Florida, Homestead, showed typical Tahiti lime characters vegetatively and in the fruit, except for long thorns on the trunk and branches.

This lime has been customarily budded onto rough lemon, but in recent years more commonly on the alemow, C. macrophylla. Many sweet orange and grapefruit trees have been successfully topworked to the Tahiti lime. Today, 40% of the commercial Tahiti lime trees have been grown from air-layers.

Culture

In Dade County's limestone, the trees are planted at the intersection of mechanically-cut trenches 16 in (40.5 cm) deep, or on mounds of crushed limestone and soil on scarified ground. The Tahiti lime tree is less vigorous than the Mexican lime and accordingly lends itself to close-planting. Spacing may be as close as 10 or 15 ft (3-4.5 m) in rows 20 ft (6 m) apart, which permits about 150 to 200 trees per acre (60-80/ha). When the trees overlap, they are mechanically hedged and topped. Greater yields will result if the trees are spaced at 20 ft (6 m) and hedging and topping are performed at 2- to 3-year intervals. The tree produces few water sprouts. A 12-month study in Cuba showed that hedging does not affect yield a year later, and does not alter the normal growth of the tree.

Air-layered trees begin to bear a year before budded trees but, as they mature, they generally do not yield as well. Because of their year-around growth, lime trees demand more fertilization and irrigation than other Citrus species. In commercial groves, irrigation is provided by overhead sprinklers, portable or stationery.

In early days, many trees were afflicted with bark lesions and even girdling, killing the affected branches or the entire tree if on the trunk. Splitting high-nitrogen fertilizer applications into 4 applications annually instead of 2 seemed to eliminate the problem. More recently, it has been recommended that a 4-6-6 formula of NPK be applied every 60 days. Potash is particularly important in relation to yield. In California, experimental spraying with gibberellic acid (10 ppm) delayed maturity and increased fruit size. The fruit stayed green longer in the packinghouse.

Harvesting

Tahiti limes are harvested 8 to 12 times a year — once a month in winter, but 70% of the crop matures from May to fall. The peak period is July to September. The demand persists year-around and off-season fruits sell at premium prices. Most harvesting is by hand but some use a "gig". If picked too immature, the fruits will be deficient in juice. Since 1955, a Federal Marketing Order has prevented the harvesting of immature fruit and has provided for the industry's setting of standards of quality, grade and size. The minimum permissible juice content is 42%. If left too long on the tree, the fruits will be subject to stylar-end-breakdown and are apt to turn yellowish before they reach distant markets.

The limes are collected in wooden field boxes and conveyed by truck to packinghouses where they are graded, washed, waxed, and packed in 10-, 20-, 40-, or 55-lb (4.5-, 9-, 18-, or 25-kg) corrugated cartons for shipment to retailers. About 40% of the crop is processed locally for lime juice concentrate. Cull limes are shipped to out-of-state manufacturers of citrus juices and peel oil extractors. Limes for shipment to Hawaii and Arizona must be fumigated with methyl bromide because of possible infestation by Caribbean fruit fly.

Yield

The yield from 7 ft (2.13 m) trees grafted on alemow rootstock has averaged 90 lbs (41 kg), while trees of the same size on rough lemon yielded 63 lbs (29 kg). Under advanced methods of management, Florida lime groves produce 600 bushels per acre (243 bu/ha) annually.

Storage

The Tahiti lime requires no curing. The fresh fruits remain in good condition for 6 to 8 weeks under refrigeration.

Pests and Diseases

The citrus red mite (purple mite, red spider, spider mite), and the broad mite may heavily infest Tahiti lime leaves and fruits.

Formerly, the trees and fruits commonly evidenced lime blotch (yellow areas on leaves and fruits) but the replacing of susceptible trees has largely eliminated this problem. The tree is immune to withertip, moderately susceptible to scab and greasy spot. Red alga is a major problem, causing bark splitting and dieback of branches. It can be prevented by regular and thorough spraying with copper or other suitable fungicides. The tree is subject to several viruses: crinkly leaf, psorosis, tatterleaf, tristeza, exocortis and xyloporosis.

The fruits are highly subject to oil spotting (oleocellosis), which occurs most frequently during rainy seasons and when limes are harvested when wet with dew. Stylar-end-breakdown, or stylar-end-rot, has been a very serious post-harvest disorder in the summer. It may develop within 2 hours after picking or several days later. It is apparently induced in oversize fruits, larger than 2½ in (6.25 cm) picked early in the morning when internal pressure is high and left too long in the hot sun in the field boxes. The effect is an expansion and rupturing of juice vesicles and the development of a brown, soft area at the apex of the fruit, occasionally at the base also. Fruit losses have been as high as 40%. Precooling the fruits for 24 hours greatly reduces the incidence of this disease.

Food Uses

The Tahiti lime is utilized for making limeade and otherwise for the same purposes as the Mexican lime. In Florida, a wedge of lime is commonly served with avocado, and lime juice is frequently used as an alternative to vinegar in dressings and sauces.

It was formerly held that the oil from the peel of the Tahiti lime was of inferior quality. Since the late 1960's, it has been accepted by the trade and produced in quantity as a by-product of the juice-extraction process. It is utilized for enhancing lime juice and for most of the other purposes for which Mexican lime peel oil is employed.

Toxicity

Excessive exposure to the peel oil of the Tahiti lime may cause dermatitis. Rolling the limes between the hands before squeezing in order to extract more of the juice will coat the hands with oil and this will be transferred to whatever parts of the body are touched before washing the hands. Subsequent exposure to sunlight often results in brown or red areas that itch intensely, and sometimes severe blistering. The sap of the tree and scratches by the thorns may cause rash in sensitive individuals.

Other Uses

Lime juice is employed as a rinse after shampooing the hair. Light streaks have been bleached in the hair by applying lime juice and then going out into the sun for a time. One should be sure that there is no peel oil on the hands when doing this. Lime juice has been applied on the face as a freshening lotion. Some Florida housewives use lime juice for cleaning the inside of coffeepots, and grind a whole lime in the electric garbage-disposal to eliminate unpleasant odor. Dilute lime juice will dissolve, overnight, calcium deposits in teakettles.

Medicinal Uses: Lime juice, given quickly, is an effective antidote for the painful oral irritation and inflammation that result from biting into aroids such as *Dieffenbachia* spp., *Xanthosoma* spp., *Philodendron* spp., and their allies. Lime juice has also been applied to relieve the effects of stinging corals.

Sweet Lime

The sweet lime, *Citrus limettioides* Tan. (syn. *C. lumia* Risso et Poit.), is called *limettier doux* in French; *lima dulce* in Spanish; *mitha limbu, mitha nimbu,* or *mitha nebu,* in India (*mitha* meaning "sweet"); *quit giây* in Vietnam; *limûn helou,* or *succari* in Egypt; *laymûn-helo* in Syria and Palestine. It is often confused with the sweet lemon, *C. limetta* Tan., (q.v. under LEMON) which, in certain areas, is referred to as "sweet lime". In some of the literature, it is impossible to tell which fruit is under discussion.

Description

The tree, its foliage, and the form and size of the fruit resemble the Tahiti lime; the leaves are serrated and the petioles nearly wingless. The fruit is not at all similar to the Mexican lime. The flowers are borne singly in the leaf axils or in terminal clusters of 2 to 10; the fruits may be solitary or in bunches of 2 to 5.

Origin and Distribution

It is not known where or how the sweet lime originated, but it is thought to be a hybrid between a Mexican-type lime and a sweet lemon or sweet citron. Mediterranean botanists refer to it as native to India. Central and northern India, northern Vietnam, Egypt and other countries around the coasts of Mediterranean, and tropical America, are the chief areas of cultivation. It came to the United States from Saharanpur, India, in 1904 (S.P.I. #10365).

There is very limited culture in California where the fruits produced by desert-grown trees differ markedly from those in cooler coastal regions. It is not grown for its fruit nor used as a rootstock in Florida because of its high susceptibility to viruses. In India and Israel it is much utilized as a rootstock for the sweet orange and other *Citrus* species.

Varieties

There are said to be several strains in India differing in fruit shape and tree productivity.

'Indian' ('Palestine')—oblong, ovoid or nearly round, with rounded base and small nipple at apex, occasionally slightly ribbed; peel aromatic, greenish to orange-yellow when ripe, smooth, with conspicuous oil glands, thin; pulp pale-yellow, usually in 10 segments, tender, very juicy, non-acid, bland, faintly bitter. The tree may be large or shrubby; is spreading, irregular, thorny, with leaves resembling those of the orange but paler and with more prominent oil glands, their petioles faintly winged. Buds and flowers are white. The tree is hardier than that of the acid lime; bears late in the rainy season in India when other citrus fruits are out-of-season.

'Columbia' — a clonal selection mentioned by Reuther *et al.* (*Citrus Industry*, Vol. I, rev'd, 1967).

'Soh Synteng' — a strongly acid variation in Assam with new shoots and flower buds briefly pinkish.

Pollination

The sweet lime is self-compatible. In studies aimed at improving yield, Indian scientists found that self-pollination results in maximum fruit set, while cross-pollination with sweet orange or grapefruit results in greater fruit retention, at the same time increasing fruit size and seed count. Therefore, the practice of interplanting with sweet orange and grapefruit has been adopted in commercial orchards.

Propagation

In India, the sweet lime is grown from cuttings.

Food Uses

In the West Indies and Central America, the fruits are commonly enjoyed out-of-hand. The stem-end is cut off, the core is pierced with a knife, and the juice is sucked out. The fruit is eaten fresh in India as well as cooked and preserved.

The hand-pressed peel oil has a strong lemon odor. It contains pinene, limonene, linalool, linalyl acetate and possibly dipentene and citral.

Medicinal Uses

In India the sweet lime is therapeutically valued for its cooling effect in cases of fever and jaundice.

Calamondin

Prized for its ornamental value more widely than for its fruit, the calamondin was formerly identified as *Citrus mitis* Blanco (syn. *C. microcarpa* Bunge); more recently in *Citrus* circles, erroneously, as *C. madurensis* Lour.; now it has been given the hybrid name: X *Citrofortunella mitis* J. Ingram & H.E. Moore. Among alternate common names are: calamondin orange; Chinese, or China, orange; Panama orange; golden lime; scarlet lime; and, in the Philippines, *kalamondin, kalamunding, kalamansi, calamansi, limonsito,* or *agridulce.* Malayan names are *limau kesturi* ("musk lime") and *limau chuit.* In Thailand it is *ma–nao–wan.*

Description

The calamondin tree, ranging from 6½ to 25 ft (2–7.5 m) high, is erect, slender, often quite cylindrical, densely branched beginning close to the ground, slightly thorny, and develops an extraordinarily deep taproot. The evergreen leaves (technically single leaflets) are alternate, aromatic, broad-oval, dark-green, glossy on the upper surface, yellowish-green beneath, 1½ to 3 in (4–7.5 cm) long, faintly toothed at the apex, with short, narrowly-winged petioles. The richly and sweetly fragrant flowers, having 5 elliptic-oblong, pure-white petals, are about 1 in (2.5 cm) wide and borne singly or in 2's or 3's terminally or in the leaf axils near the branch tips. The showy fruits are round or oblate and to 1¾ in (4.5 cm) wide, with very aromatic, orange-red peel, glossy, and dotted with numerous small oil glands; tender, thin, easily-removed, sweet, and edible. The pulp, in 6 to 10 segments, is orange, very juicy, highly acid, seedless or with 1 to 5 small, obovoid seeds, green within.

Origin and Distribution

The calamondin is believed native to China and thought to have been taken in early times to Indonesia and the Philippines. It became the most important *Citrus* juice source in the Philippine Islands and is widely grown in India and throughout southern Asia and Malaysia. It is a common ornamental dooryard tree in Hawaii, the Bahamas, some islands of the West Indies, and parts of Central America. Dr. David Fairchild introduced it into Florida from Panama in 1899. It quickly became popular in Florida and Texas. The California climate is not as favorable but a variegated form ('Peters') is cultivated there.

Fig. 44: The calamondin (X *Citrofortunella mitis*), a showy ornamental, makes excellent marmalade.

Since 1960, thousands of potted specimens have been shipped from southern Florida to all parts of the United States for use as house plants. Israel is now similarly raising such plants for the European market. The calamondin is also valued as a rootstock for the oval kumquat (q.v.) for pot culture.

At the Agricultural Experiment Station of the University of Florida in Gainesville, the calamondin is much utilized for greenhouse research on the various aspects of flowering and fruiting in *Citrus*.

Climate

The calamondin is as cold-hardy as the Satsuma orange and can be grown all along the Gulf Coast of the southern United States. It is moderately drought-tolerant.

Soil

The tree seems able to tolerate a wide range of soils from clay-loam in the Philippines to limestone or sand in Florida.

Propagation

Calamondin trees may be easily grown from seeds, which are polyembryonic with 3 to 5 embryos each. For commercial fruit production in the Philippines, the trees are budded onto calamondin seedlings. In Florida, propagation by cuttings rooted under constant mist is the more common commercial procedure for pot culture. Even leaf-cuttings will root readily.

Culture

Plants grown from cuttings fruit during the rooting period and will reach 18 to 24 in (45-60 cm) in height in 10½ months. The flowers are self-fertile and require no cross-pollination. Transplanted into a large container and well cared for, a calamondin will grow at the rate of 1 ft (30 cm) per year; will produce an abundant crop of fruit at the age of 2 years and will continue to bear the year around. Potted plants for shipment can be stored in the dark for 2 weeks at 53.6°F (12°C) without loss of leaves or fruits in storage or in subsequent transit and marketing.

In orchard plantings, Philippine workers have established that a complete commercial fertilizer with a 1:1 nitrogen to potassium ratio gives the best growth. There are 2 applications: one prior to the onset of the rainy season and the second just before the cessation of rains. Adequate moisture is the principal factor in yield, size and quality of the fruit. Drought and dehydrating winds often lead to mesophyll collapse.

Harvesting

Calamondins are harvested by clipping the stems as they become fully colored throughout the year. In the Philippines the peak season is mid-August through October.

Storage

The fruits will keep in good condition for 2 weeks at 48° to 50°F (8.89°-10°C) and 90% relative humidity. Weight loss will be only 6.5%. Waxing retards ascorbic acid loss for 2 weeks in storage but not thereafter.

Pests and Diseases

The calamondin is a prime host of the Mediterranean and Caribbean fruit flies, and for this reason is much less planted in Florida than formerly. It may be attacked by other pests and diseases that affect the lemon and lime including the viruses: crinkly leaf, exocortis, psorosis, xyloporosis and tristeza, but it is immune to canker and scab.

Food Uses

Calamondin halves or quarters may be served with iced tea, seafood and meats, to be squeezed for the acid juice. They were commonly so used in Florida before limes became plentiful. Some people boil the sliced fruits with cranberries to make a tart sauce. Calamondins are also preserved whole in sugar sirup, or made into sweet pickles, or marmalade. A superior marmalade is made by using equal quantities of calamondins and kumquats. In Hawaii, a calamondin-papaya marmalade is popular. In Malaya, the calamondin is an ingredient in chutney. Whole fruits, fried in coconut oil with various seasonings, are eaten with curry. The preserved peel is added as flavoring to other fruits stewed or preserved.

The juice is primarily valued for making acid beverages. It is often employed like lime or lemon juice to make gelatin salads or desserts, custard pie or chiffon pie. In the Philippines, the extracted juice, with the addition of gum tragacanth as an emulsifier, is pasteurized and bottled commercially. This product must be stored at low temperature to keep well. Pectin is recovered from the peel as a by-product of juice production.

Food Value Per 100 g of Edible Portion*		
	Whole Fruit (%)	*Juice* (%)
Calories/lb	173 (380/kg)	
Moisture	87.08-87.12	89.66
Protein	0.86	0.01
Fat	2.41	0.53
Carbohydrates	3.27	
Ash	0.54-0.64	0.62
Calcium	0.14	
Phosphorus	0.07	
Iron	0.003	
Citric Acid	2.81	5.52

*The chemistry of the calamondin has received only moderate attention. Wester (1924) and Marañon (1935) reported the above constituents from Philippine analyses. Mustard found the ascorbic acid content of the *whole fruit* to be, 88.4-111.3 mg/100 g; of the *juice*, 30-31.5 mg; and of the *peel*, 130-173.9 mg.

Other Uses

The fruit juice is used in the Philippines to bleach ink stains from fabrics. It also serves as a body deodorant.

Medicinal Uses: The fruits may be crushed with the saponaceous bark of *Entada phaseoloides* Merr. for shampooing the hair, or the fruit juice applied to the scalp after shampooing. It eliminates itching and promotes hair growth. Rubbing calamondin juice on insect bites banishes the itching and irritation. It bleaches freckles and helps to clear up *acne vulgaris* and *pruritus vulvae*. It is taken orally as a cough remedy and antiphlogistic. Slightly diluted and drunk warm, it serves as a laxative. Combined with pepper, it is prescribed in Malaya to expel phlegm. The root enters into a treatment given at childbirth. The distilled oil of the leaves serves as a carminative with more potency than peppermint oil. The volatile oil content of the leaves is 0.90% to 1.06%.

Mandarin Lime

This is a group name embracing three more or less similar fruits:

1) **Rangpur** (*Citrus* X *limonia* Osbeck) is also called rangpur lime, *rungpur,* marmalade lime, lemandarin; Canton lemon in southern China, *hime* lemon in Japan; *Japanche citroen* in Indonesia; *sylhet* lime, *surkh nimboo* and *shabati* in India; *limao cravo* in Brazil. It is probably a lemon X mandarin orange hybrid originating in India.

Sir Joseph Hooker recorded this as a small, slender tree in the very bottom of valleys, along the foot of the Himalayas, from Gurhwal to the Khasia Hills. The Reasoner Brothers, nurserymen, at Oneco, Florida, introduced seeds from northwestern India and catalogued the tree as a lime.

The fruit resembles a mandarin orange; is round, oblate, or obovate, of irregular surface, the base becoming furrowed and slightly necked with age, the apex rounded or faintly nippled; 1¾ to 2½ in (4.5-6.25 cm) wide, 1⅝ to 2¼ in (4.1-5.7 cm) high; peel is reddish-orange, with large oil glands, thin, easily removed; pulp has limelike aroma, is deep-orange, in 8 to 10 segments having tender walls and separating readily from each other; melting, very juicy; flavor exceedingly sour but suggestive of orange; there may be 6-18 seeds, small, green within.

The tree is fast-growing, more or less spreading, reaching 15 to 20 ft (4.5-6 m); has short thorns; the flower buds and petals are purple-tinted. It is more cold-tolerant than the lime and in California has endured freezes better than the lemon. Unfortunately, it is highly subject to scab. It bears abundantly, from November through winter, and the fruits remain on the tree in good condition. It is a casual dooryard tree in Florida and a minor commercial fruit tree in California. Until the late 1930s, it was much used in Brazil and Argentina as a root-stock but trees budded onto it proved to be short-lived It is grown to some extent in Australia and the Hawaiian Islands, rarely in Trinidad where it was introduced from Montserrat in 1920.

In India, mandarin orange juice is improved by adding 20-40% Rangpur juice. Small, whole fruits can be candied or pickled, but the Rangpur is not fully appreciated until it is made into marmalade. This product is superb and rivals or excels that made from the sour orange.

2) **Kusiae** or kusiae lime is presumably a form of the Rangpur though it is even more limelike in aroma. It is believed to have evolved in India where virtually identical fruits are called *nasaran* and *nemu tenga*. Hawaiians believe that early Spanish settlers planted it on Kusiae, or Strongs Island, in the Caroline Islands of Micronesia. In 1885, Henry Swinton introduced it into Hawaii where it was described and pictured by Gerrit Wilder in 1911. Budwood was taken from Wilder's garden in Honolulu to the Citrus Experiment Station at Riverside, California, in 1914.

The fruit is oval, oblate or round, furrowed and sometimes faintly necked at the base, the apex rounded or with a slight pointed nipple; 1½ to 2½ in (4-6.25 cm) wide; the peel is deep-yellow with prominent oil glands,

medium-thick to thin, leathery, easily removed; pulp is honey-yellow, in 8 or 9 segments having tender walls; melting, somewhat less acid than the true lime and not so rich in flavor; contains 6 to 10 small seeds; the abundant juice is colorless, transparent.

The tree is vigorous, of bushy habit, branched to the ground, but reaching 10 to 20 ft (4.5-6 m) in height; has only a few small thorns and oval to lanceolate leaves; new growth is pale-green; sends up many root sprouts, forming thickets. It is generally grown from seeds and seedlings may be less thorny and seedy than their parents; can be grafted onto sour orange or other non-sprouting citrus rootstocks to avoid root suckers. Fruiting begins in 1½ to 3 years and the tree is nearly everbearing and prolific. In Hawaii, 11-year-old trees have borne 2,000 fruits, nearly 200 lbs (90.5 kg) per tree. The Kusiae lime is cold-tolerant, immune to withertip but prone to scab and root-rot. It is a common dooryard fruit tree in Hawaii and also grown in Trinidad, little-known elsewhere.

3) **Otaheite,** or Otaite, orange, or Otaheite Rangpur, formerly known as *C. otaitensis* Risso & Poit. (syn. *C. taitensis* Risso), is now thought to be a non-acid form of the Rangpur. Its origin is unknown. It was introduced into France from Tahiti by way of England in 1813; was being grown in Paris by the botanist Noisitte in 1915. It was catalogued by a San Francisco nurseryman in 1882.

The fruit is oblate to spherical, 1½ to 2 in (4-5 cm) wide, furrowed and rounded or slightly necked at the base, the apex rounded or with a flat nipple; peel is orange with small oil glands; thin; pulp is orange, in 7 to 10 segments, juicy, slightly limelike in aroma and flavor but bland with scarcely any acidity; seedless, or with 3 to 6 small, abortive seeds.

The tree is a dwarf, spreading, thornless, with oblong to elliptic, finely-toothed leaves having narrowly-winged petioles; the new growth is deep-purple; flowers are fragrant and purple outside. Grown from cuttings or air-layers, the tree is widely sold in the United States as a potted "miniature orange", especially in the Christmas season when it bears flowers and fruits concurrently.

Citron (Plate XXI)

A fruit better known to most consumers in its preserved rather than in its natural form, the citron, *Citrus medica* Linn., is called in French, *cedrat, cidratier, citronnier des Juifs;* in Spanish, *cidra, poncil, poncidre, cedro limón, limón cidra, limón Francés,* though in Central America it is often referred to as *toronja,* the popular Spanish name for grapefruit. In Portuguese, it is *cidrao;* in Italian, *cedro* or *cedrone;* in German, *cedratzitrone* or *ceder-* *appelen;* in Dutch, *citroen;* in India, *citron, beg-poora,* or *leemoo;* in Malaya, *limau susu, limau mata kerbau, limau kerat lingtang;* in Thailand, *som-mu, som manão* or *som ma-nguâ;* in Laos, *manao ripon, mak vo* or *mak nao;* in Vietnam, *thank-yen* or *chanh;* in Samoa, *tipolo* or *moli-apatupatu;* in China, *kou-yuan.* Theophrastus wrote of it as the Persian, or Median, Apple, and it was later called the Citrus Apple.

Description

The citron is borne by a slow-growing shrub or small tree reaching 8 to 15 ft (2.4-4.5 m) high with stiff branches and stiff twigs and short or long spines in the leaf axils. The leaflets are evergreen, lemon-scented, ovate-lanceolate or ovate elliptic, 2½ to 7 in (6.25-18 cm) long; leathery, with short, wingless or nearly wingless petioles; the flower buds are large and white or purplish; the fragrant flowers about 1½ in (4 cm) wide, in short clusters, are mostly perfect but some male because of pistil abortion; 4- to 5-petalled, often pinkish or purplish on the outside, with 30 to 60 stamens. The fruit is fragrant, mostly oblong, obovoid or oval, occasionally pyriform, but highly variable; various shapes and smooth or rough fruits sometimes occurring on the same branch; one form is deeply divided from the apex into slender sections; frequently there is a protruding style; size also varies greatly from 3½ to 9 in or even 1 ft (9-22.8 or 30 cm) long; peel is yellow when fully ripe; usually rough and bumpy but sometimes smooth; mostly very thick, fleshy, tightly clinging; pulp pale-yellow or greenish divided into as many as 14 or 15 segments, firm, not very juicy, acid or sweet; contains numerous monoembryonic seeds, ovoid, smooth, white within.

Origin and Distribution

The citron's place of origin is unknown but seeds were found in Mesopotamian excavations dating back to 4000 B.C. The armies of Alexander the Great are thought to have carried the citron to the Mediterranean region about 300 B.C. A Jewish coin struck in 136 B.C. bore a representation of the citron on one side. A Chinese writer in AD 300 spoke of a gift of "40 Chinese bushels of citrons from Ta-ch'in" in AD 284. Ta-ch'in is understood to mean the Roman Empire. The citron was a staple, commercial food item in Rome in AD 301. There are wild citron trees in Chittagong, Sitakund Hill, Khasi and Garo hills of northern India. Dioscorides mentioned citron in the 1st Century AD and Pliny called it *malus medica, malus Assyria* and *citrus* in AD 177. The fruit was imported into Greece from Persia (now Iran). Greek colonists began growing the citron in Palestine about 200 B.C. The tree is assumed to have been successfully introduced into Italy in the 3rd Century. The trees were mostly destroyed by barbarians in the 4th Century but those in the "Kingdom of Naples" and in Sardinia and Sicily survived. By the year 1003, the citron was commonly cultivated at Salerno and fruits (called *poma cedrina*) were presented as a token of gratitude to Norman lords. For centuries, this area supplied citron to the Jews in Italy, France and Germany for their Feast of the Tabernacles (*sukkot*) ceremony. Moses had specified the cone of the cedar, *kadar* (*kedros* in Greek) and when it fell into disfavor it was replaced by the citron, and the Palestine Greeks called the latter *kedromelon* (cedar apple). *Kedros* was Latinized as *cedrus* and this evolved into *citrus,* and subsequently into citron. For many years, most *Citrus* species were identified as botanical varieties of *Citrus medica.*

Spaniards probably brought the citron with other *Citrus* species to St. Augustine, Florida, though it could have survived there only in greenhouses. The tree was introduced into Puerto Rico in 1640. Commercial citron culture and processing began in California in 1880. The trees suffered severe cold damage in 1913 and, within a few years, the project was abandoned. From 1926 to 1936, there were scattered small plantings of citron in Florida, and particularly one on Terra Ceia Island, supplying fruits to the Hills Brothers Canning Company. The groves eventually succumbed to cold and today the citron is grown in southern Florida only occasionally as a curiosity. The main producing areas of citron for food use are Sicily, Corsica and Crete and other islands off the coasts of Italy, Greece and France, and the neighboring mainland. Citron is also grown commercially in the central, mountainous coffee regions of Puerto Rico. Some is candied locally but most is shipped in brine to the United States and Europe. Citron is casually grown in several other islands of the Caribbean and in Central and South America. It has been rather commonly grown in Brazil for many years. There have long been scattered citron trees in the Cauca Valley of Colombia. After 5 years of study, horticulturists decided in 1964 that commercial culture could be profitable. Citron trees are not uncommon in some of the Pacific Islands but are rare in the Philippines.

Varieties

Citron cultivars are mainly of two types: 1) those with pinkish new growth, purple flower buds and purple-tinted petals, acid pulp and dark inner seed coat and chalazal spot; 2) those with no pink or purple tint in the new growth nor the flowers, with non-acid pulp, colorless inner seed coat, and pale-yellow chalazal spot. Among the better-known cultivars are:

'Corsican' —origin unknown but the leading citron of Corsica; introduced into the United States around 1891 and apparently the cultivar grown in California; ellipsoid or faintly obovate, furrowed at base; large; peel yellow, rough, lumpy, very thick, fleshy; pulp crisp, non-juicy, non-acid, seedy. Tree small, spreading, moderately thorny with some large spines.

'Diamante' ('Cedro Liscio'; possibly the same as 'Italian' and 'Sicilian')—of unknown origin but the leading cultivar in Italy and preferred by processors elsewhere; long-oval or ellipsoid, furrowed at base, broadly nippled at apex; peel yellow, smooth or faintly ribbed; very thick, fleshy; pulp crisp, non-juicy, acid; seedy. Tree small, spreading, thorny as 'Corsican'. Very similar is a cultivar called "Earle" in Cuba.

'Etrog' ('Ethrog', 'Atrog'; *C. medica* var. *Ethrog* Engl.)— the leading cultivar in Israel; ellipsoid, spindle-shaped or lemon-like with moderate neck and often with persistent style at base; usually with prominent nipple at apex; medium-small as harvested; if not picked early, it will remain on the tree, continuing to enlarge for years until the branch cannot support it. For ritual use, the fruit should be about 5 oz (142 g) and not oblong in form. Peel is yellow, semi-rough and bumpy, faintly ribbed, thick, fleshy; flesh is crisp, firm, with little juice; acid; seedy. Tree is small, not vigorous; leaves rounded at apex and cupped. This cultivar has been the official citron for use in the Feast of the Tabernacles ritual but if unavailable any yellow,

Plate XXI
CITRON
Citrus medica

Plate XXII
FINGERED CITRON
Citrus medica var. *sarcodactylus*

Plate XXIII
SANTOL
Sandoricum koetjape

Plate XXIV
LANGSAT
Lansium domesticum

Plate XXV
BARBADOS CHERRY
Malpighia punicifolia

Plate XXVI
BIGNAY
Antidesma bunius

Plate XXVII
MANGO
Mangifera indica
'Cambodiana'

Plate XXVIII
MANGO
Mangifera indica
'Kent', 'Tommy Atkins', and 'Irwin'

Plate XXIX
GANDARIA
Bouea gandaria

Plate XXX
CASHEW APPLE
Anacardium occidentale

Plate XXXI
AMBARELLA
Spondias dulcis

unblemished, lemon-sized citron with adhering style can be substituted.

'Fingered Citron', Plate XXI, ('Buddha's Hand', or 'Buddha's Fingers'); *C. medica* var. *sarcodactylus* Swing.); called *fu shou* in China, *bushukon* in Japan, *limau jari, jeruk tangan, limau kerat lingtang,* in Malaya; *djerook tangan* in Indonesia; *som–mu* in Thailand; *phât thu* in Vietnam. The fruit is corrugated, wholly or partly split into about 5 finger-like segments, with little or no flesh; seedless or with loose seeds. The fruit is highly fragrant and is placed as an offering on temple altars. It is commonly grown in China and Japan; is candied in China.

In India, there are several named types, in addition to the 'Fingered', in the northwest:

'Bajoura'—small, with thin peel, much acid juice.
'Chhangura'—believed to be the wild form and commonly found in a natural state; fruit rough, small, without pulp.
'Madhankri' or **'Madhkunkur'**—fruit large with sweetish pulp.
'Turunj'—fruit large, with thick peel, the white inner part sweet and edible; pulp scant, dry, acid. Leaves are oblong and distinctly notched at the apex.

Climate

The citron tree is highly sensitive to frost; does not enter winter dormancy as early as other *Citrus* species. Foliage and fruit easily damaged by very intense heat and drought. Best citron locations are those where there are no extremes of temperature.

Soil

The soils where the citron is grown vary considerably, but the tree requires good aeration.

Propagation

Citron trees are grown readily from cuttings taken from branches 2 to 4 years old and quickly buried deeply in soil without defoliation. For quicker growth, the citron may be budded onto rough lemon, grapefruit, sour orange or sweet orange but the fruits do not attain the size of those produced from cuttings, and the citron tends to overgrow the rootstock. Rough lemon has been found too susceptible to gummosis to be employed as a rootstock for citron in Colombia. The 'Etrog', to be acceptable for ritual use, must not be budded or grafted.

Culture

The citron tree tends to put out water sprouts that should be eliminated, and the grower should prune branches hanging so low that they touch the ground with the weight of the fruit. Italian producers keep the tree low and stake the branches, and may even trim off the thorns, to avoid scarring of the fruits. The trees begin to bear when 3 years old and reach peak production in 15 years; die in about 25 years.

In 'Etrog' orchards, the Israeli growers are careful to take every precaution to protect the fruit, tying the fruiting branch securely in place and trimming away any twigs that might touch the fruit. To avoid moving irrigation equipment through the groves, the trees are manually watered and frequently sprayed to eliminate destructive insects.

If citrons are allowed to fully ripen on the tree they will be very aromatic and the peel yellow, the inner peel very tender. In India, a fruiting branch may be bent down and the immature fruit put into a jar shaped like a human head (or other form) so that the mature fruit will be of the same shape. These are sold as curiosities and are said to be intensely fragrant.

Harvesting

The citron tree blooms nearly all year, but mostly in spring and the spring blooms produce the major part of the crop. The fruit is dark-green when young, takes 3 months to turn yellow. To retain the green color, firmness and uniformity desired by the dealers in candied citron, the fruit must be picked when only 5 to 6 in (12.5-15 cm) long and 3 to 4 in (7.5-10 cm) wide. Mature trees yield an average of 66 lbs (30 kg) per year but exceptional trees have borne as much as 150 to 220 lbs (68-100 kg). 'Etrog' fruits are wrapped in hemp fiber immediately after picking. Those for local use are inspected by rabbis, and those for export by agents of the Ministry of Agriculture.

Pests and Diseases

The citron tree is undoubtedly subject to most of the pests that attack other *Citrus* species. The citrus bud mite (*Eriophyes sheldoni*), citrus rust mite (*Phyllocoptruta oleivora*), and snow scale (*Unaspis citri*) are among its major enemies.

Horticulturists in Florida report that citron trees in this state are nearly always unthrifty, are subject to gummosis, and usually in a state of decline and dieback, and are accordingly poor bearers.

Branch knot, caused by the fungus *Sphaeropsis tumefaciens,* was first noticed on citron trees in Puerto Rico in 1977. By 1983, it had become a serious threat to the local citron industry. The deformations become large and necrotic, lead to witches' broom, dieback and breaking of branches.

Food Uses

The most important part of the citron is the peel which is a fairly important article in international trade. The fruits are halved, depulped, immersed in seawater or ordinary salt water to ferment for about 40 days, the brine being changed every 2 weeks; rinsed, put in denser brine in wooden barrels for storage and for export. After partial de-salting and boiling to soften the peel, it is candied in a strong sucrose/glucose solution. The candied peel is sun-dried or put up in jars for future use. Candying is done mainly in England, France and the United States. The candied peel is widely employed in the food industry, especially as an ingredient in fruit cake, plum pudding, buns, sweet rolls and candy.

Puerto Rican food technologists reported in 1970 that the desalted citron could be dehydrated in a hot air tray dryer at 108°F (42.22°C), reducing the weight by 95% to lower costs of shipment, then stored in polyethylene bags and later reconstituted and candied. In 1979, after further experiments, it was announced that fresh citron cubes, blanched for ½ minute in water at 170°F (76.7°C) can be candied and the product is equal in quality to the brined and candied peel, and this procedure saves the costs of salt, storage, and shipping of heavy barrels. If the citron lacks flavor, a few orange or lemon leaves may be added to the sirup.

The fruit of the wild 'Chhangura' is pickled in India. In Indonesia, citron peel is eaten raw with rice. The entire fruit of the 'Fingered citron' is eaten.

If there is sufficient juice in the better cultivars, it is utilized for beverages and to make desserts. In Guatemala, it is used as flavoring for carbonated soft-drinks. In Malaya, citron juice is used as a substitute for the juice of imported, expensive lemons. A product called "citron water" is made in Barbados and shipped to France for flavoring wine and vermouth.

In order to expand the market for citron, Puerto Rican workers have established that the green-mature fruits can be peeled by immersing in a boiling lye solution to save the labor of hand-peeling and then the fruits can be made into marmalade, jelly, and fruit bars that are crusty on the outside, soft within.

In Spain, a sirup made from the peel is used to flavor unpalatable medical preparations.

Food Value Per 100 g of Edible Portion*	
Moisture	87.1 g
Protein	0.081 g
Fat	0.04 g
Fiber	1.1 g
Ash	0.41 g
Calcium	36.5 mg
Phosphorus	16.0 mg
Iron	0.55 mg
Carotene	0.009 mg
Thiamine	0.052 mg
Riboflavin	0.029 mg
Niacin	0.125 mg
Ascorbic Acid	368 mg
*According to analyses made in Central America.	

Other Uses

Fruit: Chinese and Japanese people prize the citron for its fragrance and it is a common practice in central and northern China to carry a ripe fruit in the hand or place the fruit in a dish on a table to perfume the air of a room. The dried fruits are put with stored clothing to repel moths. In southern China, the juice is used to wash fine linen. Formerly, the essential oil was distilled from the peel for use in perfumery.

Leaves and twigs: In some of the South Pacific islands, "Cedrat Petitgrain Oil" is distilled from the leaves and twigs of citron trees for the French perfume industry.

Flowers: The flowers have been distilled for essential oil which has limited use in scent manufacturing.

Wood: Branches of the citron tree are used as walking-sticks in India. The wood is white, rather hard and heavy, and of fine grain. In India, it is used for agricultural implements.

Medicinal Uses: In ancient times and in the Middle Ages, the 'Etrog' was employed as a remedy for seasickness, pulmonary troubles, intestinal ailments and other disorders. Citron juice with wine was considered an effective purgative to rid the system of poison. In India, the peel is a remedy for dysentery and is eaten to overcome halitosis. The distilled juice is given as a sedative. The candied peel is sold in China as a stomachic, stimulant, expectorant and tonic. In West Tropical Africa, the citron is used only as a medicine, particularly against rheumatism. The flowers are used medicinally by the Chinese. In Malaya, a decoction of the fruit is taken to drive off evil spirits. A decoction of the shoots of wild plants is administered to improve appetite, relieve stomachache and expel intestinal worms. The leaf juice, combined with that of *Polygonum* and *Indigofera* is taken after childbirth. A leaf infusion is given as an antispasmodic. In Southeast Asia, citron seeds are given as a vermifuge. In Panama, they are ground up and combined with other ingredients and given as an antidote for poison. The essential oil of the peel is regarded as an antibiotic.

Kumquat

Kumquats have been called "the little gems of the citrus family". They were included in the genus *Citrus* until about 1915 when Dr. Walter T. Swingle set them apart in the genus *Fortunella*, which embraces six Asiatic species. The common name, which has been spelled cumquat, or comquot, means "gold orange" in China. The Japanese equivalent is *kin kan* or *kin kit* for the round type, *too kin kan,* for the oval type. In Southeast Asia, the round is called *kin, kin kuit,* or *kuit xu,* and the oval, *chu tsu* or *chantu*. In Brazil, the trade name may be kumquat, kunquat, or *laranja de ouro dos orientais*.

Description

The kumquat tree is slow-growing, shrubby, compact, 8 to 15 ft (2.4-4.5 m) tall, the branches light-green and angled when young, thornless or with a few spines.

The apparently simple leaves are alternate, lanceolate, 1¼ to 3⅜ in (3.25-8.6 cm) long, finely toothed from the apex to the middle, dark-green, glossy above, lighter beneath. Sweetly fragrant, 5-parted, white flowers are borne singly or 1 to 4 together in the leaf axils. The fruit is oval-oblong or round, ⅝ to 1½ in (1.6-4 cm) wide; peel is golden-yellow to reddish-orange, with large, conspicuous oil glands, fleshy, thick, tightly clinging, edible, the outer layer spicy, the inner layer sweet; the pulp is scant, in 3 to 6 segments, not very juicy, acid to subacid; contains small, pointed seeds or sometimes none; they are green within.

Origin and Distribution

Kumquats are believed native to China. They were described in Chinese literature in 1178 A.D. A European writer in 1646 mentioned the fruit as having been described to him by a Portuguese missionary who had labored 22 years in China. In 1712, kumquats were included in a list of plants cultivated in Japan. They have been grown in Europe and North America since the mid-19th Century, mainly as ornamental dooryard trees and as potted specimens in patios and greenhouses. They are grown mainly in California, Florida and Texas; to a lesser extent in Puerto Rico, Guatemala, Surinam, Colombia and Brazil. In South India, they can be grown only at high elevations. There is limited cultivation in Australia and South Africa.

Varieties

The various kumquats are distinguished as botanical species rather than as cultivars. The following are those most utilized for food:

'Hong Kong', or Hong Kong Wild (*F. Hindsii* Swing.), called *chin chü, shan chin kan,* and *chin tou* by the Chinese—native to Hong Kong and adjacent hilly and mountainous regions of Kwantung and Chekiang Provinces of China; nearly round, ⅝ to ¾ in (1.6-2 cm) wide; peel orange or scarlet when ripe, thin, not very fleshy; pulp in only 3 or 4 small segments; seeds plump. Chinese people flock to the foothills to gather the fruits in season. In the western world, the very thorny shrub is grown only as an ornamental pot plant.

'Marumi', or Round Kumquat (*F. japonica* Swing., syn. *Citrus madurensis* Lour.)—fully described for the first time in 1784; introduced into Florida from Japan by Glen St. Mary and Royal Palm nurseries in 1885; fruit is round, slightly oblate or obovate; to 1¼ in (3.2 cm) long; peel is golden-yellow, smooth, with large oil glands, thin, aromatic and spicy; pulp, in 4 to 7 segments, is scant and acid, with 1 to 3 seeds which are smaller than those of 'Nagami'. The tree reaches 9 ft (2.75 m); is otherwise similar to that of 'Nagami' except that it is slightly thorny, has somewhat smaller leaves and is considerably more cold-tolerant; bears at the same season.

'Meiwa', or Large Round Kumquat (*F. crassifolia* Swing.), called *ninpo* or *neiha kinkan* in Japan—possibly a hybrid between 'Nagami' and 'Marumi'; introduced from Japan by the United States Department of Agriculture between 1910 and 1912; short-oblong to round, about 1½ in (4 cm) wide; peel orange-yellow, very thick, sweet; pulp usually in 7 segments, relatively sweet or subacid; often seedless or with few seeds. The tree is a dwarf, frequently thornless or having short, stout spines; the leaves differ from those of other kumquats in being very thick and rigid and partly folded lengthwise; they are pitted with numerous dark-green oil glands. Extensively grown in Chekiang Province, China, and less commonly in Fukuoka Prefecture, Japan. There is an ornamental form with variegated fruits in Japan. This kumquat is the best for eating fresh; is still somewhat rare in the United States.

'Nagami', or Oval, Kumquat (*F. margarita* Swing.)—plants introduced from China into London in 1846 by Robert Fortune, plant explorer for the Royal Horticultural Society; was reported in North America in 1850; introduced into Florida from Japan by Glen St. Mary and Royal Palm nurseries in 1885; obovate or oblong; up to 1¾ in (4.5 cm) long and 1³⁄₁₆ in (3 cm) wide; pulp divided into 4 or 5 segments, contains 2 to 5 seeds. In season October to January. Tree to 15 ft (4.5 m) tall. A mature specimen on rough lemon rootstock at Oneco, Florida, in 1901, bore a crop of 3,000 to 3,500 fruits. This is the most often cultivated kumquat in the United States.

Climate

Robert Fortune reported that the 'Nagami' kumquat required a hot summer, ranging from 80° to 100°F (26.67°-37.78°C), but could withstand 10 to 15 degrees of frost without injury. It grows in the tea regions of China where the climate is too cold for other citrus fruits, even the Satsuma orange. The trees differ also from other *Citrus* species in that they enter into a period of winter dormancy so profound that they will remain through several weeks of subsequent warm weather without putting out new shoots or blossoms. Despite their ability to survive low temperatures, as in the vicinity of San Francisco, California, the kumquat trees grow better and produce larger and sweeter fruits in warmer regions.

Propagation

Kumquats are rarely grown from seed as they do not do well on their own roots. In China and Japan they are grafted onto the trifoliate orange (*Poncirus trifoliata*). This has been found the best rootstock for kumquats in northern Florida and California and for dwarfing for pot culture. Sour orange and grapefruit are suitable rootstocks for southern Florida. Rough lemon is unsatisfactory in moist soils and tends to be too vigorous for the slow-growing kumquats.

Culture

In orchard plantings, kumquats on trifoliate orange can be set 8 to 12 ft (2.4-3.65 m) apart, or they may be spaced at 5 ft (1.5 m) in hedged rows 12 ft (3.65 m) apart. For pot culture, they must be dwarfed; must not be allowed to become pot-bound, and need faithful watering to avoid dehydration and also need regular feeding.

Harvesting

For the fresh fruit market, it has been customary to clip the fruits individually with 2 or 3 leaves attached to the stem. For decorating gift packs of other citrus fruits, or for use as table decorations, leafy branches bearing several fruits are clipped. This practice has been common in Florida but in cooler California the tree is not sufficiently vigorous to stand much depletion.

Fig. 45: Nagami, or Oval, kumquat (*Fortunella margarita*) (left); and Marumi, or Round, kumquat (*F. japonica*) (right).

Keeping Quality

Because of the thick peel, the kumquat has good keeping quality and stands handling and shipment well.

Food Value Per 100 g of Edible Portion (*raw*)*	
Calories	274
Protein	3.8 g
Fat	0.4 g
Carbohydrates	72.1 g
Calcium	266 mg
Phosphorus	97 mg
Iron	1.7 mg
Sodium	30 mg
Potassium	995 mg
Vitamin A	2,530 I.U.
Thiamine	0.35 mg
Riboflavin	0.40 mg
Niacin	
Ascorbic Acid	151 mg

*According to analyses published by the United States Department of Agriculture.

Pests and Diseases

Potted kumquats are subject to mealybug infestations. Dooryard and orchard trees may be attacked by most of the common citrus pests. They are highly resistant or even immune to citrus canker. The following diseases are recorded by the Florida Department of Agriculture as observed on kumquats: scab (*Elsinoë fawcetti* and its conidial stage, *Sphaceloma fawcetti;* algal leaf spot, or green scurf (*Cephaleuros virescens*); greasy spot (*Cercospora citri–grisea*); anthracnose (*Colletotrichum gloeosporioides*); fruit rot, melanose (*Diaporthe citri*); stem-end rot and gummosis (*Physalospora rhodina*).

Food Uses

Fresh kumquats, especially the 'Meiwa', can be eaten raw, whole. For preserving, they should be left until they lose some of their moisture and acquire richer flavor. The fruits are easily preserved whole in sugar sirup. Canned kumquats are exported from Taiwan and often served as dessert in Chinese restaurants. For candying, the fruits are soaked in hot water with baking soda, next day cut open and cooked briefly each day for 3 days in heavy sirup, then dried and sugared. Kumquats are excellent for making marmalade, either alone or half-and-half with calamondins. The fruit may be pickled by merely packing in jars of water, vinegar, and salt, partially sealing for 4 to 5 days, changing the brine, sealing and letting stand for 6 to 8 weeks. To make sweet pickles, halved fruits are boiled until tender, drained, boiled again in a mixture of corn sirup, vinegar, water and sugar, with added cloves and cinnamon, and then baked until the product is thick and transparent. Kumquat sauce is made by cooking chopped, seeded fruits with honey, orange juice, salt and butter.

Sundry Hybrids and Rootstocks

TRIFOLIATE ORANGE (*Poncirus trifoliata* Raf., syn. *Citrus trifoliata* Linn.) grown for thousands of years in central and northern China; from the 8th Century in Japan if not earlier; a small, fast-growing, deciduous tree, with palmate leaves usually having 3 leaflets, rarely 4 or 5; flowers showy, white, 5-petalled; fruits round to pear-shaped, 1¼ to 2 in (3.2-5 cm) wide; peel fragrant, dull-yellow, minutely downy, rough, with numerous oil glands, thick; pulp scant, sour, with a little acrid oil in the center; seeds ovoid, plump, numerous. Immature fruits and dried mature fruits used medicinally in China. In southern Germany, fruit juice after 2 weeks' storage used to make a flavoring sirup, the peel is candied and used as a spice, and is a source of pectin. The plant is much grown as an ornamental in cool areas of Europe, Asia and North America. In Brazil, it is valued as a protective hedge against animals and human trespassers. Seedlings are important in most citrus-growing areas as rootstocks for various *Citrus* and related species.

CITRANGE (X *Citroncirus Webberi* J. Ingram & H.E. Moore); a trifoliate orange X sweet orange hybrid created by Dr. Walter Swingle or under his direction, beginning in 1897. Tree is evergreen or semi-deciduous, usually trifoliolate, deciduous; not as cold-resistant as the trifoliate orange. Fruits more or less aromatic, outwardly orange-like; 2 to 3 in (5-7.5 cm) wide; peel yellow to deep-orange, may be hairy or non-hairy, wrinkled, ribbed, or smooth; thin; pulp often very juicy and tender, richly flavored, highly acid, slightly bitter; seedless or with a few, mostly polyembryonic, seeds. Certain cultivars, 'Coleman', 'Morton', 'Rusk' and 'Savage', especially 'Rusk', yield juice valued for ade and mixed drinks. They are also desirable for pie, jams and marmalade. 'Troyer' ('Carrizo'), a 'Washington Navel' X trifoliate orange hybrid created by Dr. Walter Reuther in 1909, named 'Troyer' by Swingle in 1934 and renamed 'Carrizo' in 1938, has become a very important rootstock, particularly in California. When budded onto trifoliate orange, can be grown in Georgia.

In early 1985, citrange hybrids 'C35' and 'C32' ('Ruby' orange X trifoliate orange) were released by the Citrus Research Center, Riverside, California, for trial as rootstocks because of their resistance to the citrus nematode, also to *Phytophthora* spp. and the tristeza virus.

CITRANGEQUAT (*Fortunella* sp. X citrange). The first crosses were made by Dr. Swingle at Eustis, Florida, in 1909. Tree is vigorous, erect, thorny or thornless, with mostly trifoliolate leaves; highly cold-resistant. Fruit resembles the oval kumquat, mostly very acid. One cultivar, 'Thomasville', becomes edible when fully mature, though it is relatively seedy. It is very juicy, valued for eating out-of-hand, for ade and marmalade. The tree is strongly resistant to citrus canker and is very ornamental. Two other cultivars, 'Swinton' and 'Telfair', have few seeds, but are less desirable; have limited use for juice and as ornamentals.

LIMEQUAT (X *Citrofortunella* spp.) — Mexican lime X kumquat hybrids made by Dr. Swingle in 1909, described and named in 1913. Tree vigorous, evergreen, the single leaflets having narrowly-winged petioles; nearly spineless or with a few short thorns; more cold-tolerant than the lime but not as hardy as the kumquat; very resistant to withertip. Fruit much like the Mexican lime. There are three named cultivars:

'Eustis' (X *C. floridana* J. Ingram & H.E. Moore) — Mexican lime crossed with round kumquat; oval or round, 1⅛ to 1½ in (2.8-4 cm) wide; peel pale-yellow, smooth, glossy, with prominent oil glands, thin, edible; pulp light greenish in 6 to 9 segments, tender, juicy, very acid, with 5 to 12 small seeds. Of excellent quality, nearly everbearing but mainly in fall-to-winter. Tree has small spines and pure-white buds and flowers; prolific.
'Lakeland' (different seed from same hybrid parent) — oval, 1¼ to 2¼ in (4.5-7 cm) wide; peel bright-yellow, smooth, thin; pulp in 5 to 8 segments, pale-yellow, juicy, pleasantly acid, with 2 to 9 large seeds. Tree nearly spineless; flowers white with pink streaks.
'Tavares' (X *C. Swinglei* J. Ingram & H. E. Moore) — a Mexican lime X oval kumquat hybrid; obovate to oval, about 1¼ to 1⅞ in (3.2-4.75 cm) wide; peel pale orange-yellow, smooth, thin, tender, edible; pulp buff-yellow, in 7 to 8 segments, juicy, very acid, with 6 to 11 large seeds. Tree vigorous with short spines and pink flower buds.

Limequats are cultivated as dooryard trees to a limited extent in central Florida; are more commonly grown in California as potted ornamentals.

VOLKAMER LEMON is described and illustrated in great detail by H. Chapot as *Citrus volkameriana* Pas-

Fig. 46: 'Eustis' limequat (X *Citrofortunella floridana*), a cross between a Mexican lime and the Marumi kumquat.

quale, though the author views it as a hybrid between the lemon and possibly the sour orange. Tanaka and others suggest that it may be a variety of mandarin lime.

The tree is a little smaller than the average lemon tree. Young seedlings bear a few spines ½ to ⅗ in (12.5-15 mm) long, but these disappear with age and are produced only occasionally on older specimens. The leaves are short-petioled, ellipsoid, more or less toothed, 3¾ to 6 in (9.5-15 cm) long. The flowers, only slightly fragrant, short-stalked, 3- to 6-petalled, 1⅜ in (3.5 cm) wide, are borne in small clusters all along the branches and at the tips. The fruit, borne profusely, is lemon-shaped, 2¼ in (5.7 cm) long, 2⅛ in (5.4 cm) wide, rough, bright-reddish-orange. The yellow-orange pulp, in 7 to 11 segments, is very juicy, acid, faintly bitter, of agreeable odor and flavor, with few seeds. The fruiting tree is exceptionally ornamental and the fruit can be used as a substitute for the lemon.

The Volkamer lemon has been known for more than 3 centuries. In the mid-1950's, it was reported in Italy to be a promising rootstock for lemon because of its high resistance to *mal secco* (*Deuterophoma tracheiphila*) and foot-rot (*Phytophthora* sp.). Trials in Morocco in 1972-1973 with scions of sour orange, sweet orange, mandarin orange, grapefruit, lemon and rough lemon, and inoculated Volkamer rootstock, showed it to be highly susceptible to gummosis caused by *Phytophthora citrophthora* in contrast to 'Carrizo' citrange rootstock's high resistance. The degree of necrosis varied somewhat with the scion. (See Chapot in Bibliography).

During tristeza studies on Reunion, workers noted on several trunks of the Volkamer lemon woody galls associated with a wood-bark-socket stem-pitting, according to Aubert *et al.* Protopapadakis and Zambettakis have reported that, in Crete, Volkamer lemon has proved to be second only to sour orange in resistance to *mal secco*.

Bael Fruit

Though more prized for its medicinal virtues than its edible quality, this interesting member of the family Rutaceae is, nevertheless, of sufficient importance as an edible fruit to be included here. The bael fruit, *Aegle marmelos* Correa (syns. *Feronia pellucida* Roth., *Crataeva marmelos* L.), is also called Bengal quince, Indian quince, golden apple, holy fruit, stone apple, *bel, bela, sirphal, maredoo* and other dialectal names in India; *matum* and *mapin* in Thailand; *phneou* or *pnoi* in Cambodia; *bau nau* in Vietnam; *bilak,* or *maja pahit* in Malaya; *modjo* in Java; *oranger du Malabar* in French; *marmelos* in Portuguese. Sometimes it is called elephant apple, which causes confusion with a related fruit of that name, *Feronia limonia* Swingle (q.v.).

Description

The bael fruit tree is slow-growing, of medium size, up to 40 or 50 ft (12-15 m) tall with short trunk, thick, soft, flaking bark, and spreading, sometimes spiny branches, the lower ones drooping. Young suckers bear many stiff, straight spines. A clear, gummy sap, resembling gum arabic, exudes from wounded branches and hangs down in long strands, becoming gradually solid. It is sweet at first taste and then irritating to the throat. The deciduous, alternate leaves, borne singly or in 2's or 3's, are composed of 3 to 5 oval, pointed, shallowly toothed leaflets, 1½ to 4 in (4-10 cm) long, ¾ to 2 in (2-5 cm) wide, the terminal one with a long petiole. New foliage is glossy and pinkish-maroon. Mature leaves emit a disagreeable odor when bruised. Fragrant flowers, in clusters of 4 to 7 along the young branchlets, have 4 recurved, fleshy petals, green outside, yellowish inside, and 50 or more greenish-yellow stamens. The fruit, round, pyriform, oval, or oblong, 2 to 8 in (5-20 cm) in diameter, may have a thin, hard, woody shell or a more or less soft rind, gray-green until the fruit is fully ripe, when it turns yellowish. It is dotted with aromatic, minute oil glands. Inside, there is a hard central core and 8 to 20 faintly defined triangular segments, with thin, dark-orange walls, filled with aromatic, pale-orange, pasty, sweet, resinous, more or less astringent, pulp. Embedded in the pulp are 10 to 15 seeds, flattened-oblong, about ⅜ in (1 cm) long, bearing woolly hairs and each enclosed in a sac of adhesive, transparent mucilage that solidifies on drying.

Origin and Distribution

The tree grows wild in dry forests on hills and plains of central and southern India and Burma, Pakistan and Bangladesh, also in mixed deciduous and dry dipterocarp forests of former French Indochina. Mention has been found in writings dating back to 800 B.C. It is cultivated throughout India, mainly in temple gardens, because of its status as a sacred tree; also in Ceylon and northern Malaya, the drier areas of Java, and to a limited extent on northern Luzon in the Philippine Islands where it first fruited in 1914. It is grown in some Egyptian gardens, and in Surinam and Trinidad. Seeds were sent from Lahore to Dr. Walter T. Swingle in 1909 (P.I. No. 24450). Specimens have been maintained in citrus collections in Florida and in agriculture research stations but the tree has never been grown for its fruit in this state except by Dr. David Fairchild at his home, the "Kampong", in Coconut Grove, after he acquired a taste for it, served with jaggery (palm sugar), in Ceylon.

Climate

The bael fruit tree is a subtropical species. In the Punjab, it grows up to an altitude of 4,000 ft (1,200 m) where the temperature rises to 120°F (48.89°C) in the shade in summer and descends to 20°F (-6.67°C) in the winter, and prolonged droughts occur. It will not fruit where there is no long, dry season, as in southern Malaya.

Soil

The bael fruit is said to do best on rich, well-drained soil, but it has grown well and fruited on the oolitic limestone of southern Florida. According to L.B. Singh (1961), it "grows well in swampy, alkaline or stony soils". . . "grows luxuriantly in the soils having pH range from 5 to 8". In India it has the reputation of thriving where other fruit trees cannot survive.

Varieties

One esteemed, large cultivar with thin rind and few seeds is known as 'Kaghzi'. Dr. L.B. Singh and co-workers at the Horticultural Research Institute, Saharanpur, India, surveyed bael fruit trees in Uttar Padesh, screened about 100 seedlings, selected as the most promising for commercial planting: 'Mitzapuri', 'Darogaji', 'Ojha', 'Rampuri', 'Azamati', 'Khamaria'. Rated the best was 'Mitzapuri', with very thin rind, breakable with slight pressure of the thumb, pulp of fine texture, free of gum, of excellent flavor, and containing few seeds.

S.K. Roy, in 1975, reported on the extreme variability of 24 cultivars collected in Agra, Calcutta, Delhi and Varanasi. He decided that selections should be made for high sugar content and low levels of mucilage, tannin and other phenolics.

Fig. 47: A hard-shelled bael fruit (*Aegle marmelos*), of the type valued more for medicinal purposes than for eating.

Only the small, hard-shelled type is known in Florida and this has to be sawed open, cracked with a hammer, or flung forcefully against a rock. Fruits of this type are standard for medicinal uses rather than for consuming as normal food.

Propagation

The bael fruit is commonly grown from seed in nurseries and transplanted into the field. Seedlings show great variation in form, size, texture of rind, quantity and quality of pulp and number of seeds. The flavor ranges from disagreeable to pleasant. Therefore, superior types must be multiplied vegetatively. L.B. Singh achieved 80% to 95% success in 1954 when he budded 1-month-old shoots onto 2-year-old seedling bael rootstocks in the month of June. Experimental shield-budding onto related species of *Afraegle* and onto *Swinglea glutinosa* Merr. has been successful. Occasionally, air-layers or root cuttings have been used for propagation.

Culture

The tree has no exacting cultural requirements, doing well with a minimum of fertilizer and irrigation. The spacing in orchards is 25 to 30 ft (6-9 m) between trees. Seed-

lings begin to bear in 6 to 7 years, vegetatively propagated trees in 5 years. Full production is reached in 15 years. In India flowering occurs in April and May soon after the new leaves appear and the fruit ripens in 10 to 11 months from bloom — March to June of the following year.

Harvesting

Normally, the fruit is harvested when yellowish-green and kept for 8 days while it loses its green tint. Then the stem readily separates from the fruit. The fruits can be harvested in January (2 to 3 months before full maturity) and ripened artificially in 18 to 24 days by treatment with 1,000 to 1,500 ppm ethrel (2-chloroethane phosphonic acid) and storage at 86°F (30°C). Care is needed in harvesting and handling to avoid causing cracks in the rind.

A tree may yield as many as 800 fruits in a season but an average crop is 150 to 200, or, in the better cultivars, up to 400.

Keeping Quality

Normally-harvested bael fruits can be held for 2 weeks at 86°F (30°C), 4 months at 48.2°F (9°C). Thereafter, mold is likely to develop at the stem-end and any crack in the rind.

Pests and Diseases

The bael fruit seems to be relatively free from pests and diseases except for the fungi causing deterioration in storage.

Food Uses

Bael fruits may be cut in half, or the soft types broken open, and the pulp, dressed with palm sugar, eaten for breakfast, as is a common practice in Indonesia. The pulp is often processed as nectar or "squash" (diluted nectar). A popular drink (called "sherbet" in India) is made by beating the seeded pulp together with milk and sugar. A beverage is also made by combining bael fruit pulp with that of tamarind. These drinks are consumed perhaps less as food or refreshment than for their medicinal effects.

Mature but still unripe fruits are made into jam, with the addition of citric acid. The pulp is also converted into marmalade or sirup, likewise for both food and therapeutic use, the marmalade being eaten at breakfast by those convalescing from diarrhea and dysentery. A firm jelly is made from the pulp alone, or, better still, combined with guava to modify the astringent flavor. The pulp is also pickled.

Bael pulp is steeped in water, strained, preserved with 350 ppm SO_2, blended with 30% sugar, then dehydrated for 15 hrs at 120°F (48.89°C) and pulverized. The powder is enriched with 66 mg per 100 g ascorbic acid and can be stored for 3 months for use in making cold drinks ("squashes"). A confection, bael fruit toffee, is prepared by combining the pulp with sugar, glucose, skim milk powder and hydrogenated fat. Indian food technologists view the prospects for expanded bael fruit processing as highly promising.

The young leaves and shoots are eaten as a vegetable in Thailand and used to season food in Indonesia. They are said to reduce the appetite. An infusion of the flowers is a cooling drink.

Toxicity

The leaves are said to cause abortion and sterility in women. The bark is used as a fish poison in the Celebes. Tannin, ingested frequently and in quantity over a long period of time, is antinutrient and carcinogenic.

Other Uses

Fruit: The fruit pulp has detergent action and has been used for washing clothes. Quisumbing says that bael fruit is employed to eliminate scum in vinegar-making. The gum enveloping the seeds is most abundant in wild fruits and especially when they are unripe. It is commonly used as a household glue and is employed as an adhesive by jewelers. Sometimes it is resorted to as a soap-substitute. It is mixed with lime plaster for waterproofing wells and is added to cement when building walls. Artists add it to their watercolors, and it may be applied as a protective coating on paintings.

The limonene-rich oil has been distilled from the rind for scenting hair oil. The shell of hard fruits has been fashioned into pill- and snuff-boxes, sometimes decorated with gold and silver. The rind of the unripe fruit is

Food Value Per 100 g of Edible Portion*	
Water	54.96–61.5 g
Protein	1.8–2.62 g
Fat	0.2–0.39 g
Carbohydrates	28.11–31.8 g
Ash	1.04–1.7 g
Carotene	55 mg
Thiamine	0.13 mg
Riboflavin	1.19 mg
Niacin	1.1 mg
Ascorbic Acid	8–60 mg
Tartaric Acid	2.11 mg
*Fresh bael fruit, as analyzed in India and in the Philippines.	

The pulp also contains a balsam-like substance, and 2 furocoumarins — psoralen and marmelosin ($C_{13}H_{12}O_3$), highest in the pulp of the large, cultivated forms.

There is as much as 9% tannin in the pulp of wild fruits, less in the cultivated types. The rind contains up to 20%. Tannin is also present in the leaves, as is skimmianine.

The essential oil of the leaves contains d–limonene, 56% a-d-phellandrene, cineol, citronellal, citral; 17% p-cymene, 5% cumin aldehyde. The leaves contain the alkaloids O-(3,3-dimethylallyl)-halfordinol, N-2-ethoxy-2-(4-methoxyphenyl) ethylcinnamide, N-2-methoxy-2-[4-(3',3'-dimethalloxy) phenyl]]ethylcinnamide, and N-2-methoxy-2-(4-methoxyphenyl)-ethylcinnamamide.

employed in tanning and also yields a yellow dye for calico and silk fabrics.

Leaves: In the Hindu culture, the leaves are indispensable offerings to the 'Lord Shiva'. The leaves and twigs are lopped for fodder.

Flowers: A cologne is obtained by distillation from the flowers.

Wood: The wood is strongly aromatic when freshly cut. It is gray-white, hard, but not durable; has been used for carts and construction, though it is inclined to warp and crack during curing. It is best utilized for carving, small-scale turnery, tool and knife handles, pestles and combs, taking a fine polish.

Medicinal Uses: The fresh ripe pulp of the higher quality cultivars, and the "sherbet" made from it, are taken for their mild laxative, tonic and digestive effects. A decoction of the unripe fruit, with fennel and ginger, is prescribed in cases of hemorrhoids. It has been surmised that the psoralen in the pulp increases tolerance of sunlight and aids in the maintaining of normal skin color. It is employed in the treatment of leucoderma. Marmelosin derived from the pulp is given as a laxative and diuretic. In large doses, it lowers the rate of respiration, depresses heart action and causes sleepiness.

For medicinal use, the young fruits, while still tender, are commonly sliced horizontally and sun-dried and sold in local markets. They are much exported to Malaya and Europe. Because of the astringency, especially of the wild fruits, the unripe bael is most prized as a means of halting diarrhea and dysentery, which are prevalent in India in the summer months. Bael fruit was resorted to by the Portuguese in the East Indies in the 1500's and by the British colonials in later times.

A bitter, light-yellow oil extracted from the seeds is given in 1.5 g doses as a purgative. It contains 15.6% palmitic acid, 8.3% stearic acid, 28.7% linoleic and 7.6% linolenic acid. The seed residue contains 70% protein.

The bitter, pungent leaf juice, mixed with honey, is given to allay catarrh and fever. With black pepper added, it is taken to relieve jaundice and constipation accompanied by edema. The leaf decoction is said to alleviate asthma. A hot poultice of the leaves is considered an effective treatment for ophthalmia and various inflamma-tions, also febrile delirium and acute bronchitis.

A decoction of the flowers is used as eye lotion and given as an antiemetic. The bark contains tannin and the coumarin, aegelinol; also the furocourmarin, marmesin; umbelliferone, a hydroxy coumarin; and the alkaloids, fagarine and skimmianine. The bark decoction is administered in cases of malaria. Decoctions of the root are taken to relieve palpitations of the heart, indigestion, and bowel inflammations; also to overcome vomiting.

The fruit, roots and leaves have antibiotic activity. The root, leaves and bark are used in treating snakebite. Chemical studies have revealed the following properties in the roots: psoralen, xanthotoxin, O-methylscopoletin, scopoletin, tembamide, and skimmin; also decursinol, haplopine and aegelinol, in the root bark.

Wood–Apple

The wood-apple, *Feronia limonia* Swingle (syns. *F. elephantum* Correa; *Limonia acidissima* L.; *Schinus limonia* L.) is the only species of its genus, in the family Rutaceae. Besides wood-apple, it may be called elephant apple, monkey fruit, curd fruit, *kath bel* and other dialectal names in India. In Malaya it is *gelinggai* or *belinggai;* in Thailand, *ma-khwit;* in Cambodia, *kramsang;* in Laos, *ma–fit.* In French, it is *pomme d'elephant, pomme de bois,* or *citron des mois.*

Description

The slow-growing tree is erect, with a few upward-reaching branches bending outward near the summit where they are subdivided into slender branchlets drooping at the tips. The bark is ridged, fissured and scaly and there are sharp spines ¾ to 2 in (2-5 cm) long on some of the zigzag twigs. The deciduous, alternate leaves, 3 to 5 in (7.5-12.5 cm) long, dark-green, leathery, often minutely toothed, blunt or notched at the apex, are dotted with oil glands and slightly lemon-scented when crushed. Dull-red or greenish flowers to ½ in (1.25 cm) wide are borne in small, loose, terminal or lateral panicles. They are usually bisexual. The fruit is round to oval, 2 to 5 in (5-12.5 cm) wide, with a hard, woody, grayish-white, scurfy rind about ¼ in (6 mm) thick. The pulp is brown, mealy, odorous, resinous, astringent, acid or sweetish, with numerous small, white seeds scattered through it.

Origin and Distribution

The wood-apple is native and common in the wild in dry plains of India and Ceylon and cultivated along roads and edges of fields and occasionally in orchards. It is also frequently grown throughout Southeast Asia, in northern Malaya and on Penang Island. In India, the fruit was traditionally a "poor man's food" until processing techniques were developed in the mid-1950's.

Varieties

There are 2 forms, one with large, sweetish fruits; one with small, acid fruits.

Climate

The tree grows up to an elevation of 1,500 ft (450 m) in the western Himalayas. It is said to require a monsoon climate with a distinct dry season.

Soil

Throughout its range there is a diversity of soil types, but it is best adapted to light soils.

Propagation

The wood-apple is generally grown from seeds though seedlings will not bear fruit until at least 15 years old. Multiplication may also be by root cuttings, air-layers, or by budding onto self-seedlings to induce dwarfing and precociousness.

Season

In Malaya, the leaves are shed in January, flowering occurs in February and March, and the fruit matures in October and November. In India, the fruit ripens from early October through March.

Harvesting

The fruit is tested for maturity by dropping onto a hard surface from a height of 1 ft (30 cm). Immature fruits bounce, while mature fruits do not. After harvest, the fruit is kept in the sun for 2 weeks to fully ripen.

Food Uses

The rind must be cracked with a hammer. The scooped-out pulp, though sticky, is eaten raw with or without sugar, or is blended with coconut milk and palm-sugar sirup and drunk as a beverage, or frozen as an ice cream. It is also used in chutneys and for making jelly and jam. The jelly is purple and much like that made from black currants.

A bottled nectar is made by diluting the pulp with water, passing through a pulper to remove seeds and fiber, further diluting, straining, and pasteurizing. A clear juice for blending with other fruit juices, has been obtained by clarifying the nectar with Pectinol R-10. Pulp sweetened with sirup of cane or palm sugar, has been canned and sterilized. The pulp can be freeze-dried for future use but it has not been satisfactorily dried by other methods.

Food Value Per 100 g of Edible Pulp*		
	Pulp (ripe)	*Seeds*
Moisture	74.0%	4.0%
Protein	8.00%	26.18%
Fat	1.45%	27%
Carbohydrates	7.45%	35.49%
Ash	5.0%	5.03%
Calcium	0.17%	1.58%
Phosphorus	0.08%	1.43%
Iron	0.07%	0.03%
Tannins	1.03%	0.08%
*According to analyses made in India.		

The pulp represents 36% of the whole fruit. The pectin content of the pulp is 3 to 5% (16% yield on dry-weight basis). The seeds contain a bland, non-bitter, oil high in unsaturated fatty acids.

Other Uses

Pectin: The pectin has potential for multiple uses in pectin-short India, but it is reddish and requires purification.

Rind: The fruit shell is fashioned into snuffboxes and other small containers.

Gum: The trunk and branches exude a white, transparent gum especially following the rainy season. It is utilized as a substitute for, or adulterant of, gum arabic, and is also used in making artists' watercolors, ink, dyes and varnish. It consists of 35.5% arabinose and xylose, 42.7% d-galactose, and traces of rhamnose and glucuronic acid.

Wood: The wood is yellow-gray or whitish, hard, heavy, durable, and valued for construction, pattern-making, agricultural implements, rollers for mills, carving, rulers, and other products. It also serves as fuel.

The heartwood contains ursolic acid and a flavanone glycoside, 7-methylporiol-β-D-xylopyranosyl-D-glucopyranoside.

Medicinal Uses: The fruit is much used in India as a liver and cardiac tonic, and, when unripe, as an astringent means of halting diarrhea and dysentery and effective treatment for hiccough, sore throat and diseases of the gums. The pulp is poulticed onto bites and stings of venomous insects, as is the powdered rind.

Juice of young leaves is mixed with milk and sugar candy and given as a remedy for biliousness and intestinal troubles of children. The powdered gum, mixed with honey, is given to overcome dysentery and diarrhea in children.

Oil derived from the crushed leaves is applied on itch and the leaf decoction is given to children as an aid to digestion. Leaves, bark, roots and fruit pulp are all used against snakebite. The spines are crushed with those of other trees and an infusion taken as a remedy for menorrhagia. The bark is chewed with that of *Barringtonia* and applied on venomous wounds.

The unripe fruits contain 0.015% stigmasterol. Leaves contain stigmasterol (0.012%) and bergapten (0.01%). The bark contains 0.016% marmesin. Root bark contains aurapten, bergapten, isopimpinellin and other coumarins.

White Sapote

The genus *Casimiroa* of the family Rutaceae was named in honor of Cardinal Casimiro Gomez de Ortega, a Spanish botanist of the 18th Century. It embraces 5 or 6 species of shrubs or trees. Of these, 3 shrubby species, *C. pubescens* Ramirez, *C. pringlei* Engl. and *C. watsonii* Engl., are apparently confined to Mexico and have received scant attention. An additional species, *C. emarginata* Standl. & Steyerm., was described in 1944, based on a single specimen in Guatemala. It may be merely a form of *C. sapota,* below.

Of the 3 larger-growing forms, the best known is the common white sapote, called *zapote blanco* by Spanish-speaking people, *abché* or *ahache* by Guatemalan Indians, and Mexican apple in South Africa, and widely identified as *C. edulis* Llave & Lex. The *matasano* (or *matazano*), *C. sapota* Oerst., is often not distinguished from *C.*

Fig. 48: A seedless white sapote, natural size, photographed by Dr. David Fairchild at Orange, California, in October 1919. In his notes accompanying the picture in *Inventory of Seeds and Plants Imported,* No. 60, he says: "It is not rare for trees of this species…may often be due to defective pollination." (Bureau of Plant Industry, United States Department of Agriculture)

edulis in the literature and the name *matasano* has been applied to other species in various localities. The woolly-leaved white sapote, known to the Maya as *yuy* and set apart in Guatemala as *matasano de mico,* has been commonly considered a distinct species, *C. tetrameria* Millsp., but it may be only a variant of *C. edulis.*

Description

White sapote trees range from 15 to 20 ft (4.5-6 m) up to 30 to 60 ft (9-18 m) in height. They have light-gray, thick, warty bark and often develop long, drooping branches. The leaves, mostly evergreen are alternate, palmately compound, with 3 to 7 lanceolate leaflets, smooth or hairy on the underside. The odorless flowers, small and greenish-yellow, are 4- or 5-parted, and borne in terminal and axillary panicles. They are hermaphrodite or occasionally unisexual because of aborted stigmas.

The fruit is round, oval or ovoid, symmetrical or irregular, more or less distinctly 5-lobed; 2½ to 4½ in (6.25-11.25 cm) wide and up to 4¾ in (12 cm) in length; with thin green, yellowish or golden skin coated with a very thin bloom, tender but inedible; and creamy-white or yellow flesh glinting with many tiny, conspicuous,

yellow oil glands. The flavor is sweet with a hint or more of bitterness and sometimes distinctly resinous. There may be 1 to 6 plump, oval, hard, white seeds, 1 to 2 in (2.5-5 cm) long and ½ to 1 in (1.25-2.5 cm) thick, but often some seeds are under-developed (aborted) and very thin. The kernels are bitter and narcotic.

C. edulis has leaves that are usually composed of 5 leaflets, glabrous to slightly pubescent on the underside, and 5-parted flowers. The fruit is somewhat apple-like externally, generally smooth, fairly symmetrical and 2½ to 3 in (6.25-7.5 cm) wide. *C. sapota* is very similar but the leaves usually have only 3, somewhat smaller, leaflets. The woolly-leaved white sapote usually has 5 leaflets, larger and thicker than those of *C. edulis* and velvety-white on the underside, and all the parts of the flowers are in 4's. The fruits are usually 4 to 4½ in (10-11.25 cm) wide, ovoid, irregular and knobby, with rough, pitted skin, and there are often gritty particles in the flesh.

Origin and Distribution

The common white sapote occurs both wild and cultivated in central Mexico. It is planted frequently in Guatemala, El Salvador and Costa Rica and is occasionally grown in northern South America, the Bahamas, West Indies, along the Riviera and other parts of the Mediterranean region, India and the East Indies. It is grown commercially in the Gisborne district of New Zealand and to some extent in South Africa. Horticulturists in Israel took serious interest in white sapotes around 1935 and planted a number of varieties. The trees grew well and produced little in the coastal plain; bore good crops in the interior and commercial prospects seemed bright but the fruit did not appeal to consumers and was too attractive to fruit flies. White sapotes have not done well in the Philippines. The common species was introduced into California by Franciscan monks about 1810, and it is still cultivated on a limited scale in the southern part of that state. In Florida, it was first planted with enthusiasm. Today it is seldom seen outside of fruit tree collections. Of course, many of the trees planted have been seedlings bearing fruits of inferior size and quality, but even the best have never attained popularity in this country.

C. sapota is wild in southern Mexico and Nicaragua, commonly cultivated in Oaxaca and Chiapas. The woolly-leaved white sapote is native from Yucatan to Costa Rica and has not been widely distributed in cultivation. According to Chandler, the fruits are objectionably bitter in California. In southern Florida, the woolly-leaved is sometimes planted in preference to *C. edulis*.

White sapote trees often are grown strictly as ornamentals in California. They are planted as shade for coffee plantations in Central America.

Varieties

Clonal selections were made in California from about 1924 to 1954, and several also in Florida. Some of these may actually be chance hybrids. A surprising number have been named and propagated: 'Blumenthal', 'Chapman', 'Coleman', 'Dade', 'Flournoy', 'Galloway', 'Gillespie', 'Golden' or 'Max Golden', 'Johnston's Golden', 'Harvey', 'Lenz', 'Lomita', 'Maechtlen', 'Maltby' or 'Nancy Maltby', 'Nies', 'Page', 'Parroquia', 'Pike', 'Sarah Jones', 'Suebelle', or 'Hubbell', 'Walton', 'Whatley', 'Wilson', 'Wood', 'Yellow'.

'Coleman'—was one of the first named in California; fruit is oblate, somewhat lobed, furrowed at apex; to 3 in (7.5 cm) wide; skin is yellow-green; flesh of good flavor (22% sugar) but resinous; seeds small. Fruit ripens from late fall to summer. Tree somewhat dwarf; leaflets small and tend to twist. Difficult to propagate.

'Dade'—grown at the Agricultural Research and Education Center, Homestead, Florida from a seed of a selected fruit of a local seedling tree. It was planted in 1935 and fruited in 1939. Round; skin golden-yellow tinged with green, thin; flesh of good, non-bitter flavor. There are 4 to 5 seeds. Ripens in June-July. The tree is low-growing and spreading, with smooth leaflets.

'Gillespie'—originated in California; fruit is round, 3 in (7.5 cm) wide; skin is light-green with russet cheek, fairly tough, rough; flesh is white, of very good flavor. Tree is prolific bearer.

'Golden', or 'Max Golden'—woolly-leaved; fruit conical, depressed at apex; up to 4½ in (11.25 cm) wide; skin yellow-green, fairly tough; flesh has strong flavor, somewhat bitter; few seeds.

'Harvey'—originated in California; round; 3½ in (9 cm) wide; skin smooth, yellow-green with bright orange cheek; flesh cream-colored to pale-yellow; not of the best flavor. Tree is a prolific bearer.

'Maechtlen—named for the parent, an old tree on property owned by the Maechtlen family in Covina, California. Propagated by budding and sold by nurserymen in the 1940's.

'Maltby', or 'Nancy Maltby'—originated in California; round, faintly furrowed, blunt-pointed at apex, base slightly tapered; large; skin yellow-green, smooth, of good flavor but slightly bitter. Tree bears well.

'Parroquia'—originated in California; oval, 2½ in (6.25 cm) wide, 3 in (7.5 cm) long; skin yellow-green, smooth, thin; flesh ivory, of very good flavor. A fairly prolific bearer.

'Pike'—originated in California; rounded or oblate, slightly 5-lobed; to 4 in (10 cm) wide; skin green, very fragile; flesh white to yellowish, of rich, non-bitter, flavor. The tree bears regularly and heavily in California and South Africa.

'Suebelle', or 'Hubbell'—originated in California; round; medium to small; skin green or yellowish-green; of excellent flavor (22% sugar). Tree is precocious and blooms and fruits all year. Fairly widely planted in California.

'Wilson'—originated in California; round to oblate; medium to large; skin smooth, medium thick; flesh of high quality and excellent flavor. Fruit ripens in fall and winter or more or less all year. Tree bears heavily and has been rather widely planted in California.

'Yellow'—originated in California; oval with pointed apex, furrowed; skin is bright-yellow and fairly tough; flesh is firm. Fruit keeps well. Tree bears regularly and heavily in California.

Pollination

There is a great variation in the amount of pollen produced by seedlings and grafted cultivars. Some flowers bear no pollen grains; others have an abundance. Sterile pollen or lack of cross-pollination are suggested causes of aborted seeds and heavy shedding of immature fruits. In Florida, flowers of some heavy-bearing, double-cropping, trees have been observed so heavily worked by bees that their humming is heard several feet away.

Climate

The white sapotes can be classed as subtropical rather than tropical. *C. edulis* is usually found growing naturally at elevations between 2,000 and 3,000 ft (600-900 m) and occasionally in Guatemala up to a maximum of 9,000 ft (2,700 m) in areas not subject to heavy rainfall.

In California, light frosts cause some leaf shedding but otherwise do not harm the tree. Mature trees have withstood temperature drops to 20°F (-6.67°C) in California and 26°F (-3.33°C) in Florida without injury.

The trees prosper near the coast of southern California where the mean temperature from April to October is about 65°F (18°C). They do poorly and often fail to survive further north near San Francisco where the mean temperature for the same period is 57° to 58°F (13.89°-14.44°C). The woolly-leaved is somewhat less hardy than the common white sapote.

Soil

As long as there is good drainage, the trees will do very well on sandy loam or even on clay. In California, some of the early plantings were on light, decomposed granite soil, and they were fruitful for many years. In Florida, the trees grow and fruit well on deep sand and on oolitic limestone, though, on the latter, they may become chlorotic. They are fairly drought-resistant.

Propagation

White sapotes are commonly grown from seeds and seedlings usually begin to bear in 7 or 8 years. Grafting is a common practice in California and Florida in mid-summer. Seedlings of 'Pike', being vigorous growers, are preferred as rootstock. Shield-budding and side-grafting in spring onto stocks up to ¾ in (2 cm) thick give good results. Cleft grafts and slot grafts are made on larger rootstocks and when topworking mature trees. Grafted trees will start bearing in 3 or 4 years. Commercial growers in New Zealand have had success with air-layers. Cuttings are very difficult to root.

Culture

In California, the young trees are cut back to 3 ft (0.9 m) when planted out, in order to encourage low-branching. As the branches elongate, some pruning is done to induce lateral growth.

Fertilizer formulas should vary with the nature of the soil, but, in general, the grower is advised to follow procedures suitable for citrus trees. Many white sapote trees have received little or no care and yet have been long-lived. One of the original trees in Santa Barbara, California, was said to be over 100 years old in 1915.

Season

In the Bahamas, the fruits ripen from late May through August. In Mexico, flowering occurs in January and February and the fruits mature from June to October. In Florida there is usually just a spring-summer crop, but a heavy-bearing woolly-leaved tree in Miami blooms in December, fruits in the spring, blooms again and produces a second crop in the fall. In California, 'Pike' and 'Yellow' bloom in the spring and again in late summer and fall, the fruits from late blooms maturing gradually over the winter. 'Suebelle' blooms for 6 to 8 weeks in spring and again in midsummer and fruits ripen in September and October.

Harvesting

Mature fruits must be clipped from the branches leaving a short piece of stem attached. This stub will fall off naturally when the fruits become eating-ripe. If plucked by hand, the fruits will separate from the stem if given a slight twist but they will soon show a soft bruised spot at the stem-end which quickly spreads over much of the fruit, becoming watery and decayed. The fruits must be handled with care even when unripe as they bruise so easily and any bruised skin will blacken and the flesh beneath turns bitter. If picked just a few days before fully ripe and ready to fall, the fruits turn soft quickly but they can be picked several weeks in advance of the falling stage and most will develop full flavor. 'Pike', however, if picked a month early, will take 2 weeks to ripen and will be substandard in flavor. Fruits that have ripened on hand will keep in good condition in the home refrigerator for at least 2 weeks. Fruits from commercial orchards are graded for size, wrapped individually to retard full ripening, packed in wooden boxes, and well-padded for transportation under refrigeration.

Pests and Diseases

The white sapote has few natural enemies but the fruits of some cultivars are attacked by fruit flies. Black scale often occurs on nursery stock and occasionally on mature trees in California.

Food Uses

Within its native range, the white sapote is commonly eaten out-of-hand. The flesh of ripe fruits may be added to fruit cups and salads or served alone as dessert, but it is best cut into sections and served with cream and sugar. Sometimes it is added to ice cream mix or milk shakes, or made into marmalade. Even in their countries of origin, where the fruits may at times appear in markets, their repute is due largely to a belief in their therapeutic value, while, at the same time, there prevails a fear that over-indulgence may be harmful. The epithet "matasano" (interpreted as "kill health") has a sinister connotation. Dr. J.B. Londoño, in his *Frutas de Antioquia*, published in Medellín, Colombia, in 1934, referred to the white sapote as disagreeable and indigestible. Some years ago in Central America there were unsuccessful efforts to manufacture from the pulp an acceptable preserve. In processing trials at the Western Regional Research Laboratory of the United States Department of Agriculture, Albany, California, technologists decided that white sapotes are not suitable for either canning in sirup or freezing as a puree.

Food Value Per 100 g of Fresh Pulp*	
Moisture	78.3 g
Protein	0.143 g
Fat	0.03 g
Fiber	0.9 g
Ash	0.48 g
Calcium	9.9 mg
Phosphorus	20.4 mg
Iron	0.33 mg
Carotene	0.053 mg
Thiamine	0.042 mg
Riboflavin	0.043 mg
Niacin	0.472 mg
Ascorbic Acid	30.3 mg

*According to analyses made in El Salvador.

As bearers of edible fruits, the white sapotes, despite their prolificacy, will doubtless continue to occupy the minor position which they now hold in subtropical horticulture.

Toxicity

The seed is said to be fatally toxic if eaten raw by humans or animals.

Other Uses

Seeds: In 1959, Dr. Everette Burdick, Consulting Chemist, of Coral Gables, Florida, made several extractions from the kernels, securing small amounts of needle-like yellow crystals. From one process, a yellow resinous mass resulted which functioned as an attractive and lethal bait for American cockroaches, having the advantage of killing on the spot rather than at some distance after ingestion of the poison. The United States Department of Agriculture's Agricultural Handbook 154, *Insecticides from Plants,* mentions no experiments with *Casimiroa* seed extracts but reports that extracts from branches and leaves of *C. edulis* are non-toxic to both American and German roaches.

Wood: The wood is yellow, fine-grained, compact, moderately dense and heavy, medium strong and resistant, but not durable for long. It is occasionally employed in carpentry and for domestic furniture in Central America.

Medicinal Uses: The ancient Nahuatl name for the fruits, "cochiztzapotl", is translated "sleepy sapote" or "sleep-producing sapote", and it is widely claimed in Mexico and Central America that consumption of the fruit relieves the pains of arthritis and rheumatism. This belief may stem only from the oft-quoted statement to this effect by Dr. Leopoldo Flores in *Manual Terapeutica de Plantas Mexicanas,* published in 1907, although the Mexican National Commission has received frequent reports of anti-arthritic, anti-rheumatic effects from physicians and their patients.

The eminent Francisco Hernandez, in his writings during the period 1570-1575 (translated and published as *Rerum Medicarum Novae Hispaniae* in 1651), noted that eating the fruit produced drowsiness. He referred to the seeds as "deadly poison" but efficacious, when crushed and roasted, in healing putrid sores. This vulnerary use of the seeds is cited in the obsolete *Farmacopea Mexicana,* where the fruit is mentioned as a vermifuge. For many years, extracts from the leaves, bark, and especially the seeds have been employed in Mexico as sedatives, soporifics and tranquilizers.

The narcotic property of the seeds was first identified as an alkaloid by Dr. Jesus Sanchez of Mexico in his thesis, *Breve estudio sobre la almendra del zapote blanco,* in 1893; and, in 1898, it was made the subject of chemical study by an especially appointed commission. One of the investigators, Alfonso Altimirano, reported the isolation of a glucoside as a pale yellow, amorphous mass, at first sweet but with a prolonged bitter aftertaste. White sapote derivatives were among the medicinal plant products displayed at the St. Louis Exposition in 1904 and explained in the slender book, *Materia Medica Mexicana: A Manual of Mexican Medicinal Herbs,* prepared by the Mexican National Commission for that occasion.

In 1900, a quantity of white sapote seeds was sent from Mexico to F.H. Worlee & Co., in Hamburg, Germany, with an accompanying explanation that both the fruit and the seeds possessed sleep-inducing principles but without the undesirable after-effects of opium. This material came to the attention of W. Bickern. He proceeded to work on the seeds, from which he obtained a substance which he called an alkaloidal glycoside, *casimirin.* In France, several investigators confirmed the narcotic nature of the seeds. Subsequently, Frederick Power and Thomas Callan of the Wellcome Chemical Research Laboratories in London, declared that, though they isolated 6 substances including 2 alkaloids, *casimiroine* and *casimiroedine,* there was "no evidence of the presence of a definite glucoside or a so-called glucoalkaloid . . . and physiological tests conducted with animals . . . likewise failed to confirm . . . reported hypnotic or toxic properties." Meanwhile, the seed extracts, in liquid, capsule, or tablet form, continued in use in Mexico, one product bearing the trade name "Rutelina".

In 1934, José de Lille proceeded to test the effect on blood pressure of dogs. He found a dose of .20 g per kilo of animal weight to be definitely hypotensive. A large dose (1 g) administered to a dog weighing 11 lbs (5 kg) produced a drastic lowering of blood pressure which persisted even after a brief rise induced by injecting adrenalin. In 1936, M. Mendez described the preparation of a tincture of "a clear yellow color with neither special odor nor taste" which produces "a state of depression in the entire nervous system, especially in the sensory sphere, and sleep." Dr. Faustino Miranda reported that an infusion of the leaves of *Casimiroa sapota* is used for similar purposes, and he assumed that this species has the same properties as *C. edulis.* According to *Materia Medica Mexicana,* the extracts from the leaves and bark are half as strong as those from the seeds and can be safely administered to children. In Costa Rica, the leaf decoction is taken as a treatment for diabetes.

In 1956, four chemists, F. Kinel, J. Rosso, O. Rosenkranz and F. Sondheimer, on the staff of the Mexican branch of the pharmaceutical company, Syntex, undertook chemical studies of the seeds. They did not find the "gluco-alkaloid" casimirin, but isolated 13 substances, 6 of which coincided with those reported by Power and Callan. One of these, *casimirolid,* was later found by F. Sondheimer, A Meisels and F. Kinel, to be identical with *obacunone,* an attribute of citrus oil. Of the 7 additional compounds, one *palmitamide,* had not previously been noted in the plant kingdom. Another, *N-benzoyltyramine,* they suggested might have much to do with the reputed potency of the seed, for *tyramine* is one of the active principles of ergot (is also found in mistletoe and thistle) and is well known for its physiological action. The main alkaloid of the seeds, *casimiroedine,* representing 0.143%, was crystallized in the form of needles.

Investigations of the bark from the trunk and roots of *C. edulis* were undertaken for Syntex by J. Iriarte, F. Kinel, O. Rosenkranz and F. Sondheimer. No *casimiroe-dine* was found but 12 substances were identified, only 2 of which, *zapotin* and *casimiroin*, occur in the seeds. The root bark contained .22% of the latter, while the seeds yielded only 0.0076%. In 1957, Meisels and Sondheimer announced that one of the bark alkaloids, *edulein*, which they had considered new, is identical with an alkaloid found in the bark and leaves of *Lunaria amara* Blanco, a citrus relative of Malayan origin. In 1958, R.T. Major and F. Dürsch, of the Cobb Chemical Laboratory, University of Virginia, working under a grant from Merck & Co., isolated from *C. edulis* seeds a compound which they identified as N^a, N^a-*dimethylhistamine*, formerly found in nature only in the sponge, *Geodia gigas*. J.S.L. Ling, S.Y. P'an and F.A. Hockstein, of Chas. Pfizer & Co. Research Laboratories, Brooklyn, New York, in experimental work with this compound in rabbits, dogs and cats, observed strong vasodepressive action. Dr. Hockstein suggested that all of the hypotensive properties and at least part of the sedative and pain-relieving qualities could be attributed to this compound, which "is not considered acceptable in man".

In early July of 1960, the writer furnished approx-imately 2 bushels of largely overripe, fallen fruits of *C. edulis* and the woolly-leaved white sapote to Delta Pharmaceuticals of Hialeah, Florida. They readily extracted from the seeds a soporific substance, 50 mg of which, taken by humans, induced sound sleep within 2 hours, with no apparent ill effects. The extract also acted as a narcotic on goldfish.

The following statement (translated from Spanish) is made in a communication received in 1961 from the Sección Administrativa, Dirección de Control de Medicamentos, Secretaria de Salubridad y Asistencia, Mexico City: "In Mexico, the white sapote is not used other than in folk medicine and not in any way by pharmacists nor doctors; neither is it an official drug in the Pharmacopoeia".

In India, extensive studies have been made of the seeds, roots and bark, which contain histamine derivatives with strong hypotensive activity, as well as furoquinoline alkaloids and 2-quinolones and 4-quinolones, including *edulein, edulitin, edulinine* and *casimiroin*. Also present are coumarins, flavonoids, and limonoids, including *zapoterin, zapotin, zapotinin, casimirolid, deacetyl-nomilin*, and *7-a-obacunol*. Leaves and twigs yield *isoplimpinellin* (diuretic) and *n-hentriacontane* (anti-inflammatory).

Fig. 49: The common white sapote (*Casimiroa edulis*) (left) and the woolly-leaved white sapote, often called *C. tetrameria* (right). The latter may be only a variant of *C. edulis*.

Wampee

A minor member of the Rutaceae and distant relative of the citrus fruits, the wampee, *Clausena lansium* Skeels (syns. *C. wampi* (Blanco), D. Oliver; *C. punctata* (Sonn.), Rehd. & E.H. Wils.; *Cookia punctata* Sonn.; *Cookia wampi* Blanco; *Quinaria lansium* Lour.), has not traveled sufficiently to acquire many vernacular names and most are derived from the Chinese *huang-p'i-kuo, huang p'i ho, huang p'i kan,* or *huang-p'i-tzu.* In Malaya, it is known as *wampi, wampoi,* or *wang-pei;* in the Philippines, *uampi, uampit, huampit* or *galumpi;* in Vietnam, *hong bi,* or *hoang bi.* In Thailand it is *som-ma-fai.*

Description

The tree is fairly fast-growing or rather slow, depending on its situation; attractive, reaching 20 ft (6 m), with long, upward-slanting, flexible branches, and gray-brown bark rough to the touch. Its evergreen, spirally-arranged, resinous leaves are 4 to 12 in (10-30 cm) long, pinnate, with 7 to 15 alternate, elliptic or elliptic-ovate leaflets 2¾ to 4 in (7-10 cm) long, oblique at the base, wavy-margined and shallowly toothed; thin, minutely hairy on the veins above and with yellow, warty midrib prominent on the underside. The petiole also is warty and hairy. The sweet-scented, 4- to 5-parted flowers are whitish or yellowish-green, about ½ in (1.25 cm) wide, and borne in slender, hairy panicles 4 to 20 in (10-50 cm) long. The fruits, on ¼ to ½ in (0.6-1.25 cm) stalks, hang in showy, loose clusters of several strands. The wampee may be round, or conical-oblong, up to 1 in (2.5 cm) long, with 5 faint, pale ridges extending a short distance down from the apex. The thin, pliable but tough rind is light brownish-yellow, minutely hairy and dotted with tiny, raised, brown oil glands. It is easily peeled and too resinous to be eaten. The flesh, faintly divided into 5 segments, is yellowish-white or colorless, grapelike, mucilaginous, juicy, pleasantly sweet, subacid, or sour. There may be 1 to 5 oblong, thickish seeds ½ to ⅝ in (1.25-1.6 cm) long, bright-green with one brown tip.

Origin and Distribution

The wampee is native and commonly cultivated in southern China and the northern part of former French Indochina, especially from North to Central Vietnam. It was growing in the Philippines before 1837 and was reintroduced in 1912. It is only occasionally grown in India and Ceylon. Chinese people in southern Malaya, Singapore and elsewhere in the Malaysian Archipelago grow the tree in home gardens. It is cultivated to a limited extent in Queensland, Australia and Hawaii. In 1908, it was said to have been growing in a few Hawaiian gardens for many years but was not in general cultivation. It was brought to Florida as an unidentified species in 1908. The United States Department of Agriculture received seed from Hong Kong in 1914 (P.I. 39176); from Canton in 1917 (P.I. 45328), and from Hawaii in 1922 (P.I. 55598). Dr. David Fairchild was pleased with a wampee tree he grew at his 'Kampong' in Coconut Grove, Miami, and a small cottage near it was named the **'Wamperi'**.

A few other specimens have been growing in southern Florida for some years, mostly in experimental collections, but the fruit is unknown to most residents despite some efforts to arouse interest in it. The wampee was growing in Jamaica in 1913. Two trees were thriving at the Federal Experimental Station, Mayaguez, Puerto Rico, and there were specimens on St. Croix, in the 1920's. Seeds from a Chinese grower in Panama were planted at the Lancetilla Experimental Garden, Tela, Honduras, in 1944. The tree does well in greenhouses in England.

Varieties

A Chinese work translated and published in 1936, mentioned 7 varieties of Foochow, describing and illustrating 6 of them. They vary somewhat in form and size, number of seeds, season of ripening, as well as in flavor:

'Niu Shen' ("cow's kidney")—sour in flavor;
'Yuan Chung' ("globular variety")—sweet-subacid;
'Yeh Sheng' ("wild growing")—sour;
'Suan Tsao' ("sour jujube")—is very sour, of poor quality;
'Hsiao Chi Hsien' ("small chicken heart")—sweet subacid;
'Chi Hsin' ("chicken heart")—sweet; "best flavor of all";
'Kua Pan' ("melon section")—sweet-subacid.

A professor at Sun Yat-sen University in Canton listed 8 varieties of Kwangtung with, as Dr. Swingle stated, long, descriptive names such as "white-hairy-chicken-heart-sweet-wampee" and "long-chicken-heart-sour-wampee".

Climate

The wampee is subtropical to tropical, and young and mature trees have been scarcely hurt by brief exposure to 28° to 30°F (-2.22° to -1.11°C) in Florida, but they have been killed at temperatures of 20°F (-6.667°C) and lower.

Soil

The tree seems quite tolerant of a range of soils, including the deep sand and the oolitic limestone of southern Florida but thrives best in rich loam. It requires watering in dry periods though good drainage is essential.

Fig. 50: The wampee (*Clausena lansium*) is an attractive tree with somewhat grapelike fruits, but the pulp is scant and the seeds large.

Propagation

The wampee grows readily from seeds which germinate in a few days. It can also be grown from softwood cuttings and air-layers, and can be veneer-grafted onto wampee seedlings. Dr. Swingle said it could be grafted onto grapefruit. However, trials on various *Citrus* rootstocks in Florida have shown various degrees of incompatibility and few, if any, can be said to have been really successful in the long run. The wampee is not a first-class fruit and the tree is of only casual interest, even as an ornamental, except in Asia.

Cultivation

No particular cultural requirements have been noted in the literature, except that the wampee is subject to chlorosis on limestone soils and needs applications of manganese and zinc as well as organic fertilizer and mulch to overcome this condition. Sturrock recommends thinning of the crown to avoid overcrowding.

Season and Yield

The fruits ripen in July and August in Florida; from June to October in Southeast Asia; in November and December in Queensland. Seedlings begin to bear when 5 to 8 years of age or sometimes older. Mature trees may yield 100 lbs (45 kg) of fruits in a season.

Food Uses

A fully ripe, peeled wampee, of the sweet or subacid types, is agreeable to eat out-of-hand, discarding the large seed or seeds. The seeded pulp can be added to fruit cups, gelatins or other desserts, or made into pie or jam. Jelly can be made only from the acid types when underripe. The Chinese serve the seeded fruits with meat dishes.

In Southeast Asia, a bottled, carbonated beverage resembling champagne is made by fermenting the fruit with sugar and straining off the juice.

Food Value

Florida-grown fruits have shown 28.8 to 29.2 mg/100 g ascorbic acid.

Medicinal Uses

The fruit is said to have stomachic and cooling effects and to act as a vermifuge. The Chinese say that if one has eaten too many lychees, eating the wampee "will counteract the bad effects. Lychees should be eaten when one is hungry, and wampees only on a full stomach".

The halved, sun-dried, immature fruit is a Vietnamese and Chinese remedy for bronchitis. Thin slices of the dried roots are sold in Oriental pharmacies for the same purpose. The leaf decoction is used as a hair wash to remove dandruff and preserve the color of the hair.

Santol (Plate XXIII)

Perhaps the only important edible fruit in the family Meliaceae, the santol, *Sandoricum koetjape* Merr. (syns. *S. indicum* Cav., *S. nervosum* Blume, *Melia koetjape* Burm. f.), is also known as *sentieh, sentol, setol, sentul, setul, setui, kechapi* or *ketapi*, in Malaya; *saton, satawn, katon*, or *ka–thon* in Thailand; *kompem reach* in Cambodia; *tong* in Laos; *sau chua, sau tia, sau do, mangoustanier sauvage*, or *faux mangoustanier* in North Vietnam. In the Philippines, it is *santor* or *katul;* in Indonesia, *ketjapi* or *sentool;* on Sarawak and Brunei, it is *klampu*. In India, it may be called *sayai, sevai, sevamanu* or *visayan*. In Guam, it is *santor* or wild mangosteen.

Description

The santol is a fast-growing, straight-trunked, pale-barked tree 50 to 150 ft (15-45 m) tall, branched close to the ground and buttressed when old. Young branchlets are densely brown-hairy. The evergreen, or very briefly deciduous, spirally-arranged leaves are compound, with 3 leaflets, elliptic to oblong-ovate, 4 to 10 in (20-25 cm) long, blunt at the base and pointed at the apex. The greenish, yellowish, or pinkish-yellow, 5-petalled flowers, about 3/8 in (1 cm) long are borne on the young branchlets in loose, stalked panicles 6 to 12 in (15-30 cm) in length. The fruit (technically a capsule) is globose or oblate, with wrinkles extending a short distance from the base; 1½ to 3 in (4-7.5 cm) wide; yellowish to golden, sometimes blushed with pink. The downy rind may be thin or thick and contains a thin, milky juice. It is edible, as is the white, translucent, juicy pulp (aril), sweet, subacid or sour, surrounding the 3 to 5 brown, inedible seeds which are up to ¾ in (2 cm) long, tightly clinging or sometimes free from the pulp.

Origin and Distribution

The santol is believed native to former Indochina (especially Cambodia and southern Laos) and Malaya, and to have been long ago introduced into India, the Andaman Islands, Malaysia, Indonesia, the Moluccas, Mauritius, and the Philippines where it has become naturalized. It is commonly cultivated throughout these regions and the fruits are abundant in the local markets.

Only a few specimens are known in the western hemisphere: one in the Lancetilla Experimental Garden at Tela, Honduras, and one or more in Costa Rica. Seeds have been introduced into Florida several times since 1931. Most of the seedlings have succumbed to cold injury. At least 3 have survived to bearing age in special collections. Grafted plants from the Philippines have fruited well at Fairchild Tropical Garden, Miami.

In Asia and Malaysia, the tree is valued not just for its fruit, but for its timber and as a shade tree for roadsides, being wind-resistant and non-littering.

Varieties

There are two general types of santol: the **Yellow** (formerly *S. indicum* or *S. nervosum*); and the **Red** (formerly *S. koetjape*). The leaflets of the Yellow, to 6 in (15 cm) long, turn yellow when old; the flowers are pinkish-yellow in panicles to 6 in (15 cm) long; the fruit has a thin rind and the pulp is ¼ to ½ in (0.6-1.25 cm) thick around the seeds and typically sweet. The fruit may not fall when ripe. Only the Yellow is now found wild in Malayan forests.

The leaflets of the Red, to 12 in (30 cm) long, velvety beneath, turn red when old; the flowers are greenish or ivory, in panicles to 12 in (30 cm) long; the fruit has a thick rind, frequently to ½ in (1.25 cm); there is less pulp around the seeds, and it is sour. The fruit falls when ripe.

However, Corner says that these distinctions are not always clear-cut except as to the dying leaf color, and the fruit may not correspond to the classifications. There are sweet and acid strains of both the Yellow and Red types and much variation in rind thickness.

Climate

The santol is tropical and cannot be grown above 3,280 ft (1,000 m) in Java. It flourishes in dry as well as moist areas of the Philippine lowlands.

Soil

The tree has grown well in Florida in acid sandy soil and oolitic limestone, but in the latter the foliage becomes chlorotic.

Propagation

The santol is reproduced by seeds, air-layering, inarching, or by budding onto self rootstocks.

Season

The fruit ripens in Malaya in June and July; in Florida, August and September; in the Philippines, from July to October.

Fig. 51: Santol (*Sandoricum indicum*) marmalade made in the Philippines is sometimes imported into the United States.

Food Value Per 100 g of Edible Pulp

	Yellow*	Red**	Fruits (unspecified type)***
Moisture	87.0 g	83.07–85.50%	85.4 g
Protein	0.118 g	0.89%	0.06 g
Carbohydrates		11.43%	–
Fat	0.10 g	1.43%	0.52 g
Fiber	0.1 g	2.30%	1.26 g
Ash	0.31 g	0.65–0.88%	0.39 g
Calcium	4.3 mg	0.01%	5.38 mg
Phosphorus	17.4 mg	0.03%	12.57 mg
Iron	0.42 mg	0.002%	0.86 mg
Carotene	0.003 mg	–	
Thiamine	0.045 mg	0.037 mg	
Niacin	0.741 mg	0.016 mg	
Ascorbic Acid	86.0 mg	0.78 mg	
Pectin	–	–	14.89 mg
			17.01 g

*According to analyses of yellow, thick-skinned, acid fruits in Honduras.

**According to analyses of the red type in the Philippines.

***According to analyses of unspecified type in India. The pericarp contains glucose, sucrose, malic acid, tartaric acid and much pectin.

Pests

The Caribbean fruit fly (*Anastrepha suspensa*) causes freckle-like blemishes on the surface of the fruit but cannot penetrate the rind.

Food Uses

The fruit is usually consumed raw without peeling. In India, it is eaten with spices. With the seeds removed, it is made into jam or jelly. Pared and quartered, it is cooked in sirup and preserved in jars. Young fruits are candied in Malaysia by paring, removing the seeds, boiling in water, then boiling a second time with sugar. In the Philippines, santols are peeled chemically by dipping in hot water for 2 minutes or more, then into a lye solution at 200°F (93.33°C) for 3 to 5 minutes. Subsequent washing in cool water removes the outer skin. Then the fruits are cut open, seeded and commercially preserved in sirup. Santol marmalade in glass jars is exported from the Philippines to Oriental food dealers in the United States and probably elsewhere. Very ripe fruits are naturally vinous and are fermented with rice to make an alcoholic drink.

Other Uses

Wood: The sapwood is gray, merging into the heartwood which is reddish-brown when dry, imparting the color to water. It is fairly hard, moderately heavy, close-grained and polishes well, but is not always of good quality. It is not durable in contact with moisture and is subject to borers. However, it is plentiful, easy to saw and work, and accordingly popular. If carefully seasoned, it can be employed for house-posts, interior construction, light-framing, barrels, cabinetwork, boats, carts, sandals, butcher's blocks, household utensils and carvings. When burned, the wood emits an aromatic scent.

Fig. 52: Santol fruits photographed by Dr. Walter T. Swingle, Plant Explorer for the United States Department of Agriculture.

The dried heartwood yields 2 triterpenes—katonic acid and indicic acid—and an acidic resin.

Bark: In the Philippines, the bark is used in tanning fishing lines.

Medicinal Uses: The preserved pulp is employed medicinally as an astringent, as is the quince in Europe. Crushed leaves are poulticed on itching skin.

In cases of fever in the Philippines, fresh leaves are placed on the body to cause sweating and the leaf decoction is used to bathe the patient. The bitter bark, containing the slightly toxic sandoricum acid, an unnamed, toxic alkaloid, and a steroidal sapogenin, is applied on ringworm and also enters into a potion given a woman after childbirth. The aromatic, astringent root also serves the latter purpose, and is a potent remedy for diarrhea. An infusion of the fresh or dried root, or the bark, may be taken to relieve colic and stitch in the side. The root is a stomachic and antispasmodic and prized as a tonic. It may be crushed in a blend of vinegar and water which is then given as a carminative and remedy for diarrhea and dysentery. Mixed with the bark of *Carapa obovata* Blume, it is much used in Java to combat leucorrhea.

Langsat (Plate XXIV)

A somewhat less edible fruit of the family Meliaceae, the langsat, *Lansium domesticum* Corr., is also known as *lansa, langseh, langsep, lanzon, lanzone, lansone,* or *kokosan,* and by various other names in the dialects of the Old World tropics.

Description

The tree is erect, short-trunked, slender or spreading; reaching 35 to 50 ft (10.5 to 15 m) in height, with red-brown or yellow-brown, furrowed bark. Its leaves are pinnate, 9 to 20 in (22.5-50 cm) long, with 5 to 7 alternate leaflets, obovate or elliptic-oblong, pointed at both ends, 2¾ to 8 in (7-20 cm) long, slightly leathery, dark-green and glossy on the upper surface, paler and dull beneath, and with prominent midrib. Small, white or pale-yellow, fleshy, mostly bisexual, flowers are borne in simple or branched racemes which may be solitary or in hairy clusters on the trunk and oldest branches, at first standing erect and finally pendant, and 4 to 12 in (10-30 cm) in length.

The fruit, borne 2 to 30 in a cluster, is oval, ovoid-oblong or nearly round, 1 to 2 in (2.5-5 cm) in diameter, and has light grayish-yellow to pale brownish or pink, velvety skin, leathery, thin or thick, and containing milky latex. There are 5 or 6 segments of aromatic, white, translucent, juicy flesh (arils), acid to subacid in flavor. Seeds, which adhere more or less to the flesh, are usually present in 1 to 3 of the segments. They are green, relatively large— ¾ to 1 in (2-2.5 cm) long and ½ to ¾ in (1.25-2 cm) wide, very bitter, and sometimes, if the flesh clings tightly to the seed, it may acquire some of its bitterness.

Origin and Distribution

The langsat originated in western Malaysia and is common both wild and cultivated throughout the Archipelago and on the island of Luzon in the Philippines where the fruits are very popular and the tree is being utilized in reforestation of hilly areas. It is much grown, too, in southern Thailand and Vietnam and flourishes in the Nilgiris and other humid areas of South India and the fruits are plentiful on local markets. The langsat was introduced into Hawaii before 1930 and is frequently grown at low elevations. An occasional tree may be found on other Pacific islands.

The species is little known in the American tropics except in Surinam. There it is commercially grown on a small scale. Seeds were sent from Java to the Lancetilla Experimental Garden at Tela, Honduras, in 1926 and plants arrived from the same source in 1927. The trees have grown well but are usually unfruitful, occasionally having a small number of fruits. There are bearing trees in Trinidad, where the langsat was established in 1938, and a few around Mayaguez, Puerto Rico, that have been bearing well for about 60 years. There were young specimens growing on St. Croix in 1930.

Southern Florida does not have climatic and soil conditions favorable to the langsat, but the rare-fruit fancier, William Whitman, has managed to raise two bearing trees in special soil and tented for the first several years. Winter cold has caused complete defoliation and near-girdling at the base of the trunks, but the trees made good recovery. Other specimens have survived on the Lower Keys in pits prepared with non-alkaline soil. There have been attempts to maintain langsats at the University of Florida's Agricultural Research and Education Center in Homestead, but the trees have succumbed either to the limestone terrain or low temperatures.

Varieties

There are two distinct botanical varieties: 1) *L. domesticum* var. *pubescens,* the typical wild langsat which is a rather slender, open tree with hairy branchlets and nearly round, thick-skinned fruits having much milky latex; 2) var. *domesticum,* called the *duku, doekoe,* or *dookoo,* which is a more robust tree, broad-topped and densely foliaged with conspicuously-veined leaflets; the fruits, borne few to a cluster, are oblong-ovoid or ellipsoid, with thin, brownish skin, only faintly aromatic and containing little or no milky latex. The former is often referred to as the "wild" type but both varieties are cultivated and show considerable range of form, size and quality. There are desirable types in both groups. Some small fruits are completely seedless and fairly sweet.

'Conception' is a sweet cultivar from the Philippines; 'Uttaradit" is a popular selection in Thailand; 'Paete' is a leading cultivar in the Philippines.

Climate

The langsat is ultra-tropical. Even in its native territory it cannot be grown at an altitude over 2,100 to 2,500 ft (650-750 m). It needs a humid atmosphere, plenty of moisture and will not tolerate long dry seasons. Some shade is beneficial especially during the early years.

Soil

The tree does best on deep, rich, well-drained, sandy loam or other soils that are slightly acid to neutral and high in organic matter. It is inclined to do poorly on clay that dries and cracks during rainless periods, and is not at all adapted to alkaline soils. It will not endure even a few days of water-logging.

Propagation

Langsats are commonly grown from seeds which must be planted within 1 or 2 days after removal from the fruit. Viability is totally lost in 8 days unless the seeds are stored in polyethylene bags at 39.2°-42.8°F (4°-6°C) where they will remain viable for 14 days.

Seedlings will bear in 12 to 20 years. Air-layering is discouraging, as the root system is weak and the survival rate is poor after planting out. Shield-budding has a low rate of success. Cleft- and side-grafting and approach-grafting give good results. The budwood should be mature but not old, 2½ to 3½ in (6.5-9 cm) long, ¼ to ¾ in (6-20 mm) thick, and it is joined to rootstock of the same diameter about 2½ to 4 in (6.5-10 cm) above the soil. Some preliminary experiments have been conducted in Puerto Rico with hormone-treated cuttings under intermittent mist. Whitman found that a potted cutting 3 to 4 in (7.5-10 cm) long, will root if covered with a clear plastic bag.

Culture

The trees are spaced 25 to 33 ft (8-10 m) apart in orchards. In the Philippines they are frequently planted around the edges of coconut plantations. Generally, the langsat is casually grown in dooryards and on roadsides and receives no cultural attention. Regular irrigation results in better fruit size and heavier crops. Whitman has demonstrated that thrice-yearly applications of a 6-6-6 fertilizer formula with added minor elements result in good growth, productivity and high quality fruits even in an adverse environment.

Season and Harvesting

Langsats in Malaya generally bear twice a year—in June and July and again in December and January or even until February. In India, the fruits ripen from April to September but in the Philippines the season is short and most of the fruits are off the market in less than one month.

Yield

Trees in the Nilgiris average 30 lbs (13.5 kg) of fruits annually. In the Philippines, a productive tree averages 1,000 fruits per year.

Keeping Quality

Langsats are perishable and spoil after 4 days at room temperature. They can be kept in cold storage for 2 weeks at 52° to 55°F (11.11°-12.78°C) and relative humidity of 85-90%. Sugar content increases over this period, while acidity rises only up to the 7th day and then gradually declines.

Fruits treated with fungicide and held at 5% 0 and zero CO_2 and 58°F (14.44°C) with 85% to 90% humidity, have remained in good condition for more than 2 weeks. High CO_2 promotes browning and elevates acidity.

Waxing reduces weight loss, increases sweetness, but causes browning over at least half the surface within 5 days in storage.

Pests and Diseases

In Puerto Rico, young langsat trees have been defoliated by the sugarcane root borer, *Diaprepes abbreviatus*. Scale insects, especially *Pseudaonidia articulatus* and *Pseudaulacaspis pentagona*, and the red spider mite, *Tetranychus bimaculatus*, are sometimes found attacking the foliage, and sooty mold is apt to develop on the honeydew deposited by the scales. Rats gnaw on the branchlets and branches and the mature fruits.

Anthracnose caused by *Colletotrichum gloeosporioides* is evidenced by brown spots and other blemishes on the fruit and peduncle and leads to premature shedding of fruits.

Canker which makes the bark become rough and corky and flake off has appeared on langsats in Florida, Hawaii and Tahiti. It was believed to be caused by a fungus, *Cephalosporium* sp., and larvae of a member of the Tineidae have been observed feeding under the loosened bark. However, other fungi, *Nectria* sp. (perfect stage of *Volutella* sp.) and *Phomopsis* sp. are officially recorded as causes of stem gall canker on the langsat in Florida.

Food Uses

The peel of the langsat is easily removed and the flesh is commonly eaten out-of-hand or served as dessert, and may be cooked in various ways.

Varieties with much latex are best dipped into boiling water to eliminate the gumminess before peeling.

The peeled, seedless or seeded fruits are canned in sirup or sometimes candied.

Toxicity

An arrow poison has been made from the fruit peel and the bark of the tree. Both possess a toxic property, lansium acid, which, on injection, arrests heartbeat in frogs. The peel is reportedly high in tannin. The seed contains a minute amount of an unnamed alkaloid, 1% of an alcohol-soluble resin, and 2 bitter, toxic principles.

Food Value Per 100 g of Edible Portion*	
Moisture	86.5 g
Protein	0.8 g
Carbohydrates	9.5 g
Fiber	2.3 g
Calcium	20.0 mg
Phosphorus	30.0 mg
Carotene (Vitamin A)	13.0 I.U.
Thiamine	89 mcg
Riboflavin	124 mcg
Ascorbic Acid	1.0 mg
Phytin	1.1 mg (dry weight)

*According to analyses made in India.

The edible flesh may constitute 60% of the fruit.

Other Uses

Peel: The dried peel is burned in Java, the aromatic smoke serving as a mosquito repellent and as incense in the rooms of sick people.

Wood: The wood is light-brown, medium-hard, fine-grained, tough, elastic and durable and weighs 52.3 lbs/cu ft. It is utilized in Java for house posts, rafters, tool handles and small utensils. Wood-tar, derived by distillation, is employed to blacken the teeth.

Medicinal Uses: The fresh peel contains 0.2% of a light-yellow volatile oil, a brown resin and reducing acids. From the dried peel, there is obtained a dark, semi-liquid oleoresin composed of 0.17% volatile oil and 22% resin. The resin is non-toxic and administered to halt diarrhea and intestinal spasms; contracts rabbit intestine *in vitro*.

The pulverized seed is employed as a febrifuge and vermifuge. The bark is poulticed on scorpion stings. An astringent bark decoction is taken as a treatment for dysentery and malaria. Leaves may be combined with the bark in preparing the decoction. The leaf juice is used as eye-drops to dispel inflammation.

Fig. 53: The langsat, photographed by Dr. Walter T. Swingle, Plant Explorer for the United States Department of Agriculture.

MALPIGHIACEAE

Barbados Cherry (Plate XXV)

The Barbados cherry, a member of the Malpighiaceae, is an interesting example of a fruit that rose, like Cinderella, from relative obscurity about 40 years ago. It was at that time the subject of much taxonomic confusion, having been described and discussed previously under the binomial *Malpighia glabra* L., which properly belongs to a wild relative inhabiting the West Indies, tropical America and the lowlands of Mexico to southern Texas, and having smaller, pointed leaves, smaller flowers in peduncled umbels, styles nearly equal, and smaller fruits. *M. punicifolia* L. (*M. glabra* Millsp. NOT Linn.) has been generally approved as the correct botanical name for the Barbados cherry, which is also called West Indian cherry, native cherry, garden cherry, French cherry; in Spanish, *acerola, cereza, cereza colorada, cereza de la sabana,* or *grosella;* in French, *cerisier, cerise de St. Domingue;* in Portuguese, *cerejeira.* The name in Venezuela is *semeruco,* or *cemeruco;* in the Netherlands Antilles, *shimarucu;* in the Philippines, *malpi* (an abbreviation of the generic name).

Description

The Barbados cherry is a large, bushy shrub or small tree attaining up to 20 ft (6 m) in height and an equal breadth; with more or less erect or spreading and drooping, minutely hairy branches, and a short trunk to 4 in (10 cm) in diameter. Its evergreen leaves are elliptic, oblong, obovate, or narrowly oblanceolate, somewhat wavy, ¾ to 2¾ in (2-7 cm) long, ⅜ to 1⅝ in (9.5-40 mm) wide, obtuse or rounded at the apex, acute or cuneate at the base; bearing white, silky, irritating hairs when very young; hairless, dark green, and glossy when mature. The flowers, in sessile or short-peduncled cymes, have 5 pink or lavender, spoon-shaped, fringed petals. The fruits, borne singly or in 2's or 3's in the leaf axils, are oblate to round, cherry-like but more or less obviously 3-lobed; ½ to 1 in (1.25-2.5 cm) wide; bright-red, with thin, glossy skin and orange-colored, very juicy, acid to subacid, pulp. The 3 small, rounded seeds each have 2 large and 1 small fluted wings, thus forming what are generally conceived to be 3 triangular, yellowish, leathery-coated, corrugated inedible "stones".

Origin and Distribution

The Barbados cherry is native to the Lesser Antilles from St. Croix to Trinidad, also Curacao and Margarita and neighboring northern South America as far south as Brazil. It has become naturalized in Cuba, Jamaica and Puerto Rico after cultivation, and is commonly grown in dooryards in the Bahamas and Bermuda, and to some extent in Central and South America.

The plant is thought to have been first brought to Florida from Cuba by Pliny Reasoner because it appeared in the catalog of the Royal Palm Nursery for 1887-1888. It was carried abroad rather early for it is known to have borne fruit for the first time in the Philippines in 1916. In 1917, H.M. Curran brought seeds from Curacao to the United States Department of Agriculture. (S.P.I. #44458). The plant was casually grown in southern and central Florida until after World War II when it became more commonly planted. In Puerto Rico, just prior to that war, the Federal Soil Conservation Department planted Barbados cherry trees to control erosion on terraces at the Rio Piedras Experiment Station. During the war, 312 seedlings from the trees with the largest and most agreeably-flavored fruits were distributed to families to raise in their Victory Gardens. Later, several thousand trees were provided for planting in school yards to increase the vitamin intake of children, who are naturally partial to the fruits.

An explosion of interest occurred as a result of some food analyses being conducted at the School of Medicine, University of Puerto Rico, in Rio Piedras in 1945. The emblic (*Emblica officinalis* L.) was found to be extremely high in ascorbic acid. This inspired one of the laboratory assistants to bring in some Barbados cherries which the local people were accustomed to eating when they had colds. These fruits were found to contain far more ascorbic acid than the emblic, and, because of their attractiveness and superior eating quality, interest quickly switched from the emblic to the Barbados cherry. Much publicity ensued, featuring the fruit under the Puerto Rican name of *acerola.* A plantation of 400 trees was established at Rio Piedras in 1947 and, from 1951 to 1953, 238 trees were set out at the Isabela Substation. By 1954, there were 30,000 trees in commercial groves on the island. Several plantings had been made in Florida and a 2,000-acre (833-ha) plantation in Hawaii. There was a great flurry of activity. Horticulturists were busy making selections of high-ascorbic-acid clones and improving methods of vegetative propagation, and agronomists were studying the effects of cultural practices. Smaller plantings were being developed in Jamaica, Venezuela, Guatemala, Ghana, India, the Philippines and Queensland, Australia, and even in Israel. Many so-called "natural food" outlets promoted various "vitamin C" products from the fruits — powder, tablets, capsules, juice, sirup.

At length, enthusiasm subsided when it was realized that a fruit could not become a superstar because of its ascorbic acid content alone; that ascorbic acid from a natural source could not economically compete with the much cheaper synthetic product, inasmuch as research proved that the ascorbic acid of the Barbados cherry is metabolized in a manner identical to the assimilation of crystalline ascorbic acid.

The large plantation of the Hawaiian Acerola Company (a subsidiary of Nutrilite Products Company) was abandoned for this reason, and low fruit yields; and, so it is said, the low ascorbic acid content because of the high copper levels in the soil. Puerto Rican production was directed thereafter mainly to the use of the fruit in specialty baby foods.

Frozen fruits are shipped to the United States for processing.

Varieties

In 1956, workers at the University of Florida's Agricultural Research and Education Center in Homestead, after making preliminary evaluations and selections, chose as superior and named the **'Florida Sweet',** a clone that was observed to have an upright habit of growth, large fruits, thick skin, apple-like, semi-sweet flavor, and high yield.

The first promising selections in Puerto Rico, on the bases of fruit size, yield and vitamin content, were identified as 'A-1' and 'B-17', but these were later found to be inferior to 'B-15' in ascorbic acid level and productivity. Yields of 10 clones ('A-1', 'A-2', 'A-4', 'A-10', 'A-21', 'B-2', 'B-9', 'B-15', 'B-17', and 'K-7') were compared over a 2-year period (1955-56) in Puerto Rico and 'B-15' far exceeded the others in both years.

A horticultural variety in St. Croix, formerly known as *M. thompsonii* Britton & Small, has displayed unusually large leaves and fruits and more abundant flowers than the common strain of Barbados cherry.

Climate

The Barbados cherry can be classed as tropical and subtropical, for mature trees can survive brief exposure to 28°F (-2.22°C). Young plants are killed by any drop below 30°F (-1.11°C). It is naturally adapted to both medium- and low-rainfall regions; can tolerate long periods of drought, though it may not fruit until the coming of rain.

Soil

The tree does well on limestone, marl and clay, as long as they are well drained. The pH should be at least 5.5. Elevation to 6.5 significantly improves root development. Acid soils require the addition of lime to avoid calcium deficiency and increase yield. The lime should be worked into the soil to a depth of 8 in (20 cm) or more.

Propagation

If seeds are used for planting, they should be selected from desirable clones not exposed to cross-pollination by inferior types. They should be cleaned, dried, and dusted with a fungicide. It should also be realized that the seeds in an individual fruit develop unevenly and only those that are fully developed when the fruit is ripe will germinate satisfactorily. Germination rates may be only 50% or as low as 5%. Seedlings should be transferred from flats to containers when 2 to 3 in (5-7.5 cm) high.

Air-layering (in summer) and side-veneer, cleft, or modified crown grafting are feasible but not popular because it is so much easier to raise the tree from cuttings. Cuttings of branches ¼ to ½ in (6-12.5 mm) thick and 8 to 10 in (20-25 cm) long, with 2 or 3 leaves attached, hormone-treated and set in sand or other suitable media under constant or intermittent mist, will root in 60 days. They are then transplanted to nursery rows or containers and held in shade for 6 months or a year before being set out in the field. Some fruits will be borne a year after planting but a good crop cannot be expected until the 3rd or 4th year. The tree will continue bearing well for about 15 years. There is a lapse of only 22 days between flowering and complete fruit maturity.

Grafting is generally practiced only when cuttings of a desired clone are scarce or if a nematode-resistant rootstock is available on which to graft a preferred cultivar; or when top-working a tree that bears fruits of low quality.

Culture

The Barbados cherry tree will grow and fruit fairly well with little care. For best performance, Puerto Rican agronomists have recommended a fertilizer formula of 8-8-13 twice annually for the first 4 years at the rate of ½ to 1 lb (0.22-0.45 kg). Older trees should have 3 to 5 lbs (1.35-2.25 kg) per tree. In addition, organic material should be worked into the planting hole and also supplied in amounts of 10 to 20 lbs (4.5-9 kg) per tree. Under Florida conditions, a 10-10-10 formula is given in February, 1 lb (0.22 kg) for each year of growth. In May, July and September, a 4-7-5-3 formula is recommended, 1 lb (0.22 kg) for each year of age up to the 10th year. Thereafter, a 6-4-6-3 mixture is given — 5 lbs (2.25 kg) per tree in late winter and 10 lbs (4.5 kg) per tree for each of the summer feedings. On limestone soils, sprays of minor elements — copper, zinc, and sometimes manganese — will enhance growth and productivity. Young trees need regular irrigation until well established; older trees require watering only during droughts. Mature plants will bear better if thinned out by judicious pruning after the late crop and then fertilized once more.

Pollination and Fruit Set

In Florida, bees visit Barbados cherry flowers in great numbers and are the principal pollinators. Maintenance of hives near Barbados cherry trees substantially improves fruit set. In Hawaii, there was found to be very little transport of pollen by wind, and insect pollination is inadequate. Consequently, fruits are often seedless. Investigations have shown that growth regulators (IBA at 100 ppm; PCA at 50 ppm) induce much higher fruit set but these chemicals may be too costly to buy and apply.

Season

In Florida, the Bahamas, Puerto Rico and Hawaii the fruiting season varies with the weather. There may be a spring crop ripening in May and then successive small crops off and on until December, but sometimes, if spring rains are lacking, there may be no fruits at all until December and then a heavy crop. In Zanzibar, the bearing season is said to be just the months of December and January.

Harvesting

For home use, as dessert, the fruits are picked when fully ripe. For processing or preserving, they can be harvested when slightly immature, when they are turning from yellow to red. As there is continuous fruiting over long periods, picking is done every day, every other day, or every 3 days to avoid loss by falling.

The fruits are usually picked manually in the cool of the early morning, and must be handled with care. For immediate processing, some growers shake the tree and allow the ripe fruits to fall onto sheets spread on the ground. Harvested fruits should be kept in the shade until transferred from the field, which ought to be done within 3 hours, and collecting lugs are best covered with heavy canvas to retard loss of ascorbic acid.

Yield

There is great variation in productivity. Individual trees may yield 30 to 62 lbs (13.5-28 kg) in Puerto Rico. In Jamaica, maximum yield in the 6th year is about 80 lbs (36 kg) per tree; 24,000 lbs/acre (24,000 kg/ha). Venezuelan growers have reported 10 to 15 tons/ha; the average in Puerto Rico is 25 tons/ha/yr. 'Florida Sweet' in Florida has yielded 65 tons/ha. A plot of 300 trees of 'Florida Sweet' has borne crops of 6,300 to 51,300 lbs (2,858-23,270 kg) of fruit from March to November, in Homestead, Florida.

In Puerto Rico, a planting of 200 trees may be expected to produce 3,600 to 5,400 lbs (1,636-2,455 kg) of juice. From the juice there can be extracted at least 120 lbs (54.5 kg) of vitamin C expressed as dehydroascorbic and ascorbic acid, providing the content is determined to be 2%. In Puerto Rico, it is calculated that 10 tons of fruit should yield 435 lbs (197 kg) ascorbic acid. In a commercial operation using ion-exchange resins, the yield of ascorbic acid from Barbados cherry juice is expected to be about 88%.

Keeping Quality

Ripe Barbados cherries bruise easily and are highly perishable. Processors store them for no more than 3 days at 45°F (7.22°C). Half-ripe fruits can be maintained for a few more days. If longer storage is necessary, the fruits must be frozen and kept at 10°F (-12.22°C) and later thawed for use. At one time it was believed that the fruits could be transported to processing plants in water tanks (as is done with true cherries) but it was discovered that they lose their color and ascorbic acid content in water.

At room temperature—85°F (29.44°C) in Puerto Rico—canned Barbados cherries and also the juice lose color and fresh flavor and 53% to 80% of their ascorbic acid content in one month, and metal cans swell because of the development of CO_2. Refrigeration at 44.6°F (7°C) considerably reduces such deterioration. Juice in the home refrigerator will lose 20% of its ascorbic acid in 18 days. Therefore, the juice and the puree should be kept no longer than one week.

Pests and Diseases

One of the major obstacles to successful cultivation of the Barbados cherry is the tree's susceptibility to the root-knot nematode, *Meloidogyne incognita* var. *acrita*, especially in sandy acid soils. Soil fumigation, mulching and regular irrigation will help to keep this problem under control. The burrowing nematode, *Radopholus similis*, is also a cause of decline in otherwise healthy trees.

In Florida, the foliage is attacked by wax scale, Florida mango scale, and other scale insects, whiteflies, a leaf roller, and aphids. In Guatemala, the aphid, *Aphis spiraecola*, attacks the leaves and young, tender branches. This pest and the Hesperid caterpillar, *Ephyriades arcas*, require chemical control. In Puerto Rico, the tree is often damaged by the blue chrysomelid of acerola, *Leucocera laevicollis*. Some fruits may be malformed but not otherwise affected by the sting of stinkbugs. None of these predators is of any great importance.

The major pest in Florida is the Caribbean fruit fly, *Anastrepha suspensa*, which seems to attack all but very sour fruits and the larvae are commonly found inside. In Guatemala, a fruit worm, *Anthonomus florus*, deposits its eggs in the floral ovary and also in the fruits; the larvae feed in the fruits causing deformity and total ruin. Drastic control measures have been employed against this predator, including the incineration of all fallen, infested fruits and the elimination of all related species that serve as hosts.

Few diseases have been reported. However, in Florida, there are cases of anthracnose caused by *Colletotrichum gloeosporioides,* and leafspotting by the fungus, *Cercospora bunchosiae,* is a serious malady in Florida, Puerto Rico and Hawaii. Green scurf, identified with the alga, *Cephaleuros virescens,* occurs in Puerto Rico.

Food Uses

Barbados cherries are eaten out-of-hand, mainly by children. For dessert use, they are delicious merely stewed with whatever amount of sugar is desired to modify the acidity of the particular type available. The seeds must be separated from the pulp in the mouth and returned by spoon to the dish. Many may feel that the nuisance is compensated for by the pleasure of enjoying the flavorful pulp and juice. Otherwise, the cooked fruits must be strained to remove the seeds and the resulting sauce or puree can be utilized as a topping on cake, pudding, ice cream or sliced bananas, or used in other culinary products. Commercially prepared puree may be dried or frozen for future use. The fresh juice will prevent darkening of bananas sliced for fruit cups or salads. It can be used for gelatin desserts, punch or sherbet, and has been added as an ascorbic acid supplement to other fruit juices. The juice was dried and powdered commercially in Puerto Rico for a decade until the cost of production caused the factory to be closed down.

The fruits may be made into sirup or, with added pectin, excellent jelly, jam, and other preserves. Cooking causes the bright-red color to change to brownish-red. The pasteurization process in the canning of the juice changes the color to orange-red or yellow, and packing in tin cans brings on further color deterioration. Enamel-lined cans preserve the color better.

Wine made from Barbados cherries in Hawaii was found to retain 60% of the ascorbic acid.

Harmful Effects

Physicians in Curacao report that children often require treatment for intestinal inflammation and obstruction caused by eating quantities of the entire fruits, including seeds, from the wild Barbados cherries which abound on the island.

People who pick Barbados cherries without gloves and long sleeves may suffer skin irritation from contact with the minute stinging hairs on the leaves and petioles.

Other Uses

Bark: The bark of the tree contains 20-25% tannin and has been utilized in the leather industry.

Wood: The wood is surprisingly hard and heavy. Trials have demonstrated that it refuses to ignite even when treated with flammable fluid unless perfectly dry.

Medicinal Uses: The fruits are considered beneficial to patients with liver ailments, diarrhea and dysentery, as well as those with coughs or colds. The juice may be gargled to relieve sore throat.

Food Value Per 100 g of Edible Portion*	
Calories	59
Moisture	81.9-91.10 g
Protein	0.68-1.8 g
Ether Extract	0.19-0.09 g
Fiber	0.60-1.2 g
Fat	0.18-0.1 g
Carbohydrates	6.98-14.0 g
Ash	0.77-0.82 g
Calcium	8.2-34.6 mg
Phosphorus	16.2-37.5 mg
Iron	0.17-1.11 mg
Carotene	0.003-0.408 mg
(Vitamin A)	408-1000 I.U.
Thiamine	0.024-0.040 mg
Riboflavin	0.038-0.079 mg
Niacin	0.34-0.526 mg
Ascorbic Acid**	

*According to analyses made in Hawaii, Guatemala, and elsewhere.

**According to analyses at the Massachusetts Institute of Technology of fruits grown in Barbados: 4,500 mg (green), 3,300 mg (medium-ripe), 2,000 mg (very ripe). The ascorbic acid level of unripe fruits can range up to 4,676 mg and such ratings are exceeded only by the fruits (rose hips) of *Rosa rugosa* Thunb., which may have as much as 6,977 mg/100 g. This constituent varies as much as 25% with the clone, the locale, cultural methods and degree of exposure to sunlight during developmental stages and after harvesting. At INCAP (Instituto de Nutricion de Central America and Panama), in Guatemala assays in 1950-1955 showed distressingly low levels—an average of 17 mg/100 g, whereas fruits sent to INCAP by air and in dry ice from Florida were analyzed and contained 1,420 mg/100 g.

In field experiments, treatment of young fruits on the tree with 200 ppm gibberellic acid has brought about a marked increase in the ascorbic acid content of the mature fruits.

The ascorbic acid is not totally destroyed by heat, for the jelly may contain 499-1,900 mg/100 g. Of the total ascorbic acid in Barbados cherry juice, 0.18% is in the bound form. Other constituents include dextrose, levulose, and a little sucrose.

Nance

The fruits of a number of species of *Byrsonima* have been consumed by the Indians of Central America and northern South America. The best-known of these is the nance, *B. crassifolia* HBK. (syns. *B. cubensis* Juss.; *Malpighia crassifolia* L.), which has acquired many alternate vernacular names: *changugu, chi, nance agrio, nanche, nanchi, nancen, nanche de perro, nananche,* and *nantzin* in Mexico; *nance verde* in El Salvador; *nancito* or *crabo* in Honduras; *craboo, crapoo* and wild *craboo* in Belize; *doncela* and *maricao* in the Dominican Republic; *maricao cimaroon, maricao verde, peralejo* and *peralejo* blanco in Puerto Rico; *peralejo de sabana* in Cuba; *tapal* in Guatemala; *chaparro, chaparro manteca, maache, mantequera, nanzi, noro, peraleja hembra, yaca* or *yuco* in Colombia; *chaparro de chinche, chaparro de sabana, manero manteco, manteco merey* or *manteco sabanero* in Venezuela; *murici, mirixi, murici-do-campo,* and *muruci-da-praia* in Brazil; *hori, sabana kwari moeleidan,* and *sabana mango* in Surinam; *huria* in Guyana; *quinquina des savannes* in Guateloupe; *savanna serrette* in Trinidad; sometimes wild cherry in Panama; golden spoon in the former British West Indies.

Fig. 54: The nance (*Byrsonima crassifolia*), though a minor fruit, has culinary and beverage uses in tropical America. The flowers furnish nectar for honeybees.

Description

The nance is a slow-growing large shrub or tree to 33 ft (10 m) high, or, in certain situations, even reaching 66 ft (20 m); varying in form from round-topped and spreading to narrow and compact; the trunk short or tall, crooked or straight. Young branches are densely coated with russet hairs. The opposite leaves, ovate to elliptic or oblong-elliptic, may be 1¼ to 6½ in (3.2-17 cm) long and 1½ to 2¾ in (4-7 cm) wide, rounded or pointed at the apex, blunt or pointed at the base; leathery, usually glossy on the upper surface and more or less brown- or gray-hairy on the underside. The flowers, borne in thinly or conspicuously red-hairy, erect racemes 4 to 8 in (10-20 cm) long, are ½ to ¾ in (1.25-2 cm) wide; the 5 petals yellow at first, changing to dull orange-red. The fruit is peculiarly odorous, orange-yellow, round, 5/16 to 7/16 in (8-12 cm) wide, with thin skin and white, juicy, oily pulp varying in flavor from insipid to sweet, acid, or cheese-like. There is a single, fairly large, stone containing 1 to 3 white seeds.

Origin and Distribution

The tree is native and abundant in the wild, sometimes in extensive stands, in open pine forests and grassy savannas, from southern Mexico, through the Pacific side of Central America, to Peru and Brazil; also occurs in Trinidad, Barbados, Curacao, St. Martin, Dominica, Guadeloupe, Puerto Rico, Haiti, the Dominican Republic and throughout Cuba and the Isle of Pines.

Dr. David Fairchild brought seeds from Panama to the United States Department of Agriculture in 1899 (S.P.I. #2944). A few specimens exist in special collections in southern Florida. The species was introduced into the Philippines in 1918.

Throughout its natural range, the nance is mainly consumed by children, birds, and wild and domesticated animals. In some regions, large quantities are sold in native markets at very low prices. There is some cultivation of the tree for its fruits in Mexico and parts of Central America.

Climate

The nance is limited to tropical and subtropical climates. In Central and South America, the tree ranges from sea-level to an altitude of 6,000 ft (1,800 m). It is highly drought-tolerant.

Soil

In Mexico, the tree is often found on rocky ground. It grows well in sandy and alkaline-sandy soils. It is well suited for restoration of infertile and burned-over land.

Season

In Mexico, the tree blooms from April through July and the fruits are marketed in September and October. In Puerto Rico, the tree blooms and fruits continuously from spring to fall; in Brazil from December to April.

Keeping Quality

The fruits fall to the ground when fully ripe and are very perishable. However, they can be stored in good condition for several months by merely keeping them submerged in water.

Food Uses

The fruits are eaten raw or cooked as dessert, or may be included in soup or in stuffing for meats. J.N. Rose in

1899 wrote that he saw nances, olives and rice cooked with stewed chicken in Mexico.

The fruits are often used to prepare carbonated beverages, or an acid, oily, fermented beverage known by the standard term *chicha* applied to assorted beer-like drinks made of fruits or maize. By distillation, there is produced in Costa Rica, a rum-like liquor called *Crema de nance*.

In Magdalena, Colombia, an edible fat is extracted from the fruits with boiling water.

Food Value Per 100 g of Edible Portion*	
Moisture	79.3–83.2 g
Protein	0.109–0.124 g
Fat	0.21–1.83 g
Fiber	2.5–5.8 g
Ash	0.58–0.69 g
Calcium	23.0–36.8 mg
Phosphorus	12.6–15.7 mg
Iron	0.62–1.01 mg
Carotene	0.002–0.060 mg
Thiamine	0.009–0.014 mg
Riboflavin	0.015–0.039 mg
Niacin	0.266–0.327 mg
Ascorbic Acid	90.0–192.0 mg

*According to analyses made in Guatemala and El Salvador. The fruit is high in tannin, especially when unripe.

Other Uses

Fruit: Green fruits are sometimes used in dyeing. The fruit skin imparts a light-brown hue to cotton cloth.

Bark: The bark yields a strong fiber, and is employed in tanning, giving the leather a light-yellow tone. The bark contains 17.25–28.26% tannin and 2.73% oxalic acid.

Branches: Fresh branches are cut into small pieces and thrown into streams to stupefy fish; or they are crushed at the edge of shallow waters so that the juice spills into the water, for the same effect.

Wood: The sapwood is grayish; the heartwood reddish-brown, heavy, coarse-textured, tough, and highly prized for boat ribs though it is brittle and only medium-durable. Usually available only in small sizes, it serves for tool handles, turnery, cabinetwork and furniture and small-scale construction. In Brazil, the wood is chosen for the hot fire over which the people smoke the stimulant paste of guaraná (*Paullinia cupana* HBK.) because the burning wood has a pleasant odor. In some areas it is used for making charcoal.

Nectar: In Costa Rica, the nance provides one of the few sources of nectar for honeybees in the month of June.

Medicinal Uses: The astringent bark infusion is taken to halt diarrhea; also as a febrifuge. It is considered beneficial in pulmonary complaints, cases of leucorrhea, and allegedly tightens the teeth where the gums are diseased. In Belize, it is taken as an antidote for snakebite. In Guyana, the pounded bark is poulticed on wounds. Mexicans apply the pulverized bark on ulcers.

EUPHORBIACEAE

Bignay (Plate XXVI)

When Corner referred to this member of the Euphorbiaceae as a "shady, rather gloomy tree", he could not have been viewing it in fruit, a spectacle that has always aroused enthusiasm. The colorful bignay, *Antidesma bunius* Spreng., is called *bignai* in the Philippines; *buni* or *berunai* in Malaya; *wooni* or *hooni*, in Indonesia; *ma mao luang* in Thailand; *kho lien tu* in Laos; *choi moi* in Vietnam; *moi-kin* and *chunka* by the aborigines in Queensland. Among English names are Chinese laurel, currant tree, nigger's cord, and salamander tree.

Description

The tree may be shrubby, 10 to 26 ft (3-8 m) high, or may reach up to 50 or even 100 ft (15-30 m). It has wide-spreading branches forming a dense crown. The evergreen, alternate leaves are oblong, pointed, 4 to 9 in (10-22.5 cm) long, 2 to 3 in (5-7.5 cm) wide, dark-green, glossy, leathery, with very short petioles. The tiny, odorous, reddish male and female flowers are produced on separate trees, the male in axillary or terminal spikes, the female in terminal racemes 3 to 8 in (7.5-20 cm) long. The round or ovoid fruits, up to 1/3 in (8 mm) across, are borne in grapelike, pendent clusters (often paired) which are extremely showy because the berries ripen unevenly, the pale yellowish-green, white, bright-red and nearly black stages present at the same time. The skin is thin and tough but yields an abundance of bright-red juice which leaves a purple stain on fabrics, while the pulp, only 1/8 in (3 mm) thick is white with colorless juice. Whole fruits are very acid, much like cranberries, when unripe; are subacid, slightly sweet, when fully ripe. Some tasters detect a bitter principle or "unpleasant aftertaste" which is unnoticeable to others. There is a single, straw-colored stone, an irregular, flattened oval, ridged or fluted, very hard, 3/8 in (1 cm) long, 1/4 in (6 mm) wide.

P.J. Wester mentions a "very distinct and superior variety" as reliably reported from the Mountain Province, Philippines.

Origin and Distribution

The bignay is native and common in the wild from the lower Himalayas in India, Ceylon, and southeast Asia (but not Malaya) to the Philippines and northern Australia. It is an abundant and invasive species in the Philippines; occasionally cultivated in Malaya; grown in every village in Indonesia where the fruits are marketed in clusters.

The United States Department of Agriculture received seeds from the Philippines in 1905 (S.P.I. #18393); twice in 1913 (S.P.I. #36088 and #34691), and again in 1918 (S.P.I. #46704). Quite a few trees have been planted in southern Florida in the past and the fruits were formerly appreciated as a source of juice for jelly, commercialized in a limited way, but are rarely so used today. There are specimens in experimental stations in Cuba, Puerto Rico, Honduras and Hawaii.

Climate

The tree is not strictly tropical for it has proved to be hardy up to central Florida. It thrives in Java from sea-level to 4,000 ft (1,200 m). It grows well and flowers but does not set fruit in Israel.

Propagation

Many seeds are non-viable in Florida, perhaps because of inadequate pollination. Since seedlings may turn out to be male, and female seedlings may not bear for a number of years, vegetative propagation is preferred. The tree is readily multiplied by cuttings, grafting or air-layering. The air-layers have borne fruit in 3 years after transplanting to the field. Ochse recommends grafting in the wet season because scions will remain dormant in dry weather. Most female trees will bear some fruit without the presence of a male because many of the flowers are perfect.

Culture

The trees should be spaced 40 to 45 ft (12-14 m) apart, each way. And one male tree should be planted for every 10 to 12 females to provide cross-pollination. Wind-protection is desirable when the trees are small. Otherwise they require very little cultural attention.

Yield

Yield varies greatly from tree to tree if they are grown from seed. A mature tree in Florida has produced 15 bushels of fruit in a season. One very old tree at the home of Dr. David Fairchild produced 22 bushels yielding 72 gals (273 liters) of juice.

Season

In Indonesia, the trees flower in September and October and the fruits mature in February and March. The fruiting season is July to September in North Vietnam. In Florida it extends from late summer through fall and winter because some trees bloom much later than others.

Pests and Diseases

The tree is attacked by termites in Southeast Asia. In Florida, the leaves may be heavily attacked by mealybugs and by scale insects and sooty mold develops on their excretions. Here, also, the foliage is subject to green scurf and algal leaf spot caused by *Cephaleuros virescens*.

Food Uses

In Malaya, the fruits are eaten mostly by children. Indonesians cook the fruits with fish. Elsewhere the fruits (unripe and ripe together) are made into jam and jelly though the juice is difficult to jell and pectin must be added. Some cooks add lemon juice as well. If the extracted bignay juice is kept under refrigeration for a day or so, there will be a settling of somewhat astringent sediment which can be discarded, thus improving the flavor. For several years, the richly-colored jelly was produced on a small commercial scale in southern Florida. The juice makes an excellent sirup and has been successfully fermented into wine and brandy.

In Indonesia and the Philippines, the leaves are eaten raw or stewed with rice. They are often combined with other vegetables as flavoring.

Food Value Per 100 g of Edible Portion*	
Moisture	91.11–94.80 g
Protein	0.75 g
Ash	0.57–0.78 g
Calcium	0.12 mg
Phosphorus	0.04 mg
Iron	0.001 mg
Thiamine	0.031 mg
Riboflavin	0.072 mg
Niacin	0.53 mg

*According to analyses made in Florida and the Philippines.

Toxicity

The bark contains a toxic alkaloid. The heavy fragrance of the flowers, especially the male, is very obnoxious to some individuals.

Other Uses

Bark: The bark yields a strong fiber for rope and cordage.

Wood: The timber is reddish and hard. If soaked in water, it becomes heavy and, according to Drury, "black as iron". It has been experimentally pulped for making cardboard.

Medicinal Uses: The leaves are sudorific and employed in treating snakebite, in Asia.

Related Species

The **Herbert River cherry,** *A. dallachyanum* Baill., is a bushy tree, seldom over 25 ft (7.5 m) in height. The young shoots are slightly hairy. Mature leaves, almost hairless, are ovate to lanceolate-elliptical, 2 to 6 in (5–15 cm) long; deep-green above, bright-green beneath; thick and leathery. The odoriferous male flower spikes are hairy, generally in panicles in the leaf axils, occasionally solitary, more or less interrupted. The greenish female flowers are borne in racemes. The fruits, single or in clusters of 4 to 30, are round to obovoid, up to ¾ in (2 cm) wide, rich-red when unripe, dark purple-red (nearly black) when ripe and very acid. They ripen fairly evenly in the cluster.

The tree is native to coastal North Queensland, growing on the borders of rain forests and on the banks of streams and lagoons. Seeds were imported by the University of Florida Agricultural Research and Education Center, Homestead, Florida, in 1941 and the seedlings grew and bore well. The seeds germinate readily and seedlings begin to fruit at about 6 years of age when they may be 8 ft (2.4 m) tall. Multiplication may also be by cuttings, air-layering or grafting. One nursery in Florida offered grafted plants for sale but they did not become popular and the species is still rare.

In Australia, the trees bloom from December to February and again in September and the fruits mature in their fall and winter months. In Florida, blooming takes place from April to June and the fruit is in season in September and October.

The extracted juice is very dark-red, nearly black, but it yields, with the addition of pectin, a deep-red jelly.

The tree, like that of the bignay, is prone to infestation by mealybugs and scale insects and associated sooty mold.

The **black currant tree,** *A. ghaesembilla* Gaertn. (syn. *A. pubescens* Roxb.), called *dang kiep kdam* in Cambodia, *chop moi, choi moi, chua moi* or *chum moi* in Vietnam, is a deciduous shrub or bushy tree up to 26 or, at most, 40 ft (8–12 m), with short, russet hairs on the young branches, rosy new foliage and inflorescences. The mature leaves are broad-ovate or nearly circular, 1½ to 3 in (4–7.5 cm) long, glossy on the upper surface. Male flower spikes, purplish or light-yellow with pollen, are dense, 1 to 2 in (2.5–5 cm) long; the erect female shorter and not as compact. Both types occur in terminal panicles or rarely solitary. Some trees bear both male and female flowers but on separate branches. The trees flower off and on during the year but mostly March to May in Asia.

Fig. 55: The Herbert River Cherry of Australia (*Antidesma dallachyanum*) is less showy than the bignay but the fruits have more flesh.

The fruit is velvety, dark-red or very dark-purple, obliquely ovoid with one seed or occasionally double with 2 seeds. The seed kernels are sharply angular. When fully ripe the fruit is subacid to somewhat sweet.

This species has a wide natural range: in tropical Africa, and from the moist tropical lower Himalayas in northern India through Ceylon, southern China, Southeast Asia and Malaysia to the Walsh River region of Queensland. Generally the fruits are eaten mainly by children, but they are appreciated as thirst-quenchers by forest people of Thailand. They were made into jam by early settlers in Australia. In Malaya and Indonesia, they are made into a kind of relish, and the very young leaves are added as acid flavoring to various foods.

The wood is red, hard, close-grained, smooth and used for light rafters in huts, but for little else. Small branches are lopped twice a year for fuel. In India, the leaves are used to treat fever, headache and swollen abdomens. In Cambodia, various parts of the tree are valued in native medicine. The bark, combined with tobacco, is applied on wounds of animals. Combined with the bark of other species, it is boiled and the decoction given to halt diarrhea. The leaves and wood are similarly employed. A decoction of young branches and papaya roots is considered an effective emmenagogue. Crushed leaves are applied on the head of a newborn infant.

Emblic

This member of the Euphorbiaceae, *Phyllanthus emblica* L. (syn. *Emblica officinalis* Gaertn.) ranges in status from insignificant in the western world to highly prized in tropical Asia. Alternative English names include emblic myrobalan, Malacca tree and Indian gooseberry, though the last term is more frequently applied to the related but dissimilar Otaheite gooseberry, q.v. In Malaya the emblic is called *melaka, Asam melaka,* or *amlaka;* in Thailand, it is *ma–kham–pom;* in Laos, *mak–kham–pom;* in Cambodia, *kam lam* or *kam lam ko;* in southern Vietnam, *bong ngot;* in North Vietnam, *chu me.* In the Philippines, it is called *nelli.*

Description

The tree is a graceful ornamental, normally reaching a height of 60 ft (18 m) and, in rare instances, 100 ft (30 m). Its fairly smooth bark is a pale grayish-brown and peels off in thin flakes like that of the guava. While actually deciduous, shedding its branchlets as well as its leaves, it is seldom entirely bare and is therefore often cited as an evergreen. The miniature, oblong leaves, only 1/8 in (3 mm) wide and 1/2 to 3/4 in (1.25-2 cm) long, distichously disposed on very slender branchlets, give a misleading impression of finely pinnate foliage. Small, inconspicuous, greenish-yellow flowers are borne in compact clusters in the axils of the lower leaves. Usually, male flowers occur at the lower end of a growing branchlet, with the female flowers above them, but occasional trees are dioecious.

The nearly stemless fruit is round or oblate, indented at the base, and smooth, though 6 to 8 pale lines, sometimes faintly evident as ridges, extending from the base to the apex, give it the appearance of being divided into segments or lobes. Light-green at first, the fruit becomes whitish or a dull, greenish-yellow, or, more rarely, brick-red as it matures. It is hard and unyielding to the touch. The skin is thin, translucent and adherent to the very crisp, juicy, concolorous flesh. Tightly embedded in the center of the flesh is a slightly hexagonal stone containing 6 small seeds. Fruits collected in South Florida vary from 1 to 1¼ in (2.5-3.2 cm) in diameter but choice types in India approach 2 in (5 cm) in width. Ripe fruits are astringent, extremely acid, and some are distinctly bitter.

Origin and Distribution

The emblic tree is native to tropical southeastern Asia, particularly in central and southern India, Pakistan, Bangladesh, Ceylon, Malaya, southern China and the Mascarene Islands. It is commonly cultivated in home gardens throughout India and grown commercially in Uttar Pradesh. Many trees have been planted in southern Malaya, Singapore, and throughout Malaysia. In India, and to a lesser extent in Malaya, the emblic is important and esteemed, raw as well as preserved, and it is prominent in folk medicine. Fruits from both wild and dooryard trees and from orchards are gathered for home use and for market. In southern Thailand, fruits from wild trees are gathered for marketing.

In 1901, the United States Department of Agriculture received seeds from the Reasoner Brothers, noted nurserymen and plant importers of Oneco, Florida. Seeds were distributed to early settlers in Florida and to public gardens and experimental stations in Bermuda, Cuba, Puerto Rico, Trinidad, Panama, Hawaii and the Philippines. The fruits of these seedlings aroused no enthusiasm until 1945 when Mr. Claud Horn of the Office of Foreign Agricultural Relations in Washington, D.C., inspired by Indian ratings of the emblic as the "richest known natural source of vitamin C", asked that analyses be made in Puerto Rico. A high level of ascorbic acid was found and confirmed in Florida but interest quickly switched to the Barbados cherry (q.v.) which was casually assayed and found to be as rich or richer when underripe. The emblic was soon forgotten. Some old trees still exist in southern Florida; others have been removed in favor of housing or other developments. In 1954, the Campbell Soup Company in Camden, New Jersey, requested 5 lbs (2.25 kg) of the fruits for study. They were sent, but no further interest was evidenced. In 1982, several individuals asked for and were given seeds for planting in Australia. They did not reveal whether the tree was desired for its own sake or for its fruits.

Varieties

In India there are 3 named cultivars grown commercially:

'Banarsi'—originated in Banarsi district of Uttar Pradesh; medium to large, the 6 segments paired, giving the appearance of only 3; 1½ in (4 cm) long, 1¾ in (4.5 cm) wide; skin thin and translucent, light-green, turning whitish as the fruit ripens; flesh slightly fibrous, medium juicy, moderately astringent. Earliest in season. Tree is semi-spreading; not a heavy cropper; tends to alternate bearing unless interplanted.

'Chakaiya'—flattened at base and apex; may have 6, 7, or 8 segments; of medium size, 1¼ in (3.2 cm) long, 3¼ in (8.25 cm) wide; flesh fibrous. Tree is spreading; prolific. This cultivar is now preferred over the others because of its yield.

'Francis' ('Hathijhool')—rounded-oval, bulged at the apex; has 6 segments; large, 1⅝ in (4.3 cm) long and 2 in (5 cm) wide. The tree is a regular producer of good crops, but prone to fruit necrosis.

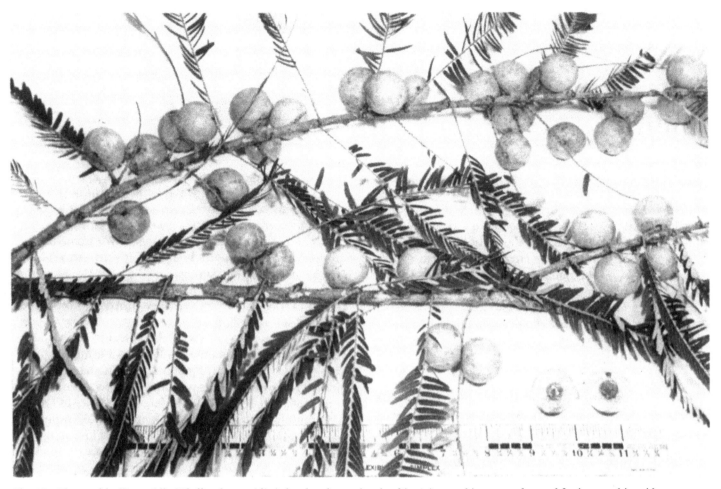

Fig. 56: The marble-like emblic (*Phyllanthus emblica*), hard and sour, is valued in Asia as a thirst-quencher and for its ascorbic acid content.

The ordinary small fruits — 5/8 to 1 in (1.5-2.5 cm) wide, with reddish skin, rarely grown commercially, are mainly used for medicinal purposes.

Pollination

Cross-pollination is desirable. 'Banarsi' bears better when interplanted with other varieties. Growers in India are beginning to scatter a few seedling trees around in their groves. Honeybees work the flowers in the morning and late evening. It is now known that lack of pollination is the cause of up to 70% shedding of flowers in the first 3 weeks after onset of blooming.

Climate

The emblic is subtropical rather than strictly tropical. In India, it flourishes from sea-level up to an altitude of 5,000 ft (1,800 m). Seeds were planted at the Agricultural Research and Education Center in Homestead, Florida, in 1955 and the seedlings were set out in the field in 1956. They survived unusually cold weather in the winter of 1957-58. That freeze damaged a tree with a trunk 1 ft (30 cm) thick at Laurel, Florida. It was set back again by cold in December 1962. It put out many shoots which, by October 2, 1963 were 10 ft (3 m) high, showing a remarkable ability to recover from cold injury. On the other hand, it is intolerant of excessive heat. In India, mature trees can stand temperatures up to 115°F (46°C) in the summer but young plants must be shaded.

Soil

The emblic seems to grow equally well under both arid and humid conditions. It is noted for being able to thrive in regions too dry and soil too poor for most other fruit crops. For maximum productivity, the tree requires deep soil ranging from sandy loam to clay, light or heavy, slightly acidic to slightly alkaline. At high pH (as much as 8.0), nutritional deficiencies are evident. Limestone is considered unsuitable but the large, old trees in southern Florida are all in oolitic limestone. Good drainage is essential. A low degree of salinity seems to be fairly well tolerated.

Propagation

The tree is often propagated by seeds taken from over-ripe fruits sun-dried to facilitate removal of the stone, or cut in half right through the stone. The extracted seeds are given the float test and 100% of those that sink will germinate. In 4 months, seedlings will have a stem diameter of 1/3 in (8 mm) and can be budded or grafted from June to September and in February and March in India. The Forkert and patch techniques have given 85% to 100% success. Chip-budding, using seedlings 1½ years old as rootstocks, is easier and 60% to 80% successful in September and October and February and March. Inarching is sometimes practiced in India but survival rate may be only 25% to 30% after separation from the stock and further losses may occur in the field. At the Experimental Farm of the University of Miami in 1955, air-layers and cuttings were unsuccessful but root sprouts grew well.

Emblic trees bearing fruits of inferior quality may be top-worked by cutting back to a height of 4 ft (1.2 m) and applying coal tar to the cut surfaces. Trials at Saharanpur showed that this is best done in March when the trees are not in active growth. Budding of the new shoots can be done successfully any time from June to September.

Culture

While the emblic has long been established as an important and remunerative crop in India, the systematic culture of high-quality fruit is a modern development actively promoted by the Indian Government. It is recommended that the trees be spaced 30 to 40 ft (9-12 m) apart and planted in well-prepared holes enriched with a composted manure and soil mixture, and well-watered. Thereafter, watering is done only in the dry season. Seedlings in Florida have attained 8 to 9 ft (2.4-2.7 m) in height in 5 years. They usually begin to bear when 5 to 6 years old and normally bear for about 50 years.

There are no standard practices for fertilizing the emblic but 1 to 1½ oz (28-42 g) of nitrogen per tree for each year of age up to 10 years has been suggested. After 10 years the nitrogen is increased and potash and superphosphate are added. Half of the fertilizer should be given after fruit-set and the other half 4 months later.

The branches are brittle and judicious pruning to develop a strong framework is advocated to avoid branch breakage from heavy loads of fruit.

Season and Harvesting

The emblic is sensitive to day-length. In northern India, flowering takes place from March to May. In Madras, the tree blooms in June-July and again in February-March, the second flowering producing only a small crop. In Florida flowering occurs during the summer months, the main crop maturing during the winter and early spring. A few fruits developed from late blooms are found in summer and fall.

In India, people shake down the fruits that are ready to fall and gather from the ground those that have already fallen, and take them to market. They stand handling well. The yield varies a great deal as many young fruits are shed throughout the period of fruit development, and there is considerable difference in the productivity of seedlings and cultivars. P.N. Bajpai, in a study of the fruiting habits of four 15-year-old emblic trees, found an average yield of 415 fruits, which weighed approximately 24½ lbs (11 kg). 'Banarsi' trees 10 years old have yielded 35.2 lbs (16 kg). 'Chakaiya' trees of the same age have yielded 39.6 lbs (18 kg). Mature 'Chakaiya' trees may bear 55 lbs (25 kg) per year.

Pests and Diseases

The chief pest of this tree in India is the bark-eating caterpillar, *Indarbela* sp., which tunnels into the branches and trunk. A secondary enemy produces shoot galls. A non-pathogenic problem, especially in 'Francis', is called "fruit necrosis" in India. It is evidenced by internal browning which gradually extends to the surface where dark spots become corky and gummy. It can be overcome by bi-monthly sprays of borax in September and October. There are few serious diseases but the fungi, *Bestonea stylophora*, *Phakospora phyllanthi* and *Ravenelia emblicae*, cause ring rust, leaf rust and fruit rot.

Fresh emblics on the market or in storage are subject to blue mold and rotting caused by *Penicillium islandicum*. Rinsing with very dilute borax or sodium chloride solutions helps retard such spoilage. Emblic preserves on the market have been found contaminated with yeasts, molds and bacteria. Pre-processing treatment with 0.01% sulfur dioxide or sodium benzoate prolongs keeping quality.

Food Uses

Rural folk in India claim that the highly acid, fresh, raw fruit, followed by water, produces a sweet and refreshing aftertaste. Wood-cutters in Southeast Asia eat the emblic to avoid thirst, as the fruit stimulates the flow of saliva. This is the one tree left standing when forests are clear-cut in Thailand, and busses stop along highways to let thirsty travelers run to the tree to get the fruits. The emblic is regarded as sacred by many Hindus and the Hindu religion prescribes that ripe fruits be eaten for 40 days after a fast in order to restore health and vitality. It is a common practice in Indian homes to cook the fruits whole with sugar and saffron and give one or two to a child every morning.

Fresh emblics are baked in tarts, added to other foods as seasoning during cooking, and the juice is used to flavor vinegar. Both ripe and half-ripe fruits are candied whole and also made into jam and other preserves, sweetmeats, pickles and relishes. They are combined with other fruits in making chutney. In Indonesia, emblics are added to impart acidity to many dishes, often as a substitute for tamarinds.

When necessary, bitterness is overcome by soaking the fruits in a salt solution or by adding citrus fruit, unripe mango or tamarind. In preserving emblics whole, the fruit is first brined, washed and pricked, blanched in an alum solution, layered with sugar until a sirup is formed, and then boiled. It is finally packed in enameled cans or crystallized as a confection. In India, a sauce is made from the dried, chipped flesh. In its preparation, the chips are cooked in water, mashed in a mortar with caraway seeds, and further seasoned with salt and yogurt. This, also, is commonly eaten after fasting. During World War II, emblic powder, tablets and candies were issued to Indian military personnel as vitamin C rations. Drs. Rama Rao, Balakushnan and Rajagopalan, of the Institute of Science at Bangalore, describe a method of spray-drying emblic juice to produce a special powder for fortifying salt as a means of increasing vitamin C intake.

In Thailand, where the tree is common in the forests, the fruits are favored by deer, especially the tiny barking deer.

Food Value Per 100 g of Edible Portion*	
Moisture	77.1 g
Protein	0.07 g
Fat	0.2 g
Carbohydrates	21.8 g
Fiber	1.9 g
Ash	0.5 g
Calcium	12.5 mg
Phosphorus	26.0 mg
Iron	0.48 mg
Carotene	0.01 mg
Thiamine	0.03 mg
Riboflavin	0.05 mg
Niacin	0.18 mg
Tryptophan	3.0 mg
Methionine	2.0 mg
Lysine	17.0 mg
Ascorbic Acid**	625 mg

*As reported by the Finlay Institute Laboratory, Havana.

**The ascorbic acid ratings vary immensely. Analyses in Puerto Rico, showed 625 mg; fruits from one tree in Avon Park, Florida, showed only 467 mg, while 2 adjacent trees in Homestead, Florida, showed 1,130 and 1,325 mg; and Dr. Margaret Mustard reported an average of 1,561.0 and a high of 1,814 mg in 7 samples analyzed.

The ascorbic acid in the emblic is considered highly stable, apparently protected by tannins (or leucoanthocyanins) which retard oxidation. Biochemical studies at the Central Drug Research Institute, Lucknow, India, show 13 tannins plus 3 or 4 colloidal complexes. In juice extracted from the fresh fruit, the ascorbic acid is stable for at least a week. Fresh juice stored at 35.6°F (2°C) loses only 14% ascorbic acid after 45 days. Only 30% is lost in evaporation over open flame at 149°F (65°C), but the product loses 40% during a week in a refrigerator and 100% in 20 days.

Efforts in India to prepare a stable ascorbic acid concentrate from the dried fruit have been frustrating because sun-drying loses 65% ascorbic acid. Artificial drying at 185°F (85°C) loses 34%; and at 212°F (100°C), 72%. Once dried, there is negligible loss. However, vacuum-drying (27 in. Hg) at 140-176°F (60-80°C) retains the original ascorbic acid levels, the dried product containing 2,000 to 3,500 mg per 100 g, depending on the content of the fresh fruit. Even after 14 months of refrigerated storage, there is a loss of only 15 to 20%.

Separation of tannins from expressed juice by precipitation with neutral lead acetate and ion exchange chromatographic purification has yielded crystalline ascorbic acid amounting to 70-72% of that in the juice.

The dry, powdered fruit contains 6.3% phyllembic acid, 6% fatty matter, 5% gallic acid, ellagic acid, emblicol (a crystalline phenolic product) and other constituents. Phyllemblin (ethyl gallate) isolated from dried fruit, acts as a mild CNS depressant and has spasmolytic activity.

Other Uses

Other uses of the fruit and parts of the tree are numerous:

Fruit: The dried fruit yields ink and hair-dye and, having detergent properties, is sometimes used as a shampoo. A fixed oil derived from the fruit allegedly acts as a hair-restorer and is used in shampoos in India. This oil is the main ingredient in an "Amla Conditioner" currently sold by Shikai Products of Santa Rosa, California, by mail and through "health food" stores and other "natural" product outlets. A most curious custom is the making of simulated pottery jars from a paste of the boiled fruit, the surface being decorated with impressed colored seeds. Dyes from the fruit and leaves impart an appealing light-brown or yellow-brown hue to silk and wool. When sulfate of iron is added as a mordant, the color becomes black.

Bark: The tannin-rich bark, as well as the fruit and leaves, is highly valued and widely employed in conjunction with other so-called myrobalans, especially fruits of various species of *Terminalia*. The twig bark is particularly esteemed for tanning leather and is often used with leaves of *Carissa spinarum* A. DC. and *Anogeissus latifolia* Wall.

Leaves: The foliage furnishes fodder for cattle and branches are lopped for green manure. They are said to correct excessively alkaline soils.

Wood: The hard but flexible red wood, though highly subject to warping and splitting, is used for minor construction, furniture, implements, gunstocks, hookas and ordinary pipes. Durable when submerged and believed to clarify water, it is utilized for crude aqueducts and inner braces for wells, and branches and chips of the wood are thrown into muddy streams for clarification and to impart a pleasant flavor. The wood serves also as fuel and a source of charcoal.

Medicinal Uses: The emblic is of great importance in Asiatic medicine, not only as an antiscorbutic, but in the treatment of diverse ailments, especially those associated with the digestive organs. For such use, the fruit juice is prepared in the form of a sherbet or is fermented. In the latter state, it is prescribed in jaundice, dyspepsia and coughs. The dried chips of flesh are dispensed by apothecaries and often are mixed with grape juice and honey for dosage. The fruit is considered diuretic and laxative. *Triphala*, a decoction of emblic with *Terminalia chebula* Retz. and *T. bellerica* Roxb. is given for chronic dysentery, biliousness, hemorrhoids, enlarged liver, and other disorders. A powder prepared from the dried fruit is an effective expectorant as it stimulates the bronchial glands. The juice that exudes when the fruit is scored while still on the tree is valued as an eyewash and an application for inflamed eyes. An infusion made by steeping dried fruit overnight in water also serves as an eyewash, as does an infusion of the seeds. A liquor made from the fermented fruits is prescribed as a treatment for indigestion, anemia, jaundice, some cardiac problems, nasal congestion and retention of urine.

Emblic leaves, too, are taken internally for indigestion and diarrhea or dysentery, especially in combination with

Fig. 57: Emblics, heavily sugared, are sold in the native markets of Southeast Asia.

buttermilk, sour milk or fenugreek. The milky sap of the tree is applied on foul sores. The plant is considered an effective antiseptic in cleaning wounds, and it is also one of the many plant palliatives for snakebite and scorpion stings. A decoction of the leaves is used as a mouthwash and as a lotion for sore eyes.

The flowers, considered refrigerant and aperient, and roots, emetic, are also variously employed. The root bark, mixed with honey, is applied to inflammations of the mouth. The bark is strongly astringent and used in the treatment of diarrhea and as a stomachic for elephants. The juice of the fresh bark is mixed with honey and turmeric and given in cases of gonorrhea. It is clear that the majority of the applications of the fruit and other parts are based on the astringent action of the tannins they contain. The short-term effects of tannins appear beneficial, but habitual indulgence can be highly detrimental, inasmuch as tannin is antinutrient and carcinogenic.

An ointment made from the burnt seeds and oil is applied to skin afflictions. The seeds are used in treating asthma, bronchitis, diabetes and fevers. They contain proteolytic and lipolytic enzymes, phosphatides and a small amount of essential oil. Approximately 16% consists of a brownish-yellow fixed oil.

Otaheite Gooseberry

Totally unlike a gooseberry except for its acidity, the Otaheite gooseberry, *Phyllanthus acidus* Skeels (syns. *P. distichus* Muell. Arg.; *Cicca acida* Merr.; *C. disticha* L.), is another of the few members of the family Euphorbiaceae having edible fruit. It has been widely distributed and is variously known as Malay gooseberry, country gooseberry, *cheremai, chermela, chamin–chamin,* or *kemangor* (Malaya); *cherme, tjerme,* or *tjareme* (Java); *cherim-billier, tam duot, chum ruot* (Vietnam); *mayom* (Thailand); *mak–nhom* (Laos); star gooseberry, West India

Fig. 58: No fruit is borne in greater abundance than the crisp, sour, pale-yellow Otaheite gooseberry (*Phyllanthus acidus*). When cooked in sugar, the fruit and juice turn ruby-red. In: K. & J. Morton, *Fifty Tropical Fruits of Nassau,* 1946.

gooseberry, *jimbling, chalmeri, harpharori* (India); *iba* (Philippines); *ciruela corteña, manzana estrella* (Mexico); *pimienta* or *guinda* (El Salvador); *grosella* (Costa Rica, Cuba, Guatemala, Nicaragua); *groselha* (Brazil); *groseillier des Antilles* (French West Indies); *cereza amarilla, cerezo comun, cerezo de la tierra* (Puerto Rico); *cerezo agrio* (Venezuela); *cerezo occidental* (Cuba); wild plum (Belize, Yucatan); *cheramina, jimbling,* short *jimbelin* (Jamaica).

Description

This is a curious and ornamental shrub or tree, 6½ to 30 ft (2-9 m) high, with spreading, dense, bushy crown of thickish, rough, main branches, in general aspect resembling the **Bilimbi** (q.v.). At the branch tips are clusters of deciduous, greenish or pinkish branchlets 6 to 12 in (15-30 cm) long, bearing alternate, short-petioled, ovate or ovate-lanceolate, pointed leaves ¾ to 3 in (2-7.5 cm) long, thin, green and smooth on the upper surface, blue-green with a bloom on the underside; altogether giving the impression of pinnate leaves with numerous leaflets. There are 2 tiny, pointed stipules at the base of each leaf. Small, male, female, and some hermaphrodite, 4-parted, rosy flowers, are borne together in little clusters arranged in panicles 2 to 5 in (5-12.5 cm) long, hanging directly from leafless lengths of the main branches and the upper trunk, and the fruits develop so densely that they form spectacular masses. The fruit is oblate with 6 to 8 ribs; is ⅜ to 1 in (1-2.5 cm) wide; pale-yellow to nearly white when fully ripe; waxy, fleshy, crisp, juicy and highly acid. Tightly embedded in the center is a hard, ribbed stone containing 4 to 6 seeds.

Origin and Distribution

This species is believed to have originated in Madagascar and to have been carried to the East Indies. Quisumbing says that it was introduced into the Philippines in prehistoric times and is cultivated throughout those islands but not extensively. It is more commonly grown in Indonesia, South Vietnam and Laos, and frequently in northern Malaya, and in India in home gardens. The tree is a familiar one in villages and on farms in Guam, where the fruit is favored by children, and occurs in Hawaii and some other Pacific Islands.

It was introduced into Jamaica from Timor in 1793 and has been casually spread throughout the Caribbean islands and to the Bahamas and Bermuda. It has long been naturalized in southern Mexico and the lowlands of Central America, and is occasionally grown in Colombia, Venezuela, Surinam, Peru and Brazil. Formerly an escape from cultivation in South Florida, there are now only scattered specimens remaining here as curiosities.

Climate

The Otaheite gooseberry is subtropical to tropical, being sufficiently hardy to survive and fruit in Tampa, Florida, where cold spells are more severe than in the southeastern part of the state. It thrives up to an elevation of 3,000 ft (914 m) in El Salvador.

Soil

The tree grows on a wide range of soils but prefers rather moist sites.

Propagation

The tree is generally grown from seed but may also be multiplied by budding, greenwood cuttings, or air-layers. Seedlings will produce a substantial crop in 4 years.

Pests

The Otaheite gooseberry is prone to attack by the phyllanthus caterpiller in Florida. This pest eats the bark and also the young leaves, causing total defoliation in a few days if not controlled by pesticides.

Season

The tree often bears two crops a year in South India, the first in April and May, and the second in August and September. In other areas, the main crop is in January with scattered fruiting throughout the year.

Food Uses

The flesh must be sliced from the stone, or the fruits must be cooked and then pressed through a sieve to separate the stones. The sliced raw flesh can be covered with sugar and let stand in the refrigerator for a day. The sugar draws out the juice and modifies the acidity so that the flesh and juice can be used as a sauce. If left longer, the flesh shrivels and the juice can be strained off as a clear, pale-yellow sirup. In Indonesia, the tart flesh is added to many dishes as a flavoring. The juice is used in cold drinks in the Philippines. Bahamian cooks soak the whole fruits in salty water overnight to reduce the acidity, then rinse, boil once or twice, discarding the water, then boil with equal amount of sugar until thick, and put up in sterilized jars without removing seeds. The repeated processing results in considerable loss of flavor. Fully ripe fruits do not really require this treatment. If cooked long enough with plenty of sugar, the fruit and juice turn ruby-red and yield a sprightly jelly. In Malaya, the ripe or unripe Otaheite gooseberry is cooked and served as a relish, or made into a thick sirup or sweet preserve. It is also combined with other fruits in making chutney and

Food Value Per 100 g of Edible Portion*	
Moisture	91.9 g
Protein	0.155 g
Fat	0.52 g
Fiber	0.8 g
Ash	0.51 g
Calcium	5.4 mg
Phosphorus	17.9 mg
Iron	3.25 mg
Carotene	0.019 mg
Thiamine	0.025 mg
Riboflavin	0.013 mg
Niacin	0.292 mg
Ascorbic Acid	4.6 mg

*According to analyses made in El Salvador.

jam because it helps these products to "set". Often, the fruits are candied, or pickled in salt. In the Philippines, they are used to make vinegar.

The young leaves are cooked as greens in India and Indonesia.

Other Uses

Wood: The wood is light-brown, fine-grained, attractive, fairly hard, strong, tough, durable if seasoned, but scarce, as the tree is seldom cut down.

Root bark: The root bark has limited use in tanning in India.

Medicinal Uses: In India, the fruits are taken as liver tonic, to enrich the blood. The sirup is prescribed as a stomachic; and the seeds are cathartic. The leaves, with added pepper, are poulticed on sciatica, lumbago or rheumatism. A decoction of the leaves is given as a sudorific. Because of the mucilaginous nature of the leaves, they are taken as a demulcent in cases of gonorrhea.

The root is drastically purgative and regarded as toxic in Malaya but is boiled and the steam inhaled to relieve coughs and headache. The root infusion is taken in very small doses to alleviate asthma. Externally, the root is used to treat psoriasis of the soles of feet. The juice of the root bark, which contains saponin, gallic acid, tannin and a crystalline substance which may be lupeol, has been employed in criminal poisoning.

The acrid latex of various parts of the tree is emetic and purgative.

Rambai

A fruit somewhat resembling the **langsat** (q.v.) but belonging to a different family, Euphorbiaceae, is the rambai, *Baccaurea motleyana* Hook. f., called *rambi* in the Philippines, *mai–fai farang* in Thailand.

Description

The slow-growing tree, ordinarily to 30 or 40 ft (9-12 m), occasionally up to 60 ft (18 m), has a short, thick trunk, broad, dense, rounded crown and silky-hairy new branchlets. The leaves are evergreen, spiralled, 6 to 13 in (15-33 cm) long, 3 to 6 in (7.5-15 cm) wide; dark-green, glossy, with conspicuously indented veins on the upper surface; greenish-brown and hairy below. The small, fragrant male and female flowers are borne on separate trees. They are petalless, with 4 to 6 chartreuse, velvety sepals, the female arranged in racemes 10 to 30 in (25-75 cm) long; the male in racemes 3 to 6 in (7.5-15 cm) long. The fruits, in showy strands dangling from the older branches and trunk, are oval, 1 to 1¾ in (2.5-4.5 cm) long and 1 in (2.5 cm) thick, with thin, salmon-colored or brownish-yellow, velvety skin becoming wrinkled after ripening. The translucent, white, sweet-to-acid pulp is in 3 to 5 segments which separate readily, each segment containing a brown, flat seed about ½ in (1.25 cm) long, adherent to the pulp.

The rambai is native and commonly cultivated in the lowlands of Malaya, grows wild in Bangha and Borneo and is occasionally cultivated in Java. It is valued for its shade as well as its fruits, which are eaten raw, stewed or made into jam or wine.

The wood is of low quality but used for posts. The bark serves as a mordant for dyes and is employed to relieve eye inflammation.

The very similar *kapoendoeng, B. racemosa* Muell. Arg., native to West, Central and East Java, is commonly cultivated and is budded onto its own rootstocks or those of *B. motleyana*.

A lesser-known species, the so-called Burmese grape, *B. sapida* Muell.-Art., called *tempui* in Malaya, *lutqua* in India, and *mai fai* in Thailand, grows to 30 or even 70 ft (9-21 m). The leaves are rarely, and then only slightly, hairy; the fruit, in strands 6 to 12 in (15-30 cm) long, is smooth, nearly round or oval, 1 to 1¼ in (2.5-3.2 cm) long. The skin turns from ivory to yellowish or pinkish-buff or sometimes bright-red. The pulp is not translucent; is whitish, occasionally deep-pink near the seeds; varies from acid to sweet.

The tree grows wild from southern China, Thailand and Cambodia to Malacca and it is occasionally cultivated in northern Malaya and Thailand.

B. dulcis Muell.-Arg., the *tjoepa, toepa* or *ketoepa* of southern Sumatra, has relatively large, sweet fruits which are abundant on local markets. It is sometimes cultivated in West Java.

Mango (Plates XXVII and XXVIII)

It is a matter of astonishment to many that the luscious mango, *Mangifera indica* L., one of the most celebrated of tropical fruits, is a member of the family Anacardiaceae—notorious for embracing a number of highly poisonous plants. The extent to which the mango tree shares some of the characteristics of its relatives will be explained further on. The universality of its renown is attested by the wide usage of the name, mango in English and Spanish and, with only slight variations in French (*mangot, mangue, manguier*), Portuguese (*manga, mangueira*), and Dutch (*manja*). In some parts of Africa, it is called *mangou,* or *mangoro.* There are dissimilar terms only in certain tribal dialects.

Description

The mango tree is erect, 30 to 100 ft (roughly 10-30 m) high, with a broad, rounded canopy which may, with age, attain 100 to 125 ft (30-38 m) in width, or a more upright, oval, relatively slender crown. In deep soil, the taproot descends to a depth of 20 ft (6 m), the profuse, wide-spreading, feeder root system also sends down many anchor roots which penetrate for several feet. The tree is long-lived, some specimens being known to be 300 years old and still fruiting.

Nearly evergreen, alternate leaves are borne mainly in rosettes at the tips of the branches and numerous twigs from which they droop like ribbons on slender petioles 1 to 4 in (2.5-10 cm) long. The new leaves, appearing periodically and irregularly on a few branches at a time, are yellowish, pink, deep-rose or wine-red, becoming dark-green and glossy above, lighter beneath. The midrib is pale and conspicuous and the many horizontal veins distinct. Full-grown leaves may be 4 to 12.5 in (10-32 cm) long and ¾ to 2⅛ in (2-5.4 cm) wide. Hundreds and even as many as 3,000 to 4,000 small, yellowish or reddish flowers, 25% to 98% male, the rest hermaphroditic, are borne in profuse, showy, erect, pyramidal, branched clusters 2½ to 15½ in (6-40 cm) high. There is great variation in the form, size, color and quality of the fruits. They may be nearly round, oval, ovoid-oblong, or somewhat kidney-shaped, often with a break at the apex, and are usually more or less lop-sided. They range from 2½ to 10 in (6.25-25 cm) in length and from a few ounces to 4 to 5 lbs (1.8-2.26 kg). The skin is leathery, waxy, smooth, fairly thick, aromatic and ranges from light- or dark-green to clear yellow, yellow-orange, yellow and reddish-pink, or more or less blushed with bright- or dark-red or purple-red, with fine yellow, greenish or reddish dots, and thin or thick whitish, gray or purplish bloom, when fully ripe. Some have a "turpentine" odor and flavor, while others are richly and pleasantly fragrant. The flesh ranges from pale-yellow to deep-orange. It is essentially peach-like but much more fibrous (in some seedlings excessively so—actually "stringy"); is extremely juicy, with a flavor range from very sweet to subacid to tart.

There is a single, longitudinally ribbed, pale yellowish-white, somewhat woody stone, flattened, oval or kidney-shaped, sometimes rather elongated. It may have along one side a beard of short or long fibers clinging to the flesh cavity, or it may be nearly fiberless and free. Within the stone is the starchy seed, monoembryonic (usually single-sprouting) or polyembryonic (usually producing more than one seedling).

Origin and Distribution

Native to southern Asia, especially eastern India, Burma, and the Andaman Islands, the mango has been cultivated, praised and even revered in its homeland since ancient times. Buddhist monks are believed to have taken the mango on voyages to Malaya and eastern Asia in the 4th and 5th Centuries B.C. The Persians are said to have carried it to East Africa about the 10th Century A.D. It was commonly grown in the East Indies before the earliest visits of the Portuguese who apparently introduced it to West Africa early in the 16th Century and also into Brazil. After becoming established in Brazil, the mango was carried to the West Indies, being first planted in Barbados about 1742 and later in the Dominican Republic. It reached Jamaica about 1782 and, early in the 19th Century, reached Mexico from the Philippines and the West Indies.

In 1833, Dr. Henry Perrine shipped seedling mango plants from Yucatan to Cape Sable at the southern tip of mainland Florida but these died after he was killed by Indians. Seeds were imported into Miami from the West Indies by a Dr. Fletcher in 1862 or 1863. From these, two trees grew to large size and one was still fruiting in 1910 and is believed to have been the parent of the 'No. 11' which was commonly planted for many years thereafter.

Fig. 59: Some mangoes (*Mangifera indica*) more or less commonly grown in dooryards of southern Florida in the mid–1940's.

In 1868 or 1869, seeds were planted south of Coconut Grove and the resultant trees prospered at least until 1909, producing the so-called 'Peach' or 'Turpentine' mango which became fairly common. In 1872, a seedling of 'No. 11' from Cuba was planted in Bradenton. In 1877 and 1879, W.P. Neeld made successful plantings on the west coast but these and most others north of Ft. Myers were killed in the January freeze of 1886.

In 1885, seeds of the excellent 'Bombay' mango of India were brought from Key West to Miami and resulted in two trees which flourished until 1909. Plants of grafted varieties were brought in from India by a west coast resident, Rev. D.G. Watt, in 1885 but only two survived the trip and they were soon frozen in a cold spell. Another unsuccessful importation of inarched trees from Calcutta was made in 1888. Of six grafted trees that arrived from Bombay in 1889, through the efforts of the United States Department of Agriculture, only one lived to fruit nine years later. The tree shipped is believed to have been a 'Mulgoa' (erroneously labeled 'Mulgoba', a name unknown in India except as originating in Florida). However, the fruit produced did not correspond to 'Mulgoa' descriptions. It was beautiful, crimson-blushed, just under 1 lb (454 g) with golden-yellow flesh. No Indian visitor has recognized it as matching any Indian variety. Some sug-

gest that it was the fruit of the rootstock if the scion had been frozen in the freeze of 1894-95. At any rate, it continued to be known as 'Mulgoba', and it fostered many offspring along the southeastern coast of the State and in Cuba and Puerto Rico, though it proved to be very susceptible to the disease, anthracnose, in this climate. Seeds from this tree were obtained and planted by a Captain Haden in Miami. The trees fruited some years after his death and his widow gave the name 'Haden' to the tree that bore the best fruit. This variety was regarded as the standard of excellence locally for many decades thereafter and was popular for shipping because of its tough skin.

George B. Cellon started extensive vegetative propagation (patch-budding) of the 'Haden' in 1900 and shipped the fruits to northern markets. P.J. Wester conducted many experiments in budding, grafting and inarching from 1904 to 1908 with less success. Shield-budding on a commercial scale was achieved by Mr. Orange Pound of Coconut Grove in 1909 and this was a pioneer breakthrough which gave strong impetus to mango growing, breeding, and dissemination.

Enthusiastic introduction of other varieties by the U.S. Department of Agriculture's Bureau of Plant Industry, by nurserymen, and other individuals followed, and the mango grew steadily in popularity and importance. The

Reasoner Brothers Nursery, on the west coast, imported many mango varieties and was largely responsible for the ultimate establishment of the mango in that area, together with a Mr. J.W. Barney of Palma Sola who had a large collection of varieties and had worked out a feasible technique of propagation which he called "slot grafting".

Dr. Wilson Popenoe, one of the early Plant Explorers of the U.S. Department of Agriculture, became Director of the Escuela Agricola Panamericana, Tegucigalpa, Honduras. For more than a quarter of a century, he was a leader in the introduction and propagation of outstanding mangos from India and the East Indies, had them planted at the school and at the Lancetilla Experiment Station at Tela, Honduras, and distributed around tropical America.

In time, the mango became one of the most familiar domesticated trees in dooryards or in small or large commercial plantings throughout the humid and semi-arid lowlands of the tropical world and in certain areas of the near-tropics such as the Mediterranean area (Madeira and the Canary Islands), Egypt, southern Africa, and southern Florida. Local markets throughout its range are heaped high with the fragrant fruits in season and large quantities are exported to non-producing countries.

Altogether, the U.S. Department of Agriculture made 528 introductions from India, the Philippines, the West Indies and other sources from 1899 to 1937. Selection, naming and propagation of new varieties by government agencies and individual growers has been going on ever since. The Mango Form was created in 1938 through the joint efforts of the Broward County Home Demonstration Office of the University of Florida's Cooperative Extension Service and the Fort Lauderdale Garden Club, with encouragement and direction from the University of Florida's Subtropical Experiment Station (now the Agricultural Research and Education Center) in Homestead, and Mrs. William J. Krome, a pioneer tropical fruit grower. Meetings were held annually, whenever possible, for the exhibiting and judging of promising seedlings, and exchanging and publication of descriptions and cultural information.

Meanwhile, a reverse flow of varieties was going on. Improved mangos developed in Florida have been of great value in upgrading the mango industry in tropical America and elsewhere.

With such intense interest in this crop, mango acreage advanced in Florida despite occasional setbacks from cold spells and hurricanes. But with the expanding population, increased land values and cost and shortage of agricultural labor after World War II, a number of large groves were subdivided into real estate developments given names such as "Mango Heights" and "Mango Terrace". There were estimated to be 7,000 acres (2,917 ha) in 27 Florida counties in 1954, over half in commercial groves. There were 4,000 acres (1,619 ha) in 1961. Today, mango production in Florida, on approximately 1,700 acres (688 ha), is about 8,818 tons (8,000 MT) annually in "good" years, and valued at $3 million. Fruits are shipped not only to northern markets but also to the United Kingdom, Nether-

lands, France and Saudi Arabia. In advance of the local season, quantities are imported into the USA from Haiti and the Dominican Republic, and, throughout the summer, Mexican sources supply mangos to the Pacific Coast consumer. Supplies also come in from India and Taiwan.

A mango seed from Guatemala was planted in California about 1880 and a few trees have borne fruit in the warmest locations of that state, with careful protection when extremely low temperatures occur.

Mangos have been grown in Puerto Rico since about 1750 but mostly of indifferent quality. A program of mango improvement began in 1948 with the introduction and testing of over 150 superior cultivars by the University of Puerto Rico. The south coast of the island, having a dry atmosphere, is best suited for mango culture and substantial quantities of mangos are produced there without the need to spray for anthracnose control. The fruits are plentiful on local markets and shipments are made to New York City where there are many Puerto Rican residents. A study of 16 cultivars was undertaken in 1960 to determine those best suited to more intense commercial production. Productivity evaluations started in 1965 and continued to 1972.

The earliest record of the mango in Hawaii is the introduction of several small plants from Manila in 1824. Three plants were brought from Chile in 1825. In 1899, grafted trees of a number of Indian varieties, including 'Pairi', were imported. Seedlings became widely distributed over the six major islands. In 1930, the 'Haden' was introduced from Florida and became established in commercial plantations. The local industry began to develop seriously after the importation of a series of monoembryonic cultivars from Florida. But Hawaiian mangos are prohibited from entry into mainland USA, Australia, Japan and some other countries, because of the prevalence of the mango seed weevil in the islands.

In Brazil, most mangos are produced in the state of Minas Gerais where the crop amounts to 243,018 tons (22,000 MT) annually on 24,710 acres (10,000 ha). These are mainly seedlings, as are those of the other states with major mango crops—Ceará, Paraibá, Goias, Pernambuco, and Maranhao. Sao Paulo raises about 63,382 tons (57,500 MT) per year on 9,884 acres (4,000 ha). The bulk of the crop is for domestic consumption. In 1973, Brazil exported 47.4 tons (43 MT) of mangos to Europe.

Mango growing began with the earliest settlers in North Queensland, Australia, with seeds brought casually from India, Ceylon, the East Indies and the Philippines. In 1875, 40 varieties from India were set out in a single plantation. Over the years, selections have been made for commercial production and culture has extended to subtropical Western Australia.

There is no record of the introduction of the mango into South Africa but a plantation was set out in Durban about 1860. Production today probably has reached about 16,535 tons (15,000 MT) annually, and South Africa exports fresh mangos by air to Europe.

Kenya exports mature mangos to France and Germany and both mature and immature to the United Kingdom, the latter for chutney-making. Egypt produces 110,230 tons (100,000 MT) of mangos annually and exports moderate amounts to 20 countries in the Near East and Europe. Mango culture in the Sudan occupies about 24,710 acres (10,000 ha) producing a total of 66,138 tons (60,000 MT) per year.

India, with 2,471,000 acres (1,000,000 ha) of mangos (70% of its fruit-growing area) produces 65% of the world's mango crop—9,920,700 tons (9,000,000 MT). In 1985, mango growers around Hyderabad sought government protection against terrorists who cut down mango orchards unless the owners paid ransom (50,000 rupees in one case). India far outranks all other countries as an exporter of processed mangos, shipping ⅔ of the total 22,046 tons (20,000 MT). Mango preserves go to the same countries receiving the fresh fruit and also to Hong Kong, Iraq, Canada and the United States. Following India in volume of exports are Thailand, 774,365 tons (702,500 MT), Pakistan and Bangladesh, followed by Brazil. Mexico ranks 5th with about 100,800 acres (42,000 ha) and an annual yield of approximately 640,000 tons (580,000 MT). The Philippines have risen to 6th place. Tanzania is 7th, the Dominican Republic, 8th and Colombia, 9th.

Leading exporters of fresh mangos are: the Philippines, shipping to Hong Kong, Singapore and Japan; Thailand, shipping to Singapore and Malaysia; Mexico, shipping mostly 'Haden' to the United States, 2,204 tons (2,000 MT), annually, also to Japan and Paris; India, shipping mainly 'Alphonso' and 'Bombay' to Europe, Malaya, Saudi Arabia and Kuwait; Indonesia, shipping to Hong Kong and Singapore; and South Africa shipping (60% 'Haden' and 'Kent') by air to Europe and London in mid-winter.

Chief importers are England and France, absorbing 82% of all mango shipments. Mango consumers in England are mostly residents of Indian origin, or English people who formerly lived in India.

The first International *Symposium on Mango and Mango Culture,* of the International Society for Horticultural Science, was held in New Delhi, India, in 1969 with a view to assembling a collection of germplasm from around the world and encouraging cooperative research on rootstocks and bearing behavior, hybridization, disease, storage and transport problems, and other areas of study.

Varieties

The original wild mangos were small fruits with scant, fibrous flesh, and it is believed that natural hybridization has taken place between *M. indica* and *M. sylvatica* Roxb. in Southeast Asia. Selection for higher quality has been carried on for 4,000 to 6,000 years and vegetative propagation for 400 years.

Over 500 named varieties (some say 1,000) have evolved and have been described in India. Perhaps some are duplicates by different names, but at least 350 are propagated in commercial nurseries. In 1949, K.C. Naik described 82 varieties grown in South India. L.B. and R.N. Singh presented and illustrated 150 in their monograph on the mangos of Uttar Pradesh (1956). In 1958, 24 were des-

cribed as among the important commercial types in India as a whole, though in the various climatic zones other cultivars may be prominent locally. Of the 24, the majority are classed as early or mid-season:

Early:
    'Bombay Yellow' ('Bombai') — high quality
    'Malda' ('Bombay Green')
    'Olour' (polyembryonic) — a heavy bearer.
    'Pairi' ('Paheri', 'Pirie', 'Peter', 'Nadusalai', 'grape', 'Raspuri', 'Goha bunder')
    'Safdar Pasand'
    'Suvarnarekha' ('Sundri')

Early to Mid–Season:
    'Langra'
    'Rajapuri'

Mid–Season:
    'Alampur Baneshan' — high quality but shy bearer
    'Alphonso' ('Badami', 'gundu', 'appas', 'khader') — high quality
    'Bangalora' ('Totapuri', 'collection', 'kili-mukku', abu Samada' in the Sudan) — of highest quality, best keeping, regular bearer, but most susceptible to seed weevil.
    'Banganapally' ('Baneshan', 'chaptai', 'Safeda') — of high quality but shy bearer
    'Dusehri' ('Dashehari aman', 'nirali aman', 'kamyab') — high quality
    'Gulab Khas'
    'Zardalu'
    'K.O. 11'

Mid– to Late–Season:
    'Rumani' (often bearing an off-season crop)
    'Samarbehist' ('Chowsa', 'Chausa', 'Khajri') — high quality
    'Vanraj'
    'K.O. 7/5' ('Himayuddin' X 'Neelum')

Late:
    'Fazli' ('Fazli malda') — high quality
    'Safeda Lucknow'

Often Late:
    'Mulgoa' — high quality but a shy bearer
    'Neelum' (sometimes twice a year) somewhat dwarf, of indifferent quality, and anthracnose-susceptible.

Most of the leading Indian cultivars are seedling selections. Over 50,000 crosses were made over a period of 20 years in India and 750 hybrids were raised and screened. Of these, **'Mallika',** a cross of 'Neelum' (female parent) with **'Dashehari'** (male parent) was released for cultivation in 1972. The hybrid tends toward regular bearing, the fruits are showier and are thicker of flesh than either parent, the flavor is superior and keeping quality better. The season is nearly a month later than 'Dashehari'. Another new hybrid, **'Amrapali',** of which 'Dashehari' was the female parent and 'Neelum' the male, is definitely dwarf, precocious, a regular and heavy bearer, and late in the season. The fruit is only medium in size; flesh is rich orange, fiberless, sweet and 2 to 3 times as high in carotene as either parent.

The Central Food Technological Research Institute Experiment Station in Hyderabad has evaluated 9 "table varieties" (firm-fleshed), 4 "juicy" varieties, and 5 hybrids as to suitability for processing. 'Baneshan', 'Suvarnarekha' and '5/5 Rajapuri' X 'Langra' were deemed suitable for slicing and canning. 'Baneshan', 'Navaneetam', 'Goabunder', 'Royal Special', 'Hydersaheb' and '9/4 Neelum Baneshan', for canned juice; and 'Baneshan', 'Navaneetam', 'Goabunder', 'K.O. 7' and 'Sharbatgadi' for canned nectar.

It is interesting to note that all but four of the leading Indian cultivars are yellow-skinned. The exceptions are: two yellow with a red blush on shoulders, one red-yellow with a blush of red, and one green. In Thailand, there is a popular mango called 'Tong dum' ('Black Gold') marketed when the skin is very dark-green and usually displayed with the skin at the stem end cut into points and spread outward to show the golden flesh in the manner that red radishes are fashioned into "radish roses" in American culinary art.

European consumers prefer a deep-yellow mango that develops a reddish-pink tinge. In Florida, the color of the mango is an important factor and everyone admires a handsome mango more or less generously overlaid with red. Red skin is considered a necessity in mangos shipped to northern markets, even though the quality may be inferior to that of non-showy cultivars. Also, dependable bearing and shippability are rated above internal qualities for practical reasons. And a shipping mango must be one that can be picked 2 weeks before full maturity without appreciable loss of flavor. Too, there must be several varieties to extend the season over at least 3 months.

Florida mangos are classed in 4 groups:

1—Indian varieties, mainly monoembryonic, introduced in the past and maintained mostly in collections; typically of somewhat "turpentine" character.

2—Philippine and Indo-Chinese types, largely polyembryonic, non-turpentiney, fiberless, fairly anthracnose-resistant. Scattered in dooryard plantings.

3—West Indian/South American mangos, especially 'Turpentine' and 'No. 11' and the superior 'Julie' from Trinidad, 'Madame Francis' from Haiti, 'Itamaraca' from Brazil. These are non-commercial.

4—Florida-originated selections or cultivars, of which many have risen and declined over the decades.

In general, mangos from the Philippines ('Carabao') and Thailand ('Saigon', 'Cambodiana') behave better in Florida's humidity than the Indian varieties.

The much-prized 'Haden' was being recognized in the late 1930's and early 1940's as anthracnose-prone, a light and irregular bearer, and was being replaced by more disease-resistant and prolific cultivars. The present-day leaders for commercial production and shipping are 'Tommy Atkins', 'Keitt', 'Kent', 'Van Dyke' and 'Jubilee'. The first 2 represent 50% of the commercial crop.

'Tommy Atkins' (from a seed planted early in the 1920's at Fort Lauderdale, Florida; commercially adopted in the late 1950's); oblong-oval; medium to large; skin thick, orange-yellow, largely overlaid with bright- to dark-red and heavy purplish bloom, and dotted with many large, yellow-green lenticels. Flesh medium- to dark-yellow, firm, juicy, with medium fiber, of fair to good quality; flavor poor if over-fertilized and irrigated. Seed small. Season: mid-May to early July, or late June through July, depending on spring weather; can be picked early, developing good color and usually has long shelf-life. Sometimes there is an open space in the flesh at the stem-end. Interior softening near the seed occurs in some years. Anthracnose-resistant.

'Keitt'—rounded-oval to ovate; large; skin medium-thick, yellow with light-red blush and a lavender bloom; the many lenticels small, yellow to red. Flesh orange-yellow, firm, fiberless except near the seed; of rich, sweet flavor; very good quality. Seed small, or medium to large. Season: early July through August or August and September, depending on spring weather. Tree small to medium, erect, open, rather scraggly but very productive. For market acceptance, requires post-harvest ethylene treatment to enhance color.

'Kent'—ovate, thick; large; skin greenish-yellow with dark-red blush and gray bloom; many small, yellow lenticels. Flesh fiberless, juicy, sweet; very good to excellent. Seed small. Season: July and August and often into September, but if left on too long the seed tends to sprout in the fruit—a condition called ovipary. Subject to black spot. Tree is of erect, slender habit, of moderate size, precocious; bears very well and fruit ships well, but, for the market, needs ethylene treatment to enrich color.

'Van Dyke' and 'Jubilee' are relatively new cultivars maturing from late June through July. 'Van Dyke' is of superior color and excellent quality but subject to anthracnose and may not hold its place for long.

Two cultivars that have stood the test of time and have been shipped north on a lesser scale are:

'Sensation' (originated in North Miami; tree moved to Carmichael grove near Perrine and propagated and grown commercially since 1949). Oval, oblique, and faintly beaked; medium to medium-small; skin thin, adherent; basically yellow to yellow-orange overlaid with dark plum-red, and with tiny, pale-yellow lenticels. Flesh pale-yellow, firm, with very little fiber, faintly aromatic, of mild, slightly sweet flavor; of good quality. Monoembryonic. Tree bears heavily in August.

'Palmer'—oblong-ovate, plump; large; skin medium-thick, orange-yellow with red blush and pale bloom and many large lenticels. Flesh dull-yellow, firm, with very little or no fiber; of fair to good quality. Seed long, of medium size. Season: July and August, sometimes into September. Tree is medium to large; precocious; usually bears well.

The leading cultivar for local market at present is:

'Irwin' (a seedling of 'Lippens', planted by F.D. Irwin of Miami in 1939; bore its first fruits in 1945); oblong-ovate, one shoulder oblique; of medium size; skin orange to pink with extensive dark-red blush and small, white lenticels. Seed of medium size. Flesh yellow, almost fiberless, with mild, sweet flavor; good to very good quality. Seed small. Season: mid-May to early July; or June through July. Tree somewhat dwarf; bears heavy crops of fruits in clusters. Fruit no longer shipped because if picked before full maturity ripens with a mottled appearance which is not acceptable on the market.

Fig. 60: The tiny, colorful 'Azucar' mango of Santa Marta and Baranquilla, Colombia, is sweet and freestone.

Non-colorful or not high-yielding cultivars of excellent quality recommended for Florida homeowners include:

'Carrie' (somewhat dwarf); 'Edward' ('Haden' seedling); 'Florigon'; 'Jacquelin'; 'Cambodiana'; 'Cecil'; 'Saigon'.

Among cultivars formerly commercial but largely top-worked to others favored for various reasons: 'Davis-Haden' (a 'Haden' seedling); 'Fascell'; 'Lippens' (a 'Haden' seedling); 'Smith' (a 'Haden' seedling); 'Spring-fels'; 'Dixon'; 'Sunset'; 'Zill' (a 'Haden' seedling).

Many cultivars that have lost popularity in Florida have become of importance elsewhere. 'Sandersha', for example, has proved remarkably resistant to most mango fruit diseases in South Africa.

The histories and descriptions of 46 cultivars growing in Brazil were published in 1955. These included 'Brooks', 'Cacipura', 'Cambodiana', 'Goa–Alphonso', 'Haden', 'Mulgoba', 'Pairi', 'Pico', 'Sandersha', 'Singapore', 'White Langra', all brought in from Florida. The rest are mostly local seedlings. 'Haden' was introduced from Florida in 1931 and has been widely cultivated. It is still included among the cultivars of major importance, the others being 'Extrema', 'Non–Plus–Ultra', 'Carlota'; but in 1977 the leading cultivar in Brazil was reported to be 'Bourbon', also known as 'Espada'. It is found especially in northeastern Brazil but is recommended for all other

mango areas. A collection of 53 cultivars is maintained at Piricicaba and another of 82 at Bahia.

Of Mexican mangos, 65% are Florida selections; 35% are of the type commonly grown in the Philippines. Over a period of 3 years detailed studies have been made of the commercial cultivars in Culiacan, Sinaloa, Mexico, with a view to determining the most profitable for export. Results indicated that propagation of 'Purple Irwin', 'Red Irwin', 'Sensation' and 'Zill' should be discontinued, and that 'Haden', 'Kent' and 'Keitt' will continue to be planted, the first two because of their color and quality, and the third in spite of its deficiency in color.

'Manila', a Philippine mango, early-ripening, is much grown in Veracruz. 'Manzanillo–Nunez', a chance seedling first noticed in 1972, is gaining in popularity because of its regular bearing, skin color (75% red), nearly fiberless flesh, good quality, high yield and resistance to anthracnose.

'Julie' is the main mango exported from the West Indies to Europe. The fruit is somewhat flattened on one side, of medium size; the flesh is not completely fiberless but is of good flavor. It came to Florida from Trinidad but has long been popular in Jamaica. The tree is somewhat dwarf, has 30% to 50% her-maphrodite flowers; bears well and regularly. It is adaptable to humid environments and disease-resistant and the fruit is resistant to the fruit fly. 'Julie' has been grown in Ghana since the early 1920's. From 'Julie', the well-known mango breeder,

Lawrence Zill, developed 'Carrie', but 'Julie' has not been planted in Florida for many years.

Grafted plants of the 'Bombay Green', so popular in Jamaica, were brought there from India in 1869 by the then governor, Sir John Peter Grant, but were planted in Castleton gardens where the trees flourished but failed to fruit in the humid atmosphere. Years later, a Director of Agriculture had budwood from these trees transferred to rootstocks at Hope Gardens. The results were so successful that the 'Bombay Green' became commonly planted on the island. The author brought six grafted trees from Jamaica to Miami in 1951 and, after they were released from quarantine, distributed them to the Subtropical Experiment Station in Homestead, the Newcomb Nursery, and a private grower, but all succumbed to the cold in succeeding winters. The fruit is completely fiberless and freestone so that it is frequently served cut in half and eaten with a spoon. The seed is pierced with a mango fork and served also so that the luscious flesh that adheres to it may be enjoyed as well.

One of the best-known mangos peculiar to the West Indies is 'Madame Francis' which is produced abundantly in Haiti. It is a large, flattened, kidney-shaped mango, light-green, slightly yellowish when ripe, with orange, low-fiber, richly flavored flesh. This mango has been regularly exported to Florida in late spring after fumigation against the fruit fly.

Ghana received more than a dozen cultivars back in the early 1920's. In 1973, it was found that only three of these—'Julie', 'Jaffna' and 'Rupee'—could be recognized with certainty. More than a dozen other cultivars were brought in much later from Florida and India. An effort was begun in 1967 to classify the seedlings (from 10 to 50 years of age) in the Ejura district, the Ejura Agricultural Station, and the plantation of the Faculty of Agriculture, University of Science and Technology, Kumasi, in order to eliminate confusion and have identifiable cultivars marked for future research. After checking with available published material on other cultivars for possible resemblances, descriptions and photographs of 21 newly named cultivars were published in 1973. Of these, 12 are fibrous and 9 fiberless. (See Godfrey-Sam-Aggrey and Arbutiste in the Bibliography). One of the fibrous cultivars, named 'Tee-Vee-Dee', is so well flavored and aromatic that it is locally extremely popular.

Until the mid-1960's mangos were grown only in dooryards in Surinam and the few varieties were largely polyembryonic types from Indonesia, and these have given rise to many chance seedlings. In order to discover the best for commercial planting, mango exhibits were sponsored and budwood of the best selections has been grafted onto various rootstocks at the Paramaribo Agricultural Experiment Station. The two most important local mangos are:

'Golek' (from Java; also grown in Queensland) long-oblong; skin dull-green or yellowish-green even when ripe, leathery; flesh pale yellow, thick, fiberless, sweet, rich, of excellent quality. Keeps well in cold storage for 3 weeks. Season: early (December in Queensland). Tree bears moderately to heavily.

This cultivar is considered the most promising for large-scale culture and export. In Queensland it tends to crack longitudinally as it matures.

'Roodborstje'—medium to large; skin deep-red; flesh sweet, juicy, with very little fiber. Not a good keeper. Season: early to midseason. Tree is a heavy bearer.

In Venezuela, eleven cultivars were evaluated by food technologists for processing suitability—'Blackman', 'Glenn', 'Irwin', 'Kent', 'Lippens', 'Martinica', 'Sensation', 'Smith', 'Selection 80', 'Selection 85', and 'Zill'. The most appropriate, because of physicochemical characteristics and productivity were determined to be: 'Glenn', 'Irwin', 'Kent' and 'Zill'.

In Hawaii, **'Haden'** has represented 90% of all commercial production. 'Pairi' is more prized for home use but is a shy bearer, a poor keeper, not as colorful as 'Haden', so it never attained commercial status. In a search for earlier and later varieties of commercial potential, over 125 varieties were collected and tested between 1934 and 1969. In 1956, one of the winning entries in a mango contest attracted much attention. After propagation and due observation it was named 'Gouveia' in 1969 and described as: ovate-oblong, of medium size, with medium-thick, ochre-yellow skin blushed with blood-red over ⅔ of the surface. Flesh is orange, nearly fiberless, sweet, juicy. Seed is small, slender, monoembryonic. Season: late. Tree is of medium size, a consistent but not heavy bearer. In quality tests 'Gouveia' received top scoring over 'Haden', 'Pairi', and several other cultivars. Florida mangos rated as promising for Hawaii were 'Pope', 'Kent', 'Keitt' and 'Brooks' (later than 'Haden') and 'Earlygold' and 'Zill' (earlier than 'Haden').

In Queensland, **'Kensington Pride'** is the leading commercial cultivar in the drier areas. In humid regions it is anthracnose-prone and requires spraying. It is thought to have been introduced by traders in Bowen who were shipping horses for military use in India. It may be called 'Kensington', 'Bowen', or, because of its color, 'Apple' or 'Strawberry'. The fruit is distinctly beaked when immature, with a groove extending from the stem to the beak. It is medium-large; the skin is bright orange-yellow with red-pink blush overlying areas exposed to the sun. Flesh is orange, thick, nearly fiberless, juicy, of rich flavor. This cultivar is classified as mid-season. The fruit matures from early to mid-November at latitude 13°S; 6 weeks later at Bowen (20°S) and 1 week later for each degree of latitude from Bowen to Brisbane. But at 17°S and an altitude of 1,148 ft (350 m) peak maturity is in mid- to late-January. Polyembryonic. The fruit ships well but the tree is not a dependable nor heavy bearer. It has an oval crown and unusually sweet-scented leaves.

In 1981, after evaluating 43 accessions seeking to lengthen the mango season in Queensland, 9 that mature between 2 weeks earlier and 4 weeks later than 'Kensington Pride' were chosen for commercial testing. Only one, 'Banana-1', was a Queensland selection. The other 8 were introductions from Florida—'Smith', 'Palmer', 'Haden', 'Zill', 'Carrie', 'Irwin', 'Kent', 'Keitt'. 'Kent' and 'Haden' have proved to be highly susceptible to black spot in Queensland; 'Keitt', 'Smith', and 'Zill' less so; and 'Palmer' and 'Kensington Pride' resistant.

In the Philippines, the **'Carabao'** constitutes 66% of the crop and **'Pico'** 26%. These cultivars, apparently of Southeast Asian origin have remained the most commonly grown and exported for many years.

In Israel, 'Haden' has been popular for a long time though it is sensitive to low temperatures in spring. An Egyptian introduction, **'Mabroka'** is later in season and escapes the early frosts. **'Maya'**, a local seedling of 'Haden' has done well. Perhaps the most promising today is **'Nimrod',** a seedling of 'Maya', open pollinated, perhaps by 'Haden', planted in 1943, observed for 20 years and budded progeny for another 9 years; named and released in 1970. The fruit is round-ovate, large; skin is fairly thin, olive-green to yellow-green, blushed with red; attractive. Flesh is deep-yellow, nearly fiberless, of fair flavor. Seed is large, monoembryonic. Matures in mid-season (all August to mid-September in Israel). Tree is large, upright, very cold-resistant. Average yield is 480 lbs (218 kg) per tree over 10 years.

It is impressive to see how the early favorite, 'Haden', has influenced mango culture in many parts of the world. Today, the Subtropical Horticulture Research Unit of the U.S. Department of Agriculture and the Agricultural Research and Education Center of the University of Florida, together maintain 125 mango cultivars as a resource for mango growers and breeders in many countries.

Blooming and Pollination

Mango trees less than 10 years old may flower and fruit regularly every year. Thereafter, most mangos tend toward alternate, or biennial, bearing. A great deal of research has been done on this problem which may involve the entire tree or only a portion of the branches. Branches that fruit one year may rest the next, while branches on the other side of the tree will bear.

Blooming is strongly affected by weather, dryness stimulating flowering and rainy weather discouraging it. In most of India, flowering occurs in December and January; in northern India, in January and February or as late as March. There are some varieties called "Baramasi" that flower and fruit irregularly throughout the year. The cultivar **'Sam Ru Du'** of Thailand bears 3 crops a year—in January, June and October. In the drier islands of the Lesser Antilles, there are mango trees that flower and fruit more or less continuously all year around but never heavily at any time. Some of these are cultivars introduced from Florida where they flower and fruit only once a year. In southern Florida, mango trees begin to bloom in late November and continue until February or March, inasmuch as there are early, medium, and late varieties. During exceptionally warm winters, mango trees have been known to bloom 3 times in succession, each time setting and maturing fruit.

In the Philippines, various methods are employed to promote flowering: smudging (smoking), exposing the roots, pruning, girdling, withholding nitrogen and irrigation, and even applying salt. In the West Indies, there is a common folk practice of slashing the trunk with a machete to make the tree bloom and bear in "off" years. Deblossoming (removing half the flower clusters) in an "on" year will induce at least a small crop in the next "off" year. Almost any treatment or condition that retards vegetative growth will have this effect. Spraying with growth-retardant chemicals has been tried, with inconsistent results. Potassium nitrate has been effective in the Philippines.

In India, the cultivar 'Dasheri', which is self incompatible, tends to begin blooming very early (December and January) when no other cultivars are in flower. And the early panicles show a low percentage of hermaphrodite flowers and a high incidence of floral malformation. Furthermore, early blooms are often damaged by frost. It has been found that a single mechanical deblossoming in the first bud-burst stage, induces subsequent development of panicles with less malformation, more hermaphrodite flowers, and, as a result, a much higher yield of fruits.

There is one cultivar, 'Neelum', in South India that bears heavily every year, apparently because of its high rate (16%) of hermaphrodite flowers. (The average for 'Alphonso' is 10%.) However, Indian horticulturists report great tree-to-tree variation in seedlings of this cultivar; in some surveys as much as 84% of the trees were rated as poor bearers. Over 92% of 'Bangalora' seedlings have been found bearing light crops.

Mango flowers are visited by fruit bats, flies, wasps, wild bees, butterflies, moths, beetles, ants and various bugs seeking the nectar and some transfer the pollen but a certain amount of self-pollination also occurs. Honeybees do not especially favor mango flowers and it has been found that effective pollination by honeybees would require 3 to 6 colonies per acre (6-12 per ha). Many of the unpollinated flowers are shed or fail to set fruit, or the fruit is set but is shed when very young. Heavy rains wash off pollen and thus prevent fruit setting. Some cultivars tend to produce a high percentage of small fruits without a fully developed seed because of unfavorable weather during the fruit-setting period.

Shy-bearing cultivars of otherwise desirable characteristics are hybridized with heavy bearers in order to obtain better crops. For example: shy-bearing 'Himayuddin' X heavy-bearing 'Neelum'. Breeders usually hand-pollinate all the flowers that are open in a cluster, remove the rest, and cover the inflorescence with a plastic bag. But researchers in India have found that there is very little chance of contamination and that omitting the covering gives as much as 3.85% fruit set in place of 0.23% to 1.57% when bagged. Thus large populations of hybrids may be raised for study. One of the latest techniques involves grafting the male and female parents onto a chosen tree, then covering the panicles with a polyethylene bag, and introducing house flies as pollinators.

Indian scientists have found that pollen for crossbreeding can be stored at 32°F (0°C) for 10 hours. If not separated from the flowers, it remains viable for 50 hours in a humid atmosphere at 65° to 75°F (18.33°-23.09°C). The stigma is receptive 18 hours before full flower opening and, some say, for 72 hours after.

Climate

The mango is naturally adapted to tropical lowlands between 25°N and 25°S of the Equator and up to elevations of 3,000 ft (915 m). It is grown as a dooryard tree at slightly cooler altitudes but is apt to suffer cold damage. The *amount* of rainfall is not as critical as *when it occurs.* The best climate for mango has rainfall of 30 to 100 in (75-250 cm) in the four summer months (June to September) followed by 8 months of dry season. This crop is well suited to irrigated regions bordering the desert frontier in Egypt. Nevertheless, the tree flourishes in southern Florida's approximately 5 months of intermittent, scattered rains (October to February), 3 months of drought (usually March to May) and 4 months of frequently heavy rains (June to September).

Rain, heavy dews or fog during the blooming season (November to March in Florida) are deleterious, stimulating tree growth but interfering with flower production and encouraging fungus diseases of the inflorescence and fruit. In Queensland, dry areas with rainfall of 40 in (100 cm), 75% of which occurs from January to March, are favored for mango growing because vegetative growth is inhibited and the fruits are well exposed to the sun from August to December, become well colored, and are relatively free of disease. Strong winds during the fruiting season cause many fruits to fall prematurely.

Soil

The mango tree is not too particular as to soil type, providing it has good drainage. Rich, deep loam certainly contributes to maximum growth, but if the soil is too rich and moist and too well fertilized, the tree will respond vegetatively but will be deficient in flowering and fruiting. The mango performs very well in sand, gravel, and even oolitic limestone (as in southern Florida and the Bahamas).

A polyembryonic seedling, 'No. 13-1', introduced into Israel from Egypt in 1931, has been tested since the early 1960's in various regions of the country for tolerance of calcareous soils and saline conditions. It has done so well in sand with a medium (15%) lime content and highly saline irrigation water (over 600 ppm) that it has been adopted as the standard rootstock in commercial plantings in salty, limestone districts of Israel. Where the lime content is above 30%, iron chelates are added.

Propagation

Mango trees grow readily from seed. Germination rate and vigor of seedlings are highest when seeds are taken from fruits that are fully ripe, not still firm. Also, the seed should be fresh, not dried. If the seed cannot be planted within a few days after its removal from the fruit, it can be covered with moist earth, sand, or sawdust in a container until it can be planted, or kept in charcoal dust in a dessicator with 50% relative humidity. Seeds stored in the latter manner have shown 80% viability even after 70 days. High rates of germination are obtained if seeds are stored in polyethylene bags but the seedling behavior may be poor. Inclusion of sphagnum moss in the sack has no benefit and shows inferior rates of germination over 2- to 4-week periods, and none at all at 6 weeks.

The flesh should be completely removed. Then the husk is opened by carefully paring around the convex edge with a sharp knife and taking care not to cut the kernel, which will readily slide out. Husk removal speeds germination and avoids cramping of roots, and also permits discovery and removal of the larva of the seed weevil in areas where this pest is prevalent. Finally, the husked kernels are treated with fungicide and planted without delay. The beds must have solid bottoms to prevent excessive taproot growth, otherwise the taproot will become 18 to 24 in (45-60 cm) long while the top will be only one-third to a half as high, and the seedling will be difficult to transplant with any assurance of survival. The seed is placed on its ventral (concave) edge with 1/4 protruding above the sand. Sprouting occurs in 8 to 14 days in a warm, tropical climate; 3 weeks in cooler climates. Seedlings generally take 6 years to fruit and 15 years to attain optimum yield for evaluation.

However, the fruits of seedlings may not resemble those of the parent tree. Most Indian mangos are monoembryonic; that is, the embryo usually produces a single sprout, a natural hybrid from accidental crossing, and the resulting fruit may be inferior, superior, or equal to that of the tree from which the seed came. Mangos of Southeast Asia are mostly polyembryonic. In these, generally, one of the embryos in the seed is a hybrid; the others (up to 4) are vegetative growths which faithfully reproduce the characteristics of the parent. The distinction is not absolute, and occasionally a seed supposedly of one class may behave like the other.

Seeds of polyembryonic mangos are most convenient for local and international distribution of desirable varieties. However, in order to reproduce and share the superior monoembryonic selections, vegetative propagation is necessary. Inarching and approach-grafting are traditional in India. Tongue-, saddle-, and root-grafting (stooling) are also common Indian practices. Shield- and patch-grafting have given up to 70% success but the Forkert system of budding has been found even more practical. After many systems were tried, veneer grafting was adopted in Florida in the mid-1950's. Choice of rootstock is important. Use of seedlings of unknown parentage has resulted in great variability in a single cultivar. Some have believed that polyembryonic rootstocks are better than monoembryonic, but this is not necessarily so. In trials at Tamil Nadu Agricultural University, 10-year-old trees of 'Neelum' grafted on polyembryonic 'Bapakkai' showed vigor and spread of tree and productivity far superior to those grafted on 'Olour' which is also polyembryonic. Those grafted on monoembryonic rootstock also showed better growth and yield than those on 'Olour'. In 1981, experimenters at Lucknow, India, reported the economic advantage of "stone-grafting", which requires less space in the nursery and results in greater uniformity. Scions from the spring flush of selected cultivars are defoliated and, after a 10-day delay, are cleft-grafted on 5-day-old seedlings which must thereafter be kept in the shade and protected from drastic changes in the weather.

Old trees of inferior types are top-worked to better cultivars by either side-grafting or crown-grafting the beheaded trunk or beheaded main branches. Such trees need protection from sunburn until the graft affords shade. In South Africa, the trunks are whitewashed and bunches of dry grass are tied onto cut branch ends. The trees will bear in 2 to 3 years. Attempts to grow 3 or 4 varieties on one rootstock may appear to succeed for a while but the strongest always outgrows the others.

Cuttings, even when treated with growth regulators, are only 40% successful. Best results are obtained with cuttings of mature trees, ringed 40 days before detachment, treated, and rooted under mist. But neither cuttings nor air layers develop good root systems and are not practical for establishing plantations. Clonal propagation through tissue culture is in the experimental stage.

In spite of vegetative propagation, mutations arise in the form of bud sports. The fruit may differ radically from the others on a grafted tree—perhaps larger and superior—and the foliage on the branch may be quite unlike that on other branches.

Dwarfing

Reduction in the size of mango trees would be a most desirable goal for the commercial and private planter. It would greatly assist harvesting and also would make it possible for the homeowner to maintain trees of different fruiting seasons in limited space.

In India, double-grafting has been found to dwarf mango trees and induce early fruiting. Naturally dwarf hybrids such as 'Julie' have been developed. The polyembryonic Indian cultivars, 'Olour' and 'Vellai Colamban', when used as rootstocks, have a dwarfing effect; so has the polyembryonic 'Sabre' in experiments in Israel and South Africa.

In Peru, the polyembryonic 'Manzo de Ica', is used as rootstock; in Colombia, 'Hilaza' and 'Puerco'. 'Kaew' is utilized in Thailand.

Culture

About 6 weeks before transplanting either a seedling or a grafted tree, the taproot should be cut back to about 12 in (30 cm). This encourages feeder-root development in the field. For a week before setting out, the plants should be exposed to full morning sun.

Inasmuch as mango trees vary in lateral dimensions, spacing depends on the habit of the cultivar and the type of soil, and may vary from 34 to 60 ft (10.5-18 m) between trees. Closer planting will ultimately reduce the crop. A spacing of 34 x 34 ft (10.5x10.5 m) allows 35 trees per acre (86 per ha); 50 x 50 ft (15.2x15.2 m) allows only 18 trees per acre (44.5 per ha). In Florida's limestone, one commercial grower maintains 100 trees per acre (247 per ha), controlling size by hedging and topping.

The young trees should be placed in prepared and enriched holes at least 2 ft (60 cm) deep and wide, and ¾ of the top should be cut off. In commercial groves in southern Florida, the trees are set at the intersection of cross trenches mechanically cut through the limestone.

Mangos require high nitrogen fertilization in the early years but after they begin to bear, the fertilizer should be higher in phosphate and potash. A 5-8-10 fertilizer mix is recommended and applied 2 or 3, or possibly even 4, times a year at the rate of 1 lb (454 g) per year of age at each dressing. Fertilizer formulas will vary with the type of soil. In sandy acid soils, excess nitrogen contributes to "soft nose" breakdown of the fruits. This can be counteracted by adding calcium. On organic soils (muck and peat), nitrogen may be omitted entirely. In India, fertilizer is applied at an increasing rate until the tree is rather old, and then it is discontinued. Ground fertilizers are supplemented by foliar nutrients including zinc, manganese and copper. Iron deficiency is corrected by small applications of chelated iron.

Indian growers generally irrigate the trees only the first 3 or 4 years while the taproot is developing and before it has reached the water table. However, in commercial plantations, irrigation of bearing trees is withheld only for the 2 or 3 months prior to flowering. When the blooms appear, the tree is given a heavy watering and this is repeated monthly until the rains begin. In Florida groves, irrigation is by means of overhead sprinklers which also provide frost protection when needed.

Usually no pruning is done until the 4th year, and then only to improve the form and this is done right after the fruiting season. If topping is practiced, the trees are cut at 14 ft (4.25 m) to facilitate both spraying and harvesting. Grafted mangos may set fruit within a year or two from planting. The trees are then too weak to bear a full crop and the fruits should be thinned or completely removed.

Harvesting

Mangos normally reach maturity in 4 to 5 months from flowering. Fruits of "smudged" trees ripen several months before those of untreated trees. Experts in the Philippines have demonstrated that 'Carabao' mangos sprayed with ethephon (200 ppm) 54 days after full bloom can be harvested 2 weeks later at recommended minimum maturity. The fruits will be larger and heavier even though harvested 2 weeks before untreated fruits. If sprayed at 68 days after full bloom and harvested 2 weeks after spraying, there will be an improvement in quality in regard to soluble solids and titratable acidity.

When the mango is full-grown and ready for picking, the stem will snap easily with a slight pull. If a strong pull is necessary, the fruit is still somewhat immature and should not be harvested. In the more or less red types of mangos, an additional indication of maturity is the development of a purplish-red blush at the base of the fruit. A long-poled picking bag which holds no more than 4 fruits is commonly used by pickers. Falling causes bruising and later spoiling. When low fruits are harvested with clippers, it is desirable to leave a 4-inch (10 cm) stem to avoid the spurt of milky/resinous sap that exudes if the stem is initially cut close. Before packing, the stem is cut off ¼ in (6 mm) from the base of the fruit. In Queensland, after final clipping of the stem, the fruits are placed stem-end-down to drain.

In a sophisticated Florida operation, harvested fruits are put into tubs of water on trucks in order to wash off the sap that exudes from the stem end. At the packing house, the fruits are transferred from the tubs to bins, graded and sized and packed in cartons ("lugs") of 8 to 20 each depending on size. The cartons are made mechanically at the packing house and hold 14 lbs (6.35 kg) of fruit. The filled cartons are stacked on pallets and forklifted into refrigerated trucks with temperature set at no less than 55°F (12.78°C) for transport to distribution centers in major cities throughout the USA and Canada.

Yield

The yield varies with the cultivar and the age of the tree. At 10 to 20 years, a good annual crop may be 200 to 300 fruits per tree. At twice that age and over, the crop will be doubled. In Java, old trees have been known to bear 1,000 to 1,500 fruits in a season. Some cultivars in India bear 800 to 3,000 fruits in "on" years and, with good cultural attention, yields of 5,000 fruits have been reported. There is a famous mango, 'Pane Ka Aam' of Maharashtra and Khamgaon, India, with "paper-thin" skin and fiberless flesh. One of the oldest of these trees, well over 100 years of age, bears heavily 5 years out of 10 with 2 years of low yield. Average annual yield is 6,500 fruits; the highest record is 29,000.

Reported annual yields for 6 cultivars in Puerto Rico are:

'Lippens'	67,079 lbs per acre
'Keitt'	45,608 lbs per acre
'Earlygold'	42,310 lbs per acre
'Parvin'	38,369 lbs per acre
'Haden'	32,732 lbs per acre
'Palmer'	28,868 lbs per acre

The number of lbs per acre is roughly the equivalent of kg per hectare.

Average mango yield in Florida is said to be about 30,000 lbs/acre. One leading commercial grower has reported his annual crop as 22,000 to 27,500 lbs/acre. One grower who has hedged and topped trees close-planted at the rate of 100 per acre (41/ha) averages 14,000 to 19.000 lbs/acre.

Ripening

In India, mangos are picked quite green to avoid bird damage and the dealers layer them with rice straw in ventilated storage rooms over a period of one week. Quality is improved by controlled temperatures between 60° and 70°F (15°-21°C). In ripening trials in Puerto Rico, the 'Edward' mango was harvested while deep-green, dipped in hot water at 124°F (51°C) to control anthracnose, sorted as to size, then stored for 15 days at 70°F (21°C) with relative humidity of 85% to 90%. Those picked when more than 3 in (7.5 cm) in diameter ripened satisfactorily and were of excellent quality.

Ethylene treatment causes green mangos to develop full color in 7 to 10 days depending on the degree of maturity, whereas untreated fruits require 10 to 15 days. One of the advantages is that there can be fewer pickings and the fruit color after treatment is more uniform. Therefore, ethylene treatment is a common practice in Israel for ripening fruits for the local market. Some growers in Florida depend on ethylene treatment. Generally, 24 hours of exposure is sufficient if the fruits are picked at the proper stage. It has been determined that mangos have been picked prematurely if they require more than 48 hours of ethylene treatment and are not fit for market.

Keeping Quality and Storage

Washing the fruits immediately after harvest is essential, as the sap which leaks from the stem burns the skin of the fruit making black lesions which lead to rotting.

Some cultivars, especially 'Bangalora', 'Alphonso', and 'Neelum' in India, have much better keeping quality than others. In Bombay, 'Alphonso' has kept well for 4 weeks at 52°F (11.11°C); 6 to 7 weeks at 45°F (7.22°C). Storage at lower temperatures is detrimental inasmuch as mangos are very susceptible to chilling injury. Any temperature below 55.4°F (13°C) is damaging to 'Kent'. In Florida, this is regarded as the optimum for 2 to 3 weeks storage. The best ripening temperatures are 70° to 75°F (21.11°-23.89°C).

Experiments in Florida have demonstrated that 'Irwin', 'Tommy Atkins' and 'Kent' mangos, held for 3 weeks at storage temperature of 55.4°F (13°C), 98% to 100% relative humidity and atmospheric pressure of 76 or 152 mmHg, ripened thereafter with less decay at 69.8°F (21°C) under normal atmospheric pressure, as compared with fruits stored at the same temperature with normal atmospheric pressure. Those stored at 152 mmHg took 3 to 5 days longer to ripen than those stored at 76 mmHg. Decay rates were 20% for 'Tommy Atkins' and 40% for 'Irwin'. Spoilage from anthracnose has been reduced by immersion for 15 min in water at 125°F (51.67°C) or for 5 min at 132°F (55.56°C). Dipping in 500 ppm maleic hydrazide for 1 min and storing at 89.6°F (32°C) also retards decay but not loss of moisture. In South Africa, mangos are submerged immediately after picking in a suspension of benomyl for 5 min at 131°F (55°C) to control soft brown rot.

In Australia, mature-green 'Kensington Pride' mangos have been dipped in a 4% solution of calcium chloride under reduced pressure (250 mm Hg) and then stored in containers at 77°F (25°C) in ethylene-free atmosphere. Ripening was retarded by a week; that is, the treated fruits ripened in 20 to 22 days whereas controls ripened in 12 to 14 days. Eating quality was equal except that the calcium-treated fruits were found slightly higher in ascorbic acid.

Wrapping fruits individually in heat-shrinkable plastic film has not retarded decay in storage. The only benefit has been 3% less weight loss. Coating with paraffin wax or fungicidal wax and storing at 68° to 89.6°F (20°-32°C) delays ripening 1 to 2 weeks and prevents shriveling but interferes with full development of color.

Gamma irradiation (30 Krad) causes ripening delay of 7 days in mangos stored at room temperature. The irradiated fruits ripen normally and show no adverse effect on quality. Irradiation has not yet been approved for this purpose.

In India, large quantities of mangos are transported to distant markets by rail. To avoid excessive heat buildup and consequent spoilage, the fruits, padded with paper shavings, are packed in ventilated wooden crates and loaded into ventilated wooden boxcars. Relative humidity varies from 24% to 85% and temperature from 88° to 115°F (31.6°-46.6°C). These improved conditions have proved superior to the conventional packing of the fruits in *Phoenix*-palm-midrib or bamboo, or the newer pigeonpea-stem, baskets padded with rice straw and mango leaves and transported in steel boxcars, which has resulted in 20% to 30% losses from shriveling, unshapeliness and spoilage.

Green seedling mangos, harvested in India for commercial preparation of chutneys and pickles as well as for table use, are stored for as long as 40 days at 42° to 45°F (5.56°-7.22°C) with relative humidity of 85% to 99%. Some of these may be diverted for table use after a 2-week ripening period at 62° to 65°F (16.67°-18.13°C).

Pests and Diseases

The fruit flies, *Dacus ferrugineus* and *D. zonatus*, attack the mango in India; *D. tryoni* (now *Strumeta tryoni*) in Queensland, and *D. dorsalis* in the Philippines; *Pardalaspis cosyra* in Kenya; and the fruit fly is the greatest enemy of the mango in Central America. Because of the presence of the Caribbean fruit fly, *Anastrepha suspensa*, in Florida, all Florida mangos for interstate shipment or for export must be fumigated or immersed in hot water at 115°F (46.11°C) for 65 minutes.

In India, South Africa and Hawaii, mango seed weevils, *Sternochetus* (*Cryptorhynchus*) *mangiferae* and *S. gravis*, are major pests, undetectable until the larvae tunnel their way out. The leading predators of the tree in India are jassid hoppers (*Idiocerus* spp.) variously attacking trunk and branches or foliage and flowers, and causing shedding of young fruits. The honeydew they excrete on leaves and flowers gives rise to sooty mold.

The mango-leaf webber, or "tent caterpillar", *Orthaga euadrusalis*, has become a major problem in North India, especially in old, crowded orchards where there is excessive shade. Around Lucknow, 'Dashehari' is heavily infested by this pest; 'Samarbehist' ('Chausa') less. In South Africa, 11 species of scales have been recorded on the fruits. *Coccus mangiferae* and *C. acuminatus* are the most common scale insects giving rise to the sooty mold that grows on the honeydew excreted by the pests. In some areas, there are occasional outbreaks of the scales, *Pulvinaria psidii*, *P. polygonata*, *Aulacaspis cinnamoni*, *A. tubercularis*, *Aspidiotus destructor* and *Leucaspis indica*. In Florida, pyriform scale, *Protopulvinaria pyriformis*, and Florida wax scale, *Ceroplastes floridensis*, are common, and the lesser snow scale, *Pinnaspis strachani*, infests the trunks of small trees and lower branches of large trees. Heavy attacks may result in cracking of the bark and oozing of sap.

The citrus thrips, *Scirtothrips aurantii*, blemishes the fruit in some mango-growing areas. The red-banded thrips, *Selenothrips rubrocinctus*, at times heavily infests mango foliage in Florida, killing young leaves and causing shedding of mature leaves. Mealybugs, *Phenacoccus citri* and *P. mangiferae*, and *Drosicha stebbingi* and *D. mangiferae* may infest young leaves, shoots and fruits. The mango stem borer, *Batocera rufomaculata* invades the trunk. Leaves and shoots are preyed on by the caterpillars of *Parasa lepida*, *Chlumetia transversa* and *Orthaga exvinacea*. Mites feed on mango leaves, flowers and young fruits. In Florida, the most common is the avocado red mite, *Paratetranychus yothersii*.

Mistletoe (*Loranthus* and *Viscum* spp.) parasitizes and kills mango branches in India and tropical America. Dr. B. Reddy, Regional Plant Production and Protection Officer, FAO, Bangkok, compiled an extensive roster of insects, mites, nematodes, other pests, fungi, bacteria and phanerogamic parasites in Southeast Asia and the Pacific Region (1975).

One of the most serious diseases of the mango is powdery mildew (*Oidium mangiferae*), which is common in most growing areas of India, occurs mostly in March and April in Florida. The fungus affects the flowers and causes young fruits to dehydrate and fall, and 20% of the crop may be lost. It is controllable by regular spraying. In humid climates, anthracnose caused by *Colletotrichum gloeosporioides* (*Glomerella cingulata*) affects flowers, leaves, twigs, fruits, both young and mature. The latter show black spots externally and the corresponding flesh area is affected. Control measures must be taken in advance of flowering and regularly during dry spells. In Florida, mango growers apply up to 20 sprayings up to the cut-off point before harvesting. The black spots are similar to those produced by *Alternaria* sp. often associated with anthracnose in cold storage in India. Inside the fruits attacked by *Alternaria* there are corresponding areas of hard, corky, spongy lesions. Inasmuch as the fungus enters the stem-end of the fruit, it is combatted by applying Fungicopper paste in linseed oil to the cut stem and also by sterilizing the storage compartment with Formalin 1:20. A pre-harvest dry stem-end rot was first noticed on 'Tommy Atkins' in Mexico in 1973, and it has spread to all Mexican plantings of this cultivar causing losses of 10-80% especially in wet weather. *Fusarium*, *Alternaria* and *Cladosporium* spp. were prominent among associated fungi.

Malformation of inflorescence and vegetative buds is attributed to the combined action of *Fusarium moniliforme* and any of the mites, *Aceria mangifera*, *Eriophyes* sp., *Tyrophagus castellanii*, or *Typhlodromus asiaticus*. This grave problem occurs in Pakistan, India, South Africa and Egypt, El Salvador, Nicaragua, Mexico, Brazil and Venezuela, but not as yet in the Philippines. It is on the increase in India. Removing and burning the inflorescence has been the only remedy, but it has been found that malformation can be reduced by a single spray of NAA (200 mg in 50 ml alcohol with water added to make 1 liter) in October, and deblooming in early January.

There are 14 types of mango galls in India, 12 occurring on the leaves. The most serious is the axillary bud gall caused by *Apsylla cistellata* of the family Psyllidae.

In Florida, leaf spot is caused by *Pestalotia mangiferae, Phyllosticta mortoni,* and *Septoria* sp.; algal leaf spot, or green scurf by *Cephaleuros virescens*. In 1983, a new disease, crusty leaf spot, caused by the fungus, *Zimmermaniella trispora,* was reported as common on neglected mango trees in Malaya. Twig dieback and dieback are from infection by *Phomopsis* sp., *Physalospora abdita,* and *P. rhodina*. Wilt is caused by *Verticillium alboatrum;* brown felt by *Septobasidium pilosum* and *S. pseudopedicellatum;* wood rot, by *Polyporus sanguineus;* and scab by *Elsinoe mangiferae (Sphaceloma mangiferae)*. *Cercospora mangiferae* attacks the fruits in the Congo.

A number of organisms in India cause white sap, heart rot, gray blight, leaf blight, white pocket rot, white spongy rot, sap rot, black bark and red rust. In South Africa, *Aspergillus* attacks young shoots and fruit rot is caused by *A. niger. Gloeosporium mangiferae* causes black spotting of fruits. *Erwinia mangiferae* and *Pseudomonas mangiferaeindicae* are sources of bacterial black spot in South Africa and Queensland. *Bacterium carotovorus* is a source of soft rot. Stem-end rot is a major problem in India and Puerto Rico from infection by *Physalospora rhodina (Diplodia natalensis)*. Soft brown rot develops during prolonged cold storage in South Africa.

Leaf tip burn may be a sign of excess chlorides. Manganese deficiency is indicated by paleness and limpness of foliage followed by yellowing, with distinct green veins and midrib, fine brown spots and browning of leaf tips. Inadequate zinc is evident in less noticeable paleness of foliage, distortion of new shoots, small leaves, necrosis, and stunting of the tree and its roots. In boron deficiency, there is reduced size and distortion of new leaves and browning of the midrib. Copper deficiency is seen in paleness of foliage and severe tip-burn with gray-brown patches on old leaves; abnormally large leaves; also dieback of terminal shoots; sometimes gummosis of twigs and branches. Magnesium is needed when young trees are stunted and pale, new leaves have yellow-white areas between the main veins and prominent yellow specks on both sides of the midrib. There may also be browning of the leaf tips and margins. Lack of iron produces chlorosis in young trees.

Food Uses

Mangos should always be washed to remove any sap residue, before handling. Some seedling mangos are so fibrous that they cannot be sliced; instead, they are massaged, the stem-end is cut off, and the juice squeezed from the fruit into the mouth. Non-fibrous mangos may be cut in half to the stone, the two halves twisted in opposite directions to free the stone which is then removed, and the halves served for eating as appetizers or dessert. Or the two "cheeks" may be cut off, following the contour of the stone, for similar use; then the remaining side "fingers" of flesh are cut off for use in fruit cups, etc.

Most people enjoy eating the residual flesh from the seed

Fig. 61: Low-fiber mangoes are easily prepared for the table by first cutting off the "cheeks" which can then be served for eating by spooning the flesh from the "shell".

and this is done most neatly by piercing the stem-end of the seed with the long central tine of a mango fork, commonly sold in Mexico, and holding the seed upright like a lollypop. Small mangos can be peeled and mounted on the fork and eaten in the same manner. If the fruit is slightly fibrous especially near the stone, it is best to peel and slice the flesh and serve it as dessert, in fruit salad, on dry cereal, or in gelatin or custards, or on ice cream. The ripe flesh may be spiced and preserved in jars. Surplus ripe mangos are peeled, sliced and canned in sirup, or made into jam, marmalade, jelly or nectar. The extracted pulpy juice of fibrous types is used for making mango halva and mango leather. Sometimes corn flour and tamarind seed jellose are mixed in. Mango juice may be spray-dried and powdered and used in infant and invalid foods, or reconstituted and drunk as a beverage. The dried juice, blended with wheat flour has been made into "cereal" flakes. A dehydrated mango custard powder has also been developed in India, especially for use in baby foods.

Ripe mangos may be frozen whole or peeled, sliced and packed in sugar (1 part sugar to 10 parts mango by weight) and quick-frozen in moisture-proof containers. The diced flesh of ripe mangos, bathed in sweetened or unsweetened lime juice, to prevent discoloration, can be quick-frozen, as can sweetened ripe or green mango puree. Immature mangos are often blown down by spring winds. Half-ripe or green mangos are peeled and sliced as filling for pie, used for jelly, or made into sauce which, with added milk and egg whites, can be converted into mango sherbet. Green mangos are peeled, sliced, parboiled, then combined with sugar, salt, various spices and cooked, sometimes with raisins or other fruits, to make chutney; or they may be salted, sun-dried and kept for use in chutney and pickles. Thin slices, seasoned with turmeric, are dried, and sometimes powdered, and used to impart an acid flavor to chutneys, vegetables and soup. Green or ripe mangos may be used to make relish.

In Thailand, green-skinned mangos of a class called "keo", with sweet, nearly fiberless flesh and very commonly grown and inexpensive on the market, are soaked whole for 15 days in salted water before peeling, slicing and serving with sugar.

Processing of mangos for export is of great importance in Hawaii in view of the restrictions on exporting the fresh fruits. Hawaiian technologists have developed methods for steam- and lye-peeling, also devices for removing peel from unpeeled fruits in the preparation of nectar. Choice of suitable cultivars is an essential factor in processing mangos for different purposes.

The Food Research Institute of the Canada Department of Agriculture has developed methods of preserving ripe or green mango slices by osmotic dehydration.

The fresh kernel of the mango seed (stone) constitutes 13% of the weight of the fruit, 55% to 65% of the weight of the stone. The kernel is a major by-product of the mango-processing industry. In times of food scarcity in India, the kernels are roasted or boiled and eaten. After soaking to dispel the astringency (tannins), the kernels are dried and ground to flour which is mixed with wheat or rice flour to make bread and it is also used in puddings.

The fat extracted from the kernel is white, solid like cocoa butter and tallow, edible, and has been proposed as a substitute for cocoa butter in chocolate.

The peel constitutes 20% to 25% of the total weight of the fruit. Researchers in India have shown that the peel can be utilized as a source of pectin. Average yield on a dry-weight basis is 13%.

Immature mango leaves are cooked and eaten in Indonesia and the Philippines.

Food Value Per 100 g of Ripe Mango Flesh*	
Fruit	
Calories	62.1-63.7
Moisture	78.9-82.8 g
Protein	0.36-0.40 g
Fat	0.30-0.53 g
Carbohydrates	16.20-17.18 g
Fiber	0.85-1.06 g
Ash	0.34-0.52 g
Calcium	6.1-12.8 mg
Phosphorus	5.5-17.9 mg
Iron	0.20-0.63 mg
Vitamin A (carotene)	0.135-1.872 mg
Thiamine	0.020-0.073 mg
Riboflavin	0.025-0.068 mg
Niacin	0.025-0.707 mg
Ascorbic Acid	7.8-172.0 mg
Tryptophan	3-6 mg
Methionine	4 mg
Lysine	32-37 mg

*Minimum and maximum levels of food constituents derived from various analyses made in Cuba, Central America, Africa and India.

Puerto Rican analyses of 30 cultivars showed β-carotene as ranging from a low of 4,171 I.U./100 g in 'Stringless Peach' to a high of 7,900 I.U. in 'Carrie'. Ascorbic acid ranged from 3.43 mg/100 g in 'Keitt' to 62.96 in 'Julie'.

Seed Kernel**	
Moisture	10.55-11.35%
Protein	4.76-8.5%
Fat	6-15%
Starch	40-72%
Sugar	1.07%
Fiber	1.17-2.6%
Ash	1.72-3.66%
Silica	0.41%
Iron	0.03%
Calcium	0.11-0.23%
Magnesium	0.34%
Phosphorus	0.21-0.66%
Sodium	0.28%
Potassium	1.31%
Sulfur	0.23%
Carbonate	0.09%

**According to analyses made in India and Cuba.

Indian analyses of the mango kernel reveal the amino acids—alanine, arginine, aspartic acid, cystine, glutamic acid, glycine, histidine, isoleucine, leucine, lysine, methionine, phenylalanine, proline, serine, threonine, tyrosine, valine, at levels lower than in wheat and gluten. Tannin content may be 0.12-0.18% or much higher in certain cultivars.

Fig. 62: The long center tine of the mango fork is designed for piercing the base of the center section and right through the seed. With the strip of peel removed, the most flavorful flesh around the seed can be enjoyed like a lollipop.

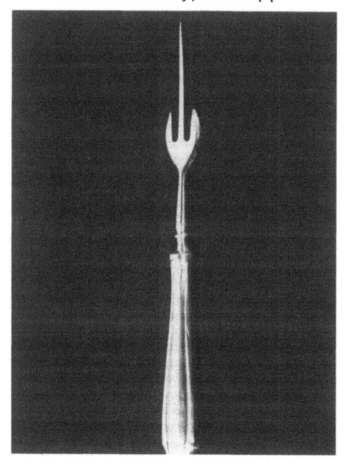

Fig. 63: 'Black Gold' mangoes, dark-green externally when ripe, are partly peeled like "radish roses" on the Bangkok market to show their yellow, fiberless flesh.

Food Value (continued)

Kernel Flour*

Protein	5.56%
Fat	16.17%
Carbohydrates	69.2%
Ash (minerals)	0.35%

*It is said to be equal to rice in food value, if tannin-free.

Kernel Fat

Fully saturated glycerides	14.2%
Mono-oleoglycerides	24.2%
Di-oleoglycerides	60.8%
Tri-unsaturated glycerides	0.8%

   Fatty Acids:

Myristic	0.69%
Palmitic	4.4-8.83%
Stearic	33.96-47.8%
Arachidic	2.7-6.74%
Oleic	38.2-49.78%
Linoleic	4.4-5.4%
Linolenic	0.5%

Leaves (immature)

Moisture	78.2%
Protein	3.0%
Fat	0.4%
Carbohydrates	16.5%
Fiber	1.6%
Ash	1.9%
Calcium	29 mg/100 g
Phosphorus	72 mg
Iron	6.2 mg
Vitamin A (carotene) β	1,490 I.U.
Thiamine	0.04 mg
Riboflavin	0.06 mg
Niacin	2.2 mg
Ascorbic Acid**	53 mg/100 g

**According to various analyses made in India.

Toxicity

The sap which exudes from the stalk close to the base of the fruit is somewhat milky at first, also yellowish-resinous. It becomes pale-yellow and translucent when dried. It contains mangiferen, resinous acid, mangiferic acid, and the resinol, mangiferol. It, like the sap of the trunk and branches and the skin of the unripe fruit, is a potent skin irritant, and capable of blistering the skin of the normal individual. As with poison ivy, there is typically a delayed reaction. Hypersensitive persons may react with considerable swelling of the eyelids, the face, and other parts of the body. They may not be able to handle, peel, or eat mangos or any food containing mango flesh or juice. A good precaution is to use one knife to peel the mango, and a clean knife to slice the flesh to avoid contaminating the flesh with any of the resin in the peel.

The leaves contain the glucoside, mangiferine. In India, cows were formerly fed mango leaves to obtain from their urine euxanthic acid which is rich yellow and has been used as a dye. Since continuous intake of the leaves may be fatal, the practice has been outlawed.

When mango trees are in bloom, it is not uncommon for people to suffer itching around the eyes, facial swelling and respiratory difficulty, even though there is no airborne pollen. The few pollen grains are large and they tend to adhere to each other even in dry weather. The stigma is small and not designed to catch windborne pollen. The irritant is probably the vaporized essential oil of the flowers which contains the sesquiterpene alcohol, mangiferol, and the ketone, mangiferone.

Mango wood should never be used in fireplaces or for cooking fuel, as its smoke is highly irritant.

Other Uses

Seed kernels: After soaking and drying to 10% moisture content, the kernels are fed to poultry and cattle. Without the removal of tannins, the feeding value is low. Cuban scientists declare that the mineral levels are so low mineral supplementation is needed if the kernel is used for poultry feed, for which purpose it is recommended mainly because it has little crude fiber.

Seed fat: Having high stearic acid content, the fat is desirable for soap-making. The seed residue after fat extraction is usable for cattle feed and soil enrichment.

A mango stone decorticator has been designed and successfully operated by the Agricultural Engineering Department of Pantnagar University, India.

Wood: The wood is kiln-dried or seasoned in saltwater. It is gray or greenish-brown, coarse-textured, medium-strong, hard, durable in water but not in the ground;

Fig. 64: A low-fiber, green-skinned mango on the market in Merida, Yucatan, is mounted on a lollipop stick. The fruit may be peeled and the flesh deeply cut to resemble the petals of a flower.

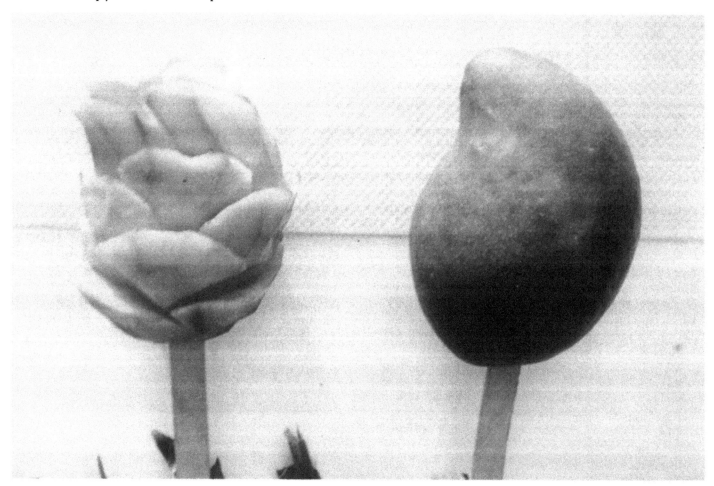

easy to work and finishes well. In India, after preservative treatment, it is used for rafters and joists, window frames, agricultural implements, boats, plywood, shoe heels and boxes, including crates for shipping tins of cashew kernels. It makes excellent charcoal.

Bark: The bark possesses 16% to 20% tannin and has been employed for tanning hides. It yields a yellow dye, or, with turmeric and lime, a bright rose-pink.

Gum: A somewhat resinous, red-brown gum from the trunk is used for mending crockery in tropical Africa. In India, it is sold as a substitute for gum arabic.

Medicinal Uses: Dried mango flowers, containing 15% tannin, serve as astringents in cases of diarrhea, chronic dysentery, catarrh of the bladder and chronic urethritis resulting from gonorrhea. The bark contains mangiferine and is astringent and employed against rheumatism and diphtheria in India. The resinous gum from the trunk is applied on cracks in the skin of the feet and on scabies, and is believed helpful in cases of syphilis.

Mango kernel decoction and powder (not tannin-free) are used as vermifuges and as astringents in diarrhea, hemorrhages and bleeding hemorrhoids. The fat is administered in cases of stomatitis. Extracts of unripe fruits and of bark, stems and leaves have shown antibiotic activity. In some of the islands of the Caribbean, the leaf decoction is taken as a remedy for diarrhea, fever, chest complaints, diabetes, hypertension and other ills. A combined decoction of mango and other leaves is taken after childbirth.

Related Species

Of approximately 40 other species of *Mangifera,* a few are cultivated for their fruits and several have been employed as rootstocks for the mango in Malaya.

M. sylvatica Roxb., is a large tree to 150 ft (45 m) growing wild in the eastern Himalayas, Nepal and the Andaman Islands, from 980 to 4,200 ft (300-1,300 m). The elliptic fruit, 3¼ to 4 in (8-10 cm) long, has yellow skin and fiberless, though rather thin, flesh. It is mostly utilized while still unripe for pickles and other preserves. The tree is valued mainly for its timber which is largely sapwood, light in weight and easily worked but medium-hard and strong.

M. foetida Lour., the horse mango, is a handsome, well-formed tree, 60 to 80 ft (18-24 m) tall with very stiff leaves and showy panicles of pink-red, odorless flowers. The fruit is oblong, 3 to 5½ in (7.5-16 cm) long, plump, with yellowish- or grayish-green skin when ripe. The flesh is variable, in some types orange, acid, strongly turpentine-scented; in others, pale-yellow, sweet in flavor and mildly aromatic. All types are fibrous and the stone has much fiber. Sweet types are eaten raw when ripe; others are used for pickles, chutneys and in curries. The sap of the tree and the immature fruit is highly irritating.

M. caesia Jack, ranging from 65 to 150 ft (20-45 m) at low altitudes in Malaysia and the Philippines, is frequently cultivated in Indonesia. The flowers are blue or lavender. Strongly and, to some people, unpleasantly aromatic, the fruit is oval to pear-shaped, 4¼ to 6 in (11-15 cm) long, with thin, pale-green or light-brown, scurfy skin which clings to the white or pale-yellow, juicy, fibrous flesh. Quality is highly variable; some types being subacid to sweet and agreeable and these are commonly eaten in Malaya. The seed is large and pink, enclosed in matted fibers; edible; monoembryonic. Young leaves are eaten raw. The sap of the tree and immature fruits is exceedingly irritant.

M. odorata Griff. is a medium to large tree, 60 to 80 ft (15-24 m) high, better suited than the mango to humid regions and much cultivated from Malaya to the Philippines where it is more familiar than the mango in eastern Mindanao. The flowers are whitish to yellowish and very fragrant. The fruit is round-oblique, somewhat oblate; to 5 in (12.5 cm) long, plump, with green or yellow-green, thick, tough skin. When ripe the flesh is pale-orange or yellowish, fibrous and resinous but juicy and sweet, though most types are distinctly turpentine-flavored. Nevertheless, all types are popular for curries and pickles. The stone is large with many coarse fibers. The sap of this tree is said to be fairly mild, but the milky sap of the immature fruit extremely acrid.

In addition to the above, Malayan villagers occasionally cultivate some lesser-known species: *M. longipetiolata* King, *M. maingayi* Hook f., *M. kemanga* Blume, and *M. pentandra* Hook f.

The gandaria, Plate XXIX, *Bouea gandaria* Blume (syn. *B. macrophylla* Griff.), is also called *kundangan, kundang, setar, star* and *rumia* in Malaya; *gandareed* in Java; *ma-prang* in Thailand. The tree, usually to 30 ft (9 m), sometimes to 60 ft (18 m), is short-trunked with resinous sap, drooping branches and evergreen, opposite, resinous, leathery, downward-pointing leaves 4 to 12 in (10-30 cm) long, 2 to 4½ in (5-11.25 cm) wide. They are purple-red and silky when they first appear. Small, greenish flowers are borne in pendent panicles to 5 in (12.5 cm) in length. The fruit, like a miniature mango, is oval, round or oblong-ovoid, 1½ to 2½ in (4-6.25 cm) long, with thin, smooth, brittle, edible skin, yellow or apricot-colored when ripe. The yellow or orange pulp is juicy, varies from acid to sweet, and adheres to the leathery, whiskered stone. There is great variation in the fruits of seedling trees, especially in the degree of "turpentine" odor. The tree is native to Malaya and Sumatra; is frequently cultivated, either from seed or air-layers, in its natural range and also rather widely through Malaysia and the fruits are sold in markets. They are made into jam and chutney. When still immature, they are pickled

Fig. 65: Mango trees produce massive sprays of reddish or yellowish flowers but only a few fruits develop from each spray.

in brine and used in curries. In Indonesia, the young leaves are marketed and eaten raw with rice. Budwood of a cultivar named 'Wan', meaning "sweet", was obtained by William F. Whitman from an orchard near Bangkok in 1967. His resulting grafted tree, in a protected location in South Florida, fruited in 1974. Earlier introductions (1935, 1936 and 1938) by the Agricultural Research and Education Center in Homestead failed to survive.

A lesser species, *B. oppositifolia* Adelb. (syn. *B. microphylla* Griff.), is called plum mango, *rembunia*, *gemis*, or *rumia* in Malaya; *ma–pring* in Thailand. The tree is similar but deciduous, smaller in all its parts, and the fruit is orange or yellow and only 1 in (2.5 cm) long, acid and usually cooked when half-ripe. This species is abundant wild in lowland forests of Malaya and much cultivated as a shade tree. The wood is hard and very heavy, sinks in water, and is used for houseposts.

Food Value

Fruits from a 20-year-old gandaria tree (*Bouea gandaria* Blume) in the Lancetilla Experimental Garden, Tela, Honduras, were analyzed in 1950 and the following values were reported:

Food Value Per 100 g of Edible Portion	
Moisture	85.2 g
Protein	0.112 g
Fat	0.04 g
Fiber	0.6 g
Ash	0.23 g
Calcium	6.0 mg
Phosphorus	10.8 mg
Iron	0.31 mg
Carotene	0.043 mg
Thiamine	0.031 mg
Riboflavin	0.025 mg
Niacin	0.286 mg
Ascorbic Acid	75.0 mg

Cashew Apple (Plate XXX)

This pseudofruit (or "false fruit") is a by-product of the cashew nut industry. The cashew tree, *Anacardium occidentale* L., is called *marañon* in most Spanish-speaking countries, but *merey* in Venezuela; and *caju* or *cajueiro* in Portuguese. It is generally bushy, low-branched and spreading; may reach 35 ft (10.6 m) in height and width. Its leaves, mainly in terminal clusters, are oblong-oval or obovate, 4 to 8 in (10-20 cm) long and 2 to 4 in (5-10 cm) wide, and leathery. Yellowish-pink, 5-petalled flowers are borne in 6 to 10-in (15-25 cm) terminal panicles of mixed male, female and bisexual. The true fruit of the tree is the cashew nut resembling a miniature boxing-glove; consisting of a double shell containing a caustic phenolic resin in honeycomb-like cells, enclosing the edible kidney-shaped kernel. An interesting feature of the cashew is that the nut develops first and when it is full-grown but not yet ripe, its peduncle or, more technically, receptacle, fills out, becomes plump, fleshy, pear-shaped or rhomboid-to-ovate, 2 to 4½ in (5-11.25 cm) in length, with waxy, yellow, red, or red-and-yellow skin and spongy, fibrous, very juicy, astringent, acid to subacid, yellow pulp. Thus is formed the conspicuous, so-called cashew apple.

The cashew is native to arid northeast Brazil and, in the 16th Century, Portuguese traders introduced it to Mozambique and coastal India, but only as a soil retainer to stop erosion on the coasts. It flourished and ran wild and formed extensive forests in these locations and on nearby islands, and eventually it also became dispersed in East Africa and throughout the tropical lowlands of northern South America, Central America and the West Indies. It has been more or less casually planted in all warm regions and a few fruiting specimens are found in experimental stations and private gardens in southern Florida.

The production and processing of cashew nuts are complex and difficult problems. Because of the great handi-cap of the toxic shell oil, Latin Americans and West Indians over the years have been most enthusiastic about the succulent cashew apple and have generally thrown the nut away or processed it crudely on a limited scale, except in Brazil, where there is a highly developed cashew nut processing industry, especially in Ceara. In Mozambique, also, the apple reigned supreme for decades. Attention then focused on the nut, but, in 1972, the industrial potential of the juice and sirup from the estimated 2 million tons of surplus cashew apples was being investigated. In India, on the other hand, vast tonnages of cashew apples have largely gone to waste while that country pioneered in the utilization and promotion of the nut.

Fig. 66: The so-called "cashew apple", a pseudofruit — actually the swollen stalk of the true fruit of *Anacardium occidentale*, the cashew nut — is fibrous but juicy and locally popular preserved in sirup.

Fig. 67: Food technologists in Mysore, India, developed a candied cashew apple product, more appealing than the canned. A similar confection is made and sold in the Dominican Republic.

have found that good condition can be maintained for 5 weeks at 32° to 35°F (0°-1.67°C) and relative humidity of 85% to 90%. Inasmuch as the juice is astringent and somewhat acrid due to 35% tannin content (in the red: less in the yellow) and 3% of an oily substance, the fruit is pressure-steamed for 5 to 15 minutes before candying or making into jam or chutney or extracting the juice for carbonated beverages, sirup or wine. Efforts are made to retain as much as possible of the ascorbic acid. Food technologists in Costa Rica recently worked out an improved process for producing the locally popular candied, sun-dried cashew apples. Failure to remove the tannin from the juice may account for the nutritional deficiency in heavy imbibers of cashew apple wine in Mozambique, for tannin prevents the body's full assimilation of protein.

Food Value Per 100 g of Fresh Cashew Apple*	
Moisture	84.4-88.7 g
Protein	0.101-0.162 g
Fat	0.05-0.50 g
Carbohydrates	9.08-9.75 g
Fiber	0.4-1.0 g
Ash	0.19-0.34 g
Calcium	0.9-5.4 mg
Phosphorus	6.1-21.4 mg
Iron	0.19-0.71 mg
Carotene	0.03-0.742 mg
Thiamine	0.023-0.03 mg
Riboflavin	0.13-0.4 mg
Niacin	0.13-0.539 mg
Ascorbic Acid	146.6-372.0 mg
*Analyses made in Central America and Cuba.	

The apple and nut fall together when both are ripe and, in commercial nut plantations, it is most practical to twist off the nut and leave the apple on the ground for later grazing by cattle or pigs. But, where labor costs are very low, the apples may be gathered up and taken to markets or processing plants. In Goa, India, the apples are still trampled by foot to extract the juice for the locally famous distilled liquor, feni. In Brazil, great heaps are displayed by fruit vendors, and the juice is used as a fresh beverage and for wine.

In the field, the fruits are picked up and chewed for refreshment, the juice swallowed, and the fibrous residue discarded. In the home and, in a limited way for commercial purposes, the cashew apples are preserved in sirup in glass jars. Fresh apples are highly perishable. Various species of yeast and fungi cause spoilage after the first day at room temperature. Food technologists in India

Medicinal Uses: Cashew apple juice, without removal of tannin, is prescribed as a remedy for sore throat and chronic dysentery in Cuba and Brazil. Fresh or distilled, it is a potent diuretic and is said to possess sudorific properties. The brandy is applied as a liniment to relieve the pain of rheumatism and neuralgia.

Ambarella (Plate XXXI)

An under-appreciated member of the Anacardiaceae, but deserving of improvement, is the ambarella, *Spondias dulcis* Forst. (syn. *S. cytherea* Sonn.). Among various colloquial names are Otaheite apple, Tahitian quince, Polynesian plum, Jew plum and golden apple. In Malaya it is called great hog plum or *kedondong;* in Indonesia, *kedongdong;* in Thailand, *ma-kok-farang;* in Cambodia, *mokak;* in Vietnam, *coc, pomme cythere* or *pommier de cythere.* In Costa Rica, it is known as *juplón;* in Colombia, *hobo de racimos;* in Venezuela, *jobo de la India, jobo de Indio,* or *mango jobo;* in Ecuador, *manzana de oro;* in Brazil, *caja-manga.*

Description

The tree is rapid-growing, attaining a height of 60 ft (18 m) in its homeland; generally not more than 30 or 40 ft (9-12 m) in other areas. Upright and rather rigid and symmetrical, it is a stately ornamental with deciduous, handsome, pinnate leaves, 8 to 24 in (20-60 cm) in length, composed of 9 to 25 glossy, elliptic or obovate-oblong leaflets 2½ to 4 in (6.25-10 cm) long, finely toothed toward the apex. At the beginning of the dry, cool season, the leaves turn bright-yellow and fall, but the tree with its nearly smooth, light gray-brown bark and graceful, rounded branches is not unattractive during the few weeks that it remains bare. Small, inconspicuous, whitish flowers are borne in large terminal panicles. They are assorted, male, female and perfect in each cluster. Long-stalked fruits dangle in bunches of a dozen or more; oval or somewhat irregular or knobby, and 2½ to 3½ in (6.25-9 cm) long, with thin but tough skin, often russetted. While still green and hard, the fruits fall to the ground, a few at a time, over a period of several weeks. As they ripen, the skin and flesh turn golden-yellow. While the fruit is still firm, the flesh is crisp, juicy and subacid, and has a somewhat pineapple-like fragrance and flavor. If allowed to soften, the aroma and flavor become musky and the flesh difficult to slice because of conspicuous and tough fibers extending from the rough ridges of the 5-celled, woody core containing 1 to 5 flat seeds. Some fruits in the South Sea Islands weigh over 1 lb (0.45 kg) each.

Origin and Distribution

The ambarella is native from Melanesia through Polynesia and has been introduced into tropical areas of both the Old and New World. It is common in Malayan gardens and fairly frequent in India and Ceylon. The fruits are sold in markets in Vietnam and elsewhere in former Indochina. It first fruited in the Philippines in 1915. It is cultivated in Queensland, Australia, and grown on a small scale in Gabon and Zanzibar.

It was introduced into Jamaica in 1782 and again 10 years later by Captain Bligh, probably from Hawaii where it has been grown for many years. It is cultivated in Cuba, Haiti, the Dominican Republic, and from Puerto Rico to Trinidad; also in Central America, Venezuela, and Surinam; is rare in Brazil and other parts of tropical America. Popenoe said there were only a few trees in the Province of Guayas, Ecuador, in 1924.

The United States Department of Agriculture received seeds from Liberia in 1909, though Wester reported at that time that the tree had already been fruiting for 4 years in Miami, Florida. In 1911, additional seeds reached Washington from Queensland, Australia. A number of specimens are scattered around the tip of Florida, from Palm Beach southward, but the tree has never become common here. Some that were planted in the past have disappeared.

Climate

The tree flourishes in humid tropical and subtropical areas, being only a trifle tenderer than its close relative, the mango. It succeeds up to an altitude of 2,300 ft (700 m). In Israel, the tree does not thrive, remaining small and bearing only a few, inferior fruits.

Soil

The ambarella grows on all types of soil, including oolitic limestone in Florida, as long as they are well-drained.

Propagation

The tree is easily propagated by seeds, which germinate in about 4 weeks, or by large hardwood cuttings, or air-layers. It can be grafted on its own rootstock, but Firminger says that in India it is usually grafted on the native *S. pinnata* Kurz (see below). Wester advised: "Use non-petioled, slender, mature, but green and smooth budwood; cut large buds with ample wood-shield, 1½ to 1¾ in (4-4.5 cm) long; insert the buds in the stock at a point of approximately the same age and appearance as the scion."

Culture

Seedlings may fruit when only 4 years old. Ochse recommends that the young trees be given light shade. Mature trees are somewhat brittle and apt to be damaged by strong winds; therefore, sheltered locations are preferred.

Season

In Hawaii, the fruit ripens from November to April; in Tahiti, from May to July. In Florida, a single tree provides a steady supply for a family from fall to midwinter, at a time when mangos and many other popular fruits are out of season.

Pests and Diseases

Ochse says that in Indonesia the leaves are severely attacked by the larvae of the kedongdong spring-beetle, *Podontia affinis*. In Costa Rica, the bark is eaten by a wasp ("Congo"), causing necrosis which leads to death. No particular insects or diseases have been reported in Florida. In Jamaica, the tree is subject to gummosis and is consequently short-lived.

Food Uses

The ambarella has suffered by comparison with the mango and by repetition in literature of its inferior quality. However, taken at the proper stage, while still firm, it is relished by many out-of-hand, and it yields a delicious juice for cold beverages. If the crisp sliced flesh is stewed with a little water and sugar and then strained through a wire sieve, it makes a most acceptable product, much like traditional applesauce but with a richer flavor. With the addition of cinnamon or any other spices desired, this sauce can be slowly cooked down to a thick consistency to make a preserve very similar to apple butter. Unripe fruits can be made into jelly, pickles or relishes, or used for flavoring sauces, soups and stews.

Young ambarella leaves are appealingly acid and consumed raw in southeast Asia. In Indonesia, they are steamed and eaten as a vegetable with salted fish and rice, and also used as seasoning for various dishes. They are sometimes cooked with meat to tenderize it.

Food Value Per 100 g of Edible Portion*	
Calories	157.30
Total Solids	14.53–40.35%
Moisture	59.65–85.47%
Protein	0.50–0.80%
Fat	0.28–1.79%
Sugar (sucrose)	8.05–10.54%
Acid	0.47%
Crude Fiber	0.85–3.60%
Ash	0.44–0.65%
*According to analyses made in the Philippines and Hawaii.	

Miller, Louis and Yanazawa in Hawaii reported an ascorbic acid content of 42 mg per 100 g of raw pulp. It is a good source of iron. Unripe fruits contain 9.76% of pectin.

Other Uses
Wood: The wood is light-brown and buoyant and in the Society Islands has been used for canoes.

Medicinal Uses: In Cambodia, the astringent bark is used with various species of *Terminalia* as a remedy for diarrhea.

Related Species

The amra, *S. pinnata* Kurz (syns. *Mangifera pinnata* L. f.; *Pourpartia pinnata* Blanco), which some botanists consider merely a wild form of *S. dulcis*, is wild and cultivated from the Himalayas of northern India to the Andaman Islands and is commonly cultivated throughout southeast Asia and Malaysia. The twigs are smooth and the leaves are not toothed; the fruit is smaller than the ambarella and inferior in quality but has the same uses. The aromatic, acidulous leaves and flowers are employed as flavoring and consumed raw or cooked, especially in curries. The wood is used for making boats, floats, matches, etc. There are several medicinal applications of the bark, root, and the gum that exudes from the trunk.

Purple Mombin

One of the most popular small fruits of the American tropics, the purple mombin, *Spondias purpurea* L., has acquired many other colloquial names: in English, red mombin, Spanish plum, hog plum, scarlet plum; purple plum in the Virgin Islands; Jamaica plum in Trinidad; Chile plum in Barbados; wild plum in Costa Rica and Panama; red plum, as well as *noba* and *makka pruim* in the Netherlands Antilles. Spanish names include: *ajuela ciruela; chiabal; cirguelo; ciruela; ciruela agria; ciruela calentana; ciruela campechana; ciruela colorada; ciruela de coyote; ciruela de hueso; ciruela del país; ciruela de Mexico; ciruela morada; ciruela roja; ciruela sanjuanera; hobo; hobo colorado; ismoyo; jobillo; jobito; jobo; jobo colorado; jobo francés; jocote; jocote agrio; jocote amarillo* (yellow form); *jocote común; jocote de corona; jocote de iguana; jocote iguanero; jocote tronador; jocotillo; pitarillo; sineguelas* (Philippines); *sismoyo*. In Portuguese, it is called *ambu; ambuzeiro; ameixa da Espanha; cajá vermelha* (yellow form); *ciriguela; ciroela; imbu; imbuzeiro; umbu*, or *umbuzeiro*. In French, it is *cirouelle, mombin rouge, prune du Chili, prune d'Espagne, prune jaune* (yellow form) or *prune rouge*.

Description
The purple mombin may be a shrub or low-branched, small tree in lowlands, or a spreading, thick-trunked tree reaching 25 or even 50 ft (7.5–15 m) in highlands. The branches are thickish and brittle. The deciduous, alternate, compound leaves bright-red or purple when young; 4¾ to 10 in (12–25 cm) long when mature; have 5 to 19 nearly sessile, obovate to lanceolate or oblong-elliptic leaflets ¾ to 1½ in (2–4 cm) long; oblique toward the base and faintly toothed toward the apex. The tiny, 4- to 5-petalled flowers, male, female and bisexual, are red or purple and borne in short, hairy panicles along the branches before the leaves appear. Somewhat plumlike, the fruits, borne singly or in groups of 2 or 3, may be purple, dark- or bright-red, orange, yellow, or red-and-yellow. They vary from 1 to 2 in (2.5–5 cm) in length and may be oblong, oval, obovoid or pear-shaped, with small indentations and often a knob at the apex. The skin is glossy and firm; the flesh aromatic, yellow, fibrous, very juicy, with a rich, plum-like, subacid to acid flavor, sometimes a trifle turpentiney; and it adheres to the rough, fibrous, hard, oblong, knobby, thick, pale stone, which is ½ to ¾ in (1.25–2 cm) long and contains up to 5 small seeds.

Origin and Distribution
The purple mombin is native and common both wild and cultivated from southern Mexico through northern Peru and Brazil, particularly in arid zones. There are some recent commercial plantings in Mexico and Venezuela. It is commonly planted in most of the islands of the West Indies and the Bahamas. Everywhere the fruits are sold along the roads and streets as well as in the native markets. Spanish explorers carried this species to the

Fig. 68: The purple, or red, mombin (*Spondias purpurea*), despite its large seed, is popular for casual nibbling. In: K. & J. Morton, *Fifty Tropical Fruits of Nassau*, 1946.

Philippines, where it has been widely adopted. The tree is naturalized throughout much of Nigeria and occasionally cultivated for its fruit. It has been infrequently planted in southern Florida, mainly as a curiosity.

Varieties

The fruit is highly variable. The yellow form (uncommon) has been identified by some botanists as *S. purpurea* forma *lutea* F. & R., or even as a separate species, *S. cirouella* Tussac. It has been confused with the true yellow mombin, *S. mombin* L. (syn. *S. lutea* L.), q.v.

In Guatemala, the variety called *jocote de corona,* which is flattened and somewhat shouldered at the apex, is said to be of superior quality, and *jocote tronador* is nearly its equal.

Climate

The tree is tropical, ranging from sea-level to 5,500 or 6,000 ft (1,700-1,800 m) in Mexico and Central America; to 2,500 ft (760 m) in Jamaica, in either dry or humid regions. It flowers but does not fruit in Israel; is cold-sensitive in Florida.

Soil

The tree is found growing naturally on a great diversity of soils throughout Latin America—sand, gravel, heavy clay loam, or limestone.

Propagation and Culture

The purple mombin, including its yellow form, is grown very easily and quickly by setting large cuttings upright in the ground. It is one of the trees most used to create "living fences". It grows very slowly from seed.

Season

There are flowers and fruits of the red form nearly all year in Jamaica, but mainly in July and August, while the yellow variant fruits only from September to November. In the Bahamas, the fruiting season of the red type is brief, just May and June; the yellow ripens from August to early October.

Pests and Diseases

Fruit flies commonly infest the ripe fruits. In Florida, the foliage is subject to spot anthracnose caused by *Sphaceloma spondiadis.*

Food Uses

The ripe fruits are commonly eaten out-of-hand. While not of high quality, they are popular with people who have enjoyed them from childhood, and they serve a useful purpose in the absence of "snackbars". In the home, they are stewed whole, with sugar, and consumed as dessert. They can be preserved for future use merely by boiling and drying, which keeps them in good condition for several months. The strained juice of cooked fruits yields an excellent jelly and is also used for making wine and vinegar. It is a pleasant addition to other fruit beverages.

In Mexico, unripe fruits are made into a tart, green sauce, or are pickled in vinegar and eaten with salt and chili peppers.

The new shoots and leaves are acid and eaten raw or cooked as greens in northern Central America.

Toxicity

In the Philippines, it is said that eating a large quantity of the fruits on an empty stomach may cause stomachache.

Other Uses

Gum: The tree exudes a gum that has served in Central America as a glue.

Wood: The wood is light and soft; has been found to be suitable for paper pulp in Brazil. It is sometimes burned to ashes which are employed in making soap.

Leaves and fruits: The leaves are readily grazed by cattle and the fruits are fed to hogs.

Lac: Lac insects have been raised on the red mombin in Mexico.

Food Value Per 100 g of Edible Portion*	
Moisture	65.9–86.6 g
Protein	0.096–0.261 g
Fat	0.03–0.17 g
Fiber	0.2–0.6 g
Ash	0.47–1.13 g
Calcium	6.1–23.9 mg
Phosphorus	31.5–55.7 mg
Iron	0.09–1.22 mg
Carotene	0.004–0.089 mg
Thiamine	0.033–0.103 mg
Riboflavin	0.014–0.049 mg
Niacin	0.540–1.770 mg
Ascorbic Acid	26.4–73.0 mg

Amino Acids** (mg per g nitrogen [N = 6.25])

Lysine	316 mg
Methionine	178 mg
Threonine	219 mg
Tryptophan	57 mg

*Analyses made in Central America and Ecuador.
**Brazilian analyses.

Fig. 69: The yellow form of the purple mombin, which has been called S. *purpurea* var. *lutea,* is smaller, less irregular in form.

Plate XXXII
LYCHEE
Litchi chinensis

Plate XXXIII
LYCHEE
Litchi chinensis: dried

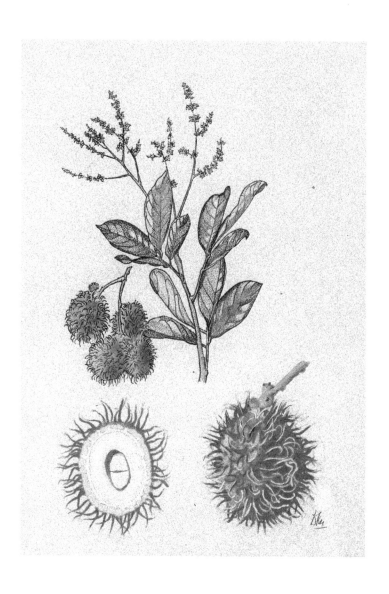

Plate XXXIV
RAMBUTAN
Nephelium lappaceum
Painted by Dr. M.J. Dijkman

Plate XXXV
INDIAN JUJUBE
Zizyphus mauritiana

Plate XXXVI
ROSELLE
Hibiscus sabdariffa

Plate XXXVII
ROSELLE
Hibiscus sabdariffa
(calyces raw and cooked)

Plate XXXVIII
DURIAN
Durio zibethinus

Plate XXXIX
CHUPA-CHUPA
Quararibea cordata

Plate XL
KIWIFRUIT
Actinidia chinensis

Medicinal Uses: In Mexico, the fruits are regarded as diuretic and antispasmodic. The fruit decoction is used to bathe wounds and heal sores in the mouth. A sirup prepared from the fruit is taken to overcome chronic diarrhea. The astringent bark decoction is a remedy for mange, ulcers, dysentery and for bloating caused by intestinal gas in infants. In the Philippines, the sap of the bark is used to treat stomatitis in infants.

The juice of the fresh leaves is a remedy for thrush. A decoction of the leaves and bark is employed as a febrifuge. In southwestern Nigeria, an infusion of shredded leaves is valued for washing cuts, sores and burns. Researchers at the University of Ife have found that an aqueous extract of the leaves has antibacterial action, and an alcoholic extract is even more effective. The gum-resin of the tree is blended with pineapple or soursop juice for treating jaundice. Most of the other uses indicate that the fruits, leaves and bark are fairly rich in tannin.

Yellow Mombin

The true yellow mombin, *S. mombin* L. (syn. *S. lutea* L.) is most often called hog plum in the Caribbean Islands. In Jamaica, it is also known as Spanish plum, or gully plum. In Malaya, it is distinguished as thorny hog plum; in Ghana, it is hog plum or Ashanti plum. Among its Spanish names are *caimito, chupandilla, ciruela agria, ciruela amarilla, ciruela de jobo, ciruela del pais, ciruela de monte, ciruela loca, ciruela mango, ciruela obo, cuajo, guama zapotero, hobo de monte, hubu, jobillo, jobito, jobo, jobo arisco, joboban, jobo blanco, jobo de Castilla, jobo de perro, jobo de puerco, jobo espino, jobo espinoso, jobo gusanero, jobo hembra, jobo jocote, jobo negro, jobo roñoso, jobo vano, jocote, jocote amarillo, jocote de chanche, jocote de jobo, jocote jobo, jocote montanero, jocote montero, jovo, marapa, obo de zopilote, palo de mulato, noma, tobo de montana, obo* and *uvo*. In Portuguese, it is called *acaiba, acaimiri, acaja, acajaiba, caja, caja mirim, caja pequeno, cajazeiro,* and *caja miudo*. In French, it is *mombin franc, mombin fruits jaunes, mombinier, myrobalane, prune mombin, prune myrobalan,* or *prunier mombin*. Local names in Surinam are *hoeboe, mompe, monbe, mopé* and *moppé*. Amazonian Indians call it *taperiba* or *tapiriba* (fruit of the tapir).

Description

The yellow mombin tree, unlike that of the purple mombin, is erect, stately, to 65 ft (20 m) tall, with trunk to 2 or 2½ ft (60-75 cm) in diameter, somewhat buttressed, and thick, fissured bark, often, in young trees, bearing many blunt-pointed spines or knobs up to ¾ in (2 cm) long. Generally, its lower branches are whorled. Its deciduous, alternate, pinnate leaves, 8 to 18 in (20-45 cm) long, have hairy, often pinkish, petioles and 9 to 19 sub-opposite, ovate or lanceolate, pointed leaflets, 2 to 6 in (5-15 cm) long, inequilateral and oblique at the base. Small, fragrant, whitish, male, female and bisexual flowers are borne, after the new leaves, in panicles 6 to 12 in (15-30 cm) long. The fruit, hanging in numerous, branched, terminal clusters of a dozen or more, is aromatic, ovoid or oblong, 1¼ to 1½ in (3.2-4 cm) long and up to 1 in (2.5 cm) wide; golden-yellow; with thin, tough skin, and scant, medium-yellow, translucent, fibrous, very juicy pulp, somewhat musky, very acid, often with a hint of turpentine, clinging to the white, fibrous or "corky" stone.

Origin and Distribution

The tree is native and common in moist lowland forests from southern Mexico to Peru and Brazil, and in many of the West Indies. It has been planted in Bermuda; is grown to a limited extent in India and Indonesia; is rare in Malaya, but widely cultivated and naturalized in tropical Africa.

The United States Department of Agriculture received seeds from Colombia in 1914 (S.P.I. #39563); more seeds arrived in 1917 (S.P.I. #45086); and Dr. David Fairchild collected seeds in Panama in 1921 (S.P.I. #54632). Still, only a few specimens exist in special collections in southern Florida.

Climate

This is a strictly tropical tree, not growing above an elevation of 3,200 ft (1,000 m) in South America. It is well-adapted to arid as well as humid zones.

Propagation

The tree may be propagated by seeds but it is usually grown from large cuttings which root quickly.

Culture

The tree is fast-growing in full sun and in the American tropics and Africa is extensively planted as a living fence-post, as well as for shade and for its fruits.

Season

In Costa Rica, the tree blooms in November and December and again in March, and the fruits ripen in August, and in December/January. Blooming occurs in Jamaica in April, May and June and the crop matures in July and August. The fruits are in season in Mexico from July to October; in Florida from August to November. They fall to the ground when fully ripe, but children throw sticks up into the trees to bring them down sooner.

Fig. 70: The true yellow mombin (*Spondias mombin*) is borne in dangling clusters. It is eaten mostly by children and livestock.

Pests

The fruits are commonly infested with fruit-fly larvae.

Food Uses

The yellow mombin is less desirable than the purple mombin and is appreciated mostly by children and way-farers as a means of alleviating thirst. Ripe fruits are eaten out-of-hand, or stewed with sugar. The extracted juice is used to prepare ice cream, cool beverages and jelly. Some people make those of fair quality into jam and various other preserves.

In Amazonas, the fruit is used mainly to produce wine sold as "*Vinho de Taperiba*". In Guatemala, the fruit is made into a cider-like drink.

Mexicans pickle the green fruits in vinegar and eat them like olives with salt and chili, as they do with the unripe purple mombin.

Young leaves are cooked as greens.

Toxicity

According to Altschul, E.L. Little recorded on an herbarium specimen collected in Colombia: ". . . fruit edible, but said to be bad for the throat." In tropical Africa, excessive indulgence in the fruits is said to cause dysentery.

Food Value Per 100 g of Edible Portion*	
Calories	21.8–48.1
Moisture	72.8–88.53 g
Protein	1.28–1.38 g
Fat	0.1–0.56 g
Fiber	1.16–1.18 g
Carbohydrates	8.70–10.0 g
Ash	0.65–0.66 g
Calcium	31.4 mg
Iron	2.8 mg
Carotene (Vitamin A)	71 I.U.
Thiamine	95 mcg
Riboflavin	50 mcg
Ascorbic Acid	46.4 mg

*Analyses made in Guatemala, Africa and the Philippines.

Other Uses

Fruits: The fruits are widely valued as feed for cattle and pigs.

Gum: The tree exudes a gum that is used as a glue.

Wood: The wood is yellow or yellowish-brown with darker markings; light in weight, buoyant, flexible,

Fig. 71: The imbu (*Spondias tuberosa*) of northeastern Brazil is an appreciated wild source of juice in that semi-arid land. Photo'd by the plant explorer, P.H. Dorsett in 1914, for the U.S. Dept. of Agriculture.

strong; prone to attack by termites and other pests. It is much used in carpentry, also for matchsticks, matchboxes, physician's spatulas, sticks for sweetmeats, pencils, pen-holders, packing cases, interior sheathing of houses and boats and as a substitute for cork. It is not suited for turnery and does not polish well. In Brazil, the woody tubercles on the trunk are cut off and used for bottle stoppers and to make seals for stamping sealing wax. In tropical Africa, saplings serve as poles for huts; branches for garden poles and for axe and hoe handles. In Costa Rica and Puerto Rico the wood is employed only as fuel. Ashes from the burned wood are utilized in indigo-dyeing in Africa.

Bark: The bark, because of its tannin content, is used in tanning and dyeing. It is so thick that it is popular for carving amulets, statuettes, cigarette holders, and various ornamental objects.

Roots: Potable water can be derived from the roots in emergency.

Nectar: The flowers are worked intensively by honeybees early in the morning.

Medicinal Uses: The fruit juice is drunk as a diuretic and febrifuge. The decoction of the astringent bark serves as an emetic, a remedy for diarrhea, dysentery, hemorrhoids and a treatment for gonorrhea and leucorrhea; and, in Mexico, it is believed to expel calcifications from the bladder. The powdered bark is applied on wounds. A tea of the flowers and leaves is taken to relieve stomachache, biliousness, urethritis, cystitis and eye and throat inflammation. In Belize, a decoction of the young leaves is a remedy for diarrhea and dysentery. The juice of crushed leaves and the powder of dried leaves are used as poultices on wounds and inflammations. The gum is employed as an expectorant and to expel tapeworms.

Related Species

The imbu, or umbu, *S. tuberosa* Arruda, is a low-branching tree to 13 or 16 ft (4–5 m) high, spreading to a width of 30 ft (9 m). It has a shallow system of soft, tuberous roots called *cunca,* which store much water. The pinnate leaves have 5 to 9 oblong-ovate leaflets, 1 to 1¾ in (2.5–4.5 cm) long, sometimes faintly toothed. Flowers, small, white and 4- to 5-petalled, are produced in panicles 4 to 6 in (10-15 cm) in length. The fruit, borne in great abundance, exhibits minor seedling variations; is usually more or less oval, 1½ in (4 cm) long, with greenish-yellow, fairly thick, tough skin and tender, melting pulp, acid unripe, sweet when ripe, and adherent to the single stone, ¾ in (2 cm) long.

The tree thrives in very dry soil, gravelly loam, sandy or partly clay, throughout much of subtropical, semi-arid northeastern Brazil. It is rarely cultivated. It is a much-appreciated, bountiful, wild food resource of rural people. The fruits are gathered from the ground and sold in village markets. They are eaten out-of-hand, or the juice is blended with boiled milk and sugar, or made into ice cream or jelly. The roots have been consumed in emergency and they readily yield potable water.

Introductions into Florida and Malaya have been unsuccessful.

Fig. 72: The imbu (*Spondias tuberosa*) from The Navel Orange of Bahia, with notes on some little-known Brazilian fruits, by P.H. Dorsett, A.D. Shamel and W. Popenoe. Bull. 445, Bureau of Plant Industry, U.S. Department of Agriculture, Washington, D.C. 1917.

Lychee (Plates XXXII and XXXIII)

The lychee is the most renowned of a group of edible fruits of the soapberry family, Sapindaceae. It is botanically designated *Litchi chinensis* Sonn. (*Nephelium litchi* Cambess) and widely known as litchi and regionally as *lichi, lichee, laichi, leechee* or lychee. Professor G. Weidman Groff, an influential authority of the recent past, urged the adoption of the latter as approximating the pronunciation of the local name in Canton, China, the leading center of lychee production. I am giving it preference here because the spelling best indicates the desired pronunciation and helps to standardize English usage. Spanish- and Portuguese-speaking people call the fruit *lechia;* the French, *litchi,* or, in French-speaking Haiti, *quenepe chinois,* distinguishing it from the *quenepe, genip* or mamoncillo of the West Indies, *Melicoccus bijugatus,* q.v. The German word is *litschi.*

The lychee tree is handsome, dense, round-topped, slow-growing, 30 to 100 ft (9-30 m) high and equally broad. Its evergreen leaves, 5 to 8 in (12.5-20 cm) long, are pinnate, having 4 to 8 alternate, elliptic-oblong to lanceolate, abruptly pointed, leaflets, somewhat leathery, smooth, glossy, dark-green on the upper surface and grayish-green beneath, and 2 to 3 in (5-7.5 cm) long. The tiny petalless, greenish-white to yellowish flowers are borne in terminal clusters to 30 in (75 cm) long. Showy fruits, in loose, pendent clusters of 2 to 30 are usually strawberry-red, sometimes rose, pinkish or amber, and some types tinged with green. Most are aromatic, oval, heart-shaped or nearly round, about 1 in (2.5 cm) wide and 1½ in (4 cm) long; have a thin, leathery, rough or minutely warty skin, flexible and easily peeled when fresh. Immediately beneath the skin of some varieties is a small amount of clear, delicious juice. The glossy, succulent, thick, translucent-white to grayish or pinkish fleshy aril which usually separates readily from the seed, suggests a large, luscious grape. The flavor of the flesh is subacid and distinctive. There is much variation in the size and form of the seed. Normally, it is oblong, up to ¾ in (20 mm) long, hard, with a shiny, dark-brown coat and is white internally. Through faulty pollination, many fruits have shrunken, only partially developed seeds (called "chicken tongue") and such fruits are prized because of the greater proportion of flesh. In a few days, the fruit naturally dehydrates, the skin turns brown and brittle and the flesh becomes dry, shriveled, dark-brown and

raisin-like, richer and somewhat musky in flavor. Because of the firmness of the shell of the dried fruits, they came to be nicknamed "lychee, or litchi, nuts" by the uninitiated and this erroneous name has led to much misunderstanding of the nature of this highly desirable fruit. It is definitely not a "nut", and the seed is inedible.

Origin and Distribution

The lychee is native to low elevations of the provinces of Kwangtung and Fukien in southern China, where it flourishes especially along rivers and near the seacoast. It has a long and illustrious history having been praised and pictured in Chinese literature from the earliest known record in 1059 A.D. Cultivation spread over the years through neighboring areas of southeastern Asia and offshore islands. Late in the 17th Century, it was carried to Burma and, 100 years later, to India. It arrived in the West Indies in 1775, was being planted in greenhouses in England and France early in the 19th Century, and Europeans took it to the East Indies. It reached Hawaii in 1873, and Florida in 1883, and was conveyed from Florida to California in 1897. It first fruited at Santa Barbara in 1914. In the 1920's, China's annual crop was 30 million lbs (13.6 million kg). In 1937 (before WW II) the crop of Fukien Province alone was over 35 million lbs (16 million kg). In time, India became second to China in lychee production, total plantings covering about 30,000 acres (12,500 ha). There are also extensive plantings in Pakistan, Bangladesh, Burma, former Indochina, Taiwan, Japan, the Philippines, Queensland, Madagascar, Brazil and South Africa. Lychees are grown mostly in dooryards from northern Queensland to New South Wales, but commercial orchards have been established in the past 20 years, some consisting of 5,000 trees.

Madagascar began experimental refrigerated shipments of lychees to France in 1960. It is recorded that there were 2 trees about 6 years old in Natal, South Africa, in 1875. Others were introduced from Mauritius in 1876. Layers from these latter trees were distributed by the Durban Botanical Gardens and lychee-growing expanded steadily until in 1947 there were 5,000 bearing trees on one estate and 5,000 newly planted on another property, a total of 40,000 in all.

In Hawaii, there are many dooryard trees but commercial plantings are small. The fruit appears on local markets and small quantities are exported to the mainland but the lychee is too undependable to be classed as a crop of serious economic potential there. Rather, it is regarded as a combination ornamental and fruit tree.

There are only a few scattered trees in the West Indies and Central America apart from some groves in Cuba, Honduras and Guatemala. In California, the lychee will grow and fruit only in protected locations and the climate is generally too dry for it. There are a few very old trees and one small commercial grove. In the early 1960's, interest in this crop was renewed and some new plantings were being made on irrigated land.

At first it was believed that the lychee was not well suited to Florida because of the lack of winter dormancy, exposing successive flushes of tender new growth to the occasional periods of low temperature from December to March. The earliest plantings at Sanford and Oviedo were killed by severe freezes. A step forward came with the importation of young lychee trees from Fukien, China, by the Rev. W.M. Brewster between 1903 and 1906. This cultivar, the centuries-old 'Chen-Tze' or 'Royal Chen Purple', renamed 'Brewster' in Florida, from the northern limit of the lychee-growing area in China, withstands light frost and proved to be very successful in the Lake Placid area—the "Ridge" section of Central Florida.

Layered trees were available from Reasoner's Royal Palm Nurseries in the early 1920's, and the Reasoner's and the U.S. Department of Agriculture made many new introductions for trial. But there were no large plantings until an improved method of propagation was developed by Col. William R. Grove who became acquainted with the lychee during military service in the Orient, retired from the Army, made his home at Laurel (14 miles south of Sarasota, Florida) and was encouraged by knowledgeable Prof. G. Weidman Groff, who had spent 20 years at Canton Christian College. Col. Grove made arrangements to air-layer hundreds of branches on some of the old, flourishing 'Brewster' trees in Sebring and Babson Park and thus acquired the stock to establish his lychee grove. He planted the first tree in 1938, and by 1940 was selling lychee plants and promoting the lychee as a commercial crop. Many small orchards were planted from Merritt's Island to Homestead and the Florida Lychee Growers' Association was founded in 1952, especially to organize cooperative marketing. The spelling "lychee" was officially adopted by the association upon the strong recommendation of Professor Groff.

In 1960, over 6,000 lbs (2,720 kg) were shipped to New York, 4,000 lbs (1,814 kg) to California, nearly 6,000 lbs (2,720 kg) to Canada, and 3,900 lbs (1,769 kg) were consumed in Florida, though this was far from a record year. The commercial lychee crop in Florida has fluctuated with weather conditions, being affected not only by freezes but also by drought and strong winds. Production was greatly reduced in 1959, to a lesser extent in 1963, fell drastically in 1965, reached a high of 50,770 lbs (22,727 kg) in 1970, and a low of 7,200 lbs (3,273 kg) in 1974. Some growers lost up to 70% of their crop because of severe cold in the winter of 1979-80. Of course, there are many bearing trees in home gardens that are not represented in production figures. The fruit from these trees may be merely for household consumption or may be purchased at the site by Chinese grocers or restaurant operators, or sold at roadside stands.

Though the Florida lychee industry is small, mainly because of weather hazards, irregular bearing and labor of hand-harvesting, it has attracted much attention to the crop and has contributed to the dissemination of planting material to other areas of the Western Hemisphere. Escalating land values will probably limit the expansion of lychee plantings in this rapidly developing state. Another limiting factor is that much land suitable for lychee culture is already devoted to citrus groves.

Varieties

Professor Groff, in his book, *The lychee and the lungan,* tells us that the production of superior types of lychee is a matter of great family pride and local rivalry in China, where the fruit is esteemed as no other. In 1492, a list of 40 lychee varieties, mostly named for families, was published in the *Annals of Fukien.* In the Kwang provinces there were 22 types, 30 were listed in the Annals of Kwangtung, and 70 were tallied as varieties of Ling Nam. The Chinese claim that the lychee is highly variable under different cultural and soil conditions. Professor Groff concluded that one could catalog 40 or 50 varieties as recognized in Kwangtung, but there were only 15 distinct, widely-known and commercial varieties grown in that province, half of them marketed in season in the City of Canton. Some of these are classed as "mountain" types; the majority are "water types" (grown in low, well-irrigated land). There is a special distinction between the kinds of lychee that leak juice when the skin is broken and those that retain the juice within the flesh. The latter are called "dry-and-clean" and are highly prized. There is much variation in form (round, egg-shaped or heart-shaped), skin color and texture, the fragrance and flavor and even the color, of the flesh; and the amount of "rag" in the seed cavity; and, of prime importance, the size and form of the seed.

The following are the 15 cultivars recognized by Professor Groff:

'No Mai Tsze', or 'No mi ts 'z' (glutinous rice) is the leading variety in China; large, red, "dry-and-clean"; seeds often small and shriveled. It is one of the best for drying, and is late in season. It does best when grafted onto the 'Mountain' lychee.

'Kwa luk' or 'Kua lu' (hanging green) is a famous lychee; large, red with a green tip and a typical green line; "dry-and-clean"; of outstanding flavor and fragrance. It was, in olden times, a special fruit for presentation to high officials and other persons in positions of honor. Professor Groff was given a single fruit in a little red box!

'Kwai mi' or 'Kuei Wei', (cinnamon flavor) which came to be called 'Mauritius' is smaller, heart-shaped, with rough red skin tinged with green on the shoulders and usually having a thin line running around the fruit. The seed is small and the flesh very sweet and fragrant. The branches of the tree curve upward at the tips and the leaflets curl inward from the midrib.

'Hsiang li', or 'Heung lai' (fragrant lychee) is borne by a tree with distinctive erect habit having upward-pointing leaves. The fruit is small, very rough and prickly, deep-red, with the smallest seeds of all, and the flesh is of superior flavor and fragrance. It is late in season. Those grown in Sin Hsing are better than those grown in other locations.

'Hsi Chio tsu', or 'Sai kok tsz' (rhinoceros horn) is borne by a large-growing tree. The fruit is large, rough, broad at the base and narrow at the apex; has somewhat tough and fibrous, but fragrant, sweet, flesh. It ripens early.

'Hak ip', or 'Hei yeh', (black leaf) is borne by a densely-branched tree with large, pointed, slightly curled, dark-green leaflets. The fruit is medium-red, sometimes with green tinges, broad-shouldered, with thin, soft skin and the flesh, occasionally pinkish, is crisp and sweet. This is rated as "one of the best 'water' lychees."

'Fei tsu hsiao', or 'Fi tsz siu' (imperial concubine's laugh, or smile) is large, amber-colored, thin-skinned, with very sweet, very fragrant flesh. Seeds vary from large to very small. It ripens early.

'T' ang po', or 'T' ong pok' (pond embankment) is from a small-leaved tree. The fruit is small, red, rough, with thin, juicy acid flesh and very little rag. It is a very early variety.

'Sheung shu wai' or 'Shang hou huai', (President of a Board's embrace) is borne on a small-leaved tree. The fruit is large, rounded, red, with many dark spots. It has sweet flesh with little scent and the seed size is variable. It is rather late in season.

'Ch'u ma lsu', or 'Chu ma lsz' (China grass fiber) has distinctive, lush foliage. The leaves are large, overlapping, with long petioles. The fruits are large with prominent shoulders and rough skin, deep red inside. While very fragrant, the flesh is of inferior flavor and clings to the seed which varies from large to small.

'Ta tsao', or 'Tai tso' (large crop) is widely grown around Canton; somewhat egg-shaped; skin rough, bright-red with many small, dense dots; flesh firm, crisp, sweet, faintly streaked with yellow near the large seed. The juice leaks when the skin is broken. The fruit ripens early.

'Huai chih', or 'Wai chi' (the Wai River lychee) has medium-sized, blunt leaves. The fruit is round with medium-smooth skin, a rich red outside, pink inside; and leaking juice. This is not a high class variety but the most commonly grown, high yielding, and late in season.

'San yueh hung', or 'Sam ut hung' (third month red), also called 'Ma yuen', 'Ma un', 'Tsao kuo', 'Tso kwo', 'Tsao li', or 'Tsoli' (early lychee) is grown along dykes. The branches are brittle and break readily; the leaves are long, pointed, and thick. The fruit is very large, with red, thick, tough skin and thick, medium-sweet flesh with much rag. The seeds are long but aborted. This variety is popular mainly because it comes into season very early.

'Pai la li chih', or 'Pak lap lai chi' (white wax lychee), also called 'Po le tzu', or 'Pak lik tsz (white fragrant plant), is large, pink, rough, with pinkish, fibrous, not very sweet flesh and large seeds. It ripens very late, after 'Huai chih'.

'Shan chi', or 'Shan chih' (mountain lychee), also called 'Suan chih', or 'Sun chi' (sour lychee) grows wild in the hills and is often planted as a rootstock for better varieties. The tree is of erect habit with erect twigs and large, pointed, short-petioled leaves. The fruit is bright-red, elongated, very rough, with thin flesh, acid flavor and large seed.

'T'im ngam', or 'T' ien yeh' (sweet cliff) is a common variety of lychee which Professor Groff reported to be quite widely grown in Kwantung, but not really on a commercial basis.

In his book, *The Litchi,* Dr. Lal Behari Singh wrote that Bihar is the center of lychee culture in India, producing 33 selected varieties classified into 15 groups. His extremely detailed descriptions of the 10 cultivars recommended for large-scale cultivation I have abbreviated (with a few bracketed additions from other sources):

'Early Seedless', or 'Early Bedana'. Fruit 1⅓ in (3.4 cm) long, heart-shaped to oval; rough, red, with green interspaces; skin firm and leathery; flesh [ivory] to white, soft, sweet; seed shrunken, like a dog's tooth. Of good quality. The tree bears a moderate crop, early in season.

'Rose-scented'. Fruit 1¼ in (3.2 cm) long; rounded-heart-shaped; slightly rough, purplish-rose, slightly firm skin; flesh gray-white, soft, very sweet. Seed round-ovate, fully developed. Of good quality. [Tree bears a moderate crop] in midseason.

'Early Large Red'. Fruit slightly more than 1⅓ in (3.4 cm) long, usually obliquely heart-shaped; crimson [to carmine], with green interspaces; very rough; skin very firm and leathery, adhering slightly to the flesh. Flesh grayish-white, firm, sweet and flavorful. Of very good quality. [Tree is a moderate bearer], early in season.

'Dehra Dun', [or 'Dehra Dhun']. Fruit less than 1½ in (4 cm) long; obliquely heart-shaped to conical; a blend of red and orange-red; skin rough, leathery; flesh gray-white, soft, of good, sweet flavor. Seed often shrunken, occasionally very small. Of good quality; midseason. [This is grown extensively in Uttar Pradesh and is the most satisfactory lychee in Pakistan.]

'Late Long Red', or 'Muzaffarpur'. Fruit less than 1½ in (4 cm) long; usually oblong-conical; dark-red with greenish interspaces; skin rough, firm and leathery, slightly adhering to the flesh; flesh grayish-white, soft, of good, sweet flavor. Seed cylindrical, fully developed. Of good quality. [Tree is a heavy bearer], late in season.

'Pyazi'. Fruit 1⅓ in (3.4 cm) long; oblong-conical to heart-shaped; a blend of orange and orange-red, with yellowish-red, not very prominent, tubercles. Skin leathery, adhering; flesh gray-white, firm, slightly sweet, with flavor reminiscent of "boiled onion". Seed cylindrical, fully developed. Of poor quality. Early in season.

'Extra Early Green'. Fruit 1¼ in (3.2 cm) long; mostly heart-shaped, rarely rounded or oblong; yellowish-red with green interspaces; skin slightly rough, leathery, slightly adhering; flesh creamy-white, [firm, of good, slightly acid flavor]; seed oblong, cylindrical or flat. Of indifferent quality. Very early in season.

'Kalkattia', ['Calcuttia', or 'Calcutta']. Fruit 1½ in (4 cm) long; oblong or lopsided; rose-red with darker tubercles; skin very rough, leathery, slightly adhering; flesh grayish ivory, firm, of very sweet, good flavor. Seed oblong or concave. Of very good quality. [A heavy bearer; withstands hot winds]. Very late in season.

'Gulabi'. Fruit 1⅓ in (3.4 cm) long; heart-shaped, oval or oblong; pink-red to carmine with orange-red tubercles; skin very rough, leathery, non-adherent; flesh gray-white, firm, of good subacid flavor; seed oblong-cylindrical, fully developed. Of very good quality. Late in season.

'Late Seedless', or 'Late Bedana'. Fruit less than 1⅜ in (3.65 cm) long; mainly conical, rarely ovate; orange-red to carmine with blackish-brown tubercles; skin rough, firm, non-adherent; flesh creamy-white, soft; very sweet, of very good flavor except for slight bitterness near the seed. Seed slightly spindle-shaped, or like a dog's tooth; underdeveloped. Of very good quality. [Tree bears heavily. Withstands hot winds.] Late in season.

There are numerous lychee orchards in the submontane region of the Punjab. The leading variety is:

'Panjore common'. Fruit is large, heart-shaped, deep-orange to pink; skin is rough, very thin, apt to split. Tree bears heavily and has the longest fruiting season—for an entire month beginning near the end of May. Six other varieties commonly grown there are: 'Rose-scented', 'Bhadwari', 'Seedless No. 1', 'Seedless No. 2', 'Dehra Dun', and 'Kalkattia'.

In South Africa, only one variety is produced commercially. It is the 'Kwai Mi' but it is locally called 'Mauritius' because nearly all of the trees are descendants of those brought in from that island. In South Africa, the fruit is of medium size, nearly round but slightly oval, reddish-brown. Flesh is firm, of good quality and usually contains a medium-sized seed, but certain fruits with broad, flat shoulders and shortened form tend to have "chicken-tongue" seeds.

There have been many other introductions into South Africa from China and India but most failed to survive. In 1928, 16 varieties from India were planted at Lowe's Orchards, Southport, Natal, but the records were lost and they remained unnamed. A Litchi Variety Orchard of 26 cultivars from India, China, Taiwan and elsewhere was established at the Subtropical Horticulture Research Station in Nelspruit. Tentative classifications grouped these into 3 distinct types—'Kwai Mi' ['Mauritius'], 'Hak Ip' (of high quality and small seed but a shy bearer in the Low-veld), and the 'Madras', a heavy bearer of choice fruits, bright-red, very rough, and with large seeds, but very sweet, luscious flesh.

The first lychee introduced into Hawaii was the 'Kwai Mi', as was the second introduction several years later. The high quality of this variety (sometimes locally called 'Charlie Long') caused the lychee to become extremely popular and widely planted. The Hawaiian Agricultural Experiment Station imported 3 'Brewster' trees in 1907, and various efforts were made to bring other types from China but not all survived. A total of 16 varieties became well established in Hawaii, including 'Hak Ip' which has become second to 'Kwai Mi' in importance.

In 1942, the Agricultural Experiment Station set out a collection of 500 seedlings of 'Kwai Mi', 'Hak Ip' and 'Brewster' with a view to selecting the trees showing the best performance. One tree of outstanding character (a seedling of 'Hak Ip') was first designated H.A.E.S. Selection 1-18-3 and was given the name 'Groff' in 1953. It is a consistent bearer, late in season. The fruit is of medium size, dark rose-red with green or yellowish tinges on the apex of each tubercle. The flesh is white and firm; there is no leaking juice; the flavor is excellent, sweet and sub-acid; most of the fruits have abortive, "chicken-tongue" seeds and, accordingly have 20% more flesh than if the seeds were fully developed.

'No Mai Tsze' has been growing in Hawaii for over 40 years but has produced very few fruits. 'Pat Po Heung' (eight precious fragrances), erroneously called 'Pat Po Hung' (eight precious red), somewhat resembles 'No Mai Tsze' but is smaller; the skin is purplish-red, thin and pliable; the juice leaks when the skin is broken; the flesh is soft, juicy, sweet even when slightly unripe; the seed varies from medium to large. The tree is slow-growing and of weak, spreading habit; it bears well in Hawaii. Nevertheless, it is not commonly planted.

'Kaimana', or 'Poamoho', an open-pollinated seedling of 'Hak Ip', developed by Dr. R.A. Hamilton at the Poamoho Experiment Station of the University of Hawaii, was released in 1982. The fruit resembled 'Kwai Mi' but is twice as large, deep-red, of high quality, and the tree is a regular bearer.

'Brewster' is large, conical or wedge-shaped, red, with soft flesh, more acid than that of 'Kwai mi', and the seeds are very often fully formed and large. The leaflets are flat with slightly recurved margins and taper to a sharp point.

There were many other introductions of seeds, seedlings, cuttings or air-layers into the United States, from 1902 to 1924, mostly from China; also from India and Hawaii, and a few from Java, Cuba, and Trinidad; and these were distributed to experimenters in Florida and California, and some to botanical gardens in other states, and to Cuba, Puerto Rico, Panama, Honduras, Costa Rica and Brazil. Many were killed by cold weather in California and Florida.

In 1908, the United States Department of Agriculture brought in 27 plants of 'Kwai mi'. At the same time, 20 plants of 'Hak ip' were imported and these were sent to George B. Cellon in Miami in 1918. A tree of the 'Bedana' was introduced from India in 1913. In 1920, Professor Groff obtained seedlings of 'Shan Chi' (mountain lychee) from Kwantung Province, together with air-layers of 'Sheung shu wai', 'No mai ts 'z', and 'T' im ngam' (sweet cliff). The latter was found to bear more regularly than 'Brewster' but exhibited nutritional deficiencies in limestone soil.

Most of the various plants and rooted cuttings from them were distributed for trial; the rest were kept in U.S. Department of Agriculture greenhouses in Maryland.

'Bengal'—In 1929, the U.S. Department of Agriculture received a small lychee plant, supposedly a seedling of 'Rose-scented', from Calcutta. It was planted at the Plant Introduction Station in Miami and began bearing in 1940. The fruits resembled 'Brewster' but were more elongated, were borne in large clusters, and the flesh was firm, not leaking juice when peeled. All the fruits had fully developed seeds but smaller in proportion to flesh than those of 'Brewster'. The habit of the tree is more spreading than that of 'Brewster'; it has larger, more leathery, darker green leaves, and the bark is smoother and paler. The original tree and its air-layered progeny have shown no chlorosis on limestone in contrast to 'Brewster' trees growing nearby.

'Peerless', believed to be a seedling of 'Brewster', originated at the Royal Palm Nursery at Oneco; was transplanted to the T.R. Palmer Estate in Belleair where C.E. Ware noticed from 1936 to 1938 that it bore fruit of larger size, brighter color and higher percentage of abortive seed than 'Brewster'. In 1938, Ware air-layered and removed 200 branches, purchased the tree and moved it to his property in Clearwater. It resumed fruiting in 1940 and annual crops recorded to 1956 showed good productivity—averaging 383.4 lbs (174 kg) per year, and the rate of abortive seeds ranged from 62% to 85%. The 200 air-layers were planted out by Ware in 1942 and began bearing in 1946. Most of the fruits had fully developed seeds but the rate of abortive seeds increased year by year and in 1950 was 61%

to 70%. The cultivar was named with the approval of the Florida Lychee Growers Association. Two seedling selections by Col. Grove, 'Yellow Red' and 'Late Globe', Prof. Groff believed to be natural hybrids of 'Brewster' X 'Mountain'.

In northern Queensland, 'Kwai Mi' is the earliest cultivar grown, and about 10% of the fruits have "chicken tongue" seeds. 'Brewster' bears in mid-season and is important though the seed is nearly always fully formed and large. 'Hak Ip' is also midseason and large-seeded there. 'Bedana' is grown only in home gardens and the fruits have large seeds unlike the usual "chicken tongue" seeds of the fruits of this cultivar borne in India.

'Wai Chi' is late in season (December), has small, round fruits, basically yellow overlaid with red; the seed is small and oval. The tree is very compact with upright branches, and prefers a cooler climate than that of coastal north Queensland where it does not fruit heavily. The leaflets are concave like those of 'Kwai Mi'.

A very similar, perhaps identical, cultivar called 'Hong Kong' is grown in South Queensland. 'No Mai' bears poorly in Queensland and seems better adapted to cooler areas.

Blooming and Pollination

There are 3 types of flowers appearing in irregular sequence or, at times, simultaneously, in the lychee inflorescence: a) male; b) hermaphrodite, fruiting as female (about 30% of the total); c) hermaphrodite fruiting as male. The latter tend to possess the most viable pollen. Many of the flowers have defective pollen and this fact probably is the main cause of the abortive seeds and also the common problem of shedding of young fruits. The flowers require transfer of pollen by insects.

In India, L.B. Singh recorded 11 species of bees, flies, wasps and other insects as visiting lychee flowers for nectar. But honeybees, mostly *Apis cerana indica*, *A. dorsata* and *A. florea,* constitute 78% of the lychee-pollinating insects and they work the flowers for pollen and nectar from sunrise to sundown. *A. cerana* is the only hive bee and is essential in commercial orchards for maximum fruit production.

A 6-week survey in Florida revealed 27 species of lychee-flower visitors, representing 6 different insect Orders. Most abundant, morning and afternoon, was the secondary screw-worm fly (*Callitroga macellaria*), an undesirable pest. Next was the imported honeybee (*Apis mellifera*) seeking nectar daily but only during the morning and apparently not interested in the pollen. No wild bees were seen on the lychee flowers, though wild bees were found in large numbers collecting pollen in an adjacent fruit-tree planting a few weeks later. Third in order, but not abundant, was the soldier beetle (*Chauliognathus marginatus*). The rest of the insect visitors were present only in insignificant number. Maintenance of bee hives in Florida lychee groves is necessary to enhance fruit set and development. The fruits mature 2 months after flowering.

In India and Hawaii, there has been some interest in possible cross-breeding of the lychee and pollen storage tests have been conducted. Lychee pollen has remained viable at room temperature for 10 to 30 days in petri dishes; for 3 to 5 months in desiccators; 15 months at 32°F (0°C) and 25% relative humidity in desiccators; and 31 months under deep-freeze, -9.4°F (-23°C). There is considerable variation in the germination rates of pollen from different cultivars. In India, 'Rose Scented' has shown mean viability of 61.99% compared with 42.52% in 'Khatti'.

Climate

Groff provided a clear view of the climatic requirements of the lychee. He said that it thrives best in regions "not subject to heavy frost but cool and dry enough in the winter months to provide a period of rest." In China and India, it is grown between 15° and 30° N. "The Canton delta . . . is crossed by the Tropic of Cancer and is a subtropical area of considerable range in climate. Great fluctuations of temperature are common throughout the fall and winter months. In the winter sudden rises of temperature will at times cause the lychee . . . to flush forth . . . new growth. This new growth is seldom subject to a freeze about Canton. On the higher elevations of the mountain regions which are subject to frost the lychee is seldom grown . . . The more hardy mountainous types of the lychee are very sour and those grown near salt water are said to be likewise. The lychee thrives best on the lower plains where the summer months are hot and wet and the winter months are dry and cool."

Heavy frosts will kill young trees but mature trees can withstand light frosts. Cold tolerance of the lychee is intermediate between that of the sweet orange on one hand and mango and avocado on the other. Location, land slope, and proximity to bodies of water can make a great difference in degree of damage by freezing weather. In the severe low temperature crisis during the winter of 1957-58, the effects ranged from minimal to total throughout central and southern Florida. A grove of 12- to 14-year-old trees south of Sanford was killed back nearly to the ground; on Merritt Island trees of the same age were virtually undamaged, while a commercial mango planting was totally destroyed. L.B. Singh resists the common belief that the lychee needs winter cold spells that provide periods of temperature between 30° and 40°F (-1.11° and 4.44°C) because it does well in Mauritius where the temperature is never below 40°F (-1.11°C). However, lychee trees in Panama, Jamaica, and other tropical areas set fruit only occasionally or not at all.

Heavy rain or fog during the flowering period is detrimental, as are hot, dry, strong winds which cause shedding of flowers, also splitting of the fruit skin. Splitting occurs, too, during spells of alternating rain and hot, dry periods, especially on the sunny side of the tree. Spraying with Ethephon at 10 ppm reduced splitting in 'Early Large Red' in experiments in Nepal.

Soil

The lychee grows well on a wide range of soils. In China it is cultivated in sandy or clayey loam, "river mud", moist sandy clay, and even heavy clay. The pH should be between 6 and 7. If the soil is deficient in lime, this must be added. However, in an early experiment in a greenhouse in Washington, D.C., seedlings planted in acid soil showed superior growth and the roots had many nodules filled with mycorrhizal fungi. This caused some to speculate that inoculation might be desirable. Later, in Florida, profuse nodulation was observed on roots of lychee seedlings that had not been inoculated but merely grown in pots of sphagnum moss and given a well-balanced nutrient solution.

The lychee attains maximum growth and productivity on deep alluvial loam but flourishes in extreme southern Florida on oolitic limestone providing it is put in an adequate hole and irrigated in dry seasons.

The Chinese often plant the lychee on the banks of ponds and streams. In low, wet land, they dig ditches 10 to 15 ft (3-4.5 m) wide and 30 to 40 ft (9-12 m) apart, using the excavated soil to form raised beds on which they plant lychee trees, so that they have perfect drainage but the soil is always moist. Though the lychee has a high water requirement, it cannot stand water-logging. The water table should be at least 4 to 6 ft (1.2-1.8 m) below the surface and the underground water should be moving inasmuch as stagnant water induces root rot. The lychee can stand occasionally brief flooding better than citrus. It will not thrive under saline conditions.

Propagation

Lychees do not reproduce faithfully from seed, and the choicest have abortive, not viable, seed. Furthermore, lychee seeds remain viable only 4 to 5 days, and seedling trees will not bear until they are 5 to 12, or even 25, years old. For these reasons, seeds are planted mostly for selection and breeding purposes or for rootstock.

Attempts to grow the lychee from cuttings have been generally discouraging, though 80% success has been claimed with spring cuttings in full sun, under constant mist and given weekly liquid nutrients. Ground-layering has been practiced to some extent. In China, air-layering (marcotting, or gootee) is the most popular means of propagation and has been practiced for ages. By their method, a branch of a chosen tree is girdled, allowed to callus for 1 to 2 days and then is enclosed in a ball of sticky mud mixed with chopped straw or dry leaves and wrapped with burlap. With frequent watering, roots develop in the mud and, in about 100 days, the branch is cut off, the ball of earth is increased to about 12 in (30 cm) in width, and the air-layer is kept in a sheltered nursery for a little over a year, then gradually exposed to full sun before it is set out in the orchard. Some air-layers are planted in large clay pots and grown as ornamentals.

The Chinese method of air-layering has many variations. In fact, 92 modifications have been recorded and experimented with in Hawaii. Inarching is also an ancient custom, selected cultivars being joined to 'Mountain' lychee rootstock.

In order to make air-layering less labor-intensive, to eliminate the watering, and also to produce portable, shippable layers, Colonel Grove, after much experimentation, developed the technique of packing the girdle with wet sphagnum moss and soil, wrapping it in moisture-proof clear plastic that permits exchange of air and gasses, and

tightly securing it above and below. In about 6 weeks, sufficient roots are formed to permit detaching of the layer, removal of the plastic wrap, and planting in soil in nursery containers. It is possible to air-layer branches up to 4 in (10 cm) thick, and to take 200 to 300 layers from a large tree.

Studies in Mexico have led to the conclusion that, for maximum root formation, branches to be air-layered should not be less than ⅝ in (15 mm) in diameter, and, to avoid undue defoliation of the parent tree, should not exceed ¾ in (20 mm). The branches, of any age, around the periphery of the canopy and exposed to the sun, make better air-layers with greater root development than branches taken from shaded positions on the tree. The application of growth regulators, at various rates, has shown no significant effect on root development in the Mexican experiments. In India, certain of the various auxins tried stimulated root formation, forced early maturity of the layers, but contributed to high mortality. South African horticulturists believe that tying the branch up so that it is nearly vertical induces vigorous rooting.

The new trees, with about half of the top trimmed off and supported by stakes, are kept in a shadehouse for 6 weeks before setting out. Improvements in Colonel Grove's system later included the use of constant mist in the shadehouse. Also, it was found that birds pecked at the young roots showing through the transparent wrapping, made holes in the plastic and caused dehydration. It became necessary to shield the air-layers with a cylinder of newspaper or aluminum foil. As time went on, some people switched to foil in place of plastic for wrapping the air-layers.

The air-layered trees will fruit in 2 to 5 years after planting, Professor Groff said that a lychee tree is not in its prime until it is 20 to 40 years old; will continue bearing a good crop for 100 years or longer. One disadvantage of air-layering is that the resultant trees have weak root systems. In China, a crude method of cleft-grafting has long been employed for special purposes, but, generally speaking, the lychee has been considered very difficult to graft. Bark, tongue, cleft, and side-veneer grafting, also chip- and shield-budding, have been tried by various experimenters in Florida, Hawaii, South Africa and elsewhere with varing degrees of success. The lychee is peculiar in that the entire cambium is active only during the earliest phases of secondary growth. The use of very young rootstocks, only ¼ in (6 mm) in diameter and wrapping the union with strips of vinyl plastic film, have given good results. A 70% success rate has been achieved in splice-grafting in South Africa. Hardened-off, not terminal, wood of young branches ¼ in (6 mm) thick is first ringed and the bark-ring removed. After a delay of 21 days, the branch is cut off at the ring, defoliated but leaving the base of each petiole, then a slanting cut is made in the rootstock 1 ft (30 cm) above the soil, at the point where it matches the thickness of the graftwood (scion), and retaining as many leaves as possible. The cut is trimmed to a perfectly smooth surface 1 in (2.5 cm) long; the scion is then trimmed to 4 in (10 cm) long, making a slanting cut to match that on the rootstock. The scion should have 2 slightly swollen buds. After joining the scion and the rootstock, the union is wrapped with plastic grafting tape and the scion is completely covered with grafting strips to prevent dehydration. In 6 weeks the buds begin to swell, and the plastic is slit just above the bud to permit sprouting. When the new growth has hardened off, all the grafting tape is removed. The grafting is performed in a moist, warm atmosphere. The grafted plants are maintained in containers for 2 years or more before planting out, and they develop strong taproots.

In India, a more recent development is propagation by stooling, which has been found "simpler, quicker and more economical" there than air-layering. First, air-layers from superior trees are planted 4 ft (1.2 m) apart in "stool beds" where enriched holes have been prepared and left open for 2 weeks. Fertilizer is applied when planting (at the beginning of September) and the air-layers are well established by mid-October and putting out new flushes of growth in November. Fertilizer is applied again in February-March and June-July. Shallow cultivation is performed to keep the plot weed-free. At the end of 2½ years, in mid-February, the plants are cut back to 10 in (25 cm) from the ground. New shoots from the trunk are allowed to grow for 4 months. In mid-June, a ring of bark is removed from all shoots except one on each plant and lanolin paste containing IBA (2,500 ppm) is applied to the upper portion of the ringed area. Ten days later, earth is heaped up to cover 4 to 6 in (10-15 cm) of the stem above the ring. This causes the shoots to root profusely in 2 months. The rooted shoots are separated from the plant and are immediately planted in nursery beds or pots. Those which do not wilt in 3 weeks are judged suitable for setting out in the field. The earth around the parent plants is leveled and the process of fertilization, cultivation, ringing and earthing-up and harvesting of stools is repeated over and over for years until the parent plants have lost their vitality. It is reported that the transplanted shoots have a survival rate of 81-82% as compared with 40% to 50% in air-layers.

Culture

Spacing: For a permanent orchard, the trees are best spaced 40 ft (12 m) apart each way. In India, a 30-ft spacing is considered adequate, probably because the drier climate limits the overall growth. Portions of the tree shaded by other trees will not bear fruit. For maximum productivity, there must be full exposure to light on all sides.

In the Cook Islands, the trees are planted on a 40 x 20 ft (12x6 m) spacing — 56 trees per acre (134 per ha) — but in the 15th year, the plantation is thinned to 40 x 40 ft (12x12 m).

Wind protection: Young trees benefit greatly by wind protection. This can be provided by placing stakes around each small tree and stretching cloth around them as a windscreen. In very windy locations, the entire plantation may be protected by trees planted as windbreaks but these should not be so close as to shade the lychees. The lychee tree is structurally highly wind-resistant, having withstood typhoons, but shelter may be needed to safeguard the crop. During dry, hot months, lychee trees of any age will benefit from overhead sprinkling; they are seriously retarded by water stress.

Fertilization: Newly planted trees must be watered but not fertilized beyond the enrichment of the hole well in advance of planting. In China, lychee trees are fertilized only twice a year and only organic material is used, principally night soil, sometimes with the addition of soybean or peanut residue after oil extraction, or mud from canals and fish ponds. There is no great emphasis on fertilization in India. It has been established that a harvest of 1,000 lbs (454.5 kg) removes approximately 3 lbs (1,361 g) K_2O, 1 lb (454 g) P_2O_5, 1 lb (454 g) N, ¾ lb (340 g) CaO, and ½ lb (228 g) MgO from the soil. It is judged, therefore, that applications of potash, phosphate, lime and magnesium should be made to restore these elements.

Fertilizer experiments on fine sand in central Florida have shown that medium rates of N (either sulfate of ammonia or ammonium nitrate), P_2O_5, K_2O, and MgO, together with one application of dolomite limestone at 2 tons/acre (4.8 tons/ha) are beneficial in counteracting chlorosis and promoting growth, flowering and fruit-set and reducing early fruit shedding. Excessive use of nitrogen suppresses growth and interferes with the uptake of other nutrients. If vegetative dormancy is to be encouraged in bearing trees, fertilizer should be withheld in fall and early winter.

In limestone soil, it may be necessary to spread chelated iron 2 or 3 times a year to avoid chlorosis. Zinc deficiency is evidenced by bronzing of the leaves. It is corrected by a foliar spray of 8 lbs (3.5 kg) zinc sulphate and 4 lbs (1.8 kg) hydrated lime in 48 qts (45 liters) of water. Because of the very shallow root system of the lychee, a surface mulch is very beneficial in hot weather.

Pruning: Ordinarily, the tree is not pruned after the judicious shaping of the young plant, because the clipping off of a branch tip with each cluster of fruits is sufficient to promote new growth for the next crop. Severe pruning of old trees may be done to increase fruit size and yield for at least a few years.

Girdling: The Indian farmer may girdle the branches or trunk of his lychee trees in September to enhance flowering and fruiting. Tests on 'Brewster' in Hawaii confirmed the much higher yield obtained from branches girdled in September. Girdling of trees that begin to flush in October and November is ineffective. Similar trials in Florida showed increased yield of trees that had poor crops the previous year, but there was no significant increase in trees that had been heavy bearers. Furthermore, many branches were weakened or killed by girdling. Repeated girdling as a regular practice would probably seriously interfere with overall growth and productivity.

Indian horticulturists warn that girdling in alternate years, or girdling just half of the tree, may be preferable to annual girdling and that, in any case, heavy fertilization and irrigation should precede girdling. Fall spraying of growth inhibitors has not been found to increase yields.

Harvesting

For home use or for local markets, lychees are harvested when fully colored; for shipment, when only partly colored. The final swelling of the fruit causes the protuberances on the skin to be less crowded and to slightly flatten out, thus an experienced picker will recognize the stage of full maturity. The fruits are rarely picked singly except for immediate eating out-of-hand, because the stem does not normally detach without breaking the skin and that causes the fruit to spoil quickly. The clusters are usually clipped with a portion of stem and a few leaves attached to prolong freshness. Individual fruits are later clipped from the cluster leaving a stub of stem attached. Harvesting may need to be done every 3 to 4 days over a period of 3-4 weeks. It is never done right after rain, as the wet fruit is very perishable. The lychee tree is not very suitable for the use of ladders. High clusters are usually harvested by metal or bamboo pruning poles. A worker can harvest 55 lbs (25 kg) of fruits per hour.

Yield

The yield varies with the cultivar, age, weather, presence of pollinators, and cultural practices. In India, a 5-year-old tree may produce 500 fruits, a 20-year-old tree 4,000 to 5,000 fruits—160 to 330 lbs (72.5-149.6 kg). Exceptional trees have borne 1,000 lbs (455 kg) of fruit per year. One tree in Florida has borne 1,200 lbs (544 kg). In China, there are reports of 1,500-lb crops (680 kg). In South Africa, trees 25 years old have averaged 600 lbs (272 kg) each in good years; and an average yield per acre is approximately 10,000 lbs annually (roughly equivalent to 10,000 kg per hectare).

Keeping Quality, Storage and Shipping

Freshly picked lychees keep their color and quality only 3 to 5 days at room temperature. If pre-treated with 0.5% copper sulphate solution and kept in perforated polyethylene bags, they will remain fresh somewhat longer.

Fresh fruits, picked individually by snapping the stems and later de-stemmed during grading, and packed in shallow, ventilated cartons with shredded-paper cushioning, have been successfully shipped by air from Florida to markets throughout the United States and also to Canada. In South Africa, freshly picked lychees have been placed on trays in ventilated sheds, dusted with sulphur and left overnight, and then allowed to "wilt" in lugs for 24 to 48 hours to permit any infested or injured fruits to become conspicuous before grading and packing. It is said that fruits so treated retain their fresh color and are unaffected by fungi or pests for several weeks.

In China and India, lychees are packed in baskets or crates lined with leaves or other cushioning. The clusters or loose fruits are best packed in trays with protective sheets between the layers and no more than 5 single layers or 3 double layers are joined together. The pack should not be too tight. Containers for stacked trays or fruits not so arranged, must be fairly shallow to avoid too much weight and crushing. Spoilage may be retarded by moistening the fruits with a salt solution.

In the Cook Islands, the fruits are removed from the clusters, dipped in Benlate to control fungal growth, dried on racks, then packed in cartons for shipment to New Zealand. South African shippers immerse the fruits for 10 minutes in a suspension of 0.375 dicloran 50% wp plus 0.625 g benomyl 50% wp per liter of water warmed to 125.6°F (52°C). Tests at CSIRO, Div. of Food Research, New South Wales, Australia, in 1982, showed good color retention, retardation of weight loss and fungal spoilage in lychees dipped in hot benomyl 0.05% at 125.6°F (52°C) for two minutes and packed in trays with PVC "skrink" film covering. The chemical treatment had not yet been approved by health authorities.

Lychee clusters shipped to France by air from Madagascar have arrived in fresh condition when packed 13 lbs (6 kg) to the carton and cushioned with leaves of the traveler's tree (*Ravenala madagascariensis* Sonn.).

Boat shipment requires hydrocooling at the plantation at 32°-35.6°F (0°-2°C), packing in sealed polyethylene bags, storing and conveying to the port at -4° to -13°F (-20°--25°C) and shipping at 32° to 35.6°F (0°-2°C).

In Florida, fresh lychees in sealed, heavy-gauge polyethylene bags keep their color for 7 days in storage or transit at 35° to 50°F (1.67°-10°C). Each bag should contain no more than 15 lbs (6.8 kg) of fruit.

Lychees placed in polyethylene bags with moss, leaves, paper shavings or cotton packing have retained fresh color and quality for 2 weeks in storage at 45°F (7.22°C); for a month at 40°F (4.44°C). At 32° to 35°F (0°-1.67°C) and 85% to 90% relative humidity, untreated lychees can be stored for 10 weeks; the skin will turn brown but the flesh will be virtually in fresh condition but sweeter.

Frozen, peeled or unpeeled, lychees in moisture-vapor-proof containers keep for 2 years.

Drying of Lychees

Lychees dehydrate naturally. The skin loses its original color, becomes cinnamon-brown, and turns brittle. The flesh turns dark-brown to nearly black as it shrivels and becomes very much like a raisin. The skin of 'Kwai Mi' becomes very tough when dried; that of 'Madras' less so. The fruits will dry perfectly if clusters are merely hung in a closed, air-conditioned room.

In China, lychees are preferably dried in the sun on hanging wire trays and brought inside at night and during showers. Some are dried by means of brick stoves during humid weather.

When exports of dried fruits from China to the United States were suspended, India welcomed the opportunity to supply the market. Experimental drying involved preliminary disinfection by immersing the fruits in 0.5% copper sulphate solution for 2 minutes. Sun-drying on coir-mesh trays took 15 days and the results were good except that thin-skinned fruits tended to crack. It was found that shade-drying for 2 days before full exposure to the sun prevented cracking.

Electric-oven drying of single layers arranged in tiers, at 122° to 140°F (50°-65°C), requires only 4 days. Hot-airblast at 160°F (70°C) dries seedless fruits in 48 hours. Fire-oven and vacuum-oven drying were found unsatisfactory. Florida researchers have demonstrated the feasibility of drying untreated lychees at 120°F (48.8°C) with free-stream air flow rates above 35 CMF/f². Drying at higher temperatures gave the fruits a bitter flavor.

The best quality and light color of flesh instead of dark-brown is achieved by first blanching in boiling water for 5 minutes, immersing in a solution of 2% potassium metabisulphite for 48 hours, and dipping in citric acid prior to drying.

Dried fruits can be stored in tins at room temperature for about a year with no change in texture or flavor.

Pests

In most areas where lychees are grown, the most serious foliage pest is the erinose, or leaf-curl, mite, *Aceria litchii*, which attacks the new growth causing hairy, blister-like galls on the upperside of the leaves, thickening, wrinkling and distorting them, and brown, felt-like wool on the underside. The mite apparently came to Florida on plants from Hawaii in 1953 but has been effectively eradicated. A leaf-webber, *Dudua aprobola*, attacks the new growth of all lychee trees in the Punjab.

The most destructive enemy of the lychee in China is a stinkbug (*Tessaratoma papillosa*) with bright-red markings. It sucks the sap from young twigs and they often die; at least there is a high rate of fruit-shedding. This pest is combatted by shaking the trees in winter, collecting the bugs and dropping them into kerosene. Without such efforts, it works havoc. A stinkbug (*Banasa lenticularis*) has been found on lychee foliage in Florida. The leaf-eating false-unicorn caterpillar (*Schizura ipomeae*), which is parasitized by a tachinid fly (*Thorocera floridensis*) feeds on the leaves. The foliage is sometimes infested with red spider mites (*Paratetranychus hawaiiensis*). The citrus aphid (*Toxoptera aurantii*) preys on flush foliage. Two leaf rollers, *Argyroploce leucaspis*, and *A. aprobola*, are active on lychee trees in India. Thrips (*Dolicothrips idicus*) attack the foliage and *Megalurothrips* (*Taeniothrips*) *distalis* and *Lymantria mathura* damage the flowers.

A twig-pruner, *Hypermallus villosus*, has damaged lychee trees in Florida and a twig borer, *Proteoteras implicata*, has killed twigs of new growth on Florida lychees. The larvae of a native leaf beetle, *Exema nodulosa*, has been found puncturing and girdling lychee branchlets 1/8 to 1/4 in (3-6 mm) thick. Ambrosia beetles bore into the stems of young trees and fungi enter through their holes. A shoot-borer, *Chlumetia transversa*, is found on lychee trees all over India. Two bark-boring caterpillars, *Indarbela quadrinotata* and *I. tetraonis*, bore rings around the trunk underneath the bark of older trees. The larvae of a small moth, *Acrocerops cramerella*, eat developing seeds and the pith of young twigs. A small parasitic wasp helps to control this predator, as does the sanitary practice of burning the fallen lychee leaves.

The aphid (*Aphis spiraecola*) occurs on young plants in shaded nurseries, as does the armored scale, or lychee bark scale, *Pseudaulacaspis major*, and white peach scale,

P. pentagona. The Florida red scale, *Chrysomphalus aonidum*, has been seen on lychee trees, also the banana-shaped scale, *Coccus acutissimus*, and green-shield scale, *Pulvinaria psidii*. The latter is the second most serious pest in Florida. Others are the six-spotted mite, *Eotetranychus sexmaculatus*, the leaf-footed bug, *Leptoglossus phyllopus*, and less troublesome creatures such as the several species of Scarabaeidae (related to June bugs) which attack leaves and flower buds.

In South Africa, the parasitic nematode *Hemicricone-moides mangiferae* and *Xiphinema brevicolle* cause dieback, decline and ultimately death of lychee trees, sometimes devastating orchards. The root-knot nematode, *Meloidogyne javanica*, also attacks the lychee in South Africa but is less prevalent.

In Florida, the southern green stinkbug, *Nezara viridula*, and the larvae of the cotton square borer, *Strymon metinus*, attack the fruit. Seed-feeding Lepidoptera, especially *Cryptophlebia ombrodelta* and *Lobesia* sp. cause much fruit damage and falling in northern Queensland. Carbaryl sprays considerably reduce the losses. In South Africa, a moth, *Argyroploce peltastica*, lays eggs on the surface of the fruit and the larvae may penetrate weak areas of the skin and infest the flesh. The fruit flies, *Ceratites capitata* and *Pterandrus rosa* make minute holes and cracks in the skin and cause internal decay. These pests are so detrimental that growers have adopted the practice of enclosing bunches of clusters (with most of the leaves removed) in bags made of "wet-strength" paper or unbleached calico 6 to 8 weeks before harvest-time. The Caribbean fruit fly, *Anastrepha suspensa*, has attacked lychee fruits in Florida.

Birds, bats and bees damage ripe fruits on the trees in China and sometimes a stilt house is built beside a choice lychee tree for a watchman to keep guard and ward off these predators, or a large net may be thrown over the tree. In Florida, birds, squirrels, raccoons and rats are prime enemies. Birds have been repelled by hanging on the branches thin metallic ribbons which move, gleam and rattle in the wind. Grasshoppers, crickets, and katydids may, at times, feed heavily on the foliage.

Diseases

Few diseases have been reported from any lychee-growing locality. The glossy leaves are very resistant to fungi. In Florida, lychee trees are occasionally subject to green scurf, or algal leaf spot (*Cephaleuros virescens*), leaf blight (*Gleosporium* sp.), die-back, caused by *Phomopsis* sp., and mushroom root rot (*Clitocybe tabescens*) which is most likely to attack lychee trees planted where oak trees formerly stood. Old oak roots and stumps have been found thoroughly infected with the fungus.

In India, leaf spot caused by *Pestalotia pauciseta* may be prevalent in December and can be controlled by lime-sulphur sprays. Leaf spots caused by *Botryodiplodia theobromae* and *Colletotrichum gloeosporioides*, which begin at the tip of the leaflet, were first noticed in India in 1962.

Lichens and algae commonly grow on the trunks and branches of lychee trees.

The main post-harvest problem is spoilage by the yeast-like organism, which is quick to attack warm, moist fruits. It is important to keep the fruits dry and cool, with good circulation of air. When conditions favor rotting, dusting with fungicide will be necessary.

Food Uses

Lychees are most relished fresh, out-of-hand. Peeled and pitted, they are commonly added to fruit cups and fruit salads. Lychees stuffed with cottage cheese are served as salad topped with dressing and pecans. Or the fruit may be stuffed with a blend of cream cheese and mayonnaise, or stuffed with pecan meats, and garnished with whipped cream. Sliced lychees, congealed in lime gelatin, are served on lettuce with whipped cream or mayonnaise. The fruits may be layered with pistachio ice cream and whipped cream in parfait glasses, as dessert. Halved lychees have been placed on top of ham during the last hour of baking, or grilled on top of steak. Pureed lychees are added to ice cream mix. Sherbet is made by extracting the juice from fresh, seeded lychees and adding it to a mixture of prepared plain gelatin, hot milk, light cream, sugar and a little lemon juice, and freezing.

Peeled, seeded lychees are canned in sugar sirup in India and China and have been exported from China for many years. Browning, or pink discoloration, of the flesh is prevented by the addition of 4% tartaric acid solution, or by using 30° Brix sirup containing 0.1% to 0.15% citric acid to achieve a pH of about 4.5, processing for a maximum of 10 minutes in boiling water, and chilling immediately.

Fig. 73: Peeled, seeded, lychees (*Litchi chinensis*) are canned in sirup in the Orient and exported to the United States and other countries.

Food Value Per 100 g of Edible Portion*		
	Fresh	*Dried*
Calories	63-64	277
Moisture	81.9-84.83%	17.90-22.3%
Protein	0.68-1.0 g	2.90-3.8 g
Fat	0.3-0.58 g	0.20-1.2 g
Carbohydrates	13.31-16.4 g	70.7-77.5 g
Fiber	0.23-0.4 g	1.4 g
Ash	0.37-0.5 g	1.5-2.0 g
Calcium	8-10 mg	33 mg
Phosphorus	30-42 mg	–
Iron	0.4 mg	1.7 mg
Sodium	3 mg	3 mg
Potassium	170 mg	1,100 mg
Thiamine	28 mcg	–
Nicotinic Acid	0.4 mg	–
Riboflavin	0.05 mg	0.05 mg
Ascorbic Acid	24-60 mg	42 mg

*According to analyses made in China, India and the Philippines.

The lychee is low in phenols and non-astringent in all stages of maturity.

To a small extent, lychees are also spiced or pickled, or made into sauce, preserves or wine. Lychee jelly has been made from blanched, minced lychees and their accompanying juice, with 1% pectin, and combined phosphoric and citric acid added to enhance the flavor.

The flesh of dried lychees is eaten like raisins. Chinese people enjoy using the dried flesh in their tea as a sweetener in place of sugar.

Whole frozen lychees are thawed in tepid water. They must be consumed very soon, as they discolor and spoil quickly.

Other Uses

In China, great quantities of honey are harvested from hives near lychee trees. Honey from bee colonies in lychee groves in Florida is light amber, of the highest quality, with a rich, delicious flavor like that of the juice which leaks when the fruit is peeled, and the honey does not granulate.

Medicinal Uses: Ingested in moderate amounts, the lychee is said to relieve coughing and to have a beneficial effect on gastralgia, tumors and enlargements of the glands. One stomach-ulcer patient in Florida, has reported that, after eating several fresh lychees he was able to enjoy a large meal that, ordinarily, would have caused great discomfort. Chinese people believe that excessive consumption of raw lychees causes fever and nosebleed. According to legends, ancient devotees have consumed from 300 to 1,000 per day.

In China, the seeds are credited with an analgesic action and they are given in neuralgia and orchitis. A tea of the fruit peel is taken to overcome smallpox eruptions and diarrhea. In India, the seeds are powdered and, because of their astringency, administered in intestinal troubles, and they have the reputation there, as in China, of relieving neuralgic pains. Decoctions of the root, bark and flowers are gargled to alleviate ailments of the throat.

Lychee roots have shown activity against one type of tumor in experimental animals in the United States Department of Agriculture/National Cancer Institute Cancer Chemotherapy Screening Program.

Longan

Closely allied to the glamorous lychee, in the family Sapindaceae, the longan, or lungan, also known as dragon's eye or eyeball, and as *mamoncillo chino* in Cuba, has been referred to as the "little brother of the lychee", or *li–chih-nu,* "slave of the lychee". Botanically, it is placed in a separate genus, and is currently designated *Dimocarpus longan* Lour. (syns. *Euphoria longan* Steud.; *E. longana* Lam.; *Nephelium longana* Cambess.). According to the esteemed scholar, Prof. G. Weidman Groff, the longan is less important to the Chinese as an edible fruit, more widely used than the lychee in Oriental medicine.

Description

The longan tree is handsome, erect, to 30 or 40 ft (9–12 m) in height and to 45 ft (14 m) in width, with rough-barked trunk to 2½ ft (76.2 cm) thick and long, spreading, slightly drooping, heavily foliaged branches. The evergreen, alternate, paripinnate leaves have 4 to 10 opposite leaflets, elliptic, ovate-oblong or lanceolate, blunt-tipped; 4 to 8 in (10–20 cm) long and 1⅜ to 2 in (3.5–5 cm) wide; leathery, wavy, glossy-green on the upper surface, minutely hairy and grayish-green beneath. New growth is wine-colored and showy. The pale-yellow, 5- to 6-petalled, hairy-stalked flowers, larger than those of the lychee, are borne in upright terminal panicles, male and female mingled. The fruits, in drooping clusters, are globose, ½ to 1 in (1.25–2.5 cm) in diameter, with thin, brittle, yellow-brown to light reddish-brown rind, more or less rough (pebbled), the protuberances much less prominent than those of the lychee. The flesh (aril) is mucilaginous, whitish, translucent, somewhat musky, sweet, but not as sweet as that of the lychee and with less "bouquet". The seed is round, jet-black, shining, with a circular white spot at the base, giving it the aspect of an eye.

Origin and Distribution

The longan is native to southern China, in the provinces of Kwangtung, Kwangsi, Schezwan and Fukien, between elevations of 500 and 1,500 ft (150–450 m). Groff wrote: "The lungan, not so highly prized as the lychee, is nevertheless usually found contiguous to it. . . . It thrives much better on higher ground than the lychee and endures more frost. It is rarely found growing along the dykes of streams as is the lychee but does especially well on high ground near ponds. . . . The lungan is more seldom grown under orchard conditions than is the lychee. There is not so large a demand for the fruit and the trees therefore more scattered although one often finds attractive groups of lungan." Groff says that the longan was introduced into India in 1798 but, in Indian literature, it is averred that the longan is native not only to China but also to southwestern India and the forests of upper Assam and the Garo hills, and is cultivated in Bengal and elsewhere as an ornamental and shade tree. It is commonly grown in former Indochina (Thailand, Cambodia, Laos and Vietnam and in Taiwan). The tree grows but does not fruit in Malaya and the Philippines. There are many of the trees in Reúnion and Mauritius.

The longan was introduced into Florida from southern China by the United States Department of Agriculture in 1903 and has flourished in a few locations but never became popular. There was a young tree growing at the Agricultural Station in Bermuda in 1913. A tree planted at the Federal Experiment Station in Mayaguez, Puerto Rico, was 10 ft (3 m) high in 1926, 23 ft (7 m) in 1929. A longan tree flourished in the Atkins Garden in Cuba and seedlings were distributed but found to fruit irregularly and came to be valued mostly for their shade and ornamental quality. In Hawaii, the longan was found to grow faster and more vigorously than the lychee but the fruit is regarded there as less flavorful than the lychee.

Varieties

It seems that the type of longan originally brought to the New World was not one of the best, having aroused so little interest in the fruit. Groff stated that the leading variety of Fukien was the round-fruited 'Shih hsia', the "Stone Gorge Lungan" from P'ing Chou. There were 2 types, one, 'Hei ho shih hsia', black-seeded, and 'Chin ch' i ho shih hsia', brown-seeded. This variety did not excel in size but the flesh was crisp, sweeter than in other varieties, the seed small and the dried flesh, after soaking in water, was restored almost to fresh condition.

None of the other 4 varieties described by Groff has any great merit.

'Wu Yuan' ("black ball") has small, sour fruit used for canning. The tree is vigorous and seedlings are valued as rootstocks. 'Kao Yuan' is believed to be a slightly better type of this variety and is widely canned.

'Tsao ho' ('Early Rice') is the earliest variety and a form called 'Ch'i chin tsao ho' precedes it by 2 weeks. In quality, both are inferior to 'Wu Yuan'.

Fig. 74: The brown-skinned longan (*Euphoria longan*), less luscious than the lychee, is hardier, bears heavily and later in the year.

'She p' i' ('Snake skin') has the largest fruit, as big as a small lychee and slightly elongated. The skin is rough, the seed large, some of the juice is between the rind and the flesh, and the quality is low. Its only advantage is that it is very late in season.

'Hua Kioh' ('Flower Skin'), slightly elongated, has thin, nearly tasteless flesh, some of the juice is between the rind and the flesh, and the overall quality is poor. It is seldom propagated vegetatively.

There are no "chicken-tongue" (aborted seed) varieties in China.

There are 2 improved cultivars grown extensively in Taiwan—'Fukien Lungan' ('Fukugan') was introduced from Fukien Province in mainland China. The other, very similar and possibly a mutant of 'Fukien', is 'Lungan Late', which matures a month later than 'Fukien'.

In 1954, William Whitman of Miami introduced a superior variety of longan, the 'Kohala', from Hawaii. It began to bear in 1958. The fruit is large for the species, the seed is small, and the flesh is aromatic, sweet and spicy. The tree produces fairly good crops in midsummer. One hundred or more air-layers have been brought by air from Hawaii and planted at various locations in southern Florida and in the Bahamas. A seedling planting and selection program was started in 1962 at the USDA Subtropical Horticulture Research Unit, Miami. The plants were all open-pollinated seedlings of the canning variety, 'Wu Yuan', brought in from Canton in 1930 as P. I. #89409. Some set fruit in 1966 and 1967 but more of them in 1968. Evaluation of these and other acquisitions continues. Included in the study are M-17886, 'Chom Poo Nuch', and M-17887, 'E-Haw'.

Climate

Professor Groff wrote that "the lungan . . . is found growing at higher latitudes and higher altitudes than the lychee." Also: "On the higher elevations of the mountainous regions which are subject to frost the lychee is seldom grown. The longan appears in these regions more often but it, too, cannot stand heavy frosts." The longan's range in Florida extends north to Tampa on the west coast and to Merritt Island on the east coast. Still, small trees suffer leaf- and twig-damage if the temperature falls to 31° or 30°F (-0.56°--1.11°C) and are killed at just a few degrees lower. Larger trees show leaf injury at 27° to 28°F (-2.78°--2.22°C), small branch injury at 25° to 26°F (-3.89°--3.33°C), large branch and trunk symptoms at 24°F (-4.44°C) and sometimes fail to recover.

On the other hand, after a long period of cool weather over the 3 winter months, with no frost, longan trees bloom well. Blooming is poor after a warm winter.

Soil

The longan thrives best on a rich sandy loam and nearly as well on moderately acid, somewhat organic, sand. It also grows to a large size and bears heavily in oolitic limestone. In organic muck soils, blooming and fruiting are deficient.

Propagation

Most longan trees have been grown from seed. The seeds lose viability quickly. After drying in the shade for ½ day, they should be planted without delay, but no more than ¾ in (2 cm) deep, otherwise they may send up more than one sprout. Germination takes place within a week or 10 days. The seedlings are transplanted to shaded nursery rows the following spring and set in the field 2-3 years later during winter dormancy.

In Kwangtung Province, when vegetative propagation is undertaken, it is mostly by means of inarching, nearly always onto 'Wu Yuan' trees 3-5 years old and 5 to 6 ft (1.5-1.8 m) high. The union is made no less than 4 ft (1.2 m) from the ground because it is most convenient. Nevertheless, the point of attachment remains weak and needs to be braced with bamboo to avoid breaking in high winds.

Grafting is uncommon and when it is done, it is a sandwich graft on longan rootstock, 3 or 4 grafts being made successively, one onto the beheaded top of the preceding one, in the belief that it makes the graft wind-resistant and that it induces better size and quality in the fruit.

Conventional modes of grafting have not been successful in Florida, but whip-grafting has given 80% success in Taiwan. Air-layering is frequently done in Fukien Province and was found a feasible means of distributing the 'Kohala' from Hawaii. Air-layers bear in 2 to 3 years after planting. A tree can be converted to a preferred cultivar by cutting it drastically back and veneer-grafting the new shoots.

Culture

In China, if the longan is raised on the lowlands it is always put on the edges of raised beds. On high ground, the trees are placed in pre-enriched holes on the surface. The trees are fertilized after the fruit harvest and during the blooming season, at which time the proportion of nitrogen is reduced. Fresh, rich soil is added around the base of the trees year after year. The longan needs an adequate supply of water and can even stand brief flooding, but not prolonged drought. Irrigation is necessary in dry periods.

An important operation is the pruning of many flower-bearing twigs — ¾ of the flower spikes in the cluster being removed. Later, the fruit clusters are also thinned, in order to increase the size and quality of the fruits.

Generally, the trees are planted too close together, seriously inhibiting productivity when they become overcrowded. In China, full-grown trees given sufficient room — at least 40 ft (12 m) apart — may yield 400 to 500 lbs (180-225 kg) in good years. Crops in Florida from trees 20 ft (6 m) tall and broad, have varied from light — 50-100 lbs (22.5-45 kg) — to medium — 150-250 lbs (68-113 kg), and heavy — 300-500 lbs (135-225 kg). Rarely such trees may produce 600-700 lbs (272-317 kg). Larger trees have larger crops but if the trees become too tall harvesting is too difficult, and they should be topped. Harvesters, working manually from ladders, or using pruning poles cut the entire cluster of fruit with leaves attached.

A serious problem with the longan is its irregular bearing — often one good year followed by 1 or 2 poor years. Another handicap is the ripening season — early to mid-August in China, which is the time of typhoons; August and September in Florida which is during the hurricane season. Rain is a major nuisance in harvesting and in conveying the fruit to market or to drying sheds or processing plants.

Keeping Quality

At room temperature, longans remain in good condition for several days. Because of the firmer rind, the fruit is less perishable than the lychee.

Preliminary tests in Florida indicate that the fruit can be frozen and will not break down as quickly as the lychee when thawed.

Pests and Diseases

The longan is relatively free of pests and diseases. At times, there may be signs of mineral deficiency which can be readily corrected by supplying minor elements in the fertilization program.

Food Uses

Longans are much eaten fresh, out-of-hand, but some have maintained that the fruit is improved by cooking. In China, the majority are canned in sirup or dried. The canned fruits were regularly shipped from Shanghai to the United States in the past. Today, they are exported from Hong Kong and Taiwan.

For drying, the fruits are first heated to shrink the flesh and facilitate peeling of the rind. Then the seeds are removed and the flesh dried over a slow fire. The dried product is black, leathery and smoky in flavor and is mainly used to prepare an infusion drunk for refreshment.

A liqueur is made by macerating the longan flesh in alcohol.

Food Value Per 100 g of Edible Portion		
	Fresh	*Dried*
Calories	61	286
Moisture	82.4 g	17.6 g
Protein	1.0 g	4.9 g
Fat	0.1 g	0.4 g
Carbohydrates	15.8 g	74.0 g
Fiber	0.4 g	2.0 g
Ash	0.7 g	3.1 g
Calcium	10 mg	45 mg
Phosphorus	42 mg	196 mg
Iron	1.2 mg	5.4 mg
Thiamine		0.04 mg
Ascorbic Acid	6 mg (possibly)	28 mg

Other Uses

Seeds and rind: The seeds, because of their saponin content, are used like soapberries (*Sapindus saponaria* L.) for shampooing the hair. The seeds and the rind are burned for fuel and are part of the payment of the Chinese women who attend to the drying operation.

Wood: While the tree is not often cut for timber, the wood is used for posts, agricultural implements, furniture and construction. The heartwood is red, hard, and takes a fine polish. It is not highly valued for fuel.

Medicinal Uses: The flesh of the fruit is administered as a stomachic, febrifuge and vermifuge, and is regarded as an antidote for poison. A decoction of the dried flesh is taken as a tonic and treatment for insomnia and neurasthenic neurosis. In both North and South Vietnam, the "eye" of the longan seed is pressed against a snakebite in the belief that it will absorb the venom.

Leaves and flowers are sold in Chinese herb markets but are not a part of ancient traditional medicine. The leaves contain quercetin and quercitrin. Burkill says that the dried flowers are exported to Malaysia for medicinal purposes. The seeds are administered to counteract heavy sweating and the pulverized kernel, which contains saponin, tannin and fat, serves as a styptic.

Rambutan (Plate XXXIV)

Though a close relative of the lychee and an equally desirable fruit, this member of the Sapindaceae is not nearly as well-known. Botanically, it is *Nephelium lappaceum* L. (syns. *Euphoria nephelium* DC.; *Dimocarpus crinita* Lour.). In the vernacular, it is generally called rambutan (in French, *ramboutan* or *ramboutanier;* in Dutch, *ramboetan*); occasionally in India, *ramboostan*. To the Chinese it is *shao tzu,* to Vietnamese, *chom chom* or *vai thieu;* to Kampucheans, *ser mon,* or *chle sao mao.* There are other local names in the various dialects of southeast Asia and the East Indies.

Description

The rambutan tree reaches 50 to 80 ft (15-25 m) in height, has a straight trunk to 2 ft (60 cm) wide, and a dense, usually spreading crown. The evergreen leaves are alternate, pinnately compound, 2¾ to 12 in (7-30 cm) long, with reddish rachis, hairy when young, and 1 to 4 pairs of leaflets, subopposite or alternate, elliptic to oblong-elliptic, or rather obovate, sometimes oblique at the base; slightly leathery; yellowish-green to dark-green and somewhat dull on the upper surface, yellowish or bluish-green beneath; 2 to 8 in (5-20 cm) long, 1 to 4⅓ in (2.5-11 cm) wide, the 6 to 15 pairs of principal veins prominent on the underside. The small, petalless flowers, of three kinds: males, hermaphrodite functioning as males, and hermaphrodite functioning as females, are borne in axillary or pseudo-terminal, much branched, hairy panicles. The fruit is ovoid, or ellipsoid, pinkish-red, bright- or deep-red, orange-red, maroon or dark-purple, yellowish-red, or all yellow or orange-yellow; 1⅓ to 3⅛ in (3.4-8 cm) long. Its thin, leathery rind is covered with tubercles from each of which extends a soft, fleshy, red, pinkish, or yellow spine ⅕ to ¾ in (0.5-2 cm) long, the tips deciduous in some types. The somewhat hairlike covering is responsible for the common name of the fruit, which is based on the Malay word *"rambut",* meaning "hair". Within is the white or rose-tinted, translucent, juicy, acid, subacid or sweet flesh, ⅙ to ⅓ in (0.4-0.8 cm) thick, adhering more or less to the ovoid or oblong, somewhat flattened seed, which is 1 to 1⅓ in (2.5-3.4 cm) long and ⅖ to ⅗ in (1-1.5 cm) wide. There may be 1 or 2 small undeveloped fruits nestled close to the stem of a mature fruit.

Origin and Distribution

The rambutan is native to Malaysia and commonly cultivated throughout the archipelago and southeast Asia. Many years ago, Arab traders introduced it into Zanzibar and Pemba. There are limited plantings in India, a few trees in Surinam, and in the coastal lowlands of Colombia, Ecuador, Honduras, Costa Rica, Trinidad and Cuba. Some fruits are being marketed in Costa Rica. The rambutan was taken to the Philippines from Indonesia in 1912. Further introductions were made in 1920 (from Indonesia) and 1930 (from Malaya), but until the 1950's its distribution was rather limited. Then popular demand brought about systematic efforts to improve the crop and resulted in the establishment of many commercial plantations in the provinces of Batangas, Cavite, Davan, Iloilo, Laguna, Oriental Mindoro and Zamboanga. Seeds were imported into the United States from Java in 1906 (SPI #17515) but the species is not grown in this country.

Varieties

Popular varieties in Malaya include 'Chooi Ang', 'Peng Thing Bee', 'Ya Tow', 'Azimat', and 'Ayer Mas'. Dr. J.J. Ochse described 6 named varieties in Indonesia:

'Lebakbooloos'—a broad-topped tree with dark-red fruits having uncrowded spines ⅗ in (1.5 cm) long, and grayish-white, tough, subacid flesh ⅕ in (0.5 cm) thick, frequently difficult to separate from the seed and often takes pieces of the testa with it. Ships well over long distances. (Cultivated also in India).

'Seematjan'—Tree has an open crown and long, flexible branches. Fruits are dark-red with spines to ¾ in (2 cm) long.

In Java the tree is especially prone to attack by various insects. It is cultivated also in India and in the Philippines where it has averaged 16 lbs/acre (16 kg/ha). There are 2 forms: 1) 'Seematjan besar' with small fruit, thin rind, spines fairly far apart; very sweet, somewhat coarse, fairly juicy flesh to which the coarse, fibrous testa tightly adheres; 2) 'Seematjan ketjil' (or 'Koombang')—the fruit has soft, tough, and less sweet flesh to which the seed coat does not tightly adhere.

'Seenjonja'—Tree low-growing; has a drooping crown. Fruit nearly ovoid, about 1½ in (4 cm) long and 1⅕ in (3 cm) wide; dark wine-red with slender, flexible spines about ⅖ in (1 cm) long. Flesh clings firmly to the seed. In the Philippines has yielded on the average 41 lbs/acre (41 kg/ha).

'Seetangkooweh'—Tree broad-topped. Fruit flattened ellipsoid, about 2 in (5 cm) long, 1½ in (4 cm) wide with slim spines ⅖ in (1 cm) long. Rind is thin, pliable, tough. Flesh yellowish-white, sweet, clings tightly to the thick testa which separates from the seed. Fruits stand long-distance shipment.

'Seelengkeng'—Tree low-growing with drooping crown. Fruit ovoid, 1⅕ in (3 cm) long, ¾ in (2 cm) wide, with very fine, soft spines. Flesh slightly glossy, tough, moderately sweet, and separates from the seed with a few particles of testa clinging to it. Air-layers are unsatisfactory, so it is rare in cultivation and expensive on the market. Much favored by Chinese because of its resemblance to the lychee. (Cultivated also in India.)

'Seekonto'—Tree has broad crown; is fast-growing. Fruits ellipsoid, faintly flattened, about 2 in (5 cm) long, 1½ in (4 cm) wide. Spines are thick and short. Flesh is dull, grayish-white, somewhat coarse and dry; clings to the testa which separates readily from the seed.

'Maharlika' (no description available) has yielded 21 lbs/acre (21 kg/ha) in the Philippines.

Yellow-fruited rambutans are called 'Atjeh koonig' in Batavia. In Malaya, 'Rambutan gading' indicates a yellow type.

Among the many "races" of rambutan in Malaya, the best "freestone" types are found in Penang. One race with a partly free stone is known as 'rambutan lejang'. Burkill says that some rambutans are so sour that monkeys are reluctant to eat them.

In 1950, Philippine agriculturists undertook a program of selection and the creation of a Testing Plot at the Provincial Nursery, Victoria, Oriental Mindoro. There they assembled 360 trees of which 140 were found to be bearing in 1960 and 196 (mostly males) were non-bearing. Observations of the bearing trees there and at the Arago Farm not far away, resulted in the selection of 21 clones which they classified into 4 groups according to fruit size: 1) very large, 14 or less per lb (31 or less/kg); 2) large, 15 to 16 per lb (32–36/kg); 3) medium, 17 to 19 per lb (37–41/kg); 4) small, 20 or more per lb (42 or more/kg).

The main characteristics of the 21 named selections are here summarized:

'Queen Zaida'—Dark-red, oblong, medium-size; flesh thick (38.76% of fruit), sweet, juicy; freestone; 60% of fruits kept well for 2 weeks in cold storage. Yield: 275 lbs (125 kg) per tree at 20 years of age.

'Baby Eulie'—Light-red, very large, flesh thick (39.92% of fruit), soft, freestone. Kept well only 1 week at 60°F (15.56°C). Yield: 352 lbs (160 kg) per tree at 8 years of age.

'Princess Caroline'—Dark-red, small, rind pliable; flesh thick (44.14% of fruit); seeds small. Kept well for 2 weeks at 60°F (15.56°C). Yield; 440 lbs (200 kg) per tree at 8 years of age.

'Quezon'—Yellowish-red, small to medium; rind pliable; flesh thick (38.24% of fruit); sweet, slightly acid, juicy. Yield: 343 lbs (156 kg) per tree at 8 years of age.

'Roxas'—Dark-red; medium-sized; flesh thick (42.97% of fruit); juicy, sweet, adheres to seed. Yield: 429 lbs (195 kg) per tree at 8 years of age.

'Zamora'—Yellowish rind with pale-pink spines; oblong; small; rind hard; flesh thick (38.29% of fruit), juicy and sweet. Yield: 330 lbs (150 kg) per tree at 7 years of age. Ripens mid- to late October. After 2 weeks of refrigeration at 60°F (15.56°C) 80% of the fruits were still in good condition.

'Quirino'—Yellowish with pinkish-red spines; small; flesh thick (32.78% of fruit), juicy and sweet. Borne in large clusters of up to 85 fruits each.

'Magsaysay'—Dark-red to near-black with dark-red spines; oblong, large; rind pliable; flesh thick (42.68% of fruit); juicy, sweet; freestone. Yield: 176 lbs (80 kg) per tree at 6 years of age.

'Santo Tomas'—Yellowish-pink with reddish-pink, soft spines. Nearly round; rind hard; flesh thick (43.25% of fruit); seed small. Yield: 352 lbs (160 kg) per tree at 8 years of age.

'Victoria'—Yellowish with red spines; rind thick; flesh thick, juicy, sweet, freestone. Yield: 132 lbs (60 kg) per tree at 6 years of age. Early in season (mid-July).

'Baby Christie'—Yellowish-red with soft, silvery-pink spines; large. Flesh thick (36.41% of fruit).

'Governor Infantada'—Oblong, very large; rind pliable; flesh thick (39.28% of fruit), juicy, sweet and slightly acid; adheres tightly to seed. Yield: 330 lbs (150 kg) per tree at 6 years of age. Fruits keep only 1 week at 60°F (15.56°C).

'Laurel, Sr.'—Pinkish-red, small; flesh thick (39.76% of fruit). Tree very low-growing, spreading.

'Fortich'—Yellowish-red; medium-sized; flesh thick (40.95% of fruit); juicy, sweet; freestone. Early in season.

'Osmeña, Sr.'—Purple-red; medium-sized; flesh thick (38.90% of fruit); juicy, sweet; freestone. Ripens late in season.

'Ponderosa Ferreras' (from Arago Farm)—Crimson red with very prominent spines; very large; flesh thick (35.73% of fruit); juicy, sweet, freestone. Early in season. Yield: 303 lbs (138 kg) per tree at 6 years of age.

'Rodrigas' (from Arago Farm)—Medium-sized; flesh thick (38.46% of fruit).

'Manahan' (from Arago Farm)—Medium-sized; flesh thick (37.37% of fruit).

'Santan' (from Arago Farm)—Flesh thick (34.26% of fruit).

'Arago' (from Arago Farm)—flesh very thick (41.42% of fruit).

'Cruz' or 'Cruzas' (from Arago Farm)—flesh medium-thick (26.15% of fruit).

About 1960, 10 outstanding rambutans were selected in an evaluation of 100 seedling trees of the unsurpassed Indonesian 'Seematjan', also 'Seenjonja', 'Maharlika', 'Divata', 'Marikit', 'Dalisay', 'Marilag', 'Bituin', 'Alindog', and 'Paraluman'.

Climate

The rambutan flourishes from sea-level to 1,600 or even 1,800 ft (500–600 m), in tropical, humid regions having well-distributed rainfall. In the ideal environment of Oriental Mindora, Philippines, the average temperature year-round is about 81°F (27.3°C), relative humidity is 82%, rainfall 71 in (180 cm)—about 165 rainy days. The dry season should not last much over 3 months.

Soil

The tree does best on deep, clay-loam or rich sandy loam rich in organic matter, or in deep peat. It needs good drainage.

Propagation

Rambutan seeds, after removal from the fruit and thorough washing, should be planted horizontally with the flattened side downward in order that the seedling will grow straight and have a normal, strong root system. Seeds will germinate in 9 to 25 days, the earlier, the more vigor in the seedling. The rate of germination of 2-day-old seeds is 87% to 95%. A week after seed removal from the fruit, there may be only 50% to 65% germination. Sun-drying for 8 hours and oven-drying at 86°F (30°C) kills seeds within a week. Washed seeds will remain viable in moist sawdust, sphagnum moss or charcoal for 3-4 weeks, and some will even sprout in storage. The juice of the flesh inhibits germination. Accordingly, unwashed seeds or seeds treated with the juice can be held for a month in moist sawdust without sprouting.

Rambutan seedlings bear in 5-6 years, but the ratio of female to male trees is 4 or 5 to 7. One Philippine seedling orchard was found to have 67% male trees. Then, too, hardly 5% of female trees give a profitable yield. Vegetative propagation is essential.

Cuttings have been rooted experimentally under mist and with the use of growth-promoting hormones, but this technique is not being practiced. Air-layering may at first appear successful, but many air-layers die after being transplanted into 5-gal containers, or, later, in the field, long after separation from the mother tree.

Inarching is very effective onto 5- to 9-month-old seedlings of rambutan or of pulasan (*N. mutabile* L.) or *N. intermedium* Radlk., but is a rather cumbersome procedure. After 2 or 3 months, the scion is notched 3 times over a period of 2 weeks and then severed from the parent tree. Cleft-, splice-, and side-grafting are not too satisfactory. Patch-budding is preferred as having a much greater rate of success. Seedlings for use as rootstocks are taken from the seedbed after 45 days and transplanted into 1-quart cans with a mixture of 50% cured manure and later transferred to 5 gal containers. In Oriental Mindoro Province, if the budding is done in the month of May, they can achieve 83.6% success; if done in June and July, 82%. Budded trees flower 2½ to 3 years after planting in the field.

Culture

In the Philippines, it is recommended that the trees be planted at least 33 ft (10 m) apart each way, though 40 ft (12 m) is not too much in rich soil. If the trees are set too close to each other, they will become overcrowded in a few years and production will be seriously affected.

Philippine agronomists apply 2.2 lbs (1 kg) ammonium sulfate together with 2.2 lbs (1 kg) complete fertilizer (12-24-12) per tree immediately after harvest and give the same amount of ammonium sulfate to each tree near the end of the rainy season. Studies in Malaya show that a harvest of 6,000 lbs/acre (6,720 kg/ha) of rambutan fruits removes from the soil 15 lbs/acre (approximately 15 kg/ha) nitrogen, 2 lbs/acre (2 kg/ha) phosphorus, 11.5 lbs/acre (11.5 kg/ha) potassium, 5.9 lbs/acre (5.9 kg/ha) calcium, and 2.67 lbs/acre (2.67 kg/ha) magnesium.

Irrigation is given as needed in dry seasons. Light pruning is done only to improve the form of the tree and strengthen it. Rambutan trees should be sheltered from strong winds which do much damage during the flowering and fruiting periods.

Harvesting

In Malaya, the rambutan generally fruits twice a year, the first, main crop in June and a lesser one in December. In the Philippines, flowering occurs from late March to early May and the fruits mature from July to October or occasionally to November.

The entire fruit cluster is cut from the branch by harvesters. If single fruits are picked, they should be snapped off with a piece of the stem attached, so as not to rupture the rind. The fruits must be handled carefully to avoid bruising and crushing, and kept dry, cool, and well-ventilated to delay spoilage.

Yield

Generally, shoots that bear fruit one year will put out new growth and will bloom and fruit the next year, so that biennial bearing is rare in the rambutan. However, yield may vary from year to year. Individual trees 8 years old or older have borne as much as 440 lbs (200 kg) one season and only 132 lbs (60 kg) the next. In the Philippines, the average production per tree of 21 selections was 264 lbs (120 kg) over a 4-year period, while the general average is only 106 lbs (48 kg).

From 1965 to 1967, agronomists at the College of Agriculture, University of the Philippines, studied the growth, flowering habits and yield of the Indonesian cultivars, 'Seematjan', 'Seenjonja', and 'Maharlika'. They found that all the 'Seematjan' flowers were hermaphrodite functioning as female (h.f.f.) and that it is necessary to plant male trees with this cultivar. 'Seenjonja' and 'Maharlika' flowers were mostly h.f.f. with a very few hermaphrodite functioning as males (h.f.m.) in the same panicles, and concluded that, though self-pollination is possible, planting of male trees with these cultivars should improve production.

Keeping Quality

Ordinarily, the fruits must be gotten to local markets within 3 days of picking before shriveling and decay begin. Fungicidal applications and packing in perforated polyethylene bags have extended fresh life somewhat. Weight loss has been reduced by packing in sawdust, or coating with a wax emulsion. Storing in sealed polyethylene bags at 40°F (10°C) and 95% relative humidity has preserved the fruits in fresh condition for 12 days. Some cultivars, as noted, keep better than others.

Pests and Diseases

Few pests or diseases have been reported by rambutan growers. Leaf-eating insects, the mealybug, *Pseudococcus lilacinus,* and the giant bug, *Tessaratoma longicorne,* may require control measures. The mango twig-borer, *Nipho-noclea albata,* occasionally appears on rambutan trees. The Oriental fruit fly attacks very ripe fruits. Birds and flying foxes (fruit-eating bats) consume many of the fruits, probably considerably reducing yield figures.

There are several pathogens that attack the fruits and cause rotting under warm, moist conditions. Powdery mildew, caused by *Oidium* sp., may affect the foliage or other parts of the tree. A serious disease, stem canker, caused by *Fomes lignosus* in the Philippines and *Ophio-ceras* sp. in Malaya, can be fatal to rambutan trees if not controlled at the outset.

Food Uses

Rambutans are most commonly eaten out-of-hand after merely tearing the rind open, or cutting it around the middle and pulling it off. It does not cling to the flesh. The peeled fruits are occasionally stewed as dessert. They are canned in sirup on a limited scale. In Malaya a preserve is made by first boiling the peeled fruit to separate the flesh from the seeds. After cooling, the testa is discarded and the seeds are boiled alone until soft. They are combined with the flesh and plenty of sugar for about 20 minutes, and 3 cloves may be added before sealing in jars. The seeds are sometimes roasted and eaten in the Philippines, although they are reputedly poisonous when raw.

Toxicity

There are traces of an alkaloid in the seed, and the testa contains saponin and tannin. The seeds are said to be bitter and narcotic. The fruit rind also is said to contain a toxic saponin and tannin.

Food Value Per 100 g of Edible Portion*	
Moisture	82.3 g
Protein	0.46 g
Total Carbohydrates	16.02 g
Reducing Sugars	2.9 g
Sucrose	5.8 g
Fiber	0.24 g
Calcium	10.6 mg
Phosphorus	12.9 mg
Ascorbic Acid	30 mg

*Analyses made in Ceylon.

Other Uses

Seed fat: the seed kernel yields 37-43% of a solid, white fat or tallow resembling cacao butter. When heated, it becomes a yellow oil having an agreeable scent. Its fatty acids are: palmitic, 2.0%; stearic, 13.8%; arachidic, 34.7%; oleic, 45.3%; and ericosenoic, 4.2%. Fully saturated glycerides amount to 1.4%. The oil could be used in making soap and candles if it were available in greater quantity.

Wood: The tree is seldom felled. However, the wood—red, reddish-white, or brownish—is suitable for construction though apt to split unless carefully dried.

Medicinal Uses: The fruit (perhaps unripe) is astringent, stomachic; acts as a vermifuge, febrifuge, and is taken to relieve diarrhea and dysentery. The leaves are poulticed on the temples to alleviate headache. In Malaya the dried fruit rind is sold in drugstores and employed in local medicine. The astringent bark decoction is a remedy for thrush. A decoction of the roots is taken as a febrifuge.

Pulasan

The pulasan, or poolasan, *Nephelium mutabile* Blume (family, Sapindaceae), is closely allied to the rambutan and sometimes confused with it. One of its local names in Malaya is *rambutan–kafri* (negro's rambutan); another is *rambutan paroh.* In Malacca it is sometimes called *pen-ing–pening–ramboetan.* The Dutch name in Java is *kap-oelasan.* In the Philippines it is mostly known as *bulala.* There are numerous tribal names for this species throughout Malaysia.

Description

The pulasan tree is a handsome ornamental; attains 33 to 50 ft (10-15 m); has a short trunk to 12 to 16 in (30-40 cm) thick; and the branchlets are brown-hairy when young. The alternate leaves, pinnate or odd-pinnate, and 6¾ to 18 in (17-45 cm) long, have 2 to 5 pairs of opposite or nearly opposite leaflets, oblong- or elliptic-lanceolate, 2½ to 7 in (6.25-17.5 cm) long and up to 2 in (5 cm) wide; slightly wavy, dark-green and barely glossy on the upper surface; pale, somewhat bluish, with a few short, silky hairs on the underside. Very small, greenish, petalless flowers with 4-5 hairy sepals, are borne singly or in clusters on the branches of the erect, axillary or terminal, panicles clothed with fine yellowish or brownish hairs. The fruit is ovoid, 2 or 3 in (5-7.5 cm) long, dark- or light-red, or yellow, its thick, leathery rind closely set with conical, blunt-tipped tubercles or thick, fleshy, straight spines, to ⅜ in (1 cm) long. There may be 1 or 2 small, undeveloped fruits nestled close to the stem. Within is the glistening, white or yellowish-white flesh (aril) to ⅜ in (1 cm) thick, more or less clinging to the thin, grayish-brown seedcoat (testa) which separates from the seed. The flavor is generally much sweeter than that of the rambutan. The seed is ovoid, oblong or ellipsoid, light-brown, somewhat flattened on one side, ¾ to 1⅓ in (2-3.5 cm) long.

Origin and Distribution

The pulasan is native to Western Malaysia. Wild trees are infrequent in lowland forests around Perak, Malaya but abundant in the Philippines at low elevations from Luzon to Mindanao. The tree has long been cultivated in Malaya and Thailand; is rarely domesticated in the Philippines. Ochse reported that there were extensive plantings in Java only around Bogor and the villages along the railway between Boger and Djakarta.

The tree was planted at the Trujillo Plant Propagation Station in Puerto Rico in 1926 and young trees from Java were sent to the Lancetilla Experimental Garden, Tela, Honduras, in 1927. The latter were said in 1945 to be doing well at Tela and fruiting moderately. The pulasan is little-known elsewhere in the New World except in Costa Rica where it is occasionally grown and the fruits sometimes appear on the market.

Varieties

Ochse refers to 2 forms of pulasan in Java: in one group, distinguished as "Seebabat' or 'Kapoolasan seebabat', the fruit is mostly dark-red, the tubercles are crowded together, the flesh is very sweet and juicy and separates easily from the seed. In the other group, the fruit is light-red and smaller, the tubercles are not so closely set, and the flesh adheres firmly to the seed.

Wester mentions a fine variety growing in Jolo. The plants introduced into Honduras were 2 superior varieties called 'Asmerah Tjoplok' and 'Kapoelasan mera tjoplok'. There are some trees in Malaya and in Thailand that bear seedless fruits and these are being vegetatively propagated.

Climate

The pulasan is ultra-tropical and thrives only in very humid regions between 360 and 1,150 ft (110-350 m) of altitude. In Malaya, it is said that the tree bears best after a long, dry season.

Soil

There is little information on the soil requirements of the pulasan but Ochse says it must be constantly moist. He was of the opinion that the richer soil around Bogor contributed to the superior quality of the fruits grown in that area.

Propagation

Planting of seeds is not favored because the seedlings may be male or female. As with the rambutan, air-layers are very short-lived. Budding is successful if it is done in the rainy season on rootstocks already set out in the field so that they will not be subject to transplanting which causes many fatalities, particularly during dry weather.

Culture

The trees require less space than rambutan trees and can be 26 to 33 ft (8 to 10 m) apart each way. As a rule, they receive little or no fertilizer or other cultural attention.

Food Uses

The flesh of ripe fruits is eaten raw or made into jam. Boiled or roasted seeds are used to prepare a cocoa-like beverage.

Food Value Per 100 g of Edible Portion*	
Moisture	84.54-90.87 g
Protein	0.82 g
Carbohydrates	12.86 g
Fiber	0.14 g
Fat	0.55 g
Ash	0.43-0.45 g
Calcium	0.01-0.05 mg
Iron	0.002 mg

*Analyses made in the Philippines.

Toxicity

Hydrocyanic acid has been detected in the bark and leaves.

Other Uses

Oil: The dried seed kernels yield 74.9% of a solid, white fat, melting at 104° to 107.6°F (40°-42°C), to a faintly perfumed oil. Presumably, this could be utilized in soap-making.

Wood: The wood is light-red, harder and heavier than that of the rambutan and of excellent quality but rarely available.

Medicinal Uses: The leaves and roots are employed in poultices. The root decoction is administered as a febrifuge and vermifuge. Burkill says that the roots are boiled with *Gleichenia linearis* Clarke, and the decoction is used for bathing fever patients.

Mamoncillo

One of the minor fruits of the family Sapindaceae, the mamoncillo (*Melicoccus bijugatus* Jacq., syn. *Melicocca bijuga* L.) has, nevertheless acquired an assortment of regional names, such as: ackee (Barbados only; not to be confused with *Blighia sapida,* q.v.); genip, ginep, ginepe, guenepa, guinep (Barbados, Jamaica, Bahamas, Puerto Rico, Trinidad and Tobago); *grosella de miel* (Mexico); *guayo* (Mexico); honeyberry (Guyana); Jamaica bullace plum, kanappy (Puerto Rico); *kenet* (French Guiana); *knepa* (Surinam); *knepe* (French West Indies); *knippa* (Surinam); *limoncillo* (Dominican Republic); *macao* (Colombia, Venezuela); *maco* (Venezuela); *mamon* (Colombia, Venezuela, El Salvador, Nicaragua, Costa Rica, Panama, Argentina); *mamon de Cartagena* (Costa Rica); marmalade box (Guyana); *mauco* (Venezuela); *muco* (Colombia, Venezuela); *quenepa* (Dominican Republic, Puerto Rico, Colombia); *quenepe* (Haiti); *quenett* (French Guiana); *sensiboom* (Surinam); Spanish lime (Florida); *tapaljocote* (El Salvador).

Description

The mamoncillo tree is slow-growing, erect, stately, attractive; to 85 ft (25 m) high, with trunk to 5½ ft (1.7 m) thick; smooth, gray bark, and spreading branches. Young branchlets are reddish. The leaves are briefly deciduous, alternate, compound, having 4 opposite, elliptic, sharp-pointed leaflets 2 to 5 in (5-12.5 cm) long and 1¼ to 2½ in (3.25-6.25 cm) wide, the rachis frequently conspicuously winged as is that of the related soapberry (*Sapindus saponaria* L.). The flowers, in slender racemes 2⅓ to 4 in (6-10 cm) long, often clustered in terminal panicles, are fragrant, white, ⅕ to ⅓ in (5-8 mm) wide, with 4 petals and 8 stamens. Male and female are usually borne on separate trees but some trees are partly polygamous. The fruit clusters are branched, compact and heavy with nearly round, green fruits tipped with a small protrusion, and suggesting at first glance small unripe limes, but there the resemblance ends. The skin is smooth, thin but leathery and brittle. The glistening pulp (aril) is salmon-colored or yellowish, translucent, gelatinous, juicy but very scant and somewhat fibrous, usually clinging tenaciously to the seed. When fully ripe, the pulp is pleasantly acid-sweet but if unripe acidity predominates. In most fruits there is a single, large, yellowish-white, hard-shelled seed, while some have 2 hemispherical seeds. The kernel is white, crisp, starchy, and astringent.

Origin and Distribution

The mamoncillo is native to Colombia, Venezuela, and the island of Margarita, also French Guiana, Guyana and Surinam. It is commonly cultivated and spontaneous in those countries, also in coastal Ecuador, the lowlands of Central America, the West Indies and in the Bahamas. In Florida, it is occasionally grown as far north as Ft. Myers on the West Coast and Palm Beach on the east; is much more plentiful in Key West, especially as a street tree. There are some specimens in California and in botanical gardens in the Philippines, Zanzibar, Hawaii and elsewhere. According to Britton, there was a tree about 30 ft (9 m) tall in Bermuda in 1914 but it had never bloomed. There are a few trees in Israel but none has flowered before 10 years of age.

Varieties

Little horticultural attention has been given this fruit. In the 1950's, a large-fruited, sweet type was found in Key West. Air-layers and inarchings were made in order to permit trial of this type on the mainland. In the 1960's, horticulturist George Jackson evaluated the fruits of 54 trees in southern Puerto Rico. Fruits with less than 45% edible pulp and 20% total sugars were disregarded. He rated 9 trees as meriting further testing. Of these, 4 were selected as having the most desirable qualities. Their main characters were listed as follows:

'Puerto Rico #1'—round, of medium size, 28 to the lb (62/kg); flesh firm, semi-dry, separating easily from the seed; sweet, with 26.0% sugars.
'Puerto Rico #2'—round, of medium size, 27 to the pound (60/kg); rind medium-thick; flesh firm, semi-dry, separating easily from seed; sweet, 24.1% total sugars.
'Puerto Rico #3'—round-oblong, small, 49 to the pound (108/kg); rind thin, pliable; flesh firm, semi-dry, separating easily from the seed; very sweet, 24.1% total sugars.
'Puerto Rico #4'—round, medium-small, 40 to the lb (88/kg); rind medium-thin, flesh firm, semi-dry, separating easily from seed; agreeably acid and slightly sweet; 22.7% total sugar.

The percentage of edible matter by fruit-weight ranged from 46.6% to 48.6%.

In 1976, Dr. Carl Campbell of the University of Florida's Agricultural Research and Education Center in Homestead, Florida, reported on his comparison of 3 selections made by interested individuals and an ordinary seedling growing at the Center. The latter, labeled 'No. 1', was graded as: small, 49.1% pulp, but of only fair flavor, and poor annual yield.

'No. 2', or 'Queen', brought by W.F. Whitman from Key West; large, 55.6% pulp, only fair in flavor, and medium in yield.
'No. 3', brought by R.G. Newcomb from Key West; of good

size, 48.2% pulp; of good flavor and borne heavily in most years.

'No. 4', or 'Montgomery', from the Montgomery (later, Jennings) Estate in Coral Gables; large, with sometimes 18% of crop having 2 seeds; 51.5% pulp; of good flavor, and borne heavily in most years.

Pollination

Generally, the presence of a male tree is necessary to pollinate the flowers of trees that are predominantly female (or hermaphrodite functioning as female). However, in Cuba, some trees have sufficient numbers of flowers of both sexes to yield regularly large crops without interplanting.

Climate

The mamoncillo is not strictly tropical, for it ascends up to 3,300 ft (1,000 m) above sea-level in South America. It can stand several degrees of frost in Florida. Nevertheless, it is too tender to fruit in California though it has been planted there on various occasions. It is well adapted to areas of low rainfall. That of Key West ranges from 30 to 50 in (75-125 cm) annually. The tree can tolerate long periods of drought.

Soil

In Cuba, the tree is said to flourish in nearly all types of terrain but particularly in deep, rich soil of calcareous origin. It seems perfectly at home in the oolitic-limestone of southern Florida and the Florida Keys. In Colombia, it has been observed to grow on such poor soils that it has been adopted for planting in soil reclamation efforts. It is spontaneous especially in dry, coastal districts.

Propagation

The mamoncillo is usually grown from seed but superior types should be vegetatively reproduced. Air-layering of fairly large branches, at least 2 in (5 cm) in diameter, is successful in the summer and there will be adequate root development in 5 to 6 weeks. Approach-grafting is feasible provided the rootstocks are raised in a lightweight medium, in plastic bags to facilitate attachment to the selected tree. Attempts to veneer-graft or chip-bud have generally failed.

Culture

Ordinarily, the mamoncillo tree is given no care except for watering and fertilizing when first planted. Vegetatively propagated trees bear earlier than seedlings.

Season and Harvesting

In Florida, the fruits ripen from June to September. In the Bahamas, the season extends from July to October. Ladders and picking poles equipped with cutters are necessary in harvesting fruits from tall trees. The entire cluster is clipped from the branch when sampling indicates that the fruits are fully ripe. At this stage, the rind becomes brittle but does not change color. If picked prematurely, the rind turns blackish, a sign of deterioration.

Keeping Quality

Because of the leathery skin, the fruit remains fresh for a long time and ships and markets well. The tropical horticulturist, David Sturrock, related that horsemen in Cuba often hung branches of mamoncillos on the saddle horn to enjoy and relieve thirst during long rides.

Pests and Diseases

The tree is a host of the Citrus black fly, *Aleurocanthus woglumi*. There are several parasites (*Prospaltella* spp., *Eretmocerus serius*, and *Amitus hesperidium*) which provide effective control of this pest. In Florida, *Armillariella (Clitocybe) tabescens* causes mushroom root rot; *Fusarium* and *Phyllosticta* cause leaf spot; and *Cephaleuros virescens*, algal leaf spot and green scurf.

Food Uses

For eating out-of-hand, the rind is merely torn open at the stem end and the pulp-coated seed is squeezed into the mouth, the juice being sucked from the pulp until there is nothing left of it but the fiber. With fruits that have non-adherent pulp, the latter may be scraped from the seed and utilized to make pie-filling, jam, marmalade or jelly, but this entails much work for the small amount of edible material realized. More commonly, the peeled fruits are boiled and the resulting juice is prized for cold drinks. In Colombia, the juice is canned commercially.

The seeds are eaten after roasting. Indians of the Orinoco consume the cooked seeds as a substitute for cassava.

Food Value Per 100 g of Edible Portion*	
Calories	58.11-73
Moisture	68.8-82.5 g
Protein	0.50-1.0 g
Fat	0.08-0.2 g
Carbohydrates	13.5-19.2 g
Fiber	0.07-2.60 g
Ash	0.34-0.74 g
Calcium	3.4-15 mg
Phosphorus	9.8-23.9 mg
Iron	0.47-1.19 mg
Carotene	0.02-0.44 mg (70 I.U.)
Thiamine	0.03-0.21 mg
Riboflavin	0.01-0.20 mg
Niacin	0.15-0.90 mg
Ascorbic Acid	0.8-10 mg
Tannin	1.88 g
Amino Acids	
Tryptophan	14 mg
Methionine	0
Lysine	17 mg
*Analyses made in Cuba, Central America and Colombia.	

Seed Hazard

It has been said that the pulp fibers coat the lining of the stomach, adversely affecting the health, but this has been denied by the Government Chemist of the Department of Science and Agriculture in Jamaica who declares that fatalities in children are the result of choking on the seed. When coated with pulp, it is very slippery, is accidentally swallowed and, because of its size, lodges in the throat, causing suffocation or strangulation.

Other Uses

Juice: A dye has been experimentally made from the juice of the raw fruit which makes an indelible stain.

Flowers: The flowers are rich in nectar and highly appealing to hummingbirds and honeybees. The honey is somewhat dark in color but of agreeable flavor. The tree is esteemed by Jamaican beekeepers though the flowering season (March/April) is short.

Leaves: In Panama, the leaves are scattered in houses where there are many fleas. It is claimed that the fleas are attracted to the leaves and are cast out with the swept-up foliage. Some believe that the leaves actually kill the fleas.

Wood: The heartwood is yellow with dark lines, compact, hard, heavy, fine-grained; inclined to decay out of doors, but valued for rafters, indoor framing, and cabinetwork.

Fig. 75: The mamoncillo (*Melicoccus bijugatus*), with its large seeds and thin layer of adhering flesh, provides little but juice.

Medicinal Uses: In Venezuela, the astringent roasted seed kernels are pulverized, mixed with honey and given to halt diarrhea. The astringent leaf decoction is given as an enema for intestinal complaints.

Akee

More widely known for its poisonous properties than as an edible fruit, the akee, *Blighia sapida* K. Konig (syn. *Cupania sapida* Voigt.), of the family Sapindaceae, is sometimes called ackee, akee apple, or vegetable brain (*seso vegetal* in Spanish). Other Spanish names are *arbol de seso, palo de seso* (Cuba); *huevo vegetal* and *fruto de huevo* (Guatemala and Panama); *arbol del huevo* and *pera roja* (Mexico); *merey del diablo* (Venezuela); *bien me sabe* or *pan y quesito* (Colombia); *akí* (Costa Rica). In Portuguese, it is *castanha* or *castanheiro de Africa*. In French, it is *arbre fricassé* or *arbre a fricasser* (Haiti); *yeux de crabe* or *ris de veau* (Martinique). In Surinam it is known as *akie*. On the Ivory Coast of West Africa, it is called *kaka* or *finzan;* in the Sudan, *finza*. Elsewhere in Africa it is generally known as *akye, akyen* or *ishin,* though it has many other dialectal names. In the timber trade, the wood is marketed as *achin.*

It should be noted that the name "akee" may refer to the mamoncillo, q.v., in Barbados. As a colloquial term for the mamoncillo it may be a corruption of the Mayan "acche" which was applied to several plants whose flowers attract honeybees.

Description

The tree, reaching 33 to 40 ft (10-12 m), is rather handsome, usually with a short trunk to 6 ft (1.8 m) in circumference, and a dense crown of spreading branches. Its bark is gray and nearly smooth. The evergreen (rarely deciduous), alternate leaves are compound with 3 to 5 pairs of oblong, obovate-oblong, or elliptic leaflets, 6 to 12 in (15-30 cm) long, rounded at the base, short-pointed at the apex; bright-green and glossy on the upper surface, dull and paler and finely hairy on the veins on the underside. Bisexual and male flowers, borne together in simple racemes 3 to 7 in (7.5-17.5 cm) long, are fragrant, 5-petalled, white and hairy. The fruit is a leathery, pear-shaped, more or less distinctly 3-lobed capsule 2¾ to 4 in (7-10 cm) long; basically yellow, more or less flushed with bright-scarlet. When it is fully mature, it splits open revealing 3 cream-colored, fleshy, glossy arils, crisp, somewhat nutty-flavored, attached to the large, black, nearly round, smooth, hard, shining seeds—normally 3; often 1 or 2 may be aborted. The base of each aril is attached to the inside of the stem-end of the "jacket" by pink or orange-red membranes.

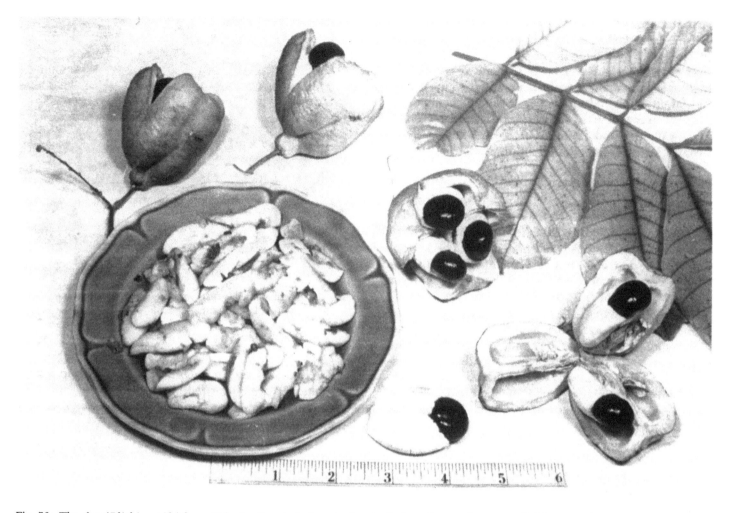

Fig. 76: The akee (*Blighia sapida*) from Africa is a favorite in Jamaica but the fleshy arils are poisonous until fully exposed to light. The seeds are always poisonous.

Origin and Distribution

The akee is indigenous to the forests of the Ivory Coast and Gold Coast of West tropical Africa where it is little eaten but various parts have domestic uses. In Ghana, the fruiting tree is admired as an ornamental and is planted in villages and along streets for shade. The akee was brought to Jamaica in 1793 by the renowned Captain Bligh to furnish food for the slaves. It was readily adopted and became commonly grown in dooryards and along roadsides and, to some extent, naturalized. The arils still constitute a favorite food of the island and the fruit is featured in a calypso despite the health hazards associated with it. Canned arils are exported to the United Kingdom where they are welcomed by Jamaican immigrants. Importation has been banned by the United States Food and Drug Administration.

The akee was planted also in Trinidad and Haiti and some other islands of the West Indies and the Bahamas and apparently was carried by Jamaican slaves to Panama and the Atlantic Coast of Guatemala and Costa Rica. In 1900 it was outlawed in Trinidad after it had caused some fatalities. There are scattered trees in Surinam, Venezuela, Colombia, Ecuador and Brazil, quite a number maintained as curiosities in southern Florida; and some

planted around Calcutta, India. The tree has been tried in the warm, moist climate of Guyana and Malaya but has never survived. At Lamao in the Philippines it first bore fruit in 1919.

Climate

The akee tree is tropical to subtropical; flourishes from sea-level to an elevation of 3,000 ft (900 m) in Jamaica. It does not bear fruit in Guatemala City; fruits heavily in southern Florida where young trees have been killed by winter cold but mature trees have escaped serious injury during brief periods of 26°F (-3.33°C).

Soil

The tree does very well on oolitic limestone and on sand in southern Florida and the Bahamas, though it grows faster in more fertile soils.

Propagation and Culture

Akee trees are grown from seeds or by shield-budding, and show very little variation. In European greenhouses, cuttings of ripe shoots are rooted in sand and raised in a mixture of peat and loam. In warm climates, the tree grows fast and requires little cultural attention.

Season

There is some flowering and fruiting all year in Jamaica. In Florida, flowers appear in spring and the fruits in midsummer and there may be a light blooming period in the fall. In the Bahamas, there are 2 distinct crops a year, one from February through April and the second from July to October.

Food Uses

The akee must be allowed to open fully or at least partly before it is detached from the tree. When it has "yawned", the seeds are discarded and the arils, while still fresh and firm, are best parboiled in salted water or milk and then lightly fried in butter. Then they are really delicious. In Jamaica, they are often cooked with codfish, onions and tomatoes. After parboiling, they are added to a stew of beef, salt-pork and scallions, thyme and other seasonings. Sometimes they are curried and eaten with rice. They are served, not only in the home, but also in hotel dining rooms and other restaurants. In Africa, they may be eaten raw or in soup, or after frying in oil.

Food Value Per 100 g of Raw Arils*	
Moisture	57.60 g
Protein	8.75 g
Fat	18.78 g
Fiber	3.45 g
Carbohydrates	9.55 g
Ash	1.87 g
Calcium	83 mg
Phosphorus	98 mg
Iron	5.52 mg
Carotene	—
Thiamine	0.10 mg
Riboflavin	0.18 mg
Niacin	3.74 mg
Ascorbic Acid	65 mg
*Analyses made in Mexico.	

Toxicity

The toxicity of the akee was long misunderstood and believed to reside in the membranes attaching the arils to the jacket, or only in the overripe and decomposing arils. There have been intensive clinical and chemical studies of the akee and its effects since 1940, and it is now known that the unripe arils contain hypoglycin, a-amino-B-(2-methylenecyclopropyl) propionic acid, formerly called hypoglycin A. This toxic property is largely dispelled by light as the jacket opens. When fully ripe, the arils still possess $1/12$ of the amount in the unripe. The seeds are always poisonous. They contain hypoglycin and its y-glutamyl derivative, y-L-glutamyl-a-amino-B-(2-methylenecyclopropyl) propionic acid, formerly called hypoglycin B. The latter is $1/2$ as toxic as the former.

In feeding experiments at the University of Miami, Dr. Edward Larson found that the membrane of open fruits was harmless; rabbits were readily killed by the unripe arils; rats were resistant and had to be force-fed to be fatally poisoned. I have found that squirrels will make holes in the unopened fruits on the tree to consume the unripe arils but they leave the seeds untouched.

Akee poisoning in humans is evidenced by acute vomiting, sometimes repeated, without diarrhea (called "vomiting sickness" in Jamaica), followed by drowsiness, convulsions, coma and, too often, death. Because of hypoglycaemic effects, administration of sugar solutions have been found helpful. Most cases occur in winter in Jamaica when 30% to 50% of the arils have small, underdeveloped seeds, often not apparent externally. Ingestion of such arils, raw or cooked, is hazardous. For more information on the toxicity of the akee, one may consult Kean, *Hypoglycin* (1975), and Morton, *Forensic Medicine,* Vol. III, Chap. 71 (1977).

Other Uses

Fruit: In West Africa, the green fruits, which produce lather in water, are used for laundering. Crushed fruits are employed as fish poison. The seeds, because of their oil content, and the jacket because of its potash content, are burned and the ashes used in making soap.

Flowers: In Cuba an extract of the flowers is appreciated as cologne.

Bark: On the Gold Coast, a mixture of the pulverized bark and ground hot peppers is rubbed on the body as a stimulant.

Wood: The sapwood is white or light greenish-brown. The heartwood is reddish-brown, hard, coarse-grained, durable, immune to termites. It is used locally for construction and pilings and has been recommended for railway sleepers. It is also fashioned into oars, paddles and casks.

Medicinal Uses: In Brazil, repeated small doses of an aqueous extract of the seed has been administered to expel parasites. The treatment is followed by a saline or oily purative. Cubans blend the ripe arils with sugar and cinnamon and give the mixture as a febrifuge and as a treatment for dysentery. On the Ivory Coast, the bark is mixed with pungent spices in an ointment applied to relieve pain. The crushed new foliage is applied on the forehead to relieve severe headache. The leaves, crushed with salt, are poulticed on ulcers. The leaf juice is employed as eye drops in ophthalmia and conjunctivitis. In Colombia, the leaves and bark are considered stomachic. Various preparations are made for treatment of epilepsy and yellow fever.

RHAMNACEAE

Indian Jujube (Plate XXXV)

While the better-known, smooth-leaved Chinese jujube (*Ziziphus jujuba* Mill.) of the family Rhamnaceae, is of ancient culture in northern China and is widely grown in mild-temperate, rather dry areas, of both hemispheres, the Indian jujube, *Z. mauritiana* Lam. (syn. *Z. jujuba* L.) is adapted to warm climates. It is often called merely jujube, or Chinese date, which leads to confusion with the hardier species. Other English names are Indian Plum, Indian cherry and Malay jujube. In Jamaica it may be called coolie plum or crabapple; in Barbados, dunk or mangustine; in Trinidad and Tropical Africa, dunks; in Queensland, Chinee apple. In Venezuela it is *ponsigne* or *yuyubo;* in Puerto Rico, *aprin* or *yuyubi;* in the Dominican Republic, *perita haitiana;* in the French-speaking West Indies, *pomme malcadi, pomme surette, petit pomme, liane croc chien, gingeolier* or *dindoulier*. In the Philippines it is called *manzana* or *manzanita* ("apple" or "little apple"); in Malaya, *bedara;* in Indonesia and Surinam, *widara;* in Thailand, *phutsa* or *ma–tan;* in Cambodia, *putrea;* in Vietnam, *tao* or *tao nhuc*. In India it is most commonly known as *ber,* or *bor*.

Description

The plant is a vigorous grower and has a rapidly-developing taproot. It may be a bushy shrub 4 to 6 ft (1.2-1.8 m) high, or a tree 10 to 30 or even 40 ft (3-9 or 12 m) tall; erect or wide-spreading, with gracefully drooping branches and downy, zigzag branchlets, thornless or set with short, sharp straight or hooked spines. It may be evergreen, or leafless for several weeks in hot summers. The leaves are alternate, ovate- or oblong-elliptic, 1 to 2½ in (2.5-6.25 cm) long, ¾ to 1½ in (2-4 cm) wide; distinguished from those of the Chinese jujube by the dense, silky, whitish or brownish hairs on the underside and the short, downy petioles. On the upper surface, they are very glossy, dark-green, with 3 conspicuous, depressed, longitudinal veins, and there are very fine teeth on the margins.

The 5-petalled flowers are yellow, tiny, in 2's or 3's in the leaf axils. The fruit of wild trees is ½ to 1 in (1.25-2.5 cm) long. With sophisticated cultivation, the fruit reaches 2½ in (6.25 cm) in length and 1¾ in (4.5 cm) in width. The form may be oval, obovate, round or oblong; the skin smooth or rough, glossy, thin but tough, turns from light-green to yellow, later becomes partially or wholly burnt-orange or red-brown or all-red. When slightly underripe,

the flesh is white, crisp, juicy, acid or subacid to sweet, somewhat astringent, much like that of a crabapple. Fully ripe fruits are less crisp and somewhat mealy; overripe fruits are wrinkled, the flesh buff-colored, soft, spongy and musky. At first the aroma is applelike and pleasant but it becomes peculiarly musky as the fruit ages. There is a single, hard, oval or oblate, rough central stone which contains 2 elliptic, brown seeds, ¼ in (6 mm) long.

Origin and Distribution

The Indian jujube is native from the Province of Yunnan in southern China to Afghanistan, Malaysia and Queensland, Australia. It is cultivated to some extent throughout its natural range but mostly in India where it is grown commercially and has received much horticultural attention and refinement despite the fact that it frequently escapes and becomes a pest. It was introduced into Guam about 1850 but is not often planted there or in Hawaii except as an ornamental. Specimens are scattered about the drier parts of the West Indies, the Bahamas, Colombia and Venezuela, Guatemala, Belize, and southern Florida. In Barbados, Jamaica and Puerto Rico the tree is naturalized and forms thickets in uncultivated areas. In 1939, 6 trees from Malaysia were introduced into Israel and flourished there. They bore very light crops of fruit heavily infested with fruit flies and were therefore destroyed to protect other fruit trees.

Varieties

In India, there are 90 or more cultivars differing in the habit of the tree, leaf shape, fruit form, size, color, flavor, keeping quality, and fruiting season. Among the important cultivars, eleven are described in the encyclopaedic *Wealth of India:* '**Banarasi** (or Banarsi) **Pewandi**', '**Dandan**', '**Kaithli**' ('Patham'), '**Muria Mahrara**', '**Narikelee**', '**Nazùk**', '**Sanauri 1**', '**Sanauri 5**', '**Thornless**' and '**Umran**' ('Umri'). The skin of most is smooth and greenish-yellow to yellow.

At Haryana Agricultural University, a study was made of 70 cultivars collected from all jujube-growing areas of northern India and set out in an experimental orchard in 1967-68. In 1980, 16 midseason selections from these were evaluated. '**Banarasi Karaka**' (poor-flavored) gave the highest yield—286 lbs (130 kg) per tree—followed by 'Mudia Murhara' and 'Kaithli' (both of good flavor), and

'Sanauri 5' and **'Desi Alwar'** (both of medium flavor). It was decided that 'Mudia Murhara', 'Kaithli' and 'Sanauri 5' were worthy of commercial cultivation. For breeding purposes, 'Banarasi Karaka' and 'Desi Alwar' could contribute high pulp content; 'Mudia Murhara', total soluble solids; 'Kaithli', high ascorbic acid content and good flavor, in efforts to develop a superior midseason cultivar.

In 1982, 4 were singled-out as the most promising cultivars:

'Umran'—large, golden-yellow turning chocolate-brown when fully ripe; sweet; 19% TSS; 0.12% acidity; average fruit weight, 30-89 g; yield, 380-440 lbs (150-200 kg) per tree; late-ripening; of good keeping and shipping quality.

'Gola'—medium to large (average, 14-17 g); 17-19% TSS; 0.46-0.51% acidity; golden-yellow, juicy, of good flavor; yield, 175-220 lbs (80-100 kg) per tree. Earliest to ripen; sells at a high price.

'Kaithli'—of medium size (average 180.0 g); 18% TSS; 0.5% acidity; pulp soft and sweet. Average yield, 220-330 lbs (100-150 kg).

'Katha phal'—small to medium (average 10.0 g); greenish blushed on one cheek with reddish-yellow; 23% TSS; 0.77% acidity; yield, medium, 175-220 lbs (80-100 kg) per tree. Late in season.

In addition to these, 5 cultivars have been described at the Indian Agricultural Research Institute, New Delhi. All are grown in Delhi, the southeastern Punjab and neighboring Uttar Pradesh. Their special features are, briefly, as follows:

'Dandan'—non-spiny; fruit medium to large; of fairly good quality; keeps well. Late in season.

'Gular Bashi'—fruit of medium size, juicy, sweet, non-acrid; of excellent quality when fresh, musky after storage. TSS 18.8% when yellow, 22.4% after turning brown. Stone medium to thin, funnel-shaped, easily separated from the flesh. Late in season. Keeps well.

'Kheera'—medium to large, oval with a beak; pulp soft, juicy, of good, sweet flavor. TSS 19.8%. Late; a heavy bearer; of fairly good keeping quality.

'Nazuk'—medium to small, elliptic-oblong; pulp slimy, fairly juicy; of good, sweet flavor, nearly without astringency. TSS 17.4%. Midseason. A moderate bearer. Of poor keeping quality.

'Seo ber' ('Seb')—medium to large; skin thick; pulp moderately juicy, astringent unless peeled or not eaten until light-brown, when it is very sweet and excellent. TSS 19%. Stone large, thick, pitted. Late in season. Keeps very well.

In Assam 5 wild or cultivated types, collected from various parts of the state, have been described by S. Dutta:

'Var. 1'—a very thorny wild shrub, with small, round, inferior fruits; grown as a fence to protect crops.

'Var. 2'—a wild, thorny tree to 30 ft (9 m) with red-brown, tough-skinned fruit having slimy, acid-sweet pulp. Much eaten by children and rural folk. Commonly used in cooking and preserving.

'Var. 3'—a very thorny, spreading tree. Fruit dark-red or brown, with sour pulp. Bears heavily. Planted for shade.

'Var. 4' ('Bali bogri')—a wild, thornless tree, with greenish-yellow fruits blushed with red; pulp slightly slimy, mealy, sweet-and-acid, of good flavor. Bears heavily.

'Var. 5' ('Tenga-mitha-bogri')—A wild, thorny tree, with oblong, brownish fruit; pulp slightly slimy, sweet-and-acid, with very pleasant flavor. Bears heavily. A choice jujube recommended for vegetative propagation and commercial cultivation.

Pollination

Pollen of the Indian jujube is thick and heavy. It is not airborne but is transferred from flower to flower by honeybees (*Apis* spp.), a yellow wasp (*Polister hebraeus*), and the house fly (*Musca domestica*).

The cultivars 'Banarasi Karaka', 'Banarasi Pewandi' and 'Thornless' are self-incompatible. 'Banarasi Karaka' and 'Thornless' are reciprocally cross-incompatible.

Climate

In China and India, wild trees are found up to an elevation of 5,400 ft (1,650 m) but commercial cultivation extends only up to 3,280 ft (1,000 m). In northern Florida, it is sensitive to frost. Young trees may be frozen to the ground but will recover. Mature trees have withstood occasional short periods of freezing temperatures without damage. In India, the minimum shade temperature for survival is 44.6° to 55.4°F (7°-13°C); the maximum, 98.6° to 118°F (37°-48°C). The tree requires a fairly dry climate with an annual rainfall of 6 to 88.5 in (15-225 cm), being unsuited to the lower, wetter parts of Malaysia. For high fruit production, the tree needs full sun.

Soil

In India, the tree does best on sandy loam, neutral or slightly alkaline. It also grows well on laterite, medium black soils with good drainage, or sandy, gravelly, alluvial soil of dry river-beds where it is vigorously spontaneous. Even moderately saline soils are tolerated. The tree is remarkable in its ability to tolerate water-logging as well as drought.

Propagation

The Indian jujube is widely grown from seeds, which may remain viable for 2½ years but the rate of germination declines with age. Superior selections are grafted or budded onto seedlings of wild types. Vegetative propagation of highly prized varieties was practiced near Bombay about 1835 but kept secret until 1904, and then was quickly adopted by many people. Ring-budding has been popular in the past but has been largely superseded by shield-budding or T-budding. Grafted plants are less thorny than seedlings.

To select seeds for growing rootstocks, the stones must be taken from fruits that have fully ripened on the tree. They are put into a 17 to 18% salt solution and all that float are discarded. The stones that sink are dipped in 500 ppm thiourea for 4 hours, then cracked and the separated seeds will germinate in 7 days. Seeds in uncracked stones require 21 to 28 days. If seeds are sown in spring, the seedlings will be ready for budding in 4 months. Great care must be taken in transplanting nursery stock to the field because of the taproot. Therefore, the rootstocks may be raised directly in the field and budding done *in situ.* Inferior seedling trees, including wild trees, can be topworked to preferred cultivars in June and some fruit will be borne a year later. From 1935 to 1939, the Punjab Department of Agriculture top-worked 50,000 trees without cost to the growers. Air-layers will root if treated with IBA and NAA at 5,000 to 7,500 ppm and given 100 ppm boron. Cuttings of mature wood at least 2 years old can be rooted and result in better yields than those taken at a younger stage.

At Punjab University, horticulturists have experimented with stooling as a means of propagation. They transplanted one-year-old seedlings into stool beds, cut them back to 4 in (10 cm), found that the shoots would root only if ringed and treated with IBA, preferably at 12,000 ppm.

Culture

Untrimmed trees must be spaced at 36 to 40 ft (11-12 m), but carefully pruned trees can be set at 23 to 26 ft (7-8 m). Pruning should be done during the first year of growth to reduce the plant to one healthy shoot, and branches lower than 30 in (75 cm) should be removed. At the end of the year, the plant is topped. During the 2nd and 3rd years, the tree is carefully shaped. Thereafter, the tree should be pruned immediately after harvesting at the beginning of dormancy and 25 to 50% of the previous year's growth may be removed. Sometimes a second lighter pruning is performed just before flowering. There will be great improvement in size, quality and number of fruits the following season.

In India, it has been traditional to apply manure and ash as fertilizer, but, in recent years, each tree has been given annual treatments of 22 lbs (10 kg) manure with 1.1 lbs (0.5 kg) ammonium sulphate for every year of age up to the 5th year. More advanced farmers utilize only commercial fertilizer (NPK) in larger amounts, twice annually, the first at the rate of 110 lbs/acre (about 110 kg/ha) and the second at 172 lbs/acre (about 172 kg/ha). Growth regulators are now being utilized to bring about early and heavier blooming, enhance fruit setting, prevent fruit drop, and increase fruit size, and promote uniform ripening. These practices have demonstrated that an improved crop can bring in 2 to 3 times the revenue of that achieved by conventional practices.

During hot weather and also in the period of fruit development, irrigation is highly beneficial. Water-stress will cause immature fruit drop. In India, water has been applied as many as 35 times during the winter months. Zinc and boron sprays are sometimes applied to enhance glossiness of the fruits.

Season and Harvesting

In India, some types ripen as early as October, others from mid-February to mid-March, others in March, or mid-March, to the end of April. In the Assiut Governorate, there are 2 crops a year, the main in early spring, the second in the fall. In India, 2 or 3 pickings are done by hand from ladders, a worker being capable of manually harvesting about 110 lbs (50 kg) per day. The fruits remaining on the tree are shaken down. After wrapping in white cloth, the fruits are put into paper-lined burlap bags holding 110 lbs (50 kg) for long trips to markets throughout the country.

Yield

Seedling trees bear 5,000 to 10,000 small fruits per year in India. Superior grafted trees may yield as many as 30,000 fruits. The best cultivar in India, with fruits normally averaging 30 to the lb (66 to the kg), yields 175 lbs (77 kg) annually. Special cultural treatment increases both fruit size and yield.

Keeping Quality

The Indian jujube stands handling, shipment and marketing very well. Storage experiments in India showed that slightly underripe fruits ripen and keep for 8 days under wheat straw, 7 days under leaves, and 4 days in carbide (50 to 60 g).

Pests and Diseases

The greatest enemies of the jujube in India are fruit flies, *Carpomyia vesuviana* and *C. incompleta.* Some cultivars are more susceptible than others, the flies preferring the largest, sweetest fruits, 100% of which may be attacked while on a neighboring tree, bearing a smaller, less-sweet type, only 2% of the crop may be damaged. The larvae pupate in the soil and it has been found that treatment of the ground beneath the tree helps reduce the problem. Control is possible with regular and effective spraying of insecticide.

A leaf-eating caterpillar, *Porthmologa paraclina,* and the green slug caterpillar, *Thosea* sp., attack the foliage. A mite, *Larvacarus transitans,* forms scale-like galls on twigs retarding growth and reducing the fruit crop.

Lesser pests include a small caterpillar, *Meridarches scyrodes,* that bores into the fruit; the gray-hairy caterpillar, *Thiacidas postica,* also *Tarucus theophrastus, Myllocerus transmarinus,* and *Xanthochelus superciliosus.*

The tree is subject to shrouding by a parasitic vine (*Cuscuta* spp.). Powdery mildew (*Oidium* sp.) causes defoliation and fruit-drop. Sooty mold (*Cladosporium zizyphi*) causes leaves to fall. Leafspot results from infestation by *Cercospora* spp. and *Isariopsis indica* var. *zizyphi.* In 1973, a witches'-broom disease caused by a mycoplasma-like organism was found in jujube plants near Poona University. It proved to be transmitted by grafting or budding diseased scions onto healthy *Z. mauritiana* seedlings. Leaf rust, caused by *Phakopsora zizyphivulgaris,* ranges from mild to severe on all commercial cultivars in the Punjab.

Fruits on the tree are attacked by *Alternaria chartarum, Aspergillus nanus, A. parasiticus, Helminthosporium atroolivaceum, Phoma hessarensis,* and *Stemphyliomma valparadisiacum.* Twigs and branches may be affected by *Entypella zizyphi, Hypoxylon hypomiltum,* and *Patellaria atrata.* In storage, the fruits may be spotted by the fungi, *Alternaria brassicicola, Phoma* spp., *Curvularia lunata, Cladosporium herbarum.* Fruit rots are caused by *Fusarium* spp., *Nigrospora oryzae, Epicoccum nigrum,* and *Glomerella cingulata.*

Food Uses

In India, the ripe fruits are mostly consumed raw, but are sometimes stewed. Slightly underripe fruits are candied by a process of pricking, immersing in a salt solution gradually raised from 2 to 8%, draining, immersing in another solution of 8% salt and 0.2% potassium metabisulphite, storing for 1 to 3 months, rinsing and cooking in sugar sirup with citric acid. Residents of Southeast Asia eat the unripe fruits with salt. Ripe fruits crushed in water form a very popular cold drink. Ripe fruits are preserved by sun-drying and a powder is prepared for out-of-season purposes. Acid types are used for pickling or for chutneys. In Africa, the dried and fermented pulp is pressed into cakes resembling gingerbread.

Young leaves are cooked and eaten in Indonesia. In Venezuela, a jujube liqueur is made and sold as *Crema de ponsigue*. Seed kernels are eaten in times of famine.

Food Value Per 100 g of Edible Portion	
Fruits, fresh:	
Moisture	81.6-83.0 g
Protein	0.8 g
Fat	0.07 g
Fiber	0.60 g
Carbohydrates	17.0 g
Total Sugars	5.4-10.5 g
Reducing Sugars	1.4-6.2 g
Non-Reducing Sugars	3.2-8.0 g
Ash	0.3-0.59 g
Calcium	25.6 mg
Phosphorus	26.8 mg
Iron	0.76-1.8 mg
Carotene	0.021 mg
Thiamine	0.02-0.024 mg
Riboflavin	0.02-0.038 mg
Niacin	0.7-0.873 mg
Citric Acid	0.2-1.1 mg
Ascorbic Acid	65.8-76.0 mg
Fluoride	0.1-0.2 ppm
Pectin (dry basis)	2.2-3.4%

The fresh fruits also contain some malic and oxalic acid and quercetin.

**Fruits, dried:	
Calories	473/lb (1,041/kg)
Moisture	68.10 g
Protein	1.44 g
Fat	0.21 g
Carbohydrates	2.47 g
Sugar	21.66 g
Fiber	1.28 g

*Analyses made in India and Honduras.
**Analyses made in the Philippines.

Toxicity

In Ethiopia, the fruits are used to stupefy fish (possibly there is sufficient saponin for this purpose). The leaves contain saponin because they are known to produce lather if rubbed in water.

Other Uses

Wood: The wood is reddish, close-grained, fine-textured, hard, tough, durable, planing and polishing well. It has been used to line wells, to make legs for bedsteads, boat ribs, agricultural implements, house poles, tool handles, yokes, gunstocks, saddle trees, sandals, golf clubs, household utensils, toys and general turnery. It is also valued as firewood; is a good source of charcoal and activated carbon. In tropical Africa, the flexible branches are wrapped as retaining bands around conical thatched roofs of huts, and are twined together to form thorny corral walls to retain livestock.

Leaves: The leaves are readily eaten by camels, cattle and goats and are considered nutritious. Analyses show the following constituents (% dry weight): crude protein, 12.9-16.9; fat, 1.5-2.7; fiber, 13.5-17.1; N-free extract, 55.3-56.7; ash, 10.2-11.7; calcium, 1.42-3.74; phosphorus, 0.17-0.33; magnesium, 0.46-0.83; potassium, 0.47-1.57; sodium, 0.02-0.05; chlorine, 0.14-0.38; sulphur, 0.13-0.33%. They also contain ceryl alcohol and the alkaloids, protopine and berberine.

The leaves are gathered as food for silkworms.

Dye: In Burma, the fruit is used in dyeing silk. The bark yields a non-fading, cinnamon-colored dye in Kenya.

Nectar: In India and Queensland, the flowers are rated as a minor source of nectar for honeybees. The honey is light and of fair flavor.

Lac: The Indian jujube is one of several trees grown in India as a host for the lac insect, *Kerria lacca*, which sucks the juice from the leaves and encrusts them with an orange-red resinous substance. Long ago, the lac was used for dyeing, but now the purified resin is the shellac of commerce. Low grades of shellac are made into sealing wax and varnish; higher grades are used for fine lacquer work, lithograph-ink, polishes and other products. The trees are grown around peasant huts and heavily inoculated with broodlac in October and November every year, and the resin is harvested in April and May. The trees must be pruned systematically to provide an adequate number of young shoots for inoculation.

Medicinal Uses: The fruits are applied on cuts and ulcers; are employed in pulmonary ailments and fevers; and, mixed with salt and chili peppers, are given in indigestion and biliousness. The dried ripe fruit is a mild laxative. The seeds are sedative and are taken, sometimes with buttermilk, to halt nausea, vomiting, and abdominal pains in pregnancy. They check diarrhea, and are poulticed on wounds. Mixed with oil, they are rubbed on rheumatic areas.

The leaves are applied as poultices and are helpful in liver troubles, asthma and fever and, together with catechu, are administered when an astringent is needed, as on wounds. The bitter, astringent bark decoction is taken to halt diarrhea and dysentery and relieve gingivitis. The bark paste is applied on sores. The root is purgative. A root decoction is given as a febrifuge, taenicide and emmenagogue, and the powdered root is dusted on wounds. Juice of the root bark is said to alleviate gout and rheumatism. Strong doses of the bark or root may be toxic. An infusion of the flowers serves as an eye lotion.

TILIACEAE

Phalsa

In the family Tiliaceae, only one genus, *Grewia*, yields edible fruit. The only species of any importance is *G. subinaequalis* DC. (syns. *G. asiatica* Mast. in part, NOT L.; *G. hainesiana* Hole), long referred to in literature as *G. asiatica* L. Phalsa is the most used vernacular name in India where there are a number of dialectal names. The plant is called *falsa* in Pakistan.

Description

A large, scraggly shrub or small tree to 15 ft (4.5 m) or more, the phalsa has long, slender, drooping branches, the young branchlets densely coated with hairs. The alternate, deciduous, widely spaced leaves are broadly heart-shaped or ovate, pointed at the apex, oblique at the base, up to 8 in (20 cm) long and 6½ in (16.25 cm) wide, and coarsely toothed, with a light, whitish bloom on the underside. Small, orange-yellow flowers are borne in dense cymes in the leaf axils. The round fruits, on 1-in (2.5 cm) peduncles are produced in great numbers in open, branched clusters. Largest fruits are ½ to ⅝ in (1.25-1.6 cm) wide. The skin turns from green to purplish-red and finally dark-purple or nearly black. It is covered with a thin, whitish bloom and is thin, soft and tender. The soft, fibrous flesh is greenish-white stained with purplish-red near the skin and becoming suffused with this color as it progresses to overripeness. The flavor is pleasantly acid, somewhat grapelike. Large fruits have 2 hemispherical, hard, buff-colored seeds 3/16 in (5 mm) wide. Small fruits are single-seeded.

Origin and Distribution

The phalsa is indigenous throughout much of India and Southeast Asia. It is cultivated commercially mainly in the Punjab and around Bombay. It was introduced into the Philippines before 1914 and is naturalized at low elevations in dry zones of the island of Luzon. Only a few specimens have been planted in the New World, for example, at the former Federal Experiment Station, Mayaguez, Puerto Rico, and the Agricultural Research and Education Center, Homestead, Florida.

Varieties

The tall-growing wild plants bear acid fruits which are not relished. The dwarf, shrubby type, with a blend of sweet-and-acid in the best fruits, is cultivated.

Climate

In India, the phalsa grows well up to an elevation of 3,000 ft (914 m). It can stand light frosts which cause only shedding of leaves.

Soil

The phalsa grows in most any soil — sand, clay or limestone — but rich loam improves fruit production, as does irrigation during the fruiting season and in dry periods, even though the tree is drought-tolerant. Generally, it is grown in marginal land close to city markets.

Propagation

Seeds are the usual means of propagation and they germinate in 15 days. Ground-layers, treated with hormones, have been 50% successful; air-layers, 85%. Cuttings are difficult to root. Only 20% of semi-hardwood cuttings from spring flush, treated with 1,000 ppm NAA, and planted in July (in India) rooted and grew normally.

Culture

Seedlings are transplanted from seedbeds into well-prepared holes when a year old and are usually spaced 10 to 15 ft (3-4.5 m) apart, though some experiments have favored 6 x 6 ft (1.8x1.8 m) or 8 x 8 ft (2.4x2.4 m) to maximize efficiency in harvesting. Fruiting will commence in 13 to 15 months. Annual pruning to a height of 3 to 4 ft (0.9-1.2 m) encourages new shoots and better yields than more drastic trimming.

Sprays of 10 ppm gibberellic acid have increased fruit-set. At 40 ppm, there is increased fruit size but decreased fruit-set. In fertilizer experiments, the plant has shown good vegetative response to applications of nitrogen. High levels of phosphorus increase sugar content, while potassium decreases sugar and elevates acidity.

Harvesting and Yield

Summer is the fruiting season. Only a few fruits in a cluster ripen at any one time, so continuous harvesting is necessary. The fruits keep poorly and must be marketed within 24 hours. Average yield per plant is 20 to 25 lbs (9-11 kg) in a season.

Pests and Diseases

Leaf-cutting caterpillars attack the foliage at night. A

Fig. 77: The phalsa (*Grewia asiatica*) is primarily a beverage fruit in India. Uneven ripening requires many pickings.

blackish caterpillar causes galls on the growing shoots. Termites often damage the roots. In some areas, leaf spot is caused by *Cercospora grewiae*.

Food Uses

The fruits are eaten fresh as dessert, are made into sirup, and extensively employed in the manufacture of soft drinks. The juice ferments so readily that sodium benzoate must be added as a preservative.

Food Value

Analyses made long ago in the Philippines show the following values: calories, 329 per lb (724 per kg); moisture, 81.13%; protein, 1.58%; fat, 1.82%; crude fiber, 1.77%; sugar, 10.27%.

Other Uses

Leaves: The fresh leaves are valued as fodder.

Bark: The bark is used as a soap substitute in Burma. A mucilaginous extract of the bark is useful in clarifying sugar. Fiber extracted from the bark is made into rope.

Wood: The wood is yellow-white, fine-grained, strong and flexible. It is used for archers' bows, spear handles, shingles and poles for carrying loads on the shoulders. Stems that are pruned off serve as garden poles and for basket-making.

Medicinal Uses: The fruit is astringent and stomachic. When unripe, it alleviates inflammation and is administered in respiratory, cardiac and blood disorders, as well as in fever.

An infusion of the bark is given as a demulcent, febrifuge and treatment for diarrhea. The root bark is employed in treating rheumatism. The leaves are applied on skin eruptions and they are known to have antibiotic action.

Sundry Chemistry

The flowers have been found to contain grewinol, a long chain keto alcohol, tetratricontane-22-ol-13-one. The seeds contain 5% of a bright-yellow oil containing 8.3% palmitic acid, 11.0% stearic acid, 13.4% oleic acid, 64.5% linoleic acid; 2.8% unsaponifiable.

ELAEOCARPACEAE

Jamaica Cherry

This is a minor but well-known and wholesome fruit, borne by a multipurpose tree and therefore merits inclusion. The Jamaica cherry, *Muntingia calabura* L., is a member of the family Elaeocarpaceae. It has acquired a wide assortment of vernacular names, among them *capuli* or *capulin* which are better limited to *Prunus salicifolia* (q.v.). In Florida, it has been nicknamed strawberry tree because its blooms resemble strawberry blossoms, but strawberry tree is a well-established name for the European ornamental and fruit tree, *Arbutus unedo* L., often cultivated in the western and southern United States, and should not be transferred to the Jamaica cherry.

In Mexico, local names for the latter are *capolin, palman, bersilana, jonote* and *puan;* in Guatemala and Costa Rica, *Muntingia calabura* is called *capulin blanco;* in El Salvador, *capulin de comer;* in Panama, *pasito* or *majagüillo;* in Colombia, *chitató, majaguito, chirriador, acuruco, tapabotija* and *nigua;* in Venezuela, *majagua, majaguillo, mahaujo, guácimo hembra, cedrillo, niguo, niguito;* in Ecuador, *niguito;* in Peru, *bolina, iumanasa, yumanaza, guinda yunanasa,* or *mullacahuayo;* in Brazil, *calabura* or *pau de seda;* in Argentina, *cedrillo majagua;* in Cuba, *capulina, chapuli;* in Haiti, *bois d'orme; bois de soie marron;* in the Dominican Republic, *memiso* or *memizo;* in Guadeloupe, *bois ramier* or *bois de soie;* in the Philippines, *datiles, ratiles, latires, cereza* or *seresa;* in Thailand, *takop farang* or *ta kob farang;* in Cambodia, *kakhop;* in Vietnam, *cay trung ca;* in Malaya, *buah cheri; kerukup siam* or Japanese cherry; in India, Chinese cherry or Japanese cherry; in Ceylon, jam fruit.

Description

This is a very fast-growing tree of slender proportions, reaching 25 to 40 ft (7.5-12 m) in height, with spreading, nearly horizontal branches. The leaves are evergreen, alternate, lanceolate or oblong, long-pointed at the apex, oblique at the base; 2 to 5 in (5-12.5 cm) long, dark-green and minutely hairy on the upper surface, gray- or brown-hairy on the underside; and irregularly toothed. The flowers, borne singly or in 2's or 3's in the leaf axils, are ½ to ¾ in (1.25-2 cm) wide with 5 green sepals and 5 white petals and many prominent yellow stamens. They last only one day, the petals falling in the afternoon. The abundant fruits are round, ⅜ to ½ in (1-1.25 cm) wide, with red or sometimes yellow, smooth, thin, tender skin and light-brown, soft, juicy pulp, with very sweet, musky, somewhat fig-like flavor, filled with exceedingly minute, yellowish seeds, too fine to be noticed in eating.

Origin and Distribution

The Jamaica cherry is indigenous to southern Mexico, Central America, tropical South America, the Greater Antilles, St. Vincent and Trinidad. The type specimen was collected in Jamaica. It is widely cultivated in warm areas of the New World and in India, southeast Asia, Malaya, Indonesia, and the Philippines, in many places so thoroughly naturalized that it is thought by the local people to be native.

Macmillan says that it was first planted in Ceylon about 1912. Several trees were introduced into Hawaii by the United States Department of Agriculture in 1922. Dr. David Fairchild collected seeds of a yellow-fruited form in the Peradeniya Botanic Gardens, Ceylon, in 1926 (S.P.I. #67936). The tree has been grown in southern Florida for its fruits and as quick shade for nursery plants. It is seldom planted at present. Volunteers from bird-distributed seeds spring up in disturbed hammocks and pinelands. The author supplied seeds requested by the Kenya Agriculture Research Institute, Kihuyu, in 1982. The Jamaica cherry is said to grow better than any other tree in the polluted air of Metropolitan Manila. It runs wild on denuded mountainsides and on cliffs and is being evaluated for reforestation in the Philippines where other trees have failed to grow and also for wildlife sanctuaries since birds and bats are partial to the fruits.

The fruits are sold in Mexican markets. In Brazil, they are considered too small to be of commercial value but it is recommended that the tree be planted on river banks so that the abundance of flowers and fruits falling into the water will serve as bait, attracting fish for the benefit of fishermen. In Malaya, the tree is considered a nuisance in the home garden because fruit-bats consume the fruits and then spend the day under the eaves of houses and disfigure the porch and terrace with their pink, seedy droppings.

Climate

The Jamaica cherry is tropical to near-tropical. The mid-19th Century botanist, Richard Spruce saw it in Ecuador "in the plains on both sides of the Cordillera"

Fig. 78: The Jamaica cherry (*Muntingia calabura*) is a fast-growing, useful tree and the sweet fruit is popular in tropical America and Southeast Asia.

growing "abundantly by the Rio San Antonio, up to 2,500 ft" (760 m). It is found up to 4,000 ft (1,300 m) in Colombia. When well-established, it is not harmed by occasional low winter temperatures in southern Florida.

Soil

The tree has the reputation of thriving with no care in poor soils and it does well in both acid and alkaline locations, and even on old tin tailings in Malaya. It is drought-resistant but not salt-tolerant.

Propagation

Brazilian planters sow directly into the field fresh seeds mixed with the sweet juice of the fruit. To prepare seeds for future planting, water is added repeatedly to the squeezed-out seeds and juice and, as the seeds sink to the bottom of the container, the water is poured off several times until the seeds are clean enough for drying in the shade.

Culture

The planting hole is prepared with a mixture of organic fertilizer and soil and with a fungicidal solution to prevent the young seedlings from damping-off. To assure good distribution of the seeds, they are mixed with water and sown with a sprinkling can. When well fertilized and watered, the seedlings will begin fruiting in 18 months and will be 13 ft (4 m) high in 2 years.

Season

Wherever it grows, fruits are borne nearly all year, though flowering and fruiting are interrupted in Florida and Sao Paulo, Brazil, during the 4 coolest months. Ripe fruits can easily be shaken from the branches and caught on cloth or plastic sheets.

Pests and Diseases

In Florida, in recent years, the fruits are infested with the larvae of the Caribbean fruit fly and are accordingly rarely fit to eat.

The foliage is subject to leaf spot caused by *Phyllosticta* sp. and *Pseudocercospora muntingiae* (formerly *Cercospora muntingiae*), and the tree is subject to crown gall caused by *Agrobacterium tumefaciens*.

Food Uses

The Jamaica cherry is widely eaten by children out-of-hand, though it is somewhat sticky to handle. It is often cooked in tarts and made into jam.

The leaf infusion is drunk as a tea-like beverage.

Food Value Per 100 g of Edible Portion	
Moisture	77.8 g
Protein	0.324 g
Fat	1.56 g
Fiber	4.6 g
Ash	1.14 g
Calcium	124.6 mg
Phosphorus	84.0 mg
Iron	1.18 mg
Carotene	0.019 mg
Thiamine	0.065 mg
Riboflavin	0.037 mg
Niacin	0.554 mg
Ascorbic Acid	80.5 mg

*Analyses made in El Salvador.

Other Uses

Wood: The sapwood is yellowish, the heartwood reddish-brown, firm, compact, fine-grained, moderately strong, light in weight, durable indoors, easily worked, and useful for interior sheathing, small boxes, casks, and general carpentry. It is valued mostly as fuel, for it ignites quickly, burns with intense heat and gives off very little smoke. Jamaicans seek out trees blown down by storms, let them dry for a while and then cut them up, preferring this to any other wood for cooking. It is being evaluated in Brazil as a source of paper pulp.

Bark: The bark is commonly used for lashing together the supports of rural houses. It yields a very strong, soft fiber for twine and large ropes.

Medicinal Uses: The flowers are said to possess antiseptic properties. An infusion of the flowers is valued as an antispasmodic. It is taken to relieve headache and the first symptoms of a cold.

Roselle (Plates XXXVI and XXXVII)

True roselle is *Hibiscus sabdariffa* L. (family Malvaceae) and there are 2 main types. The more important economically is *H. sabdariffa* var. *altissima* Wester, an erect, sparsely-branched annual to 16 ft (4.8 m) high, which is cultivated for its jute-like fiber in India, the East Indies, Nigeria and to some extent in tropical America. The stems of this variety are green or red and the leaves are green, sometimes with red veins. Its flowers are yellow and calyces red or green, non-fleshy, spiny and not used for food. This type at times has been confused with kenaf, *H. cannabinus* L., a somewhat similar but more widely exploited fiber source.

The other distinct type of roselle, *H. sabdariffa* var. *sabdariffa*, embraces shorter, bushy forms which have been described as races: *bhagalpuriensis, intermedius, albus,* and *ruber,* all breeding true from seed. The first has green, red-streaked, inedible calyces; the second and third have yellow-green edible calyces and also yield fiber. We are dealing here primarily with the race *ruber* and its named cultivars with edible calyces; secondarily, the green-fruited strains which have similar uses and which may belong to race *albus*.

Vernacular names, in addition to roselle, in English-speaking regions are rozelle, sorrel, red sorrel, Jamaica sorrel, Indian sorrel, Guinea sorrel, sour-sour, Queensland jelly plant, jelly okra, lemon bush, and Florida cranberry. In French, roselle is called *oseille rouge,* or *oseille de Guinée;* in Spanish, *quimbombó chino, sereni, rosa de Jamaica, flor de Jamaica, Jamaica, agria, agrio de Guinea, quetmia ácida, viña* and *viñuela;* in Portuguese, *vinagreira, azeda de Guiné, cururú azédo,* and *quiabeiro azédo;* in Dutch (Surinam), *zuring.* In North Africa and the Near East roselle is called *karkadé* or *carcadé* and it is known by these names in the pharmaceutical and food-flavoring trades in Europe. In Senegal, the common name is *bisap.* The names *flor de Jamaica* and hibiscus *flores* (the latter employed by "health food" vendors), are misleading because the calyces are sold, not the flowers.

Description

H. sabdariffa var. *sabdariffa* race *ruber* is an annual, erect, bushy, herbaceous subshrub to 8 ft (2.4 m) tall, with smooth or nearly smooth, cylindrical, typically red stems. The leaves are alternate, 3 to 5 in (7.5-12.5 cm) long, green with reddish veins and long or short petioles.

Leaves of young seedlings and upper leaves of older plants are simple; lower leaves are deeply 3- to 5- or even 7-lobed; the margins are toothed. Flowers, borne singly in the leaf axils, are up to 5 in (12.5 cm) wide, yellow or buff with a rose or maroon eye, and turn pink as they wither at the end of the day. At this time, the typically red calyx, consisting of 5 large sepals with a collar (epicalyx) of 8 to 12 slim, pointed bracts (or bracteoles) around the base, begins to enlarge, becomes fleshy, crisp but juicy, 1¼ to 2¼ in (3.2-5.7 cm) long and fully encloses the velvety capsule, ½ to ¾ in (1.25-2 cm) long, which is green when immature, 5-valved, with each valve containing 3 to 4 kidney-shaped, light-brown seeds, ⅛ to 3/16 in (3-5 mm) long and minutely downy. The capsule turns brown and splits open when mature and dry. The calyx, stems and leaves are acid and closely resemble the cranberry (*Vaccinium* spp.) in flavor.

A minor ornamental in Florida and elsewhere is the red-leaf hibiscus, *H. acetosella* Welw. (syn. *H. eetveldeanus* Wildem. & Th.) of tropical Africa, which has red stems to 8 ft (2.4 m) high, 5-lobed, red or bronze leaves, and mauve, or red-striped yellow, flowers with a dark-red eye, succeeded by a hairy seed pod enclosed in a red, ribbed calyx bearing a basal fringe of slender, forked bracts. This plant has been often confused with roselle, though its calyx is not fleshy and only the young leaves are used for culinary purposes—usually cooked with rice or vegetables because of their acid flavor.

Origin and Distribution

Roselle is native from India to Malaysia, where it is commonly cultivated, and must have been carried at an early date to Africa. It has been widely distributed in the Tropics and Subtropics of both hemispheres, and in many areas of the West Indies and Central America has become naturalized.

The Flemish botanist, M. de L'Obel, published his observations of the plant in 1576, and the edibility of the leaves was recorded in Java in 1687. Seeds are said to have been brought to the New World by African slaves. Roselle was grown in Brazil in the 17th Century and in Jamaica in 1707. The plant was being cultivated for food use in Guatemala before 1840. J.N. Rose, in 1899, saw large baskets of dried calyces in the markets of Guadalajara, Mexico.

Fig. 79: Seedpods of roselle (*Hibiscus sabdariffa*), enclosed in their red, fleshy, acid calyces, are piled high in the markets of Panama in January.

In 1892, there were 2 factories producing roselle jam in Queensland, Australia, and exporting considerable quantities to Europe. This was a short-lived enterprise. In 1909, there were no more than 4 acres (1.6 ha) of edible roselle in Queensland. A Mr. Neustadt of San Francisco imported seeds from Australia about 1895 and shared them with the California State Agricultural Experiment Station for test plantings and subsequent seed distribution. It was probably about the same time that Australian seeds reached Hawaii. In 1904, the Hawaiian Agricultural Experiment Station received seeds from Puerto Rico. In 1913 there was much interest in interplanting roselle with Ceara rubber (*Manihot glaziovii* Muell. Arg.) on the island of Maui and there were some plantations established also on the island of Hawaii, altogether totaling over 200 acres (81 ha). The anticipated jelly industry failed to materialize and promotional efforts were abandoned by 1929.

P.J. Wester believed that roselle was brought to Florida from Jamaica about 1887. Plants were grown by Dr. H.J. Webber at the United States Department of Agriculture's Subtropical Laboratory at Eustis, Florida, in the early 1890's, but all the roselle was killed there by a severe freeze in 1895. Cook and Collins reported that roselle was commonly cultivated in southern Florida in 1903. In 1904,

Wester acquired seeds from Mr. W.A. Hobbs of Coconut Grove and planted them at the United States Department of Agriculture's Subtropical Garden in Miami. He was enthusiastic about roselle's potential as a southern substitute for the cranberry. In 1907, he stated that the fresh calyces were being sold by the quart in South Florida markets. He introduced 3 edible cultivars into the Philippines in 1905. In 1920, he declared: "No plant that has ever been brought into the Philippines is more at home and few grow with so little care as the roselle, or are so productive. Still, like so many other new introductions, the roselle has been slow to gain hold in the popular taste though here and there it is now found in the provincial markets."

In 1928, Paul C. Standley wrote: "roselle . . . is grown in large quantities in Panama, especially by the West Indians. So much of the plant is seen in the markets and on the roads that one would think the market oversupplied." This situation has not changed. I saw great quantities of the whole fruits and the calyces in Panama markets in January of 1976.

Roselle became and remained a common home garden crop throughout southern and central Florida until after World War II when this area began to develop rapidly and home gardening and preserving declined. Mrs. Edith Trebell of Estero, Florida, was one of the last remaining suppliers of roselle jelly. In February, 1961, I purchased the last 2 jars made from the small crop salvaged following the 1960 hurricane and before frost killed all her plants.

In 1954, roselle was still being grown by individuals in the Midwest for its edible herbage. By 1959 and 1960, when there was widespread alarm concerning coal-tar food dyes, it was easy to arouse interest in roselle as a coloring source but difficult to obtain seeds in Florida. At that time, I purchased them from Gleckler's Seedsmen in Metamora, Ohio. Roselle had by then become nearly extinct in Puerto Rico also. From time to time over the next dozen years I was able to obtain a few seeds from oldtimers in Central Florida. In 1973, roselle was featured in the catalog of John Brudy's Rare Plant House, Cocoa Beach (now John Brudy Exotics, Brandon, Florida and no longer listing the seed). Reasoner's Tropical Nurseries in Bradenton was selling plants in containers and giving to purchasers a sheet of recipes. From Lawrence Adams of Arcadia, I obtained seeds which came from the Virgin Islands where this particular strain is said to mature its fruit a month early. These seeds and seeds purchased by John G. DuPuis, Jr., from Brudy were the basis of a large planting at DuPuis' Bar D Ranch in Martin County. Many packets of seeds were distributed to home growers during the following winter.

Today, roselle is attracting the attention of food and beverage manufacturers and pharmaceutical concerns who feel it may have exploitable possibilities as a natural food product and as a colorant to replace some synthetic dyes.

In 1962, Sharaf referred to the cultivation of roselle as a "recent" crop in Egypt, where interest is centered more on its pharmaceutical than its food potential. In 1971, it was reported that roselle calyces, produced and dried in Senegal, particularly around Bambey, were being shipped to Europe (Germany, Switzerland, France and Italy) at the rate of 10 to 25 tons annually.

Varieties

In 1920, Wester described 3 named, edible cultivars as being grown at that time in the Philippines:

'Rico' (named in 1912)—plant relatively low-growing, spreading, with simple leaves borne over a long period and the lobed leaves mostly 3-parted. Flower has dark-red eye and golden-yellow pollen. Mature calyx to 2 in (5 cm) long and to 1¼ in (3.2 cm) wide; bracts plump and stiffly horizontal. Highest yielder of calyces per plant. Juice and preserves of calyx and herbage rich-red.

'Victor'—a superior selection from seedlings grown at the Subtropical Garden in Miami in 1906. Plant taller—to 7 ft (2.13 m), more erect and robust. Flower has dark-red eye and golden-brown pollen. It blooms somewhat earlier than 'Rico'. Calyces as long as those of 'Rico' but slenderer and more pointed at apex; bracts longer, slenderer and curved upward. Juice and preserves of calyx and herbage rich-red.

'Archer' (sometimes called "white sorrel") resulted from seed sent to Wester by A.S. Archer of the island of Antigua. It is believed to be of the race albus. Edward Long referred to "white" as well as red roselle as being grown in most gardens of Jamaica in 1774. Plant is as tall and robust as 'Victor' but has green stems. Flower is yellow with deeper yellow eye and pale-brown pollen. Calyx is green or greenish-white and smaller than in the 2 preceding, but the yield per plant is much greater. Juice and other products are nearly colorless to amber. Green-fruited roselle is grown throughout Senegal, but especially in the Cape Vert region, mainly for use as a vegetable.

Another roselle selection which originated in 1914 at the Lamao experiment station and was named 'Temprano' because of its early flowering, Wester reported as no longer grown, the plant being less robust and less productive than the others.

A strain with dark-red, plump but stubby calyces (the sepals scarcely longer than the seed capsule) is grown in the Bahamas.

Climate

Roselle is very sensitive to frost. It succeeds best in tropical and subtropical regions from sea-level up to 3,000 ft (900 m) with a rainfall of about 72 in (182 cm) during its growing season. Where rainfall is inadequate, irrigation has given good results. It can be grown as a summer crop in temperate regions. The fruits will not ripen, but the herbage is usable.

Soil

While deep, fairly fertile sandy loam is preferable, roselle grew and produced well over many years in the oölitic limestone of Dade County. Wester observed that the high pinelands were far more suitable than low-lying muck soils. The plants tended to reseed themselves and on some properties they spread so extensively they became a nuisance and were eradicated.

Propagation

Roselle is usually propagated by seed but grows readily from cuttings. The latter method results in shorter plants preferred in India for interplanting with tree crops but the yield of calyces is relatively low.

Culture

Seedlings may be raised in nursery beds and transplanted when 3 to 4 in (7.5-10 cm) high, but seeds are usually set directly in the field, 4 to 6 to a hill, the hills 3 to 6 ft (0.9-1.8 m) apart in rows 5 to 10 ft (1.5-3 m) apart. When 2 or 3 leaves have developed, the seedlings are thinned out by 50%. If grown mainly for herbage, the seed can be sown as early as March, and no early thinning is done.

Roselle is a short-day plant and photoperiodic. Unlike kenaf, roselle crops cannot be grown successively throughout the year.

If intended solely for the production of calyces, the ideal planting time in southern Florida is mid-May. Blooming will occur in September and October and calyces will be ready to harvest in November and December. Harvesting causes latent buds to develop and extends the flowering life of the plant to late February. When the fruit is not gathered but left to mature, the plants will die in January.

Fig. 80: Dried roselle calyces are sold in plastic bags in Mexico, labeled "Flor de Jamaica", leading many to believe that they are flower petals. Actually, the flower falls before the red calyx enlarges and becomes fit for food use.

Rolfs recommended whatever fertilizer would be ordinarily used for vegetables but warned that only ¼ to ½ the usual amount should be applied. He wryly remarked: "As a whole, the plants are rather more vigorous than need be; consequently no attention need be paid in the direction of vigor." An excess of ammonia encourages vegetative growth and reduces fruit production. Commercial fertilizer of the formula 4-6-7 NPK has proved satisfactory.

Weeding is necessary at first, but after the plants reach 1½ to 2 ft (45-60 cm) in height, weeds will be shaded out and no longer a problem. Early pruning will increase branching and development of more flowering shoots.

Harvesting

For herbage purposes, the plants may be cut off 6 weeks after transplanting, leaving only 3 to 4 in (7.5-10 cm) of stem in the field. A second cutting is made 4 weeks later and a third after another 4 weeks. Then the shorn plants are thinned out—2 of every 3 rows removed—and the remaining plants left to grow and develop fruit as a second product.

The fruits are harvested when full-grown but still tender and, at this stage, are easily snapped off by hand. They are easier to break off in the morning than at the end of the day. If harvesting is overdue and the stems have toughened, clippers must be used.

The fruits of roselle ripen progressively from the lowest to the highest. Harvesting of seeds takes place when the lower and middle tiers of the last of the fruits are allowed to mature, at which time the plants are cut down, stacked for a few days, then threshed between canvas sheets.

Yield

Calyx production per plant has ranged from 3 lbs (1.3 kg) in California to 4 lbs (1.8 kg) in Puerto Rico and 16 lbs (7.25 kg) in southern Florida. In Hawaii, roselle intercropped with rubber yielded 16,800 lbs per acre (roughly 16,800 kg/ha) when planted alone. Dual-purpose plantings can yield 19,000 lbs (17,000 kg) of herbage in 3 cuttings and, later, 13,860 lbs (6,300 kg) of calyces.

Pests and Diseases

Roselle's major enemy is the root-knot nematode, *Heterodera rudicicola*. Mealybugs may be very troublesome. In Australia, 3 beetles, *Nisotra breweri, Lagris cyanea,* and *Rhyparida discopunctulata,* attack the leaves. The "white" roselle has been found heavily infested with the cocoa beetle, *Steirastoma breve* in Trinidad, with a lighter infestation of the red roselle in an intermixed planting. Occasional minor pests are scales, *Coccus hesperidum* and *Hemichionaspis aspidistrae,* on stems and branches; yellow aphid, *Aphis gossypii,* on leaves and flower buds; and the cotton stainer, *Dysdercus suturellus,* on ripening calyces.

In Florida, mildew (*Oidium*) may require control. Late in the season, leaves on some Philippine plants have appeared soft and shriveled; and *Phoma sabdariffae* has also done minimal damage.

Keeping Quality

Rolfs, in 1929, reported that fresh roselle calyces, as harvested, were successfully shipped by rail to Washington for retail sale and he judged that they could stand rail transport to any markets east of the Mississippi.

Food Uses

Roselle fruits are best prepared for use by washing, then making an incision around the tough base of the calyx below the bracts to free and remove it with the seed capsule attached. The calyces are then ready for immediate use. They may be merely chopped and added to fruit salads. In Africa, they are frequently cooked as a side-dish eaten with pulverized peanuts. For stewing as sauce or filling for tarts or pies, they may be left intact, if tender, and cooked with sugar. The product will be almost indistinguishable from cranberry sauce in taste and appearance. For making a finer-textured sauce or juice, sirup, jam, marmalade, relish, chutney or jelly, the calyces may be first chopped in a wooden bowl or passed through a meat grinder. Or the calyces, after cooking, may be pressed through a sieve. Some cooks steam the roselle with a little water until soft before adding the sugar, then boil for 15 minutes.

Roselle sauce or sirup may be added to puddings, cake frosting, gelatins and salad dressings, also poured over gingerbread, pancakes, waffles or ice cream. It is not necessary to add pectin to make a firm jelly. In fact, the calyces possess 3.19% pectin and, in Pakistan, roselle has been recommended as a source of pectin for the fruit-preserving industry.

Juice made by cooking a quantity of calyces with ¼ water in ratio to amount of calyces, is used for cold drinks and may be frozen or bottled if not for immediate needs. In sterilized, sealed bottles or jars, it keeps well providing no sugar has been added. In the West Indies and tropical America, roselle is prized primarily for the cooling, lemonade-like beverage made from the calyces. This is still "one of the most popular summer drinks of Mexico", as Rose observed in 1899. In Egypt, roselle "ade" is consumed cold in the summer, hot in winter. In Jamaica, a traditional Christmas drink is prepared by putting roselle into an earthenware jug with a little grated ginger and sugar as desired, pouring boiling water over it and letting it stand overnight. The liquid is drained off and served with ice and often with a dash of rum. A similar spiced drink has long been made by natives of West Tropical Africa. The juice makes a very colorful wine.

John Ripperton of the Hawaiian Experiment Station maintained that, for jelly and wine-making, it is unnecessary to take out the seed capsule, but neglecting to do so may result in a "stringy" product which would be contaminated with the minute hairs from the surface of the capsule and these hairs are quite likely to be injurious unless carefully filtered out.

The calyces are either frozen or dried in the sun or artificially for out-of-season supply, marketing or export. In Mexico today, the dried calyces are packed for sale in imprinted, plastic bags. It is calculated that 11 lbs (5 kg) of fresh calyces dehydrate to 1 lb (0.45 kg) of dried roselle, which is equal to the fresh for most culinary purposes. However, dried calyces as sold for "tea" do not yield high color and flavor if merely steeped; they must be boiled.

For retailing in Africa, dried roselle is pressed into solid cakes or balls. In Senegal, the dried calyces are squeezed into great balls weighing 175 lbs (80 kg) for shipment to Europe, where they are utilized to make extracts for flavoring liqueurs. In the United States, Food and Drug Administration regulations permit the use of the extracts in alcoholic beverages.

The young leaves and tender stems of roselle are eaten raw in salads or cooked as greens alone or in combination with other vegetables or with meat or fish. They are also added to curries as seasoning. The leaves of green roselle are marketed in large quantities in Dakar, West Africa. The juice of the boiled and strained leaves and stems is utilized for the same purposes as the juice extracted from the calyces. The herbage is apparently mostly utilized in the fresh state though Wester proposed that it be evaporated and compressed for export from the Philippines.

The seeds are somewhat bitter but have been ground to a meal for human food in Africa and have also been roasted as a substitute for coffee. The residue remaining after extraction of oil by parching, soaking in water containing ashes for 3 or 4 days, and then pounding the seeds, or by crushing and boiling them, is eaten in soup or blended with bean meal in patties. It is high in protein.

Other Uses

The seeds are considered excellent feed for chickens. The residue after oil extraction is valued as cattle feed when available in quantity.

Medicinal Uses: In India, Africa and Mexico, all above-ground parts of the roselle plant are valued in native medicine. Infusions of the leaves or calyces are regarded as diuretic, cholerectic, febrifugal and hypotensive, decreasing the viscosity of the blood and stimulating intestinal peristalsis. Pharmacognosists in Senegal recommend roselle extract for lowering blood pressure. In 1962, Sharaf confirmed the hypotensive activity of the calyces and found them antispasmodic, anthelmintic and antibacterial as well. In 1964, the aqueous extract was found effective against *Ascaris gallinarum* in poultry. Three years later, Sharaf and co-workers showed that both the aqueous extract and the coloring matter of the calyces are lethal to *Mycobacterium tuberculosis*. In experiments with domestic fowl, roselle extract decreased the rate of absorption of alcohol and so lessened its effect on the system. In Guatemala, roselle "ade" is a favorite remedy for the aftereffects of drunkenness.

Food Value Per 100 g of Edible Portion					
Calyces, fresh*		*Leaves, fresh***			
Moisture	9.2 g	Moisture	86.2%		
Protein	1.145 g	Protein	1.7-3.2%	*Seeds*	
Fat	2.61 g	Fat	1.1%	Moisture	12.9%
Fiber	12.0 g	Carbohydrates	10%	Protein	3.29%
Ash	6.90 g	Ash	1%	Fatty Oil	16.8%
Calcium	1,263 mg	Calcium	0.18%	Cellulose	16.8%
Phosphorus	273.2 mg	Phosphorus	0.04%	Pentosans	15.8%
Iron	8.98 mg	Iron	0.0054%	Starch	11.1%
Carotene	0.029 mg	Malic Acid	1.25%		
Thiamine	0.117 mg				
Riboflavin	0.277 mg				
Niacin	3.765 mg	*Analyses made in Guatemala.			
Ascorbic Acid	6.7 mg	**Analyses made in the Philippines.			

Food Value cont'd	
Amino acids (N = 16 p. 100 According to Busson)*	
Arginine	3.6
Cystine	1.3
Histidine	1.5
Isoleucine	3.0
Leucine	5.0
Lysine	3.9
Methionine	1.0
Phenylalanine	3.2
Threonine	3.0
Tryptophan	–
Tyrosine	2.2
Valine	3.8
Aspartic Acid	16.3
Glutamic Acid	7.2
Alanine	3.7
Glycine	3.8
Proline	5.6
Serine	3.5

*Calyces, fresh

The dried calyces contain the flavonoids gossypetine, hibiscetine and sabdaretine. The major pigment, formerly reported as hibiscin, has been identified as daphniphylline. Small amounts of delphinidin 3-monoglucoside, cyanidin 3-monoglucoside (chrysanthenin), and delphinidin are also present. Toxicity is slight.

In East Africa, the calyx infusion, called "Sudan tea", is taken to relieve coughs. Roselle juice, with salt, pepper, asafetida and molasses, is taken as a remedy for biliousness.

The heated leaves are applied to cracks in the feet and on boils and ulcers to speed maturation. A lotion made from leaves is used on sores and wounds. The seeds are said to be diuretic and tonic in action and the brownish-yellow seed oil is claimed to heal sores on camels. In India, a decoction of the seeds is given to relieve dysuria, strangury and mild cases of dyspepsia and debility. Brazilians attribute stomachic, emollient and resolutive properties to the bitter roots.

Food Value

Nutritionists have found roselle calyces as sold in Central American markets to be high in calcium, niacin, riboflavin and iron.

Durian (Plate XXXVIII)

The family Bombacaceae is best known for showy flowers and woody or thin-shelled pods filled with small seeds and silky or cottonlike fiber. The durian, *Durio zibethinus* L., is one member that differs radically in having large seeds surrounded by fleshy arils. Apart from variants of the word "durian" in native dialects, there are few other vernacular names, though the notorious odor has given rise to the unflattering terms, "civet cat tree", and "civet fruit" in India and *"stinkvrucht"* in Dutch. Nevertheless the durian is the most important native fruit of southeastern Asia and neighboring islands.

Description

The durian tree, reaching 90 to 130 ft (27-40 m) in height in tropical forests, is usually erect with short, straight, rough, peeling trunk to 4 ft (1.2 m) in diameter, and irregular dense or open crown of rough branches, and thin branchlets coated with coppery or gray scales when young. The evergreen, alternate leaves are oblong-lance-olate, or elliptic-obovate, rounded at the base, abruptly pointed at the apex; leathery, dark-green and glossy above, silvery or pale-yellow, and densely covered with gray or reddish-brown, hairy scales on the underside; 2½ to 10 in (6.25-25 cm) long, 1 to 3½ in (2.5-9 cm) wide. Malodorous, whitish to golden-brown, 3-petalled flowers, 2 to 3 in (5-7.5 cm) wide, with 5-lobed, bell-shaped calyx, are borne in pendant clusters of 3 to 30 directly from the old, thick branches or trunk.

The fruits are ovoid or ovoid-oblong to nearly round, 6 to 12 in (15-30 cm) long, 5 to 6 in (12.5-15 cm) wide, and up to 18 lbs (8 kg) in weight. The yellow or yellowish-green rind is thick, tough, semi-woody, and densely set with stout, sharply pointed spines, 3- to 7-sided at the base. Handling without gloves can be painful. Inside there are 5 compartments containing the creamy-white, yellowish, pinkish or orange-colored flesh and 1 to 7 chestnut-like seeds, ¾ to 2¼ in (2-6 cm) long with glossy, red-brown seedcoat. In the best fruits, most seeds are abortive. There are some odorless cultivars but the flesh of the common durian has a powerful odor which reminded the plant explorer, Otis W. Barrett, of combined cheese, decayed onion and turpentine, or "garlic, Limburger cheese and some spicy sort of resin" but he said that after eating a bit of the pulp "the odor is scarcely noticed." The nature of the flesh is more complex—in the words of Alfred Russel Wallace (much-quoted), it is "a rich custard highly flavored with almonds . . . but there are occasional wafts of flavour that call to mind cream cheese, onion-sauce, sherry wine and other incongruous dishes. Then there is a rich glutinous smoothness in the pulp which nothing else possesses, but which adds to its delicacy. It is neither acid, nor sweet, nor juicy; yet it wants none of these qualities, for it is in itself perfect. It produces no nausea or other bad effect, and the more you eat of it the less you feel inclined to stop." (*The Treasury of Botany*, Vol. I, p. 435). Barrett described the flavor as "triplex in effect, first a strong aromatic taste, followed by a delicious sweet flavor, then a strange resinous or balsam-like taste of exquisite but persistent savor." An American chemist working at the U.S. Rubber Plantations in Sumatra in modern times, was at first reluctant to try eating durian, was finally persuaded and became enthusiastic, declaring it to be "absolutely delicious", something like "a concoction of ice cream, onions, spices, and bananas, all mixed together."

Some fruits split into 5 segments, others do not split, but all fall to the ground when mature.

Origin and Distribution

The durian is believed to be native to Borneo and Sumatra. It is found wild or semi-wild in South Tenasserim, Lower Burma, and around villages in peninsular Malaya, and is commonly cultivated along roads or in orchards from southeastern India and Ceylon to New Guinea. Four hundred years ago, there was a lively trade in durians between Lower Burma to Upper Burma where they were prized in the Royal Palace. Thailand and South Vietnam are important producers of durians. The Association of Durian Growers and Sellers was formed in 1959 to standardize quality and marketing practices. The durian is grown to a limited extent in the southern Philippines, particularly in the Provinces of Mindanao and Sulu. The tree grows splendidly but generally produces few fruits in the Visayas Islands and on the island of Luzon. There are many bearing trees in Zanzibar, a few in Pemba and Hawaii. The durian is not included in the latest *Flora of Guam* (1970) which covers both indigenous and exotic species. It has been introduced into New Guinea, Tahiti, and Ponape.

The durian is rare in the New World. Seeds from Java were planted at the Federal Experiment Station in Mayaguez, Puerto Rico in 1920. The single resulting tree bloomed heavily in February and March in 1944 but only one fruit matured in July and it had but 3 normal carpels. Nevertheless, there were 6 fully developed seeds which germinated and were planted. The tree has fruited in Dominica and Jamaica. There have been specimens in the Royal Botanic Gardens, Port-au-Spain, Trinidad, for many years though they are not very much at home there. Young trees and seeds were introduced into Honduras from Java in 1926 and 1927, and the trees have grown well at the Lancetilla Experimental Garden at Tela, but they bear poorly to moderately. Seedlings have lived only briefly in southern Florida.

Varieties

Much variation occurs in seedlings. There are over 300 named varieties of durian in Thailand. Only a few of these are in commercial cultivation. In Malaysia, 100 types are graded for size and quality. In peninsular Malaya, there are 44 clones with small differences in time and extent of flowering, floral and fruit morphology, productivity and edible quality.

Pollination

There is no evidence that the durian is wind-pollinated and it is believed that bats (mainly *Eoncyteris spelea*) transfer pollen when they visit the flowers for nectar. Honeybees are seen on the flowers too early in the afternoon to serve as pollinators. Natural pollination is possible only at night, the heavily fragrant flowers opening in late afternoon and being receptive from 5 P.M. until 6 A.M., but pollen begins to shed at 7 P.M. and other floral parts gradually fall, only the pistil remaining at 11 P.M.

The durian has a high rate of self-incompatibility. In peninsular Malaya, the norm is 20% to 25% fruit-set, and it is realized that cross-pollination is essential to obtaining good crops. Hand-pollination performed during the day on buds that would open in 24 to 36 hours gives a much higher percentage of fruit-set than pollination of opened flowers. In unopened flowers the style is 1/3 as long as in fully opened flowers and the pollen reaches the ovules more quickly.

Climate

The durian is ultra-tropical and cannot be grown above an altitude of 2,000 ft (600 m) in Ceylon; 2,300 ft (700 m) in the Philippines, 2,600 ft (800 m) in Malaysia. The tree needs abundant rainfall. In India, it flourishes on the banks of streams, where the roots can reach water.

Soil

Best growth is achieved on deep alluvial or loamy soil.

Propagation

Durian seeds lose viability quickly, especially if exposed even briefly to sunlight. Even in cool storage they can be kept only 7 days. Viability can be maintained for as long as 32 days if the seeds are surface-sterilized and placed in air-tight containers and held at 68°F (20°C).

They have been successfully shipped to tropical America packed in a barely moist mixture of coconut husk fiber and charcoal. Ideally, they should be planted fresh, flat-side down, and they will then germinate in 3 to 8 days. Seeds washed, dried for 1 or 2 days and planted have shown 77-80% germination. It is reported that, in some countries, seedling durian trees have borne fruit at 5 years of age. In India, generally, they come into bearing 9 to 12 years after planting, but in South India they will not produce fruit until they are 13 to 21 years old. In Malaya, seedlings will bloom in 7 years; grafted trees in 4 years or earlier.

Neither air-layers nor cuttings will root satisfactorily. Inarching can be accomplished with 50% success but is not a popular method because the grafts must be left on the trees for many months. Selected cultivars are propagated by patch-budding (a modified Forkert method) onto rootstocks 2 months old and pencil-thick, and the union should be permanent within 25 to 30 days. The plants can be set out in the field within 14 to 16 months. Grafted trees never grow as tall as seedlings; they are usually between 26 to 32 ft (8-10 m) tall; rarely 40 ft (12 m).

Culture

Generally, durian trees receive little or no horticultural attention in the Far East. Young grafted plants, however, need good care. They should be staked, irrigated daily in the dry season, given monthly feedings of about 1/5 oz (5 g) of a 6-6-6 fertilizer formula, and the rootstock should be pruned gradually as leaves develop on the scion. When set out in the field, the trees should be 30 to 40 ft (9 to 12 m) apart each way.

Studies in Malaya have shown that a harvest of 6,000 lbs of fruits from an acre (6,720 kg from a hectare) removes the following nutrients from the soil: N, 16.1 lbs/acre (roughly equal kg/ha); P, 2.72 lbs/acre (roughly equal kg/ha); K, 27.9 lbs/acre (roughly equal kg/ha); Ca, 1.99 lbs/acre (roughly equal kg/ha); Mg, 3.26 lbs/acre (roughly equal kg/ha).

Season

In Ceylon, the durian generally blooms in March and April and the fruits mature in July and August, but these periods may shift considerably with the weather. Malaya has two fruiting seasons: early, in March and April; late, in September and October. Nearly all cultivars mature within the very short season during which the fruits are present in great numbers in local markets.

Harvesting

In rural areas, villagers clear the ground beneath the durian tree. They build grass huts nearby at harvest time and camp there for 6 or 8 weeks in order to be ready to collect each fruit as soon as it falls. Caution is necessary when approaching a durian tree during the ripening season, for the falling fruits can cause serious injury. Hunters place traps in the surrounding area because the fallen fruits attract game animals and all kinds of birds. The fruit is also placed as bait for game in the forests.

Fig. 81: The heavy, spiny durian (*Durio zibethinus*) is prized in Southeast Asia and Malaysia for its custard-like, odorous flesh.

Yield

Durians mature in 3½ to 4½ months from the time of fruit-set. Seedling trees in India may bear 40 to 50 fruits annually. Well-grown, high-yielding cultivars should bear 6,000 lbs of fruit per acre (6,720 kg/ha).

Keeping Quality

Durians are highly perishable. They are fully ripe 2 to 4 days after falling and lose eating quality in 5 or 6 days.

Pests and Diseases

Minor pests in the Philippines are the white mealybug (*Pseudococcus lilacinus*) and the giant mealybug (*Drosicha townsendi*) which infest young and developing fruits.

Very few diseases have been reported. In West Malaysia, patch canker caused by *Phytophthora palmivora* was first noted in 1934. It is becoming increasingly common on roots and stems of durian seedlings. Infection in the field begins at the collar with oozing of brownish-red gum and extends up the trunk and down to the roots. Sometimes a tree is completely girdled at the base and dies. Testing of 13 clones showed that all but 2 were susceptible. The 2 resistant clones succumbed after the stems were wounded and inoculated. It is evident that pruning injuries have provided access for the organism. The disease is encouraged by close-planting which shades the soil and promotes dampness. Weeds, grass and mulch around the collar are also contributing factors. Budded trees are particularly susceptible because of their habit of putting forth low branches and the occurrence of cracks where these join the main stem. When these low branches are pruned, the wound must be immediately treated with a fungicide.

Food Uses

Durians are sold whole, or cut open and divided into segments, which are wrapped in clear plastic. The flesh is mostly eaten fresh, often out-of-hand. It is best after being well chilled in a refrigerator. Sometimes it is simply boiled with sugar or cooked in coconut water, and it is a popular flavoring for ice cream. Javanese prepare the flesh as a sauce to be served with rice; they also combine the minced flesh with minced onion, salt and diluted vinegar as a kind of relish; and they add half-ripe arils to certain dishes. Arabian residents prefer to mix the flesh with ice and sirup. In Palembang, the flesh is fermented in earthen pots, sometimes smoked, and eaten as a special sidedish.

Durian flesh is canned in sirup for export. It is also dried for local use and export. Blocks of durian paste are sold in the markets. In Bangkok much of the paste is adulterated with pumpkin. Malays preserve the flesh in

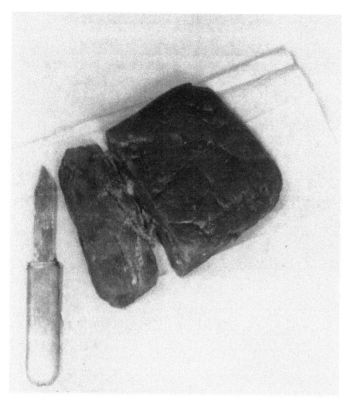

Fig. 82: Blocks of preserved durian paste are sold in the Bangkok market.

Food Value Per 100 g of Edible Portion*		
	Fresh Arils	*Dried Arils*
Calories	144	
Moisture	58.0–62.9 g	18.0 g
Protein	2.5–2.8 g	
Fat	3.1–3.9 g	3.0–6.0 g
Sugars	(approx.) 12.0 g	37.0–43.0 g
Starch	(approx.) 12.0 g	8.0–13.0 g
Total Carbohydrates	30.4–34.1 g	
Fiber	1.7 g	
Ash	1.1–1.2 g	3.0 g
Calcium	7.6–9.0 mg	
Phosphorus	37.8–44.0 mg	
Iron	0.73–1.0 mg	
Carotene	0.018 mg	
(as Vitamin A)	20–30 I.U.	
Thiamine	0.24–0.352 mg	
Riboflavin	0.20 mg	
Niacin	0.683–0.70 mg	
Ascorbic Acid	23.9–25.0 mg	
Vitamin E	"high"	

*Analyses made in Malaya, Honduras and elsewhere.

salt in order to keep it on hand the year around to eat with rice, even though it acquires a very strong and, to outsiders, most disagreeable odor. The unripe fruit is boiled whole and eaten as a vegetable.

The seeds are eaten after boiling, drying, and frying or roasting. In Java, the seeds may be sliced thin and cooked with sugar as a confection; or dried and fried in coconut oil with spices for serving as a side-dish.

Young leaves and shoots are occasionally cooked as greens. Sometimes the ash of the burned rind is added to special cakes.

Toxicity

The seeds are believed to possess a toxic property that causes shortness of breath.

Other Uses

Rind: The dried or half-dried rinds are burned as fuel and fish may be hung in the smoke to acquire a strong flavor. The ash is used to bleach silk.

Wood: The sapwood is white, the heartwood light red-brown, soft, coarse, not durable nor termite-resistant. It is used for masts and interiors of huts in Malaya.

Medicinal Uses: The flesh is said to serve as a vermifuge. In Malaya, a decoction of the leaves and roots is prescribed as a febrifuge. The leaf juice is applied on the head of a fever patient. The leaves are employed in medicinal baths for people with jaundice. Decoctions of the leaves and fruits are applied to swellings and skin diseases. The ash of the burned rind is taken after childbirth. The leaves probably contain hydroxy-tryptamines and mustard oils.

The odor of the flesh is believed to be linked to indole compounds which are bacteriostatic. Eating durian is alleged to restore the health of ailing humans and animals. The flesh is widely believed to act as an aphrodisiac. In the late 1920's, Durian Fruit Products, Inc., of New York City, launched a product called "Dur-India" as a "health-food accessory" in tablet form, selling at $9 for a dozen bottles, each containing 63 tablets—a 3-months' supply. The tablets reputedly contained durian and a species of *Allium* from India, as well as a considerable amount of vitamin E. They were claimed to provide "more concentrated healthful energy in food form than any other product the world affords"—to keep the body vigorous and tireless; the mind alert with faculties undimmed; the spirit youthful.

A toothpaste flavored with durian is currently marketed for durian fanciers.

Related Species

There are estimated to be 28 species in the genus *Durio* in Malaysia. Only 5 species in addition to the durian bear edible fruits. These are *D. dulcis* Becc., in Sabah and Indonesian Borneo; *D. grandiflorus* Kost., in Sabah, Sarawak, and Indonesian Borneo; *D. graveolens* Becc., in peninsular Malaya and all of Borneo and Sumatra; *D. kutejensis* Becc., all over Borneo, and ranked second to the durian in edibility; and *D. oxleyanus* Griff., in peninsular Malaya and all of Borneo and Sumatra. All five are cultivated in Brunei and a few to some extent in Malaysian Borneo.

It is believed that some of the other species, especially *D. malaccensis* Planch. and *D. Wyatt-Smithii* Kost., which are very closely allied to *D. zibethinus*, may be useful in breeding for pest- and disease-resistance and other characters.

There is evidence that natural interspecific cross-pollination is going on because a hybrid of *D. zibethinus* and *D. graveolens* has been found in northeastern Indonesian Borneo, and some trees of normally white-flowered *D. malaccensis* have been discovered in Johore State with reddish flowers, perhaps from cross-pollination by the pink or red-flowered *D. lowianus* King and *D. pinangianus*.

Chupa-Chupa (Plate XXXIX)

Little-known outside its natural range, this member of the Bombacaceae has nomenclatural problems. Its current botanical designation is *Quararibea cordata* Vischer (syn. *Matisia cordata* Humb. & Bonpl.), though it still is being dealt with in Brazil and Colombia under the latter binomial, and there are taxonomists who prefer not to merge *Matisia* with *Quararibea*. In addition, there is no generally accepted vernacular name. "Sapote" and "zapote" predominate in native countries but these terms, derived from the Nahuatl word for "soft, sweet", are applied to several other fruits and to one in particular, the sapote, *Pouteria sapota*, q.v. To distinguish *Quararibea cordata*, one writer proposed "South American sapote", and this has been repeated, but it is cumbersome and strictly artificial, not a name in use in any country of origin. Therefore, I have chosen *chupa-chupa*, which is a valid colloquial name in Colombia and Peru, certainly euphonius, and, as Dr. Victor Patino has stated, descriptive of the manner in which the flesh is chewed from the large seeds. In Peru and Colombia, the species may also be called *zapote chupachupa, zapote chupa, sapote de monte,* or *sapotillo*. In Brazil, it is known as *sapota, sapote-do-peru,* or *sapota-do-solimões*, in reference to the Solimões River.

Description

The chupa-chupa tree is fast-growing, erect, to 130 or even 145 ft (40-45 m) high in the wild, though often no more than 40 ft (12 m) in cultivation. It is sometimes buttressed; has stiff branches in tiered whorls of 5; and copious gummy yellow latex. The semi-deciduous, alternate, long-petioled leaves, clustered in rosettes near the ends of the branches, are broadly heart-shaped, normally 6 to 12 in (15-30 cm) long and nearly as wide. Short-stalked, yellowish-white or rose-tinted, 5-petalled flowers, about 1 in (2.5 cm) wide, with 5 conspicuous, protruding stamens and pistil, are borne in masses along the lesser branches and on the trunk. The fruit is rounded, ovoid or elliptic with a prominent, rounded knob at the apex and is capped with a 2- to 5-lobed, velvety, leathery, strongly persistent calyx at the base; 4 to 5¾ in (10-14.5 cm) long and to 3³⁄₁₆ in (8 cm) wide, and may weigh as much as 28 oz (800 g). The rind is thick, leathery, greenish-brown, and downy. The flesh, orange-yellow, soft, juicy, sweet and of agreeable flavor surrounds 2 to 5 seeds, to 1½ in (4 cm) long and 1 in (2.5 cm) wide, from which long fibers extend through the flesh.

Origin and Distribution

The tree grows wild in lowland rainforests of Peru, Ecuador and adjacent areas of Brazil, especially around the mouth of the Javari River. It is common in the western part of Amazonas, southwestern Venezuela, and in the Cauca and Magdalena Valleys of Colombia. It flourishes and produces especially well near the sea at Tumaco, Colombia. The fruits are plentiful in the markets of Antioquia, Buenaventura and Bogotá, Colombia; Puerto Viejo, Ecuador; the Brazilian towns of Tefé, Esperanca, Sao Paulo de Olivenca, Tabetinga, Benjamin Constant and Atalaia do Norte; and elsewhere.

There were only 3 trees in gardens in Belém in 1979. The Instituto Nacional de Pesquisas da Amazonia had 150 fruits sent there for evaluation and 80 to 90% of the samplers rated them as of excellent flavor and expressed interest in obtaining trees. However, it is recognized that there is need for horticultural improvement. In 1964, William Whitman obtained seeds from Iquitos, Peru, raised seedlings; planted one on his own property at Bal

Harbour, Florida, and distributed the rest to private experimenters. The first to fruit was that grown by B.C. Bowker, Miami, in 1973. Whitman's tree and several others have also borne fruit.

Varieties

Some of the fruits borne in Florida appear to be of better than average quality. In northern Peru, there is reportedly a type with little fiber and superior flavor.

Pollination

The flowers are pollinated by hummingbirds, bees and wasps. In the afternoon some trees become self-compatible.

Climate

The chupa-chupa is a tropical to subtropical species. In Ecuador, it ranges from sea-level to 4,000 or even 6,500 ft (1,200-2,000 m). In Florida, young trees need protection from winter cold. For best performance, the tree needs full sun and plenty of moisture.

Soil

The tree attains maximum dimensions in the low, wet, deep soils of South American forests, yet it does well in cultivation on the slopes of the Andes and seems to tolerate the dry, oolitic limestone of South Florida's coastal ridge when enriched with topsoil and fertilizer.

Propagation

The tree is commonly grown from seed but superior types should be vegetatively propagated. Side-veneer grafting can be easily done. Budding is not feasible.

Season and Harvesting

In Brazil, the tree blooms from August to November and fruits mature from February to May. Trees in Florida bloom in midwinter and ripen their fruits in November. The fruit will stay on the tree until it rots. It must be harvested with a knife or a long cutting-pole. Light color around the edge of the calyx is a sign of ripeness.

Yield

Whitman's tree bore 58 fruits in 1976. A normal crop may be 3,000. One tree in Tefé, Brazil, produced an estimated crop of 6,000 or more fruits in a season.

Pests and Diseases

The chupa-chupa is very prone to attack by fruit flies and in some locations in South America is commonly infested with their larvae. In Florida, the Keys whitefly, *Aleurodicus dispersus,* and the Cuban May beetle, *Phyllophaga bruneri,* attack the foliage.

Food Uses

This is a fruit that has always been eaten out-of-hand. Those that have the least fibrous flesh may be utilized for juice or in other ways.

Food Value Per 100 g of Edible Portion*	
Moisture	85.3 g
Protein	0.129 g
Fat	0.10 g
Fiber	0.5 g
Ash	0.38 g
Calcium	18.4 mg
Phosphorus	28.5 mg
Iron	0.44 mg
Carotene	1.056 mg
Thiamine	0.031 mg
Riboflavin	0.023 mg
Niacin	0.33 mg
Ascorbic Acid	9.7 mg

*Analyses made in Ecuador.

ACTINIDIACEAE

Kiwifruit (Plate XL)

A late-comer on the international market, the kiwifruit long identified as (*Actinidia chinensis* Planch.), was formerly placed in the family Dilleniaceae; is now set apart in Actinidiaceae which includes only two other genera. In the August 1986 issue of *HortScience* (Vol. 21 #4 : 927), there appears an announcement that China's leading authority on this fruit has renamed the stiff-haired form (which includes the kiwifruit) *A. deliciosa* (A. Chevalier) C.F. Liang et A.R. Ferguson var. *deliciosa*, and has retained *A. chinensis* for the smooth-skinned form. The Chinese name, *yang tao*, meaning "strawberry peach", was replaced by Europeans with the descriptive term, Chinese gooseberry (because of the flavor and color of the flesh). In 1962, New Zealand growers began calling it "kiwifruit" to give it more market appeal, and this name has been widely accepted and publicized despite the fact that it is strictly artificial and non-traditional. It was commercially adopted as the trade name in 1974. There are a few little-used colloquial names such as Ichang gooseberry, monkey peach and sheep peach.

Description

The kiwifruit is borne on a vigorous, woody, twining vine or climbing shrub reaching 30 ft (9 m). Its alternate, long-petioled, deciduous leaves are oval to nearly circular, cordate at the base, 3 to 5 in (7.5-12.5 cm) long. Young leaves and shoots are coated with red hairs; mature leaves are dark-green and hairless on the upper surface, downy-white with prominent, light-colored veins beneath. The fragrant, dioecious or bisexual flowers, borne singly or in 3's in the leaf axils, are 5- to 6-petalled, white at first, changing to buff-yellow, 1 to 2 in (2.5-5 cm) broad, and both sexes have central tufts of many stamens though those of the female flowers bear no viable pollen. The oval, ovoid, or oblong fruit, up to 2½ in (6.25 cm) long, with russet-brown skin densely covered with short, stiff brown hairs, is capped at the base with a prominent, 5-pointed calyx when young but this shrivels and dehisce from the mature fruit while 5 small sepals persist at the apex. The flesh, firm until fully ripe, is glistening, juicy and luscious, bright-green, or sometimes yellow, brownish or off-white, except for the white, succulent center from which radiate many fine, pale lines. Between these lines are scattered minute dark-purple or nearly black seeds, unnoticeable in eating. Cross-sections are very at-tractive. In some inferior types, the central core is fibrous or even woody. The flavor is subacid to quite acid, somewhat like that of the gooseberry with a suggestion of strawberry.

Origin and Distribution

This interesting species is native to the provinces of Hupeh, Szechuan, Kiangsi and Fukien in the Yangtze Valley of northern China—latitude 31°N—and Zhejiang Province on the coast of eastern China. It was cultivated on a small scale at least 300 years ago, but still today most of the 1,000-ton crop is derived from wild vines scattered over 33 of the 48 counties of Zhejiang. The plants may be seen climbing tall trees or, near Lung to ping, Hupeh, sprawling over low scrub or rocks exposed to strong northeast winds and bearing heavily. The Chinese have never shown much interest in exploiting the fruit. Because of the dense population, there is little room for expansion of the industry. Nevertheless, trial shipments of canned fruits were made to West Germany in 1980.

Specimens of the plant were collected by the agent for the Royal Horticultural Society, London, in 1847 and described from his dried material. In 1900, seeds gathered in Hupeh were sent to England by E.H. Wilson. The resulting plants flourished and bloomed in 1909. When both male and female vines were planted together, fruits were produced but usually only solitary vines were grown as ornamentals. Seeds from China were introduced into New Zealand in 1906 and some vines bore fruits in 1910. Several growers raised numerous seedlings (many of which were males) and selected the best fruiting types, which were propagated around 1930. By 1940 there were many plantings, one with 200 vines, especially on the eastern coast of the North Island. The fruits were being marketed and were very popular with American servicemen stationed in New Zealand during World War II. Commercial exporting was launched in 1953, the fruits going mainly to Japan, North America and Europe, with small quantities to Australia, the United Kingdom and Scandinavia. In 1981, a survey of smallholders in the Auckland suburbs revealed that the great majority of them intended to plant kiwifruit for the local market. Today, West Germany is New Zealand's biggest customer for kiwifruit. Production in 1983 was reportedly 40,000 tons as compared with 300 tons in 1937. New Zealand supplies 99% of the world production of kiwifruit and

Fig. 83: Kiwifruits on the vine and harvested fruits *en route* to packing house. Courtesy Blue Anchor, Inc., Sacramento, California.

95% of the crop is harvested within 35 miles (56 km) of the little town of Te Puke, Bay of Plenty—38° S latitude. The small industry was greatly assisted in 1971 by an arrangement with the Bay of Plenty Co-operative Dairy Company for the use of cool storage facilities and the construction of a cooperative central packing house. In 1984, there were 2,500 growers, more than 400 packing sheds and 200 "coolstores" with a capacity of 1.9 million tons. A $10,000 prize was offered for the design of a new package for export that would accommodate fruits of varying shapes and sizes.

Plants and seeds have been distributed from New Zealand to the United States of America (including Hawaii), and to Australia, South Africa, Germany, the Netherlands and Denmark. In 1981, plant exports amounted to $430,000 NZ. But in 1982 the New Zealand Kiwifruit Authority issued an appeal to cease exporting plants to reduce the likelihood of competition for foreign markets.

The United States Department of Agriculture received seeds from Consul-General Wilcox in Hankow in 1904 (P.I. 11629, 11630) and the resulting vines were fruiting at the Plant Introduction Field Station at Chico, California in 1910. In 1905, a Rev. Hugh White sent in seeds from Kiangsi (P.I. 18535). E.H. Wilson supplied seeds from western Hupeh and Szechuan (P.I. 21781). In 1917, the agricultural explorer, Frank Meyer, sent back to Washington seeds from fruits he found growing near Lung to ping, Hupeh, ranging in size from "that of a gooseberry to a good-sized plum" (P.I. 45946). A plant from this introduction was given to Mr. William Hertrich of San Gabriel, California. It had perfect flowers and bore fruit "of good size and quality." Mr. Hertrich reproduced it by cuttings and in 1919 supplied some of the plants to the

Station at Chico (P.I. 46864). In 1935, a New Zealand grower sent plants of a large-fruited kiwifruit (later named 'Hayward' in New Zealand). One of the plants was reported as still flourishing and fruiting—400 lbs (160 kg) annually—in 1982. After cultural techniques were developed in the 1960's, two California growers imported several thousands of plants from New Zealand. Special kiwifruit nurseries were established in 1966 and, by 1970, there were 40 acres (20.25 ha) devoted to this crop. By 1977 there were over 2,000 acres (800 ha) planted with kiwifruit vines but only 10% of the plants had reached bearing age. In 1982, there were about 1,000 small commercial farms in the state. In 1984, kiwifruit groves in California totalled 6,000 acres (2,040 ha). Most of the crop, worth $18,000,000 to the growers, is sold locally, but some has been shipped to Japan and to the Netherlands. The trade association, Kiwi Growers of California, was organized in 1972 and incorporated in 1975 to sponsor research and exchange and publish information. Nationwide publicity and marketing is handled by Blue Anchor, Inc., the California Fruit Exchange, greatly stimulating demand despite the high retail price of the fruits. The California Kiwifruit Administration Committee has set rigid quality standards, preventing the shipping of "unclassified" grade.

The Fruit and Fruit Technology Research Institute of Stellenbosch, South Africa, obtained budwood of New Zealand cultivars in 1960 and experimental plantings were made in a number of areas around the country. The success of the vines in the northeastern Transvaal inspired the installment of a large plantation of mostly seedlings, some plants from cuttings, at Chiremba in the lower Vumba based on New Zealand and California selections. At this location, the altitude is 3,280 ft (1,000 m) and the annual rainfall is 60 in (152 cm). The mean temperatures in southern Cape areas are close to those at Sacramento, California. However, there are great extremes in South African weather and occasionally very high day temperatures which may cause sunburn on exposed fruits. Nevertheless, the South African Kiwifruit Association was formed in 1981 at the University of Natal with expectations of developing successful cultivation.

The kiwifruit was already being grown in Cambodia, Vietnam and southern Laos, France, Spain, Belgium, and Italy where plantings were first made in the late 1960's and commercial growing started in the late 1970's. Italy advanced to third place in world production by 1983, with a crop of 6,000 tons from 4,800 acres (2,000 ha). Over one-half of Italy's crop is exported to France and other European countries.

French interest in the kiwifruit has been stimulated by the low returns from apple-growing. By 1971, there were small plantings scattered around southwestern and southeastern areas of the country—valleys of the Garonne, Dordogne, Rhone and Loire rivers—totalling about 123 acres (50 ha). Greece is now producing kiwifruits for export to other European countries, filling the seasonal gap when fruits from New Zealand are not available. A recent development is the raising of kiwifruits in greenhouses in the Channel Islands, especially as an alternative to tomatoes suffering from European competition.

The vine was introduced into the Philippines at Baguio, in 1923. It succeeds there only above 3,280 ft (1,000 m) and has not been exploited. Large plantings are being made in Chile, not far from Santiago.

Varieties

There are 4 main Chinese classes of kiwifruit:

1) **'Zhong Hua'** ("Chinese gooseberry")—round to oval, or oblate; weight varies from 6.5 to 80 g, averaging 30 to 40 g. Sugar content is 4.6 to 13.1%; ascorbic acid, 25.5 to 139.7 mg per 100 g. This is the most commonly grown.
Three subvarieties are: "Yellow flesh"—average weight, 30.2 g; sugar content, 9.0%; ascorbic acid 101.9 mg per 100 g. "Green flesh"—average weight 18.4 g; sugar content, 5.4%; ascorbic acid, 55.7 mg per 100 g. "Yellow-green" and "Green-yellow"—average weight 31 to 48 g; sugar content 5.4%; ascorbic acid 85.5 mg per 100 g. Not suitable for canning sliced or for jam.
2) **'Jing Li'** ("northern pear gooseberry")—elongated oval with green flesh. Leaves usually hairless.
3) **'Ruan Zao'** ("Soft date gooseberry")—small, with green flesh; quite sweet. Good for jam. Usually grows in the hills.
4) **'Mao Hua'**—may be tight- or loose-haired; has green, sweet flesh. The leaves are elongated oval, relatively broad and thick.

Selections made by growers for fresh fruit market:

1) **'Qing Yuan #17'**—fruit weighs a maximum of 70.3 g; skin is yellow-brown, smooth, thin; flesh is juicy and of excellent flavor; sugar content 8.2%; ascorbic acid, 169.7 mg per 100 g. Rated as of superior quality.

2) **'Qing Yuan #22'**—fruit has maximum weight of 67 g; average is 47.3 g; skin is yellow-brown, smooth, thin; sugar content 7.9%; ascorbic acid, 11.42 mg per 100 g. Of high quality.
3) **'Qing Yuan #28'**—fruit cylindrical; weighs a maximum of 46 g; averages 40.6 g. Skin is smooth; flesh fine-textured and juicy. Sugar content 9.1%; ascorbic acid 103.2 mg per 100 g. Of medium quality.
4) **'Qing Yuan #18'**—fruit cylindrical; maximum weight 56 g, average 36 g; flesh very tender, medium juicy, of good flavor. Ascorbic acid content 178.9 mg per 100 g. Good fresh and for processing.
5) **'Qing Yuan #20'**—small, elongated cylindrical; maximum weight 26 g; average 21.5 g. Sugar content 12.4%; ascorbic acid, 189.2 mg per 100 g. Excellent quality.
6) **'Long Quan #3'**—oblate; average weight 31 g. Flesh yellow, fine-textured, juicy, and of good flavor. Sugar content, 9.5%; ascorbic acid, 99.7 mg per 100 g. Above average quality.

Selections made by growers for processing because of uniform shape and size, yellow or reddish-brown flesh, minimum woodiness at base, high ascorbic acid content:

1) **'Qing Yuan #27'**—cylindrical; average weight, 27.9 g; flesh yellow, fine-textured; seeds few; core small. Good for processing.
2) **'Qing Yuan #29'**—average weight, 27 g; flesh yellow, fine-textured, with small core.
3) **'Qing Yuan #6'**—average weight 27.3 g. Flesh pale-yellow and fine-textured. Sugar content 7.6%; ascorbic acid 140 mg per 100 g. Of superior quality for processing.
4) **'Huang Yan'**—yellow-skinned, cylindrical; average weight 21.9 g; flesh yellow-white, fine-textured, and of good flavor, with medium-large core. Sugar content 7.4%; ascorbic acid 170.8 mg per 100 g. Above average quality for processing.

The leading cultivars in New Zealand are:

'Abbott' ('Green's'; 'Rounds')—a chance seedling, discovered in the 1920's; introduced into cultivation in the 1930's. Fruit oblong, of medium size, with brownish skin and especially dense, long, soft, hairs; flesh is light-green and of good flavor. Of good keeping quality. Resembles 'Allison'. Ripens in early May. Vine is vigorous, precocious, productive. Petals do not overlap; styles are horizontal. Most exports to the United Kingdom have been of this cultivar.

'Allison' ('Large-fruited')—a chance seedling discovered in 1920's; introduced in early 1930's. Fruit oblong, slightly broader than 'Abbott'; of medium size, with densely hairy, brownish-skin; flesh is light-green, of good flavor. Fruit is of good keeping quality. Vine very vigorous, prolific; blooms later than 'Abbott'; fruits ripen early May. Flowers have broader, more overlapping petals than 'Abbott' and they are crinkled on the margins. Styles elevate to 30 or 60° angle as flower ages. Formerly very popular but has lost ground to 'Hayward'.

'Bruno' ('McLoughlin'; 'Longs'; 'Long-fruited'; 'Te puke')—a chance seedling; discovered in the 1920's; introduced in the 1930's. Fruit large, elongated cylindrical, broadest at apex; has darker-brown skin than other cultivars and dense, short, bristly hairs. Flesh is light-green, of good flavor. Ripens in early May. Vine is vigorous and productive, blooms with or slightly after 'Allison'. Sometimes exported. Flowers borne singly or sometimes in pairs. Petals narrower and overlap less; styles longer and stouter than those of 'Abbott', more regularly arranged than those of 'Allison'.

'Hayward' ('Giant'; 'Hooper's Giant'; 'McWhannel's')—chance seedling in Auckland; discovered in 1920's; introduced into cultivation in early 1930's; introduced into the United States as P.I. 112053 before being named in New Zealand and was called 'Chico' in California. Fruit exceptionally large, broad-oval, with slightly flattened sides; skin light greenish-brown with dense, fine, silky hairs. Flesh light green; of superior flavor and fruit is of good keeping quality. Ripens in early May. Vine is moderately vigorous, blooms very late; is moderately prolific, partly because of scanty pollination and late-blooming males must be planted with it. Flowers borne singly or, rarely, in pairs. The petals are broad, overlapping, cupped, and the styles more erect than those of other cultivars though they vary from horizontal to vertical. This is the leading cultivar in New Zealand; the only commercial cultivar in California; produces 72% of Italy's crop.

'Monty' ('Montgomery')—a chance seedling in New Zealand, discovered in the early 1950's; introduced into cultivation about 1957. Fruit oblong, somewhat angular, widest at apex; of medium size; skin brownish with dense hairs. Flesh is light-green. Fruit ripens in early May. Vine is highly vigorous and productive, sometimes excessively so. Petals overlap only slightly at the base.

'Greensill'—a more recent selection; it is the most cylindrical of all, flattened on both ends, slightly wider at base than at apex; a little shorter than 'Allison' but thicker. Petals narrow, constricted, do not overlap at the base; styles are mostly erect.

Plant breeders are endeavoring to develop an acceptable hairless kiwifruit and several thousand seedlings of a promising clone were set out in an experimental plot in Pukekohe, New Zealand, in 1980.

Male plants commonly used for pollination are:

'Matua', with short hairs on peduncles and flowers in groups of 1 to 5, usually 3.

'Tomuri', with long hairs on peduncles, flowers in groups of 1 to 7, usually 5.

Climate

The kiwifruit vine grows naturally at altitudes between 2,000 and 6,500 ft (600-2,000 m). The Kwangsi latitude is approximately that of Galveston, Texas; the climate has been likened to that of Virginia or North Carolina, with heavy rainfall and an abundance of snow and ice in the winter.

In the Bay of Plenty region the winter mean minimum daily temperatures are from 40° to 42°F (4.44°-5.56°C); mean maximum, 57° to 60°F (13.89°-15.56°C); in summer, mean minimum is 56° to 57°F (13.33°-13.89°C); mean maximum, 75° to 77°F (23.89°-25°C). Annual rainfall is 51 to 64 in (130-163 cm) and relative humidity 76 to 78%.

In California, the kiwifruit is an appropriate crop wherever citrus fruits, peaches and almonds are successful, though the leaves and flowers are more sensitive to cold than those of orange and peach trees. Autumn frosts retard new growth and kill developing flower buds, or, if they occur after the flowers have opened, will prevent the setting of fruits. Late winter frosts are said to improve the flavor of full-grown fruits.

Kiwifruit vines in leaf are killed by drops in temperature below 29°F (-1.67°C), while dormant mature vines can survive temperatures down to 10°F (-12.22°C). In France, 1-year-old plants have been killed to the ground by frosts. California growers report that the kiwifruit requires a temperature drop to 32°F (0°C) to cause it to drop its leaves and then 400 hours of dormancy, or 40 days of 40°F (4.44°C), in order to set fruit properly. At Pietermaritzburg, South Africa, where there are only 150 to 200 hours of chilling weather, the vines are slow to put out new spring leaves.

Alternating warm and cold spells during the winter will reduce flowering. A seedling selection at the Citrus and Subtropical Fruit Research Institute, Nelspruit, has borne well and appears to be more tolerant of mild winters than other cultivars which are not successful in this warm region of the eastern Transvaal. There have been several attempts to grow kiwifruits in northern and central Florida, and a few vines are growing experimentally in the southern part of the state and even on the Florida Keys but, so far, only the plants at Tallahassee have fruited to any extent.

Soil

For good growth, the vine needs deep, fertile, moist but well-drained soil, preferably a friable, sandy loam. Heavy soils subject to waterlogging are completely unsuitable. In Kiangsi Province, China, the wild plants flourish in a shallow layer of "black wood earth" on top of stony, red subsoil.

Pollination

The flowers are mostly insect-pollinated. For small, single-row plantings, one male vine to every 5 females is necessary. In commercial plantings, 10 to 12% of the vines must be males, that is, about 1 male for every 8 or 9 female vines, and the males should be staggered evenly throughout the block plantations. The time of flowering must be ascertained so that the male and female plants will coincide. The female plants yield no nectar. It is recommended that there be 3⅓ beehives per acre (8 per ha) when 10 to 15% of the flowers are open in order to assure adequate pollination. In anticipation of a shortage of hives for expanding culture, work was begun in New Zealand about 1980 to perfect means of collecting and drying pollen and preparing a suspension for spraying onto the blooming vines by tractor-drawn equipment. Pollen is commercially available in California also for artificial pollination.

Propagation

Inasmuch as seedlings show great variation, it is not recommended that the vine be grown from seed except in experimental plots for clone selection or to produce rootstocks for budding or grafting. To obtain the small seeds, ripe fruits are pulped in an electric blender and then the pulp is strained through a fine screen. The seeds, mixed with moist sand, are placed in a plastic bag, plastic box or other covered container, and kept in a refrigerator (below freezing temperature) for 2 weeks. Then the seed/sand mixture can be planted in nursery flats of sterilized soil, or directly in the garden or field, no deeper than ⅛ in (3 mm) and kept moist. Germination will take 2 to 3 weeks. The seedlings should be thinned out to prevent overcrowding and can be successfully transplanted when 3 in (7.5 cm) high if the soil is taken up with the root system intact. If intended for rootstocks, they should be set 12 to 15 in (30-45 cm) apart in nursery rows. When 1-year-old, the plants are ready for budding.

Budwood is taken from the current season's growth and defoliated, leaving only ½ in (1.25 cm) of the petiole of each leaf, and is inserted in the rootstock about 4 in (10 cm) above the ground, using the "T" or shield method. When the buds have "taken", the stock is cut back to just above the union.

For grafting, scions are taken from a parent vine while it is dormant and should be trimmed at both ends, leaving 2 or 3 buds. The scion is joined to the stock by either the whip or tongue process about 4 in (10 cm) from the ground.

Soft-wood cuttings, trimmed to leave only 2 leaves, are treated with hormones and rooted under intermittent mist. Dormant cuttings have a low percentage of success. In New Zealand, cuttings are not popular because they do not develop a strong root system and are prone to attack by crown-gall. Root-grafting was formerly practiced but abandoned because of susceptibility to crown-gall at the graft union.

Old vines bearing inferior fruits can be reworked by budding or, preferably cleft-grafting, which must be done before new growth begins or the vine will bleed sap. Some growers graft a branch of a compatible male onto a female vine to promote pollination. The increasing demand for plants of cv. 'Hayward' in South Africa has led to *in vitro* propagation using vegetative buds of female plants.

Culture

The kiwifruit is alleged to be a difficult crop to establish, and many new plantations in California have been costly failures. The soil should be well worked to a fine tilth for easy penetration by the shallow, fibrous root system. It is important to fumigate in advance of planting. The land should be level to give all plants equal moisture. There should be good drainage and protection from strong winds which severely damage tender spring shoots. The vines are set not opposite each other but alternated and a generally used spacing has been 18 to 20 ft (5.56-6 m) apart in rows 15 ft (4.5 m) apart. In 1983 it was announced that between-plant spacing was being reduced to 8.2 ft (2.5 m). It has been customary to train the vines to grow on strong horizontal trellises with wood "T" supports 6 to 7 ft (1.8-2.1 m) high, holding 3 wires 2 ft (60 cm) apart. One New Zealand grower has developed a metal arch system which provides headroom under the canopy for pruning and harvesting, and also provides frost protection by allowing cold air to flow downward and settle on the ground, and this air movement helps reduce the frequency of disease. Also, it has been found that A-frame pergolas are producing 3 times as much fruit as the traditional flat trellises.

By the common method, the plants are staked until they reach the wire and, as they develop, they must be kept under control, otherwise a tangled mass of unwieldy vegetation will result. Training of the vines is very important. There should be a single leader and fruiting arms every 18 to 28 in (45-71 cm). Summer pruning is for the purpose of heading the fruiting arms and suppressing shoots. Shoots from summer pruning will not bear fruit until the following year after dormancy. Male plants will yield more pollen in the spring if new shoots are topped to leave 5 to 7 buds during the summer. Renewing of fruiting arms is done every 4 years, in the winter. The vines should be trained to fruit above the foliage instead of beneath it because excessive shading from the canopy results in poor shoot development, delayed blooming, dehydration and dying of flowerbuds, reduced size of fruits. This is more critical in New Zealand than in California where the light is more intense and penetrating.

The mature plants require a minimum of 150 lbs nitrogen per acre (about 150 kg/ha). In New Zealand, they are usually fertilized twice a year, once in spring and once in early summer, using a total of 500 lbs (225 kg) nitrogen, 220 lbs (100 kg) P_2O_5, 121 lbs (55 kg) K_2O, per hectare—equivalent to 202 lbs (92 kg) nitrogen, 89 lbs (4.5 kg) P_2O_5, 49 lbs (22.2 kg) K_2O per acre.

Apart from land cost, it takes a minimum of $3,500 to bring each acre into production. The first 2 years are the most critical, coping with the variable growth habits of individual plants, but the vines become more manageable with age. One producer in California, who also raises and sells grafted plants, believes that many people have set out plants that are too young. He sells only 2-year-old vines, bare-root for planting in the dormant season, which gives the roots maximum freedom unlike those which develop in containers. In France, where cuttings from New Zealand are kept in cool storage during the winter and planted out in the spring, vines have made 5 to 6 ft (1.5-1.8 m) of growth in the first 2 months.

Kiwifruit vines can stand wet seasons that destroy peach orchards. Drip irrigation is now being used in California plus overhead sprinklers which have the additional value of plant protection during cold spells and protection from heat in dry seasons. A mature orchard is said to require 40 in (1,000 mm) of water during the 8-month growing season, more than ½ of it in the 3 summer months. Some growers plant a permanent cover crop of inoculated clover to control dust, aid water penetration and provide additional nitrogen for the kiwifruit crop. However, clover must be mowed at pollinating time to prevent the flowers from attracting the bees away from the kiwi vines.

Season

New Zealand production begins in May and the fresh or stored fruits are exported through November. In California, the vines put out new leaves in mid-March, bloom in early May and the fruit ripens in November after the leaves have fallen. The marketing season extends from November through April because the fruits hold so well in storage. The French season corresponds to that of California.

Harvesting and Packing

In New Zealand, a minimum picking-maturity standard is 6.25% soluble solids. California kiwifruits are harvested when they attain 6.5 to 8% soluble solids. They are picked by hand, either by breaking the stalk at its natural abscission point or are clipped very close to the base of the fruit to avoid stem punctures. They are carried in field boxes to packing stations. In well-equipped packing plants, the fruits are mechanically conveyed across a brushing machine that removes the hairs and, in some plants, the styles and sepals as well. The fruits are graded for size (25 to 54 per flat). For shipment, about 7 lbs (3¼ kg) of fruits are arranged in a plastic tray covered with perforated polyethylene and packed in a fiberboard or wooden box.

Yield

In California, 4-year-old vines have yielded 14,000 lbs per acre (15.7 MT/ha). Vines 8 years old have yielded 18,000 lbs per acre (20 MT/ha), which is nearly the maximum for mature plants (8 to 10 years old).

The bearing habits of the vine are variable—a light crop one season is likely to be followed by a heavy crop the next season, and *vice versa*.

Keeping Quality

Firm fruits can be kept 8 weeks at room temperature, 65° to 70°F (18.33°-21.11°C). Fully ripe fruits can be kept for a week or more in the home refrigerator. Fruits harvested at the firm stage will keep for long periods at 31° to 32°F (-0.56°-0°C) and at least 90% relative humidity, wrapped in unsealed polyethylene in containers. Lower relative humidity, even 85%, will cause a weight loss of as much as 4.5% in 6 weeks. Fruits that are cooled to a temperature of 32°F (0°C) within 12 hours after harvesting, will keep in good condition for as long as 6 months under commercial refrigeration. Experiments have shown that an atmosphere modified with 10 to 14% CO_2 will increase cold storage life by 2 months, providing the fruits enter storage within a week after harvest and are removed from the controlled atmosphere shortly in advance of marketing. Some studies by Arpaia *et al.*, indicate that optimum storage atmosphere may be obtained with 5% CO_2 and 2% O_2, with C_2H_4 excluded and/or removed to keep it below 0.05 mcg per liter. Kiwifruits freeze at storage temperatures between 28° and 30°F (-1.8° and -2.1°C).

For consumption fresh or for processing, kiwifruits are customarily kept refrigerated for at least 2 weeks to induce softening and then allowed to further soften at room temperature to improve flavor. The fruits will ripen too rapidly and lose quality if stored with other fruits, such as apples, pears, peaches, plums, etc., because of the ethylene these fruits emit.

Pests and Diseases

Kiwifruit vines are subject to attack by rootknot nematodes—*Meloidogyne hapla* and, to a lesser extent, *Heterodera marioni*—in New Zealand. Because of the surface hairs, the fruit is not damaged by fruit flies. The leaf roller, *Ctenopseustis obliquana*, which scars the surface of the fruit, sometimes eats holes where 2 or more fruits touch each other. In New Zealand, crawlers of the greedy scale insect, *Hemiberlesia repac*, have been conveyed to the plants by wind. This pest infests the leaves and fruit and kills the growing tips of the vines. The passionvine hopper sucks the sap of the vine and deposits honeydew on the fruit, and sooty mold growing on this sticky substance renders the fruit unmarketable. A small moth native to New Zealand—*Stathmopoda skellone*—may occur in abundance some seasons and do damage to the fruit under the sepals or where fruits touch each other. Silvering and browning of the leaves may occur in late summer or early fall because of infestation by thrips (*Heliothrips haemorrhoidalis*). Other pests in New Zealand include the salt marsh caterpillar and mites. In Chiremba, South Africa, red scale has been observed but it is easily controlled by spraying. In 1984, the New Zealand Pesticides Board approved Ivon Watkins-Dow's Lorsban insecticides for spraying on kiwifruit crops for export, and also cleared 4 herbicides for kiwifruit orchards.

A major disease of the vine is crown gall caused by *Agrobacterium tumefaciens*, but many suspected cases have turned out to be merely natural callousing. Crown gall can be avoided in budded or grafted plants by leaving the upper roots exposed. The roots may be attacked by *Phytophthora cactorum* and *P. cinnamomi*, and also by oak root fungus (*Armillaria mellea*) which is fatal. In humid climates, *Botrytis cinerea* infects the flowers and contaminates the young fruits. New Zealand growers may apply 8 or 9 sprays during the dormant period to achieve control of pests and diseases.

Post-harvest fruit decay is caused by *Alternaria* spp. and *Botrytis* spp. The greatest enemy is gray mold rot arising from *Botrytis cinerea* which enters through even minute scratches on the skin during storage at high humidity. *Alternaria alternata* mold is superficial and can be avoided if styles and sepals are completely removed during the brushing operation. *Alternaria*-caused hard, dry rot often is found on stored fruits that have been sunburned in the orchard. Such fruits should be culled during grading. Blue mold, resulting from infection by *Penicillium expansum*, may occur on injured fruits.

Leaf scorch results from hot dry winds in summer and early fall.

Food Uses

The Chinese have never been overly fond of the kiwifruit, regarding it mainly as a tonic for growing children and for women after childbirth. It is ripe for eating when it yields to slight pressure. For home use, the fruits are hand-picked. In addition to eating out-of-hand, they are served as appetizers, in salads, in fish, fowl and meat dishes, in pies, puddings, and prepared as cake-filling. Ice cream may be topped with kiwifruit sauce or slices, and the fruit is used in breads and various beverages. Kiwifruit cannot be blended with yogurt because an enzyme conflicts with the yogurt process. A cookbook, *Kiwifruit Recipes*, is published by the Kiwi Growers of California.

For commercial canning, the partly softened fruits are peeled by a mechanical steam peeler or by immersing in a boiling 15% lye solution for 90 seconds. Then they are washed in cold water, trimmed by hand, rinsed, and cooked in sirup in standard #2½ vacuum-sealed cans.

For preservation by freezing, the fruits are similarly peeled, sliced and immersed for 3 minutes in a solution of 12% sucrose, 1% ascorbic acid, and 0.25% malic acid, quick-frozen, then put into polyethylene bags and stored at 0°F (-17.78°C). Experiments have shown drying to be practical if the lye-peeled whole fruits are first dipped in a sugar solution to improve flavor, then dehydrated at temperatures below 150°F (65.56°C).

Only overripe or poorly shaped fruits are utilized for flavoring ice cream and for commercial juice production blended with apple to reduce acidity. The fruits so used are not peeled but put through a processing machine that removes the hairs, skin and seeds. In 1983, 2,378 gals (9,000 liters) of kiwifruit concentrate from 1,000,000 fruits were sold in Germany, and 13,210 gals (50,000 liters) were to be provided in 1984.

Slightly underripe fruits, which are high in pectin, must be chosen for making jelly, jam and chutney. Freeze-dried kiwifruit slices are shipped to health food outlets in Sweden and Japan. In the latter country, they are sometimes coated with chocolate. The peeled whole fruits may be pickled with vinegar, brown sugar and spices. Cull fruits can be made into wine. The Kiwifruit Wine Company of New Zealand, Ltd., has a contract to sell "Durham Light", a medium-sweet wine, throughout Japan. The Gibson Wine Company in Elk Grove, California, is making kiwifruit wine with an 11.5% alcohol content.

In the home kitchen, meat can be tenderized by placing slices of kiwifruit over it or by rubbing the meat with the flesh. After 10 minutes the fruit must be lifted or scraped off, otherwise the enzymatic action will be excessive. The meat should then be cooked immediately.

Toxicity

The hairs on the skin can cause throat irritation if ingested. It might be wise to avoid excessive consumption of

Food Value Per 100 g of Edible Portion*			
	Fresh	*Canned*	*Frozen*
Calories	66		66
Moisture	81.2 g	73.0 g	80.7 g
Protein	0.79 g	0.89 g	0.95 g
Fat	0.07 g	0.06 g	0.08 g
Carbohydrates	17.5 g	25.5 g	17.6 g
Ash	0.45 g	0.45 g	0.53 g
Calcium	16 mg	23 mg	18 mg
Iron	0.51 mg	0.40 mg	0.51 mg
Magnesium	30 mg	30 mg	27 mg
Phosphorus	64 mg	48 mg	67 mg
Thiamine	0.02 mg	0.02 mg	0.01 mg
Niacin	0.50 mg	0.40 mg	0.22 mg
Riboflavin	0.05 mg	0.02 mg	0.03 mg
Vitamin A	175 I.U.	155 I.U.	117 I.U.
Ascorbic Acid	105 mg	103 mg	218 mg (natural and added by pre-dip)

*Analyses made at the University of California.

Quinic acid predominates in young fruits, disappears with the formation of ascorbic acid. Boiling for 2 hours reduces ascorbic acid content by 20%. The same amount is lost when frozen fruits are thawed at room temperature.

Kiwifruits, even when ripe, contain the proteolytic enzyme actinidin, which is said to aid digestion. It can be extracted and purified as a powder for tenderizing meat. The tannin content is low, 0.95%, in mature fruits. According to a recent report from New Zealand, the kiwifruit is rich in folic acid, potassium, chromium and Vitamin E.

raw kiwifruits until more is known of the body's reaction to actinidin.

Medicinal Uses

The branches and leaves are boiled in water and the liquid used for treating mange in dogs. In China, the fruit and the juice of the stalk are esteemed for expelling "gravel". The scraped stems of the vine are used as rope in China, and paper has been made from the leaves and bark. If the bark at the base of the vine, close to the roots, is removed in one piece and placed in hot ashes, it will roll into a firm tube which can be used as a pencil.

Related Species

The United States Department of Agriculture has, in the past, made various introductions of other species, especially *A. arguta* Planch. ex Miq., *A. kolomikta* Maxim., and *A. polygama* Maxim., which are often grown as ornamental vines in the northern states.

A. arguta, KOKUWA, or TARA VINE, from Japan, Korea and Manchuria, has greenish-yellow fruit, or sometimes dark-green blushed with red; oblong or oval, about 1 in (2.5 cm) long, tipped with the persistent style. The skin is smooth and very thin; the flesh is green and sweet when fully ripe, and the seeds are minute. The fruits are edible but somewhat purgative.

The vine was growing in a private garden in Marblehead, Massachusetts, in 1888. The United States Department of Agriculture received seeds from that vine in 1908; but had been sent seeds from Germany in 1901; and more seeds came from Korea in 1909. This species has been cultivated as an ornamental and screening vine in subtemperate zones of this country since these early dates. Currently, Henry Field's Seed and Nursery Company, Shenandoah, Iowa, is glamorizing the "Hardy Kiwi" as a new "tropical fruit . . . surviving down to 25° below zero". The Richard Owen Nursery in Bloomington, Illinois, is advertising "*A. arguta annasnaja*" as a "Hardy Kiwi" ¾ to 1½ in (2-4 cm) in diameter, ripening in late September or early October.

It is true that the wild fruits of *A. arguta* are gathered and sold in northern China, and the success of the kiwifruit has aroused some interest in the fruits of *A. arguta* in cool areas of the United States. However, most seedlings are non-fruiting males, and female or bisexual specimens are rare and may be unreliable bearers. Much experi-

mental work may be necessary to determine whether or not the kokuwa can be developed into a practical fruit source.

In June, 1923, Dr. David Fairchild applied pollen of the kiwifruit on the flowers of a vine of *A. arguta* in a garden in Maryland and he harvested some fruits in October of that year. He had hopes for the future of his hybrid and distributed cuttings and seeds. Later, he sadly reported that all his "hybrid plants made poor growths and never bore."

A. kolomikta, ranging from Japan to Manchuria and western China, has blue, oblong-ovoid fruits, of sweet flavor. Cats are very partial to the plant.

A. polygama, SILVER VINE, is native from Japan to western China. It has beaked, yellow, bitter fruits to 1½ in (4 cm) long. The Japanese eat the salted fruits and the leaves. This species is prized in horticulture for the silvery tone of the young growth of male plants. The bark, twigs and leaves contain actinidine and also metatabilacetone, similar to catnip oil, and they lure and intoxicate cats. They are said to be used for taming lions and tigers in captivity.

Fig. 83-a: A by-product of kiwifruit culture: California growers have found the vine trimmings unsuitable for mulch or disposal by burning. They are shipping them to florists. Being naturally coiled and curiously twisted, they are attractive and useful in enhancing flower arrangements. Some that are fairly fresh may put out a temporary flurry of downy green leaves and tendrils. (Stems courtesy Flower Wagon, Miami, FL).

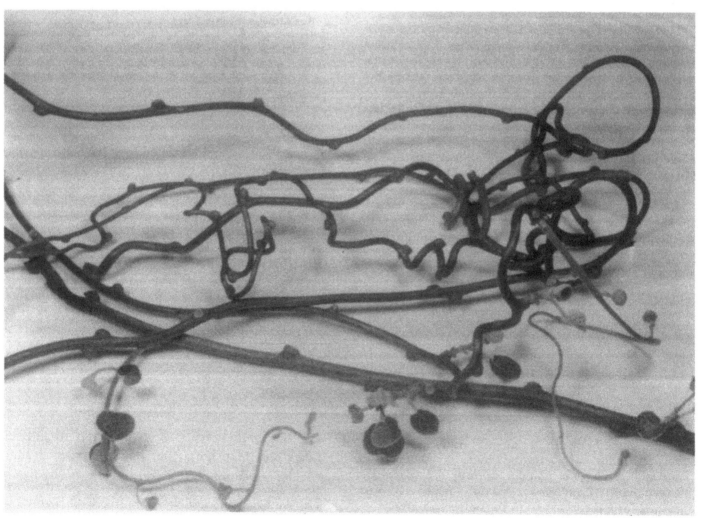

Mangosteen (Plate XLI)

One of the most praised of tropical fruits, and certainly the most esteemed fruit in the family Guttiferae, the mangosteen, *Garcinia mangostana* L., is almost universally known or heard of by this name. There are numerous variations in nomenclature: among Spanish-speaking people, it is called *mangostan;* to the French, it is *mangostanier, mangoustanier, mangouste* or *mangostier;* in Portuguese, it is *mangostao, mangosta* or *mangusta;* in Dutch, it is *manggis* or *manggistan;* in Vietnamese, *mang cut;* in Malaya, it may be referred to in any of these languages or by the local terms, *mesetor, semetah,* or *sementah;* in the Philippines, it is *mangis* or *mangostan.* Throughout the Malay Archipelago, there are many different spellings of names similar to most of the above.

Description

The mangosteen tree is very slow-growing, erect, with a pyramidal crown; attains 20 to 82 ft (6-25 m) in height, has dark-brown or nearly black, flaking bark, the inner bark containing much yellow, gummy, bitter latex. The evergreen, opposite, short-stalked leaves are ovate-oblong or elliptic, leathery and thick, dark-green, slightly glossy above, yellowish-green and dull beneath; 3½ to 10 in (9-25 cm) long, 1¾ to 4 in (4.5-10 cm) wide, with conspicuous, pale midrib. New leaves are rosy. Flowers, 1½ to 2 in (4-5 cm) wide and fleshy, may be male or hermaphrodite on the same tree. The former are in clusters of 3-9 at the branch tips; there are 4 sepals and 4 ovate, thick, fleshy petals, green with red spots on the outside, yellowish-red inside, and many stamens though the aborted anthers bear no pollen. The hermaphrodite are borne singly or in pairs at the tips of young branchlets; their petals may be yellowish-green edged with red or mostly red, and are quickly shed.

The fruit, capped by the prominent calyx at the stem end and with 4 to 8 triangular, flat remnants of the stigma in a rosette at the apex, is round, dark-purple to red-purple and smooth externally; 1⅓ to 3 in (3.4-7.5 cm) in diameter. The rind is ¼ to ⅜ in (6-10 mm) thick, red in cross-section, purplish-white on the inside. It contains bitter yellow latex and a purple, staining juice. There are 4 to 8 triangular segments of snow-white, juicy, soft flesh (actually the arils of the seeds). The fruit may be seedless or have 1 to 5 fully developed seeds, ovoid-oblong, somewhat flattened, 1 in (2.5 cm) long and ⅝ in (1.6 cm) wide, that cling to the flesh. The flesh is slightly acid and mild to distinctly acid in flavor and is acclaimed as exquisitely luscious and delicious.

Origin and Distribution

The place of origin of the mangosteen is unknown but is believed to be the Sunda Islands and the Moluccas; still, there are wild trees in the forests of Kemaman, Malaya. Corner suggests that the tree may have been first domesticated in Thailand, or Burma. It is much cultivated in Thailand—where there were 9,700 acres (4,000 ha) in 1965—also in Kampuchea, southern Vietnam and Burma, throughout Malaya and Singapore. The tree was planted in Ceylon about 1800 and in India in 1881. There it succeeds in 4 limited areas—the Nilgiri Hills, the Tinnevelly district of southern Madras, the Kanyakumani district at the southernmost tip of the Madras peninsula, and in Kerala State in southwestern India. The tree is fairly common only in the provinces of Mindanao and Sulu (or Jolo) in the Philippines. It is rare in Queensland, where it has been tried many times since 1854, and poorly represented in tropical Africa (Zanzibar, Ghana, Gabon and Liberia). There were fruiting trees in greenhouses in England in 1855. The mangosteen was introduced into Trinidad from the Royal Botanic Garden at Kew, England, between 1850 and 1860 and the first fruit was borne in 1875. It reached the Panama Canal Zone and Puerto Rico in 1903 but there are only a few trees in these areas, in Jamaica, Dominica and Cuba, and some scattered around other parts of the West Indies. The United States Department of Agriculture received seeds from Java in 1906 (S.P.I. #17146). A large test block of productive trees has been maintained at the Lancetilla Experimental Station at Tela, Honduras, for many years. Quite a few trees distributed by the United Fruit Company long ago have done well on the Atlantic coast of Guatemala. In 1924, Dr. Wilson Popenoe saw the mangosteen growing at one site in Ecuador. In 1939, 15,000 seeds were distributed by the Canal Zone Experiment Gardens to many areas of tropical America. It is probable that only a relatively few seedlings survived. It is known that many die during the first year. Dr. Victor Patiño has observed flourishing mangosteen trees at the site of an old mining settlement in Mariquita, Colombia, in the Magdalena Valley and the fruits are sold on local markets. Dierberger Agricola Ltda., of Sao Paulo, included the mangosteen in their nursery catalog in 1949.

Despite early trials in Hawaii, the tree has not become well acclimatized and is still rare in those islands. Neither has it been successful in California. It encounters very unfavorable soil and climate in Florida. Some plants have been grown for a time in containers in greenhouses. One tree in a very protected coastal location and special soil lived to produce a single fruit and then succumbed to winter cold.

Despite the oft-repeated Old World enthusiasm for this fruit, it is not always viewed as worth the trouble to produce. In Jamaica, it is regarded as nice but over-rated; not comparable to a good field-ripe pineapple or a choice mango.

Varieties

According to Corner, the fruit from seedling trees is fairly uniform; only one distinct variation is known and that is in the Sulu Islands. The fruit is larger, the rind thicker than normal, and the flesh more acid; the flavor more pronounced. In North Borneo, a seemingly wild form has only 4 carpels, each containing a fully-developed seed, and this is probably not unique.

Climate

The mangosteen is ultra-tropical. It cannot tolerate temperatures below 40°F (4.44°C), nor above 100°F (37.78°C). Nursery seedlings are killed at 45°F (7.22°C).

It is limited in Malaya to elevations below 1,500 ft (450 m). In Madras it grows from 250 to 5,000 ft (76–1,500 m) above sea-level. Attempts to establish it north of 20° latitude have all failed.

It ordinarily requires high atmospheric humidity and an annual rainfall of at least 50 in (127 cm), and no long periods of drought. In Dominica, mangosteens growing in an area having 80 in (200 cm) of rain yearly required special care, but those in another locality with 105 in (255 cm) and soil with better moisture-holding capacity, flourished.

Soil

The tree is not adapted to limestone and does best in deep, rich organic soil, especially sandy loam or laterite. In India, the most productive specimens are on clay containing much coarse material and a little silt. Sandy alluvial soils are unsuitable and sand low in humus contributes to low yields. The tree needs good drainage and the water table ought to be about 6 ft (1.8 m) below ground level. However, in the Canal Zone, productive mangosteen groves have been established where it is too wet for other fruit trees—in swamps requiring drainage ditches between rows and in situations where the roots were bathed with flowing water most of the year, in spite of the fact that standing water in nursery beds will kill seedlings. The mangosteen must be sheltered from strong winds and salt spray, as well as saline soil or water.

Propagation

Technically, the so-called "seeds" are not true seeds but adventitious embryos, or hypocotyl tubercles, inasmuch as there has been no sexual fertilization. When growth begins, a shoot emerges from one end of the seed and a root from the other end. But this root is short-lived and is replaced by roots which develop at the base of the shoot. The process of reproduction being vegetative, there is naturally little variation in the resulting trees and their fruits. Some of the seeds are polyembryonic, producing more than one shoot. The individual nucellar embryos can be separated, if desired, before planting.

Inasmuch as the percentage of germination is directly related to the weight of the seed, only plump, fully developed seeds should be chosen for planting. Even these will lose viability in 5 days after removal from the fruit, though they are viable for 3 to 5 weeks in the fruit. Seeds packed in lightly dampened peat moss, sphagnum moss or coconut fiber in airtight containers have remained viable for 3 months. Only 22% germination has been realized in seeds packed in ground charcoal for 15 days. Soaking in water for 24 hours expedites and enhances the rate of germination. Generally, sprouting occurs in 20 to 22 days and is complete in 43 days.

Because of the long, delicate taproot and poor lateral root development, transplanting is notoriously difficult. It must not be attempted after the plants reach 2 ft (60 cm). At that time the depth of the taproot may exceed that height. There is greater seedling survival if seeds are planted directly in the nursery row than if first grown in containers and then transplanted to the nursery. The nursery soil should be 3 ft (1 m) deep, at least. The young plants take 2 years or more to reach a height of 12 in (30 cm), when they can be taken up with a deep ball of earth and set out. Fruiting may take place in 7 to 9 years from planting but usually not for 10 or even 20 years.

Conventional vegetative propagation of the mangosteen is difficult. Various methods of grafting have failed. Cuttings and air-layers, with or without growth-promoting chemicals, usually fail to root or result in deformed, short-lived plants. Inarching on different rootstocks has appeared promising at first but later incompatibility has been evident with all except *G. xanthochymus* Hook. f. (*G tinctoria* Dunn.) or *G. lateriflora* Bl., now commonly employed in the Philippines.

In Florida, approach-grafting has succeeded only by planting a seed of *G. xanthochymus* about 1¼ in (3 cm) from the base of a mangosteen seedling in a container and, when the stem of the *G. xanthochymus* seedling has become ⅛ in (3 mm) thick, joining it onto the 3/16 to ¼ in (5–6 mm) thick stem of the mangosteen at a point about 4 in (10 cm) above the soil. When the graft has healed, the *G. xanthochymus* seedling is beheaded. The mangosteen will make good progress having both root systems to grow on, while the *G. xanthochymus* rootstock will develop very little.

Culture

A spacing of 35 to 40 ft (10.7–12 m) is recommended. Planting is preferably done at the beginning of the rainy season. Pits 4 x 4 x 4½ ft (1.2x1.2x1.3 m) are prepared at least 30 days in advance, enriched with organic matter and topsoil and left to weather. The young tree is put in place very carefully so as not to injure the root and given a heavy watering. Partial shading with palm fronds or by other means should be maintained for 3 to 5 years. Indian growers give each tree regular feeding with well-rotted manure — 100 to 200 lbs (45-90 kg) — and peanut meal — 10 to 15 lbs (4.5-6.8 kg) total, per year.

Some of the most fruitful mangosteen trees are growing on the banks of streams, lakes, ponds or canals where the roots are almost constantly wet. However, dry weather just before blooming time and during flowering induces a good fruit-set. Where a moist planting site is not available, irrigation ditches should be dug to make it possible to maintain an adequate water supply and the trees are irrigated almost daily during the dry season.

In Malaya and Ceylon, it is a common practice to spread a mulch of coconut husks or fronds to retain moisture. A 16-in (40-cm) mulch of grass restored trees that had begun dehydrating in Liberia. It has been suggested that small inner branches be pruned from old, unproductive trees to stimulate bearing. In Thailand, the tree is said to take 12 to 20 years to fruit. In Panama and Puerto Rico trees grown from large seed and given good culture have borne in six years.

Season and Harvesting

At low altitudes in Ceylon the fruit ripens from May to July; at higher elevations, in July and August or August and September. In India, there are 2 distinct fruiting seasons, one in the monsoon period (July-October) and another from April through June. Puerto Rican trees in full sun fruit in July and August; shaded trees, in November and December.

Cropping is irregular and the yield varies from tree to tree and from season to season. The first crop may be 200 to 300 fruits. Average yield of a full-grown tree is about 500 fruits. The yield steadily increases up to the 30th year of bearing when crops of 1,000 to 2,000 fruits may be obtained. In Madras, individual trees between the ages of 20 and 45 years have borne 2,000 to 3,000 fruits. Productivity gradually declines thereafter, though the tree will still be fruiting at 100 years of age.

Ripeness is gauged by the full development of color and slight softening. Picking may be done when the fruits are slightly underripe but they must be fully mature (developed) or they will not ripen after picking. The fruits must be harvested by hand from ladders or by means of a cutting pole and not be allowed to fall.

Keeping Quality

In dry, warm, closed storage, mangosteens can be held 20 to 25 days. Longer periods cause the outer skin to toughen and the rind to become rubbery; later, the rind hardens and becomes difficult to open and the flesh turns dry.

Ripe mangosteens keep well for 3 to 4 weeks in storage at 40° to 55°F (4.44°-12.78°C). Trials in India have shown that optimum conditions for cold storage are temperatures of 39° to 42°F (3.89°-5.56°C) and relative humidity of 85 to 90%, which maintain quality for 49 days. It is recommended that the fruits be wrapped in tissue paper and packed 25-to-the-box in light wooden crates with excelsior padding. Fruits picked slightly unripe have been shipped from Burma to the United Kingdom at 50° to 55°F (10°-12.78°C). From 1927 to 1929, trial shipments were made from Java to Holland at 37.4°F (approximately 2.38°C) and the fruits kept in good condition for 24 days.

Pests and Diseases

Few pests have been reported. A leaf-eating caterpillar in India may perhaps be the same as that which attacks new shoots in the Philippines and which has been identified as *Orgyra* sp. of the tussock moth family, Lymantridae. A small ant, *Myrnelachista ramulorum*, in Puerto Rico, colonizes the tree, tunnels into the trunk and branches, and damages the new growth. Mites sometimes deface the fruits with small bites and scratches. Fully ripe fruits are attacked by monkeys, bats and rats in Asia.

In Puerto Rico, thread blight caused by the fungus, *Pellicularia koleroga,* is often seen on branchlets, foliage and fruits of trees in shaded, humid areas. The fruits may become coated with webbing and ruined. In Malaya, the fungus, *Zignoella garcineae,* gives rise to "canker" — tuberous growths on the branches, causing a fatal dying-back of foliage, branches and eventually the entire tree. Breakdown in storage is caused by the fungi *Diplodia gossypina, Pestalotia* sp., *Phomopsis* sp., *Gloeosporium* sp., and *Rhizopus nigricans.*

A major physiological problem called "gamboge" is evidenced by the oozing of latex onto the outer surface of the fruits and on the branches during periods of heavy and continuous rains. It does not affect eating quality. Fruit-cracking may occur because of excessive absorption of moisture. In cracked fruits the flesh will be swollen and mushy. Bruising caused by the force of storms may be an important factor in both of these abnormalities. Fruits exposed to strong sun may also exude latex. Mangosteens produced in Honduras often have crystal-like "stones" in the flesh and they may render the fruit completely inedible.

Food Uses

To select the best table fruits, choose those with the highest number of stigma lobes at the apex, for these have the highest number of fleshy segments and accordingly the fewest seeds. The numbers always correspond. Mangosteens are usually eaten fresh as dessert. One need only hold the fruit with the stem-end downward, take a sharp knife and cut around the middle completely through the rind, and lift off the top half, which leaves the fleshy segments exposed in the colorful "cup" — the bottom half of the rind. The segments are lifted out by fork.

The fleshy segments are sometimes canned, but they are said to lose their delicate flavor in canning, especially if pasteurized for as much as 10 minutes. Tests have shown that it is best to use a 40% sirup and sterilize for only 5 minutes. The more acid fruits are best for preserving. To make jam, in Malaya, seedless segments are boiled with an equal amount of sugar and a few cloves for 15 to 20 minutes and then put into glass jars. In the Philippines, a preserve is made by simply boiling the segments in brown sugar, and the seeds may be included to enrich the flavor.

The seeds are sometimes eaten alone after boiling or roasting.

The rind is rich in pectin. After treatment with 6% sodium chloride to eliminate astringency, the rind is made into a purplish jelly.

Food Value Per 100 g of Edible Portion*	
Calories	60–63
Moisture	80.2–84.9 g
Protein	0.50–0.60 g
Fat	0.1–0.6 g
Total Carbohydrates	14.3–15.6 g
Total Sugars	16.42–16.82 g
(sucrose, glucose and fructose)	
Fiber	5.0–5.1 g
Ash	0.2–0.23 g
Calcium	0.01–8.0 mg
Phosphorus	0.02–12.0 mg
Iron	0.20–0.80 mg
Thiamine	0.03 mg
Ascorbic Acid	1.0–2.0 mg

*Minimum/maximum values from analyses made in the Philippines and Washington, D.C.

Phytin (an organic phosphorus compound) constitutes up to 0.68% on a dry-weight basis. The flesh amounts to 31% of the whole fruit.

Other Uses

Mangosteen twigs are used as chewsticks in Ghana. The fruit rind contains 7 to 14% catechin tannin and rosin, and is used for tanning leather in China. It also yields a black dye.

Wood: In Thailand, all non-bearing trees are felled, so the wood is available but usually only in small dimensions. It is dark-brown, heavy, almost sinks in water, and is moderately durable. It has been used to make handles for spears, also rice pounders, and is employed in construction and cabinetwork.

Medicinal Uses: Dried fruits are shipped from Singapore to Calcutta and to China for medicinal use. The sliced and dried rind is powdered and administered to overcome dysentery. Made into an ointment, it is applied on eczema and other skin disorders. The rind decoction is taken to relieve diarrhea and cystitis, gonorrhea and gleet and is applied externally as an astringent lotion. A portion of the rind is steeped in water overnight and the infusion given as a remedy for chronic diarrhea in adults and children. Filipinos employ a decoction of the leaves and bark as a febrifuge and to treat thrush, diarrhea, dysentery and urinary disorders. In Malaya, an infusion of the leaves, combined with unripe banana and a little benzoin is applied to the wound of circumcision. A root decoction is taken to regulate menstruation. A bark extract called "amibiasine", has been marketed for the treatment of amoebic dysentery.

The rind of partially ripe fruits yields a polyhydroxy-xanthone derivative termed mangostin, also β-mangostin. That of fully ripe fruits contains the xanthones, gartanin, 8-disoxygartanin, and normangostin. A derivative of mangostin, mangostin-e,6-di-O-glucoside, is a central nervous system depressant and causes a rise in blood pressure.

Mamey (Plate XLII)

The mamey stands almost midway between "major" and "minor" tropical fruits and is unique in remaining virtually static in the past 40 years, receiving little attention at home or abroad. Botanically, it is identified as *Mammea americana* L., of the family Guttiferae, and therefore related to the mangosteen, q.v. Among alternative names in English are mammee, mammee apple, St. Domingo apricot and South American apricot. To Spanish-speaking people, it is known as *mamey de Santo Domingo*, *mamey amarillo*, *mamey de Cartagena*, *mata serrano*, *zapote mamey*, or *zapote de Santo Domingo*. In Portuguese it is called *abricote*, *abrico do Pará*, *abrico selvagem*, or *pecego de Sao Domingos*. In French, it is *abricot d' Amerique*, *abricot des Antilles*, *abricot pays*, *abricot de Saint-Dominque* or *abricotier sauvage*.

This species is often confused with the sapote, or *mamey colorado*, *Pouteria sapota*, q.v., which is commonly called *mamey* in Cuba; and reports of its occurring wild in Africa are due to confusion with the African mamey, *M. africana* Sabine (syn. *Ochrocarpus africana* Oliv.).

Description

The mamey tree, handsome and greatly resembling the southern magnolia, reaches 60 to 70 ft (18-21 m) in height, has a short trunk which may attain 3 or 4 ft (0.9-1.2 m) in diameter, and ascending branches forming an erect, oval head, densely foliaged with evergreen, opposite, glossy, leathery, dark-green, broadly elliptic leaves, up to 8 in (20 cm) long and 4 in (10 cm) wide. The fragrant flowers, with 4 to 6 white petals and with orange stamens or pistils or both, are 1 to 1½ in (2.5-4 cm) wide when fully open and borne singly or in groups of 2 or 3 on short stalks. They appear during and after the fruiting season: male, female and hermaphrodite together or on separate trees.

The fruit, nearly round or somewhat irregular, with a short, thick stem and a more or less distinct tip or merely a bristle-like floral remnant at the apex, ranges from 4 to 8 in (10-20 cm) in diameter, is heavy and hard until fully ripe when it softens slightly. The skin is light-brown or grayish-brown with small, scattered, warty or scurfy areas, leathery, about ⅛ in (3 mm) thick and bitter. Beneath it, a thin, dry, whitish membrane, or "rag", astringent and often bitter, adheres to the flesh. The latter is light- or golden-yellow to orange, non-fibrous, varies from firm and crisp and sometimes dry to tender, melting and juicy. It is more or less free from the seed though bits of the seed-covering, which may be bitter, usually adhere to the immediately surrounding wall of flesh. The ripe flesh is appetizingly fragrant and, in the best varieties, pleasantly subacid, resembling the apricot or red raspberry in flavor. Fruits of poor quality may be too sour or mawkishly sweet. Small fruits are usually single-seeded; larger fruits may have 2, 3 or 4 seeds. The seed is russet-brown, rough, ovoid or ellipsoid and about 2½ in (6.25 cm) long. The juice of the seed leaves an indelible stain.

Origin and Distribution

The mamey is native to the West Indies and northern South America. It was recorded as growing near Darién, Panama, in 1514, and in 1529 was included by Oviedo in his review of the fruits of the New World. It has been nurtured as a specimen in English greenhouses since 1735. It grows well in Bermuda and is quite commonly cultivated in the Bahama Islands and the Greater and Lesser Antilles. In St. Croix it is spontaneous along the roadsides where seeds have been tossed. In southern Mexico and Central America, it is sparingly grown except in the lowlands of Costa Rica, El Salvador and in Guatemala where it may be seen planted as a windbreak and ornamental shade tree along city streets, and is frequently grown for its fruit on the plains and foothills of the Pacific coast. Cultivation is scattered in Colombia, Venezuela, Guyana, Surinam and French Guiana, Ecuador and northern Brazil.

Introduced into the tropics of the Old World, it is of very limited occurrence in West Africa (particularly Sierra Leone), Zanzibar, southeastern Asia, Java, the Philippines, and Hawaii. All seedlings planted in Israel have died in the first or second year. From time to time, seedlings have been planted in California, but most have succumbed the first winter. Dr. Robert Hodson, of the University of California, stated in 1940: "I know of only one large and old tree of *Mammea americana* growing out of doors in southern California, and it has never fruited."

The mamey may have been brought to Florida first from the Bahamas, but the United States Department of Agriculture received seeds from Ecuador in 1919 (S.P.I.

#47425). One of the largest fruiting specimens in Florida is at the Fairchild Tropical Garden, Miami, standing on a site formerly part of an early nursery, and thought to be over 60 years of age. Another, as old or older, on a private estate in Palm Beach, was fruiting heavily before 1940. The most northerly reached 30 feet (9 m) and fruited in Dr. Talmadge Wilson's garden at Stuart but was killed by lightning about 1956. There was a 35-foot (10.5 m) fruiting tree in the Edison Botanical Garden at Fort Myers, its trunk at least 20 in (50 cm) thick, but it was removed after severe hurricane damage in 1960 and replaced by a young one. A number of fruiting trees on private property in the Miami area have been destroyed to make room for construction. The Fairchild Tropical Garden has distributed numerous seedlings from their large tree but most apparently fail to survive the winter in the hands of new owners Many seeds were planted as nursery stock by Robert Newcomb of Homestead who offered grafted plants for sale from 1953 to 1956 and then, discouraged by winter-killing, gave his remaining plants to a garden club on Key Largo. Hurricane "Donna" of 1960 doubtless eliminated most of these.

Climate

The mamey is limited to tropical or near-tropical climates. In Central America, it thrives from near sea-level to 3,300 ft (1,000 m). Three trees at the Agricultural Research and Education Center, Homestead, in southern Florida, were killed by a temperature drop to 28°F (-2.22°C) in January 1940.

Soil

The mamey tree favors deep, rich, well-drained soil, but is apparently quite adaptable to even shallow, sandy terrain, and it grows naturally in limestone areas of Jamaica, also does well in the oolitic limestone of the Bahamas and southeastern Florida.

Propagation

Seeds are the usual means of dissemination and they germinate in 2 months or less and sprout readily in leaf-mulch under the tree. Seedlings bear in 6 to 8 years in Mexico, 8 to 10 years in the Bahamas. Vegetative propagation is preferable to avoid disappointment in raising male trees and to achieve earlier fruiting. In English greenhouse culture, half-ripe cuttings with lower leaves attached are employed. Both Robert Newcomb and Albert Caves of Palm Lodge Tropical Grove, Homestead, successfully grafted the mamey onto self-seedlings.

Culture

The mamey generally receives little or no cultural attention, apart from protection from cold during the first few winters in other than strictly tropical climates. It seems remarkably resistant to pests and diseases.

Season

In Barbados, the fruits begin to ripen in April and continue for several weeks. The season extends from May through July in the Bahamas, some fruits being offered in the Nassau native market and on roadside stands. In southern Florida, mameys ripen from late June through July and August. In Puerto Rico, some trees produce two crops a year. Central Colombia has two crops occurring in June and December.

Harvesting

Ripeness may be indicated by a slight yellowing of the skin or, if this is not apparent, one can scratch the surface very lightly with a fingernail. If green beneath, the fruit should not be picked, but, if yellow, it is fully mature. If fruits are allowed to fall when ripe, they will bruise and spoil. They should be clipped, leaving a small portion of stem attached.

Yield

The productivity of individual trees varies considerably. In Puerto Rico, high-yielding trees may bear 150 to 200 fruits per crop, totalling 300 to 400 fruits per year.

Food Uses

To facilitate peeling, the skin is scored from the stem to the apex and removed in strips. The rag must be thoroughly scraped from the flesh which is then cut off in slices, leaving any part which may adhere to the seed, and trimming off any particles of seed-covering from the roughened inner surface of the flesh.

The flesh of tender varieties is delicious raw, either plain, in fruit salads, or served with cream and sugar or wine. In Jamaica, it may be steeped in wine and sugar for a while prior to eating. In the Bahamas, some prefer to let the flesh stand in lightly salted water "to remove the bitterness" before cooking with much sugar to a jam-like consistency. I have often stewed the flesh, without pre-treatment, adding a little sugar and possibly a dash of lime or lemon juice. Once, some of the pulp, stewed without citrus juice, was left in a covered plastic container in a refrigerator for one month. At the end of this time, there was no loss of flavor, no fermentation or other evidence of spoilage; and the fruit was eaten with no ill effect. In this connection, it is interesting to note that an antibiotic principle in the mamey was reported by the Agricultural Experiment Station, Rio Piedras, Puerto Rico, in 1951.

Sliced mamey flesh may also be cooked in pies or tarts, and may be seasoned with cinnamon or ginger. Canned, sliced mamey has in the past been exported from Cuba. The mamey is widely made into preserves such as spiced marmalade and pastes (resembling guava paste) and used as a filler for products made of other fruits. Slightly underripe fruits, rich in pectin, are made into jelly. Wine is made from the fruit and fermented "toddy" from the sap of the tree in Brazil.

In the Dominican Republic, the uncooked flesh, blended with sugar, is made into frozen sherbet. The juice or sirup of stewed flesh, is seasoned with sugar and lemon juice to make "ade". When cooking the flesh for any purpose, one is advised to skim off any foam that forms on the surface of the water, as this is usually bitter.

Food Value Per 100 g of Fresh Pulp*	
Calories	44.5–45.3
Moisture	85.5–87.6 g
Protein	0.470–0.088 g
Fat	0.15–0.99 g
Total Carbohydrates	11.52–12.67 g
Fiber	0.80–1.07 g
Ash	0.17–0.29 g
Calcium	4.0–19.5 mg
Phosphorus	7.8–14.5 mg
Iron	0.15–2.51 mg
Vitamin A (β-Carotene)	0.043–0.37 mg
Thiamine	0.017–0.030 mg
Riboflavin	0.025–0.068 mg
Niacin	0.160–0.738 mg
Ascorbic Acid	10.2–22.0 mg
Amino Acids:	
Tryptophan	5 mg
Methionine	5–6 mg
Lysine	14–35 mg

*Analyses made in Cuba and Central America.

Toxicity

Rural folk in the Dominican Republic have some doubt of the wholesomeness of mamey flesh. In the *Description and History of Vegetable Substances Used in the Arts and Domestic Economy*, published in London in 1829, it is stated: "To people with weak stomachs, it is said to be more delicious than healthful." The Bahamian practice of soaking the pulp in salted water may be a safety precaution inasmuch as bitterness is not only disliked but distrusted. The old Jamaican custom of steeping in wine might also be considered a safeguard. Kennard and Winters observe that, in Puerto Rico, "Although the fruit is widely eaten, it is recommended that only moderate amounts be consumed." A former Spanish professor at the University of Miami related that, when he was about 19 in Mayaguez, Puerto Rico, he ate half of a large mamey from a tree in his home yard, after peeling and scraping off the rag but not removing any adherent seed-covering. Then he ate the pulp of one star apple. An hour later, he had stomach cramps and, later, his abdomen was reddened and oddly reticulated. He attributed this reaction to the mamey and was convinced there was "something poisonous about it."

Morris *et al.* (1952) commented that, while the delicious mamey "has formed part of the diet of the inhabitants of the Caribbean Islands for many generations, it is well known that this fruit produces discomfort, especially in the digestive system, in some persons." They reported also that "a concentrated extract of the fresh fruit" proved fatally toxic to guinea pigs, and was also found poisonous to dogs and cats. The extract was made from the edible portion only. The authors likened the mamey to the akee

(*Blighia sapida*), q.v., as a human hazard, and Djerassi, *et al.*, aver that "reports of poisoning in humans are known."

Other Uses

Insecticidal value: That various parts of the mamey tree contain toxic properties has been long recognized and was first reported by Grosourdy in *El Médico Botanico Criollo* in 1864. A Colombian decoction of mamey resin was displayed at the Paris Exposition in 1867. It is significant that in the United States Department of Agriculture's record of mamey seed introduction from Ecuador in 1919, only the insecticidal and medicinal uses of the species were noted. There was no comment on edible uses of the fruit.

In Puerto Rico, there is a time-honored practice of wrapping a mamey leaf like a collar around young tomato plants when setting them in the ground to protect them from mole crickets and cutworms. The leaf must be placed at just the right height, half above ground and half below.

In Mexico and Jamaica, the thick, yellow gum from the bark is melted with fat and applied to the feet to combat chiggers and used to rid animals of fleas and ticks. A greenish-yellow, gummy resin from the skin of immature fruits, and an infusion of half-ripe fruits are similarly employed. The bark is strongly astringent and a decoction is effective against chiggers. In El Salvador, a paste made of the ground seeds is used against poultry lice, mites and head lice. In the Dominican Republic, mamey seeds, avocado seeds, and *Zamia* seeds fried in oil, are mashed and applied to the head as a "therapeutic shampoo", probably to eliminate lice.

At the Federal Experiment Station, Mayaguez, Puerto Rico, the insecticidal activity of various parts of the mamey tree and the fruit have been under active investigation. The seed kernel, most potent, was found, in feeding experiments and when tested as a contact poison applied as a dust or spray, to be effective in varying degree against armyworms, melonworms, cockroaches, ants, drywood termites, mosquitoes and their larvae, flies, larvae of diamond-back moth, and aphids. In certain tests, mamey seeds appeared to be 1/5 as toxic as pyrethrum and less toxic to plant pests than nicotine sulfate and DDT. When powdered seeds and sliced unripe fruit infusions, 1 lb (0.45 kg) in a gallon (3.78 liters) of water, were tested on dogs, both products were as effective as DDT and faster in killing fleas and ticks but not as long-lasting in regard to reinfestation. None of the dogs was poisoned despite the presence of healing sores and minor abrasions of the skin, but, after similar trials on mice, 4 out of 70 died. The active ingredients of the infusion are the resin from the unripe skin and the developing seeds. In Ecuador, animals with mange or sheep ticks are washed with a decoction made by boiling the seed but, in one instance, a dog with mange and ulcers died 48 hours after two applications.

The dried and powdered immature fruit, the bark, wood, roots and flowers have shown poor insecticidal activity; the seed hulls appeared inert. The powdered leaves were found 59% effective against fall armyworms and 75% against the melonworm. Various extracts from the fruit, bark, leaves or roots are toxic to webbing clothes moths, black carpet beetle larvae and also to milkweed bugs.

In fish-poisoning experiments, Pagan and Morris reported mamey seed extracts to be 1/30 as toxic as rotenone; 1/60 to 1/80 as potent as powdered dried derris root. Feeding trails have shown the seeds to be very toxic to chicks and they are considered a hazard to hogs in the Virgin Islands.

The crude resinous extract from powdered mamey seeds, given orally, has produced symptoms of poisoning in dogs and cats and a dose of 200 mg per km weight has caused death in guinea pigs within 8 hours. The crystalline insecticidal principle from the dried and ground seeds, potent even after several months of storage, has been named *mammein* and assigned the formula $C_{22}H_{28}O_5$. The stability of this principle was demonstrated by M.P. Morris who found no significant difference in toxicity of powdered fresh mamey fruit and mamey powder stored for 6 years in steel drums. Neither was the potency of mamey extract destroyed by subjection to 392°F (200°C).

Extensive chemical experiments with the extracted compound are reported by S.P. Marfey who considered the mamey a potential substitute for pyrethrum and rotenone.

The main constituent of a wax isolated from the seed oil is the symmetrical C_{48} homolog, tetracosanyl tetracosanoate.

Wood: In Central America, the tree is protected because the fruit is valued. Elsewhere, if the mamey is common, it may be felled for its timber. The heartwood is reddish- or purple-brown; the sapwood much lighter in color. The wood is heavy, hard, but not difficult to work, fine-grained and strong; has an attractive grain and polishes well. It is useful in cabinetwork, valued for pillars, rafters, decorative features of fine houses, interior sheathing, turnery and for fenceposts since it is fairly decay-resistant. It is, however, highly susceptible to termites. Some of the wood is consumed as fuel.

Bark: The tannin from the bark is sometimes used for home treatment of leather in the Virgin Islands.

Medicinal Uses: In Venezuela, the powdered seeds are employed in the treatment of parasitic skin diseases. In Brazil, the ground seeds, minus the embryo, which is considered convulsant, are stirred into hot water and the infusion employed as an anthelmintic for adults only.

In the French West Indies, an aromatic liqueur called *Eau de Creole*, or *Creme de Creole*, is distilled from the flowers and said to act as a tonic or digestive.

An infusion of the fresh or dry leaves (one handful in a pint [0.47 liter] of water) is given by the cupful over a period of several days in cases of intermittent fever and it is claimed to have been effective where quinine has failed.

Bakuri

A relatively obscure member of the Guttiferae, the bakuri, *Platonia insignis* Mart. (syn. *Aristoclesia esculenta* Stuntz), is also called *bacuri, bacuri assu, bacuri do Pará, bacury, pacuri* or *pacoury–uva* in Brazil; *pakuri, pakouri* or *maniballi* in Guyana; *pacouri* in French Guiana; *packoeri, pakoeri* or *geelhart* in Surinam; *goherica* or *ko* by the Indians in Amazonian Colombia. It is, unfortunately, sometimes referred to as *bakupari*, a name better limited to *Rheedia brasiliensis*, q.v. In Brazil the tree is called *bacurizeiro*.

Description

The tree is erect, to 80 ft (25 m) high, with pyramidal crown and copious yellow latex in the bark. The leaves are deciduous, opposite, oblong or elliptic, to 6 in (15 cm) long, dark-green and glossy above; leathery, with wavy margins. Borne singly or in 3's, the flowers are 2¾ in (7 cm) long, rose-colored, 5-petalled, with many stamens. The fruit is nearly round or ovoid, 3 to 5 in (7.5-12.5 cm) wide, weighing up to 32 oz (900 g); yellow when ripe. The rind is yellow, hard, fleshy on the inside, ⅜ to ¾ in (1-2 cm) thick, and contains gummy, yellow, resinous latex. The white, pithy pulp, of pleasant odor and agreeable, subacid flavor, contains 1 to 4, rarely 5, oblong, angular seeds, dark-brown and 2 to 2⅜ in (5-6 cm) long. The infertile seed compartments are filled with pulp called *"filho"* which is the part preferred.

Origin and Distribution

The bakuri was first reported in European literature in 1614. The tree is common, wild, in the Amazon region of northern Brazil from Maranhao, Goias to Paraguay. It is abundant in the State of Para, especially around Marajo and Salgado. Its native territory extends across the border into Colombia and northeast to the humid forests of Guyana. It is seldom cultivated but when the Indians clear the land for planting or pastures, they always leave this tree standing for the sake of its fruits. In Marajo, it is viewed as a weed because it proliferates from fallen seeds and, if felled, produces abundant suckers from the roots. In the district of Marapanim, there is a hamlet called Bacurteua because of its many bakuri trees.

Climate

The bakuri requires a moist, lowland, tropical habitat.

Season

In Brazil, the tree flowers in June and July, after the shedding of the leaves. The first fruits mature in early December and the season extends to the following May, the peak of the crop ripening in February and March.

Food Uses

The pulp is much eaten raw but is mainly used to make sherbet, ice cream, marmalade or jelly.

Food Value Per 100 g of Edible Portion*	
Calories	105
Moisture	72.3 g
Protein	1.9 g
Lipids	2.0 g
Glycerides	22.8 g
Fiber	7.4 g
Ash	1.0 g
Calcium	20.0 mg
Phosphorus	36.0 mg
Iron	2.2 mg
Vitamin B$_1$	0.04 mg
Vitamin B$_2$	0.04 mg
Niacin	0.50 mg
Ascorbic Acid	33.0 mg
Amino Acids (mg per g of nitrogen [N = 6.25])	
Lysine	316 mg
Methionine	178 mg
Threonine	219 mg
Tryptophan	57 mg
*Analyses made in Brazil.	

Other Uses

The sapwood is yellowish-white; the heartwood dull-yellow to orange-brown with many fine, dark, often black streaks. It is hard but easy to work and fairly durable. It is valued for construction, furniture, flooring, ship-building and general carpentry.

Medicinal Uses: The latex derived from the bark is used in veterinary practice in Guyana. The seeds contain 6 to 11% of an oil that is mixed with sweet almond oil and used to treat eczema and herpes.

Plate XLI
MANGOSTEEN
Garcinia mangostana
Painted by Dr. M.J. Dijkman

Plate XLII
MAMEY
Mammea americana

Plate XLIII
YELLOW PASSIONFRUIT
Passiflora edulis var. *flavicarpa*

Plate XLIV
YELLOW PASSIONFRUIT
Passiflora edulis var. *flavicarpa*

Plate XLV
GIANT GRANDADILLA
Passiflora quadrangularis

Plate XLVI
SWEET GRANADILLA
Passiflora ligularis

Plate XLVII
PAPAYA
Carica papaya

Plate XLVIII
PITAHAYA AMARILLA
possibly yellow form of *Hylocereus undatus*

Bakupari

Of the approximately 45 species of *Rheedia* (family Guttiferae), several have edible fruits. Perhaps the best-known is the bakupari, *R. brasiliensis* Planch. & Triana, which is also known as *bacupary* or *bacoropary* in Brazil; as *guapomo* in Bolivia.

The very attractive tree is pyramidal like that of the bakuri but smaller; is equally rich in yellow latex. The leaves are short-petioled, ovate, oblong-ovate or lanceolate, narrowed at the base, blunt or slightly pointed at the apex, and leathery. The flowers, profuse in axillary clusters, are polygamous. The fruit, ovate, pointed at the apex, may be 1¼ to 1½ in (3.2-4 cm) long, with orange-yellow, pliable, leathery, tough skin, ⅛ in (3 mm) thick and easily removed. The aril-like pulp is white, translu-cent, soft, subacid, of excellent flavor, and encloses 2 rounded seeds.

The tree grows wild in the state of Rio de Janeiro in southeastern Brazil and adjacent Paraguay; is rarely cultivated. It blooms in December and matures its fruit in January and February. The ripe fruit is mostly used in making sweetmeats or jam.

The seeds contain 8 to 9% oil (by weight) which is used in Brazil in poultices on wounds, whitlows, tumors and, externally, over an enlarged liver. An infusion of the pulp has a narcotic action with an effect like that of nicotine. The root bark extract contains rheediaxanthone and a polyprenylated benzophenone, other lesser constituents, and 3 new prenylated xanthones.

Related Species

The **mameyito**, *R. edulis* Triana & Planch. (syn. *Calophyllum edule* Seem.), is also known as *arrayan* and *palo de frutilla* in Guatemala; waiki plum in Belize; *chaparrón* in El Salvador; *caimito* or *caimito de montaña* in Honduras; *jorco* in Costa Rica; *sastra* in Panama; *berba* in the Philippines.

The elegant, erect tree, ranging up to 100 ft (30 m), has copious gummy, yellow latex and opposite, short-petioled, thick, leathery, elliptic-oblong or elliptic-lanceolate leaves, 3³/₁₆ to 6 in (8-15 cm) long, ¾ to 2 in (2-5 cm) wide, or much larger, with numerous lateral veins conspicuous on both surfaces; dark-green above, pale or brownish on the underside. Young foliage is reddish. The small, greenish-white or ivory flowers, densely clustered below the leaves, are 4-petalled, the male with 25 to 30 stamens, the perfect with 10 to 12. The fruit is oval or oblong, ¾ to 1¼ in (2-3.2 cm) long, smooth, orange or yellow, the thin, soft skin easily peeled. There is a little flesh, sweet or acid, adhering to the 1 or 2 seeds.

The tree is native and common in humid forests on both the Atlantic and Pacific sides of Central America, from southern Mexico to Panama, up to an elevation of 4,000 ft (1,200 m). It is often planted in Central America as a shade or ornamental tree. It has been grown in the Philippines, Puerto Rico and California. The fruits mature from late January to March in Costa Rica.

The heartwood is rose-yellow, hard, medium-heavy, coarse-textured, with numerous gum ducts, but tough, strong, easy to work, fairly durable, and valued for construction because it is nearly immune to insects. It is also used for tool handles, fenceposts, and temporary railroad ties. The bark is rich in tannin.

The **bacuripari**, *R. macrophylla* Planch. & Triana, is also called *bacury-pary* in Brazil; *charichuela* in Peru.

It is a pyramidal tree, 26 to 40 ft (8-12 m) tall, with stiff, leathery, lanceolate-oblong or broad-lanceolate leaves, 12 to 18 in (30-45 cm) long and 3 to 7 in (8-18 cm) wide, pointed at both ends, with numerous lateral, nearly horizontal veins. New foliage is maroon. The 4-petalled, male and female flowers are borne in small axillary clusters on separate trees, the male on delicate stalks to 1½ in (4 cm) long and having numerous stamens, the female on thick, short stalks and sometimes having a few stamens with sterile anthers.

The fruit is rounded-conical, pointed at one or both ends, about 3³/₁₆ in (8 cm) wide, with thick yellow rind, usually smooth, sometimes rough, containing gummy yellow latex. The white, aril-like pulp, agreeably subacid, encloses 3 to 4 oblong seeds.

The tree is native to humid forests of Surinam and Brazil to northern Peru. The fruit is not much esteemed but widely eaten and sold in native markets. The bacuripari was introduced into Florida in 1962 and planted at the Agricultural Research and Education Center in Homestead, at Fairchild Tropical Garden and in several private gardens. One tree fruited in 1970, another in 1972, and the latter has continued to bear. Young specimens have been killed by drops in temperature to 29° to 30°F (-1.67°--1.11°C). Older trees have been little harmed by 27° to 28°F (-2.78°--2.22°C). The tree is accustomed to light-to moderate-shade. Seeds have remained viable for 2 to 3 weeks but require several weeks to germinate.

In Brazil, the tree blooms from August to November and the fruits mature from December to May. In Florida, flowers appear in April and May and a second time in August and September, and the fruits are in season from May to August and again in October and November. Some 15- to 20-year-old trees have produced 100 to 200 fruits when there have been no adverse weather conditions.

The **madroño,** *R. madruno* Planch. & Triana, may be called *machari* or *fruta de mono* in Panama; *cerillo* in Costa Rica; *cozoiba* in Venezuela; *kamururu* in Bolivia.

The tree is erect, lush, compact, with pyramidal or nearly round crown, 20 to 65 ft (6-20 m) high, and has much gummy yellow latex. The opposite leaves are elliptic to oblong, wedge-shaped at the base, rounded or pointed at the apex, 2⅜ to 8 in (6-20 cm) long, ¾ to 3 in (2-7.5 cm) wide; dark green above, paler beneath, with numerous veins conspicuous on both surfaces and merging into a thick marginal vein. The fragrant male and female flowers are borne on separate trees in clusters of up to 14 in the leaf axils; have 4 reflexed, pale-yellow petals; the male, 25 to 30 light-yellow stamens. The fruit is round or ellipsoidal, sometimes with a prominent nipple at each end; 2 to 3 in (5-7.5 cm) long, with thick, leathery, warty, greenish-yellow rind containing a deep-yellow, resinous latex. The white, translucent, juicy, sweet-acid, aromatic pulp adheres tightly to the 1 to 3 ovate or oblong seeds which are about ¾ in (2 cm) long.

The tree is native to the Golfo Dulce region of Costa Rica, the Atlantic slope of Panama, and northern South America—Colombia and Ecuador through Venezuela to Guyana and Bolivia. It is particularly common in the Cauca Valley of Colombia where the fruits are marketed in quantity. It is limited to elevations below 4,000 ft (1,200 m). Dr. Wilson Popenoe collected seeds for the United States Department of Agriculture near Palmira, Colombia, in 1921 (S.P.I. #52301). The tree was introduced into Puerto Rico in 1923 and into the Philippines at about the same time. A few old trees have been fruiting more or less in southern Florida for many years, in midsummer. In Costa Rica, flowers are borne from December to February and fruits from May to August.

The yellow latex of the tree is used in Panama to treat ulcers and other sores. The wood is pinkish and hard but not commonly used.

Fig. 83-b: Peeled mangosteens, in light sirup, canned in Thailand, are appearing in Asiatic food outlets in the United States.[1] According to the *Wall Street Journal*, April 7, 1987, fresh fruits, cut open, inspected, sealed with tape, and quick-frozen, are exported from Malaysia to Japan where they sell readily at nearly $4 each. They are defrosted in boiling water for 2 minutes before eating.

Ketembilla

Somewhat better-known than the kei apple, q.v., the ketembilla, *Dovyalis hebecarpa* Warb. (syns. *Aberia gardneri* Clos.; *Roumea hebecarpa* Gard.), is often called Ceylon gooseberry; sometimes ketambilla, or kitembilla; and it is known as *aberia* in Cuba and Central America.

Description

The shrub or small tree reaches no more than 15-20 ft (4.5-6 m) in height but its long, slender, arching, wide-spreading branches may cover 30 ft (9 m) of ground. Sharp spines to 1½ in (4 cm) long, are plentiful on the trunk and lower branches. The alternate leaves are elliptical to ovate, pointed, 2¾ to 4 in (7-10 cm) long, wavy-margined, gray-green, finely velvety, with pinkish, woolly petioles, and thin in texture. Male, female and hermaphrodite flowers are borne on separate trees. They are petalless, greenish-yellow, nearly ½ in (1.25 cm) wide and clustered in the leaf axils. The fruit, borne in great abundance, is globose, ½ to 1 in (1.25-2.5 cm) wide. Its thin, bitter skin turns from somewhat orange to dark purple on ripening and is coated with short, grayish-green, velvety hairs, unpleasant in the mouth. The pulp is very juicy, extremely acid, purple-red, enclosing 9 to 12 hairy seeds about ¼ in (6 mm) long.

Fig. 84: Ripe fruits of the ketembilla are furry-skinned, extremely acid and slightly bitter.

Origin and Distribution

The ketembilla is native to Ceylon. It was introduced into the United States by Dr. David Fairchild and was one of the few fruits he admitted he never liked very much. The first fruiting specimens in the western hemisphere were apparently those growing in southern Florida. P.J. Wester carried seeds to the northern islands of the Philippines where it began fruiting in 1916. From Florida, also, the plant was introduced into the Atkins Garden of Harvard University at Cienfuegos, Cuba. Seeds from the Garden were shipped to the Hawaiian Sugar Planters' Association in 1920, and to the Lancetilla Experimental Garden at Tela, Honduras, in 1927. Seeds from Florida were supplied to the Mayaguez and Trujillo Experimental Stations in Puerto Rico where the plants were 16 ft (5 m) high by 1929 and 1930. Plants were distributed widely throughout the Hawaiian Islands and use of the fruits was officially encouraged.

Fig. 85: Formerly grown for jelly-making, the too-vigorous, productive ketembilla or Ceylon gooseberry (*Dovyalis hebecarpa*) is no longer planted in southern Florida.

Florida pioneers grew the species and utilized the fruits until the plants took up too much space. When South Florida began to develop rapidly after World War II, most people had no room for such an aggressive plant. One enthusiast maintained a small commercial plot in West Palm Beach for juice production.

In 1935, horticulturists in Israel imported seeds from Ceylon and plants grew and fruited well in a variety of locations. Commercial exploitation was anticipated but was suspended during World War II because of the shortage of sugar for preserving.

Climate

In the Philippines, the ketembilla flourishes from sea-level to 2,600 ft (800 m). In Malaya, it is found from near-sea-level up to 4,000 ft (1,200 m). It has never survived at Singapore. Fruiting is not consistent at Tela, Honduras. However, it does do well planted at appropriate elevations in either dry or moist climates.

Soil

In Florida, the plant grows entirely too vigorously on sand or limestone, but a rich soil is best for maximum fruit production and plenty of water is desirable during fruit development.

Season

In Israel, fruit ripens from winter to spring. In Florida, there are two crops a year—spring and fall, but the fruits may be infested with the larvae of the Caribbean fruit fly, *Anastrepha suspensa,* and unusable.

Food Uses

In Florida, in the past, the ketembilla was used primarily for jelly. Recipes developed in Hawaii include juice, spiced jelly, ketembilla-papaya jam, ketembilla-guava jelly, and ketembilla-apple butter. In Israel, the fruit is valued mainly as a source of jelly for export.

Food Value Per 100 g of Edible Portion*	
Moisture	81.9-83.6 g
Protein	0.174-0.206 g
Fat	0.64-1.02 g
Crude Fiber	1.7-1.9 g
Ash	0.61-0.63 g
Calcium	12.6-13.3 mg
Phosphorus	24.5-26.8 mg
Iron	0.91-1.41 mg
Carotene	0.125-0.356 mg
Thiamine	0.017 mg
Riboflavin	0.033-0.042 mg
Niacin	0.261-0.316 mg
Ascorbic Acid	91.7-102.5 mg
(Slightly unripe fruits are high in pectin.)	
*Analyses made in Honduras.	

Other Uses

In the West Indies and Central America, honeybees are seen to work the blossoms eagerly from July to December.

Related Species

The **Abyssinian gooseberry,** *D. abyssinica* Warb. (syns. *D. engleri* Gilg; *Aberia abyssinica* Clos.) is a bushy, more or less thorny, shrub or tree to 30 ft (9 m) high, with alternate leaves, ovate-lanceolate to oblong, 1 to 3½ in (2.5-9 cm) long, ¾ to 1½ in (2-4 cm) wide; glabrous or slightly hairy, light-green, glossy, wavy, and sometimes finely toothed. Male and female flowers are borne on separate plants. They are small, greenish-white, and emerge at the leaf axils, the male clustered, the female singly. The fruits are oblate, ½ to 1 in (1.25-2.5 cm) wide, with thin, tender, apricot-colored skin and concolorous, apricot-flavored, juicy, melting, astringent, acid pulp containing several flat seeds.

This species is native and common in forests of East Africa (Ethiopia, Kenya, Uganda) at elevations between 6,000 and 8,000 ft (1,800-2,400 m). Seeds were obtained by the United States Department of Agriculture from the Atkins Garden in Cuba in 1935 (S.P.I. #112086) and planted at the then Plant Introduction Station in Miami. Three seedlings were supplied to the University of Florida's experiment station in Homestead, two of which died and the survivor was a male. Two plants remaining at the United States Department of Agriculture showed considerable hardiness with only minor injury in cold spells just below freezing. Some die-back was attributed to infestation by scale insects or root damage by nematodes. These plants had female flowers but never bore fruit until there occurred accidental pollination by a ketembilla 50 to 60 ft (15-18 m) distant. A heavy crop of fruits was borne in 1951. A dozen seedlings were sent to Homestead. A scion from one of the 2 female plants was grafted onto the male plant at the Homestead station and bore fruit less than a year later. The attractive fruits caused considerable interest, grafted plants were sold by nurseries and someone proceeded to invent the frivolous term, "Florida apricot".

The seedlings from the 1951 crop planted at Homestead fruited in October 1953. Both foliage and fruit suggest that hybridization had taken place between the ketembilla and the Abyssinian gooseberry. One of the seedlings bore perfect flowers in small clusters.

The hybrid fruit is oblate, ¾ to 1⅜ in (2-3.5 cm) across, with a velvety skin, brownish-orange or burnt-orange, dappled with many flecks of yellow. The flesh is burnt-orange or orange-yellow, juicy, very sour, more or less acrid, the flavor modifying somewhat when the fruit becomes extra-ripe and dark-red in color. There are 3 to

9 flat, pointed, nearly white seeds to 5/16 in (8 mm) long, mostly underdeveloped and not very noticeable when the fruit is eaten. Plants reproduced by cuttings or air-layers (though producing strong, spiny shoots) were soon being offered by local nurserymen as "*Dovyalis* hybrid", no other name having been adopted. In 1960, I proposed "ketcot" as concisely representing its 2 parents and Dr. George H. Lawrence, then Director of the Bailey Hortorium wanted to record this in *Hortus* as soon as it became popularized, which it never was.

The hybrid proved to be remarkably hardy, more stalwart and vigorous than either parent, forming massive, formidable mounds to 15 ft (4.5 m) high, the branches weighed down with excessive crops. One practical disadvantage is that the green, 6-pointed calyx, 3/8 in (1 cm) wide, remains on the plant as the fruit is picked, leaving a cavity in the base of the fruit. It is, therefore, not marketable as a fresh fruit but can be used to make sirup, jam or other preserves.

I was informed in 1962 that a hybrid of *D. abyssinica* and the ketembilla had originated in the Kitchen Door Nursery, North Miami. It was given the name "Kandy" after a village in Ceylon, and had survived several winters in Winter Haven.

Despite productivity and hardiness and the promotion of less-spiny, less rampant plants grafted on ketembilla, few homeowners have welcomed the "*Dovyalis* hybrid" and its position has remained static over the past 25 years.

Fig. 86: The Abyssinian gooseberry (*Dovyalis abyssinica*), more attractive in color and of more pleasing flavor than the ketembilla, is still too astringent to be popular.

Fig. 87: An apparent chance cross between the ketembilla and the Abyssinian gooseberry, known only as "Dovyalis hybrid", was briefly promoted in southern Florida. The fruits are large but astringent.

Kei Apple

The kei apple, *Dovyalis caffra* Warb. (syn. *Aberia caffra* Harv. & Sond.) is also known as *umkokolo* in Africa and this is abbreviated to *umkolo* in the Philippines. The generic name has been rendered *Doryalis* by many writers but botanists now agree that this form was not the original spelling.

Description

The shrub or small tree, growing to a height of 30 ft (9 m) with a spread of 25 ft (7.5 m), usually has many sharp spines 1 to 3 in (2.5-7.5 cm) long, though it is often entirely spineless if not trimmed. The leaves, often clustered on short spurs, are oblong-obovate, 1 to 3 in (2.5-7.5 cm) long, glossy and short-petioled. Pale-yellow male and female flowers are usually borne on separate trees. They are small, petalless, and clustered in the leaf axils. The aromatic fruit is oblate or nearly round, 1 to 1½ in

(2.5-4 cm) long, with bright-yellow, smooth but minutely downy, somewhat tough skin, and mealy, apricot-textured, juicy, highly acid flesh. There are 5 to 15 seeds arranged in double rings in the center. They are flat, pointed and surrounded by threadlike fibers. The tree is spectacular when its branches are laden with these showy fruits.

Origin and Distribution

The kei apple is native to the Kei River area of southwest Africa and abundant in the wild around the eastern Cape, Kaffraria and Natal. It is cultivated in the Transvaal. In 1838, it was introduced into England and from there distributed to Egypt, Algeria, southern France and Italy, the Philippines, northwestern Australia, Jamaica, southern California and Florida. The United States Department of Agriculture obtained plants from Reasoner Bros., Oneco, Florida in 1901 (S.P.I. #6857); seeds from

Fig. 88: The kei apple tree (*Dovyalis caffra*) is drought-tolerant, salt-resistant and strikingly fruitful, but the fruit is intensely acid.

South Africa in 1901 (S.P.I. #7955 & #7956); seeds from the Cape Town Public Gardens in 1906 (S.P.I. #18667); seeds from the Middle Egypt Botanic Gardens in 1912 (S.P.I. #34250); and seeds from Hubert Buckley, St. Petersburg, Florida (S.P.I. #145592) and the resulting seedlings were being distributed from the Plant Introduction Garden, Coconut Grove, in 1942 and 1943. A few specimens were planted in experimental stations in Puerto Rico and St. Croix, and in private gardens in southern and central Florida, and the plant was adopted as a coastal, rough hedge in southern California. It has been grown as a hedge and for its fruit in some parts of Costa Rica. It was in the past extensively cultivated as a hedge around citrus groves in Israel, but the fruits were not liked, they accumulated on the ground and became breeding places for the Mediterranean fruit fly. Therefore, nearly all the plants were destroyed.

Climate

The kei apple is subtropical; does poorly at sea-level in the Philippines but thrives at and above 2,600 ft (800 m). Introductions have failed to survive in Malaya. In Florida, the plant has been grown in a small way as far north as

Gainesville, enduring brief drops in temperature to 20°F (-6.67°C) but descents to 16°F (-8.80°C) have been lethal in this state and in California.

Soil

The kei apple does well in almost any soil that does not have a high water table. It is extremely drought-resistant and tolerates saline soil and salt spray and is accordingly valued as a coastal hedge in the Mediterranean region and in California.

Propagation

Propagation is ordinarily by seeds, though layering is successfully done in Australia. Seeds germinate readily when fresh and seedlings begin to bear in 4 or 5 years. For fruit production, Wilson Popenoe recommended a spacing of no less than 12 to 15 ft (3.5-4.5 m). Hedge plants can be set 3 to 5 ft (0.9-1.5 m) apart. According to Popenoe there should be 1 male for every 20 or 30 females. However, certain female trees have borne profusely in the absence of male pollinators. A kei apple hedge must be trimmed twice a year. If neglected and allowed to become leggy, it can be cut to the ground and given a new start. Weeding should not be a problem, for the kei apple exhibits allelopathy, that is, its roots excrete growth inhibitors which prevent the occurrence of other plants in its vicinity. Investigators in Egypt have demonstrated that the roots, stem and fruit, but not the leaves and branches, possess antibiotic properties.

Season

Generally, the plants bloom in spring and the fruits ripen from August to October. The thorns make harvesting difficult. The top may have to be thinned out in order to facilitate fruit-picking.

Food Uses

Most people consider the fruit too acid for eating out-of-hand even when fully ripe. It is best cut in half, peeled, seeded, sprinkled with sugar and allowed to stand for a few hours before serving as dessert or in fruit salads. The halves can stand only a few minutes of cooking before they turn into sauce. Simmered briefly in sirup, they make excellent shortcake. Kei apples are customarily made into jam and jelly, and, when underripe, pickles.

Food Value

Fresh ripe fruits contain 83 mg ascorbic acid per 100 g and 3.7% pectin. Scientists in Egypt have reported 15 amino acids: alanine, 0.41%; arginine, 0.36%; aspartic acid, 0.96%; glutamic acid, 2.00%; glycine, 0.39%; histidine, 0.10%; isoleucine, 0.25%; leucine, 0.75%; lysine, 0.36%; methionine + valine, 0.28%; phenylalanine, 0.40%; proline, trace; serine, 0.48%; threonine, 0.34%.

Related Species

In the family Flacourtiaceae, there are several species of *Flacourtia* that have been distributed as fruit producers. None has any great merit, and four shall be treated as minor subjects here.

The **louvi**, *F. inermis* Roxb., is called *rukam masam, rokam masam, lovi-lovi, lobeh-lobeh, tomi* and thornless *rukam* in Malaya. The tree is short-trunked, bushy, to 25 or 30 ft (7.6-9 m) tall, and thornless. The evergreen, alternate leaves, bright-red when young, are glossy on the upper surface, dull beneath; 3½ to 10 in (9-25 cm) long and 2 to 5 in (5-12.5 cm) wide. Unlike other species, the tree has bisexual flowers. They are petalless with green sepals and many yellow stamens and borne in small clusters along the branches. The fruit is round but slightly flattened at the apex, ¾ to 1 in (2-2.5 cm) wide, smooth, bright-red, thin-skinned. The flesh is whitish tinged with pink, astringent, acid or occasionally sweet. There are 4 to 14 hard, sharp, irregular seeds under ¼ in (6 mm) wide.

The tree is of unknown origin; cultivated in Ceylon, Malaya and Indonesia. Its lifespan is said to be about 20 years. It is propagated by seed in Malaya, by air-layering or budding in Java. Flowering occurs several times a year. Yield from dooryard trees varies from 81 to 241 lbs (36.8-109.5 kg) a year. Those given good cultural attention may bear a total of 374-576 lbs (170-261.8 kg). The fruits are not favored raw but are seeded and cooked with apples to add color, or are made into pie, jam, jelly, sirup, chutney and pickles.

The **paniala**, *F. jangomas* Raeusch. (syn. *F. cataphracta* Roxb.) is also called *puneala, puneala* plum, *jaggam,* Chinese plum, Indian plum; in Malaya, *kerkup, kerkup besar, kerkup bakoh*; in Thailand, *ta-khop-thai*; in Vietnam, *mu cuon, mung quan, bo quan,* and *prunier malgache.* The shrub or erect, low-branched tree, 20 to 40 ft (6-12 m) high, has flaking bark and sharp spines on the trunk. The leaves are alternate, deciduous, pale pink when young, spirally arranged, oval-lanceolate, long-pointed, toothed, very thin, glossy on both surfaces; 2 to 4 in (5-10 cm) long, ½ to 2 in (1.25-5 cm) wide. Male and female flowers are on separate trees. They are greenish, heavily fragrant, borne in small clusters on new branchlets. The fruits are round or slightly oval, ¾ to 1 in (2-2.5 cm) long, dark-maroon to nearly black; the flesh greenish to white or amber, varying from acid to sweet, and containing 7 to 12 flat, hard, pale-yellow seeds.

The tree is native to North Bengal, East Bengal and Chittagong in India; commonly cultivated throughout Southeast Asia, eastern Malaya, and also in the Philippines. It has been planted in a very limited way in Surinam, Trinidad, Puerto Rico and southern Florida. The seeds are slow to germinate, therefore propagation is usually by inarching or budding onto self-seedlings.

For eating out-of-hand, the fruit is rolled between the hands to reduce astringency, and is better-liked than that of other species. It is stewed as dessert, made into juice, sirup, jam, marmalade and pickles and also used in chutneys. When slightly underripe, it is used to make jelly. The acid young shoots are eaten in Indonesia.

Philippine analyses show: moisture, 78.28%; protein, 0.03%; fat, 0.39%; sugar, 4.86%; ash, 0.94%; acidity, 1.16%. The fruit is fairly rich in pectin; contains 9.9% tannin on a dry-weight basis.

The wood, red or scarlet, is close-grained, hard, brittle, durable and polishes well. It is used for agricultural implements.

The fruits are eaten to overcome biliousness, nausea and diarrhea. The leaf decoction is taken to halt diarrhea. Powdered, dried leaves are employed to relieve bronchitis and coughs. The leaves and bark are applied on bleeding gums and aching teeth, and the bark infusion is gargled to alleviate hoarseness. Pulverized roots are poulticed on sores and skin eruptions and held in the mouth to soothe toothache.

Fig. 89: The paniala (*Flacourtia jangomas*) of southern Asia and the Philippines, has wine-red, plumlike fruits, unfortunately very astringent.

The **ramontchi** is *F. ramontchi* L'Her. *F. indica* has been frequently recorded as a synonym but Indian botanists disagree and treat *F. indica* Merr. as a distinct species. The common name for the ramontchi in India is governor's plum; in Malaya, it is *kerkup kechil* or lesser *kerkup;* in Thailand, *ta-khop-pa;* in the Philippines, *bitongol, bolong,* or *palutan;* in Africa, it is called *kokowi,* Madagascar plum or Indian plum.

The tree is bushy and spreading but may reach 50 ft (15 m) and usually has sharp spines on the trunk and on main branches which tend to arch and droop at the tips. The evergreen, alternate leaves, red when young, are obovate to oblong-obovate, 1 to 2 in (2.5-5 cm) long and finely toothed. Male and female flowers are borne on separate trees. They are white, about 3/16 in (5 mm) wide, and appear singly or paired in the leaf axils. The fruit is round, ½ to 1 in (1.25-2.5 cm) thick, smooth, glossy, dark red-purple, with light-brown, acid to sweet, astringent, slightly bitter, flesh and 6 to 10 small, flat seeds.

The ramontchi is native to tropical Africa, Madagascar, India, parts of Malaya and Southeast Asia, and much of Malaysia including the Philippines. It has been planted in Florida, Puerto Rico, Trinidad, Guatemala, Honduras and Venezuela and advocated as a source of fruit. It has never become popular anywhere, but jelly can be made from it by not squeezing the jelly bag and thus avoiding excessive astringency. The fruit is usually infested by fruit flies. Analyses made in the Philippines show: moisture, 66.42%; protein, 0.69%; fat, 1.67%; sugar, 7.68%; ash, 1.09%; acidity, 1.78%.

In Florida, birds scatter the seeds and volunteers invade natural areas. In Puerto Rico, the tree is considered useful as a tall barrier hedge or windbreak. Farmers in India lop the branches for fodder. The wood is used only for fuel.

The leaves and roots are believed to be effective against snakebite and the pulverized bark, mixed with sesame oil, is applied on rheumatic parts. Filipinos use the bark infusion as a gargle. A root infusion is taken in cases of pneumonia. The leaf juice is given as a febrifuge and remedy for coughs, dysentery and diarrhea. The dried leaves are regarded as carminative, expectorant, tonic and astringent.

The **rukam,** *F. rukam* Zoll. & Mor., also called *rukam manis, rukam gajah* and Indian prune in Malaya; *khrop-dong* in Thailand, is a much-branched, crooked tree to 40 or even 65 ft (12-20 m), sometimes thornless in cultivation but usually heavily armed with forked, woody spines on the trunk and old branches. The leaves are evergreen, spiralled, red when young, elliptic-oblong, 3 to 6 in (7.5-15 cm) long, 1¼ to 2½ in (3.2-6.25 cm) wide, coarsely toothed, slightly shiny. Flowers are in small clusters in the leaf axils. Male and female are usually on separate trees; occasionally both occur on the same plant. There are no petals; the male have many stamens.

The fruits, borne on old branches or on the trunk, are nearly round, slightly flattened at the apex, ½ to 1 in (1.25-2.5 cm) wide, dark purple-red, smooth, with whitish, juicy, acid flesh. There are 4 to 7 flat seeds.

The tree is native to India, Southeast Asia, Malaysia and Oceania; cultivated in southern Malaya and Indonesia. It is adapted to elevations up to 5,200 ft (1,600 m). Seeds came to the USDA from Bangkok in 1920 (S.P.I. #51772). A few specimens have been grown in Florida.

The fruits are eaten raw, especially after rolling them between the palms to reduce astringency. They are also cooked, made into pie, jam and chutney. The young shoots are marketed and eaten raw in Java.

Analyses made in the Philippines show: calories 82.80 per 100 g; moisture, 76.93%; protein, 1.72%; fat, 1.26%; reducing sugars, 4.32%; fiber, 3.71%; other carbohydrates, 11.29%; ash, 0.771%; acidity, 1.29%.

The heavy, strong wood is made into rice pounders in Java; pestles in the Philippines; and clubs in Samoa.

The juice of immature fruit is taken to halt diarrhea and dysentery. Leaf juice is applied on inflamed eyelids, and dried, pulverized leaves are spread on wounds. The root decoction is given to women after childbirth. The inner bark is used against filariasis in Samoa.

Fig. 90: The ramontchi, or governor's plum (*F. ramontchi*), closely resembles the paniala. The fruit is sweet but astringent and slightly bitter. The leaves are useful as fodder.

PASSIFLORACEAE

Passionfruit (Plates XLIII and XLIV)

Of the estimated 500 species of *Passiflora,* in the family Passifloraceae, only one, *P. edulis* Sims, has the exclusive designation of passionfruit, without qualification. Within this species, there are two distinct forms, the standard purple, and the yellow, distinguished as *P. edulis* f. *flavicarpa* Deg., and differing not only in color but in certain other features as will be noted further on.

General names for both in Spanish are *granadilla, parcha, parchita, parchita maracuyá,* or *ceibey* (Cuba); in Portuguese, *maracuja peroba;* in French, *grenadille,* or *couzou.* The purple form may be called purple, red, or black granadilla, or, in Hawaii, *lilikoi;* in Jamaica, mountain sweet cup; in Thailand, *linmangkon.* The yellow form is widely known as yellow passionfruit; is called yellow *lilikoi* in Hawaii; golden passionfruit in Australia; *parcha amarilla* in Venezuela.

Description

The passionfruit vine is a shallow-rooted, woody, perennial, climbing by means of tendrils. The alternate, evergreen leaves, deeply 3-lobed when mature, are finely toothed, 3 to 8 in (7.5-20 cm) long, deep-green and glossy above, paler and dull beneath, and, like the young stems and tendrils, tinged with red or purple, especially in the yellow form. A single, fragrant flower, 2 to 3 in (5-7.5 cm) wide, is borne at each node on the new growth. The bloom, clasped by 3 large, green, leaflike bracts, consists of 5 greenish-white sepals, 5 white petals, a fringelike corona of straight, white-tipped rays, rich purple at the base, also 5 stamens with large anthers, the ovary, and triple-branched style forming a prominent central structure. The flower of the yellow is the more showy, with more intense color. The nearly round or ovoid fruit, 1½ to 3 in (4-7.5 cm) wide, has a tough rind, smooth, waxy, ranging in hue from dark-purple with faint, fine white specks, to light-yellow or pumpkin-color. It is 1/8 in (3 mm) thick, adhering to a ¼ in (6 mm) layer of white pith. Within is a cavity more or less filled with an aromatic mass of double-walled, membranous sacs filled with orange-colored, pulpy juice and as many as 250 small, hard, dark-brown or black, pitted seeds. The flavor is appealing, musky, guava-like, subacid to acid.

Origin and Distribution

The purple passionfruit is native from southern Brazil through Paraguay to northern Argentina. It has been stated that the yellow form is of unknown origin, or perhaps native to the Amazon region of Brazil, or is a hybrid between *P. edulis* and *P. ligularis* (q.v.). Cytological studies have not borne out the hybrid theory. Speculation as to Australian origin arose through the introduction of seeds from that country into Hawaii and the mainland United States by E.N. Reasoner in 1923. Seeds of a yellow-fruited form were sent from Argentina to the United States Department of Agriculture in 1915 (S.P.I. No. 40852) with the explanation that the vine was grown at the Guemes Agricultural Experiment Station from seeds taken from fruits purchased in Covent Garden, London. Some now think the yellow is a chance mutant that occurred in Australia. However, E.P. Killip, in 1938, described *P. edulis* in its natural range as having purple or yellow fruits.

Brazil has long had a well-established passionfruit industry with large-scale juice extraction plants. The purple passionfruit is there preferred for consuming fresh; the yellow for juice processing and the making of preserves.

In Australia, the purple passionfruit was flourishing and partially naturalized in coastal areas of Queensland before 1900. Its cultivation, especially on abandoned banana plantations, attained great importance and the crop was considered relatively disease-free and easily managed. Then, about 1943, a widespread invasion of *Fusarium* wilt killed the vines and forced the undertaking of research to find fungus-resistant substitutes. It was discovered that the neglected yellow passionfruit is both wilt- and nematode-resistant and does not sucker from the roots. It was adopted as a rootstock and plants propagated by grafting were soon made available to planters in Queensland and northern New South Wales.

The Australian taste is strongly prejudiced in favor of the purple passionfruit and growers have been reluctant to relinquish it altogether. Only in the last few decades have they begun to adopt hybrids of the purple and yellow which have shown some ability to withstand the serious virus disease called "woodiness".

New Zealand, in the early 1930's, had a small but thriving purple passionfruit industry in Auckland Province but in a few years the disease-susceptibility of this type brought about its decline. Good local marketing and export prospects have brought about a revival of efforts to control infestations and increase acreage, mostly in the Bay of Plenty region. Today, fruits and juice are ex-

Fig. 91: Purple passionfruit (*Passiflora edulis*) is subtropical, important in some countries, while the more tropical yellow passionfruit excels in others. Both yield delicious juice.

ported. A profitable purple passionfruit industry has developed also in New Guinea.

In Hawaii, seeds of the purple passionfruit, brought from Australia, were first planted in 1880 and the vine came to be popular in home gardens. It quickly became naturalized in the lower forests and, by 1930, could be found wild on all the islands of the Hawaiian chain. In the 1940's, a Mr. Haley attempted to market canned passionfruit juice in a small way but the product was unsatisfactory and his effort was terminated by World War II. A processor on Kauai produced a concentrate in glass jars and this project, though small, proved successful. In 1951, when Hawaiian passionfruit plantings totalled less than 5 acres (2 ha), the University of Hawaii chose this fruit as the most promising crop for development and undertook to create an industry based on quick-frozen passionfruit juice concentrate. From among Mr. Haley's vines, choice strains of yellow passionfruit were selected. These gave four times the yield of the purple passionfruit and had a higher juice content. By 1958, 1,200 acres (486 ha) were devoted to yellow passionfruit production and the industry was firmly established on a satisfactory economic level.

Commercial culture of purple passionfruit was begun in Kenya in 1933 and was expanded in 1960, when the crop was also introduced into Uganda for commercial production. In both countries, the large plantations were devastated several times by easily-spread diseases and pests. It became necessary to abandon them in favor of small and isolated plantings which could be better protected.

South Africa in 1947 produced 2,000 tons of purple passionfruit for domestic consumption. Production was doubled by 1950. In 1965, passionfruit plantations were initiated over large areas of the Transvaal to meet the market demand and apparently there have been no serious setbacks as yet, from disease or other causes.

India, for many years, has enjoyed a moderate harvest of purple passionfruit in the Nilgiris in the south and in various parts of northern India. In many areas, the vine has run wild. The yellow form was unknown in India until just a few decades ago when it was introduced from Ceylon and proved well adapted to low elevations around Madras and Kerala. It was quickly approved as having a more pronounced flavor than the purple and producing within a year of planting heavier and more regular crops.

The purple passionfruit was introduced into Israel from Australia early in the 20th Century and is commonly grown in home gardens all around the coastal plain, with small quantities being supplied to processing factories.

Passionfruit vines are found wild and cultivated to some extent in many other parts of the Old World—including the highlands of Java, Sumatra, Malaya, Western Samoa, Norfolk Islands, Cook Islands, Solomon Islands, Guam, the Philippines, the Ivory Coast, Zimbabwe and Taiwan. From several of these sources, considerable quantities of yellow passionfruit juice and pulp are exported to Australia, causing some protests from Queensland growers. The yellow passionfruit was introduced into Fiji from Hawaii in 1950, was distributed to farmers in 1960 and became the basis of a small juice-processing industry. Fiji has exported to Australia, New Zealand, and Canada as well as to nearby islands.

In South America, interest in yellow passionfruit culture intensified in Colombia and Venezuela in the mid-1950's and in Surinam in 1975. In Colombia, there are commercial plantations mainly in the Cauca Valley.

Since the introduction of the yellow passionfruit from Brazil into Venezuela in 1954, it has achieved industrial status and national popularity. Much effort is being devoted to improving the yield to better meet the demand for the extracted juice, passionfruit ice cream, and other appealing products such as bottled passionfruit-and-rum cocktail.

The purple passionfruit was naturalized in the Blue Mountains of Jamaica by 1913, and both the purple and the yellow are planted to some extent in Puerto Rico.

Various species of *Passiflora* have reached the United States Plant Introduction Station (now the Subtropical Horticulture Research Unit) in Miami, Florida, in the routine course of plant accession. Some vines were known to exist and bear fruit year after year here and there in the southern and central areas of the state since 1887 or earlier. In 1953, I requested seeds of good strains of the purple and yellow forms from the Queensland Department of Agriculture and Stock and gave seeds to experimenters. In 1955, one yellow-fruited vine from these seeds was flourishing at Pinecrest and, from the reports of hunters camping beyond that locality, it appears that bird-transported seeds have produced fruiting vines in outlying Everglades hammocks. In 1957, a very fruitful specimen was thriving at the home of Benjamin Blumberg in Coconut Grove, and an escape was bearing unusually large fruits in the treetops of a natural hammock a few miles away. At this time, the purple passionfruit was being grown successfully by a homeowner further north, at Land

Fig. 92: Flowers of the purple passionfruit are fragrant and lovely, though those of the yellow are richer in color.

O'Lakes, Pasco County, and the seeds were advertised for sale. There were small plantations of purple passionfruit in San Diego County, California, the fruits being sold on the fresh fruit market and also processed for juice. However, there was little interest in developing either form as a crop in the United States. At the University of Florida's Subtropical Experiment Station in Homestead, Florida, limited trials with the purple and yellow forms resulted in words of discouragement, the purple vine in particular having proved so susceptible to disease. Certain vines at the Plant Introduction Station had died from *Fusarium* attack and the survivors showed poor fruiting performance.

Dr. Robert Knight and Harold F. Winters of the United States Department of Agriculture prepared two reports on the pollination of the yellow passionfruit and the problems affecting yield. They expressed a dim view of economical juice production and the need for extensive field studies. They offered plant material to anyone qualified to undertake such work. The Minute Maid Company established a test plot of the yellow form at Indiantown in 1965. They found the fruit entirely satisfactory for processing but abandoned the project 2 years later, stating: "The yields are not as large as in more tropical areas where the plant remains productive all year round. Our plants went out of production during the winter season. During the windy spring months of March and April, the vines are badly damaged and no flowers are set until sometime in May. We also found that the passionfruit were expensive to harvest. The fruit has to fall on the ground and sometimes it gets hung up in the vines. There is a continual collection of small quantities of fruit throughout the [bearing] year. Special equipment is needed to obtain the juice from the fruit without bits of the calyx showing up as objectionable black specks. This equipment is costly and can only be justified when a large volume of fruit is being processed."

In 1965, the Laboratorie de Recherche des Produits Nestlé, Vevey, Switzerland, placed the passionfruit among the three insufficiently-known tropical fruits having the greatest potential for nectar processing for the European market. It is obvious, then, that in spite of the handicaps of passionfruit culture, the crop offers revenue-earning opportunities for developing countries with low labor costs.

Varieties

The yellow form has a more vigorous vine and generally larger fruit than the purple, but the pulp of the purple is less acid, richer in aroma and flavor, and has a higher proportion of juice—35-38%. The purple form has black seeds, the yellow, brown seeds.

The following are some of the older cultivars as well as some of the more recent:

'Australian Purple', or 'Nelly Kelly'—a purple selection of mild, sweet flavor, grown in Australia and Hawaii.

'Common Purple'—the form growing naturalized in Hawaii; thick-skinned, with small seed cavity, but of fine flavor and low acidity.

'Kapoho Selection'—a cross of 'Sevcik' and other yellow strains in Hawaii. A heavy bearer of large fruits but subject to brown rot; many fruits contain little or no pulp and the juice has the off-flavor of 'Sevcik' though not as pronounced.

'Pratt Hybrid'—apparently a natural cross between the 'Common Purple' and a yellow strain; subject to rot, but juice is of fine color and flavor, low in acid.

'Sevcik Selection'—a golden form of the yellow selected in Hawaii; a heavy bearer, but subject to brown rot and the juice has a peculiar woody flavor.

'University Round Selection'—Hawaiian crosses of 'Waimanalo' and 'Yee'—fruit smaller than 'Yee'; not as attractive but yields 10% more juice of very good flavor.

'University Selection No. B-74'—a Hawaiian hybrid between 'Pratt' and 'C-77', usually yellow, occasionally with red tinges; resembles 'Waimanalo'; has good juice yield and very good flavor.

'Waimanalo Selection'—consists of 4 strains: 'C-54', 'C-77', 'C-80', of similar size, shape, color and very good flavor, and 'C-39' as pollinator.

'Yee Selection'—yellow, round, very attractive, highly disease-resistant, but fruit has thick rind and low yield of juice which is of very good flavor.

What may be a great improvement over any of the above is the cultivar known as **'Noel's Special'**. It is a yellow passionfruit selected in 1968 from open-pollinated seedlings of a vine discovered at an abandoned farm on Hilo, Hawaii, by Noel Fujimoto in the early 1950's. The fruit is round, averages 3.17 oz (90 g); the cavity is filled with dark-orange pulp yielding 43 to 56% bright-orange, richly flavored juice. The vine is vigorous, begins to bear in one year, and is tolerant to brown spot. It produces 88% marketable fruit in a season—a higher proportion than any other cultivar.

In 1967, two purple X yellow hybrids—'3-1' and '3-26', developed at the Redlands Horticulture Research Station, Queensland, had nearly replaced the purple passionfruit in commercial plantations on the coast of southern Queensland and New South Wales. They have a longer fruiting season than the purple, are high-yielding, with high pulp content, keep very well, and meet with little market resistance. Australian breeders continued to strive for a type that would have the needed characteristics and reproduce true from seed. Hybrid '23-E' followed. By 1981, hybrid '3-1' had succumbed to a new, more virulent strain of "woodiness" virus and had to be abandoned. Other popular hybrids are **'Lacey'** and **'Purple-gold'**.

In early 1980, several purple passionfruit hybrids, all insect-pollinated, were introduced into the island of Niue, as possible substitutes for the yellow form cultivated commercially there for export since 1955, with the view of eliminating the labor of hand-pollination required by the yellow for top production. However, the hybrids are more susceptible to mealybug infestation.

One New Zealand grower has exported purple passionfruits to the United States under the trade name of **'Bali Hai'**.

Commercial cultivars of the purple form in Brazil include **'Ouropretano'**, **'Muico'**, **'Peroba'**, and **'Pintado'**; of the yellow form, **'Mirim'** or **'Redondo'**, and **'Guassu'** or **'Grande'**.

In the Cauca Valley of Colombia, the best-performing yellow passionfruit is the **'Hawaiiana'**. Venezuelan growers favor the 'Hawaiiana', 'Brasilera amarilla', and the purple-fruited 'Brasilera rosada'.

A highly promising hybrid, 'M-21471A' has been developed by Dr. R.J. Knight at the United States Department of Agriculture's Subtropical Horticulture Research Station, Miami. The fruit is maroon, weighs about 3 oz (85 g); is close to the purple parent in quality; is self-compatible and resists soil-borne diseases like its yellow parent. F_1 hybrids may be reddish-purple with more conspicuous white dots than on the purple parent, and sometimes there is a tinge of yellow in the background. F_2 hybrids show three variations of purple and are difficult to distinguish from the purple parent.

Pollination

Yellow passionfruit flowers are perfect but self-sterile. In controlled pollination studies at the College of Agriculture of Jaboticabal, Sao Paulo, Brazil, it was found that the yellow passionfruit has three types of flowers according to the curvature of the style: TC (totally curved), PC (partially curved), and SC (upright-styled). TC flowers are most prevalent. Carpenter bees (*Xylocopa megaxylocopa frontalis* and *X. neoxylocopa*) efficiently pollinated TC and PC flowers. Honey bees (*Apis mellifera adansonii*) were much less efficient. Wind is ineffective because of the heaviness and stickiness of the pollen. SC flowers have fertile pollen but do not set fruit. To assure the presence of carpenter bees, it is wise to have decaying logs among the vines to provide nesting places. Carpenter bees will not work the flowers if the nectary is wet. If rain occurs in 1½ hrs after pollination, there will be no fruit set, but if 2 hrs pass before rain falls, it will have no detrimental effect. In the absence of carpenter bees in Fiji, farmers cross-pollinate by hand, treating 600 flowers an hour, with 70% fruit set and 60% of fruit reaching maturity.

The purple form blooms in spring and early summer (July-November) in Queensland and again for a shorter period in fall and early winter (February-April). In Florida, blooming occurs from mid-March through April. The flowers open early in the morning (about dawn) and close before noon, and are self-compatible. The yellow form has one flowering season in Queensland (October-June). In Florida, blooming has occurred from mid-April to mid-November. The flowers open around noon and close about 9 to 10 PM and are self-incompatible.

In crossing the yellow and purple forms, it is necessary to use the purple as the seed parent because the flowers of the yellow are not receptive to the pollen of the purple, and an early-blooming yellow must be utilized in order to have a sufficient overlapping period for pollen transfer. Dr. R.J. Knight has suggested lengthening the overlap by exposing the yellow to artificial light for 6 weeks before the normal flowering season. However, despite the seasonal and hourly differences, natural hybrids between the two forms occur in South Africa, Queensland and in Hawaii. Growers of purple passionfruit in South Africa are warned not to take seed from any vine in proximity to a planting of yellow passionfruit, otherwise the seedlings are apt to produce hybrid fruit of inferior quality.

In some areas, trellis-grown vines of the yellow passionfruit require hand-pollination to assist fruit set. In the home garden, at least two vines of different parentage should be planted and allowed to intertwine for cross-pollination.

Climate

The purple passionfruit is subtropical. It grows and produces well between altitudes of 2,000 and 4,000 ft (650-1,300 m) in India. In Java, it grows well in lowlands but will flower and fruit only above 3,200 ft (1,000 m). In west-central Florida, at 28° N latitude and slightly above sea-level, 3-year-old vines have survived freezing temperatures with the lower 3 ft (.9 m) of the stems wrapped in fiberglass 4 in (10 cm) thick. The upper parts suffered cold injury, were cut back, the vines were heavily fertilized, recovered rapidly and fruited heavily the second summer thereafter.

The yellow passionfruit is tropical or near-tropical. In Western Samoa, it is grown from near sea-level up to an elevation of 2,000 ft (600 m).

Both forms need protection from wind. Generally, annual rainfall should be at least 35 in (90 cm), but in the Northern Transvaal, in South Africa, there is reduced transpiration because of high atmospheric humidity and commercial culture is carried on with precipitation of only 24 in (60 cm). It is reported that annual rainfall in passionfruit-growing areas of India ranges between 40 and 100 in (100-250 cm).

Soil

Passionfruit vines are grown on many soil types but light to heavy sandy loams of medium texture are most suitable, and pH should be from 6.5 to 7.5. If the soil is too acid, lime must be applied. Good drainage is essential to minimize the incidence of collar rot.

Propagation

Passionfruit vines are usually grown from seeds. With the yellow form, seedling variation provides cross-pollination and helps overcome the problem of self-sterility. Some say that the fruits should be stored for a week or two to allow them to shrivel and become perfectly ripe before seeds are extracted. If planted soon after removal from the fruit, seeds will germinate in 2 to 3 weeks. Cleaned and stored seeds have a lower and slower rate of germination. Sprouting may be hastened by allowing the pulp to ferment for a few days before separating the seeds, or by chipping the seeds or rubbing them with fine sandpaper. Soaking, often recommended, has not proved helpful. Seeds are planted ½ in (1.25 cm) deep in beds, and seedlings may be transplanted when 10 in (25 cm) high. If taller—up to 3 ft (.9 m)—the tops should be cut back and the plants heavily watered.

Some growers prefer layers or cuttings of matured wood with 3 to 4 nodes. Cuttings should be well rooted and ready for setting out in 90 days. Rooting may be hastened by hormone treatment. Grafting is an important means of perpetuating hybrids and reducing nematode damage and diseases by utilizing the resistant yellow passionfruit rootstock. If seeds are available in the early spring, seedlings for rootstocks can be raised 4 in (10 cm) apart in rows 24 in (60 cm) apart and the grafted plants will be ready to set out in late summer. If seeds cannot be obtained until late summer, the seedlings are raised and grafted in pots and set out in the spring. Scions from healthy young vines are preferred to those from mature plants. The diameter of the selected scion should match that of the rootstock. Either a cleft graft, whip graft, or side-wedge graft may be made.

If approach-grafting is to be done, a row of potted scions must be placed close alongside the row of rootstocks so that the union can be made at about ¾ of the height of the plant.

Culture

Root-pruning should precede transplanting of seedlings by 2 weeks. Transplanting is best done on a cool, overcast day. The soil should be prepared and enriched organically a month in advance if possible. Grafted vines must be planted with the union well above ground, not covered by soil or mulch, otherwise the disease resistance will be lost. Mounding of the rows greatly facilitates fruit collection.

In plantations, the vines are set at various distances, but studies in Venezuela indicate that highest yields in yellow passionfruit are obtained when the vines are set 10 ft (3 m) apart each way. In South Africa, purple passionfruit vines are set 8 ft (2½ m) apart in cool areas, and 12 to 15 ft (3½–4½ m) apart in warm areas. Spacing of purple passionfruit in Kenya has been 10 ft (3 m) between vines and 6 ft (1.8 m) between rows. Recent 3-year trials of 4 ft (1.2 m) between rows, with light pruning the 2nd and 3rd years, resulted in the highest yield (50% of the crop being borne the first year). But it is recognized that such close planting can lead to disease problems and replanting after the 3rd year.

Commercially, vines are trained to strongly-supported wire trellises at least 7 ft (2.13 m) high. However, for the benefit of the homeowner, it should be pointed out that the yellow passionfruit is more productive and less subject to pests and diseases if allowed to climb a tall tree.

After a vine of either the yellow or purple passionfruit attains 2 years of age, pruning once a year will stimulate new growth and consequently more flower and fruit production. The average life of a plantation in Fiji is only 3 years. Judicious pruning of lateral branches after fruiting aids in disease control and can extend plantation life to 5 or 6 years. In South Africa, at elevations between 4,000 and 4,800 ft (1,200–1,460 m), plantations are kept in full production for as long as 8 years.

Regular watering will keep a vine flowering and fruiting almost continuously. Least flowers develop during the winter season due to short day length. Water requirement is high when fruits are approaching maturity. If soil is dry, fruits may shrivel and fall prematurely. Fertilizer (10-5-20 NPK) should be applied at the rate of 3 lbs (1.36 kg) per plant 4 times a year, under normal conditions. In India, trials of purple passionfruit on red sandy loam with a pH of 6.5 and high organic content, the optimum fertilizer treatment was found to be 290 lbs (132 kg) N and 69½ lbs (31.6 kg) P per ha per year. French horticulturists have reported that, in plantations on the Ivory Coast, annual supplements of 8 oz (220 g) urea and 7½ oz (210 g) potassium sulfate per plant per year of age will have a highly favorable effect on production. It is said that 32 to 36 oz (900–1,000 g) of nitrogen are required to produce 66 lbs (30 kg) of fruits, but excessive nitrogen will cause premature fruit drop. Passionfruit vines should always be watched for deficiencies, particularly in potassium and calcium, and of less importance, magnesium.

The passionfruit vine, especially the yellow, is fast-growing and will begin to bear in 1 to 3 years. Ripening occurs 70 to 80 days after pollination. Injuries to the base of the vine, which allow entrance of disease organisms, can be avoided by hand-weeding or the application of herbicides around the main stems. These practices will also protect the shallow root system. In Surinam, good weed control under trellises has been achieved by covering the soil with black plastic.

Seasons and Harvesting

The different flowering seasons of the purple and yellow passionfruits have been mentioned under "Pollination". In some areas, as in India, the vines bear throughout the year but peak periods are, first, August to December, and, second, March to May. At the latter time, the fruits are somewhat smaller, with less juice. In Hawaii, passionfruits mature from June through January, with heaviest crops in July and August and October and November. With variations according to cultivar, and with commercial cultivation both above and below the Equator, there need never be a shortage of raw material for processing.

Ripe fruits fall to the ground and will roll in between mounded rows. They do not attract flies or ants but should be collected daily to avoid spoilage from soil organisms. In South Africa, they are subject to sunburn damage on the ground and, for that reason, are picked from the vines 2 or 3 times a week in the summertime before they are fully ripe, that is, when they are light-purple. At this stage, they will reach the fresh fruit market before they wrinkle. In winter, only one picking per week is necessary. For juice processing, the fruit is allowed to attain a deep-purple color. In India and Israel the fruits are always picked from the vine rather than being allowed to fall. It has been found that fallen fruits are lower in soluble solids, sugar content, acidity and ascorbic acid content.

The fruits should be collected in lugs or boxes, not in bags which will cause "sweating". If not sent immediately to processing plants, the fruits should be spread out on wire racks where there will be good air circulation.

Yield

Many factors influence the yield of passionfruit vines. In general, yields of commercial plantations range from

20,000 to 35,000 lbs per acre (roughly the same number of kg per ha). In Fiji, with hand pollination, 173 acres (70 ha) will yield 33 tons (30 MT) of fruits. Hybrids in Australia have raised yields far beyond those obtained with the purple passionfruit.

On the average, a bushel of passionfruits in Australia weighs 36 lbs (16 kg); yields 13⅓ lbs (6 kg) of pulp from which is obtained 1 gal (3.785 liters) — that is 10.7 lbs (4.5 kg) of juice, and 2.6 lbs (1.18 kg) of seeds. With some strains, the juice yield is much higher.

Storage

Underripe yellow passionfruits can be ripened and stored at 68°F (20°C) with relative humidity of 85 to 90%. Ripening is too rapid at 86°F (30°C). Ripe fruits keep for one week at 36° to 45°F (2.22°-7.22°C). Fruits stored in unperforated, sealed, polyethylene bags at 74°F (23.1°C), have remained in good condition for 2 weeks. Coating with paraffin and storing at 41° to 44.6°F (5° to 7°C) and relative humidity of 85 to 90%, has prevented wrinkling and preserved quality for 30 days.

Pests and Diseases

In Hawaii and Australia, infestations of the passion vine mite (*Brevipalpus phoenicis*) occur during dry weather in the warm season, defoliate the younger portions of the vines but not the terminus, and make brown blemishes on the fruits. The passion vine bug (*Leptoglossus australis*) feeds on flowers and young, green fruits in Queensland. The green vegetable bug, or stinkbug, (*Nezara viridula*) is a similar but lesser menace to the plant and young fruits. Both the immature and the adult stages suck the sap of the growing tips, as do the brown stinkbug (*Boerias maculata*), the large black stinkbug (*Anoplocnemis* sp.) and the small black stinkbug (*Leptoglossus membranaceus*). In Florida, the yellow passionfruit is commonly found to be superficially punctured by a stinkbug (*Chrondrocera laticornis*), affecting only its appearance. Thrips (*Thysanoptera* sp.) injure and cause stunting of young seedlings in nurseries. In dry weather, they also feed on leaves and fruits, leaving them defaced and prone to shrivel and fall prematurely. In East Africa, injury from the tobacco white fly (*Bemisia tabaci*) may lead to galls on the leaves. Leaf beetles (*Haltica* sp.) and weevils (*Systates* spp.) chew the foliage, and cutworms behead seedlings in nurseries. Two lepidopterous pests, *Dione*, or *Agraulis*, *vanillae* and *Mechanitis variabilis* are common in Colombia.

Among scales attacking the vine and petioles, white peach scale (*Pseudaulacaspis pentagona*) is most troublesome in Queensland. Not as prevalent are round purple scale (*Chrysomphalus ficus*) and granadilla purple scale (*Parasaissetia nigra*). These pests may cause dieback of the entire plant if not controlled. Red scale (*Aonidiella aurantii*) is common on mature passion vines in Queensland. Soft brown scale (*Coccus hesperidum*) is occasionally troublesome. The passion vine leaf hopper (*Scolypopa australis*) requires protective measures. The citrus mealybug (*Planococcus citri*) is a major Queensland pest in summer. Spraying, unfortunately, kills its chief predator, the mealybug ladybird, *Cryptolaemus montrouzieri*. The aphids, *Aphis gossypii* and *Myzus persicae*, transmit the virus which causes "woodiness" (see below).

There has been no report of attack by the Caribbean fruit fly (*Anastrepha suspensa*) in Florida, though *Anastrepha* infestation was on one occasion observed by Curtis Dowling in *Passiflora* fruits in Costa Rica. In Brazil, fruit flies of the genus *Anastrepha*, and in Hawaii the Oriental fruit fly and the melon fly, deposit eggs in the very young, tender fruits. In these, the larvae seem able to develop and cause the immature fruits to shrivel and fall. If fruits are punctured when nearly mature, the only effect is an external scar. The same is reported concerning the dominant Queensland fruit fly (*Dacus tryoni*) and the less common Mediterranean fruit fly (*Ceratitis capitata*) in Australia.

In South Africa, purple passionfruit vines are damaged by several species of nematodes. The most important, which causes extreme thickening of the roots, is the root-knot nematode, *Meloidogyne javanica*. Others include the spiral nematode (*Scutellonema truncatum* and *Helicotylenchus* sp.), and the lesion nematode (*Pratylenchus* sp.). The yellow passionfruit is nematode-resistant.

The main diseases of purple passion fruit in Australia are brown spot, *Septoria* spot and base rot, *Phytophthora* blight, *Fusarium* wilt, woodiness, and damping-off. Brown spot, caused by *Alternaria passiflorae* in warm weather, is a major affliction of the purple passionfruit also in New Zealand and East Africa. In Hawaii, brown spot is the leading disease of the yellow passionfruit and *A. tenuis* was found to be the dominant species associated with the disease in 1969. *A. macrospora* has occasioned severe leaf spot and branch lesions in India. A similar disease causing spotting and crinkling of leaves and fruit first appeared in Ceylon in 1970. Septoria spot, from the fungus *Septoria passiflorae*, most common in summer and fall, is evidenced by more numerous and smaller spots than brown spot, on all parts of the vine and on the fruits, and it is spread by rain, dew and overhead irrigation. Some believe this fungus to be also the source of base rot, often induced by injury from mowers or other mechanical equipment.

Phytophthora cinnamoni, the source of collar rot in Fiji, makes it necessary to replace yellow passionfruit plantings there every 30 to 35 months. *P. nicotinae* var. *parasitica* has been linked to fatal blight, or stem rot, and fruit rot in purple passionfruit vine, but not in the yellow, in wet periods of summer and fall in Queensland and South Africa. *P. cinnamoni* and *P. nicotinae* are responsible for root rot in New Zealand and Western Australia, and the latter is identified with wilt in South Africa and Sarawak, and with damping-off and leaf blight in both the purple and the yellow passionfruits in India.

Fusarium wilt, arising from the soil-borne fungus, *Fusarium oxysporum* f. sp. *passiflorae*, can be reduced only by grafting the purple, or, better still, purple-yellow hybrids, onto the *Fusarium*-resistant yellow passionfruit rootstock. However, Bedoya *et al.* have reported that, in the zones of Palmira, Cerrito and Ginebra of the Cauca

Valley of Colombia, but not in the zone of Unión, collar rot limits the life of yellow passionfruit plantations to 3 years, and they found, in inoculation experiments, that *Fusarium solani* produced the symptoms. The first signs are chlorosis, necrosis and defoliation; next there is splitting of the trunk and separation of the bark. The root becomes progressively discolored and red rays extend to the surface of the soil.

Nectria haematococca, or *Hypomyces solani*, the ascogenous state of *Fusarium solani*, has been determined to be the organism girdling the collar zone and bringing on sudden wilt of the purple passionfruit vine in Uganda.

The virus disease, "woodiness", or "bullet", appearing as small misshapen fruits with thick rind and small pulp cavity, has been the most serious plague of the purple passionfruit in Australia and East Africa, but it has little effect on the yellow form. The "woodiness" virus (PWV) is also the source of tip blight in the coastal districts of central Queensland. This virus has a wide host range, not only in the genus *Passiflora*, but also weedy species in the families Amaranthaceae, Chenopodiaceae, Cucurbitaceae and Solanaceae.

There are a number of different strains of the "woodiness" virus. For many years, inoculation of passionfruit vines with mild strains protected them from further infection, and commercial hybrids containing small doses of mild strains were released to farmers. But, in 1978, a new, more virulent, strain of virus appeared and overcame the "mild strain protection". The New South Wales Passionfruit Growers Association, in response to this new threat, established, in 1979, a Passionfruit Scion Accreditation Scheme to "improve the quality of planting material by field selection and provide scionwood free of the severe strain of woodiness virus", for a standard fee. Generally, 100 scions can be taken from each accredited vine in a season. By 1981, 16,000 scions had been supplied to commercial growers.

In 1973, two mosaic viruses—PPMV-K and PFMV-MY—said to differ from other reported *Passiflora* viruses, were found to be prevalent in commercial plantings of the yellow passionfruit in the Bantung district of Selangor, Malaya. Damping-off is caused by *Rhizoctonia solani* and *Pythium* spp. in Queensland. Thread blight of yellow passionfruit vine in Fiji and Western Samoa, seen as patches of black, papery, shredded leaves with gray to tan layer of merged "threads" beneath, has been attributed to *Rhizoctonia solani* (also called *Thanatephorus cucumeris*). It may invade the entire vine.

Food Uses

The fruit is of easy preparation. One needs only cut it in half lengthwise and scoop out the seedy pulp with a spoon. For home use, Australians do not trouble to remove the seeds but eat the pulp with cream and sugar or use it in fruit salads or in beverages, seeds and all. Elsewhere it is usually squeezed through two thicknesses of cheesecloth or pressed through a strainer to remove the seeds. Mechanical extractors are, of course, used industrially. The resulting rich juice, which has been called a

natural concentrate, can be sweetened and diluted with water or other juices (especially orange or pineapple), to make cold drinks. In South Africa, passionfruit juice is blended with milk and an alginate; in Australia the pulp is added to yogurt. After primary juice extraction, some processors employ an enzymatic process to obtain supplementary "secondary" juice from the double juice sacs surrounding each seed. The high starch content of the juice gives it exceptional viscosity. To produce a freeflowing concentrate, it is desirable to remove the starch by centrifugal separation in the processing operation.

Passionfruit juice can be boiled down to a sirup which is used in making sauce, gelatin desserts, candy, ice cream, sherbet, cake icing, cake filling, meringue or chiffon pie, cold fruit soup, or in cocktails. The seeded pulp is made into jelly or is combined with pineapple or tomato in making jam. The flavor of passionfruit juice is impaired by heat preservation unless it is done by agitated or "spin" pasteurization in the can. The frozen juice can be kept without deterioration for 1 year at 0°F (-17.78°C) and is a very appealing product. The juice can also be "vacuum-puff" dried or freeze-dried. Swiss processors have marketed a passionfruit-based soft drink called "Passaia" for a number of years in Western Europe. Costa Rica produces a wine sold as "Parchita Seco."

Food Value Per 100 g of Edible Portion (Purple passionfruit, pulp and seeds)*	
Calories	90
Moisture	75.1 g
Protein	2.2 g
Fat	0.7 g
Carbohydrates	21.2 g
Fiber	?
Ash	0.8 g
Calcium	13 mg
Phosphorus	64 mg
Iron	1.6 mg
Sodium	28 mg
Potassium	348 mg
Vitamin A	700 I.U.
Thiamine	Trace
Riboflavin	0.13 mg
Niacin	1.5 mg
Ascorbic Acid	30 mg
*According to U.S. Dept. Agr., ARS.	

The yellow passionfruit has somewhat less ascorbic acid than the purple but is richer in total acid (mainly citric) and in carotene content. It is an excellent source of niacin and a good source of riboflavin. Free amino acids in purple passionfruit juice are: arginine, aspartic acid, glycine, leucine, lysine, proline, threonine, tyrosine and valine. Carotenoids in the purple form constitute 1.160%; in the yellow, 0.058%; flavonoids in the purple, 1.060%; in the yellow, 1.000%; alkaloids in the purple, 0.012%; in the yellow, 0.700% (mainly harman), and the juice is slightly sedative. Starch content of purple passionfruit juice is 0.74%; of the yellow, 0.06%.

Toxicity

A cyanogenic glycoside is found in the pulp of passionfruits at all stages of development, but is highest in very young, unripe fruits and lowest in fallen, wrinkled fruits, the level in the latter being so low that it is of no toxicological significance.

Other Uses

Commercial processing of the yellow passionfruit yields 36% juice, 51% rinds, and 11% seeds.

Rind: The rinds have a very low pectin content—only 2.4% (14% on a dry weight basis). Nevertheless, it has been determined in Fiji that extraction of pectin from the rinds—up to 5 tons (4.5 MT) annually—reduces the otherwise burdensome problem of waste disposal. The rind residue contains about 5 to 6% protein and could be used as a filler in poultry and stock feed. In Brazil, pectin is extracted from the purple form which has a better quality pectin than that in the yellow. In Hawaii, the pectin is not extracted. Instead, the rinds are chopped, dried, and combined with molasses as cattle or pig feed. They can also be converted into silage.

Seeds: The seeds yield 23% oil which is similar to sunflower and soybean oil and accordingly has edible as well as industrial uses. Up to 3,400 gallons (13,000 liters) can be obtained per year in Fiji. The seed meal contains about 12% protein and 50 to 55% fiber. It has been judged unsuitable for cattle feed.

Analyses of the fresh rind show: moisture, 78.43-85.24%; crude protein, 2.04-2.84%; fat, 0.05-0.16%; crude starch, 0.75-1.36%; sugars (sucrose, glucose, fructose), 1.64%; crude fiber, 4.57-7.13%; phosphorus, 0.03-0.06%; silica, 0.01-0.04%; potassium, 0.60-0.78%; organic acids (citric and malic), 0.15%; ascorbic acid, 78.3-166.2%. The outer skin of the purple form contains 1.4 mg per 100 g of the anthocyanin pigment, pelargonidin 3-diglucoside. There is also some tannin.

The composition of the air-dried seeds is reported as: moisture, 5.4%; fat, 23.8%; crude fiber, 53.7%; protein, 11.1%; N-free extract, 5.1%; total ash, 1.84%; ash insoluble in HC1, 0.35%; calcium, 80 mg; iron, 18 mg; phosphorus, 640 mg per 100 g.

The seed oil contains 8.90% saturated fatty acids; 84.09% unsaturated fatty acids. The fatty acids consist of: palmitic, 6.78%; stearic, 1.76%; arachidic, 0.34%; oleic, 19.0%; linoleic, 59.9%; linolenic, 5.4%.

Medicinal Uses: There is currently a revival of interest in the pharmaceutical industry, especially in Europe, in the use of the glycoside, *passiflorine*, especially from *P. incarnata* L., as a sedative or tranquilizer. Italian chemists have extracted *passiflorine* from the air-dried leaves of *P. edulis*.

In Madeira, the juice of passionfruits is given as a digestive stimulant and treatment for gastric cancer.

Giant Granadilla (Plate XLV)

The largest fruit in its genus, the giant granadilla, *Passiflora quadrangularis* L. (syn. *P. macrocarpa* M.T. Mast.), is often called merely *granadilla*, or *parcha*, Spanish names loosely applied to various related species; or it may be distinguished as *granadilla real, grandadilla grande, parcha granadina* or *parcha de Guinea*. In El Salvador, it is known as *granadilla de fresco* or *granadilla para refrescos;* in parts of Colombia, it is *badea* or *corvejo;* in the State of Tachira, Venezuela, *badea;* in Bolivia, *granadilla real* or *sandia de pasión*. In Brazil, it is *maracuya–acu, maracuja-assu, maracuja silvestre, maracuya grande, maracuja suspiro, maracuja mamao,* or *maracuja de caiena*. In Surinam, it is *grote* or *groote markoesa;* in Peru and Ecuador, *tumbo* or *tambo*. In the Philippines, its local names are *parola, kasaflora,* and square-stemmed passion flower. To Indonesians, it is familiar as *markiza, markoesa, markeesa,* or *manesa*, and to the Malays, *timun belanda, marquesa* or *mentimun*. In Thailand, it is *su–khontha–rot;* in Vietnam, *dua gan tay,* or *barbadine,* the French name.

Description

The vine is fast-growing, large, coarse, herbaceous but woody at the base, arising from a fleshy root that becomes enlarged with age, and climbing trees to a height of 33 to 50 ft (10-15 m) or even 150 ft (45 m) in Java. It has thick, 4-angled stems prominently winged on the angles, and axillary tendrils to 12 in (30 cm) long, flanked by leaflike, ovate or ovate-lanceolate stipules ¾ to 1⅜ in (2-3.5 cm) long, sometimes faintly toothed. The alternate leaves are broad-ovate or oblong-ovate, 3¼ to 6 in (8.25-15 cm) wide, 4 to 8 in (10-20 cm) long; rounded or cordate at the base, abruptly pointed at the apex, sometimes toothed near the base; thin, with conspicuous veins sunken on the upper surface, prominent beneath. The solitary, fragrant flowers, up to 4¾ or 5 in (12-12.5 cm) wide, have a bell-shaped calyx, the 5 sepals greenish or reddish-green on the outside, white, pink or purple inside; the 5 petals, to 1¾ in (4.5 cm) long, white-and-pink; the corona filaments 2-ranked, to 2⅜ in (6 cm) long, purple-and-white below, blue in the middle, and pinkish-blue above, around the typical complex of pistil, style and stigmas.

The pleasantly aromatic, melon-like fruit is oblong-ovoid, 4¾ to 6 in (12-15 cm) wide, and 8 to 12 in (10-30 cm) long; may be faintly ribbed or longitudinally 3-lobed; has a thin, delicate skin, greenish-white to pale- or deep-yellow, often blushed with pink. Beneath it is a layer of firm, mealy, white or pink flesh, 1 to 1½ in (2.5-4 cm) thick, of very mild flavor, and coated with a parchment-like material on the inner surface. The central cavity contains some juice and masses of whitish, yellowish, partly yellow or purple-pink, sweet-acid arils (commonly referred to as the pulp), enclosing flattened-oval, purplish-brown seeds to ½ in (1.25 cm) long.

Origin and Distribution

The giant granadilla is generally agreed to be a native of tropical America, though the actual place of origin is unknown. It was growing in Barbados in 1750 and is present in several other Caribbean Islands and in Bermuda. It is commonly cultivated, and sometimes an escape from cultivation or truly wild, from Mexico to Brazil and Peru. At some point in the 18th Century, it was introduced into Malaya, where it thrives in both the north and the south. In Vietnam, it is limited to the southern half of the country. Perhaps it had reached Indonesia earlier, for it is more common and even naturalized there. It is also cultivated in the lowlands of India, Ceylon and the Philippines; in tropical Africa, and throughout Queensland, Australia. In tropical North Queensland it has run wild, growing lushly in jungle areas. It flourishes and fruits heavily especially in the Cairns district. It was being grown in Hawaii in 1888 and by 1931 had become naturalized in moist places. The United States Department of Agriculture received seeds from Trinidad in 1909 and the vine is very occasionally planted in southern Florida, but is too cold-sensitive to survive in California.

Varieties

There are various strains producing fruits of different sizes and quality. Wester stated that some are insipid, while one of superior flavor had originated at Cotabato. One strain with especially large fruits and good flavor was formerly considered a separate species (*P. macrocarpa*), but it hybridizes readily with smaller strains and there are intermediate types. An ornamental form, 'Variegata', has leaves splashed with yellow.

Pollination

The vine may produce few or no fruits in a dry atmosphere, or in the absence of insect pollinators. Also the pollen may ripen before the stigma is ready to receive it, and, at times, bees may steal the pollen too early in the morning. Hand-pollination is regularly practiced in Queensland and has been successful in limited experiments in Florida. It should be done in the late morning, no later than 4 to 6 hours after the flowers open.

Climate

The ideal climate for the giant granadilla is one that is truly tropical, warm both day and night, with little fluctuation, and with high humidity. It is grown between 700 and 1,500 ft (213 and 457 m) elevations in Jamaica and Hawaii, and up to 3,000 ft (914 m) in India; to 5,000 or, at most, 7,200 ft (1,800 or 2,200 m) in Ecuador. Vines several years old have been killed by winter cold on the Riviera.

Soil

For maximum growth and productivity, the vine requires deep, fertile, moist but well-drained soil. Australians have observed good growth on volcanic, alluvial, and sandy soil, and even decomposed granite. Vines planted in highly alkaline situations in Israel have died after evidencing acute chlorosis.

Propagation

The giant granadilla grows readily from seeds, which germinate in 2 to 3 weeks and the seedlings can be set out when 6 to 12 in (15-30 cm) high. Cuttings of mature wood 10 to 12 in (25-30 cm) or even 2 to 3 ft (.6-.9 m) long, are partially defoliated and deeply planted in well-watered sand. There will be sufficient vegetative growth and root development to permit transplanting in 30 days. Air- or ground-layers are also satisfactory.

Culture

In commercial plantings in Indonesia, the vines are set 6.5 to 10 ft (2-3 m) apart each way. When the plants reach about 6.5 ft (2 m) in height, they must be trained to a strong, horizontal trellis. Pruning may be necessary if the growth becomes too dense. Regular applications of fertilizer high in organic matter, and copious watering are necessary.

Harvesting and Yield

In Indonesia and Queensland, a productive vine will fruit more or less continually all year and the annual yield may range from 25 to 35 fruits in the larger types to 70 to 120 fruits in medium to small types. Venezuelan horticulturists report that their main blooming period is May to October and the fruits ripen in 62 to 85 days from flower-opening, the crop being harvested mainly from July through October. The yield of 2- to 3-year-old vines varies from 16 to 50 fruits. The fruits are ready for harvesting when the skin becomes translucent and glossy and is beginning to turn yellowish at the apex. It is clipped from the vine. Very careful handling and packing are essential.

Pests and Diseases

Young plants in nurseries may be severely defoliated by *Disonycha glabrata* in Venezuela.

In Queensland, the principal pest of the giant granadilla is the green vegetable bug, *Nezara viridula*, which punctures young fruits and sucks out the juice, causing them to wither and fall; or hard lumps will form in the flesh. To avoid damage by fruit flies, the fruits are sometimes bagged.

Leaf spot, from fungal infection, occurs occasionally in Queensland but it is considered of little importance. Stem-end rot in East Africa has been attributed to the fungus, *Botryodiplodia theobromae*.

Food Uses

The flesh of the ripe fruit, with the inner skin removed, is cut up and added to papaya, pineapple and banana slices in fruit salads, seasoned with lemon or lime juice. It is cooked with sugar and eaten as dessert, or is canned in sirup; sometimes candied; but it is so bland that it needs added flavoring. In Indonesia, the flesh and arils are eaten together with sugar and shaved ice. Australians add a little orange juice and usually serve the dish with cream. They also use the stewed flesh and raw arils together as pie filling. The whole arils can be eaten raw without removing the seeds.

Food Value Per 100 g of Edible Portion*		
	Thick Flesh	*Arils and Seeds*
Moisture	94.4 g	78.4 g
Protein	0.112 g	0.299 g
Fat	0.15 g	1.29 g
Crude Fiber	0.7 g	3.6 g
Ash	0.41 g	0.80 g
Calcium	13.8 mg	9.2 g
Phosphorus	17.1 mg	39.3 mg
Iron	0.80 mg	2.93 mg
Carotene	0.004 mg	0.019 mg
Thiamine	–	0.003 mg
Riboflavin	0.033 mg	0.120 mg
Niacin	0.378 mg	15.3 mg
Ascorbic Acid	14.3 mg	

*According to analyses made in El Salvador.

Jelly can be made from the unpeeled flesh boiled for 2 hours and the pulp simmered separately. The juice strained from both is combined and, with added sugar and lemon juice, is boiled until it jells.

The pulp (arils) yields a most agreeable juice for cold drinks. It is bottled in Indonesia and served in restaurants. Wine is made in Australia by mashing several of the whole ripe fruits, adding sugar and warm water and allowing the mix to ferment for 3 weeks, adding 2 pints of brandy, and letting stand for 9 to 12 months.

The young, unripe fruit may be steamed or boiled and served as a vegetable, or may be cut up, breaded and cooked in butter with milk, pepper and nutmeg. In Java ripe fruits are scarce because of squirrels and other predators.

The root of old vines is baked and eaten in Jamaica as a substitute for yam.

Toxicity

The leaves, skin and immature seeds contain a cyanogenic glycoside. The pulp contains passiflorine and, if indulged in excessively, causes lethargy and somnolence. The raw root is said to be emetic, narcotic and poisonous.

Medicinal Uses

The fruit is valued in the tropics as antiscorbutic and stomachic. In Brazil, the flesh is prescribed as a sedative to relieve nervous headache, asthma, diarrhea, dysentery, neurasthenia and insomnia. The seeds contain a cardiotonic principle, are sedative, and, in large doses, narcotic. The leaf decoction is a vermifuge and is used for bathing skin afflictions. Leaf poultices are applied in liver complaints. The root is employed as an emetic, diuretic and vermifuge. Powdered and mixed with oil, it is applied as a soothing poultice.

Sweet Granadilla (Plate XLVI)

Ranking close to *Passiflora edulis* in popular appeal and potential, the sweet granadilla, *P. ligularis* Juss., is also known as *granadilla* (Bolivia, Costa Rica, Ecuador, Mexico, Peru); *granadilla común* (Guatemala); *granadilla de China* or *parchita amarilla* (Venezuela); and *granadita* (Jamaica).

Description

The vine is a vigorous, strong grower, woody at the base, climbing by tendrils, topping the highest trees, shading out and killing the understory. Its leaves are broadly heart-shaped, pointed at the apex, 3³/₁₆ to 8 in (8-20 cm) long, 2³/₈ to 6 in (6-15 cm) wide, conspicuously veined, medium-green on the upper surface, pale-green with a bloom on the underside. Spaced along the petiole, are 3 pairs of hairlike glands about ³/₈ in (1 cm) long. At the leaf axils, there are paired, leaflike stipules, ovate-oblong and about 1 in (2.5 cm) long and a little over ½ in (1.25 cm) wide; more or less finely toothed.

The flowers, sweet and musky in odor, usually 2 to a node, may be 4 in (10 cm) across, on a 1½ in (4 cm) peduncle bearing 3 leaflike, ovate-oblong, pointed bracts, 1½ in (4 cm) long and 1 in (2.5 cm) wide, faintly toothed. The sepals are greenish-white, lanceolate; the petals pinkish white; the filaments, in 2 rows, white, horizontally striped with purple-blue.

The fruit is broad-elliptic, 2³/₈ to 3 in (6-7.5 cm) long, green with purple blush on sunny side and minutely dotted when unripe, orange-yellow with white specks when ripe. The rind is smooth, thin, hard and brittle externally, white and soft on the inside. The pulp (arils) is whitish-yellow or more or less orange, mucilaginous, very juicy, of sprightly, aromatic flavor, and encloses numerous black, flat, pitted, crisp but fairly tender seeds.

Origin and Distribution

The sweet granadilla is the common species of *Passiflora* ranging from central Mexico through Central America and western South America, through western Bolivia to south-central Peru. Throughout this region, it is popular and abundant in the markets.

It has been grown in Hawaii since late in the 19th Century. In 1916, the United States Department of Agriculture received seeds from Quito, Ecuador. The vine is not suited to California, has been grown in greenhouses in Florida but has never survived for long. Northern gardeners sometimes plant it as a summer ornamental. It is not reported in Guam; may be grown to some extent in New Guinea. Trial plantings in Israel were killed by cold weather. It is cultivated and naturalized in Jamaica and, in recent years, has been blooming and fruiting prolifically in mountainous Haiti.

Climate

The sweet granadilla is subtropical, not tropical. In its natural range, it is wild and cultivated at elevations of 3,000 to 8,850 ft (900-2,700 m). In Hawaii, it finds sufficiently cool temperatures at 3,000 ft (900 m). In Jamaica, the vine volunteers freely at altitudes between 3,500 and 4,000 ft (1,000-1,200 m). At 5,000 to 8,200 ft (1,500-2,500 m) in Colombia, the vine fruits well. At higher altitudes, it flourishes and blooms but will not fruit. An elevation of 6,000 ft (1,828 m), where the clouds descend on peaks in the afternoon, has proven ideal in Haiti. The vine is intolerant of heat. It will do well over the winter in Florida but declines with the onset of hot weather.

Soil

Thin, volcanic soils do not discourage the sweet granadilla, providing they are moist. It is naturally adapted to high rainforests.

Propagation

The sweet granadilla can be grown from seeds or cuttings.

Season and Keeping Quality

There is but one crop per year. In Bolivia, the fruits ripen in May and June. The fruit, despite its hard shell, has poor keeping quality, deteriorating soon after the harvest.

Pests

In Haiti, the planted seeds are often devoured by rodents, though the seeds of *P. edulis* in the same situation have never been disturbed. Squirrels ravage the crop in the forests of Ecuador.

Food Uses

Usually, the fruit is cracked open and the pulp and seeds consumed out-of-hand. For the table, the fruit is cut in half and the contents are eaten with a spoon. The strained juice is much used for making cold drinks and sherbet (ice).

Food Value Per 100 g of Edible Portion*	
	Pulp and Seeds Combined
Moisture	69.9-79.1 g
Protein	0.340-0.474 g
Fat	1.50-3.18 g
Crude Fiber	3.2-5.6 g
Ash	0.87-1.36 g
Calcium	5.6-13.7 mg
Phosphorus	44.0-78.0 mg
Iron	0.58-1.56 mg
Carotene	0.00-0.035 mg
Thiamine	0.00-0.002 mg
Riboflavin	0.063-0.125 mg
Niacin	1.42-1.813 mg
Ascorbic Acid	10.8-28.1 mg

*Analyses made in Ecuador, El Salvador, Costa Rica and Guatemala.

Water Lemon

One of the best of the lesser-known passionfruit relatives, the water lemon, *Passiflora laurifolia* L., is also known as bell-apple, sweet cup, yellow granadilla, Jamaica honeysuckle, vinegar pear, golden apple, where English is spoken; as *pomme d'or, pomme liane,* or *pomme de liane, Marie-Tambour,* or *maritambou,* in the French West Indies; as *parcha, parcha de culebra,* or *pasionaria con hojas de laurel* in Spanish. In the Portuguese language, in Brazil, it is called *maracuja comum* or *maracuja laranja.* It is *paramarkoesa* in Surinam. In Malaya, it is *markusa leutik, buah susu, buah belebar,* or *buah selaseh;* in Thailand, *sa-wa-rot;* in Vietnam, *guoi tay.*

Description

The water lemon vine is a moderately vigorous climber, to 32 ft (10 m) or more, its twining, more or less woody or wiry stems longitudinally grooved and bearing slender, tough tendrils in the leaf axils flanked by 2 slim, green stipules. The alternate leaves are oblong-ovate or elliptical, rounded at the base, abruptly pointed at the apex; 6 to 8 in (15-20 cm) long, 1⅓ to 3⅛ in (3.4-8 cm) wide; thick and leathery. The fragrant, solitary, 5-petalled flowers, 3 to 4 in (7.4-10 cm) across, have a bell-shaped calyx, oblong, red or purple-red sepals and petals, and corona filaments 6-ranked, banded with red, blue, purple and white. The fruit is ellipsoidal or ovoid, 2 to 3⅛ in (5-8 cm) long, 1½ to 2⅜ in (4-6 cm) wide; orange-yellow; clasped at the base by 3 large, green, leaflike bracts, toothed and edged with conspicuous glands. The rind is leathery, to ⅛ in (3 mm) thick, white and spongy within; becomes hard when dry. Pleasantly rose-scented, the translucent, nearly white pulp is juicy, mucilaginous and of agreeable, subacid flavor, and encloses numerous seeds, flat and minutely ribbed.

Origin and Distribution

The water lemon is native to tropical America and common, wild and cultivated from southern Venezuela, Surinam, Guyana and French Guiana down through the Amazon region of Brazil to Peru. In the dry season, the fruits are regularly sold in local markets. The vine is cultivated and naturalized from Trinidad and Barbados to Jamaica, Puerto Rico, Hispaniola and Cuba. In Bermuda, it is only occasionally grown. It was introduced into Malaya in the 18th Century; is commonly cultivated in the lowlands and naturalized in Singapore and Penang. According to Petelot, the water lemon is grown in Thailand and throughout the southern half of Vietnam. In India, Ceylon and Hawaii, the vine is grown as an ornamental but rarely fruits except in hot, dry situations where the pollen is dry enough to be naturally transmitted. There are only a few specimens in Florida.

Pollination

The water lemon flowers open only in the afternoon, and apparently are not self-pollinated, or only slightly so. Cross-pollination is required for good crops. If carpenter bees are not present at the right time, the pollen must be transferred by hand.

Climate

A warm, dry atmosphere is essential for early ripening of the stigmas. On Oahu, Hawaii, best yields have been obtained at sea-level, though the vine grows vigorously up to 1,500 ft (457 m).

Soil

The vine has grown and flowered well on sand and on limestone in Florida.

Propagation

The water lemon grows readily from seeds or cuttings.

Pests

Trials have shown that the vine is fairly resistant to rootknot nematodes in Florida.

Food Uses

Children and adults make a hole in one end of the fruit and suck out the pulp and seeds for refreshment. The juice of the strained pulp makes an excellent beverage.

Food Value

The pulp contains 1.55 mg of pantothenic acid per 100 g; the rind, 1.87 mg. This element belongs to the vitamin B complex group and is sometimes called vitamin B_5.

Toxicity

The rind, leaves and seeds contain a cyanogenic glycoside. On the other hand, the leaves possess 387 mg, per 100 g, ascorbic acid. The leaf decoction is taken as a vermifuge. The seeds have a sedative action on the nervous system and heart and, in strong doses, are hypnotic. The root acts as a very potent vermifuge.

Banana Passionfruit

A distinctive and much admired passionfruit relative, *Passiflora mollissima* Bailey (syns. *P. tomentosa* var. *mollissima* Tr. & Planch.; *Tacsonia mollissima* HBK.), was given this appealing and appropriate English name in New Zealand. In Hawaii, it is called banana *poka.* In its Latin American homeland, it is known as *curuba, curuba de Castilla,* or *curuba sabanera blanco* (Colombia); *tacso, tagso, tauso* (Ecuador); *parcha* (Venezuela), *tumbo* or *curuba* (Bolivia); *tacso, tumbo, tumbo del norte, trompos,* or *tintin* (Peru).

Description

The vine is a vigorous climber to 20 or 23 ft (6-7 m), its nearly cylindrical stems densely coated with yellow hairs. Its deeply 3-lobed leaves, 3 to 4 in (7.5-10 cm) long and 2⅜ to 4¾ in (6-12 cm) wide, are finely toothed and downy above, grayish- or yellowish-velvety beneath. The stipules are short, slender and curved. The attractive blossom has a tube 3 to 4 in (7.5-10 cm) long, gray-green, frequently blushed with red, rarely downy; corolla with 5 oblong sepals and deep-pink petals flaring to a width of 2 to 3 in (5-7.5 cm); and a rippled, tuberculated, purple corona. The fruit is oblong or oblong-ovoid, 2 to 4¾ in (5-12 cm) long, 1¼ to 1½ in (3.2-4 cm) wide. The rind is thick, leathery, whitish-yellow or, in one form, dark-green, and minutely downy. Very aromatic pulp (arils), salmon-colored, subacid to acid and rich in flavor, surrounds the small, black, flat, elliptic, reticulated seeds.

Origin and Distribution

The banana passionfruit is native and commonly found in the wild in Andean valleys from Venezuela and eastern Colombia to Bolivia and Peru. It is believed to have been domesticated only shortly before the Spanish Conquest. Today it is commonly cultivated and the fruits, which are highly favored, are regularly sold in local markets. In

1920, the United States Department of Agriculture received seeds from Guayaquil, Ecuador (S.P.I. No. 51205), and from Bogotá, Colombia (S.P.I. No. 54399). The vine is grown in California as an ornamental under the name "softleaf passionflower". It has never succeeded in Florida; is grown to some extent in Hawaii and the State of Madras, India. The climate of New Zealand seems highly suitable for it and it has been grown there, more or less commercially, for several decades.

Varieties

In general, the fruit is smaller in Peru than in Colombia and Ecuador. There are said to be several varieties. A form called *curuba quiteña* in Colombia is dark-green externally even when fully ripe, the apex is abruptly pointed and furrowed; the pulp is dark-orange or orange-brown.

Climate

This species is at home at elevations between 6,000 and 7,200 ft (1,800-3,200 m) in the Andes, and has adapted well to altitudes of 4,000 to 6,000 ft (1,200-1,800 m) in Hawaii and New Zealand. It can tolerate brief drops in temperature to 28.4°F (-2°C).

Propagation

The vine can be propagated from cuttings but is usually grown from seeds which normally germinate in 10 weeks. The time can be shortened to 5 weeks by preliminary soaking in lukewarm water.

Culture

The seedlings can be transplanted when 3 months old and need to be trained onto a horizontal trellis 6½ ft (2 m) high with crosswires 16 in (40 cm) apart. At a vine spacing of 6.5 ft (2 m) each way, there will be 607 plants per acre (1,500 plants/ha). Less dense planting, allowing 10 ft (3 m) each way between vines, and 20 in (50 cm) between crosswires, will result in 445 vines per acre (1,100/ha). The first crop will be produced in 2 years. At dense spacing, and with good weed control and adequate fertilization, the annual harvest in Colombia will be 200 to 300 fruits per vine, amounting to 200,000 to 303,000 fruits per acre (500,000-750,000 fruits per ha), or about 31,000 to 47,000 lbs per acre (roughly the same number of kg per ha). The individual fruits range from 2 to 5½ oz each (approximately 50-150 g). Some growers have practiced pruning, which improves air-flow, reducing disease, and facilitates weeding, irrigation, spraying and harvesting. It produces larger fruits but fewer and therefore is generally viewed as not practical as size is not important to the consumer. In India, the average yield is said to be 40 to 50 fruits per vine beginning with the 6th year from planting.

Season

There is more or less continuous fruiting the year around in Colombia. In New Zealand, the crop ripens from late March or early April to September or October.

Keeping Quality

The fruit stands shipment well and will keep in good condition in a dry and not too cold atmosphere for a reasonable length of time.

Pests and Diseases

In humid and poorly drained situations, some plantations suffer from nematodes (*Meloidogyne* sp.). Leaves and shoots may be attacked by leafhoppers (*Empoasca* sp.) and by *Dione* or *Agraulis, vanillae;* leaves and fruits may be plagued by mites (*Tetranychus* sp.); larvae of *Hepialus* sp. invade the flowerbud; stems may be bored and tunneled by *Heteractes* sp. and *Nyssodrys* sp. Occasionally the fruits are attacked by fruit flies. Young shoots are prone to powdery mildew (*Asterinia* sp.) and anthracnose (*Colletotrichum* sp.) may affect the vine and fruits. Boron deficiency causes cracking of fruits. Sometimes, for physiological reasons not yet fully understood, 50 to 60% of the fruits may drop prematurely.

Food Uses

The pulp is eaten out-of-hand or is strained for its juice which is not consumed alone but employed in refreshing mixed cold beverages. In Bolivia, the juice, combined with aguardiente and sugar, is served as a pre-dinner cocktail. Colombians strain out the seeds and serve the pulp with milk and sugar, or use it in gelatin desserts. In Ecuador, the pulp is made into ice cream.

The New Zealand Department of Agriculture has developed enticing recipes to encourage the growing and utilization of the seeded pulp as pie filling, and also for making meringue pie, sauce, spiced relish, jelly, jam and other preserves. It is also advocated as an ingredient in fruit salad, especially with pineapple, and for blending with whipped cream as a pudding, and for cooking and preserving as an ice-cream topping.

Canning the juice with benzoate of soda as a preservative loses much of the quality and, therefore, there is as yet no commercial processing.

Food Value Per 100 g of Edible Portion*	
Calories	25
Moisture	92.0 g
Protein	0.6 g
Fat	0.1 g
Carbohydrates	6.3 g
Fiber	0.3 g
Ash	0.7 g
Calcium	4 mg
Phosphorus	20 mg
Iron	0.4 mg
Riboflavin	0.03 mg
Niacin	2.5 mg
Ascorbic Acid	70 mg
*Analyses made in Colombia.	

Sweet Calabash

Of minor status among the cultivated species of *Passiflora*, the sweet calabash, *P. maliformis* L., has been called water lemon (Bermuda); *ceibey cimarron* (Cuba), *callebassie* (Haiti), *calabacito de Indio* (Dominican Republic); sweet cup, conch apple, conch nut (Jamaica); *parcha cimarrona* (Puerto Rico); *pomme calabas, liane a agouti* (Guadeloupe); *pomme-liane de la Guadeloupe* (Martinique); *culupa, granadilla, curuba* or *kuruba* (Colombia); *granadilla de hueso* or *granadilla de mono* (Ecuador); *guerito* (Cuba).

Description

The vine is woody but slender, climbing to 33 ft (10 m) or more by means of tendrils in the leaf axils, and draping trees, walls and small buildings. The evergreen leaves are ovate-cordate, or ovate-oblong, with a short, recurved point at the apex; fairly thin, light-green; 2 3/8 to 6 in (6-15 cm) long, with 2 round, flat glands at about the middle of the petiole. The peduncle bears 3 thin, ovate, pointed bracts, to 2 in (5 cm) long which enclose the unopened bud and form an ivory-hued background for the opened flower, which is fragrant, 2 to 2 3/8 in (5-6 cm) wide, with keeled, green, maroon-dotted sepals and 5 small petals, greenish-white, dotted with red or purple. The corona is 3-ranked and variegated white, purple and blue.

The fruit is oblate to nearly round-oval, the specific name implying "apple-shaped", being derived from *Malus*, the apple genus. It is 1 3/4 to 2 in (4.5-5 cm) long, 1 3/8 to 1 1/2 in (3.5-4 cm) wide. The rind is yellow to brownish when fully ripe, thin; varies from rather flexible and leath-

Fig. 93: The sweet calabash (*Passiflora maliformis*) is light-yellow with a very hard shell. Photographed at the experimental station, Palmira, Colombia, in 1969.

ery to hard and brittle. The pulp is grayish or pale orange-yellow, juicy, sweet or subacid and pleasingly aromatic, containing many black, flat, ovate, pitted seeds.

Origin and Distribution

This species is native and common in the wild in Cuba, Puerto Rico, the Dominican Republic, Jamaica, and from Saba to Barbados and Trinidad; also Venezuela, Colombia and northern Ecuador. It is cultivated in Jamaica, Brazil and Ecuador for its fruits, and in Hawaii as an ornamental in private gardens and in experimental stations for use in breeding work. The United States Department of Agriculture received seeds from Trinidad in 1909 (P.I. No. 26269); seeds of 4 varieties from Colombia in September 1914 (P.I. Nos. 39223-226); and more seeds from Colombia in November 1914 (P.I. No. 39383). However, the species has not been successful in Florida or California.

Climate

The vine grows and fruits at cool altitudes—up to 5,500 ft (1,700 m)—in South America; in Jamaica, be-tween 500 and 1,200 ft (152-366 m). Lefroy saw it in Bermuda in 1871 but the climate apparently did not favor survival.

Season

The fruits ripen from September to December in Jamaica.

Pests and Diseases

This species is noted for its resistance to pests and diseases that affect its relatives.

Food Uses

The fruit, whether leathery or hard-shelled, is difficult to open but the seedy pulp is much enjoyed locally. In Jamaica, it is scooped from the shell and served with wine and sugar. The strained juice is excellent for making cold drinks.

Other Uses

Snuff boxes have been made of the shell of the hard type.

CARICACEAE

The Papaya (Plate XLVII)

The papaya, *Carica papaya* L., is a member of the small family Caricaceae allied to the Passifloraceae. As a dual- or multi-purpose, early-bearing, space-conserving, herbaceous crop, it is widely acclaimed, despite its susceptibility to natural enemies.

In some parts of the world, especially Australia and some islands of the West Indies, it is known as papaw, or pawpaw, names which are better limited to the very different, mainly wild *Asimina triloba* Dunal, belonging to the Annonaceae. While the name papaya is widely recognized, it has been corrupted to *kapaya, kepaya, lapaya* or *tapaya* in southern Asia and the East Indies. In French, it is *papaye* (the fruit) and *papayer* (the plant), or sometimes *figuier des Iles*. Spanish-speaking people employ the names *melón zapote, lechosa, payaya* (fruit), *papayo* or *papayero* (the plant), *fruta bomba, mamón* or *mamona*, depending on the country. In Brazil, the usual name is *mamao*. When first encountered by Europeans it was quite naturally nicknamed "tree melon".

Description

Commonly and erroneously referred to as a "tree", the plant is properly a large herb growing at the rate of 6 to 10 ft (1.8-3 m) the first year and reaching 20 or even 30 ft (6-9 m) in height, with a hollow green or deep-purple stem becoming 12 to 16 in (30-40 cm) or more thick at the base and roughened by leaf scars. The leaves emerge directly from the upper part of the stem in a spiral on nearly horizontal petioles 1 to 3½ ft (30-105 cm) long, hollow, succulent, green or more or less dark purple. The blade, deeply divided into 5 to 9 main segments, each irregularly subdivided, varies from 1 to 2 ft (30-60 cm) in width and has prominent yellowish ribs and veins. The life of a leaf is 4 to 6 months. Both the stem and leaves contain copious white milky latex.

The 5-petalled flowers are fleshy, waxy and slightly fragrant. Some plants bear only short-stalked pistillate (female) flowers, waxy and ivory-white; or hermaprodite (perfect) flowers (having female and male organs), ivory-white with bright-yellow anthers and borne on short stalks; while others may bear only staminate (male) flowers, clustered on panicles to 5 or 6 ft (1.5-1.8 m) long. There may even be monoecious plants having both male and female flowers. Some plants at certain seasons produce short-stalked male flowers, at other times perfect flowers. This change of sex may occur temporarily during high temperatures in midsummer. Some "all-male" plants occasionally bear, at the tip of the spray, small flowers with perfect pistils and these produce abnormally slender fruits. Male or hermaphrodite plants may change completely to female plants after being beheaded.

Generally, the fruit is melon-like, oval to nearly round, somewhat pyriform, or elongated club-shaped, 6 to 20 in (15-50 cm) long and 4 to 8 in (10-20 cm) thick; weighing up to 20 lbs (9 kg). Semi-wild (naturalized) plants bear miniature fruits 1 to 6 in (2.5-15 cm) long. The skin is waxy and thin but fairly tough. When the fruit is green and hard it is rich in white latex. As it ripens, it becomes light- or deep-yellow externally and the thick wall of succulent flesh becomes aromatic, yellow, orange or various shades of salmon or red. It is then juicy, sweetish and somewhat like a cantaloupe in flavor; in some types quite musky. Attached lightly to the wall by soft, white, fibrous tissue, are usually numerous small, black, ovoid, corrugated, peppery seeds about 3/16 in (5 mm) long, each coated with a transparent, gelatinous aril.

Origin and Distribution

Though the exact area of origin is unknown, the papaya is believed native to tropical America, perhaps in southern Mexico and neighboring Central America. It is recorded that seeds were taken to Panama and then the Dominican Republic before 1525 and cultivation spread to warm elevations throughout South and Central America, southern Mexico, the West Indies and Bahamas, and to Bermuda in 1616. Spaniards carried seeds to the Philippines about 1550 and the papaya traveled from there to Malacca and India. Seeds were sent from India to Naples in 1626. Now the papaya is familiar in nearly all tropical regions of the Old World and the Pacific Islands and has become naturalized in many areas. Seeds were probably brought to Florida from the Bahamas. Up to about 1959, the papaya was commonly grown in southern and central Florida in home gardens and on a small commercial scale. Thereafter, natural enemies seriously reduced the plantings. There was a similar decline in Puerto Rico about 10 years prior to the setback of the industry in Florida. While isolated plants and a few commercial plots may be fruitful and long-lived, plants in some fields may reach 5 or 6 ft, yield one picking of undersized and misshapen fruits and then are so affected by virus and other diseases that they must be destroyed.

In the 1950's an Italian entrepreneur, Albert Santo, imported papayas into Miami by air from Santa Marta, Colombia, Puerto Rico and Cuba for sale locally as well as shipping fresh to New York, and he also processed quantities into juice or preserves in his own Miami factory.

Since there is no longer such importation, there is a severe shortage of papayas in Florida. The influx of Latin American residents has increased the demand and new growers are trying to fill it with relatively virus-resistant strains selected by the University of Florida Agricultural Research and Education Center in Homestead.

Successful commercial production today is primarily in Hawaii, tropical Africa, the Philippines, India, Ceylon, Malaya and Australia, apart from the widespread but smaller scale production in South Africa, and Latin America.

Annual papaya consumption in Hawaii is 15 lbs (6.8 kg) per capita, yet 26 million lbs (11,838,700 kg) of fresh fruits were shipped by air freight to mainland USA in 1974, mainly direct from Hilo or via Honolulu.

Puerto Rican production does not meet the local demand and fruits are imported from the Dominican Republic for processing.

The papaya is one of the leading fruits of southern Mexico and 40% of that country's crop is produced in the state of Veracruz on 14,800 acres (6,000 ha) yielding 120,000 tons annually.

Fruits from bisexual plants are usually cylindrical or pyriform with small seed cavity and thick wall of firm flesh which stands handling and shipping well. In contrast, fruits from female flowers are nearly round or oval and thin-walled. In some areas, bisexual types are in greatest demand. In South Africa, round or oval papayas are preferred.

Fig. 94: A healthy papaya (*Carica papaya*) in Homestead, Florida, in 1946, when virus diseases were not prevalent.

Varieties

Despite the great variability in size, quality and other characteristics of the papaya, there were few prominent, selected and named cultivars before the introduction into Hawaii of the dioecious, small-fruited papaya from Barbados in 1911. It was named 'Solo' in 1919 and by 1936 was the only commercial papaya in the islands. 'Solo' produces no male plants; just female (with round, shallowly furrowed fruits) and bisexual (with pear-shaped fruits) in equal proportions. The fruits weigh 1.1 to 2.2 lbs (½ - 1 kg) and are of excellent quality. When the fruit is fully ripe the thin skin is orange-yellow and the flesh golden-orange and very sweet.

'**Kapoho Solo**' or 'Puna Solo' was discovered and became popular with growers on Kauai before 1950. In 1955 a '**Dwarf Solo**' (a back-cross of Florida's 'Betty' and 'Solo') was introduced to aid harvesting, and this became the leading commercial papaya on the island of Oahu. It was, up to 1974, the only export cultivar. It is pear-shaped, 14 to 28 oz (400-800 g) in weight in high rainfall areas, and has yellow skin and pale-orange flesh.

'**Waimanalo**' (formerly 'Solo' Line 77) was selected in 1960 and released by the Hawaii Agricultural Experiment Station in 1968 and soon superseded Line 8 'Solo' on Oahu for the fresh fruit market because of its firmness and quality, but there it is

usually too large for export. It has long storage life and is recommended for sale fresh and for processing. Since 1974 this cultivar has been produced commercially on the low-rainfall island of Maui where it ripens at a greener color than on the island of Hawaii and is exported to cities in the northwestern and central USA. The growers raised only bisexual plants; they say that the fruits of female plants are too rough in appearance.

'**Higgins**' (formerly Line 17A), the result of crosses in 1960, was introduced to Hawaiian growers in 1974. It is of high quality, pear-shaped, with orange-yellow skin, deep-yellow flesh, and averages 1 lb (0.45 kg) when grown under irrigation. In arid territory or seasons of low rainfall, the fruit is undersized.

'**Wilder**' (formerly Line 25) is a cultivar admired for its uniformity of size, firmness and small cavity and it is now popular for export.

'**Hortus Gold**', a South African cultivar, launched in the early 1950's, is dioecious, early-maturing, with round-oval, golden-yellow fruits, 2 to 3 lbs (0.9-1.36 kg) in weight. From 200 female 'Hortus Gold' seedlings planted at the University of Natal's Ukulinga Research Farm in 1960, selections were made of the plants showing the highest yield. Of these, one clone having the best sugar content and disease resistance was chosen and named '**Honey Gold**' in 1976. This cultivar has a slight beak at the apex, golden-yellow skin; is of sweet flavor and good texture but becomes mushy when overripe. It averages 2.2 lbs (1 kg) per fruit except for those at the end of the season which are much smaller. It does not reproduce true from seed and is therefore propagated by cuttings. It is late in season and late-maturing (10 months from fruit set to maturity) and therefore brings nearly double the price of other cultivars.

'Bettina' and 'Petersen', long-standing cultivars in Queensland, Australia, were inbred for several generations to obtain pure lines. 'Bettina', a hybrid of Florida's 'Betty' and a Queensland strain, is a low, shrubby, dioecious plant producing well-colored, round-oval fruits weighing 3 to 5 lbs (1.36-2.27 kg).

'Improved Petersen', of local origin, is dioecious, tall-growing, with fruits deficient in external color and indifferent as to keeping quality but noted for the fine color and flavor of the flesh. In 1947 'Bettina 100A' was crossed with 'Petersen 170' to produce the superior, semi-dwarf 'Hybrid No. 5', smooth, yellow, rounded-oval, 3 lbs (1.36 kg) in weight, thick-fleshed, of excellent flavor and prized for marketing fresh and for canning. It bore more heavily than either of its parents and remained a preferred cultivar for more than 20 years. 'Solo' and 'Hortus Gold' are often grown but most plantations are open-pollinated mixtures.

In Western Australia, after trials of 9 cultivars—'Hybrid No. 5', 'Petersen', 'Yarwun Yellow', 'Gold Cross', 'Goldy', 'Hong Kong', 'Guinea Gold', 'Golden Surprise' and 'Sunnybank'—only 'Sunnybank' and 'Guinea Gold' were chosen as having sufficient yield and quality to be worth cultivating commercially. 'Sunnybank' fruits average 1.39 lbs (0.63 kg), and ripen over 11 months. 'Guinea Gold' averages 2.4 lbs (a little over 1 kg) and ripens over a period of 18 months.

The Universidad Agraria, La Molina, Peru, began to assemble papaya strains in 1964, collecting 40 from various parts of the country and introducing 3 from Brazil, 1 from Puerto Rico, 3 from Mexico and 2 lines of 'Solo' from Hawaii, and embarked on an evaluation and breeding program and the creation of a germplasm bank.

In Ghana, dioecious cultivars such as 'Solo', 'Golden Surprise', 'Hawaii', and 'No. 5595', were introduced and commonly cultivated by farmers but they hybridized with local types and lost their identities after several generations. A number of types were collected at the Agricultural Research Station at Kade from 1966 to 1970 and classified according to sex type, fruit form, weight, skin and flesh color, flesh thickness, texture and flavor, number of seeds, and various plant factors. It was determined that preference should be given female plants with short, stout stems, early maturing, and bearing heavily all year medium-size fruits of bright color, thick-flesh and with few seeds.

The Instituto Colombiano Agropecuario, at Palmira, Colombia, began a papaya breeding program in 1963 by bringing together Colombian-grown cultivars—'Campo Grande', 'Tocaimera', 'Zapote', 'Solo',—with some from Brazil—'Betty', 'Bettina' and '43-A-3'—South Africa—'Hortus Gold'—and Puerto Rico, and representatives of related species: *C. candamarcensis* Hook. F., *C. pentagona* Heilborn, *C. goudotiana* Tr. & Pl. (one type yellow with green peduncles and another red with purple peduncles), *C. cauliflora* Jacq. of Colombia and *C. monoica* Desf. and *Jacaratia dodecaphylla* A. DC. from Peru.

The first two of these species were not suited to conditions at Palmira.

The progeny of crosses with *C. cauliflora* were the only hybrids showing some virus resistance but they were unfruitful when attacked. There were no viable seeds and 30% of the fruits were seedless. *C. monoica* proved well adapted to Palmira, bore small, yellow fruits, but succumbed to virus. The introductions from Brazil were by far the most promising. 'Zapote', with rich, red flesh is much grown on the Atlantic coast of Colombia.

In India, papaya breeding and selection work has been carried on for over 30 years beginning with 100 introduced strains and 16 local variations. A well-known cultivar is 'Coorg Honey Dew', a selection from 'Honey Dew' at Chethalli Station of the Indian Institute of Horticultural Research. There are no male plants; female and bisexual occur in equal proportions. The plant is low-bearing and prolific. The fruit is long to oval, weighs 4.4 to 7.7 lbs (2-3½ kg); has yellow flesh with a large cavity, and keeps fairly well. 'Washington', popular in Bombay, has dark-red petioles and yellow flowers. The fruits are of medium size with excellent, sweet flavor. 'Burliar Long' is prolific, bearing as many as 103 fruits the first year, mostly in pairs densely packed along the stem down to 18 in (45 cm) from the ground. Seedlings are 70% females and bloom 3 months after transplanting.

'Co. 1' and 'Co. 2' were developed at Tamil Nadu Agricultural University. Both are dioecious and dwarf, the first fruits being borne 3 ft (1 m) from the ground. 'Co. 1' is valued for eating fresh; 'Co. 2' is grown for table use and for papain extraction. The fruits are of medium size—3.3 to 5.5 lbs (1½-2½ kg), with yellow, sweet flesh.

The Regional Research Station at Pusa has introduced some promising selections:

'Pusa Delicious' ('Pusa 1-15')—medium size; flesh deep-orange, of excellent flavor; female and hermaphrodite plants; high-yielding.

'Pusa Majesty' ('Pusa 22-3')—round, of medium size; flesh yellowish, solid; keeps well and ships well; virus resistant; hermaphrodite plants higher-yielding than the female.

'Pusa Giant' ('Pusa 1-45V')—large fruits suitable for marketing ripe, or green for use as a vegetable, also for canning. Plant dioecious, fast-growing; tall; trunk thick, wind-resistant.

'Pusa Dwarf' ('Pusa 1-45')—fruit oval, of medium size. Plant is dwarf; begins bearing fruit at 10 to 12 in (25-30 cm) above the ground. In much demand for home and commercial culture; suitable for high-density plantings.

In 1965, a program of papaya improvement was undertaken in Trinidad and Tobago utilizing promising selections from local types, including 'Santa Cruz Grant', a vigorous plant mainly bisexual (having both male and female flowers), very large fruits weighing 10 to 15 lbs (4.5-6.8 kg), with firm, yellow flesh of agreeable flavor. The fruit is too large for marketing fresh but is processed both green and ripe. 'Cedro' is dioecious, rarely bisexual, a heavy bearer and highly resistant to anthracnose. The fruits weigh from 3 to 8 lbs (1.37-3.6 kg) but average 6 lbs (2.7 kg); have firm, yellow, melon-like flesh and are suitable for sale fresh or for processing.

In **'Singapore Pink'**, the plants are mainly bisexual, producing cylindrical fruit. The minority are female with round fruit. Average weight of fruit is 5 lbs (2.27 kg) though there is variation from 2 to 7 lbs (1–3 kg). The flesh is pink. The fruit surface is prone to anthracnose in rainy periods, so, at such times, the fruits must be picked and sold in the green state. Two smaller-fruited types, 2 to 3 lbs (1–1.37 kg) in weight, with bright-yellow skin and thick, firm flesh, were selected for marketing fresh.

The **'Solo'** of Hawaii has performed unsatisfactorily in Florida, producing low yields of small fruits. Scott Stambaugh, a papaya specialist, began his papaya breeding with a strain designated USDA Bureau of Plant Industry #28533 obtained from the then Plant Introduction Station in Miami. From offspring of this he made a selection which he named **'Norton'**. When he acquired seed of a type called **'Purplestem'**, later **'Bluestem'**, he crossed it with 'Norton' and the hybrid yielded fruits 10 lbs (4.5 kg) in weight and was named **'Big Bluestem'**. The latter was crossed with 'Solo' and the hybrid was called **'Bluestem Solo'** or **'Blue Solo'**. The 'Blue Solo' has been well regarded in Florida for its low growth, dependable yields of good quality fruits, 2 to 4 lbs (1–2 kg) in weight, orange-fleshed and rich in flavor.

'Cariflora' is a new cultivar developed at the recently renamed Tropical Research and Education Center of the University of Florida at Homestead. It is nearly round, about the size of a cantaloupe, with thick, dark-yellow to light-orange flesh; tolerant of papaya ringspot virus, but not resistant to papaya mosaic virus or papaya apical necrosis virus. Yield is good in southern Florida and warm lowlands of tropical America but not at elevations above 2625 ft (800 m).

'Sunrise Solo' (formerly HAES 63–22) was introduced from Hawaii into Puerto Rico. The fruit has pink flesh with high total solid content. In Puerto Rican trials, seeds were planted in mid-November, seedlings were transplanted to the field 2 months later, flowering occurred in April and mature fruits were harvested from early August to January. Recent selections from Puerto Rican breeding programs are **'P.R. 6-65'** (early), **'P.R. 7-65'** (late), and **'P.R. 8-65'**.

Venezuelan papayas are usually long and large, ranging in weight from 2 to 13 lbs (1–6 kg) and mostly for domestic consumption or shipment by boat to nearby islands.

Pollination

If a papaya plant is inadequately pollinated, it will bear a light crop of fruits lacking uniformity in size and shape. Therefore, hand-pollination is advisable in commercial plantations that are not entirely bisexual.

Bags are tied over bisexual blossoms for several days to assure that they are self-pollinated. The progeny of self-pollinated bisexual flowers are 67% bisexual, the rest being female.

To cross-pollinate, one or 2 stamens from a bisexual flower are placed on the pistil of a female flower about to open and a bag is tied over the flower for a few days. Most of such cross-pollinated blooms should set fruit. Resulting seeds will produce ½ female and ½ bisexual plants.

By another method, all but the apical female flower bud are removed from a stalk and the apical bud is bagged 1–2 days before opening. At full opening, the stigma is dusted with pollen from a selected male bloom and the bag quickly resealed and it remains so for 7 days.

Plants from female flowers crossed with male flowers are 50–50 male and female. Bisexual flowers pollinated by males give rise to ⅓ female, ⅓ bisexual and ⅓ male plants.

South African growers have long been urged to practice hand-pollination in order to maintain a selected strain and, in breeding, to incorporate factors such as purple stem, yellow flowers and reddish flesh so that the improved selection will be distinguishable from ordinary strains with non-purple stems, white flowers and yellow flesh.

Climate

The papaya is a tropical and near-tropical species, very sensitive to frost and limited to the region between 32° north and 32° south of the Equator. It needs plentiful rainfall or irrigation but must have good drainage. Flooding for 48 hours is fatal. Brief exposure to 32°F (-0.56°C) is damaging; prolonged cold without overhead sprinkling will kill the plants.

Soil

While doing best in light, porous soils rich in organic matter, the plant will grow in scarified limestone, marl, or various other soils if it is given adequate care. Optimum pH ranges from 5.5 to 6.7. Overly acid soils are corrected by working in lime at the rate of 1–2 tons/acre (2.4–4.8 tons/ha). On rich organic soils the papaya makes lush growth and bears heavily but the fruits are of low quality.

Propagation

Papayas are generally grown from seed. Germination may take 3 to 5 weeks. It is expedited to 2 to 3 weeks and percentage of germination increased by washing off the aril. Then the seeds need to be dried and dusted with fungicide to avoid damping-off, a common cause of loss of seedlings. Well-prepared seeds can be stored for as long as 3 years but the percentage of germination declines with age. Dipping for 15 seconds in hot water at 158°F (70°C) and then soaking for 24 hrs in distilled water after removal from storage will improve the germination rate. If germination is slow at some seasons, treatment with gibberellic acid may be needed to get quicker results.

To reproduce the characteristics of a preferred strain, air-layering has been successfully practiced on a small scale. All offshoots except the lowest one are girdled and layered after the parent plant has produced the first crop of fruit. Later, when the parent has grown too tall for convenient harvesting the top is cut off and new buds in the crown are pricked off until offshoots from the trunk appear and develop over a period of 4 to 6 weeks. These are layered and removed and the trunk cut off above the originally retained lowest sprout which is then allowed to

Fig. 95: Papaya fruits vary in form, size, thickness, color and flavor of flesh. Favored types have little, if any, muskiness of odor.

grow as the main stem. Thereafter the layering of off-shoots may be continued until the plant is exhausted.

Rooting of cuttings has been practiced in South Africa, especially to eliminate variability in certain clones so that their performance can be more accurately compared in evaluation studies. Softwood cuttings made in midsummer rooted quickly and fruited well the following summer. Cuttings taken in fall and spring were slow to root and deficient in root formation. The commercial cultivar 'Honey Gold' is grown entirely from cuttings. Once rooted, the cuttings are planted in plastic bags and kept under mist for 10 days, and then put in a shadehouse for hardening before setting in the field.

Hawaiian workers have found that large branches 2-3 ft (60-90 cm) long rooted more readily than small cuttings. Planted 1 ft (30 cm) deep in the rainy season, they began fruiting in a few months very close to the ground.

In budding experiments both Forkert and chip methods have proved satisfactory in Trinidad. However, it is reported that a vegetatively propagated selected strain deteriorates steadily and is worthless after 3 or 4 generations.

In Hawaii, 'Solo' grafted onto 'Dwarf Solo' was reduced in vigor and productivity, but 'Dwarf Solo' grafted onto 'Solo' showed improved performance.

In recent years, the potential of rapid propagation of papaya selections by tissue culture is being explored and promises to be feasible even for the establishment of commercial plantations of superior strains.

Efforts have been made to determine the sex of seedlings in the nursery, Indian scientists making colorimetric tests of leaf extracts have had 87% success in identifying seedlings as female; 67% in classifying males/bisexuals grouped together.

Variable Season

Planting may be done at any time of year and local conditions determine when it is best for the crop to come in. Papayas mature in 6 to 9 months from seed in the hotter areas of South Africa; in 9 to 11 months where it is cooler, providing an opportunity to supply markets in the off-season when prices are high. Seeds planted in early summer or midsummer will produce the first crop in the second winter. Thereafter, the same plants will mature fruit from spring to early summer. Spring fruits are apt to be sunburned because of winter leaf loss; are also subject to fruit spot and have a low sugar content. Sunburn can be avoided by advance whitewashing of sides exposed to the afternoon sun. Some growers manipulate the harvest season by stripping off 6 of the newly set fruits, thus forcing the plant to bloom again and produce fruits 6 to 8 weeks later than they normally would.

In southern Florida, plants set out in March or April will ripen their fruits in November and December and have the advantage of a "tourist" market. July plantings will be slowed down by winter and will not fruit for 10 months or more. Some growers advocate planting in September and October so that the crop will be ready for harvest before the onset of the main hurricane season. Further north in the state, papayas must be set out in March or April in order to have the required growing season before frost.

Spacing

Puerto Rican trials have shown that papaya plants set in the field on 6 ft (1.8 m) centers made stronger, stouter growth and were more fruitful than those at closer spacings. Some growers insist on an 8 x 8 ft (2.4 x 2.4 m) area per plant. In India, 'Co. 1' and 'Co. 2' and 'Solo' are set on 6 ft (1.8 m) centers; 'Coorg Honey Dew' and 'Washington' on 8 ft (2.4 m) centers. Princess Orchards on Maui, Hawaii, plant in double rows with an alley between each pair providing room for cultural and harvesting operations. In Queensland, plants may be set only 3 ft (1 m) apart on level ground and then thinned out by removal of unwanted plants after flowering.

Culture

Seeds may be planted directly in the field, or seedlings raised in beds or pots may be transplanted when 6 weeks old or even up to 6 months of age, though there must be great care in handling and the longer the delay the greater the risk of dehydrated or twisted roots; also, transplanting often results in trunk-curvature in windy locations.

Experiments in Hawaii indicate that direct seeding results in deeper tap-roots, erect and more vigorous growth, earlier flowering and larger yields.

In Puerto Rico, it is customary to set 2 plants per hole. In El Salvador planters place 5 to 6 seeds, separated from each other, in each hole at a depth of 3/8 in (1 cm). When the plants bloom, 90% of the males are removed, preferably by cutting off at ground level. Pulling up disturbs the roots of the remaining plants. If the plantation is isolated and there is no chance of cross-pollination by males, all the seed will become female or hermaphrodite plants. Fruits should mature 5 to 8 months later.

In India, seeds are usually treated with fungicide and planted in beds 6 in (15 cm) above ground level that have been organically enriched and fumigated. The seeds are sown 2 in (5 cm) apart and 3/4 to 1 1/8 in (2-3 cm) deep in rows 6 in (15 cm) apart. They are watered daily and transplanted in 2 1/2 months when 6 to 8 in (15-20 cm) high. Transplanting is more successful if polyethylene bags of enriched soil are used instead of raised beds. Two seeds are planted in each bag but only the stronger seedling is maintained. Transplanting is best done in the evening or on cloudy, damp days. On hot, dry days, each plant must be protected with a leafy branch or palm leaf stuck in the soil. Except for 'Coorg Honey Dew' and 'Solo', the plants are set out in 3's, 6 in (15 cm) apart in enriched pits. After flowering, one female or hermaphrodite plant is retained, the other two removed. But one male is kept for every 10 females. 'Coorg Honey Dew' and 'Solo' are planted one to a pit and no males are necessary. Watering is done every day until the plants are well established, but overwatering is detrimental to young plants. Double rows of *Sesbania aegyptiaca* are planted as a windbreak.

The installation of constant drip irrigation (12 gals per day) has made possible papaya cultivation on mountain slopes on the relatively dry island of Maui which averages 10 in (25 cm) of rain annually.

Papaya plants require frequent fertilization for satisfactory production. In India, best results have been obtained by giving 9 oz (250 g) of nitrogen, 9 oz (250 g) of phosphorus, and 18 oz (500 g) potash to each plant each year, divided into 6 applications.

Because of the need to expedite growth and production before the onslaught of diseases, Puerto Rican agronomists recommend treating the predominantly clay soil with a nematicide before planting, giving each plant 4 oz (113 g) of 15-15-15 fertilizer at the end of the first week, and each month thereafter increasing the dose by 1 oz (28 g) until the beginning of flowering, then applying .227 g per plant as a final treatment. In trials, this program has permitted 6 harvests of green fruits for processing, each over 1 lb (1/2 kg) in weight, spanning a period of 13 months. The roots usually extend out beyond the leaves and it is advisable to spread fertilizer over the entire root area.

In late fertilizer applications of a crop destined for canning, nitrogen should be omitted because it renders the fruit undesirable for processing. High nitrate content in canned papaya (as with several common vegetables) removes the tin from the can. To avoid nitrogen deficiency at the beginning of flowering for the next crop, 1 or 2% urea sprays can be applied.

In southern Florida, on oolitic limestone, experts have prescribed liquid fertilizer weekly for the first 10 weeks and then 1 lb (1/2 kg) of 4-8-6 dry fertilizer mixture (with added minor elements) per plant weekly until flowering. Here a heavy organic mulch is desirable to conserve moisture, control weeds, keep the soil cool, and help repel nematodes.

Mechanical cultivation between rows is apt to disturb the shallow roots. Judicious use of herbicides is preferable.

Overcrowded fruits should be thinned out when young to provide room for good form development and avoid pressure injury. Cold weather may interfere with pollination and cause shedding of unfertilized female flowers. Spraying the inflorescence with growth regulators stops flower drop and significantly enhances fruit set. After the first crop, the terminal growth may be nipped off to induce branching which tends to dwarf the plant and facilitates harvesting. However, unless the plants are strong growers, fruiting branches may need to be propped to avoid collapse.

Harvesting

Studies in Hawaii have shown that papaya flavor is at its peak when the skin is 80% colored. For the local market, in winter months, papayas may be allowed to color fairly well before picking, but for local market in summer and for shipment, only the first indication of yellow is permissible. The fruits must be handled with great care to avoid scratching and leaking of latex which stains the fruit skin. Home growers may twist the fruit to break the stem, but in commercial operations it is preferable to use a sharp knife to cut the stem and then trim it level with the base of the fruit. However, to expedite harvesting of high fruits, most Hawaiian growers furnish their pickers with a bamboo pole with a rubber suction cup (from the well-known "plumber's helper") at the tip. With the cup held against the lower end of the fruit, the pole is thrust upward to snap the stem and the falling fruit is caught by hand. One man can thus gather 800–1,000 lbs (363–454 kg) daily.

In Hawaii, it has been calculated that manual picking and field sorting constitute 40% of the labor cost of the crop (1,702 man-hours per acre to pick and pack). Therefore, in 1970, an experimental mechanical aid was tested and results indicated that a machine with one operator and 2 pickers could harvest 1,000 lbs (454 kg) of fruit per hour, the equivalent of 8 men hand-picking. Many factors, such as investment, operation and repair costs, useful life, and so forth must be considered before such a machine could be determined to be feasible. On the island of Maui, harvesting is aided by hydraulic lifts, each operated by a single worker. Picking starts when the plants are 11 months of age and continues for 48 months when the trees are 25 ft (7.5 m) high, too tall for further usefulness.

The fruits are best packed in single layers and padded to avoid bruising. The latex oozing from the stem may irritate the skin and workers should be required to wear gloves and protective clothing.

Yield

In the usual papaya plantation, each plant may ripen 2 to 4 fruits per week over the fruiting season. Healthy plants, if well cared for, may average 75 lbs (34 kg) of fruit per plant per year, though individual plants have borne as much as 300 lbs (136 kg). In South Africa, branched 'Honey Gold' plants set 20 ft (6 m) apart in rows 10 ft (3 m) apart have produced 45 lbs (100 kg) of fruit each in their 4th year. A field of 1,000 plants occupying 2½ acres (1 ha) gave 30 tons of fruit. In the Hilo area of the island of Hawaii, production averages 15 tons per acre (37 tons/ha). From 250 acres (100 ha), Princess Orchards on Maui harvests 150,000 lbs (68,180 kg) weekly during the season.

In the Kapoho region of the island of Hawaii, yields average 38,000 lbs/acre (roughly 38,000 kg/ha) the first year, 25,000 lbs (11,339 kg) the second year. Papaya plants bear well for 2 years and then productivity declines and commercial plantings are generally replaced after 3–4 years. By that time they have attained heights which make harvesting difficult.

Renovation of Plantings

In Trinidad and Tobago, plants that have become too tall are cut to the ground and side shoots are allowed to grow and bear. In El Salvador, after the 3rd year of bear-

ing, the main stem is cut off about 3 ft (1 m) from the ground at the beginning of winter and is covered with a plastic bag to protect it from rain and subsequent rotting. Several side shoots will emerge within a few days. When these reach 8 in to 1 ft (20–30 cm) in height, all are cut off except the most vigorous one which replaces the original top.

Postharvest Treatment

Fruits can be held at 85°F (29.64°C) and high atmospheric humidity for 48 hours to enhance coloring before packing. Standard decay control has been a 20-minute submersion in water at 120°F (49°C) followed by a cool rinse. In India, dipping in 1,000 ppm of aureofungin has been shown to be effective in controlling postharvest rots. In Philippine trials, thiabendazole reduced fruit rot by 50%. In 1979, Hawaiian workers demonstrated that spreading an aqueous solution of carnauba wax and thiabendazole over harvested fruits gives good protection from postharvest diseases and can eliminate the hot-water bath.

In Puerto Rico, fruits of 'P.R. 8-65', picked green, were ripened successfully by 6-7 days treatment with ethylene gas in airtight chambers at 77°F (25°C) and 85 to 95% humidity, following the hot-water bath.

Hawaiian papayas must be sanitized before shipment to the mainland USA to avoid introduction of fruit flies. Fruits picked ¼ ripe are prewarmed in water at 110°F (43.33°C) for about 40 min, then quickly immersed for 20 min at 119°F (48.33°C). This double-dipping may be replaced by irradiation. One little-used method is a vapor-heat treatment following dry heat at 110°F (43.33°C) and 40% relative humidity.

Fruits that have had hot water treatment and EDB fumigation and then have been stored in 1.5% oxygen at 55°F (13°C) for 12 days will have a shelf life of about 3½ days at room temperature. Fruits that have had hot water treatment when ¼ colored, followed by irradiation at 75-100 krad, and storage at 2-4% oxygen and 60°F (16°C) for 6 days will have a market life of 8 days. Those held for 12 days will be saleable thereafter for 5 days.

In Puerto Rico, gamma irradiation (25-50 krads) delayed ripening up to 7 days. Treatment at 100 krads slightly accelerated ripening in storage. Even at the lowest level irradiation inhibited fungal growth. Carotenoid content was unaffected but ascorbic acid was slightly reduced at all exposures.

Partly ripe papayas stored below 50°F (10°C) will never fully ripen. This is the lowest temperature at which ripe papayas can be held without chilling injury.

'Solo 62/3' fruits harvested in Trinidad at the first sign of yellow, treated with fungicide, placed in perforated polyethylene bags and packed in individual compartments in cartons, have been shipped to England by air (2 days' flight), ripened at 68°F (20°C), and found to be of excellent quality and flavor.

The same cultivar, similarly handled, withstood transport in the refrigerated hold of a ship for 21 days. Immediately ripened on arrival, the fruits were well accepted on the London market. Sea shipment proved to be the more economical.

Hypobaric (low pressure) containers have made possible satisfactory sea shipment (18-21 days) of hot-water treated and fungicidal-waxed papayas from Hilo, Hawaii, to Los Angeles and New York.

Pests

A major hazard to papayas in Florida and Venezuela is the wasp-like papaya fruit fly, *Toxotrypana curvicauda*. The female deposits eggs in the fruit which will later be found infested with the larvae. Only thick-fleshed fruits are safe from this enemy. Control on a commercial scale is very difficult. Home gardeners often protect the fruit from attack by covering with paper bags, but this must be done early, soon after the flower parts have fallen, and the bags must be replaced every 10 days or 2 weeks as the fruits develop. Rolled newspaper may be utilized instead of bags and is more economical. India has no fruit fly with ovipositor long enough to lay eggs inside papayas.

An important and widespread pest is the papaya webworm, or fruit cluster worm, *Homolapalpia dalera*, harbored between the main stem and the fruit and also between the fruits. It eats into the fruit and the stem and makes way for the entrance of anthracnose. Damage can be prevented if spraying is begun at the beginning of fruit set, or at least at the first sign of webs.

The tiny papaya whitefly, *Trialeuroides variabilis*, is a sucking insect and it coats the leaves with honeydew which forms the basis for sooty mold development. Shaking young leaves will often reveal the presence of whiteflies. Spraying or dusting should begin when many adults are noticed. Hornworms (immature state of the sphinx moth— *Erinnyis obscura* in Jamaica, *E. ello* in Venezuela, *E. alope* in Florida) feed on the leaves, as do the small, light-green leafhoppers.

Mention is made later on of the aphids that transmit virus diseases and other infections.

Other pests requiring control measures in Australia include the red spider, or red spider mite, *Tetranychus seximaculatus*, which sucks the juice from the leaves. In India and on the island of Maui, plant and fruit infestation by red spider has been a major problem. This pest and the cucumber fly and fruit-spotting bugs feed on the very young fruits and cause them to drop. In Hawaii, the red-and-black-flat mite feeds on the stem and leaves and scars the fruit. The broad mite damages young plants especially during cool weather.

In the Virgin Islands scale has been most troublesome, apart from rats and fruit-bats that attack ripe fruits. In Australia, 5 species of scale insects have been found on papayas, the most serious being oriental scale, *Aonidiella orientalis*, which occurs on both the fruit and the stem. So far, it is confined to limited areas. In Florida, the scale insects *Aspidiotus destructor* and *Coccus hesperidium* may infest bagged fruit more than unbagged fruit. Another scale, *Philaphedra* sp., has recently been reported here.

Indian scientists have observed that immature earthworms, *Megascolex insignis*, are attracted by and feed on rotting tissue of papaya plants. They hasten the demise of plants afflicted with stem rot from *Pythium aphanidermatum* and may act as vectors for this fungus.

Root-knot nematodes, *Meloidogyne incognita acrita*, and reniform nematodes, *Rotylenchulus reniformis*, are detrimental to the growth and productivity of papaya plants and should be combatted by pre-planting soil fumigation if the nematode population is high.

Diseases

Hawaii, partly because of its distance from other papaya-growing areas, is less afflicted with disease problems than Florida and Puerto Rico, but still has to combat a number of major and minor maladies. Most serious of all is the mosaic virus, on plant and fruit, which is common in Florida, Cuba, Puerto Rico, Trinidad, and first seen in Hawaii in 1959. It is transmitted mechanically or by the green peach aphid, *Myzus persicae*, and other aphids including the green citrus aphid, *Aphis spiraecola*, in Puerto Rico. Two forms of mosaic virus are reported in Puerto Rico: the long-known "southern coast papaya mosaic virus", the symptoms of which include extreme leaf deformation, and the relatively recent "Isabela mosaic virus" on the northern coast which is similar but without leaf distortion. Both forms occur in some north-coast plantations. There is no remedy, but measures to avoid spread include the destruction of affected plants, control of aphids by pesticides, and elimination of all members of the Cucurbitaceae from the vicinity. Mosaic is sporadic and scattered and not of great concern in Queensland.

Papaya ringspot virus, prevalent in Florida, the Dominican Republic and Venezuela, is occasionally serious in the Waianae area on the dry leeward side of Oahu. It is transmitted by the same vectors. Mosaic and ringspot viruses are the main limiting factors in papaya production in the Cauca Valley of Colombia.

In Florida, virus diseases were recognized as the greatest threat to the papaya industry in the early 1950's. The first signs are irregular mottling of young leaves, then yellowing with transparent areas, leaf distortion, and rings on the fruit. If affected plants are not removed, the condition spreads throughout the plantation. Fruits borne 2 or 3 months after the first symptoms will have a disagreeable, bitter flavor.

At the Agricultural Research and Education Center of the University of Florida in Homestead, the late Dr. Robert Conover established a test plot of papayas grown from seed of 95 accessions from a number of countries and 94 collections in Florida in the hope of finding some virus-free strains. Most of the introductions were highly susceptible to papaya ringspot virus; local strains showed some resistance. Highest tolerance was shown by a dioecious, round-fruited, yellow-fleshed strain brought from Colombia by Dr. S.E. Malo several years ago. The fruits weigh 3-5 lbs (1.36-2.27 kg).

It is thought that at least 3 virus diseases are involved in papaya decline in East Africa and it has been suggested that the diseases are spread in part by the tapping of green fruits for their latex (the source of papain).

Bunchy top is a common, controllable mycoplasma disease transmitted by a leafhopper, *Empoasca papayae* in Puerto Rico, the Dominican Republic, Haiti, and Jamaica; by that species and *E. dilitara* in Cuba; and by *E. stevensi* in Trinidad. Bunchy top can be distinguished from boron deficiency by the fact that the tops of affected plants do not ooze latex when pricked.

In the subtropical part of Queensland, but not in the tropical, wet climate of northern Queensland, papaya plants are subject to die-back, a malady of unknown origin, which begins with shortening of the petioles and bunching of inner crown leaves. Then the larger crown leaves quickly turn yellow. Affected plants can be cut back at the first sign of the disease and if the cut stem is covered to avoid rotting, the top will be replaced by healthy side branches. The problem occurs mainly in the hot, dry spring after a season of heavy rains.

Anthracnose, which usually attacks the ripe fruits and is caused by the fungus *Colletotrichum gloeosporioides*, was formerly the most important papaya disease in Hawaii, Mexico and India, but it is controllable by spraying every 10 days, or every week in hot, humid seasons, and hot-water treatment of harvested fruits. A strain of this fungus produces "chocolate spot" (small, angular, superficial lesions). A disease resembling anthracnose but which attacks papayas just beginning to ripen, was reported from the Philippines in 1974 and the causal agent was identified as *Fusarium solani*.

A major disease in wet weather is phytophthora blight. *Phytophthora parasitica* attacks and rots the stem and roots of the plant and infects and spoils the fruit surface and the stem-end, inducing fruit fall and mummification. Fungicidal sprays and removal of diseased plants and fruits will reduce the incidence. *P. palmivora* has been identified as the chief cause of root-rot in Hawaii and Costa Rica. In Hawaii, the strains, 'Waimanalo-23' and -24, 'Line 8' and 'Line 40', are resistant to this fungus. 'Kapoho Solo' and '45-T_{22}' are moderately resistant, and 'Higgins' is susceptible.

Root-rot by *Pythium* sp. is very damaging to papayas in Africa and India. *P. ultimum* causes trunk rot in Queensland. Collar rot in 8- to 10-month old seedlings, evidenced by stunting, leaf-yellowing and shedding, and total loss of roots, was first observed in Hawaii in 1970, and was attributed to attack by *Calonectria* sp. Collar rot is sometimes so severe in India as to cause growers to abandon their plantations.

Powdery mildew, caused by *Oidium caricae* (the imperfect state of *Erysiphe cruciferarum* the source of mildew in the Cruciferae) often affects papaya plants in Hawaii and both plants and fruits elsewhere. Sulfur, judiciously applied, is an effective control. Powdery mildew is caused by *Sphaerotheca humili* in Queensland and by *Ovulariopsis papayae* in East Africa. Angular leaf spot, a form of powdery mildew, is linked in Queensland to the fungus *Oidiopsis taurica*.

Corynespora leaf spot, or brown leaf spot, greasy spot or "papaya decline" (spotting of leaves and petioles and defoliation) in St. Croix, Puerto Rico, Florida and Queensland, is caused by *Corynespora cassiicola*, which is controllable with fungicides.

A new papaya disease, yellow strap leaf, similar to YSL of chrysanthemums, appeared in Florida during the summer in 1978 and 1979.

Black spot, resulting from infection by *Cercospora papayae*, has plagued Hawaiian growers since the winter of 1952-53. It causes defoliation, reduces yield, blemishes the fruit, and is unaffected by the hot-water dip. It can be prevented by field use of fungicides.

Rhizopus oryzae is most commonly linked with rotting fruits on Pakistan markets. *R. nigricans* is the usual source of fruit rot in Queensland. Injured fruits are prone to fungal rotting caused by *R. stolonifer* and *Phytophthora palmivora*. Stem-end rot occurs when fruits are pulled, not cut, from the plant and the fungus, *Ascochyta caricae*, is permitted entrance. This fungus attacks very young and older fruits in Queensland and also causes trunk rot. In South Africa, it affects cv 'Honey Gold' which is also subject to spotting by *Asperisporium caricae* on the fruits and leaves. Both of these diseases are controllable by fungicidal sprays.

Infection at the apex by *Cladosporium* sp. is manifested by internal blight. A pre-harvest fruit rot caused by *Phomopsis caricae papayae* is troublesome in Queensland and was announced from India in 1971. A new disease, papaya apical necrosis, caused by a rhabdovirus, was reported in Florida in 1981.

Papayas are frequently blemished by a condition called "freckles", of unknown origin; and mysterious hard lumps of varying size and form may be found in ripe fruits. Star spot (grayish-white, star-shaped superficial markings) appears on immature fruits in Queensland after exposure to cold winter winds. In Uttar Pradesh, an alga, *Cephaleuros mycoidea*, often disfigures the fruit surface.

In Brazil, Hawaii and other areas, a fungus, *Botryodiplodia theobromae*, causes severe stem rot and fruit rot. Trichothecium rot (*T. roseum*) is evidenced by sunken spots soon covered by pink mold on fruits in India. Charcoal rot, *Macrophomina phaseoli*, is reported in Pakistan.

Young papaya seedlings are highly susceptible to damping-off, a disease caused by soil-borne fungi—*Pythium aphanidermatum, P. ultimum, Phytophthora palmivora*, and *Rhizoctonia* sp.,—especially in warm, humid weather. Pre-planting treatment of the soil is the only means of prevention.

Papayas generally do poorly on land previously planted with papayas and this is usually the result of soil infestation by *Pythium aphanidermatum* and *Phytophthora palmivora*. Plant refuse from previous plantings should never be incorporated into the soil. Soil fumigation is necessary before replanting papayas in the same field.

Food Uses

Ripe papayas are most commonly eaten fresh, merely peeled, seeded, cut in wedges and served with a half or quarter of lime or lemon. Sometimes a few seeds are left attached for those who enjoy their peppery flavor but not many should be eaten. The flesh is often cubed or shaped into balls and served in fruit salad or fruit cup. Firm-ripe papaya may be seasoned and baked for consumption as a vegetable. Ripe flesh is commonly made into sauce for shortcake or ice cream sundaes, or is added to ice cream just before freezing; or is cooked in pie, pickled, or preserved as marmalade or jam. Papaya and pineapple cubes, covered with sugar sirup, may be quick-frozen for later serving as dessert. Half-ripe fruits are sliced and crystallized as a sweetmeat.

Papaya juice and nectar may be prepared from peeled or unpeeled fruit and are sold fresh in bottles or canned. In Hawaii, papayas are reduced to puree with sucrose added to retard gelling and the puree is frozen for later use locally or in mainland USA in fruit juice blending or for making jam.

Unripe papaya is never eaten raw because of its latex content. Even for use in salads, it must first be peeled, seeded, and boiled until tender, then chilled. Green papaya is frequently boiled and served as a vegetable. Cubed green papaya is cooked in mixed vegetable soup. Green papaya is commonly canned in sugar sirup in Puerto Rico for local consumption and for export. Green papayas for canning in Queensland must be checked for nitrate levels. High nitrate content causes detinning of ordinary cans, and all papayas with over 30 ppm nitrate must be packed in cans lacquered on the inside. Australian growers are hopeful that the papaya can be bred for low nitrate uptake.

A lye process for batch peeling of green papayas has proven feasible in Puerto Rico. The fruits may be immersed in boiling 10% lye solution for 6 minutes, in a 15% solution for 4 minutes, or a 20% solution for 3 minutes. They are then rapidly cooled by a cold water bath and then sprayed with water to remove all softened tissue. Best proportions are 1 lb (.45 kg) of fruit for every gallon (3.8 liters) of solution.

Young leaves are cooked and eaten like spinach in the East Indies. Mature leaves are bitter and must be boiled with a change of water to eliminate much of the bitterness. Papaya leaves contain the bitter alkaloids, carpaine and pseudocarpaine, which act on the heart and respiration like digitalis, but are destroyed by heat. In addition, two previously undiscovered major Δ^1-piperideine alkaloids, dehydrocarpaine I and II, more potent than carpaine, were reported from the University of Hawaii in 1979. Sprays of male flowers are sold in Asian and Indonesian markets and in New Guinea for boiling with several changes of water to remove bitterness and then eating as a vegetable. In Indonesia, the flowers are sometimes candied. Young stems are cooked and served in Africa. Older stems, after peeling, are grated, the bitter juice squeezed out, and the mash mixed with sugar and salt.

In India, **papaya seeds** are sometimes found as an adulterant of whole black pepper. Collaborating chemists in Italy and Somalia identified 18 amino acids in papaya seeds, principally, in descending order of abundance, glutamic acid, arginine, proline, and aspartic acid in the endosperm; and proline, tyrosine, lysine, aspartic acid, and glutamic acid in the sarcotesta. A yellow to brown, faintly scented oil was extracted from the sundried, powdered seeds of unripe papayas at the Central Food Technological Research Institute, Mysore, India.

White seeds yielded 16.1% and black seeds 26.8% and it was suggested that the oil might have edible and industrial uses.

Food Value

The papaya is regarded as a fair source of iron and calcium; a good source of vitamins A, B and G and an excellent source of vitamin C (ascorbic acid). The following figures represent the minimum and maximum levels of constituents as reported from Central America and Cuba.

Food Value Per 100 g of Edible Portion		
	Fruit	*Leaves**
Calories	23.1–25.8	
Moisture	85.9–92.6 g	83.3%
Protein	.081–.34 g	5.6%
Fat	.05–.96 g	0.4%
Carbohydrates	6.17–6.75 g	8.3%
Crude Fiber	0.5–1.3 g	1.0%
Ash	.31–.66 g	1.4%
Calcium	12.9–40.8 mg	0.406% (CO)
Phosphorus	5.3–22.0 mg	
Iron	0.25–0.78 mg	0.00636%
Carotene	.0045–.676 mg	28,900 I.U.
Thiamine	.021–.036 mg	
Riboflavin	.024–.058 mg	
Niacin	.227–.555 mg	
Ascorbic Acid	35.5–71.3 mg	38.6%
Tryptophan	4–5 mg	
Methionine	1 mg	
Lysine	15–16 mg	
Magnesium		0.035%
Phosphoric Acid		0.225%
	*Analyses made in Malaya.	

Carotenoid content of papaya (13.8 mg/100 g dry pulp) is low compared to mango, carrot and tomato. The major carotenoid is cryptoxanthin.

Papain

The latex of the papaya plant and its green fruits contains two proteolytic enzymes, papain and chymopapain. The latter is most abundant but papain is twice as potent. In 1933, Ceylon (Sri Lanka) was the leading commercial source of papain but it has been surpassed by East Africa where large-scale production began in 1937.

The latex is obtained by making incisions on the surface of the green fruits early in the morning and repeating every 4 or 5 days until the latex ceases to flow. The tool is of bone, glass, sharp-edged bamboo or stainless steel (knife or razor blade). Ordinary steel stains the latex. Tappers hold a coconut shell, clay cup, or glass, porcelain or enamel pan beneath the fruit to catch the latex, or a container like an "inverted umbrella" is clamped around the stem. The latex coagulates quickly and, for best results, is spread on fabric and oven-dried at a low temperature, then ground to powder and packed in tins. Sun-drying tends to discolor the product. One must tap 1,500 average-size fruits to gain 1½ lbs (.68 kg) of papain.

The lanced fruits may be allowed to ripen and can be eaten locally, or they can be employed for making dried papaya "leather" or powdered papaya, or may be utilized as a source of pectin.

Because of its papain content, a piece of green papaya can be rubbed on a portion of tough meat to tenderize it. Sometimes a chunk of green papaya is cooked with meat for the same purpose.

One of the best known uses of papain is in commercial products marketed as meat tenderizers, especially for home use. A modern development is the injection of papain into beef cattle a half-hour before slaughtering to tenderize more of the meat than would normally be tender. Papain-treated meat should never be eaten "rare" but should be cooked sufficiently to inactivate the enzyme. The tongue, liver and kidneys of injected animals must be consumed quickly after cooking or utilized immediately in food or feed products, as they are highly perishable.

Papain has many other practical applications. It is used to clarify beer, also to treat wool and silk before dyeing, to de-hair hides before tanning, and it serves as an adjunct in rubber manufacturing. It is applied on tuna liver before extraction of the oil which is thereby made richer in vitamins A and D_2. It enters into toothpastes, cosmetics and detergents, as well as pharmaceutical preparations to aid digestion.

Papain has been employed to treat ulcers, dissolve membranes in diphtheria, and reduce swelling, fever and adhesions after surgery. With considerable risk, it has been applied on meat impacted in the gullet. Chemo-papain is sometimes injected in cases of slipped spinal discs or pinched nerves. Precautions should be taken because some individuals are allergic to papain in any form and even to meat tenderized with papain.

Folk Uses

In tropical folk medicine, the fresh latex is smeared on boils, warts and freckles and given as a vermifuge. In India, it is applied on the uterus as an irritant to cause abortion. The unripe fruit is sometimes hazardously in-gested to achieve abortion. Seeds, too, may bring on abortion. They are often taken as an emmenagogue and given as a vermifuge. The root is ground to a paste with salt, diluted with water and given as an enema to induce abortion. A root decoction is claimed to expel roundworms. Roots are also used to make salt.

Crushed leaves wrapped around tough meat will tenderize it overnight. The leaf also functions as a vermifuge and as a primitive soap substitute in laundering. Dried leaves have been smoked to relieve asthma or as a tobacco substitute. Packages of dried, pulverized leaves are sold by "health food" stores for making tea, despite the fact that the leaf decoction is administered as a purgative for horses in Ghana and in the Ivory Coast it is a treatment for genito-urinary ailments. The dried leaf infusion is taken for stomach troubles in Ghana and they say it is purgative and may cause abortion.

Antibiotic Activity

Studies at the University of Nigeria have revealed that extracts of ripe and unripe papaya fruits and of the seeds are active against gram-positive bacteria. Strong doses are effective against gram-negative bacteria. The substance has protein-like properties. The fresh crushed seeds yield the aglycone of glucotropaeolin benzyl isothiocyanate (BITC) which is bacteriostatic, bactericidal and fungicidal. A single effective does is 4-5 g seeds (25-30 mg BITC).

In a London hospital in 1977, a post-operative infection in a kidney-transplant patient was cured by strips of papaya which were laid on the wound and left for 48 hours, after all modern medications had failed.

Papaya Allergy

Mention has already been made of skin irritation in papaya harvesters because of the action of fresh papaya latex, and of the possible hazard of consuming undercooked meat tenderized with papain. It must be added that the pollen of papaya flowers has induced severe respiratory reactions in sensitive individuals. Thereafter, such people react to contact with any part of the plant and to eating ripe papaya or any food containing papaya, or meat tenderized with papain.

Related Species

The mountain papaya (*C. candamarcencis* Hook. f.), is native to Andean regions from Venezuela to Chile at altitudes between 6,000 and 10,000 ft (1,800-3,000 m). The plant is stout and tall but bears a small, yellow, conical, 5-angled fruit of sweet flavor. It is cultivated in climates too cold for the papaya, including northern Chile where it thrives mainly in and around the towns of Coquimbo and La Serena at near-sea-level. The fruit (borne all year) is too rich in papain for eating raw but is popular cooked, and is canned for domestic consumption and for export. The plant grows on mountains in Ceylon and South India; does well at 1800 ft (549 m) in Puerto Rico. Its high resistance to papaya viruses is of great interest to plant breeders there and elsewhere.

The babaco, or *chamburo* (*C. pentagona* Heilborn), is commonly cultivated in mountain valleys of Ecuador. The plant is slender and no more than 10 ft (3 m) high, but the 5-angled fruits reach a foot (30 cm) in length. Usually seedless, or with only a few seeds at most, the fruits are locally eaten only after cooking. The plant is not known in the wild and botanists have suggested that it may be a hybrid. It is propagated by cuttings and is grown on a small scale in Australia and New Zealand primarily for export.

Strawberry Pear

This is one of the most beautiful and widespread members of the family Cactaceae, with one common name for its fruit, strawberry pear, and another for the plant, night-blooming cereus. *Hylocereus undatus* Britt. & Rose (syn. *Cereus undatus* Haw.), has been often misnamed *H. triangularis*, a binomial restricted today to a very similar cactus, *H. triangularis* Britt. & Rose (syns. *Cereus triangularis* Haw.; *Cactus triangularis* L.), endemic in Jamaica.

The Spanish terms *pitaya, pitajaya, pitahaya*, are applied to the strawberry pear in Latin America, in common with the edible fruits of several other species of cacti; but *pitahaya roja* and *pitahaya blanca* are applied specifically to *H. undatus* in Mexico; *pitahaya de cardón* in Guatemala.

Description

This cactus may be terrestrial or epiphytic. Its heavy, 3-sided, green, fleshy, much-branched stems with flat, wavy wings having horny margins, may reach 20 ft (6 m) in length. They arch over rocks or bushes, climb and form dense masses in trees, and cling to walls, by means of numerous, strong aerial roots. There are 2 to 5 short, sharp spines at each areole. The magnificent, night-blooming, very fragrant, bell-shaped, white flowers, up to 14 in (35 cm) long and 9 in (22.5 cm) wide, have a thick tube bearing several linear, green scales 1½ to 3 in (4-7.5 cm) long, above which is a circle of recurved, greenish-yellow, linear segments 4⅜ in (11 cm) long and ⅜ to ⅝ in (1-1.6 cm) wide, and an inner circle of about 20 white, oblong-lanceolate segments 4 in (10 cm) long and 1¼ to 1½ in (3.2-4 cm) wide. Very numerous, cream-colored stamens form a showy fringe in the center and at the apex of the thick perianth tube. The non-spiny fruit is oblong-oval, to 4 in (10 cm) long, 2½ in (6.25 cm) thick, coated with the bright-red, fleshy or yellow, ovate bases of scales. Within is white, juicy, sweet pulp containing innumerable tiny black, partly hollow seeds.

Origin and Distribution

The strawberry pear is believed native to southern Mexico, the Pacific side of Guatemala and Costa Rica, and El Salvador. It is commonly cultivated and naturalized throughout tropical American lowlands, the West Indies, the Bahamas, Bermuda, southern Florida and the tropics of the Old World.

Degener tells how this species reached Hawaii in 1830 in a shipment of plants loaded at a Mexican port aboard a ship en route from Boston to Canton, China. He says most of the plants died and were being discarded during a stopover in Hawaii, but the Captain noticed that the strawberry pear was still partly alive. Cuttings were planted and flourished and the cactus became a common ornamental in the islands. It blooms there spectacularly but rarely sets fruit. This species is often used as a rootstock on which to graft various ornamental cacti including *Zygocactus, Epiphyllum* and *Rhipsalis*.

It blooms and fruits mainly in August and September.

Varieties

It is not clear whether the *pitahaya amarilla* of Colombia is the same as the yellow form of *H. undatus* which occurs in Mexico. Perez-Arbelaez describes it under *Cereus triangularis* Haw. but expresses doubt as to its true identity. The attractive and delicious fruit is served whole or halved as dessert in hotels in Bogotá. (see Plate XLVIII).

Food Uses

The ripe strawberry pear is much appreciated, especially if chilled and cut in half so that the flesh can be eaten with a spoon. The juice is enjoyed as a cool drink. A sirup made of the whole fruit is used to color pastries and candy. The unopened flowerbud can be cooked and eaten as a vegetable.

Food Value

We have only Aguilar Giron's assay of the pulp: water, 92.20; protein, 0.48-0.50; carbohydrates, 4.33-4.98; fat, 0.17-0.18; fiber, 1.12; ash, 1.10%.

Analyses made in Guatemala were published under the heading *"Hylocereus undatus"*. However, the pulp is described in accompanying notes as being "a bright, clear cerise", and the fruits analyzed were accordingly those of *H. guatemalensis* Britt. & Rose which is very much like *H. undatus*, but has smaller, red-fleshed fruits instead of white-fleshed. A large vine of the Guatemalan species has festooned a tree at the Agricultural Research and Education Center, Homestead, Florida, for many years. The composition of this species, analyzed by Munsell, *et al.* (1950), is tabulated here in lieu of comparable data on *H. undatus*.

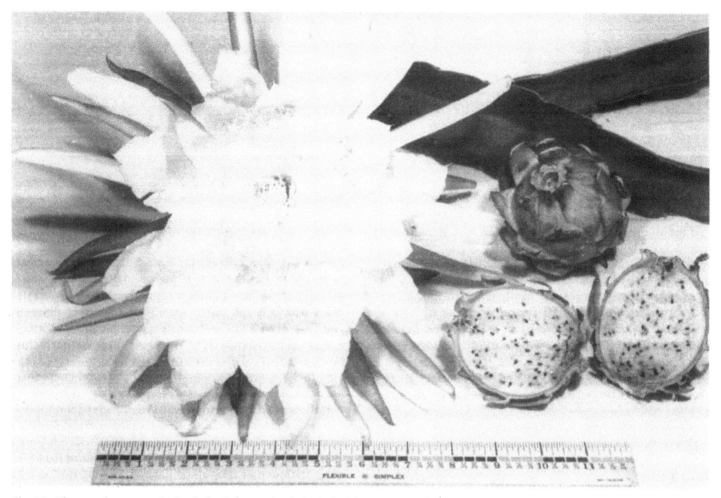

Fig. 96: The strawberry pear is the fruit of the much-admired climbing cactus (*Hylocereus undatus*), one of several species called "night-blooming cereus".

Food Value Per 100 g of Edible Portion*	
Moisture	82.5–83.0 g
Protein	0.159–0.229 g
Fat	0.21–0.61 g
Crude Fiber	0.7–0.9 g
Ash	0.54–0.68 g
Calcium	6.3–8.8 mg
Phosphorus	30.2–36.1 mg
Iron	0.55–0.65 mg
Carotene	0.005–0.012 mg
Thiamine	0.28–0.043 mg
Riboflavin	0.043–0.045 mg
Niacin	0.297–0.430 mg
Ascorbic Acid	8.0–9.0 mg

*Analyses of *H. guatemalensis*.

Medicinal Uses

The sap of the stems of *H. undatus* has been utilized as a vermifuge but it is said to be caustic and hazardous. The air-dried, powdered stems contain *B*-sitosterol.

Related Species

H. ocamponis Britt. & Rose (syn. *Cereus ocamponis* Salm-Dyck) is a similar cactus cultivated in Guatemala, Colombia, Bolivia and Puerto Rico. It has more deeply undulate wings bordered with brown, and longer spines. The fruit is wine-red outside and inside and the pulp is sweet.

The so-called apple cactus is *Cereus peruvianus* Mill., a striking, large, erect, multiple-stemmed, ribbed, spiny columnar species from South America, much grown as an ornamental in southern Florida and Hawaii. The fruit is oval, to 4 in (10 cm) long, deep-pink externally and white internally, sweet, juicy and desirable.

Barbados Gooseberry

A climbing, leafy cactus, the Barbados gooseberry, *Pereskia aculeata* Mill., (syn. *P. pereskia* Karst.; *Cactus pereskia* L.), has various English names: West Indian gooseberry, Spanish gooseberry, lemon vine, sweet Mary, leaf cactus, blade apple, and gooseberry shrub—the latter in Barbados. It is known as *grosellero* or *ramo de novia* in Cuba; *buganvilla blanca* in Chiapas, Mexico; *guamacho* in Venezuela; *ora–pro–nobis* (pray for us) in Brazil; *bladappel* in Surinam. The generic name is sometimes spelled *Peireskia*, especially in Europe, for it was adopted in honor of Nicholas Peiresk, a senator of Aix in Provence, France, and a patron of botany.

Description

The plant is an erect woody shrub when young, becoming, with age, scrambling or climbing and vinelike, with branches up to 33 ft (10 m) long that may shroud a large tree. Spines on the trunk are long, slender, in groups; those on the branches are short, recurved, usually in pairs, rarely solitary or in 3's, in the leaf axils. The deciduous, alternate, short-petioled, waxy leaves are elliptic, oblong or ovate, with a short point at the apex; 1¼ to 4 in (3.2-10 cm) long, sometimes fleshy. To some people, the flowers are lemon-scented; others say sweet and pungent in odor; still others, of unpleasant or repulsive odor. They are borne profusely in panicles or corymbs; are white, yellowish or pink-tinted; 1 to 1¾ in (2.5-4.5 cm) across and the calyx tube is prickly. The fruit is round, oval or pyriform, lemon- or orange-yellow or reddish; ⅜ to ¾ in (1-2 cm) wide, with thin, smooth, somewhat leathery skin. It is beset with the curling, leafy sepals of the calyx and often a few spines, until fully ripe, when it is juicy

Fig. 97: A leafy, spiny, climbing shrub, the Barbados gooseberry (*Pereskia aculeata*) is an atypical cactus.

Fig. 98: The pecular yellow or reddish fruits of the Barbados gooseberry bear recurved, leafy sepals until fully ripe.

and subacid to tart. There are only a few flat, thin, brown or black, soft seeds about ⅙ in (4 mm) long.

Origin and Distribution

The Barbados gooseberry is believed to be indigenous to the West Indies, coastal northern South America and Panama. It is seldom found truly wild but is frequently grown as an ornamental or occasionally for its fruits in the American tropics, Bermuda, California, Hawaii, Israel, the Philippines, India and Australia. In many areas it has escaped from cultivation and become thoroughly naturalized. It was growing at the Agricultural Research and Education Center in Homestead in the early 1940's and running wild to some extent in the Redlands, but has since disappeared, possibly destroyed by winter cold or excessive rainfall. At least one nursery in Winter Haven, Florida, is now growing the plant in quantity. Gardeners had to give up the plant in South Africa in 1979 when it was banned as an illegal weed because it had been invading and overwhelming natural vegetation. It is frequently grown in greenhouses and as a house plant in temperate regions of both hemispheres. Horticulturists often use this species as a rootstock on which to graft other less vigorous cacti.

Varieties

There are 2 cultivars in the ornamental-plant trade:

'**Godseffiana**' — bushy, with broad leaves basically yellow-green variegated with scarlet and copper on the upper surface, purplish or rosy-red on the underside.
'**Rubescens**' — the leaves variegated with red.

Climate

The Barbados gooseberry is tropical and suited only to low elevations. In greenhouses, the favorable temperature range is from 68°F (20°C) at night to 99°F (37.22°C) in daytime. Chilling causes the leaves to fall.

Propagation

The plant is easily grown from seeds or cuttings of half-ripe wood.

Culture

Flourishing with little or no care, the plant is drought-tolerant and suffers from over-watering. In greenhouse experiments, it has been found highly responsive to light. Under high light intensity, it can be kept erect and compact; under low light, it grows higher, with ascending stems and the leaves are larger and thinner.

Season

In Jamaica, the plant blooms in June and again in October and November; fruits mature in March and October.

Food Uses

The fruits are generally stewed or preserved with sugar, or made into jam. Young shoots and leaves are cooked and eaten as greens. In rural Brazil, they are important as food for humans and livestock.

Other Uses

In Israel, the flowers are said to be of great value in apiculture.

Medicinal Uses: In Brazil, the leaves are valued for their emollient nature and are applied on inflammations and tumors.

Food Value Per 100 g of Edible Portion					
	Fruit	*Leaves*		*Fruit*	*Leaves*
Moisture	91.4 g		Ascorbic Acid	2 mg	
Protein	1.0 g		Magnesium		1.2–1.5 mg
Fat	0.7 g	6.8–11.7 g			
Carbohydrates	6.3 g		*Amino acid per 100 g protein:*		
Fiber	0.7 g	9.1–9.6 g	Arginine		5.00–5.36 g
Ash	0.6 g	20.1–21.7 g	Histidine		2.49–2.54 g
Calcium	174 mg	2.8–3.4 mg	Isoleucine		3.78–4.23 g
Phosphorus	26 mg	1.8–2.0 mg	Leucine		6.99–8.03 g
Iron	Trace		Lysine		5.32–5.43 g
Vitamin A	3,215 I.U.		Methionine		1.72–2.03 g
Thiamine	0.03 mg		Phenylanine		5.06–5.08 g
Riboflavin	0.03 mg		Threonine		3.09–3.60 g
Niacin	0.9 mg		Valine		4.78–5.52 g

Studies of the leaves in Brazil show a protein content of 17.4–25.5% and a mean digestibility of 85.0%.
Protein, lysine, calcium, phosphorus and magnesium levels are higher than in cabbage, lettuce and spinach.

PUNICACEAE

Pomegranate (Plate XLIX)

Steeped in history and romance and almost in a class by itself, the pomegranate, *Punica granatum* L., belongs to the family Punicaceae which includes only one genus and two species, the other one, little-known, being *P. proto-punica* Balf. peculiar to the island of Socotra.

Despite its ancient background, the pomegranate has acquired only a relatively few commonly recognized vernacular names apart from its many regional epithets in India, most of which are variations on the Sanskrit *dadima* or *dalim*, and the Persian *dulim* or *dulima*. By the French it is called *grenade*; by the Spanish, *granada* (the fruit), *granado* (the plant); by the Dutch, *granaatappel*, and Germans, *granatapfel*; by the Italians, *melogranato, melograno granato, pomo granato*, or *pomo punico*. In Indonesia, it is *gangsalan;* in Thailand, *tab tim;* and in Malaya, *delima*. Brazilians know it as *roma, romeira* or *romazeira*. The Quecchi Indian name in Guatemala is *granad*. The Samoan name is *limoni*. The generic term, *Punica,* was the Roman name for Carthage from whence the best pomegranates came to Italy.

Description

An attractive shrub or small tree, to 20 or 30 ft (6 or 10 m) high, the pomegranate is much-branched, more or less spiny, and extremely long-lived, some specimens at Versailles known to have survived two centuries. It has a strong tendency to sucker from the base. The leaves are evergreen or deciduous, opposite or in whorls of 5 or 6, short-stemmed, oblong-lanceolate, 3/8 to 4 in (1–10 cm) long, leathery. Showy flowers are borne on the branch tips singly or as many as 5 in a cluster. They are 1¼ in (3 cm) wide and characterized by the thick, tubular, red calyx having 5 to 8 fleshy, pointed sepals forming a vase from which emerge the 3 to 7 crinkled, red, white or variegated petals enclosing the numerous stamens. Nearly round, but crowned at the base by the prominent calyx, the fruit, 2½ to 5 in (6.25–12.5 cm) wide, has a tough, leathery skin or rind, basically yellow more or less overlaid with light or deep pink or rich red. The interior is separated by membranous walls and white spongy tissue (rag) into compartments packed with transparent sacs filled with tart, flavorful, fleshy, juicy, red, pink or whitish pulp (technically the aril). In each sac, there is one white or red, angular, soft or hard seed. The seeds represent about 52% of the weight of the whole fruit.

Origin and Distribution

The pomegranate tree is native from Iran to the Himalayas in northern India and has been cultivated since ancient times throughout the Mediterranean region of Asia, Africa and Europe. The fruit was used in many ways as it is today and was featured in Egyptian mythology and art, praised in the Old Testament of the Bible and in the Babylonian Talmud, and it was carried by desert caravans for the sake of its thirst-quenching juice. It traveled to central and southern India from Iran about the first century A.D. and was reported growing in Indonesia in 1416. It has been widely cultivated throughout India and drier parts of southeast Asia, Malaya, the East Indies and tropical Africa. The most important growing regions are Egypt, China, Afghanistan, Pakistan, Bangladesh, Iran, Iraq, India, Burma and Saudi Arabia. There are some commercial orchards in Israel on the coastal plain and in the Jordan Valley.

It is rather commonly planted and has become naturalized in Bermuda where it was first recorded in 1621, but only occasionally seen in the Bahamas, West Indies and warm areas of South and Central America. Many people grow it at cool altitudes in the interior of Honduras. In Mexico it is frequently planted, and it is sometimes found in gardens in Hawaii. The tree was introduced in California by Spanish settlers in 1769. It is grown for its fruit mostly in the dry zones of that state and Arizona. In California, commercial pomegranate cultivation is concentrated in Tulare, Fresno and Kern counties, with small plantings in Imperial and Riverside counties. There were 2,000 acres (810 ha) of bearing trees in these areas in the 1920's. Production declined from lack of demand in the 1930's but new plantings were made when demand increased in the 1960's.

Cultivars

There is little information available on the types grown in the Near East, except that the cultivars 'Ahmar', 'Aswad', 'Halwa' are important in Iraq, and 'Mangulati' in Saudi Arabia. 'Wonderful' and 'Red Loufani' are often grown in the Jewish sector of Israel, while the sweeter, less tangy 'Malissi' and 'Ras el Baghl', are favored in the Arab sector.

In India there are several named cultivars. Preference is usually given those with fleshy, juicy pulp around the seeds. Types with relatively soft seeds are often classed as "seedless". Among the best are **'Bedana'** and **'Kandhari'**. 'Bedana' is medium to large, with brownish or whitish rind, pulp pinkish-white, sweet, seeds soft. 'Kandhari' is large, deep-red, with deep-pink or blood-red, subacid pulp and hard seeds. Others include:

'Alandi' ('Vadki')—medium-sized, with fleshy red or pink, subacid pulp, very hard seeds.

'Dholka'—large, yellow-red, with patches of dark-pink and purple at base, or all-over greenish-white; thick rind, fleshy, purplish-white or white, sweet, pulp; hard seeds. The plant is evergreen, non-suckering, desirable for commercial purposes in Delhi.

'Kabul'—large, with dark-red and pale-yellow rind; fleshy, dark-red, sweet, slightly bitter pulp.

'Muscat Red'—small to medium, with thin or fairly thick rind, fleshy, juicy, medium-sweet pulp, soft or medium-hard seeds. The plant is a moderately prolific bearer.

'Paper Shell'—round, medium to large, pale-yellow blushed with pink; with very thin rind, fleshy, reddish or pink, sweet, very juicy pulp and soft seeds. Bears heavily.

'Poona'—large, with dark-red, gray or grayish-green rind, sometimes spotted, and orange-red or pink-and-red pulp.

'Spanish Ruby'—round, small to medium or large; bright-red, with thin rind, fleshy, rose-colored, sweet, aromatic pulp, and small to medium, fairly soft seeds. Considered medium in quality.

'Vellodu'—medium to large, with medium-thick rind, fleshy, juicy pulp and medium-hard seeds.

'Muscat White'—large, creamy-white tinged with pink; thin rind; fleshy, cream-colored, sweet pulp; seeds medium-hard. Bears well. Desirable for commercial planting in Delhi.

'Wonderful'—originated as a cutting in Florida and propagated in California in 1896. The fruit is oblate, very large, dark purple-red, with medium-thick rind; deep-red, juicy, winey pulp; medium-hard seeds. Plant is vigorous and productive.

In California, 'Spanish Ruby' and 'Sweet Fruited' were the leading cultivars in the past century, but were superseded by 'Wonderful'. In recent years 'Wonderful' is losing ground to the more colorful 'Grenada'.

Mexicans take especial pride in the pomegranates of Tehuacan, Puebla. Many cultivars are grown, including **'Granada de China'** and **'Granada Agria'**.

The Japanese dwarf pomegranate, *P. granatum* var. *nana,* is especially hardy and widely grown as an ornamental in pots. The flowers are scarlet, the fruit only 2 in (5 cm) wide but borne abundantly. Among other ornamental cultivars are **'Multiplex'** with double, creamy white blooms; **'Chico'**, double, orange-red; **'Pleniflora'**, double, red; **'Rubra Plena'**, double, red; **'Mme. Legrelle'** and **'Variegata'**, double, scarlet bordered and streaked with yellowish-white.

Pollination

The pomegranate is both self-pollinated and cross-pollinated by insects. There is very little wind dispersal of pollen. Self-pollination of bagged flowers has resulted in 45% fruit set. Cross-pollination has increased yield to 68%. In hermaphrodite flowers, 6 to 20% of the pollen may be infertile; in male, 14 to 28%. The size and fertility of the pollen vary with the cultivar and season.

Climate

The species is primarily mild-temperate to subtropical and naturally adapted to regions with cool winters and hot summers, but certain types are grown in home dooryards in tropical areas, such as various islands of the Bahamas and West Indies. In southern Florida, fruit development is enhanced after a cold winter. Elsewhere in the United States, the pomegranate can be grown outdoors as far north as Washington County, Utah, and Washington, D.C., though it doesn't fruit in the latter locations. It can be severely injured by temperatures below 12°F (-11.11°C). The plant favors a semi-arid climate and is extremely drought-tolerant.

Soil

The pomegranate thrives on calcareous, alkaline soil and on deep, acidic loam and a wide range of soils in between these extremes. In northern India, it is spontaneous on rockstrewn gravel.

Propagation

Pomegranate seeds germinate readily even when merely thrown onto the surface of loose soil and the seedlings spring up with vigor. However, to avoid seedling variation, selected cultivars are usually reproduced by means of hardwood cuttings 10 to 20 in (25-50 cm) long. Treatment with 50 ppm indole-butyric acid and planting at a moisture level of 15.95% grealty enhances root development and survival. The cuttings are set in beds with 1 or 2 buds above the soil for 1 year, and then transplanted to the field. Grafting has never been successful but branches may be air-layered and suckers from a parent plant can be taken up and transplanted.

Culture

Rooted cuttings or seedlings are set out in pre-fertilized pits 2 ft (60 cm) deep and wide and are spaced 12 to 18 ft (3.5-5.5 m) apart, depending on the fertility of the soil. Initially, the plants are cut back to 24 to 30 in (60-75 cm) in height and after they branch out the lower branches are pruned to provide a clear main stem. Inasmuch as fruits are borne only at the tips of new growth, it is recommended that, for the first 3 years, the branches be judiciously shortened annually to encourage the maximum number of new shoots on all sides, prevent straggly development, and achieve a strong, well-framed plant. After the 3rd year, only suckers and dead branches are removed.

For good fruit production, the plant must be irrigated. In Israel, brackish water is utilized with no adverse effect. In California, irrigation water is supplied by overhead sprinklers which also provide frost protection during cold spells. The pomegranate may begin to bear in 1 year after planting out, but 2½ to 3 years is more common.

Harvesting and Yield

The fruits ripen 6 to 7 months after flowering. In Israel, cultivar 'Wonderful' is deemed ready for harvest when the soluble solids (SSC) reach 15%. In California, maturity has been equated with 1.8% titratable acidity

(TA) and SSC of 17% or more. The fruit cannot be ripened off the tree even with ethylene treatment. Growers generally consider the fruit ready for harvest if it makes a metallic sound when tapped. The fruit must be picked before overmaturity when it tends to crack open if rained upon or under certain conditions of atmospheric humidity, dehydration by winds, or insufficient irrigation. Of course, one might assume that ultimate splitting is the natural means of seed release and dispersal.

The fruits should not be pulled off but clipped close to the base so as to leave no stem to cause damage in handling and shipping. Appearance is important, especially in the United States where pomegranates may be purchased primarily to enhance table arrangements and other fall (harvest-time) decorations. Too much sun exposure causes sunscald—brown, russeted blemishes and roughening of the rind.

The fruit ships well, cushioned with paper or straw, in wooden crates or, for nearby markets, in baskets. Commercial California growers grade the fruits into 8 sizes, pack in layers, unwrapped but topped with shredded plastic, in covered wood boxes, precool rapidly, and ship in refrigerated trucks.

Keeping Quality and Storage

The pomegranate is equal to the apple in having a long storage life. It is best maintained at a temperature of 32° to 41°F (0°–5°C). The fruits improve in storage, become juicier and more flavorful; may be kept for a period of 7 months within this temperature range and at 80 to 85% relative humidity, without shrinking or spoiling. At 95% relative humidity, the fruit can be kept only 2 months at 41°F (5°C); for longer periods at 50°F (10°C). After prolonged storage, internal breakdown is evidenced by faded, streaky pulp of flat flavor. 'Wonderful' pomegranates, stored in Israel for Christmas shipment to Europe, are subject to superficial browning ("husk scald"). Control has been achieved by delaying harvest and storing in 2% O_2 at 35.6°F (2°C). Subsequent transfer to 68°F (20°C) dispels off-flavor from ethanol accumulation.

Pests and Diseases

The pomegranate butterfly, *Virachola isocrates*, lays eggs on flower-buds and the calyx of developing fruits; in a few days the caterpillars enter the fruit by way of the calyx. These fruit borers may cause loss of an entire crop unless the flowers are sprayed 2 times 30 days apart. A stem borer sometimes makes holes right through the branches. Twig dieback may be caused by either *Pleuroplaconema* or *Ceuthospora phyllosticta*. Discoloration of fruits and seeds results from infestation by *Aspergillus castaneus*. The fruits may be sometimes disfigured by *Sphaceloma punicae*. Dry rot from *Phomopsis* sp. or *Zythia versoniana* may destroy as much as 80% of the crop unless these organisms are controlled by appropriate spraying measures. Excessive rain during the ripening season may induce soft rot. A post-harvest rot caused by *Alternaria solani* was observed in India in 1974. It is particularly prevalent in cracked fruits.

Minor problems are leaf and fruit spot caused by *Cercospora*, *Gloeosporium* and *Pestalotia* sp.; also foliar damage by whitefly, thrips, mealybugs and scale insects; and defoliation by *Euproctis* spp. and *Archyophora dentula*. Termites may infest the trunk. In India, paper or plastic bags or other covers may be put over the fruits to protect them from borers, birds, bats and squirrels.

Food Uses

For enjoying out-of-hand or at the table, the **fruit** is deeply scored several times vertically and then broken apart; then the clusters of juice sacs can be lifted out of the rind and eaten. Italians and other pomegranate fanciers consider this not a laborious handicap but a social, family or group activity, prolonging the pleasure of dining.

In some countries, such as Iran, the **juice** is a very popular beverage. Most simply, the juice sacs are removed from the fruit and put through a basket press. Otherwise, the fruits are quartered and crushed, or the whole fruits may be pressed and the juice strained out. In Iran, the cut-open fruits may be stomped by a person wearing special shoes in a clay tub and the juice runs through outlets into clay troughs. Hydraulic extraction of juice should be at a pressure of less than 100 psi to avoid undue yield of tannin. The juice from crushed whole fruits contains excess tannin from the rind (as much as .175%) and this is precipitated out by a gelatin process. After filtering, the juice may be preserved by adding sodium benzoate or it may be pasteurized for 30 minutes, allowed to settle for 2 days, then strained and bottled. For beverage purposes, it is usually sweetened. Housewives in South Carolina make pomegranate jelly by adding 7½ cups of sugar and 1 bottle of liquid pectin for every 4 cups of juice. In Saudi Arabia, the juice sacs may be frozen intact or the

Food Value Per 100 g of Edible Portion*	
Calories	63–78
Moisture	72.6–86.4 g
Protein	0.05–1.6 g
Fat	Trace only to 0.9 g
Carbohydrates	15.4–19.6 g
Fiber	3.4–5.0 g
Ash	0.36–0.73 g
Calcium	3–12 mg
Phosphorus	8–37 mg
Iron	0.3–1.2 mg
Sodium	3 mg
Potassium	259 mg
Carotene	None to Trace
Thiamine	0.003 mg
Riboflavin	0.012–0.03 mg
Niacin	0.180–0.3 mg
Ascorbic Acid	4–4.2 mg
Citric Acid	0.46–3.6 mg
Boric Acid	0.005 mg

*Analyses of fresh juice sacs made by various investigators.

extracted juice may be concentrated and frozen, for future use. Pomegranate juice is widely made into grenadine for use in mixed drinks. In the Asiatic countries it may be made into a thick sirup for use as a sauce. It is also often converted into wine.

In the home kitchen, the juice can be easily extracted by reaming the halved fruits on an ordinary orange-juice squeezer.

In northern India, a major use of the wild fruits is for the preparation of "anardana"—the juice sacs being dried in the sun for 10 to 15 days and then sold as a spice.

Toxicity

A tannin content of no more than 0.25% in the edible portion is the desideratum. Many studies have shown that tannin is carcinogenic and excessive ingestion of tannin from one or more sources, over a prolonged period, is detrimental to health. (See also "Medicinal Uses" regarding overdoses of bark.)

Other Uses

All parts of the tree have been utilized as sources of tannin for curing leather. The **trunk bark** contains 10 to 25% tannin and was formerly important in the production of Morocco leather. The **root bark** has a 28% tannin content, the leaves, 11%, and the fruit rind as much as 26%. The latter is a by-product of the "anardana" industry. Both the **rind** and the **flowers** yield dyes for textiles. Ink can be made by steeping the **leaves** in vinegar. In Japan, an insecticide is derived from the bark. The pale-yellow **wood** is very hard and, while available only in small dimensions, is used for walking-sticks and in woodcrafts.

Medicinal Uses: The juice of wild pomegranates yields citric acid and sodium citrate for pharmaceutical purposes. Pomegranate juice enters into preparations for treating dyspepsia and is considered beneficial in leprosy.

The bark of the stem and root contains several alkaloids including *iso*pelletierine which is active against tapeworms. Either a decoction of the bark, which is very bitter, or the safer, insoluble Pelletierine Tannate may be employed. Overdoses are emetic and purgative, produce dilation of pupila, dimness of sight, muscular weakness and paralysis.

Because of their tannin content, extracts of the bark, leaves, immature fruit and fruit rind have been given as astringents to halt diarrhea, dysentery and hemorrhages. Dried, pulverized flower buds are employed as a remedy for bronchitis. In Mexico, a decoction of the flowers is gargled to relieve oral and throat inflammation. Leaves, seeds, roots and bark have displayed hypotensive, antispasmodic and anthelmintic activity in bioassay.

MYRTACEAE

Guava (Plate L)

One of the most gregarious of fruit trees, the guava, *Psidium guajava* L., of the myrtle family (Myrtaceae), is almost universally known by its common English name or its equivalent in other languages. In Spanish, the tree is *guayabo*, or *guayavo*, the fruit *guayaba* or *guyava*. The French call it *goyave* or *goyavier;* the Dutch, *guyaba, goeajaaba;* the Surinamese, *guave* or *goejaba;* and the Portuguese, *goiaba* or *goaibeira*. Hawaiians call it guava or *kuawa*. In Guam it is *abas*. In Malaya, it is generally known either as *guava* or *jambu batu*, but has also numerous dialectal names as it does in India, tropical Africa and the Philippines where the corruption, *bayabas*, is often applied. Various tribal names—*pichi, posh, enandi*, etc.— are employed among the Indians of Mexico and Central and South America.

Description

A small tree to 33 ft (10 m) high, with spreading branches, the guava is easy to recognize because of its smooth, thin, copper-colored bark that flakes off, showing the greenish layer beneath; and also because of the attractive, "bony" aspect of its trunk which may in time attain a diameter of 10 in (25 cm). Young twigs are quadrangular and downy. The leaves, aromatic when crushed, are evergreen, opposite, short-petioled, oval or oblong-elliptic, somewhat irregular in outline; 2¾ to 6 in (7–15 cm) long, 1⅕ to 2 in (3–5 cm) wide, leathery, with conspicuous parallel veins, and more or less downy on the underside. Faintly fragrant, the white flowers, borne singly or in small clusters in the leaf axils, are 1 in (2.5 cm) wide, with 4 or 5 white petals which are quickly shed, and a prominent tuft of perhaps 250 white stamens tipped with pale-yellow anthers.

The fruit, exuding a strong, sweet, musky odor when ripe, may be round, ovoid, or pear-shaped, 2 to 4 in (5–10 cm) long, with 4 or 5 protruding floral remnants (sepals) at the apex; and thin, light-yellow skin, frequently blushed with pink. Next to the skin is a layer of somewhat granular flesh, ⅛ to ½ in (3–12.5 mm) thick, white, yellowish, light- or dark-pink, or near-red, juicy, acid, subacid, or sweet and flavorful. The central pulp, concolorous or slightly darker in tone, is juicy and normally filled with very hard, yellowish seeds, ⅛ in (3 mm) long, though some rare types have soft, chewable seeds. Actual seed counts have ranged from 112 to 535 but some guavas are seedless or nearly so.

When immature and until a very short time before ripening, the fruit is green, hard, gummy within and very astringent.

Origin and Distribution

The guava has been cultivated and distributed by man, by birds, and sundry 4-footed animals for so long that its place of origin is uncertain, but it is believed to be an area extending from southern Mexico into or through Central America. It is common throughout all warm areas of tropical America and in the West Indies (since 1526), the Bahamas, Bermuda and southern Florida where it was reportedly introduced in 1847 and was common over more than half the State by 1886. Early Spanish and Portuguese colonizers were quick to carry it from the New World to the East Indies and Guam. It was soon adopted as a crop in Asia and in warm parts of Africa. Egyptians have grown it for a long time and it may have traveled from Egypt to Palestine. It is occasionally seen in Algeria and on the Mediterranean coast of France. In India, guava cultivation has been estimated at 125,327 acres (50,720 ha) yielding 27,319 tons annually.

Apparently it did not arrive in Hawaii until the early 1800's. Now it occurs throughout the Pacific islands. Generally, it is a home fruit tree or planted in small groves, except in India where it is a major commercial resource. A guava research and improvement program was launched by the government of Colombia in 1961. In 1968, it was estimated that there were about 10 million wild trees (around Santander, Boyacá, Antioquia, Palmira, Buga, Cali and Cartago) bearing 88 lbs (40 kg) each per year and that only 10% of the fruit was being utilized in processing. Bogotá absorbs 40% of the production and preserved products are exported to markets in Venezuela and Panama.

Brazil's modern guava industry is based on seeds of an Australian selection grown in the botanical garden of the Sao Paulo Railway Company at Tatu. Plantations were developed by Japanese farmers at Itaquera and this has become the leading guava-producing area in Brazil. The guava is one of the leading fruits of Mexico where the annual crop from 36,447 acres (14,750 ha) of seedling trees totals 192,850 tons (175,500 MT). Only in recent years has there been a research program designed to evaluate and select superior types for vegetative propagation and large-scale cultivation.

In Florida, the first commercial guava planting was established around 1912 in Palma Sola. Others appeared at Punta Gorda and Opalocka. A 40-acre (16 ha) guava grove was planted by Miami Fruit Industries at Indiantown in 1946. There have been more than two dozen guava jelly manufacturers throughout the state. A Sarasota concern was processing 250 bushels of guavas per day and a Pinellas County processor was operating a 150-bushel capacity plant in 1946. There has always been a steady market for guava products in Florida and the demand has increased in recent years with the influx of Caribbean and Latin American people.

The guava succumbs to frost in California except in a few favorable locations. Even if summers are too cool—a mean of 60°F (15.56°C)—in the coastal southern part of the state, the tree will die back and it cannot stand the intense daytime heat of interior valleys.

In many parts of the world, the guava runs wild and forms extensive thickets—called "guayabales" in Spanish—and it overruns pastures, fields and roadsides so vigorously in Hawaii, Malaysia, New Caledonia, Fiji, the U.S. Virgin Islands, Puerto Rico, Cuba and southern Florida that it is classed as a noxious weed subject to eradication. Nevertheless, wild guavas have constituted the bulk of the commercial supply. In 1972, Hawaii processed, for domestic use and export, more than 2,500 tons (2,274 MT) of guavas, over 90% from wild trees. During the period of high demand in World War II, the wild guava crop in Cuba was said to be 10,000 tons (9,000 MT), and over 6,500 tons (6,000 MT) of guava products were exported.

Cultivars

Formerly, round and pear-shaped guavas were considered separate species—P. pomiferum L. and P. pyriferum L.—but they are now recognized as mere variations. Small, sour guavas predominate in the wild and are valued for processing.

'Redland', the first named cultivar in Florida, was developed at the University of Florida Agricultural Research and Education Center, Homestead, and described in 1941. Very large, with little odor, white-fleshed and with relatively few seeds, it was at first considered promising but because of its excessively mild flavor, low ascorbic acid content, and susceptibility to algal spotting, it was abandoned in favor of better selections.

'Supreme' came next, of faint odor, thick, white flesh, relatively few, small seeds, high ascorbic acid content and ability to produce heavy crops over a period of 8 months from late fall to early spring.

'Red Indian', of strong odor, medium to large size, round but slightly flattened at the base and apex, yellow skin often with pink blush; with medium thick, red flesh of sweet flavor; numerous but small seeds; agreeable for eating fresh; fairly productive in fall and early winter.

'Ruby', with pungent odor, medium to large size; ovate; with thick, red flesh, sweet flavor, relatively few seeds. An excellent guava for eating fresh and for canning; fairly productive, mainly in fall and early winter.

'Blitch' (a seedling which originated in West Palm Beach and was planted at Homestead)—of strong odor, medium size, oval, with light-pink flesh, numerous, small seeds; tart, pleasant flavor; good for jelly.

'Patillo' (a seedling selection at DeLand propagated by a root sucker and from that by air-layer and planted at Homestead)—of very mild odor, medium size, ovate to obovate, with pink flesh, moderate number of small seeds; subacid, agreeable flavor; good for general cooking. (As grown in Hawaii it is highly acid and best used for processing).

'Miami Red' and 'Miami White', large, nearly odorless and thick-fleshed, were released by the University of Miami's Experimental Farm in 1954.

In early 1952, Dr. J.J. Ochse imported into Florida air-layers of a seedless guava from Java. All died. In September 1953, the writer received air-layers from Saharanpur, India. One survived and was turned over to the Agricultural Research and Education Center, Homestead. Four more were ordered from Coimbatore but arrived dead. Willim Whitman brought in a grafted plant from Java in 1954 which grew well, fruited and was the source of propagating material. In 1955, Whitman obtained a plant of a seedless guava from Cuba and it bore its first fruit in 1957. Seedless guavas are the result of low fertility of pollen grains and self-incompatibility. The fruits tend to be malformed and the trees are scant bearers. Applications of gibberellic acid increase fruit size, weight and ascorbic acid content but induce prominent ridges on the surface.

Among early California cultivars were:

'Webber' (formerly 'Riverside'), of medium-large size, pale-yellowish flesh, good flavor and 9.5% sugar.

'Rolfs', of medium size with pink flesh; of good quality and containing 9% sugar.

'Hart', fairly large, with pale-yellow flesh, and 8% sugar content.

Currently, some rare fruit fanciers grow the Florida-developed 'Red Indian' and 'White Indian'; also 'Detwiler' and 'Turnbull'.

In 1975, a guava trial project was undertaken at the Maroochy Horticultural Research Station in southeastern Queensland, beginning with 5 strains from Hawaii. By 1981, 4 selections (GA9-39R1T2', 'GA11-56T7', GA11-56R5T2' and 'GA11-564T1') seemed to hold promise for processing and 2 selections ('GA11-56T3' and 'GA11-56R1T1') for marketing fresh. They were all vegetatively propagated and tested as to performance. The green-skinned, acid, 'GA11-56' and another Hawaiian selection, '1050', yellow-skinned and mild in flavor and odor, are being grown commercially for processing in New South Wales.

In India much attention is given the characteristics of local and introduced guava cultivars and their suitability for various purposes. Among common white-fleshed cultivars are:

'Apple Colour'—of medium size, slightly oblate; deep-pink skin, creamy-white flesh, moderate amount of seeds, very sweet flavor (0.34-2.12% acid, 9 to 11.36% sugar); heavy bearer; good keeping quality; good for canning.

'Behat Coconut'—large, with thick white flesh, few seeds; poor for canning.

'Chittidar'—medium to large, round-ovate, white-fleshed, mild acid-sweet flavor; bears moderately well; keeps well; good for canning.

'Habshi'—of medium size with thick, white flesh, few seeds; halves good for canning.

'Lucknow 42'—of medium size, roundish, with creamy-white, soft flesh; sweet, pleasant flavor; very few seeds; good quality; bears heavily; keeps fairly well; not suitable for canning.

'Lucknow 49'—medium-large with cream-white, thick flesh, few seeds; acid-sweet; good quality; heavy bearer; high in pectin and good for jelly; halves good for canning.

'Safeda'—of medium size, with very thin skin, thick, white flesh, few seeds. Outstanding quality for canning. A famous guava, widely planted, but susceptible to wilt and branches are brittle and break readily.

'Smooth Green'—of medium size, with thick white flesh, few, small, hard seeds. Halves are firm, good for canning.

'Allahabad'—large, white-fleshed, with few, medium-sized, fairly hard seeds.

'Karela'—medium-large, pear-shaped, furrowed, rough-skinned, with soft, granular, white flesh; sweet, rich, pleasant flavor. Poor bearer. Not popular.

'Nagpur Seedless'—small to medium, often irregular in shape; white-fleshed.

'Seedless' (from Allahabad)—medium to large, pear-shaped to ovoid; with thick white flesh, firm to soft, sweet. Light bearer; poor keeper.

A seedless type at Poona, India, was found to be a triploid with 33 chromosomes in place of the usual 22.

Other white-fleshed guavas with poor canning qualities are: 'Dharwar', 'Mirzapuri', 'Nasik', 'Sindh', and 'White Supreme X Ruby'.

Among red-fleshed cultivars in India there are:

'Anakapalle'—small, with thin, red flesh, many seeds; not suitable for canning.

'Florida Seedling'—small, with thin, red, acid flesh; many seeds; not suitable for canning.

'Hapi'—medium to large, with red flesh.

'Hybrid Red Supreme'—large, with thin, red, acid flesh; moderate amount of seeds; not suitable for canning.

'Kothrud'—of medium size with medium thick, red flesh; moderate amount of seeds; not suitable for canning.

'Red-fleshed'—of medium size with many (about 567) fairly soft seeds; high in pectin and good for jelly; not suitable for canning.

Among other Indian cultivars are: 'Banaras', 'Dholka', 'Hasijka', 'Kaffree', and 'Wickramasekara'. The latter is a small fruit and poor bearer.

Indian breeders have crossed the guava with its dwarf, small-fruited relative, *P. guineense* Sw., with a view to reducing tree size and enhancing hardiness and yield.

In Egypt, a cultivar named 'Bassateen El Sabahia' has long been the standard commercial guava. Efforts have been made to improve quality and yield and to this end selections were made from 300 seedlings. The most promising selection was tested and introduced into cultivation in 1975 under the name 'Bassateen Edfina'. It is pear-shaped, of medium size, sometimes pink-blushed, with thick, white flesh, few seeds, good flavor and higher ascorbic acid content than the parent. It bears well over a long season.

In Puerto Rico, over 100 promising selections were under observation in 1963.

Numerous cultivated clones identified only by number have been evaluated for processing characters. Others have been tested and rated for resistance to *Glomerella* disease. Among the few named cultivars are 'Corozal Mixta', 'Corriente', and 'Seedling 57–6–79'.

In Trinidad, a large, white-fleshed type is known as 'Cayenne'.

In 1967, French horticulturists made a detailed evaluation of 11 guava cultivars grown at the Neufchateau Station in Guadeloupe:

'Elisabeth'—large, round, pink-fleshed, very acid; good for processing.

'Red' X 'Supreme' X 'Ruby'—large, ovoid, with deep-pink flesh; agreeable for eating fresh.

'Large White'—large, round, white-fleshed; low sugar content, astringent; can be useful as filler in preserves.

'Acid Speer'—large, round, with pale-yellow flesh; acid; recommended only as source of pectin.

'Red' X 'Supreme' X 'Ruby' X 'White'—large to very large, pear-shaped, with creamy-white flesh; good for eating fresh and for juice and nectar.

'Pink Indian'—of medium size, red-fleshed; agreeably acid; good for eating fresh and for processing.

'Red Hybrid'—medium, sub-ovoid, red-fleshed; medium quality.

'Supreme' X 'Ruby'—medium, sub-ovoid, white-fleshed; unremarkable except for high productivity.

'Stone'—small, ovoid, with deep-pink flesh; attractive and of agreeable flavor for eating fresh.

'Supreme'—small, ovoid, with pale-yellow, pink-tinged flesh; sweet; good for sherbet and paste; very productive.

'Patricia'—very small, ovoid, salmon-fleshed; attractive; good to eat fresh but quickly loses its distinct strawberry flavor; good for sirup; very productive.

Between 1948 and 1969, 21 guava cultivars from 7 countries were introduced into Hawaii. Some have been test planted and evaluated at the Waimanalo Experimental Farm. Four sweet, white-fleshed, thick-walled cultivars were rated as commercially desirable: 'Indonesian White', 'Indonesian Seedless', 'Lucknow 49', and 'No. 6363' (a 'Ruby' X 'Supreme' hybrid from Florida). Lower ratings were given four others of this group: 'Apple' (too musky and seedy); 'Allahabad Safeda' (too bumpy of surface); 'Burma' (too seedy) and 'Hong Kong White' (too seedy). Of the sweet, pink-fleshed, thick-walled cultivars examined, 'Hong Kong Pink' was preferred. Second-choice was 'No. 6362' (a seedling of a 'Ruby' X 'Supreme' cross in Florida). 'No. 7199', a seedling of a 'Stone Acid' X 'Ruby' cross in Florida, was considered too musky. Among acid, non-musky, thick-walled guavas, 'Beaumont', a Hawaiian selection, is large and pink-fleshed. 'Pink Acid' (#7198), from a Florida cross of 'Speer' and 'Stone Acid', has dark-pink flesh and few seeds. These cultivars are employed in breeding programs in Hawaii. In 1978, a new cultivar, 'Ka Hua Kula', selected from 1,200 seedlings of 'Beaumont', was released and recommended for commercial guava puree. The fruit is large, with thick, deep-pink flesh, and fewer seeds than 'Beaumont', and is less acid. It is also a heavier bearer.

In Colombia, the cultivars **'Puerto Rico'**, **'Rojo Africano'**, and **'Agrio'**, all yield over 2,200 fruits annually. Other high-yielding cultivars being evaluated are **'White'**, **'Red'**, **'D–13'**, **'D–14'**, and **'Trujillo 2'**.

Collecting guava cultivars is a hobby of Mr. Arthur Stockdale, Finca Catalina, Zitacuaro, Mexico. He is said to have some very superior selections in his grove.

Pollination

The chief pollinator of guavas is the honeybee (*Apis mellifera*). The amount of cross-pollination ranges from 25.7 to 41.3%.

Climate

The guava thrives in both humid and dry climates. In India, it flourishes up to an altitude of 3,280 ft (1,000 m); in Jamaica, up to 3,900 ft (1,200 m); in Costa Rica, to 4,590 ft (1,400 m); in Ecuador, to 7,540 ft (2,300 m). It can survive only a few degrees of frost. Young trees have been damaged or killed in cold spells at Allahabad, India, in California and in Florida. Older trees, killed to the ground, have sent up new shoots which fruited 2 years later. The guava requires an annual rainfall between 40 and 80 in (1,000-2,000 mm); is said to bear more heavily in areas with a distinct winter season than in the deep Tropics.

Soil

The guava seems indiscriminate as to soil, doing equally well on heavy clay, marl, light sand, gravel bars near streams, or on limestone; and tolerating a pH range from 4.5 to 9.4. It is somewhat salt-resistant. Good drainage is recommended but guavas are seen growing spontaneously on land with a high water table—too wet for most other fruit trees.

Propagation

Guava seeds remain viable for many months. They often germinate in 2 to 3 weeks but may take as long as 8 weeks. Pretreatment with sulfuric acid, or boiling for 5 minutes, or soaking for 2 weeks, will hasten germination. Seedlings are transplanted when 2 to 30 in (5-75 cm) high and set out in the field when 1 or 2 years old. Inasmuch as guava trees cannot be depended upon to come true from seed, vegetative propagation is widely practiced.

In Hawaii, India and elsewhere, the tree has been grown from root cuttings. Pieces of any roots except the smallest and the very large, cut into 5 to 10 in (12.5-20 cm) lengths, are placed flat in a prepared bed and covered with 2 to 4 in (5-10 cm) of soil which must be kept moist. Or one can merely cut through roots in the ground 2 to 3 ft (0.6-0.9 m) away from the tree trunk; the cut ends will sprout and can be dug up and transplanted.

By another method, air-layers of selected clones are allowed to grow 3 to 5 years and are then sawn off close to the ground. Then a ring of bark is removed from each new shoot; root-inducing chemical is applied. Ten days later, the shoots are banked with soil to a height 4 to 5 in (10-12.5 cm) above the ring. After 2 months, the shoots are separated and planted out.

Pruned branches may serve as propagating material. Cuttings of half-ripened wood, ¼ to ½ in (6-12.5 mm) thick will root with bottom heat or rooting-hormone treatment. Using both, 87% success has been achieved. Treated softwood cuttings will also root well in intermittent mist. In Trinidad, softwood, treated cuttings have been rooted in 18 days in coconut fiber dust or sand in shaded bins sprayed 2 or 3 times daily to keep humidity above 90%. Over 100,000 plants were produced by this method over a 2-year period. Under tropical conditions (high heat and high humidity), mature wood ¾ to 1 in (2-2.5 cm) thick and 1½ to 2 ft (45-60 cm) long, stuck into 1-ft (30-cm) high black plastic bags filled with soil, readily roots without chemical treatment.

In India, air-layering and inarching have been practiced for many years. However, trees grown from cuttings or air-layers have no taproot and are apt to be blown down in the first 2 or 3 years. For this reason, budding and grafting are preferred.

Approach grafting yields 85 to 95% success. Trials have been made of the shield, patch and Forkert methods of budding. The latter always gives the best results (88 to 100%). Vigorous seedlings ½ to 1 in (1.25-2.5 cm) thick are used as rootstocks. The bark should slip easily to facilitate insertion of the bud, which is then tightly bound in place with a plastic strip and the rootstock is beheaded, leaving only 6 to 8 leaves above the bud. About a month later, an incision is made halfway through 2 or 3 in (5-7.5 cm) above the bud and the plant is bent over to force the bud to grow. When the bud has put up several inches of growth, the top of the rootstock is cut off immediately above the bud. Sprouting of the bud is expedited in the rainy season.

At the Horticultural Experiment and Training Center, Basti, India, a system of patch budding has been demonstrated as commercially feasible. A swollen but unsprouted, dormant bud is taken as a ¾ x ⅜ in (2x1 cm) patch from a leaf axil of previous season's growth and taped onto a space of the same size cut 6 to 8 in (15-20 cm) above the ground on a 1-year-old, pencil-thick seedling during the period April-June. After the bud has "taken", ⅓ is cut from the top of the seedling; 2–3 weeks later, the rest of the top is cut off leaving only ¾ to 1¼ in (2-3.2 cm) of stem above the bud. This method gives 80 to 90% success. If done in July, only 70%. In Hawaii, old seedling orchards have been topworked to superior selections by patch budding on stump shoots.

Culture

Guava trees are frequently planted too close. Optimum distance between the trees should be at least 33 ft (10 m). Planting 16½ ft (5 m) apart is possible if the trees are "hedged". The yield per tree will be less but the total yield per land area will be higher than at the wider spacing. Some recommend setting the trees 8 ft (2.4 m) apart in rows 24 ft (7.3 m) apart and removing every other tree as soon as there is overcrowding. Where mass production is not desired and space is limited, guava trees can be grown as cordons on a wire fence. Rows should always run north and south so that each tree receives the maximum sunlight. Exudates from the roots of guava trees tend to inhibit the growth of weeds over the root system.

Light pruning is always recommended to develop a strong framework, and suckers should also be eliminated around the base. Experimental heading-back has increased yield in some cultivars in Puerto Rico. In Palestine, the trees are cut back to 6½ ft (2 m) every other spring to facilitate harvesting without ladders. Fruits are borne by new shoots from mature wood. If trees bear too heavily, the branches may break. Therefore, thinning is recommended and results in larger fruits.

Guava trees grow rapidly and fruit in 2 to 4 years from seed. They live 30 to 40 years but productivity declines after the 15th year. Orchards may be rejuvenated by drastic pruning.

The tree is drought-tolerant but in dry regions lack of irrigation during the period of fruit development will cause the fruits to be deficient in size. In areas receiving only 15 to 20 in (38-50 cm) rainfall annually, the guava will benefit from an additional 2,460 cm (2 acre feet) applied by means of 8 to 10 irrigations, one every 15-20 days in summer and one each month in winter.

Guava trees respond to a complete fertilizer mix applied once a month during the first year and every other month the second year (except from mid-November to mid-January) at the rate of 8 oz (227 g) per tree initially with a gradual increase to 24 oz (680 g) by the end of the second year. Nutritional sprays providing copper and zinc are recommended thrice annually for the first 2 years and once a year thereafter. In India, flavor and quality of guavas has been somewhat improved by spraying the foliage with an aqueous solution of potassium sulfate weekly for 7 weeks after fruit set.

Control of Wild Trees

Large trees that have overrun pastures are killed in Fiji with 2,4-D dicamba or 2,4,5-T in diesel fuel or old engine oil. Extensive wild stands of young trees are best burned. Cutting results in regrowth with multiple stems.

Cropping and Yield

The fruit matures 90 to 150 days after flowering. Generally, there are 2 crops per year in southern Puerto Rico; the heaviest, with small fruits, in late summer and early fall; another, with larger fruits, in late winter and early spring. In northern India, the main crop ripens in midwinter and the fruits are of the best quality. A second crop is borne in the rainy season but the fruits are less abundant and watery. Growers usually withhold irrigation after December or January or root-prune the trees in order to avoid a second crop. The trees will shed many leaves and any fruits set will drop. An average winter crop in northern India is about 450 fruits per tree. Trees may bear only 100-300 fruits in the rainy season but the price is higher because of relative scarcity despite the lower quality. Of course, yields vary with the cultivar and cultural treatment. Experiments have shown that spraying young guava trees with 25% urea plus a wetting agent will bring them into production early and shorten the harvest period from the usual 15 weeks to 4 weeks.

Handling and Keeping Quality

Ripe guavas bruise easily and are highly perishable. Fruits for processing may be harvested by mechanical tree-shakers and plastic nets. For fresh marketing and shipping, the fruits must be clipped when full grown but underripe, and handled with great care. After grading for size, the fruits should be wrapped individually in tissue and packed in 1 to 4 padded layers with extra padding on top before the cover is put on. They have been successfully shipped from Miami to wholesalers in major northern cities in refrigerated trucks at temperatures of 45° to 55°F (7.22°-12.78°C). It is commonly said that guavas must be tree-ripened to attain prime quality, but the cost of protecting the crop from birds makes early picking necessary. It has been demonstrated that fruits picked when yellow-green and artificially ripened for 6 days in straw at room temperature developed superior color and sugar content.

Guavas kept at room temperature in India are normally overripe and mealy by the 6th day, but if wrapped in pliofilm will keep in good condition for 9 days. In cold storage, pliofilm-wrapped fruits remain unchanged for more than 12 days. Wrapping checks weight loss and preserves glossiness. Unwrapped 'Safeda' guavas, just turned yellow, have kept well for 4 weeks in cold storage at 47° to 50°F (8.33°-10°C) and relative humidity of 85-95%, and were in good condition for 3 days thereafter at room temperature of 76° to 87°F (24°-44°C).

Fruits coated with a 3% wax emulsion will keep well for 8 days at 72° to 86°F (22.2°-30°C) and 40 to 60% relative humidity, and for 21 days at 47° to 50°F (8.3°-10°C) and relative humidity of 85-90%. Storage life of mature green guavas is prolonged at 68°F (20°C), relative humidity of 85%, less than 10% carbon dioxide, and complete removal of ethylene.

Researchers at Kurukshetra University, India, have shown that treatment of harvested guavas with 100 ppm morphactin (chlorflurenol methyl ester 74050) increases the storage life of guavas by controlling fungal decay, and reducing loss of color, weight, sugars, ascorbic acid and non-volatile organic acids. Combined fungicidal and double-wax coating has increased marketability by 30 days.

Australian workers report prolonged life and reduced rotting in storage after a hot water dip, but better results were achieved by dipping in an aqueous benomyl suspension at 122°F (50°C). Higher temperatures cause some skin injury, as does a guazatine dip which is also a less effective fungicide.

Fruits sprayed on the tree with gibberellic acid 20-35 days before normal ripening, were retarded nearly a week as compared with the untreated fruits. Also, mature guavas soaked in gibberellic acid off the tree showed a prolonged storage life.

Trials at Haryana Agricultural University, Hissar, India, showed that weekly spraying with 1.0% potassium sulfate—1.6 gals (6 liters) per tree—beginning 7 days after fruit set and ending just before harvesting at the pale-green stage, delays yellowing, retains firmness and flavor beyond normal storage life.

Food technologists in India found that bottled guava juice (strained from sliced guavas boiled 35 minutes), preserved with 700 ppm SO_2, lost much ascorbic acid but little pectin when stored for 3 months without refrigeration, and it made perfectly set jelly.

Pests and Diseses

Guava trees are seriously damaged by the citrus flat mite, *Brevipalpus californicus* in Egypt. In India, the tree is attacked by 80 insect species, including 3 bark-eating caterpillars (*Indarbella* spp.) and the guava scale, but this and other scale insects are generally kept under control by their natural enemies. The green shield scale, *Pulvinaria psidii*, requires chemical measures in Florida, as does the guava white fly, *Trialeurodes floridensis*, and a weevil, *Anthonomus irroratus*, which bores holes in the newly forming fruits.

The red-banded thrips feed on leaves and the fruit surface. In India, cockchafer beetles feed on the leaves at the end of the rainy season and their grubs, hatched in the soil, attack the roots. The larvae of the guava shoot borer penetrates the tender twigs, killing the shoots. Sometimes aphids are prevalent, sucking the sap from the underside of the leaves of new shoots and excreting honeydew on which sooty mold develops.

The guava fruit worm, *Argyresthia eugeniella*, invisibly infiltrates hard green fruits, and the citron plant bug, *Theognis gonagia*, the yellow beetle, *Costalimaita ferruginea*, and the fruit-sucking bug, *Helopeltis antonii*, feed on ripe fruits. A false spider mite, *Brevipalpus phoenicis*, causes surface russeting beginning when the fruits are half-grown. Fruit russeting and defoliation result also from infestations of red-banded thrips, *Selenothrips rubrocinctus*. The coconut mealybug, *Pseudococcus nipae*, has been a serious problem in Puerto Rico but has been effectively combatted by the introduction of its parasitic enemy, *Pseudaphycus utilis*.

Soil-inhabiting white grubs require plowing-in of an approved and effective pesticide during field preparation in Puerto Rico. There are other minor pests, but the great problems wherever the guava is grown are fruit flies.

The guava is a prime host of the Mediterranean, Oriental, Mexican, and Caribbean fruit flies, and the melon fly—*Ceratitis capitata, Dacus dorsalis, Anastrepha ludens, A. suspensa*, and *Dacus cucurbitae*. Ripe fruits will be found infested with the larvae and totally unusable except as feed for cattle and swine. To avoid fruit fly damage, fruits must be picked before full maturity and this requires harvesting at least 3 times a week. In Brazil, choice, undamaged guavas are produced by covering the fruits with paper sacks when young (the size of an olive). Infested fruits should be burned or otherwise destroyed. In recent years, the Cooperative Extension Service in Dade County,

Florida, has distributed wasps that attack the larvae and pupae of the Caribbean fruit fly and have somewhat reduced the menace.

In Puerto Rico, up to 50% of the guava crop (mainly from wild trees) may be ruined by the uncontrollable fungus, *Glomerella cingulata*, which mummifies and blackens immature fruits and rots mature fruits. *Diplodia natalensis* may similarly affect 40% of the crop on some trees in South India.

Fruits punctured by insects are subject to mucor rot (caused by the fungus, *Mucor hiemalis*) in Hawaii. On some trees, 80% of the mature green fruits may be ruined.

Algal spotting of leaves and fruits (caused by *Cephaleuros virescens*) occurs in some cultivars in humid southern Florida but can be controlled with copper fungicides. During the rainy season in India, and the Province of Sancti Spiritus, Cuba, the fungus, *Phytophthora parasitica*, is responsible for much infectious fruit rot. *Botryodiplodia* sp. and *Dothiorella* sp. cause stem-end rot in fruits damaged during harvesting. *Macrophomina* sp. has been linked to fruit rot in Venezuela and *Gliocladium roseum* has been identified on rotting fruits on the market in India.

In Bahia, Brazil, severe deficiency symptoms of guava trees was attributed to nematodes and nematicide treatment of the soil in a circle 3 ft (0.9 m) out from the base restored the trees to normal in 5 months. Zinc deficiency may be conspicuous when the guava is grown on light soils. It is corrected by two summer sprayings 60 days apart with zinc sulphate.

Wilt, associated with the fungi *Fusarium solani* and *Macrophomina phaseoli*, brings about gradual decline and death of undernourished 1- to 5-year-old guava trees in West Bengal. A wilt disease brought about by the wound parasite, *Myxosporium psidii*, causes the death of many guava trees, especially in summer, throughout Taiwan. Wilt is also caused by *Fusarium oxysporum* f. *psidii* which invades the trunk and roots through tunnels bored by the larvae of *Coelosterna* beetles. Anthracnose (*Colletotrichum gloeosporioides*) may attack the fruits in the rainy season. *Pestalotia psidii* sometimes causes canker on green guavas in India and rots fruits in storage.

Severe losses are occasioned in India by birds and bats and some efforts are made to protect the crop by nets or noisemakers.

Food Uses

Raw guavas are eaten out-of-hand, but are preferred seeded and served sliced as dessert or in salads. More commonly, the fruit is cooked and cooking eliminates the strong odor. A standard dessert throughout Latin America and the Spanish-speaking islands of the West Indies is stewed guava shells (*cascos de guayaba*), that is, guava halves with the central seed pulp removed, strained and added to the shells while cooking to enrich the sirup. The canned product is widely sold and the shells can also be quick-frozen. They are often served with cream cheese. Sometimes guavas are canned whole or cut in half without seed removal.

Bars of thick, rich guava paste and guava cheese are staple sweets, and guava jelly is almost universally marketed. Guava juice, made by boiling sliced, unseeded guavas and straining, is much used in Hawaii in punch and ice cream sodas. A clear guava juice with all the ascorbic acid and other properties undamaged by excessive heat, is made in South Africa by trimming and mincing guavas, mixing with a natural fungal enzyme (now available under various trade names), letting stand for 18 hours at 120° to 130°F (49°-54°C) and filtering. It is made into sirup for use on waffles, ice cream, puddings and in milkshakes. Guava juice and nectar are among the numerous popular canned or bottled fruit beverages of the Caribbean area. After washing and trimming of the floral remnants, whole guavas in sirup or merely sprinkled with sugar can be put into plastic bags and quick-frozen.

There are innumerable recipes for utilizing guavas in pies, cakes, puddings, sauce, ice cream, jam, butter, marmalade, chutney, relish, catsup, and other products. In India, discoloration in canned guavas has been overcome by adding 0.06% citric acid and 0.125% ascorbic acid to the sirup. For pink sherbet, French researchers recommend 2 parts of the cultivar 'Acid Speer' and 6 parts 'Stone'. For white or pale-yellow sherbet, 2 parts 'Supreme' and 4 parts 'Large White'. In South Africa, a baby-food manufacturer markets a guava-tapioca product, and a guava extract prepared from small and over-ripe fruits is used as an ascorbic-acid enrichment for soft drinks and various foods.

Dehydrated guavas may be reduced to a powder which can be used to flavor ice cream, confections and fruit juices, or boiled with sugar to make jelly, or utilized as pectin to make jelly of low-pectin fruits. India finds it practical to dehydrate guavas during the seasonal glut for jelly-manufacture in the off-season. In 1947, Hawaii began sea shipment of frozen guava juice and puree in 5-gallon cans to processors on the mainland of the United States. Since 1975, Brazil has been exporting large quantities of guava paste, concentrated guava pulp, and guava shells not only to the United States but to Europe, the Middle East, Africa and Japan.

Canned, frozen guava nectar is an important product in Hawaii and Puerto Rico but may be excessively gritty unless stone cells from the outer flesh and skin are reduced by use of a stone mill or removed by centrifuging.

In South Africa, guavas are mixed with cornmeal and other ingredients to make breakfast-food flakes.

Green mature guavas can be utilized as a source of pectin, yielding somewhat more and higher quality pectin than ripe fruits.

Food Value Per 100 g of Edible Portion*	
Calories	36-50
Moisture	77-86 g
Crude Fiber	2.8-5.5 g
Protein	0.9-1.0 g
Fat	0.1-0.5 g
Ash	0.43-0.7 g
Carbohydrates	9.5-10 g
Calcium	9.1-17 mg
Phosphorus	17.8-30 mg
Iron	0.30-0.70 mg
Carotene (Vitamin A)	200-400 I.U.
Thiamine	0.046 mg
Riboflavin	0.03-0.04 mg
Niacin	0.6-1.068 mg
Vitamin B_3	40 I.U.
Vitamin G_4	35 I.U.
*Analyses of whole ripe guavas.	

Ascorbic acid—mainly in the skin, secondly in the firm flesh, and little in the central pulp—varies from 56 to 600 mg. It may range up to 350-450 mg in nearly ripe fruit. When specimens of the same lot of fruits are fully ripe and soft, it may decline to 50-100 mg. Canning or other heat processing destroys about 50% of the ascorbic acid. Guava powder containing 2,500-3,000 mg ascorbic acid was commonly added to military rations in World War II. Guava seeds contain 14% of an aromatic oil, 15% protein and 13% starch. The strong odor of the fruit is attributed to carbonyl compounds.

Other Uses

Wood: The wood is yellow to reddish, fine-grained, compact, moderately strong, weighs 650-750 kg per cubic meter; is durable indoors; used in carpentry and turnery. Though it may warp on seasoning, it is much in demand in Malaya for handles; in India, it is valued for engravings. Guatemalans use guava wood to make spinning tops, and in El Salvador it is fashioned into hair combs which are perishable when wet. It is good fuelwood and also a source of charcoal.

Leaves and bark: The leaves and bark are rich in tannin (10% in the leaves on a dry weight basis, 11-30% in the bark). The bark is used in Central America for tanning hides. Malayans use the leaves with other plant materials to make a black dye for silk. In southeast Asia, the leaves are employed to give a black color to cotton; and in Indonesia, they serve to dye matting.

Wood flowers: In Mexico, the tree may be parasitized by the mistletoe, *Psittacanthus calyculatus* Don, producing the rosette-like malformations called "wood flowers" which are sold as ornamental curiosities.

Medicinal Uses: The roots, bark, leaves and immature fruits, because of their astringency, are commonly employed to halt gastroenteritis, diarrhea and dysentery, throughout the tropics. Crushed leaves are applied on wounds, ulcers and rheumatic places, and leaves are chewed to relieve toothache. The leaf decoction is taken as a remedy for coughs, throat and chest ailments, gargled to relieve oral ulcers and inflamed gums; and also taken as an emmenagogue and vermifuge, and treatment for leucorrhea. It has been effective in halting vomiting and diarrhea in cholera patients. It is also applied on skin diseases. A decoction of the new shoots is taken as a febrifuge. The leaf infusion is prescribed in India in cerebral ailments, nephritis and cachexia. An extract is given in epilepsy and chorea and a tincture is rubbed on the spine of children in convulsions. A combined decoction of leaves and bark is given to expel the placenta after childbirth.

The leaves, in addition to tannin, possess essential oil containing the sesquiterpene hydrocarbons caryophyl-lene, β-bisabolene, aromadendrene, β-selinene, nero-lidiol, caryophyllene oxide and sel-11-en-4x -ol, also some triterpenoids and β-sitosterol. The bark contains tannin, crystals of calcium oxalate, ellagic acid and starch. The young fruits are rich in tannin.

Cattley Guava

Much more attractive in foliage and fruit than the common guava, the cattley guava, *Psidium cattleianum* Sabine (syns. *P. littorale* Raddi; *P. chinense* Hort.), is also known as the strawberry or purple guava, Chinese guava, Calcutta guava, *araca da praia* (Brazil), *araza* (Uruguay), *cas dulce* (Costa Rica), *guayaba japonesa* (Guatemala), and *guayaba peruana* (Venezuela). In Hawaii, the yellow-fruited is called *waiawi,* and the red-fruited *waiawi ulaula.*

Description

A fairly slow-growing shrub or small tree, the cattley guava generally ranges from 6.5 to 14 ft (2-4 m) tall but the yellow-fruited may attain 40 ft (12 m). Both have slender, smooth, brown-barked stems and branches, and alternate, evergreen, obovate, dark, smooth, glossy, some-what leathery leaves 1⅓ to 4¾ in (3.4-12 cm) long and ⅝ to 2⅓ in (1.6-6 cm) wide. The fragrant flowers, ⅝ to 2⅓ in (1.5-6 cm) wide are white with prominent stamens about ¾ in (2 cm) long, and are borne singly or in 3's in the leaf axils. The fruit is round or obovoid, 1 to 1½ in (2.5-4 cm) long, tipped with the protruding 4- to 5-parted calyx; thin-skinned, dark-red or purple-red or, in variety *lucidum,* lemon-yellow. Red-skinned fruits have white flesh more or less reddish near the skin. Yellow-skinned fruits have faintly yellowish flesh. In both types, the flesh is aromatic, about ⅛ in (3 mm) thick, surrounding the central juicy, somewhat translucent pulp filled with hard, flattened-triangular seeds ³⁄₃₂ in (2.5 mm) long. Free of the muskiness of the common guava, the flavor is somewhat strawberry-like, spicy, subacid.

Origin and Distribution

The cattley guava is believed native to the lowlands of eastern Brazil, especially near the coast. It is cultivated to a limited extent in other areas of South America and Central America and in the West Indies, Bermuda, the Bahamas, southern and central Florida and southern California. A commercial planting of about 3,000 trees was established at La Mesa, California, around 1884 and the trees were still producing heavily a half century later. Today there is much more use of the cattley guava as an ornamental hedge than as a fruit tree. It is grown occa-sionally in subtropical Africa, and in highlands of the Philippines at elevations up to 5,000 ft (1,500 m), India, Ceylon and Malaya. It was introduced into Singapore in 1877 and at various times thereafter but failed to survive at low altitudes. In Hawaii, it has become naturalized in moist areas, forming dense, solid stands, and is subject to eradication in range lands. It is one of the major "weed trees" of Norfolk Island; has escaped into pastures and woods at elevations between 1,500 and 3,000 ft (457-914 m) in Jamaica.

Cultivars

No named cultivars are reported but there is consid-erable variation, apart from the distinct botanical variety *lucidum*. Types with pubescent foliage are seen in culti-vation in tropical America.

Climate

The red cattley guava is hardier than the common guava and can survive temperatures as low as 22°F (-5.56°C). It can succeed wherever the orange is grown without artificial heating. The yellow is tenderer and its climatic requirements are similar to those of the lemon. Both kinds flourish in full sun.

Soil

The cattley guava does well in limestone and poor soils that would barely support other fruit trees. It is shallow-rooted but the red type is fairly drought tolerant. The yellow is able to endure flooding for short periods.

Propagation

The tree is not easily multiplied by budding or grafting because of its thin bark. It can be propagated by layering or rooting of soft tip cuttings or root cuttings, but is usually grown from seed even though seedlings of the red type vary in habit of growth, fruit size and seediness, also bear-ing season. The yellow comes fairly true from seed.

Culture

Cultural information is scant except that irrigation is necessary to obtain full-size fruits on poor soil, and the tree benefits from mulching when grown in limestone. Seedlings are set out 10 ft (3 m) apart in rows 10 ft (3 m) apart.

Cropping and Yield

On good soil and under irrigation, the cattley guava has yielded 30 tons from 5 acres (2 ha). In India, it bears

Fig. 99: Red Cattley guava (*Psidium cattleianum*) (left) and the yellow (var. *lucidum*) are flavorful but seedy. The trees are very ornamental.

two crops a year, one in July and August and another in January and February. Near the coast in California, fruits ripen continuously from August to March; inland the season is shorter, October to December.

Keeping Quality

The fresh fruit is very perishable when fully ripe and can be kept only 3 to 4 days at room temperature. For shipping, the fruit must be picked slightly unripe, handled carefully and refrigerated during transit. Generally it is sent to local processors instead of to fresh fruit markets. Hawaiian-grown fruits, slightly underripe, were stored at 32° to 36°F (0°-2.22°C) for a month and were found shriveled and decomposed. Accordingly, much higher temperatures are recommended.

Pests and Diseases

The cattley guava is usually reported as disease- and pest-free. In California, there are occasional infestations of the greenhouse thrips (*Heliothrips haemorrhoidalis*). The Caribbean fruit fly attacks the fruits in southern Florida and wherever this pest abounds. In India, birds compete with humans for the ripe fruits.

Food Uses

Cattley guavas are eaten out-of-hand without preparation except the removal of the calyx. A delicious puree or tart-filling can be made by trimming and cooking 6 cups of red cattleys with 1 cup water and 2 cups granulated sugar and pressing through a sieve. The resulting 3 cups of puree will be subacid, spicy and a dull, old-rose in color. Commercial growers ship to factories which convert the fruits into jelly, jam, butter, paste and sherbet. In Hawaii, either half-ripe or full-ripe cattleys are cut in half, boiled, and the juice strained to make ade or punch.

Food Value

Analyses of ripe fruits in the Philippines, Hawaii and Florida have shown the following constituents:

Red: seeds, 6%; water, 81.73-84.9%; ash, 0.74-1.50%; crude fiber, 6.14%; protein, 0.75-1.03%; fat, 0.55%; total sugar, 4.42-4.46%.

Yellow: seeds, 10.3%; water, 84.2%; ash, 0.63-0.75%; crude fiber, 3.87%; protein, 0.80%; fat, 0.42%; total sugar, 4.32-10.01%.

Red or *Yellow:* ascorbic acid, 22-50 mg/100 g. Calories per 2.2 lbs (1 kg), 268.

Costa Rican Guava

Perhaps the most noteworthy of the lesser species of *Psidium* is *P. friedrichsthalianum* Ndz., known variously in Latin America as *cas* or *cas ácida* (Costa Rica), *guayaba ácida* (Guatemala), *guayaba agria* (Colombia), *guayaba de danto* (Honduras), *guayaba de agua* (Panama), *guayaba del Choco* (Ecuador), *guayaba montes* (Mexico), *guayaba* (Nicaragua), and *arrayan* (El Salvador).

Description
An attractive, shapely tree, 20 to 35 ft (6-10 m) high, it has wiry, quadrangular, or 4-winged, branchlets which are dark reddish and minutely hairy. The trunk bark is red-brown with grayish patches. The evergreen leaves are 2 to 4¾ in (5-12 cm) long, 1 to 2 in (2.5-5 cm) wide, elliptic or oval, pointed, gland-dotted, thin; dark and smooth above, pale beneath. Flowers, usually borne singly, are fragrant, white, 1 in (2.5 cm) wide, with 5 waxy petals and about 300 stamens up to ½ in (1.25 cm) long. The fruit is round or oval, 1¼ to 2½ in (3-6 cm) long, with yellow skin and soft, white, very acid flesh, and a few flattened seeds ³⁄₁₆ in (5 mm) long. There is no musky odor.

Distribution
This tree grows naturally in Colombia (especially in the Cauca and Magdalena valleys), throughout Central America and around Oaxaca in southern Mexico, usually bordering streams and in swampy woods along the coast and inland. It is commonly cultivated in home gardens in temperate highlands of Costa Rica, occasionally in El Salvador, Guatemala and northern Ecuador. It thrives in the Philippines at medium and low elevations. Introductions into California and Florida have not been very successful, the tree bearing poorly and eventually succumbing to cold spells.

Food Uses
Because of its acidity, the fruit is mostly used for ade, jelly and jam. It makes fine filling for pies. Early Spaniards complained that eating the raw fruits "set the teeth on edge".

Food Value
Analyses in Guatemala show: moisture, 83.15%; protein, 0.78-0.88%; carbohydrates, 5.75-6.75%; fat, 0.39-0.52%; fiber, 7.90%; ash, 0.80%. The fruit is rich in pectin even when fully ripe.

Other Uses
The wood is fine-grained and durable, with specific gravity of 0.650-0.700. Weight per cubic meter is 650-700 kg.

Brazilian Guava

This guava relative has been the subject of much confusion, beginning with its scientific name, *Psidium guineense* Sw., based on the botanist Swartz' belief that it originated on the Guinea Coast of Africa. For a long time it was considered distinct from the guisaro, *P. molle* Bertol (syn. *P. schiedeanum* Berg.), but now these names as well as *P. aracá* Raddi, are treated as synonyms of *P. guineense*, and all the corresponding colloquial names should be applied to this one confirmed species.

In Brazil the popular names are *aracá*, *aracá do campo*, or *aracahy;* in the Guianas it is called wild guava or *wilde guave*. Among other regional names are: *guabillo, huayava, guayaba brava* and *sacha guayaba* (Peru); *allpa guayaba* (Ecuador); *guayaba de sabana, guayaba sabanera* and *guayaba agria* (Venezuela); *guayaba,* or *guayaba acida, guayaba hedionda, chamach, chamacch, pataj* and *pichippul* (Guatemala); *guisaro,* or *cas extranjero* (Costa Rica); *guayabita, guayaba arraijan,* and *guayabita de sabana* (Panama); *guayabillo* (El Salvador). The name, *guayaba agria*, seems to be the only one employed in Mexico. In California it is called either Brazilian or Castilian guava.

Description
The Brazilian guava is a relatively slow-growing shrub 3 to 10 ft (1-3 m) tall; sometimes a tree to 23 ft (7 m); with grayish bark, hairy young shoots and cylindrical or slightly flattened branchlets. The evergreen, grayish

leaves, 1⅓ to 5½ in (3.5-14 cm) long and 1 to 3⅛ (2.5-8 cm wide), are stiff, oblong, elliptic, ovate or obovate, sometimes finely toothed; scantily hairy on the upperside but coated beneath with pale or rusty hairs and distinctly dotted with glands. Flowers, borne singly or in clusters of 3 in the leaf axils, are white and have 150 to 200 prominent stamens. The fruit, round or pear-shaped, is from ⅜ to 1 in (1-2.5 cm) wide, with yellow skin, thick, pale-yellowish flesh surrounding the white central pulp, and of acid, resinous, slightly strawberry-like flavor. It contains numerous small, hard seeds and is quite firm even when fully ripe.

Distribution

The most wide-ranging guava relative, *P. guineense* occurs naturally from northern Argentina and Peru to southern Mexico, and in Trinidad, Martinique, Jamaica and Cuba, at medium elevations. It is cultivated to a limited extent in Martinique, Guadeloupe, the Dominican Republic and southern California. Trials in Florida have not been encouraging. At Agartala in Tripura, northeast India, this plant has become thoroughly naturalized and runs wild.

Cultivars

While no named cultivars have been reported, this species has been crossed with the common guava and the hybrids are dwarf, hardy and bear heavy crops.

Soil

The plant will not develop satisfactorily on light sandy soil.

Food Uses

This guava is suitable for baking and preserving. It makes a distinctive jelly which some consider superior to common guava jelly.

Other Uses

The **wood** is strong and used for tool handles, beams, planks and agricultural instruments. The **bark**, rich in tannin, is used for curing hides.

Medicinal Uses: In the interior of Brazil, a decoction of the bark or of the roots is employed to treat urinary diseases, diarrhea and dysentery. In Costa Rica, it is said to reduce varicose veins and ulcers on the legs. A leaf decoction is taken to relieve colds and bronchitis.

Related Species

The Pará guava has been known as *Britoa acida* Berg. Calvacante now shows this binomial as a synonym of *Psidium acutangulum* DC. and gives the Brazilian vernacular name as *aracá–pera*. Cruz (1965) calls it *araca piranga, aracandiva, aracanduba* and *goiabarana*. Le Cointe shows it as *araca comum do Pará* and he describes *P. aracá* Raddi as a separate species. In Bolivia, *P. acutangulum* is known as guabira; in Peru, as *ampi yacu, puca yacu, guayava del agua*.

The shrub or tree ranges in height from 26 to 40 ft (8-12 m). Its branchlets are quadrangular and winged near the leaf base. New growth is finely hairy. The leaves, with very short petioles, are elliptical, 4 to 5½ in (10-14 cm) long, 1½ to 2⅜ in (4-6 cm) wide, rounded at the base, pointed at the apex. The long-stalked, white, 5-petalled flowers, with more than 300 stamens, are borne singly or in 2's or 3's in the leaf axils. The fruit is round, pear-shaped or ellipsoid, 1¼ to 3³⁄₁₆ in (3-8 cm) wide, pale-yellow, with yellowish-white, very acid but well-flavored pulp containing a few hard, triangular seeds. The crop ripens in the spring.

The tree occurs wild and cultivated at low and medium elevations throughout Amazonia and from Peru to Colombia, Bolivia, Venezuela and the Guianas. Some specimens have been grown in southern Florida in the past under the name *P. aracá*. The fruit is eaten mixed with honey or made into acid drinks or preserves.

Of recent interest as a possible new crop is *Eugenia stipitata* McVaugh, treated by Calvacante as a variable species, but separated by McVaugh (*Flora of Peru*, Vol. XIII, Pt. 4, No. 2, 1958) into 2 subspecies, as follows:

E. stipitata subsp. *stipitata* McVaugh, called *pichi* in Peru, *araca–boi* in Brazil, is a tree to 40 or 50 ft (12-15 m) tall, with short-petioled, opposite, broad-elliptic leaves, pointed at the apex, 3 to 7 in (7.5-18 cm) long and 1⅓ to 3¼ in (3.4-8.25 cm) wide, with indented veins on the upper surface, densely hairy on the underside, faintly dotted with oil glands on both sides. The flowers, in compound, axillary racemes, are white, hairy, ¾ in (2 cm) wide, with numerous prominent stamens.

According to horticulturists and Calvacante, the fruit is somewhat like a small guava; very aromatic, round to oblate, less than 2 oz (56 g) in the wild, up to 4¾ in (12 cm) wide under cultivation and weighing as much as 14½ oz (420 g) or even 28 oz (800 g). The skin is thin and delicate; the pulp soft, juicy, very acid, containing 8 to 10 irregular-oblong or kidney-shaped seeds to 1 in (2.5 cm) long and ⅝ in (1.5 cm) wide. Ascorbic acid content has been reported as 38 to 40 mg per 100 g of edible portion. The fruiting season is February to May around Belem, Brazil. There may be 4 crops a year in Peru and Ecuador. The tree is native and abundant in the wild in Amazonian regions of Peru, Ecuador and Brazil. The fruit is eaten by the Indians and the tree is being cultivated experimentally in Peru and Ecuador and a collection of 360 seedlings has been established at Manaus. Seeds germinate in 4-12 months.

Seedlings grow slowly at first, are transplanted in about 6 months. They begin to fruit 18 months later. Yields of 12.7 tons per acre (28 T/ha) have been obtained in Peru. The tree is subject to leafspot and the fruit is prone to attack by fruit flies. The fruit loses flavor when cooked; is quick-boiled for jam. A Peruvian grower is exporting the frozen pulp to Europe.

Subspecies *sororia,* called *rupina caspi* in Peru, is a shrub or small tree to 10 ft (3 m) high with elliptic leaves 3½ to 5 in (9-12.5 cm) long, 1 to 1¾ in (2.5-4.5 cm) wide with barely visible veins; minutely hairy beneath or hairless when fully mature; and having a few dark dots. The flowers are ½ in (1.25 cm) wide with 75 stamens. The fruit is oblate, ⅝ in (1.6 cm) wide, velvety, acid, with numerous kidney-shaped seeds, ⅛ to a little over ¼ in (3-7 mm) long. McVaugh shows as native to Peru, Ecuador, Bolivia and Colombia.

Feijoa

Few fruit bearers have received as much initial high-level attention and yet have amounted to so little as this member of the Myrtaceae, *Feijoa sellowiana* Berg. It is the best known of only 3 species in the genus which the German botanist, Ernst Berger, named after Don da Silva Feijoa, a botanist of San Sebastian, Spain. The specific name honors F. Sellow, a German who collected specimens in the province of Rio Grande do Sul in southern Brazil. The paucity of vernacular names is indicative of its lack of popularity. In Uruguay, it is called, in Spanish, *guayabo del pais.* It has been nicknamed "pineapple guava", "Brazilian guava" and "fig guava". The term "guavasteen" has been adopted in Hawaii. The most unlikely term, "New Zealand banana", has shown up in agricultural literature from that country.

Description

The plant is a bushy shrub 3 to 20 ft (0.9-6 m) or more in height with pale gray bark; the spreading branches swollen at the nodes and white-hairy when young. The evergreen, opposite, short-petioled, bluntly elliptical leaves are thick, leathery, 1⅛ to 2½ in (2.8-6.25 cm) long, ⅝ to 1⅛ in (1.6-2.8 cm) wide; smooth and glossy on the upper surface, finely veiny and silvery-hairy beneath. Conspicuous, bisexual flowers, 1½ in (4 cm) wide, borne singly or in clusters, have 4 fleshy, oval, concave petals, white outside, purplish-red inside; ⅝ to ¾ in (1.6-2 cm) long, and a cluster of numerous, erect, purple stamens with round, golden-yellow anthers. The fruit is oblong or ovoid or slightly pear-shaped, 1-1½ to 2½ in (4-6 cm) long and 1⅛ to 2 in (2.8-5 cm) wide, with the persistent calyx segments adhering to the apex. The thin skin is coated with a "bloom" of fine whitish hairs until maturity, when it remains dull-green or yellow-green, sometimes with a red or orange blush. The fruit emits a strong long-lasting perfume, even before it is fully ripe. The thick, white, granular, watery flesh and the translucent central pulp enclosing the seeds are sweet or subacid, suggesting a combination of pineapple and guava or pineapple and strawberry in flavor. There are usually 20 to 40, occasionally as many as 100, very small, oblong seeds hardly noticeable when the fruit is eaten.

Origin and Distribution

The feijoa is native to extreme southern Brazil, northern Argentina, western Paraguay and Uruguay where it is common wild in the mountains. It is believed that the plant was first grown in Europe by M. de Wette in Switzerland and, a little later, about 1887, it was known to be in the Botanic Garden at Basle. In 1890, the renowned French botanist and horticulturist, Dr. Edouard Andre, brought an air-layered plant from La Plata, Brazil and planted it in his garden on the Riviera. It fruited in 1897. Dr. Andre published a description with color plates of the leaves, flowers and fruit, in the *Revue Horticole* in 1898, praising the fruit and recommending cultivation in southern France and all around the Mediterranean area.

A nurseryman in Lyons distributed air-layers from the Andre plant in 1899 and many were planted on the Riviera, some in Italy and Spain and some in greenhouses further north. That same year, the prominent nurserymen, Besson Freres, obtained seeds from Montevideo and raised thousands of plants which were widely sold and proved to be of a different type than Dr. Andre's plant. Seeds were imported by one or two other French nurserymen, and then, in 1901, seedlings from Dr. Andre's plant were obtained by Dr. F. Franceschi of Santa Barbara, California, from M. Naudin of Antibes. These were planted at several different California locations. In 1903, Dr. Franceschi acquired, through F. Morel of Lyons, several air layers from Dr. Andre's plant. He planted 1 or 2 at Santa Barbara and most of the rest were sent to Florida. The plant did not succeed in southern Florida but became quite popular in northern Florida, primarily as an ornamental and particularly as a clipped hedge. Dr. Henry Nehrling had two plants growing well in a shed in half-shade at Gotha in central Florida, in 1911. They flowered and fruited but the fruit dropped before maturity and rotted quickly. In recent years, the cultivar 'Coolidge', vegetatively propagated, has borne well in Florida. In California, the feijoa is grown in a limited way for its fruit, especially in cool coastal locations, mainly around San Francisco. At the Experimental Station in Honolulu a plant flourished for 15 years without bearing fruit. Later plantings have succeeded at higher elevations.

Fig. 100: The feijoa, or pineapple guava (*Feijoa sellowiana*) which thrives best in areas too cool for the common guava, is not fully ripe until it falls to the ground.

The feijoa is sometimes cultivated in the highlands of Chile and other South American countries and in the Caribbean area. Jamaica received a few plants from California in 1912 and planted them at various altitudes. I have seen occasional plants on roadsides and in private gardens in the Bahamas, but they do not fruit and often fail to flower. In southern India, the feijoa is grown for its fruit in home gardens at temperate elevations—about 3,500 ft (1,067 m).

Nowhere has the feijoa received more attention than in New Zealand. An Auckland nurseryman introduced 3 cultivars from Australia—'Coolidge', 'Choiceana', and 'Superba'—about 1908. They remained little known until 1930 when the feijoa was advertised as an ornamental plant. Later, after improvement by selection and naming of types with large, superior fruits and their vegetative propagation, small commercial plantings were made in citrus-growing areas of the North Island. The New Zealand Feijoa Growers' Association was formed in 1983 and some fruit is being exported to the United States, United Kingdom, Germany, Netherlands, France and Japan. New Zealanders also plant the feijoa as a windbreak around wind-sensitive crops. It is planted as an ornamental and for its fruit in southern Africa. Following

WW II, feijoa plantations were established in North Africa, the Caucasian region of southern Russia, as well as in Sicily, Portugal and Italy.

In England, the feijoa is much appreciated as a wall shrub, though it flowers profusely only in sunny locations. Planting of feijoas has been officially discouraged in New South Wales and Victoria, Australia, because the fruit is a prime host of the fruit fly.

Varieties

As stated, right at the outset seedlings from different sources showed distinct characteristics. It is reported that a man named H. Hehre of Los Angeles got seeds from Argentina and among the seedlings he raised there was one that seemed superior to the others and was earlier bearing. It became known as the 'Hehre' variety. The fruit is large, slender-pyriform, sometimes curved; yellow-green, with thin skin, finely granular flesh, abundant, very juicy pulp, fairly numerous and larger than ordinary seeds, sweet but not aromatic flavor; seedlings erect, compact, vigorous, with lush foliage but only moderately fruitful.

'Andre' (the original air-layer from Brazil), has a medium to large, oblong to round fruit, rough-surfaced, light-green, thick-fleshed, few-seeded; richly flavored and very aromatic.

Seedlings are upright, spreading to intermediate. Self-fertile; bears heavily.

'Besson' (seeds from Uruguay in 1899) has small to medium, oval, smooth fruits with red or maroon cheek; thin-skinned, with medium-thick, fine-grained flesh, very juicy pulp, numerous seeds, and rich, aromatic flavor. Seedlings are upright or spreading. This is the type grown in southern India. Both 'Andre' and 'Besson' have long been prominent in France.

'Coolidge', most commonly grown in California, has fruit varying from pyriform to oblong or elongated, of medium size, with somewhat crinkled skin. It is of indifferent flavor but is a dependable bearer being 100% self-fertile. The plant is upright and strong growing.

'Choiceana', next in favor, has round to oval, fairly smooth, medium sized to small fruit, 2 to 3½ in (5-9 cm) long, of good flavor; almost always or no less than 42% self-fertile; the plant of spreading habit and medium vigor.

'Superba' has round to slightly oval, medium smooth, medium to small fruits of good flavor; it is partially (33%) self-incompatible. The plant is spreading, straggly in habit and of medium vigor.

The two leading New Zealand cultivars are selections made there from 'Choiceana' seedlings: 'Triumph' has oval, short, plump fruits, not as pointed as those of 'Coolidge'; medium to large; smooth. The plant is upright, of medium vigor.

'Mammoth' has oval fruits resembling those of 'Coolidge'; large, to 8½ oz (240 g); somewhat wrinkled. The plant is of upright habit, and strong-growing. In 1979, 'Mammoth', 'Coolidge', and 'Triumph' grown from cuttings were being advertised in the New Zealand Journal of Agriculture as suitable for export.

Two new New Zealand cultivars, of which 20,000 plants had been sold in 1983, are 'Apollo', with thin skin subject to bruising and purpling; and 'Gemini', having very small fruits with thin skin. The Association recommends that growers plant the tried and true 'Triumph'.

Among Australian selections are 'Large Oval' and 'Chapman'.

'David' has round or oval fruits with skin of sweet and agreeable flavor; matures in November in Europe.

'Roundjon' has oval or rounded fruits, somewhat rough-skinned and red-blushed; of agreeable flavor; matures in November in Europe.

'Magnifica' is a selected seedling with very large fruits of inferior quality.

'Robert' has oval fruits with grainy flesh, and undesirable brownish leaves.

'Hirschvogel' is highly self-incompatible. 'Bliss' is partially self-incompatible.

The botanical variety *variegata* has variegated foliage.

Pollination

It has been said that feijoa pollen is transferred by birds that are attracted to and eat the flowers, but bees are the chief pollinators. Most flowers pollinated with compatible pollen show 60 to 90% fruit-set. Hand-pollination is nearly 100% effective. One should plant 2 or more bushes together for cross-pollination unless the cultivar is known to be self-compatible. Poor bearing is usually the result of inadequate pollination.

Climate

The feijoa needs a subtropical climate with low humidity. The optimum annual rainfall is 30 to 40 in (762-1,016 mm). The plant thrives where the weather is cool part of the year and it can withstand temperatures as low as 12° to 15°F (-11.11°-9.44°C). The flavor of the fruit is much better in cool than in warm regions.

Soil

While the shrub is often said to be adapted to a wide range of soil types and in England does well even where there is a high chalk content, it actually prefers rich organic soil and is not very thrifty on light or sandy terrain. Some believe that an acid soil is best but the feijoa has done well on soil with a pH of 6.2. It is drought-resistant but needs adequate water for fruit production. The site must be well-drained. The feijoa can tolerate partial shade and slight exposure to salt spray.

Propagation

The feijoa is generally grown from seed and reproduces fairly, but not absolutely, true to type. Seeds are separated by squeezing the seedy pulp into a container, covering with water, and letting the liquid stand for 4 days to ferment. Seeds are then strained out and dried before sowing. The seeds will retain viability for a year or more if kept dry. Germination takes place in 3 weeks. Soil in nursery flats must be sterile, otherwise there will be much loss of seedlings from damping-off. The young plants are transplanted to pots when they have produced their second leaves and later transferred to the field without difficulty. The plant fruits in 3 to 5 years from seed. To reproduce a special selection, vegetative propagation is, of course, necessary. In France and New Zealand, ground-layering is practiced and rooting occurs in 6 months. Air-layering is usually successful and the layers will fruit the second year.

Whip-, tongue-, and veneer-grafting on own rootstock the thickness of a pencil (about 2 years old) gives a low percentage of "takes" but grafted plants will bear in 2 years. Feijoa cuttings are said to be hard to root, but in England and Auckland cuttings are preferred. Young wood from branch tips will root in 1 to 2 months with bottom heat. If placed in sand in a glass-covered box in full sun and kept well watered, they will root in 10 days. In New Zealand, growers are advised to take 4 to 6 in (10-15 cm) cuttings of side shoots in late summer, cutting close to the firm base or pulling off with a heel of older wood which is then trimmed off; and a hormone rooting agent is applied.

Culture

A 20-year-old plant on the Riviera was reported to be 15 ft (4.5 m) high and 18 ft (5.5 m) in diameter with a trunk 8 in (20 cm) thick at the base. Because of the spreading habit of such types, 15 to 18 ft (4.5-5.5 m) should be allowed between plants for good fruit production. As the fruit is borne on young wood, pruning reduces the crop, but all shoots below 12 in (30 cm) from the ground should be removed. Some seedlings have a more erect habit and these should be chosen where space is limited. The shrubs may be set 5 ft (1.5 m) apart to form a barrier hedge; 3 ft (1 m) apart in a compact foundation planting. A 15 x 15 ft (4.5x4.5 m) spacing requires 190 plants per acre (468 per hectare).

The feijoa requires little care beyond good soil preparation before planting. Subsequent cultivation is inadvisable because of the plant's shallow, fibrous root system which should be left undisturbed. If planted for its fruit, fertilizer should be low in nitrogen to avoid excessive vegetative growth. It should be watered liberally during hot, dry spells.

Season and Yield

Flowering occurs in November in Uruguay, in late April in northern Florida, May in southern California, early June in the San Francisco Bay area and July in England. In southern California the fruits ripen 4½ to 6 months after flowers appear, in the San Francisco Bay area, 5½ to 7 months. In New Zealand fruits are borne from early February to May. The fruits fall when mature and are collected daily from the ground and kept cool until slightly soft to the touch. Straw mulch beneath the plants helps avoid bruising. If picked from the tree before they are ready to fall or if eaten before they are fully ripe, the fruits will not have their full richness of flavor.

The 20-year-old Riviera plant referred to above is said to have borne a crop of 2,000 fruits. The yield is poor in India where the maximum crop per season is 100 fruits per plant, probably due to inadequate pollination or flower damage by birds.

New Zealand test plantings have given the following yields: 3rd year, 13.2 lbs (6 kg) per plant; 4,000 lbs/acre, (4,000 kg/ha); 4th year, 26.5 lbs (12 kg) per plant; 8,000 lbs/acre, (8,000 kg/ha); 5th year, 39.7 lbs (18 kg) per plant; 12,000 lbs/acre (12,000 kg/ha). The growers now foresee 66 lbs (30 kg) per plant—25 tons per hectare. In 1978, New Zealand produced 333 tons of feijoas—149 tons to be sold fresh, and 184 tons to be processed.

In New Zealand, flat tomato boxes are employed for shipping feijoas. A case 4½ in (11.25 cm) deep and 12 in (30 cm) to 16 in (40 cm) long and wide holds about 20 lbs (9.07 kg).

Keeping Quality

If the atmosphere is too warm, the interior of the fruit turns brown and decays in 3 to 4 days even though the fruit may appear intact on the surface. In cool storage, undamaged fruits will remain in good condition for one month or longer. In France, fruits harvested in November and December have been kept till spring at a cool temperature and with sufficient humidity. In the early days of its introduction, feijoa shipments were successfully made from France to California despite being 30 days at sea. Today, air transport is essential for New Zealand feijoas en route to Europe. They can be held 1 mo at 32°F (0°C) and then have only a week's life on the market.

Pests and Diseases

The shrub is remarkably pest-resistant. Occasionally it may be attacked by hard wax scale (*Ceroplastes sinensis*) and associated sooty mold in New Zealand and Florida, also greedy scale in New Zealand, by black scale (*Saissetia oleae*) in California and southern Europe. In New Zealand, the larvae of a leaf-rolling caterpillar (*Tortrix spp.*) and of a bagworm moth may eat holes in the leaves but they are effectively controlled with suitable sprays. Fruit flies attack the ripe fruits. A leaf-spotting fungus (*Sphaceloma* sp.) occasionally requires control measures. In Florida, leaf spot is caused by the fungi *Cercospora* sp., *Cylindrocladium scoparium*, and *Phyllosticta* sp.; algal leaf spot by *Cephaleuros virescens*. Thread blight (*Corticium stevensii* Burt. and *Rhizoctonia ramicola*), and mushroom root rot (*Clytocybe tabescens*).

Food Uses

When preparing feijoas for eating or preserving, peeling should be immediately followed by dipping into a weak salt solution or into water containing fresh lemon juice. Both of these methods will prevent the flesh from oxidizing (turning brown). The flesh and pulp (with seeds) are eaten raw as dessert or in salads, or are cooked in puddings, pastry fillings, fritters, dumplings, fruit-sponge-cake, pies or tarts, or employed as flavoring for ice cream or soft drinks. Surplus fruits may be peeled, halved and preserved in sirup in glass jars, or sliced and crystallized, or made into chutney, jam, jelly, conserve, relish, sauce or sparkling wine.

The thick petals are spicy and are eaten fresh by children and sometimes by adults. The petals may be plucked without interfering with fruit set.

Food Value Per 100 g of Edible Portion*	
Moisture	84%
Protein	0.9%
Fat	0.2%
Carbohydrates**	10%
Ash	0.5%
Minerals:	
Potassium	166 mg
Sodium	5 mg
Calcium	4 mg
Magnesium	8 mg
Phosphorus	10 mg
Iron	0.05 mg
Ascorbic Acid	28-35 mg
*Analyses reported in the literature.	

**Sugar 6% compared to 13% in the orange.

The fruit is rich in water-soluble iodine compounds. The percentage varies with locality and from year to year but the usual range is 1.65 to 3.90 mg/kg of fresh fruit. Most types are high in pectin, so that 3 lbs (1.4 kg) of jelly can be made from 1 lb (.45 kg) of fruit.

Jaboticabas (Plate LI)

Little known outside their natural range, these members of the myrtle family, Myrtaceae, are perhaps the most popular native fruit-bearers of Brazil. Generally identified as *Myrciaria cauliflora* Berg. (syn. *Eugenia cauliflora* DC.), the names *jaboticaba, jabuticaba* or *yabuticaba* (for the fruit; *jaboticabeira* for the tree) actually embrace 4 species of very similar trees and fruits: *M. cauliflora, sabará jaboticaba,* also known as *jabuticaba sabará, jabuticaba de Campinas, guapuru, guaperu, hivapuru,* or *ybapuru; M. jaboticaba* Berg., great jaboticaba, also known as *jaboticaba de Sao Paulo, jaboticaba do mato, jaboticaba batuba, jaboticaba grauda; M. tenella* Berg., *jaboticaba macia,* also known as *guayabo colorado, cambui preto, murta do campo, camboinzinho; M. trunciflora* Berg., long-stemmed jaboticaba, also called *jaboticaba de Cabinho,* or *jaboticaba do Pará.*

The word "jaboticaba" is said to have been derived from the Tupi term, *jabotim,* for turtle, and means "like turtle fat", presumably referring to the fruit pulp.

Description

Jaboticaba trees are slow-growing, in *M. tenella,* shrubby, 3½ to 4½ ft (1-1.35 m) high; in *M. trunciflora,* 13 to 23 or rarely 40 ft (4-7 or 12 m); in the other species usually reaching 35 to 40 ft (10.5-12 m). They are profusely branched, beginning close to the ground and slanting upward and outward so that the dense, rounded crown may attain an ultimate spread of 45 ft (13.7 m). The thin outer bark, like that of the guava, flakes off, leaving light patches. Young foliage and branchlets are hairy.

The evergreen, opposite leaves, on very short, downy petioles, are lanceolate or elliptic, rounded at the base, sharply or bluntly pointed at the apex; 1 to 4 in (2.5-10 cm) long, ½ to ¾ in (1.25-2 cm) in width; leathery, dark-green, and glossy. Spectacularly emerging from the multiple trunks and branches in groups of 4, on very short, thick pedicels, the flowers have 4 hairy, white petals and about 60 stamens to ⅙ in (4 mm) long. The fruit, borne in abundance, singly or in clusters, on short stalks, is largely hidden by the foliage and the shade of the canopy, but conspicuous on the lower portions of the trunks. Round, slightly oblate, broad-pyriform, or ellipsoid, with a small disk and vestiges of the 4 sepals at the apex, the fruits vary in size with the species and variety, ranging from ¼ in (6 mm) in *M. tenella* and from ⅝ to 1½ in (1.6-4 cm) in diameter in the other species. The smooth, tough skin is very glossy, bright-green, red-purple, maroon-purple, or so dark a purple as to appear nearly black, slightly acid and faintly spicy in taste; encloses a gelatinous, juicy, translucent, all-white or rose-tinted pulp that clings firmly to the seeds. The fruit has an overall subacid to sweet, grapelike flavor, mildly to disagreeably resinous, and is sometimes quite astringent. There may be 1 to 5 oval to nearly round but flattened, hard to tender, light-brown seeds, ¼ to ½ in (6-12.5 mm) long, but often some are abortive. The fruit has been well likened to a muscadine grape except for the larger seeds.

Origin and Distribution

M. cauliflora is native to the hilly region around Rio de Janeiro and Minas Gerais, Brazil, also around Santa Cruz, Bolivia, Asunción, Paraguay, and northeastern Argentina. *M. jaboticaba* grows wild in the forest around Sao Paulo and Rio de Janeiro; *M. tenella* occurs in the arid zone of Bahia and the mountains of Minas Gerais; in the states of Sao Paulo, Pernambuco and Rio Grande do Sul; also around Yaguarón, Uruguay, and San Martin, Peru. *M. trunciflora* is indigenous to the vicinity of Minas Gerais.

Jaboticabas are cultivated from the southern city of Rio Grande to Bahia, and from the seacoast to Goyaz and Matto Grosso in the west, not only for the fruits but also as ornamental trees. They are most common in parks and gardens throughout Rio de Janeiro and in small orchards all around Minas Gerais. Many cultivated forms are believed to be interspecific hybrids.

An early "hearsay" account of the jaboticabas of Brazil was published in Amsterdam in 1658. The jaboticaba was introduced into California (at Santa Barbara) about 1904. A few of the trees were still living in 1912 but all were gone by 1939. In 1908, Brazil's National Society of Agriculture sent to the United States Department of Agriculture plants of 3 varieties, **'Coroa', 'Murta',** and **'Paulista'.** The first 2 died soon but 'Paulista' lived until 1917. A Dr. W. Hentz bought 6 small inarched plants in Rio Janeiro in 1911 and planted them in City Point, Brevard County, Florida. Only one, variety 'Murta', survived and he moved it to Winter Haven in 1918. It began fruiting in 1932 and continued to bear in great abundance. Another introduction was made by the U.S. Department of Agriculture in 1913 in the form of seeds collected by the plant explorers, P.H. Dorsett, A.D. Shamel, and W. Popenoe from marketed fruits in Rio de Janeiro, the best of which was described as 1½ in (3.8 cm) thick. In 1914, the U.S. Department of Agriculture received seeds from 40 lbs (28 kg) of fruit purchased in the public market in Rio de Janeiro, which appeared different from previous introductions being purple-maroon, round or slightly oblate, and, at most, not quite 1 in (2.5 cm) in diameter. Plants

Fig. 101: A jaboticaba tree in full bloom in Brazil is a striking example of cauliflory (flowers arising from axillary buds on main trunks or older branches).

grown from these seeds, believed to represent more than one species, were distributed to Florida, California and Cuba. A seedling of *M. trunciflora* from this lot was, up until 1928, grown at the Charles Deering estate, Buena Vista, Florida, and then transferred to the then U.S.D.A. Plant Introduction Station (now the Subtropical Horti-culture Research Unit) on Old Cutler Road. It made poor growth in the limestone, but survived.

In 1918, seeds were presented to the U.S. Department of Agriculture by the Director of the Escola Agricola de Lavras in Minas Gerais, and most of the resulting trees were growing at the Brickell Avenue Garden until 1926 when they were killed by the 3 ft (1 m) of salt water pushed over the garden by the disastrous hurricane of that year. Dr. David Fairchild rejoiced that, in 1923, he had set out two of the seedlings at his home, "The Kampong", in Coco-

Plate XLIX
POMEGRANATE
Punica granatum

Plate L
GUAVA
Psidium guajava

Plate LI
JABOTICABA
Myrciaria cauliflora

Plate LII
JAMBOLAN
Syzygium cumini

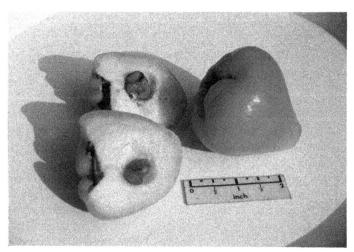

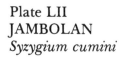

Plate LIII
JAVA APPLE
Syzygium samarangense

Plate LIV
RUMBERRY
Myrciaria floribunda

Plate LV
SAPOTE
Pouteria sapota

Plate LVI
CANISTEL
Pouteria campechiana

Plate LVII
ABIU
Pouteria caimito

Plate LVIII
STAR APPLE
Chrysophyllum cainito

nut Grove and these lived; one fruiting for the first time in 1935. Seedlings of the same lot were successfully grown and fruited heavily at the Atkins Garden of Harvard University at Soledad, near Cienfuegos, Cuba.

In 1920, Dr. Fairchild and P.H. Dorsett took several young trees to Panama and planted them at Juan Mina at sea-level where they grew well and fruited for many years. Later, jaboticabas were set out in the new Summit Botanic Garden. Between 1930 and 1940, plants presumably from the Summit Garden, were installed at the Estacion Agricola de Palmira, in southern Colombia.

Seeds were sent from Washington to the Philippines in 1924. Plants were sent to Puerto Arturo, Honduras, and transferred to the Lancetilla Experimental Garden, at Tela, in 1926 and again in 1929. Other plants were transferred from the Summit Garden in 1928. The trees flourished and fruited well in Honduras. Dr. Hamilton P. Traub, of the Orlando, Florida, branch of the U.S. Department of Agriculture, was establishing a 2½ acre (nearly 1 ha) experimental block of jaboticabas in 1940 for testing and study. At that time there were only a few bearing trees in the state. Soon, nurseries began selling grafted trees and they began appearing in home gardens.

Varieties

M. cauliflora differs mainly from the other species in the large size of the tree and of the fruits. The well-known variety 'Coroa' is believed to belong to this species, also 'Murta' which has smaller leaves and larger fruits. The latter was among those sent to California in 1904.

Among commercial sorts in Brazil are:

'Sabará', a form of *M. cauliflora,* is the most prized and most often planted. The fruit is small, thin-skinned and sweet. The tree is of medium size, precocious, and very productive. Early in season; bears 4 crops a year. Susceptible to rust on flowers and fruits.

'Paulista'—fruit is very large, with thick, leathery skin. The tree is a strong grower and highly productive though it bears a single crop. Later in season than 'Sabará'. Fruits are resistant to rust. Was introduced into California in 1904.

'Rajada'—fruit very large, skin green-bronze, thinner than that of 'Paulista'. Flavor is sweet and very good. The tree is much like that of 'Paulista'. Midseason.

'Branca'—fruit is large, not white, but bright-green; delicious. Tree is of medium size and prolific; recommended for home gardens.

'Ponhema'—fruit is turnip-shaped with pointed apex; large; with somewhat leathery skin. Must be fully ripe for eating raw; is most used for jelly and other preserves. Tree is very large and extremely productive.

'Rujada'—fruit is striped white and purple.

'Roxa'—an old type mentioned by Popenoe as being more reddish than purple, as the name (meaning "red") implies.

'Sao Paulo' (probably *M. jaboticaba*)—tree is large-leaved.

'Mineira'—was introduced into California in 1904.

Pollination

It has been reported from Brazil that solitary jaboticaba trees bear poorly compared with those planted in groups, which indicates that cross-pollination enhances productivity.

Climate

In Brazil, jaboticabas grow from sea-level to elevations of more than 3,000 ft (910 m). At Minas Gerais, the temperature rarely falls below 33°F (0.56°C). Trees in central Florida have lived through freezing weather. In 1917, one very young jaboticaba tree at Brooksville survived a drop in temperature to 18°F (-7.78°C), only the foliage and branches being killed back. In southern Florida, jaboticabas have not been damaged by brief periods of 26°F (-3.33°C).

Soil

Jaboticaba trees grow best on deep, rich, well-drained soil, but have grown and borne well on sand in central Florida and have been fairly satisfactory in the southern part of the state on oolitic limestone.

Propagation

Jaboticabas are usually grown from seeds in South America. These are nearly always polyembryonic, producing 4 to 6 plants per seed. They germinate in 20 to 40 days.

Selected strains can be reproduced by inarching (approach-grafting) or air-layering. Budding is not easily accomplished because of the thinness of the bark and hardness of the wood. Side-veneer grafting is fairly successful. And experimental work has shown that propagation by tissue culture may be feasible.

At the Agricultural Research and Education Center in Homestead, Florida, 6 related genera, including 10 species, were tried as rootstocks in grafting experiments but none was successful. However, *M. cauliflora* scions were satisfactorily joined to rootstock of the same species ⅛ to ¼ in (3-6 mm) thick, bound with parafilm and grown in plastic bags under mist.

Culture

Jaboticaba trees in plantations should be spaced at least 30 ft (9 m) apart each way. Dr. Wilson Popenoe wrote that in Brazil they were nearly always planted too close—about 15 ft (4.5 m) apart, greatly restricting normal development.

Growth is so slow that a seedling may take 3 years to reach 18 in (45 cm) in height. However, a seedling tree in sand at Orlando, Florida, was 15 ft (4.5 m) high when 10 years old. Others on limestone at the United States Department of Agriculture's Subtropical Horticulture Research Unit were shrubby and only 5 to 6 ft (1.5-1.8 m) high when 10 and 11 years old. Seedlings may not bear fruit until 8 to 15 years of age, though one seedling selection flowered in 4 to 5 years. Grafted trees have fruited in 7 years. One planted near Bradenton, Florida, in bagasse-enriched soil started bearing the 6th year. The fruit develops quickly, in 1 to 3 months, after flowering.

Traditionally, jaboticabas have not been given fertilizer in Brazil, the belief prevailing that it might be prejudicial rather than beneficial because of the sensitivity of the root system. Some agronomists have advocated digging a series of pits around the base of the tree and filling them with organic matter enriched with 1 part ammonium sulfate, 2 parts superphosphate, and 1 part potassium chlorate. The pits store and gradually release the nutrients and the water from the fall rains.

In 1978, E.A. Ackerman of the Rare Fruit Council International, Inc., reported on fertilizer experiments with 63 one-year-old and 48 two- and three-year-old seedlings in containers. Better growth was obtained with plants in a mixture of equal amounts of acid sandy muck, vermiculite, and peat, given feedings of 32 g of 14-14-14 slow-release fertilizer (Osmocote), roughly every 2½ months, and 3 gallons (11.4 liters) of well water (pH 7.20) by a drip system every 2 days over a period of 18 months, than plants given other treatments. The addition of chelated iron was of no advantage; chelated zinc retarded growth rate, chelated manganese stopped growth and caused defoliation. Abundant water was found to be essential to survival. Irrigation to promote flowering in the dry season is recommended in Brazil to avoid the detrimental effects of flowering in the rainy season.

Season

The time of fruiting varies with the species and/or cultivar and, of course, the locale. In Rio de Janeiro, *M. cauliflora* fruits in May and *M. jaboticaba* in September. If the trees are heavily irrigated in the dry season, they may bear several crops a year. Trees in southern Florida usually produce 2 crops a year.

Harvesting and Packing

In Brazil, jaboticabas harvested in the interior are shipped crudely in second-hand wooden boxes to urban markets. The toughness of the skin prevents serious bruising if the boxes are handled with some care.

Keeping Quality

Jaboticabas, once harvested, ferment quickly at ordinary temperatures.

Pests and Diseases

If the jaboticaba blooms during a period of drought, many flowers desiccate. If blooming occurs during heavy rains, many flowers will be affected by rust caused by a fungus. The variety 'Sabara' is particularly susceptible to attacks of rust on the flowers and fruits. This is the most serious disease of the jaboticaba in Brazil. The initial signs are circular spots, at first yellow then dark-brown.

Fruit-eating birds are very troublesome to jaboticaba growers in Brazil. To protect the crop, double-folded newspaper pages are placed around individual clusters and tied at the top. If birds are very aggressive, or if there are high winds, the paper must be secured with string at the bottom also. To facilitate this operation, it may be necessary in winter or early spring to do some pruning to make it easier to climb the trees and this will result in protecting a larger portion of the crop. Furthermore, reducing the number of fruits has the effect of increasing the size of those that remain. In Florida, raccoons and opossums make raids on jaboticabas.

Food Uses

Jaboticabas are mostly eaten out-of-hand in South America. By squeezing the fruit between the thumb and forefinger, one can cause the skin to split and the pulp to slip into the mouth. The plant explorers, Dorsett, Shamel and Popenoe, wrote that children in Brazil spend hours "searching out and devouring the ripe fruits." Boys swallow the seeds with the pulp, but, properly, the seeds should be discarded.

The fruits are often used for making jelly and marmalade, with the addition of pectin. It has been recommended that the skin be removed from at least half the fruits to avoid a strong tannin flavor. In view of the undesirability of tannin in the diet, it would be better to peel most of them. The same should apply to the preparation of juice for beverage purposes, fresh or fermented. The aborigines made wine of the jaboticabas, and wine is still made to a limited extent in Brazil.

Food Value Per 100 g of Edible Portion*	
Calories	45.7
Moisture	87.1 g
Protein	0.11 g
Fat	0.01 g
Carbohydrates	12.58 g
Fiber	0.08 g
Ash	0.20 g
Calcium	6.3 mg
Phosphorus	9.2 mg
Iron	0.49 mg
Carotene	–
Thiamine	0.02 mg
Riboflavin	0.02 mg
Niacin	0.21 mg
Ascorbic Acid**	22.7 mg
Amino Acids:	
Tryptophan	1 mg
Methionine	–
Lysine	7 mg

*Analyses made in 1955 at the Laboratories FIM de Nutricion, Havana, Cuba.

**Others have shown 30.7 mg.

Toxicity

Regular, quantity consumption of the skins should be avoided because of the high tannin content, inasmuch as tannin is antinutrient and carcinogenic if intake is frequent and over a long period of time.

Medicinal Uses

The astringent decoction of the sun-dried skins is prescribed in Brazil as a treatment for hemoptysis, asthma, diarrhea and dysentery; also as a gargle for chronic inflammation of the tonsils. Such use also may lead to excessive consumption of tannin.

Jambolan (Plate LII)

This member of the Myrtaceae is of wider interest for its medicinal applications than for its edible fruit. Botanically it is *Syzygium cumini* Skeels (syns. *S. jambolanum* DC., *Eugenia cumini* Druce, *E. jambolana* Lam., *E. djouat* Perr., *Myrtus cumini* L., *Calyptranthes jambolana* Willd.). Among its many colloquial names are Java plum, Portuguese plum, Malabar plum, black plum, purple plum, and, in Jamaica, damson plum; also Indian blackberry. In India and Malaya it is variously known as *jaman, jambu, jambul, jambool, jambhool, jamelong, jamelongue, jamblang, jiwat, salam,* or *koriang*. In Thailand, it is *wa,* or *ma–ha;* in Laos, *va;* Cambodia, *pring bai* or *pring das krebey;* in Vietnam, *voi rung;* in the Philippines, *duhat, lomboy, lunaboy* or other dialectal appelations; in Java, *djoowet,* or *doowet*. In Venezuela, local names are *pésjua extranjera* or *guayabo pésjua;* in Surinam, *koeli, jamoen,* or *druif* (Dutch for "grape"); in Brazil, *jambuláo, jaláo, jameláo* or *jambol*.

Description

The jambolan is fast-growing, reaching full size in 40 years. It ranges up to 100 ft (30 m) in India and Oceania; up to 40 or 50 ft (12–15 m) in Florida; and it may attain a spread of 36 ft (11 m) and a trunk diameter of 2 or 3 ft (0.6–0.9 m). It usually forks into multiple trunks a short distance from the ground. The bark on the lower part of the tree is rough, cracked, flaking and discolored; further up it is smooth and light-gray. The turpentine-scented evergreen leaves are opposite, 2 to 10 in (5–25 cm) long, 1 to 4 in (2.5–10 cm) wide; oblong-oval or elliptic, blunt or tapering to a point at the apex; pinkish when young; when mature, leathery, glossy, dark-green above, lighter beneath, with conspicuous, yellowish midrib. The fragrant flowers, in 1- to 4-in (2.5–10 cm) clusters, are ½ in (1.25 cm) wide, 1 in (2.5 cm) or more in length; have a funnel-shaped calyx and 4 to 5 united petals, white at first, then rose-pink, quickly shed leaving only the numerous stamens.

The fruit, in clusters of just a few or 10 to 40, is round or oblong, often curved; ½ to 2 in (1.25–5 m) long, and usually turns from green to light-magenta, then dark-purple or nearly black as it ripens. A white-fruited form has been reported in Indonesia. The skin is thin, smooth, glossy, and adherent. The pulp is purple or white, very juicy, and normally encloses a single, oblong, green or brown seed, up to 1½ in (4 cm) in length, though some fruits have 2 to 5 seeds tightly compressed within a leathery coat, and some are seedless. The fruit is usually astringent, sometimes unpalatably so, and the flavor varies from acid to fairly sweet.

Origin and Distribution

The jambolan is native in India, Burma, Ceylon and the Andaman Islands. It was long ago introduced into and became naturalized in Malaya. In southern Asia, the tree is venerated by Buddhists, and it is commonly planted near Hindu temples because it is considered sacred to Krishna. The leaves and fruits are employed in worshipping the elephant-headed god, Ganesha or Vinaijaka, the personification of "Pravana" or "Om", the apex of Hindu religion and philosophy.

The tree is thought to be of prehistoric introduction into the Philippines where it is widely planted and naturalized, as it is in Java and elsewhere in the East Indies, and in Queensland and New South Wales, also on the islands of Zanzibar and Pemba and Mombasa and adjacent coast of Kenya. In Ghana, it is found only in gardens. Introduced into Israel perhaps about 1940, it grows vigorously there but bears scantily, the fruit is considered valueless but the tree is valued as an ornamental and for forestry in humid zones. It is grown to some extent in Algiers.

By 1870, it had become established in Hawaii and, because of seed dispersal by mynah birds, it occurs in a semiwild state on all the Hawaiian islands in moist areas below 2,000 ft (600 m). There are vigorous efforts to exterminate it with herbicides because it shades out desirable forage plants. It is planted in most of the inhabited valleys in the Marquesas. It was in cultivation in Bermuda, Cuba, Haiti, Jamaica, the French Islands of the Lesser Antilles and Trinidad in the early 20th Century; was introduced into Puerto Rico in 1920; but still has remained little-known in the Caribbean region. At the Lancetilla Experimental Garden at Tela, Honduras, it grows and fruits well. It is seldom planted elsewhere in tropical America but is occasionally seen in Guatemala, Belize, Surinam, Venezuela and Brazil.

The Bureau of Plant Industry of the United States Department of Agriculture received jambolan seeds from the Philippines in 1911, from Java in 1912, from Zanzibar and again from the Philippines in 1920. The tree flourishes in California, especially in the vicinity of Santa Barbara, though the climate is not congenial for production or ripening of fruit. In southern Florida, the tree was rather commonly planted in the past. Here, as in Hawaii, fruiting is heavy, only a small amount of the crop has been utilized in home preserving. The jambolan has lost popularity, as it has in Malaya where it used to be frequently grown in gardens. Heavy crops litter streets, sidewalks and lawns, attracting insects, rapidly fermenting and creating a foul atmosphere. People are eager to have the trees cut down. Where conditions favor spontaneous growth, the seedlings become a nuisance, as well.

Varieties

The common types of jambolan in India are: 1) *Ra Jaman*, with large, oblong fruits, dark-purple or bluish, with pink, sweet pulp and small seeds; 2) *Kaatha*, with small, acid fruits. Among named cultivars are, mainly, 'Early Wild', 'Late Wild', 'Pharenda'; and, secondarily, 'Small Jaman' and 'Dabka' ('Dubaka'). In Java, the small form is called *Djoowet kreekil;* a seedless form is *Djoowet booten.* In southern Malaya, the trees are small-leaved with small flower clusters. Farther north, the variety called 'Krian Duat' has larger, thicker leaves and red inner bark. Fruits with purple flesh are more astringent than the white-fleshed types.

Climate

The jambolan tree grows well from sea-level to 6,000 ft (1,800 m) but, above 2,000 ft (600 m) it does not fruit but can be grown for its timber. It develops most luxuriantly in regions of heavy rainfall, as much as 400 in (1,000 cm) annually. It prospers on river banks and has been known to withstand prolonged flooding. Yet it is tolerant of drought after it has made some growth. Dry weather is desirable during the flowering and fruiting periods. It is sensitive to frost when young but mature trees have been undamaged by brief below-freezing temperatures in southern Florida.

Soil

Despite its ability to thrive in low, wet areas, the tree does well on higher, well-drained land whether it be in loam, marl, sand or oolitic limestone.

Propagation

Jambolan seeds lose viability quickly. They are the most common means of dissemination, are sown during the rainy season in India, and germinate in approximately 2 weeks. Semi-hardwood cuttings, treated with growth-promoting hormones have given 20% success and have grown well. Budding onto seedlings of the same species has also been successful. Veneer-grafting of scions from the spring flush has yielded 31% survivors. The modified Forkert method of budding may be more feasible. When a small-fruited, seedless variety in the Philippines was budded onto a seeded stock, the scion produced large fruits, some with seeds and some without. Approach-grafting and inarching are also practiced in India. Air-layers treated with 500 ppm indolebutyric acid have rooted well in the spring (60% of them) but have died in containers in the summer.

Culture

Seedlings grow slowly the first year, rapidly thereafter, and may reach 12 ft (3.65 m) in 2 years, and begin bearing in 8 to 10 years. Grafted trees bear in 4 to 7 years. No particular cultural attention seems to be required, apart from

frost protection when young and control measures for insect infestations. In India, organic fertilizer is applied after harvest but withheld in advance of flowering and fruiting to assure a good crop. If a tree does not bear heavily, it may be girdled or root-pruned to slow down vegetative growth.

The tree is grown as shade for coffee in India. It is wind-resistant and sometimes is closely planted in rows as a windbreak. If topped regularly, such plantings form a dense, massive hedge. Trees are set 20 ft (6 m) apart in a windbreak; 40 ft (12 m) apart along roadsides and avenues.

Fruiting Season

The fruit is in season in the Marquesas in April; in the Philippines, from mid-May to mid-June. In Hawaii, the crop ripens in late summer and fall. Flowering occurs in Java in July and August and the fruits ripen in September and October. In Ceylon, the tree blooms from May to August and the fruit is harvested in November and December. The main fruiting season in India and southern Florida (where the tree blooms principally in February and March) extends through late May, June and July. Small second crops from late blooms have been observed in October. Individual trees may habitually bear later than others.

Harvesting and Yield

In India, the fruits are harvested by hand as they ripen and this requires several pickings over the season. Indian horticulturists have reported a crop of 700 fruits from a 5-year-old tree. The production of a large tree may be overwhelming to the average homeowner.

Pests and Diseases

In Florida, some jambolan trees are very susceptible to scale insects. The whitefly, *Dialeurodes eugeniae*, is common on jambolans throughout India. Of several insect enemies in South India, the most troublesome are leaf-eating caterpillars: *Carea subtilis, Chrysocraspeda olearia, Phlegetonia delatrix, Oenospila flavifuscata, Metanastria hyrtaca,* and *Euproctis fraterna.* These pests may cause total defoliation. The leafminer, *Acrocercops phaeospora,* may be a major problem at times. *Idiocerus atkinsoni* sucks the sap of flowering shoots, buds and flower clusters, causing them to fall.

The fruits are attacked by fruit flies (*Dacus diversus* in India), and are avidly eaten by birds and four-footed animals (jackals and civets). In Australia, they are a favorite food of the large bat called "flying fox."

Diseases recorded as found on the jambolan by inspectors of the Florida Department of Agriculture are: black leaf spot (*Asterinella puiggarii*); green scurf or algal leaf spot (*Cephaleuros virescens*); mushroom root rot (*Clitocybe tabescens*); anthracnose (*Colletotrichum gloeosporioides*); and leaf spot caused by *Phyllosticta eugeniae.*

Food Uses

Jambolans of good size and quality, having a sweet or subacid flavor and a minimum of astringency, are eaten raw and may be made into tarts, sauces and jam. Astringent fruits are improved in palatability by soaking them in salt water or pricking them, rubbing them with a little salt, and letting them stand for an hour. All but decidedly inferior fruits have been utilized for juice which is much like grape juice. When extracting juice from cooked jambolans, it is recommended that it be allowed to drain out without squeezing the fruit and it will thus be less astringent. The white-fleshed jambolan has adequate pectin and makes a very stiff jelly unless cooking is brief. The more common purple-fleshed yields richly colored jelly but is deficient in pectin and requires the addition of a commercial jelling agent or must be combined with pectin-rich fruits such as unripe or sour guavas, or ketembillas.

Good quality jambolan juice is excellent for sherbet, sirup and "squash". In India, the latter is a bottled drink prepared by cooking the crushed fruits, pressing out the juice, combining it with sugar and water and adding citric acid and sodium benzoate as a preservative.

Food Value Per 100 g of Edible Portion*	
Moisture	83.7-85.8 g
Protein	0.7-0.129 g
Fat	0.15-0.3 g
Crude Fiber	0.3-0.9 g
Carbohydrates	14.0 g
Ash	0.32-0.4 g
Calcium	8.3-15 mg
Magnesium	35 mg
Phosphorus	15-16.2 mg
Iron	1.2-1.62 mg
Sodium	26.2 mg
Potassium	55 mg
Copper	0.23 mg
Sulfur	13 mg
Chlorine	8 mg
Vitamin A	80 I.U.
Thiamine	0.008-0.03 mg
Riboflavin	0.009-0.01 mg
Niacin	0.2-0.29 mg
Ascorbic Acid	5.7-18 mg
Choline	7 mg
Folic Acid	3 mcg

*Values reported from Asian and tropical American analyses.

Also present are gallic acid and tannin and a trace of oxalic acid.

In Goa and the Philippines, jambolans are an important source of wine, somewhat like Port, and the distilled liquors, brandy and "jambava" have also been made from the fermented fruit. Jambolan vinegar, extensively made throughout India, is an attractive, clear purple, with a pleasant aroma and mild flavor.

Virmani gives the following vinegar analysis: specific gravity, 1.0184; total acidity (as acetic acid), 5.33 per 100 cc; volatile acid (as ascetic acid), 5.072 per 100 cc; fixed acidity, as citric, .275%; total solids, 4.12 per 100 cc; ash, .42; alkalinity of ash, 32.5 (N/10 alkali); nitrogen, .66131; total sugars, .995; reducing sugars, .995; non-volatile reducing sugars, .995; alcohol, .159% by weight; oxidation value, (K MnO₁), 186.4; iodine value, 183.7; ester value, 40.42.

Other Uses

Nectar: The jambolan tree is of real value in apiculture. The flowers have abundant nectar and are visited by bees (*Apis dorsata*) throughout the day, furnishing most of the honey in the Western Ghats at an elevation of 4,500 ft (1,370 m) where the annual rainfall is 300 to 400 in (750-1,000 cm). The honey is of fine quality but ferments in a few months unless treated.

Leaves: The leaves have served as fodder for livestock and as food for tassar silkworms in India. In Zanzibar and Pemba, the natives use young jambolan shoots for cleaning their teeth. Analyses of the leaves show: crude protein, 9.1%; fat, 4.3%; crude fiber, 17.0%; ash, 6.0%; calcium, 1.3%; phosphorus, 0.19%. They are rich in tannin and contain the enzymes esterase and galloyl carboxylase which are presumed to be active in the biosynthesis of the tannins.

The essential oil distilled from the leaves is used to scent soap and is blended with other materials in making inexpensive perfume. Its chemical composition has been reported by Craveiro et al. in Brazil. It consists mainly of mono- or sesqui-terpene hydrocarbons which are "very common in essential oils."

Bark: Jambolan bark yields durable brown dyes of various shades depending on the mordant and the strength of the extract. The bark contains 8 to 19% tannin and is much used in tanning leather and preserving fishing nets.

Wood: The wood is red, reddish-gray or brownish-gray, with close, straight grain. The very small, oval pores are often connected by waxy belts of loose tissue. The medullary rays are so fine as to be clearly visible only when greatly magnified. When fresh, the sapwood is attacked by powerpost beetles, pinhole borers and ambrosia beetles. Both sapwood and heartwood are perforated by the borer, *Aeolesthes holosericea*, if the bark is left on for as long as 10 months. Air-dried wood is apt to crack and split. When kiln dried, the heartwood is hard, difficult to work but polishes well. It is durable in water and resistant to borers and termites; tends to warp slightly. In India, it is commonly used for beams and rafters, posts, bridges, boats, oars, masts, troughs, well-lining, agricultural implements, carts, solid cart wheels, railway sleepers and the bottoms of railroad cars. It is sometimes made into furniture but has no special virtues to recommend it for cabinetwork. It is a fairly satisfactory fuel.

Medicinal Uses: The jambolan has received far more recognition in folk medicine and in the pharmaceutical trade than in any other field. Medicinally, the fruit is stated to be astringent, stomachic, carminative, antiscorbutic and diuretic. Cooked to a thick jam, it is eaten to allay acute diarrhea. The juice of the ripe fruit, or a decoction of the fruit, or jambolan vinegar, may be administered in India in cases of enlargement of the spleen, chronic diarrhea and urine retention. Water-diluted juice is used as a gargle for sore throat and as a lotion for ringworm of the scalp.

The seeds, marketed in ¼ inch (7 mm) lengths, and the bark are much used in tropical medicine and are shipped from India, Malaya and Polynesia, and, to a small extent, from the West Indies, to pharmaceutical supply houses in Europe and England. Extracts of both, but especially the seeds, in liquid or powdered form, are freely given orally, 2 to 3 times a day, to patients with diabetes mellitus or glycosuria. In many cases, the blood sugar level reportedly is quickly reduced and there are no ill effects. However, in some quarters, the hypoglycemic value of jambolan extracts is disclaimed. Mercier, in 1940, found that the aqueous extract of the seeds, injected into dogs, lowered the blood sugar for long periods, but did not do so when given orally. Reduction of blood sugar was obtained in alloxan diabetes in rabbits. In experiments at the Central Drug Research Institute, Lucknow, the dried alcoholic extract of jambolan seeds, given orally, reduced blood sugar and glycosuria in patients.

The seeds are claimed by some to contain an alkaloid, jambosine, and a glycoside, jambolin or antimellin, which halts the diastatic conversion of starch into sugar. The seed extract has lowered blood pressure by 34.6% and this action is attributed to the ellagic acid content. This and 34 other polyphenols in the seeds and bark have been isolated and identified by Bhatia and Bajaj.

Other reported constituents of the seeds are: protein, 6.3-8.5%; fat, 1.18%; crude fiber, 16.9%; ash, 21.72%; calcium, 0.41%; phosphorus, 0.17%; fatty acids (palmitic, stearic, oleic and linoleic); starch, 41%; dextrin, 6.1%; a trace of phytosterol; and 6 to 19% tannin.

The leaves, steeped in alcohol, are prescribed in diabetes. The leaf juice is effective in the treatment of dysentery, either alone or in combination with the juice of mango or emblic leaves. Jambolan leaves may be helpful as poultices on skin diseases. They yield 12 to 13% tannin (by dry weight).

The leaves, stems, flowerbuds, opened blossoms, and bark have some antibiotic activity. A decoction of the bark is taken internally for dyspepsia, dysentery, and diarrhea and also serves as an enema. The root bark is similarly employed. Bark decoctions are taken in cases of asthma and bronchitis and are gargled or used as mouthwash for the astringent effect on mouth ulcerations, spongy gums, and stomatitis. Ashes of the bark, mixed with water, are spread over local inflammations, or, blended with oil, applied to burns. In modern therapy, tannin is no longer approved on burned tissue because it is absorbed and can cause cancer. Excessive oral intake of tannin-rich plant products can also be dangerous to health.

Malay Apple

A delight to the eye in every respect, the Malay apple is much admired for the beauty of the tree, its flowers and its colorful, glistening fruits, without parallel in the family Myrtaceae. Botanically identified as *Syzygium malaccense* Merr. & Perry (syns. *Eugenia malaccensis* L., *Jambos malaccensis* DC.), this species has earned a few alternate English names including Malay rose-apple, mountain apple, water apple, and, unfortunately, Otaheite apple, which is better limited to the ambarella, *Spondias dulcis* Park., and cashew, or French cashew (Guyana) or Otaheite cashew (India) because of its resemblance to the cashew apple, the pseudofruit or swollen fruit-stalk of the cashew nut.

In Malaya there are many local names including *jambu merah, jambu bar, jambu bol, jambu melaka, jambu kling* and *jambu kapal*. In Thailand, it is *chom-phu-sa-raek* or *chom-phu-daeng;* in Cambodia, *chompuh kraham;* in Vietnam, *man hurong tau;* in Indonesia, *darsana, jambu tersana,* or *djamboo bol;* in the Philippines, *makopang-kalabau* or *tersana;* in Guam, *makupa;* in Tahiti, *ahia;* in Hawaii, *ohia*. In the French language it is *jambosier rouge, poire de Malaque, pomme Malac* (corrupted to pomerac), *pomme de Malaisie,* and *pomme de Tahiti*. Among Spanish names are: *pomarosa,* or *pomarrosa, Malaya* (Puerto Rico); *manzana* (Costa Rica), *marañon japonés* (El Salvador), *pomarosa de Malaca* (Colombia); *pera de agua* or *pomagás* (Venezuela); and *marañon de Curacao* (Panama), though the somewhat similar plant in Curacao is *S. samarangense* Merr. & Perry, locally called *cashu di Surinam,* in Papiamento, *Curacaose appel,* in Dutch. The latter species has yellowish-white flowers and light-red, greenish-white or cream-colored fruits. (See Java apple pp. 381-2.)

Description

The Malay apple tree is rather fast-growing, reaching 40 to 60 ft (12-18 m) in height, and has an erect trunk to 15 ft (4.5 m) in circumference and a pyramidal or cylindrical crown. Its evergreen leaves are opposite, short-petioled, elliptic-lanceolate or oblanceolate; soft-leathery, dark-green and fairly glossy on the upper surface, paler beneath; 6 to 18 in (15-45 cm) long, 3½ to 8 in (9-20 cm) wide. The veins are indistinct above, but they and the pale midrib are prominent on the underside. New growth is wine-red at first, changing to pink-buff. The abundant flowers, only mildly fragrant, and borne on the upper

Fig. 102: Glossy, red, juicy, Malay apples (*Syzygium malaccense*) are sold in markets and along streets in warm areas of the Old and New World.

trunk and along leafless portions of mature branches in short-stalked clusters of 2 to 8, are 2 to 3 in (5-7.5 cm) wide, and composed of a funnel-like base topped by 5 thick, green sepals, 4 usually pinkish-purple to dark-red (sometimes white, yellow or orange) petals, and numerous concolorous stamens to 1½ in (4 cm) long tipped with yellow anthers. Though showy, the flowers are hidden by the foliage until they fall and form a lovely carpet on the ground. The fruit, oblong, obovoid, or bell-shaped, 2 to 4 in (5-10 cm) long, 1 to 3 in (2.5-7.5 cm) wide at the apex, has thin, smooth, waxy skin, rose-red or crimson or sometimes white with streaks of red or pink, and white, crisp or spongy, juicy flesh of very mild, sweetish flavor. There may be a single oblate or nearly round seed or 2 hemispherical seeds, ⅝ to ¾ in (1.6-2 cm) in width, light-brown externally, green internally and somewhat meaty in texture. The fruits of some trees are entirely seedless.

Origin and Distribution

The Malay apple is presumed to be a native of Malaysia. It is commonly cultivated from Java to the Philippines and Vietnam, also in Bengal and South India. Portuguese voyagers carried it from Malacca to Goa and from there it was introduced into East Africa. It must have spread throughout the Pacific Islands in very early times for it is featured in Fijian mythology and the wood was used by ancient Hawaiians to make idols. Indeed, it has been recorded that, before the arrival of missionaries in Hawaii, there were no fruits except bananas, coconuts and the Malay apple. The flowers are considered sacred to Pele, the fiery volcano goddess. Captain Bligh conveyed small trees of 3 varieties from the islands of Timor and Tahiti to Jamaica in 1793. The tree was growing under glass in Cambridge, Massachusetts, in 1839, and specimens were fruiting in Bermuda in 1878.

Eggers, who studied the flora of St. Croix, reported seeing naturalized trees in shaded valleys during his stay on the island from 1870 to 1876. The Malay apple was unknown in Puerto Rico in 1903 but must have arrived soon after. Britton and Wilson observed 2 trees 43 ft (13 m) high at Happy Hollow in 1924. Thereafter, the tree was rather frequently planted as an ornamental or windbreak. Perhaps the Portuguese were responsible for its introduction into Brazil, for it is cultivated there, as it is also in Surinam and Panama. Dr. David Fairchild sent seeds from Panama to the United States Department of Agriculture in 1921. In 1929, young trees from the Canal Zone were transported to the Lancetilla Experimental Gardens at Tela, Honduras, where they flourished and fruited. The Malay apple is sometimes seen in other parts of Central America, including Belize, El Salvador and Costa Rica, much more frequently in parks and gardens in Venezuela. The fruits are sold in local markets and along the streets wherever the tree is grown.

Varieties

Ochse mentions an oblong to pear-shaped, white form called *djamboo pootih, djamboo bodas,* or *djamboo kemang,* which, in Java, is less flavorful than the red type. He says that there are many forms because of seedling variation. A large, especially sweet and juicy clone was introduced into the Philippines from Hawaii in 1922.

Climate

The Malay apple is strictly tropical, too tender for Florida and California except under very unusual conditions. It is naturalized and cultivated from sea-level to 9,000 ft (2,740 m) in valleys and on mountain slopes of the lowest forest zone of the Hawaiian Islands, and is grown up to 2,000 ft (610 m) in Ceylon and Puerto Rico. The tree needs a humid climate, with an annual rainfall of 60 in (152 cm) or more.

Soil

The tree grows vigorously on a range of soil types from sand to heavy clay. It tolerates moderately acid soil, reacts unfavorably to highly alkaline situations. In India, it grows best on the banks of ponds, lakes and streams where there is good drainage and no standing water. It is reported to be one of the first trees to spring up in new lava flows in Hawaii.

Propagation

Malay apple seeds germinate readily. Many sprout on the ground under the tree. While seed propagation is common, superior types are multiplied by budding onto their own seedlings. Air-layering has been successful and cuttings have been rooted in sand in Hawaii. Seeds are planted no more than 1½ in (4 cm) deep in nurseries or directly in the field. They will germinate in 2 to 4 weeks and, if in nurseries, the seedlings are transplanted to the field when 8 months old. Cuttings are ready for transplanting in 6 weeks after rooting.

Culture

In India, Malay apple trees are spaced 26 to 32 feet (8-10 m) apart in fields prepared and enriched as for any other crop, and thereafter they require little care except for elimination of weeds and periodic fertilization and plentiful irrigation in very dry weather.

Pests and Diseases

Young Malay apple trees are frequently attacked by termites in India. It is reported that sap-feeders, defoliators, miners and borers have been found on the foliage and on dead stems.

Season

In Java, the tree flowers in May and June and the fruits ripen in August and September. The fruiting season is about the same around Castleton Gardens in Jamaica but at the lower level of Kingston it is earlier and ends during the first week of June. In India, the main crop occurs from May to July and there is often a second crop in November and December. In Puerto Rico, the tree may flower 2 or 3 times a year, in spring, summer and fall, the blooming season covering 40 to 60 days. The spring and fall flowering seasons produce the biggest crops. Fruits mature in 60 days from the full opening of the flowers and they fall quickly after they become fully ripe and deteriorate rapidly. For marketing, they must be hand-picked to avoid damage and to have longer shelf-life.

Yield

The yield varies from 48 to 188 lbs (21-85 kg) per tree.

Food Uses

The ripe fruit is eaten raw though many people consider it insipid. It is best stewed with cloves or other flavoring and served with cream as dessert. Asiatic people in Guyana stew the peeled fruit, cooking the skin separately to make a sirup which they add to the cooked fruit. Malayan people may add the petals of the red-flowered hibiscus (*Hibiscus rosa-sinensis* L.) to make the product more colorful. Malay apples are often cooked with acid fruits to the benefit of both. They are sometimes made into sauce or preserves. The slightly unripe fruits are used for making jelly and pickles.

In Puerto Rico, both red and white table wines are made from the Malay apple. The fruits are picked as soon as they are fully colored (not allowed to fall) and immediately dipped in boiling water for one minute to destroy surface bacteria and fungi. The seeds are removed and, for red wine, the fruits are passed through a meat grinder and the resulting juice and pulp weighed. To this material, they add twice the amount of water and 1½ lbs (680 g) of white sugar per gallon, and pour into sterilized barrels with the mouth covered soon with cheesecloth. Yeast is added and a coil inserted to maintain circulation of the water. The barrels are kept in the coolest place possible for 6 months to 1 year, then the wine is filtered. It will be of a pale-rose color so artificial color is added to give it a rich-red hue. In making white wine, the fruits are peeled, the only liquid is the fruit juice, and less sugar is used, only 1¼ lbs (565 g) per gallon, so as to limit alcohol formation over a fermenting period of 3 to 6 months.

In Indonesia, the flowers are eaten in salads or are preserved in sirup. Young leaves and shoots, before turning green, are consumed raw with rice or are cooked and eaten as greens.

Food Value Per 100 g of Edible Portion*	
Moisture	90.3–91.6 g
Protein	0.5–0.7 g
Fat	0.1–0.2 g
Fiber	0.6–0.8 g
Ash	0.26–0.39 g
Calcium	5.6–5.9 mg
Phosphorus	11.6–17.9 mg
Iron	0.2–0.82 mg
Carotene	0.003–0.008 mg
(Vitamin A)	3–10 I.U.
Thiamine	15–39 mcg
Riboflavin	20–39 mcg
Niacin	0.21–0.40 mg
Ascorbic Acid	6.5–17.0 mg

*According to analyses made in Hawaii, El Salvador and Ghana.

Other Uses

Wood: The timber is reddish, soft to hard, tough and heavy, but inclined to warp. It is difficult to work, but is employed for construction, railway ties, and for fashioning bowls and poi-boards in Hawaii.

Medicinal Uses: According to Akana's translation of *Hawaiian Herbs of Medicinal Value*, the astringent bark has been much used in local remedies. It is pounded together with salt, the crushed material is strained through coconut husk fiber, and the juice poured into a deep cut. "The patient must exercise absolute self-control as the liquid burns its way into the flesh and nerves."

In the Molucca, or Spice, Islands, a decoction of the bark is used to treat thrush. Malayans apply a powder of the dried leaves on a cracked tongue. A preparation of the root is a remedy for itching. The root acts as a diuretic and is given to alleviate edema. The root bark is useful against dysentery, also serves as an emmenagogue and abortifacient. Cambodians take a decoction of the fruit, leaves or seeds as a febrifuge. The juice of crushed leaves is applied as a skin lotion and is added to baths. In Brazil, various parts of the plant are used as remedies for constipation, diabetes, coughs, pulmonary catarrh, headache and other ailments. Seeded fruits, seeds, bark and leaves have shown antibiotic activity and have some effect on blood pressure and respiration.

Java Apple (Plate LIII)

Much less known than the Malay Apple, this member of the Myrtaceae is botanically identified as *Syzygium samarangense* Merr. & Perry (syns. *S. javanicum* Miq.; *Eugenia javanica* Lam. in part; *E. alba* Roxb.). Among its various vernacular names are: samarang rose apple, *djamboe semarang* (Indonesia); *jambu ayer rhio* (Malaya); *pini jambu* (Ceylon); *jumrool, jamrul,* or *amrool* (India); *chom pu kao,* or *chom pu kio* (Thailand); *makopa* (Philippines); *cashu di Surinam,* or *Curacaose appel* (Curacao); wax apple, wax jambu and water apple, generally.

Description

The tree, 16 to 50 ft (5–15 m) tall, has a short trunk 10 to 12 in (25–30 cm) thick, and open, widespreading crown, and pinkish-gray, flaking bark. The opposite leaves are nearly sessile, elliptic-oblong, rounded or slightly cordate at the base; yellowish to dark bluish-green; 4 to 10 in (10–25 cm) long and 2 to 4¾ in (5–12 cm) wide; very aromatic when crushed. Flowers, borne in drooping panicles of 3 to 30 at the branch tips or in smaller clusters in the axils of fallen leaves, are fragrant, yellowish-white, ¾ to 1½ in (2–4 cm) broad, 4-petalled, with numerous stamens

⅗ to 1 in (1.5–2.5 cm) long. The waxy fruit, usually light-red, sometimes greenish-white or cream-colored, is pear-shaped, narrow at the base, very broad, flattened, indented and adorned with the 4 fleshy calyx lobes at the apex; 1⅓ to 2 in (3.4–5 cm) long, 1¾ to 2⅛ in (4.5–5.4 cm) wide. The skin is very thin, the flesh white, spongy, dry to juicy, subacid and very bland in flavor. There may be 1 or 2 somewhat rounded seeds 3/16 to 5/16 in (0.5–0.8 cm) wide, or none.

Origin and Distribution

The tree is indigenous from Malaya to the Andaman and Nicobar Islands where there are wild trees in the coastal forests. It was introduced into the Philippines in prehistoric times and is widely grown throughout those islands. It is common in Thailand, Cambodia, Laos, Vietnam and Taiwan, frequently cultivated in India and in Zanzibar and Pemba, but primarily as an ornamental, seldom for its fruits which are little valued. It was introduced into Jamaica before 1903 and also into Surinam and the islands of Curacao, Aruba and Bonaire. A few trees have been grown in Israel but have borne sparsely.

Climate

The Java apple is extra-tropical, growing only at the lower altitudes—up to 4,000 ft (1,220 m)—in India. It does best in parts of the Philippines that have a long dry season.

Soil

The soil must be fertile, or the crops will be small and the fruit quality poor.

Propagation

The trees grow spontaneously from seed. Preferred types are reproduced by layering, budding onto their own rootstocks, or onto seedlings of *S. densiflorum* A. DC., (the beautiful Wild Rose Apple of Malaya, which has edible flowers, undesirable fruits, but is not attacked by termites). Sometimes the Java apple is grafted onto the cultivated Rose Apple (q.v.).

Culture

If planted in orchards, the trees are spaced 26 to 32 ft (8-10 m) apart and are given a minimum of attention.

Season

In Ceylon, the fruits are ripe from March to May; in India, the tree blooms in March and April and the fruit ripens in May and June; in Java, flowering occurs from April to June and fruiting from June to August.

Yield

The Java apple is a heavy bearer on good soil. When 5 years old it may yield a crop of 700 fruits.

Food Uses

In Malaya, the greenish fruits are eaten raw with salt or may be cooked as a sauce. They are also stewed with true apples. The pink fruits are juicier and more flavorful and suitable for eating out-of-hand or cooking without accompaniments except sugar.

Food Value Per 100 g of Edible Portion*	
Moisture	91.40-92.96 g
Protein	0.50 g
Sugar	6.56 g
Iron	0.001 g
Ash	0.21-0.27 g
Calcium	0.01 g
Phosphorus	0.03 g
Sulphuric Acid	0.17%
Citric Acid	0.15%

*Analyses made in the Philippines.

Other Uses

Wood: The wood is red, coarse, hard; used for constructing huts in the Andaman and Nicobar Islands.

Medicinal Uses: The flowers are astringent and used in Taiwan to treat fever and halt diarrhea. Investigators have found their principal constituent to be tannin. They also contain desmethoxymatteucinol, 5-*O*-methyl-4'-desmethoxymatteucinol, oleanic acid and *B*-sitosterol. They show weak antibiotic action against *Staphylococcus aureus, Mycobacterium smegmatis,* and *Candida albicans.*

Water Apple

The water apple is the least of the small group of somewhat similar fruits of the genus *Syzygium* (family Myrtaceae). This species, *S. aqueum* Alst. (syn. *Eugenia aquea* Burm. f.), also known as watery rose apple, is distinguished in Malaya as *jambu chili, jambu ayer, jambu ayer mawar,* or *jambu penawar;* in Indonesia as *djamboo aer, djamboo wer,* or *djamboo wir.* In the Philippines, it is called *tambis;* in Thailand, it is *chom-phu-pa.*

Description

The tree may reach 10 or even 32 ft (3-10 m); has a short, crooked trunk branching close to the ground, and a non-symmetrical, open crown. The opposite leaves, on very short, thick petioles, are obovate- or elliptic-oblong, cordate at the base and clasping the twig; blunt and notched or short-pointed at the apex; 2 to 10 in (5-25 cm) long, 1 to 6⅜ in (2.5-16 cm) wide; dull, light-green above, yellowish-green beneath; leathery; not aromatic or only slightly so when crushed. Flowers, faintly fragrant, are borne in loose terminal or axillary clusters of 3 to 7, mostly hidden by the foliage. The 4-parted calyx and 4 petals are pale-yellow, yellowish-white or pinkish and there are numerous concolorous stamens to ¾ in (2 cm) long. Thin-skinned and shining, the fruit varies from white, to light-red or red, is pear-shaped with a narrow neck and broad apex; ⅝ to ¾ in (1.6-2 cm) long, 1 to 1⅓ in (2.5-3.4 cm) wide. The apex is concave; bears the thick calyx segments and the protruding, slender, bristle-like style. The flesh is white or pink, mildly fragrant, dry or juicy, crisp or spongy, and usually of sweetish but faint flavor. There may be 3 to 6 small seeds, frequently only 1 or 2, but generally the fruits are seedless.

Origin and Distribution

The water apple occurs naturally from southern India to eastern Malaysia. It is commonly cultivated in India, southeastern Asia, and Indonesia. In the Philippines, it grows as though wild in the Provinces of Mindanao, Basilan, Dinagat and Samar. It has never been widely distributed but is occasionally grown in Trinidad and Hawaii. It was introduced into Puerto Rico in 1927 but survived only a few years.

Varieties

In Indonesia, two forms are recognized—one white-fruited and the other red, the color of the latter developing from the base upward. Much variation is seen in the fruits from different trees in Malaya and the flavor of some types is quite acid.

Climate

The water apple is suited only to low altitudes in the tropics and areas where there is rainfall fairly well spaced throughout the year.

Propagation

The tree may be air-layered or budded onto rootstocks of *Eugenia javanica* Lam. or *E. densiflora* A. DC. Experiments in Hawaii proved that cuttings can be successfully rooted.

Culture

Little cultural attention has been given the water apple. In Indonesia, when it is set out in orchards, it is spaced at a 20 to 26 ft (6-8 m) distance from tree-to-tree.

Season

In Malaya there are two crops a year, one in the spring and a second in the fall. In Indonesia, the tree frequently blooms in July and again in September, the fruits ripening in August and November.

Food Uses

The water apple is mainly consumed by children, the appeal being largely its thirst-relieving character. In Indonesia, the fruits are sold in markets in piles or skewered on slender bamboo sticks. Superior types are sometimes served sliced in salads. According to early writings, a water apple salad is a ceremonial dish for new mothers.

Other Uses

Wood: The wood is hard and is fashioned into small pieces of handicraft.

Medicinal Uses: A decoction of the astringent bark is a local application on thrush.

Rose Apple

Like many other fruits to which the word "apple" has been attached, the rose apple in no way resembles an apple, neither in the tree nor in its fruit. It is a member of the myrtle family, Myrtaceae, and is technically known as *Syzygium jambos* Alston (syn. *Eugenia jambos* L.; *Jambosa jambos* Millsp.; *Jambosa vulgaris* DC.; *Caryophyllus jambos* Stokes).

The term "rose apple" (in French, *pomme rose, pommier rose;* in Spanish, *poma rosa, pomarrosa, manzana rosa,* or *manzanita de rosa*) is so widely employed that the species has few alternate names apart from those in the many local dialects of Africa, India, Malaya, southeastern Asia, the East Indies and Oceania. It is sometimes called *jambosier* by French-speaking people, plum rose or malabar plum in the English-speaking West Indies, *pommeroos* or *appelroos* in Surinam, and *jambeiro* or *jambo amarelo* in Brazil; *jaman* in India, and *yambo* in the Philippines.

Description

The rose apple tree may be merely a shrub but is generally a tree reaching 25 or even 40 ft (7.5-12 m) in height, and has a dense crown of slender, wide-spreading branches, often the overall width exceeding the height. The evergreen leaves are opposite, lanceolate or narrow-elliptic, tapering to a point; 4 to 9 in (10-22 cm) long, and from 1 to 2½ in (2.5-6.25 cm) wide; somewhat leathery, glossy, dark-green when mature, rosy when young. The flowers are creamy-white or greenish-white, 2 to 4 in (5-10 cm) wide, consisting mostly of about 300 conspicuous stamens to 1½ in (4 cm) long, a 4-lobed calyx, and 4 greenish-white, concave petals. There are usually 4 or 5 flowers together in terminal clusters. Capped with the prominent, green, tough calyx, the fruit is nearly round, oval, or slightly pear-shaped, 1½ to 2 in (4-5 cm) long, with smooth, thin, pale-yellow or whitish skin, sometimes pink-blushed, covering a crisp, mealy, dry to juicy layer of yellowish flesh, sweet and resembling the scent of a rose in flavor. In the hollow center, there are 1 to 4 brown, rough-coated, medium-hard, more or less rounded seeds, ⅜ to ⅝ in (1-1.6 cm) thick, which loosen from the inner wall and rattle when the fruit is shaken. Fragments of the seedcoat may be found in the cavity.

Origin and Distribution

The rose apple is native to the East Indies and Malaya and is cultivated and naturalized in many parts of India, Ceylon and former Indochina and the Pacific Islands. It was introduced into Jamaica in 1762 and became well distributed in Bermuda, the Bahamas, the West Indies and,

Fig. 103: The rose apple (*Syzygium jambos*) is a minor fruit, but the tree is a quick-growing source of fuel and other products.

at low and medium elevations, from southern Mexico to Peru. In Guatemala, the tree may be planted as a living fencepost or in hedgerows around coffee plantations. For this purpose, it is drastically pruned to promote dense growth. It grows wild abundantly, forming solid stands and thickets, in Puerto Rico, the Virgin Islands, Guatemala, Honduras and Panama.

In 1825, eight young trees were taken from Rio de Janeiro to Hawaii by ship, and, in 1853, a United States warship delivered avocado and rose apple trees from Central America to the island of Hilo. The rose apple became naturalized on the islands of Kauai, Molokai, Oahu, Maui and Hawaii. In 1893, it was reported as already cultivated in Ghana. It is semi-naturalized in some areas of West Tropical Africa and on the islands of Zanzibar, Pemba and Reunion. It is believed to have been first planted in Queensland, Australia, about 1896. A tree obtained from an Italian nursery has grown and borne well on the coastal plain of Israel. However, it is not of interest there as a fruit tree but rather as an ornamental.

The rose apple was introduced into Florida, at Jacksonville, before 1877, but, as a fruit tree, it is suited only to the central and southern parts of the state. In California, it is planted as far north as San Francisco for its ornamental

foliage and flowers. Because the tree occupies considerable space and the fruit is little valued, the rose apple has not been planted in Florida in recent years, though there are quite a number of specimens remaining from former times.

Climate

The rose apple flourishes in the tropical and near-tropical climates only. In Jamaica, it is naturalized from near sea-level up to an altitude of 3,000 ft (915 m); in Hawaii, from sea-level to 4,000 ft (1,200 m). In India, it ranges up to 4,400 ft (1,350 m); in Ecuador, to 7,500 ft (2,300 m). At the upper limits, as in California, the tree grows vigorously but will not bear fruit.

In India, it does best on the banks of canals and streams and yet tolerates semi-arid conditions. Prolonged dry spells, however, are detrimental.

Soil

A deep, loamy soil is considered ideal for the rose apple but it is not too exacting, for it flourishes also on sand and limestone with very little organic matter.

Propagation

Most rose apple trees are grown from seeds, which are

polyembryonic (producing 1 to 3 sprouts), but the seedlings are not uniform in character nor behavior. In India, vegetative propagation has been undertaken with a view to standardizing the crop and also to select and perpetuate dwarf types. Using cuttings, it was found that hardwood does not root even with chemical growth promoters. Treated semihard wood gave 20% success. Air-layers taken in the spring and treated with 1,000 ppm NAA gave 60% success. Air-layers did not root in the rainy season. In budding experiments, neither chip nor "T" buds would take. Veneer grafting in July of spring-flush scions on 1-year-old rootstocks was satisfactory in 31% of the plants. In West Bengal, air-layering is commonly performed in July and the layers are planted in October and November. Fruiting can be expected within 4 years. Sometimes the rose apple is inarched onto its own seedlings.

Culture

Rarely do rose apple trees receive any cultural attention. Some experimental work has shown that seedless, thick-fleshed fruits can be produced by treating opened flowers with growth regulators—naphthoxy acetic acid (NOA), 2,4,5-T, or naphthalene acetic acid.

Season

In Jamaica and Puerto Rico, the rose apple trees bloom and fruit sporadically nearly all year, though somewhat less in summer than at other times. The main season in the Bahamas and in Florida is May through July. The fruiting period varies in different parts of India. In South India, blooming usually occurs in January, with fruit ripening in March and April, whereas in the Circars, ripening takes place in April and May. In the central part of the country, flowering occurs in February, March and April and the fruits ripen from June through July. Then again, it is reported that there are varieties that produce fruit in February and March.

Yield

In India, they say that a mature rose apple tree will yield 5 lbs (2 kg) of fruit each season. The fruits are, of course, very light in weight because they are hollow, but this is a very small return for a tree that occupies so much space.

Keeping Quality

Rose apples bruise easily and are highly perishable. They must be freshly picked to be crisp. Some studies of respiration rate and ethylene production in storage have been made in Hawaii. The fruit is non-climacteric.

Pests and Diseases

The rose apple tree has few insect enemies. In humid climates, the leaves are often coated with sooty mold growing on the honeydew excreted by aphids. They are also prone to leaf spot caused by *Cercospora* sp., *Gloeosporium* sp., and *Phyllosticta eugeniae*; algal leaf spot (*Cephaleuros virescens*); black leaf spot (*Asterinella puiggarii*); and anthracnose (*Glomerella cingulata*). Root rot caused by *Fusarium* sp., and mushroom root rot (*Armillariella* (*Clitocybe*) *tabescens*) attack the tree.

Food Uses

Around the tropical world, rose apples are mostly eaten out-of-hand by children. They are seldom marketed. In the home, they are sometimes stewed with some sugar and served as dessert. Culinary experimenters have devised other modes of using the cuplike halved fruits. One stuffs them with a rice-and-meat mixture, covers them with a tomato sauce seasoned with minced garlic, and bakes them for about 20 minutes. Possible variations are limitless. The fruit is made into jam or jelly with lemon juice added, or more frequently preserved in combination with other fruits of more pronounced flavor. It is also made into a sirup for use as a sauce or to flavor cold drinks. In Jamaica, the halved or sliced fruits are candied by stewing them in very heavy sugar sirup with cinnamon.

Food Value Per 100 g of Edible Portion*	
Calories	56
Moisture	84.5-89.1 g
Protein	0.5-0.7 g
Fat	0.2-0.3 g
Carbohydrates	14.2 g
Fiber	1.1-1.9 g
Ash	0.4-0.44 g
Calcium	29-45.2 mg
Magnesium	4 mg
Phosphorus	11.7-30 mg
Iron	0.45-1.2 mg
Sodium	34.1 mg
Potassium	50 mg
Copper	0.01 mg
Sulfur	13 mg
Chlorine	4 mg
Carotene	123-235 I.U.
Thiamine	0.01-0.19 mg
Riboflavin	0.028-0.05 mg
Niacin	0.521-0.8 mg
Ascorbic Acid	3-37 mg

*Analyses made in Central America and elsewhere.

Toxicity

The seeds are said to be poisonous. An unknown amount of hydrocyanic acid has been reported in the roots, stems and leaves. An alkaloid, jambosine, has been found in the bark of the tree and of the roots, and the roots are considered poisonous.

Other Uses

Fruit: In 1849, it was announced in Bengal that the ripe fruits, with seeds removed, could be distilled 4 times to make a "rosewater" equal to the best obtained from rose petals.

Branches: The flexible branches have been employed in Puerto Rico to make hoops for large sugar casks, and also are valued for weaving large baskets.

Bark: The bark has been used for tanning and yields a brown dye.

Wood: The sapwood is white. The heartwood is dark-red or brown, fibrous, close-grained, medium-heavy to heavy, strong; and has been used to make furniture, spokes for wheels, arms for easy chairs, knees for all kinds of boats, beams for construction, frames for musical instruments (violins, guitars, etc.), and packing cases. It is also popular for general turnery. It is not durable in the ground and is prone to attack by drywood termites.

The tree grows back rapidly after cutting to a stump and consequently yields a continuous supply of small wood for fuel. Rose apple wood makes very good charcoal.

Leaves: A yellow essential oil, distilled from the leaves, contains, among other properties, 26.84% *dl-a*-pinene and 23.84% *l*-limonene, and can be resorted to as a source of these elements for use in the perfume industry.

Flowers: The flowers are a rich source of nectar for honeybees and the honey is a good amber color. Much comes from the San Cristobal River Valley in Cuba.

Medicinal Uses: In India, the fruit is regarded as a tonic for the brain and liver. An infusion of the fruit acts as a diuretic.

A sweetened preparation of the flowers is believed to reduce fever. The seeds are employed against diarrhea, dysentery and catarrh. In Nicaragua, it has been claimed that an infusion of roasted, powdered seeds is beneficial to diabetics. They say in Colombia that the seeds have an anesthetic property.

The leaf decoction is applied to sore eyes, also serves as a diuretic and expectorant and treatment for rheumatism. The juice of macerated leaves is taken as a febrifuge. Powdered leaves have been rubbed on the bodies of smallpox patients for the cooling effect.

The bark contains 7-12.4% tannin. It is emetic and cathartic. The decoction is administered to relieve asthma, bronchitis and hoarseness. Cuban people believe that the root is an effective remedy for epilepsy.

Surinam Cherry

The most widely known of the edible-fruited *Eugenia* species, because of its great adaptability, the Surinam cherry, *E. uniflora* L. (syns. *E. Michelii* Lam.; *Stenocalyx Michelii* Berg; *Plinia rubra* Vell.), is also called Brazil or Brazilian cherry, Cayenne cherry, pitanga, and, unfortunately, Florida cherry. In Spanish it is generally *cereza de cayena;* but *pendanga* in Venezuela; *guinda* in El Salvador; *ñanga-pirĕ* in Argentina; *cereza quadrada* in Colombia. In Guadeloupe and Martinique it is called *cerese à côtes* or *cerises-cotes;* in French Guiana, *cerise de Cayenne, cerise de pays,* or *cerise carĕe;* in Surinam, *Surinaamsche kersh, zoete kers,* or *monkie monkie kersie.*

Description

The shrub or tree, to 25 ft (7.5 m) high, has slender, spreading branches and resinously aromatic foliage. The opposite leaves, bronze when young, are deep-green and glossy when mature; turn red in cold, dry winter weather. They are ovate to ovate-lanceolate, blunt- to sharp-pointed, 1½ to 2½ in (4-6.25 cm) long. Long-stalked flowers, borne singly or as many as 4 together in the leaf axils, have 4 delicate, recurved, white petals and a tuft of 50 to 60 prominent white stamens with pale-yellow anthers. The 7- to 8-ribbed fruit, oblate, ¾ to 1½ in (2-4 cm) wide, turns from green to orange as it develops and, when mature, bright-red to deep-scarlet or dark, purplish maroon ("black") when fully ripe. The skin is thin, the flesh orange-red, melting and very juicy; acid to sweet, with a touch of resin and slight bitterness. There may be 1 fairly large, round seed or 2 or 3 smaller seeds each with a flattened side, more or less attached to the flesh by a few slender fibers.

Origin and Distribution

The plant is native from Surinam, Guyana and French Guiana to southern Brazil (especially the states of Rio de Janeiro, Paraña, Santa Catharina and Rio Grande do Sul), and to northern, eastern and central Uruguay. It grows wild in thickets on the banks of the Pilcomayo River in Paraguay. It was first described botanically from a plant growing in a garden at Pisa, Italy, which is believed to have been introduced from Goa, India. Portuguese voyagers are said to have carried the seed from Brazil to India, as they did the cashew. It is cultivated and naturalized in Argentina, Venezuela and Colombia; also along the Atlantic coast of Central America; and in some islands of the West Indies—the Cayman Islands, Jamaica, St. Thomas, St. Croix, Puerto Rico, Cuba, Haiti, the Dominican Republic, and in the Bahamas and Bermuda. In 1918, Britton wrote, in the *Flora of Bermuda*, that ". . .as it harbors the fruit fly, the tree has been largely cut out in recent years." It is frequently grown in Hawaii, Samoa, India and Ceylon as an ornamental plant and occasionally in tropical Africa, southern China and in the Philippines where it first fruited in 1911. It was long ago planted on the Mediterranean coast of Africa and the European Riviera. The first Surinam cherry was introduced into coastal Israel in 1922 and aroused considerable interest because it produced fruit in May when other fruits are scarce, and it requires so little care; but over 10 years of observation, the yields recorded were disappointingly small.

In Florida, the Surinam cherry is one of the most common hedge plants throughout the central and southern parts of the state and the Florida Keys. The fruits are today mostly eaten by children. In the past, many people al-

Fig. 104: The Surinam cherry (*Eugenia uniflora*) is primarily grown as a hedge, the showy fruits being eaten mainly by children.

lowed the tree to grow naturally and harvested the fruits for culinary use. For a while, small quantities were sold in Miami markets. In temperate zones, the plant is grown in pots for its attractive foliage and bright fruits.

Varieties

There are 2 distinct types: the common bright-red and the rarer dark-crimson to nearly black, which tends to be sweeter and less resinous.

Climate

The Surinam cherry is adapted to tropical and sub-tropical regions. In the Philippines, it thrives from sea-level to 3,300 ft (1,000 m); in Guatemala, up to 6,000 ft (1,800 m). Young plants are damaged by temperatures below 28°F (-2.22°C), but well-established plants have suffered only superficial injury at 22°F (-5.56°C). The plant revels in full sun. It requires only moderate rainfall and, being deep-rooted, can stand a long dry season.

Soil

The Surinam cherry grows in almost any type of soil—sand, sandy loam, stiff clay, soft limestone—and can even stand waterlogging for a time, but it is intolerant of salt.

Propagation

Seeds are the usual means of propagation. They remain viable for not much longer than a month and germinate in 3 to 4 weeks. Volunteer seedlings can be taken up and successfully transplanted. Layering has been successful in India. The seedlings can be topworked to superior selections by side- or cleft-grafting but they tend to sucker below the graft.

Culture

Surinam cherry seedlings grow slowly; some begin to fruit when 2 years old; some may delay fruiting for 5 or 6 years, or even 10 if in unfavorable situations. They are most productive if unpruned, but still produce a great many fruits when close-clipped in hedges. Quarterly feeding with a complete fertilizer formula promotes fruiting. The plant responds quickly to irrigation, the fruit rapidly becoming larger and sweeter in flavor after a good watering.

Season and Harvesting

The fruits develop and ripen quickly, only 3 weeks after the flowers open. In Brazil, the plants bloom in September and fruits ripen in October; they bloom again in December and January. In Florida and the Bahamas, there is a spring crop, March or April through May or June; and a second crop, September through November, coinciding with the spring and fall rains.

The fruits should be picked only when they are so ripe as to fall into the hand at the lightest touch, otherwise they will be undesirably resinous. Gathering must be done daily or even twice a day.

Yield

In India, pruned bushes yield an average of 6 to 8 lbs (2.7–3.6 kg) per plant. The highest yield obtained in Israel was 2,700 fruits weighing about 24 lbs (11 kg) from one untrimmed plant.

Pests and Diseases

Surinam cherries are highly attractive to Caribbean and Mediterranean fruit flies, but the incidence of infestation was found to vary greatly in Israel from location to location, some plants being unmolested.

The foliage is occasionally attacked by scale insects and caterpillars. A large, extensive hedge along a canal in Dade County blew down in September 1982. Examination showed that the roots had been chewed off and there were about a dozen white grubs up to 2 in (5 cm) long under each plant. These were identified as the larvae of a sugar cane pest that is common in Haiti.

Among diseases encountered in Florida are leaf spot caused by *Cercospora eugeniae*, *Helminthosporium* sp., and *Phyllostica eugeniae;* thread blight from infection by *Corticium stevensii;* anthracnose from *Colletotrichum gloeosporioides;* twig dieback and root rot caused by *Rhizoctonia solani;* and mushroom root rot, *Armillariella* (*Clitocybe*) *tabescens.*

Food Uses

Children enjoy the ripe fruits out-of-hand. For table use, they are best slit vertically on one side, spread open to release the seed(s), and kept chilled for 2 or 3 hours to dispel most of their resinously aromatic character. If seeded and sprinkled with sugar before placing in the refrigerator, they will become mild and sweet and will exude much juice and serve very well instead of strawberries on shortcake and topped with whipped cream. They are an excellent addition to fruit cups, salads and custard pudding; also ice cream; and can be made into pie or sauce

or preserved whole in sirup. They are often made into jam, jelly, relish or pickles. Brazilians ferment the juice into vinegar or wine, and sometimes prepare a distilled liquor.

Food Value Per 100 g of Edible Portion*	
Calories	43–51 g
Moisture	85.4–90.70 g
Protein	0.84–1.01 g
Fat	0.4–0.88 g
Carbohydrates	7.93–12.5 g
Fiber	0.34–0.6 g
Ash	0.34–0.5 g
Calcium	9 mg
Phosphorus	11 mg
Iron	0.2 mg
Carotene (Vitamin A)	1,200–2,000 I.U.
Thiamine	0.03 mg
Riboflavin	0.04 mg
Niacin	0.03 mg
Ascorbic Acid**	20–30 mg
*A composite of analyses made in Hawaii, Africa, Florida.	

**Dr. Margaret Mustard found 33.9–43.9 mg in ripe red fruits; 25.3 in the "black" type.

Toxicity

The seeds are extremely resinous and should not be eaten. Diarrhea has occurred in dogs that have been fed the whole fruits by children. The strong, spicy emanation from bushes being pruned irritates the respiratory passages of sensitive persons.

Other Uses

The **leaves** have been spread over the floors of Brazilian homes. When walked upon, they release their pungent oil which repels flies. The **bark** contains 20 to 28.5% tannin and can be used for treating leather. The **flowers** are a rich source of pollen for honeybees but yield little or no nectar.

Medicinal Uses: In Brazil the leaf infusion is taken as a stomachic, febrifuge and astringent. In Surinam, the leaf decoction is drunk as a cold remedy and, in combination with lemongrass, as a febrifuge. The leaves yield essential oil containing citronellal, geranyl acetate, geraniol, cineole, terpinene, sesquiterpenes and polyterpenes.

Rumberry (Plate LIV)

A tiny fruit, formerly in demand, the rumberry, *Myrciaria floribunda* Berg. (syns. *M. protracta* Berg.; *Eugenia floribunda* West ex Willd.), is also called guavaberry, *mirto* or *murta* in Puerto Rico; guaveberry in St. Martin and St. Eustatius; *guayabillo* in Guatemala; *coco-carette, merisier-cerise,* or *bois de basse batard* in Guadeloupe and Martinique; *cabo de chivo* in El Salvador; *escobillo* in Nicaragua; *mije* or *mije colorado* in Cuba; *mijo* in the

Dominican Republic; *bois mulatre* in Haiti; *roode bosch guave, saitjaberan,* or *kakrioe hariraroe tataroe* in Surinam. In Venezuela the names *guayabito* and *guayabillo blanco* are applied to the related species, *M. caurensis* Steyerm, as well as to some other plants.

Description

This is an attractive shrub or slender tree reaching 33 or even 50 ft (10-15 m) in height, with reddish-brown branchlets, downy when young, and flaking bark. The evergreen, opposite leaves are ovate, elliptical, or oblong-lanceolate, pointed at the apex; 1 to 3³⁄₁₆ in (2.5-8 cm) long, ⅓ to 1³⁄₁₆ in (0.8-3 cm) wide; glossy, slightly leathery, minutely dotted with oil glands. The flowers, borne in small axillary or lateral clusters, are white, silky-hairy with about 75 prominent white stamens. The fruit is round or oblate, ⁵⁄₁₆ to ⅝ in (8-16 mm) in diameter; dark-red (nearly black) or yellow-orange; highly aromatic and of bittersweet, balsam-like flavor; with one globular seed. In Surinam, according to Pulle, there are sometimes deformed fruits, rounded, flattened, leathery, dehiscent, and to ¾ in (2 cm) across.

Origin and Distribution

The rumberry occurs wild over a broad territory—Cuba, Hispaniola, Jamaica, Puerto Rico (including Vieques), the Virgin Islands, St. Martin, St. Eustatius, St. Kitts, Guadeloupe, Martinique, Trinidad, southern Mexico, Belize, Guatemala, El Salvador to northern Colombia; also Guyana, Surinam and French Guiana, and eastern Brazil. It has been occasionally cultivated in Bermuda, rarely elsewhere, but, throughout its natural range, when land is cleared for pastures, the tree is left standing for the sake of its fruits. The plant was introduced into the Philippines in the early 1900's and has been included in propagation experiments in Hawaii. There is a healthy fruiting specimen at Fairchild Tropical Garden, Miami.

Varieties

O.W. Barrett wrote in 1928: "There are 3 or 4 varieties in the dry hills of St. Croix; these vary as to size and color, but all are intensely aromatic." In St. John, they say the fruits produced by wild trees on Bordeaux Mountain and along Reef Bay Trail are "unusually good".

Climate and Soil

In Puerto Rico, the rumberry grows naturally in dry and moist coastal forests from sea-level to an elevation of 700 ft (220 m). In Vieques and the Virgin Islands, it abounds in dry forests up to 1,000 ft (300 m). In South Florida it is growing well, but as a small tree, on oolitic limestone.

Food Uses

In Cuba, the fruits are relished out-of-hand and are made into jam, and the fermented juice is rated as *"una bebida exquisita"* (an exquisite beverage). People on the island of St. John use the preserved fruits in tarts. The local "guavaberry liqueur" is made from the fruits "with pure grain alcohol, rum, raw sugar and spices" and it is a special treat at Christmastime. In the past, a strong wine and a heavy liqueur were exported from St. Thomas to Denmark in "large quantities".

Other Uses

In Camaguey, Cuba, the rumberry is included among the nectar sources visited by honeybees.

Medicinal Uses: The fruits are sold by herbalists in Camaguey for the purpose of making a depurative sirup; and the decoction is taken as a treatment for liver complaints.

Related Species

The **camu-camu**, *Myrciaria dubia* McVaugh (syns. *M. paraensis* Berg; *M. spruceana* Berg), is also called *camocamo* in Peru and *cacari* in Brazil. It is a shrub, or bushy tree, to 43 ft (13 m) high with minute prickly hairs on the young branchlets and petioles. The opposite leaves are broad- or narrow-ovate, or elliptic, often lop-sided; 1¾ to 4 in (4.5-10 cm) long and ⅝ to 1¾ in (1.6-4.5 cm) wide, pointed at the apex, rounded at the base where the margins curve inward to the petiole, forming winglike appendages. Fragrant flowers, nearly sessile, are borne in 4's in or near the leaf axils; have tiny, white petals and about 125 stamens ¼ to ⅜ in (6-10 mm) long. The fruit is nearly round, ⅜ to 1 in (1-2.5 cm) wide, yellow at first, becoming maroon to purple-black and soft and juicy when ripe. It is of acid or sweet flavor and contains 3 seeds. Locally it is considered good fish food.

This species occurs abundantly wild in swamps along rivers and lakes, especially the Rio Mazán near Iquitos, Peru, and in Amazonian Brazil and Venezuela, often with the base of the trunk under water, and, during the rainy season, the lower branches are also submerged for long periods.

Seeds were brought to Florida by William F. Whitman in 1964, and plants were raised, he says, in an "acid hammock sand soil" and regularly watered. One plant bore rather heavily in 1972, mainly in late summer with a few scattered fruits the following winter. One plant was 12 ft (3.65 m) tall and equally broad in 1974. In Brazil, the fruit is borne mainly from November to March.

Half-ripe fruits have been found to contain 1,950 to 2,700 mg of ascorbic acid per 100 g edible portion, values comparable to the high ranges of the Barbados cherry, q.v. These findings led to a certain amount of exploitation of the fruit, which must be harvested by boat. There is a trial plot at Manaus, Brazil, and some experimental plantings in Peru and the juice is frozen or bottled and exported to the United States for the production of "vitamin C" tablets for the "health food" market. In plantations, in non-flooded land a single plant may bear 400 to 500 fruits. On flooded land, the per-plant harvest has been 1,000 fruits.

Though there are still people who can be persuaded to believe that "natural vitamin C" is superior to synthetic, the commercial prospects for the camu-camu are no brighter than those for the Barbados cherry. In 1969, V.L.S. Charley, Consultant to Beecham Products, Brentford,

England, in assessing prospects for the camu-camu, with its "very slight flavour characteristics", declared that the idea that natural vitamin C, *per se*, had some magical quality is not now acceptable . . . there is little doubt that the presence of a full, clean, well-balanced flavour is more commercially important than the possession of a high ascorbic acid content."

Grumichama

An often admired but still very minor fruiting member of the Myrtaceae, the grumichama, *Eugenia brasiliensis* Lam. (syn. *E. dombeyi* Skeels), is also called *grumixama, grumichameira,* or *grumixameira* in Brazil, and sometimes Brazil cherry elsewhere.

Description

The highly ornamental tree is slender, erect, usually to 25 or 35 ft (7.5–10.5 m) high, short-trunked and heavily foliaged with opposite, oblong-oval leaves 3½ to 5 in (9–16 cm) long, 2⅜ in (5–6 cm) wide, with recurved margin; glossy, thick, leathery, and minutely pitted on both surfaces. They persist for 2 years. New shoots are rosy. The flowers, borne singly in the leaf axils, are 1 in (2.5 cm) wide; have 4 green sepals and 4 white petals, and about 100 white stamens with pale-yellow anthers. The long-stalked fruit is oblate, ½ to ¾ in (1.25–2 cm) wide; turns from green to bright-red and finally dark-purple to nearly

Fig. 105: The grumichama (*Eugenia brasiliensis*) is more cherry-like than many so-called "cherries" but handicapped by small size, apical sepals and large seeds.

black as it ripens, and bears the persistent, purple- or red-tinted sepals, to ½ in (1.25 cm) long, at its apex. The skin is thin, firm and exudes dark-red juice. The red or white pulp is juicy and tastes much like a true subacid or sweet cherry except for a touch of aromatic resin. There may be 1 more or less round, or 2 to 3 hemispherical, hard, light-tan or greenish-gray seeds to ½ in (1.25 cm) wide and half as thick.

Origin and Distribution

The grumichama is native and wild in coastal southern Brazil, especially in the states of Parana and Santa Catarina. It is cultivated in and around Rio de Janeiro, also in Paraguay. A specimen was growing in Hope Gardens, Jamaica, in 1880 and a tree was planted in the Botanical Gardens, Singapore, in 1888, fruited in 1903. It has long since vanished from both of these locations. An attempt to grow it in the Philippines in the early 1920's did not meet with success. Neither did a trial in Israel. An early introduction, perhaps by Don Francisco de Paula Marin in 1791, was made in Hawaii and the tree was adopted into numerous local gardens.

The United States Department of Agriculture received seeds from Mauritius in 1911 (S.P.I. #30040); plants and seeds from Bahia, Brazil, in 1914 (S.P.I. #36968), and more seeds from Mauritius in 1922 (S.P.I. #54797). Plants were set out at the Plant Introduction Station in Miami and prospered. Other plantings were made in California where it seemed even better adapted but has apparently disappeared. The United States Department of Agriculture raised seedlings at Puerto Arturo, Honduras, and transferred some plants to the Lancetilla Experimental Garden at Tela in 1926. They flourished there and flowered and fruited well.

Over the years there have been mild efforts to encourage interest in the virtues of the grumichama in Florida, mainly because of the beauty and hardiness of the tree and the pleasant flavor of the fruit but the sepals are a nuisance and there is too little flesh in proportion to seed for the fruit to be taken seriously.

Varieties

Variety *leucocarpus* Berg. in Brazil becomes a large tree to 65 ft (20 m) high and has fruits with white flesh. It is not as common as the red-fleshed type.

Climate

The grumichama is subtropical, surviving temperatures of 26°F (-3.33°C) in Brazil. It is better suited to Palm Beach than to southern Florida. In Hawaii, the tree fruits best from sea-level to an altitude of no more than 300 ft (90 m).

Soil

The grumichama does better on acid sand in Central Florida than it does on limestone in the south. It is reported to prefer deep, fertile, sandy loam. Sturrock says it grows well in rich clay in Cuba but is adversely affected by the long dry season.

Propagation

Wilson Popenoe stated that propagation in Brazil is entirely by seeds which remain viable for several weeks and germinate in about a month. Fenzi says that seeds, cuttings and air-layers are employed, and Sturrock has mentioned that grafting is easy.

Culture

The grumichama is of slow growth when young unless raised in a mixture of peat moss and sand and then given a thick layer of peat moss around the roots when setting out, and kept heavily fertilized. In Hawaii, it has taken 7 years to reach 7 ft. Fruiting begins when the plants are 4 to 5 years old.

Season

The tree is regarded as remarkable for the short period from flowering to fruiting. In Florida, it has been in full bloom in late April and loaded with fruits 30 days later. The crop ripens quickly over just a few days. In Hawaii, the trees bloom and fruit from July to December, with the main crop in the fall. Trees in Brazil vary considerably in time of flowering and fruiting so that the overall season extends from November to February.

Pests

In Hawaii, the fruits are heavily attacked by the Mediterranean fruit fly.

Food Uses

Fully ripe grumichamas are pleasant to nibble out-of-hand. In Hawaii, half-ripe fruits are made into pie, jam or jelly.

Food Value Per 100 g of Edible Portion*	
Moisture	83.5 g
Protein	0.102 g
Fiber	0.6 g
Ash	0.43 g
Calcium	39.5 mg
Phosphorus	13.6 mg
Iron	0.45 mg
Carotene	0.039 mg
Thiamine	0.044 mg
Riboflavin	0.031 mg
Niacin	0.336 mg
Ascorbic Acid	18.8 mg
*Analyses made in Honduras.	

Medicinal Uses

The bark and leaves contain 1.5% of essential oil. The leaf or bark infusion—⅓ oz (10 g) of plant material in 10½ oz (300 g) water—is aromatic, astringent, diuretic and taken as a treatment for rheumatism at the rate of 2 to 4 cups daily, in Brazil.

Pitomba

The pitomba, *Eugenia luschnathiana* Klotzsch ex O. Berg. (syn. *Phyllocalyx luschnathianus* Berg.) is also called *uvalha do campo, ubaia do campo,* or *uvalheira* in Brazil.

It is an attractive, slow-growing tree to 20 or 30 ft (6-9 m) high, with dense foliage. The evergreen, opposite, short-petioled, oblong-lanceolate leaves, 1 to 3 in (2.5-7.5 cm) long, are glossy, dark-green on the upper surface, paler beneath. New growth is temporarily coated with bronze hairs on the underside. The long-stalked, 4-petalled, white flowers are borne singly in the leaf axils.

The fruit, broad-obovate, faintly 4-lobed, 1 to 1¼ in (2.5-3.2 cm) long, is bright orange-yellow with 4 or 5 green sepals ½ in (1.25 cm) long protruding from the apex. The skin is thin, tender, and the pulp golden-yellow, apricot-like in texture, soft, melting, juicy, aromatic and slightly acid, faintly resinous in flavor. In the central cavity there may be one round seed or 2 to 4 irregular, angular seeds, light-tan and ⅜ to ⅝ in (1-1.6 cm) in diameter.

This little-known species is native to the State of Bahia, Brazil, is cultivated to a limited extent locally and is grown in the botanical garden in Rio de Janeiro. Seeds were brought to the United States from Brazil by plant explorers for the federal Department of Agriculture in 1914 (S.P.I. #37017). A very few specimens, scarcely more than shrubs, have been grown to the fruiting stage in southern Florida. The pitomba was at first considered promising for this area but has made no progress at all in the last 40 years.

When in good soil, well-fertilized and frequently and heavily watered, the tree begins to bear when less than 3½ ft (a little over 1 m) high. There is much variation in the size of fruits produced by seedlings. Sturrock made some selections and grafted them successfully. Flowers appear in late spring and early summer in Florida and the fruiting season is in midsummer. In Brazil the fruits ripen in November and December. The fruits are there used mainly for jelly, preserves, and carbonated beverages.

Fig. 106: Little-known, the orange-yellow pitomba (*Eugenia luschnathiana*) is of fair size and thick-fleshed when well-irrigated and fertilized.

Sapodilla

One of the most interesting and desirable of all tropical fruit trees, the sapodilla, a member of the family Sapotaceae, is now known botanically as *Manilkara zapota* van Royen (syns. *M. achras* Fosb., *M. zapotilla* Gilly; *Achras sapota* L., *A. zapota* L.; *Sapota achras* Mill.).

Among numerous vernacular names, some of the most common are: *baramasi* (Bengal and Bihar, India); *buah chiku* (Malaya); *chicle* (Mexico); *chico* (Philippines, Guatemala, Mexico); *chicozapote* (Guatemala, Mexico, Venezuela); *chikoo* (India); *chiku* (Malaya, India); dilly (Bahamas; British West Indies); *korob* (Costa Rica); *mespil* (Virgin Islands); *mispel, mispu* (Netherlands Antilles, Surinam); *muy* (Guatemala); *muyozapot* (El Salvador); naseberry (Jamaica; British West Indies); neeseberry (British West Indies; *nispero* (Puerto Rico, Central America, Venezuela); *nispero quitense* (Ecuador); sapodilla plum (India); *sapota* (India); *sapoti* (Brazil); *sapotille* (French West Indies); tree potato (India); *Ya* (Guatemala; Yucatan); *zapota* (Venezuela); *zapote* (Cuba); *zapote chico* (Mexico; Guatemala); *zapote morado* (Belize); *zapotillo* (Mexico).

Description

The sapodilla is a fairly slow-growing, long-lived tree, upright and elegant, distinctly pyramidal when young; to 60 ft (18 m) high in the open but reaching 100 ft (30 m) when crowded in a forest. It is strong and wind-resistant, rich in white, gummy latex. Its leaves are highly ornamental, evergreen, glossy, alternate, spirally clustered at the tips of the forked twigs; elliptic, pointed at both ends, firm, 3 to 4½ in (7.5-11.25 cm) long and 1 to 1½ in (2.5-4 cm) wide. Flowers are small and bell-like, with 3 brown-hairy outer sepals and 3 inner sepals enclosing the pale-green corolla and 6 stamens. They are borne on slender stalks at the leaf bases. The fruit may be nearly round, oblate, oval, ellipsoidal, or conical; varies from 2 to 4 in (5-10 cm) in width. When immature it is hard, gummy and very astringent. Though smooth-skinned it is coated with a sandy brown scurf until fully ripe. The flesh ranges in color from yellowish to light- or dark-brown or sometimes reddish-brown; may be coarse and somewhat grainy or smooth; becomes soft and very juicy, with a sweet flavor resembling that of a pear. Some fruits are seedless, but normally there may be from 3 to 12 seeds which are easily removed as they are loosely held in a whorl of slots in the center of the fruit. They are brown or black, with one white margin; hard, glossy; long-oval, flat, with usually a distinct curved hook on one margin; and about ¾ in (2 cm) long.

Origin and Distribution

The sapodilla is believed native to Yucatan and possibly other nearby parts of southern Mexico, as well as northern Belize and Northeastern Guatemala. In this region there were once 100,000,000 trees. The species is found in forests throughout Central America where it has apparently been cultivated since ancient times. It was introduced long ago throughout tropical America and the West Indies, the Bahamas, Bermuda, the Florida Keys and the southern part of the Florida mainland. Early in colonial times, it was carried to the Philippines and later was adopted everywhere in the Old World tropics. It reached Ceylon in 1802.

Cultivation is most extensive in coastal India (Maharastra, Gujarat, Andhra Pradesh, Madras and Bengal States), where plantations are estimated to cover 4,942 acres (2,000 ha), while Mexico has 3,733.5 acres (1,511 ha) devoted to the production of fruit (mainly in the states of Campeche and Veracruz) and 8,192 acres (4,000 ha) primarily for extraction of chicle (see under "Other Uses") as well as many dooryard and wild trees. Commercial plantings prosper in Sri Lanka, the Philippines, the interior valleys of Palestine, as well as in various countries of South and Central America, including Venezuela and Guatemala.

Cultivars

In most areas, types are distinguished merely by shape, as 'Round' and 'Oval' in Saharanpur, India. Several named cultivars are grown for commercial or home use in western and southern India: **'Kalipatti'**, small, early, high quality; **'Calcutta Special'**, large, late; **'Pilipatti'**, small, midseason to late; **'Bhuripatti'**, small, midseason; **'Jumakhia'**, small, in clusters, late; **'Mohan Gooti'**, small, midseason, not very sweet; **'Kittubarti'**, very small, ridged, very sweet; **'Kittubarti Big'**, large, but of inferior quality; **'Cricket Ball'**, very large, with crisp, granular, very sweet flesh but not distinctive in flavor; **'Dwarapudi'**, similar, but not quite as big, sweet and very popular; **'Bangalore'**, large, ridged, and **'Vavivalasa'** are oval and popular in the Circars but are only medium-sweet and bear poorly.

Fig. 107: The sapodilla (*Manilkara zapota*) is sweet, luscious, practical and borne abundantly by a handsome, drought- and wind-resistant tree.

Other prominent cultivars in India are **'Jonnava-losa–I'**, of medium size, pale-fleshed, sweet; **'Jonnava-losa–II'**, of medium size, ridged, with yellowish-pink flesh, sweet but not agreeable in flavor; **'Jonnavalosa Round'**, large, ridged, with cream-colored flesh, very sweet; **'Gauranga'**, small, lop-sided, ridged, very sweet, bears heavily; **'Ayyangar'**, large, very thick-skinned, sweet, rose-scented; **'Thagarampudi'**, of medium size, thin-skinned, very sweet; **'Oaka'**, small, rounded to oval, of good flavor and popular. Among the lesser-known are **'Badam'**, **'Bhuri'**, **'Calcutta Round'**, **'CO. 1'** ('Cricket Ball' X 'Long Oval'), **'Dhola diwani'**, **'Fingar'**, **'Gava-rayya'**, **'Guthi'**, **'Kali'**, and **'Vanjet'**.

A dwarf type called **'Pot'** bears early and can be maintained as a pot specimen for 10 years.

Henry Pittier, in 1914, described what he deemed a "remarkable variety" called *nispero de monte* at Patiño, Panama. The trees do not exceed 26 ft (8 m) in height and bear small, oblate fruits in dense clusters.

In Indonesia, sapodillas are classed in two main groups: 1) Sawo maneela, normal-size trees having narrow, pointed leaves; and 2) Sawo apel, low, shrublike trees, with oblong leaves broadest above the middle. Belonging to group #1

are the common cultivars **'Sawo betawi'** (fruit large, in clusters of 2-4, popular, perishable, ripening in 3 days from picking); **'Sawo koolon'** (fruit large, solitary, thick-skinned, with firm flesh, shipping well); **'Sawo madja'** (large, with persistent scurf, pulp of fine texture, sweet with an acid tang). Belonging to group #2 are **'Sawo apel bener'** (fruits small in clusters of 3-6, thick-skinned); **'Sawo apel klapa'** (fruits medium-size, with persistent scurf). Some others are little grown because the fruits are either very small, too sandy, too gummy, or too dry.

In Mexico, some superior selections are known merely as **'SCH–02'**, **'SCH–03'**, **'SCH–07'**, **SCH–08'**, and **'SCH–28'**.

In Florida, seedling selections of high quality have been named and vegetatively reproduced. The first of these was **'Russell'** from Islamorada in the Florida Keys, named and propagated by R.H. Fitzpatrick. It is nearly round, up to 4 in (10 cm) in diameter and length, brown-scurfy with gray patches, and luscious, reddish flesh. It is not a dependable bearer. The second, **'Prolific'**, a seedling grown at the Agricultural Research and Education Center, Homestead, and released in 1941, is round-conical, 2½ to 3½ in (6.25-9 cm) long and broad, with smooth, pinkish-tan flesh. The skin is lighter than that of the 'Russell' and tends to lose much of the scurf as it ripens. The tree bears early, consistently and heavily. Of later selection, 'Modello' is a good quality fruit but not a heavy producer; 'Seedless' yields poorly; 'Brown Sugar' is a good, regular, high yielder; handles and keeps well.

Some introduced cultivars being tested in Florida include: 'Boetzberg', 'Larsen', 'Morning Star', 'Jamaica 8', and 'Jamaica 10'. 'Tikal', a recent seedling selection, seems very promising. It is light-brown, elliptic to conical, much smaller than 'Prolific', but of excellent flavor and comes into season very early. Several cultivars not recommended because of low yield in southern Florida are 'Addley', 'Adelaide', 'Big Pine Key', 'Black', 'Jamaica No. 4', 'Jamaica No. 5', 'Martin' and 'Saunders'.

In 1951, in Jamaica, I visited an English gentleman who had a very special sapodilla tree which bore great quantities of tiny sapodillas, no more than 1½ in (4 cm) in diameter. They were all seedless and he served them chilled, whole.

In the Philippines, selected cultivars, 'Ponderosa', 'Java', 'Sao Manila', 'Native', 'Formosa', 'Rangel', and the 'Prolific' from Florida are maintained by the Bureau of Plant Industry for propagation and distribution to farmers. 'Sao Manila' fruits mature in 190 days and ripen 3 to 5 days after picking.

Hybridization studies have been conducted in India.

Climate

The sapodilla grows from sea level to 1,500 ft (457 m) in the Philippines, up to 4,000 ft (1,220 m) in India, to 3,937 ft (1,200 m) in Venezuela, and is common around Quito, Ecuador, at 9,186 ft (2,800 m). It is not strictly tropical, for mature trees can withstand temperatures of 26° to 28°F (-3.33° to -2.2°C) for several hours. Young trees are tenderer and apt to be killed by 30°F (-1.11°C) unless the stem is banked with sand or wrapped with straw and burlap during the cold spell. A number of sapodilla trees have lived for a few years in California without fruiting and then have succumbed to cold. Cool nights are considered a constant limiting factor. However, I have learned of one tree in a protected location in the Sacramento Valley that has survived for many years, reaching a large size and fruiting regularly. The sapodilla seems equally at home in humid and relatively dry atmospheres.

Soil

The sapodilla grows naturally in the calcareous marl and disintegrated limestone of its homeland, therefore it should not be surprising that it is so well adapted to southern Florida and the Florida Keys. Nevertheless, it flourishes also in deep, loose, organic soil, or on light clay, diabase, sand or lateritic gravel. Good drainage is essential, the tree bearing poorly in low, wet locations. It is highly drought-resistant, can stand salt spray, and approaches the date palm in its tolerance of soil salinity, rated as ECe 14.20.

Propagation

Seeds remain viable for several years if kept dry. The best seeds are large ones from large fruits. They germinate readily but growth is slow and the trees take 5 to 8 years to bear. Since there is great variation in the form, quality and yield of fruits from seedling trees, vegetative propagation has long been considered desirable but has been hampered by the gummy latex. In India, several methods are practiced: grafting, inarching, ground-layering and air-layering. Grafts have been successful on several rootstocks: sapodilla, *Bassia latifolia*, *B. longifolia*, *Sideroxylon dulcificum* and *Mimusops hexandra*. The last has been particularly successful, the grafts growing vigorously and fruiting heavily.

In Florida, shield-budding, cleft-grafting and side-grafting were moderately successful but too slow for large-scale production. An improved method of side-grafting was developed using year-old seedlings with stems ¼ in (6 mm) thick. The scion (young terminal shoot) was prepared 6 weeks to several months in advance by girdling and defoliating. Just before grafting the rootstock was scored just above the grafting site and the latex "bled" for several minutes. After the stock was notched and the scion set in, it was bound with rubber and given a protective coating of wax or asphalt. The scion started growing in 30 days and the rootstock was then beheaded. Some years later, further experiments showed that better results were obtained by omitting the pre-conditioning of the scion and the bleeding of the latex. The operator must work fast and clean his knife frequently. The scions are veneer-grafted and then completely covered with plastic, allowing free gas exchange while preventing dehydration. Success is deemed most dependent on season: the 2 or 3 months of late summer and early fall.

In the Philippines, terminal shoots are completely defoliated 2 to 3 weeks before grafting onto rootstock which has been kept in partial shade for 2 months. However, inarching is there considered superior to grafting, giving a greater percentage of success. Homeowners often find air-layering easier and more successful than grafting, and air-layered trees often begin bearing within 2 years after planting.

In India, 50% success has been realized in top-working 20-year-old trees—cutting back to 3½ ft (1 m) from the ground and inserting scions of superior cultivars.

Culture

Seedlings for grafting are best grown in full sun, kept moist and fertilized with 8-4-8 N P K every 45 days.

Trees set out in commercial groves should be spaced 30 to 45 ft (9-13.5 m) apart each way.

In India, the plants are placed in deep, pre-fertilized pits and manured twice a year, sometimes with the addition of castor bean meal or residue of neem seed (*Azadirachta indica* A. Juss.), wood ash and/or ammonium sulfate. In an experiment at Marathwada Agricultural University, Parbhani, India, with 8-year-old trees planted at 12 m, application of 28 oz (800 g) N/tree increased trunk size and number and weight of fruits. Combined application of this amount of N plus 6¼ oz (176 g) P and 5¾ oz (166 g) K/tree gave the highest fruit yield. Fertilizer experiments over a period of 25 years at Gujarat Agricultural University revealed that N alone increases yield by 70%, a combination of N and P elevates yield by 90%, and combined N and K, 128%, over that of control (unfertilized) trees. Of course, optimum nutrient formulas depend on the character of the soil. In South Florida's limestone a mixed fertilizer of N, P, K, Mg in a 4-7-5-3 ratio is recommended in spring, summer and fall.

Most mature sapodilla trees receive no watering, but irrigation in dry seasons will increase productivity. In some parts of India, brackish or saline water is sometimes used to reduce vegetative growth and promote fruiting.

Season

The fruits mature 4 to 6 months after flowering. In the tropics, some cultivars bear almost continuously. In India, the main season is from December to March. The trees bear from May to September in Florida, with the peak of the crop in June and July. In Mexico, there are two peak seasons: February–April and October–December.

Harvesting

Most people find it difficult to tell when a sapodilla is ready to pick. With types that shed much of the "sand" on maturity, it is relatively easy to observe the slight yellow or peach color of the ripe skin, but with other types it is necessary to rub the scurf to see if it loosens readily and then scratch the fruit to make sure the skin is not green beneath the scurf. If the skin is brown and the fruit separates from the stem easily without leaking of the latex, it is fully mature though still hard and must be kept at room temperature for a few days to soften. It is best to wash off the sandy scurf before putting the fruit aside to ripen. It should be eaten when firm-soft, not mushy.

In the Bahamas, children bury their "dillies" in potholes in the limestone to ripen, or the fruits may be wrapped in sweaters or other thick material and put in drawers to hasten softening. Fruits picked immature will shrivel as they soften and will be of inferior quality, sometimes with small pockets of gummy latex.

In commercial groves, it is judged that when a few fruits have softened and fallen from the tree, all the full-grown fruits may be harvested for marketing. If in any doubt, the grower should cut open a few fruits to make sure the seeds are black (or very dark-brown). Pickers should use clippers or picking poles with bag and sharp notch at the peak of the metal frame to cut the fruit stem.

In India, the fruits are spread out in the shade to allow any latex at the stem end to dry before packing. The fruits ship well with minimal packing.

Yield

The 'Prolific' sapodilla yields 6 to 9 bushels per tree annually; or, 200 to 450 lbs (90 to 180 kg). 'Brown Sugar' yields 5 to 8 bushels. In India, it is said that a productive tree will bear 1,000 fruits in its 10th year and the yield increases steadily. At 30-35 years of age, the tree should produce 2,500 to 3,000 fruits annually. A great deal depends on the cultivar. A 10-year-old 'Oval' tree gave 1,158 fruits weighing 184 lbs (128.8 kg), while a 10-year-old 'Cricket Ball' bore 353 fruits weighing 112 lbs (50 kg). Hand-pollination has been found to increase fruit set.

Keeping Quality and Storage

Mature, hard sapodillas will ripen in 9 to 10 days and rot in 2 weeks at normal summer temperature and relative humidity. More than 50 years ago, sapodillas were shipped from Java to Holland, held at 40°-50°F (4.44-10°C) for 3 days, and they ripened satisfactorily after arrival. They were smoked over burning straw for a few hours before packing. Storage trials in Malaya demonstrated that mature, hard sapodillas stored at 68°F (20°C) will ripen in 10 days and remain in good condition for another 5 days. In Venezuela, mature fruits held at 68°F (20°C) and 90% relative humidity were in excellent condition at the end of 23 days. Lower temperatures, in efforts to prolong storage life, seriously retard ripening and lower fruit quality. Low relative humidity causes shriveling and wrinkling. Humid conditions promote sogginess. If long storage is necessary, the fruits may be kept at 59°-68°F (15°-20°C) in a controlled atmosphere of 85-90% relative humidity, 5-10% (v/v) CO_2, with total removal of C_2H_4 to delay ripening.

Firm-ripe sapodillas may be kept for several days in good condition in the home refrigerator. At 35°F (1.67°C), they can be kept for 6 weeks. Fully ripe fruits frozen at 32°F (0°C) keep perfectly for 33 days.

Pests and Diseases

In general, the sapodilla tree remains supremely healthy with little or no care. In India, it is sometimes attacked by a bark-borer, *Indarbela* (*Arbela*) *tetraonis*. Mealybugs may infest tender shoots and deface the fruits. A galechid caterpillar (*Anarsia*) has caused flower buds and flowers to dry up and fall. In Indonesia, caterpillars of *Tarsolepis remicauda* may completely defoliate the tree. A caterpillar, *Nephopteryx engraphella*, feeds on the leaves, flower buds and young fruits in parts of India. The ripening and overripe fruits are favorite hosts of the Mediterranean, Caribbean, Mexican and other fruit flies.

Various scales, including *Howardia biclavis*, *Pulvinaria* (or *Chloropulvinaria*) *psidii*, *Rastrococcus iceryoides*, and pustule scale, *Asterolecanium pustulans* Ckll., may lead to black sooty mold caused by the fungus *Capnodium* sp. on stems, foliage and fruits. In some years, during winter and spring in Florida, a rust (possibly *Uredo sapotae*) may affect the foliage of some cultivars. A leaf spot (*Septoria* sp.) has caused defoliation in a few locations. The moth of a leaf miner (*Acrocercops gemoniella*) is active on young leaves. Other minor enemies have been occasionally observed.

In India, it may be necessary to spread nets over the tree to protect the fruits from fruit bats.

Food Uses

Generally, the ripe sapodilla, unchilled or preferably chilled, is merely cut in half and the flesh is eaten with a spoon. It is an ideal dessert fruit as the skin, which is not eaten, remains firm enough to serve as a "shell". Care must be taken not to swallow a seed, as the protruding hook might cause lodging in the throat. The flesh, of course, may be scooped out and added to fruit cups or salads. A dessert sauce is made by peeling and seeding ripe sapodillas, pressing the flesh through a colander, adding orange juice, and topping with whipped cream. Sapodilla flesh may also be blended into an egg custard mix before baking.

It was long proclaimed that the fruit could not be cooked or preserved in any way, but it is sometimes fried in Indonesia and, in Malaya, is stewed with lime juice or ginger. I found that Bahamians often crush the ripe fruits, strain, boil and preserve the juice as a sirup. They also add mashed sapodilla pulp to pancake batter and to ordinary bread mix before baking. My own experiments showed that a fine jam could be made by peeling and stewing cut-up ripe fruits in water and skimming off a green scum that rises to the surface and appears to be dissolved latex, then adding sugar to improve texture and sour orange juice and a strip of peel to offset the increased sweetness. Skimming until all latex scum is gone is the only way to avoid gumminess. Cooking with sugar changes the brown color of the flesh to a pleasing red.

One lady in Florida developed a recipe for sapodilla pie. She peeled the ripe fruits, cut them into pieces as apples are cut, and filled the raw lower crust, sprinkled ½ cup of raisins over the fruit, poured over evenly ½ cup of 50-50 lime and lemon juice to prevent the sapodilla pieces from becoming rubbery, and then sprinkled evenly ½ cup of granulated sugar. After covering with the top crust and making a center hole to release steam, she baked for 40 minutes at 350°F (176.67°C). In India, it has been shown that ripe fruits can be peeled and sliced, packed in metal cans, heated for 10 minutes at 158°F (70°C), then treated for 6 minutes at a vacuum of 28 in Hg, vacuum double-seamed, and irradiated with a total dose of 4 x 10^5 rads at room temperature. This process provides an acceptable canned product.

Ripe sapodillas have been successfully dried by pre-treatment with a 60% sugar solution and osmotic dehydration for 5 hours, and the product has retained acceptable quality for 2 months.

Mr. Edward Smith of Crescent Place, Trinidad, made sapodilla wine and told me that it was very good. Young leafy shoots are eaten raw or steamed with rice in Indonesia, after washing to eliminate the sticky sap.

Food Value

Immature sapodillas are rich in tannin (proanthocyanadins) and very astringent. Ripening eliminates the tannin except for a low level remaining in the skin.

Analyses of 9 selections of sapodillas from southern Mexico showed great variation in total soluble solids, sugars and ascorbic acid content. Unfortunately, the fruits were not peeled and therefore the results show abnormal amounts of tannin contributed by the skin:

Moisture ranged from 69.0 to 75.7%; ascorbic acid from 8.9 to 41.4 mg/100 g; total acid, 0.09 to 0.15%; pH, 5.0 to 5.3; total soluble solids, 17.4° to 23.7° Brix; as for carbohydrates, glucose ranged from 5.84 to 9.23%, fructose, 4.47 to 7.13%, sucrose, 1.48 to 8.75%, total sugars, 11.14 to 20.43%, starch, 2.98 to 6.40%. Tannin content, because of the skins, varied from 3.16 to 6.45%.

Toxicity

The seed kernel (50% of the whole seed) contains 1% saponin and 0.08% of a bitter principle, sapotinin. Ingestion of more than 6 seeds causes abdominal pain and vomiting.

Other Uses

Chicle: A major by-product of the sapodilla tree is the gummy latex called "chicle", containing 15% rubber and 38% resin. For many years it has been employed as the chief ingredient in chewing gum but it is now in some degree diluted or replaced by latex from other species and by synthetic gums.

Chicle is tasteless and harmless and is obtained by repeated tapping of wild and cultivated trees in Yucatan, Belize and Guatemala. It is coagulated by stirring over low fires, then poured into molds to form blocks for export. Processing consists of drying, melting, elimination of foreign matter, combining with other gums and resins, sweeteners and flavoring, then rolling into sheets and cutting into desired units.

The dried latex was chewed by the Mayas and was introduced into the United States by General Antonio Lopez de Santa Ana about 1866 while he was on Staten Island awaiting clearance to enter this country. He had a supply in his pocket for chewing and gave a piece to the son of Thomas Adams. The latter at first considered the possibility of using it to make dentures, then decided it was useful only as a masticatory. He found he could easily incorporate flavoring and thus soon launched the chicle-based chewing-gum industry. In 1930, at the peak of production, nearly 14,000,000 lbs (6,363,636 kg) of chicle were imported.

Efforts have been made to extract chicle from the leaves and unripe fruit but the yield is insufficient. It has been estimated that 3,200 leaves would be needed to produce one pound (0.4535 kg) of gum.

Among miscellaneous uses: the latex is employed as birdlime, as an adhesive in mending small articles in India; it has been utilized in dental surgery, and as a substitute for gutta percha. The Aztecs used it for modeling figurines.

Timber: Sapodilla wood is strong and durable and timbers which formed lintels and supporting beams in Mayan temples have been found intact in the ruins. It has also been used for railway crossties, flooring, native carts, tool handles, shuttles and rulers. The red heartwood is valued for archer's bows, furniture, bannisters, and cabinetwork but the sawdust irritates the nostrils. Felling of the tree is prohibited in Yucatan because of its value as a source of chicle.

Bark: The tannin-rich bark is used by Philippine fishermen to tint their sails and fishing lines.

Medicinal Uses: Because of the tannin content, young fruits are boiled and the decoction taken to stop diarrhea. An infusion of the young fruits and the flowers is drunk to relieve pulmonary complaints. A decoction of old, yellowed leaves is drunk as a remedy for coughs, colds and diarrhea. A "tea" of the bark is regarded as a febrifuge and is said to halt diarrhea and dysentery. The crushed seeds have a diuretic action and are claimed to expel bladder and kidney stones. A fluid extract of the crushed seeds is employed in Yucatan as a sedative and soporific. A combined decoction of sapodilla and chayote leaves is sweetened and taken daily to lower blood pressure. A paste of the seeds is applied on stings and bites from venomous animals. The latex is used in the tropics as a crude filling for tooth cavities.

Sapote (Plate LV)

The word "sapote" is believed to have been derived from the Aztec "tzapotl", a general term applied to all soft, sweet fruits. It has long been utilized as a common name for *Pouteria sapota* (Jacq.) H.E. Moore & Stearn (syns. *P. mammosa* (L.) Cronquist, *Lucuma mammosa* Gaertn., *Achradelpha mammosa* Cook, *Vitellaria mammosa* Radlk., *Calocarpum mammosum* Pierre, *C. sapota* Merrill, *Sideroxylon sapota* Jacq.). Alternate vernacular names include *sapota, zapote, zapote colorado, zapote mamey, lava-zapote, zapotillo, mamey sapote, mamee sapote, mamee zapote, mamey colorado, mamey rojo, mammee* or *mammee apple* or red sapote. In El Salvador, it is known as *zapote grande*, in Colombia as *zapote de carne;* in Cuba, it is *mamey*, which tends to confuse it with *Mammea americana* L., a quite different fruit widely known by that name. The usual name in Panama is *mamey de la tierra;* in Haiti, *sapotier jaune d'oeuf,* or *grand sapotillier;* in Guadeloupe, *sapote à creme;* in Martinique, *grosse sapote;* in Jamaica, it is marmalade fruit or marmalade plum; in Nicaragua, it may be called *guaicume;* in Mexico, *chachaas* or *chachalhaas* or *tezon-zapote;* in Malaya and the Philippines, *chico-mamei,* or *chico-mamey.*

The sapote belongs to the family Sapotaceae, the same family as the sapodilla (*Manilkara zapota* van Royen) which has also been called *sapote, zapote,* or *zapote chico* to distinguish it from the larger fruit.

Description
The sapote tree is erect, frequently to 60 ft (18 m) sometimes to 100 or 130 ft (30 or 40 m) with short or tall trunk to 3 ft (1 m) thick, often narrowly buttressed, a narrow or spreading crown, and white, gummy latex. The evergreen or deciduous leaves, clustered at the branch tips, on petioles ¾ to 2 in (2-5 cm) long, are obovate, 4 to 12 in (10-30 cm) long, and 1½ to 4 in (4-10 cm) wide, pointed at both ends. The small, white, to pale-yellow 5-parted flowers emerge in clusters of 6 to 12 in the axils of fallen leaves along the branches. The fruit may be round, ovoid or elliptic, often bluntly pointed at the apex, varies from 3 to 9 in (7.5-22.8 cm) long, and ranges in weight from ½ lb to 5 lbs (227 g-2.3 kg). It has rough, dark-brown, firm, leathery, semi-woody skin or rind to ¹/₁₆ in (1.5 mm) thick, and salmon-pink to deep-red, soft flesh, sweet and pumpkin-like in flavor, enclosing 1 to 4 large, slick, spindle-shaped, pointed seeds, hard, glossy-brown, with a whitish, slightly rough hilum on the ventral side. The large kernel is oily, bitter, and has a strong bitter-almond odor.

Origin and Distribution
The sapote occurs naturally at low elevations from southern Mexico to northern Nicaragua. It is much cultivated and possibly also naturalized up to 2,000 ft (600 m) and occasionally found up to 5,000 ft (1,500 m) throughout Central America and tropical South America. It is abundant in Guatemala. In the West Indies, it is planted to a limited extent from Trinidad to Guadeloupe, and in Puerto Rico, Haiti and Jamaica, but mainly in Cuba where it is often grown in home gardens and along streets and for shading coffee because it loses its leaves at the period when coffee plants need sun, and the fruit is extremely popular. It is grown only occasionally in Colombia, Ecuador, Venezuela and Brazil. It was introduced into the Philippines by the early Spaniards but is grown only around Cavite and Laguna on Luzon and Cagayan on Mindanao. From the Philippines, it was carried to southern Vietnam where the fruit is eaten when very ripe.

The sapote has existed in Florida for at least a century. The prominent horticulturist, Pliny Reasoner, included it in his report in the U.S. Department of Agriculture's Pomological Bulletin in 1887. Subsequently, seeds were brought into the United States on various occasions. In 1914, the Office of Foreign Seed and Plant Introduction received seeds from the Costa Rican National Museum, San José (P.I. #39357). Mr. Ramon Arias-Feraud supplied seeds from Panama in 1918 (P.I. #46236). In July, 1919, seeds from Laguna, Philippines, were sent by the Bureau of Agriculture, Manila (P.I. #47516). More seeds from Costa Rica were presented by Mr. Carlos Werckle in October, 1919 (P.I. #47956). Seeds of a superior selection were obtained and planted at the Federal Experiment Station, Mayaguez, Puerto Rico, in 1939.

Despite the favorable comments that accompanied these and other introductions, the sapote was represented by only a few scattered trees in southern Florida for a long time. One of the discouraging factors was the tree's slowness in coming into bearing. William J. Krome, a leading pioneer, planted a seedling on his property in Homestead in 1907 and it bore its first fruits in 1949, after having suffered repeated setbacks from freezes and hurricanes over the years, and it was then only 18 ft (5½ m) high. Other trees in more protected locations had fared much better.

The arrival of many Cubans in Dade County during the past 2 decades has created an active demand for the fruits and for the trees for home planting, and some commercial orchards of 5 to 20 acres (2-8⅓ ha) or more have been established. In 1983, one man with 15 trees in his backyard was selling the fruits to Cuban people and bringing in seedlings 5 ft (1½ m) high from the Dominican Republic at $100 each. Such enthusiasm has spurred efforts to develop practical methods of vegetative propagation and one expert propagator is now selling grafted trees at $10.50 each, wholesale. In the fall of 1984, a nursery had acquired a stock of 1,000 of these trees and one customer bought them all. Thus has the status of the sapote risen dramatically in southern Florida because of an ethnic change in the population.

Varieties

There is much seedling variation in the sapote. Superior selections have been made in Cuba, Central America and in Florida in recent years. The following named cultivars are being cultivated domestically or commercially, or merely being tested in Florida:

'AREC No. 3' — Seed received from Isle of Pines, Cuba, 1940. Seedling grown at AREC, Homestead. Grafted trees planted later. Fruit medium to large, 14 to 26 oz (400-740 g). Flesh pink; of poor to good quality; contains 3-4 seeds. Ripens July-Sept. Tree of medium size, a fair bearer; probably useful source of seeds for rootstocks.

'Cayo Hueso' — A selection from the Dominican Republic; favored by Cubans.

'Chenox' — Obtained by Lawrence Zill from Belize. Grafted trees being tested at AREC, Homestead, and elsewhere.

'Copan' ('AREC No. 1') — Seed received from Cuba in 1938. Seedling set out in field at AREC, Homestead, 1940. Grafted trees planted out in 1975; later propagated by nurseries. Fruit of medium size; 15-32 oz (425-900 g). Flesh red, of excellent quality; contains 1 seed. Fruit ripens in July-Aug. Tree is of spreading habit and medium in size. Leaves turn red in Dec., then become brown and are shed in spring.

'Cuban No. 1' — Believed to have originated in Cuba but introduced from El Salvador. Fruit large; 9 in (22.8 cm) long; weighs 2.2 lbs (1 kg).

'Flores' — A Guatemalan selection introduced by Tom Economou of Miami and being tested at AREC, Homestead.

'Francisco Fernandez' — A Cuban selection named for the Miami man who introduced it into Florida.

'Magana' — Introduced from El Salvador in 1961. Seedling set in field at AREC, Homestead, in 1952. Grafted trees planted in 1975. Later propagated by nurseries. Fruit large to very large; 26 to 85 oz (740-2,400 g). Flesh pink, of good to excellent quality; contains 1 seed. Fruit matures in less than 12 mos (Apr.-May). Tree is small, slow-growing; may fruit 1 yr. after planting. Bears poorly in Puerto Rico; very well in Florida. Evergreen.

'Mayapan' ('AREC No. 2') — Seed sent from Isle of Pines, Cuba, in 1940. Fruit a little above medium size; 18 to 40 oz (510-1,135 g). Skin very scurfy. Flesh red, of good quality though slightly fibrous; contains 1 seed. Tree is erect and tall. Grafted trees slow to fruit but produce well after the lapse of a few years.

'Pantin' (or 'Key West') — In 1956, Pantin family in Miami provided budwood from a seedling tree in Key West. Fruit of medium size; 14 to 40 oz (400-1,130 g). Flesh pink to red, of excellent quality, fiberless; contains 1 seed. Tree is tall. Grafted trees grow slowly at first, bear little or no fruit for 2-3 years, then growth rate increases and yield is good. Leaves become brown in winter. Grafted trees sold by nurseries.

'Progreso' — Obtained by Lawrence Zill from Belize. Grafted trees being tested at AREC, Homestead, and elsewhere.

'Tazumal' — AREC, Homestead, received seedling tree from El Salvador in 1949. Grafted and planted several trees in 1975. Fruit is of medium size, 14 to 30 oz (400-850 g). Flesh pink, of good quality; contains 1-2 seeds. First crop ripens Jan.-Feb.; second crop, July-Aug. Tree is of medium size, fast-growing, bears regularly and heavily. Grafted trees sold by nurseries. Usually evergreen.

In western Puerto Rico, there are some high-yielding trees producing large fruits 2.2 lbs (1 kg) or more in weight having dark-red flesh.

Climate

The sapote tree is limited to tropical or near-tropical climates. In Central America, it flourishes from sea-level up to 2,000 ft (610 m); it is less common at 3,000 ft (914 m); and rare at 4,000 ft (1,220 m). Occasional trees have survived at 5,000 ft (1,500 m) but these grow slowly and fruit maturity is considerably delayed. Young specimens are highly cold-sensitive and the large leaves of the tree are subject to damage by cold winds. The sapote has been found too tender for California. It thrives in regions of moderate rainfall — about 70 in (178 cm) annually — and is intolerant of prolonged drought. Even a short dry spell may induce shedding of leaves.

Soil

The tree makes its best growth on the heavy soils — deep clay and clay loam — of Guatemala but it does well on a wide range of soil types, even infertile, porous sand. It was originally believed unsuited to the oolitic limestone soils of southern Florida. However, with adequate planting holes, it has proved to be long-lived and fruitful in Dade County. The tree will not thrive where there is poor drainage, a high water table, or impermeable subsoil restricting root development.

Propagation

Sapote seeds lose viability quickly and must be planted soon after removal from the fruit. They normally germinate in 2 to 4 weeks. Removal of the hard outer coat will speed germination. The seeds must be planted with the more pointed end upward and protruding ½ in (1.25 cm) above the soil in order to assure good form in the seedling. Rodents are attracted to the seeds and cause considerable losses in Cuba. Seedlings should be grown only in experimental plantings intended for selection of desirable characters, or for use as rootstocks. Normally seedlings will not bear until they are 8 to 10 years old and they do not necessarily come true from seed. In Cuba, seeds are taken only from esteemed trees that are isolated from those of low quality in order to avoid any detrimental influence through cross-pollination. For fruit production, the sapote is best propagated vegetatively and it will then produce fruit in 1 to 4 years, depending on the cultivar. Air-layering is seldom successful. Cuttings treated with indolebutyric acid fail to root. Various methods of grafting have been tried. Approach-grafting has been commonly practiced in Cuba and is a reliable but somewhat cumbersome technique. Chip-budding has given good results at times. Side-veneer grafting is considered most feasible in Mexico and Florida. It has been achieved with 80 to 98% success utilizing 1-yr-old defoliated trees in the February-May dry season, but still presents difficulties. Ing. Filiberto Lazo, a horticulturist of long experience in Cuba, has provided detailed instructions for tip-grafting which he proved to be practical. The seedlings for use as rootstocks are first grown in 1-quart (.94 liter) containers and, when the first tender leaves appear, are transplanted into gallon (3.8 liter) containers and kept in semi-shade until the leaves are full-grown and dark-green. At this stage they are given more sun and are fertilized and watered faithfully. Within a year the stem will be ¾ in (2 cm) thick and ready for grafting. An important point is to select budwood (scion) that is not as thick as the rootstock. The scion may be prepared by one of two methods: a) select from a tree that you wish to propagate a branch that has flowered; cut off the tip just below the leaves. About 10 to 12 days later the lateral buds of the beheaded branch begin to swell and this is the time to clip off the scion, 8 in (20 cm) in length, wrap it in a damp cloth, and proceed to graft as soon as possible; or b) clip off the terminal 8 in (20 cm) or more of a branch that has flowered, then immediately cut off the apex with the leaves, wrap the decapitated scion in a damp cloth and keep in the nursery until you see the lateral buds of the scion begin to swell; then proceed to graft.

The first cut in the rootstock should be a transverse one with pruning shears, leaving the stem about 1 ft (30 cm) high. Because of the copious latex, one must wait for it to drain out before going ahead. When the flow stops, take the scion (prepared either way), clip off 2 in (5 cm) or more from the base, leaving the scion about 6 in (15 cm) long. Using the budding knife, make a diagonal cut from 2½ in (6.25 cm) below the tip downward, the slant terminating at the side opposite the side where it was begun. A reverse cut of the same length is made in the tip of the rootstock so that the base of the scion and the tip of the rootstock will fit together perfectly and the bark will match up. The scion must then be tightly bound to the rootstock with polyethylene ribbon, leaving no air-space, and covering all of the scion up to 2½ in (6.25 cm) above the rootstock. A rubber band is put around over the polyethylene to make sure the wrapping is completely secure. When the scion has developed mature leaves, this is a sign that the graft has taken. The plastic is removed from the scion except for the part covering the graft which is left on until the scion has developed a quantity of leaves and displays distinct vigor. The grafted plant is ready to set out in the field one year later. Inferior cultivars, or grafted trees that have been frozen back, can be topworked by veneer-grafting mature or "juvenile-like" scions onto interstocks (seedling tops prepared for the purpose).

Spacing

Planting distances may vary with the fertility of the soil and the form and growth habit of the cultivar. On rich soil, sapote trees of spreading habit should be no less than 30 ft (9 m) apart each way. Lazo preferred a spacing of 40 ft (12 m) on an equilateral triangle. Where the soil is less fertile and the cultivar is fairly compact, the distance may be reduced to 25 ft (7.5 m).

Culture

Sapote trees do not require elaborate care, but should be given the advantage of adequate holes, pre-enriched, and routine fertilizer applications, at first high in nitrogen to stimulate vegetative growth. When nearing fruiting age, the tree will benefit from applications of a balanced fertilizer in spring and fall, the amount increasing each year. In dry seasons, frequent watering is desirable until the tree is well established. Grafted trees grow more slowly than seedlings and do not grow as tall, which is a distinct advantage in harvesting.

Harvesting and Yield

It is not easy to determine when the sapote is sufficiently mature to harvest. Some say the fruits are picked when they show a reddish tinge. Actually, in Cuba, 10 or 12 fruits from each tree are sampled by removing a small part of the rind and judging the color of the flesh. If it has achieved maximum color for that particular cultivar, the entire crop is deemed ready to pick. Fruits are not harvested from trees in active vegetative growth (a state called "primavera"), because they will never ripen completely. Harvesting of large trees requires a picking pole with a cutter and a basket to catch the fruits; or workers must use ladders and twist the fruit until the stem breaks. Trees that become too tall may be topped so that the crop will be within reach. After picking, the stem is close-clipped and the fruits are packed in boxes or baskets to avoid injury. There are no available figures on productivity but it is said in Cuba that trees on fertile soil will live for at least 100 years and bear abundantly throughout their lives.

Keeping Quality

A fully mature sapote will ripen in a few days. If shipped right after picking, the fruits can be sent to distant markets. In the past, they were exported from Mexico and Cuba to the United States.

Pests and Diseases

Sapote leaves and roots are attacked by the West Indian sugar cane root borer, *Diaprepes abbreviatus*, in Puerto Rico. The red spider mite, *Tetranychus bimaculatus*, may infest the leaves.

The fungus, *Colletotrichum gloeosporioides*, causes anthracnose on the leaves and fruit stalks in rainy seasons and causes fruits to fall prematurely. Leafspot resulting from attack by the fungus *Phyllosticta sapotae* occurs in Cuba and the Bahamas but seldom in Puerto Rico. In addition, black leaf spot (*Phyllachora* sp.) and root rot (*Pythium* sp.) may occur in Florida.

Food Uses

The sapote is credited with sustaining Cortez and his army in their historic march from Mexico City to Honduras. The fruit is of such importance to the Indians of Central America and Mexico that they usually leave this tree standing when clearing land for coffee plantations or other purposes. They generally eat the fruit out-of-hand or spooned from the half-shell. In urban areas, the pulp is made into jam or frozen as sherbet. In Cuba, fibrous types are set aside for processing.

A prominent dairy in Miami has for many years imported sapote pulp from Central America to prepare and distribute commercially as "Spanish sherbet". In Cuba, a thick preserve called *"crema de mamey colorado"* is very popular. The pulp is sometimes employed as a filler in making guava cheese.

The decorticated seeds, called *zapoyotas, sapuyules,* or *sapuyulos,* strung on sticks or cords, are marketed in the Isthmus of Tehuantepec, Mexico, and in Central America. The kernel is boiled, roasted and mixed with cacao in making chocolate—some say to improve the flavor, others say to increase the bulk, in which case it is actually an adulterant. In Costa Rica, it is finely ground and made into a special confection. Around Oaxaca, in southern Mexico, the ground-up kernel is mixed with parched corn, or cornmeal, sugar and cinnamon and prepared as a nutritious beverage called "pozol".

Food Value Per 100 g of Edible Portion*	
Calories	114.5
Moisture	55.3–73.1 g
Protein	0.188–1.97 g
Fat	0.09–0.25 g
Carbohydrates	1.41–29.7 g
Fiber	1.21–3.20 g
Ash	0.89–1.32 g
Calcium	28.2–121.0 mg
Phosphorus	22.9–33.1 mg
Iron	0.52–2.62 mg
Carotene	0.045–0.665 mg
Thiamine	0.002–0.025 mg
Riboflavin	0.006–0.046 mg
Niacin	1.574–2.580 mg
Ascorbic Acid	8.8–40.0 mg
Amino Acids:	
Tryptophan	19 mg
Methionine	12 mg
Lysine	90 mg

*Analyses made in Cuba and Central America.

Toxicity

De la Maza, in 1893, reported that the seed has stupefying properties, and this may be due to its HCN content. One is cautioned not to rub the eyes after handling the green fruit because of the sap exuding from the cut or broken stalk. The milky sap of the tree is highly irritant to the eyes and caustic and vesicant on the skin. The leaves are reportedly poisonous.

Other Uses

Seeds: Early in the 19th Century, the seeds were used in Costa Rica to iron starched fine linen. The seed kernel yields 45 to 60% of a white, semi-solid, vaseline-like oil which is edible when freshly extracted and refined. It is sometimes used in soap and considered to have a greater potential in the soap industry, in cosmetics and pharmaceutical products. It was used in olden times to fix the colors on painted gourds and other articles of handicraft. The seeds have served as a source of Noyeau scent in perfumery. The nectar of the flowers is gathered by honeybees.

Trees: The trees are seldom cut for timber, unless they bear poor quality fruit. There is very little sapwood. The heartwood is buff or brown when fresh, becoming reddish with age; sometimes resembles mahogany but is redder and more or less mottled with darker tones. It is fine-grained, compact, generally hard and fairly heavy, strong, easy to work and fairly durable. It is rated as suitable for cabinetwork and is made into furniture, but mostly serves for building carts, and for shelving and house frames.

Medicinal Uses: In Santo Domingo, the seed kernel oil is used as a skin ointment and as a hair dressing believed to stop falling hair. In Mexico, 2 or 3 pulverized kernels are combined with 10 oz (300 g) castor oil for application to the hair. In 1970, clinical tests at the University of California at Los Angeles failed to reveal any hair-growth promoting activity but confirmed that the oil of sapote seed is effective in stopping hair-fall caused by seborrheic dermatitis. The oil is employed as a sedative in eye and ear ailments. The seed residue after oil extraction is applied as a poultice on painful skin afflictions.

A seed infusion is used as an eyewash in Cuba. In Mexico, the pulverized seed coat is reported to be a remedy for coronary trouble and, taken with wine, is said to be helpful against kidney stones and rheumatism. The Aztecs employed it against epilepsy. The seed kernel is regarded as a digestive; the oil is said to be diuretic. The bark is bitter and astringent and contains lucumin, a cyanogenic glycoside. A decoction of the bark is taken as a pectoral. In Costa Rica a "tea" of the bark and leaves is administered in arteriosclerosis and hypertension. The milky sap is emetic and anthelmintic and has been used to remove warts and fungal growths on the skin.

Related Species

The **green sapote,** *Pouteria viridis* Cronq., (syns. *Calocarpum viride* Pitt.; *Achradelpha viridis* O.F. Cook), is called *injerto, injerto verde* or *raxtul* in Guatemala; *zapote injerto* in Costa Rica; white faisan or red faisan in Belize. The tree is erect, to 40 or even 80 ft (12-24 m) in height, its young branches densely brown-hairy. It possesses an abundance of white, gummy latex. The leaves are clustered at the tips of flowering branches and irregularly alternate along non-fruiting limbs. They are oblanceolate, pointed, 4 to 10 in (10-25 cm) long, 2 to 2¾ in (5-7 cm) wide; hairy on the upper midrib and downy-white beneath. Flowers, borne in groups of 2 to 5 in the leaf axils and massed along leafless branches, are tubular, 5-lobed, pinkish or ivory and silky-hairy. The fruit varies from nearly round to ovoid, pointed at the apex and sometimes at the base; may be 3½ to 5 in (9-12.5 cm) long and 2½ to 3 in (6.25-7.5 cm) thick, with thin, olive-green or yellow-green skin dotted with red-brown and clinging tightly to the flesh. The flesh is light-russet, of fine texture, melting, fairly juicy and sweet; of better flavor than the sapote. There may be 1 or 2 dark-brown, shiny, elliptic or ovate seeds to 2 in (5 cm) long, with a large, dull, grayish hilum on one surface. The fruit is picked while hard and held until soft. The flesh is generally eaten raw, spooned from the skin, but a preserve is made from it in Guatemala.

The tree is native and common in the wild in Guatemala and Honduras; rarer in Costa Rica and southward to Panama; at elevations between 3,000 and 7,000 ft (900-2,100 m). The fruits are commonly marketed.

In 1916, 50 seeds from fruits on the market in Guatemala were introduced by the United States Department of Agriculture (S.P.I. #43788). Experimental plantings were made in California and Florida. More seeds were sent by Dr. Wilson Popenoe from the Lancetilla Experimental Garden at Tela, Honduras, in 1929 (S.P.I. #80383). Other introductions followed. There were no survivors in California or Florida in 1940. Trees 8 to 10 ft (2.4-3 m) high at the Agricultural Research and Education Center, Homestead, Florida, were killed by a flood in 1948. A private experimenter, William Whitman, obtained budwood

from Honduras in 1954 and grafted it onto sapote rootstock. Other such grafts were made by a commercial fruit grower and the first fruits were borne in 1961. Subsequently, grafted trees were offered for sale by the Brooks-Tower Nursery and various seedlings have been distributed to private growers. The tree seems to flourish with little care on rich hammock soil but needs regular fertilizing on limestone. The Cuban May beetle feeds on the leaves. Seedlings begin to bear when 8 to 10 years old. The crop ripens in fall and winter.

Food Value Per 100 g of Edible Portion*	
Moisture	68.1-69.5 g
Protein	0.152-0.283 g
Fat	0.24-0.28 g
Fiber	1.2-1.6 g
Ash	0.69-1.38 g
Calcium	18.6-35.7 mg
Phosphorus	22.1-23.6 mg
Iron	0.57-0.74 mg
Carotene	0.031-0.069 mg
Thiamine	0.009-0.011 mg
Riboflavin	0.027 mg
Niacin	1.88-1.189 mg
Ascorbic Acid	49.9-62.3 mg
*Analyses made in Guatemala.	

The latex (chicle) has been commercially collected and marketed like that from the sapodilla for use in chewing gum. The wood is reddish, fine-grained, compact, strong, durable; occasionally used in construction, carpentry, turnery, and for furniture and paneling in Guatemala.

Canistel (Plate LVI)

The canistel, *Pouteria campechiana* Baehni, has been the subject of much botanical confusion as is evidenced by its many synonyms: *P. campechiana* var. *nervosa* Baehni; *P. campechiana* var. *palmeri* Baehni; *P. campechiana* var. *salicifolia* Baehni; *Lucuma campechiana* HBK.; *L. Heyderi* Standl.; *L. laeteviridis* Pittier; *L. multiflora* Millsp. NOT A. DC.; *L. nervosa* A. DC.; *L. palmeri* Fernald; *L. rivicoa* Gaertn.; *L. rivicoa* var. *angustifolia* Miq.; *L. salicifolia* HBK.; *Richardella salicifolia* Pierre; *Sideroxylon campestre* T.S. Brandeg.; *Vitellaria campechiana* Engl.; *V. salicifolia* Engl.

It is the showiest fruit of the family Sapotaceae but generally underevaluated in horticultural literature and by those who have only a casual acquaintance with it.

Colloquial names applied to this species include: eggfruit, canistel, ti-es, yellow sapote (Cuba, Hawaii, Jamaica,

Puerto Rico, Bahamas, Florida); canistel, *siguapa*, *zapotillo* (Costa Rica); *costiczapotl, custiczapotl, fruta de huevo, zapote amarillo* (Colombia); *cakixo, canizte, kanis, kaniste, kantzé, kantez, limoncillo, mamee ciruela, zapotillo de montana* (Guatemala); *huevo vegetal* (Puerto Rico, Venezuela); *mammee sapota*, eggfruit, *ti-es* (Bahamas); *mamey cerera, mamey cerilla, mamee ciruela, kanizte* (Belize); *atzapotl* (the fruit), *atzapolquahuitl* (the tree), *caca de niño, cozticzapotl, cucumu, mamey de Campechi, mamey de Cartagena, huicumo, huicon, kan 'iste', kanixte, kanizte, palo huicon, zapote amarillo, zapote de niño, zapote borracho* (drunken sapote, perhaps because the fallen fruits ferment on the ground); *zapote mante, zubul* (Mexico); *guaicume, guicume, zapotillo, zapotillo amarillo* (El Salvador); *zapote amarillo* (Nicaragua); *boracho*, canistel, *toesa* (Philippines).

Fig. 108: Glossy, yellow, long-keeping, highly nutritious, the canistel (*Pouteria campechiana*) deserves wider recognition as a good food.

Description

The canistel tree is erect and generally no more than 25 ft (8 m) tall, but it may, in favorable situations, reach a height of 90 to 100 ft (27-30 m) and the trunk may attain a diameter of 3 ft (1 m). Slender in habit or with a spreading crown, it has brown, furrowed bark and abundant white, gummy latex. Young branches are velvety brown. The evergreen leaves, alternate but mostly grouped at the branch tips, are relatively thin, glossy, short- to long-stemmed, oblanceolate, lanceolate-oblong, or obovate, bluntly pointed at the apex, more sharply tapered at the base; 4½ to 11 in (11.25-28 cm) long, 1½ to 3 in (4-7.5 cm) wide. Fragrant, bisexual flowers, solitary or in small clusters, are borne in the leaf axils or at leafless nodes on slender pedicels. They are 5- or 6-lobed, cream-colored, silky-hairy, about 5/16 to 7/16 in (8-11 mm) long.

The fruit, extremely variable in form and size, may be nearly round, with or without a pointed apex or curved beak, or may be somewhat oval, ovoid, or spindle-shaped. It is often bulged on one side and there is a 5-pointed calyx at the base which may be rounded or with a distinct depression. Length varies from 3 to 5 in (7.5-12.5 cm) and width from 2 to 3 in (5-7.5 cm), except in the shrubby form, var. *palmeri*, called *huicon*—4 to 9 ft (1.5-3 m) high—which has nearly round fruits only 1 in (2.5 cm) long. When unripe the fruit is green-skinned, hard and gummy internally. On ripening, the skin turns lemon-yellow, golden-yellow or pale orange-yellow, is very smooth and glossy except where occasionally coated with light-brown or reddish-brown russetting.

Immediately beneath the skin the yellow flesh is relatively firm and mealy with a few fine fibers. Toward the center of the fruit it is softer and more pasty. It has been often likened in texture to the yolk of a hard-boiled egg. The flavor is sweet, more or less musky, and somewhat like that of a baked sweet potato. There may be 1 to 4 hard, freestone seeds, ¾ to 2⅛ in (2-5.3 cm) long and ½ to 1¼ in (1.25-3.2 cm) wide, near-oval or oblong-oval, glossy and chestnut-brown except for the straight or curved ventral side which is dull light-brown, tan or grayish-white. Both ends are sharp-tipped.

Origin and Distribution

The canistel is sometimes erroneously recorded as native to northern South America where related, somewhat similar species are indigenous. Apparently, it occurs wild only in southern Mexico (including Yucatan), Belize, Guatemala and El Salvador. It is cultivated in these countries and in Costa Rica (where it has never been found wild), Nicaragua and Panama, Puerto Rico, Jamaica, Cuba (where it is most popular and commercialized in

Pinar del Rio), the Bahamas, southern Florida and the Florida Keys. Some writers have reported the canistel as naturalized on the Florida Keys, in the Bahamas and Cuba, but specimens that appear to be growing in the wild are probably on the sites of former homesteads. Oris Russell, who has explored hundreds of acres of coppices in the Bahamas, has never seen the canistel or its close relative, *P. domingensis* Baehni, in a wild state. He says that abandoned plantings can be completely overgrown by coppice in 3 to 4 years. Also, it is possible that a seedling might arise from the seed of a fruit carried into the woods by an animal or tossed away by a human. Mango trees are sometimes unintentionally planted in this way in southern Florida, especially if the seed lands in a hedge which provides a moist and shady site and physical protection.

Seeds from Cuba were planted at the Lancetilla Experimental Garden, La Lima, Honduras, in 1927. Dr. Victor M. Patiño bought fruits in a Cuban market in 1957 and had the seeds planted at the Estacion Agricola Experimental de Palmira, Colombia. He reported that several trees were growing well there in 1963. The canistel is included in experimental collections in Venezuela. The tree was introduced at low and medium elevations in the Philippines before 1924 and it reached Hawaii probably around the same time. Attempts to grow it in Singapore were not successful. In 1949 there were a few canistel trees growing in East Africa.

Varieties

There are apparently no named cultivars but certain types are so distinct as to have been recorded as different species in the past. The spindle-shaped form (called mammee sapota or eggfruit) was the common strain in the Bahamas for many years, at least as far back as the 1920's. The rounded, broader form began to appear in special gardens in the 1940's, and the larger types were introduced from Florida in the 1950's

In 1945, large, handsome, symmetrical fruits were being grown under the names *Lucuma salicifolia* and yellow sapote at the Agricultural Research and Education Center and at Palm Lodge Tropical Grove, Homestead, Florida, but these were soon classified as superior strains of canistel. Some fruits are muskier in odor and flavor than others, some are undesirably dry and mealy, some excessively sweet. An excellent, non-musky, fine-textured, rounded type of medium size has been selected and grown by Mr. John G. DuPuis, Jr., at his Bar-D Ranch in Martin County. It is well worthy of dissemination. There is considerable variation as to time of flowering and fruiting among seedling trees.

Climate

The canistel needs a tropical or subtropical climate. In Guatemala, it is found at or below 4,600 ft (1,400 m) elevation. In Florida, it survives winter cold as far north as Palm Beach and Punta Gorda and in protected areas of St. Petersburg. It has never reached fruiting age in California. It requires no more than moderate precipitation; does well in regions with a long dry season.

Soil

The canistel is tolerant of a diversity of soils—calcareous, lateritic, acid-sandy, heavy clay. It makes best vegetative growth in deep, fertile, well-drained soil but is said to be more fruitful on shallow soil. It can be cultivated on soil considered too thin and poor for most other fruit trees.

Propagation

Canistel seeds lose viability quickly and should be planted within a few days after removal from the fruit. If decorticated, seeds will germinate within 2 weeks; otherwise there may be a delay of 3 to 5 months before they sprout. The seedlings grow rapidly and begin to bear in 3 to 6 years. There is considerable variation in yield and in size and quality of fruits. Vegetative propagation is preferred in order to hasten bearing and to reproduce the best selections. Side-veneer grafting, cleft grafting, patch budding and air-layering are usually successful. Cuttings take a long time to root.

Culture

Mulching is beneficial in the early years. A balanced fertilizer applied at time of planting and during periods of rapid growth is advisable though the tree does not demand special care. Outstanding branches should be pruned back to avoid wind damage and shape the crown.

Pests and Diseases

Few pests and diseases attack the canistel. In Florida only scale insects and the fungi, *Acrotelium lucumae* (rust); *Colletotrichum gloeosporioides* (fruit spot); *Elsinoë lepagei* (leaf spot and scab); and *Gloeosporium* (leaf necrosis) have been recorded for this species. The tree is nearly always vigorous and healthy.

Fruiting Season and Harvesting

Blooming extends from January to June in Mexico (26). In Cuba, flowers are borne mostly in April and May though some trees flower all year. The canistel has the advantage of coming into season in late fall and winter, when few other tropical fruits are available. The fruits generally mature from September to January or February in the Bahamas, from November or December to February or March in Florida. In Cuba, the main fruiting season is from October to February but some trees produce more or less continuously throughout the year. The mature but still firm fruits should be clipped to avoid tearing the skin. When left to ripen on the tree, the fruits split at the stem end and fall. A severe drop in temperature will cause firm-mature fruits to split and drop to the ground.

Storage and Shipment

If kept at room temperature, the fruits will soften to eating-ripe in 3 to 10 days. They should not be allowed to become too soft and mushy before eating. Ripe fruits can be kept in good condition in the vegetable tray of a home refrigerator for several days.

Freshly picked, hard fruits have been successfully shipped from Florida to fruiterers and other special customers in

New York City and Philadelphia by Palm Lodge Tropical Grove, Homestead.

Unfortunately, no studies have been made to determine optimum temperature and humidity levels for long-term storage and long-distant shipment. This is an ideal fruit for export to European markets where its bright color, smoothness and appealing form would be especially welcome in the winter season.

Food Uses

The fact that the canistel is not crisp and juicy like so many other fruits seems to dismay many who sample it casually. Some take to it immediately. During World War II when RAF pilots and crewmen were under training in the Bahamas, they showed great fondness for the canistel and bought all they could find in the Nassau market.

Some Floridians enjoy the fruit with salt, pepper and lime or lemon juice or mayonnaise, either fresh or after light baking. The pureed flesh may be used in custards or added to ice cream mix just before freezing. A rich milkshake, or "eggfruit nog", is made by combining ripe canistel pulp, milk, sugar, vanilla, nutmeg or other seasoning in an electric blender.

The late Mrs. Phyllis Storey of Homestead made superb "mock-pumpkin" pie with 1½ cups mashed canistel pulp, ⅔ cup brown sugar, ½ teaspoon salt, ¼ teaspoon nutmeg, 1 teaspoon lime juice, 2 beaten eggs, 2 cups evaporated milk or light cream. The mixture is poured into one crust and baked for 1 hr at 250°F (121°C).

Others have prepared canistel pancakes, cupcakes, jam, and marmalade. Mrs. Gladys Wilbur made canistel "butter" by beating the ripe pulp in an electric blender, adding sugar, and cooking to a paste, with or without lemon juice. She used it as a spread on toast. The fruit could also be dehydrated and reduced to a nutritious powder as is being done with the lucmo (q.v.) and this might well have commercial use in pudding mixes.

Food Value

Canistels are rich in niacin and carotene (provitamin A) and have a fair level of ascorbic acid. The following analyses show that the canistel excels the glamorized carambola (*Averrhoa carambola* L.) in every respect except in moisture and fiber content, and riboflavin.

Food Value Per 100 g of Edible Portion*	
Calories	138.8
Moisture	60.6 g
Protein	1.68 g
Fat	0.13 g
Carbohydrates	36.69 g
Fiber	0.10 g
Ash	0.90 g
Calcium	26.5 mg
Phosphorus	37.3 mg
Iron	0.92 mg
Carotene	0.32 mg
Thiamine	0.17 mg
Riboflavin	0.01 mg
Niacin	3.72 mg
Ascorbic Acid	58.1 mg
Amino Acids:	
Tryptophan	28 mg
Methionine	13 mg
Lysine	84 mg

*According to analyses made at the Laboratorio FIM de Nutricion in Havana.

Other Uses

Latex extracted from the tree in Central America has been used to adulterate chicle. The **timber** is fine-grained, compact, strong, moderately to very heavy and hard, and valued especially for planks and rafters in construction. The **heartwood** is grayish-brown to reddish-brown and blends into the sapwood which is somewhat lighter in color. The darker the color, the more resistant to decay.

Medicinal Uses: A decoction of the astringent bark is taken as a febrifuge in Mexico and applied on skin eruptions in Cuba. A preparation of the seeds has been employed as a remedy for ulcers.

In 1971, a pharmaceutical company in California was exploring a derivative of the seed of *Pouteria sapota* (mamey, q.v.) which seemed to be active against seborrheic dermatitis of the scalp. Since they were having difficulty in procuring sufficient seeds for study, I suggested that they test the more readily available seeds of the canistel. They found these acceptable and were pursuing the investigation when last heard from.

Lucmo

This is a rare case of a species of ancient cultivation, little-known outside its homeland, that has recently found a place in modern food processing. The lucmo, *Pouteria lucuma* O. Ktze. (syns. *P. insignis* Baehni, *Lucuma obovata* HBK. and perhaps *L. bifera* Mol.; also *Richardella lucuma* Aubr.; *Achras lucuma* Ruiz & Pavón), is called *lucumo* in Chile and Peru; *lucma* in Ecuador; *lucuma* or *rucma* in Colombia; and *mamón* in Costa Rica.

Description

This attractive tree ranges from 25 to 50 ft (8-15 m) in height, has a dense, rounded crown, velvety hairs on its young branchlets, and copious milky latex. The evergreen leaves, clustered at the tips of small branches, are obovate, oval or elliptic, blunt at the apex, pointed at the base, 5 to 10 in (12.5-25 cm) long; thin or slightly leathery; dark-green on the upper surface, pale and sometimes brown-hairy on the underside. The profuse flowers, borne singly or 2 or 3 together in the leaf axils, are tubular, yellowish-green, with hairy sepals and 5- to 7-lobed mouth about ½ in (1.25 cm) across. The fruit is oblate, ovate or elliptic, pointed or depressed at the apex; 3 to 4 in (7.5-10 cm) long, with thin, delicate skin, brownish-green more or less overlaid with russet, and bright-yellow, firm, dry, mealy, very sweet pulp, permeated with latex until almost over-ripe. There may be 1 to 5, usually 2, rounded or broad-oval, dark-brown, glossy seeds with a whitish hilum on one flattish side.

Origin and Distribution

The lucmo was first seen and reported by Europeans in Ecuador in 1531. Archaeologists have found it frequently depicted on ceramics at burial sites of the indigenous people of coastal Peru. It is native and cultivated in the highlands of western Chile and Peru and possibly south-eastern Ecuador where it is known to have been cultivated since ancient times. It is grown also, to a limited extent, in the Andes of eastern Bolivia and the fruit is sold in the markets of La Paz. It is most popular in central Chile, less so in Ecuador. In 1776, it was reported as planted only in the warmest parts of northern Chile. In 1912, there were a few trees growing in gardens around San José, Costa Rica where the lucmo was introduced by returning exiles in the first half of the 19th Century. In 1915, O.F. Cook col-lected seeds at Ollantaytambo, Peru, for the United States Department of Agriculture (S.P.I. #41332). In January of 1922, Wilson Popenoe introduced seeds from Santiago, Chile (S.P.I. #54653). There have been several attempts to grow the tree in southern Florida. It has not lived long. One specimen actually bore fruit at the Fairchild Tropical Garden, developed galls, and eventually succumbed. The lucmo grows well in parts of Mexico and Hawaii but the fruit is not widely favored.

Climate

This species is not tropical, but grows at temperate elevations—between 9,000 and 10,000 ft (2,700-3,000 m) in Peru. It is adapted to fairly dry locations.

Season

The tree blooms and fruits all year. Mature fruits fall to the ground but they are not edible until they have been kept on hand for several days. Peruvian Indians bury them in stored grain, cured hay, chaff, dry leaves or other ma-terials until they become soft.

Food Uses

The fruit is eaten raw, out-of-hand, when fully ripe but Costa Ricans find that, though the flavor is appealing at first, one soon finds it repulsive because of the peculiar aftertaste. The lucmo has been stewed in sirup, used as pie-filling, and made into preserves. Currently, some fruits are being shipped from Chile to England where they are being used in making ice cream. A dehydrated, powdered product is being produced by a tomato cannery in Peru.

Other Uses

The wood is pale, compact, durable, and used for con-struction in Peru.

Abiu (Plate LVII)

A minor member of the Sapotaceae, the abiu, *Pouteria caimito* Radlk. (syns. *Lucuma caimito* Roem. & Schult.; *Achras caimito* Ruiz & Pavón), has acquired few vernac-ular names. In Colombia, it is called *caimito, caimito amarilla, caimo* or *madura verde;* in Ecuador, *luma* or *cauje;* in Venezuela, *temare;* in Brazil, *abiu, abi, abio, abieiro* or *caimito.* It is called yellow star apple in Trinidad.

Description

The tree has a pyramidal or rounded crown; is gen-erally about 33 ft (10 m) high but may reach 115 ft (35 m) in favorable situations. A gummy latex, white or reddish, exudes from wounds in the bark. The leaves are alternate and highly variable; may be ovate-oblong, obovate or el-liptic; 4 to 8 in (10-20 cm) long, 1¼ to 2⅜ in (3-6 cm) wide; short-pointed at the apex, sometimes long-tapering at the base; smooth or with a few scattered hairs. The flowers, borne singly or in groups of 2 to 5 in the leaf axils, are cylindrical, 4- to 5-lobed, white or greenish; ⅙ to ⅓ in (4-8 mm) long. The fruit, downy when young, is ovoid, elliptical or round; 1½ to 4 in (4-10 cm) long, sometimes having a short nipple at the apex; with smooth, tough, pale-yellow skin when ripe and fragrant, white, mucilag-inous, translucent, mild-flavored, sweet or insipid pulp containing 1 to 4 oblong seeds, brown, with a pale hilum on one side. Until fully ripe, the fruit is permeated with latex and is very gummy and astringent.

Origin and Distribution

The abiu is a denizen of the headwaters of the Amazon. It grows wild on the lower eastern slopes of the Andes from southwestern Venezuela to Peru. It is often cultivated

Fig. 109: The pale-yellow abiu (*Pouteria caimito*) as sold in the native market of Buenaventura, Colombia. The fruit is gummy with latex until it becomes fully ripe.

around Iquitos, Peru. In Ecuador, it is common in the Province of Guayas and the fruits are sold in the markets of Guayaquil. It is much grown around Pará, Brazil; less frequently near Rio de Janeiro, and to a limited extent at Bahia. In Colombia, it is fairly common in the regions of Caquetá, Meta and Vaupés and it abounds in the adjacent areas of Amazonas, Venezuela. It has been growing for many years in Trinidad.

The plant explorers, Dorsett, Shamel and Popenoe, collected seeds for the United States Department of Agriculture in Bahia in 1914 (S.P.I. #37929). In 1915, seeds were received from Lavoras, Minas, Brazil (S.P.I. #41003). This species has been planted several times at the Agricultural Research and Education Center, Homestead, Florida, but most of the young plants have been killed by winter cold. A few trees planted in 1953 fruited in 1962.

Varieties

There is much variation in the form, size and quality of the fruits of seedling trees, some having firm flesh, some soft; and some are insipid, while others have agreeable flavor. At Puerto Ospina, along the Putamayo River in Colombia, there is a type that fruits in 4 years. The fruit is round and large. Near the River Inirida, in Vaupés,

Colombia, there is a type that bears in one year from seed, but the fruits are small with little pulp.

Climate

The abiu is strictly tropical or near-tropical. It thrives best in a year-around warm and moist climate, yet Popenoe noted that it does well in somewhat cooler Rio de Janeiro. In Peru it has not been found above 2,000 ft (650 m), though in Colombia, it can be grown up to an elevation of 6,000 ft (1,900 m).

Soil

The tree is naturally suited to fertile, wet soil. It is subject to chlorosis in the limestone of southern Florida.

Season

The fruits are in season in March and April in Ecuador. They are sold in some Brazilian markets from September to April but only a few are seen in the much shorter season of February and March at Bahia. Fruits have matured in October in Florida. The abiu can be picked while underripe and firm for transport to markets.

Propagation and Culture

In Brazil, the washed seeds are dried in the shade and

then planted, 3 together and 2 in (5 cm) deep in enriched soil. They will germinate in 15 to 20 days. When the seedlings are 4 in (10 cm) high, the 2 weakest are removed. The strong one is set out when 12 to 16 in (30-40 cm) high. Spacing is 17 x 20 ft (6x5 m). One year later, the lower branches are pruned. Fruiting will begin in 3 years; will be substantial in 5 years.

Pests and Diseases

Actually, the fruit has little value commercially because it is commonly damaged by small insects (*bichos* in Spanish and Portuguese). In Brazil, the chief pests are said to be fruit flies.

Food Uses

In Colombia, people who wish to eat the abiu are advised to grease their lips beforehand to keep the gummy latex from clinging to them. It is mostly eaten out-of-hand but, in Pará, some types are used to make ices and ice cream.

Other Uses

Wood: The wood is dense and heavy, hard, and valued for construction.

Medicinal Uses: In Brazil, the pulp, because of its mucilaginous nature, is eaten to relieve coughs, bronchitis and other pulmonary complaints. The latex is given as a vermifuge and purge and is applied on abscesses.

Food Value Per 100 g of Edible Portion*	
Calories	95
Moisture	74.1 g
Protein	2.1 g
Lipids	1.1 g
Glycerides	22.0 g
Fiber	3.0 g
Ash	0.7 g
Calcium	96.0 mg
Phosphorus	45.0 mg
Iron	1.8 mg
Vitamin B$_1$	0.2 mg
Vitamin B$_2$	0.2 mg
Niacin	3.4 mg
Ascorbic Acid	49.0 mg
Amino Acids (mg per g of nitrogen [N = 6.25])	
Lysine	316 mg
Methionine	178 mg
Threonine	219 mg
Tryptophan	57 mg

*According to analyses made in Brazil.

Star Apple (Plate LVIII)

One of the relatively minor fruits of the family Sapotaceae, the star apple or goldenleaf tree, *Chrysophyllum cainito* L. (syn. *Achras caimito* Ruiz & Pavon), has acquired a moderate assortment of regional names. In Spanish, it is usually *caimito* or *estrella;* in Portuguese, *cainito* or *ajara;* in French, generally, *caimite* or *caimitier;* in Haiti, *pied caimite* or *caimitier a feuilles d'or;* in the French West Indies, *pomme surette,* or *buis;* in the Virgin Islands, cainit; in Trinidad and Tobago, it is caimite or kaimit; in Barbados, star-plum; in Colombia, it may be *caimo, caimo morado* (purple variety) or *caimito maduraverde* (green variety); in Bolivia, *caimitero,* or *murucuja;* in Surinam, *sterappel, apra* or *goudblad boom;* in French Guiana, *macoucou;* in Belize, damsel; in El Salvador, *guayabillo;* in Argentina, *aguay* or *olivoa.* The Chinese in Singapore call it "chicle durian".

Description

The star apple tree is erect, 25 to 100 ft (8-30 m) tall, with a short trunk to 3 ft (1 m) thick, and a dense, broad crown, brown-hairy branchlets, and white, gummy latex. The alternate, nearly evergreen, leaves are elliptic or oblong-elliptic, 2 to 6 in (5-15 cm) long, slightly leathery, rich green and glossy on the upper surface, coated with silky, golden-brown pubescence beneath when mature, though silvery when young. Small, inconspicuous flowers, clustered in the leaf axils, are greenish-yellow, yellow, or purplish-white with tubular, 5-lobed corolla and 5 or 6 sepals. The fruit, round, oblate, ellipsoid or somewhat pear-shaped, 2 to 4 in (5-10 cm) in diameter, may be red-purple, dark-purple, or pale-green. It feels in the hand like a rubber ball. The glossy, smooth, thin, leathery skin adheres tightly to the inner rind which, in purple fruits, is dark-purple and ¼ to ½ in (6-12.5 mm) thick; in green fruits, white and ⅛ to 3/16 in (3-5 mm) thick. Both have soft, white, milky, sweet pulp surrounding the 6 to 11 gelatinous, somewhat rubbery, seed cells in the center which, when cut through transversely, are seen to radiate from the central core like an asterisk or many-pointed star, giving the fruit its common English name. The fruit may have up to 10 flattened, nearly oval, pointed, hard seeds, ¾ in (2 cm) long, nearly ½ in (1.25 cm) wide, and up to ¼ in (6 mm) thick, but usually several of the cells are not occupied and the best fruits have as few as 3 seeds. They appear black at first, with a light area on the ventral side, but they dry to a light-brown.

Origin and Distribution

It is commonly stated that the star apple is indigenous to Central America but the eminent botanists Paul Standley and Louis Williams have declared that it is not native to that area, no Nahuatl name has been found, and the tree may properly belong to the West Indies. However, it is more or less naturalized at low and medium altitudes from southern Mexico to Panama, is especially abundant on the Pacific side of Guatemala, and frequently cultivated as far south as northern Argentina and Peru. It was recorded by Ciezo de Leon as growing in Peru during his travels between 1532 and 1550. It is common throughout most of the Caribbean Islands and in Bermuda. In Haiti, the star apple was the favorite fruit of King Christophe and he held court under the shade of a very large specimen at Milot. The United States Department of Agriculture received seeds from Jamaica in 1904 (S.P.I. #17093). The star apple is grown occasionally in southern Florida and in Hawaii where it was introduced before 1901. There are some trees in Samoa and in Malaya though they do not bear regularly. The tree is grown in southern Vietnam and in Kampuchea for its fruits but more for its ornamental value in West Tropical Africa, Zanzibar, and the warmer parts of India. It was introduced into Ceylon in 1802, reached the Philippines much later but has become very common there as a roadside tree and the fruit is appreciated.

Varieties

Apart from the two distinct color types, there is little evidence of such pronounced variation that growers would be stimulated to make vigorous efforts to select and propagate superior clones. William Whitman of Miami observed a tree yielding heavy crops of well-formed, high quality fruits in Port-au-Prince, Haiti, from late January to the end of June. He brought budwood to Florida in 1953. Grafted progeny and trees grown from air-layers have borne well here even prior to reaching 10 ft (3 m) in height. This introduction, named the "Haitian Star Apple", is propagated commercially for dooryard culture. Seeds of the Port-au-Prince tree have produced seedlings that have performed poorly in Florida.

Climate

The star apple tree is a tropical or near-tropical species ranging only up to 1,400 ft (425 m) elevation in Jamaica. It does well only in the warmest locations of southern Florida and on the Florida Keys. Mature trees are seriously injured by temperatures below 28°F (-2.22°C) and recover slowly. Young trees may be killed by even short exposure to 31°F (-0.56°C).

Soil

The tree is not particular as to soil, growing well in deep, rich earth, clayey loam, sand, or limestone, but it needs perfect drainage.

Propagation

Star apple trees are most widely grown from seeds which retain viability for several months and germinate readily. The seedlings bear in 5 to 10 years. Vegetative propagation hastens production and should be more commonly practiced. Cuttings of mature wood root well. Air-layers can be produced in 4 to 7 months and bear early. Budded or grafted trees have been known to fruit one year after being set in the ground. In India, the star apple is sometimes inarched on star apple seedlings. Grafting on the related satinleaf tree (*C. oliviforme* L.) has had the effect of slowing and stunting the growth.

Culture

During the first 6 months, the young trees should be watered weekly. Later irrigation may be infrequent except during the flowering season when watering will increase fruit-set. Most star apple trees in tropical America and the West Indies are never fertilized but a complete, well-balanced fertilizer will greatly improve performance in limestone and other infertile soils.

Harvesting

Star apples are generally in season from late winter or early spring to early summer. They do not fall when ripe but must be hand-picked by clipping the stem. Care must be taken to make sure that they are fully mature. Otherwise the fruits will be gummy, astringent and inedible. When fully ripe, the skin is dull, a trifle wrinkled, and the fruit is slightly soft to the touch.

Yield

In India, a mature star apple tree may bear 150 lbs (60 kg) of fruits in the short fruiting season of February and March.

Keeping Quality

Ripe fruits remain in good condition for 3 weeks at 37.4° to 42.8°F (3°-6°C) and 90% relative humidity.

Pests and Diseases

Larvae of small insects are sometimes found in the ripe fruits.

The main disease problem in the Philippines is stem-end decay caused by species of *Pestalotia* and *Diplodia*. In Florida, some fruits may mummify before they are full-grown.

The foliage is subject to leaf spots from attack by *Phomopsis* sp., *Phyllosticta* sp., and *Cephaleuros virescens*, the latter known as algal leaf spot or green scurf.

Birds and squirrels attack the fruits if they are left to fully ripen on the tree.

Food Uses

Star apples must not be bitten into. The skin and rind (constituting approximately 33% of the total) are inedible. When opening a star apple, one should not allow any of the bitter latex of the skin to contact the edible flesh. The ripe fruit, preferably chilled, may be merely cut in half and the flesh spooned out, leaving the seed cells and core. A combination of the chopped flesh with that of mango, citrus, pineapple, other fruits and coconut water is frozen and served as Jamaica Fruit Salad Ice. An attractive way to serve the fruit is to cut around the middle completely through the rind and then, holding the fruit stem-end down, twisting the top gently back and forth. As this is done, the flesh will be felt to free itself from the downward half of the rind, and the latter will pull away, taking with it the greater part of the core.

In Jamaica, the flesh is often eaten with sour orange juice, a combination called "matrimony"; or it is mixed with orange juice, a little sugar, grated nutmeg and a spoonful of sherry and eaten as dessert called "strawberries-and-cream". Bolivians parboil the edible portion, and also prepare it as a decoction. An emulsion of the slightly bitter seed kernels is used to make imitation milk-of-almonds, also nougats and other confections.

Food Value Per 100 g of Edible Portion*	
Calories	67.2
Moisture	78.4–85.7 g
Protein	0.72–2.33 g
Carbohydrates	14.65 g
Fiber	0.55–3.30 g
Ash	0.35–0.72 g
Calcium	7.4–17.3 mg
Phosphorus	15.9–22.0 mg
Iron	0.30–0.68 mg
Carotene	0.004–0.039 mg
Thiamine	0.018–0.08 mg
Riboflavin	0.013–0.04 mg
Niacin	0.935–1.340 mg
Ascorbic Acid	3.0–15.2 mg
Amino Acids:	
Tryptophan	4 mg
Methionine	2 mg
Lysine	22 mg
*Analyses made in Cuba and Central America.	

Toxicity

The seeds contain 1.2% of the bitter, cyanogenic glycoside, lucumin; 0.0037% pouterin; 6.6% of a fixed oil; 0.19% saponin; 2.4% dextrose and 3.75% ash. The leaves possess an alkaloid, also resin, resinic acid, and a bitter substance.

Other Uses

Wood: The tree is seldom felled for timber unless there is a particular need for it. The heartwood is pinkish or red-brown, violet, or dark-purple; fine-grained, compact, heavy, hard, strong, tough but not difficult to work; durable indoors but not outside in humid conditions. It has been utilized for heavy construction and for deluxe furniture, cabinetwork and balustrades.

Latex: The latex obtained by making incisions in the bark coagulates readily and has been utilized as an adulterant of gutta percha. It was formerly proposed as a substitute for wax on the shelves of wardrobes and closets.

Medicinal Uses: The ripe fruit, because of its mucilaginous character, is eaten to sooth inflammation in laryngitis and pneumonia. It is given as a treatment for diabetes mellitus, and as a decoction is gargled to relieve angina. In Venezuela, the slightly unripe fruits are eaten to overcome intestinal disturbances. In excess, they cause constipation. A decoction of the rind, or of the leaves, is taken as a pectoral. A decoction of the tannin-rich, astringent bark is drunk as a tonic and stimulant, and is taken to halt diarrhea, dysentery and hemorrhages, and as a treatment for gonorrhea and "catarrh of the bladder". The bitter, pulverized seed is taken as a tonic, diuretic and febrifuge. Cuban residents in Miami are known to seek the leaves in order to administer the decoction as a cancer remedy. Many high-tannin plant materials are believed by Latin Americans to be carcinostatic. In Brazil, the latex of the tree is applied on abscesses and, when dried and powdered, is given as a potent vermifuge. Elsewhere, it is taken as a diuretic, febrifuge and remedy for dysentery.

EBENACEAE

Japanese Persimmon (Plates LIX and LX)

In great contrast to the native American persimmon, *Diospyros virginiana* L., which has never advanced beyond the status of a minor fruit, an oriental member of the family Ebenaceae, *D. kaki* L. f., is prominent in horticulture. Perhaps best-known in America as the Japanese, or Oriental, persimmon, it is also called kaki (in Spanish, *caqui*), Chinese plum or, when dried, Chinese fig.

Description

The tree, reaching 15 to 60 ft (4.5-18 m) is long-lived and typically round-topped, fairly open, erect or semi-erect, sometimes crooked or willowy; seldom with a spread of more than 15 to 20 ft (4.5-6 m). The leaves are deciduous, alternate, with brown-hairy petioles ¾ in (2 cm) long; are ovate-elliptic, oblong-ovate, or obovate, 3 to 10 in (7.5-25 cm) long, 2 to 4 in (5-10 cm) wide, leathery, glossy on the upper surface, brown-silky beneath; bluish-green, turning in the fall to rich yellow, orange or red. Male and female flowers are usually borne on separate trees; sometimes perfect or female flowers are found on male trees, and occasionally male flowers on female trees. Male flowers, in groups of 3 in the leaf axils, have 4-parted calyx and corolla and 24 stamens in 2 rows. Female flowers, solitary, have a large leaflike calyx, a 4-parted, pale-yellow corolla, 8 undeveloped stamens and oblate or rounded ovary bearing the style and stigma. Perfect flowers are intermediate between the two. The fruit, capped by the persistent calyx, may be round, conical, oblate, or nearly square, has thin, smooth, glossy, yellow, orange, red or brownish-red skin, yellow, orange, or dark-brown, juicy, gelatinous flesh, seedless or containing 4 to 8 flat, oblong, brown seeds ¾ in (2 cm) long. Generally, the flesh is bitter and astringent until fully ripe, when it becomes soft, sweet and pleasant, but dark-fleshed types may be non-astringent, crisp, sweet and edible even before full ripening.

Origin and Distribution

The tree is native to Japan, China, Burma and the Himalayas and Khasi Hills of northern India. In China it is found wild at altitudes up to 6,000-8,000 ft (1,830-2,500 m) and it is cultivated from Manchuria southward to Kwangtung. Early in the 14th Century, Marco Polo recorded the Chinese trade in persimmons. Korea has long-established ceremonies that feature the persimmon. Culture in India began in the Nilgiris. The tree has been grown for a long time in North Vietnam, in the mountains of Indonesia above 3,500 ft (1,000 m) and in the Philippines. It was introduced into Queensland, Australia, about 1885.

It has been cultivated on the Mediterranean coast of France, Italy, and other European countries, and in southern Russia and Algeria for more than a century. The first trees were introduced into Palestine in 1912 and others were later brought in from Sicily and America.

Seeds first reached the United States in 1856 when they were sent from Japan by Commodore Perry. Grafted trees were imported in 1870 by the U.S. Department of Agriculture and distributed to California and the southern states. Other importations were made by private interests until 1919. Seeds, cuttings, budwood and live trees of numerous types were brought into the United States at various times from 1911 to 1923 by government plant explorers and the tree has been found best adapted to central and southern California, Arizona, Texas, Louisiana, Mississippi, Georgia, Alabama, southeastern Virginia, and northern Florida. A few specimens have been grown in southern Maryland, eastern Tennessee, Illinois, Indiana, Pennsylvania, New York, Michigan and Oregon.

By 1930, California had over 98,000 bearing trees and nearly 97,000 non-bearing, on 3,000 acres (1,214 ha). California production in 1965 amounted to 2,100 tons. Real estate development reduced persimmon groves to 540 acres by 1968. In 1970, California produced 1,600 tons—92% of the total U.S. crop.

In parts of Central America, Japanese persimmons have been planted from sea-level to 5,000 ft (1,524 m). The tree was first grown in Brazil by Japanese immigrants. By 1961, the total crop was 2,271,046,000 fruits, mainly in the State of Ceará, followed by Pernambuco and Piaui, with Bahia far behind. At present, the largest orchards are mainly in the States of Sao Paulo, Parana and Rio Grande do Sul, with lesser groves in Minais Gerais and Espirtu Santo. Of 111,412 acres (45,088 ha) all told, 60,336 acres (24,418 ha) are in Ceará. Israel and Italy have developed commercial plantings, and cultivar trials began in 1976 with a view to establishing persimmon-growing for export in southeastern Queensland.

Cultivars

Of the 2,000 cultivars known in China, cuttings of 52, from the provinces of Honan, Shensi and Shansi, were brought into the United States in 1914. J. Russell Smith, an esteemed economic-geographer, collected a number of types near the Great Wall of China in 1925 and some of the trees still survive in his derelict orchard in the Blue Ridge Mountains of southern Virginia. Over 800 kinds are grown in Japan but less than 100 are considered important. Among prominent cultivars are the non-astringent 'Fuyu', 'Jiro', 'Gosho' and 'Suruga'; the astringent 'Hiratanenashi', 'Hachiya', 'Aizumishirazu', 'Yotsumizo' and 'Yokono'. It was formerly believed that the flesh color and astringency can vary considerably depending on whether or not the flowers were effectively pollinated, and cultivars were classed as: 1) Pollination Constants; and 2) Pollination Variants.

It has been recently discovered that there are two different mechanisms affecting astringency; one is degree of pollination, the other is the amount of ethanol produced in the seeds and accumulated in the flesh. Pollination Variant fruits with naturally high levels of ethanol lose astringency on the tree. So does Pollination Constant 'Fuyu' but other non-astringent Pollination Constant cultivars have been found to have low levels of ethanol. Pomologists at Kyoto University, Japan, have classified 40 cultivars into 4 types depending upon the ways or degrees their fruits lose astringency on the tree and upon flesh color — Pollination Constant Non-astringent (PCNA), Pollination Variant Non-astringent (PVNA), Pollination Variant Astringent (PVA) and Pollination Constant Astringent (PCA). They evidently have not studied seedless cultivars.

Dr. H.H. Hume, of the University of Florida, separated 13 seeded and seedless (or nearly seedless) cultivars according to the earlier pollination classification, and Drs. Camp and Mowry added 'Fuyu'. The following 8 comprise *Group 1*:

'Costata' — conical, pointed, somewhat 4-sided, 2⅝ in (6.5 cm) long, 2⅛ in (5.4 cm) wide, with salmon-yellow skin, light-yellow flesh, with no seeds; or dark flesh and a few seeds. Astringent until fully ripe, then sweet; late (Oct.-Nov. in Florida). Keeps very well.

'Fuyu' (or 'Fuyugaki') — oblate, faintly 4-sided, 2 in (5 cm) long; 2¾ in (7 cm) wide; skin deep-orange; flesh light-orange; firm when ripe; non-astringent even when unripe; with few seeds or none. Keeps well; excellent packer and shipper. It is the most popular non-astringent persimmon in Florida. 'Matsumoto Early Fuyu' ripens three weeks earlier.

'Hachiya' — oblong-conical, 3¾ in (9.5 cm) long, 3¼ in (8.25 cm) wide; skin glossy, deep orange-red; flesh dark-yellow with occasional black streaks; astringent until fully ripe and soft, then sweet and rich. Seedless or with a few seeds. Midseason to late. Much used in Japan for drying. Tree vigorous, well-formed and prolific in Kulu Valley, India. Scanty bearer in southeastern United States; does well on *D. virginiana* in Florida, but tends to growth-ring cracking; often prolific in California.

'Ormond' — oblong-conical, 2⅝ in (6.5 cm) long, 1⅞ in (4.7 cm) thick. Skin reddish-yellow with thin bloom; flesh orange-red, moderately juicy; seeds large. Very late (Nov. and Dec. in Florida). Keeps well.

'Tamopan' — Introduced from China in 1905, again in 1916 (S.P.I. Nos. 16912, 16921, 26773). Broad oblate, somewhat 4-sided; indented around the middle or closer to the base; 3 to 5 in (7.5-12.5 cm) wide; skin thick, orange-red; flesh light-orange, usually astringent until fully ripe, then sweet and rich. In some parts of China and Japan said to be non-astringent. Seedless or nearly so. Of medium quality; late (Nov.) in Florida; midseason in California. Was being grown commercially in North Carolina and at Glen St. Mary, Florida, in 1916.

'Tanenashi' — round-conical, 3⅓ in (8.3 cm) long, 3⅜ in (8.5 cm) wide; skin light-yellow or orange, turning orange-red; thick; flesh yellow, astringent until soft, then sweet; seedless. Early; prolific. Much esteemed. Much used for drying in Japan. Leading cultivar in southeastern United States without pollination. In California tends to bear in alternate years.

'Triumph' — oblate, faintly 4-sided; of small to medium size; skin yellowish to dark orange-red. Flesh yellowish-red, translucent, soft, juicy; seedless or with 5 to 8 seeds; astringent until fully ripe, then sweet. Of high quality. Medium-late. In Florida begins in September and lasts until mid-November.

'Tsuru' — long-conical, pointed; 3⅜ in (8.5 cm) long, 2⅜ in (6 cm) wide; skin bright orange-red, turning red with purple bloom when mature; flesh orange-yellow or dark-yellow, granular; astringent until fully ripe; with few or no seeds. Very late.

Group 2:

'Gailey' — roundish to conical with rounded apex; small; skin dull-red, pebbled; flesh dark, firm, juicy, of good flavor. Bears many male flowers regularly and is planted for cross-pollination.

'Hyakume' — round-oblong to round-oblate, somewhat 4-angled and flat at both ends; 2¾ in (7 cm) long, 3⅛ in (8 cm) wide; skin pale dull-yellow to light-orange, with brown russeting when ripe; flesh dark-brown, crisp, sweet, non-astringent whether hard or ripe. Midseason. Fairly good quality; somewhat unattractive externally. Stores and ships well.

'Okame' — round-oblate, 2⅜ in (6 cm) long, 3⅛ in (8 cm) wide; skin orange-yellow turning to bright-red with waxy bloom; flesh light but brownish around the seeds; sometimes seedless; sweet, of excellent quality. Fairly early, beginning about Sept. 1 in Florida. Productive.

'Yeddo-ichi' — oblate, 2½ in (6.25 cm) long, 3 in (7.5 cm) wide; skin dark orange-red with a bloom; flesh dark-brown with purplish tint; sweet, rich, non-astringent whether hard or ripe. Of high quality.

'Yemon' — oblate, 4-sided; 2¼ in (5.7 cm) long, 3¼ in (8.25 cm) wide; skin light-yellow becoming reddish with orange-yellow mottling; flesh red-brown or light-colored, astringent at first, sweet after softening; seedless or with few seeds and then dark around the seeds. Of high quality, but becomes too soft for shipping.

'Zengi' ('Zengimaru') — round or round-oblate, 1¾ in (4.5 cm) long, 2¼ in (5.6 cm) wide. Skin dark orange-red or yellow-red; flesh dark with black streaks; sweet even when hard; with some seeds. Early, prolific; of medium quality.

Cultivars that are especially hardy in Maryland, Pennsylvania and Virginia include: 'Atome', 'Benigaki', 'Delicious', 'Eureka', 'Great Wall', 'Manerh', 'Okame', 'Peiping', 'Pen', 'Shaumopan', 'Sheng', 'Tsurushigaki', 'Yokono', etc.

'Delicious' is oblate, medium to large; skin is smooth, light-red; flesh light-yellow, non-astringent when hard, but more flavorful when soft; contains a few seeds; tree is vigorous and a regular bearer.

'Eureka' (from Texas) is oblate, medium to large, puckered at calyx, bright orange-red, astringent; of good quality; drought- and frost-resistant; late (Nov. in Florida). One of the most satisfactory in Florida.

'Great Wall' is small, flat, 4-sided with fine black stripes extending from the calyx; astringent, dry-fleshed; tree is vigorous, a biennial bearer; does well in Florida.

'Hanafuyu' is oblate, non-astringent and usually seedless; late-midseason; tree is small, bears regularly but yield is low; prone to premature shedding of fruit; fairly common in northern Florida.

'Ichikikeijiro' is medium-large, orange, non-astringent; early-ripening; tree is not vigorous but still this cultivar is among the best of the non-astringent class in Florida.

'Jumbu' resembles 'Fuyu' but is somewhat more conical and larger; non-astringent; edible either firm or soft. Ripens a little later than 'Fuyu'; of good quality.

'Ogasha' is oblate, non-astringent and usually seedless; prone to immature shedding of fruit; fairly common in northern Florida.

'Sheng' is large, ribbed, puckered at calyx, astringent; popular in Florida; bears annually when pollinated.

'Shogatsu' is flattened, non-astringent, of fair quality; bears an abundance of male flowers. Does well in Florida.

'Siajo' is small, astringent, of good quality and flavor; performs well in Florida.

'Taber No. 23' is round to oblate with flat apex; fairly small; skin is dark-red, stippled. Begins to ripen in September in Florida.

'Yamato Hyakume' is large, with red skin; has little tannin when seed content is low; tends to growth-ring cracking; is a heavy bearer in Florida.

'Yokono' is large, orange-red, astringent, of good quality; bears well but tends to shed fruit; keeps well.

Maru is a group name for several roundish types of Japanese persimmon with brilliant orange-red skin, cinnamon-colored flesh; medium to small in size; flesh is juicy, sweet, richly flavored; they have excellent keeping quality after ripening, store and ship well and are very decorative.

At the Pomological Station, Coonor, India, an unnamed type and a named cultivar, 'Dai Dai Maru' have performed well. The unnamed cultivar is broad at the base, large, attractive, deep-red, astringent until fully ripe, then very sweet; bears well regularly. The tree is semi-erect.

'Dai Dai Maru' has a broadly rounded apex, is of medium size; orange-red, glossy, with a slight bloom; has dark flesh, is not edible until fully cured; seedless unless cross-pollinated; bears good crops regularly. The tree is of semi-erect habit.

In Brazil, cultivars are sorted into 3 groups. *Group 1*, 'Sibugaki', includes those that are yellow-fleshed, always astringent whether seedless or not ('Taubaté', 'Hachiya', 'Trakoukaki', 'Hatemya', etc.).

'Taubaté', the most popular of this group, is round, slightly flattened, large, yellow-fleshed, very astringent; highly perishable, lasting only 3 to 4 days after ripening.

Group 2, 'Amagaki', includes those that are yellow-fleshed, never astringent whether seedless or not ('Jiro', 'Fuyu', 'Hannagosho').

'Hannagosho' is of excellent quality but in Florida is slow in losing astringency and the tree is deficient in male flowers.

'Jiro' is second to 'Fuyu' in importance in Japan; is of high quality and ships well. The fruit is colorful and the tree vigorous in Florida.

Group 3, 'Variavel', or 'Variaveis', includes those that are astringent when they have several seeds, and partially or totally non-astringent when they have only one or a few seeds. The flesh is yellow when there are no seeds and dark when seeds are present ('Rama Forte', 'Guiombo', 'Luiz de Queiroz', 'Hyakume', 'Chocolate', etc.).

'Guiombo' (perhaps the same as 'Korean') is one of the best in Florida, with thin skin; but it is a biennial bearer when young.

'Rama Forte', the most popular of this group is oblate, medium to large, with dark-yellow flesh, or dark-brown when there are many seeds; keeps well—8 to 10 days at room temperature after ripening; yields 30% more than 'Taubaté' and its branches are less apt to break under a heavy crop.

The Instituto Agronomico do Estado de Sao Paulo has developed various promising hybrids.

In 1922, seeds of 'Kai Sam T'sz' (chicken-heart persimmon) from Canton, China, were sent to the United States Department of Agriculture as a subtropical cultivar which might be appropriate for southern Florida and the West Indies in contrast to the hardier types brought in from Japan and northern and central China, but it seems to have soon dropped out of sight.

Among commercial cultivars in Japan not already mentioned are:

'Suruga' (distributed in 1959); orange-red, non-astringent, very sweet, keeps well.

'Gosho', orange-red, non-astringent, sweet, of high quality but giving a low yield because of excessive shedding of immature fruits.

'Hiratanenashi', oblate, somewhat 4-sided, astringent, thick-skinned; seedless; of high quality, but keeps only a short time after curing; mostly used for drying.

'Aizumishirazu', rounded, astringent, black-spotted around seeds; of fair quality; bears well.

'Yotsumizo', small, astringent, usually seedless, sweet after curing; bears well; often dried.

Of six cultivars tested in Queensland ('Tanenashi', 'Hyakume', 'Dai Dai Maru', 'Tsuru Magri', 'Flat Seedless', and 'Nightingale'), all grafted on *D. lotus*, only 'Nightingale' proved satisfactory in fruit quality and yield in an assessment made after 3 years of fruiting.

'Nightingale' is classed as PCA (pollination constant, astringent); is conical, 3½ in (9 cm) long; red; of distinctive flavor; with an average of 2½ seeds per fruit. The tree is semi-dwarf and fairly precocious.

Pollination

Some cultivars in certain locations and under some conditions, will fruit abundantly without cross-pollination, but this trait is not dependable. In commercial groves, the cultivar known as 'Gailey', which regularly produces many male flowers, is interplanted to insure adequate pollination. The formula is one male for every 8 female trees, uniformly dispersed throughout the grove; or 12 to 24 pollinating trees per acre (30-60 per ha). Japanese farmers sometimes plant the pollinating trees as a hedge around the grove. If hand-pollination of early cultivars is necessary, unopened male buds are collected, dried, opened and the pollen separated and stored. When needed, it is mixed with skimmed milk or club moss (*Lycopodium*) and applied at 1/7 to 2/7 oz per acre (10-20 g per ha).

If the flowers are not effectively pollinated, the entire crop of fruit may fall prematurely. This is a fault of the cultivar 'Isu' in Japan. Losses can be reduced by girdling the tree after flowering but the practice has the effect of retarding growth. If the weather is hot and dry at blooming time, pollination will be inadequate and very few fruits will be set. The maintenance of bee colonies (1 or 2 hives for every 2½ acres, or per ha) in persimmon orchards will enhance pollination, especially in cultivar 'Fuyu'.

Climate

The Japanese persimmon needs a subtropical to mild-temperate climate. It will not fruit in tropical lowlands. In Brazil, the tree is considered suitable for all zones favorable to *Citrus*, but those zones with the coldest winters induce the highest yields. The atmosphere may range from semi-arid to one of high humidity.

Trees in the Middle Atlantic States have been known to have withstood temperatures as low as 20°F (-6.67°C) and to have remained in excellent condition and fruitful after 40 years.

Soil

The tree is not particular as to soil, and does well on any moderately fertile land with deep friable subsoil. In Florida, a sandy loam with clay subsoil promotes good growth. While the young tree needs plentiful watering, good drainage is essential.

Propagation

Indonesians propagate the tree by means of root suckers. In the Orient, selected cultivars are raised from seed or grafted onto wild rootstocks of the same species, or onto the close relative, *D. lotus* L. In the eastern United States, the trees are grafted onto the native American persimmon, *D. virginiana*. This rootstock significantly contributes to cold-resistance. California growers have found *D. kaki* the most satisfactory rootstock, *D. lotus* rootstock resulting in much lower yields.

Seeds for the production of rootstocks need no pretreatment. They are planted in seedbeds or directly in the nursery row 8 to 12 in (20-30 cm) apart with 3 to 3½ ft (0.9-1.06 m) between the rows. After a season of growth, they may be whip-grafted close to the surface of the soil, using freshly cut scions or scions from dormant trees kept moist in sphagnum moss.

Cleft-grafting is preferred on larger stock and for top-working old trees. In India, cleft-grafting on stem has been 88.9% successful; while cleft-grafting on crown and tongue-grafting on stem have been 73.4% successful when the grafted plants were left for 2 weeks at about 77°F (25°C) and relative humidity of 75% for 2 weeks before planting.

In the Kulu Valley, India, scions are grafted onto 2-year-old *D. lotus* seedlings which are mounded with earth to cover the graft until it begins to sprout. At the Fruit Research Station, Kandaghat, 2-year-old *D. lotus* seedlings were used as rootstock for veneer and tongue grafts from cv 'Hachiya' between late June and the third week of August. Success rates ranged from 80 to 100%.

In Palestine, trees grafted on *D. lotus* and grown on light soil are dwarfish, fruit heavily at first, but are weak and short-lived. Those grafted on *D. virginiana* are larger and vigorous and bear heavily consistently. The only disadvantage is that the shallow root system fans out to 65 ft (20 m) from the base of the tree and wherever the roots are injured by cultivation, suckers spring up and become a nuisance.

Culture

The soil should be well prepared—deeply plowed and enriched with organic matter. Trees should be set out at spacings ranging from 15 x 5 ft (4.5x1.5 m) to 20 x 20 ft (6x6 m), depending on the habit of the cultivar. In Japan, 404.7 plants per acre (1,000 per ha) may be installed at the outset, to be thinned down to 85 trees per acre (200 per ha) in 10-15 years.

Good results have been obtained with a fertilizer mixture of 4 to 6% N, 8 to 10% P and 3 to 6% K at the rate of 1 lb (.45 kg) per tree per year of age. Generally the application is made in spring, but some growers apply half in the spring, half in July. Over-fertilization or excessive amounts of nitrogen fertilizers will cause shedding of fruits.

Young trees are pruned back to 2½ ft to 3 ft (.74-.91 m) when planted and later the new shoots are thinned with a view to forming a well-shaped tree. Some cultivars tend to develop a willowy growth and require cutting back occasionally to avoid the development of weak branches which break when heavy with fruit. Annual pruning during the first 4 to 5 winters is desirable in some cultivars. If a tree tends to overbear and shows signs of decline, it should be drastically cut back to give it a fresh start.

After flowering, the trees should be irrigated every 3 weeks on light soil, every month on heavier soil, until time for harvest. One California grower, with trees on deep river loam, has provided furrow irrigation every 2 weeks from April through September. Branches are fragile and must be propped when heavily laden with fruits.

Cropping and Yield

Many cultivars begin to bear 3-4 years after planting out; others after 5-6 years. Shedding of many blossoms, immature and nearly mature fruits is characteristic of the Japanese persimmon as well as the tendency toward alternate bearing. The annual yield of a young tree ranges from 50 to 96 lbs (22.6-40.8 kg); of a full-grown tree, 330 to 550 lbs (150-250 kg). Estimated yield in Brazil is 6.5 tons per acre (15 tons per ha), but yields will vary with the cultivar and cultural practices.

Harvesting takes place in fall and early winter. Late ripening cultivars may be picked after hard frosts or light-snowfall. Japan produces about 300,000 tons per year.

Japanese growers use color charts to determine when each cultivar is ready for harvest. Astringent cultivars are picked when fully mature but hard and are cured before marketing.

Curing

In the Orient, much of the crop is left in piles covered by bamboo mats to cure (near-freeze) naturally and is marketed throughout the winter. In some parts of China, the fruit is cured in covered pits by introducing the smoke from burning dung. There are several other methods of curing: soaking in vinegar or immersing in boiling water and letting stand for 12 hours. 'Hachiya' fruits kept in warm water—104°F (40°C)—for 24 hours will be firm and non-astringent 2 days after treatment. One practice is to leave the astringent fruits in lime water for 2 days but tests have shown no advantage of a lime solution over pure water except that lime disinfects and can prevent the rotting that might follow soaking.

In Japan, the fruits may be sprayed with ethanol, or stored for 10 days to 2 weeks in kegs which previously contained *sake;* or they may be stored in air-tight containers with ethylene gas for 3 days. Carbon dioxide is widely employed and the treatment consists of storing in a 95% CO_2 atmosphere for 24 hours at 68° to 77°F (20°-25°C), but the fruit softens very quickly thereafter. In Brazil, successful curing has been achieved by immersing 'Taubate' persimmons in 1,000 ppm solution of ethephon (an ethylene generator) for 1 hour and then storing at room temperature for 4 days. Large quantities are cured by exposure to the fumes of alcohol (aguardiente), acetylene gas from combustion of calcium carbonate, or gas from burning sawdust, in hermetically sealed chambers at temperatures between 68° and 82.4° F (20° and 28°C) at relative humidity of 80%. Various other chemical processes and gamma radiation have been successfully employed in other countries.

A simple method was discovered in California some years ago. The newly picked fruits were merely pierced once at the apex with a needle dipped in alcohol, then the fruits were layered with straw in a tightly closed box for 10 days. The homeowner may merely keep the fruits at room temperature in a closed vessel or plastic bag for 2-4 days with bananas, pears, tomatoes, apples, or other fruits which give off ethylene gas. In India, the persimmons are individually paper-wrapped and placed in alternate rows with 'Kieffer' pears in a closed container and are edible in 3 days. Non-astringent cultivars need no curing.

Packing, Keeping Quality and Storage

In California, persimmons are graded by size, then tissue-wrapped and packed in peach boxes for rail shipment in refrigerated cars. Packing in other areas is similar. Astringent types soften in 2 or 3 days after treatment and quickly become overripe. Non-astringent types are usually harder than astringent types when picked, and they therefore ship and keep better. Persimmons have been kept for 2 months at 30°F (-1.11°C) and 85-90% relative humidity. 'Triumph' is frequently stored in Israel for as long as 4 months at 30°F (-1.11°C). Persimmons have been kept in good condition for several months in sealed 0.06 mm polyethylene bags at 32°F (0°C).

Spraying the bearing branches with gibberellic acid 3 days before harvest has retarded maturity on the tree; has doubled the storage life of astringent types after curing.

Pests and Diseases

In Brazil, premature fall of 'Fuyu' is partly linked to heavy infestation by the mite, *Aceria diospyri*. Spraying with Sevin 85 ppm 3 times at 30-day intervals right after petal fall controls the mite and increases yield. *Retithrips syriacus* feeds on and blemishes the leaves and fruit skin in Palestine but has been controlled by spraying with nicotine sulfate. The greenhouse thrips (*Heliothrips haemorrhoidalis*) blemishes fruits in Queensland. San José scale is combatted by a dormant application of Bordeaux in diesel emulsion in India. In Florida, white peach scale, *Pseudaulacaspis pentagona*, has required control and a twig girdler, *Onsideres cingulatus*, has been troublesome. Also, a flat-headed borer drills into the bark and the wood causing oozing of gum and decline in vigor. The main enemies in the eastern United States are mealybugs which distort young shoots and kill all new growth unless controlled. They do not seriously affect mature trees.

In Brazil and Queensland, fruit flies may attack the fruits, especially in dry years. Tree-ripe persimmons are sought by all kinds of birds, especially by parrots and crows in India, where flying foxes are a nocturnal menace. The less astringent types seem to be preferred by all of these predators. Bird-repellent sprays have given good control in Queensland. There, sunburn affects marketability especially of 'Tanenashi' and 'Tsuru magri'.

In India, low germination rates of planted seeds has been traced to dry rot caused by *Penicillium* sp. It can be controlled by pretreatment with an appropriate fungicide.

D. lotus rootstock is subject to root rot and crown gall in Florida but resistant to wilt caused by *Cephalosporium diospyri* which induces severe defoliation and has killed trees on *D. virginiana* rootstock. In Brazil, *Cercospora* may spot the leaves, and a virus causes "mosaic"—mottling of leaves and premature leaf fall, shedding of flowers, and necrotic spots on fruits; also a different necrosis on the tree and the bark of shoots, twigs and branches that causes die-back. Anthracnose occurs on fruits that have slightly cracked or have been pierced by insects. In Florida, leaf spot, algal leaf spot, twig blight, twig dieback, root rot, thread blight and other fungal diseases may occur.

Food Uses

Fully ripe Japanese persimmons are usually eaten out-of-hand or cut in half and served with a spoon, preferably after chilling. Some people prefer to add lemon juice or cream and a little sugar. The flesh may be added to salads, blended with ice cream mix or yogurt, used in pancakes, cakes, gingerbread, cookies, gelatin desserts, puddings, mousse, or made into jam or marmalade. The pureed pulp can be blended with cream cheese, orange juice, honey and a pinch of salt to make an unusual dressing.

Ripe fruits can be frozen whole or pulped and frozen in the home freezer. Large quantities of 'Tamopan' are preserved by drying. Drying is commonly practiced in Brazil and the dried fruit is popular throughout the country. Some California growers dry the 'Hachiya' by a Chinese method. The fruits are picked when mature but firm, are peeled and hung up by their stems for 30-50 days to dry in the sun. Kneading every 4-5 days is necessary to give uniform texture and improve flavor. Then they are taken down and sweated for 10 days in heaps under mats. Sugar crystals form on the surface. Lastly, they are hung up again to dry in the wind. In the Orient, the peelings are dried separately and are mixed in with fruits when packed for sale. An inferior product is made by slitting the skin with a knife, then spreading the fruits out on mats to dry for several weeks, then sweating them in piles, and the product is sold at a very low price.

In Indonesia, ripe fruits are stewed until soft, then pressed flat and dried in the sun. Early travelers called such fruits "red figs". Intestinal compaction from consumption of persimmons in Israel has been eliminated by drying the fruits before marketing, and some dried fruits are now being exported to Europe. Surplus persimmons may be converted into molasses, cider, beer and wine. Roasted seeds have served as a coffee substitute.

Other Uses

Tannin from unripe Japanese persimmons has been employed in brewing *sake*, also in dyeing and as a wood preservative. Juice of small, inedible wild persimmons, crushed whole, calyx, seeds and all, is diluted with water and painted on paper or cloth as an insect- and moisture-repellent.

The **wood** of the tree is fairly hard and heavy, black with streaks of orange-yellow, salmon, brown or gray; close-

Food Value Per 100 g of Edible Portion*	
Calories	77
Moisture	78.6 g
Protein	0.7 g
Fat	0.4 g
Carbohydrates	19.6 g
Calcium	6 mg
Phosphorus	26 mg
Iron	0.3 mg
Sodium	6 mg
Potassium	174 mg
Magnesium	8 mg
Carotene	2,710 I.U.
Thiamine	0.03 mg
Riboflavin	0.02 mg
Niacin	0.1 mg
Ascorbic Acid	11 mg
*Average values.	

The astringent substance in the persimmon, generally called "tannin", has been much studied and variously defined as knowledge of tannins and other phenols has unfolded. To put it simply, it is classed as a condensed tannin (proanthocyanidin) of complex structure.

One would be wise to eat only fully ripe persimmons from which the tannin has been almost entirely eliminated. The skin, which retains some tannin, should not be eaten.

grained; takes a smooth finish and is prized in Japan for fancy inlays, though it has an unpleasant odor.

Medicinal Uses: A decoction of the calyx and fruit stem is sometimes taken to relieve hiccups, coughs and labored respiration.

Black Sapote (Plate LXI)

The black sapote is not, as might be assumed, allied to either the sapote (*Pouteria sapota* H.E. Moore & Stearn) or the white sapote (*Casimiroa edulis* Llave & Lex.). Instead, it is closely related to the persimmon in the family Ebenaceae. For many years it has been widely misidentified as *Diospyros ebenaster* Retz., a name confusingly applied also to a strictly wild species of the West Indies now distinguished as *D. revoluta* Poir. The presently accepted binomial for the black sapote is *D. digyna* Jacq. (syn. *D. obtusifolia* Humb. & Bonpl. ex Willd.).

In Spanish, it is known variously as sapote, *sapote negro, zapote, zapote negro, zapote prieto, zapote de mico, matasano* (or *matazano*) *de mico,* or *ebano.* It has been called black persimmon in Hawaii.

Description

The tree is handsome, broad-topped, slow-growing, to 80 ft (25 m) in height, with furrowed trunk to 30 in (75 cm) in diameter, and black bark. The evergreen, alternate leaves, elliptic-oblong to oblong-lanceolate, tapered at both ends or rounded at the base and bluntly acute at the apex, are leathery, glossy, 4 to 12 in (10-30 cm) long. The flowers, borne singly or in groups of 3 to 7 in the leaf axils, are tubular, lobed, white, 3/8 to 5/8 in (1-1.6 cm) wide, with persistent green calyx. Some have both male and female organs, large calyx lobes and are faintly fragrant; others are solely male and have a pronounced gardenia-like scent and a few black specks in the throat of the corolla. The fruit is bright-green and shiny at first; oblate or nearly round; 2 to 5 in (5-12.5 cm) wide; with a prominent, 4-lobed, undulate calyx, 1½ to 2 in (4-5 cm) across, clasping the base. On ripening, the smooth, thin skin becomes olive-green and then rather muddy-green. Within is a mass of glossy, brown to very dark-brown, almost black, somewhat jelly-like pulp, soft, sweet and mild in flavor. In the center, there may be 1 to 10 flat, smooth, brown seeds, ¾ to 1 in (2-2.5 cm) long, but the fruits are often seedless.

Origin and Distribution

The black sapote is native along both coasts of Mexico from Jalisco to Chiapas, Veracruz and Yucatan and in the forested lowlands of Central America, and it is frequently cultivated throughout this range. It was apparently carried by the Spaniards to Amboina before 1692, and to the Philippines long before 1776, and eventually reached Malacca, Mauritius, Hawaii, Brazil, Cuba, Puerto Rico and the Dominican Republic. In 1919, seeds from Guadalajara, Mexico, were sent to the Bureau of Plant Industry of the United States Department of Agriculture; cuttings and seeds were received from the Isle of Pines, Cuba, in 1915; seeds arrived from Hawaii in 1916 and 1917; others from Oaxaca, Mexico, in 1920. Numerous seedlings have been grown in southern California but all have been killed by low temperatures. The tree does very well in southern Florida, though it has been grown mainly as a curiosity. Outside of its homeland, the fruit has not achieved any great popularity. In Mexico, the fruits are regularly marketed.

Varieties

Certain trees tend to bear very large, seedless or nearly seedless fruits maturing in summer instead of winter as most do, but no varietal names have been attached to them in Florida.

Climate

The black sapote is not strictly tropical inasmuch as it is hardy as far north as Palm Beach County, Florida, if protected from frost during the first few years. Trees that have become well established have withstood occasional brief exposures to 28° or 30°F (-2.22° or -1.11°C). In Mexico, the tree is cultivated up to elevations of 5,000 or even 6,000 ft (1,500-1,800 m).

Soil

The tree has a broad adaptability as to terrain. In Mexico it grows naturally in dry forests or on alluvial clay near streams or lagoons where it is frequently subject to flooding. Nevertheless, it thrives on moist sandy loam, on well-drained sand or oolitic limestone with very little topsoil in southern Florida. It is said to flourish on all the soils of Cuba.

Propagation

The black sapote is usually grown from seeds, which remain viable for several months in dry storage and germinate in about 30 days after planting in flats. Vegetative propagation is not commonly practiced but the tree has been successfully air-layered and also shield-budded using mature scions.

Culture

Seedlings are best transplanted to pots when about 3 in (7.5 cm) high and they are set in the field when 1 to 2 years old, at which time they are 1 to 2 ft (30-60 cm) in height. They should be spaced at least 40 ft (12 m) apart. Most begin to bear in 5 to 6 years but some trees may take somewhat longer. The tree is naturally vigorous and receives little or no cultural attention in Florida though it has been noted that it benefits from fertilization.

Season

In Mexico, the fruits are common in the markets from August to January. Most black sapotes in Florida ripen in December, January or February. Certain trees, especially the large-fruited types, regularly come into season in June, others in July and August.

Harvesting

It is difficult to detect the slight color change of mature fruits amid the dense foliage of the black sapote tree. Many black sapotes ripen, fall and smash on the ground before one has the chance to pick them, and this is one reason why the tree is not favored for landscaping in urban areas. An experienced picker can harvest the fruits at the green-mature or olive-green stage with a cutting pole equipped with a cloth sack.

Yield

No yield figures are available but the tree is noted for bearing well. In 1899, the annual crop in Mexico was valued at $27,000, a considerable sum at that time.

Keeping Quality

Fruits picked when full-grown but unripe (bright-green) have ripened in 10 days at room temperature. Therefore it is at this stage that they must be picked for marketing and shipping. Firm, olive-green fruits will ripen in 2 to 6 days. Fruits displayed on markets in Mexico are somewhat shriveled and wrinkled. The black sapote is very soft when fully ripe. Though it may remain fit for eating if held for a few days in cold storage, it is too soft to stand handling.

Food Uses

Unkind writers have employed unflattering phrases in describing the flesh of the black sapote and have probably hindered its acceptance. This seems quite unreasonable because the color and texture of the pulp closely match stewed prunes, to which there seems to be no aesthetic objection. In the Philippines, the seeded pulp is served as dessert with a little milk or orange juice poured over it. The addition of lemon or lime juice makes the pulp desirable as a filling for pies and other pastry. It is also made into ice cream. In Mexico, the pulp may be mashed, beaten or passed through a colander and mixed with orange juice or brandy, and then served with or without whipped cream.

Also, they sometimes mix the pulp with wine, cinnamon and sugar and serve as dessert. Some Floridians use an eggbeater to blend the pulp with milk and ground nutmeg. A foamy, delicious beverage is made by mixing the pulp with canned pineapple juice in an electric blender. In Central America, the fermented fruits are made into a liqueur somewhat like brandy.

Food Value Per 100 g of Edible Portion*	
Moisture	79.46-83.1 g
Protein	0.62-0.69 g
Carbohydrates	12.85-15.11 g
Fat	0.01 g
Ash	0.37-0.6 g
Calcium	22.0 mg
Phosphorus	23.0 mg
Iron	0.36 mg
Carotene	0.19 mg
Thiamine	–
Riboflavin	0.03 mg
Niacin	0.20 mg
Ascorbic Acid**	191.7 mg
*According to analyses in Mexico and Guatemala.	

**The ascorbic acid content is said to be about twice that of the average orange.

Toxicity

Unripe black sapotes are very astringent, irritant, caustic and bitter, and have been used as fish poison in the Philippines.

Other Uses

Wood: The wood is yellowish to deep-yellow with black markings near the heart of old trunks; compact and suitable for cabinetwork but little used. Reports of dark wood utilized for furniture are probably the result of confusion with other species of *Diospyros*.

Medicinal Uses: The crushed bark and leaves are applied as a blistering poultice in the Philippines. In Yucatan, the leaf decoction is employed as an astringent and is taken internally as a febrifuge. Various preparations are used against leprosy, ringworm and itching skin conditions.

Note: The rare, wild relative *D. revoluta* Poir., mentioned at the beginning, has not only been included with the black sapote under the erroneous *D. ebanaster*, but has also been dealt with as *D. nigra* Perr. and under at least 8 other binomials. In Puerto Rico, the Dominican Republic, Montserrat, Dominica and Guadeloupe it is variously called black apple, *barbara*, *bambarat*, *barbequois*, *bois noir*, *bois negresse*, *ebene*, *guayabota*, *plaqueminier*, and *zapote negro*. It has smaller, thicker leaves and smaller fruits than the black sapote and the calyx is square. Little, Woodbury and Wadsworth say the fruits are poisonous and, with the bark, used as fish poison.

Mabolo (Plate LXII)

A minor member of the family Ebenaceae, more admired for its ornamental than its edible value, the mabolo has appeared in literature for many years under the illegitimate binomial *Diospyros discolor* Willd. In 1968, Dr. Richard Howard, Director of the Arnold Arboretum, Harvard University, proposed the adoption of *D. blancoi* A. DC., and this is now regarded as the correct botanical designation for this species. The fruit is sometimes called velvet apple, or, in India, peach bloom. In Malaya, it is *buah mantega* (butter fruit)—a term now often applied to the avocado—, or *buah sakhlat*, or *sagalat* (scarlet fruit). Mabolo (or mabulo) is the most common of the several Philippine dialectal names. Another, *kamagon*, is rendered *camagon* in Spanish.

Description

The mabolo varies in form from a small straggly tree with drooping branches, to an erect, straight tree to 60 or even 100 ft (18-33 m), with stout, black, furrowed trunk to 50 in (80 cm) thick. It is rather slow-growing. The evergreen, alternate leaves, oblong, pointed at the apex, rounded or pointed at the base, are 6 to 9 in (15-22.8 cm) long, 2 to 3½ in (5-9 cm) wide; leathery, dark-green, smooth and glossy on the upper surface, silvery-hairy underneath. New leaves are showy, pale-green or pink and silky-hairy. The tubular, 4-lobed, waxy, faintly fragrant blooms are short-stalked, creamy-white, downy. Male flowers ¼ in (6 mm) wide, in small clusters, and female flowers, ½ in (12.5 mm) wide, and solitary, are borne on separate trees. Attractive and curious, the oval or oblate fruit, 2 to 4 in (5-10 cm) wide, has thin, pink, brownish, yellow, orange or purple-red skin, densely coated with short, golden-brown or coppery hairs, and is capped at the base with a dull-green, stiff calyx. The fruits are often borne in pairs, very close together on opposite sides of a branch. A strong, unpleasant, cheese-like odor is given off by the whole fruit but emanates from the skin, for it is absent in the peeled flesh, which is whitish, firm, mealy, somewhat like that of an overripe apple; moist but not very juicy; of mild, more or less sweet flavor, suggesting a banana-flavored apple. There may be 4 to 8 brown, smooth, wedge-shaped seeds, about 1½ in (4 cm) long and 1 in (2.5 cm) wide, standing in a circle around the central core, though the fruits are often completely seedless. Each seed is covered with a whitish membrane that is transparent when fresh, opaque when dried.

Origin and Distribution

The mabolo is indigenous to the low and medium altitude forests of the Philippine Islands from the island of Luzon to the southernmost of the Sulu Islands, and is commonly cultivated for its fruit and even more as a shade tree for roadsides. The tree was introduced into Java and Malaya, and, in 1881, into Calcutta and the Botanical Garden in Singapore, though it existed in Singapore before that date. In recent times, it has been decreasing in numbers in Malaya. It is only occasionally planted in India and then mainly as an ornamental because of the attractiveness of the foliage and the fruits.

Seeds were sent to the United States Department of Agriculture by W.S. Lyon, of the Philippine Bureau of Agriculture, in 1906, with a note of admiration for the tree and the exterior of the fruit but not the interior; still, more seeds were sent in 1909 and the seedlings thrived at the Plant Introduction Station in Miami. There are occasional specimens grown elsewhere in southern Florida and some scattered around the Caribbean area, in Jamaica, Puerto Rico, Cuba, Trinidad and the Lancetilla Experimental Garden in Honduras where plants were received from the Philippines in 1926 and seeds from Cuba in 1927. There are a few in Bermuda and in Hawaii where the mabolo first fruited in 1928. Nowhere has the mabolo gained the favor it enjoys in its homeland.

Varieties

Mabolo trees vary in the degree of hairiness on the twigs and leaves. Burkill (in Malaya) and Mendiola (in the Philippines) refer to mabolos with red and copper-colored skin as distinct races. A race with purplish-red skin and unusually sweet flavor was long ago introduced into Malaya. In 1921, budded trees of a superior seedless cultivar called 'Manila' were shipped to the United States Department of Agriculture by P.J. Wester, who was then Horticulturist in charge of the Manila Experiment Station. The parent tree in the Philippines had a history of bearing crops of oblate, sweet, juicy fruits, 80% of them seedless, 20% having 1 to 3 seeds. Another seedless Philippine cultivar was named 'Valesca'.

Mendiola (1926) wrote that seedless mabolos "are easily distinguished from the seedy ones as they are flatter. It is believed by some horticulturists and growers that these seedless fruits come from branches that are bud sports . . . it is impossible to confirm or deny this claim until it is known how much parthenocarpy has to do with . . . these seedless forms . . . the genus *Diospyros* is, in a number of cases, parthenocarpic."

Propagation

The tree is generally grown from seeds. Shield-budding has been successfully practiced in the Philippines and is the preferred means of perpetuating superior types.

Cultivation

Male trees must be planted near the female trees for effective pollination and fruit production. The tree does best in loam but flourishes very well in almost any soil with little care. It is rarely fertilized and seems to need no protective spraying.

Season

In India, the mabolo blooms in March and April and the fruits ripen in July and August. The main season in Florida is June to September but occasional fruits may be found on the tree at almost any time of the year.

Keeping Quality

Investigators in Hawaii studied carbon dioxide and ethylene production of mature green and 5% red-colored mabolos. Mature-green fruits reached the climacteric peak stage in 9 days; the slightly ripe fruits, in 5 days.

Food Uses

The surface fuzz adheres tightly even when the fruit is ripe. Also, the skin, though thin and pliable, is tough and papery when chewed. Therefore, the fruits should be peeled before eating, and then kept in the refrigerator for a few hours before serving. Then the odor, which is mainly in the skin, will have largely dissipated.

Some people slice or quarter the flesh, season with lime or lemon juice or Grenadine sirup and serve fresh as dessert. The flesh is also diced and combined with that of other fruits in salads. If stewed in sirup, the flesh becomes fibrous and tough. Cut into strips and fried in butter, it is crisp and fairly agreeable as a vegetable of the dasheen or taro type appropriate for serving with ham, sausage or other spicy meat.

Food Value Per 100 g of Edible Portion*		
	Ordinary type	*Seedless type*
Calories	–	504
Moisture	77.80 g	71.95–86.04 g
Protein	0.75 g	0.82–2.79 g
Fat	–	0.22–0.38 g
Carbohydrates	–	(other) 5.49–6.12 g
Sugar	11.47 g	(reducing) 6.25–18.52 g
Fiber	–	0.74–1.76 g
Ash	0.83 g	0.43–1.08 g
Sulphuric Acid	0.11 g	–
Malic Acid	0.16 g	–
Phytin	–	3.26% (on dry basis)
*Analyses made in the Philippines and India.		

The fruit is considered a fairly good source of iron and calcium and a good source of vitamin B.

Toxicity

The hairs may be somewhat irritating to sensitive skin.

Other Uses

Mabolo seedlings: Useful as rootstock on which to graft the Japanese persimmon.

Wood: The sapwood is pinkish or reddish; may have gray markings. The heartwood is streaked and mottled with gray and is sometimes all-black. In the Philippines, it is carved into highly prized hair combs.

APOCYNACEAE

Carissa

Two species of the notorious family Apocynaceae are noteworthy because of their edible fruits and innocuous milky latex. The more attractive of these is the carissa, *Carissa macrocarpa* A. DC. (syn. *C. grandiflora* A. DC.), also called Natal plum and *amantungula*.

Description

A vigorous, spreading, woody shrub with abundant white, gummy sap, the carissa may reach a height of 15 to 18 ft (4.5–5.5 m) and an equal breadth. The branches are armed with formidable stout, double-pronged thorns to 2 in (5 cm) long. The handsome, evergreen, opposite leaves are broad-ovate, 1 to 2 in (2.5–5 cm) long, dark-green, glossy, leathery. Sweetly fragrant, white, 5-lobed, tubular flowers to 2 in (5 cm) broad are borne singly or a few together at the tips of branchlets all year. Some plants bear flowers that are functionally male, larger than normal and with larger anthers, and stamens much longer than the style. Functionally female flowers have stamens the same length as the style and small anthers without pollen.

The round, oval or oblong fruit, to 2½ in (6.25 cm) long and up to 1½ in (4 cm) across, is green and rich in latex when unripe. As it ripens, the tender, smooth skin turns to a bright magenta-red coated with a thin, whitish bloom, and finally dark-crimson. The flesh is tender, very juicy, strawberry-colored and -flavored, with flecks of milky sap. Massed in the center are 6 to 16 small, thin, flat, brown seeds, not objectionable when eaten.

Origin and Distribution

The carissa is native to the coastal region of Natal, South Africa, and is cultivated far inland in the Transvaal. It was first introduced into the United States in 1886 by the horticulturist Theodore L. Meade. Then, in 1903, Dr. David Fairchild, heading the Office of Foreign Seed and Plant Introduction of the United States Department of Agriculture, brought in from the Botanical Garden at Durban, a large quantity of seeds. Several thousand seedlings were raised at the then Plant Introduction Garden at Miami and distributed for testing in Florida, the Gulf States and California, and much effort was devoted to following up on the fate of the plants in different climatic zones. The carissa was introduced into Hawaii in 1905 and over the next few years was extensively distributed throughout the islands. It was planted in the Bahamas in 1913. It first fruited in the Philippines in 1924; is grown to a limited extent in India and East Africa. It was widely planted in Israel, flourished and flowered freely but rarely set fruit. Elsewhere, it is valued mainly as a protective hedge and the fruit is a more-or-less-welcomed by-product.

Varieties

Horticulturists in South Africa, California and Florida have selected and named some types that tend to bear more reliably than others:

'Fancy', selected in California in the 1950's, was an erect form bearing an abundance of large fruits with few seeds.
'Torrey Pines' produces good crops of fruit and pollen.
'Gifford' is one of the best fruit bearers in Florida.
'Extra Sweet' was advertised in Florida in the early 1960's.
'Alles' ('Chesley') produces few fruits in California.
'Frank' is a light bearer though it has a good supply of pollen.

As space for massive barrier hedges has diminished and interest in the fruits declined, efforts have been directed to the development of dwarf, compact, less spiny types for landscape use. Some of the popular ornamental cultivars include: 'Bonsai', 'Boxwood Beauty', 'Dainty Princess', 'Grandiflora', 'Green Carpet', 'Horizontalis', 'Linkii', 'Low Boy', 'Minima', 'Nana', 'Nana Compacta', 'Prostrata' and 'Tuttlei'.

Pollination

In its homeland, the carissa is pollinated by small beetles and hawk-moths and other night-flying insects. Various degrees of unfruitfulness in America has been attributed to inadequate pollination. Some seedlings are light-croppers, but others never bear at all. It has been found that unproductive plants, apparently self-infertile, will bear fruits after cross-pollination by hand.

Climate

The carissa is subtropical to near-tropical, thriving throughout the state of Florida and enduring temperatures as low as 25°F (-3.89°C) when well-established. Young plants need protection when the temperature drops below 29°F (-1.67°C). Best growth is obtained in full sun.

Soil

The shrub thrives in dry, rocky terrain in Hawaii; in red clay or sandy loam in California, and in sandy or alkaline soils in Florida, though the latter may induce defi-

Fig. 110: Beautiful of foliage, flower and fruit, the thorny carissa (*Carissa grandiflora*) is primarily an ornamental but the fruits are edible and enjoyable.

ciencies in trace elements. The plant has moderate drought tolerance and high resistance to soil salinity and salt spray. It cannot stand water-logging.

Propagation

Seeds germinate in 2 weeks but the seedlings grow very slowly at first and are highly variable. Vegetative propagation is preferred and can be done easily by air-layering, ground-layering, or shield-budding. Cuttings root poorly unless the tip of a young branchlet is cut half-way through and left attached to the plant for 2 months. After removal and planting in sand, it will root in about 30 days. Grafting onto seedlings of the karanda (q.v.) has considerably increased fruit yield.

Culture

Seedlings may begin to produce fruit in 2 years; cuttings earlier. A standard, well-balanced fertilizer suffices except on limestone where trace elements must be added. Dwarf cultivars must be kept under control, otherwise they are apt to revert to the ordinary type. Vigorous shoots will develop and outgrow the compact form.

Season

While the carissa flowers and fruits all year, the peak period for blooming and fruiting is May through September. The 5-pointed calyx remains attached to the plant when the fruit is picked.

Pests and Diseases

Spider mites, thrips and whiteflies, and occasionally scale insects, attack young plants, especially in nurseries and in the shade.

A number of fungus diseases have been recorded in Florida; algal leaf spot and green scurf caused by *Cephaleuros virescens;* leaf spot from *Alternaria* sp., *Botryosphaeria querquum, Fusarium* sp., *Gloeosporium* sp., *Phyllosticta* sp. and *Colletotrichum gloeosporioides* which also is responsible for anthracnose; stem gall from *Macrophoma* sp., *Nectria* sp., *Phoma* sp., *Phomopsis* sp., and both galls and cankers from *Sphaeropsis tumefaciens*; dieback caused by *Diplodia natalensis* and *Rhizoctonia solani*; thread blight from *Rhizoctonia ramicola*; root rot resulting from infection by *Phytophthora parasitica* and *Pythium* sp.

Food Uses

The carissa must be fully ripe, dark-red and slightly soft to the touch to be eaten raw. It is enjoyed whole, without peeling or seeding, out-of-hand. Halved or quartered and seeded it is suitable for fruit salads, adding to gelatins and using as topping for cakes, puddings and ice cream. Carissas can be cooked to a sauce or used in pies and tarts. Stewing or boiling causes the latex to leave the fruit and adhere to the pot (which must not be aluminum), but this can be easily removed by rubbing with cooking oil.

Carissas are preserved whole by pricking, cooking briefly in a sugar sirup and sterilizing in jars. Peeled or unpeeled, they are made into jam, other preserves, sirup or sweet pickles. Jelly is made from slightly underripe fruits, or a combination of ripe and unripe to enhance the color.

Food Value

Analyses made in the Philippines show the following values: calories, 270/lb (594/kg); moisture, 78.45%; protein, 0.56%; fat, 1.03%; sugar, 12.00%; fiber, 0.91%; ash, 0.43%. Ascorbic acid content has been calculated as 10 mg/100 g in India.

Karanda

Less showy than the carissa, q.v., the karanda has attracted more interest as a source of fruit and as a medicinal plant than as an ornamental. Its botanical name was in recent years changed to *Carissa congesta* Wight (syn. *C. carandas* Auct., formerly widely shown as *C. carandas* L.). It is called *kerenda* in Malaya, *karaunda* in Malaya and India; Bengal currant or Christ's thorn in South India; *nam phrom,* or *namdaeng* in Thailand; *caramba, caranda, caraunda* and *perunkila* in the Philippines.

Description

This species is a rank-growing, straggly, woody, climbing shrub, usually growing to 10 or 15 ft (3-5 m) high, sometimes ascending to the tops of tall trees; and rich in white, gummy latex. The branches, numerous and spreading, forming dense masses, are set with sharp thorns, simple or forked, up to 2 in (5 cm) long, in pairs in the axils of the leaves. The leaves are evergreen, opposite, oval or elliptic, 1 to 3 in (2.5-7.5 cm) long; dark-green, leathery, glossy on the upper surface, lighter green and dull on the underside. The fragrant flowers are tubular with 5 hairy lobes which are twisted to the left in the bud instead of to the right as in other species. They are white, often tinged with pink, and borne in terminal clusters of 2 to 12. The fruit, in clusters of 3 to 10, is oblong, broad-ovoid or round, ½ to 1 in (1.25-2.5 cm) long; has fairly thin but tough, purplish-red skin turning dark-purple or nearly black when ripe; smooth, glossy; enclosing very acid to fairly sweet, often bitter, juicy, red or pink, juicy pulp, exuding flecks of latex. There may be 2 to 8 small, flat, brown seeds.

Varieties

Formerly there were believed to be 2 distinct varieties: *C. carandas* var. *amara*—with oval, dark-purple, red-fleshed fruits, of acid flavor; and var. *dulcis*—round, maroon, with pink flesh and sweet-subacid flavor. However, David Sturrock, a Florida horticulturist who took a special interest in the karanda, observed these and other variations throughout seedling populations.

Origin and Distribution

The karanda is native and common throughout much of India, Burma and Malacca and dry areas of Ceylon; is rather commonly cultivated in these areas as a hedge and for its fruit and the fruit is marketed in villages. It is rare in Malaya except as a potted plant in the north; often grown in Thailand, Cambodia, South Vietnam and in East Africa. It was introduced into Java long ago as a hedge and has run wild around Djakarta. The karanda first fruited in the Philippines in 1915 and P.J. Wester described it in 1918 as "one of the best small fruits introduced into the Philippines within recent years."

The United States Department of Agriculture received seeds from the Middle Egypt Botanic Garden in 1912 (S.P.I. #34364); from P.J. Wester in the Philippines in 1918 (S.P.I. #46636) and again in 1920 (S.P.I. #51005); and a third time in 1925 (S.P.I. #65334). The shrub has been cultivated in a limited way in Florida and California and in some experimental gardens in Trinidad and Puerto Rico.

Climate

The karanda is more cold-tolerant than the carissa. It grows from sea-level to 2,000 ft (600 m) in the Philippines; but up to an altitude of 6,000 ft (1,800 m) in the Himalayas. Burkill says it is not really suited to the humid climate of Malaya. Like the carissa, its chief requirement is full exposure to sun.

Soil

The plant grows vigorously in Florida on sand or limestone. In India, it grows wild on the poorest and rockiest soils and is grown as a hedge plant in dry, sandy or rocky soils. It is most fruitful on deep, fertile, well-drained soil but if the soil is too wet, there will be excessive vegetative growth and lower fruit production.

Propagation

Propagation is usually by seed because cuttings have never rooted readily. Experimental work in India has

Fig. 111: The karanda (*Carissa carandas*) is small and gummy but yields colorful, tart juice.

shown that cuttings from mature plants may not root at all; 20% of hardwood cuttings from trimmed hedges have rooted in November but not when planted earlier. Cuttings from nursery stock gave best results: 10% rooted in late September; 20% in early October; 30% in late October; and 50% in early November. In all cases, cuttings were pre-treated with indolebutyric acid at 500 ppm in 50% alcohol. Sturrock found that tender tip cuttings could be rooted under constant mist; also that the karanda can be grafted onto self-seedlings. It has proved to be a good rootstock for carissa.

Culture

The plant grows slowly when young. Once well-established, it grows more vigorously and becomes difficult to control. If kept trimmed to encourage new shoots, it will bloom and fruit profusely.

Season

The karanda may bloom and fruit off and on throughout the year. For use unripe, the fruits are harvested from mid-May to mid-July. The main ripening season is August and September. The 5-pointed calyx remains attached to the plant when the fruit is picked, leaving a gummy aperture at the base.

Keeping Quality

Freshly-picked ripe fruits can be kept at room temperature only 3 or 4 days before they begin to shrivel.

Pests and Diseases

Nursery plants are probably prone to the same pests that attack young carissas.

Fungus diseases recorded on the karanda in Florida are algal leaf spot and green scurf caused by *Cephaleuros virescens;* twig dieback from *Diplodia natalensis;* and stem canker induced by *Dithiorella* sp.

Food Uses

The sweeter types may be eaten raw out-of-hand but the more acid ones are best stewed with plenty of sugar. Even so, the skin may be found tough and slightly bitter. The fruit exudes much gummy latex when being cooked but the rich-red juice becomes clear and is much used in cold beverages. The sirup has been successfully utilized on a small scale by at least one soda-fountain operator in Florida. In Asia, the ripe fruits are utilized in curries, tarts, puddings and chutney. When only slightly underripe, they are made into jelly. Green, sour fruits are made into pickles in India. With skin and seeds removed and seasoned with sugar and cloves, they have been popular as a substitute

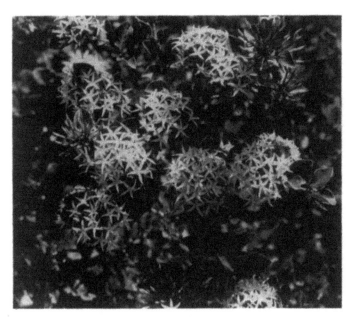

Fig. 112: The karanda, shrubby or climbing, is conspicuous when in starry bloom.

for apple in tarts. British residents in India undoubtedly favored the karanda as being reminiscent of gooseberries.

Food Value

Analyses made in India and the Philippines show the following values for the ripe karanda: calories, 338 to 342/lb (745-753/kg); moisture, 83.17-83.24%; protein, 0.39-0.66%; fat, 2.57-4.63%; carbohydrate, 0.51-0.94%; sugar, 7.35-11.58%; fiber, 0.62-1.81%; ash, 0.66-0.78%. Ascorbic acid content has been reported as 9 to 11 mg per 100 g.

Other Uses

Fruit: The fruits have been employed as agents in tanning and dyeing.

Leaves: Karanda leaves have furnished fodder for the tussar silkworm.

Root: A paste of the pounded roots serves as a fly repellent.

Wood: The white or yellow wood is hard, smooth and useful for fashioning spoons, combs, household utensils and miscellaneous products of turnery. It is sometimes burned as fuel.

Medicinal Uses: The unripe fruit is used medicinally as an astringent. The ripe fruit is taken as an antiscorbutic and remedy for biliousness. The leaf decoction is valued in cases of intermittent fever, diarrhea, oral inflammation and earache. The root is employed as a bitter stomachic and vermifuge and it is an ingredient in a remedy for itches. The roots contain salycylic acid and cardiac glycosides causing a slight decrease in blood pressure. Also reported are carissone; the D-glycoside of B-sitosterol; glucosides of odoroside H; carindone, a terpenoid; lupeol; ursolic acid and its methyl ester; also carinol, a phenolic lignan. Bark, leaves and fruit contain an unnamed alkaloid.

Naranjilla (Plate LXIII)

An intriguing and highly appealing member of the nightshade family, Solanaceae, the naranjilla, *Solanum quitoense* Lam. (syn. *S, angulatum* Lam.), acquired its Spanish name, meaning "little orange" because it is round, and is bright-orange when fully ripe. In Ecuador it is called *naranjilla de Quito*, or *nuqui;* in Peru, *naranjita de Quito.* The Incas called it *lulum.* In Mexico, it is *lulun;* in Colombia, *lulo, naranjilla* or *toronja.* Variety *septentrionale* Schultes & Cuatr. is called *lulo de castilla, lulo de perro,* or *lulo morado.*

Description

The naranjilla plant is a spreading, herbaceous shrub to 8 ft (2.5 m) high with thick stems that become somewhat woody with age; spiny in the wild, spineless in cultivated plants. The alternate leaves are oblong-ovate, to 2 ft (60 cm) long and 18 in (45 cm) wide, soft and woolly. There may be few or many spines on petioles, midrib and lateral veins, above and below, or the leaves may be completely spineless. Young leaves, young stems and petioles are coated with richly purple stellate hairs. Hairs on other parts may appear simple. Borne in short axillary clusters of as many as 10, the fragrant flowers, about 1⅕ in (3 cm) wide, have 5 petals, white on the upper surface, purple hairy beneath, and 5 prominent yellow stamens. The unopened buds are likewise covered with purple hairs. A brown, hairy coat protects the fruit until it is fully ripe, when the hairs can be easily rubbed off, showing the bright-orange, smooth, leathery, fairly thick peel. The fruit, crowned with the persistent, 5-pointed calyx, is round or round-ovate, to 2½ in (6.25 cm) across and contains 4 compartments separated by membranous partitions and filled with translucent green or yellowish, very juicy, slightly acid to acid, pulp of delicious flavor which has been likened to pineapple-and-lemon. There are numerous pale-buff seeds, thin, flat, hard and ⅛ in (3 mm) in diameter.

Origin and Distribution

The usually spineless naranjilla is believed to be indigenous and most abundant in Peru, Ecuador and southern Colombia. The forms found in the rest of Colombia and in the central and northern Andes of Venezuela and interior mountain ranges of Costa Rica may vary from partly to very spiny. Some botanists have suggested that these spiny forms belong to the botanical variety *septentrionale.*

In Ecuador, 90% of commercial naranjilla cultivation is in a 15-mile area in the valley and adjacent hillsides of the Pastaza River, a tributary of the Amazon.

Seeds were first sent to the United States Department of Agriculture from Colombia in 1913; from Ecuador in 1914 and 1916. Many other introductions were made but the resulting plantings in California, Florida and northern greenhouses flourished only briefly, some set fruit, and all died. Trial plantings were made in the Philippines about 1922. The exhibition of fruits and 1,500 gallons of freshly made juice of Ecuadorian naranjillas at the New York World's Fair in 1939 roused a great deal of interest. In February, 1948, 20 naranjilla plants were set out in a field at the University of Florida's Agricultural Research and Education Center in Homestead, Florida. They flourished and were beginning to fruit when nearly all were destroyed by hurricanes. Dr. Milton Cobin tried grafting the naranjilla on the so-called "potato tree", *Solanum macranthum* Dunal of Brazil, hoping to give it wind-resistance. The grafted plants were set out in 1949 and fruited well. Seeds of acid and sweet strains were obtained from the United States Department of Agriculture in 1950. Some of the resulting plants were grafted onto *S. macranthum* and did well; others, set out on their own roots, became severely infested with rootknot nematodes and died. In 1951, the naranjilla was grafted onto *S. erianthum* D. Don but the plants were dwarfed by this rootstock and short-lived. A number of fruit fanciers took up the growing of grafted naranjilla plants in home gardens. Interest was aroused in Caribbean horticulturists and other visitors to the Homestead station. In the early 1950's, plantings were made in Puerto Rico, Jamaica, Panama, Hawaii and Queensland, and in the Meseta Central of Costa Rica where one of several growers set out 70,000 plants of the local wild variety which bears a larger fruit than the nonspiny South American type.

In 1962, a commercial plantation owned by Frederic Zeuner, proprietor of Cia Procesadora de Naranjilla Ltda, of San José, covered 1,200 acres (511 ha) and a $55,000 factory was built to process the fruits. The pulp was being shipped to the United States in No. 10 cans. It was blended with apple or pineapple juice, put up in small cans and frozen for retail sale. In 1966, I was advised by the U.S. Agricultural Attaché in San José that this pilot effort failed because the canned product was not properly processed and had a metallic taste, also because of the collapse of

Fig. 113: Naranjilla (*Solanum quitoense*) juice is most prized fresh or preserved, but some is made into wine in Colombia.

the canners' contracts with farmers. Production of a better product with proper cooling and storage continued on a local scale. In 1963, the naranjilla was a relatively new crop in Guatemala and there was an experimental plantation and others that were semi-commercial.

The naranjilla is much admired as an ornamental foliage plant in northern conservatories but it will not fruit in temperate latitudes.

Varieties

The botanical variety *septentrionale* already referred to is found in Valle, Cundinamarca, Magdalena, Santanderes and Tolima, in central and northern Colombia, and also in Ecuador and Venezuela. It is said to differ from the typical form, var. *quitoense,* of Ecuador, Peru and southern Colombia, only in having spines on the stem, branches, petioles, and principal veins of the leaves.

There is a sweet, but not very juicy strain around the Andean town of Baza, about 50 miles (80 km) east of Quito, Ecuador.

A wild, spiny form in Costa Rica, called *berenjena de olor* ("fragrant eggplant"), has woodier stem and branches and unusually large fruits to 2½ in (6.25 cm) in diameter.

The fruit of seedling plants shows much variation. However, there seems to be little or no effort to select and name superior cultivars.

Climate

In Colombia, the naranjilla flourishes in humid regions at elevations between 3,600 and 7,900 ft (1,600 and 2,400 m) where the annual rainfall is about 60 in (150 cm). Precipitation up to 120 in (250 cm) is tolerable if well distributed throughout the year. In Panama, the naranjilla has made good growth at altitudes from 4,000 to 6,000 ft (1,200-1,800 m). It is grown in southern Florida at near sea-level. The best plantations in Ecuador are between 5,000 and 6,000 ft (1,500-1,800 m), where the mean temperature is 62.6° to 66.2°F (17°-19°C). The naranjilla cannot tolerate temperatures over 85°F (29.4°C). It is not adapted to full sun but favors semi-shade.

Soil

The plant does best in a rich, organic soil; also grows well on poor, stony ground, and on scarified limestone. It must have good drainage. In Latin America, naranjillas are planted on virgin soil in tracts where the large trees have been felled and the undergrowth burned off. The remaining trees provide semi-shade and wind protection.

Propagation

The naranjilla can be propagated by air-layering or by cuttings of mature wood. In Latin America, it is commonly grown from seeds which must first be spread out in the shade to ferment slightly to eliminate the mucilage, then washed, air-dried, and dusted with fungicide. There are about 140,000 seeds to the pound (.5 kg); 9,000 to the ounce (28 g). Seedlings are raised in nurseries by the same methods appropriate for tomato seedlings, and are ready for transplanting in 2 to 3 months.

In Florida, the naranjilla is easily cleft-grafted onto *S. macranthum* seedlings that have grown 2 ft (60 cm) tall and have been cut back to 1 ft (30 cm) from the ground, then split down the center for a distance of 1 to 2 in (2.5-5 cm). Selected scions 2 to 3 in (5-7.5 cm) long are inserted in the slit and tightly bound in place. It takes 2 to 3 weeks for the scion to fully unite with the stock. The plants are not set out until the scion has grown about 2 ft (30 cm). Other grafting methods—saddle, side, and whip—have also been successful.

Trials on tree tomato (*Cyphomandra betacea* Sendt.) seemed promising in 1952. In tropical Africa, the naranjilla has done well on the nematode-resistant relative, *S. torvum* Sw.

Culture

Naranjilla plants should be set 6 to 8 ft (1.8-2.4 m) apart each way, which provides 1,250 plants per acre (3,000/ha). Colombians transplant young seedlings from the nursery bed into polyethylene bags containing 5½ lbs. (2.5 kg) of soil, keep them in semi-shade, give them ½ oz (14 g) of super-phosphate and frequent irrigation. When 14 in (35 cm) high, they are set out in holes enriched with 8.8 lbs (4 kg) of organic compost, breaking the plastic bag as they place the plant in the hole. In Latin America, generally, the naranjilla is planted out in the afternoon of a cloudy day at the beginning of the rainy season. The planting hole is 12 x 12 x 12 in (30x30x30 cm) and a circle at least 3 ft (1 m) in diameter is kept free of weeds. The plant is a heavy feeder and growth is rapid if fertilizer is given once a month, though most plantations are given no such nutritional care. A 12-12-20 mixture of NPK at the rate of 3 oz (85 g) per plant every 2 months has been recommended. In the coffee zone of Caldas, Colombia, where the soil is organically rich but low in phosphorus, the addition of urea, superphosphate and potassium sulphate, has been found to double productivity.

Seedlings flower 4 to 5 months after transplanting. Fruiting begins 10 to 12 months from seed and is continuous for 3 years in Panama. When the plants reach 4 years of age, productivity declines and they begin to die. In Costa Rica, they are said to bear until 4 to 7 years old. Grafted plants begin to bear about 1 year from planting in the field. In Florida, they continue fruiting for 2 years, then they die back and are replaced by young ones. Watering is essential in dry periods.

Harvesting and Yield

Though everbearing in its natural habitat, the naranjilla fruits mainly in the winter in Florida; rarely, or very lightly, in the summer. For eating out-of-hand, the fruits are picked fully ripe, at which stage the calyx naturally separates from the fruit, leaving a circular depression. In the field, workers remove the hairs by stooping down and rubbing the fruit in dry grass. For marketing, the fruits must be picked when half-colored to avoid falling and bruising and to assure they are firm enough to withstand handling and packing. They are individually cleaned with a dry cloth and then packed in wooden boxes holding 400 fruits—about 70 lbs (32 kg).

In large-scale processing operations, there are mechanized devices for inspection and grading of fruits, washing off the hairy coat, drying, and removing the peduncle and calyx. For underripe fruits with firmly adhering hairs, the machine must be equipped with brushes. Because of the continuous bearing, fruits must be collected every 7 to 10 days. In Ecuador, long trains of mules and burros make weekly trips with sacks and boxes of naranjillas down the trails to central market places.

A healthy plant bears 100 to 150 fruits a year. A good annual yield is 135 fruits—20 lbs (9 kg)—per plant. This results in 25,000 lbs (10,417 kg) per acre, 60,000 lbs (27,273 kg) per hectare.

Keeping Quality

Fully ripe naranjillas soften and ferment very quickly. Fruit picked when half colored will remain in good condition at ordinary temperatures for 8 days. They can be stored for 1 or 2 months at 45°-50°F (7.22°-10°C) and relative humidity of 70 to 80%.

Pests and Diseases

The chief enemies of the naranjilla are the rootknot nematodes (*Meloidogyne* sp.) and grafting on nematode-resistant rootstock is essential to fruit production in south-ern Florida. In the Chinchiná coffee-growing region of Caldas, Colombia, nematicide-treatment of the soil each time it is invaded is considered too expensive, and the plants can therefore be kept in production only one year before they succumb to nematode damage. Nematodes are causing a drop in naranjilla production in various parts of the country and Dr. Charles Heiser of Indiana University is studying the possibility of hybridization with nematode-resistant wild relatives in order to save the industry. Measures to reduce nematode populations in Guatemalan fields include discarding nursery seedlings and adult plants that show typical symptoms (chlorosis, dwarfing, rachitic appearance), mulching, or frequent plowing during hot, dry spells. In Panama, the main stem and branches, and sometimes even the fruits, of mature plants are attacked by the *cochinilla blanca* (white, or West Indian, peach scale, *Pseudaulacaspis pentagona*). A number of other pests and diseases affect naranjilla plants in Colombia. Bacterial wilt is a serious problem in Puerto Rico.

Food Uses

Ripe naranjillas, freed of hairs, may be casually consumed out-of-hand by cutting in half and squeezing the contents of each half into the mouth. The empty shells are discarded. The flesh, complete with seeds, may be squeezed out and added to ice cream mix, made into sauce for native dishes, or utilized in making pie and various other cooked desserts. The shells may be stuffed with a mixture of banana and other ingredients and baked. But the most popular use of the naranjilla is in the form of juice. For home preparation, the fruits are washed, the hairs are rubbed off, the fruits cut in half, the pulp squeezed into an electric blender and processed briefly; then the green juice is strained, sweetened, and served with ice cubes as a cool, foamy drink. A dozen fruits will yield 8 oz (227 g) of juice. Commercially, the juice is extracted mechanically from the cleaned and chopped fruits,

Food Value Per 100 g of Edible Portion*	
Calories	23
Moisture	85.8-92.5 g
Protein	0.107-0.6 g
Carbohydrates	5.7 g
Fat	0.1-0.24 g
Fiber	0.3-4.6 g
Ash	0.61-0.8 g
Calcium	5.9-12.4 mg
Phosphorus	12.0-43.7 mg
Iron	0.34-0.64 mg
Carotene	0.071-0.232 mg (600 I.U.)
Thiamine	0.04-0.094 mg
Riboflavin	0.03-0.047 mg
Niacin	1.19-1.76 mg
Ascorbic Acid	31.2-83.7 mg

*According to analyses of fresh fruits in Colombia and Ecuador.

strained, concentrated and canned or put into plastic bags and frozen.

Sherbet is made in the home by mixing naranjilla juice with corn sirup, sugar, water, and a little lime juice, partially freezing, then beating to a froth and freezing. Naranjilla jelly and marmalade are produced on a small scale in Cali, Colombia.

Toxicity

People with very sensitive skin may find the hairs on the fruits irritating and should protect the hands when rubbing off the fuzz.

Closely Related Species

Dr. Charles Heiser has made a survey of wild relatives of the naranjilla in the hope that one or more of them may be used in cross-breeding to incorporate nematode-resistance without adversely affecting the fruit quality, productivity and other desirable characteristics. He found *S. tequilense* A. Gray most like *S. quitoense*. It is native from central Mexico to central Ecuador, usually between 3,200 and 6,200 ft (1,000-1,900 m) of elevation, and its fruit is sometimes eaten though its hairy coat is more persistent than that of the naranjilla. Fertile hybrids of the two species have been achieved.

Among other wild species reported by Heiser as having edible, naranjilla-like fruits: *S. pseudolulo* Heiser, of Colombia, with cream-colored flesh and short hairs which are readily shed. The fruits are gathered and sold by local vendors. This species, also, has made fertile hybrids with *S. quitoense*.

S. candidum occurring in lowland areas from Mexico to northern Peru and called *huevo de gato*. The juice is less flavorful than that of the naranjilla and the hairs do not detach readily.

S. pectinatum Dunal (syn. *S. hirsutissimum* Standl.), often a small tree, ranges from Mexico to Venezuela and Peru, is known variously as *lulita*, *lulo de la tierra fria*, *toronja*, or *tumo*. It has juice of fine flavor but is handicapped by persistent hairs and the fruit reportedly contains alkaloids which may hinder its exploitation. The spiny plant is a local folk-remedy for hypertension.

The inedible *S. hirtum* Vahl., *huevo de gato*, found wild in Trinidad and Tobago, Yucatan, Central America, Colombia and Venezuela, is nematode-resistant and hybrids of this species and *S. quitoense* retain this character and have moderately good fruits. Dr. Heiser is encouraging further efforts at cross-breeding in Colombia and Costa Rica.

Cocona (Plate LXIV)

Closely allied to the naranjilla, and similar vegetatively but with a quite different fruit, the cocona is much less known outside its natural range. At one time it was erroneously identified as *Solanum hyporhodium* A. Br. & Bouché. This binomial was dropped in favor of *S. topiro* HBK., which is now replaced by *S. sessiliflorum* Dunal. The Amazonian Indian name, *cubiyú*, is a term applied to several species of *Solanum*, but around Manaus, Brazil, *cubiu* pertains specifically to *S. sessiliflorum*. The Indians of the Upper Orinoco call it *tupiro* or *topiro*. Some Colombians refer to it as *coconilla*, or as *lulo*, a name more often given to the naranjilla. It has been casually dubbed "turkey berry", "peach tomato", or "Orinoco apple".

Description

The cocona plant is a much-branched, herbaceous shrub 6½ ft (2 m) high, with downy stem, densely white-hairy twigs, and ovate leaves, oblique at the base, scalloped on the margins, downy on the upper surface, prominently

veined beneath; 18 in (45 cm) long and 15 in (38 cm) wide. New shoots are rusty-hairy on the underside. The wild variety *georgicum* has spines on stem, branches and leaves. The flowers, in clusters of 2 or more in the leaf axils, are 1 in (2.5 cm) wide, with 5 pale greenish-yellow petals, 5 yellow stamens, and a dark-green, 5-pointed calyx. Borne singly or in compact clusters on very short peduncles, and capped with the persistent calyx, the fruit may be round, oblate, oblong or conical-oval, with bluntly rounded apex; 1 in (2.5 cm) to 4 in (10 cm) long, and up to 2⅓ in (6 cm) wide at the base. The thin, tough skin is coated with a slightly prickly, peach-like fuzz until the fruit is fully ripe, then it is smooth, golden- to orange-yellow, burnt-orange, red, red-brown or deep purple-red, and has a bitter taste. Within is a ¼ to ⅜ in (6-10 mm) layer of cream-colored, firm flesh enclosing the yellow, jelly-like central pulp. The cut-open fruit has a faint, tomato-like aroma. The flesh has a mild flavor faintly suggestive of tomato, while the pulp has a pleasant, lime-like acidity.

Abundant throughout the central pulp are the thin, flat, oval, cream-colored seeds, 3/32 to 3/16 in (2-4 mm) in length and unnoticeable in eating.

Origin and Distribution

The spineless cocona is apparently unknown in the wild, having been observed by botanists only in cultivation from Peru and Colombia to Venezuela and bordering regions of Brazil. In 1760, a Spanish surveyor, Apolinar Diez de la Fuente, found the cocona with maize and beans in an Indian garden between Guaharibos Falls and the juncture of the Casiquiare and Orinoco rivers. In 1800, Humboldt and Bonpland, traveling up the Orinoco, observed that the cocona was one of the common plants in the region between the Javita and Pimichin rivers, and they collected specimens on which the first technical description was based. In the mid-1940's, seeds from the upper Amazon were planted at the Experiment Station in Tingo Maria, Peru, and, later on, the plant was grown at the Instituto Interamericano de Agricultura at Turrialba, Costa Rica. Seeds sent from Natal, South Africa, were planted at the University of Florida's Agricultural Research and Education Center, Homestead, Florida, in 1948. By 1950, all the resulting plants had succumbed to nematode damage. The seeds sent to Medellin, Colombia, in 1948 could have been from these plants. Dr. J.J. Ochse grew specimens in a plot outside the then Botany Building at the University of Miami, Coral Gables, Florida, in 1953.

Dr. Niilo Virkki of Cupey, Puerto Rico, bought one fruit from a street vendor in Manaus, Brazil, in June 1964 and planted the seeds when he returned home. The seedlings grew vigorously and began fruiting in March 1965. Plant breeders studied the plant and fruits in view of its possible potential for hybridizing with the naranjilla. They determined the chromosome number of the cocona to be $2n = 24$.

The fruits are much eaten by the Indians and commonly marketed throughout the producing areas of Latin America. In Colombia and Brazil, the cocona is a domestic product, in Peru it is the basis of an industry. Cultivation is being encouraged by Gerber's Baby Foods and farmers are guaranteed a good price. Canned juice is being exported to Europe.

Varieties

The wild variety, *S. topiro* var. *georgicum* Heiser, of the lowlands of eastern Ecuador and Colombia, is a smaller plant with smaller fruits and with spines on the stem, branches and leaves. It spontaneously hybridizes with the typical var. *topiro*, and Dr. Charles Heiser of Indiana University views it as the ancestor of the cultivated cocona.

In Peru, 4 types are distinguished: a) small, purple-red; b) medium, yellow; c) round, resembling an apple, yellow; d) pear-shaped. The medium-sized cocona is in greatest demand in Peru and especially for juice.

The Divisão de Ciencias Agronomicas of INPA in Amazonia, made a collection of 35 strains of cocona from Belem do Pará, Brazil, and Iquitos, Peru, and established an experimental block of 149 plants in pure sand for evaluation. The range of variation indicated that seedling coconas represent a great reservoir of characters to be utilized in improvement of the crop, to enhance nematode resistance, reduce seed count, and increase sweetness.

Climate

In Florida and Trinidad, the cocona is grown at near sea-level. In Colombia, it is grown from sea-level to an elevation of 2,000 ft (610 m), while elsewhere in South America it thrives at altitudes up to 3,000 or 4,000 ft (910-1,200 m). Unlike the naranjilla, the plant needs full sun.

Soil

The cocona grows in soil of medium fertility on Peruvian mountain slopes; in Amazonian Brazil, on latisols or pure sand. In Puerto Rico, it has done well on clay; in southern Florida on scarified limestone. Good drainage is essential.

Pollination

The cocona is self-fertile. Bees are always visiting the flowers and carrying pollen, and natural crosses are common. Fruits mature about 8 weeks after pollination.

Propagation

There are from 800 to 2,000 seeds in each fruit. New plants spring up voluntarily from seeds clinging to discarded rinds in full sun on disturbed ground in northern South America. For planting, seeds extracted from the ripe fruits are placed in the shade for 2 days to ferment a little and break down the mucilage. Then they are washed and dried briefly out of the direct sun, and finally dusted with fungicide—2¼ g per lb (5 g per kg) of seeds. The seeds are planted 3/8 in (1 cm) deep in nursery beds in rows 8 in (20 cm) apart; or in polyethylene bags containing a 50-50 mixture of potting soil and sand. In each bag, or each hole, one puts 4 to 5 seeds expecting the emergence of 1 or 2 sturdy seedlings. Germination time varies from 15 to 40 days.

Vegetative propagation is possible, in order to perpetuate a particular cultivar. Air-layers and cuttings of mature wood have been rooted successfully.

Culture

Seedlings are transplanted to the field when 8 to 12 in (20-30 cm) high and they are spaced 5 to 7 ft (1.5-2.5 m) apart each way, depending on the fertility of the soil. Flowering commences 2 to 3 months after transplanting. The plants usually begin fruiting in 6 to 7 months from seed and will continue fruiting for several months.

A fertilizer formula of 10-8-10 NPK is applied 6 times during the year at the rate of 1.8 to 2.5 oz (50-70 g) per plant. If the soil is low in phosphorus, the formula should be 10-20-10. Productivity has been greatly enhanced in field trials at Manaus on pure sand, by applying organic fertilizer—104 tons per acre (250 tons/ha), with the addition of appropriate amounts of triple super-phosphate, urea and chlorate of potassium.

Yield

Average annual yield in Colombia is 22 to 44 lbs (10-20 kg) per plant. In Costa Rica, cocona plants have yielded 40 to 60 lbs (18-27 kg) of fruit. In variety trials at Manaus, productivity per plant varied from 5½ to 30 lbs (2.5-14 kg). An unfertilized plantation may provide 20 to 30 fruits per plant—12 tons per acre (29 tons/ha). With a high-yielding selection and a well-fertilized field, one can realize up to 136 fruits per plant—61 tons per acre (146 tons/ha). The fresh fruit keeps well for 5 to 10 days at normal temperature.

Processing studies have shown that 22 lbs (10 kg) of fruit will yield about 6½ pints (3 liters) of preserved flesh and 3¼ lbs (1½ liters) of jelly, or 2 gallons (7½ liters) of juice. A plantation providing 30 tons fruit per acre (70 tons/ha) will yield 5,548 gallons preserved flesh and 2,774 gallons of jelly, or 13,738 gallons (52,000 liters) of juice.

Pests and Diseases

The cocona is prone to attack by rootknot nematodes (*Meloidogyne* sp.). In 1973, it was decided, after test plantings at the Universidad Central de Venezuela, that it was impossible to cultivate the cocona commercially in that country because of its susceptibility to nematodes, but the experimenters at Manaus believe that they have demonstrated that selection for nematode-resistance and soil-enrichment can give the farmer good returns.

In Puerto Rico, a mealybug, *Pseudococcus* sp., infests the new growth but causes little harm. However, *Psara periosalis* has been very damaging in the fall. Cutworms and leaf-eating insects require control. In Brazil, a hemipterous bug of the family Tingidae colonizes the underside of the leaves, causing them to discolor and fall. A fungal disease (*Sclerotium* sp.) has been identified with wilting.

Food Uses

The ripe fruit is peeled and eaten out-of-hand by South American Indians. More sophisticated people use the fruit in salads, cook it with fish and also in meat stews. Sweetened, it is used to make sauce and pie-filling. It is prized for making jam, marmalade, paste, and jelly, and is sometimes pickled or candied. It is often processed as a nectar or juice which, sweetened with sugar, is a popular cold beverage. Dr. Victor Patiño of Cali, Colombia, states that a 50-50 cocona-naranjilla juice mixture is superior to naranjilla alone.

In Brazil, the leaves are cooked and eaten as well.

Food Value Per 100 g of Edible Portion*	
Protein	0.6 g
Fiber	0.4 g
Carbohydrates	5.7 g
Calcium	12 mg
Phosphorus	14 mg
Iron	0.6 mg
Carotene	140 mcg
Thiamine	25 mcg
Riboflavin	-
Niacin	500 mcg

*Analyses made in Brazil.

The fruit has a high level of citric acid, about 0.8%. Venezuelan studies reveal 142 mg tannin.

Toxicity

The cocona is utilized by Indians of eastern Peru to rid the head of lice.

Cape Gooseberry

The genus *Physalis*, of the family Solanaceae, includes annual and perennial herbs bearing globular fruits, each enclosed in a bladderlike husk which becomes papery on maturity. Of the more than 70 species, only a very few are of economic value. One is the strawberry tomato, husk tomato or ground cherry, *P. pruinosa* L., grown for its small yellow fruits used for sauce, pies and preserves in mild-temperate climates. Though more popular with former generations than at present, it is still offered by seedsmen. Various species of *Physalis* have been subject to much confusion in literature and in the trade. A species which bears a superior fruit and has become widely known is the cape gooseberry, *P. peruviana* L. (*P. edulis* Sims). It has many colloquial names in Latin America: *capuli, aguaymanto,* *tomate sylvestre*, or *uchuba*, in Peru; *capuli* or *motojobobo embolsado* in Bolivia; *uvilla* in Ecuador; *uvilla, uchuva, vejigón* or *guchavo* in Colombia; *topotopo*, or *chuchuva* in Venezuela; *capuli, amor en bolsa*, or *bolsa de amor*, in Chile; *cereza del Peru* in Mexico. It is called cape gooseberry, golden berry, *pompelmoes* or *apelliefie* in South Africa; *alkekengi* or *coqueret* in Gabon; *lobo-lobohan* in the Philippines; *teparee, tiparee, makowi,* etc., in India; cape gooseberry or *poha* in Hawaii.

Description

This herbaceous or soft-wooded, perennial plant usually reaches 2 to 3 ft (1.6-0.9 m) in height but occasionally may attain 6 ft (1.8) m. It has ribbed, often purplish, spreading

Fig. 114: The golden cape gooseberry (*Physalis peruviana*) keeps well and makes excellent preserves. The canned fruits have been exported from South Africa and the jam from England.

branches, and nearly opposite, velvety, heart-shaped, pointed, randomly-toothed leaves 2 3/8 to 6 in (6-15 cm) long and 1 1/2 to 4 in (4-10 cm) wide, and, in the leaf axils, bell-shaped, nodding flowers to 3/4 in (2 cm) wide, yellow with 5 dark purple-brown spots in the throat, and cupped by a purplish-green, hairy, 5-pointed calyx. After the flower falls, the calyx expands, ultimately forming a straw-colored husk much larger than the fruit it encloses. The berry is globose, 1/2 to 3/4 in (1.25-2 cm) wide, with smooth, glossy, orange-yellow skin and juicy pulp containing numerous very small yellowish seeds. When fully ripe, the fruit is sweet but with a pleasing grape-like tang. The husk is bitter and inedible.

Origin and Distribution

Reportedly native to Peru and Chile, where the fruits are casually eaten and occasionally sold in markets but the plant is still not an important crop, it has been widely introduced into cultivation in other tropical, subtropical and even temperate areas. It is said to succeed wherever tomatoes can be grown. The plant was grown by early settlers at the Cape of Good Hope before 1807. In South Africa it is commercially cultivated and common as an escape and the jam and canned whole fruits are staple commodities, often exported. It is cultivated and naturalized on a small scale in Gabon and other parts of Central Africa.

Soon after its adoption in the Cape of Good Hope it was carried to Australia and there acquired its common English name. It was one of the few fresh fruits of the early settlers in New South Wales. There it has long been grown on a large scale and is abundantly naturalized, as it is also in Queensland, Victoria, South Australia, Western Australia and Northern Tasmania. It was welcomed in New Zealand where it is said that "the housewife is sometimes embarrassed by the quantity of berries [cape gooseberries] in the garden," and government agencies actively promote increased culinary use.

In China, India and Malaya, the cape gooseberry is commonly grown but on a lesser scale. In India, it is often interplanted with vegetables. It is naturalized on the island of Luzon in the Philippines. Seeds were taken to Hawaii before 1825 and the plant is naturalized on all the islands at medium and somewhat higher elevations. It was at one time extensively cultivated in Hawaii. By 1966, commercial culture had nearly disappeared and processors had to buy the fruit from backyard growers at high prices. It is widespread as an exotic weed in the South Sea Islands but not seriously cultivated. The first seeds were planted in Israel in 1933. The plants grew and bore very well in cultivation and soon spread as escapes, but the fruit did not appeal to consumers, either fresh or preserved, and promotional efforts ceased.

In England, the cape gooseberry was first reported in 1774. Since that time, it has been grown there in a small way in home gardens, and after World War II was canned commercially to a limited extent. Despite this background, early in 1952, the Stanford Nursery, of Sussex, announced the "Cape Gooseberry, the wonderful new fruit, especially developed in Britain by Richard I. Cahn." Concurrently, jars of cape gooseberry jam from England appeared in South Florida markets and the product was found to be attractive and delicious. It is surprising that this useful little fruit has received so little attention in the United States in view of its having been reported on with enthusiasm by the late Dr. David Fairchild in his well-loved book, *The World Was My Garden.* He there tells of its fruiting "enormously" in the garden of his home, "In The Woods", in Maryland, and of the cook's putting up over a hundred jars of what he called "Inca Conserve" which "met with universal favor." It is also remarkable that it is so little known in the Caribbean islands, though naturalized plants were growing profusely along roadsides in the Blue Mountains of Jamaica before 1913.

With a view to encouraging cape gooseberry culture in Florida, the Bahamas, and the West Indies, seeds have been repeatedly purchased from the Stanford Nursery and distributed for trial. Good crops have been obtained. Nevertheless there was no incentive to make further plantings.

Pollination

In England, growers shake the flowers gently in summer to improve distribution of the pollen, or they will give the plants a very light spraying with water.

Climate

The cape gooseberry is an annual in temperate regions and a perennial in the tropics. In Venezuela, it grows wild in the Andes and the coastal range between 2,500 and 10,000 ft (800-3,000 m). It grows wild in Hawaii at 1,000 to 8,000 ft (300-2,400 m). In northern India, it is not possible to cultivate it above 4,000 ft (1,200 m), but in South India it thrives up to 6,000 ft (1,800 m).

In England, the plants have been undamaged by 3 degrees of frost. In South Africa, plants have been killed to the ground and failed to recover after a temperature drop to 30.5°F (-0.75°C).

The plant needs full sun but protection from strong winds; plenty of rain throughout its growing season, very little when the fruits are maturing.

Soil

The cape gooseberry will grow in any well-drained soil but does best on sandy to gravelly loam. On highly fertile alluvial soil, there is much vegetative growth and the fruits fail to color properly. Very good crops are obtained on rather poor sandy ground. Where drainage is a problem, the plantings should be on gentle slopes or the rows should be mounded. The plants become dormant in drought.

Propagation

The plant is widely grown from seed. There are 5,000 to 8,000 seeds to the ounce (28 g) and, since germination rate is low, this amount is needed to raise enough plants for an acre—2½ oz (70 g) for a hectare. In India, the seeds are mixed with wood ash or pulverized soil for uniform sowing.

Sometimes propagation is done by means of 1-year-old stem cuttings treated with hormones to promote rooting, and 37.7% success has been achieved. The plants thus grown flower early and yield well but are less vigorous than seedlings. Air-layering is also successful but not often practiced.

Culture

It is necessary to determine the time of planting for each area. In India, seeds are broadcast from March through May. In Hong Kong, planting in seedbeds is done in September/October and again in March/April. In the Bahamas the first seeds planted in late summer of 1952 produced healthy plants and a continuous crop of fruits for 3 months during the following winter. Additional seeds procured from England were planted in April of 1953. The plants started to blossom in mid-July and from September on continued to flower and set fruit, although no fruits remained on the plants to maturity until the cooler months of winter when a good yield was obtained. Seeds were again planted the following November. Thirteen weeks later, the first fruits were ripening, and by mid-May of the following year a heavy crop was harvested. In late June, the plants were still growing and flowering profusely but only a few fruits were being set and these failed to develop to maturity. This condition continued into September, by which time some of the more robust plants had reached 6 ft (1.8 m) in height with much lateral growth.

In Jamaica, the initial planting of cape gooseberries in late January of 1954 made slow growth until June when development accelerated. By mid-August the plants had reached 15 in (37.5 cm) in height with much lateral growth, and were flowering and setting fruit. It would appear that the heat of summer is unfavorable for fruit development and, therefore, the best time to plant the cape gooseberry is in the fall so that fruit can be set during the cooler weather and harvested in late spring or early summer. In California, the plants do not fruit heavily until the second year unless started early in greenhouses.

Some growers have kept plants in production for as long as 4 years by cutting back after each harvest, but these plants have been found more susceptible to pests and diseases.

In India, plants 6 to 8 in (15-20 cm) high are set out 18 in (45 cm) apart in rows 3 ft (0.9 m) apart. Farmers in South Africa space the plants 2 to 3 ft (0.6-0.9 m) apart in rows 4 to 6 ft (1.2-1.8 m) or even 8 ft (2.4 m) apart in very rich soil. They apply 200 to 400 lbs (90-180 kg) of complete fertilizer per acre (approx. = kg/ha) on sandy loam. Foliar spraying of 1% potassium chloride solution before and just after blooming enhances fruit quality.

In dry seasons, irrigation is necessary to keep the cape gooseberry plant in production.

Season

In parts of India, the fruits ripen in February, but, in the South, the main crop extends from January to May. In Central and southern Africa, the crop extends from the beginning of April to the end of June. In England, plants from seeds sown in spring begin to fruit in August and continue until there is a strong frost.

Harvesting and Yield

In rainy or dewy weather, the fruit is not picked until the plants are dry. Berries that are already wet need to be lightly dried in the sun. The fruits are usually picked from the plants by hand every 2 to 3 weeks, although some growers prefer to shake the plants and gather the fallen fruits from the ground in order to obtain those of more uniform maturity. At the peak of the season, a worker can pick 2½ bushels (90 liters) a day, but at the beginning and end of the season, when the crop is light, only ½ bushel (18 liters).

A single plant may yield 300 fruits. Seedlings set 1,800 to 2,150 to the acre (228-900/ha) yield approximately 3,000 lbs of fruit per acre (approx. = kg/ha). The fruits are usually dehusked before delivery to markets or processors. Manual workers can produce only 10 to 12 lbs. (4.5-5.5 kg) of husked fruits per hour. Therefore, a mechanical husker, 4 to 5 times more efficient, has been designed at the University of Hawaii.

Keeping Quality

Cape gooseberries are long-lasting. The fresh fruits can be stored in a sealed container and kept in a dry atmosphere for several months. They will still be in good condition. If the fresh fruits are to be shipped, it is best to leave the husk on for protection.

Fig. 115: The cape gooseberry is a useful small fruit crop for the home garden; is labor-intensive in commercial plantings.

Pests and Diseases

In South Africa, the most important of the many insect pests that attack the cape gooseberry are cutworms, in seedbeds; red spider after plants have been established in the field; the potato tuber moth if the cape gooseberry is in the vicinity of potato fields. Hares damage young plants and birds (francolins) devour the fruits if not repelled. In India, mites may cause defoliation. In Jamaica, the leaves were suddenly riddled by what were apparently flea beetles of the family Chrysomelidae. In the Bahamas, whitefly attacks on the very young plants and flea beetles on the flowering plants required control.

In South Africa, the most troublesome diseases are powdery mildew and soft brown scale. The plants are prone to root rots and viruses if on poorly-drained soil or if carried over to a second year. Therefore, farmers favor biennial plantings. Bacterial leaf spot (*Xanthomonas* spp.) occurs in Queensland. A strain of tobacco mosaic may affect plants in India.

Food Uses

In addition to being canned whole and preserved as jam, the cape gooseberry is made into sauce, used in pies, puddings, chutneys and ice cream, and eaten fresh in fruit salads and fruit cocktails. In Colombia, the fruits are stewed with honey and eaten as dessert. The British use the husk as a handle for dipping the fruit in icing.

Toxicity

Unripe fruits are poisonous. The plant is believed to have caused illness and death in cattle in Australia.

Food Value Per 100 g of Edible Portion*	
Moisture	78.9 g
Protein	0.054 g
Fat	0.16 g
Fiber	4.9 g
Ash	1.01 g
Calcium	8.0 mg
Phosphorus	55.3 mg
Iron	1.23 mg
Carotene	1.613 mg
Thiamine	0.101 mg
Riboflavin	0.032 mg
Niacin	1.73 mg
Ascorbic Acid	43.0 mg

*According to analyses of husked fruits made in Ecuador.

The ripe fruits are considered a good source of Vitamin P and are rich in pectin.

Other Uses

Fruits: In the 18th Century, the fruits were perfumed and worn for adornment by native women in Peru.

Medicinal Uses: In Colombia, the leaf decoction is taken as a diuretic and antiasthmatic. In South Africa, the heated leaves are applied as poultices on inflammations and the Zulus administer the leaf infusion as an enema to relieve abdominal ailments in children.

Indian chemists have isolated from the leaves a minor steroidal constituent, *physalolactone* C.

Mexican Husk Tomato

Somewhat suggesting a miniature tomato, the Mexican, or Mayan, husk tomato, *Physalis ixocarpa* Brot. (syn. *P. aequata* Jacq.), is also called *tomate de cáscara, tomate verde, tomate Mexicano, tomate de fresadilla, tomate de culebra, tomatillo, miltomate* and *farolito.*

Description

The plant, which is a semi-woody annual, may attain a height of 4 to 5 ft (1.2-1.5 m), but is often prostrate and spreading. Its branches and leaves are smooth, not downy. The leaves are ovate, pointed at apex, wedge-shaped at base, sometimes wavy-margined; 2½ in (6.25 cm) long, 1¼ in (3.2 cm) wide. Borne singly in the leaf axils, the flowers, clasped halfway by a 5-toothed, green calyx, are ½ to ¾ in (1.25-2 cm) long and wide; yellow with dark-brown spots in the throat. As the fruit develops, the calyx enlarges to more or less enclose it and finally becomes straw-colored and papery. It is so tight-fitting that it often bursts. The berry is slightly oblate, 1 to 2½ in (2.5-6.25 cm) wide. When ripe, its thin skin may be yellow, purple, or, more rarely reddish, or still green. The flesh is pale-yellow, crisp or soft, and acid, subacid, sweet, or insipid, and contains many tiny seeds.

Origin and Distribution

The Mexican husk tomato was a prominent staple in Aztec and Mayan economy. The plant abounds in Mexico and the highlands of Guatemala and the fruits are commonly seen in native markets. Nevertheless, this species has not been as widely distributed abroad as the Cape gooseberry. It was introduced into India in the 1950's and is cultivated in the northwest desert region of Rajasthan. In Queensland, Australia, and in South Africa it has fruited prolifically. There is some commercial cultivation in Pietersburg, South Africa, for processing. It was too-successfully introduced into East Africa, for, in 1967, it was reported to be the most important weed of agricultural fields in the highlands of Kenya.

Fig. 116: The Mexican husk tomato, (*Physalis ixocarpa*), page-green, yellow, purple or reddish when ripe, is a staple food in Mexico and Guatemala and commonly marketed.

Before 1863, it was thoroughly naturalized and commonly growing in abundance in the far west of the United States. Mr. Sun Jue cultivated some 20 acres (8 ha) of Mexican husk tomatoes near Los Angeles, California, from 1930 to about 1939, supplying the fruits to Mexican and Italian markets. In 1945, the American Fruit Grower publicized this species under the concocted name "Jamberry", as a new fruit introduced by scientists at Iowa State College. Dr. I.E. Melhus, Director of the Iowa State College-Guatemala Tropical Research Center reported in 1953 that, as a result of 6 years' testing of hundreds of selections, only a few were found suitable for the American Midwest. They were then sending out a strain to which they had given the name "Mayan husk tomato"; 4,000 packets of seed were distributed in Iowa and adjoining states. Sampling data from 200 people that grew the plant showed that over 60% were successful and liked the fruit. Later, that strain was offered by the Earl May Seed Company of Shenandoah, Iowa. An apparently independent introduction was made by Glecklers, Seedsmen, of Metamora, Ohio, and first offered by them as "Jumbo husk tomato" in 1952. Seeds obtained from these sources and from fruits purchased in the Mexican markets were given by the writer to experimenters in the Bahamas, Puerto Rico, Jamaica and Florida.

Plantings were successful in the Bahamas and Puerto Rico but did not arouse enough interest to cause further cultivation. Florida and Jamaica trials were failures. In recent years, test plantings have been made in Trinidad and Taiwan, and plants have fruited well in greenhouse culture in England. The principal areas of production in Mexico are the States of Morelos and Hidalgo. The former has about 32,000 acres (13,000 ha) with a total production of 101,366 tons.

Varieties

There is great variation, not only in color and flavor of the numerous strains of Mexican husk tomato. Some require long days and others short days. Some mature early, others late. The husk may be long or short. The flesh may be soft and spongy or firm and crisp. A large number of selections has been made at the Campo Agrícola Experimental de Zacatepec, in the State of Morelos, Mexico. The most promising, **'Rendidora',** is more erect than the common type, the fruit is large, green, ripens 15 days earlier than others and gives 80% greater yield. Horticulturists at the Universidad de San Simón in Bolivia have long maintained a collection of various types received from Mexico.

The "Mayan husk tomato" selection at Iowa is semiprostrate, vigorous, branching at a height of 4 to 6 in (10-15 cm); the stems are pale-green, smooth and succulent when young. The fruit is round, yellow, with light-yellow, firm flesh and mild-acid flavor. According to Dr. Margaret Menzel, an authority on the genus, the so-called *Physalis macrocarpa*, or "Golden Nugget Cape Gooseberry", offered by seedsmen in Australia, is really a yellow-fruited form of the Mexican husk tomato, *P. ixocarpa*.

Pollination

The Mexican husk tomato is highly self-incompatible. When the flowering plants are bagged, no fruits are set. K.K. Pandey, while at the University of Ohio, studied this problem. He reported that only a few seedlings in a group produce rare fruits by natural-selfing and such fruits usually contain no seeds or only a small number. An occasional fruit may have 100 or more.

Climate and Soil

This species is not ultra-tropical but tropical and, like the tomato, is grown in summer in temperate regions. The plant needs full sun. It will grow in any soil suitable for tomatoes but not in wet situations.

Propagation

The Mexican husk tomato is usually raised from seed and it takes about 2¼ oz (60 g) to plant an acre; 5¼ oz (150 g) to plant a hectare. In Puerto Rico, seeds saved from the first crop and kept for 6 months without refrigeration were planted and 80% germinated.

Cuttings should root easily. Heavy rains cause the plants to bend down to the ground and it has been observed that tips that touch the soil take root and the new shoots grow vigorously.

Culture

Ideal spacing for cv. 'Rendidora' is 16 in (40 cm) between plants and 4 ft (1.25 m) between rows.

From 4 to 6 seeds are planted ½ in (1.25 cm) deep in hills 2 ft (60 cm) apart in rows 5 ft (1.5 m) apart. When 4 to 5

in (10-12.5 cm) high, the seedlings are thinned to 1 plant per hill. In the midwestern United States, seedlings are raised in greenhouses and are transplanted when about 3 weeks old as soon as all likelihood of spring frosts is past. They will begin to bear 6 to 18 weeks later and continue for about 1½ months.

In Bahamian trials, seeds were planted in mid-April. By mid-September, the plants were fruiting heavily. They reseeded themselves and a healthy clump of "volunteers" sprang up on the site. In Puerto Rico, seeds planted at Mayaguez produced an abundant crop in the winter of 1953-54. The plot was fertilized at the rate of 2 oz (56 g) per plant, side dressing, of 9-8-8 fertilizer. The plants were staked and tied twice and grew to a height of 5 ft (1.5 m).

Season

Wild plants in Mexico flower from June to October. In the midwestern United States, flowering takes place in mid-June and fruits start to ripen in late July and fruiting continues until fall frosts. The plants bear during the summer months in South Africa; in northern India, both summer and winter.

Harvesting

With the Mexican husk tomato, falling of fruits before ripening is not uncommon, and, according to Dr. Melhus, they may be allowed to remain on the ground until fully colored. Collecting must be done every day. The green-skinned variety grown commercially by Mr. Jue was harvested as soon as it burst its husk, and the crop was then kept on hand 2 to 4 weeks for the husk to dry before the fruit was considered acceptable to the consumer. If left too long on the plant, there is much loss of flavor.

Yield

Individual plants may produce 64 to 200 fruits in a season. In test plantings at Ames, Iowa, the fruit yield averaged 2½ lbs (1.1 kg) per plant; equal to approximately 9 tons per acre (20.2 MT/ha). In Mexico and India, yields of 7.5 to 10 tons per acre (17-22.5 MT/ha) have been reported.

Keeping Quality

The unhusked fresh fruits can be stored in single layers in a cool, dry atmosphere for several months. Mexican and Central American people may pull up the entire plant with fruits attached and hang it upside-down in a dry place until the fruits are needed.

Pests and Diseases

The Mexican husk tomato is subject to few pests and diseases. In Mexico, the main pest is the so-called *mosquita blanca* (see below). The larvae of *Heliothis virescens* attack the fruits. It has been found that various species of *Trichogramma* parasitize the eggs, found mainly on the underside of the leaves, though only in certain localities at certain seasons. In India, fruit and stem borers are troublesome during the rainy season but not in the winter.

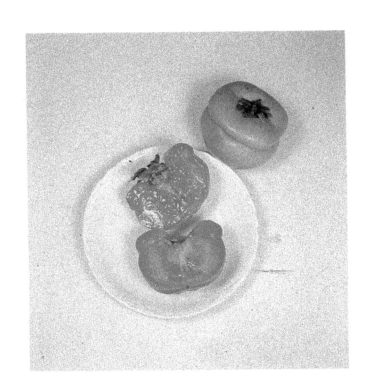

Plate LIX
JAPANESE PERSIMMON
Diospyros kaki
'Tamopan'

Plate LX
JAPANESE PERSIMMON
Diospyros kaki
'Tanenashi'

Plate LXI
BLACK SAPOTE
Diospyros digyna

Plate LXII
MABOLO
Diospyros blancoi

Plate LXIII
NARANJILLA
Solanum quitoense

Plate LXIV
COCONA
Solanum sessiliflorum

Plate LXV
TREE TOMATO
Cyphomandra betacea

Plate LXVI
TREE TOMATO
Cyphomandra betacea

Plate LXVII
GENIPAP
Genipa americana

Plate LXVIII
CASSABANANA
Sicana odorifera

No insects attacked the plant in Puerto Rico. The two trials in Florida were at first promising, the plants flowering and setting fruit satisfactorily. However, as the fruits began to mature, they were attacked within the husk by a species of cutworm and only a few mature fruits were harvested. In Jamaica, seeds planted in late January produced vigorous and precocious plants which flowered when only 4 in (10 cm) high. Fruit-setting began in May and a high yield was expected but nearly all of the fruits were damaged by caterpillars before reaching maturity.

In Puerto Rico, no diseases were evident. In the Bahamas, only a slight incidence of leaf spot was observed. In Mexico, the husk tomato and the common tomato are both subject to a disease called *chino* or *chahuixtle* which occurs in irrigated plantings in Morelos. It is transmitted by the *mosquita blanca, Trialeurodes vaporariorum.*

Food Uses

This species, in contrast with the cape gooseberry, is used more largely as a vegetable than as a dessert fruit, though it is often consumed ripe, raw, out-of-hand. In Mexico, it is generally made into a sauce, *salsa verde,* for meats, alone or together with green chili peppers. Suggestions for use distributed by Iowa State College include recipes for stewing, frying, baking, cooking with chopped meat, making into soup, marmalade and dessert sauce. The fruit is an excellent addition to salads and curries. It has been utilized commercially for jam in Australia but the product is there considered inferior to that made from the cape gooseberry. The fruits, canned whole in Mexico, are sold domestically and in the western United States.

Food Value Per 100 g of Edible Portion*	
Moisture	90.4-91.7 g
Protein	0.171-0.7 g
Fat	0.6 g
Carbohydrates	5.8 g
Fiber	0.6-1.7 g
Ash	0.6-0.69 g
Calcium	6.3-10.9 mg
Magnesium	23 mg
Phosphorus	21.9-40 mg
Phytin Phosphorus	7 mg
Iron	0.57-1.4 mg
Ionisable Iron	1.0 mg
Sodium	0.4 mg
Potassium	243 mg
Copper	0.09 mg
Sulfur	27 mg
Chloride	14 mg
Carotene (Vitamin A)	80 I.U. or 0.061-0.074 mg
Thiamine	0.054-0.106 mg
Riboflavin	0.023-0.057 mg
Niacin	2.1-2.7 mg
Ascorbic Acid	2-4.8 mg

*According to analyses of the husked fruit made in Guatemala and India.

Medicinal Use

It is said in Mexico that a decoction of the calyces will cure diabetes.

Tree Tomato (Plates LXV and LXVI)

The tree tomato, *Cyphomandra betacea* Sendt. (*C. hartwegi* Sendt.; *Solanum betaceum* Cav.) is the best-known of about 30 species of *Cyphomandra* (family Solanaceae). Among its various regional names are: *tomate, tomate extranjero, tomate de arbol, tomate granadilla, granadilla, pix,* and *caxlan pix* (Guatemala); *tomate de palo* (Honduras); *arvore do tomate, tomate de arvore* (Brazil); *lima tomate, tomate de monte, sima* (Bolivia); *pepino de arbol* (Colombia); *tomate dulce* (Ecuador); *tomate cimarron* (Costa Rica); and *tomate francés* (Venezuela, Brazil). In 1970, or shortly before, the construed name "tamarillo" was adopted in New Zealand and has become the standard commercial designation for the fruit.

Description

The plant is a small, half-woody, attractive, fast-growing, brittle tree; shallow-rooted; reaching 10 to 18 ft (3-5.5 m) in height; rarely as much as 25 ft (7.5 m). The leaves are muskily odorous, evergreen, alternate, more or less heart-shaped at the base, ovate, pointed at the apex, 4 to 13½ in (10-35 cm) long and 1½ to 4¾ in (4-12 cm) broad, thin, softly hairy, with conspicuous coarse veins. Borne in small, loose clusters near the branch tips, the fragrant flowers, ½ to ¾ in (1.25-2 cm) wide, have 5 pale-pink or lavender, pointed lobes, 5 prominent yellow stamens, and green-purple calyx. The long-stalked, pendent fruit, borne singly, or in clusters of 3 to 12, is smooth, egg-shaped but pointed at both ends and capped with the persistent conical calyx. In size it ranges from 2 to 4 in (5-10 cm) long and 1½ to 2 in (4-5 cm) in width. Skin color may be solid deep-purple, blood-red, orange or yellow, or red-and-yellow, and may have faint dark, longitudinal stripes. Flesh color varies accordingly from orange-red or orange to yellow or cream-yellow. While the skin is somewhat tough and unpleasant in flavor, the outer layer of flesh is slightly firm, succulent and bland, and the pulp surrounding the seeds in the two lengthwise compartments is soft, juicy, subacid to sweet; it is black in dark-purple and red fruits, yellow in yellow and orange fruits. The seeds are thin, nearly flat, circular, larger and harder than those of the true tomato and distinctly bitter. The fruit has a slightly resinous aroma and the flavor suggests a mild or underripe tomato with a faintly resinous aftertaste.

Origin and Distribution

Although its place of origin is not certain, the tree tomato is generally believed to be native to the Andes of Peru and probably also Chile, Ecuador and Bolivia where it is extensively grown, as it is also in Argentina, Brazil and Colombia. It is cultivated and naturalized in Venezuela and grown in the highlands of Costa Rica, Guatemala, Jamaica, Puerto Rico and Haiti.

It must have been carried at an early date to East Africa, Asia and the East Indies, as it is well established in the Nilgiri heights and the hills of Assam in southern India, and in the mountains of Malaya, and was popular in Ceylon and the Dutch East Indies before 1903. It has been grown in Queensland, Australia, in home gardens, for many years and is a practical crop in the highlands of the Australian part of New Guinea.

D. Hay & Sons, nurserymen, introduced the tree tomato into New Zealand in 1891 and commercial growing on a small scale began about 1920. Shortages of tropical fruits in World War II justified an increased level of production. A promotional campaign was launched in 1961; window banners and 100,000 recipe leaflets were distributed. This small industry prospered until 1967 when annual production reached a peak of 2,000 tons. There was a heavy loss of trees at Kerikeri in 1968. Replanting took place there and at the Bay of Plenty and cultivation of this crop continues to expand. In 1970, there were 209,110 trees on 476 acres (130 ha) in New Zealand. Shipment of the fresh fruits to Australia has not been very successful and the surplus crop is being delivered to processors for the making of preserves.

The United States Department of Agriculture received seeds from Argentina in 1913; from Sumatra and Ceylon in 1926. The plant was fruiting at the United States Department of Agriculture's Plant Introduction Station at Chico, California, in 1915. It is still grown casually in California and occasionally in Florida. It is frequently advertised and sold throughout the United States for growing indoors in pots as a curiosity. It fruits satisfactorily in northern greenhouses.

Varieties

There are apparently no named cultivars, but there are local preferences according to fruit color. Red fruits are chosen for the fresh fruit markets because of their appealing color. The dark-red strain (called "black") now leading in commercial plantings in New Zealand was obtained by selection around 1920 as a variation from the yellow and purple types grown up to that time. It was propagated and reselection thereafter resulted in this large, higher quality, red variety.

Yellow fruits are considered best for preserving because of their superior flavor.

Climate

The tree tomato is not tropical but subtropical. It flourishes between 5,000 and 10,000 ft (1,525-3,050 m) in Ecuador; between 1,000 and 3,000 ft (305-915 m) in Puerto Rico; 1,000 to 7,500 ft (305-2,288 m) in India. In Haiti it grows and fruits to perfection at 6,000 ft (1,830 m). In cooler climates, it succeeds at lower elevations. It does best where the temperature remains above 50°F (10°C). Frost at 28°F (-2.2°C) kills the small branches and foliage of mature trees but not the largest branches and main stem. The tree will recover if such frosts are not prolonged or frequent. However, seedlings and cuttings are readily killed by frost during their first year.

Protection from wind is necessary as the tree is shallow-rooted and easily blown over. It is also brittle and its branches are easily broken by gusts, especially when laden with fruit. It is suggested that windbreaks be established for each ½ acre (⅕ ha) before setting out the plantation in order to protect the young plants. Hedges of *Albizia lophantha* Benth. and of *Hakea saligna* R. Br., kept trimmed and narrow, are popular in the North Auckland area of New Zealand.

Soil

The tree tomato cannot tolerate tightly compacted soil with low oxygen content. It requires fertile, light soil. It grows well on deep lateritic soil in Haiti. Perfect drainage is necessary. Water standing for even a few days may kill the tree.

Propagation

Seeds or cuttings may be used for propagation. Seeds produce a high-branched, erect tree, ideal for sheltered locations. Cuttings develop into a shorter, bushy plant with low-lying branches, suitable for exposed, windy sites. The tree does not always come true from seed, but is most likely to if one is careful to take seed from red fruits with black seed pulp or yellow fruits with yellow seed pulp.

In Brazil, seeds for planting are first washed, dried in the shade, and then placed in a freezer for 24 hours to accelerate germination. They are then planted in boxes of rich soil—12 in (30 cm) between plants and 24 in (60 cm) between rows—and virtually 100% will germinate in 4 to 6 days.

Culture

The seedlings are set out in the field when 2 to 2¾ in (5-7 cm) high, spaced 32 in (80 cm) apart in rows 6½ ft (2 m) apart. In New Zealand, the trees are set 8 to 10 ft (2.5-3 m) apart in paired rows 8 ft (2.5 m) apart with 14 ft (4.25 m) between each pair. If the soil is very rich, 9 ft (2.75 m) is allowed between the rows and 16 ft (5 m) between the pairs. Closer planting is recommended in windy, unprotected locations—5 to 6 ft (1.5-1.8 m) between the plants and 8 to 10 ft (2.5-3 m) between the rows, and the trees may be staked to prevent swaying and disturbing the roots. In India, the trees are set out in pits 4 to 5 ft (1.2-1.5 m) apart.

Cuttings should be of 1- to 2-year-old wood ⅜ to 1 in (10-25 mm) thick and 18 to 30 in (45-75 cm) long; the leaves are removed and the base cut square below a node. They can be planted directly in the field and, while precocious, should not be permitted to fruit in the first year.

Recommended fertilizer application is 0.5 to 2.2 lbs (0.25-1.0 kg) per tree of NPK 5:6:6, half in early spring and half in midsummer. In the 5th or 6th year, the grower is advised to give a special feeding of 2 parts superphosphate, 1½ parts nitrate of soda, 1 part sulphite of potash, in late winter or early spring, at the rate of 2 to 3 lbs (1-1.5 kg) per plant—approximately 10 to 16 cwt per acre, or 100 kg per hectare.

Because of the shallow root system, deep cultivation is not possible, but light cultivation is desirable to eliminate weeds until there is sufficient vegetative growth to shade them out.

Seedling trees are pruned back the first year after planting to a height of 3 or 4 ft (0.9-1.2 m) to encourage branching. Annual pruning thereafter is advisable to eliminate branches that have already fruited and induce ample new shoots close to the main branches, inasmuch as fruit is produced on new growth. Otherwise, the tree will develop a broad top with fruits only on the outer fringe. And wide-spreading branches are subject to wind damage. Pruning facilitates harvesting and, if timed appropriately, can extend the total fruiting period. Early spring pruning of some of the owners' trees brings about early maturity; fall pruning of other trees delays fruit maturity to the following fall.

Irrigation

The tree tomato cannot tolerate prolonged drought and must have an ample water supply during extremely dry periods. A mulch is very beneficial in conserving moisture at such times.

Pollination

Tree tomato flowers are normally self-pollinating. If wind is completely cut off so as not to stir the branches, this may adversely affect pollination unless there are bees to transfer the pollen. Unpollinated flowers will drop prematurely.

Cropping and Yield

The tree usually begins to bear when 1½ to 2 years old and continues to be productive for 5 or 6 years. If then adequately nourished, it may keep on fruiting for 11 to 12 years. In Brazil, each tree is expected to yield 44 to 66 lbs (20-30 kg) of fruit annually.

The crop does not ripen simultaneously and several pickings are necessary. The fruits are clipped, leaving about ½ in (12.5 cm) of stem attached. They are collected in bags worn by the harvesters.

In New Zealand, the fruits are sorted by size—small, medium and large—and packed in paper-lined wooden boxes for marketing. Because of its firm flesh and tough skin, the fruit can be shipped long distances without bruising.

However, it deteriorates rather rapidly under ordinary storage conditions.

Pests and Diseases

The tree tomato is generally regarded as fairly pest-resistant. A looper caterpillar makes large holes in the leaves of young plants in the nursery but causes little damage to trees in the field. Occasionally the plants are attacked by the green aphis.

In South America and the Caribbean, the fruits are subject to attack by fruit flies—*Anastrepha* sp. and *Carpolonchaea pendula* (syn. *Silba pendula*). In Colombia, the tree tomato has been found to be the preferred host of the tree tomato worm (*Neoleucinodes* sp.) which infests also the tomato and the eggplant. The larvae feed on the fruits and cause heavy losses. Rigorous spraying and sanitary measures are required to reduce losses and means of biological control are being sought.

The principal disease is powdery mildew (both *Erysiphe* sp. and *Oidium* sp.), which may cause serious defoliation if not controlled. Minor problems include Sclerotinia disease (*Sclerotinia sclerotiorum*), the black lesions of which girdle stems and cause terminal wilting; and Ascochyta disease (*Ascochyta* sp.) which is evidenced by small, round, black, dead areas on leaves, especially mature leaves. Tree tomato mosaic virus causes pale mottling on leaves and sometimes on the fruits which has not been considered a serious disadvantage. Another virus disorder, called "bootlace virus", distorts the leaf, especially on young plants, reducing it to little more than the midrib. Affected plants are pulled up and destroyed.

The tree tomato is noted for its resistance to tobacco mosaic virus, though it is susceptible to cucumber mosaic virus and potato virus. Die-back, of unknown origin, at times is lethal to the flowers, fruit cluster, twigs and new shoots. A strain of Arabis mosaic virus (which, in combination with two other unidentified viruses, causes sunken necrotic rings on the fruit surface) was reported in two plantations in the TePuke-Tauranga area of New Zealand in 1971, together with the identification of its vector, the nematode *Xiphinema diversicaudatum*.

Abnormality: In Haiti and New Zealand, small, hard, irregular, semi-transparent "stones" occur in the flesh of tree tomatoes and must be strained out in the process of jam-making. It is not known if these are similar to the "two gritty lumps in the wall of the fruit (on opposite sides)" mentioned by E.J.H. Corner as observed in Malaya. Samples of the stones were examined at the Division of Plant Industry, Florida State Department of Agriculture, and were found to contain "large amounts of sodium and calcium, probably as silicates, borates, aluminum-magnesium-oxygen complexes, or aluminates or magnesium oxides. In addition, small amounts of tin, copper, chromium, iron and phosphorus were found." It is well known that plants may accumulate minerals from mineral-rich soils, but such stony accretions are found in the leaves, not in the fruits. At Tela, Honduras, concretions occur in mangosteens, often rendering the fruit inedible. The cause has not been determined.

Food Uses

Ripe tree tomatoes may be merely cut in half lengthwise, sprinkled with sugar and served for eating by scooping out the flesh and pulp. Or the halves may be seasoned and grilled or baked for 15 minutes for service as a vegetable. The fruit should not be cut on a wooden or other permeable surface, as the juice will make an indelible stain. For other purposes, the skin must be removed and this is easily done by pouring boiling water over the fruit and letting it stand for 4 minutes, then peeling is begun at the stem end. The peeled fruit can then be sliced and the slices added to stews or soups, or served with a sprinkling of sugar and perhaps with a scoop of vanilla ice cream. Seasoned with salt and pepper, the slices can serve as sandwich-filling or may be used in salads. Chopped slices are blended with cream cheese and used as sandwich spread.

Peeled, diced fruits, with diced onion, breadcrumbs, butter and appropriate seasonings are employed as stuffing for roast lamb. Tree tomato slices, alone, or combined with sliced apple, are cooked in pies. They may be packed in preserving jars with water or sugar sirup and cooked for 55 minutes, or may be put into plastic containers with a 50% sirup and quick-frozen for future use in pies or puddings. The peeled fruits can be pureéd in a blender or by cooking, strained to remove the seeds and then packed in plastic containers and frozen. Lemon juice may be added to the puree to enhance flavor. The peeled, stewed fruits are combined with gelatin, milk, sugar and lemon juice to make a dessert which is then garnished with fresh tree tomato slices. Peeled, sliced and seeded tree tomatoes, with lemon rind, lemon juice and sugar, are cooked to a jam; or, with onions and apples, are made into chutney.

Chutney is prepared commercially in a factory in Auckland, New Zealand. Being high in pectin, the fruit is easily made into jelly but the fruit oxidizes and discolors without special treatment during processing. Whole, peeled fruits, with sugar, are cooked to a sauce for use on ice cream. The peeled fruits may be pickled whole, or may be substituted for tomatoes in a hot chili sauce.

Food Value Per 100 g of Edible Portion*	
Moisture	82.7–87.8 g
Protein	1.5 g
Carbohydrates	10.3 g
Fat (ether extract)	0.06–1.28 g
Fiber	1.4–4.2 g
Nitrogen	0.223–0.445 g
Ash	0.61–0.84 g
Calcium	3.9–11.3 mg
Phosphorus (with seeds)	52.5–65.5 mg
(without seeds)	13.1 mg
Iron	0.66–0.94 mg
Carotene	0.371–0.653 mg
(or calculated as Vitamin A)	540 I.U.
Thiamine	0.038–0.137 mg
Riboflavin	0.035–0.048 mg
Niacin (with seeds)	1.10–1.38 mg
(without seeds)	1.011 mg
Ascorbic Acid**	23.3–33.9 mg

*Analyses made in Ecuador, Guatemala and India.

**Most of the ascorbic acid is lost in cooking.

Genipap (Plate LXVII)

Rating low as an edible fruit but popular as a source of beverages, the genipap, *Genipa americana* L. (syns. *G. americana* var. *caruto* Schum.; *G. caruto* HBK.), of the family Rubiaceae, has a number of colloquial names: marmalade box in former British West Indies; *genipa, jagua* or *caruto* in Puerto Rico and several other Spanish-speaking countries; *genipapo* or *jenipapo* in parts of Colombia and Brazil; *chipara* or *chibara* or *guanapay* among Colombian Indians; *carcarutoto, caruto rebalsero,* or *guaricha* in Venezuela; *tapoeripa* in Surinam; *lana* in Guyana; *bi, bicito* or *totumillo* in Bolivia; *huitoc, vito, vitu* or *palo colorado* in Peru; *maluco* in Mexico; *crayo, irayol de montaña,* or *guali* in Guatemala; *guaitil* or *tapaculo* in Costa Rica; *irayol, tambor* or *tiñe-dientes* in El Salvador; *guayatil colorado* or *jagua blanca* in Panama.

Description

The tree is erect, to 60 or even 110 ft (18-33 m), with a tall, slender trunk and spreading branches. One form with a dense coating of soft hairs on the young branchlets and underside of the leaves has been separated by some botanists as a distinct species, *G. caruto* or *G. americana* var. *caruto*, though most botanists now view this as just a variation of *G. americana*. The leaves are abundant, deciduous, short-petioled, opposite but mostly clustered at the branch tips; oblong-obovate, 4 to 13 in (10-33 cm) long, 1½ to 5¼ in (4-13 cm) wide; sometimes faintly toothed, and with prominent whitish midrib. The faintly fragrant pale-yellow or white, tubular, 5-petalled flowers, to 1½ in (4 cm) wide, are borne in short, branched, terminal clusters.

The fruit, 3½ to 6 in (9-15 cm) long, 2¾ to 3½ in (7-9 cm) wide, weighing 7 to 14 oz (200-400 g), is elliptic or rounded-oval tapering briefly at the stem end, and having a short hollow tube at the apex. It has a thin leathery, yellow-brown, scurfy skin adherent to a ¼ to ½ in (6-12.5 mm) layer of muskily odorous, rubbery, whitish flesh (turning yellowish on exposure). The central cavity is filled with flat, circular, yellowish or brown seeds ⅜ to ½ in (1-1.25 cm) long, enclosed in grayish-yellow, mucilaginous membranes arranged in rows around a central fleshy core. The fruit, like the European medlar (*Mespilus germanica* L.) is edible only when overripe and soft to the touch, when the flavor, acid to subacid, resembles that of dried apples or quinces.

Origin and Distribution

The genipap is native to wet or moist areas of Cuba, Hispaniola, Puerto Rico, the Virgin Islands, and from Guadeloupe to Trinidad; also from southern Mexico to Panama, and from Colombia and Venezuela to Peru, Bolivia and Argentina. Its usefulness to the Indians was reported by several European writers in Brazil in the 16th Century. It is widely cultivated in dooryards as an ornamental tree and for its fruits, but Patiño stated in 1967 that it was no longer as commonly grown in the Cauca Valley of Colombia as it had been in the past. In Trinidad, the tree is occasionally planted as a living fencepost for pasture fences. In 1965, a program was launched to utilize the genipap for reforestation in northeastern Brazil and plantations were established with a view to the exploitation of the fruit for liquor manufacture and the timber and other products for local use and possible export.

The tree first fruited in the Philippines in 1913 and is occasionally planted there. Otherwise, it is virtually unknown in the Old World. Burkill wrote that it had been tried in the southern part of the Malay Peninsula several times but without success.

P.J. Wester sent seeds from the Philippines to the United States Department of Agriculture in 1917 (S.P.I. #44090). A tree at the Plant Introduction Station, Miami, was 20 ft (6 m) tall in 1951 but had never bloomed. It is still alive and well today. There was a large tree at the Agricultural Research and Education Center, Homestead. It did not bear fruit and was killed by a freeze. A tree at the Fairchild Tropical Garden, Coral Gables, bloomed for the first time in the spring of 1980 but did not produce fruit. A few small seedlings were distributed by the Rare Fruit Council in 1980.

Varieties

It is reported in Brazil that there are varieties that bear all year. There is a shrubby form, *jenipaporana,* or *jenipapo-bravo*, no more than 10 to 13 ft (3-4 m) high, that grows in swamps along the edges of rivers and lakes in Brazil. The fruit is small and inedible.

Climate and Soil

The tree is strictly tropical; is limited to elevations below 3,300 ft (100 m) in Peru and has been killed by low temperatures in Florida. The genipap flourishes best in a humid atmosphere and deep, rich, loamy, moist soil.

Propagation

The genipap is mostly grown from seed but P.J. Wester determined that it can be propagated by shield-budding, using mature, non-petioled scions.

Culture

Seeds germinate in 25 to 30 days. The seedlings reach 4¾ in (12 cm) in 3 to 4 months and are transplanted when 6 to 12 months old at a height of 8 in (20 cm). The tree requires little cultural attention and thrives even in arid situations. For fruit production, the trees are spaced 33 to 50 ft (10-15 m) apart. Temporary crops such as cassava or cotton are interplanted to provide shade for the young trees and income for the farmer. For purposes of reforestation, the spacing may be 5 x 10 ft (1.5x3 m) or up to 10 x 10 ft (3x3 m). The heavy leaf fall of the genipap is important in improving the soil of the plantation.

Season

Flowers and fruits appear continuously from spring to fall in Puerto Rico. In Brazil, the tree flowers in November and the fruits appear in the markets in February and March.

Fig. 117: The genipap (*Genipa americana L.*) photographed by P. H. Dorsett, plant explorer for the U. S. Department of Agriculture, Bureau of Plant Industry, in Bahia, Brazil, November 12, 1913.

Food Uses

In Puerto Rico, the fruit is cut up and put in a pitcher of water with sugar added to make a summer drink like lemonade. Sometimes it is allowed to ferment slightly. A bottled concentrate is served with shaved ice by street vendors. In the Philippines, also, the fruit is used to make cool drinks, as well as jelly, sherbet and ice cream. The flesh is sometimes added as a substitute for commercial pectin to aid the jelling of low-pectin fruit juices. Rural Brazilians prepare sweet preserves, sirup, a soft drink, *genipapada*, wine, and a potent liqueur from the fruits.

Food Value

Analyses made in the Philippines many years ago show the following values for the edible portion (70%) of the fruit: protein, 0.51%; carbohydrates, 11.21%; sugar, 4.30%; ash, 0.20%; malic acid, 0.63%.

Recent analyses made at the Fundacion Instituto Brasileiro de Geografia e Estatistica show:

Food Value per 100 g of Edible Portion	
Calories	113
Moisture	67.6 g
Protein	5.2 g
Lipids	0.3 g
Glycerides	25.7 g
Fiber	9.4 g
Ash	1.2 g
Calcium	40.0 mg
Phosphorus	58.0 mg
Iron	3.6 mg
Vitamin B$_1$	0.04 mg
Vitamin B$_2$	0.04 mg
Niacin	0.50 mg
Ascorbic Acid	33.0 mg
Amino Acids (per g of nitrogen [N = 6.25])	
Lysine	316 mg
Methionine	178 mg
Threonine	219 mg
Tryptophan	57 mg

The fruit contains too much tannin to be a desirable article of diet.

Other Uses

Fruit: In Guyana, the ripe fruit is used mainly as fish bait. The fallen, astringent fruits are much eaten by wild and domestic animals. The juice of the unripe fruit is colorless but oxidizes on exposure to the air and gradually turns light brown, then blue-black, and finally jet black. It has been commonly employed by South American Indians to paint their faces and bodies for adornment and to repel insects; and to dye clothing, hammocks, utensils and basket materials a bluish-purple. The dye is indelible on the skin for 15 to 20 days. Oviedo wrote that the Indian men sometimes playfully sprinkled the women with the fresh juice mixed with perfume so that mysterious spots would appear on their bodies and alarm them. Cardenas tells of seeing the robe of a Franciscan monk which was dyed a very dark purple with this juice.

Leaves: The foliage is readily eaten by cattle.

Bark: The bark, rich in tannin, has been used for treating leather. It also yields a fiber employed in making rough clothing.

Wood: When 5 or 6 years old, saplings can be harvested for firewood, poles or fenceposts. Ten-year-old trees can be cut for timber. The wood is yellowish-white or sometimes slightly pinkish or lavender, with light, reddish-brown streaks. It is fibrous, compact, hard, elastic, strong but not durable, being subject to attack by termites, borers and fungi. It has been used for spears, rifle stocks, shoe lasts, frames for sieves, barrel hoops, ammunition chests and other boxes, packing cases, plows, tool handles, boards for flooring, door frames and cabinetwork.

Flowers: The flowers yield nectar for honeybees.

Medicinal Uses: The fruit is eaten as a remedy for jaundice in El Salvador. Ingested in quantity, it is said to act as a vermifuge. The fruit juice is given as a diuretic. It is a common practice in Puerto Rico to cut up the fruits, steep them in water until there is a little fermentation, then add flavoring and drink the infusion as a cold remedy.

Because the fruit and its infusion have unusually good keeping quality, Puerto Rican scientists investigated the possible presence of antibiotic principles and proved the existence of antibiotic activity in all parts of the fruit. In 1964, Dr. W.H. Tallent of G.D. Searle & Company in Chicago, isolated and identified 2 new antibiotic cyclopentoid monoterpenes, primarily genipic acid and secondarily genipinic acid, its carbomethoxyl derivative.

The crushed green fruit and the bark decoction are applied on venereal sores and pharyngitis. The root decoction is a strong purgative.

The bark exudes when cut a whitish, sweetish gum which is diluted and used as an eyewash and is claimed to alleviate corneal opacities. The juice expressed from the leaves is commonly given as a febrifuge in Central America. The pulverized seeds are emetic and caustic. The flower decoction is taken as a tonic and febrifuge.

Superstitious Uses: Guatemalan Indians carry the fruits in their hands in the belief that this will provide protection from disease and ill-fortune.

CUCURBITACEAE

Cassabanana (Plate LXVIII)

A handsome and interesting member of the Cucurbitaceae, the cassabanana, *Sicana odorifera* Naud. (syn. *Cucurbita odorifera* Vell.), is also called sikana or musk cucumber. It is known as *melocotonero, calabaza de olor, calabaza melón, pérsico* or *alberchigo* in Mexico; *melocotón* or *melón de olor* in El Salvador and Guatemala; *calabaza de chila* in Costa Rica; *cojombro* in Nicaragua; *chila* in Panama; *paví* in Bolivia; *padea, olerero, secana* or *upe* in Peru; *calabaza de Paraguay, curuba,* or *pepino melocoton* in Colombia; *cura, coróa, curua, curuba, cruatina, melão caboclo* or *melão macã* in Brazil; *cajú, cajuba, cajua, cagua, calabaza de Guinea* in Venezuela; *pepino, pepino angolo* or *pepino socato* in Puerto Rico; *cohombro* in Cuba.

Description

The vine is perennial, herbaceous, fast-growing, heavy, requiring a strong trellis; climbing trees to 50 ft (15 m) or more by means of 4-parted tendrils equipped with adhesive discs that can adhere tightly to the smoothest surface. Young stems are hairy. The leaves are gray-hairy, rounded-cordate or rounded kidney-shaped, to 1 ft (30 cm) wide, deeply indented at the base, 3-lobed, with wavy or toothed margins, on petioles 1½ to 4¾ in (4-12 cm) long. Flowers are white or yellow, urn-shaped, 5-lobed, solitary, the male ¾ in (2 cm) long, the female about 2 in (5 cm) long. Renowned for its strong, sweet, agreeable, melon-like odor, the striking fruit is ellipsoid or nearly cylindrical, sometimes slightly curved; 12 to 24 in (30-60 cm) in length, 2¾ to 4½ in (7-11.25 cm) thick, hard-shelled, orange-red, maroon, dark-purple with tinges of violet, or entirely jet-black; smooth and glossy when ripe, with firm, orange-yellow or yellow, cantaloupe-like, tough, juicy flesh, ¾ in (2 cm) thick. In the central cavity, there is softer pulp, a soft, fleshy core, and numerous flat, oval seeds, ⅝ in (16 mm) long and ¼ in (6 mm) wide, light-brown bordered with a dark-brown stripe, in tightly-packed rows extending the entire length of the fruit.

Origin and Distribution

The cassabanana is believed native to Brazil but it has been spread throughout tropical America. Historians have evidence that it was cultivated in Ecuador in pre-Hispanic times. It was first mentioned by European writers in 1658 as cultivated and popular in Peru. It is grown near sea-level in Central America but the fruit is carried to markets even up in the highlands. Venezuelans and Brazilians are partial to the vine as an ornamental, but in Cuba, Puerto Rico and Mexico it is grown for the usefulness of the fruit.

In 1903, O.F. Cook saw one fruit in a market in Washington, D.C. The United States Department of Agriculture received seeds from C.A. Miller, the American Consul in Tampico, Mexico, in 1913 (S.P.I. #35136). H.M. Curran collected seeds in Brazil in 1915 (S.P.I. #41665). Wilson Popenoe introduced seeds from Guatemala in

Fig. 118: The cassabanana (*Sicana odorifera* Naud.), photographed by Wilson Popenoe, plant explorer for the U.S.D.A. Bureau of Plant Industry, in Guatemala on September 23, 1916.

1916 (S.P.I. #43427). The author brought seeds from Rio Piedras, Puerto Rico, to the Agricultural Research and Education Center, Homestead, in 1951. A resulting vine grew to large size but produced a single 2 ft (60 cm) fruit. Dr. John Thieret, formerly Professor of Botany at Southwestern Louisiana University, says that the Cajuns in the southern part of that state grow the cassabanana for making preserves. Verrill stated in 1937, "The fruit is now on sale in New York markets."

According to Burkill, the vine was tried in the Botanic Gardens in Singapore but lived for only a short time. Wester wrote that it fruited at Lamao in the Philippines in 1916 and became heavily attacked by a destructive fly (*Dacus* sp.).

Culture

Fenzi says that the cassabanana is grown from seeds or cuttings. A high temperature during the fruiting season is needed to assure perfect ripening. Brazilians train the vine to grow over arbors or they may plant it close to a tree. However, if it is allowed to climb too high up the tree there is the risk that it may smother and kill it.

Keeping Quality and Marketing

The cassabanana remains in good condition for several months if kept dry and out of the sun.

The fruit has high market value in Puerto Rico. It is cut up and sold by the piece, the price being determined by weight.

Food Uses

The ripe flesh, sliced thin, is eaten raw, especially in the summer when it is appreciated as cooling and refreshing. However, it is mainly used in the kitchen for making jam or other preserves. The immature fruit is cooked as a vegetable or in soup and stews.

Food Value Per 100 g of Edible Portion		
	Analyses of ripe fruit made in Guatemala (without peel, seeds, or soft central pulp)	*Analyses of peeled green fruit made in Nicaragua* (including seeds)
Moisture	85.1 g	92.7 g
Protein	0.145 g	0.093 g
Fat	0.02 g	0.21 g
Fiber	1.1 g	0.6 g
Ash	0.70 g	0.38 g
Calcium	21.1 mg	8.2 mg
Phosphorus	24.5 mg	24.2 mg
Iron	0.33 mg	0.87 mg
Carotene	0.11 mg	0.003 mg
Thiamine	0.058 mg	0.038 mg
Riboflavin	0.035 mg	–
Niacin	0.767 mg	0.647 mg
Ascorbic Acid	13.9 mg	10.0 mg

Other Uses

Fruit: People like to keep the fruit around the house, and especially in linen- and clothes-closets, because of its long-lasting fragrance, and they believe that it repels moths. It is also placed on church altars during Holy Week.

Medicinal Uses: In Puerto Rico, the flesh is cut up and steeped in water, with added sugar, overnight at room temperature so that it will ferment slightly. The resultant liquor is sipped frequently and strips of the flesh are eaten, too, to relieve sore throat. It is believed beneficial also to, at the same time, wear a necklace of the seeds around the neck.

The seed infusion is taken in Brazil as a febrifuge, vermifuge, purgative and emmenagogue. The leaves are employed in treating uterine hemorrhages and venereal diseases. In Yucatan, a decoction of leaves and flowers (2 g in 180cc water) is prescribed as a laxative, emmenagogue and vermifuge, with a warning not to make a stronger preparation inasmuch as the seeds and flowers yield a certain amount of hydrocyanic acid.

BIBLIOGRAPHY

Abdel Akher, M., Abdel Moneim Youssef, and M.S. Ezzat. 1964. The chemical composition and utilization of palm tree leaves. Alexandria J. Agr. Res. 12 (1): 179-189.

Abdel-Fattah, A.F., D.A. Zaki, and M. Edrees. 1975. Some investigations on the pectin and amino acid composition of *Dovyalis caffra* W. fruit. Qualitas Plant. — Pl. Foods & Human Nutr. 24 (3/4): 311-316.

Abou Aziz, A.B., S.M. El-Nabarvy and H.A. Zaki. 1975. Effect of different temperatures on the storage of papaya fruits and respirational activity during storage. Scientia Hort. 3: 173-177.

Abraham, E.A.V. 1912. Materia Medica Guian. Brit. Temehri. J. Roy. Agr. Soc. Brit. Guiana 2 (3rd Ser.) (1): 179-196.

Abrams, R. and G.C. Jackson. 1959. Influence of root-inducing substances and time intervals on the rooting of acerola cuttings. J. Agr. U. of Puerto Rico 43 (3): 152-158.

Abusin, A.M. and B.H. Hamid. 1971. Evaluation of the major mango varieties for processing. Sudan J. of Food Sci. & Tech. Vol. 3: 61-63.

Abutiate, W.S. and N.Y. Nakasone. 1972. Studies of vegetative propagation of the lychee (*Litchi chinensis* Sonn.) with special reference to graftage. Ghana J. Agr. Sci. 5: 201-211.

Ackerman, E.D. 1978. Jaboticaba nutrition experiment. Proc. Fl. St. Hort. Soc. 91: 187-189.

Adams, C.D. 1972. Flowering Plants of Jamaica. Univ. of the West Indes, Mona, Jamaica.

Adams, C.S. 1929. Growing melons on trees. [papaya] Fla. Grower, Nov. 16-17.

Adams, M.R. 1978. Small-scale vinegar production from bananas. Trop. Sci. 20 (1): 11-19.

Adamson, R.S. and T.M. Salter. 1950. Flora of the Cape Peninsula. Juta & Co., Ltd., Cape Town & Johannesburg, S. Africa.

Adcock, R.E. 1968. Malaysia increases its pineapple profits. For. Agr. 6 (32): 11.

Adesogan, E.K. and J.I. Durodola. 1976. Antitumour and antibiotic principles of *Annona senegalensis*. Phytochem. 15: 1311-1312.

Adsuar, J. 1972. A new virus disease of papaya (*Carica papaya*) in Puerto Rico. J. Agr. U. Puerto Rico 56 (4): 397-402.

Adsuar, J. 1947. Studies on virus diseases of papaya (*Carica papaya*) in Puerto Rico, I, II, and III. J. Agr. U. Puerto Rico 31 (3): 248-264.

Adsuar, J. 1950. Studies on virus diseases of papaya (*Carica papaya*) in Puerto Rico. Tech. Paper #6. U. of Puerto Rico Agr. Expt. Sta., Rio Piedras.

Adsule, P.G., M.A. Ismael and P.J. Fellers. 1984. Quality of citrus fruit following cold treatment as a method of disinfestation against the Caribbean fruit fly. J. Amer. Soc. Hort. Sci. 109 (6): 851-854.

Agarwal, K.D. 1970. Table-cum-juicy mango — a promising variety. Indian Hort. 14 (2): 15, 32.

Agena, U., J.E. Dunn, Sr., and Jaw-Kai Wang. 1969. Current status of papaya mechanization. Hawaii Farm Sci. 18 (2): 1-3.

Agnew, G.W.J. 1968. Growing quality papaws in Queensland. Queensland Agr. J. 94 (1): 24-36.

Agnihotri, B.N. 1950. The bael, *Aegle marmelos* Correa. Indian Food Packer 4 (9): 29-31.

Aguilar Giron, J.I. 1966. Relacion de unos aspectos de la flora util de Guatemala. 2nd ed. Tipografia Nacional de Guatemala, for Associacion Amigos del Bosque.

Aguilar-Santos, G., J.R. Librea, and A.C. Santos. 1967. The alkaloids of *Annona muricata* Linn. Phil. J. Sci. 96 (4): 399-407.

Agusti, M., V. Almela, and J.L. Guardiola. 1981. The regulation of fruit cropping in mandarins through the use of growth regulators. Proc. Internat. Soc. Citriculture, Nov. 9-12, 1981, Tokyo. Vol. I: 216-220.

Ahlawat, V.P. and R. Yamdagni. 1981. Effect of potassium sprays on quality of guava (*Psidium guajava* L.) cv. Sardar (L. 49). Agr. Sci. Digest 1 (4): 213-214.

Ahmad, M., M.H. Naqvi, A. Hussain, M. Mohyuddin, A. Sattar, and M. Ali. 1972. Induction of ripening delay in mangoes (var. Dushhri) by gamma irradiation and refrigeration. Pakistan J. Sci. Ind. Res. 15 (4-5): 314-319.

Ahmad, M.N., Nasir-ud-Din Zahid, M. Rafio and I. Ahmad. 1964. Composition of the oil of *Grewia asiatica* (phalsa) seeds. Pakistan J. Sci. & Indus. Res. 7 (2): 145-146.

Ahmed, S. 1961. The litchi. Agr. Pakistan 12 (4): 769-775.

Ahmed, S. and M.A. Ghafoor. 1961. Falsa (*Grewia asiatica*) cultivation in West Pakistan. Agr. Pakistan 12 (4): 762-768.

Aiyappa, K.M. 1969. Higher pineapple yields possible. Indian Hort. 14 (1): 5-6 & inside back cover.

Ajao, A.O., O. Shonukan and B. Femi-Onadeko. 1985. Antibacterial effect of aqueous and alcohol extracts of *Spondias mombin* and *Alchornea cordifolia* — two local antimicrobial remedies. Internat. J. Crude Drug Res. 23 (2): 67-72.

Akamine, E.K. 1971. Controlled atmosphere storage of papayas. Misc. Pub. 64. (Proc. 5th Ann. Mtg. Haw. Papaya Indus. Assn.). U. Hawaii Coop. Exten. Serv., Honolulu. Pp.23-24.

Akamine, E.K. 1965. Fumigation of bananas with special reference to ripening. Hawaii Farm Sci. 14 (2): 1-3.

Akamine, E.K. 1960. Preventing the darkening of fresh lychees prepared for export. Tech. Progress Rpt. 127. U. Hawaii, Agr. Expt. Sta., Honolulu.

Akamine, E.K. and G. Girolami. 1959. Pollination and fruit set in the yellow passion fruit. Tech. Bul. 39. Hawaii Agr. Expt. Sta., Honolulu.

Akamine, E.K. and T. Goo. 1981. Carbon dioxide and ethylene production in *Diospyros discolor* Willd. HortSci. 16 (4): 519.

Akamine, E.K. and T. Goo. 1979. Concentrations of carbon dioxide and ethylene in the cavity of attacked papaya fruit. HortSci. 14 (2): 138-139.

Akamine, E.K. and T. Goo. 1971. Relationship between surface color development and total soluble solids in papaya. HortSci. 6 (6): 567-568.

Akamine, E.K. and T. Goo. 1979. Respiration and ethylene production in fruits of species and cultivars of *Psidium* and species of *Eugenia*. J. Amer. Soc. Hort. Sci. 104 (5): 632-635.

Akamine, E.K., T. Goo, T. Steefy, T. Greidanus, and N. Irvaoka. 1975. Control of endogenous brown spot of fresh pineapple in postharvest handling. J. Amer. Soc. Hort. Sci. 100 (1): 60-65.

Akamine, E.K., R.A. Hamilton, T. Nishida, G.D. Sherman, W.B. Storey, W.J. Yee and T.N. Shaw. 1954. Passion Fruit Culture. Exten. Cir. 345. U. Hawaii Agr. Exten. Serv., Honolulu.

Albizu de Santiago, C. 1968. Karyotype of the blue chrysomelid of acerola, *Leucoceia laevicollis* Weise (Coleoptera), Chrysomelidae. J. Agr. U. Puerto Rico 52 (1): 74-75.

Alcorn, J.L. 1968. Cucurbit powdery mildew on papaw. Queensland J. Agr. & Anim. Sci. 25 (3): 161-164.

Aldrich, W.W. and H.Y. Nakasone. 1975. Day versus night application of calcium carbide for flower induction in pineapple. J. Amer. Soc. Hort. Sci. 100 (4): 410-413.

Ali, A., M. Sahai and A.B. Ray. 1984. Physalolactone C, a new withazolide from *Physalis peruviana*. J. Nat. Prod. 47 (4): 648-651.

Ali Dinar, H.M. and A.H. Krezdorn. 1976. Extending the grapefruit harvest season with growth regulators. Proc. Fl. St. Hort. Soc. 89: 4-6.

Allan, P. 1981. Clonal "Honey Gold" pawpaws a horticultural and commercial success. Citrus and Subtrop. Fruit J. 575: 19-23.

Allan, P. 1979. Cultivation practices for Kiwi fruit with special reference to South African conditions. Hortus 26: 33-40.

Allan, P. 1967. Papaw research at Pietermaritzburg. 1. Production from seedlings. Farming S. Afr. 41 (10): 25, 29, 31.

Allan, P. 1963. Pollen studies in *Carica papaya*. I. Formation, development, morphology and production of pollen. S. Afr. J. Agr. Sci. 6: 517-530.

Allan, P. 1963. Pollen studies in *Carica papaya*. II. Germination and storage of pollen. S. Afr. J. Agr. Sci. 6: 613-624.

Allan, P. 1964. Papajas van steggies gekweek. Boerderj in S. Afr. 39

(11): Feb.: 35-40.

Allan, P. 1967. Papaw research at Pietermaritzburg. 2. Production from cuttings. Farming in So. Afr. 42 (11): 15-21.

Allen, B.M. 1967. Malayan fruits: an introduction to the cultivated species. Donald Moore Press, Ltd., Singapore.

Allen, D.J. 1970. Botryodia fruit rot of giant granadilla in East Africa. Pl. Dis. Reporter 54 (9): 809-811.

Allen, J.L. 1970. Lime juice and lime oil production and markets. Report G-45. Trop. Prod. Inst., London.

Allen, P.H. 1943. Poisonous and injurious plants of Panama. Supp. to Amer. J. Trop. Med. Vol. 23, No. 1. Williams & Wilkins Co., Baltimore, Maryland.

Allen, P.H. 1977. The rain forests of Golfo Dulce. U. Fla. Press, Gainesville, Fl.

Almeyda, N. and F.W. Martin. 1976. Cultivation of neglected tropical fruits with promise. Pt. 1. The mangosteen. ARS-S-155. U.S. Dept. Agr., Agr. Res. Serv., Washington, D.C.

Almeyda, N. and F.W. Martin. 1976. Cultivation of neglected tropical fruits with promise. Pt. 2. The Mamey Sapote. ARS-S-156. U.S. Dept. Agr., Agr. Res. Serv., Washington, D.C.

Almeyda, N. and F.W. Martin. 1977. Cultivation of neglected tropical fruits with promise. Pt. 4. The Lanson. ARS-S-171. U.S. Dept. Agr., Agr. Res. Serv., Washington, D.C.

Almeyda, N. and F.W. Martin. 1980. Cultivation of neglected tropical fruits with promise. Pt. 8. The pejibaye. U.S. Dept. Agr., Sci. & Educ. Admin., Washington, D.C.

Alston, A.H.G. 1938. The Kandy flora. Ceylon Gov't Press, Colombo.

Altman, L.J. and S.W. Zito. 1976. Sterols and triterpenes from the fruit of *Artocarpus altilis*. Phytochem. 15: 829-830.

Altschul, S. v. R. 1973. Drugs and foods from little-known plants. Harvard U. Press, Cambridge, Mass.

Alvarez, A.M. 1980. Improved marketability of fresh papaya by shipment in hypobaric containers. HortSci. 15 (4): 517-518.

Alvarez-Garcia, L.A. and J. Lopez-Garcia. 1971. Gummosis, die-back and fruit rot disease of mango (*Mangifera indica* L.) caused by *Physalospora rhodina* (B. & C.) Cke. in Puerto Rico. J. Agr. U. of Puerto Rico 55 (4): 435-450.

Ammer, K., A. Afifi, and A.E.A. Fahmy. 1925. Wine production from bitter orange juice. Egypt J. Food Sci. 3 (1-2): 17-23.

Ammerman, C.B., D.A. Hansen, F.G. Martin and L.R. Arrington. 1976. Nutrient composition of dried citrus pulp as influenced by season of production and production source. Proc. Fl. St. Hort. Soc. 89: 168-170.

Anand, J.C. 1960. Efficacy of sodium benzoate to control yeast fermentation in phalsa (*Grewia asiatica* L.) juice. Indian J. Hort. 17 (2): 138-141.

Andam, C.J. 1983. Response of maturing 'Carabao' mango fruits to pre-harvest ethephon application. NSTA Tech. J. July-Sept. 4-13.

Anderson, D.L., M. Sedgley, J.R.T. Short and A.J. Allwood. 1982. Insect pollination of mango in Northern Australia. Aust. J. Agr. Res. 33: 541-548.

Anelli, G., R. Fiorentini, and A.A. Lepidi. 1984. Food and feed from banana by-products. Revista di Agricoltura Subtrop. e Trop. 76 (1-2): 67-75.

Angeles, E.R. 1962. A survey of patch-budding of rambutan (*Nephelium lappaceum* Linn.) at the provincial nursery in Oriental Mindoro. Arenata J. Agr. 9 (4): 267-273.

Anon. 1965. A Climbing Mandarin. Queensland Fruit & Veg. News. June. p. 519.

Anon. 1974. A frutifera calabura. O Estado de Sao Paulo: Supp. Agricola. Mar. 24. P. 15.

Anon. 1977. Alkop Farms—California's largest kiwi planting. Blue Anchor 54 (1): 23-26.

Anon. 1981. All out alert to stop citrus enemy #1. Plant Indus. News 23 (2): 3-5, 15.

Anon. 1979. Annotated Bibliographies: Loquat (*Eriobotrya japonica*), 1967-1978. Plant Ser. ISSN 0141-5921. Commonwealth Agr. Bur., Farnham House, Farnham Royal Slough, UK.

Anon. 1963. A 'Peach Belt' for Florida? Fla. Grower & Rancher, June, pp. 9, 16.

Anon. 1952. Avocado oil in toilet preparations and pharmaceuticals.

Perfum. & Essen. Oil Rec. 43 (11): 399-400.

Anon. 1979. Avocados: $40 million business. Israel Economist No. 35. P. 21.

Anon. 1982. Babaco, mountain papaya and papaya: all are susceptible to powdery mildew. N. Zeal. J. Agr. 145 (4): 28.

Anon. 1950. Big lychee planting started. Fla. Grower. Oct. Pp. 19, 22.

Anon. 1980. Big retail profit opportunity with record new California avocado crop. Blue Anchor 57 (5): 34.

Anon. 1962. Brazil product [banana flakes] suits many purposes. Braz. Bull. Dec. 15. P. 6.

Anon. 1978. Breadfruit reconnaisance study in the Caribbean region. CIAT, Inter-Amer. Develop. Bank. 65 pp.

Anon. 1972. California mechanizes desert grapefruit harvest. Fla. Grower & Rancher–South Central. July-Aug. P. 12.

Anon. 1970. Caqui fruit could reward investors. Brazilian Bull. Feb. P. 7.

Anon. 1963. Caquis. Brasil. 19: 24.

Anon. 1965. Caqui śo detesta mismo é o vento forte na maturarao. O Dirigente Rural 4 (9): 18, 20.

Anon. 1977. Carta Informativa #97: [cherimoya]. Sociedad Amer. de Ciencias Horticolas—Region Tropical.

Anon. 1977. C B F [Citrus blackfly] update: Parasite Lab. Proposed and quarantine line extended. Pl. Indus. News 19 (2): 3-5.

Anon. 1977. Certified Star Ruby is (finally!) here. Plant Indus. News 18 (3): 6-7.

Anon. 1951. Ceylon jack fruit can be grown in India. Indian Food Packer 5 (12): 17-18.

Anon. 1942. Chemistry investigations . . . The non-volatile acids and flavor of the soursop were studied. Rpt. of Puerto Rico Exp. Sta. 1940. U.S. Dept. Agr., Washington, D.C. Pp. 88-91.

Anon. 1967. Choko pickles. Queensland Agr. J. 93 (4): 251.

Anon. 1982. Chymopapain approved. FDA Drug Bull. 12 (3): 17-18.

Anon. 1977. Citrus blackfly—eradication or containment. Plant Indus. News 18 (4): 3-5, 15-18.

Anon. 1983. Citrus Budwood Registration—a model program for quality control. Plant Indus. News 24 (4): 3-7, 15.

Anon. 1965. Collar rot and shell bark in lemon trees. Queensland Fr. & Veg. News. Apr. P. 343.

Anon. 1978. Cultivar and germplasm releases: papaya. HortSci. 13 (5): 527.

Anon. 1962. Cultivo de la curuba [*Passiflora mollissima*] Rev. Nac. de Agr. 55(691): 54.

Anon. 1959. Dehydration of litchis. Ann. Rpt. for 1957-59. Fruit Res. Sta., Hort. Res. Inst., Dept. of Agr., Uttar Pradesh, Saharanpur, India. P. 17.

Anon. 1975. Deno Giannandrea and his kiwifruit. Blue Anchor 52 (4): 17-20.

Anon. 1975. Development and prospects of pineapple production industry. J. Phil. Statist. 26 (2): ix-xxix.

Anon. 1982. Disk surgery the easy way [papaya]. Amer. Health (5): 84.

Anon. 1962. Drug research and development of new drugs. J. Sci. & Indus. Res. 21A (5): 227.

Anon. 1968. Drying lychee fruit. Fla Grower & Rancher 61: 11.

Anon. 1964. Estudio economico de pre-inversion en plantas frutales. Inst. Latinoamericano de Mercadeo Agricola, Bogotá, Colombia. Pp. 1-13 & tables.

Anon. 1981. Estudo nacional da despesa familiar. Tabelas de composiciao de alimentos. 2nd ed. Fundacion Instituto Brasileiro de Geografia e Estatistica. Rio de Janeiro, Brazil.

Anon. 1979. Exotic plants declared weeds [*Pereskia*]. So. Afr. Dig. June 1. P. 19.

Anon. 1983. Feijoas forge ahead. New Zeal. J. Agr. 147 (2): 58.

Anon. 1986. Flordaprince peach now available. Amer. Fruit Grower 106 (1): 36.

Anon. 1980. Florida Christmas merry and green [lychee]. Fla. Market Bull. 23 (24): 1.

Anon. 1982. Florida exports sunshine. Fla. Market Bull. 25 (6): 1, 4.

Anon. 1953. Grapefruit born of scalawag tree. Fla. Grower & Rancher. Oct. P. 5.

Anon. 1963. Guavas. The Farmer (Jamaica) 68: 2-4, 57-59, 84, 97.

Anon. 1967. Hybrid No. 5 papaw seed unavailable. Queensland Fr. &

Veg. News 31 (9): 215.

Anon. 1965. Hybrid papayas gaining popularity. Queensland Fr. & Veg. News. Apr. P. 399.

Anon. 1978. In search of the perfect peeler. Amer. Fr. Grower 98 (6): 8.

Anon. 1982. In search of the perfect pine. So. Afr. Dig. Jan. 15, p. 16 (from Sci. Prog. 14 (3)).

Anon. 1965. Jenipapeiro é boa essência e vai bem no Nordeste. O Dirigente Rural 4 (8): 28.

Anon. 1977. Kiwifruit conquers new lands. World Farming 19 (1): 14, 18-19, 31.

Anon. 1986. Kiwifruit is nutritional food. The Orchardist of New Zealand 58 (4): 177; 1985; reproduced in HortSci. 21 (1): 13: 1986.

Anon. 1977. Kiwifruit pioneers Tanimotos still see bright future. Blue Anchor 54 (1): 21-22.

Anon. 1971. Le palmier Dattier en Tunisie. Fiche Technique #29. Ministere de l'Agriculture, Direction de la Production Agricole, Div. de la Vulgarisation Agricole. 16 pp.

Anon. 1954. Lemonade concentrate — the rush is on. Fla Grower & Rancher, Aug. Pp. 11, 28.

Anon. 1973. Lemons — cultivation lacks organization. Rev. of the economic situation of Mexico 49 (569): 152-157.

Anon. 1829. Library of entertaining knowledge: Description and listing of vegetable substances used in the arts and in domestic economy: timber trees, fruits. Charles Knight, London.

Anon. 1958. Lychee Varieties. Fruit Varieties & Hort. Dig. 13 (2).

Anon. 1975. Mainland papaya consumption looking up. Blue Anchor 52 (1): 22-24.

Anon. 1904. Materia Medica Mexicana: A manual of Mexican medicinal herbs. Mexican Nat'l Comm., St. Louis, Mo.

Anon. 1954. Mayan husk tomato is well received in Iowa. Better Iowa 40 (24): 2.

Anon. 1982. Mexican canker investigated. Plant Indus. News 24 (2): 6-7.

Anon. 1958. *Monstera deliciosa.* Bull. #7. National Botanical Garden, Lucknow, India.

Anon. 1979. Mystery disease threat to Valley's date palms. Houston Chronicle Dec. 23, P. 18. Sec. 1.

Anon. 1965. Nematode fumigation lifts pineapple yield. Queensland Agr. J. 91 (12): 712-714.

Anon. 1970. New pineapple variety for 1975. Queensland Agr. J. 96 (5): 317.

Anon. 1966. One ton pines — picked and loaded in 8 minutes. Queensland Fr. & Veg. News 29 (19): 444-445.

Anon. 1964. Oriental persimmons, California grown. Blue Anchor 40 (4): 8-12.

Anon. 1981. Oroblanco grapefruit. Amer. Fruit Grower 101 (6): 26.

Anon. 1974. Papaw dieback and calcium deficiency. Queensland Agr. J. 100 (7): 316.

Anon. 1967. Papaya insect control, pests to watch, what to do about them. Fla. Grower & Rancher 75 (8): 27.

Anon. 1977. Pawpaw cure. So. Afr. Dig. Apr. 29, p. 28.

Anon. 1963. Pectin and papain from raw papaya. Ann. Rpt. Central Food Tech. Res. Inst. Mysore, India. 1961-1962. Pp. 66-68.

Anon. 1968. Persimmons. Blue Anchor 45 (4): 42.

Anon. 1980. Picture caption: "Unusual bunch of 12 pineapples". So. Afr. Dig. Mar. 28. (from The Star).

Anon. 1975. Pomegranates a specialty for the Don Miller family. Blue Anchor 52 (4): 29-30.

Anon. 1977. Poor fruiting on tamarillo. New Zeal. J. Agr. 135 (5): 63.

Anon. 1962. Possible use for lychee roots. Fla. Dept. Agr., Div. of Pl. Indus. 4 (1): 4.

Anon. 1974. Problems in Paradise. Amer. Fr. Grower 94 (1): 14.

Anon. 1977. Production of Maui papayas moving up at Princess Orchards. Blue Anchor 54 (4): 40-44.

Anon. 1965. Pruning breadfruit trees. Agr. Exten. Leaflet No. 4. Div. of Agr., Dept. of Resources & Dev., Trust Terr. of the Pacific Islands, Saipan, Marianna Islands.

Anon. 1973. Pruning of tamarillo. New Zeal. J. Agr. 127 (4): 9.

Anon. 1962. Rambutan (condensed from Tech. Bull. #7, Coll. of Agr., U. of Phil.). Farmers' J 4 (2): 10, 12-15.

Anon. 1891. Report on the condition of tropical and semitropical fruits in the United States in 1887. Bull. #1. U.S. Dept. Agr., Div. of Pomology, Washington, D.C.

Anon. 1974. Scientific literature reviews on generally recognized as safe (GRAS) food ingredients — papain. PB-228-540. Prepared for FDA by Food & Drug Res. Lab., Inc., East Orange, N.J. Distributed by Nat'l Tech. Info. Serv., U.S. Dept. Commerce, Washington, D.C. 86 pp.

Anon. 1983. Slipped disc drug gains approval. AARP News Bull. 24 (1).

Anon. 1978. 'Star Ruby' case upheld. Plant Indus. News 20 (1): 9, 19-20.

Anon. 1977. 'Star Ruby' distributed. Plant Indus. News. 18 (4).

Anon. 1977. Steady kiwi expansion. Blue Anchor 54 (5): 30.

Anon. 1978. "Strange fruit" no longer strange to eastern retailers. Blue Anchor 55 (5): 21-23.

Anon. 1970. Talking over tamarillos. New Zeal. J. Agr. 120 (6): 111.

Anon. 1979. (based on report by P.J. Ferree, U.S. Agr. Attaché, Bangkok). Thailand brakes growth of pineapple industry. For. Agr. 17 (10): 10.

Anon. 1953. The Cape gooseberry. Queensland Agr. J. 76 (5): 288-290.

Anon. 1949. The feijoa: host for fruit fly in New South Wales. The Seed & Nursery Trader, Sept. 20. P. 13.

Anon. 1942. The nonvolatile acids of the carambola were investigated. Report of the Puerto Rico Exper. Sta., U.S. Dept. Agr. 1940 (issued May 1942). Pp. 91-93.

Anon. 1961. The Queensland Agricultural and Pastoral Handbook. Vol. II. 2nd ed. Fruits and Vegetables. Dept. of Agr. & Stock, Brisbane, Australia.

Anon. 1971. The world banana economy. Commodity Bull. Ser. #50. Food & Agr. Org. of the United Nations, Rome. 65 pp.

Anon. 1966. The world's largest persimmon ranch. Blue Anchor 43 (4): 20-21.

Anon. 1972. Transporte de banana por barco volador resulta rápido y eficiente. La Hacienda 67 (10): 20-21.

Anon. 1961. Transvaal lowveld farmers now 'milk' their pawpaws. So. Afr. Sugar J. 45 (10): 886.

Anon. 1961. Tree tomatoes popular in New Zealand. Queensland Fr. & Veg. News. 19 (13): 307 (reproduced from N.Z. Orchardist).

Anon. 1953. Uses of New Zealand grapefruit. New Zeal. J. Agr. 87 (5): 491-496.

Anon. 1970. Vegetative propagation of jaman. Indian Horticult. 14 (2): 17.

Apelbaum, A., G. Zauberman and Y. Fuchs. 1977. Prolonging storage life of avocado fruits by subatmospheric pressure. HortSci. 12 (2): 115-117.

Aponte, C.E. 1963. El cultivo de guayaba en Puerto Rico. Carib. Agr. 1 (3): 199-215.

Aponte, C.E. 1974. El cultivo del agaucate, Pt. I. La Hacienda 69 (12): 23-26.

Appleman, D. 1944. Preliminary report on toxicity of avocado leaves. Calif. Avocado Soc. Yearbook. P. 37.

Aragaki, M., S. Nakata and C. Long. 1969. The etiology, including nutritional considerations, of passion fruit brown spot in Hawaii. Pl. Dis. Rep. 53 (10): 789-792.

Aranha, C. 1970. A árvore do tomate. O Estado de Sao Paulo, Supp. Agricola, Nov. 25, p. 4.

Araque, R. 1963. El higo. Serie de Cultivos #7. Consejo de Bienestar Rural, Caracas, Venez. 12 pp.

Araque, R. 1963. La piña. Serie de Cultivos #5. Consejo de Bienestar Rural, Caracas, Venez. 33 pp.

Araque, R. 1967. Produccion de frutales en Venezuela (Mimeo). Consejo de Bienestar Rural, Caracas. 28 pp.

Araujo, J. 1965. Seleccione sus papayas por medio de la flor. Agr. en el Salvador 6 (4,5,6): 3-5.

Archapong, A. and F. Wieneke. 1977. Processing of plantain in simple operations. Proc. Int'l Conf. on Rural Develop. Tech.: An International Approach. Asian Inst. Tech., Bangkok. Pp. 227-242.

Arctander, S. 1960. Perfume and flavor materials of natural origin. Steffen Arctander, Elizabeth, N.J. 370 pp.

Arechavaleta, J. 1902. Flora Uruguay Vol. II (Part 1). Establecimiento Tipo-litografico "Oriental", Montevideo, Uruguay. 160 pp.

Arias, O. 1979. Efecto del flurenol en la produccion de hijos laterales de pejibaye (*Bactris gasipaes* HBK). Agron. Costarr. 3 (1): 47-52.

Ariosa Terry, M.D. 1982. Una nueva enfermedad de la guayaba (*Psidium guajava* L.) en la provincia de Sancti Spiritus. Centro Agricola 9 (2): 3-7.

Arostegui, F., C.F. Asenjo and A.I. Munez. 1954. Studies on the West Indian Cherry, *Malpighia punicifolia* L., observation and data on a promising selection. Proc. Fl. St. Hort. Soc. 67: 250-255.

Arostegui, F., C.F. Asenjo, A.I. Muniz and L. Alemany. 1955. Observations and data on a promising selection of the West Indian Cherry, *Malpighia punicifolia* L. J. Agr. U. Puerto Rico 39 (2): 51-56.

Arnoldo, M. and P.W. Hummelinck. 1936. Over *Hylocereus undatus*. Succulenta 8: 1-5.

Arroyo, R. 1943. Preparacion de vinos en el hogar usando la manza Malaya, tambien conocido como pomarrosa americana. Agr. Exper. 3 (3): 9-12.

Arpaia, M.L., F.G. Mitchell, A.A. Kader and G. Mayer. 1986. Ethylene and temperature effects on softening and white core inclusions of kiwifruit stored in air controlled atmospheres. J. Amer. Soc. Hort. Sci. 111 (1): 149-153.

Arshad Ali, M. and Tarikul Alam, M. 1963. Quantitative survey of *Holarrhena antidysenterica* (Kurchi) and *Emblica officinalis* plants in the andhermanik forests of the Chittagong Forest Division. Pakistan J. Sci. 15 (4): 161-165.

Arteaga, P.H. 1957. Papayo (*Carica papaya* L.). Cafe de el Salvador 27 (302/3): 33-34.

Arumugam, S. and K.G. Shanmugavelu. 1975. Studies on the effect of Sarcotesta on the seed germination of papaya (*Carica papaya* L.). Seed Research 3 (2): 77-80.

Asenjo, C.F. 1953. The story of the West Indian Cherry. Bol. del Colegio de Quimicos 10 (8-11): 1-11.

Asenjo, C.F. 1959. Vitamin C in acerola and rose hips. J. Agr. U. of Puerto Rico 43 (3): 212-213.

Asenjo, C.F. and A.R. Freise de Guzman. 1946. The high ascorbic acid content of the West Indian Cherry. Science 103: 219.

Asiedu, J.J. 1980. Foodstuff technology: the processing of plantain, an example from Ghana. GATE 3: 19-20.

Asprey, G.F. and P. Thornton. 1953. Medicinal plants from Jamaica. Pt. I. West Indies Med. J. 2 (4): 233-252.

Assaf, R. and P. Rivals. 1977. Neoculture du néflier du Japon (*Eriobotrya japonica* Lindl.). Fruits 32 (4): 237-251.

Astridge, S.J. 1975. Cultivars of Chinese gooseberry (*Actinidia chinensis*) in New Zealand. Econ. Bot. 29: 357-360.

Atchley, J. and P.A. Cox. 1984. Breadfruit fermentation in Micronesia. Econ. Bot. 39 (3): 326-335.

Aubert, B., R. Vogel, J.M. Bove, and C. Bove. 1982. Virus infections of *Citrus* transmitted by aphids in the island of Reunion. [Volkamer lemon]. Fruits 37 (7-8): 441-465.

Audas, J.W. 1919. The litchi. J. Dept. Agr. Victoria 17 (6): 371-373.

Aurnhammer, G., H. Wagner, L. Hörhammer and L. Farkas. 1971. Synthese des 7-B—neohesperidosyl-4'-B-D-glucopyranosyl-naringenins, eines Flavanontriglykosides aus citrusfrüchten. Chem. Ber. 104: 473-478.

Awad, N.M. and H. Amenomori. 1972. Astringency removal in persimmon fruits with ethephon. HortSci. 7 (2): 174.

Awada, M. 1977. Relations of nitrogen, phosphorus and potassium fertilization to nutrient composition of the petiole and growth of papaya. J. Amer. Soc. Hort. Sci. 102 (4): 413-418.

Awada, M. and C. Long. 1978. Relation of nitrogen and phosphorus fertilization to fruiting and petiole composition of 'Solo' papaya. J. Amer. Soc. Hort. Sci. 103 (2): 217-219.

Awada, M. and C. Long. 1971. Relation of petiole nitrogen levels to nitrogen fertilization and yield of papaya. J. Amer. Soc. Hort. Sci. 96 (6): 745-749.

Awada, M., R. Suchisa, and Y. Kanchiro. 1975. Effects of lime and phosphorus on yield, growth, and petiole composition of papaya. J. Amer. Soc. Hort. Sci. 100 (3): 294-298.

Awadallab, A., S. Fouds, O.F. Selim, and A.G. Hashem. 1975. The chemical control of the *Ziziphus* fruit fly, *Carpomyia incompleta*, Becker, in Assiut area. Agr. Res. Rev. 53 (1): 123-125.

Ayala, A. 1968. Nematode problems of pineapple in Puerto Rico. In: Tropical Nematology; G.C. Smart and V.G. Perry, editors. Pp. 49-60.

Ayers, E.L. 1957. Erinose. Fla. Lychee Growers Assn. Year Book and Proc. 5th Ann. Mtg., Winter Haven, Fla. Pp. 15-17.

Azam, B., F. Lafitte, F. Obry and J.L. Paulet. 1981. Le Feijoa en Novelle-Zélande. Fruits 36 (6): 361-384.

Bafna, A.M., N.M. Parikh, G.B. Shah and P.M. Bhatt. 1983. Effect of long term fertilisation on yield of sapota (*Achras sapota* L.) [sapodilla] and N fractions in soil. So. Indian Hort. 31 (2/3): 66-69.

Bailey, F.L. 1952. Culture of feijoa trees. New Zeal. J. Agr. 84 (4): 291-293, 295-296.

Bailey, F.L. and E. Topping. 1951. Chinese gooseberries, their culture and uses. Bull. 349. New Zeal. Dept. Agr., Wellington, N.Z. 23 pp.

Bailey, L.H. 1947. The standard cyclopedia of horticulture. 3 vols. (Reprint of 1942 ed.). The Macmillan Co., N.Y. 3639 pp.

Bajpai, P.N. 1957. Blossom biology and fruit set in *Phyllanthus emblica*. Indian J. Hort. 14 (2): 99-102.

Bajpai, P.N. and R.K. Trivedi. 1961. Storage of mango seedstone. Hort. Adv. 7: 228-229.

Bakhshi, J.C. and N.S. Randhawa. 1967. Bright prospects for guava cultivation in the Punjab. Indian J. Hort. 11 (3): 3-4, 23.

Bakker, W. 1974. Notes on East African plant virus diseases VI: Virus diseases of *Passiflora* in Kenya. E. Afr. Agr. & For. J. 40 (1): 11-23.

Baldry, J. and F.D. Dempster. 1976. Green bananas for cooking: a study of taste-panel preferences between certain clones. Trop. Sci. 18 (4): 219-225.

Balls, A.K., R.R. Thompson and M.W. Kies. 1941. Bromelin-properties and commercial production. Indus. & Engrg. Chem. 33 (7): 350-353.

Bandyopadhyay, C. and A.S. Gholap. 1979. On the chemical composition of mango kernel fat (*Mangifera indica* L.) Curr. Sci. 48 (21): 935-936.

Banks, N.F. 1964. Tree tomatoes, a rich source of vitamin C for winter. New Zeal. Agr. 108 (4): 413.

Banta, E.D. 1952. Behold! The lychee. Amer. Fr. Grower, Oct. 10-11, 20-21.

Barker, W.G. 1959. A system of maximum multiplication of the banana plant. Trop. Agr. (Trin.) 36: 275-284.

Barmore, C.R. and E.F. Mitchell. 1975. Ethylene preripening of mangos prior to shipment. Proc. Fl. St. Hort. Soc. Vol. 88: 469-471.

Barmore, C.R. and A.H. Rouse. 1976. Pectinerase activity in controlled atmosphere stored avocados. J. Amer. Soc. Hort. Sci. 101 (3): 294-296.

Barnett, G.B. 1947. The papaw or papaya (*Carica papaya*). Leaflet 542; reprinted from J. Dept. Agr. W. Aust. 15, 2nd Ser. (1): 21-29. W. Aust. Dept. Agr., Perth. 9 pp.

Barnett, G.B. 1931. The pineapple. Bull. (reprinted from Agr. Gaz. Feb. 1931). New So. Wales Dept. Agr. 5 pp.

Barrau, J. 1959. Investigation to extend season of breadfruit yield. So. Pacific Quart. Bull. 9 (2): 41, 43.

Barrett, M.F. 1956. Common exotic trees of South Florida. Univ. Fla. Press, Gainesville, Fl.

Barrett, O.W. 1928. The tropical crops. The Macmillan Co., N.Y.

Barreveld, W.H. 1941. Industrial use of second quality dates. World Crops 23 (4): 209-210.

Barron, R.W. and R.W. Olsen. 1960. Processed products from Murcott orange. Pt. II. Characteristics of processed products. Proc. Fl. St. Hort. Soc. 73: 279-283.

Baslas, K.K. 1961. Essential oil from the flowers of *Mangifera indica*. Perf. & Ess. Oil Record. Mar.: 156-158.

Basu, D. and R. Sen. 1974. Alkaloids and coumarins from root-bark of *Aegle marmelos*. Phytochem. 13: 2329-2330.

Bates, R.P. 1964. Factors affecting foam production and stabilization of tropical fruit products. Food Tech. 18 (1): 93-96.

Bates, R.P. 1968. The retardation of enzymatic browning in avocado puree and guacamole. Proc. Fl. St. Hort. Soc. 81: 230-235.

Beal, P.R. 1975. Hybridization of *Passiflora edulis* Sims and *P. edulis* Sims f. *flavicarpa* Degener. Bull. 723. Queensland J. Agr. & Anim. Sci. 32 (1): 101-111.

Beal, P.R. 1981. Screening of mango varieties at Bowen, Queensland. Queensland J. Agr. & Anim. Sci. 38 (1): 71-85.

Becker, S. 1958. The production of papain — an agricultural industry for tropical America. Econ. Bot. 12 (1): 62-79.

Bedoya, L., J., O. Medina L., R.D. Zarate R., and R. Torres M. 1983. Etiologia de la pudricion radicular de Maracuya Amarillo *Passiflora edulis* var. *flavicarpa* Degener. Acta Agron. 33 (4): 54-60.

Beerh, O.P. and V.R. Rane. 1964. Utilization of Rangpur limes. Indian Food Packer 18 (2): 22-24.

Behl, P.N., R.M. Captain, B.M.S. Bedi and S. Gupta. 1966. Skin-irritating and sensitizing plants found in India. Dr. P.N. Behl, Dept. Derm., Irwin Hosp. & M.A. Med. Coll., New Delhi. 206 pp.

Bembower, W. and A. Champoopho. 1955. Mango varieties of Thailand. Indian J. Hort. 12 (2): 58-62.

Ben-Aire, R. and S. Guelfat-Reich. 1976. Softening effects of CO_2 treatment for removal of astringency from stored persimmon fruits. J. Amer. Soc. Hort. Sci. 101 (2): 179-181.

Ben-Aire, R., N. Segal, and S. Guelfat-Reich. 1984. The maturation and ripening of the 'Wonderful' pomegranate. J. Amer. Soc. Hort. Sci. 109 (6): 898-902.

Benero, J.R., L.E. Cancel, and A.L. Rivero. 1970. Dehydration of fermented citron. J. Agr. U. Puerto Rico 54 (3): 508-516.

Benero, J.R., A.L. Collazo de Rivera and L.M.I. de George. 1974. Studies on the preparation and shelf-life of soursop, tamarind, and blended soursop-tamarind soft drinks. J. Agr. U. Puerto Rico 58 (1): 99-104.

Benero, J.R., and A.J. Rodriguez. 1971. Mango pulp extracting method. J. Agr. U. Puerto Rico 60 (4): 513-515.

Benero, J.R., A.A. Rodriguez and A. Collazo de Rivera. 1972. A mechanical method for extracting tamarind pulp. J. Agr. U. Puerto Rico 56 (2): 185-186.

Benero, J.R., A.J. Rodriguez, and A. Roman de Sandoval. 1971. A soursop pulp extraction procedure. J. Agr. U. Puerto Rico 55 (4): 518-519.

Ben-Jaacov, J., B. Steinitz and Y. Tendler. 1984. Dark storage of calamondin. HortSci. 19 (2): 263-264.

Benschoter, C.A., R.J. King and P.C. Witherell. 1984. Large chamber fumigations with methyl bromide to destroy Caribbean fruit fly in grapefruit. Proc. Fl. St. Hort. Soc. 97: 123-125.

Benthall, A.P. 1946. Trees of Calcutta and its neighborhood. Thacker Spink & Co. (1933) Ltd., Calcutta. 513 pp.

Bento de Oliveira, J. 1959. Tamareiras — da viabilidad desta cultura em Cabo Verde. V. Cabo Verde 11 (122): 12-15.

Bento de Oliveira, J. 1960. Tamareiras — da viabilidade desta cultura em Cabo Verde. VI. Cabo Verde 11 (124): 29-34.

Ben-Yehoshua, S. 1969. Gas exchange transpiration and the commercial deterioration in storage of orange fruit. J. Amer. Soc. Hort. Sci. 94: 524-528.

Ben-Yehoshua, S., B. Shapiro, and I. Kobiler. 1982. New method of degreening lemons by a combined treatment of ethylene-releasing agents and seal-packaging in high-density polyethylene film. J. Amer. Soc. Hort. Sci. 107 (3): 365-368.

Berk, Z. 1973. Soybean proteins for bananas. Israel J. Res. & Devel. 1 (2): 29-30.

Bernhardt, L.W., Z. de Martin, and E. Angolucci. 1971/72. Industrializacao do pseudofruto do cajueiro. Coletanea do Inst. de Tech. de Alimentos 4: 47-61.

Berrill, F.W. 1966. New chemical flower induction in pines. Queensland Fr. & Veg. News 29 (1): 9-10.

Berrill, F.W. 1959. Plant sword-leaved banana suckers. Queensland Agr. J. 85 (1): 15-17.

Bertin, Y. 1976. La taille de l'avocatier a la Martinique. Fruits 31 (6): 391-399.

Bertin, Y. 1970. L' Avocatier en Martinique. Proc. 7th Ann. Mtg. C.F.C.S., pp. 101-104.

Bettai Gowder, R. and I. Irulappan. 1971. Performance of 'Neelum' variety of mango (*Mangifera indica* L.) on polyembryonic rootstocks as compared to that on monoembryonic rootstock. Madras Agr. J. 58 (3): 183-189.

Beutel, J.A., F.H. Winter, S.C. Mannars and M.W. Miller. 1976. A new crop for California: kiwifruit. Calif. Agr. 30 (10): 5-7. (reprinted under the title "Growing, Processing and Marketing kiwifruit", Blue Anchor 54 (1): 18-21; Jan.-Feb. 1977).

Beutel, M. 1975. Kiwifruit recipes. Kiwi Growers of Calif., P.O. Box 922, Gridley, Calif. 95948. 64 pp.

Bhaduri, A.P., R.P. Rastogi, and N.M. Khanna. 1968. Biologically active carissone derivatives. Indian J. Chem. 6 (7): 405-406.

Bhakuni, D.S., S. Tewari and M.M. Dhar. 1972. Aporphine alkaloids of *Annona squamosa*. Phytochem. 11: 1819-1822.

Bhan, K.C. and P.K. Mazumder. 1956. Propagation trials on banana. II. Butts and bits of rhizomes as planting materials. Indian J. Hort. 13 (3): 141-148.

Bhan, K.C. and P.K. Mazumder. 1961. Propagation trials on banana. III. Effect of age of suckers and season of planting on banana production. Ind. J. Hort. 18 (3): 187-197.

Bharath, S. 1969. Cultivation of papaw in Trinidad and Tobago. Crop Bull. #15. Min. of Agr., Lands and Fish., (Trin. & Tobago); pp. 3-7.

Bharath, S. 1970. Guava propagation from stem cuttings in Trinidad. J. Agr. Soc. Trin. and Tobago. 70 (4): 396-399.

Bhatia, B.S. and G. Lal. 1956. Development of products from jack fruit. Pt. VI. Canned green jack fruit in brine and in curry. Indian Food Packer 10 (8): 7-8.

Bhatia, B.S. and G. Lal. 1956. Development of products from jack fruit. Pt. VII: Dried green jack fruit and jack pickle. Indian Food Packer 10 (9): 13-14.

Bhatia, B.S., G.L. Mehta, and K.G. Nair. 1963. Dehydration of litchi (*Litchi chinensis*). Food Sci. 12 (11): 313-315.

Bhatia, B.S., G.S. Siddappa and G. Lal. 1955. Development of products from jack fruit. Pt. I. Frozen canned jack fruit and jack fruit jam. Indian Food Packer 9 (9): 8-12.

Bhatia, B.S., G.S. Siddappa and G. Lal. 1955. Development of products from jack fruit. Pt. II. Jack fruit preserve, candy, chutney and dried bulbs. Indian Food Packer 9 (11): 7-9.

Bhatia, B.S., G.S. Siddappa and G. Lal. 1956. Development of products from jack fruit. Pt. III. Canned jack seeds in brine, tomato sauce and curry. Indian Food Packer 10 (1): 9-10, 12.

Bhatia, B.S., G.S. Siddappa and G. Lal. 1956. Development of products from jack fruit. Pt. IV. Jack fruit concentrate, powder, pulp, toffees and squash. Indian Food Packer 10 (6): 11-12, 18.

Bhatia, B.S., G.S. Siddappa and G. Lal. 1956. Development of products from jack fruit. Pt. V. Dried jack seeds and flower, roasted nut and jack "papad" (papai). Indian Food Packer 10 (7): 9.

Bhatia, B.S., G.S. Siddappa and G. Lal. 1960. Preparation of pectin, pectin extract and syrup from jack fruit rind. Food Sci. 9 (12): 421.

Bhatia, B.S., G.S. Siddappa and G. Lal. 1956. Some physico-chemical changes in canned jack-fruit during storage. J. Sci. & Indus. Res. 15C (4): 91-95.

Bhatia, I.S. and K.L. Bajaj. 1975. Chemical constituents of the seeds and bark of *Syzygium cumini*. Planta Med. 28: 346-352.

Bhatia, I.S., S.K. Sharma and K.L. Bajaj. 1974. Esterase and galloyl carboxylase from *Eugenia jambolana* (Lam.) leaves. Indian J. Exper. Biol. 12: 550-552.

Bhatnagar, S.S., H. Santapan, F. Fernandez, V.N. Kamat, N.J. Dastoor, and T.S.N. Rao. 1961. Physiological activity of Indian medicinal plants [jambolan]. J. Sci. & Indus. Res. 20A (8): 1-24.

Bhutiani, R.C. 1956. Fruit and vegetable preservation industry in India. [jackfruit]. Cent. Food Tech. Res. Inst., Mysore, India.

Bhutiani, R.C., J.V. Shankar and P.G. Kutty Menon. 1963. Papaya. Cent. Food Tech. Res. Inst., Mysore, India.

Bickern, W. 1903. Bietrag zur kenntnis der *Casimiroa edulis* La Llave. Arch. de Pharmacie 241: 166-176.

Bielorai, H., J. Shalhevet, and Y. Levy. 1983. The effect of high sodium irrigation water on soil salinity and yield of mature grapefruit orchard. Irrig. Sci. 4: 255-266.

Bierrenbach de Castro, J. 1973. As goiabas australianas. O Estado de Sao Paulo, 19 (944) Supp. Pp. 8-10.

Bierrenbach de Castro, J. 1979. Cultura da lichia na regiao de Bauru,

S.P. O Estado de Sao Paulo, Supp. Agr. Jan. 31, pp. 1 & 3.

Biggs, R.H. and S.V. Kossuth. 1980. Modeling the effectiveness of Release® as a citrus harvest aid for 'Valencia' fruits. Proc. Fl. St. Hort. Soc. 93: 301-304.

Bird, J., A. Krochmal, G. Zentmyer and J. Adsuar. 1966. Fungus diseases of papaya in the U.S. Virgin Islands. J. Agr. U. Puerto Rico 50 (3): 186-200.

Bischoff, W.H. 1950. Autos-Arias-Avocados. Miami Sunday News Mag. Oct. 22, p. 20.

Bischoff, W.H. 1949. Florida pineapple comes back. Miami Sunday News Mag. July 31. p. 20.

Bischoff, W.H. 1949. Lychee fever: Florida west coast residents rush to plant trees yielding big profit. Miami Sunday News Mag. Oct. 23: p. 20.

Bischoff, W.H. 1949. Tailor-made papaya tree. Miami Sunday News Mag. July 24, p. 20.

Bishop, E.J.B., J.B. Gradwell, J.A.G. Nell and D.M. Bradfield. 1965. Pineapple plants, a good drought fodder for cattle. Farming in So. Afr. 41 (9): 6-7, 9.

Blaak, G. 1976. Pejibaye. Abs. on Trop. Agr. 2 (9): 9-17.

Blaak, G. 1980. Vegetative propagation of pejibaye (*Bactris gasipaes* HBK). Turrialba 30 (3): 258-261.

Black, R.F. and P.E. Page. 1969. Pineapple growth and nutrition over a plant crop cycle in south-eastern Queensland. 2. Uptake and concentrations of nitrogen, phosphorus and potassium. Queensland J. Agr. & Anim. Sci. 26 (3): 385-405.

Blacker, G.W.J. 1960. Grafting passion vines. New Zeal. J. Agr. 101 (4): 402, 403, 405.

Blazquez, C.H. 1967. Ortanique, a new orange-tangerine cross. Proc. Fl. St. Hort. Soc. 80: 331-337.

Blazquez, C.H., A.G. Naylor and D. Hastings. 1966. Sphaeropsis knot of lime. Proc. Fl. St. Hort. Soc. 79: 344-350.

Bloch, F. and R.J. Binder. 1960. Report on processing trials with fruits and nuts not common in United States. Chemurgic Dig. 19 (10): 13-14.

Blohm, H. 1962. Poisonous plants of Venezuela. Harvard U. Press, Cambridge, Mass. 136 pp.

Blondeau, J.P. and Y. Bertin. 1978. Carences minérales chez la grenadille (*Passiflora edulis* Sims var. *flavicarpa*). I. Carences totales en N, P, K, Ca, Mg. Croissance et symptomes. Fruits 33 (6): 433-443.

Blumenfeld, A. 1980. Fruit growth of loquat. J. Amer. Soc. Hort. Sci. 105 (5): 747-750.

Bolt, L.C. and A.J. Joubert. 1968. Litchi grafting trials. Farming in So. Africa 44 (2): 11, 13.

Bombardelli, A., A. Bonati, B. Gabetta, E.M. Martinelli and G. Mustich. 1975. Passiflorine, a new glycoside from *Passiflora edulis*. Phytochem. 14: 2661-2665.

Bonavia, E. 1890. The cultivated oranges and lemons, etc. of India and Ceylon. Vol. I, 384 pp. & Vol. II, 259 pls. W.H. Allen & Co., 13, Waterloo Place, Pall Mall, S.W., London.

Bondad, N.D. 1982. Mango and its relatives in the Philippines. Phil. Geog. J. 26 (2): 88-100.

Bondad, N.D. 1982. Origin and distribution of mango. Phil. Geog. J. 26 (1): 44-52.

Bondad, N.D. 1980. World mango production and trade. World Crops (Nov.): 160-168.

Bopaiah, B.M. 1985. Microbial spoilage of cashew apples and its prevention. Indian Cashew J. 16 (2): 15-17.

Borg, J. 1937. Cacti—a gardener's handbook for their identification and cultivation. Blandford Press, Ltd., London.

Borrel, M. and A. Diaz. 1982. Effects of mechanical pruning on yield of citrus trees. Proc. Internat. Soc. of Citriculture. Vol. I. 190-194.

Bowden, R.P. 1968. Processing quality of oranges grown in the Near North Coast area of Queensland. Queensland J. Agr. & Anim. Sci. 25 (3): 93-119.

Bowers, R. 1981. Breadfruit—a low energy requirement source of carbohydrate for the wet tropics. Entevicklung & landlicher raum 15 (2): 11-13.

Bowman, G.F. 1980. Utilizacion de desperdicios de banano. La Hacienda 75 (3): 34-36.

Braddock, R.J. 1974. Citrus seeds: a potential food source. Span 17 (2): 86-87.

Braddock, R.J. and J.W. Kesterson. 1973. Dried citrus juice sacs for the food and beverage industries. Proc. Fl. St. Hort. Soc. 86: 261-263.

Brekke, J.E. and L. Allen. 1967. Dehydrated bananas. Food Technol. 21 (139): 101-105.

Brekke, J.E., C.G. Cavaletto, T.O.M. Nakayama and R.H. Suehina. 1976. Effects of storage temperature and container lining on some quality attributes of papaya nectar. J. Agr. Food Chem. 24 (2): 341-343.

Brekke, J.E., C.G. Cavaletto, A.E. Stafford & H.T. Chan, Jr. 1975. Mango: processed products. U.S. Dept. Agr., Agr. Res. Serv., in coop. with Hawaii Agr. Exper. Sta. 26 pp.

Brimblecombe, A.R. 1961. Scale insects on papaws. Queensland Agr. J. 87 (3): 163-164.

Brite, J.R. 1951. A plantation with a future—pineapple production soars in Miami area. Miami Herald. Aug. 12, p. 6-G.

British War Charities. 1951. Buen Provecho: Caracas Cookery. Rev'd. Caracas J., Venez., p. 151, 161. [mamey].

Britton, N.L. 1918. Flora of Bermuda. Charles Scribner's Sons, N.Y.

Britton, N.L. and C.F. Millspaugh. 1920. The Bahama flora. Pub'd by authors, N.Y.

Britton, N.L. and P. Wilson. 1925-6; 1930. Botany of Porto Rico and the Virgin Islands (Sci. Surv. P.R. and V.I.) Vol. 6, Pts. 1-4. N.Y. Acad. Sci., N.Y.

Broadley, R.A. 1974. Nematode control in banana planting material. Queensland Agr. J. 100 (7): 279-280.

Brocklehurst, K., J. Carlsson, P. Marek, P.J. Kierstan, and E.M. Crook. 1973. Covalent chromatography: Preparation of fully active papain from dried papaya latex. Biochem. J. 133: 573-584.

Brodrick, H.T. 1971. Mango diseases. Leaflet 72. Subtrop. Fr. Ser. 12; Mango Ser. #3. Dept. Agr. & Tech. Serv., Pretoria, So. Africa.

Brogdon, J., F.G. Butcher, G.W. Dekle, E.G. Kelsheimer, and D.O. Wolfenberger. 1966. Lychee insect control. Cir. 131. Coop. Exten. Work, Agr. Exten. Serv., U. of Fla., Gainesville. 4 pp.

Brooks, R.F. and J.L. Knapp. 1983. Control of chaff scale on 'Dancy' tangerine. Proc. Fl. St. Hort. Soc. 96: 367-369.

Brooks, R.M. and H.P. Olmo. 1952. Register of new fruit and nut varieties, 1920-1950. [white sapote]. Berkeley, Calif.

Brooks, R.M. and H.P. Olmo. 1972. Register of new fruit and nut varieties, List 27. HortSci. 7 (5): 455-458. [Chinese gooseberry]

Brooks, R.M. and H.P. Olmo. 1982. Register of new fruit and nut varieties. List 32. HortSci. 17 (1): 19. [grapefruit]

Brooks, R.M. and H.P. Olmo. 1984. Register of new fruit and nut varieties. List 34. Lychee: Kaimana. HortSci. 19 (3): 361.

Broughton, W.J. and B.E. Chan. 1979. Maturation of Malaysian fruits. IV. Storage conditions and ripening of passion fruit (*Passiflora edulis* L. var. *flavicarpa*). MARDI Res. Bull. 7 (2): 27-34.

Broughton, W.J. and S.F. Leong. 1979. Maturation of Malaysian fruits. III. Storage conditions and ripening of guava (*Psidium guajava* L. var. GU3 and GU4). MARDI Res. Bull. 7 (2): 12-26.

Broughton, W.J. and Ten Guat. 1979. Storage conditions and ripening of the custard apple *Annona squamosa* L. Scientia Hort. 10: 73-82.

Broughton, W.J. and H.C. Wong. 1979. Storage conditions and ripening of chiku fruits, *Achras sapota* L. Scientia Hort. 10: 377-385.

Brown, B.I. 1966. Observations on physical and chemical properties of acerola fruit and puree. Queensland J. of Agr. & Anim. Sci. 23 (4): 599-604.

Brown, F.B.H. 1935. Flora of southeastern Polynesia. Bull. 130. Bernice P. Bishop Mus., Honolulu.

Brown, M.B. 1974. Kiwi culture in the Vumba. Hortus 21: 24-26.

Brown, R.L., C.S. Tang and R.K. Nishimoto. 1983. Growth inhibition from guava root exudates. HortSci. 18 (3): 316-318.

Brown, W.H. 1921. Minor products of Philippine forests. Vol. III. Bull. 22. Bur. Forestry, Dept. Agr. & Nat. Res., Manila.

Brown, W.H. 1951. Useful plants of the Philippines. Vol. I (Tech. Bull. 10). Phil. Dept. Agr. & Nat. Res., Manila.

Brown, W.H. 1946. Useful plants of the Philippines. Vol. 3 (Tech. Bull. 10) Dept. of Agr. & Commerce, Manila.

Browne, F.G. 1955. Forest trees of Sarawak and Brunei and their products. Gov't Ptg. Office, Kiching, Sarawak.

Burden, O.J. 1968. A wilt of papaw caused by *Verticillium dahliae* Kleb. Queensland J. Agr. & Anim. Sci. 25 (3): 153-164.

Burden, O.J. 1968. Reduction of banana anthracnose following hot-water treatment of the green fruit. Queensland J. Agr. & Anim. Sci. 25 (3): 135-144.

Burdick, E.M. 1971. Carpaine: an alkaloid of *Carica papaya*—its chemistry and pharmacology. Econ. Bot. 25 (4): 363-365.

Burg, S.P. and E.A. Burg. 1966. Auxin induced ethylene formation: its relation to flowering in the pineapple. Science Mag. 27. P. 1269.

Burkill, I.H. 1935. Dictionary of the economic products of the Malay Peninsula (2 vols.). Crown Agents for the Colonies, London, Eng.

Burney, B. 1980. Exotics: Second part of our guide to some that may be grown here. New Zeal. J. Agr. 140 (3): 56.

Burns, R.M., S.B. Boswell, C.D. McCarty and B.W. Lee. 1972. Experimental growing of lemons on trellises. Calif. Agr. 26 (9): 10-15.

Burns, R.M., D.O. Rosedale, J.E. Pehrson and C.W. Coggins. 1964. Preliminary trials indicate gibberellin delays lime maturity. Calif. Citrograph 49 (12): 488-490.

Busson, F. 1965. Plantes Alimentaires de l'Ouest Africain. Leconte, Marseille, France.

Butani, D.K. 1975. Insect pests of fruit crops and their control: Sapota. Pesticides 9 (11): 37-39.

Butani, D.K. 1976. Insect pests of fruit crops and their control: 21. Pomegranate. Pesticides 10 (6): 23-26.

Butani, D.K. 1978. Insect pests of tamarind and their control. Pesticides. 12 (11): 34-41.

Butani, D.K. 1975. Papaya (In: "Insect pests of fruit crops and their control"). Pesticides. Feb. P. 23.

Butani, D.K. 1978. Pests and diseases of jackfruit in India and their control. Fruits 33 (5): 351-357.

Butani, D.K. 1974. Pests of fruit crops in India and their control. #12. Loquat. Pesticides 3 (12): 17-18.

Butani, D.K. 1977. Pests of litchi in India and their control. Fruits 32 (4): 269-273.

Butcher, F.G. 1955. A new caterpillar on lychees. Fl. Lychee Growers Assn. Year Book and Proc. 3rd. Ann. Mtg., Winter Haven, Fl. Pp. 19-20.

Butcher, F.G. 1956. Bees pollinate lychee blooms. Proc. Fl. Lychee Growers Assn., Winter Haven, Fl. Nov. 14; 59-60.

Butcher, F.G. 1957. Pollinating insects on lychee blossoms. Proc. Fl. St. Hort. Soc. 70: 326-328.

Butcher, F.G. 1955. Some insect problems on guavas. Proc. Fl. St. Hort. Soc. 68: 292-293.

Butcher, F.G. 1957. Tung borer injury in the lychee. Proc. Fl. Lychee Growers Assn., Winter Haven, Fl. Nov. 11, pp. 21-22.

Butterfield, H.M. 1970. History of the persimmon, quince and pomegranate industries in California. Blue Anchor 47 (4): 18-19, 22.

By-Products Dept., TSC. 1975. A study on production of alcohol from pineapple skin-juice. Taiwan Sugar 22 (1): 14-16.

Cadillat, R.M. 1968. L'ananas en Amérique Centrale. Fruits 23 (2): 119-122.

Cailes, R.L. 1952. The cultivation of the Cape gooseberry. Leaflet 2001. Reprint from J. Agr. of Western Australia.

California Fruit Exchange. 1964. Fresh persimmon recipes. Blue Anchor 40 (4): 21-22.

Calvacante, P.B. 1972, 1974, 1979. Frutas comestiveis da Amazonia: I, II, III. Museu Paraense Emilio Goeldi, Belem, Brazil.

Calzada Benza, J., V. Bautista Carrasco and J. Bermudez Rodriguez. 1967. Status del mejoramiento del papayo en la Molina, Peru. Agron. Trop. Peru 17 (4): 381-389.

Camacho, E. and J. Soria V. 1970. Palmito de pejibaye. Proc. Trop. Reg. Amer. Soc. Hort. Sci. 14: 122-132.

Cameron, J.W. and R.K. Soost. 1986. 'C35' and 'C32': citrange rootstocks for citrus. HortSci. 21 (1): 157-158.

Camp, W.H. 1947. Some additional comments on the naranjilla. J. N.Y. Bot. Gard. 48 (571): 159-160.

Campbell, C.W. 1985. Carambola industry in Florida. HortSci. 20 (1): 16.

Campbell, C.W. 1979. Effect of gibberellin treatment and hand pollination on fruit set of atemoya (*Annona* hybrid). Proc. Trop. Reg. Amer. Soc. Hort. Sci. 23: 122-124.

Campbell, C.W. 1966. Growing the mangosteen in southern Florida. Proc. Fl. St. Hort. Soc. 79: 399-401.

Campbell, C.W. 1965. Performance of peach selections at Homestead, Florida. Proc. Fl. St. Hort. Soc. 78: 324-327.

Campbell, C.W. 1963. Promising new guava varieties. Proc. Fl. St. Hort. Soc. 76: 363-365.

Campbell, C.W. 1976. Selection and propagation of the Spanish lime in Florida. Proc. Fl. St. Hort. Soc. 89: 227-229.

Campbell, C.W. 1959. Storage behavior of fresh Brewster and Bengal lychees. Proc. Fl. St. Hort. Soc. 72: 356-360.

Campbell, C.W. 1983. The bacuripari: a shade-tolerant tropical fruit tree for southern Florida. Proc. Fl. St. Hort. Soc. 96: 219-220.

Campbell, C.W. 1965. The Golden Star carambola. Cir. S-173. U. of Florida, Agr. Exp. Sta., Gainesville.

Campbell, C.W. 1984. The kwai muk, a tropical fruit tree for southern Florida. Proc. Fl. St. Hort. Soc. 97: 318-319.

Campbell, C.W. 1965. The naranjilla. [bulletin] Vol. 2. Rare Fruit Council of S. Fla., Inc., Miami.

Campbell, C.W. 1975. Ten new 'Tahiti' lime selections. Proc. Fl. St. Hort. Soc. 88:455-458.

Campbell, C.W. 1973. The 'Tommy Atkins' mango. Proc. Fl. St. Hort. Soc. 86: 348-350.

Campbell, C.W., R.J. Knight, Jr., and R. Olszack. 1985. Carambola production in Florida. Proc. Fla. State Hort. Soc. 98: 145-149.

Campbell, C.W. and S.P. Lara. 1982. Mamey sapote cultivars in Florida. Proc. Fl. St. Hort. Soc. 95: 114-115.

Campbell, C.W. and S.E. Malo. 1981. Evaluation of the longan as a potential crop for Florida. Proc. Fl. St. Hort. Soc. 94: 307-309.

Campbell, C.W. and S.E. Malo. 1972. The carambola. Fruit Crops Fact Sheet 12. U. of Florida, Coop. Exten. Serv., IFAS and USDA, Cooperating.

Campbell, C.W. and J. Popenoe. 1967. Effect of gibberellic acid on seed dormancy of *Annona diversifolia* Saff. Proc. Trop. Reg. Amer. Soc. Hort. Sci. XV Ann. Mtg. Pp. 33-36.

Campbell, C.W., J. Popenoe and H.Y. Ozaki. 1962. Adaptation trials of tropical and subtropical fruits on sandy soils in Broward County, Florida. Proc. Fl. St. Hort. Soc. 75: 361-363.

Cancel, H.L. 1974. Harvesting and storage conditions for pineapples of the Red Spanish variety. J. Agr. U. Puerto Rico 58 (2): 162-169.

Cancel, H.L. and T. Garcia de Perez. 1979. Effect of fruit size on the ripening of Edward mangoes. J. Agr. U. of Puerto Rico 63 (4): 460-463.

Cancel, L.E., M.A. Gonzales and F. Sanchez Nieva. 1962. Elaboracion del platanutre. Pub. #6. Lab. de Tecnol. de Alimentos, U. de Puerto Rico, Estac. Exper. Agr., Rio Piedras, P.R.

Cancel, L.E., M.A. Gonzalez, and F. Sanchez Nieva. 1964. Utilizacion de los residuos del enlatado de la piña—preparacion de un sirope adecuado para enlatar piñas en rebanadas. Bull. 185. U. de Puerto Rico, Estac. Exper. Agr., Rio Piedras.

Cancel, L.E. and E.R. de Hernandez. 1979. Effect of blanching and freezing on the texture and color of candied citron. J. Agr. U. Puerto Rico 63 (3): 309-314.

Cancel, L.E., I. Hernandez, and E. Rodriguez-Sosa. 1970. Lye-peeling of green papaya (*Carica papaya* L.). J. Agr. U. Puerto Rico 54 (1): 10-27.

Cancel, L.E., J.M. Rivera-Ortiz, and M.R. de Montalvo. 1972. Lye peeling of citron (*Citrus medica* L.). J. Agr. U. Puerto Rico 56 (2): 154-161.

Cañizares Zayas, J. 1944. Los frutales en Cuba. 1944. Rev. de Agr. July, Aug.-Sept. Pp. 23-140.

Cann, H.J. 1956. Split corm in bananas. Queensland Fr. & Veg. News 9 (13): 388.

Cann, H.J. 1967. The custard apple [atemoya]. Agr. Gaz. New So. Wales 18 (2): 85-90.

Cann, H.J. 1966. The papaw. Agr. Gaz. New So. Wales 77: 347-350.

Cannon, R.C. 1963. Bunchy top—a threat to bananas. Queensland Fr. & Veg. News 23 (13): 296-297.

Cannon, R.C. 1957. Closer spacing of pineapples. Queensland Agr. J. 83 (10): 575-578.

Cardenas, M. 1969. Manual de plantas economicas de Bolivia. Imprenta Ichthus, Cochabamba, Bolivia.

Cary, P.R. 1982. Citrus tree density and Pruning practices for the 21st Century. Proc. Internat. Soc. Citriculture. Vol. I: 165-168.

Casimir, D.J., J.F. Keffod and F.B. Whitfield. 1981. Technology and flavor chemistry of passion fruit juices and concentrates. (Pp. 243-295 in: Advances in Food Res. Vol. 27; edited by C.O. Chichester, E.M. Mrak and G.F. Stewart. Academic Press, N.Y.

Castañeda, R.R. 1961. El lulo: una fruta de importancia economica. Agr. Trop. 17 (4): 214-218.

Castañeda, R.R. 1961. Frutas silvestres de Colombia. Vol. I. Author, Bogotá, Colombia.

Castellani, J.J. 1975. Flocos de banana e soja. O Estado de Sao Paulo, Supp. Agr. 1067. Sept. 11, p. 6.

Castellar Palma, N. and A. Figueroa Escobar. 1969. Estudio biologico de dos formas de lepidopteros: *Agraulis vanillae* (Linn.) y *Mechanitis veritabilis* (Butler) en el maracuya (*Passiflora edulis* var. *flavicarpa*, D.) Acta Agron. 19 (1): 17-30.

Castle, W.S. 1983. Antitranspirant and root canopy pruning effects on mechanically transplanted eight-year-old 'Murcott' citrus trees. J. Amer. Soc. Hort. Sci. 108 (6): 981-985.

Castro, I.R. 1981. Studies on papain production. NSDB Tech. J. 6 (2): 61-66.

Ceponis, M.J. and R.A. Cappellini. 1985. Wholesale and retail losses in grapefruit marketed in Metropolitan New York. HortSci. 20 (1): 93-95.

Cereda, E. and U.A. Lima. 1976. Conservacao do maracujá amarelo para consumo "in natura". Acta Hort. 57: 145-150. (Trop. and Subtrop. Frs.)

Chadha, K.L. 1968. Litchi cultivation in India. Indian Hort. 12 (3): 13-16, 36.

Chakraborty, D.P. 1959. Chemical examination of *Feronia elephantum* Corr. J. Sci. & Indus. Res. 18B (2): 90-91.

Chakraborty, R.N., L.V.L. Sastry & G.S. Siddappa. 1962. Changes in polyphenols and ascorbic acid during the candying of cashew apples. Indian Food Packer 16 (7): 10-11.

Chakraborty, S., R. Rodriguez, S.R. Sampathu and N.K. Saha. 1974. Prevention of pink discolouration in canned litchi (*L. chinensis*). J. Food Sci. & Technol. 11 (6): 266-268.

Chalons, M.E. 1944. Naranjillas—the golden fruit of the Andes. Agr. in the Americas 4 (6): 110-112.

Chalutz, E., J. Waks, and M. Schiffman-Nadel. 1985. Reducing susceptibility of grapefruit to chilling injury during cold treatment. HortSci. 20 (2): 226-228.

Chan, H.T., Jr. 1979. Sugar composition of papayas during fruit development. HortSci. 14 (2): 140-141.

Chandler, W.H. 1950. Evergreen Orchards. Lea & Febiger, Phila.

Chandra, D. 1965. Recent advances in clonal propagation of guava. Allahabad Farmer 39 (4): 137-139.

Chandra Rao, B.R. and J.V. Ghat. 1981. Preliminary observations on the harmful effects of *Megascolex insignis* on *Carica papaya*. U.A.S. Tech. Ser. 53-60.

Chandrasekhar, N. and C.S. Vaidyanathan. 1961. Some pharmacological properties of the blood anticoagulant from *Carica papaya* Linn. J. Sci. Indus. Res. 20C (7): 213-215.

Ch'ang, T.S. 1982. Kiwifruit: the Chinese way. New Zeal. J. Agr. 144 (4): 18-19.

Chapman, H.L., Jr. 1971. Citrus feeds as substitutes for grain in livestock rations. Sunshine State Agr. Res. Report 16 (1-2): 11-14.

Chapman, K.R., J. Saranah and B. Paxton. 1979. Induction of early cropping of guava seedlings in a closely planted orchard using urea as a defoliant. Aust. J. Exp. Agr. Anim. Hubs. 19: 382-384.

Chapman, K.R., B. Paxton, J. Saranah and P.D. Scudamore-Smith. 1981. Growth, yield and preliminary selection of seedling guavas in Queensland. Aust. J. Exp. Agr. Anim. Husb. 21: 119-123.

Chapman, T. 1967. Passion fruit growing in Kenya. Econ. Bot. 17 (3): 165-168.

Chapot, H. 1965. Le *Citrus volkameriana* Pasquale. Al Awamia 14: 29-45.

Chapot, H. 1964. Les bigaradiers bruquetiers. Pt. I. Al Awamia 10: 55-95.

Charley, V.L.S. 1969. Some tropical fruit juices. Proc. Trop. & Subtrop. Fruit Conference.

Charney, P.F. and R.A. Seelig. 1967. Persimmons. 2nd rev'd ed. Fruit & Veg. Facts & Pointers, United Fresh Fruit & Veg. Assn., Washington, D.C.

Charpentier, J.M., J. Godefroy et Y. Menillet. 1970. Interet de la couverture du sol par un film de polyethylene sur bananeraie de Cote d'Ivoire. Fruits 25 (2): 77-85.

Chatterjee, A., R. Sen and D. Ganguly. 1978. Aegelinol, a minor lactonic constituent of Aegle marmelos. Phytochem. 17: 328-329.

Chattopadhyay, S.B. and S.K. Bhattacharjya. 1968. Investigation on the wilt disease of guava (*Psidium guajava* L.) in West Bengal. I & II. Ind. J. Agr. Sci. 38 (1): 65-72, 176-183.

Chauhan, K.S., M.K. Dhingra and P.K. Mehta. 1981. Ber responds to pruning. Indian Hort. 26 (2): 5-6.

Chauhan, K.S. and J. Singh. 1971. Effect of moisture stress and indolebutyric acid on rooting of stem cuttings of pomegranate (*Punica granatum* L.) Indian J. Agr. Sci. 41 (4): 293-296.

Cheema, P.S., R.S. Dixit, T. Koshi and S.L. Perti. 1958. Indigenous insecticides: Pt. II. Insecticidal properties of the seed oil of *Annona squamosa* Linn. J. Sci. & Indus. Res. 17C (8): 132-134.

Chellappan, K. and J.I. Roche. 1982. A hybrid Jack. South Indian Hort. 30 (2): 76-81.

Chen, Wen-Hsun. 1949. The culture of the lychee. Proc. Fl. St. Hort. Soc. 62: 223-226.

Cheshire, P.C. 1966. Papain: Trade and markets. TPI Rept. G. 25. Trop. Products Inst., London.

Childers, N.F., H.F. Winters, P.S. Robles and H.K. Plank. 1950. Vegetable gardening in the tropics. Cir. 32. Fed. Exper. Sta. in Puerto Rico, Mayaguez, P.R.

Chin, H.F. 1975. Germination and Storage of rambutan (*Nephelium lappaceum* L.) seeds. Mal. Agr. Res. 4: 173-180.

Chinappa, B. 1968. Studies on a leaf blight disease of custard apple (*Annona squamosa* L.). J.U. of Poona, Sci. Technol. Sect. 34: 117-118.

Chiossone V., C. 1938. Flora Medica del Estado Lara. Coop. de Artes Graficas, Caracas, Venez.

Chi-Tung, S.C. 1961. Will lungan ever come of age in Florida? Proc. Fl. St. Hort. Soc. 74: 306-309.

Chopra, R.N., R.L. Badhwar, and S. Ghosh. 1965. Poisonous Plants of India. Vol. II. Indian Coun. Agr. Res., New Delhi.

Chopra, R.N., I.C. Chopra, K.L. Handa and L.D. Kapur. 1958. Chopra's Indigenous Drugs of India. 2nd ed. U.N. Dhur & Sons Pvt. Ltd., Calcutta.

Chopra, R.N., S.I. Nayar and I.C. Chopra. 1956. Glossary of Indian Medicinal plants. Coun. Sci. & Indus. Res., New Delhi.

Christopherson, E. 1935. Flowering plants of Samoa. Bull. 128. Bernice P. Bishop Mus., Honolulu, Hawaii.

Cibes, H.R. and S. Gaztambide. 1978. Mineral-deficiency symptoms displayed by papaya plants grown under controlled conditions. J. Agr. U. Puerto Rico 62 (4): 413-423.

Cibes, H.R. and G. Samuels. 1961. Mineral-deficiency symptoms displayed by Smooth Cayenne pineapple plants grown under controlled conditions. Tech. Paper 31. U. Puerto Rico Agr. Exper. Sta., Rio Piedras.

Clark, R.K., Jr. and M.T. Ellis. 1976. The use of Release®, an abscission agent, to increase the productivity of pickers on processing oranges. Proc. Fl. St. Hort. Soc. 89: 72-73.

Claus, E.P., V.E. Tyler and L.R. Brady. 1970. Pharmacognosy 6th ed. Lea & Febiger, Phila.

Clavijo, H. and J. Maner. 1975. The use of waste bananas for swine feed. Proc. of Conference on Anim. Feeds of trop. and subtrop. origin. Trop. Prod. Inst., London.

Clement, C.R. and J. Mora Urpi. 1982. The pejibaye palm (*Bactris gasipaes*) comes of age. Principes 26 (3): 150-152.

Clemente, A.M. 1960. A chamada goiaba 'Campinas'. Supl. Agricola 6 (296): 15.

Cobin, M. 1948. Notes on the grafting of *Litchi chinensis* Sonn. Proc. Fl. St. Hort. Soc. 61: 265-268.

Cobin, M. 1945. Recent introductions and distributions [mangosteen]. In: Report of the Fed. Exper. Sta. in Puerto Rico, 1944. U.S.

Dept. Agr., Fed. Exper. Sta., Mayaguez, P.R. Pp. 23-24.

Cobin, M. 1948. The Barbados or West Indian cherry in Florida. Mimeo. Rpt. 14. U. of Florida, Subtrop. Exper. Sta., Homestead, Fla.

Cobin, M. 1961. The lychee in Florida. Bull. 176 (rev. of Bull. 546, 1954.) U. of Florida, Agr. Exten. Serv., Gainesville.

Cobin, M. 1945. Tropical fruits and vegetables: Introduction and production of tropical fruits and vegetables. Report of the Fed. Exper. Sta. in Puerto Rico. 1944. U.S. Dept. Agr., Fed. Exper. Sta., Mayaguez, P.R. P. 22.

Cobin, M. 1946. Tropical fruits [mangosteen]. In: Report of the Fed. Exper. Sta. in Puerto Rico, 1945. U.S. Dept. Agr., Fed. Exper. Sta., Mayaguez, P.R. Pp. 33-34.

Cody, R.S. 1969. Expedirán piñas de "la costa bananera". La Hacienda 64 (3): 42.

Coenen, J. and J. Barrau. 1961. The breadfruit tree in Micronesia. So. Pacific Bull. 11 (4): 37-39, 65-67.

Coetzee, W.H.K., M.M. Krynauw, F.F. Pratt and J.F. du T. Hugo. 1950. Guava juice and guava sweets. Reprint #4. Farming in So. Africa. Jan: 1-4.

Cohen, A. 1982. Recent developments in girdling of citrus trees. Proc. Internat. Soc. Citriculture. Vol. 1: 196-199.

Cohen, A., J. Lomas, and A. Rassis. 1972. Climatic effects on fruit shape and peel thickness in 'Marsh Seedless' grapefruit. J. Amer. Soc. Hort. Sci. 97 (6): 768-771.

Cohen, C.V. 1959. Key lime disappearing; revival campaign is renewed. Miami Herald Sept. 27, p. 28G.

Cohen, M. 1955. Clitocybe rot of lychee trees. Proc. Fl. St. Hort. Soc. 68: 329-332.

Coit, J.E. 1949. Carob culture in the semi-arid southwest. Walter Rittenhouse, San Diego, Calif.

Coit, J.E. 1951. Carob or St. John's bread. Econ. Bot. 5 (1): 82-96.

Coit, J.E. 1960. Carob research in California. Hort. Advance (India) 4: 16-20.

Coit, J.E. 1967. Carob varieties for the semi-arid southwest. Fr. Vars. & Hort. Dig. 21 (1): 5-9.

Colbran, R.C. 1969. Cover crops and nematode control in pineapples. Queensland Agr. J. 95 (10): 658-661.

Colbran, R.C. and G.W. Saunders. 1961. Nematode root-rot of bananas. Queensland Agr. J. 87 (1): 22-24.

Collins, J.L. 1949. History, taxonomy and culture of the pineapple. Econ. Bot. 3 (4): 335-359.

Colom-Covas, G. 1977. Effect of plant population and fertilization on growth and yield of papaya (Carica papaya L.). J. Agr. U. Puerto Rico 61 (2): 152-159.

Condit, I.J. 1947. The fig. Chronica Botanica Co., Waltham, Mass.

Conover, R.A. 1979. Yellow strap leaf, a new disease of Florida papayas. Proc. Fl. St. Hort. Soc. 92: 276-277.

Conover, R.A. and R.E. Litz. 1978. Progress in breeding papayas with tolerance to papaya ringspot virus. Proc. Fl. St. Hort. Soc. 91: 182-184.

Conover, R.A., R.E. Litz and S.E. Malo. 1986. 'Cariflora'—a Papaya Ringspot Virus-tolerant Papaya for South Florida and the Caribbean. HortScience 21 (4): 1074.

Conover, R.A. and V.H. Waddill. 1981. Permethrin as a control for the papaya fruit fly. Proc. Fl. St. Hort. Soc. 94: 353-355.

Consolmagno, E. 1968. Destaninizacao melhora o caqui. Coopercotia 25: 36-40.

Conway, T. 1963. Pruning citrus and subtropical fruits: Tree tomatoes. New Zeal. J. Agr. 105 (5): 425.

Cook, A.A. and F.W. Zettler. 1970. Susceptibility of papaya cultivars to papaya ringspot and papaya mosaic viruses. Pl. Dis. Rep. 54 (10): 893-895.

Cook, O.F. 1913. Nomenclature of the sapote and the sapodilla. Contrib. U.S. Nat. Herb. Vol. 16, Pt. 11. Smithsonian Inst., Washington, D.C.

Cook, O.F. and G.N. Collins. 1903. Economic plants of Puerto Rico. Contrib. U.S. Nat. Herb. Vol. 8, Pt. 2. Smithsonian Inst., Washington, D.C.

Cooke, P.D. and J.C. Caygill. 1974. The possible utilisation of plant proteases in cheesemaking. Trop. Sci. 16 (3): 149-153.

Cooper, J.F. 1977. Lemons require complex care. Fla. Grower & Rancher 70 (5): 10-11.

Cooper, J.F. 1982. In search of the golden apple. Vantage Press, N.Y.

Cordoba V., J.A. 1961. La chirimoya. Agr. Trop. 17 (11): 647-664.

Cordoba V., J.A. 1961. La guayaba. Agr. Trop. 17 (8): 459-479.

Corner, E.J.H. 1952. Wayside trees of Malaya (2 vols.) 2nd ed. Gov't Printing Office, Singapore.

Correll, D.S. and H.B. Correll. 1982. Flora of the Bahama Archipelago (including the Turks and Caicos Islands). J. Cramer, Vaduz, Germany.

Costantino, F. 1962. L' Anona cherimolia, a subtropical crop of great interest for the South. Agricultura (Italy) 11 (12): 37-40.

Couey, H.M. and G. Farias. 1979. Control of postharvest decay of papaya. HortSci. 14 (6): 719-721.

Courtenay, C.E. 1966. Big build-up of Malayan pineapple industry. Queensland Fr. & Veg. News 29 (7): 149, 151.

Cowen, D.V. 1965. Flowering trees and shrubs in India. Rev'd 4th ed. Thacker & Co., Ltd., Bombay.

Cox, J.E. 1961. Fusarium resistant rootstocks for passion vines. Agr. Gaz. New So. Wales 72 (6): 314-318.

Cox, P.A. 1980. Two Samoan technologies for breadfruit and banana preservation. Econ. Bot. 34 (2): 181-185.

Craigmill, A.L., R.N. Eide, T.A. Shultz and Karel Hendrick. 1984. Toxicity of avocado (Persea americana (Guatemalan var.)) leaves; review and preliminary report. Vet. & Human Toxicology 26 (5): 381-383.

Crandall, P.G. and J.W. Kesterson. 1976. Recovery of naringin and pectin from grapefruit albedo. Proc. Fl. St. Hort. Soc. 89: 189-191.

Craveiro, A.A., C.H.S. Andrade, F.J.A. Matos, J.W. Alencar and M.I.L. Machado. 1983. Essential oil of Eugenia jambolana. J. Nat. Prod. 46 (4): 591-592.

Crawford, D.L. 1937. Hawaii's Crop Parade. Advertiser Pub'g Co., Honolulu.

Crawford, M. 1978. Where do we stand on mechanical citrus harvesting? Fl. Grower & Rancher. Oct. Pp. 19-20.

Cribb, A.B. and J.W. Cribb. 1975. Wild foods in Australia. William Collins, Pub. Pty. Ltd., Sydney.

Crider, F.J. 1958. Airlayering the papaya. Trop. Living-Homemaker & Gardener. Sept. P. 12.

Crider, F.J. 1958. Airlayering the papaya (cont'd). Trop. Living-Homemaker & Gardener. Oct. P. 14.

Cronauer, S.S. and A.D. Krikorian. 1984. Rapid multiplication of bananas and plantains by in vitro shoot tip culture. HortSci. 19 (2): 234-235.

Crooks, M.R. 1978. Sparkling wine from feijoas. New Zeal. J. Agr. 136 (8): 85.

Crowther, P.C. and E.H.G. Smith. 1960. Dates from the eastern Aden Protectorate. Trop. Sci. II (1 & 2): 90-92.

Cruz, G.L. 1979. Dicionario das plantas uteis do Brasil. Editora Civilizacao Brasileira S.A., Rio de Janeiro.

Cruz, G.L. 1965. Livre Verde das Plantas medicinais e industriais do Brasil. Vols. 1 & 2. Velloso, S.A., Belo Horizonte, Brasil.

Cruz Cay, J.R. 1972. Processing chironja. J. Agr. U. Puerto Rico 56 (2): 183-184.

Cueto, C.U., A.G. Quintos, C.N. Peralta and M.S. Palmario. 1978. Pineapple fibers. The retting process. II. N.S.D.B. Technology J. 3 (2): 73-79.

Cull, B.W. and F.D. Hams. 1974. Litchi finds a new home in Queensland. Queensland J. Agr. 100 (12): 597-603.

da Cunha Parro, A. 1971. Os alcoois superiores do sumo fermentado de caju (Anacardium occidentale). Rev. Cienc. Agron. Lourenco Marques 4 (3) Ser. B.: 47-68.

Dadlani, S.A. and K.P.S. Chandal. 1970. The little-grown tree tomato. Indian Hort. 14 (2): 13-14.

Dahlgren, B.E. 1947. Tropical and subtropical fruits. Popular Ser., Botany #26. Chicago Nat. Hist. Mus., Chicago.

Dahlgren, B.E. and P.C. Standley. 1944. Edible and poisonous plants of the Caribbean Region. NAVMED. 127. U.S. Navy Dept., Bur. of Medic. and Surgery, Washington, D.C.

Dale, I.R. and P.J. Greenway. 1961. Kenya trees and shrubs. Buchanan's Kenya Estates, Ltd., Nairobi.

Dalziel, J.M. 1948. Useful plants of West Tropical Africa (Appendix to the Flora of W.T.A.). Crown Agents for the Colonies, London.

DaPonte, J.J. 1973. A cercosporiore de ateira, *Annona squamosa* L. Rev. de Agric. 48 (2-3): 121-122.

Das, R.C. and B.N. Acharya. 1969. The effect of wax emulsion on the storage life of guava fruits (*Psidium guajava* L. var. Banaras Round). Plant Sci. India 1: 233-238.

Dass, H.C., H.S. Sohi, M.C. Reddy and G.S. Prakash. 1977. Vegetative multiplication by leaf cuttings of crowns in pineapple (*Ananas comosus*). Curr. Sci. 46 (7): 241-242.

Dastur, J.F. 1952. Medicinal plants of India and Pakistan. D.B. Taraporevala Sons & Co., Ltd., Bombay.

Dastur, J.F. 1951. Useful plants of India and Pakistan. 2nd ed. D.B. Taraporevala Sons & Co., Ltd., Bombay.

Data, E.S., D.B. Mendiola and E.B. Pantastico. 1975. Note: Storage of Calamansi fruits (*Citrus mitis* Blanco). 1. Chemical changes. Phil. Agr. 59: 119-125.

Datta, S.C., V.V.S. Murti, and T.R. Seshadri. 1969. A new synthesis of zapotin, zapotinin and related flavones [white sapote]. Indian J. Chem. 7 (8): 746-750.

Daulta, B.S. and K.S. Chauhan. 1982. Ber—a fruit with rich food value. Indian Hort. 27 (3): 7, 9.

Dave, K.G., R. Mani and K. Venkataraman. 1961. The colouring matters of the wood of *Artocarpus integrifolia*. Pt. III. Constitution of artocarpin and synthesis of tetrahydroartocarpin dimethyl ether. J. Sci. & Indus. Res. 20B (3): 112-121.

Dave, K.G., S.A. Telang and K. Venkataraman. 1960. The colouring matter of the wood of *Artocarpus integrifolia:* Pt. II. Artocarpetin, a new flavone, and artocarpanone, a new flavanone. J. Sci. & Indus. Res. 19B (12): 470-476.

Dave, K.G. and K. Venkataraman. 1956. The colouring matters of the wood of *Artocarpus integrifolia*. Pt. I. Artocarpin. J. Sci. & Indus. Res. 15B (4): 183-190.

Davenport, T.L. and C.W. Campbell. 1977. Stylar-end breakdown: a pulp disorder in 'Tahiti' lime. HortSci. 12 (3): 246-248.

Davenport, T.L. and C.W. Campbell. 1977. Stylar-end breakdown in 'Tahiti' lime: aggravating effects of field heat and fruit maturity. J. Amer. Soc. Hort. Sci. 102 (4): 484-486.

Davey, J.B. 1958. Can windbreaks improve quality of papaws. Queensland Agr. J. 84 (1): 2-6.

Davies, W.N.L. 1970. The carob tree and its importance in the agricultural economy of Cyprus. Econ. Bot. 24 (4): 460-470.

Davis, T.A. and W. Charles. 1976. The pejibaye can take root in India. Indian Farming 26 (1): 13-14.

deAguiar Falcão, M., E. Lleras, W. Estevam Kerr, and L.M. Medeiros Carreira. 1981. Aspectos fenologicos e de produtividade do biriba (*Rollinia mucosa* (Jacq.) Baill.). Acta Amazonica 11 (2): 297-306.

deAlmeida Filho, J. and J. Cambraia. 1974. Estudo do valor nutritivo do "ora-pro-nobis" (*Pereskia aculeata* Mill.). Rev. Ceres 21 (114): 105-111.

deCaloni, I.B. and J.R. Cruz-Cay. 1984. Elaboration and evaluation of typical Puerto Rican dishes prepared with mixtures of plantain, cassava and tanier flours. J. Agr. U. Puerto Rico 68 (1): 67-73.

deCaloni, I.B. and F. Fernandez Coll. 1983. Elaboration, sensory and microbiological evaluation of mofongo. [plantain] J. Agr. U. Puerto Rico 67 (2): 95-99.

deCandolle, A. 1884. Origin of cultivated plants. Kegan Paul, Trench & Co., London.

DeCardenas, J. 1923. Las frutas de Cuba. Republica de Cuba, Sec. de Agr., Havana.

Dedolph, R.R., J. Hiyane and F.A.I. Bowers. 1960. A study of guava orchard renovation. Proc. Amer. Soc. Hort. Sci. 76: 262-269.

Degener, O. 1946. Flora Hawaiiensis (New illustrated flora of the Hawaiian Islands). Books 1-5. Otto Degener, Riverdale, N.Y.

Degener, O. 1946-57 - 1963. Flora Hawaiienis (New illustrated flora of the Hawaiian Islands). Book 6. Author, Mokuleia Beach, Waialua, Oahu.

Degener, O. 1945. Plants of Hawaii National Park, illustrative of plants and customs of the South Seas. Author, N.Y. Bot. Gard., Bronx Park, N.Y.

Degner, R.L. and C.B. Dunham. 1977. Economic trends in the Florida avocado industry. Proc. Fl. St. Hort. Soc. 90: 230-233.

Degner, R.L. and M.G. Rooks. 1978. Lime production in Florida: projections and economic implications for 1981-82. Proc. Fl. St. Hort. Soc. 91: 194-197.

deHernandez, E.K. and J.R. Benero. 1982. Evaluation of four mango cultivars for nectar. J. Agr. U. Puerto Rico 66 (3): 153-158.

Dei-tutu, J. 1975. Studies on the development of *tatate* mix, a plantain product. Ghana J. Agr. Sci. 8 (2): 153-157.

Dekle, G.W. 1957. A leaf beetle feeding on the stems of lychee. Proc. Fl. St. Hort. Soc. 70: 331-333.

Dekle, G.W. 1954. Insects on lychee. Fla. Subtrop. Gard. July. P. 10.

Dekle, G.W. 1955. Insects on lychees during the past year. Fla. Lychee Growers Assn. Year Book & Proc. 3rd Ann. Mtg., Winter Haven, Fla. 29-30.

Dekle, G.W. 1954. Some lychee insects of Florida. Proc. Fl. St. Hort. Soc. 67: 226-228.

deLange, J.H. and A.P. Vincent. 1972. Pollination requirements of Ortanique tangor. Agroplantae 4: 87-92.

delCampillo, A. and C.F. Asenjo. 1957. Bound ascorbic acid in acerola juice. J. Agr. U. Puerto Rico 41 (2): 134-139.

deLille, J. 1934. Nota acerca de la accion del zapote blanco sobre la tension arterial. Anales del Inst. de Biologia, Mexico 5: 45-47.

de Lima, D.C. 1971/1972. Extracao de pectina do maracuja. Coletanea do Inst. de Tech. de Alimentos [Brazil] 4: 63-69.

Della Monache, G., B. Botta, J.F. deMello, J.S. de Barros Coelho and F. Menichini. 1984. Chemical investigation of the genus Rheedia; IV: Three new xanthones from *Rheedia brasiliensis*. J. Nat. Prod. 47 (4): 620-625.

DePerello, A.O. 1955. Cocina Criolla [cookbook]. Editora del Caribe C. por A., Ciudad Trujillo, Dom. Rep.

deRamallo, N.E.V. and S. Zabala. 1966. Antracnosis del Palto (Bol. 102). Estac. Exper. Agt., Prov. Tucumán, Argentina.

deRoman, A.L. 1979. Plantana y banana: dos cultivos colombianos amenazados por enemigos presentes y ausentes (adaptado a partir de "Diseminacion de patógenos y control de enfermedades en banano", seminario presentado por el Ing. Agron. Gabriel Cubillos Z., PEG UN-1 CA, Revista ICA-INFORMA 13 (5): 23-27.

de Sousa Correa, L., C. Rugiero and J.C. Oliveira. 1979. Propagation of yellow passion fruit by graftage. Proc. Trop. Reg. Amer. Soc. Hort. Sci. 23: 149-150.

Deszyck, E.J. and S.V. Ting. 1960. Processed products from Murcott orange, Pt. 1: Availability and characteristics of fruit. Proc. Fl. St. Hort. Soc. 73: 276-279.

de Toledo Piza, Jr., C. 1963. Mudas de Caqui. Chacaras e Quintais [Sao Paulo] Apr. 15, pp. 391-392.

Deullin, R. and F. Trupin. 1965. Transport maritime des litchis. Fruits 20 (7): 341-343.

de Villiers, E.A. 1972. Which pests attack granadillas? Farming in So. Afr. Jan. 31-33, 35.

de Villiers, E.A. and D.L. Milne. 1972. Nip eelworm in the bud on granadillas. Farming in So. Afr. 48 (8): 75-77.

Dhaliwal, J.S. 1975. Insect-pollination in ber (*Zizyphus mauritiana* Lamk.). Curr. Sci. 44 (14): 527.

Dhaliwal, T.S. and J.L. Serapión. 1981. Progress in guava selection in Puerto Rico. J. Agr. U. Puerto Rico 65 (1): 74-75.

Dhar, D.C., D.L. Shrivastara, and M. Screenivasaya. 1956. Studies on *Emblica officinalis* Gaertn. Pt. I. Chromatographic study of some constituents of amla. J. Sci. & Indus. Res. 15C (9): 205-206.

Dharkar, S.D., K.A. Savagaon, U.S. Kumta, and A. Sreenivasan. 1966. Development of a radiation process for South Indian fruits: mangos and sapodillas. J. Food Sci. 31 (1): 22-28.

Dhillon, B.S., G.S. Chohan and J.S. Josan. 1982. Redblush—a grapefruit. Indian Hort. 27 (3):6.

Dhillon, J.S., K. Kirpal Singh and J.C. Bakhshi. 1963. Investigations on flowering and fruiting problems in sweet lime (*C. limettioides* Tanaka). The Punjab Hort. J. 3 (1): 46-53.

Dhingra, O.D. and M.N. Khare. 1971. A new fruit rot of papaya. Curr. Sci. 40 (22): 612-613.

Dhingra, P., R.S. Mehrotra and K.R. Aneja. 1980. A new postharvest disease of *Annona squamosa* L. Curr. Sci. 49 (12): 472-478.

Dhuria, H.S., V.P. Bhutani and B.B. Lal. 1977. A new technique or vegetative propagation of persimmons. Scientia Hort. 6: 55-58.

Diaz, A. and G. Coto. 1983. Chemical composition of two varieties of mango seed for animal feeding. Cuban J. Agr. Sci. 17: 175-182.

Diaz, N. 1976. Effect of storage at 45° F (7° C) on keeping quality of 5 Chironja clones. J. Agri. U. Puerto Rico 60 (3): 348-368.

Diaz, N., T. Rodriguez, and I.B. de Caloni. 1983. Some characteristics of the chemical composition and general quality of the Red Spanish and PR 1-67 pineapple varieties. J. Agr. U. Puerto Rico 67 (4): 507-513.

Diaz Polanco, C. and A. Rondon. 1971. Un tipo de *Macrophomina* patogeno en frutos de guayaba. Agron. Trop. Venez. 21 (2): 111-118.

Dickinson, T.A. 1962. Growing litchis in California. World Crops 14 (2): 55-56.

Dickson, F. 1973. Pineapples. Miami Herald. Sept. 28. P. 5-D.

Dierberger, J. 1949. Guia do fruticultor Brasileiro [catalog]. Dierberger Agricola Ltd. Sao Paulo, Brazil.

Dietenbeck, W. 1980. Florida avocados—$12.5 million industry—and growing. Fla. Grower & Rancher (Oct.). P. 6.

Djerassi, C., E.J. Eisenbraun, R. Gilbert, A.J. Lemin, S.P. Marfey and M.P. Morris. 1959. Naturally occurring oxygen heterocyclics. II. Characteristics of an insecticidal principle from *Mammea americana L.* J. Amer. Chem. Soc. 80 (14): 3866-3897.

Dodge, C.R. 1897. Descriptive catalogue of useful fiber plants of the world. Rpt. #9. U.S. Dept. Agr., Off. of Fiber Investigations, Washington, D.C.

Doepel, R.F. 1965. Brown spot of passion fruit. J. Agr. West. Aust. 6 (3): 4th Ser. 149-151.

Doepel, R.F. 1965. Grease spot of passion fruit. J. Agr. West. Aust. 6 (5): 4th Ser. 291.

Doepel. R.F. 1966. Loquat diseases—black spot and fleck. J. Agr. West. Aust. 7 (5): 225-227.

Dominguez Gil, O.E. 1978. Insectos perjudiciales del guanabano (*Annona muricata L.*) en el Estado Zulia, Venezuela. Rev. de la Fac. de Agron. 4 (3): 149-163.

Dominguez Gil, O.E. 1983. Insectos perjudiciales del guanabano (*Annona muricata L.*) en el estado Zulia, Venezuela. Rev. de la Fac. de Agron. 6 (2): 699-707.

Dorsett, P.H., A.D. Shamel and W. Popenoe. 1917. Navel orange of Bahia, with notes on some little-known Brazilian fruits. Bull. 445. U.S. Dept. Agr., Bur. Pl. Indus., Washington, D.C.

Doughty, D.D. and E. Larson. 1960. Tissue changes in experimental achee poisoning. Trop. Geog. Med. 12: 243-250.

Douglas, J. 1980. Persimmon and loquats—viable prospects for New Zealand? New Zeal. J. Agr. 141 (4): 11-13, 15.

Dressler, R.L. 1953. The pre-Columbian cultivated plants of Mexico. Harvard U. Bot. Mus. Leaflet 16: 128.

Drury, Col. Heber. 1873. The useful plants of India. 2nd ed. William H. Allen & Co., London.

DuCharme, E.P. and R.F. Suit. 1955. Immunity of the lychee from the burrowing nematode. Proc. Fl. St. Hort. Soc. 68: 270-272.

Dunsmore, J.R. 1957. The pineapple in Malaya (*Ananas comosus* Merr.). Malayan Agr. J. 40 (3): 159-187.

Dupaigne, P. 1975. Effets biochimiques des bromélines. Leur utilisation en therapeutique. Fruits 30 (9): 545-567.

Dupaigne, P. 1976. Le dattier, plante saccharifere. Fruits 31 (2): 111-116.

Durand, B.J., L. Orcan, U. Yanko, G. Zauberman, and Y. Fuchs. 1984. Effects of waxing on moisture loss and ripening of 'Fuerte' avocado fruit. HortSci. 19 (3): 421-422.

Durrani, S.M., V.K. Patil and B.A. Kadam. 1982. Effect of N, P and K on growth, yield, fruit quality and leaf composition of sapota [*Achras zapota*]. Indian J. Agr. Sci. 52 (4): 231-234.

Dutta, S. 1956. Cultivation of jack fruit in Assam. Indian J. Hort. 13 (4): 189-197.

Dutta, S. 1954. Jujubes of Assam. Indian J. Hort. 11 (2): 53-56.

Eaks, I.L. 1985. Effect of calcium on ripening, respiratory rate, ethylene production, and quality of avocado fruit. J. Amer. Soc. Hort. Sci. 110 (2): 145-148.

Eaks, I.L. 1972. Effects of clip vs. snap harvest of avocados on ripening and weight loss. J. Amer. Soc. Hort. Sci. 98 (1): 106-108.

Eastwood, H.W. 1942. The Monstera Deliciosa. Agr. Gaz., New So.

Wales. Jan. 1: 23.

Ebeling, W. and R.J. Pence. 1957. Orange tortrix as a pest of avocados. Calif. Citrograph. Aug. Pp. 367-368.

Edward, J.C., Z. Naim and G. Shanker. 1964. Canker and fruit-rot of guava (*Psidium guajava* L.). Allahabad Farmer 38 (2): 59-61.

Eggeling, W.J. 1951. The indigenous trees of the Uganda Protectorate. 2nd ed. Gov't of the Uganda Protectorate, Entebbe.

Egler, F.E. 1947. The role of botanical research in the chicle industry. Econ. Bot. 1 (2): 188-209.

Einset, J.W., J.L. Lyon and P. Johnson. 1981. Chemical control of abscission and degreening in stored lemon. J. Amer. Soc. Hort. Sci. 106 (5): 531-533.

Eisen, G. 1901. The fig: its history, culture and curing with a descriptive catalogue of the known varieties of figs. U.S. Dept. Agr., Div. of Pomology, Washington, D.C.

El Baradi, T.A. 1963. Date growing. Trop. Abs. 23 (8): 433-479.

El Baradi, T.A. 1975. Guava. Review article. Abs. on Trop. Agr. 1 (3): 9-16.

El Baradi, T.A. 1968. Processing and by-products of dates. Trop. Abs. 23 (9): 541-546.

El Kady, M.H., M. El Said Nassar and M. El Said El Halawany. 1977. A study of the chemical control of the citrus flat mite, *Brevipalpus californicus* (Banks) on guava (*Psidium guajava*) trees. Agr. Res. Rev. 55 (1): 139-141.

Elliott, R. 1950. Prospects for marketing Hawaiian papaya products in the United States. Agr. Econ. Bull. #1. U. Hawaii Coll. of Agr., Agr. Exper. Sta., Dept. of Agr. Economics.

El-Tayet, O., M. Kucera, V.O. Marquis and H. Kucerova. 1974. Contribution to the knowledge of Nigerian medicinal plants. III. Study on *Carica papaya* seeds as a source of a reliable antibiotic, the BITC. Planta Med. 26: 83-89.

El-Wakeel, A.T., N.A. Ali, and I.M. Ibrahim. 1967. Bearing capacity of mature orange trees in relation to rootstock effect. Agr. Res. Rev. U.A.R. 45 (4): 27-31.

El-Wakeel, A.T. and M.M. Eid. 1967. A citrus rootstock on heavy soil. Agr. Res. Rev. U.A.R. 45 (4): 21-26.

El-Wakeel, A.T., A.M. Rokba, A.M. Ezzat and M.H. Osman. 1977. Two new selected loquat cultivars. Egypt. J. Hort. 4 (1): 33-44.

Emechebe, A.M. and J. Mukiibi. 1975. Fungicidal control of brown spot of passion fruit in Uganda. Acta Horticulturae, Tropical Horticulture: 49: 281-289.

Emechebe, A.M. and J. Mukiibi. 1976. Nectria collar and root rot of passion fruit in Uganda. Pl. Dis. Rep. 60 (3): 227-231.

Emeruwa, A.C. 1982. Antibacterial substance from *Carica papaya* fruit extract. J. Nat. Prod. 45 (2): 123-127.

Emilsson, B. 1969. Problems in long-range transport of fresh avocados, mangoes and pineapples. Proc. Trop. and Subtrop. Fruits Conf. Pp. 65-69.

Ervin, R.T., D.S. Moreno, J.L. Baritelle and P.D. Gardner. 1985. Pheromone monitoring is cost-effective [California red scale on citrus]. Calif. Agr. 39 (9 & 10): 17-19.

Ewart, A. 1976. California growers turn to Kiwifruit. New Zeal. J. Agr. 132 (3): 10-11.

Ezzat, A.H., M. Naguib and S. Metwalli. 1974. Evaluation and determination of the maturity stage of the fruits of some *Annona* varieties. Agr. Res. Rev. 52 (9): 7-17.

Ezzat. A.H., A.M. Rokbal, and F.A. Khalil. 1972. Seasonal changes of the loquat fruit. Agr. Res. Rev. 50 (4): 33-38.

Fahmy, I. 1952. Grafting studies on macadamia and sapodilla in relation to carbohydrates, using pre-girdled scions. Proc. Fl. St. Hort. Soc. 65: 190-192.

Fahmy, M.A., S.A. Dewedar and M.A. Wally. 1978. Effect of calcium chloride and latex treatments on keeping quality of avocado fruits during storage. Res. Bull. 858. Ain Shams Univ., Cairo, Egypt.

Fairchild, D. 1931. Exploring for plants. The Macmillan Co., N.Y.

Fairchild, D. 1946. The alamoen—a citrus fruit of the tangelo type from Paramaribo. Proc. Fl. St. Hort. Soc. 59: 151-155.

Fairchild, D. 1946. The jack fruit (*Artocarpus integrifolia* Merrill), its planting in Coconut Grove, Fla. Paper 16. Fairchild Trop. Garden, Miami.

Fairchild, D. 1915. The mangosteen. J. Heredity. (Aug.) Pp. 339-347.

Fairchild, D. 1927. The pink-fleshed pummelo of Java. J. Heredity 18 (Oct.): 425-426.

Fairchild, D. 1947. The world grows round my door. Chas. Scribner's, N.Y.

Fairchild, G.F. and J.A. Niles. 1975. The marketing of by-products in the Florida citrus industry. Proc. Fl. St. Hort. Soc. 88: 281-285.

Farrales, J.F., A.T. Jocson and B.C. Andres. 1962. Progress report on the design and development of a cashew juice expeller. Phil. J. Agr. 23 (1-4): 71-77.

Felizardo, B.C., R.V. Valmayor, N.B. Nazareno and J.C. Hapitan, Jr. 1964. Influence of fertilizers on the growth and yield of calamondin (Citrus mitis Blanco). Phil. Agr. 47 (8): 412-418.

Felton, G. 1971. The use of pineapple by-products for livestock feeding. Misc. Pub. 75. U. Hawaii Coop. Exten. Serv., Pp. 40-42.

Fennah, R.G. 1937. Lepidopterous pests of the soursop in Trinidad (1). Trop. Agr. 14 (16): 75-78.

Fennah, R.G. 1937. Lepidopterous pests of the soursop in Trinidad (2). Trop. Agr. 14 (8): 244-245.

Fennell, J.L. 1948. "Cocona"—a desirable new fruit. Tech. Pub. 24 (A reprint from For. Agr.) Inst. Interamer. de Ciencias Agr., Turrialba, Costa Rica.

Fenzi, E.O. 1915. Fruitti Tropicali e Semitropicali (Esclusi gli agrumi). Inst. Agr. Coloniale Italiano, Firenze.

Fernandez, C.E. 1962. Pejibaye. Rev. Cafetalera 1 (Ser. 4): 5-6.

Finnegan, R.A. and E.J. Eisenbraun. 1964. Constitution of mamey wax. J. Pharm. Sci. 53 (12): 1506-1509.

Firminger, T.A. 1947. Firminger's Manual of gardening for India. 8th ed. [jackfruit]. Thacker Spink & Co. (1933), Ltd., Calcutta.

Fisher, A.E. 1968. Growers test plastic mulch in pineapples. Queensland Agr. J. 94 (2): 104-108.

Fisher, J.F. and H.E. Nordby. 1965. Isolation and spectral characterization of coumarin in Florida grapefruit peel oil. J. Food Sci. 30 (5): 869-873.

Fleming, G.D., Jr. 1953. Personal communication [lime]. Mar. 11.

Fletcher, W.A. 1952. Passionfruit culture. Bul. 135, rev'd. New Zeal. Dept. Agr., Auckland.

Fletcher, W.A. 1963. Production of subtropical fruits in 1962. [kiwifruit, passionfruit, tree tomato]. New Zeal. J. Agr. 106 (6): 489-490.

Fletcher, W.A. 1958. Subtropical fruit production and planting trends. New Zeal. J. Agr. 97 (5): 484.

Flores G., A., and D. Rivas. 1975. Estudios de maduracion y almacenamiento refrigerado de nispero (Achras sapota L.). Fitotecnia Latinoamericana 11 (1): 43-51.

Fonnesbech, A. and M. Fonnesbech. 1980. In vitro propagation of Monstera deliciosa. HortSci. 15 (6): 740-741.

Ford, H.W. 1958. The host status of lychee with reference to the burrowing nematode. Year Book & Proc. Fla. Lychee Growers Assn., Winter Haven, Fla. 10-12.

Ford, I. 1971. Chinese gooseberry production in France. New Zeal. J. Agr. 123 (4): 20-21.

Fouqué, A. 1973? Especies fruitieres d' Amerique Tropicale. Institut francais de recherches fruitieres outre-mer, Paris. (with 144 35 mm. color transparencies).

Fouqué, A. 1972. Especies fruitieres d' Amerique tropicale: genre Passiflora. Fruits 27 (5): 369-382.

Foyet, M. 1972. L'extraction de la papaine. Fruits 27 (4): 303-306.

Foyet, M. and E. Bothia. 1978. Comportement de l'ananas et qualite industrielle de la récolte en deux sites ecologiques au Cameroun. Fruits 33 (6): 425-432.

Fraire Mora, R. 1973. Evaluacion de fungicidas en la prevencion de anthracnosis, Colletotrichum gloeosporioides Penz del mango en Veracruz. Agr. Tec. en México 3 (6): 233-236.

Fraire Mora, R. 1973. Evaluacion de fungicidas en el combate de anthracnosis Colletotrichum sp. de papaya en Veracruz. Agr. Tec. en México 3 (7): 259-261.

Franzmann, B.A. 1974. Banana rust thrips in north Queensland. Queensland Agr. J. 100 (12): 595-596.

French, R.B. and O.D. Abbott. 1951. Levels of thiamine, riboflavin, and niacin in Florida-grown foods. Bull. 482. U. Fla., Agr. Exper. Sta., Gainesville.

Frossard, P. 1969. Les maladies du papayer, les maladies fongiques (Pt. 7). Fruits 24 (11-12): 473-481.

Fuchs, Y., G. Zauberman, U. Yanko and S. Homsky. 1975. Ripening of mango fruits with ethylene. Trop. Sci. 17 (4): 211-216.

Fung-Kon-Sang, W.E. 1977. Rootstocks for grapefruit in the coastal clays of Surinam. De Surinaamse Landbouw 25 (1): 14-19.

Gachanja, S.P. 1975. Training and pruning of passion fruit (Passiflora edulis Sims) in Kenya. Acta Hort. 49: 219-222.

Gaillard, J.P. 1972. Approches sur la fertilisation du papayer 'Solo' au Cameroun. Fruits 27 (5): 355-360.

Gaillard, J.P. 1971. Lutte contre le Cercospora de l'avocatier au Cameroun. Fruits 6 (3): 225-230.

Gallardo-Covas and R. Inglés Casanova. 1983. Insecticide evaluation for the control of Carpophilus humeralis F. in pineapple fields of Puerto Rico. J. Agr. U. Puerto Rico 67 (2): 174-175.

Gallardo Covas, F. and S. Medina Gaud. 1983. Conditions that affect populations of Carpophilus humeralis F. (Coleoptera: Nitidulidae) in the pineapple fields of Puerto Rico. J. Agr. U. Puerto Rico 67 (1): 11-15.

Gallego M., F.L. 1960. Gusano del tomate de arbol. Rev. Fac. Nacional del Agronomia (Medellin, Colombia) 20 (54): 39-43.

Ganapathy, K.M. and H.R. Singh. 1975. Passion-fruit for Malnad areas. Planters' Chronical 70 (12): 383-385.

Garcia Reyes, F. 1967. El cultivo de lulo en la zona cafetera colombiana. Rev. Cafetalera Colombia 17 (142): 75-77.

Garcia-Rivera, J. and M.P. Morris. 1953. Mamey toxicity. Rpt. of Fed. Exper. Sta. in Puerto Rico., U.S. Dept. Agr., Agr. Res. Admin., Washington, D.C.

Gardner, F.E. and G.E. Horanic. 1967. Poncirus trifoliata and some of its hybrids as rootstocks for Valencia sweet orange. Proc. Fl. St. Hort. Soc. 80: 85-88.

Garlick, P. 1964. More success for paper bunch covers [banana]. Queensland Fr. & Veg. News. 25 (21): 490.

Garnsey, S.M., C.O. Youtsey, G.D. Bridges and H.C. Burnett. 1976. A necrotic ringspotlike virus found in a 'Star Ruby' grapefruit tree imported without authorization into Florida. Proc. Fl. St. Hort. Soc. 89: 63-67.

Garzón Tiznado, J.A., and R.G. Alvarez. 1979. Influencia de la densidad de poblacion sobre el rendimiento y calido de fruto en el cultivar 'Rendidora' de tomate de cascara (Physalis ixocarpa Brot.). Proc. Trop. Reg. Amer. Soc. 23: 268-270.

Gatto, G. 1971. Il carrubo [carob] in Sicilia. Agri. Forum 12 (10): 369-372.

Gattoni, L.A. 1961. La naranjilla. Agr. en El Salvador 2 (1): 8-12.

Gattoni, L.A. 1961. La naranjilla o lulo. Agr. Trop. 17 (4): 218-224.

Gattoni, L.A. 1957. La naranjilla o lulun. Servicio Interamer. de Cooper. Agricola., Min. de Agr., Comercio e Industrias, Panama.

Gattoni, L.A. 1962. Naranjilla. Mexico Agricola 9 (108): 22-23.

Gattoni, L.A. 1961. Nuevo método de propagacion de la piña. Ceiba 9 (1): 13-20.

Gazit, S., I. Galon and H. Podoler. 1982. The role of nitidulid beetles in natural pollination of Annona in Israel [atemoya]. J. Amer. Soc. Hort. Sci. 107 (5): 849-852.

Gazit, S. and A. Kadman. 1980. 13-1 mango rootstock selection. HortSci. 15 (5): 669.

Gazit, S. and Y. Levy. 1963. Astringency and its removal in persimmons. Israel J. Agr. Res. 13 (3): 125-132.

Geach, W.F. 1978. High costs, low profits affect future of both rail and highway perishable carriers. Blue Anchor 55 (2): 21-22.

Generalao, M.L. 1980. Luscious lanzones for agro-forestry. Canopy. Jan. P. 3.

Genovar, C. 1961. The lychee, a prize fruit from the Orient finds a new home and future in Florida. All Florida TV Week Mag., Miami News. June 3.

Gentry, J.L., Jr. and P.C. Standley. 1974. Flora of Guatemala. Vol. 24, Pt. 10, Nos. 1-2, Pub. #1184. Field Mus. of Nat. Hist., Chicago, Ill.

George, A.P. and R.J. Nissen. 1982. Yield, growth and fruit quality of the persimmon (Diospyros kaki L.) in south-east Queensland. Queensland J. Agr. & Anim. Sci. 39 (2): 149-158.

Gerlach, W.W.P. 1980. Thread blight of passion fruit and its control in Western Samoa. Alafua Agri. Bull. 5 (2): 42-45.

Germek, E.B. 1962. A cultura da alfarrobeira [carob]. Rural (Rev. Soc. Rural Bras.) 42 (489): 33-36.

Germek, E.B. 1964. Cultura da calabura [Jamaica cherry]. Agronomico, Brasil 16 (9/10): 34-36.

Germek, E.B. 1978. Cultura experimental da pupunha no estado de Sao Paulo [pejibaye]. Agronomico, Brasil 29/30: 96-103.

Glassman, S. 1978. Seedling citrus for central Texas and points north. Texas Horticulturist 5 (3): 7-8.

Glennie, J.D. and K.R. Chapman. 1976. A review of dieback—a disorder of the papaw (Carica papaya L.) in Queensland. Queensland J. Agr. & Anim. Sci. 33 (2): 177-188.

Glennie, J.D. and R.B. Parsons. 1981. Biuret injury of pineapples. Queensland J. Agr. & Anim. Sci. 38 (1): 43-46.

Godora, N.R., K.S. Chauhan and S.S. Bisla. 1980. Evaluation of mid-season ripening ber (Zizyphus mauritiana Lamk.) germplasm. Haryana J. Hort. 9 (3-4): 101-105.

Godfrey-Sam-Aggrey, W. and W.S. Abutiate. 1973. Description of some mango cultivars in Ghana. Ghana J. Agr. Sci. 6: 33-42.

Goldweber, S. 1965. Give yourself the raspberry. Miami News, Apr. 4 [Mysore raspberry].

Goldweber, S. 1976. The use of ethylene to preripen mango fruit. (Memorandum). U. Florida, IFAS, Fla. Coop. Exten. Serv., Gainesville, Fla. June 3.

Gondwe, A.T.D. 1976. Cyanogenesis in passion fruit. 1. Detection and quantification of cyanide in passion fruit (Passiflora edulis Syms) at different stages of fruit development. E. Afr. Agr. & For. J. 42 (1): 117-120.

Gonzaga Herrera, L. and F. Lopez C. 1977. Apuntes sobre el cultivo del brevo [fig]. ICA-Informa Vol. 11, #4. Inst. Colombiano Agropecuario, Ayudantes de Técnico, Distrito ATEA Pereira. Pp. 9-20.

Gonzalez Casillas, J.A. 1970. Cultivo de plátanos y bananos. La Hacienda 65 (11): 20, 22-25.

Gonzalez-Tejera. 1970. Effects of plant density on the production of a plant crop of Red Spanish pineapple in Puerto Rico. Proc. 7th Ann. Mtg. C.F.C.S., Martinique-Guadeloupe. 1969. Pp. 72-78.

Gonzalez Villafane, E. 1982. Economic impact of modern technology on plantain production. J. Agr. U. Puerto Rico 66 (4): 250-253.

Goodin, J.R. and V.T. Stoutemyer. 1962. Salable carobs quicker with gibberellin. So. Florist & Nurseryman. Oct. P. 52.

Gooding, E.G.B., A.R. Loveless and G.R. Proctor. 1965. Flora of Barbados. Overseas Res. Pub. #7. Her Majesty's Staty. Off., London.

Goor, A. 1967. The history of the date through the ages in the Holy Land. Econ. Bot. 21 (4): 320-340.

Goor, A. 1967. The history of the pomegranate in the Holy Land. Econ. Bot. 21 (3): 215-229.

Gopalakrishna Rao, K.P. and S. Krishnamurthy. 1985. Studies on shelf-life of Coorg mandarin (Citrus reticulata Blanco). So. Indian Hort. 31 (2/3): 55-61.

Gopalkrishna, N. and P.V. Deo. 1960. Role of pouches as banana bunch covers. Hort. Adv. (India) 4: 165-167.

Gottreich, M. and Y. Halevy. 1982. Delaying ripening of the pre-harvest bananas (Dwarf Cavendish) with gibberellins. Fruits 37 (2): 97-102.

Gould, H.P. 1940. The Oriental persimmon. Leaflet 194. U.S. Dept. Agr., Bur. of Pl. Indus., Washington, D.C.

Govindachari, T.R., P.S. Kalyanaaman, N. Muthukumaraswamy and B.R. Pai. 1971. Isolation of 3 new xanthones from Garcinia mangostana Linn. Ind. J. Chem. 9 (5): 505-506.

Grant, J.A. and K. Ryugo. 1982. Influence of developing shoots on flowering potential of dormant buds of Actinidia chinensis. HortSci. 17 (6): 977-978.

Grant, J.A. and K. Ryugo. 1984. Influence of within-canopy shading on fruit size, shoot growth and return bloom in kiwifruit. J. Amer. Soc. Hort. Sci. 109 (6): 799-802.

Grant, W.G. 1960. Influence of avocados on serum cholesterol. Proc. Soc. for Ep. Biol. & Med. 104: 45-47.

Greber, R.S. 1966. Identification of the virus causing papaw yellow crinkle with tomato big bud virus by transmission tests. Queensland J. Agr. & Anim. Sci. 23 (2): 147-153.

Greber, R.S. 1966. Passion-fruit woodiness virus as the cause of passion vine tip blight disease. Queensland J. Agr. & Anim. Sci. 23 (4): 533-538.

Greening, H.G. 1965. Scalicides for use on papaws in Queensland. Queensland J. Agr. & Anim. Sci. 22 (1): 107-108.

Greenway, P.J. 1948. The papaw or papaya. E. Afr. Agr. J. 13 (4): 228-233.

Gregory, J.M. 1948. Litchi, Chinese American. Nature Mag. 41: 521-522.

Grierson, W. 1958. Finding the best lemon for Florida—a report of progress. IV. Florida lemons for fresh fruit use. Proc. Fl. St. Hort. Soc. 86: 140-146.

Grierson, W. 1982. Record grapefruit cluster. HortSci. 17 (3): 301.

Grierson, W. and S.V. Ting. 1960. Florida lemons for fresh and cannery use: Variety selection and degreening methods. Proc. Fl. St. Hort. Soc. 73: 284-288.

Grierson, W. and H.M. Vines. 1965. Carambolas for potential use in gift fruit shipments. Proc. Fl. St. Hort. Soc. 78: 349-353.

Griffith, E. and W.H. Preston, Jr. 1971. The oriental persimmon comes north. Plants & Gard. (Brooklyn Bot. Gard.) 27 (3): 22-24.

Griffith, E. and W.H. Preston, Jr. 1961. The oriental persimmon in the Middle Atlantic States. Gardeners Forum (Amer. Hort. Soc.) 4 (5): 2.

Groff, G.W. 1948. Additional notes upon the history of the 'Brewster' lychee. Proc. Fl. St. Hort. Soc. 61: 285-289.

Groff, G.W. 1927. Culture and varieties of Siamese pummelos as related to introductions into other countries. Lingnan Sci. J. 5 (3): 187-254.

Groff, G.W. 1951. Describing Florida varieties of lychee. Proc. Fl. St. Hort. Soc. 64: 276-280.

Groff, G.W. 1943. Some ecological factors involved in successful lychee culture. Proc. Fl. St. Hort. Soc. 56: 3-24.

Groff, G.W. 1921. The lychee and lungan. Orange Judd Co., N.Y.

Groff, G.W. (not dated) The papaya for South China. Bull. 12. Canton Christian College, Dept. of Agr., Canton, China.

Groff, G.W. and B.F. Galloway. 1925. List of lychees introduced and distributed by the Office of Foreign Seed and Plant Introduction and Distribution, U.S. Dept. Agr. (manuscript, unpublished).

Groff, G.W. and Su-Ying Liu. 1951. Describing Florida varieties of lychee. Proc. Fl. St. Hort. Soc. 64: 276-285.

Grove, W.R. 1950. The lychee in Florida. No. 134, New Ser. Florida Dept. Agr., Tallahassee, Fla.

Grove, W.R. Jr. and D.R. Van Sickler. 1974. The lychee in Florida today. Proc. Fl. St. Hort. Soc. 87: 338-339.

Guelfat-Reich, S. 1970. Conservation de la nifle du Japon (Eriobotrya japonica). Fruits 25 (3): 169-173.

Guerout, R. 1975. Nematodes of pineapple: a review. PANS 21 (3): 123-140.

Guillen Paiz, R. 1975. Algunas perspectivas del limon criollo en la diversification agricola. Revista Cafetalera 149: 29-35.

Gumah, A.M. and S.P. Gachanja. 1984. Spacing and pruning of purple passion fruit. Trop. Agr. (Trin.) 61: 143-147.

Gunkel, W.W., W.H. Kahl and L. Moffett. 1972. Mechanical planting of pineapples. Spec. Pub. SP-01-72. Conf. Papers. Internat. Conf. on Trop. & Subtrop. Agr., Honolulu. Apr. 11-13, 1972. Pp. 242-246.

Gupta, R.C. and R.L. Madaan. 1975. Diseases of fruits from Haryana. I. A new fruit rot of Zizyphus mauritiana Lamk. Curr. Sci. 44 (24): 908.

Gupta, R.C. and R.L. Madaan. 1977. Diseases of fruits from Haryana. A new leaf spot disease of Zizyphus mauritiana Lamk. Curr. Sci. 46 (7): 237-238.

Gupta, V.K. and D. Mukherjee. 1980. Effect of morphactin on the storage behavior of guava fruits. J. Amer. Soc. Hort. Sci. 105 (1): 115-119.

Guru Venkatesh, A.S. and Y.K. Raghunatha Rao. 1960. Minor oils—papaya seed oil. Food Sci. (Feb.): 49.

Guyot, A., A. Pinon and C. Py. 1974. L'ananas en Côte d' Ivoire. Fruits 29 (2): 85-116.

Guzman, D.J. 1947. Especies utiles de la flora Salvadoreña. Imprenta Nacional, San Salvador, El Salvador.

Guzman R., R., C. de Villaveces, and E. de Clavijo. 1977. Estudio de la composicion quimica del lulo (Solanum quitoense Lam.) y obtencion de un producto comercial a partir de este fruto. Bol. Informativo #2. Univ. Nac. Dep. de Quimica y Org. de los Estados Americanos, Bogotá, Colombia, Pp. 59-69.

Haddad G., O. 1966. Rendimientos de parchita maracuya. Fruticultura (U. de Oriente, Venez.) 1 (1): 17-25.

Haddad G., O. and M. Figueroa R. 1973. Estudios de la floracion y fructificacion en parcha granadina (*Passiflora quadrangularis* L.). Agron. Trop. 22 (5): 483-496.

Haddad G., O., A. Ordosgoitty, and J. Bechyne. 1970. Daños causados por *Disonycha glabrata* (Fabricus) en *Passiflora quadrangularis* L. Agron. Trop. 20 (5): 331-334.

Haendler, H. and R. Huet. 1965. La papaine. Fruits 20 (8): 411-415.

Haendler, L. 1965. La passiflore, su composition chimique et ses possibilites de transformation. Fruits 20 (5): 235-245.

Haendler, L. 1965. L'huile d' avocat et les produits dérivés du fruit. Fruits 20: 625-633.

Haendler, L. 1966. Produits de transformation de la banane. Fruits 21 (7): 329-342.

Hahn, B.C. 1956. Panama fishing trip brings choice recipes. Miami Sunday News July 29. [pineapple].

Hali, R. 1967. Mangosteen—a money-maker. Ind. J. Hort. 11 (3): 16.

Hall, D.J. 1983. Fungicides for postharvest decay control in loquats. Proc. Fl. St. Hort. Soc. 96: 366-367.

Hamilton, K.S. 1965. Reproduction of banana from adventitious buds. Trop. Agr. (Trin.) 42 (1): 69-73.

Hamilton, R.A. 1969. Effect of location on size and quality of papaya fruits. Misc. Pub. 64 (Proc. 5th Ann. Mtg. Hawaii papaya Indus. Assn.). U. Hawaii Coop. Exten. Serv., Honolulu. Pp. 25-27.

Hamilton, R.A. 1960. Mango varieties in Hawaii. Hort. Adv. (India). Vol. IV, p. 15.

Hamilton, R.A. and H.Y. Nakasone. 1967. Bud grafting of superior guava cultivars. Hawaii Farm Sci. 16 (2): 6-8.

Hamilton, R.A. and W. Yee. 1969. GOUVEIA, an attractive new mango. Cir. 435. U. Hawaii Coop. Exten. Serv., Honolulu.

Hamilton, R.A. and W. Yee. 1970. Lychee cultivars in Hawaii. Proc. Fl. St. Hort. Soc. 83: 322-325. [also pub'd in Proc. Trop. Reg. Amer. Soc. Hort. Sci. 14: 7-12].

Hamilton, R.G. 1948. Tree tomato culture. Bull. 306. (Reprinted from New Zeal. J. Agr. Sept. 1947). New Zeal. Dept. Agr., Wellington, N.Z.

Hampton, R.E. and P.G. Thompson. 1974. Passionfruit production in Fiji. Fiji Agr. J. 36: 23-27.

Hancock, W.G. 1944. Pineapple growing in Queensland. Bull. (reprinted from Queensland Agr. J. Oct. 1944). Queensland Dept. Agr.

Haque, S.Q. and S. Parasam. 1973. *Empoasca stevensi*, a new vector of bunchy top disease of papaya. Plant Dis. Rep. 57 (5): 412-414.

Harding, P.L. 1959. The importance and early history of the 'Temple' Orange. Proc. Fl. St. Hort. Soc. 72: 93-96.

Hargreaves, J.R. 1979. Damage to passion-fruit by the Queensland fruit fly, *Dacus tryoni* (Froggath). Bull. 805. Queensland J. Agr. & Anim. Sci. 36 (2): 147-150.

Harkness, R.W. 1954. Papaya growing in Florida. Mimeo. Rpt. 17. U. Fla. Sub-tropical Exper. Sta., Homestead, Fla.

Harrar, E.S. and J.G. Harrar. 1946. Guide to southern trees. McGraw-Hill Book Co., Inc., N.Y.

Harris, W. 1913. Notes on fruits and vegetables in Jamaica. Gov't Ptg. Off., Kingston, Jamaica.

Hart, W.J. 1964. Why not take another look at the loquat? J. Agr. West. Aust. 5 (6) (4th Ser.): 351.

Hartley, C.W.S. 1950. *Flacourtia inermis*—Rokam Masan. Malaya Agr. J. 33 (2): 93-97.

Hartwig, L. 1976. Growth regulators can put profits back in tangerines. Fla. Grower & Rancher 69 (6): 8-9, 22.

Hatton, T.T. and C.W. Campbell. 1959. Evaluation of indices for Florida avocado maturity. Proc. Fl. St. Hort. Soc. 72: 349-353.

Hatton, T.T. and R.H. Cubbedge. 1983. Preferred temperature for prestorage conditioning of 'Marsh' grapefruit to prevent chilling injury at low temperatures. HortSci. 18 (5): 721-722.

Hatton, T.T., R.H. Cubbedge, L.A. Rissee, R.W. Hale, D.H. Spalding, D. von Windeguth and V. Chew. 1984. Phytotoxic responses of Florida grapefruit to low-dose irradiation. J. Amer. Soc. Hort. Sci. 109 (5): 607-610.

Hatton, T.T. Jr. and W.F. Reeder. 1972. Quality of 'Lula' avocados stored in controlled atmospheres with or without ethylene. J. Amer. Soc. Hort. Sci. 97 (3): 339-341.

Hatton, T.T. Jr. and W.F. Reeder. 1965. Ripening and storage of Florida avocados. Marketing Res. Rpt. 697. U.S. Dept. Agr., Agr. Res. Serv., Washington, D.C.

Hatton, T.T., W.F. Reeder and C.W. Campbell. 1965. Ripening and storage of Florida mangos. Marketing Res. Rpt. 725. U.S. Dept. Agr., Agr. Res. Serv., Washington, D.C.

Hatton, T.T., W.F. Reeder and J. Kaufman. 1966. Maintaining market quality of fresh lychees during storage and transit. Marketing Res. Rpt. 770. U.S. Dept. Agr., Agr. Res. Serv., Washington, D.C.

Hawson, M.G. and D.W. Thomas. 1968. Papaw varieties for Camaroun. J. Agr. West. Aust. 9 (2) 4th Ser.: 85-88.

Hayes, W.B. 1953. Fruit growing in India. 2nd rev'd ed. Kitabistan, Allahabad, India.

Hearn, C.J. 1984. Development of seedless orange and grapefruit cultivars through seed irradiation. J. Amer. Soc. Hort. Sci. 109 (2): 270-273.

Hearn, C.J. 1979. Performance of 'Sunburst', a new citrus hybrid. Proc. Fl. St. Hort. Soc. 92: 1-3.

Heinicke, R.M. and W.A. Gortner. 1957. Stem bromelain—a new protease preparation from pineapple plants. Econ. Bot. 11 (3): 225-234.

Heiser, C.B., Jr. 1985. Ethnobotany of the naranjilla (*Solanum quitoense*) and its relatives. Econ. Bot. 39 (1): 4-11.

Heiser, C.B., Jr. 1972. The relationships of the naranjilla. Biotropica 4 (2): 77-84.

Heiser, C.B., Jr. and J. León. 1955. The pejibaye palm. Nature Mag. 48: 131-132.

Hendrickson, R. and J.W. Kesterson. 1961. Grapefruit seed oil. Proc. Fl. St. Hort. Soc. 74: 219-223.

Hendrickson, R. and J.W. Kesterson. 1954. Hesperidin, the principal glucoside of oranges. Bull. 545. U. Fla., Agr. Exper. Sta., Gainesville, Fla.

Henricksen, H.C. and M.J. Iorns. 1909. Pineapple growing in Porto Rico. Bull. 8. Porto Rico Agr. Exper. Sta., Mayaguez.

Henry, F.E., G.B. Macfie and L.H. Young. 1981. Processing Puerto Rican plantains for the school lunch program. Proc. Fl. St. Hort. Soc. 94: 309-311.

Herbas A., R. 1969. El mosaico del caqui (*Diospyros kaki*) y algunas propiedades fisicas de su agente causal. Turrialba 19 (4): 480-490.

Herklots, G.A.C. 1972. Vegetables in South-East Asia. Hafner Press Div., Macmillan Pub'g Co., Inc., N.Y.

Hernandez, I. 1973. Storage of green plantains. J. Agr. U. Puerto Rico 62 (2): 100-106.

Hernandez, J., J.M. Rivera, and C.E. Bueso. 1976. Vitamin C enrichment of banana nectar. J. Agr. U. Puerto Rico 60 (2): 250-251.

Hernandez-Medina, E. 1952. Filter press cake increases pineapple yields in Puerto Rico. Bull. 104. U. Puerto Rico Agr. Exper. Sta., Rio Piedras.

Hernandez-Medina, E. 1969. Further evidence of the need of magnesium for pineapple in oxisols. J. Agr. U. Puerto Rico 53 (4): 357-368.

Hernandez-Medina, E. 1970. Yield response of the Red Spanish pineapple in Puerto Rico as affected by different levels of magnesium. Proc. 7th Ann. Mt., C.F.C.S., Martinique-Guadeloupe. 1969.

Hernandez-Medina, E., J. Velez-Santiago and M.A. Lugo-Lopez. 1970. Root development of acerola trees as affected by liming. J. Agr. U. Puerto Rico 54 (1): 57-61.

Hernandez Roque, F. 1974. Ensayo de resistencia del jitomate y del tomate de cáscara, al "Chino" y a la mosquita blanca en el estado de Morelos. Agr. Técn. México 3 (8): 305-309.

Hernandez Roque, F. and J.L. Carrillo S. 1973. Parasitismo natural de *Trichogramma* spp. en gusano del fruto del tomate de cáscara en el Estado de Morelos Agr. Técn. en México 3 (7): 255-258.

Hernandez-Unzon, H.Y. and S. Lakshminarayana. 1982. Biochemical changes during development and ripening of tamarind fruit (*Tamarindus indica* L.). HortSci. 17 (6): 940-942.

Hernandez-Unzon, H.Y. and S. Lakshminarayana. 1982. Developmental physiology of tamarind fruit (*Tamarindus indica* L.). HortSci. 17 (6): 938-940.

Higgins, J.E. 1924. Seediness in pineapples. Phil. Agriculturist 12 (8): 333-338.

Higgs, H. 1952. Bahamian Cook Book. 8th ed. Author. Nassau, Bahamas.

Hills, L.D. 1980. The cultivation of the carob tree (*Ceratonia siliqua*). Internat. Tree Crops J. 1: 27-36.

Hine, R.B., O.V. Holtzmann and R.D. Raabe. 1965. Diseases of papaya (*Carica papaya* L.) in Hawaii. Bull. 136. Hawaii Agr. Exper. Sta., Honolulu.

Hochstein, F.A. 1951. Medical Research Lab., Chas. Pfizer & Co., Groton, Conn. Personal communication. May 26.

Hodge, W.H. 1947. Naranjillas or "Little Oranges" of the Andean highlands. J.N.Y. Bot. Gard. 48 (571): 155-159.

Hodge, W.H. 1957. *Solanum quitoense* as an ornamental. Nat. Hort. Mag. 36 (4): 365-367.

Hodge, W.H. 1960. The South American "sapote". Econ. Bot. 14 (3): 203-206.

Hodgson, R.W. 1960. Avocado industry of Chile. Calif. Citrograph. Feb: 123-128.

Hodgson, R.W. and C.A. Schroeder. 1947. On the bearing behavior of the kaki persimmon (*Diospyros kaki*). Proc. Amer. Soc. Hort. Sci. 50: 145-148.

Hoffman, R.W. and C.E. Kennett. 1985. Tracking CRS development by degree-days. Calif. Agr. 39 (9 & 10): 19-20. [Calif. red scale on Navel orange]

Hofmeyr, J.D.J. 1953. Sex reversal as a means of solving breeding problems of *Carica papaya* L. So. Afr. J. Sci. 49 (7): 228-232.

Hogarth, W.B. 1969. Blackheart in rough-leaf pineapples. Queensland Agr. J. 95 (10): 657.

Hogarth, W.B. 1966. Zinc and pineapples. Queensland Fr. & Veg. News. Oct. 20. Pp. 389-390.

Holder, E.W. Jr. 1975. The citrus industry's spreading decline program. Proc. Fl. St. Hort. Soc. 88: 85-86.

Holdridge, L.R. and A. Poveda. 1975. Arboles de Costa Rica. Vol. I. Centro Cientifico Tropical, San José, C.R.

Holtzman, O.V. and M. Ishia. 1963. Papaya mosaic virus reduces quality of papaya fruits. Hawaii Farm Sci. 12 (4): 1-2.

Hooper, D. and H. Field. 1937. Useful plants and drugs of Iran and Iraq. Pub. 387. Bot. Ser. Vol. 9, No. 3., Field Mus. Nat. Hist., Chicago, Ill.

Hope, G.W. and D.G. Vitale. 1972. Osmotic dehydration—a cheap and simple method of preserving bananas and plantains. Monograph IRDC-004e. Food Res. Inst., Canada Dept. Agr., Ottawa, Can.

Hope, T. 1963. Pineapple maturity. Queensland Agr. J. 89 (7): 429.

Hope, T. 1963. Quality tests identify best avocadoes. Queensland Agr. J. 89 (11): 657-660.

Horticulture Div., Palmerston North. 1967. Poor fruiting of feijoa. New Zeal. J. Agr. 114 (8): 5.

Horticulture Div., Auckland. 1969. Cultivating feijoa. New Zeal. J. Agr. 118 (6): 11.

Hosaka, E.Y. and A. Thistle. 1954. Noxious Plants of the Hawaiian ranges. Exten. Bull. 62. U. Hawaii Coll. Agr. with U.S. Dept. Agr. cooperating.

Hottes, A.C. 1949. Climbers and ground covers. A.T. De La Mare Co., Inc., N.Y. 2nd ptg.

Howard, A.L. 1951. A manual of the timbers of the world. 3rd ed. Macmillan & Co., Ltd., London.

Howard, R.A. 1961. The correct names for "*Diospyros ebenaster*". J. Arnold Arboretum 42 (4): 430-436.

Howell, C.W. 1976. Edible fruited *Passiflora* adapted to South Florida growing conditions. Proc. Fl. St. Hort. Soc. 89: 236-238.

Hoyt, C.R. 1958. Check lists for ornamental plants of subtropical regions. Livingston Press, San Diego, Calif.

Huet, R. and M.-C. Murail. 1972. L'huile essentielle d' orange de type Guineé. Fruits 27 (4): 297-301.

Hughes, T.J. 1960. Florida's fresh lemon deal. Fla. Grower & Rancher 68 (10): 33-34.

Hughes, T.J. 1956. Lemons loom large. Fla. Grower & Rancher Dec. P. 13.

Hume, E.P. 1951. Growing avocados in Puerto Rico. Cir. 33. U.S. Dept. Agr., Agr. Res. Serv., Washington, D.C.

Hume, E.P. 1949. Some ornamental vines for the tropics. Cir. 31. Fed. Exper. Sta. in Puerto Rico, Mayaguez.

Hume, E.P. and H.F. Winters. 1949. The "palo de tomate" or tree tomato. Econ. Bot. 3 (2): 140-142.

Hume, E.P. and H.F. Winters. 1948. Tomatoes from a tree. Foreign Agr. 12 (6): 121-122.

Hume, H.H. 1941. The cultivation of citrus fruits. The Macmillan Co., N.Y.

Hume, H.H. 1903. The kumquats. Bull. 65. Fla. Agr. Exper. Sta., Lake City, Fla.

Hundtoft, E.B. 1974. Mechanized papaya harvesting dictates continuous handling concept. Hawaii Farm Sci. 23 (2): 7-9.

Hundtoft, E.B. 1972. New concepts for papaya harvester design. Hawaii Farm Sci. 21 (1): 6-9.

Hunter, J.E. 1969. Aerial application of fungicides on papayas. Hawaii Farm Sci. 18 (2): 4-5.

Hunter, J.E. and I.W. Buddenhagen. 1972. Incidence, epidemiology and control of fruit diseases of papaya in Hawaii. Trop. Agr. (Trin.) 49 (1): 54-71.

Hunter, J.R. 1969. The lack of acceptance of the pejibaye palm and a relative comparison of its productivity to that of maize. Econ. Bot. 25 (3): 237-244.

Hurst, E. 1942. Poison plants of New South Wales. NSW Poison Plants Comm., U. Sydney & NSW, Dept. of Agr., Sidney, Australia.

Hwang, S.C., C.L. Chen, J.C. Lin and H.L. Lin. 1984. Cultivation of banana using plantlets from meristem culture. HortSci. 19 (2): 231-233.

Ibrahim, A.G., G. Singh, and H.S. King. 1979. Trapping of the fruitflies, *Dacus* spp. (Diptera: Tephritidae) with methyl eugenol in orchards. Pertanika 2 (1): 58-61.

Iguina de George, L.M., A.L. Collazo de Rivera, J.R. Benero and W. Pennock. 1969. Provitamin A and vitamin C contents of several varieties of mango (*Mangifera indica* L.) grown in Puerto Rico. J. Agr. U. Puerto Rico 53 (2): 100-105.

Ikramul Hao, M.Y. and A.F.M. Ehteshamuddin. 1971. Utilization of mango waste—stone kernels. Sci. Ind. 8 (2): 207-209.

Inch, A.J. 1978. Passionfruit diseases. Queensland Agr. J. 104 (5): 479-484.

Ireta Ojeda, A. 1969. Eliminacion de plantes "masculinas" en una plantación de papaya. Agron. en Sinaba 3 (3): 8-9.

Ireta Ojeda, A. 1975. Estudios sobre la propagacion del litchi (*Litchi chinensis* Sonn). por el método de acodo aéreo en el valle de Culiacán. Agr. Técn. en México 3 (11): 418-423.

Iriiarte, J., F.A. Kinel, G. Rosenkranz and F. Sondheimer. 1956. The constituents of *Casimiroa edulis* Llave et Lex. Pt. II. The bark. J. Chem. Soc.: 4170-4173.

Irizarry, H., J.J. Green, E. Rivera and I. Hernandez. 1978. Effect of planting season on yield and other horticultural traits of the Horn type plantain Maricongo (*Musa acuminata* X *M. balbisiana*, AAB) in North-Central Puerto Rico. J. Agr. U. Puerto Rico 62 (1): 113-118.

Irizarry, H., E. Rivera, J.A. Rodriguez and J.J. Green. 1978. Effect of planting pattern and population density on yield and quality of the Horn-type Maricongo plantain (*Musa acuminata* X *M. balbisiana*, AAB), in North-Central Puerto Rico. J. Agr. U. Puerto Rico 62 (3): 214-223.

Irizarry, H., J. Rodriguez and D. Oramas. 1979. Evaluation of four nematicides in preplant treatments of plantain (*Musa acuminata* X *M. balbisiana*, AAB cv. Maricongo) corms. J. Agr. U. Puerto Rico 63 (2): 269-271.

Irizarry, H., S. Silva and J. Vicente-Chandler. 1980. Effect of water table level on yield and root system of plantains. J. Agr. U. Puerto Rico 64 (1): 33-36.

Irvine, F.R. 1961. Woody plants of Ghana, with special reference to their uses, Oxford U. Press, London.

Ismail, K.B., J. Abdullah and Yee Soi Yin. 1974. Pineapple waste studies (Study #1) Bull. 135. Min. Agr. & Fish., Malaysia.

Ito, P.J. 1978. 'Noel's Special' passion fruit. HortSci. 13 (2): 197.

Ivens, G.W. 1967. East African weeds and their control. Oxford U. Press, Nairobi.

Iyer, C.P.S., R. Singh and M.D. Subramanyan. 1978. A simple method for rapid germination of pineapple seeds. Scientia Hort. 8: 39-41.

Jackson, G.C. 1967. Promising selections of the honeyberry (*Melicocca bijuga* L.) from Puerto Rico. J. Agr. U. Puerto Rico 51: 66-70.

Jackson, G.C. and W. Pennock. 1958. Fruit and vitamin C production of five- and six-year-old acerola trees. J. Agr. U. Puerto Rico 42 (3): 196-205.

Jackson, L.K. and W.B. Sherman. 1972. Seedlessness in 'Tahiti' lime. Proc. Fl. St. Hort. Soc. 85: 330-332.

Jacobs, C.J., H.T. Brodrick, H.D. Swarts and N.J. Mulder. 1973. Control of postharvest decay of mango fruit in So. Africa. Pl. Dis. Rep. 57 (2): 173-176.

Jacobson, M. 1958. Insecticides from plants. Agr. Handbook 154. U.S. Dept. Agr., Agr. Res. Serv., Washington, D.C.

Jadan, D. 1973. A guide to the natural history of St. John. Virgin Islands Conservation Soc., St. John, V.I.

Jahn, O.L. 1974. Degreening of Florida lemons. Proc. Fl. St. Hort. Soc. 87: 218-221.

Jahn, O.L. 1981. Effects of ethephon, gibberellin and BA on fruiting of 'Dancy' tangerines. J. Amer. Hort. Soc. 106 (5): 597-600.

Jain, N.L. 1959. A note on drying of litchi. Indian J. Hort. 16 (3).

Jain, N.L. and D.H. Barker. 1966. Preparing beverages from guava. Indian Hort. 11 (1): 5-7.

Jain, N.L., D.P. Das and G. Lal. 1958. Utilization of cashew apples. In: Fruit and vegetable preservation industry in India. Proc. of the Symposium. Cent. Food Tech. Res. Inst., Mysore.

Jain, N.L. and G. Lal. 1957. Some studies in the utilization of jack fruit wastes as a source of pectin. Indian J. Hort. 14 (4): 213-222.

Jain, N.L. and G. Lal. 1954. Studies on the preparation of amla syrup. Bull. Cent. Food Tech. Res. Inst., Mysore, India. 3:297-301.

Jain, N.L., G. Lal and G.V. Krishnamurthy. 1957. Further studies in the preparation and uses of mango cereal flakes. Indian J. Hort. 14 (3): 2-8.

Jain, S.K. 1968. Medicinal plants (India—The land and people). Nat. Book Trust, New Delhi, India.

Jamwall, K.S., I.P. Sharma and C.L. Chopra. 1959. Pharmacological investigation of the fruit of Emblica officinalis Gaertn. J. Sci. & Indus. Res. 18C (9): 180-181.

Jaramillo, R. 1982. Necesidad de obtener variedades de plátano resistentes a la Sigatoka negra. Informe Mensual (UPEB) 5 (47): 38-42.

Jaramillo C., R. 1984. Produccion del plátano en la cuenca del lago de Maracaibo, Venezuela, Informe Mensual 8 (63): 40-44.

Jarman, C.G. 1978. Banana fibre in rural industries. Appropriate Technol. 5 (1): 25.

Jarrett, F.M. 1959. Studies in Artocarpus and allied genera. III. A revision of Artocarpus subgenus Artocarpus. J. Arn. Arbor. XL (2): 113-368.

Jawanda, J.S. and S.K. Narula. 1983. Loquat cultivation in the Punjab. Progressive Farming 19 (12): 7-8.

Jernberg, D.C. and A.H. Krezdorn. 1976. Performance of commercial 'Nova' tangelo plantings. Proc. Fl. St. Hort. Soc. 89: 14-17.

Jindal, K.K. and R.N. Singh. 1976. Sex determination in vegetative seedlings of Carica papaya by phenolic tests. Scientia Hort. 4: 33-39.

Jindal, V.K. and S. Mukherjee. 1970. Structure of alkali-soluble polysaccharides from Phoenix dactylifera seeds. Indian J. Chem. 8: 417-419.

Jindal, V.K. and S. Mukherjee. 1971. Structure of alkali-soluble seed polysaccharide from Phoenix dactylifera seeds: Part 2—Isolation of oligosaccharides. Indian J. Chem. 9: 207-208.

Jiravatana, V., J. Cuevas-Ruiz and H.D. Graham. 1970. Extension of storage life of papayas grown in Puerto Rico by gamma radiation treatments. J. Agr. U. Puerto Rico. 54 (2): 314-319.

Johannessen, C.L. 1966. Pejibayes in commercial production. Turrialba, Rev. Interam. Ciencia Agr. 16 (12): 181-187.

Johannessen, C.L. 1967. Pejibaye palm: physical and chemical analysis of the fruit. Econ. Bot. 21 (4): 371-378.

Johannessen, C.L. 1966. Pejibaye palm: yields, prices and labor costs. Econ. Bot. 20 (3): 302-315.

Johannessen, C.L. 1966. The domestication process in trees reproduced by seed: the pejibaye palm in Costa Rica. Geogr. Rev. 56 (4): 363-376.

Johar, D.S. and J.C. Anand. 1952. Nature and prevention of spoilage in amla preserves. Indian Food Packer 6: 9-11.

Johar, D.S. and Y.S. Lewis. 1955. Pectin for jams and jellies from tamarind pulp. Indian Food Packer 9 (4): 28-30.

Johnson, S. 1975. Funda de polietileno com insecticida protege racimos de banano. La Hacienda 70 (3): 22-25.

Johnson, T.J. 1979. Effects of potassium on buoyancy of banana fruit. Expl. Agr. 15: 173-176.

Johnston, B. 1962. Cape gooseberries, a delicacy for many. New Zeal. J. Agr. 104 (4): 372.

Joiner, J.N. 1954. Lychee growing gains in Florida. Fla. Grower & Rancher. Sept. Pp. 12-13, 16.

Joiner, J.N. 1959. The effects of differential levels of nitrogen, potassium and magnesium on the growth of lychees. Proc. Fl. St. Hort. Soc. 72: 346-348.

Jones, M.A. and H.K. Plank. 1946. Chemical examination of mamey seed. Report of Fed. Exper. Sta., Mayaguez, P.R. 1945. U.S. Dept. Agr., Agr. Res. Serv., Washington, D.C.

Jones, M.A. and H.K. Plank. 1947. Chemical examination of mamey seed. Report of Fed. Exper. Sta., Mayaguez, P.R. 1946. U.S. Dept. Agr., Agr. Res. Serv., Washington, D.C.

Jones, W.W., T.W. Embleton, S.B. Boswell, G.E. Goodall and E.L. Barnhart. 1970. Nitrogen rate effects on lemon production, quality and leaf nitrogen. J. Amer. Soc. Hort. Sci. 95 (1): 46-49.

Jones, W.W., T.W. Embleton, M.L. Steinacker and C.B. Cree. 1970. Carbohydrates and fruiting of 'Valencia' orange trees. J. Amer. Soc. Hort. Sci. 95 (3): 380-381.

Jones, W.W., W.B. Storey, G.K. Parris, and F.G. Holdaway. 1941. Papaya production in the Hawaiian Islands. Bull. 87. Hawaii Agr. Exper. Sta., U. of Hawaii.

Jordan, B. 1980. Lime production in Niue. Alafua Agr. Bull. 5 (3): 17-20.

Jordan, B. 1980. Passionfruit production in Niue. Alafua Agr. Bull. 5 (3): 9-14.

Jordan, K.R. 1969. Comparison of four fertilizer schedules for pineapples in central Queensland. Queensland J. Agr. & Anim. Sci. 26: 495-508.

Jorgensen, K.R. 1978. Copper deficiency in pineapples. Queensland J. Agr. & Anim. Sci. 35 (2): 77-81.

Jorgensen, K.R. 1973. Fertilizing pineapples on a soil high in potassium. Queensland J. Agr. & Anim. Sci. 30 (3): 213-223.

Jorgensen, K.R. 1969. Investigation of pineapple fertilizing methods and flower induction. Queensland J. Agr. & Anim. Sci. 26 (4): 483-493.

Joseph, G.H. 1947. Citrus products—a quarter century of amazing progress. Econ. Bot. 1 (4): 415-426.

Joubert, A.J. and L.J. van Lelyveld. 1975. An investigation of preharvest browning of litchi peel. Phytophylactica 7: 9-14.

Joyner, G. 1977. Grumichama and Surinam cherries. Fla. Gardening Companion 2 (2): 12-14.

Juarez-Gutierrez, R.E. and S. Becerra Rodriguez. 1979. Comportamiento de catorce cultivares de platano (Musa spp.) en el estado de Colima, Mexico. Proc. Trop. Reg. Amer. Soc. Hort. Sci., 23: 175-178.

Judd, B.I. 1975. Goma de mascar—el gran movimiento Americano. La Hacienda 70 (10): 13-14, 16. [sapodilla].

Kaln, V. 1976. Polyphenol oxidase isoenzymes in avocado. Phytochem. 15: 267-272.

Kanehiro, Y. and G.D. Sherman. 1947. Production of a dehydrated guava-flavored pectin. U. Hawaii Agr. Exper. Sta. Biennial Rpt. 1944 & 1946.

Karikari, S.K. 1972. Plantain growing in Ghana. World Crops 24 (1): 22-24.

Karling, J.S. 1942. Collecting chicle in the American tropics. [sapodilla]. Torreya 42: 38-50, 104-113.

Katyal, S.L. 1969. Mango cultivation in Surinam. Indian Hort. 13 (3): 3-5.

Katyal, S.L. and D.V. Chugh. 1961. Problems in extensive litchi cultivation. Hort. Adv. (India) 5: 39-41.

Kay, D.E. 1967. The processing of banana products. So. Pac. Bull. 2nd quarter. 17 (2): 37-41.

Kay, D.E. 1965. The production and marketing of pineapples. TPI Rpt. G: 10. Trop. Prod. Inst., London.

Kay, D.E. and E.H.G. Smith. 1960. The production and world trade in fresh pineapples. Trop. Sci. II (1 & 2): 55-63.

Kean, E.A. (Editor). 1975. Hypoglycin—Proceedings of a Symposium Kingston, Jamaica. Academic Press, Inc., N.Y.

Kee, N.S. and S. Thamboo. 1967. Nutrient removal studies on Malayan fruits: durian and rambutan. Malayan Agr. J. 46 (2): 164-182.

Keeler, J.T., K. Mihata and W. Nakashima. 1961. Economic factors affecting the production of papayas in Waimanalo, Oahu. Agr. Econ. Rpt. 49 (rev'd). Hawaii Agr. Exper. Sta., U. Hawaii.

Kehat, M., D. Blumberg and S. Greenberg. 1976. Fruit drop and damage in dates: the role of *Coccotrypes dactyliperda* F. and nitidulid beetles, and prevention by mechanical means. Phytoparasitica 4 (2): 91-99.

Kennard, W.C. 1955. The passion fruits in Puerto Rico. Fr. Vars. & Hort. Dig. 10 (4): 50-60.

Kennard, W.C. and H.F. Winters. 1960. Some fruits and nuts for the tropics. Misc. Pub. 801. U.S. Dept. Agr., Agr. Res. Serv., Washington, D.C.

Kerharo, J. 1971. Senegal bisap (*Hibiscus sabdariffa*) or Guinea sorrel or red sorrel. Plant. Med. Phytother. 5 (4): 277-281.

Keshwal, R.L. 1979. Note on manganese and zinc deficiency disease in lemon and jack fruit trees. Allahabad Farmer 50 (3): 251-253.

Kesterson, J.W., R.J. Braddock, R.C.J. Koo and R.L. Reese. 1977. Nitrogen and potassium fertilization as related to the yield of peel oil from 'Pineapple' oranges. J. Amer. Soc. Hort. Sci. 102 (1): 3-4.

Kesterson, J.N. and R. Hendrickson. 1958. Finding the best lemon for Florida—a report of progress. III. Evaluation of coldpressed Florida lemon oil and lemon bioflavonoids. Proc. Fl. St. Hort. Soc. 86: 132-140.

Keys, J.D. 1976. Chinese herbs, their botany, chemistry and pharmacodynamics. Charles E. Tuttle Co., Rutland, Vt. and Tokyo, Japan.

Khalil, M.A. and M.B. Sial. 1974. Spray drying of mango juice powder. Mesopotamia J. Agr. (Iraq) 91 (1-2): 47-56.

Khan, M.I.H. and J. Ahmad. 1985. A pharmacognostic study of *Psidium guajava* L. Internat. J. Crude Drug Res. 23 (1) No. 2: 95-103.

Khorana, M.L., M.R. Rajarama Rao, and H.H. Siddiqi. 1960. Expectorant activity of *Emblica officinalis* Gaertn. J. Sci. & Indus. Res. 19C: 60-61.

Killip, E.P. 1938. The American species of Passifloraceae. Bot. 19. Field Mus. Natural History, Chicago.

Kinel, F.A., J. Romo, G. Rosenkranz and F. Sondheimer. 1956. The constituents of *Casimiroa edulis* Llave et Lex. Pt. I. The seed. J. Chem. Soc. 4163-4169.

King, G.S., H. Sakanashi and E. Song. 1951. High powder from papaya. Indian Food Packer 5 (12): 31.

King, K. 1964. Mechanical planting of pineapples. Queensland Agr. J. 90 (2): 72-75.

King, K. 1969. Storing pineapple tops. Queensland Agr. J. 95 (7): 448.

Kiser, L.A. 1960. Lickin' good is the luscious lychee. Fla. Grower & Rancher. June. P. 20.

Kishore, N. 1950. Preparation of jelly [emblic]. Indian Food Packer 4: 15-22.

Kitagawa, H., A. Sugiura, and M. Sugiyama. 1966. Effects of gibberellin spray on storage quality of kaki. HortSci. 1 (2): 59-60.

Kizer, L.A. 1957. Citrus gems—golden kumquats. Fla. Grower & Rancher. Jan. P. 29.

Knapp, F.F. and H.J. Nicholas. 1969. The sterols and triterpenes of banana peel. Phytochem. 8 (1): 207-214.

Knight, R.J., Jr. 1969. Edible-fruited passionvines in Florida: the history and possibilities. Proc. Trop. Reg., Amer. Soc. Hort. Sci. 13: 265-274.

Knight, R.J., Jr. 1965. Heterostyly and pollination in carambola. Proc. Fla. St. Hort. Soc. 78: 375-378.

Knight, R.J., Jr. 1982. Partial loss of self-incompatibility in 'Golden Star' carambola. HortSci. 17 (1): 72.

Knight, R.J., Jr. 1982. Response of carambola seedling populations to Dade County's oolitic limestone soil. Proc. Fl. St. Hort. Soc. 95: 121-122.

Knight, R.J., Jr. 1972. The potential for Florida of hybrids between the purple and yellow passionfruit. Proc. Fl. St. Hort. Soc. 85: 288-292.

Knight, R.J., Jr., W.E. Manis, G.W. Kosel and C.A. White. 1968. Evaluation of longan and lychee introductions. Proc. Fl. St. Hort. Soc. 82: 314-318.

Knight, R.J., Jr. and H.F. Winters. 1971. Mango and avocado evaluation in southeastern Florida. Proc. Fl. St. Hort. Soc. 84: 314-317.

Knight, R.J., Jr. and H.F. Winters. 1963. Pollination and fruit set of

yellow passionfruit in southern Florida. Proc. Fl. St. Hort. Soc. 75: 412-418.

Knobb, L.C. 1958. Finding the best lemon for Florida—a report of progress. I. The growing of lemons in Florida: historical, varietal, and cultural considerations. Proc. Fl. St. Hort. Soc. 86: 123-128.

Kok, J.B. and A.J. Joubert. 1967. Produce better litchis. Farming in So. Africa 43 (1): 6-7, 9.

Koo, R.C.J. and R.L. Reese. 1976. Influence of fertility and irrigation treatments on fruit quality of 'Temple' orange. Proc. Fl. St. Hort. Soc. 89: 45-51.

Koo, R.C.J., T.W. Young, R.L. Reese and J.W. Kesterson. 1973. Responses of 'Bearss' lemon to nitrogen, potassium and irrigation applications. Proc. Fl. St. Hort. Soc. 86: 9-12.

Koroieveibau, D. 1966. Some Fiji breadfruit varieties. Bull. 46, Dept. Agr. Fiji.

Kotalawala, J. 1971. Mass production of pineapple planting material Trop. Agriculturist 127: 199-201.

Kotalawala, J. 1968. Pineapple cultivation in coconut land in the low-country wet zone. Ceylon Coconut Planters' Rev. 5 (3): 112-117.

Kothan, K.L. 1968. Controlling fruit rot of guava. Indian Hort. 12 (3): 9-10.

Krezdorn, A.H. 1977. Citrus close-ups: Inconsistency in Minneola reproductive incompatability. Fla. Grower & Rancher 70 (10): 14, 16.

Krezdorn, A.H. 1984. Citrus Trunk Line: Control quality with cultural practices. Amer. Fr. Grower Oct. 26-27.

Krezdorn, A.H. 1983. Citrus Trunk Line: Living with blight and tristeza. Amer. Fr. Grower 103 (6): 24-25.

Krezdorn, A.H. 1982. Citrus Trunk Line: The citrus rootstock dilemma. Amer. Fr. Grower. Oct. 26-27.

Krezdorn, A.H. 1981. Citrus Trunk Line: whatever happened to mechanical harvesting? Fla. Grower & Rancher 101 (10): 8.

Krezdorn, A.H. 1980. Closer tree spacing—is it the answer? Fla. Grower & Rancher. Aug. P. 26.

Krezdorn, A.H. 1979. Grapefruit characteristics and variety selection. Fla. Grower & Rancher 72 (10): 22-24.

Krezdorn, A.H. 1980. Lemons and limes, their culture in Florida. Fla. Grower & Rancher 73 (6): 26-27.

Krezdorn, A.H. 1980. Mandarin acreage small. Fla. Grower & Rancher 73 (2): 13.

Krezdorn, A.H. 1979. Rootstocks: their relation to virus disease and blight. Fla. Grower & Rancher 72 (7): 25-26.

Krezdorn, A.H. and G.W. Adriance. 1961. Fig growing in the South. Agr. Handbook #196. U.S. Dept. Agr., Agr. Res. Serv., Washington, D.C.

Krikorian, A.D. 1968. The psychedelic properties of banana peel: an appraisal. Econ. Bot. 22 (4): 385-389.

Krikorian, A.D. and S.S. Cronauer. 1984. Aseptic culture techniques for banana and plantain improvement. Econ. Bot. 38 (3): 322-331.

Krishnamurthi, S. and N.V. Madhava Rao. 1962. Mangosteen deserves wider attention. Indian Hort. Oct.-Dec. 3-4, 8 & inside back cover.

Krishnamurthi, S. and V.N. Madhava Rao. 1965. The mangosteen (*Garcinia mangostana* L.) and its introduction and establishment in peninsular India. Adv. in Agr. Sci., Coimbatore (India). Pp. 401-409.

Krishna Murthy, G.V., N. Giridhar and B. Raghuramaiah. 1984. Suitability of some mango varieties for processing. J. Food Sci. & Technol. 21: 21-15.

Krochmal, A. 1973. Algunas enfermedades comunes de la papaya. La Hacienda 73 (3): 50, 52-53.

Krochmal, A. 1969. Low volume papaya processing in Surinam. World Crops 21 (1): 22-23.

Krochmal, A. and M.R. Henderson. 1970. Empaque y embarque aéreo. La Hacienda 65 (12): 35-36.

Krome, W.H. 1968. Economic view of lime-growing. Econ. Bot. 22 (3): 270-272.

Kruger, N.S. and R.C. Menary. 1968. Measures against high nitrate in papaws. Queensland Agr. J. 94 (4): 234-235.

Kuhn, G.D. 1962. Dehydration studies of lychee fruit. Proc. Fl. St. Hort. Soc. 75: 273-277.

Kuhn, G.D. 1962. Some factors influencing the properties of lychee jelly. Proc. Fl. St. Hort. Soc. 75: 408-410.

Kuhne, F.A. 1968. Cultivation of granadillas (1). Farming in So. Africa 43 (11): 29-32.

Kuhne, F.A. 1968. Cultivation of granadillas (2). Farming in So. Africa 43 (12): 23-28.

Kuhne, F.A. 1968. Cultivation of granadillas (3). Farming in So. Africa 43 (13): 17-19.

Kuhne, F.A. 1968. The yellow granadilla in South Africa. Farming in So. Africa 44 (5): 7.

Kuhne, F.A. and P. Allen. 1970. Seasonal variations in fruit growth of *Carica papaya*. Agr. Sci. So. Africa, Agroplanten 2 (3): 99-104.

Kumar Saha, A. 1969. Studies on fruit-set in jackfruit (*Artocarpus heterophyllus* Lam.). I. Effect [of] ringing, scoring, and shoot pruning on fruiting. Pl. Sci. India 1: 220-223.

Kunimoto, R.K., P.J. Ito and W.H. Ko. 1977. Mucor rot of guava fruits caused by *Mucor hiemalis*. Trop. Agr. 54 (2): 185-187.

Labanauskas, C.K., W.W. Jones and T.W. Embleton. 1964. Effects of foliar applications of manganese, zinc and urea on yield and fruit quality of Valencia oranges. Calif. Citrograph 49 (5): 175, 190-192.

Laemmlen, F.F. and M. Aragaki. 1971. Collar rot of papaya caused by *Calonectria* sp. Pl. Dis. Rep. 55 (8): 743-745.

Lahav, E., B. Gefen, and D. Zamet. 1972. The effect of girdling on the productivity of the avocado. Hortus 17 (Apr.): 30-33.

Lakshmi, V. and J.S. Chauhan. 1976. Grewinol, a keto-alcohol from the flowers of *Grewia asiatica*. Lloydia 39 (5): 372-374.

Lakshminarayana, S., A. Gomez-Cruz and S. Martinez-Romero. 1985. Preliminary study of a new preharvest stem-end rot and associated microflora in mango. HortSci. 20 (5): 947-948.

Lakshminarayana, S. and M.A. Moreno Rivera. 1978. Enfermedades y desordenes en la produccion y mercadeo de la guayaba mexicana. Chapingo, Nueva Epoca No. 9: 27-33.

Lakshminarayana, S. and M.A. Moreno Rivera. 1979. Promising Mexican guava selections rich in Vitamin C. Proc. Fl. St. Hort. Soc. 92: 300-303.

Lakshminarayana, S. and M.A. Moreno Rivera. 1979. Proximate characteristics and composition of sapodilla fruits grown in Mexico. Proc. Fl. St. Hort. Soc. 92: 303-305.

Lakshminarayana, S., A.R. Vijayendra Rao, N.V.N. Moorthy, B. Amandaswamy, V.B. Dalal, P. Narasimham, and H. Subramanyam. 1971. Studies on the rail shipment of mango. J. Food Sci. & Tech. 8: 121-126.

Lal, G. and D.P. Das. 1956. Studies on jelly making from papaya fruit. Ind. J. Hort. 13 (1): 38-44.

Lal, G., D.P. Das and N.L. Jain. 1958. Tamarind beverage and sauce. Indian Food Packer 12 (5): 13-14.

Lal, G., G.V. Krishnamurthy, N.L. Jain and B.S. Bhatia. 1960. Suitability of different varieties of mangoes for the preparation of mango cereal flakes. Food Sci. 9 (4): 121-123.

Lal, G., G.S. Sidappa and G.L. Tandon. 1960. Preservation of fruits and Vegetables. Indian Coun. Agr. Res., New Delhi.

Lal, G. and J. Singh Pruthi (undated). Ascorbic acid retention in pineapple products. Inc. J. Hort. 137-141.

Landrau, P., Jr. and E. Hernandez Medina. 1959. Effects of major and minor elements, lime, and soil amendments on the yield and ascorbic acid content of acerola (*Malpighia punicifolia* L.). J. Agr. U. Puerto Rico 43 (1): 19-33.

Lange, A.H. 1969. Reciprocal grafting of normal and dwarf Solo papaya on growth and yield. HortSci. 4 (4): 304-306.

Lange, A.H. 1961. Transplanting papaya versus seeding in place. Trop. Agr. (Trin.) 38 (3): 235-243.

Lara Rodriguez, E.A. and M.W. Borys. 1983. El cultivo del guayabo, *Psidium guajava* L. Chapingo 8 (41): 41-45.

Larson, E., M.F. Wynn, S.J. Lynch and D.D. Doughty. 1953. Some further studies on the akee. Quart. J. Fla. Acad. Sci. 16 (3): 151-156.

Larue, M. 1975. L'Actinidia chinensis et sa culture. Fruits 30 (1): 45-50.

Lassoudiere, A. 1969. La papaine; production, proprietes, utilisation. Fruits 24 (11-12): 503-517.

Lassoudiere, A. 1969. Le papayer Pt. III. La plante et les conditions ecologiquees. Fruits 24 (2): 3, 105-113.

Lassoudiere, A. 1969. Le papayer. Pt. 5. Analyse d'une population de papayers 'Solo' non maintenue en selection, resultats preliminaires.

Fruits 24 (4): 217-221.

Lassoudiere, A. 1969. Le papayer. Pt. 9. Recolte, conditionnement, exportation, produits transformes. Fruits 24 (11-12): 491-502.

Laumas, K.R. and T.R. Seshadri. 1958. Chemical components of the bark of *Phyllanthus emblica*. J. Sci. & Indus. Res. 17B (4): 167-168.

Laurence, G. 1968. Insect problems of papaw production in Trinidad. J. Agr. Soc. Trinidad & Tobago 68 (3): 311-312.

Lavadores Villanueva, G. 1969. Estudio de las 119 plantas medicinales mas conocidas en Yucatan, Mexico. Univ. Nac. Autonoma de Mexico, Mexico, D.F.

Lavigne, N.F., Sr. 1976. Japanese grapefruit exports continue strong. Fla. Grower & Rancher 69 (8): 22.

Lawler, F.K. 1967. Banana challenges food formulators. Pt. I. Food Engrng. 39 (5): 58-63.

Lawler, F.K. 1967. Banana challenges food formulators. Pt. II. Food Engrng. 39 (6): 62-65.

Lazo, F.R. 1965. El injerto del mamey colorado. Arroz 14 (148): 2.

Leal, F.J. 1970. Notas sobre la guanabana (*Annona muricata*) in Venezuela. Proc. Trop. Reg. Amer. Soc. Hort. Sci. 14: 118-121.

Leal, F.J., M.G. Antoni and P. Rodriguez. 1979. Descripcion de cinco variedades de piña (*Ananas comosus*) en Venezuela. Rev. Fac. Agron. (Maracay) 10 (1-4): 21-30.

Leal, F.J. and M.G. Antoni. 1979. Descripcion y clave de las variedades de piña cultivadas en Venezuela. Proc. Trop. Reg. Amer. Soc. Hort. Sci. 23: 169-172.

Leal, F.J. and J. Soule. 1977. 'Maipure', a new spineless group of pineapple cultivars. HortSci. 12 (4): 301-305.

LeBourdelles, J. and P. Estanove. 1967. La goyave aux Antilles. Fruits 22 (9): 397-412.

LeCointe, P. 1947. Arvores e plantas uteis. 2nd ed. Amazonia Brasileira III. Biblioteca Pedagogica Brasileira, Sao Paulo.

Ledin, R.B. 1956. A comparison of three clones of Barbados cherry and the importance of improved selections for commercial plantings. Proc. Fl. St. Hort. Soc. 69: 293-297.

Ledin, R.B. 1957. A note on the fruiting of the Mauritius variety of lychee. Fla. Lychee Growers Assn. Year Book & Proc. 5th Ann. Mtg., Winter Haven, Fla.

Ledin, R.B. 1953. A tropical black raspberry for South Florida. Cir. s-56. U. of Fla. Agr. Exper. Sta., Gainesville, Fla.

Ledin, R.B. 1954. Florida mango census. Fla. Subtrop. Gardener. Sept.: 20-21.

Ledin, R.B. 1958. Mango varieties in Florida. Hort. Adv. Vol. II: 16-26.

Ledin, R.B. 1952. Naranjilla (Little Orange), a new fruit for Florida. Fla. Grower. Oct. Pp. 20, 26, 27.

Ledin, R.B. 1955. Rubus trials in South Florida. Proc. Fl. St. Hort. Soc. 68: 272-274.

Ledin, R.B. 1958. The Barbados or West Indian cherry. Bull. 594. U. of Fla. Agr. Exper. Sta., Gainesville, Fla.

Ledin, R.B. 1954. The Mysore black raspberry in Florida. Fr. Vars. & Hort. Dig. 9 (1): 10-11.

Ledin, R.B. 1952. The naranjilla (*Solanum quitoense* Lam.). Proc. Fl. St. Hort. Soc. 65: 187-190.

Lee, D. 1985. The durian, a most magnificent and elusive fruit. Fairchild Trop. Garden Bull. 40 (2): 19-27.

Lee, G.R., F. Proctor and A.K. Thompson. 1973. Transport of papaya fruits from Trinidad to Britain. Trop. Agr. (Trin.) 50 (4): 303-306.

Lee, S.A. 1972. Agro-economic studies on intercropping in pineapple. Malaya Pineapple 2: 23-32.

Lee, S.A. 1977. Paraquat, diesel oil and kerosene for ratoon clearing of pineapple. Weed Res. 17 (2): 109-111.

Lee Oi Hian and Chew Tek Ann. 1974. Allocation of resources in pomelo farms. Malay. Agr. Res. 3: 119-126.

Lefevre, J.C. 1971. Revue de la litterature sur le tamarinier. Fruits 26 (10): 687-695.

Leigh, D.S. 1969. The papaw. Agr. Gaz. New So. Wales 80 (1): 6-11.

Leon, Hno. & Hno. Alain. 1953. Flora de Cuba. Vol. III. [Barbados gooseberry]. Museo de Historia Nat. de la Salle, Havana.

Leon, J. 1964. Plantas alimenticias Andinas. Bol. Tec. 6. Inst. Interamer. de Ciencias Agr. Zona Andina, Lima, Peru.

Leonard, L.Y. and P.G. Sylvain. 1930. La papaye. Bull. 20. Serv.

Tech. du Dept. de l'Agriculture et de l'Enseignement professional. Port-au-Prince, Haiti.

Leu, L.S., C.W. Kao, C.C. Wang, W.J. Liang and S.P.Y. Hsieh. 1979. *Myxosporium* wilt of guava and its control. Pl. Dis. Rep. 63 (12): 1075-1077.

Lever, R.J.A.W. 1965. The breadfruit tree. World Crops 17 (3): 63.

Leverington, R.E. 1963. Pineapple juice. Queensland J. Agr. 89 (4): 247.

Leverington, R.E. and R.C. Morgan. 1967. An experimental pineapple juice concentrate plant incorporating flavour recovery. 2. Instrumentation and control of the turbulent thin film evaporator. Queensland J. Agr. & Anim. Sci. 24 (1): 42-47.

Lewis, T. and E.F. Woodward. 1950. Papain—the valuable latex of a delicious tropical fruit. Econ. Bot. 4 (2): 192-194.

Lewis, Y.S., C.T. Dwarakanath and D.S. Johar. 1956. Acids and sugars in *Eugenia jambolana*. J. Sci. & Indus. Res. 15C (12): 280-281.

Lewis, Y.S., C.T. Dwarkanath and D.S. Johar. 1954. Utilization of tamarind pulp. J. Sci. & Indus. Res. 13A (6): 284-286.

Li, L-Y and Chou, C-Y. 1948. Notes on the Chen-tze lychee of Henghua, Fukien, China. Proc. Fl. St. Hort. Soc. 61: 283-285.

Lim, T.K. and K.C. Khoo. 1983. Crusty leaf spot disease of mango. Pertanika 6 (3): 12-14.

Lim, W.H. 1974. The etiology of fruit collapse and bacterial heart rot of pineapple. MARDI Res. Bull. (Malaya) 2 (2): 11-16.

Lind, H.Y., M.L. Bartow, and C.D. Miller. 1946. Ways to use vegetables in Hawaii. Bul. 97. 2nd ptg. U. Hawaii Agr. Exp. Sta., Honolulu.

Lindley, John and T. Moore. 1876. The Treasury of Botany: a popular dictionary of the vegetable kingdom. 2nd ed (2 vols.). Longmans, Green & Co., London.

Lindsay, W.R. 1940. Mangosteen cultivation. Proc. 8th Amer. Sci. Congress, Washington, D.C. 263-265.

Ling, J.S.L., S.Y. P'an and F.A. Hochstein. 1958. Some pharmacological properties of *N*-substituted analogs of histamine. J. Pharm. Exp. Therap. 122: 44a.

Lipitoa, S. and G.L. Robertson. 1977. The enzymatic extraction of juice from yellow passion fruit pulp. Trop. Sci. 19 (2): 105-112.

Little, E.L., Jr. and F.H. Wadsworth. 1964. Common trees of Puerto Rico and the Virgin Islands (Agr. Handbook 249). U.S. Dept. Agr., For. Serv., Washington, D.C.

Litz, R.E. 1984. *In vitro* somatic embryogenesis from callus of jaboticaba, *Myrciaria cauliflora*. HortSci. 19 (1): 62-64.

Litz, R.E. 1981. Effect of sex type, season, and other factors in *in vitro* establishment and culture of *Carica papaya* L. explants. J. Amer. Hort. Soc. 106 (6): 792-794.

Litz, R.E. and R.A. Conover. 1978. *In vitro* propagation of papaya. HortSci. 13 (3): 241-242.

Litz, R.E. and R.A. Conover. 1977. Tissue culture propagation of papaya. Proc. Fl. St. Hort. Soc. 90: 245-246.

Liu, F.W. 1976. Ethylene inhibition of senescent spots on ripe bananas. J. Amer. Soc. Hort. Sci. 101 (6): 684-686.

Lizana, L.A. and J.M. Errazuriz. 1978. Calidad de la naranja en el mercado mayoritario de Santiago. Proc. Trop. Reg. Amer. Soc. Hort. Sci. 22: 82-97.

Llorens, A.A. and R.P. Gonzalez. 1963. El mercadeo de la piña en Puerto Rico. Bull. 175. U. Puerto Rico Estac. Exper. Agr.

Lloyd, A.C. 1972. Nitrate and nitrate reductase in papaw fruit. Queensland J. Agr. & Anim. Sci. 29: 85-102.

Logan, M.D. 1960. The carob crusade, Amer. Forests 66 (6): 18-19, 63-65.

López Garcia, J. and R. Pérez Pérez. 1977. Effect of pruning and harvesting methods on guava yields. J. Agr. U. Puerto Rico 61 (2): 148-151.

Lombardo, A. 1964. Flora arborea y arborescente del Uruguay. Consejo Depart. de Montevideo, Montevideo, Uruguay.

Longe, O.G., E.O. Famojuro, and V.A. Oyenuga. 1977. Available carbohydrates and energy values of cassava, yam and plantain peels for chicks. E. Afr. Agr. J. 42 (4): 408-413.

Loo, T.G. 1969. Some aspects of the isolation of sugar from *Ceratonia siliqua* L. Koninklijk Inst. voor de Tropen, Amsterdam.

Loomis, H.F. and R.O. Nelson. 1955. Bengal, a promising large-clustered lychee. Fla. Lychee Growers Assn. Year Book and Proc.

3rd Ann. Mtg., Winter Haven, Fla.

Lopes, H.C. 1972. Composicao quimica e aproveitamento de "pera" de caju de Mocambique. Agron. Mocamb., Lourenco Marques 6 (2): 119-131.

Lopez G., J.A. and H. Garcia T. 1968. El limon: Aspectos tecnicos, cultivo, mercado, financiamiento y costo [Pt. I]. Rev. Mensual Asoc. Gen. de Agricultores 126 (3): 12-13, 16-17.

Lopez G., J.A. and H. Garcia T. 1969. El limon: Aspectos tecnicos, cultivo, mercado, financiamiento y costs [Pt. II]. Rev. Mensual Asoc. Gen. de Agricultores 128 (3): 12-13, 15-18, 20, 22-23.

Lopez G., J.A. and H. Garcia T. 1969. El limon: Aspectos technicos, cultivo, mercado, financiamiento y costo [conclusion]. Rev. Mensual Asoc. Gen. de Agricultores 129 (3): 19-22.

Lopez, H., M. Cimadevilla, E. Fernandez, C. Dubruthy, J.M. Navia, A. Valiente, I.D. Clement and R.S. Harris. 1956. Tabla provisional de la composicion nutritiva de los alimentos cubanos. Lab. FIM de Nutricion. Pub. #3. Comite Cubano Amer. pro Fundacion Invest. Med. Havana, Cuba.

Lotorto, G.F. 1978. Eradication of citrus blackfly, biological and chemical control. Proc. Fl. St. Hort. Soc. 91: 192-193.

Lovering, F.W. 1956. Citron grows mostly now in Florida gardens. Fla. Grower & Rancher. Jan. 12-18.

Loxton, P.L.D. 1970. Canned mango for export. Queensland Agr. J. 96 (4): 283.

Lunde, P. 1978. A history of date. Aramco World Mag. Mar./Apr. 21-23.

Lynch, S.J., E. Larson, and D.D. Doughty. 1951. A study on the edibility of akee (*Blighia sapida*) fruits of Florida. Proc. Fl. St. Hort. Soc. 64: 281-284.

Lynch, S.J. 1943. The Dade white sapote. Press Bull. 581. U. Fla. Agr. Exper. Sta., Gainesville, Fla.

Lynch, S.J. 1958. The effect of cold on lychees on the calcareous soils of southern Florida 1957-58. Proc. Fl. St. Hort. Soc. 71: 359-362.

Lynch, S.J. and F.J. Fuchs, Sr. 1955. A note on the propagation of *Phyllanthus emblica* L. Proc. Fl. St. Hort. Soc. 68: 301-302.

Lynn, K.R. 1973. An isolation of chymopapain. J. Chromatog. 84: 423-425.

Lyons, K. 1965. Cosechero de pina colombiano se adapta a nuevos mercados. La Hacienda 60 (4): 24-25.

Lyons, K. 1966. En Peru se obtienen piñas y aguacates muy buenos en area desertica La Hacienda 61 (5): 58, 60.

Macbride, J.F. 1937. Flora of Peru. Vol. 13, Pt. 6, #2. Bot. Ser. Pub. 393. Field Mus. Nat. Hist., Chicago.

MacDaniels, L.H. 1947. A study of the Fe'i banana and its distribution with reference to Polynesian migrations. Bull. 190. Bernice P. Bishop Mus., Honolulu.

MacLeod, Jr., W.D. 1966. Nootkatone, grapefruit flavor and the citrus industry. The Calif. Citrograph 51 (3): 120-123.

Macpherson, N. 1953. Banana passion fruit. New Zeal. J. Agr. Mar. 286-287.

Macpherson, N. 1954. Cape gooseberry recipes. New Zeal. J. Agr. 88 (4): 401-404.

Macpherson, N. 1952. Feijoas have a variety of culinary uses. New Zeal. J. Agr. 84 (4): 336-337.

Mahata, K.C. and S.M. Singh. 1960. Influence of source and time of planting on the performance of hardwood cuttings of *Carissa carandas* L. Indian J. Hort. 17 (1): 31-37.

Maiden, J.H. 1889. The useful native plants of Australia (including Tasmania). Technol. Mus. of New So. Wales, Sydney.

Major, R.T. and F. Dursch. 1958. N^a, N^a-dimethylhistamine, a hypotensive principle in *Casimiroa edulis* Llave et Lex. J. Org. Chem. 23: 1569.

Majumdar, P.K. and S.K. Mukherjee. 1968. Guava: a new vegetative propagation method. Indian Hort. 12 (2): 11-35.

Mali, V.R. and P.V. Khalikar. 1977. Passion flower mosaic virus disease, a new record for India. Curr. Sci. 46 (5): 153-154.

Mallareddy, K. and B.B. Sharma. 1983. Effect of storage conditions on germination, moisture content and some biochemical substances in citrus seeds. III. Trifoliate orange and pummelo. Seed Res. 11 (1): 56-59.

Malo, S.E. 1967. A successful method for propagating sapodilla trees.

Proc. Fl. St. Hort. Soc. 80: 373-376.

Malo, S.E. 1971. Girdling increases avocado yields in South Florida. Proc. Trop. Reg. Amer. Soc. Hort. Sci. 15: 19-25.

Malo, S.E. 1970. Mango and avocado cultivars—present status and future developments. Proc. Fl. St. Hort. Soc. 83: 357-362.

Malo, S.E. 1977. The mango in Florida. Hort. Sci. 12 (4): 286, 367.

Malo, S.E. and F.W. Martin. 1980. Tropical fruit: the durian. World Farming 22 (5): 38-39.

Manandhar, M.D., A. Shoeb, R.S. Kapil and S.P. Popli. 1978. New alkaloids from *Aegle marmelos*. Phytochem. 17: 1814-1815.

Manfred, L. 1947. 7000 recetas botanicas a base de 1300 plantas medicinales americanas. Editorial Kier, Buenos Aires.

Manis, E. 1966. Personal communication, June 16 [pineapple].

Mann, A.R. and S.A. Treaci. 1971. Comparative studies on guava jelly production. Sci. Ind. 8 (2): 164-167.

Mann, G.S. and G. Singh. 1984. Development of chemical control against the leaf-webber, *Dudua aprobola* (Meyr.) on litchi. Internat. Pest Control. May/June. p. 77.

Maor, T.G. 1958. Banana. Rev. Ser. #23. Indian Coun. Agr. Res., New Delhi.

Maranto, J. and K.D. Hake. 1985. Verdelli summer lemons: a new option for California growers. Calif. Agr. 39 (5&6): 4.

Marcelino-Ponce, J. 1979. Comportamiento de injertos de *Annona reticulata* L. sobre varios patrones. Proc. Trop. Reg. Amer. Soc. Hort. Sci. 23: 119-121.

Marchal, J., J.P. Blondeau and X. Bertin. 1978. Carences minerales chez la grenadille (*Passiflora edulis* Sims. var. *flavicarpa*). II. Carences totales en N, P, K, Ca, Mg. Influences sur la composition minerale des organes de la plante. Fruits 33 (10): 681-690.

Marchal, J. and J. Bourdeaut. 1972. Enchantillonnages foliaires de la grenadille (*Passiflora edulis* Sims var. *flavicarpa*). Fruits 27 (4): 307-311.

Martezki, A., H.J. Teas and C.F. Asenjo. 1966. Uptake and conversion of radioactive carbon dioxide and glucose in the acerola and their relationship to ascorbic acid biosynthesis. J. Agr. U. Puerto Rico 50 (1): 1-8.

Marfey, S.P. 1962. Studies in the insecticidal principal of *Mammea americana* L. Pt. 1. Doctoral thesis Wayne State U. 1955. Pub'd by Univ. Microfilms, Ann Arbor, Mich.

Marin Acosta, J.C. 1969. Insectos relacionados con la lechosa, *Carica papaya* L., en Venezuela. Agron. Trop. 19 (4): 251-267.

Marlatt, R.B. and C.W. Campbell. 1980. Incidence of algal disease (*Cephaleuros* sp.) in selections of guava (*Psidium guajava*). Proc. Fl. St. Hort. Soc. 93: 109-110.

Marloth, R.H. 1949. The litchi in South Africa. 2nd rev. ed. Bull. 286. Union of So. Africa Dept. of Agr., Pretoria.

Marriott, J., C. Perkins and B.O. Been. 1979. Some factors affecting the storage of fresh breadfruit. Scientia Hort. 10: 177-181.

Martin, F.W. and W.C. Cooper. 1977. Cultivation of neglected tropical fruits with promise, The Pummelo. ARS-S-157. U.S. Dept. Agr., Agr. Res. Serv., Washington, D.C.

Martin, F.W. and S.E. Malo. 1978. Cultivation of neglected tropical fruits with promise: Pt. 5. The canistel and its relatives. U.S. Dept. Agr., Sci. & Educ. Admin., Washington, D.C.

Martin, F.W. and H.Y. Nakasone. 1970. The edible species of *Passiflora*. Econ. Bot. 24 (3): 333-343.

Martin, M.A. 1971. Introduction a l' ethnobotanizue du Cambodge. Centre Nat. de la Recherche Scientifique, Paris.

Martindale, W.L. 1974. Tomatoes from a tree. J. Agr. Victoria. 72, Pt. 10: 347-349.

Martinez, M. 1951. Las casimiroas de Mexico y Centroamerica. An. Inst. Biol. Mexico 21: 25-81.

Martinez, M. 1959. Las plantas medicinales de Mexico. 4th ed. Ediciones Botas, Mexico, D.F.

Martinez, M. 1959. Plantas útiles de la flora Mexicana. Ediciones Botas, Mexico, D.F.

Martinez Nadal, N.G. 1964. Information Bull. 1963-64. [genipap]. Research Center, U. Puerto Rico, Coll. Agr. & Mech. Arts, Mayaguez, P.R.

Massal, E. and J. Barrau. 1956. Food plants of the South Sea islands. Tech. Paper 94. South Pac. Comm., Noumea, New Caledonia.

Matzumato, K. 1958. Normas technicas para ter mais jaboticabas. Rural, Rev. Soc. Rural Brasil 38 (450): 18-19.

Masumdar, B.C. 1979. Cape-gooseberry. World Crops Jan./Feb. 19, 23.

McCann, C. 1947. Trees of India. B.D. Taraporevala Sons & Co., Bombay.

McCann, L.P. 1947. Ecuador's naranjilla—a reluctant guest. Agr. In the Americas 7 (12): 146-149.

McCarty, C.D., S.B. Boswell & R.M. Burns. 1971. Chemically-induced sprouting of axillary buds in avocados. Calif. Agr. Dec.: 4-5.

McConnell, D.B. and J.W. Host. 1983. Watering frequency and light levels affect growth and anatomy of lemon vine. Proc. Fl. St. Hort. Soc. 96: 294-296.

McDonald, R.E., T.T. Hatton and R.H. Cubbedge. 1985. Chilling injury and decay of lemons as affected by ethylene, low temperature, and optimal storage. HortSci. 20 (1): 92-93.

McHardy, K. 1972. Feijoas are versatile. New Zeal. J. Agr. 124 (5): 59, 61.

McLennan, M. 1972. Tamarillos truly are versatile. New Zeal. J. Agr. 124 (4): 1, 53-55.

McMillan, R. 1973. Co-operation between co-operatives [kiwifruit]. New Zeal. J. Agr. 127 (1): 33-35.

McMillan, R. 1973. The kiwi fruit story—export potential. New Zeal. J. Agr. 127 (5): 59.

McMillan, R. 1973. Control of anthracnose and powdery mildew of mango with systemic and non-systemic fungicides. Trop. Agr. (Trin). 50 (3): 245-248.

McMillan, R. 1984. Control of mango anthracnose with foliage sprays. Proc. Fl. St. Hort. Soc. 97: 344-345.

McMillan, R. 1974. Rhizopus artocarpi rot of jackfruit (*Artocarpus heterophyllus*). Proc. Fl. St. Hort. Soc. 87: 392-393.

McRitchie, J.J. 1976. Div. of Plant Indus., Fla. Dept. Agr. office memorandum to C.F. Dowling, Feb. 12.

McSorley, R. and J.L. Parrado. 1983. The spiral nematode, *Helicotylenchus multicinatus* on banana in Florida and its control. Proc. Fl. St. Hort. Soc. 96: 201-207.

McSorley, R., J.L. Parrado and R.A. Conover. 1983. Population buildup and effects of the reniform nematode on papaya in southern Florida. Proc. Fl. St. Hort. Soc. 96: 198-200.

McVaugh, R. 1963. Flora of Guatemala, Botany. 24, Pt. 7, No. 3. Chicago Nat. Hist. Mus., Chicago.

McVaugh, R. 1958. Flora of Peru, 13, Pt. 4, No. 2. Bot. Ser. Pub. 861. [jaboticaba]. Field Mus. Nat. Hist., Chicago.

McVaugh, R. 1943. To make tough meat tender. Agr. in the Americas 3 (7): 134-136. [papaya].

Meisels, A. and F. Sondheimer. 1957. The constitutents of *Casimiroa edulis* Llave et Lex. III. The structure of casimiroin. J. Amer. Chem. Soc. 79: 6328-6333.

Melendez, P.L. 1968. A cercospora leaf spot of acerola in Puerto Rico. J. Agr. U. Puerto Rico 52 (1): 71-73.

Melendez, P.L. and J. Bird Pinero. 1971. *Corynespora* leaf spot of papaya (*Carica papaya* L.) in Puerto Rico. J. Agr. U. Puerto Rico 55 (4): 411-425.

Melhus, I.E., and F.O. Smith. 1953. The Mayan husk tomato: a tropical fruit comes to Iowa. Iowa Farm Sci. 7 (11): 15-16.

Mendes de Carvalho, A. 1965. Cultura do maracuja. Agron. Brasil 17 (9-10): 12-20.

Mendez, M. 1937. Pharmacologic data of some Mexican remedies. J. Amer. Inst. Homeopathy 30: 273-274.

Mendiola, N.B. 1926. A manual of plant breeding for the tropics. Bur. of Printing, Manila.

Mendoza, D.B., Jr., E.B. Pantastico and F.B. Javier. 1972. Storage and handling of rambutan (*Nephelium lappaceum* L.). The Phil. Agriculturist 4 (7 & 8): 322-332.

Mendoza-Guazon, M.P. 1930. The treatment of itching, acne vulgaris and pruritus vulva with calamansi juice (*Citrus mitis* Blanco). J. of Phil. Isls. Medical Assn. June: 233-235.

Menninger, E.A. 1966. *Actinidia chinensis*, a promising fruit and some related species. Amer. Hort. Mag. 45 (2): 253-256.

Menninger, E.A. 1959. The cultivated Eugenias in American gardens. Nat. Hort. Mag. 38 (3): 92-163.

Menzel, M.Y. 1951. The cytotaxonomy and genetics of *Physalis*. Proc.

Amer. Philosoph. Soc. 95 (2): 132-183.

Mercader, A. and E.M. Willsey. 1935. El platano: modos de preparlo para la mesa. Cir. 2. Servicio de Extension, Colegio de Agr. y Artes Mecanicas, U. Puerto Rico in cooper. with U.S. Dept. Agr.

Meredith, W.C., J.N. Joiner and R.H. Biggs. 1970. Influences of indole-e-acetic acid and kinetin on rooting and indole metabolism of *Feijoa sellowiana*. J. Amer. Soc. Hort. Sci. 95 (1): 49-52.

Merrill, E.D. 1954. The botany of Cook's voyages. Vol. 14, No. 5/6. Chronica Botanica Co., Waltham, Mass. [ambarella].

Micklem, T. 1949. Cape gooseberry culture in the Western Cape Province. Repr. 52. Reprint from Farming in So. Africa. Aug. 1-4.

Miller, C.D. 1929. Food values of breadfruit, taro leaves, coconut and sugar cane. Bull. 64. Bernice P. Bishop Mus., Honolulu.

Miller, C.D., K. Bazore and M. Bartow. 1955. Fruits of Hawaii. 2nd ed. U. Hawaii Press, Honolulu.

Miller, C.D., L. Louis and K. Yanazawa. 1947. Vitamin values of foods in Hawaii. Tech. Bull. U. Hawaii, Agr. Exten. Serv., Honolulu.

Miller, C.D., N.S. Wenkam and K.O. Fitting. 1961. Acerola, nutritive value and home use. Cir. 59. Hawaii Agr. Exp. Sta., U. Hawaii, Honolulu.

Miller, E.P. 1984. Oriental persimmons (*Diospyros kaki* L.) in Florida. Proc. Fl. St. Hort. Soc. 97: 340-344.

Miller, E.V. and A.S. Heilman. 1952. Ascorbic acid and physiological breakdown in the fruits of the pineapple (*Ananas comosus* L. Merr.). Science 116: 505-6.

Miller, W.R., P.W. Hale, D.H. Spalding and P. Davis. 1983. Quality and decay of mango fruit wrapped in heat-shrinkable film. Hort-Sci. 18 (6): 957-958.

Millspaugh, C.F. 1902. Flora of the Island of St. Croix. Pub. 68. Bot. Ser. Vol. I, No. 7. Field Columbian Mus., Chicago.

Milne, D.L. 1982. Nematode pests of litchi. In: Chap. 5, Nematology in So. Africa. Sci. Bull. #400. Pp. 38-41. Dept. Agr. & Fish, Pretoria, So. Africa.

Milne, D.L., M.R. Appleton and L.C. Holtzhausen. 1975. The anatomical reaction of *Litchi chinensis* roots attacked by three nematode species and a soil mite. Phytophylactica 7: 15-20.

Milne, D.L. and E.A. deVilliers. 1972. Campaign progressing against nematodes on litchis. Farming in So. Afr. 49 (6): 14-15.

Milne, D.L. and E.A. deVilliers. 1969. Decline in litchi orchards. Farming in So. Africa 45 (7): 32-35.

Milne, D.L. and E.A. deVilliers. 1972. Nematodes on litchis. Farming in So. Africa 49 (6): 14-15.

Milne, D.L., E.A. deVilliers, and L.C. Holtzhausen. 1971. Litchi tree decline caused by nematodes. Phytophylactica 3: 37-44.

Minessey, F.A., M.A. Barakat and E.M. El-Azab. 1970. Effect of water table on mineral content, root and shoot growth, yield and fruit quality in 'Washington Navel' orange and 'Balady Mandarin'. J. Amer. Soc. Hort. Sci. 95 (1): 81-85.

Miranda, F. 1952. La vegetacion de Chiapas. Ediciones del Gobierno del Estado, Tuxtla Guitierrez, Mexico.

Mishra, K.A. 1982. The vegetative propagation of persimmon. Scientia Hort. 17: 125-127.

Mitchell, A.R. and M.E. Nicholson. 1965. Pineapple growth and yield as influenced by urea spray schedules and potassium levels at three plant spacings. Queensland J. Agr. & Anim. Sci. 22 (4): 409-417.

Mitchell, E.F. 1971. Mango production and marketing practices. Florida. 1971. Proc. Fl. St. Hort. Soc. 84: 307-311.

Mitchell, P. and P.C. Annon. 1953. The pineapple. Advisory Leaflet 286. Queensland Dept. Agr. & Stock., Div. of Pl. Indus.

Mitscher, L.A., Wu-nan Wu, and J.L. Beal. 1973. The isolation and structural characterization of 5-O-methyldesmethoxymatteucinol from *Eugeia javanica*. Lloydia 36 (4): 422-425.

Mittal, J.P. and B. Singh. 1977. Using mango kernel. Noma (Nigeria). 2 (3): 7-8, 24.

Miyashita, R.K., H.Y. Nakasone, and C.H. Lamoureux. 1964. Reproductive morphology of acerola (*Malpighia glabra* L.). Tech. Bul. 63. Hawaii Agr. Exper. St., U. Hawaii, Honolulu.

Mobbs, J.A. 1965. Banana planting material. Queensland Agr. J. 91 (1): 16-19.

Moffett, J.O. and D.R. Rodney. 1973. Honey bee visits increase yields of 'Orlando' tangelo. HortSci. 8 (2): 100.

Moffett, M.L. and D.S. Teakle. 1966. Bacterial leaf spot of cape gooseberry in Queensland. Bull. 349. Queensland J. Agr. & Anim. Sci. 23 (2): 133-145.

Mohammed, S. and K.S. Chauhan. 1970. Vegetative propagation of phalsa (*Grewia asiatica* L.). Indian J. Agr. Sci. 40 (7): 581-586.

Mohammed, S. and H.R. Shabana. 1979. Date palm cultivation and research in Iraq. Chonica Hort. 19 (3): 49-50.

Mohammed, S. and H.R. Shabana. 1980. Effects of naphthaleneacetic acid on fruit size, quality, and ripening of 'Zahdi' date palm. Hort-Sci. 15 (6): 724-725.

Mohammed, S. and C.A. Sorhaindo. 1984. Production and rooting of etiolated cuttings of West Indian and hybrid avocado. Trop. Agr. (Trin.) 61 (3): 200-204.

Moreno, D.S., C.E. Kennett, H.S. Forster, R.W. Hoffmann and D.L. Flaherty. 1985. Predicting CRS [California red scale] infestations by trapping males. Calif. Agr. 39 (5 & 6): 10-12.

Moreuil, C. 1973. Quelques observations et essais sur le litchi. Fruits 28 (9): 637-640.

Morgan, C.N. 1944. Pineapple sunburn. Queensland Agr. J. 90 (3): 139.

Morgan, K.T. and L.W. Timmer. 1984. Effect of inoculum density, nitrogen source and saprophytic fungi on Fusarium wilt of Mexican lime. Plant & Soil 79: 203-210.

Morong, T. and N.L. Britton. 1892. Enumeration of the plants collected by Dr. Thomas Morong in Paraguay 1888-1890. (Annals N.Y. Acad. Sci. VI). N.Y. Acad. Sci., N.Y.

Morris, M.P. 1951. Chemical studies [mamey]. Report of the Fed. Exper. Sta. in Puerto Rico. 1951. U.S. Dept. Agr., Agr. Res. Admin., Washington, D.C.

Morris, M.P. 1952. Mamey chemistry. Report of the Fed. Exper. Sta. in Puerto Rico. U.S. Dept. Agr., Agr. Res. Admin., Washington, D.C.

Morris, M.P., C. Pagan and J. Garcia. 1952. Es el mamey una fruta venenosa? Rev. de Agr. de Puerto Rico Sup. Secc. Nutr. 43 (1): 288a & 288b.

Morse, J.G., M.J. Arbaugh and D.S. Moreno. 1985. Computer simulation of CRS [California red scale] populations. Calif. Agr. 39 (5 & 6): 8-10.

Mortensen, E. and E.T. Bullard. 1964. Handbook of tropical and subtropical horticulture. U.S. Dept. State, A.I.D., Washington, D.C.

Morton, J.F. 1981. Atlas of medicinal plants of Middle America. Charles C. Thomas Publisher, Springfield, Ill.

Morton, J.F. 1962. Garden Clinic [lime]. Miami News, Fla. Living Mag. Nov. 4.

Morton, J.F. 1964. Honeybee plants of South Florida. Proc. Fl. St. Hort. Soc. 77: 415-436.

Morton, J.F. 1959. Importancia economica del tamarindo. La Hacienda, June.

Morton, J.F. 1962. La naranjilla. Rev. de Alimentos 1 (1): 8-9.

Morton, J.F. 1977. Major medicinal plants: botany, culture and uses. Charles Thomas Publisher, Springfield, Ill. [pineapple].

Morton, J.F. 1955. Monstera deliciosa. Miami News, Fla. Living. Aug. 28.

Morton, J.F. 1954. New gooseberry [Dovyalis hybrid]. Miami Daily News. Feb. 21.

Morton, J.F. 1977. Poisonous and injurious higher plants and fungi. Chap. 71. in: Tedeschi, C.G., W.G. Eckert and L.G. Tedeschi, editors, Forensic Medicine Vol. III: Environmental Hazards, 1456-1567. W.B. Saunders Co., Philadelphia, Pa. [cashew; mango].

Morton, J.F. 1974. Renewed interest in roselle (*Hibiscus sabdariffa* L.), the long-forgotten "Florida cranberry". Proc. Fl. St. Hort. Soc. 87: 416-425.

Morton, J.F. 1960. Se elaboran confites con tamarindo y marañones. La Hacienda. July.

Morton, J.F. 1955. Some useful and ornamental plants of the Caribbean Gardens. Caribbean Gardens, Naples, Fla.

Morton, J.F. 1961. The cashew's brighter future. Econ. Bot. 15 (1): 57-78.

Morton, J.F. 1962. The drug aspects of the white sapotes. Econ. Bot. 16 (4): 288-294.

Morton, J.F. 1960. The emblic (*Phyllanthus emblica* L.). Econ. Bot. 14 (2): 119-128.

Morton, J.F. 1955. The emblic (*Phyllanthus emblica* L.), a rich but neglected source of vitamin C. Proc. Fl. St. Hort. Soc. 68: 315-321.

Morton, J.F. 1965. The jackfruit (*Artocarpus heterophyllus* Lam.): its culture, varieties and utilization. Proc. Fl. St. Hort. Soc. 78: 336-344.

Morton, J.F. 1963. The jambolan (*Syzygium cumini* Skeels)—its food, medicinal, ornamental and other uses. Proc. Fl. St. Hort. Soc. 76: 328-338.

Morton, J.F. 1958. The tamarind (*Tamarindus indica* L.), its food, medicinal and industrial uses. Proc. Fl. St. Hort. Soc. 71: 288-294.

Morton, J.F. 1968. Tropical fruit tree and other exotic foliage as human food. Proc. Fl. St. Hort. Soc. 81: 318-329.

Morton, J.F. 1983. Why not select and grow superior types of Canistel? Proc. Amer. Soc. for Hort. Sci., Trop. Region Vol. 27, Pt. A: 43-52.

Morton, J.F. 1961. Why not use and improve the fruitful ambarella? Hort. Adv. (India) 5: 13-16.

Morton, J.F. and O.S. Russell. 1954. The cape gooseberry and the Mexican husk tomato. Proc. Fl. St. Hort. Soc. 67: 261-266.

Morton, K. and J. Morton. 1946. Fifty tropical fruits of Nassau. Text House (Fla.) Inc., Coral Gables, Fla.

Moscoso, G.G. 1976. La chironja: una nueva fruta citrica Puertorriqueña. Bol. 248. Ext. Exper. Agr. U. Puerto Rico, Rio Piedras, P.R.

Moscoso, C.G. 1958. The Puerto Rican Chironja—new all-purpose citrus fruit. Econ. Bot. 12 (1): 87-94.

Moscoso, C.G. 1969. The Puerto Rican chironja—a new type of citrus. Proc. Trop. & Subtrop. Fruits Conf. (Trop. Prod. Inst.): 193-195.

Moscoso, C.G. 1950. West Indian cherries and the production of ascorbic acid. Misc. Pub. 2. U. Puerto Rico, Agr. Exper. Sta., Rio Piedras, P.R.

Moscoso, C.G. 1956. West Indian cherry—richest known source of natural vitamin C. Econ. Bot. 10 (3): 280-294.

Mosqueda Vazquez, R. 1969. Efecto de diversos tratamientos aplicados a la semilla papaya, sobre su poder germinativo. Agr. Técnica 2 (11): 487-491.

Mosqueda-Vasquez, R., M. Aragaki and H.Y. Nakasone. 1981. Screening of *Carica papaya* L. seedlings for resistance to root rot caused by *Phytophthora palmivora* Britt. J. Amer. Soc. Hort. Sci. 106 (4): 484-487.

Mosqueda-Vazquez, R. and H.Y. Nakasone. 1982. Diallel analysis of root rot resistance in papaya. HortSci. 17 (3): 384-5.

Mott, J. 1969. The market for passion fruit juice. Rpt. G38. Trop. Prod. Inst., London.

Motz, F.A. 1944. Gifts of the Americas. The sapodilla. Agr. in the Americas 4 (10): 198.

Motz, F.A. and L.D. Mallory. 1944. The fruit industry of Mexico. For. Agr. Rpt. #9. Off. of For. Agr. Relations, U.S. Dept. Agr., Washington, D.C.

Mouat, H.M. 1958. New Zealand varieties of yang-tao or Chinese gooseberry. New Zeal. J. Agr. 92 (2): 161-163, 165.

Mowry, H. and R.D. Dickey. 1950. Ornamental hedges for Florida. Bull. 443. U. Fla. Agr. Exper. Sta., Gainesville.

Mowry, H., L.R. Toy and H.S. Wolfe (revised by G.D. Ruehle). 1953. Miscellaneous tropical and subtropical Florida fruits. Bull. 156. U. Fla. Agr. Exten. Serv., Gainesville.

Mukerjea, T.D. and R. Govind. 1958. Studies on indigenous insecticidal plants. Pt. II. *Annona squamosa*. J. Sci. & Indus. Res. 17C (1): 9-15.

Mukerjee, P.K. 1957. Dehydration of litchis. Hort. Adv. (India) 1: 55-57.

Mukerji, B. 1951. Indigenous Indian drugs used in the treatment of diabetes. [jambolan]. J. Sci. & Indus. Res. 16A (10): Suppl 1-18.

Mukherjee, S.K. 1976. Current advances on mango research around the world. Acta Hort. 57: 37-42.

Mukherjee, S.K. 1972. Origin of mango (*Mangifera indica*). Econ. Bot. 26 (3): 260-264.

Mukherjee, S.K. and B.K. Chatterjee. 1979. Effects of forcing etiolation and indole butyric acid on rooting of cuttings of *Artrocarpus heterophyllus* Lam. Scientia Hort. 10: 295-300.

Muller, I.A. and M.J. Young. 1982. Influence of gibberellic acid and effectiveness of several carriers on growth of sour orange (*Citrus aurantium* L.) seedlings. HortSci. 17 (4): 673-674.

Mune, T.L. and J.W. Parham. 1956. The declared noxious weeds of Fiji and their control. Bul. 31. Dept. Agr., Suva, Fiji.

Munier, P. 1965. Le palmier-dattier, producteur de sucre. Fruits 20 (10): 577-579.

Munier, P. and P. Dupaigne. 1963. Un nouvel avenir pour la pate de dattes. Fruits 18 (10): 468-473.

Munier, R. 1962. La culture du lulo en Colombie. Fruits 17 (2): 91-92.

Munsell, H.E., R. Castillo, C. Zurita and J.M. Portillo. 1953. Production, uses, composition of foods of plant origin from Ecuador. Food Res. 18 (4): 319-342.

Munsell, H.E., L.O. Williams, L.P. Guild, L.T. Kelley, and R.S. Harris. 1950. Composition of food plants of Central America. VII. Honduras. Food Res. 15 (6): 421-438.

Munsell, H.E., L.O. Williams, L.P. Guild, L.T. Kelley, A.M. McNally and R.S. Harris. 1950. Composition of food plants of Central America. VI. Costa Rica. Food Res. 15 (5): 379-404.

Munsell, H.E., L.O. Williams, L.P. Guild, L.T. Kelley, A.M. McNally and R.S. Harris. 1950. Composition of food plants of Central America. VIII. Guatemala. Food Res. 15 (6): 439-453.

Munsell, H.E., L.O. Williams, L.P. Guild, C.B. Troescher and R.S. Harris. 1950. Composition of food plants of Central America. II. Guatemala. Food Res. 15 (1): 16-33.

Munsell, H.E., L.O. Williams, L.P. Guild, C.B. Troescher and R.S. Harris. 1950. Composition of food plants of Central America. V. Nicaragua. Food Res. 15 (5): 355-365.

Munsell, H.E., L.O. Williams, L.P. Guild, C.B. Troescher, G. Nightingale and R.S. Harris. 1950. Composition of food plants of Central America. III. Guatemala. Food Res. 34-52.

Munsell, H.E., L.O. Williams, L.P. Guild, C.B. Troescher, G. Nightingale, and R.S. Harris. 1949. Composition of food plants of Central America. I. Honduras. Food Res. 14 (2): 144-164.

Munsell, H.E., L.O. Williams, L.P. Guild, C.B. Troescher, G. Nightingale, L.T. Kelley and R.S. Harris. 1950. Composition of food plants of Central America. IV. El Salvador. Food Res. 15 (4): 263-296.

Murray, D.A.H. 1978. Effect of fruit fly sprays on the abundance of the citrus mealybug, *Planococcus citri* (Risso), and its predator, *Cryptolaemus montrouzieri* Mulsant, on passion-fruit in southeastern Queensland. Bull. 786. Queensland J. Agr. & Anim. Sci. 35 (2): 143-147.

Murray, D.A.H. 1982. Effects of sticky banding of custard apple [atemoya] tree trunks on ants and citrus mealybug, *Planococcus citri* (Risso) (Pseudo-coccidae (Hem.)) in south-east Queensland. Queensland J. Agr. & Anim. Sci. 39 (2): 141-146.

Murray, D.A.H. 1976. Insect pests on passion fruit. Queensland J. Agr. 102 (2): 145-151.

Murray, D.A.H. 1982. Pineapple scale (*Diaspis bromeliae*) (Kerner) distribution and seasonal history. Queensland J. Agr. & Anim. Sci. 39 (2): 125-130.

Murray, D.A.H. 1978. Population studies of the citrus mealybug, *Planococcus citri* (Risso) and its natural enemies on passion-fruit in southeastern Queensland. Bull. 785. Queensland J. Agr. & Anim. Sci. 35 (2): 139-141.

Musa, S.K. 1974. Preliminary investigations on the storage and ripening of 'Totapuri' mangoes in the Sudan. Trop. Sci. 16 (2): 65-73.

Mustard, M.J. 1952. Ascorbic acid content of some miscellaneous tropical and subtropical plants and plant products. Food Res. 17 (1): 31-35.

Mustard, M.J. 1955. Handling guavas, lychees and white sapotes for the fresh fruit market. Proc. Fl. St. Hort. Soc. 68: 267-270.

Mustard, M.J. 1964. Lychees. Fr. & Veg. Facts and Pointers. United Fresh Fruit & Veg. Assn., Washington, D.C.

Mustard, M.J. 1960. Megagametophytes of the lychee (*Litchi chinensis* Sonn.) Proc. Amer. Soc. Hort. Sci. 75: 292-304.

Mustard, M.J. 1954. Oleocellosis or rind-oil spot on Persian limes. Proc. Fl. St. Hort. Soc. 67: 224-226.

Mustard, M.J. 1954. Pollen production and seed development in the white sapote. Bot. Gaz. 116: 189-192.

Mustard, M.J. 1946. The ascorbic acid content of some Malpighia fruits and jellies. Science 104 (2697): 230-231.

Mustard, M.J. and R.O. Nelson and S. Goldweber. 1956. Exploratory study dealing with the effect of growth regulators and other factors on the fruit production of the lychee. Proc. Fla. Lychee Growers Assn. 4th Ann. Mtg., Winter Haven, Fla.

Nadarajah, M., P.A.J. Yapa, C.G. Balasingham and S. Kasinathan. 1973. The use of papain as a biological coagulant for natural rubber latex. Q.J. Rubber Res. Inst., Sri Lanka 50: 134-142.

Nagy, S., P.E. Shaw and M.K. Veldhuis. 1977. Citrus science and technology: Vol. I: Nutrition, anatomy, chemical composition and bioregulation. The AVI Pub'g Co., Westport, Conn.

Nagy, S., P.E. Shaw and M.K. Veldhuis. 1977. Citrus Science and technology. Vol. II: Fruit production, processing practices, derived products and personnel management. The AVI Pub'g Co., Westport, Conn.

Nagy, S., W.F. Wardowski and C.J. Hearn. 1982. Diphenyl absorption and decay in 'Dancy' and 'Sunburst' tangerine fruit. J. Amer. Soc. Hort. Sci. 107 (1): 154-157.

Nagy, S. and R.E. Shaw (editors). 1980. Tropical and subtropical fruits: composition, properties and uses. The AVI Pub'g Co., Westport, Conn.

Naik, K.C. 1949. South Indian fruits and their culture. P. Varadachary & Co., Madras, India.

Nair, A.G.R. and S.S. Subramanian. 1962. Evaluation of the flowers of *Eugenia jambolana*. J. Sci. & Indus. Res. 21B (9): 457-458.

Nakasone, H.Y. 1973. Guava propagation in Hawaii. Misc. Pub. 111. Coop. Exten. Serv., U. Hawaii, Honolulu.

Nakasone, H.Y. 1975. Papaya development in Hawaii. HortSci. 10 (3): 198.

Nakasone, H.Y. 1972. Production feasibility for soursop. Hawaii Farm Sci. 21 (1): 10-11.

Nakasone, H.Y., R.A. Hamilton and P. Ito. 1967. Evaluation of introduced cultivars of guava. Hawaii Farm Sci. 16 (2): 4-6.

Nakasone, H.Y., R. Hirono and P. Ito. 1967. Preliminary observations on the inheritance of several factors in the passionfruit (*Passiflora edulis* L. and forma *flavicarpa*). Tech. Prog. Rpt. 161. U. Hawaii, Agr. Exper. Sta., Honolulu.

Nakasone, H.Y. and P.J. Ito. 1978. 'Ka Hua Kula' guava. HortSci. 13 (2): 197.

Nakasone, H.Y. and C. Lamoureux. 1982. Transitional forms of hermaphroditic papaya flowers leading to complete maleness. J. Amer. Soc. Hort. Sci. 107 (4): 589-592.

Nakasone, H.Y., R.K. Miyashita and G.M. Yamane. 1966. Factors affecting ascorbic acid content of the acerola (*Malpighia glabra* L.). Proc. Amer. Soc. Hort. Sci. 89: 161-166.

Nakasone, H.Y., W. Yee, D.K. Ikehara, M.J. Doi, and R.J. Ito. 1974. Evaluation and naming of two new Hawaii papaya lines, 'Higgins' and 'Wilder'. Res. Bull. 167. Hawaii Agr. Exp. Sta., U. of Hawaii.

Nakata, S. 1953. Girdling as a means of inducing flower-bud initiation in litchi. Progress Notes #95. Hawaii Agr. Exper. Sta., U. Hawaii.

Namba, R. 1971. Aphid transmission of the papaya mosaic virus. Misc. Pub. 64 (Mimeo). Coop. Exten. Serv., U. Hawaii, Honolulu.

Namba, R. and C.Y. Kawanishi. 1963. Transmission studies on the papaya mosaic virus. Hawaii Farm Sci. 12 (4): 3.

Nanjundaswamy, A.M., L. Setty and G.S. Siddappa. 1964. Preparation and preservation of guava juice. Indian Food Packer 18 (4): 13-17.

Nanjundaswamy, A.M. and G.S. Siddappa. 1964. Preservation of the central edible core of the banana stem in the form of candy, canned curried product and dehydrated slices. Indian Food Packer 18 (6): 9-11.

Narang, D.D. and G.S. Mann. 1983. Insect pests of guava. Progressive Farming 19 (11): 12-13.

Narayanamurti, D., P. Ramarhandra Rao and R. Ram. 1957. Adhesives from tamarind seed testa. J. Sci. & Indus. Res. 168 (8): 377-378.

Nash, D.L. and J.V.A. Dieterle. 1976. Flora of Guatemala. Fieldiana: Botany V. 24, Pt. 11, #4. Field Mus. Nat. Hist., Chicago. 275-431.

Natarajan, S., D.E. Eveleigh and R.H. Dawson. 1976. A natural source for neohesperidin. Econ. Bot. 30: 38.

Natarajan, C.P., R. Balakrishnan Nair, N. Gopalakrishna Rao, C.S. Viraktamath and D.S. Bhatia. 1960. A method for the quantitative estimation of roasted date and tamarind in coffee powder. Food Sci. 9 (2): 39.

Natarajan, P.N. and R. Karunanithy. 1974. Synthetic flavour enhancers for *Artocarpus integrifolia* L. [jackfruit]. The Flavour Indus. Nov./Dec. 282-283.

Nath, N. and G.S. Randhawa. 1969. Classification and description of some varieties of *Punica granatum* L. Indian J. Hort. 16 (4): 191-201.

Nath, N. and G.S. Randhawa. 1959. Studies on floral biology in the pomegranate (*Punica granatum* L.). 1. Flowering habit, flowering season, bud development and sex-ratio in flowers. Indian J. Hort. 16 (2): 61-68.

Nath, N. and G.S. Randhawa. 1954. Studies on floral biology in the pomegranate. II. Anthesis, dehiscence, pollen studies and receptivity of stigma. Indian J. Hort. 16 (3): 121-140.

Navaratnam, S.J. 1966. Patch canker of the durian tree. Malayan Agr. J. 45 (3): 291-294.

Navia, J.M., M. Cimadevilla, E. Fernandez, I.D. Clement, A. Valiente and R.S. Harris. 1955. Tabla provisional de la composicion nutritiva de los alimentos Cubanos. Lab. FIM de Nutricion. Pub. #2. Comite Cubano Amer. pro Fundacion Invest. Med., Havana, Cuba.

Navia, J.M., H. Lopez, I.D. Clement, M. Cimadevilla, E. Fernandez, and R.S. Harris. 1954. Tabla provisional de la composicion nutritiva de los alimentos Cubanos. Lab. FIM de Nutricion. Pub. #1. Comite Cubano Amer. pro Fundacion Invest. Med., Havana, Cuba.

Navia, G., V.M. and J. Valenzuela B. 1978. Sintomatologia de deficiencias nutricionales en chirimoyo (*Annona cherimola* Mill.) cv. Bronceada. Agr. Tecn. (Chile) 38: 9-14.

Nayar, T.G., V.S. Seshadri and C.M. Bakthavathsalu. 1956. A note on mattocking practices in banana culture. Indian J. Hort. 13 (4): 210-211.

Neal, M.C. 1965. In Gardens of Hawaii. Spec. Pub. 50. Bernice P. Bishop Mus. Press, Honolulu.

Negron de Bravo, E., H.D. Graham and M. Padovani. 1983. Composition of the breadnut (seeded breadfruit). Carib. J. Sci. 19 (3-4): 27-32.

Nelson, R.O. 1953. High humidity treatment for air layers of lychee. Proc. Fl. St. Hort. Soc. 66: 198-199.

Nelson, R.O. 1954. Notes on lychee grafting. Proc. Fl. St. Hort. Soc. 67: 231-233.

Nelson, R.O. 1957. Suggested methods for top-working lychee trees. Proc. Fla. Lychee Growers Assn. Nov. 11. 27-31.

Nelson, R.O. and S. Goldweber. 1956. Further rooting trials of Barbados cherry. Proc. Fl. St. Hort. Soc. 69: 285-287.

Neville-Rolfe, E. 1897. Report on the cultivation of the carob tree. Misc. Ser. #431. Foreign Office. Her Majesty's Staty. Off., London.

Ng Siew Kee, and S. Thamboo. 1967. Nutrient removal studies on Malayan fruits: durian and rambutan. Malayan Agr. J. 46 (2): 164-182.

Nichols, M.A. 1977. Horticulture in New Zealand. HortScience 12 (6): 539.

Nicks, G.S. 1969. Pioneer promotion of Hawaiian papayas in Japan. For. Agr. 7 (22): 8.

Nieva, F.S., A.J. Rodriguez and M.A. Gonzalez. 1965. Removal of stone cells from guava nectar. J. Agr. U. Puerto Rico 49 (2): 234-243.

Nihoul, E. 1976. Le yang tao (*Actinidia chinensis* Planchon). Fruits 31 (2): 45-50.

Nirvan, R.S. 1957. Control of dry rot of persimmon (*Diospyros kaki* Linn.) seeds. Hort. Advance 1: 62-64.

Nirvan, R.S. 1960. Effect of antibiotic spray on citrus canker. Hort. Advance 4: 155-160.

Nirvan, R.S. 1961. Trial export of litchi to the United Kingdom. Ann. Rpt. Hort. Res. Unit, Saharanpur, India: 59-63.

Nishida, T. and F.G. Holdaway. 1955. The erinose mite of lychee. Cir. 48. Hawaii Agr. Exper. Sta., U. Hawaii, Honolulu.

Nisperos, M.O., L.C. Raymundo and L.B. Mabesa. 1982. Ascorbic acid, color, provitamin A and sensory qualities of calamansi (*Citrus mitis* Linn.) juice after various processing operations and lengths of storage. Phil. Agr. 65 (Oct.-Dec.): 353-361.

Nixon, R.W. 1939. Date growing in the United States. Leaflet #170. U.S. Dept. Agr., Washington, D.C.

Nixon, R.W. 1951. The date palm—"Tree of Life" in the subtropical deserts. Econ. Bot. 5 (3): 274-301.

Noonan, J.C. 1953. Review of investigations on the *Annona* species. Proc. Fl. St. Hort. Soc. 66: 205-210.

Nordby, H.E., J.F. Fisher and T.J. Kew. 1968. Apigenin 7*B*-rutinoside, a new flavonoid from the leaves of *Citrus paradisi*. Phytochemistry 7: 1653-1657.

Norman, J.C. 1972. The influence of some compounds on the growth and flowering of *Ananas comosus* (L.) Merr. cultivar Sugarloaf. Ghana J. Agr. Sci. 5: 213-219.

Northwood, P.J. 1970. A note on a spacing and nitrogen fertilizer experiment with pawpaw. E. Afr. Agr. & For. J. 36 (1): 45-48.

Nuñez L., V.R. and J. De la Cruz. 1982. Reconocimiento y descripcion de las principales insectos observados en cultivares de guanabano (*Anona muricata* L.) en el departamento del Valle. Acta Agron. 32 (1/4): 45-51.

Nuñez M., E. 1964. Plantas medicinales de Puerto Rico. Bol. 176. U. Puerto Rico Estac. Exper. Agr., Rio Piedras.

Nuñez-Elisea, R. 1984. "Manzanillo-Nuñez': a new Mexican mango cultivar. Proc. Fl. St. Hort. Soc. 97: 360-363.

Oakes, A.J. 1970. Herbicidal control of guava (*Psidium guajava* L.). Turrialba 22 (1): 30-36.

Oakes, A.J. and M.P. Morris. 1959. The West Indian Weedwoman of the United States Virgin Islands. Bul. of the History of Medic. 32 (2): 164-170.

Oatman, E.R. and G.R. Platner. 1985. Biological control of two avocado pests. Calif. Agr. 39 (11/12): 21-23.

O'Brien, M. and J.B. Smith. 1972. Bio-engineering factors in mechanically harvesting pineapple. Spec. Pub. SP-01-72. Conf. Papers. Internat. Conf. on Trop. & Subtrop. Agr., Amer. Soc. Agr. Engineers, St. Joseph, Mich.: 235-240.

Ochse, J.J. 1953. *Solanum hyporhodium*. Proc. Fl. St. Hort. Soc. 66: 211-212.

Ochse, J.J. in collab. with R.C. Bakhuizen van den Brink. 1931. Fruits and fruitculture in the Dutch East Indies. G. Kolff & Co., Batavia.

Ochse, J.J. and R.C. Bakhuizen van den Brink. 1931. Vegetables in the Dutch East Indies. Dept. Agr., Indus. & Comm. of the Netherlands East Indies, Buitenzorg, Java.

Ochse, J.J., M.J. Soule, Jr., M.J. Dijkman, and C. Wehlburg. 1961. Tropical and subtropical agriculture. 2 vols. Macmillan Co., N.Y.

Ogden, M.A.H. and C.W. Campbell. 1982. Intergeneric and interspecific rootstock trials for jaboticaba (*Myrciaria cauliflora* (Mart.) Berg.). Proc. Fl. St. Hort. Soc. 95: 119-121.

Ogden, M.A.H., C.W. Campbell and S.P. Lara. 1984. Juvenile interstocks for topworking mamey sapote (*Calocarpum sapota* (Jacq.) Mern.). Proc. Fl. St. Hort. Soc. 97: 357-358.

Ojima, M. and O. Rigitano. 1968. Cultura da nespereira. Bull. 184. Inst. Agron., Sec. Agr. Estado de Sao Paulo, Campinas, Brazil.

Ojima, M., O. Rigitano, H.J. Scaranari, F.P. Martins, F.A.C. Dall' orto and V. Nagai. 1977. Variedades e espacamento da nespereira. Bol. Tech. #46. Inst. Agron., Sec. Agr. Estado de Sao Paulo, Campinas, Brazil.

Oliver, B. 1960. Medicinal plants in Nigeria. Nigerian Coll. of Arts, Sci. & Tech., Ibadan.

Ondieki, J.J. 1975. Diseases and pests of passion fruit in Kenya. Acta Hort. 49: 291-293.

Ondieki, J.J. and J.M. Kori. 1971. Control of brown spot disease of passion fruit in Kenya using fungicides. East Africa Hort. Symp. Pp. 85-88.

Ong Ching Ang and Ting Wen Poh. 1973. Two virus diseases of passion fruit *Passiflora edulis* f. *flavicarpa*). MARDI Res. Bull. 1 (1): 33-50.

Oppenheimer, C. 1947. Acclimitisation of new tropical and subtropical fruit trees in Palestine. Bul. 44. Jewish Agency for Palestine, Agr. Res. Sta., Rehovot.

Oppenheimer, C. 1967. Nimrod—a new mango variety selected in Israel. Proc. Fl. St. Hort. Soc. 80: 358-359.

Oppenheimer, C. and Sh. Gazit. 1961. Zinc deficiency in mango groves in Israel and its correction. Hort. Advance V: 1-12.

Oppenheimer, C. and O. Reuveni. 1961. Flowering and pollination of the loquat (*Eriobotrya japonica* Lindl.) in Israel. Ind. J. Hort. 18 (2): 97-105.

Oramas, D. and J. Roman. 1982. Plant parasitic nematodes associated with plantain (*Musa acuminata* X *M. balbisiana*, AAB) in Puerto Rico. J. Agr. U. Puerto Rico 66 (1): 52-59.

Ordetx Ros, G.S. 1952. Flora apicola de la America tropical. Editorial Lex, Havana, Cuba.

Orphanos, P.I. and J. Papaconstantinou. 1969. The carob varieties of Cyprus. Tech. Bull. Cyprus Agr. Res. Inst. 5: 3-27.

Ortiz N., A.J., R.D. Cooke and R.A. Quiros M. 1982. The processing of a date-like caramel from cashew apple. Trop. Sci. 24 (1): 29-38.

Osorio Bedoya, I.A.J. 1979. Cultivo de la "Curuba de Castilla". Rec. Nac. de Agr. No. 845: 28-32.

Oste, C.A. and S. Alvarez. 1974. Tonnage: nueva variedad de palto. Estac. Exper. Agr., Tucumán, Argentina.

Ostendorf, F.W. 1962. Nuttige planten en sierplanten in Suriname. Bull. 79. Landbouwproefstation in Suriname, Paramaribo.

Ostendorf, F.W. 1963. The West Indian cherry. Trop. Abs. 18 (3): 145-150.

Oudit, D.D. & K.J. Scott. 1973. Storage of 'Hass' avocados in polyethylene bags. Trop. Agr. (Trin.) 50 (3): 241-243.

Oxenham, B.L. 1963. Pineapple tip rot and root rot. Queensland Fr. & Veg. News 24 (19): 455.

Pacini, A. 1970. Personal communication re: sapote seed oil. Dec. 29.

Pagan, C. and M.P. Morris. 1953. A comparison of the toxicity of mamey seed extract and rotenone. J. Econ. Entom. 46 (6): 1092-1093.

Page, P.E. 1970. Winter nitrogen for pineapples. Queensland Agr. J. 96 (6): 394.

Pal, R., and D.K. Kulshreshtha. 1975. A new lignan from *Carissa carandas*. Phytochem. 14: 2302-2303.

Palmario, M.S., C.V. Cueto, Z.S. Imperial, S.A. Tayco, R.P. Soriaga, R.V. Buenaventura and M.C. de Guzman. 1976. Pineapple fibers: the retting process. Sci. Rev. 17 (4): 8-16.

Palmer, G. 1956. Some aspects of the lychee as a commercial crop. Proc. Fl. St. Hort. Soc. 69: 309-312.

Pandey, K.K. 1957. Genetics of self-incompatibility in *Physalis ixocarpa* Brot. Amer. J. Bot. 44: 879-887.

Pandey, P.K., A.B. Singh, M.R. Nimbalkar and T.S. Marathe. 1976. A witches'-broom disease of jujube from India. Pl. Dis. Rep. 60 (4): 301-303.

Pandey, V.S. and P.N. Bajpai. 1969. Studies on blossom bud differentiation in litchi (*Litchi chinensis* Sonn.) var. Kalkattia and Rose-scented. Indian J. Sci. & Indus. 3 (2): 99-102.

Pantastico, E.B. 1975. Postharvest physiology, handling and utilization of tropical and subtropical fruits and vegetables. The AVI Pub'g Co., Inc., Westport, Conn.

Pantastico, E.B., D.B. Mendoza and R.M. Abcaly. 1968-69. Some chemical and physiological changes during storage of lanzones (*Lansium domesticum* Correa). Phil. Agriculturist 52 (7-8): 505-517.

Panyathorn, K.S. 1969. Thailand making gains as a canned pineapple exporter. For. Agr. 7 (48): 6.

Parham, B.E.V. 1961. Controlling "bunchy-top" in Western Samoa. So. Pacific Bull. 11 (2): 40-42, 60.

Parker, A., Vo Huu De, and L.H. Myers. 1976. The Florida grapefruit industry: an economic analysis. Proc. Fl. St. Hort. Soc. 89: 1-3.

Parsi Ros, O. 1976. The preparation of papaya jam. J. Agr. U. Puerto Rico 60 (1): 129-131.

Partridge, I. 1979. The guava threat in Fiji. So. Pacific Bull. 2nd quarter 1979. Pp. 28-30.

Passam, H.C. and G. Blunden. 1982. Experiments on the storage of limes at tropical ambient temperature. Trop. Agr. (Trin.) 59 (1): 20-24.

Passera, C. and P. Spettoli. 1981. Chemical composition of papaya seeds. Qual. Plant Foods & Human Nutr. 31: 77-83.

Patel, B. and V.J. Patel. 1979. Impact of chemicals on ber (*Zizyphus mauritiana* Lamk.). Pesticides 13 (3): 28-30.

Patil, P.K. and V.K. Patil. 1983. Studies on soil salinity tolerance of sapota [sapodilla]. So. Indian Hort. 31 (1): 3-6.

Patil, S.S., C.S. Tang and J.E. Hunter. 1973. Effect of benzyl isothiocyanate treatment on the development of postharvest rots in papayas. Pl. Dis. Rep. 57 (1): 86-89.

Patiño, V.M. 1962. Edible fruits of *Solanum* in South American historic and geographic references. Bot. Mus. Leaflets, Harvard Univ. 19 (11): 215-234.

Patiño, V.M. 1963. Plantas cultivadas y animales domesticos en America Equinoccial—Vol. I: Frutales. Imprenta Deptamental, Cali, Colombia.

Patiño, V.M. 1964. Plantas cultivadas y animales domesticos en America Equinoccial. Vol. 2: Plantas alimenticias. Imprenta Departmental, Cali, Colombia.

Patyapongan, P. 1962. Jackfruit (trans.). Kasikorn 35 (6): 549-553.

Paull, R.E. 1982. Postharvest variation in composition of soursop (*Annona muricata* L.) in relation to respiration and ethylene production. J. Amer. Soc. Hort. Sci. 107 (4): 582-585.

Paull, R.E., N.J. Chen, H. Deputy, H. Huang, G. Cheng and F. Gao. 1984. Litchi growth and compositional changes during fruit development. J. Amer. Soc. Hort. Sci. 109 (6): 817-821.

Paull, R.E., J. Deputy and N.J. Chen. 1983. Changes in organic acids, sugars, and headspace volatiles during fruit ripening of soursop (*Annona muricata* L.). J. Amer. Soc. Hort. Sci. 108 (6): 931-934.

Payumo, E.M., L.M. Pilac and P.L. Manequis. 1965. The preparation and storage properties of canned guwayabano (*Annona muricata* L.) concentrate. Phil. J. Sci. 94 (2): 161-169.

Peacock, B.C. 1980. Banana ripening—effect of temperature on fruit quality. Queensland J. Agr. & Anim. Sci. 37 (1): 39-45.

Peacock, B.C. 1972. Effect of light on preclimacteric life of bananas. Queensland J. Agr. & Anim. Sci. 29: 199-207.

Peasley, D. 1981. Passionfruit industry benefits through scionwood scheme. Agr. Gaz. of New So. Wales 92 (5): 5-8.

Pegg, K.G. 1978. Avocado root rot. Queensland Agr. J. 104 (2): 131-133.

Pegg, K.G. 1978. Disease-free avocado nursery trees. Queensland Agr. J. 104 (2): 134-136.

Pegg, K.G. 1973. Phytophthora blight in passion fruit. Queensland Agr. J. 99 (12): 655-656.

Pegg, K.G. 1969. Pineapple top rot control with chemicals. Queensland Agr. J. 95 (7): 458-459.

Peña, J.E., H. Glenn and R.M. Baranowski. 1984. Important insect pests of *Annona* spp. in Florida. Proc. Fl. St. Hort. Soc. 97: 337-340.

Peña, R. 1901. Flora Cruceña. Author. Printed by Imp. Bolivar de M. Pizarro, Sucre, Bolivia.

Penella L., J.S. 1967. El aguacate. Serie de Cultivos #11. Consejo de Bienestar Rural, Caracas, Venez.

Penella L., J.S. (undated; bet. 1963 & 1967). El platano y el cambur. Serie de Cultivos #8. Consejo de Bienestar Rural, Caracas, Venez.

Pennella L., J.S. 1968. La lechosa. Serie de Cultivos #16. Consejo de Bienestar Rural, Caracas, Venez.

Pennington, T.D., and J. Sarukhan. 1968. Arboles tropicales de Mexico. GAO & Inst. Nac. de Invest. Forest., Mexico, D.F.

Pennock, W. 1961. Doce frutas para los patios pequeños en la Costa Norte de Puerto Rico. Rev. de Agr. de Puerto Rico 48 (1): 148-152.

Pennock, W. and H. Gandia. 1975. Effect of slip size, slip storage and time of planting on yield of Red Spanish pineapple in Puerto Rico. U. Puerto Rico J. Agr. 59 (3): 141-164.

Pennock, W. and G. Maldonado. 1963. The propagation of guavas from stem cuttings. J. Agr. U. Puerto Rico 47 (4): 280-289.

Pennock, W., T. Soto, R. Abrams, R. Gandia Caro, A. Perez, and G.C. Jackson. 1963. Variedades selectas des aguacates de Puerto Rico. Bol. 172. U. Puerto Rico Estac. Exper. Agr.

Peregrine, W.T.H. 1968. A survey of pawpaw debility in Tanzania. E. Afr. Agr. & For. J. 33 (4): 316-322.

Peregrine, W.T.H. 1969. A survey of pawpaw debility in Tanzania. PANS 15 (2): 177-182.

Perez, A., E. Boneta, E. Perez, and J. Green. 1980. Behavior of ten chironja clones at three sites. II. Fruit quality. J. Agr. U. Puerto Rico 64 (3): 323-329.

Perez, A. and N.F. Childers. 1982. Growth, yield, nutrient content and fruit quality of *Carica papaya* L. under controlled conditions. I. Nitrogen effects. J. Agr. U. Puerto Rico 66 (2): 71-79.

Perez, A. and N.F. Childers. 1982. Growth, yield, nutrient content and fruit quality of *Carica papaya* L. under controlled conditions. II. Boron effects. J. Agr. U. Puerto Rico 66 (2): 80-88.

Perez, A., M.N. Reyes and J. Cuevas. 1980. Germination of two papaya varieties: effect of seed aeration, K-treatment, removing of the sarcotesta, high temperature, soaking in distilled water, and age of seeds. J. Agr. U. Puerto Rico 64 (2): 173-180.

Perez, A., C.J. Torres, E. Perez and J. Green. 1980. Behavior of ten chironja clones at three sites. I. Growth and yield. J. Agr. U. Puerto Rico 64 (3): 318-322.

Perez, A. and D. Vargas. 1977. Effect of fertilizer level and planting distance on soil pH, growth, fruit size, disease incidence, yield, and profit of two papaya varieties. J. Agr. U. Puerto Rico 61 (1): 68-76.

Perez, E., A. Perez-Lopez, C.J. Torres, and J. Green. 1981. Behavior of ten chironja clones at three sites: III. Interaction of genotype X environment. J. Agr. U. Puerto Rico 65 (3): 299-303.

Perez-Arbelaez, E. 1956. Plantas utiles de Colombia. 3rd ed. Libreria Colombiana—Camacho Roldan (Cia. Ltda.), Bogotá.

Perez-Lopez, A. 1963. Relation of maturity to some fruit characters of the West Indian cherry. J. Agr. U. Puerto Rico 47 (3): 193-200.

Perez-Lopez, A., J.R. Cruz Cay and J.R. Benero. 1982. Behavior of ten chironja clones at three sites. IV. Evaluation of fresh and processed fruits. J. Agr. U. Puerto Rico 66 (1): 65-70.

Perez-Lopez, A. and R.D. Reyes-Jurado. 1983. Effect of nitrogen and boron application on *Carica papaya* L. I. Growth and yield. J. Agr. U. Puerto Rico 67 (3): 181-187.

Perez-Lopez, A., A. Sotomayor-Rios and S. Torres-Rivera. 1980. Studies on characters of chironja seedling trees. J. Agr. U. Puerto Rico 64 (2): 232-235.

Perez-Lopez, A. and C.J. Torres. 1984. Growth, yield, efficiency and fruit quality of five Navel orange clones during 4 years. J. Agr. U. Puerto Rico 68 (4): 405-411.

Perkins, R.M. and P.F. Burkner. 1973. Mechanical pollination of date palms. Report of 50th annual Date Growers' Inst., Coachella Valley, Calif.

Permar, J.H. 1945. Catalog of plants growing in the Lancetilla Experimental Garden at Tela, Honduras. Tela Railroad Co., La Lima, Honduras.

Petelot, A. 1952, 1954. Plantes medicinales du Cambodge, du Laos et du Vietnam. Vol. I #14, Vol. 3, #22. Centre de Rech. Sci. et Tech., Arch. des Rech. Agron. au Camb., au Laos, et au Vietnam, Saigon.

Peterson, R.A., and A.J. Inch. 1980. Control of anthracnose on avocados in Queensland. Queensland J. Agr. & Anim. Sci. 37 (1): 79-83.

Phadnis, N.A. and T.R. Bagle. 1969. Rupees 10,000 from a hectare. Grow Papaya for profit. Ind. Hort. July/Sept. 12-14, 33.

Phadnis, N.A. and N. Gopalkrishna. 1958. Lest low temperature should harm the bananas. Indian Hort. 3 (1): 5-7.

Phillips, R.L. and W.S. Castle. 1977. Evaluation of twelve rootstocks for dwarfing citrus. J. Amer. Soc. Hort. Sci. 102 (5): 526-528.

Piatos, P. and R.J. Knight. 1975. Self-incompatibility in the sapodilla. Proc. Fla. St. Hort. Soc. 88: 464-465.

Pillai, N.C., G.J.S. Rao and M. Sirsi. 1957. Plant anticoagulants [emblic]. J. Sci. & Indus. Res. 16C (5): 106-107.

Pittier, H. 1957. Ensayo sobre plantas usuales de Costa Rica. Univ. de Costa Rica, San José, C.R.

Pittier, H. 1914. New and noteworthy plants from Colombia and Central America-4. Contrib. U.S. Nat. Herb. 18, Pt. 2. Smithsonian Inst., U.S. Nat. Mus., Washington, D.C.

Pittman, E.G. 1956. La chirimoya (*Annona cherimola* Mill.). Cir. 71. Estac. Exper. de 'La Molina', Min. de Agr., Lima, Peru.

Plank, H.K. 1950. Experiments with mamey for pests of man and animals. Trop. Agr. 27: 38-41.

Plank, H.K. 1945. Plant toxicology studies [mamey]. Report of the Fed. Exper. Sta. in Puerto Rico, 1944. U.S. Dept. Agr., Agr. Res. Admin., Washington, D.C.

Plank, H.K. 1948. Plant toxicology studies: mamey controlled fleas and ticks on dogs. Report of the Fed. Exper. Sta. in Puerto Rico, 1947. U.S. Dept. Agr., Agrs. Res. Admin., Washington, D.C.

Platts, P.K. 1945. Pineapple ABC's. Fla. Dept. Agr., Tallahassee, Fla.

Plucknett, D.L. and O.R. Younge. 1967. A high-lift machine for harvesting papaya. Hawaii Farm Sci. 16 (2): 1-3.

Poignant, A. 1969. Effets de deux hormones appliquées sur l' ananas pendant la formation de fruit. Fruits 21 (7-8): 353-364.

Polius, F. 1980. The agronomy of the banana plant. Windward Islands Banana Growers Assn., Res. & Dev. Div., Exten. Newsletter 11 (4): 4-9.

Pongpangan, S. and S. Poobrasert. 1972. Edible and poisonous plants in Thai forests. Sci. Soc. of Thailand, Bangkok.

Pontikis, C.A. and P. Melas. 1986. Micropropagation of *Ficus carica* HortSci. 21 (1): 153.

Pope, W.T. 1923. Acid lime fruit in Hawaii. Bull. 49. Hawaii Agr. Exper. Sta., Honolulu.

Pope, W.T. 1926. Banana culture in Hawaii. Bull. 55. Hawaii Agr. Exper. Sta., Honolulu.

Pope, W.T. 1968. Manual of wayside plants of Hawaii. Chas. E. Tuttle Co., Rutland, Vt.

Pope, W.T. 1934. Propagation of plants by cuttings in Hawaii. Cir. 9. U. of Hawaii Agr. Exper. Sta., Honolulu.

Popenoe, F.W. 1912. Feijoa sellowiana: its history, culture and varieties. Pomona Coll. J., Econ. Bot. 11 (1): 217-242.

Popenoe, F.W. 1911. The white sapote. Pomona Coll. J., Econ. Bot. 1 (2): 83-90.

Popenoe, J. 1965. Quest for the ilama. Fairchild Trop. Bull. 20 (1): 5-6.

Popenoe, J. 1974. Status of annona culture in South Florida. Proc. Fl. St. Hort. Soc. 87: 342-344.

Popenoe, J. 1962. Summer avocado varieties. Proc. Fl. St. Hort. Soc. 75: 358-360.

Popenoe, J. 1966. The 'African Pride' atemoya in Florida. Fr. Vars. & Hort. Dig. 20 (3): 44.

Popenoe, P. (edited by Henry Field). Written in 1928; pub'd 1973. The date palm. Field Res. Projects, Coconut Grove, Miami, Fla.

Popenoe, P. 1924. The date palm in antiquity. Sci. Monthly 19: 313-325.

Popenoe, W. 1952. Central American fruit culture. Ceiba 1 (5): 269-366.

Popenoe, W. 1924. Economic fruit-bearing plants of Ecuador. Contrib. U.S. Nat. Herb. Vol. 24, Pt. 5. Smithsonian Inst., Washington, D.C.

Popenoe, W. 1938. Importantes frutas tropicales (Pub. Agr. 130-1). Pan Amer. Union, Off. of Agr. Coop., Washington, D.C.

Popenoe, W. 1938. Manual of tropical and subtropical fruits. 2nd ptg. The Macmillan Co., N.Y.

Popenoe, W. 1914. The jaboticaba. J. Heredity; July: 318-326.

Popenoe, W. 1921. The native home of the cherimoya. J. Heredity, Aug.-Sept.: 331-336.

Popenoe, W. and O. Jimenez. 1921. The pejibaye, a neglected food plant of tropical America. Bull. of Pan Amer. Union, Nov.: 449-462. Same, in J. Heredity Apr.: 154-166.

Porres, M.A., P.A. Servando Rivera de Leon. [undated]. Temas Agricolas 36. Dept. de Divulgacion Agricola, DIGESA, Min. de Agr., Guatemala, C.A.

Propisil, F., S.K. Karikari, and E. Boamah-Mensah. 1972. Characterization and selection of pawpaw varieties (*Carica papaya* L.) in Ghana. Ghana J. Agr. Sci. 5 (2): 137-151.

Poucher, W.A. 1974. Perfumes, Cosmetics and Soaps. Vol. 2: The production, manufacture and application of perfumes. 8th ed. John Wiley & Sons, N.Y. [orange blossoms].

Power, F.B. and T. Callan. 1911. The constituents of the seeds of *Casimiroa edulis*. Trans. Chem. Soc. (London) 99: 1993-2010.

Prakash, V., R. Yamdagni and P.C. Jindal. 1982. Studies on the effect of pre-harvest treatments on storage behaviour of guava (*Psidium guajava* L.). Indian J. Agr. Res. 16 (3): 153-159.

Prasad, A. and J.P. Shukla. 1979. Studies on the ripening and storage behaviour of guava fruits (*Psidium guajava*). Indian J. Agr. Res. 13 (1): 39-42.

Prasad, A. and J.P. Shukla. 1978. Studies on the ripening and storage of ber (*Zizyphus mauritiana* Lamk.). Plant Sci. 10: 191-192.

Prasad, A. and J.P. Shukla. 1978. Studies on ripening and storage of custard apple [sugar apple] fruits. Plant Sci. 10: 185-196.

Prasad, J. 1982. An agro-technique for papaya production. Indian Hort. Oct.: 25-26, 28.

Prasad, J. 1980. Pectin and oil from passion fruit waste. Fiji Agr. J. 42: 45-48.

Prasad, J.S. and R.A.B. Verma. 1980. Efficacy of certain antibiotics in the control of postharvest decay of papaya fruits. Phytoparasitica 8 (2): 105-108.

Prasad, P.C. and R. Chandra. 1980. Review of passionfruit research in Fiji. Fiji Agr. J. 42 (2): 19-22.

Prasad, S.S. 1962. Two new leaf spot diseases of Nephelium litchi Cambess. Curr. Sci. 31 (7): 293.

Prest, R.L. 1955. The custard apple [atemoya]. Queensland Agr. J. Jan. 1: 17-21.

Protopadakis, E. and Ch. Zambettakis. 1982. Contribution to knowledge of the resistance of various scion/rootstock combinations against 'mal secco' [Volkamer lemon]. Fruits 37 (7-8): 467-471.

Pruthi, J.S. 1962. Chemical composition and utilization of passion fruit seed, seed oil and seed meal. Ind. Oils & Soap J. 28 (3): 55.

Pruthi, J.S. 1960. Freeze-drying of passion fruit juice. Food Sci. 9 (16): 336-338.

Pruthi, J.S. 1960. Nutritive value and utilization of passion fruit waste. (skin or rind). Food Sci. 9 (12): 397-399.

Pruthi, J.S. 1963. Spin pasteurization of canned passion fruit juice. Food Sci. 12 (1): 1-10.

Pruthi, J.S. 1965. Studies in isolation, characterization and recovery of pectin from purple passion fruit waste (rind). Chem. & Indus. 13: 555-559.

Pruthi, J.S., R.N. Chakraborty, L.V.L. Sastry and G.S. Siddappa. 1963. Studies on concentrating the juice of the cashew apple (*Anacardium occidentale*). Food Tech. 17 (11): 95-97.

Pruthi, J.S., G.V. Krishnamurthy and G. Lal. 1959. Utilization of mango waste. Indian Food Packer 13 (4): 7-15.

Pruthi, J.S. and G. Lal. 1955. Refrigerated and common storage of purple passion fruits (*Passiflora edulis* Sims.). Indian J. Hort. 12 (4): 204-211.

Pruthi, J.S., K.K. Mookerji and G. Lal. 1960. A study of factors affecting the recovery and quality of pectin from guava. Indian Food Packer 14 (7): 7-13, 19.

Pulle, A. 1951. Flora of Suriname (Netherlands Guyana). Vol. III, Pt. 2. The Royal Inst. for the Indies, Amsterdam.

Pulle, A. 1934. Flora of Surinam. Vol. III. Kon. ver Koloniaal Inst. te Amsterdam.

Purohit, A.G. 1981. Growing papaya the proper way. Ind. Hort. 25 (4): 3-5.

Purseglove, J.W. 1968. Tropical Crops—Dicotyledons 1 & 2. John Wiley & Sons, Inc., N.Y.

Purseglove, J.W. 1972. Tropical Crops—Monocotyledons 1 & 2. John Wiley & Sons, Inc., Halsted Press Div., N.Y.

Purvis, A.C. 1985. Relationship between chilling injury of grapefruit and moisture loss during storage: amelioration by polyethylene shrink film. J. Amer. Soc. Hort. Sci. 110 (3): 385-388.

Purvis, A.C. and L.G. Albrigo. 1984. Seasonal and temperature effects on seed germination in stored grapefruit. Proc. Fl. St. Hort. Soc. 97: 100-103.

Py, C. and A. Guyot in collab. with C. Garlin and J. Martial. 1970. Etude sur l' utilisation de l' ananas en conserverie. Fruits 25 (5): 349-355.

Py, C., P. Lossois and M. Karamkam. 1968. Contribution a l' etude du cycle de l' ananas. Fruits 23 (8): 403-413.

Py, C. and R. Naville. 1973. Economic production of quality pineapples. Span 16 (1): 21-23.

Pynaert, L. 1954. Le mangoustanier (*Garcinia mangostana* L.). Dir. de l' Agr. des Forets et de l' Elevage, Brussels, Belgium.

Pynaert, L. 1955. Les litchis et leurs especes fruitieres voisines. Tract #38. Dir. de l' Agr. des Forets et de l' Elevage, Brussels, Belgium.

Quijano, J. and G.J. Arango. 1979. The breadfruit from Colombia— a detailed chemical analysis. Econ. Bot. 33 (2): 199-202.

Quilantan-Carreón, J. 1979. Propagacion vegetativa del zapote mamey. Proc. Trop. Reb., Amer. Soc. Hort. Sci. 23: 180-182.

Quimio, T.H., A.N. Pordesimo and A.J. Quimio. 1975. Control of papaya fruit rots by postharvest dip in thiabenzadole. Phil. Agr. 59: 7-11.

Quimio, T.H. and A.J. Quimio. 1974. Note: a new fruit rot of papaya in the Philippines. Phil. Agriculturist 58 (7-8): 330-331.

Quiros, C.F. 1984. Overview of the genetics and breeding of husk tomato. Hort. Sci. 19 (6): 872-874.

Quisumbing, E. 1951. Medicinal plants of the Philippines. Tech. Bull. 16. Phil. Dept. Agr. Res. & Nat. Res., Manila.

Quraishi, A. and A. Nafees. 1983. Clonal propagation of mango (*Mangifera indica*) Pakistan J. Agr. Res. 4 (1): 12-16.

Raad, H.J. 1981. Economics of macadamia and lateba [litchi] cultivation in the Cook Islands. Farm Management Notes #10. Min. Agr. & Fish., Rarotonga, Cook Islands.

Raad, H.J. 1981. Economics of pawpaw cultivation on Rarotonga, Farm Management Notes #8. Min Agr. & Fish., Rarotonga, Cook Islands.

Richman, A.R. 1964. The development of plantain flakes. J. Agr. U. Puerto Rico 48 (3): 263.

Rahman, A.R., C.M. Berrocal, J.R. Cruz-Day and J.D. Rivera-Anaya. 1963. Toxicity studies on flour produced from unpeeled green plantains. J. Agr. U. Puerto Rico 47 (1): 11-13.

Rahman, M.A. 1981. Uses for mango wastes. Appropriate Tech. 7 (4): 29-30.

Rajagopalan, B. and K.I. Wilson. 1972. *Diplodia natalensis* causing dry rot of guava fruits in South India. Pl. Dis. Rep. 56 (4): 323-324.

Rajput, C.B.S. and J.E. Teskey. 1979. Loquat cultivation in India. World Crops 31 (1): 20.

Rajarama Rao, M.R. and H.H. Siddiqui. 1964. Pharmacological studies on *Emblica officinalis* Gaertn. Indian J. Exper. Biol. 2 (1): 29-31.

Ralph, W. 1980. Sunblotch disease in avocados. Rural Res. 109 (Dec.): 21-22.

Ram, B., R. Naidu and H.P. Singh. 1977. *Alternaria macrospora* Zimm. A new record on passion fruit (*Passiflora edulis* Sims.) from India. Curr. Sci. 46 (5): 165.

Ram, M. and P.K. Majumdar. 1980. Propagating litchi through stooling. Indian Hort. 25 (2): 1.

Ram, S. 1974. Aonla—a tree you must grow. Intensive Agr. 12 (10): 6-8.

Rama Rao, P.B., S. Balakrishnan and R. Rajagopalan. 1952. Spray drying of Indian gooseberry juice. Dept. Biochem., Indian Inst. of Sci., Bangalore, Typescript of paper, dated July 5.

Ramirez, C.T. and E. Gonzalez Tejera. 1983. Spacing, nitrogen and potassium on yield and quality of Cabezona pineapple. J. Agri. U. Puerto Rico 67 (1): 1-10.

Ramirez, E. and M. Rivera. 1936. Contribucion al estudio de la acción farmaco-dinamico del zapote blanco (*Casimiroa edulis*). Rev. Mensual Med. Mexico 9 (1): #3.

Ramirez, J.M. and A.H. Krezdorn. 1975. Effect of date of harvest and spot picking on yield and quality of grapefruit. Proc. Fl. St. Hort. Soc. 88: 40-44.

Ramirez, O.D. 1964. Metodo para eliminar las plantas masculinas de papaya en siembras comerciales. Bull. 180. U. de Puerto Rico, Estac. Exper. Agr., Rio Piedras.

Ramirez, O.D. and H. Gandía. 1976. Comparison of three planting distances and fertilizer applications on the yield of pineapple variety PR 1-67. J. Agr. U. Puerto Rico 60 (1): 31-35.

Ramirez, O.D. and H. Gandía. 1982. Comparison of nine planting distances on the yield of pineapple variety PR 1-67. J. Agr. U. Puerto Rico 66 (2): 130-138.

Ramirez, O.D., H. Gandía and J. Velez Fortuño. 1970. Two new pineapple varieties for Puerto Rico. J. Agr. U. Puerto Rico 54 (3): 417-428.

Ramirez Diaz., J.M. 1973. Estudio comparativo de cultivares de mango, *Mangifera indica* L., en plantaciones de Culiacan, Sinaloa. Agr. Téc. en México 3 (6): 216-222.

Ramirez Padilla, P., L. Bello Muiños and J.C. Suarez Mesa. 1982. Estudio comparativo de tres cultivares de Valencia de la Empresa Citricola Victoria de Giron. Centro Agricola 9 (2): 65-72.

Ramos Nuñez, G. 1968. La guayaba: fruto para consumo y exportacion. Rev. Nac. Agr. Colombia 62 (753): 8-11.

Randhawa, G.S. and R.R. Kohli. 1966. Sapota cultivation in India [sapodilla]. Indian Hort. 10 (4): 3-6, 26.

Randhawa, G.S., R.S. Malik and J.P. Singh. 1959. Note on clonal propagation in the phalsa (*Grewia asiatica* L.). Indian J. Hort. 16 (2): 119-120.

Randhawa, G.S., B.B. Sharma and N.L. Jain. 1961. Effect of plant regulators on fruit drop, size and quality in sweet oranges (*Citrus sinensis* Osbeck) var. Jaffa, Pineapple and Mosambi. Ind. J. Hort. 18 (3) 177-186.

Randhawa, G.S. and J.P. Singh. 1958. Grow litchi, but select the right variety. Indian Hort. 3 (1): 10-11.

Randhawa, G.S., J.P. Singh and S.S. Khanna. 1959. Effect of gibberellic acid and some other plant regulators on fruit set, size, total yield and quality in phalsa (*Grewia asiatica* L.). Indian J. Hort. 16 (4): 202-205.

Randhawa, G.S., J.P. Singh and R.S. Malik. 1958. Fruit cracking in some tree fruits with special reference to lemon (*Citrus limon*). Indian J. Hort. 15 (1): 6-9.

Randhawa, G.S. and R.K.N. Singh. 1967. Canning a loquat variety. Indian J. Hort. 11 (3): 16.

Randhawa, G.S. and R.K.N. Singh. 1967. Loquat growing is a paying proposition. Indian Hort. 11 (2): 8-12, 31.

Ranganna, S., L. Setty and K.V. Nagaraja. 1966. Discolouration in canned guava. Indian Food Packer 20 (5): 5-13.

Rao, D.P.C. and S.C. Agrawal. 1977. A new fruit rot of *Psidium guajava*. Curr. Sci. 46 (5): 162-163.

Rao, G.M. 1983. Litchi and bee-keeping. Indian Hort. 28 (2): 19-20.

Rao, P.S. 1959. Jellies and related products from tamarind seed kernels. J. Sci. & Indus. Res. 8 (9): 354-355.

Rao, P.S. 1957. Tamarind seed jellose: a new class of polysaccharides. J. Sci. & Indus. Res. 16A (3): 138-140.

Rasmussen, G.K. 1980. Relative effectiveness of dilute and concentrated abscission-chemical sprays in loosening 'Valencia' oranges from trees. J. Amer. Soc. Hort. Sci. 105 (2): 145-147.

Rastogi, R.C., R.P. Rastogi and M.L. Dhar. 1967. Studies on *Carissa carandas* Linn. Part II—the polar glycosides. Indian J. Chem. 5 (5): 215-216.

Rastogi, R.C., M.M. Vohra, R.P. Rastogi and M.L. Dhar. 1966. Studies on *Carissa carandas* Linn.: Pt. I—Isolation of the cardiac active principles. Indian J. Chem. 4 (3): 132-138.

Ratnam, C. and M. Srinivasan. 1959. Behaviour of ascorbic acid in Indian gooseberry to heat treatment. J. Sci. & Indus. Res. 18C (7): 132-133.

Rawal, R.D., N.C. Muniyappa and B.A. Ullasa. 1983. Effect of pre-harvest sprays of fungicides on the control of storage rot of papaya. Indian J. Agr. Sci. 53 (7): 614-615.

Raymond, W.D. and J.A. Squirer. 1961. Pewa or peach nuts from Trinidad. Col. Pl. & Anim. Prod. 2 (3): 203-205.

Reasoner, P.W. 1891. Report on the condition of tropical and semi-tropical fruits in the United States in 1887. U.S. Dept. Agr., Div. of Pomology, Washington, D.C.

Record, S.J. and R.W. Hess, 1947. Timbers of the New World. 3rd ptg. Yale U. Press, New Haven, Conn.

Reddy, D.B. 1975. Insects, other pests and diseases recorded in the southeast Asia and Pacific region: mango, *Mangifera indica*. Tech. Doc. #96. Plant Protect. Comm., FAO, Bangkok.

Reddy, Y.N. and P.K. Majumder, 1975. Bottom heat—a new technique for rooting hardwood cuttings of tropical fruits. Curr. Sci. 44 (12): 444-445.

Reece, P.C. and F.E. Gardner, 1959. Robinson, Osceola and Lee—new early-maturing tangerine hybrids. Proc. Fl. St. Hort. Soc. 72: 49-51.

Reece, P.C., C.J. Hearn, and F.E. Gardner. 1964. Nova tangelo—an early ripening hybrid. Proc. Fl. St. Hort. Soc. 77: 109-110.

Reece, P.C. and R.O. Register. 1961. Influence of pollinators on fruit set on Robinson and Osceola tangerine hybrids. Proc. Fl. St. Hort. Soc. 74: 104-106.

Reed, J.B. 1956. Lychee production problems solved by good soil and water practices. Fla. Grower & Rancher. Feb.: 14, 30-31.

Reeve, R.M. 1974. Histological structure and commercial dehydration potential of breadfruit. Econ. Bot. 28: 82-96.

Reinking, O.A. 1929. The double pummelo of Banda and Ambon. J. Heredity 20 (10): 448-458.

Reinking, O.A. and G.W. Groff. 1921. The Kao Pan seedless Siamese Pummelo and its culture. Phil. J. Sci. 19 (4): 389-437.

Reis de Almeida, J.M. 1974. O maracuja. Rev. Agricola 16 (171):

27-29.

Reitz, H.J. 1984. The world citrus crop, pp. 140-146. In: Outlook on Agriculture Vol. 13, No. 3. Pergamon Press, Great Britain.

Reuther, W., H.J. Webber and L.D. Batchelor. 1967. The citrus industry. Vol. I. Rev'd ed. U. California, Riverside.

Reuveni, O. and I. Adato. 1974. Endogenous carbohydrates, root promoters and root inhibitors in easy- and difficult-to-root date palm (*Phoenix dactylifera* L.) offshoots. J. Amer. Soc. Hort. Sci. 99 (4): 361-363.

Reyes, M.N., A. Perez and J. Cuevas. 1980. Detecting endogenous growth regulators on the sarcotesta, sclerotesta, endosperm, and embryo by paper chromatography on fresh and old seeds of two papaya varieties. J. Agr. U. Puerto Rico 64 (2): 164-172.

Reynhardt, J.P.K. and E.R. Dalldorf. 1968. Planting material for the Cayenne pineapple. Farming in So. Africa 44 (2): 24-25, 27, 32.

Reynhardt, J.P.K. and E.R. Dalldorf. 1968. Queen pineapple planting material. Farming in So. Africa 44 (4): 27-31.

Reynhardt, J.P.K. and J.L. van Heerden. 1968. Pruning pineapples. Farming in So. Africa 44 (4): 57-59.

Riaz, R.A. 1967. Pectin extraction from roselle sepals. Sci. & Indus. (West Pakistan) 5 (3): 435-441.

Rios-Castaño, D. 1964. Cidras Corsican e Italian, dos cítricos potencialmente económicos. Agr. Trop. (Colombia) 20 (7): 369-378.

Rios-Castaño, D. 1963. Clave para la identificacion de tres especies de kumquat (*Fortunella* spp.). Agr. Trop. (Colombia), 19 (8): 477-489.

Rivals, P. 1964. Notes biologique et culturales sur l' Actinidia de Chine (*Actinidia sinensis* Planchon). J. d' Agr. Trop. et de Bot. Appl. 11 (4): 75-83.

Rivera Lopez, C. 1975. A method to obtain relatively uniform breadfruit trees from a stockplant. J. Agr. U. Puerto Rico 59 (1): 77-78.

Rizvi, S.H., R.S. Kapil and A. Schoeb. 1985. Alkaloids and coumarins of *Casimiroa edulis*. J. Nat. Prod. 48 (1): 146.

Roberts, J.W. 1963. Feijoas. New Zeal. J. Agr. 106 (4): 363-364.

Robinson, P.W. 1970. Simple grafting methods for passion fruit. Agr. Gaz. New So. Wales 81 (11): 591-592.

Robinson, T.R. 1952. Grapefruit and pummelo. Econ. Bot. 6 (3): 228-245.

Robinson, T.R. 1921. The bud-sport origin of a new pink-fleshed grapefruit in Florida. J. Heredity 12 (3): 195-198.

Robinson, W.B. 1963. Juice from the tomato's Latin American cousin [naranjilla]. Farm Res. 29 (3): 12, 13.

Rodriguez, A.J., R. Guadalupe, and L.M. Iguina de George. 1974. The ripening of local papaya cultivars under controlled conditions. J. Agr. U. Puerto Rico 58 (2): 184-196.

Rodriguez, A.J. and L.M. Iguina de George. 1972. Evaluation of papaya nectar prepared from unpeeled papaya puree. J. Agr. U. Puerto Rico 56 (1): 79-80.

Rodriguez, A.J. and L.M. Iguina de George. 1971. Evaluation of some processing characteristics of cultivated guava clones. J. Agr. U. Puerto Rico 55 (1): 44-52

Rodriguez, A.J. and L.M. Iguina de George. 1971. Preparation and canning of a papaya drink. J. Agr. U. Puerto Rico 55 (2): 161-166.

Rodriguez, J., V.B. Dalal, N.V.N. Moorthy and H.C. Srivastava. 1963. Effect of postharvest treatment with plant growth regulators in wax emulsion on storage behaviour of limes (*Citrus aurantifolia* (Christm.) Swingle). Indian Food Packer 17 (5): 9-11.

Rodriguez, J.A. and H. Irizarry. 1979. Effect of planting material on yield and quality of two plantain cultivars (*Musa acuminata* X *M. balbisiana*, AAB). J. Agr. U. Puerto Rico 43 (3): 351-365.

Rodriguez, R. and P.L. Melendez. 1984. Occurrence of *Sphaeropsis* knot on citron (*Citrus medica* L.) in Puerto Rico. J. Agr. U. Puerto Rico 68 (2): 179-183.

Rodriguez F., R. and H. Garayar M. (undated). Cultivo de Cocona, Maracuya and Naranjilla. Oficina Tecnica de Informacion Agraria (OTIA).

Rodriguez-Sosa, E.J., I. Beauchamp de Caloni, O. Parsi-Ros and M.A. Gonzalez. 1977. The preparation of cupcakes from green banana flour. J. Agr. U. Puerto Rico 61 (4): 521-523.

Rodriguez-Sosa, E.J., M.A. Gonzalez, I.B. de Caloni and O. Parsi-Ros. 1977. The preparation of green banana flour. J. Agr. U. Puerto Rico 61 (4): 470-478.

Rodriguez-Sosa, E.J. and O. Parsi-Ros. 1984. pH affects properties of green banana starch. J. Agr. U. Puerto Rico 68 (4): 323-329.

Rogers, D.J. and A.D. Blair. 1981. Assessment of insect damage to litchi fruit in northern Queensland. Queensland J. Agr. & Anim. Sci. 38 (2): 191-194.

Rogers, D.J. and A.D. Blair. 1983. Control of the banana scab moth (*Nacoleia octasema*) (Meyrick) in south Queensland. Queensland J. Agr. 40 (1): 35-38.

Roig y Mesa, J.T. 1945. Plantas medicinales, aromaticas o venenosas de Cuba. Cultural, S.A., Havana

Rojas, H.C. 1969. Ensayos preliminares sobre la industrializacion de la fibra de hoja de piña. Rev. Univ. Indus. Santander, Invest. 2: 11-21.

Rokba, A.M., N.H. El Bakly, and N.E. Solet. 1982. Pollination studies in Annonas and their effect on fruit set, fruit maturity and fruit characteristics. Res. Bull. 2117. Fac. of Agr., Ain Shams Univ., Egypt.

Rokba, A.M., A.T. El-Wakeel and N.H. Ali. 1977. Studies and evaluation of some annona varieties. Agr. Res. Rev. 55 (3): 13-29.

Rolfs, R.H. 1929. Subtropical vegetable-gardening. The Macmillan Co., N.Y.

Roman, J., D. Oramas, J. Green and A. Torres. 1983. Control of nematodes and black weevils in plantains. J. Agr. U. Puerto Rico 67 (2): 270-277.

Rondiere, P. 1972. The pineapple gold rush. Ceres: FAO Rev. 5 (3): 39-41.

Rose, J.N. 1899. Notes on useful plants of Mexico, Vol. 5, No. 4 [roselle]. U.S. Nat. Herb., Smithsonian Inst., Washington, D.C.

Rosenzweiz, R.W. 1968. Etrog: Israel's Goodly Tree. For. Agr. 6 (49): 9-10.

Rossetto, C.J., M. Ojima and O. Rigitano. 1971. Queda dos frutos do caquizeiro, associada a infestacao de *Aceria diospyri* K. (Acarina, Eriophyidae). Bragantia 30 (1): 1-9.

Roth, G. 1963. Post harvest decay of litchi fruit (*Litchi chinensis* Sonn.). Tech. Commun. Dept. Agr. Techn. Serv. 11: 1-16.

Rouse, A.H. and P.G. Crandall. 1976. Nitric acid extraction of pectin from citrus peel. Proc. Fl. St. Hort. Soc. 89: 166-168.

Rouse, R.E. and H.K. Wutscher. 1985. Heavy soil and bud union crease with some grapefruit clones limit use of Swingle Citrumelo rootstock. HortSci. 20 (2): 259-261.

Roy, B.N., M.S. Sengupta, and T.K. Bose. 1970. Growth substances prevent flower drop in acid lime (*Citrus aurantifolia* Swingle). Allahabad Farmer 44 (5): 275-277.

Roy, R.S. and C. Sharma. 1960. Mutation in mango (*Mangifera indica* L.). Indian J. Hort. 17 (2): 142-143.

Roy, R.S. and D.R. Singh. 1952. Utilization of unripe mango fruits. Indian Food Packer. Apr.: 13-14, 31-33.

Roy, S.K. and R.N. Singh. 1979. Bael fruit (Aegle marmelos)—a potential fruit for processing. Econ. Bot. 33 (2): 203-212.

Roy, R.S. and U.P. Verma. 1950. Bael products. Indian Food Packer 4 (8): 13-14, 28.

Ruberte-Torres, R. and F.W. Martin. 1974. First-generation hybrids of edible passion fruit species. Euphytica 23: 61-70.

Ruehle, G.D. 1943. Cause and control of *Cercospora* spot and of anthracnose of the avocado. Press Bull. 583. U. Fla. Agr. Exper. Sta., Gainesville, Fla.

Ruehle, G.D. 1948. The common guava—a neglected fruit with a promising future. Econ. Bot. 2 (3): 306-325.

Ruehle, G.D. 1951. The sapodilla in Florida. Cir. S-34. U. Fla., Agr. Exper. Sta., Gainesville, Fla.

Ruehle, G.D. 1953. Two new fruits for South Florida. Proc. Fl. St. Hort. Soc. 66: 190-191.

Ruggiero, C., D.A. Bangatto and A. Lam-Sanchez. 1976. Studies on natural and controlled pollination in yellow passion fruit (*Passiflora edulis* f. *flavicarpa* Deg.). Acta Hort. 57: 121-123.

Ryerson, K. (undated). The white sapote in California (Mimeo). Cir. 3. U. California, Coll. Agr., Berkeley, Calif.

Saavedra, E. 1977. Influence of pollen grain stage at the time of hand pollination as a factor on fruit set of cherimoya. HortSci. 12 (2): 117-118.

Saavedra, E. 1979. Set and growth of *Annona cherimola* Mill. fruit obtained by hand-pollination and chemical treatments. J. Amer. Soc. Hort. Sci. 104 (5): 668-673.

Sadasivam, R., S. Muthuswamy, J.S. Sundaraj and D.D. Sundaraj 1970. Experiments with waxing of mango fruits in common storage. Madras Agr. J. 57 (11): 694-699.

Sadhu, M.K. and S.K. Ghosh. 1976. Effects of different levels of nitrogen, phosphorus and potassium on growth, flowering, fruiting and tissue composition of custard apple (*Annona squamosa* L.). Indian Agr. 20 (4): 297-301.

Sadhu, M.K., S.K. Ghosh and T.K. Bose. 1975. Mineral nutrition of fruit plants. I. Effect of different levels of nitrogen and potassium on growth, flowering, fruiting and leaf composition of phalsa (*Grewia asiatica* L.) Indian Agr. 19 (3): 319-321.

Safford, W.E. 1912. Botany: *Annona diversifolia*, a custard apple of the Aztecs. J. Washington Acad. Sci. 11 (5): 118-125.

Safford, W.E. 1914. Classification of the genus *Annona*, with descriptions of new and imperfectly known species. Contrib. U.S. Nat. Herb. Vol. 18, Pt. 1. Smithsonian Inst., Washington, D.C.

Safford, W.E. 1905. Useful plants of the island of Guam. Contrib. U.S. Nat. Herb. Vol. II. Smithsonian Inst., U.S. Nat. Mus., Washington, D.C.

Sagawa, Y. 1982. Potentials of tissue culture for the banana industry. Research & Extension Series No. 21; pp. 36-37. Proc. Ann. Hawaii Banana Indus. Conf. 1981.

Saha, A.K. 1971. Increase in fruit set of female trees of papaya (*Carica papaya* L.) by spraying 2, 4-Dichloraphenoxy acetic acid and naphthalene acetic acid. Allahabad Farmer 45 (6): 581-583.

Saha, A.K. 1969. Some investigations of vegetable propagation of jaman *Syzygium jambos* L.). Allahabad Farmer 43 (3): 187-189.

Salazar C., R. and R. Torres M. 1977. Almacenamiento de frutas de maracuya (*Passiflora edulis* var. *flavicarpa*, Degener) en bolsas de polietileno. Revista ICA 12 (1): 1-11.

Sale, P. 1981. Avocados: NZ's [New Zealand's] got all the advantages. New Zeal. J. Agr. 142 (4): 2-3.

Sale, P. 1973. The kiwi fruit story. New Zeal. J. Agr. 127 (5): 56-58.

Saleh, M.A., M.K. Mahmoud Amer, A. El-Wahab Radwan, and M. El Said Amer. 1961. Experiments on pomegranate seeds and juice preservation. Agr. Res. Rev. [U.A.R.] 42 (4): 54-64.

Salibe, A.A. 1970. Jaca; a maior fruta. O Estado do Sao Paulo. Oct. 28. P. 14.

Samarawira, I. 1983. Date palm, potential source for refined sugar. Econ. Bot. 37 (2): 181-186.

Sambamurty, K. and V. Ramalingam. 1954. A note on hybridisation in the sapota (*Achras zapota* L.). Indian J. Hort. 11 (2): 57-60.

Sambamurty, K. and V. Ramalingam. 1954. Preliminary studies in blossom biology of the jack fruit (*Artocarpus heterophyllus* Lam.) and pollination effects. Indian J. Hort. 11 (1): 24-27.

Sampalo, A. 1977. Desenvolve-se intensamente o cultivo da nespera ou ameixa-amarela. O Estado do Sao Paulo, Supl. Agricola. Dec. 12, P. 7.

Samuels, G. 1970. Pineapple cultivars, 1970. Proc. Fl. St. Hort. Soc. 83: 325-332.

Samuels, G., E. Orengo-Santiago and A. Beale. 1978. Influence of fertilizers on the production of plantains with irrigation. J. Agr. U. Puerto Rico 62 (1): 1-9.

Sanchez-Nieva, F. 1955. Extraction, processing, canning, and keeping quality of acerola juice. J. Agr. U. Puerto Rico 39 (4): 175-183.

Sanchez-Nieva, F. 1962. Nuevos procedimientos para la elaboracion de los nectares de mango, guayaba y guanabana. P. M. 6. U. Puerto Rico, Rio Piedras.

Sanchez-Nieva, F., C.E. Bueso and M. Mercado. 1983. The canning of green bananas. II. Internal corrosion of plain tin containers by the acidified green bananas. J. Agr. U. Puerto Rico 67 (4): 356-365.

Sanchez-Nieva, F. and I. Hernandez. 1967. Preparacion y conservacion por congelacion de platanos maduros en almibar. Lab. de Tech. de Alimentos. Pub. #7. U. Puerto Rico., Estac. Exper. Agr., Rio Piedras.

Sanchez-Nieva, F. and I. Hernandez. 1977. Studies on the freezing of Red Spanish and Smooth Cayenne pineapples. J. Agr. U. Puerto Rico 61 (3): 354-360.

Sanchez-Nieva, F., I. Hernandez and C. Bueso. 1975. Studies on freezing green plantains (*Musa paradisiaca*). 1. Effect of blanching treatments on the quality and storage life of raw and pre-fried slices. J. Agr. U. Puerto Rico 59 (2): 84-91.

Sanchez-Nieva, F., I. Hernandez and C.E. Bueso. 1975. Studies on the freezing of plantains (*Musa paradisiaca*). III. Effect of stage of maturity at harvest on quality of frozen products. J. Agr. U. Puerto Rico 59 (2): 107-114.

Sanchez-Nieva, F., I. Hernandez and C. Bueso de Vinas. 1970. Studies on the ripening of plantains under controlled conditions. J. Agr. U. Puerto Rico 54 (3): 517-529.

Sanchez-Nieva, F., I. Hernandez, R. Guadalupe and C. Bueso. 1971. Effect of time of planting on yields and processing characteristics of plantains. J. Agr. U. Puerto Rico 55 (4): 394-404.

Sanchez-Nieva, F., I. Hernandez and L.M. Iguina de George. 1970. Frozen soursop puree. J. Agr. U. Puerto Rico 54 (2): 220-235.

Sanchez-Nieva, F., L. Igaravidez, and B. Lopez-Ramos. 1953. The preparation of soursop nectar. Tech. Paper 11. U. Puerto Rico, Agr. Exper. Sta., Rio Piedras.

Sanchez-Nieva, F. and M. Mercado. 1981. Non-enzymatic darkening in young hot-water-peeled green bananas. J. Agr. U. Puerto Rico 65 (1): 1-7.

Sanchez-Nieva, F. and M. Mercado. 1983. The canning of green bananas. I. Processing factors affecting the acidification of hot water peeled fruit. J. Agr. U. Puerto Rico 67 (4): 339-355.

Sanchez-Nieva, F. and M. Mercado. 1983. The canning of green bananas. III. Effect of storage and processing variables on chemical composition, texture and sensory attributes. J. Agr. U. Puerto Rico 67 (4): 366-378.

Sanchez-Nieva, F., M. Mercado and C. Bueso. 1980. Effect of the stage of development at harvest on the texture, flavor, quality and yields of frozen green bananas. J. Agr. U. Puerto Rico 64 (3): 275-282.

Sanchez-Nieva, F., A.J. Rodriguez and J.R. Benero. 1959. Processing and canning mango nectars. Bull. 148. U. Puerto Rico Agr. Exper. Sta., Rio Piedras.

Sandhu, I.P.S., G.S. Dhaliwal and M.P. Singh. 1983. Effect of pruning on yield, fruit quality and maturity in ber (*Zizyphus mauritiana* Lamk.) cv. Umran. J. Res. Punjab Agr. U. 20 (2): 135-138.

Santini, R., Jr., and A. Huyke. 1956. Identification of the anthocyanin present in the acerola which produces color changes in the juice on pasteurization and canning. J. Agr. U. Puerto Rico 40 (3): 171-178.

Santini, R., Jr. and A.S. Huyke. 1956. Identification of sugar present in fruit of the acerola (*Malpighia punicifolia* L.) by paper chromatography. J. Agr. U. Puerto Rico 40 (2): 87-89.

Santini, R., Jr. and J. Navarez. 1955. Extraction of ascorbic acid from acerolas (*Malpighia punicifolia* L.). J. Agr. U. Puerto Rico 39 (4): 184-189.

Santos, E. 1962. A cultura do maracuja. Lav. e Criacao 136: 31-33.

Saras, R. 1978. Guava, the Emperor of fruits. Malaysian Panorama 8 (2): 18-23.

Sarveswara Rao, K., N.S. Kapur, H. Subramanyam, S. d'Souza and H.C. Srivastava. 1962. Gas storage of bananas. J. Sci. & Indus. Res. 21D: 331-335.

Sarwar, M. and M. Kamal. 1971. Studies on fruit rot of papaya caused by *Rhizopus oryzae* (Went. & Jeerl.). Pakistan J. Sci. Ind. Res. 14 (3): 234-236.

Sastry, L.W.L., M.N. Satyanarayona, M. Srinivasan, N. Subramanian and V. Subrahmanyan. 1956. Polyphenols in edible materials. J. Sci. & Indus. Res. 15C (3): 78-90.

Sastry, M.V. 1965. Biochemical studies in the physiology of guava. Pt. I. Physical changes. Indian Food Packer 19 (1): 11-14.

Sastry, M.V. 1966. Biochemical studies in the physiology of sapota. Pt. II. Major chemical constituents. Indian Food Packer 20 (6): 16-20.

Sastry, M.V. 1965. Biochemical studies in the physiology of guava. Pt. V. Changes during storage. Indian Food Packer 19 (6): 17-20.

Saucedo Veloz, C., F. Esparza Torres, and S. Lakshminarayana. 1977. Effect of refrigerated temperatures on the incidence of chilling injury and ripening quality of mango fruit. Proc. Fl. St. Hort. Soc. 90: 205-210.

Sauls, J.W. and C.W. Campbell. 1980. Avocado seed germination studies. Proc. Fl. St. Hort. Soc. 93: 153-154.

Sauls, J.W. and C.W. Campbell. 1980. Herbicide screening on *Carica papaya* L. Proc. Trop. Reg. Amer. Soc. Hort. Sci. 24: 93-96.

Savastano, L. 1923. Della coltivazione dell' anona in Italia. Ann. Rpt. 7. R. Stazione Sperimentale di Agrumicoltura e frutticoltura, Acireale, Italy.

Savur, G.R. 1956. Manufacture of pectin in India. Indian Food Packer 10 (5): 12-14.

Savur, G.R. 1955. Utilization of tamarind seed polyose in food industries. Indian Food Packer 9 (2): 13-16 and 9 (3): 31-32.

Sawaya, W.N., W.M. Safi, L.T. Black, A.S. Mashadi and M.M. Al-Muhammad. 1983. Physical and chemical characterisation of the major date varieties grown in Saudi Arabia. II. Sugars, tannins, vitamin A and C. Date Palm J. 2 (2): 183-196.

Schaefers, G.A. 1969. Aphid vectors of the papaya mosaic viruses in Puerto Rico. J. Agr. U. Puerto Rico 53 (1): 1-13.

Schaffer, R.S., W.E. Scott and T.D. Fontaine. 1952. Antibiotics that come from plants. Pp. 727-733. In: U.S. Dept. Agr. Yearbook 1950-1951, "Crops in Peace and War".

Schaible, C.H. 1965. "Drink that cold away with tangerine juice". Fla. Grower & Rancher. 73 (2): 13.

Schieber, C. 1963. Nemátodos que atacan a la naranjilla en Guatemala. Revista Cafetalera 46: 6-7.

Schnee, L. 1973. Plantas comunes de Venezuela. 2nd ed. Universidad Central de Venezuela, Fac. de Agron., Maracay.

Schneider, H. and J.E. Pehrson, Jr. 1985. Decline of navel orange trees with trifoliate orange rootstocks. Calif. Agr. 39 (9 & 10): 13-16.

Schroeder, C.A. 1945. Cherimoya culture in California. Cir. 15; Mimeo Cir. Series. U. Calif., Coll. of Agr., Agr. Exper. Los Angeles, Calif.

Schroeder, C.A. 1947. Cherimoya varieties in California. Fr. Vars. & Hort. Dig. 2 (3): 68-71.

Schroeder, C.A. 1981. Fruit morphology and anatomy of the cherimoya. Bot. Gaz. 112 (4): 436-446.

Schroeder, C.A. 1954. Fruit morphology and anatomy in the white sapote. Bot. Gaz. 115 (3): 248-254.

Schroeder, C.A. 1947. Hand pollination of cherimoya improves fruit set. 1947 Yearbook, Calif. Avocado Soc. Pp. 67-70.

Schroeder, C.A. 1947. Pollination requirements of the feijoa. Proc. Amer. Soc. for Hort. Sci. 49: 161-162.

Schroeder, C.A. 1946. Priority of the species *Psidium cattleianum* Sabine. J. Arn. Arbor. 27: 314-315.

Schroeder, C.A. 1947. Rootstock influence on fruit-set in the Hachiya persimmon. Proc. Amer. Soc. Hort. Sci. 50: 149-150.

Schroeder, C.A. and J.E. Coit. 1944. The Cattley — commonly known as the strawberry guava. Yearbook Calif. Avocado Soc. Pp. 44-47.

Schroeder, C.A. and W.A. Fletcher. 1967. The Chinese gooseberry (*Actinidia chinensis*) in New Zealand. Econ. Bot. 21 (1): 81-92.

Schroeder, C.A. 1949. The feijoa in California. Fr. Vars. & Hort. Dig. (Winter). P. 99.

Schroeder, C.A. 1948. The kei apple. Yearbook (1947), Calif. Avocado Soc. Pp. 71-73.

Schroeder, C.A. 1950. The loquat. Yearbook 1959, Texas Agr. Soc. Pp. 69-73.

Schroeder, C.A. 1949? The white sapote, *Casimiroa edulis*. Proc. Amer. Soc. Hort. Sci.? Pp. 60-65.

Schroeder, C.A. 1949. White sapote varieties in California. Fr. Vars. & Hort. Dig. 4 (1): 7-9.

Schultes, R.E. 1958. A little-known cultivated plant from northern South America [*Solanum topiro*]. Bot. Mus. Leaflets, Harvard U. 18 (5): 229-244.

Schultes, R.E. and J. Cuatrecasas. 1953. Notes on the cultivated lulo. Bot. Mus. Leaflets, Harvard U. 16 (5): 97-105.

Sciancalepore, V. and W. de Dorbesson. 1981. Influencia de la variedad sobre la composicion acidica y la estructura gliceridica del aciete de aguacate. Rev. Agr. Subtrop. e Trop: 109-115.

Scott, F.S., Jr. and R. Shoraka. 1974. Economic analysis of the market for guava nectar. Res. Rpt. 230. Hawaii Agr. Exper. Sta., Coll. of Agr., U. of Hawaii.

Scott, K.J., B.I. Brown, G.R. Chaplin, M.E. Wilcox and J.M. Bain. 1982. The control of rotting and browning of litchi fruit by hot benomyl and plastic film. Scientia Hort. 16: 253-262.

Scrimshaw, N.S., INCAP, Guatemala. 1955. Personal communication, June 10.

Seale, P.E. and G.D. Sherman. 1960. Commercial passion fruit processing. Cir. 58. Hawaii Agr. Exper. Sta., U. of Hawaii.

Seaman, R.J. 1978. The Australian avocado industry. Chron. Hort. 18 (1): 5.

Seaman, R.J. 1978. The Australian pineapple industry. Chron. Horticulturae 18 (1): 5-6.

Sein, F., Jr. and J. Adsuar. 1947. Transmission of the bunchy top disease of papaya (*Carica papaya* L.) by the leaf hopper *Empoasca papayae* Oman. Science 106: 130.

Sekeri-Pataryas, K.H., K.A. Mitzakos and M.K. Georgi. 1973. Yields of fungal protein from carob sugars. Econ. Bot. 27 (3): 311-319.

Sen, A.B. and Y.N. Shukla. 1968. Chemical examination of *Hylocereus undatus*. J. Indian Chem. Soc. 45 (8) [no page no.]

Sen, A.K. 1963. Furocourmarins in the treatment of leucoderma [bael fruit]. J. Sci. & Indus. Res. 22 (2): 88-91.

Sen, P.K., R.N. Basu, T.K. Bose and N. Roychoudhury. 1968. Rooting of mango cuttings under mist. Curr. Sci. 37 (5): 144-146.

Sengupta, S.C. 1972. Improved lac cultivation. Indian Farming 22 (3): 11-14, 23. [Indian jujube].

Sergent A., E. 1979. Estado actual de las zonas productoras de aguacate (*Persea americana*) en Venezuela. Rev. Fac. Agron. (Maracay) 10 (1-4): 51-56.

Seth, J.N. 1962. Cytogenetical studies in *Psidium*, I. F_1 hybrids of *P. guajava* X *P. guineense* S.W. cross. Hort. Adv. (India) 6: 173-179.

Setty, K.G.H. 1959. Blue mold of amla. Curr. Sci. 28 (5): 208.

Shaffer, R. 1982. Kiwi is growing ripe profits for California farms. Christian Sci. Monitor. Nov. 9.

Shanker, G. 1969. Aonla for your daily requirement of vitamin C. Indian Hort. July-Sept.: 9-11, 35.

Shanker, G. 1967. How to grow apple colour guava. Indian Hort. 11 (2): 22-23.

Shanker, G. 1980. Preliminary studies on propagation of aonla (*Phyllanthus emblica* L.) by chip budding. Allahabad Farmer 51 (1): 79-80.

Shanker, G. and N. Chezhiyan. 1979. A note on the effect of gibberellic acid on seedless guava (*Psidium guajava*) cultivar. Allahabad Farmer 50 (1): 29-32.

Shanker, G. and B.A. David. 1979. Guava varieties and its relatives differ in susceptibility to the guava stem girdling and boring bark caterpillar, *Indarbela* sp. Indarbelidae, Lepidoptera. Allahabad Farmer 50 (3): 327-329.

Shanmugavelu, K.G. 1971. Effect of plant growth regulators on jack (*Artocarpus heterophyllus* Lamk.). The Madras Agr. J. 58 (2): 97-103.

Sharaf, A. 1962. The pharmacological characteristics of *Hibiscus sabdariffa* L. Planta Med. 10 (1): 48-52.

Sharaf, A., M.B.E. Fayez and A.R. Negm. 1967. Pharmacological properties of *Punica granatum* L. Qual. Plant. Mater. Veg. 14 (4): 331-336.

Sharaf, A., O. Lotfy and A. Geneidy. 1967. The effect of a coloring matter separated from *Hibiscus sabdariffa* on Mycobacterium tuberculosis as compared to that of its watery extract. Qual. Plant. Mat. Veg. 15 (2): 117-122.

Sharma, B.M. and P. Singh. 1975. Diagnostic characters of seeds of *Carica papaya* L. — an adulterant of black papper of commerce. Quart. J. Crude Drug Res. 13: 69-76.

Sharma, R.D. and P.A.A. Loof. 1973. Novo tipo de enfermidade da goiabeira causada por nematoides na Bahia. Cacau Atualidades 10 (1): 20-21.

Sharma, R.D. and S.A. Sher. 1973. Occurrence of plant parasitic nematodes in avocado in Bahia, Brazil. Nematropica 3 (1): 20-23.

Sharp, J.L. and D.L. Spalding. 1984. Hot water as a quarantine treatment for Florida mangos infested with Caribbean fruit fly. Proc. Fl. St. Hort. Soc. 97: 355-356.

Sheerer, L. 1977. Intelligence Report: African folk medicine. Parade: Orlando Sentinel Star. June 5.

Shigeura, G.T. 1973. Culture and management of guava. Misc. Pub. #11. U. Hawaii Coop. Exten. Serv., Honolulu.

Shivpuri, D.N. 1963. Allergy to papaya tree (*Carica papaya* Linn.). Ann. of Allergy 21 (3): 139-144.

Shoeb, A., R.S. Kapil, and S.A. Popli. 1973. Coumarins and alkaloids of *Aegle marmelos*. Phytochem. 12: 2071-2072.

Shoemyen, J. 1969. Citrus pulp pellets formed by extrusion process. Fla. Grower & Rancher 62 (5): 4-5.

Shoji, K., M. Nakamura, and M. Matsumura. 1958. Growth and yield of papaya in relation to fertilizer applications. Prog. Notes 118 (Mimeo). Hawaii Agr. Exper. Sta., Coll. of Agr., U. Hawaii, Honolulu.

Shrestha, G.K. 1981. Effects of ethephon on fruit cracking of lychees (*Litchi chinensis* Sonn.). HortSci. 16 (4): 498.

Shrivastava, H.C., K. Kripal Singh, and P.B. Mathur. 1962. Refrigerated storage of mangosteen (*Garcinia mangostana*). Food Sci. Aug.: 226-228.

Shukla, O.P. and C.R. Krishna Murti. 1961. Bacteriolytic activity of plant latices [jackfruit]. J. Sci. & Indus. Res. 20C (7): 225-226.

Shukla, S. and R.D. Tewari. 1971. 7-methylporoil-4-*B*-D-xylopyranosyl-D-gluco-pyranoside from the heartwood of *Feronia elephantum*. Ind. J. Chem. 9 (3): 187.

Siddappa, G.S. 1957. Effect of processing on the trypsin inhibitor in jackfruit seed (*Artocarpus integrifolia*). J. Sci. & Indus. Res. 16C (10): 199-201.

Siddappa, G.S. 1959. Preparation of some useful preserved products from carambola (*Averrhoa carambola*). Indian J. Hort. 16 (1): 47-48.

Siddappa, G.S. and B.S. Bhatia. 1956. Effect of canning on the *B*-carotene content of mango, papaya and jackfruit. J. Sci. & Indus. Res. 15C (5): 118-120.

Siddappa, G.S. and B.S. Bhatia. 1952. Preparation of jelly from jackfruit rind. The Bulletin. Cent. Food Tech. Res. Inst., Mysore, India 2 (3): 70-72.

Siddappa, G.S., A.M. Nanjundaswami and S. Saroja. 1969. Utilisation of banana fruit. Indian Hort. 14 (1): 7-8, 29-30.

Silveira, A.H. da. 1958. Pequeña industria de jaboticaba. Rural, Rev. Soc. Rural Brasil. 38 (446): 19.

Silveira, A.P., M. Ojima, B.P. Bastos Cruz and S.C.P. Silveira. 1965. Controle quimico da "entomosporiose" nos viveiros de nespereira [sapodilla]. O Biologico 31 (7): 137-141.

Simao, S. 1955. Contribuicao para caracterizacao de algumas variedades de mangueira, *Mangifera indica* L. J. de Piracicaba, Piracicaba, Brazil.

Simmonds, J.H. 1959. Mild strain protection as a means of reducing losses from the Queensland woodiness virus in the passion vine. Queensland J. Agr. 16 (4): 371-380.

Simmonds, J.H. 1965. Papaw diseases. Queensland Agr. J. 91 (11): 666-678.

Simmonds, N.W. 1959. Bananas. Trop. Agr. Ser. Longmans, Green & Co., Ltd., London.

Simons, J.S. 1942. La guanabana y su cultivo. Bol. Bimestral de la Est. Exper. Agr. de la U. de Puerto Rico 2 (3): 7-8.

Simons, J.S. and R. Arroyo. 1943. La pomarrosa Malaya en la produccion de vino—(*Jambos malaccensis* (L) DC). Agr. Exper. Bol. Bimestral de la Estac. Exper. de la U. de Puerto Rico II (3): 5-7.

Sinclair, W.B. 1972. The grapefruit. U. Calif. Press, Berkeley, Calif.

Sinclair, W.B. 1961. The orange, its biochemistry and physiology. U. Calif., Div. of Agr. Sci., Riverside, Calif.

Sing Ching Tongdee. 1972. Polyethylene bags and ethylene absorbent for delaying banana ripening. Thai J. Agr. Sci. 5: 265-271.

Singh, B.B. 1962. Alternaria rot of mango in cold storage and its control. Hort. Adv. 6: 164-167.

Singh, B.B. 1959. Fruit rot of loquat (*Eriobotrya japonica* Lindl.). Hort. Adv. 3: 128-130.

Singh, H. and A.N. Rao. 1963. Seed germination and seedling morphology in *Durio zibethinus*. The Malay Forester, Apr. Pp. 98-103.

Singh, H.P. and I.S. Yadav. 1980. Ways of quick multiplication of pineapple. Indian Hort. 25 (2): 7-9, 28.

Singh, I.D. and S.C. Sirohi. 1977. Sex expression studies in papaya (*Carica papaya* L.). Pantnagar J. Res. 2 (2): 150-152.

Singh, J.P., P.S. Godara and R.P. Singh. 1961. Effect of type of wood and planting dates on the rooting of phalsa (*Grewia asiatica* L.) cuttings. Indian J. Hort. 18 (1): 46-50.

Singh, J.P. and C.B.S. Rajput. 1961. Pomological descriptions and classification of some loquat varieties. Ind. J. Hort. 18 (3): 171-176.

Singh, J.P. and G.S. Randhawa. 1958. Some interesting bud variations in citron (*Citrus medica* L.). Indian J. Hort. 15 (1): 19-21.

Singh, J.P., G.S. Randhawa and C.B.S. Rajput. 1960. Effect of plant regulators on fruit development, ripening, drop, yield, size and quality of loquat (*Eriobotrya japonica* Lindl.) var. Improved Golden Yellow and Pale Yellow. Ind. J. Hort. 17 (3-4): 156-163.

Singh, J.P. and H.C. Sharma. 1961. Effect of time and severity of pruning on growth, yield and fruit quality of phalsa (*Grewia asiatica* L.). Indian J. Hort. 28 (1): 20-28.

Singh, J.R. and R.P. Srivastava. 1963. Propagation of guava by budding. Trop. Agr. (Trin.) 40 (1): 71-73.

Singh, K.K. and P.B. Mathur. 1954. A note on investigations on the cold storage and freezing of jackfruit. Indian J. Hort. 11 (4): 149.

Singh, K.K. and P.B. Mathur. 1954. Cold storage of guavas. Indian J. Hort. 11 (1): 1-5.

Singh, K.K. and P.B. Mathur. 1953. Studies in the cold storage of cashew apples. Indian J. Hort. 10 (3): 115-121.

Singh, K.K. and P.L. Sarin. 1957. Studies on the varietal differences in fruit quality in litchi (*Litchi chinensis* Sonn.). Indian J. Hort. 14 (12): 103-107.

Singh, L.B. 1960. A new technique for inducing early bearing in jackfruit (*Artocarpus heterophyllus* Lam.). Ann. Rpt. Hort. Res. Inst., Saharanpur, India. Pp. 40-46.

Singh, L.B. 1952. A new technique for propagating aonla (*Phyllanthus emblica*). Sci. & Culture 17: 345-346.

Singh, L.B. 1960. 'Sharbati'—the wonder peach (*Prunus persica* Batsch.) for the subtropics. Ann. Rpt. 1960. Hort. Res. Inst., Saharanpur, India. Pp. 47-50.

Singh, L.B. 1961. Some promising selections of bael. Ann. Rpt. Hort. Res. Inst., Saharanpur, India. Pp. 111-119.

Singh, L.B. 1961. Studies on the rootstocks for sweet orange in the wet subtropics. I. Variety 'Mosambi' (non-blood group). Hort. Adv. 5: 156-170.

Singh, L.B. and R.N. Singh. 1956. A monograph on the mangoes of Uttar Pradesh (2 vols.). Supt., Printing & Stationery, U.P., India.

Singh, L.B. and U.P. Singh. 1954. The Litchi. Supt., Printing & Stationery, U.P., India.

Singh, L.B. and R.D. Tripathi. 1960. A new method of papain production. Reprint from Ann. Rpt. Hort. Res. Inst., Saharanpur, U.P., India. Pp. 90-92.

Singh, L.B. and R.D. Tripathi. 1960. Papaw pilot scheme. Ann. Rpt. Hort. Res. Sta., Saharanpur, U.P., India. Pp. 45-48.

Singh, L.B. and R.D. Tripathi. 1957. Studies in the preparation of papain: selection of fruits for the first lancing. Indian J. Hort. 14 (2): 77-82.

Singh, L.B. and R.D. Tripathi. 1957. Studies in the preparation of papain II. Further observations on the selection of fruits for the first lancing. Indian J. Hort. 14 (3): 141-144.

Singh, L.B. and R.D. Tripathi. 1957. Studies in the preparation of papain III. Effect of alcohol, acetone and sodium-bisulphite on the quality and recovery of the enzyme. Hort. Adv. 1: 95-101.

Singh, L.B. and R.D. Tripathi. 1960. Studies in the preparation of papain. V. Effect of different concentrations of sodium-bisulphite on the quality of the enzyme. Hort. Adv. 4: 72-77.

Singh, L.B. and R.D. Tripathi. 1962. Studies in the preparation of papain. VI. Effect of high concentrations of sodium-bisulphite and different storage temperatures of the latex on the quality of the enzyme. Hort. Adv. 6: 64-70.

Singh, M.P. 1959. Studies in pectin production from lanced papaya fruits; effect of levels of maturity and lancing on pectin content of fruits. Hort. Adv. 3: 102-107.

Singh, N. and H.S. Shukla. 1978. Response of loquat (*Eriobotrya japonica* Lindl.) fruits to GA and urea. Plant Sci. 10: 77-83.

Singh, N.P. and R.P. Srivastava, 1981. Steps to stone grafting in mango. Indian Hort. 26 (1): 7-8, 24.

Singh, O.S., R.P. Gangwas and B.S. Dhillon. 1970. Regulations of fruit ripening in guava by gibberellic acid. Trop. Agriculturist (Ceylon) 126 (2): 85-89.

Singh, P., M.S. Bajwa and R. Singh. 1973. Propagation of ber (*Zizyphus mauritiana* Lam.). I. Effect of ringing and IBA on the stooling. Plant Sci. 5: 137-139.

Singh, P., C.L. Madan and B.C. Kundu. 1963. Histological study of the roots of *Carissa carandas* and *Carissa spinarum*. Lloydia 26 (1): 49-56.

Singh, R. 1957. Improvement of packing and storage of litchi at room temperature. Indian J. Hort. 14 (4): 205-212.

Singh, R. and R.C. Khanna. 1968. Some North Indian cultivars of ber. Indian Hort. 12 (2): 23-26, 36.

Singh, R.D. 1970. Studies on flowering and pollination in sapota (*Achras sapota* L.). Indian J. Sci. & Indus. 4 (2): 65-72.

Singh, R.G. and S.N. Dafe. 1971. Micronutrient deficiency in mango (*Mangifera indica* L.) and guava (*Psidium guajava* L.). Allahabad Farmer 45 (5): 479-483.

Singh, R.L. 1954. Propagation of jackfruit (*Artocarpus heterophyllus* Lam.). Ann. Rpt. Fruit Res. Sta., Saharanpur, India.

Singh, R.L. 1961. Studies on the effect of three levels of nitrogen on the growth and cropping of mandarins and sweet oranges budded on different rootstocks. III. Effect on the fruit-set and yield. Hort. Adv. 5: 176-196.

Singh, R.L. 1962. Studies on the effect of three levels of nitrogen on the growth and cropping of mandarins and sweet oranges budded on different rootstocks. IV. Effect on the quality of fruits. Hort. Advance 6: 93-109.

Singh, R.N. 1980. Mango improvement. Indian Farming 30 (8): 56-59.

Singh, R.N., P.K. Majumdar, and D.K. Sharma. 1969. New mango hybrids of greater export potential. Indian Hort. 13 (4): 3-6, 33.

Singh, R.N., P.K. Majumdar, and D.K. Sharma. 1963. Seasonal variation in the sex expression of papaya (*Carica papaya* L.). Ind. J. Agr. Sci. 33 (4): 261-267.

Singh, R.N., P.K. Majumdar and D.K. Sharma. 1961. Sex reversal in papaya (*Carica papaya* L.). Ind. J. Hort. 18 (2): 148-149.

Singh, R.N., P.K. Majumdar, D.K. Sharma, G.C. Sinha and P.C. Bose. 1974. Effect of de-blossoming on the productivity of mango. Scientia Hort. 2: 399-403.

Singh, R.N., D.K. Sharma, and P.K. Majumdar. 1980. An efficient technique of mango hybridization. Scientia Hort. 12: 299-301.

Singh, S., D.V. Chugh and K.K. Singh. 1965. Seasonal and interaction effects in marcotting of litchi (*Litchi chinensis* Sonn.) with some plant regulators. Ind. J. Agr. Sci. 35 (2): 101-113.

Singh, S., S. Krishnamurthi and S.L. Katyal. 1963. Fruit culture in India. Indian Coun. Agr. Res., New Delhi.

Singh, S., K.K. Singh and D.V. Chugh. 1961. Marcotting with some plant regulators in loquat (*Eriobotrya japonica* Lind.). Indian J. Hort. 18 (2): 123-129.

Singh, S.M. and T. Kaisuwan. 1971. Effect of interval between extraction and sowing; and container for storage on the germination of seedstones and growth of seedlings of mango (*Mangifera indica* Linn.) Allahabad Farmer 45 (3): 287-290.

Singh, S.N. 1960. Longevity of papaya (*Carica papaya* L.) pollen. Ind. J. Hort. 17 (3-5): 238-242.

Singh, S.N. 1962. Storage of litchi pollen. Hort. Adv. 6: 78-88.

Singh, S.N. 1962. Studies on the morphology and viability of the pollen grains of litchi. Hort. Adv. 6: 28-52.

Singh, S.N. 1959. Testing viability of papaya pollen in artificial medium. Hort. Adv. 3: 111-114.

Singh, S.R., C.O. Das and K.K. Srivastava. 1967. Physico-chemical studies of three species of genus *Carissa*. Allahabad Farmer 41 (2): 83-85.

Singh, U.P. 1954. A note on hybridization technique in loquat (*Eriobotrya japonica* Lindl.) Indian J. Hort. Mar.: 30-31.

Singh Dhaliwal, T. 1959. The Mysore raspberry, a new fruit for home gardens in the central-western mountainous region of Puerto Rico. J. Agr. U. Puerto Rico 43 (2): 132-138.

Singh Dhaliwal, T. and A. Torres-Sepulveda. 1962. Performance of acerola, *Malpighia punicifolia* L., in the coffee region of Puerto Rico. J. Agr. U. Puerto Rico 46 (3): 195-204.

Singmaster, J.A. III. 1970. Comparison of four chemicals to B-hydroxyethyl hydraxine (BOH) for flower initiation in pineapple. U. Puerto Rico J. Agr. 54 (1): 184-186.

Sinha-Roy, S.P. and D.P. Chakraborty. 1976. Psoralen, a powerful germination inhibitor. Phytochem. 15: 2005-2006.

Skipworth, R.G. 1944. The commercial culture of Cape gooseberries. Bull. 1255. Reprint from Rhodesia Agr. J. 61 (1): 20-22.

Smiley, N. 1951. Pineapples can be raised in backyard if you have patience and know how. Miami Herald. Sept. 2.

Smith, D. 1973. Insect pests of avocados. Queensland Agr. J. 99 (12): 645-653.

Smith, J. 1882. A dictionary of popular names of the plants which furnish the natural and acquired wants of man, in all matters of domestic and general economy. Macmillan & Co., London.

Smith, J.H.E. 1972. Mango farmers plan in good time. II. Farming in So. Africa. Mar. Pp. 18-19, 21, 23.

Smith, J.H.E. 1972. The white sapote. Leaflet 74. Subtrop. Fruit Ser. 14. White sapote Ser. #1. Dept. Agr. Tech. Serv., Pretoria, So. Africa.

Smith, K.L. 1957. Growing and preparing guavas. Bull. 74. Dept. Agr., Tallahassee, Fla.

Smith, N.F. 1965. Banana passionfruit is useful in winter. New Zeal. J. Agr. 110 (3): 272.

Smith, P.F. 1976. Collapse of 'Murcott' tangerine trees. J. Amer. Soc. Hort. Sci. 101 (1): 23-25.

Smith, R. 1983. The Indian jujube. Living Off the Land, a Subtropic Newsletter 9 (1): 7-8.

Smith, R.A. and D.B. Fitz. 1973. Mexico's changing citrus industry finds a new star in grapefruit. For. Agr. 11 (16): 2-4.

Smith, R.J. 1982. Hawaiian milk contamination creates alarm. Science 217 July 9: 137-140.

Smith, R. and S. Siwatibau. 1975. Sesquiterpene hydrocarbons of Fijian guavas. Phytochem. 14: 2013-2015.

Sobrinho, A.P. da Cunha, O.S. Passos, W.S. Soares Filho, and Y.S. Coelho. 1981. Behaviour of citrus rootstocks under tropical conditions. Proc. Internat. Soc. Citriculture. Vol. I: 123-126.

Soepadmo, E. and B.K. Eow. 1977. The reproductive biology of *Durio zibethinus* Murr. Gardener's Bull. 29: 25-33.

Soliven, F.A., P.H. Quinitio, and L.T. Gonzalez. 1961. Chemical peeling of fruits. I. Santol (*Sandoricum koetjape* (Burm. f.) Merr.). Phil. J. Sci. 90 (1): 1-8.

Soman, R. and P.P. Pillay. 1962. Isolation of crystalline vitamin from the fruits of *Emblica officinalis* Gaertn. J. Sci. Indus. Res. 21B (7): 347.

Sondheimer, F. and A. Meisels. 1958. The constituents of *Casimiroa edulis* Llave & Lex. IV. Identification of edulein with 7 methoxy-1-methyl 2-phenyl-4-quinolone. J. Org. Chem. 23: 762-763.

Sondheimer, F., A. Meisels and F.A. Kimel. 1959. Constituents of *Casimiroa edulis* Llave & Lex. V. Identity of casimirolid and obacunone. J. Org. Chem. 24: 870.

Sookmark, S. and E.A. Tai. 1975. Vegetative propagation of papaya by budding. Acta Hort. 49: 85-90.

Soost, R.K., and J.W. Cameron. 1986. 'Melogold', a new pummelo-grapefruit hybrid. Calif. Agr. 40 (1/2): 30-31.

Soost, R.K. and J.W. Cameron. 1985. 'Melogold', a triploid pummelo-grapefruit hybrid. HortSci. 20 (6): 1134-1135.

Soost, R.K. and J.W. Cameron. 1980. 'Oroblanco', a triploid pummelo-grapefruit hybrid. HortSci. 15 (5): 667-669.

Sosa, J. 1981. Efecto de la poda en setos en lima persa (*Citrus latifolia* Tan.) a los doce meses de efectuada. Cultivos Tropicales 3 (3): 3-11.

Soto Santos, D. (undated). El moko del banano. Temas Agricoles #26. Dept. de Sanidad Vegetal. Direccion de Desarrollo Agricola, Min. de Agr., Guatemala.

Southwick, S.M. and F.S. Davies. 1982. Growth regulator effects on fruit set and fruit size in Navel Orange. J. Amer. Soc. Hort. Sci. 107 (3): 395-397.

Souza-Novelo, N. 1950. Plantes alimenticias y plantas de condimento que viven en Yucatan. Inst. Tecn. Agricola Henequenero, Mérida, Yucatan.

Spalding, D.H., R.J. Knight and W.F. Reeder. 1976. Storage of avocado seeds. Proc. Fl. St. Hort. Soc. 89: 257-258.

Spalding, D.H. and W.F. Reeder. 1977. Low pressure (hypobaric) storage of mangos. J. Amer. Soc. Hort. Sci. 102 (3): 367-369.

Spruce, R. 1908. Notes of a botanist on the Amazon and Andes, edited and condensed by Alfred Russell Wallace (2 vols.). Macmillan & Co., Ltd., London.

Spurgin, M.M. 1964. Vinegar base production from waste pineapple juice. Queensland J. Agr. Sci. 21 (2): 213-232.

Spurlock, D.H. 1960. Fruit shedding of Japanese Persimmon. So. Florist and Nurseryman. June 24. Pp. 67-68.

Sridhar, T.S. and B.A. Ullasa. 1978. Leaf blight of guava—a new record. Curr. Sci. 47 (12): 442.

Srinivasan, C., C.M. Pappiah and A. Doraipandian. 1973. Effect of gibberellic acid on ascorbic acid, sugar content and oxidative enzyme activity of West Indian cherry (*Malpighia punicifolia* L.) fruit. Ind. J. Exper. Biol. 11: 469-470.

Srinivasa Rao, N.K., S. Narayanaswamy, E.K. Chacko and R. Dore Swamy. 1981. Tissue culture of the jack tree. Curr. Sci. 50 (7): 310-312.

Srirangarajan, A.N. and A.J. Shrikhande. 1976. Mango peel waste as a source of pectin. Curr. Sci. 45 (17): 620-621.

Srivas, S.R., N.V. Subhadra, V.L. Sastry and G.S. Siddappa. 1965. Studies on the preservation of woodapple (*Feronia elephantum* Corr.). Pt. I. Physico-chemical composition of the fruit and seed. Indian Food Packer 19 (6): 5-11.

Srivastava, M.P. and R.N. Tandon. 1971. Post-harvest diseases of papaya. PANS 17 (1): 51-54.

Srivastava, R.P. and A. Verghese. 1983. Mango leaf-webber and its control. Indian Hort. 28 (2): 21-22.

Stahl, A.L. 1935. Composition of miscellaneous tropical and subtropical Florida fruits. Bull. 283. U. of Fla. Agr. Exper. Sta., Gainesville, Fla.

Stambaugh, S.U. 1955. New practices in growing pineapples in Florida. Fla. Grower & Rancher. Feb.: 14, 31-32.

Stambaugh, S.U. 1955. Old pineapple roots sliced into 1-inch discs grow healthy plants. Fla. Grower & Rancher. Dec.: 11-12, 14.

Stambaugh, S.U. 1931. Planting papaya in fall months. Fla. Grower (June): 20.

Stambaugh, S.U. 1945. The papaya. A fruit suitable for South Florida. New Ser. #90. State of Fla., Dept. of Agr., Tallahassee.

Standley, P.C. 1938. Flora of Costa Rica. Pt. III. Pub. 420. Bot. Ser. Vol. 18. Field Mus. Nat. Hist., Chicago.

Standley, P.C. 1938. Flora of Costa Rica, Pt. IV. Pub. 429. Bot. Ser. Vol. 18 Field Mus. Nat. Hist., Chicago.

Standley, P.C. 1931. Flora of the Lancetilla Valley, Honduras. Pub. 283. Bot. Ser. Vol. 10. Field Mus. Nat. Hist., Chicago.

Standley, P.C. 1928. Flora of the Panama Canal Zone. Contrib. U.S. Nat. Herb. Vol. 27. Smithsonian Inst., U.S. Nat. Mus., Washington, D.C.

Standley, P.C. 1930. Flora of Yucatan. Pub. 279. Bot. Ser. Vol. 3, No. 3. Field Mus. Nat. Hist., Chicago.

Standley, P.C. 1927. Six new trees from British Honduras and Guatemala. Trop. Woods #11: 18-22.

Standley, P.C. 1933. The flora of Barro Colorado Island, Panama. Arnold Arboretum, Harvard U., Jamaica Plain, Mass.

Standley, P.C. 1920-1926. Trees and shrubs of Mexico, Vol. 23, Pts. 1-5. Smithsonian Inst., U.S. Nat. Herb., Washington, D.C.

Standley, P.C. and S.J. Record. 1936. The forest and flora of British Honduras. Pub. 350. Vol. 12. Field Mus. Nat. Hist., Chicago.

Standley, P.C. and J.A. Steyermark. 1958. Flora of Guatemala. Vol. 24, Pt. 1. Chicago Nat. Hist., Museum, Chicago.

Standley, P.C. and J.A. Steyermark. 1946. Flora of Guatemala. Fieldiana: Botany, Vol. 24, Pt. 4. Pub. 577. Chicago Nat. Hist. Mus., Chicago.

Standley, P.C. and L.O. Williams. 1961. Flora of Guatemala. Fieldiana: Botany. Vol. 24, Pt. 7, No. 1. Chicago Nat. Hist. Mus., Chicago.

Standley, P.C. and L.O. Williams. 1967. Flora of Guatemala. Vol. 24, Pt. 8, No. 3. Field Mus. Nat. Hist., Chicago.

Stanton, W.R. 1966. The chemical composition of some tropical food plants: VI. Durian. Trop. Sci. 8 (1): 6-10.

Steinmetz, E.F. 1960. A botanical drug from the tropics used in the treatment of diabetes mellitus [jambolan]. Acta Phytotherapeutica 7 (2): 23-25.

Steinmetz, E.F. 1959. Drug Guide. Pub'd by author, Amsterdam.

Stephens, S.E. 1963. Mango varieties in tropical Queensland. Queensland J. Agr. 89 (8): 455-458.

Stephens, S.E. 1954. Rough leaf pineapples. Queensland Fr. & Veg. News 15 (6): 178-179.

Stephens, S.E. 1937. Some tropical fruits: No. 16. The granadilla. Reprint from Queensland Agr. J. Aug.

Stephens, S.E. 1935. Some tropical fruits: 1. The mangosteen. Queensland Agr. J. Sept.: 346-348.

Stephens, S.E. 1937. Some tropical fruits. No. 17. The rose apple. Queensland Agr. J. Dec.

Stephens, S.E. 1936. Some tropical fruits. No. 11: The soursop. Queensland Agr. J. Sept.: 409-412.

Stephens, S.E. 1956. The granadilla. Queensland Fr. & Veg. News 10 (20): 614-615.

Stephens, S.E. 1936. The jackfruit. Queensland Agr. J.: 67-68.

Stewart, I., G.D. Bridges, A.P. Pieringer and T.A. Wheaton. 1975. 'Rhode Red Valencia', an orange selection with improved juice color. Proc. Fl. St. Hort. Soc. 88: 17-19.

Stone, B.C. 1970-71. The flora of Guam. Micronesica: J. of U. of Guam. Vol. 6 (complete).

Storey, R.I. and D.J. Rogers. 1980. Lepidopterous pests of the litchi in North Queensland. Queensland J. Agr. & Anim. Sci. 37 (2): 207-212.

Storey, W.B., R.A. Hamilton and H.Y. Nakasone. 1953. Groff—a new variety of lychee. Cir. 39. Agr. Exper. Sta., U. Hawaii, Honolulu.

Story, G.E. and R.S. Halliwell. 1969. Identification of distortion ringspot virus disease of papaya in the Dominican Republic. Pl. Dis. Rep. 53 (9): 757-760.

Stother, J. 1971. The market for fresh mangoes in selected Western European countries. TPI Rpt. G59. Trop. Prod. Inst., London.

Stother, J. 1970. World production and trade in fresh grapefruit. Pub. G48. Trop. Prod. Inst., London.

Strain, M.B. 1966. Preserving tree tomatoes. New Zeal. J. Agr. 112 (6): 81.

Streets, R.J. 1962. Exotic forest trees in the British Commonwealth. Clarendon Press, Oxford.

Stuart, G.A. 1911. Chinese Materia Medica: vegetable kingdom. Amer. Presbyterian Mission Press, Shanghai.

Sturrock, D. 1959. Fruits for southern Florida. Southeastern Printing Co., Stuart, Fla.

Sturrock, D. 1948. The karanda as a commercial fruit. Proc. Fl. St. Hort. Soc. 61: 289-291.

Sturrock, D. 1940. Tropical fruits for southern Florida and Cuba and their uses. Harvard U., Arnold Arboretum, Jamaica Plain, Mass.

Sturtevant, E.L. 1919. Sturtevant's Notes on Edible Plants (edited by U.P. Hedrick). N.Y. Agr. Exper. Sta., Geneva, N.Y.

Subbarayan, C. and H.R. Cama. 1964. Carotenoids in *Carica papaya* (papaya fruit). Indian J. Chem. 2 (11): 451-454.

Subhadra, N.V., S.R. Srivas, V.L. Sastry and G.S. Siddappa. 1965. Studies on the preservation of woodapple (*Feronia elephantum*). Pt. II. Preservation of pulp by canning, drying and conversion into nectarlike products. Indian Food Packer 19 (6): 12-16.

Subrahmanyan, V., G.S. Sidappa and N.V.R. Iyengar. 1963. Utilization of cellulosic agricultural wastes: pulp from banana pseudostem and areca husk. Indian Pulp and Paper 17: 1.

Suguira, A. and T. Tomans. 1983. Relationships of ethanol production by seeds of different types of Japanese persimmons and their tannin content. HortSci. 18 (3): 319-321.

Sulladmath, V.V. and C.P.A. Iyer. 1982. Preliminary evaluation of some lemon (*Citrus limon* Burm.) cultivars for growth, yield and quality. So. Indian Hort. 30 (2): 69-71.

Swaine, G. 1971. Banana pests in South Queensland. Queensland Agr. J. 97 (1): 31-34.

Swaine, G. 1970. Corky scab of Cavendish banana. Queensland Agr. J. 96 (7): 474-475.

Swaine, G. and R.J. Corcoran. 1971. Skin blemishes of banana fruit. Queensland Agr. J. 97 (8): 427-430.

Swarts, D.H. and T. Anderson. 1980. Chemical control of mould growth on litchis during storage and sea shipment. Info. Bull. #98. Citrus and Subtrop. Fr. Res. Inst., Nelspruit, So. Africa. Pp. 13-15.

Swingle, W.T. 1905. A new genus, Fortunella, comprising four species of kumquat oranges. J. Washington Acad. Sci. 5 (5): 165-176.

Swingle, W.T. 1904. The date palm and its utilization in the southwestern states. Bull. 53. U.S. Dept. Agr., Bur. of Pl. Indus., Washington, D.C.

Sydenham, F. 1943. Tree-tomato culture. New Zeal. J. Agr. Feb. 15: 93-94.

Tadeo, C.B. 1962. Plants with edible fruit in the Philippines. Forestry Leaves 14 (1/2): 55-64.

Tadros, M.R., A.S. Khalifa and A. Bondok. 1983. Gouging tool for desuckering banana plants. Ann. Agr. Sci., Fac. Agric. Ain Shama U., Cairo 28 (1): 249-259.

Tai Luang Huan. 1971. Studies on Phytophthora palmivora, the causal organism of patch canker disease of durian. Malaysian Agr. J. 48 (1): 1-9.

Tai Luang Huan. 1973. Susceptibility of durian clones to patch canker disease. MARDI Res. Bull. 1 (2): 5-9.

Tainter, M.L., C.E. Alford, A. Arnold, H. Blumberg, O.H. Buchanan, T.T. Hinkel, Jr., K. Hwang, A.C. Ivy, R.K. Lager, J.R. Schmitz, J.R. Shepherd, H.S. Wyzan and C. Zippin. 1951 Papain. Ann. N.Y. Acad. Sci. 54 (2): 143-296.

Takata, R.H. and P.J. Scheuer. 1976. Isolation of glyceryl esters of caffeic and p-coumaric acids from pineapple stems. Lloydia 39 (6): 409-411.

Talapatra, S.K., M.K. Chaudhuri and B. Talapatra. 1973. Coumarins of the root bark of Feronia elephantum. Phytochem. 12: 236-237.

Tallent, W.H. 1964. Two new antibiotic cyclopentoid monoterpenes of plant origin. Tetrahedron 20 (7): 1781-1787.

Tamir, M., E. Nachtomi and E. Alumot. 1972. Urinary phenolic metabolites of Rats fed carobs (Ceratonia siliqua) and carob fractions. Internat. J. Biochem. 3: 123.

Tanchico, S.S. and C.R. Magpanlay. 1958. Analysis and composition of Artocarpus integra latex. Phil. J. Sci.: 149-157.

Tandon, G.L. 1950. Preserves and their manufacture [emblic]. The Indian Food Packer 4: 9-12 and 25-28.

Tandon, I.N. 1960. Diplodia collar rot and root rot of loquat. Hort. Adv. 4: 115-121.

Tandon, I.N. 1959. Studies in the control of damping-off of papaya (Carica papaya L.). Hort. Adv. 3 (115-122).

Tandon, I.N. 1961. Studies on the control of root rot of papaya. Ind. J. Hort. 18 (3): 223-225.

Tandon, P.L. and R.P. Srivastava. 1980. Comparative toxicity of different insecticides to mango scale Pulvinaria polygonata Cockerell. Internat. Pest Control Nov./Dec. 158-159.

Tang, Chung-Shih. 1979. New macrocyclic Δ'-piperideine alkaloids from papaya leaves: dehydrocarpaine. I & II. Phytochem. 18: 651-652

Tarr, P. 1983. A spirit of defiance moves across the southern Philippines. Christian Sci. Monitor. Dec. 30: 7-8.

Tay, T.H. and Y.C. Wee. 1976. Comparative study on the effects of different types and sizes of planting materials on pineapple yield and quality. Malaysian Agr. J. 50 (4): 502-506.

Taylor, P.V. 1962. Another fruiting indoor ornamental ready for market. So. Florist & Nurseryman. Mar. 30. Pp. 43-44.

Taylor, P.V. 1961. Citrus goes ornamental [calamondin]. So. Florist and Nurseryman. Mar. 17: 62-63.

Teakle, D.S. and G.B. Wildermuth. 1967. Host range and particle length of passionfruit woodiness virus. Bull. 411. Queensland J. Agr. & Anim. Sci. 24 (2): 173-186.

Teaotia, S.S. and R.K. Awasthi. 1967. Studies on varietal suitability of guava fruit by canning. Indian Food Packer 21 (2): 28-33.

Teaotia, S.S. and S. Bhan. 1966. Determination of maturity for harvesting pineapple fruit (Ananas comosus (L.) Merr.) variety Giant Kew. Indian Agriculturist 10 (2): 107-112.

Teaotia, S.S. and R.S. Chauhan. 1963. Flowering, pollination, fruit-set and fruit-drop studies in ber (Zizyphus mauritiana Lamk.). I. Floral biology. Punjab Hort. J. III (1): 58-70.

Teaotia, S.S., I.C. Pandey, B.N. Agnihotri and K.L. Kapur. 1962. Study of some guava varieties (Psidium guajava L.) of Uttar Pradesh. Indian Agr. 6 (1-2): 47-55.

Teaotia, S.S., R.S. Tripathi and R.N. Singh. 1972. Effect of growth substances on ripening and quality of guava (Psidium guajava L.). J. Food Sci. & Tech. 9 (1): 38-39.

Tedder, J.L.O. 1956. Breadfruit drying in the reef islands. So. Pacific Comm. Quarterly 6 (3): 21-22.

Terra, G.J.A. 1966. Tropical Vegetables: Vegetable growing in the tropics and subtropics, especially of indigenous vegetables. Comm. 54e. Royal Trop. Inst. & Neth. Org. for Internat. Assist, Amsterdam.

Terry, R.M. 1975. Use of plant growth regulators in Hawaii on pineapple. Misc. Pub. #123. U. Hawaii Coop. Exten. Serv.: 37-39.

Thomas, C.A. 1979. Jackfruit, Artocarpus heterophyllus (Moraceae), as source of food and income. Econ. Bot. 34 (2): 154-159.

Thomas, C.A. 1969. You can grow breadfruit in your garden. Indian Hort. 13 (4): 27.

Thomas, W. and C.H. Procter. 1972. Arabic mosaic virus in Cyphomandra betacea Sendt. New Zeal. J. Agr. Res. 15 (2): 395-404.

Thompson, A.K., Y.M. Abdulla and H. Silvis. 1974. Preliminary investigations into dessication and degreening of limes for export. Sudan J. Food Sci.& Tech. 6: 1-6.

Thompson, A.K., B.O. Been and C. Perkins. 1974. Storage of fresh breadfruit. Trop. Agr. 51 (3): 407-415.

Thorpe, P. 1973. Giving a lift to a lamb. New Zeal. J. Agric. 127 (4): 54. [tree tomato].

Ting, S.V., R.L. Huggart, and M.A. Ismail. 1980. Color and processing characteristics of 'Star Ruby' grapefruit. Proc. Fl. St. Hort. Soc. 93: 293-295.

Ting Wen Poh and Tai Luang Huan. 1971. Occurrence of Cercospora leaf spot of star fruit (Averrhoa carambola L.) in Selangor. Malaysian Agr. J. 148 (1): 25-27.

Tingyu, Q. 1983. Records of the larvae of longicorns injurious to the stem of jackfruit. Chinese J. of Trop. Crops 4 (1): 103-106.

Tirmazi, S.I.H. and R.B.H. Wills. 1981. Retardation of ripening of mangoes by postharvest application of calcium. Trop. Agr. 58 (2): 137-141.

Tisserat, B. 1984. Propagation of date palms by shoot tip cultures. HortSci. 19 (2): 230-231.

Toll Jubes, J. 1964. El cultivo de la cherimoya in Tucuman y sus posibilidades. Cir. 173. Estac. Exper. Agr. de Tucuman, Argentina.

Torne, S.G. and N.P. Raut Desai. 1975. Effect of ionizing radiation on seed germination of Passiflora species. Curr. Sci. 44 (4): 112-113.

Torres, J.P. and P.B. Dionido. 1962. Notes on the selection and commercial propagation of chico [sapodilla]. Coffee & Cacao J. 5 (10): 214-215.

Torres, J.P., P.B. Dionido and A. Zamora. 1962. Selection and propagation of rambutan in Oriental Mindoro, Philippines. Araneta J. Agr. 9 (2): 146-160.

Torres M., R. and D.C. Giacometti. 1966. Comportamiento del maracuja (Passiflora edulis var. flavicarpa) bajo las condiciones del Valle del Cauca. Agr. Trop. 22 (5): 247-254.

Torres M., R. and D.C. Giacometti. 1965. Virosis de la papaya (Carica papaya L.) en el Valle del Cauca. Agr. Trop. 22 (1): 27-38.

Torres, M., R. and D. Rios Castaño. 1968. Bases para un programa de mejoramiento de papaya (Carica papaya L.) in Colombia. Agr. Trop. 24 (2): 107-112.

Toutain, G. 1973. Productions du palmier dattier (Parts 1 & 2). Al Awamia 48: 73-88.

Traub, H.P., T.R. Robinson and H.E. Stevens. 1942. Papaya production in the United States. Div. of Fr. & Veg. Crops and Diseases, Bur. of Pl. Indus., U.S. Dept. Agr., Washington, D.C.

Treakle, H.C. 1965. Iraq's date industry—biggest export crop—biggest source of farm income. For. Agr. 3 (48): 7-8.

Trejo, J.A. and J.D. Interiano. 1969. Plagas y enfermedades del banano. Cir. 86. Min. de Agr. y Ganaderia, Santa Tecla, El Salvador.

Trochoulias, T. 1976. Girdling of 'Fuerte' avocado in subtropical Australia. Scientia Hort. 5: 239-242.

Trout, S.A. 1966. Dr. Trout's Notebook: the avocado. Queensland Fr. & Veg. News. Oct. 20: 392-393.

Trout, S.A. 1966. Of mangoes, nuts and pines and papayas. Queensland Fr. & Veg. News. Oct. 13. p. 361.

Trout, S.A. 1977. Queensland horticultural exports to Hawaii. Queensland Agr. J. 103 (6): 578.

Trupin, F. 1966. Bilan de deux campagnes d' exportation de litchis de Madagascar. Fruits 21 (9): 495-500.

Turnbull, J. 1951. Odd-plant fanciers admire "Tree Tomato". Floriland (magazine). Sept.

Tyler, V.C., L.R. Brady and J.E. Robbers. 1981. Pharmacognosy. 8th ed. Lea & Febiger, Phila.

Udanga, L.E. 1980. Datiles for reforestation? Canopy. May.

Udo, H. 1981. Evaluation of breadfruit as a possible energy source for poultry diets in W. Samoa. Alafua Agr. Bull. 6 (3): 49-54.

Ugas, C. 1971. La piña. Indus. Alimenticia 2 (4): 48-59.

Ullasa, B.A. and H.S. Sohi. 1975. A new phytophthora leaf blight and damping off disease of passion fruit from India. Curr. Sci. 44 (16): 593-594.

United States Dept. Agr., 1906. Yearbook - 1905. Supt. Doc., Washington, D.C.

United States Dept. Agr., Bur. Pl. Indus. 1899. Foreign seeds and plants imported. Inventory No. 7 [*Byrsonima crassifolia*] Off. For. Seed & Pl. Intro., USDA, Washington, D.C.

United States Dept. Agr., Bur. Pl. Indus. 1909. Inventory of seeds and plants imported Oct. 1 to Dec. 31, 1909. Bull. 205 [ambarella]. Off. For. Seed & Pl. Intro., USDA, Washington, D.C.

United States Dept. Agr., Bur. Pl. Indus. 1911. Inventory of seeds and plants imported Apr. 1 to June, 1911. Bull. 242. [ambarella]. Off. For. Seed & Pl. Intro., USDA, Washington, D.C.

United States Dept. Agr., Bur. Pl. Indus. 1911. Inventory of seeds and plants imported. No. 20. Bull. 248. [jambolan]. Off. For. Seed & Pl. Intro., USDA, Washington, D.C.

United States Dept. Agr., Bur. Pl. Indus. 1912. Inventory of seeds and plants imported Oct. 1-Dec. 31, 1912. No. 33. [jambolan]. Off. For. Seed & Pl. Intro., USDA, Washington, D.C.

United States Dept. Agr., Bur. Pl. Indus. 1920. Inventory of seeds and plants imported June 1-Sept. 30, 1920. No. 64 [jambolan]. Off. For. Seed & Pl. Intro., USDA, Washington, D.C.

United States Dept. Agr., Bur. Pl. Indus. 1913. New Plant Immigrants. Bull. of For. Pl. Intro. No. 84. [sansapote; tree tomato]. Off. For. Seed & Pl. Intro., USDA, Washington, D.C.

United States Dept. Agr., Bur. Pl. Indus. 1913. Plant Immigrants. No. 91 [sansapote]. Off. For. Seed & Pl. Intro., USDA, Washington, D.C.

United States Dept. Agr., Bur. Pl. Indus. 1913. Plant Immigrants. No. 92. [jaboticaba]. Off. For. Seed & Pl. Intro., USDA, Washington, D.C.

United States Dept. Agr., Bur. Pl. Indus. 1914. Plant Immigrants. No. 102. [sapote]. Off. For. Seed & Pl. Intro., USDA, Washington, D.C.

United States Dept. Agr., Bur. Pl. Indus. 1914. Plant Immigrants. No. 93 [jaboticaba]. Off. For. Seed & Pl. Intro., USDA, Washington, D.C.

United States Dept. Agr., Bur. Pl. Indus. 1914. Plant Immigrants. No. 104. [ilama] Off. For. Seed & Pl. Intro., USDA, Washington, D.C.

United States Dept. Agr., Bur. Pl. Indus. 1915. Plant Immigrants. No. 105. [atemoya]. Off. For. Seed & Pl. Intro., USDA, Washington, D.C.

United States Dept. Agr., Bur. Pl. Indus. 1915. Plant Immigrants. No. 110. [passion fruit]. Off. For. Seed & Pl. Intro., USDA, Washington, D.C.

United States Dept. Agr., Bur. Pl. Indus. 1916. Plant Immigrants. No. 120. [atemoya]. Off. For. Seed & Pl. Intro., USDA, Washington, D.C.

United States Dept. Agr., Bur. Pl. Indus. 1916. Plant Immigrants. No. 122. [Japanese persimmon; feijoa; sansapote]. Off. For. Seed & Pl. Intro., USDA, Washington, D.C.

United States Dept. Agr., Bur. Pl. Indus. 1916. Plant Immigrants. No. 125. [atemoya]. Off. For. Seed & Pl. Intro., USDA, Washington, D.C.

United States Dept. Agr., Bur. Pl. Indus. 1916. Plant Immigrants. No. 128. [green sapote]. Off. For. Seed & Pl. Intro., USDA, Washington, D.C.

United States Dept. Agr., Bur. Pl. Indus. 1917. Plant Immigrants. No. 132. [atemoya; Barbados cherry]. Off. For. Seed & Pl. Intro., USDA, Washington, D.C.

United States Dept. Agr., Bur. Pl. Indus. 1917. Plant Immigrants. No. 137. [atemoya; tree tomato]. Off. For. Seed & Pl. Intro., USDA, Washington, D.C.

United States Dept. Agr., Bur. Pl. Indus. 1917. Plant Immigrants. No. 138. [atemoya]. Off. For. Seed & Pl. Intro., USDA, Washington, D.C.

United States Dept. Agr., Bur. Pl. Indus. 1917. Plant Immigrants. Bull. 140. [ilama]. Off. For. Seed & Pl. Intro., USDA, Washington, D.C.

United States Dept. Agr., Bur. Pl. Indus. 1918. Plant Immigrants. Bull. 141. [feijoa; cassabanana]. Off. For. Seed & Pl. Intro., USDA, Washington, D.C.

United States Dept. Agr., Bur. Pl. Indus. 1918. Plant Immigrants. No. 142. [atemoya; soncoya]. Off. For. Seed & Pl. Intro., USDA, Washington, D.C.

United States Dept. Agr., Bur. Pl. Indus. 1918. Plant Immigrants. No. 143. [jaboticaba]. Off. For. Seed & Pl. Intro., USDA, Washington, D.C.

United States Dept. Agr., Bur. Pl. Indus. 1918. Plant Immigrants. No. 144. [kiwifruit; carob]. Off. For. Seed & Pl. Intro., USDA, Washington, D.C.

United States Dept. Agr., Bur. Pl. Indus. 1918. Plant Immigrants. No. 148. [sapote]. Off. For. Seed & Pl. Intro., USDA, Washington, D.C.

United States Dept. Agr., Bur. Pl. Indus. 1919. Plant Immigrants. No. 154. [ilama]. Off. For. Seed & Pl. Intro., USDA, Washington, D.C.

United States Dept. Agr., Bur. Pl. Indus. 1919. Plant Immigrants. No. 158. [mamey]. Off. For. Seed & Pl. Intro., USDA, Washington, D.C.

United States Dept. Agr., Bur. Pl. Indus. 1919. Plant Immigrants. No. 159. [sapote]. Off. For. Seed & Pl. Intro., USDA, Washington, D.C.

United States Dept. Agr., Bur. Pl. Indus. 1919. Plant Immigrants. No. 162. [sapote]. Off. For. Seed & Pl. Intro., USDA, Washington, D.C.

United States Dept. Agr., Bur. Pl. Indus. 1920. Plant Immigrants. No. 166. [ilama]. Off. For. Seed & Pl. Intro., USDA, Washington, D.C.

United States Dept. Agr., Bur. Pl. Indus. 1920. Plant Immigrants. No. 174. [jambolan]. Off. For. Seed & Pl. Intro., USDA, Washington, D.C.

United States Dept. Agr., Bur. Pl. Indus. 1921. Plant Immigrants. No. 179. [madroño]. Off. For. Seed & Pl. Intro., USDA, Washington, D.C.

United States Dept. Agr., Bur. Pl. Indus. 1921. Plant Immigrants. No. 185. [avocado; mabolo]. Off. For. Seed & Pl. Intro., USDA, Washington, D.C.

United States Dept. Agr., Bur. Pl. Indus. 1922. Plant Immigrants. No. 198. [ilama]. Off. For. Seed & Pl. Intro., USDA, Washington, D.C.

United States Dep. Agr., Bur. Pl. Indus. 1923. Plant Immigrants. No. 212. [ilama]. Off. For. Seed & Pl. Intro., USDA, Washington, D.C.

United States Dept. Agr., Bur. Pl. Indus. 1929. Plant Material Introduced by Off. For. Pl. Intro. Apr. 1-June 30, 1926. Inventory No. 87. [tree tomato]. U.S. Dept. Agr., Washington, D.C.

United States Dept. Agr., Bur. Pl. Indus. 1907. Seeds and plants imported during the period from Dec. 1903 to Dec. 1905. Inventory No. 11. Bull. 97. [mango; sweet lime]. U.S. Dept. Agr., Washington, D.C.

United States Dept. Agr., Bur. Pl. Indus. 1907. Seeds and plants imported during the period from Dec. 1905 to July 1906. Inventory No. 12, Bull. 106. [mangosteen, rambutan, star apple]. U.S. Dept. Agr., Washington, D.C.

United States Dept. Agr., Bur. Pl. Indus. 1908. Seeds and plants imported during the period from July 1906 to Dec. 31, 1907. Inventory No. 13. Bull. 132. [avocado, breadfruit, cherimoya]. U.S. Dept. Agr., Washington, D.C.

United States Dept. Agr., Bur. Pl. Indus. 1909. Seeds and plants imported during the period from Jan. 1 to Mar. 31, 1908. Inventory No. 14. Bull. 137. [kiwifruit]. U.S. Dept. Agr., Washington, D.C.

United States Dept. Agr., Bur. Pl. Indus. 1909. Seeds and plants imported during the period of Oct. 1 to Dec. 31, 1909. Inventory No. 21. Bull. 205.

United States Dept. Agr., Bur. Pl. Indus. 1923. Seeds and plants imported—June 1 to Sept. 30, 1920. Inventory No. 64. [pejibaye]. U.S. Dept. Agr., Washington, D.C.

Upadhya, M.D. and J.L. Brewbaker. 1966. Irradiation of mangoes for control of the mango seed weevil. Hawaii Farm Sci. 15 (1).

Vagholkar, B.P. 1916. Notes on propagation of the bor-tree in East Khandesh. Poona Agr. Coll. Mag. 8 (2): 1-4.

Valdez Verduzco, J. 1979. El peso del cormo en la propagacion y el desarrollo y rendimiento del platano cv. Giant Cavendish. Proc. Trop. Reg. Amer. Soc. Hort. Sci. 23: 173-175.

Valle, N. Del, O. Herrera, and A. Rios. 1982. The influence of rootstocks on the performance of 'Valencia' oranges under tropical conditions. Proc. Internat. Soc. Citriculture 1: 134-137.

Valmayor, R.V., R.E. Coronel and D.A. Ramirez. 1965. Studies in floral biology, fruit set and fruit development in durian. Phil. Agr. 48 (8-9): 355-366.

Valmayor, R.V., D.B. Mendoza, H.B. Aycardo and C.O. Palencia. 1971. Growth and flowering habits, floral biology and yield of rambutan (*Nephelium lappaceum* Linn.). Phil. Agriculturist 54 (7 & 8): 359-374.

Vanderweyen, A. 1980. Influence de la variete d' agrumes sur la sensibilite du porte-greffe a la gommose a phytophora [Volkamer lemon]. Al Awamia 60: 67-80.

Van Doesburg, P.H. 1964. Two insect pests of soursop in Surinam. Carib. Agr. 3 (1): 797-803.

Van Overbeek, J. 1945. Flower formation in the pineapple plant as controlled by 2,4-D and Naphthaleneacetic acid. Science 102: 621.

Vargas Hernandez, D. and R. Pedraza Gonzalez. 1961. Harvesting and marketing of grapefruit in Puerto Rico. 1957-58. Bull. 159. U. Puerto Rico, Agr. Exper. Sta., Rio Piedras, P.R.

Varma, U.P. and S. Ahmad. 1958. Drying bael (*Aegle marmelos* Correa) pulp. Indian Food Packer 12 (7): 7-9.

Varner, D.M. 1966. Slow, steady progress with food irradiation. Food Engrng. 38 (11): 111-113.

Velasco M., N.I., J.G. Bautista, L.N. Cruz, E.C. Gañac, V.G. Silverio and D.C. Avante. 1983. A study of the utilization of banana meal as replacement for ground corn in poultry rations. Phil. J. Anim.

Velazquez Diaz, J. 1953. El chirimoya en la costa granadina. Agr. Rev. Agropecuaria 25 (6): 432-434.

Velez Ramos, A. and J.A. Vega Lopez. 1977. Chemical control of weeds in plantains. J. Agr. U. Puerto Rico 61 (2): 259-261.

Veltkamp, H.J. 1977. Plantain growing in Surinam—a survey. De Surinaamse Landbouw 25 (1): 20-34.

Venkataraman, K. 1966. Spectroscopic methods of structure determination: applications to natural phenolic pigments [jackfruit]. J. Sci. & Indus. Res. 25 (3): 97-118.

Venkataratnam, L. and G. Satyanarayanaswamy. 1958. Studies on genetic variability in *Annona squamosa* L. In: Proc. Internat. Symp. on Orig., Cytogen. and Breeding of Trop. Fruits. Indian J. Hort. 15 (3/4): 228-238.

Venkataratnam, L. and G. Satyamaranaswamy. 1956. Vegetative propagation of sitaphal (*Annona squamosa* Linn.). Indian J. Hort. 13 (2): 80-101.

Venkatesh, A.S.G. and Y.K. Raghunatha Rao. 1960. Minor oils— papaya seed oil. Food Sci. 9 (2): 49.

Verdoorn, I.C. 1938. Edible wild fruits of the Transvaal. Bull. 85. Pl. Indus. Ser. 29. [cape gooseberry]. Dept. Agr. & For., Pretoria, So. Africa.

Verma, A.N., D.C. Srivastava, K. Ram and R.K. Sharma. 1970. Effect of plant regulator on air layering in mango (*Mangifera indica* L.), guava (*Psidium guajava*-L.), and Kagzi lime (*Citrus aurantifolia* Swingle). Allahabad Farmer 44 (3): 139-142.

Verma, K.S., S.S. Cheema and P.S. Bedi. 1983. Prevalence of ber rust (*Phakopsora zizyphi-vulgaris*) in Punjab and its control. J. Res. Punjab Agr. Univ. 20 (1): 43-46.

Verma, S.U.P. 1952. Loss of vitamin 'C' and pectin during 6 months

of storage guava extract (with water). Indian Food Packer. Apr. 5-6, 43-44.

Verma, U.P. 1950. Preservation of jackfruit. Indian Food Packer 4 (4): 15-16.

Verma, U.P. 1956. Properties and therapeutic value of papain. 1956. Indian Food Packer 10 (2): 9-26.

Verma, U.P. and S. Ahmad. 1957. Canning of litchi. Indian Food Packer 11 (2): 7-10.

Verrill, A. Hyatt. 1937. Foods America gave the world. L.C. Page & Co., Boston, Mass.

Viennot-Bourgin, G. and A. Comelli. 1959. Le Cercospora du mangier au Congo. Fruits 14 (6).

Villar, A., M. Mares, J.L. Rios and D. Cortes. 1985. Alkaloids from *Annona cherimola* leaves. J. Nat. Prod. 48 (1): 151-152.

Vines, H.M. and W. Grierson. 1966. Handling and physiological studies with the carambola. Fl. St. Hort. Soc. 79: 350-355.

Virmani, R.S. 1950. Jaman vinegar and its composition. Indian Food Packer 4 (1): 13, 14.

Vock, N.T. 1976. Diseases of bananas. Queensland Agr. J. 102 (3): back cover, both sides.

Vock, N.T. 1974. Top and root rot of pineapples; recognition and spread. Queensland Agr. J. 100 (12): back cover, both sides.

Vock, N.T. 1975. Top and root rot of pineapples: control. Queensland Agr. J. 101 (1): back cover, both sides.

Voltz, J. 1985. Guavas? You can prepare them many ways. The Miami Herald. July 27. P. 6-E.

Voltz, J. 1957. Meet Platts, the pineapple king. Miami Herald. Oct. 20: 14-E.

von der Pahlen, A. (undated). Cubiu [*Solanum topiro* (Humb. & Bonpl.)], uma fruteira da Amazonia. Acta Amazonica 7 (3): 301-307.

von Windeguth, D.L. 1984. Low dose ethylene dibromide fumigation for quarantine control of Caribbean fruit fly in grapefruit. Proc. Fl. St. Hort. Soc. 97: 120-122.

Vosbury, E.D. 1921 (reprint 1938). Pineapple culture. Reprint from USDA Farmers' Bull. 1237. Fla. St. Dept. Agr., Tallahassee.

Vosters, J.B. 1982. A pineapple in every home. Fairchild Tropical Garden Bull. 37 (2): 14-17.

Vyas, N.L. and K.S. Panwar. 1976. A new post-harvest disease of pomegranate in India. Curr. Sci. 45 (2): 76.

Wagner, C.J., W.L. Bryan, R.E. Berry and R.J. Knight, Jr. 1975. Carambola selection for commercial production. Proc. Fl. St. Hort. Soc. 88: 466-469.

Waithaka, J.H.G. and D.K. Puri. 1971. Recent research on pineapple in Kenya. World Crops July/Aug.: 190-192.

Walker, A.R. and R. Sillans. 1961. Les plantes utiles du Gabon. Encyc. Biolog. Editions Paul Le Chevalier, Paris.

Walker, W.O. and F.P. Cullinam. 1955. Notice to fruit growers and nurserymen relative to the naming and release of a new variety of lychee called Bengal, a variety adapted to soils of high pH and producing large clusters of fruit. (Mimeo.). U.S. Dept. Agr. Hort. Crops Res. Branch, Beltsville, Md. and Div. of Res. & Indus., U. of Miami, Coral Gables, Fl.

Wan, A.S.C. 1972. *Garcinia mangostana*—high resolution NMR studies of mangosteen. Planta Med. 24 (3): 297-300.

Wan, S. & R.A. Conover. 1981. A rhabdovirus associated with a new disease of Florida papayas. Proc. Fl. St. Hort. Soc. 91: 318-321.

Wang, J-K. 1966. Equipment for husking poha berries. Tech. Bull. 60. Hawaii Agr. Exper. Sta., Honolulu.

Wang, Y.P. 1975. Discovery and study on the boron deficiency of papaya. Taiwan Agr. Quart. 2 (3): 72-73.

Ward, J.F. 1961. Garden cherry [bulletin]. Min. of Agr. & Lands, Kingston, Jamaica.

Wardlaw, C.W. 1972. Banana diseases, including plantains and Abaca. 2nd ed. Longman Group Ltd., London.

Wardowski, W.F., C.R. Barmore, T.S. Smith and C.W. DuBois. 1974. Curing of Florida lemons with ethephon. Proc. Fl. St. Hort. Soc. 87: 216-218.

Ware, C.E. 1956. Peerless, an interesting new lychee. Year Book & Proc. Fla. Lychee Growers Assn. 4th Ann. Mtg., Winter Haven, Fla.: 60-62.

Warner, R.M. and R.L. Fox. 1977. Nitrogen and potassium nutrition of 'Giant Cavendish' banana in Hawaii. J. Amer. Soc. Hort. Sci. 102 (6): 739-743.

Waters, H. 1976. The Paradise Navel, ready made for the restaurant trade. Fla. Grower & Rancher 69 (9): 21.

Watkins, J.V. 1969. Florida landscape plants, native and exotic. U. Fla. Press, Gainesville, Fla.

Watkins, JV. 1952. Gardens of the Antilles. U. Fla. Press, Gainesville Fla.

Watt, B.K. and A.L. Merrill. 1963. Composition of foods: raw, processed and prepared. Agr. Handbook 8. U.S. Dept. Agr., Agr. Res. Serv., Washington, D.C.

Watt, G. 1896. Dictionary of economic products of India. 9-vols. [Durian]. Supt. of Gov't. Ptg., Calcutta. P. 198.

Watt, G. 1908. The commercial products of India. John Murray, London.

Watt, J.M. and M.G. Breyer-Brandwijk. 1962. Medicinal and poisonous plants of southern and eastern Africa. E. & S. Livingstone, Ltd., Edinburgh & London.

Webber, H.J. 1943. The 'Tristeza' disease of sour-orange rootstock. Proc. Amer. Soc. Hort. Sci. 43: 160-168.

Wee, Y.C. 1971. The effects of Planofix on the pineapple fruit. Malaya. Pineapple 1: 35-38.

Weir, C.C. 1976. Effect of various rootstocks on the growth of Valencia orange, Marsh seedless grapefruit and Ortanique trees in Jamaica. J. Agr. U. Puerto Rico 60 (4): 485-490.

Wehlburg, C., S.A. Alfieri, Jr., K.R. Langdon, and J.W. Kimbrough. 1975. Index of plant diseases in Florida. Bull. 11. Fla. Dept. Agr. & Consumer Serv., Div. of Pl. Indus., Gainesville, Fla.

Wenzel, F.W., R.W. Olsen, R.W. Barron, R.L. Higgart, R. Patrick and E.C. Hill. 1958. Finding the best lemon for Florida—a report of progress. II. Use of Florida lemons in frozen concentrate for lemonade. Proc. Fl. St. Hort. Soc. 86: 129-132.

Wester, C. 1965. Peach Industry on the move. Fla. Rural Elec. News 2 (5): 8 & 10.

Wester, P.J. 1924. Food plants of the Philippines. Bull. 39. Phil. Dept. Agr. & Nat. Res., Bur. of Agr., Manila.

Wester, P.J. 1912. Pineapple culture. Cir. 16. Gov't of Phil. Islands, Dept. of Pub. Instr., Bur. of Agr., Manila.

Wester, P.J. 1914. Propagation of the seedless breadfruit. Phil. Agr. Rev. 7 (3): 1-3.

Wester, P.J. 1907. Roselle: its culture and uses. Farmers' Bull. 307. U.S. Dept. Agr., Washington, D.C.

Wester, P.J. 1920. The breadfruit. Phil. Agr. Rev. 13 (3): 1-8.

Wester, P.J. 1920. The cultivation and uses of roselle. Cir. 20. Reprinted from the Phil. Agr. Rev. 13 (2).

Wheaton, T.A. and I. Stewart. 1965. Feruloylputrescine: isolation and identification from citrus leaves and fruit. Nature 206: 620-621.

Wheeler, L.C. 1955. The husks of the Prodigal Son. Turtox News 33 (10): 194-197. [carob].

Whitely, K.T. 1962. The cape gooseberry. J. Agr. West. Aust. 3 (1) 4th Ser. 59, 61.

Whiteside, J.O. and L.C. Knorr. 1978. Susceptibility of different rough lemon collections to foot rot, blight and Alternaria leaf spot. Proc. Fl. St. Hort. Soc. 91: 75-77.

Whitman, W.F. 1980. Growing and fruiting the langsat in Florida. Proc. Fl. St. Hort. Soc. 93: 136-140.

Whitman, W.F. 1962. New fruit varieties for southern Florida and other warm regions. Fr. Var. & Hort. Dig. 16 (3): 47-48.

Whitman, W.F. 1974. The camu camu, the 'Wan' maprang and the 'Manila' santol. Proc. Fl. St. Hort. Soc. 84: 375-379.

Whitman, W.F. 1965. The green sapote, a new fruit for South Florida. Proc. Fl. St. Hort. Soc. 78: 330-336.

Whitman, W.F. 1976. South American sapote. Proc. Fl. St. Hort. Soc. 89: 226-227.

Whitman, W.F. 1971. The olosapo, the sunsapote and the Fijian longan. Proc. Fl. St. Hort. Soc. 84: 323-325.

Whitney, C.M. (editor). 1955. The Bermuda garden. [Barbados gooseberry] Garden Club of Bermuda, Hamilton.

Whittaker, D.E. 1972. Passion fruit: agronomy, processing and marketing. Trop. Sci. 14 (1): 59-77.

Wild, B.L. and L.E. Rippon. 1973. Quality control in lemon storage. Agr. Gaz. New So. Wales 84 (3): 142-144.

Wilder, G.P. 1928. The breadfruit of Tahiti. Bull. 50. Bernice P. Bishop Mus., Honolulu, Hawaii.

Willaman, J.J. and B.G. Schubert. 1961. Alkaloid-bearing plants and their contained alkaloids. Tech. Bull. 1234. U.S. Dept. Agr., Agr. Res. Serv., Washington, D.C.

Williams, C.G. 1954. Preparing passion fruit for the fresh fruit market. Queensland Agr. J. 78 (2): 81-88.

Williams, L.O. 1977. The avocados, a synopsis of the genus *Persea*, subg. *Persea*. Econ. Bot. 31: 315-320.

Williams, R.O. 1949. Useful and ornamental plants in Zanzibar and Pemba. Zanzibar Protectorate, Gov't Printer, Zanzibar.

Williams, R.O. and R.O., Jr. 1951. Useful and ornamental plants in Trinidad and Tobago. 4th ed. Guardian Commercial Pty., Port-of-Spain, Trinidad.

Williamson, Jr. 1955. Useful plants of Nyasaland. Gov't Printer, Nyasaland.

Wills, J. McG. 1955. Tropical notes on banana growing. Queensland Fr. & Veg. News 8 (19): 589.

Wills, R.B.H., B.I. Brown, and K.J. Scott. 1982. Control of ripe fruit rots of guavas by heated benomyl and guazatine dips. Aust. J. Exper. Agr. & Anim. Husb. 22: 437-440.

Wills, R.B.H., E.E. Mulholland, B.I. Brown and K.J. Scott. 1983. Storage of two new cultivars of guava fruit for processing. Trop. Agr. (Trin.) 60 (3): 175-178.

Wills, R.B.H., A. Poi, H. Greenfield and C.J. Rigney. 1984. Postharvest changes in fruit composition of *Annona atemoya* during ripening and effects of storage temperature on ripening. Hort-Sci. 19 (1): 96-97.

Wilson, A.M. 1954. The richest known source of vitamin C. Today's Health 32 (11): 20, 21, 54-57. [Barbados cherry].

Wiltbank, W.J. 1977. Mango and avocado cultivars in Brazil. Proc. Fl. St. Hort. Soc. 90: 243-244.

Winer, N. 1980. The potential of the carob (*Ceratonia siliqua*). Internat. Tree Crops J. 1: 15-26.

Winters, H.F. 1953. The Mangosteen. Fr. Vars. & Hort. Dig. 8 (4): 57-58.

Winters, H.F. 1963. The Natal plum. Amer. Hort. Mag. 42 (2): 93-95.

Wolfe, H.S. 1962. The mango in Florida—1887-1962. Proc. Fl. St. Hort. Soc. 75: 387-391.

Wolfe, H.S. 1960. The mystery of the Mulgoba. Proc. Fl. St. Hort. Soc. 73: 309-311.

Wolfe, H.S. and S.J. Lynch. 1942. Papaya culture in Florida. Bull. 113. U. Fla., Agr. Exten. Serv., Gainesville, Fla.

Wolfenbarger, D.O. 1972. Cedar waxwing, *Bombycilla cedrorum*, feeds on avocado flowers. Proc. Fl. St. Hort. Soc. 85: 341-343.

Wolfenbarger, D.O. 1963. Controlling mango insect pests. Cir. 147A. U. Fla. Agr. Exten. Serv., Gainesville, Fla.

Wolfenbarger, D.O. 1963. Insect pests of the avocado and their control. U. Fla. Agr. Exper. Sta., Gainesville, Fla.

Wolfenbarger, D.O. 1962. Papaya fruit fly control. Proc. Fl. St. Hort. Soc. 75: 381-384.

Wolfenbarger, D.O. and H. Spencer. 1951. Insect control on pineapples. Cir. S-36. U. Fla. Agr. Exper. Sta., Gainesville, Fla.

Woods, C. 1974. Fighting young tree decline with chemotherapy. Fla. Grower & Rancher 67 (9): 6-8.

Woods, C. 1973. Florida's other citrus crop makes a comeback. Fla. Grower & Rancher 66: 10, 24-26.

Woot-Tsuen Wu Leung, F. Busson and C. Jardin. 1968. Food composition table for use in Africa. FAO, Nutr. Div., Rome, and U.S. Dept. HEW, Pub. Health Serv., Bethesda, Md.

Worthington, T.B. 1959. Ceylon trees. Colombo Apothecaries' Co., Ltd., Colombo, Ceylon.

Wren, R.W. 1970. Potter's New Cyclopaedia of botanical drugs and preparations. 7th ed. Health Sci. Press, Rustington, Sussex, England.

Wuhrmann, J.J. and A. Patron. 1965. Evaluation de quelques fruits tropicaux peu connus. Fruits 20 (11): 615-624.

WuLeung, W., R.K. Pecot and B.K. Watt. 1952. Composition of foods used in Far Eastern countries. Agr. Handbook 34. U.S. Dept. Agr., Washington, D.C.

Wutscher, H.K., N. Del Valle and A. de Bernard. 1983. Citrus blight and wood pH in Cuba and Florida. HortSci. 18 (4): 486-488.

Wutscher, H.K., N.P. Maxwell and A.V. Shull. 1975. Performance of nucellar grapefruit, *Citrus paradisi* Macf., on 13 rootstocks in South 29-32.

Wutscher, H.K. and A.V. Schull. 1976. Performance of 'Orlando' tangelo on 16 rootstocks. J. Amer. Soc. Hort. Sci. 101 (1): 88-91.

Wynne, V.A. 1977. Haiti Seed Store, Kenscoff, Haiti. Personal Communication. Oct. 10. [tree tomato].

Wynne, V.A. 1980. Terraces and better irrigation may help stop erosion in Haiti. VITA (Volunteers for Internat. Assist.) News: Oct.

Wysoki, M., E. Swirski and Y. Izhar. 1981. Biological control of avocado pests in Israel. Protection Ecology 3: 25-28.

Yamamoto, H.Y. and W. Inouye. 1963. Sucrose as a gelation inhibitor of commercially frozen papaya puree. Tech. Prog. Rpt. 137. Hawaii Agr. Exper. Sta., U. of Hawaii, Honolulu.

Yamane, G.M. and H.Y. Nakasone. 1961. Effects of growth regulators on fruit set and growth of the acerola (*Malpighia glabra* L.). Tech. Bull. 43. Hawaii Agr. Exper. Sta., U. of Hawaii, Honolulu.

Yamane, G.M. and H.Y. Nakasone. 1961. Pollination and fruit set studies of acerola, *Malpighia glabra* L., in Hawaii. Proc. Amer. Soc. Hort. Sci. 78: 141-148.

Yamdagani, R., D.S. Balyan and P.C. Jindal. 1980. A note of the effect of nitrogen on litchi (*Litchi Chinensis* Sonn.). Haryana J. Hort. Sci. 9 (3-4): 141-143.

Yan, J. 1981. Histoire d' *Actinidia chinensis* Planch. et conditions actuelles de sa production a l' estranger. J. d' Agric. Trad. et de Bot. Appl. 28 (3-4): 281-290.

Yee, W., E.K. Akamine, G.M. Aoki, F.H. Haramoto, R.B. Hine, O.V. Holtzmann, R.A. Hamilton, J.T. Ishida, J.T. Keeler and H. Nakasone. 1970. Papayas in Hawaii. Cir. 436. U. Hawaii, Coop. Exten. Serv., Honolulu.

Yong, H.S. 1979. Sukun, Nangka and Cempedak [breadfruit, jackfruit and champedak] from Mutiny on the Bounty, the fruit with the strongest and richest smell. Malaysian Panorama 9 (2): 18-23.

Young, I.N. 1935. The Hawaiian Islands and the story of pineapple. Amer. Can Co., 230 Park Ave., N.Y.

Young, R., O. Jahn, W.C. Cooper and J.J. Smoot. 1970. Preharvest sprays with 2-chloroethyllphosphonic acid to degreen 'Robinson' and 'Lee' tangerine fruits. HortScience 5 (4): 268-269.

Young, T.W. 1966. A review of the Florida lychee industry. Proc. Fl. St. Hort. Soc. 79: 395-398.

Young, T.W. 1956. Response of lychees to girdling. Proc. Fl. St. Hort. Soc. 69: 305-308.

Young, T.W. 1970. Some climatic effects on flowering and fruiting of 'Brewster' lychees in Florida. Proc. Fl. St. Hort. Soc. 83: 362-367.

Young, T.W. and R.W. Harkness. 1961. Flowering and fruiting behavior of Brewster lychees in Florida. Proc. Fl. St. Hort. Soc. 74: 358-363.

Young, T.W. and R.C.J. Koo. 1975. Effect of hedging on yield of lemon and lime trees. Proc. Fl. St. Hort. Soc. 88: 445-448.

Young, T.W. and R.C.J. Koo. 1964. Influence of nitrogen source and rate of fertilization on performance of Brewster lychees. Proc. Fl. St. Hort. Soc. 77: 406-410.

Young, T.W. and J.C. Noonan. 1958. Freeze damage to lychees. Proc. Fl. St. Hort. Soc. 71: 300-304.

Younge, O.R. and D.L. Plucknett. 1969. Estimating the number of papaw (*Carica papaya*) trees required for reliable yield data. Queensland J. Agr. & Anim. Sci. 26: 289-292.

Youngman, B.J. 1953. Chinese gooseberry. Kew Bull. 4: 567-568.

Zachrawan dan Sumarto. 1977. Percoban pengawetan buah sawo (*Achroas zapota* L.) dalam fentuk kering [a study on dehydration of sawo (*Achras zapota*)]. Bull. Penel. Hort. 5 (4): 13-19.

Zaiger, D., and G.A. Zentmyer. 1967. Epidemic decline of breadfruit in the Pacific Islands. FAO Plant Protection Bull. 15 (2): 25-29.

Zaki, D. 1975. Biological investigation of *Dovyalis caffra*. Planta Med. 27: 330-332.

Sapata, A. 1972. Pejibaye palm from the Pacific coast of Colombia (a detailed chemical analysis). Econ. Bot. 26 (2): 156-159.

Zayas, J.C. 1944. Los frutales en Cuba. Rev. de Agr. 27 (27): 23-140.

Zentmyer, G.A. 1963. Avocado root rot in the Caribbean. Carib. Agr. 1 (4): 317-324.

Zentmyer, G. 1977. California Avocado Industry. U.C. research program on production problems. The Blue Anchor 54 (5): 53.

Ziegler, L.W. and H.S. Wolfe. 1961. Citrus growing in Florida. U. Florida Press, Gainesville, Fla.

Zink, E. and M. Ojima. 1965. Influencia das condicoes de armazenagem no poder germinativo dos sementes de nespera. Bragantia 24 (3): 9-12.

Zio, S. 1975-76. Utilisation en medecine traditionnelle de quelques plantes en Hauto-Volta. Notes et documents Voltaiques 9 (1): 77-80.

INDEX OF VERNACULAR AND SCIENTIFIC NAMES
(including pests and diseases)

ERRATA

Page 28, PINEAPPLE, Col. 1, 18th l. from bottom, Kwantgung should be Kwantung.

Page 39, BANANA, Col. 2, l. 4 and l. 17: *coffaea* should be *coffeae*.

Page 78, SOURSOP, Col. 2, 6th l. from bottom: *Corticum* should be *Corticium*.

Page 171, MEXICAN LIME, Col. 1, 24th l. from bottom: *Gleosporium* should be *Gloeosporium*.

ABOUT THE AUTHOR

JULIA F. MORTON is Research Professor of Biology and Director of the Morton Collectanea, University of Miami, a research and information center devoted to economic botany. She received her D.Sc. from Florida State University in 1973 and was elected a Fellow of the Linnean Society of London in 1974. She has conducted extensive field studies for the U.S. National Institutes of Health and the Department of Defense, has served as horticultural development consultant in Florida and tropical America, and, since 1954, has been consultant for the Poison Control Centers in Florida. In 1978, she was selected as the First Distinguished Economic Botanist by the international Society for Economic Botany. She served as President of the Florida State Horticultural Society in 1979. She is a member of the Board of Trustees of Fairchild Tropical Garden, Miami, and of the Board of Directors of the Florida National Parks and Monuments Association. She is the author of 10 books and co-author of or contributor to 12 others; has written 94 scientific papers and co-authored 27 others; has produced 2 full-color wall charts of poisonous plants; a set of Survival Cards for Southeast Asia and a 157-page report on the Survival-related Flora and Fauna of the Mekong for the U.S. Department of Defense. She is well known as a lecturer on toxic, edible and otherwise useful plants.

Photo by Frank D. Venning